Empower Each Learner

Give students anytime, anywhere access with Pearson eText

Pearson eText is a simple-to-use, mobile-optimized, personalized learning experience available within MyLab Math. It allows students to easily highlight and take notes all in one place—even when offline. Seamlessly integrated videos and other rich media engage students and give them access to the help they need, when they need it.

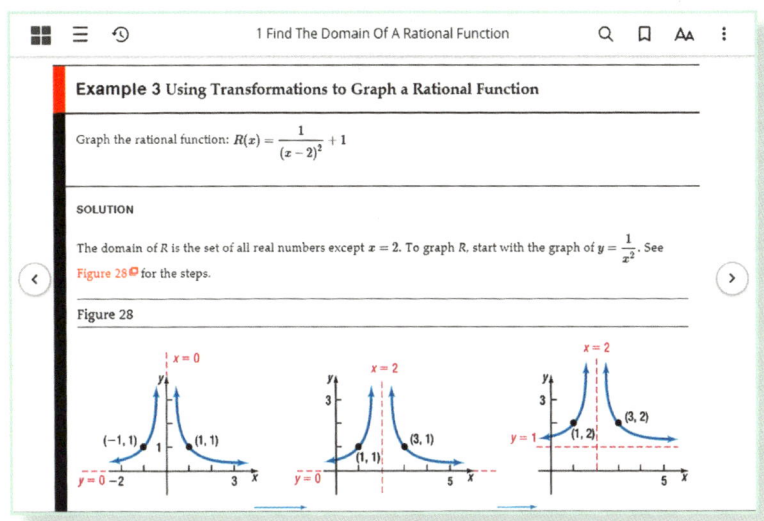

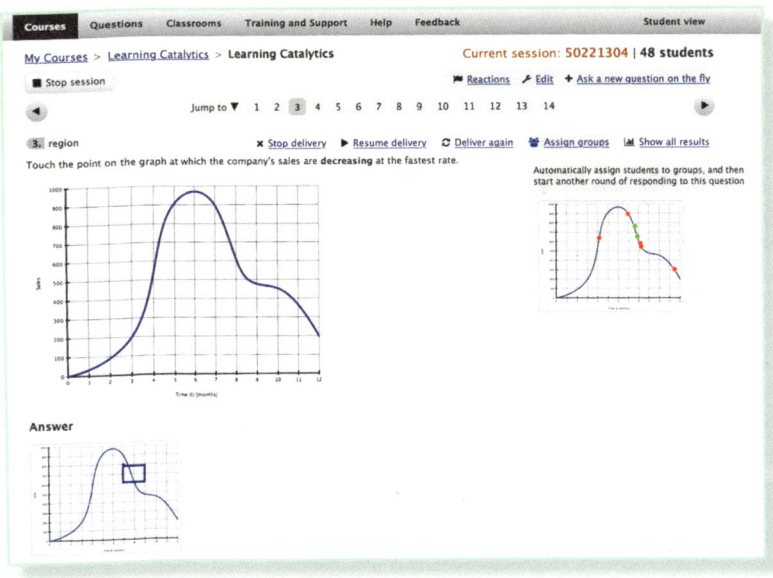

Give every student a voice with Learning Catalytics

MyLab Math and MyLab Statistics now provide Learning Catalytics™—an interactive student-response tool that uses students' smartphones, tablets, or laptops to engage them in more sophisticated tasks and thinking.

Foster student engagement, generate class discussion, guide your lecture, and promote peer-to-peer learning with real-time analytics.

Instructors, you can

- Pose a variety of open-ended questions that help your students develop critical-thinking skills.
- Monitor responses to find out where students are struggling.
- Use real-time data to adjust your instructional strategy and try other ways of engaging your students during class.
- Manage student interactions by automatically grouping students for discussion, teamwork, and peer-to-peer learning.

pearson.com/mylab/math

Annotated Instructor's Edition 6th Edition

College Algebra IN CONTEXT

with applications for the managerial, life, and social sciences

Ronald J. Harshbarger
University of South Carolina

Lisa S. Yocco
East Georgia State College

Please contact https://support.pearson.com/getsupport/s/contactsupport with any queries on this content

Microsoft and/or its respective suppliers make no representations about the suitability of the information contained in the documents and related graphics published as part of the services for any purpose. All such documents and related graphics are provided "as is" without warranty of any kind. Microsoft and/or its respective suppliers hereby disclaim all warranties and conditions with regard to this information, including all warranties and conditions of merchantability, whether express, implied or statutory, fitness for a particular purpose, title and non-infringement. In no event shall Microsoft and/or its respective suppliers be liable for any special, indirect or consequential damages or any damages whatsoever resulting from loss of use, data or profits, whether in an action of contract, negligence or other tortious action, arising out of or in connection with the use or performance of information available from the services.

The documents and related graphics contained herein could include technical inaccuracies or typographical errors. Changes are periodically added to the information herein. Microsoft and/or its respective suppliers may make improvements and/or changes in the product(s) and/or the program(s) described herein at any time. Partial screen shots may be viewed in full within the software version specified.

Microsoft® and Windows® are registered trademarks of the Microsoft Corporation in the U.S.A. and other countries. This book is not sponsored or endorsed by or affiliated with the Microsoft Corporation.

Copyright © 2021, 2017, 2013 by Pearson Education, Inc. or its affiliates, 221 River Street, Hoboken, NJ 07030. All Rights Reserved. Manufactured in the United States of America. This publication is protected by copyright, and permission should be obtained from the publisher prior to any prohibited reproduction, storage in a retrieval system, or transmission in any form or by any means, electronic, mechanical, photocopying, recording, or otherwise. For information regarding permissions, request forms, and the appropriate contacts within the Pearson Education Global Rights and Permissions department, please visit www.pearsoned.com/permissions/.

Acknowledgments of third-party content appear on page 717, which constitutes an extension of this copyright page.

PEARSON, ALWAYS LEARNING, REVEL, and MYLAB are exclusive trademarks owned by Pearson Education, Inc. or its affiliates in the U.S. and/or other countries.

Unless otherwise indicated herein, any third-party trademarks, logos, or icons that may appear in this work are the property of their respective owners, and any references to third-party trademarks, logos, icons, or other trade dress are for demonstrative or descriptive purposes only. Such references are not intended to imply any sponsorship, endorsement, authorization, or promotion of Pearson's products by the owners of such marks, or any relationship between the owner and Pearson Education, Inc., or its affiliates, authors, licensees, or distributors.

Library of Congress Cataloging-in-Publication Data
Names: Harshbarger, Ronald J., author. | Yocco, Lisa S, author.
Title: College algebra in context : with applications for the managerial, life, and social sciences / Ronald J. Harshbarger, University of South Carolina, Lisa S. Yocco, East Georgia State College.
Description: 6th edition. | Hoboken, NJ: Pearson, [2021] | Includes index.
Identifiers: LCCN 2019042866 | ISBN 9780135757765 (hardback) | ISBN 0135757762 (hardback) | ISBN 9780135757772 (annotated instructor's edition) | ISBN 0135757770 (annotated instructor's edition)
Subjects: LCSH: Algebra—Textbooks. | Social sciences—Mathematics—Textbooks.
Classification: LCC QA152.3 .H33 2021 | DDC 512.9—dc23
LC record available at https://lccn.loc.gov/2019042866

1 2019

This work is protected by United States copyright laws and is provided solely for the use of instructors in teaching their courses and assessing student learning. Dissemination or sale of any part of this work (including on the World Wide Web) will destroy the integrity of the work and is not permitted. The work and materials from it should never be made available to students except by instructors using the accompanying text in their classes. All recipients of this work are expected to abide by these restrictions and to honor the intended pedagogical purposes and the needs of other instructors who rely on these materials.

Annotated Instructor's Edition
ISBN 10: 0-13-575777-0
ISBN 13: 978-0-13-575777-2

ISBN 10: 0-13-575776-2
ISBN 13: 978-0-13-575776-5

Contents

Preface ix

CHAPTER 1 Functions, Graphs, and Models; Linear Functions 1

Algebra Toolbox 2
Sets ▪ The Real Numbers ▪ Calculating with Real Numbers ▪ Properties of Real Numbers ▪ Inequalities and Intervals on the Number Line ▪ Algebraic Expressions ▪ Evaluating Algebraic Expressions ▪ Combining Like Terms ▪ Removing Parentheses ▪ Solving Basic Linear Equations ▪ The Coordinate System ▪ Subscripts

1.1 Functions and Models 12
Function Definitions ▪ Domains and Ranges ▪ Tests for Functions ▪ Vertical Line Test ▪ Function Notation ▪ Mathematical Models ▪ Aligning Data

1.2 Graphs of Functions 31
Graphs of Functions ▪ Graphing with Technology ▪ Determining Viewing Windows ▪ Graphing Data Points

1.3 Linear Functions 45
Linear Functions ▪ Intercepts ▪ Slope of a Line ▪ Slope and y-Intercept of a Line ▪ Constant Rate of Change ▪ Revenue, Cost, and Profit ▪ Marginal Cost, Revenue, and Profit ▪ Special Linear Functions

1.4 Equations of Lines 62
Writing Equations of Lines ▪ Parallel and Perpendicular Lines ▪ Average Rate of Change ▪ Difference Quotient ▪ Approximately Linear Data

Preparing for Calculus 79
Summary 80
Key Concepts and Formulas 80
Chapter 1 Skills Check 82
Chapter 1 Review 83
Group Activities/Extended Applications 86

CHAPTER 2 Linear Models, Equations, and Inequalities 88

Algebra Toolbox 89
Multiplying and Dividing Fractions ▪ Adding and Subtracting Fractions ▪ Properties of Equations ▪ Evaluating Formulas for Given Values ▪ Properties of Inequalities ▪ Rounding Function Coefficients and Constants

2.1 Algebraic and Graphical Solution of Linear Equations 97
Algebraic Solution of Linear Equations ▪ Solutions, Zeros, and x-Intercepts ▪ Graphical Solution of Linear Equations ▪ Literal Equations; Solving an Equation for a Specified Linear Variable ▪ Direct Variation

2.2 Fitting Lines to Data Points: Modeling Linear Functions 109
Exact and Approximate Linear Models ▪ Fitting Lines to Data Points; Linear Regression ▪ Applying Models ▪ Goodness of Fit

2.3 Systems of Linear Equations in Two Variables 128
System of Equations ▪ Graphical Solution of Systems ▪ Solution by Substitution ▪ Solution by Elimination ▪ Modeling Systems of Linear Equations ▪ Dependent and Inconsistent Systems

2.4 Solutions of Linear Inequalities 141
Algebraic Solution of Linear Inequalities ▪ Graphical Solution of Linear Inequalities ▪ Intersection Method ▪ x-Intercept Method ▪ Double Inequalities

Preparing for Calculus 152
Summary 153
Key Concepts and Formulas 153
Chapter 2 Skills Check 155
Chapter 2 Review 156
Group Activities/Extended Applications 160

CHAPTER 3 Quadratic, Piecewise-Defined, and Power Functions 161

Algebra Toolbox 162
Integer Exponents ▪ Absolute Value ▪ Radicals ▪ Rational Exponents and Radicals ▪ Polynomials ▪ Operations with Polynomials ▪ Dividing Polynomials by Monomials ▪ Factoring Polynomials ▪ Complex Numbers

3.1 Quadratic Functions; Parabolas 173
Parabolas ▪ Vertex Form of a Quadratic Function

3.2 Solving Quadratic Equations 186
Factoring Methods ▪ Graphical Methods ▪ Combining Graphs and Factoring ▪ Graphical and Numerical Methods ▪ The Square Root Method ▪ Completing the Square ▪ The Quadratic Formula ▪ The Discriminant ▪ Aids for Solving Quadratic Equations ▪ Equations with Complex Solutions

3.3 Power and Root Functions 203
Power Functions ▪ Power Functions with Negative Exponents ▪ Root Functions ▪ Solution of Power and Root Equations ▪ Direct Variation as a Power

3.4 Piecewise-Defined Functions and Absolute Value Functions 215
Piecewise-Defined Functions ▪ Greatest Integer Function ▪ Absolute Value Function ▪ Solving Absolute Value Equations

3.5 Quadratic and Power Models 225
Modeling a Quadratic Function from Three Points on Its Graph ▪ Modeling with Quadratic Functions ▪ Comparison of Linear and Quadratic Models ▪ Modeling with Power Functions ▪ Comparison of Power and Quadratic Models

Preparing for Calculus 244
Summary 245
Key Concepts and Formulas 245
Chapter 3 Skills Check 247
Chapter 3 Review 248
Group Activities/Extended Applications 253

CHAPTER 4 Additional Topics with Functions 255

Algebra Toolbox 256
Linear Functions ▪ Quadratic Functions ▪ Piecewise-Defined Functions ▪ Power Functions ▪ Special Power Functions

4.1 Transformations of Graphs and Symmetry 262
Shifts of Graphs of Functions ▪ Stretching and Compressing Graphs ▪ Reflections of Graphs ▪ Symmetry; Even and Odd Functions

4.2 Combining Functions; Composite Functions 278
Operations with Functions ▪ Average Cost ▪ Composition of Functions

4.3 One-to-One and Inverse Functions 290
Inverse Functions ▪ One-to-One Functions ▪ Inverse Functions on Limited Domains

4.4 Additional Equations and Inequalities 302
Radical Equations; Equations Involving Rational Powers ▪ Equations Containing Rational Powers ▪ Equations in Quadratic Form ▪ Quadratic Inequalities ▪ Algebraic Solution of Quadratic Inequalities ▪ Graphical Solution of Quadratic Inequalities ▪ Power Inequalities ▪ Inequalities Involving Absolute Values

Preparing for Calculus 317
Summary 318
Key Concepts and Formulas 318
Chapter 4 Skills Check 320
Chapter 4 Review 321
Group Activities/Extended Applications 323

CHAPTER 5 Exponential and Logarithmic Functions 325

Algebra Toolbox 326
Additional Properties of Exponents ▪ Real Number Exponents ▪ Multiplying Radicals ▪ Dividing Radicals ▪ Adding and Subtracting Radicals ▪ Rationalizing Denominators ▪ Scientific Notation

5.1 Exponential Functions 332
Exponential Functions ▪ Transformations of Graphs of Exponential Functions ▪ Exponential Growth Models ▪ Exponential Decay Models ▪ The Number e

5.2 Logarithmic Functions; Properties of Logarithms 345
Logarithmic Functions ▪ Common Logarithms ▪ Natural Logarithms ▪ Logarithmic Properties ▪ Richter Scale

5.3 Exponential and Logarithmic Equations 362
Solving Exponential Equations Using Logarithmic Forms ▪ Change of Base ▪ Solving Exponential Equations Using Logarithmic Properties ▪ Solution of Logarithmic Equations ▪ Exponential and Logarithmic Inequalities

5.4 Exponential and Logarithmic Models 377
Modeling with Exponential Functions ▪ Finding Rates of Change in Exponential Growth and Decay Models ▪ Exponential Models ▪ Exponential Models of Growth or Decay ▪ Comparison of Models ▪ Logarithmic Models ▪ Exponents, Logarithms, and Linear Regression

5.5 Exponential Functions and Investing 394
Compound Interest ▪ Continuous Compounding and the Number e ▪ Present Value of an Investment ▪ Investment Models

5.6 Annuities; Loan Repayment 406
Future Value of an Annuity ▪ Present Value of an Annuity ▪ Loan Repayment

5.7 Logistic and Gompertz Functions 416
Logistic Functions ▪ Gompertz Functions

Preparing for Calculus 428
Summary 429
Key Concepts and Formulas 429
Chapter 5 Skills Check 432
Chapter 5 Review 433
Group Activities/Extended Applications 438

CHAPTER 6 Higher-Degree Polynomial and Rational Functions 440

Algebra Toolbox 441
Polynomials ▪ Factoring Higher-Degree Polynomials ▪ Rational Expressions ▪ Multiplying and Dividing Rational Expressions ▪ Adding and Subtracting Rational Expressions ▪ Simplifying Complex Fractions ▪ Division of Polynomials ▪ Operations with Complex Numbers ▪ Adding and Subtracting Complex Numbers ▪ Multiplying and Dividing Complex Numbers

6.1 Higher-Degree Polynomial Functions 451
Cubic Functions ▪ Quartic Functions

6.2 Modeling with Cubic and Quartic Functions 463
Modeling with Cubic Functions ▪ Modeling with Quartic Functions ▪ Model Comparisons ▪ Third and Fourth Differences

6.3 Solution of Polynomial Equations 479
Solving Polynomial Equations by Factoring ▪ Solution Using Factoring by Grouping ▪ The Root Method ▪ Estimating Solutions with Technology

6.4 Polynomial Equations Continued; Fundamental Theorem of Algebra 489
Division of Polynomials; Synthetic Division ▪ Using Synthetic Division to Solve Cubic Equations ▪ Graphs and Solutions ▪ Rational Solutions Test ▪ Fundamental Theorem of Algebra

6.5 Rational Functions and Rational Equations 501
Graphs of Rational Functions ▪ Slant Asymptotes and Missing Points ▪ Algebraic and Graphical Solution of Rational Equations ▪ Inverse Variation ▪ Combined Variation ▪ Joint Variation

6.6 Polynomial and Rational Inequalities 517
Polynomial Inequalities ▪ Rational Inequalities

Preparing for Calculus 524
Summary 525
Key Concepts and Formulas 525
Chapter 6 Skills Check 528
Chapter 6 Review 529
Group Activities/Extended Applications 532

CHAPTER 7 Systems of Equations and Matrices — 533

Algebra Toolbox 534
Proportion ▪ Proportional Triples ▪ Linear Equations in Three Variables

7.1 Systems of Linear Equations in Three Variables 537
Systems in Three Variables ▪ Left-to-Right Elimination ▪ Modeling Systems of Equations ▪ Nonunique Solutions

7.2 Matrix Solution of Systems of Linear Equations 547
Matrix Representation of Systems of Equations ▪ Echelon Forms of Matrices; Solving Systems with Matrices ▪ Gauss–Jordan Elimination ▪ Solution with Technology ▪ Nonunique Solution ▪ Dependent Systems ▪ Inconsistent Systems

7.3 Matrix Operations 560
Addition and Subtraction of Matrices ▪ Multiplication of a Matrix by a Number ▪ Matrix Multiplication ▪ Multiplication with Technology

7.4 Inverse Matrices; Matrix Equations 575
Inverse Matrices ▪ Inverses and Technology ▪ Encoding and Decoding Messages ▪ Matrix Equations ▪ Matrix Equations and Technology

7.5 Determinants and Cramer's Rule 588
The Determinant of a 2×2 Matrix ▪ Cramer's Rule ▪ The Determinant of a 3×3 Matrix ▪ Solving Systems of Equations in Three Variables with Cramer's Rule ▪ Cramer's Rule with Inconsistent and Dependent Systems

7.6 Systems of Nonlinear Equations 598
Algebraic Solution of Nonlinear Systems ▪ Graphical Solution of Nonlinear Systems

Summary 604
Key Concepts and Formulas 605
Chapter 7 Skills Check 607
Chapter 7 Review 608
Group Activities/Extended Applications 611

CHAPTER 8 Special Topics in Algebra — 613

8.1 Systems of Inequalities 614
Linear Inequalities in Two Variables ▪ Systems of Inequalities in Two Variables ▪ Systems of Nonlinear Inequalities

8.2 Linear Programming: Graphical Methods 625
Linear Programming ▪ Solution with Technology

8.3 Sequences and Discrete Functions 636
Sequences ▪ Arithmetic Sequences ▪ Geometric Sequences

8.4 Series 645
Finite and Infinite Series ▪ Arithmetic Series ▪ Geometric Series ▪ Infinite Geometric Series

8.5 The Binomial Theorem 653
Binomial Coefficients

8.6 Conic Sections: Circles and Parabolas 657
Distance and Midpoint Formulas ▪ Circles ▪ Parabolas

8.7 Conic Sections: Ellipses and Hyperbolas 665
Ellipses ▪ Hyperbolas ▪ Rectangular Hyperbolas

Summary 673
Key Concepts and Formulas 674
Chapter 8 Skills Check 677
Chapter 8 Review 678
Group Activities/Extended Applications 680

Appendix A Basic Graphing Calculator Guide 681

Appendix B Basic Guide to Excel 701

Photo Credits 717
Additional Answers A-1
Index I-1

Preface

College Algebra in Context is designed for a course in algebra that is based on data analysis, modeling, and real-life applications from the management, life, and social sciences. The text is intended to show students how to analyze, solve, and interpret problems in this course, in future courses, and in future careers. At the heart of this text is its emphasis on problem solving in meaningful contexts.

The text is application-driven and uses real data problems that motivate interest in the skills and concepts of algebra. Modeling is introduced early, in the discussion of linear functions and in the discussion of quadratic and power functions. Additional models are introduced when exponential, logarithmic, logistic, cubic, and quartic functions are discussed. Mathematical concepts are introduced informally with an emphasis on applications. Each chapter contains real data problems and extended application projects that can be solved by students working collaboratively.

The text features a constructive chapter-opening Algebra Toolbox, which reviews previously learned algebra concepts by presenting the prerequisite skills needed for successful completion of the chapter. In addition, Preparing for Calculus sections in each of the first six chapters provide problems that demonstrate how students can use their new knowledge in a calculus course.

Changes to the Sixth Edition

Based on valuable suggestions from our users and reviewers of our fifth edition, as well as our own classroom experiences and student input, we have made a number of changes in the sixth edition.

- To keep the real data applications current, more than 250 problems have been replaced with new or updated problems.

- **Algebra Toolbox** content, which opens Chapters 1–7, has been expanded and strengthened to accommodate corequisite scenarios, with more than 80 new exercises. This content has been added to the Integrated Review chapter in MyLab Math. Additional topics that have been added include the following: calculations with and properties of real numbers; operations with fractions; evaluating algebraic expressions; solving basic linear equations; evaluating formulas for given values; operations with radicals; rationalizing denominators; operations with complex numbers.

- A **Corequisite Notebook**, written by Lisa Yocco, is now available as a complement to this text. It includes all **Algebra Toolbox** topics with unique exercises for practice followed by small group activities, written by Trisha Sholar, to enhance classroom time. This resource can be used for corequisite courses or simply for students who are underprepared.

- **Preparing for Calculus** sections have been added before the Chapter Summaries in each of Chapters 1 through 6. These sections present problems that use the skills of the chapters in calculus contexts to show students how the algebra topics of each chapter apply to the development and application of calculus. This is important for those who will be taking calculus. Coverage of these exercises have also been added to MyLab Math.

- A new type of exercise call Setup & Solve has been added to MyLab Math. These multi-part exercises require students to show the setup of the solution for a particular exercise as well as the solution to gauge the students' conceptual understanding of the topic. Look for them with the label "Setup & Solve."

- Previous Section 3.3 has been divided into two new sections. New Section 3.3 now has expanded coverage of power functions and root functions, and new Section 3.4 now has expanded coverage of piecewise-defined functions and absolute value functions. These sections have many more examples and exercises than were in old Section 3.3.

- Additional modeling problems that involve decision making and critical thinking have been added throughout the text.

- Specific steps for additional topics have been added to Appendix A.

- More specific Excel steps have been added within the Spreadsheet Solutions in the text, with references to Appendix B for more details.

- Specific steps for additional topics have been added to Appendix B.

- Discussion of loan amortization using graphing calculators has been added to Chapter 5, and the detailed steps for solving loan amortization problems with graphing calculators has been added to Appendix A.

- Discussion of loan amortization using Excel has been added to Chapter 5, and the detailed steps for solving loan amortization problems with Excel has been added to Appendix B.

Continued Features

Features of the text include the following:

- The development of algebra is motivated by the need to use algebra to find the solutions to **real data–based applications**.

  Real-life problems demonstrate the need for specific algebraic concepts and techniques. Each section begins with a motivational problem couched in a real-life setting. The problem is solved after the necessary skills have been presented in that section. The aim is to prepare students to solve problems of all types by first introducing them to various functions and then encouraging them to take advantage of available technology. Special business and finance models are included to demonstrate the application of functions to the business world.

- **Technology** has been integrated into the text.

  The text discusses the use of graphing calculators and Excel to solve problems, but there are no specific technology requirements. When a new calculator or spreadsheet skill becomes useful in a section, students can find the required keystrokes or commands in the text and in the *Graphing Calculator and Excel® Manual*, located within MyLab Math, as well as in Appendixes A and B discussed below. Technology is used to enhance and support learning when appropriate—not to supplant learning.

- The text contains two technology appendixes: a **Basic Graphing Calculator Guide** and a **Basic Guide to Excel**. Footnotes throughout the text refer students to these guides for a detailed exposition when a new use of technology is introduced. Additional Excel solution procedures have been added, but, as before, they can be omitted without loss of continuity in the text.

- Each of the first seven chapters begins with an **Algebra Toolbox** section that reviews the prerequisite skills needed for successful completion of the chapter.

  Topics discussed in the Toolbox are topics that are prerequisite to a college algebra course (they are often found in a Chapter R or appendix of a college algebra text). Key objectives are listed at the beginning of each Toolbox, and topics are introduced "just in time" to be used in the chapter under consideration.

- Many problems posed in the text are **multi-part and multi-level problems**.

  Many problems require thoughtful, real-world answers adapted to varying conditions, rather than numerical answers. Questions such as "When will this model no longer be valid?" "What additional limitations must be placed on your answer?" and "Interpret your answer in the context of the application" are commonplace in the text.

- Chapter objectives are listed at the beginning of each chapter, and key objectives are given at the beginning of each section.

- Each chapter has a **Chapter Summary**, a **Chapter Skills Check**, and a **Chapter Review**.

  The Chapter Summary lists the key terms and formulas discussed in the chapter, with section references. Review problems include Chapter Skills Check, which strictly reinforce methods or procedures, and Review Exercises, which are applications.

- The text encourages **collaborative learning**.

  Each chapter ends with one or more Group Activities/Extended Applications that require students to solve multi-level problems involving real data or situations, making it desirable for students to collaborate in their solutions. These activities provide opportunities for students to work together to solve real problems that involve the use of technology and that frequently require modeling.

- The text encourages students to improve **communication skills and research skills**.

  The Group Activities/Extended Applications require written reports and frequently require use of the Internet or a library. Some Extended Applications call for students to use literature or the Internet to find a graph or table of discrete data describing an issue. They are then required to make a scatter plot of the data, determine the function type that is the best fit for the data, create the model, discuss how well the model fits the data, and discuss how it can be used to analyze the issue.

- **Answers to Selected Exercises** include answers to all Chapter Skills Checks, Chapter Reviews, and odd-numbered application exercises so students have feedback regarding the exercises they work.

- **Supplements** are provided that will help students and instructors use technology to improve the learning and teaching experience. See the instructor and student resources lists.

Acknowledgments

Many individuals contributed to the development of this text. We would like to thank the following reviewers, whose comments and suggestions were invaluable in preparing it.

Jay Abramson, *Arizona State University*
* Bader Abukhodair, *Fort Hays State University*
Ali Ahmad, *New Mexico State University*
Khadija Ahmed, *Monroe County Community College*
Janet Arnold, *Indiana University—Southeast Campus*
Jamie Ashby, *Texarkana College*
Sohrab Bakhtyari, *St. Petersburg College*
* Cindy Bales, *Oklahoma State University—Oklahoma City*
Jean Bevis, *Georgia State University*
Thomas Bird, *Austin Community College*
Len Brin, *Southern Connecticut State University*
* Judith Brummer, *Fort Hays State University*
* Lisa Buckelew, *Oklahoma City Community College*
Marc Campbell, *Daytona Beach Community College*
Rodica Cazacu, *Georgia College*
Florence Chambers, *Southern Maine Community College*
* Rebecca Diischer, *South Dakota State University*
Floyd Downs, *Arizona State University*
Aniekan Ebiefung, *University of Tennessee at Chattanooga*
Marjorie Fernandez Karwowski, *Valencia Community College*
Toni W. Fountain, *Chattanooga State Technical Community College*
Javier Gomez, *The Citadel*
John Gosselin, *University of Georgia*
David J. Graser, *Yavapai College*
Lee Graubner, *Valencia Community College East*
Linda Green, *Santa Fe Community College*
Richard Brent Griffin, *Georgia Highlands College*
Lee Hanna, *Clemson University*
Deborah Hanus, *Brookhaven College*
Steve Heath, *Southern Utah University*
* Susan Henderson, *Greenville Technical College*
Todd A. Hendricks, *Georgia Perimeter College*
Suzanne Hill, *New Mexico State University*
Sue Hitchcock, *Palm Beach Community College*
Sandee House, *Georgia Perimeter College*
Mary Hudacheck-Buswell, *Clayton State University*
David Jabon, *DePaul University*
Arlene Kleinstein, *State University of New York—Farmingdale*
Danny Lau, *Kennesaw State University*
Ann H. Lawrence, *Wake Technical Community College*
Kit Lumley, *Columbus State University*
Phoebe Lutz, *Delta College*
Antonio Magliaro, *Southern Connecticut State University*
Beverly K. Michael, *University of Pittsburgh*
Phillip Miller, *Indiana University—Southeast Campus*
Nancy R. Moseley, *University of South Carolina Aiken*
Demetria Neal, *Gwinnett Technical College*
Malissa Peery, *University of Tennessee*
Ingrid Peterson, *University of Kansas*
Beverly Reed, *Kent State University*
Jeri Rogers, *Seminole Community College—Oviedo*
Michael Rosenthal, *Florida International University*
Steven Rosin, *Delta College*
Sharon Sanders, *Georgia Perimeter College*
Carolyn Spillman, *Georgia Perimeter College*
Susan Staats, *University of Minnesota*
Marvin Stick, *University of Massachusetts Lowell*
Theresa Thomas, *Blue Ridge Community College*
Jacqueline Underwood, *Chandler-Gilbert Community College*
Erwin Walker, *Clemson University*
* Charles Warburton, *Gateway Community and Technical College*
Denise Widup, *University of Wisconsin—Parkside*
Sandi Wilbur, *University of Tennessee—Knoxville*
Jeffrey Winslow, *Yavapai College*
Asmamaw Yimer, *Chicago State University*

Many thanks to Joan Saniuk and Paul Lorczak for an outstanding job checking the accuracy of this text. Special thanks go to the Pearson team for their assistance, encouragement, and direction throughout this project: Chelsea Kharakozova, Steve Schoen, Brian Fisher, Rajinder Singh, Chere Bemelmans, Erin Carreiro, Kristina Evans, Stacey Sveum, Peggy Lucas, and Jon Krebs.

Ronald J. Harshbarger
Lisa S. Yocco

* Denotes reviewers of the sixth edition.

Get the *most* out of MyLab Math

Preparedness

Preparedness is one of the biggest challenges in many math courses. Pearson offers a variety of content and course options to support students with just-in-time remediation and key-concept review as needed.

Integrated Review in MyLab Math

Integrated Review can be used in corequisite courses or simply to help students who enter a course without a full understanding of prerequisite skills and concepts.

- Students begin each chapter by completing a Skills Check to pinpoint which topics, if any, they need to review.

- Personalized review homework provides extra support for students who need it on just the topics they didn't master in the preceding Skills Check.

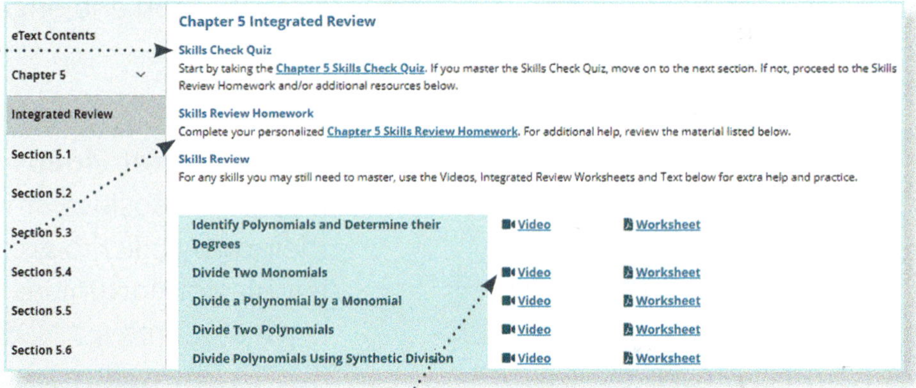

- Additional review materials, including **Updated Worksheets** and **New Videos**, are available.

Corequisite Notebook

New! The Corequisite Notebook, created by author Lisa Yocco, expands on the Algebra Toolbox topics within the text, which align to the Integrated Review videos/objectives. Students actively participate in learning the *how* and *why* of fundamental topics to ensure they are prepared for their college-level work.

- **Learning Tips** begin each chapter to set students in a learning mindset.
- **Worksheets** cover all Algebra Toolbox prerequisite topics and include unique examples and exercises for practice.
- **Small Group Activities** round out the chapter to be completed in class.

pearson.com/mylab/math

Get the *most* out of MyLab Math

MyLab Math for COLLEGE ALGEBRA IN CONTEXT 6e
by Ronald J. Harshbarger and Lisa S. Yocco
(access code required)

Motivation Resources

Motivate and inspire your students to succeed by encouraging a growth mindset and opportunities for self-reflection.

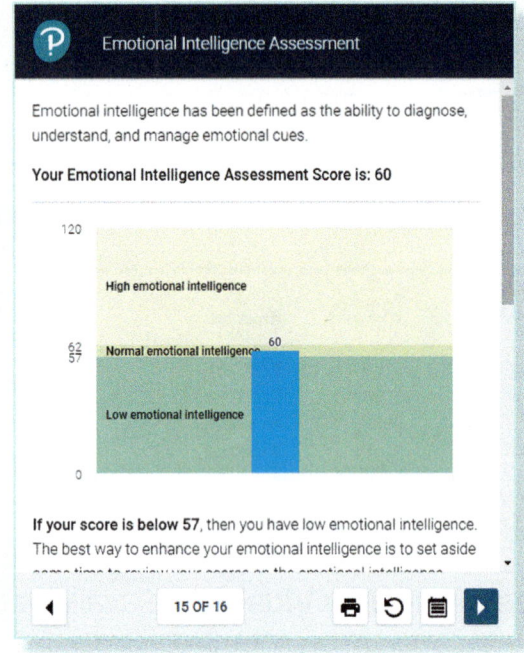

New! Personal Assessment Inventories is a collection of online activities designed to promote self-reflection and engagement in students.

New! Mindset videos encourage students to maintain a positive attitude about learning, value their own ability to grow, and view mistakes as learning opportunities—so often a hurdle for math students.

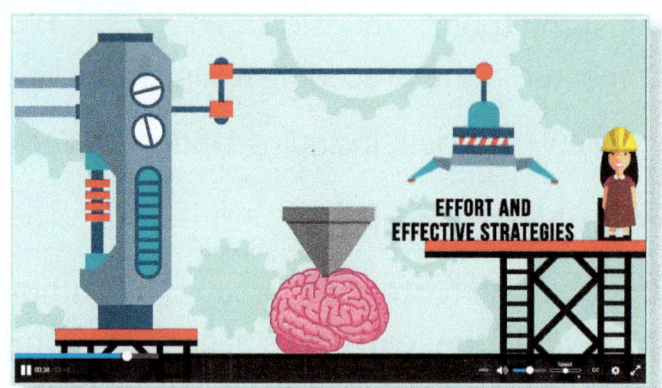

Preparing for Calculus Exercises

Mastering prerequisite content requires both a positive growth mindset *and* an understanding of why algebra is needed.

New! Preparing for Calculus Exercises place skills from each chapter in calculus contexts to demonstrate the importance of algebra for future calculus courses. Exercises are assignable in MyLab Math.

pearson.com/mylab/math

Resources for Success

Instructor Resources
Online resources can be downloaded at pearson.com/mylab/math MyLab Math or from www.pearson.com.

Instructor's Edition
ISBN - 0135757770 / 9780135757772
The instructor's edition includes all answers to the exercise sets. Shorter answers are presented on the same page as the exercise; longer answers are in the back of the text.

Instructor's Solution Manual
Includes fully worked solutions to all exercises in the text (download only).

Learning Catalytics Question Library
Questions written for this text are available to deliver through Learning Catalytics to engage students in your course. Search "Harshbarger" within Learning Catalytics to view questions.

PowerPoint® Lecture Slides
Written specifically for this text, these fully editable lecture slides provide an outline for presenting definitions, figures, and key examples. Accessible PowerPoints, which are designed to be used with screen readers, are also available.

Instructor's Testing Manual
This manual contains three alternative forms of tests per chapter with answer keys that contain more applications (download only).

TestGen®
TestGen (www.pearsoned.com/testgen) enables instructors to build, edit, print, and administer tests using a computerized bank of questions developed to cover all the objectives of the text.

Student Resources
Additional resources to enhance student success.

Corequisite Notebook:
New! The Corequisite Notebook, created by author Lisa Yocco, expands on the Algebra Toolbox topics within the text, which align to the Integrated Review videos/objectives found in MyLab Math. Students actively participate in learning the *how* and *why* of fundamental topics to ensure they are prepared for their college-level work.

Additional Skill and Exercise Manual
This manual provides additional practice and test preparation for students (download only).

Student's Solutions Manual
ISBN - 0135757665 / 9780135757666
Provides detailed worked-out solutions to odd-numbered exercises.

pearson.com/mylab/math

1

Functions, Graphs, and Models; Linear Functions

With digital TV becoming more affordable by the day, the demand for high-definition home entertainment is growing rapidly, with 180 million Americans having digital TV in 2019. The worldwide total was 553 million in 2018. Cell phone use is also on the rise, not just in the United States but throughout the world. By the middle of 2014, the number of subscribers to cell phone carriers had dramatically increased, and the number of total users had reached 4.7 billion. If the numbers continue to increase at a steady rate, the number of subscribers is expected to reach into the tens of billions over the next few years. These projections and others are made by collecting real-world data and creating mathematical models. The goal of this chapter, and future chapters, is to use real data and mathematical models to make predictions and solve meaningful problems.

sections	objectives	applications
1.1 Functions and Models	Determine graphs, tables, and equations that represent functions; find domains and ranges; evaluate functions and mathematical models; align data	Body temperature, female physicians, stock market, U.S. industrial shipments, older men in the workforce, medical 3-D printing
1.2 Graphs of Functions	Graph and evaluate functions with technology; graph mathematical models; graph data points; scale data	Global EV/PHEV sales, aging workers, cost-benefit, U.S. diabetes
1.3 Linear Functions	Identify and graph linear functions; find and interpret intercepts and slopes; find constant rates of change; model revenue, cost, and profit; find marginal revenue, marginal cost, and marginal profit; identify special linear functions	Hispanics in the United States, loan balances, revenue, cost, profit, marginal cost, marginal revenue, marginal profit
1.4 Equations of Lines	Write equations of lines; identify parallel and perpendicular lines; find average rates of change; compute the difference quotient; model approximately linear data	Service call charges, depreciation, prison population, telecom artificial intelligence, public school enrollment, average velocity

Algebra TOOLBOX

KEY OBJECTIVES

- Write sets of numbers using description or elements
- Find intersection and union of sets
- Identify sets of real numbers as being integers, rational numbers, and/or irrational numbers
- Calculate with real numbers
- Use properties of real numbers
- Express inequalities as intervals and graph inequalities
- Identify the coefficients of terms and constraints in algebraic expressions
- Evaluate algebraic expressions
- Combine like terms
- Remove parentheses and simplify expressions
- Solve basic linear equations
- Plot points on a coordinate system
- Use subscripts to represent fixed points

The Algebra Toolbox is designed to review prerequisite skills needed for success in each chapter. In this Toolbox, we discuss sets, the real numbers, the coordinate system, algebraic expressions, equations, inequalities, absolute values, and subscripts.

Sets

In this chapter we will use sets to write domains and ranges of functions, and in future chapters we will find solution sets to equations and inequalities. A **set** is a well-defined collection of objects including, but not limited to, numbers. In this section, we will discuss sets of real numbers, including natural numbers, integers, and rational numbers, and later in the text we will discuss the set of complex numbers. There are two ways to define a set. One way is by listing the **elements** (or **members**) of the set (usually between braces). For example, we may say that a set A contains 2, 3, 5, and 7 by writing $A = \{2, 3, 5, 7\}$. To say that 5 is an element of the set A, we write $5 \in A$. To indicate that 6 is not an element of the set, we write $6 \notin A$. Domains of functions and solutions to equations are sometimes given in sets with the elements listed.

If all the elements of the set can be listed, the set is said to be a **finite set**. If all elements of a set cannot be listed, the set is called an **infinite set**. To indicate that a set continues with the established pattern, we use three dots. For example, $B = \{1, 2, 3, 4, 5, \ldots, 100\}$ describes the finite set of whole numbers from 1 through 100, and $N = \{1, 2, 3, 4, 5, \ldots\}$ describes the infinite set of all whole numbers beginning with 1. This set is called the **natural numbers**.

Another way to define a set is to give its description. For example, we may write $\{x \mid x \text{ is a math book}\}$ to define the set of math books. This is read as "the set of all x such that x is a math book." $N = \{x \mid x \text{ is a natural number}\}$ defines the set of natural numbers, which was also defined by $N = \{1, 2, 3, 4, 5, \ldots\}$ above.

The set that contains no elements is called the **empty set** and is denoted by $\emptyset$.

EXAMPLE 1 ▶ Sets

Write the following sets in two ways.

a. The set A containing the natural numbers less than 7.

b. The set B of natural numbers that are at least 7.

SOLUTION

a. $A = \{1, 2, 3, 4, 5, 6\}, A = \{x \mid x \in N, x < 7\}$

b. $B = \{7, 8, 9, 10, \ldots\}, B = \{x \mid x \in N, x \geq 7\}$

The relations that can exist between two sets follow.

> **Relations Between Sets**
>
> 1. Sets X and Y are **equal** if they contain exactly the same elements.
> 2. Set A is called a **subset** of set B if each element of A is an element of B. This is denoted $A \subseteq B$.

3. If sets C and D have no elements in common, they are called **disjoint**.
4. The set containing the elements that are common to two sets is said to be the **intersection** of the two sets. The intersection of A and B is written $A \cap B$.
5. The **union** of two sets is the set that contains all the elements of both sets. The union of A and B is written $A \cup B$.

EXAMPLE 2 ▶ Relations Between Sets

For the sets $A = \{x \mid x \leq 9, x \text{ is a natural number}\}, B = \{2, 4, 6\}, C = \{3, 5, 8, 10\}$:

a. Which of the sets A, B, and C are subsets of A?

b. Which pairs of sets are disjoint?

c. Are any of these three sets equal?

SOLUTION

a. Every element of B is contained in A. Thus, set B is a subset of A. Because every element of A is contained in A, A is a subset of A.

b. Sets B and C have no elements in common, so they are disjoint.

c. None of these sets have exactly the same elements, so none are equal.

EXAMPLE 3 ▶ Intersection and Union of Sets

Given sets $A = \{5, 6, 7, 8, 9, 10, 11\}$ and $B = \{6, 8, 10, 12, 14, 16\}$, find

a. the intersection of A and B

b. the union of A and B

SOLUTION

a. The intersection of A and B, $A \cap B$, contains all elements that are common to both sets:

$$A \cap B = \{6, 8, 10\}$$

b. The union of A and B, $A \cup B$, contains all elements that are in both sets:

$$A \cup B = \{5, 6, 7, 8, 9, 10, 11, 12, 14, 16\}$$

The Real Numbers

Because most of the mathematical applications you will encounter in an applied nontechnical setting use real numbers, the emphasis in this text is the **real number system**.* Real numbers can be rational or irrational. **Rational numbers** include integers, fractions containing only integers (with no 0 in a denominator), and decimals that either terminate or repeat. Some examples of rational numbers are

$$-9, \quad \frac{1}{2}, \quad 0, \quad 12, \quad -\frac{4}{7}, \quad 6.58, \quad -7.\overline{3}$$

---

*The complex number system will be discussed in the Chapter 3 Toolbox.

Irrational numbers are real numbers that are not rational. Some examples of irrational numbers are π (a number familiar to us from the study of circles), $\sqrt{2}$, $\sqrt[3]{5}$, and $\sqrt[3]{-10}$.

The types of real numbers are described in Table 1.1.

Table 1.1

Types of Real Numbers	Descriptions
Natural numbers	1, 2, 3, 4, . . .
Integers	Natural numbers, zero, and the negatives of the natural numbers: . . . , $-3, -2, -1, 0, 1, 2, 3, \ldots$
Rational numbers	All numbers that can be written in the form $\dfrac{p}{q}$, where p and q are both integers with $q \neq 0$. Rational numbers can be written as terminating or repeating decimals.
Irrational numbers	All real numbers that are not rational numbers. Irrational numbers cannot be written as terminating or repeating decimals.

We can represent real numbers on a **real number line**. Exactly one real number is associated with each point on the line, and we say there is a one-to-one correspondence between the real numbers and the points on the line. That is, the real number line is a graph of the real numbers (see Figure 1.1).

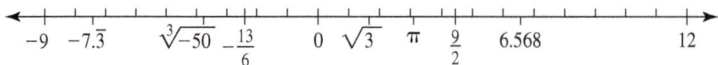

Figure 1.1

Notice the number π on the real number line in Figure 1.1. This special number, which can be approximated by 3.14, results when the circumference of (distance around) any circle is divided by the diameter of the circle. Another special real number is e; it is denoted by

$$e \approx 2.71828$$

We will discuss this number, which is important in financial and biological applications, later in the text.

Calculating with Real Numbers

We use a standard order of operations to calculate expressions when grouping symbols are not clear:

1. Perform all operations inside parentheses or other grouping symbols before removing them.

2. Raise numbers to indicated powers and take indicated roots.

3. Do all multiplications and divisions, in the order they occur from left to right.

4. Do all additions and subtractions, in the order they occur from left to right.

To *add* two signed numbers with the *same sign*, add their absolute values (the numerical values, disregarding the sign), and keep their common sign. For example, $(-4) + (-5) = -9$.

To *add* two signed numbers with *unlike signs*, subtract the smaller absolute value from the larger absolute value, and keep the sign of the number with the larger absolute value. For example, $(-7) + (3) = -(7 - 3) = -4$.

To *subtract* two signed numbers, change the sign of the number being subtracted and proceed as in addition. For example, $(-9) - (-3) = (-9) + 3 = -6$.

To *add three or more signed numbers*, add them two at a time. If no grouping symbols are present, add from left to right. If grouping symbols are present, add inside the grouping symbols first. For example, $-8 + (-6) - 4 - 3 = (-8 + (-6)) - 4 - 3 = -14 - 4 - 3 = -18 - 3 = -21$.

When *multiplying or dividing* two numbers with the *same* sign, the product or quotient is positive. For example, $(-2)(-3) = 6$; $12 \div 3 = 4$

When *multiplying or dividing* two numbers with *unlike* signs, the product or quotient is negative. For example, $(-5)(3) = -15$; $\frac{16}{-2} = -8$.

When *multiplying or dividing more than two signed numbers*, the product will be positive if there is an even number of negative signs. The product will be negative if there is an odd number of negative signs. A number **multiplied by 0** is equal to 0. **Division by 0** is undefined. Zero divided by a nonzero number is defined.

Properties of Real Numbers

1. The *Commutative Property of Addition* states that the sum of two real numbers is the same even if the order of the numbers is changed. That is, addition of real numbers is commutative: $a + b = b + a$ for all real numbers.

2. The *Commutative Property of Multiplication* states that the product of two real numbers is the same even if the order of the numbers is changed. That is, multiplication of real numbers is commutative: $a \cdot b = b \cdot a$ for all real numbers.

3. The *Associative Property of Addition* states that the sum of three numbers is the same if the first pair or the last pair is added first. That is, $a + (b + c) = (a + b) + c$.

4. The *Associative Property of Multiplication* states that the product of three numbers is the same if the first pair or the last pair is multiplied first. That is, $a \cdot (b \cdot c) = (a \cdot b) \cdot c$.

5. The *Distributive Property of Multiplication over Addition* states that multiplying a sum of two numbers by a third number gives the same result as multiplying the third number by each of the two numbers in the sum and adding the two products. That is, $a \cdot (b + c) = a \cdot b + a \cdot c$.

6. 0 is the *additive identity*. For any real number a, $a + 0 = a$.

7. 1 is the *multiplicative identity*. For any real number a, $a \cdot 1 = a$.

8. For every real number a, there exists a number b such that $a + b = 0$. The *additive inverse* of a is $-a$.

9. For every real number $a \neq 0$, there exists a number b such that $a \cdot b = 1$. The *multiplicative inverse* of a is also called the *reciprocal* of a and denoted as $\frac{1}{a}$.

Inequalities and Intervals on the Number Line

In this chapter, we will sometimes use inequalities and interval notation to describe domains and ranges of functions. An **inequality** is a statement that one quantity is greater (or less) than another quantity. We say that a is less than b (written $a < b$) if the point representing a is to the left of the point representing b on the real number line. We may indicate that the number a is greater than or equal to b by writing $a \geq b$. The subset of real numbers x that lie between a and b (excluding a and b) can be denoted by the **double inequality** $a < x < b$ or by the **open interval** (a, b). This is called an open interval because neither of the endpoints is included in the interval. The **closed interval** $[a, b]$ represents the set of all real numbers satisfying $a \leq x \leq b$. Intervals containing one endpoint, such as $[a, b)$ or $(a, b]$, are called **half-open intervals**. We can represent the inequality $x \geq a$ by the interval $[a, \infty)$, and we can represent the inequality $x < a$ by the interval $(-\infty, a)$. Note that ∞ and $-\infty$ are not numbers, but ∞ is used in $[a, \infty)$ to represent the fact that x increases without bound and $-\infty$ is used in $(-\infty, a)$ to indicate that x decreases without bound. Table 1.2 shows the graphs of different types of intervals.

Table 1.2

Interval Notation	Inequality Notation	Verbal Description	Number Line Graph
(a, ∞)	$x > a$	x is greater than a	
$[a, \infty)$	$x \geq a$	x is greater than or equal to a	
$(-\infty, b)$	$x < b$	x is less than b	
$(-\infty, b]$	$x \leq b$	x is less than or equal to b	
(a, b)	$a < x < b$	x is between a and b, not including either a or b	
$[a, b)$	$a \leq x < b$	x is between a and b, including a but not including b	
$(a, b]$	$a < x \leq b$	x is between a and b, not including a but including b	
$[a, b]$	$a \leq x \leq b$	x is between a and b, including both a and b	

Note that open circles may be used instead of parentheses, and solid circles may be used instead of brackets, in the number line graphs.

EXAMPLE 4 ▶ Intervals

Write the interval corresponding to each of the inequalities in parts (a)–(e), and then graph the inequality.

a. $-1 \leq x \leq 2$ **b.** $2 < x < 4$ **c.** $-2 < x \leq 3$ **d.** $x \geq 3$ **e.** $x < 5$

SOLUTION

a. $[-1, 2]$

b. $(2, 4)$

c. $(-2, 3]$

d. $[3, \infty)$

e. $(-\infty, 5)$

Algebraic Expressions

In algebra we deal with a combination of real numbers and letters. Generally, the letters are symbols used to represent unknown quantities or fixed but unspecified constants. Letters representing unknown quantities are usually called **variables**, and letters representing fixed but unspecified numbers are called **literal constants**. An expression created by performing additions, subtractions, or other arithmetic operations with one or more real numbers and variables is called an **algebraic expression**. Unless otherwise specified, the variables represent real numbers for which the algebraic expression is a real number. Examples of algebraic expressions include

$$5x - 2y, \quad \frac{3x - 5}{12 + 5y}, \quad \text{and} \quad 7z + 2$$

A **term** of an algebraic expression is the product of one or more variables and a real number; the real number is called a **numerical coefficient** or simply a **coefficient**. A constant is also considered a term of an algebraic expression and is called a **constant term**. For instance, the term $5yz$ is the product of the factors 5, y, and z; this term has coefficient 5.

Evaluating Algebraic Expressions

If we give a specific value to a variable, we can **evaluate an algebraic expression**. To evaluate an algebraic expression means to find its numerical value after we know the values of the variables. Algebraic expressions are often found in real-life situations. For example, $2l + 2w$ represents the perimeter of a rectangle where l is the length and w is the width.

EXAMPLE 5 ▶ Algebraic Expressions

a. State the terms of the algebraic expression $3x - 7yz + 4$.

b. State the factors of the second term in part (a).

c. Evaluate the expression in part (a) if $x = -2, y = 4$, and $z = \frac{1}{2}$.

d. If the length of a rectangle is 17 feet and its width is 9 feet, find the perimeter of the rectangle.

SOLUTION

a. The terms of the algebraic expression are $3x$, $-7yz$, and 4.

b. The factors of the term $-7yz$ are -7, y, and z.

c. Substituting -2 for x, 4 for y, and $\frac{1}{2}$ for z gives

$$3(-2) - 7(4)\left(\frac{1}{2}\right) + 4 = -6 - 14 + 4 = -16$$

d. Substituting 17 for l and 9 for w in the algebraic expression $2l + 2w$ gives

$$2(17) + 2(9) = 52$$

so the perimeter is 52 feet.

Combining Like Terms

Terms that contain exactly the same variables with exactly the same exponents are called **like terms**. For example, $3x^2y$ and $7x^2y$ are like terms, but $3x^2y$ and $3xy$ are not. We add or subtract (**combine**) algebraic expressions by combining the like terms.

EXAMPLE 6 ▶ Combining Like Terms

The sum of the expressions $5sx - 2y + 7z^3$ and $2y + 5sx - 4z^3$ is

$$(5sx - 2y + 7z^3) + (2y + 5sx - 4z^3) = 5sx - 2y + 7z^3 + 2y + 5sx - 4z^3$$
$$= 10sx + 3z^3$$

The simplified form of

$$3x^2y + 7xy^2 + 6x^2y - 4xy^2 - 5xy \text{ is } 9x^2y + 3xy^2 - 5xy$$

Removing Parentheses

We often need to remove parentheses when simplifying algebraic expressions and when solving equations. Removing parentheses frequently requires use of the **distributive property**, which says that for real numbers *a*, *b*, and *c*, $a(b + c) = ab + ac$. Care must be taken to avoid mistakes with signs when using the distributive property. Multiplying a sum in parentheses by a negative number changes the sign of each term in the parentheses.

EXAMPLE 7 ▶ Removing Parentheses and Simplifying Expressions

a. Use the distributive property to remove the parentheses in $-3(x - 2y)$ and $-(3xy - 5x^3)$.

b. Find the difference $(5sx - 2y + 7z^3) - (2y + 5sx - 4z^3)$.

SOLUTION

a. $-3(x - 2y) = -3(x) + (-3)(-2y) = -3x + 6y$
$-(3xy - 5x^3) = (-1)(3xy) + (-1)(-5x^3) = -3xy + 5x^3$

b. $(5sx - 2y + 7z^3) - (2y + 5sx - 4z^3) = 5sx - 2y + 7z^3 - 2y - 5sx + 4z^3$
$= -4y + 11z^3$

Note that removing parentheses preceded by a negative sign results in the sign of every term being changed.

Solving Basic Linear Equations

An **equation** is a mathematical statement that two expressions have equal value. An equal sign "=" is used to equate the two expressions. When an equation contains a variable, finding the value(s) of the variable that make the equation true is called **solving** the equation. A **solution** of an equation is a value for the variable that makes the equation true. For example, 4 is a solution of the equation $x + 7 = 11$ because if x is replaced with 4, the equation is true.

We can frequently solve an equation for a given variable by rewriting the equation in an equivalent form whose solution is easy to find. Two equations are **equivalent** if and only if they have the same solution. The following properties give equivalent equations:

1. *Addition Property of Equations*: Adding the same number to both sides of an equation gives an equivalent equation. For example, the equation

$$x - 8 = 7$$

is equivalent to

$$x - 8 + 8 = 7 + 8$$

or

$$x = 15$$

2. *Subtraction Property of Equations*: Subtracting the same number to both sides of an equation gives an equivalent equation. For example, the equation

$$p + 9 = -2$$

is equivalent to

$$p + 9 - 9 = -2 - 9$$

or

$$p = -11$$

3. *Multiplication Property of Equations*: Multiplying both sides of an equation by the same nonzero number gives an equivalent equation. For example, the equation

$$\frac{y}{4} = -12$$

is equivalent to

$$4\left(\frac{y}{4}\right) = 4(-12)$$

or

$$y = -48$$

4. *Division Property of Equations*: Dividing both sides of an equation by the same nonzero number gives an equivalent equation. For example, the equation

$$-3x = 48$$

is equivalent to

$$\frac{-3x}{-3} = \frac{48}{-3}$$

or

$$y = -16$$

(*Note*: Solving linear equations will be discussed in detail in Chapter 2.)

The Coordinate System

Much of our work in algebra involves graphing. To graph in two dimensions, we use a rectangular coordinate system, or **Cartesian coordinate system**. Such a system allows us to assign a unique point in a plane to each ordered pair of real numbers. We construct the coordinate system by drawing a horizontal number line and a vertical number line so that they intersect at their origins (Figure 1.2). The point of intersection is called the **origin** of the system, the number lines are called the coordinate **axes**, and the plane is divided into four parts called **quadrants**. In Figure 1.3, we call the horizontal axis the *x*-axis and the vertical axis the *y*-axis, and we denote any point in the plane as the ordered pair (x, y).

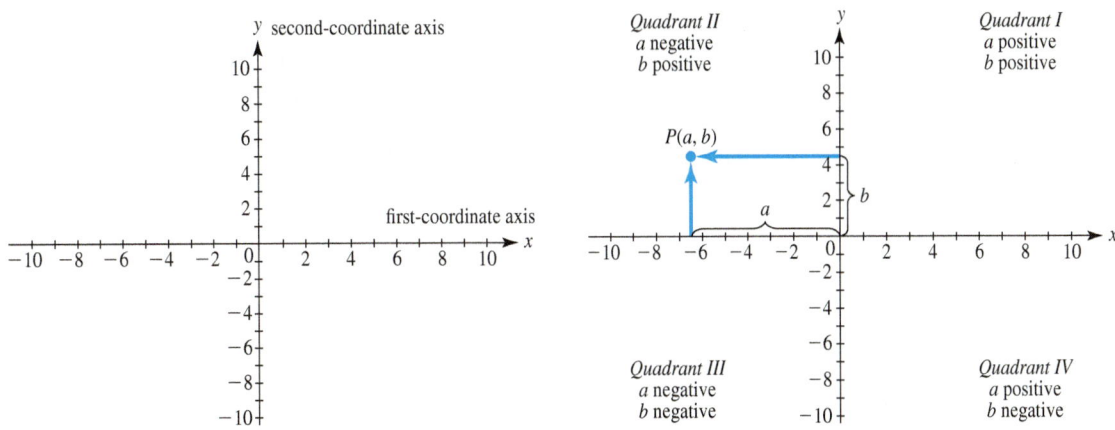

Figure 1.2 **Figure 1.3**

10 Chapter 1 Functions, Graphs, and Models; Linear Functions

The ordered pair (a, b) represents the point P that is $|a|$ units from the y-axis (right if a is positive, left if a is negative) and $|b|$ units from the x-axis (up if b is positive, down if b is negative). The values of a and b are called the **rectangular coordinates** of the point. Figure 1.3 shows point P with coordinates (a, b). The point is in the second quadrant, where $a < 0$ and $b > 0$.

Subscripts

We sometimes need to distinguish between two y-values and/or x-values in the same problem or on the same graph, or to designate literal constants. It is often convenient to do this by using **subscripts**. For example, if we have two fixed but unidentified points on a graph, we can represent one point as (x_1, y_1) and the other as (x_2, y_2). Subscripts also can be used to designate different equations entered in graphing utilities; for example, $y = 2x - 5$ may appear as $y_1 = 2x - 5$ when entered in the equation editor of a graphing calculator.

Toolbox EXERCISES

Answers that are not seen can be found in the answer section at the back of the text.

1. Write "the set of all natural numbers N less than 9" in two different ways. $\{1, 2, 3, 4, 5, 6, 7, 8\}$; $\{x \mid x \in N, x < 9\}$

2. Is it true that $3 \in \{1, 3, 4, 6, 8, 9, 10\}$? Yes

3. Is A a subset of B if $A = \{2, 3, 5, 7, 8, 9, 10\}$ and $B = \{3, 5, 8, 9\}$? No

4. Is it true that $\dfrac{1}{2} \in N$ if N is the set of natural numbers? No

5. Is the set of integers a subset of the set of rational numbers? Yes

6. Are sets of rational numbers and irrational numbers disjoint sets? Yes

7. For sets $A = \left\{-5, -3, -\dfrac{2}{3}, 0, \pi, 4, 7, \dfrac{15}{2}\right\}$ and $B = \{-3, 0, \pi, 5, 8\}$, find
   a. $A \cap B$ $A \cap B = \{-3, 0, \pi\}$
   b. $A \cup B$ $A \cup B = \left\{-5, -3, -\dfrac{2}{3}, 0, \pi, 4, 5, 7, \dfrac{15}{2}, 8\right\}$

Identify the sets of numbers in Exercises 8–10 as containing one or more of the following: integers, rational numbers, and/or irrational numbers.

8. $\{5, 2, 5, 8, -6\}$ Integers, rational

9. $\left\{\dfrac{1}{2}, -4.1, \dfrac{5}{3}, 1\dfrac{2}{3}\right\}$ Rational

10. $\left\{\sqrt{3}, \pi, \dfrac{\sqrt[3]{2}}{4}, \sqrt{5}\right\}$ Irrational

In Exercises 11–21, perform the operations, if possible, and simplify.

11. $(-6) + (-16)$ -22
12. $22 - (-13)$ 35
13. $-56 - (-34)$ -22
14. $(-12) + (-6) - (-3) - 6 + (-14)$ -70
15. $(-7)(-3)(2)$ 42
16. $18 \div (-2) + (-16)(-3)$ 39
17. $\dfrac{35}{-7} + \dfrac{-42}{-6}$ 2
18. $[(-8)(-3) + 5(-2)] \div [3(-4) - (-5)]$ -2
19. $(-3)(0)$ a. 0
20. $\dfrac{0}{-6}$ 0
21. $\dfrac{5}{0}$ Not possible; division by 0 is undefined.

22. Use a commutative property to complete each statement.
    a. $-8 + 6 = 6 + $ ___ -8
    b. $(-9)(3) = (3)$ ___ (-9)

23. Use an associative property to complete each statement.
    a. $8 + (-2) + 3 = 8 + $ ___ $3 + (-2)$
    b. $((-7)(3))(-4) = $ ___ (-4) $((3)(-7))$

24. Use the distributive property to rewrite each expression.
    a. $6(12 + 5)$ $72 + 30$
    b. $-(x + y)$ $-x - y$

25. Is the following statement an example of a commutative property or an associative property, or both?
    $$(3 + 7) + 8 = 7 + (3 + 8)$$ Both

26. Use an identity property to complete each statement.
    a. $-10 + \_\_ = -10$ 0
    b. $\_\_ \cdot (3) = 3$ 1

27. Complete the statement.
    a. $-4 + \_\_ = 0$ 4
    b. $7 \cdot \_\_ = 1$ $\frac{1}{7}$

In Exercises 28–30, express each interval or graph as an inequality.

28. [graph at -3, open] $x > -3$
29. $[-3, 3]$ $-3 \leq x \leq 3$
30. $(-\infty, 3]$ $x \leq 3$

In Exercises 31–33, express each inequality or graph in interval notation.

31. $x \leq 7$ $(-\infty, 7]$
32. $3 < x \leq 7$ $(3, 7]$
33. [graph at 4, open] $(-\infty, 4)$

In Exercises 34–36, graph the inequality or interval on a real number line.

34. $(-2, \infty)$
35. $5 > x \geq 2$
36. $x < 3$

For each algebraic expression in Exercises 37 and 38, give the coefficient of each term and give the constant term.

37. $-3x^2 - 4x + 8$ $-3x^2, -3; -4x, -4;$ constant 8
38. $5x^4 + 7x^3 - 3$ $5x^4, 5; 7x^3, 7;$ constant -3

39. Find the value of the expression $2a - 4b$ when $a = 4$ and $b = -5$. 28

40. Find the value of the expression $-2y + \frac{3}{x} - 5$ when $x = 6$ and $y = -7$. 19/2

41. Find the value of the expression $3(2a - b) + 0.6(b - 3c)$ when $a = -2, b = 4,$ and $c = 0.8$.
    -23.04

42. If $b = 4.5$ in. and $h = 6.5$ in., find the area of a triangle with these dimensions by substituting in the expression $\frac{1}{2}bh$. 14.625 sq in.

43. The expression $16t^2$ gives the distance in feet an object will fall after t seconds (neglecting air resistance). Find the distance an object will fall after 9 seconds. 1296 feet

44. Find the sum of $z^4 - 15z^2 + 20z - 6$ and $2z^4 + 4z^3 - 12z^2 - 5$. $3z^4 + 4z^3 - 27z^2 + 20z - 11$

45. Simplify the expression
    $3x + 2y^4 - 2x^3y^4 - 119 - 5x - 3y^2 + 5y^4 + 110$
    $7y^4 - 2x^3y^4 - 3y^2 - 2x - 9$

Remove the parentheses and simplify in Exercises 46–51.

46. $4(p + d)$ $4p + 4d$
47. $-2(3x - 7y)$ $-6x + 14y$
48. $-a(b + 8c)$ $-ab - 8ac$
49. $4(x - y) - (3x + 2y)$ $x - 6y$
50. $4(2x - y) + 4xy - 5(y - xy) - (2x - 4y)$
    $6x + 9xy - 5y$
51. $2x(4yz - 4) - (5xyz - 3x)$ $3xyz - 5x$

In Exercises 52–57, state the property of equations that can be used to solve each of the following equations; then use the property to solve the equation.

52. $x + 7 = 11$
    Subtraction property; 4
53. $x - 4 = -10$
    Addition property; -6
54. $5x = -20$
    Division property; -4
55. $-2x = 18$
    Division property; -9
56. $\frac{y}{6} = 3$
    Multiplication property; 18
57. $\frac{-p}{4} = -8$
    Multiplication property; 32

In Exercises 58–60, plot the points on a coordinate system.

58. $(-1, 3)$
59. $(4, -2)$
60. $(-4, 3)$

61. Plot the points $(-1, 2), (3, -1), (4, 2),$ and $(-2, -3)$ on the same coordinate system.

62. Plot the points (x_1, y_1) and (x_2, y_2) on a coordinate system if
    $$x_1 = 2, y_1 = -1, x_2 = -3, y_2 = -5$$

1.1 Functions and Models

KEY OBJECTIVES
- Determine if a table, graph, or equation defines a function
- Find the domains and ranges of functions
- Create a scatter plot of a set of ordered pairs
- Use function notation to evaluate functions
- Apply real-world information using a mathematical model

SECTION PREVIEW **Blood Alcohol Percent**

Each alcoholic drink will raise the blood alcohol percent of a 90-pound woman by 0.05. (One drink is equal to 1.25 oz of 80-proof liquor, 12 oz of regular beer, or 5 oz of table wine; many states have set 0.08% as the legal limit for driving under the influence.) Table 1.3 gives the resulting blood alcohol percent for different numbers of alcoholic drinks for such a woman. Figure 1.4(a) shows the graph of the data points from the table, and Figure 1.4(b) shows the points and a line connecting them. The equation of this line is $y = 0.05x$, for values of x from 0 through 9, and we see that the equation fits on the points. We will see that this equation gives y as a **function** of x. Because we can find or estimate the blood alcohol percent for a number of drinks or part of drinks between 0 and 9, we can say that this function is a **model** of the data points for $0 \leq x \leq 9$. See Example 8. ∎

Table 1.3

Number of Drinks	0	1	2	3	4	5	6	7	8	9
Blood Alcohol Percent	0	0.05	0.10	0.15	0.20	0.25	0.30	0.35	0.40	0.45

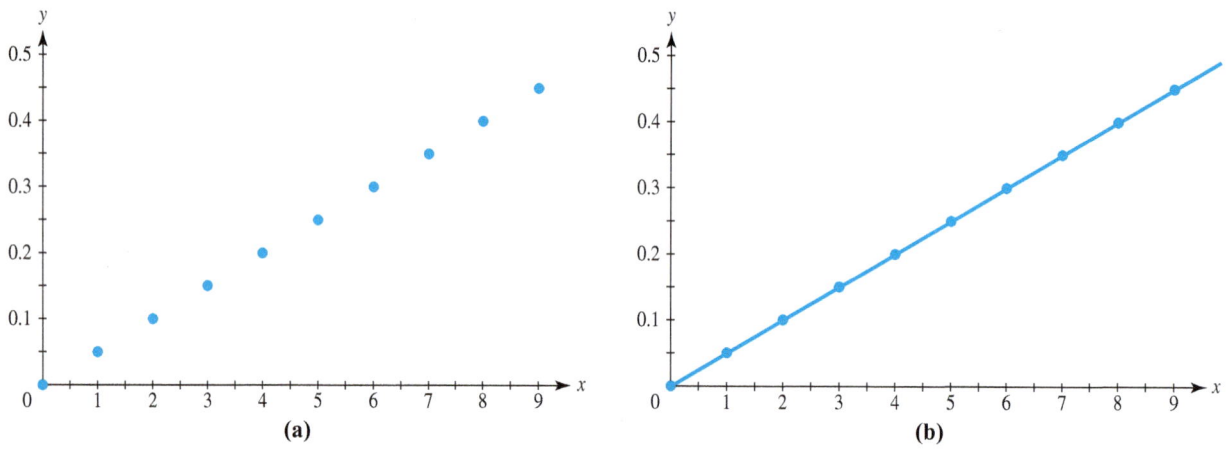

Figure 1.4

Function Definitions

There are several techniques to show how Fahrenheit degree measurements are related to Celsius degree measurements.

One way to show the relationship between Celsius and Fahrenheit degree measurements is by listing some Celsius measurements and the corresponding Fahrenheit measurements. These measurements, and any other real-world information collected in numerical form, are called **data**. These temperature measurements can be shown in a table (Table 1.4).

Table 1.4

Celsius Degrees (°C)	−20	−10	−5	0	25	50	100
Fahrenheit Degrees (°F)	−4	14	23	32	77	122	212

This relationship is also defined by the set of ordered pairs

$$\{(-20, -4), (-10, 14), (-5, 23), (0, 32), (25, 77), (50, 122), (100, 212)\}$$

We can picture the relationship between the measurements with a graph. Figure 1.5 shows a **scatter plot** of the data—that is, a graph of the ordered pairs as points. Table 1.4, the set of ordered pairs below the table, and the graph in Figure 1.5 define a **function** with a set of Celsius temperature **inputs** (called the **domain** of the function) and a set of corresponding Fahrenheit **outputs** (called the **range** of the function). A function that will give the Fahrenheit temperature measurement F that corresponds to *any* Celsius temperature measurement C between $-20°C$ and $100°C$ is the equation

$$F = \frac{9}{5}C + 32$$

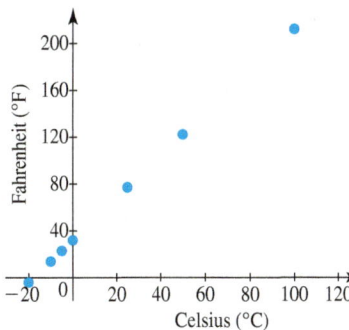

Figure 1.5

This equation defines F as a function of C because each input C results in exactly one output F. Its graph, shown in Figure 1.6, is a line that contains the points on the scatter plot in Figure 1.5 as well as other points. If we consider only Celsius temperatures from -20 to 100, then the domain of this function defined by the equation above is $-20 \leq C \leq 100$ and the resulting range is $-4 \leq F \leq 212$.

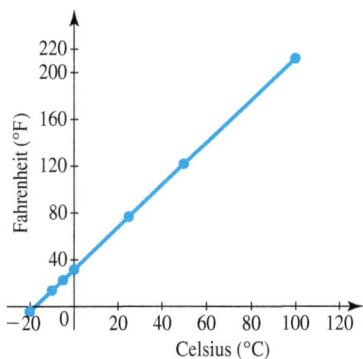

Figure 1.6

Function

A function is a rule or correspondence that assigns to each element of one set (called the domain) exactly one element of a second set (called the range).
   The function may be defined by a set of ordered pairs, a table, a graph, an equation, or a verbal description.

EXAMPLE 1 ▶ Body Temperature

Suppose a child's normal temperature is 98.6°F and the only thermometer available, which is Celsius, indicates that the child's temperature is 37°C. Does this reading indicate that the child's temperature is normal?

SOLUTION

We now have a function that will give the output temperature F that corresponds to the input temperature C. Substituting 37 for C in $F = \frac{9}{5}C + 32$ gives $F = \frac{9}{5}(37) + 32 = 98.6$. This indicates that the child's temperature is normal.

Domains and Ranges

How a function is defined determines its domain and range. For instance, the domain of the function defined by Table 1.4 or by the scatter plot in Figure 1.5 is the finite set $\{-20, -10, -5, 0, 25, 50, 100\}$ with all values measured in degrees Celsius, and the range is the set $\{-4, 14, 23, 32, 77, 122, 212\}$ with all values measured in degrees Fahrenheit. This function has a finite number of inputs in its domain.

   The inputs of the function defined by $F = \frac{9}{5}C + 32$ and graphed in Figure 1.6 are all real numbers from -20 to 100, meaning the temperature could be a number like 10.2 or -12.84. This set of numbers is the domain and can be represented by the inequality $-20 \leq C \leq 100$ or the interval $[-20, 100]$. The outputs of this function are all real numbers from -4 to 212 and can be represented by the inequality $-4 \leq F \leq 212$ or the interval $[-4, 212]$. Functions defined by equations can also be restricted by the context in which they are used. For example, if the function

14 Chapter 1 Functions, Graphs, and Models; Linear Functions

$F = \frac{9}{5}C + 32$ is used in measuring the temperature of water, its domain is limited to real numbers from 0 to 100 and its range is limited to real numbers from 32 to 212, because water changes state with other temperatures.

If x represents any element in the domain, then x is called the **independent variable**, and if y represents an output of the function from an input x, then y is called the **dependent variable**. The figure at left shows a general "function machine" in which the input is called x, the rule is denoted by f, and the output is symbolized by $f(x)$. The symbol $f(x)$ is read "f of x."

If the domain of a function is not specified or restricted by the context in which the function is used, it is assumed that the domain consists of all real number inputs that result in real number outputs in the range, and that the range is a subset of the real numbers. Sometimes domains of functions are limited to values that produce defined values, such as functions with variables in the denominator or even-root functions. These functions will be discussed later.

We can use the graph of a function to find or to verify its domain and range. We can usually see the interval(s) on which the graph exists and therefore agree on the subset of the real numbers for which the function is defined. This set is the domain of the function. We can also usually determine if the outputs of the graph form the set of all real numbers or some subset of real numbers. This set is the range of the function.

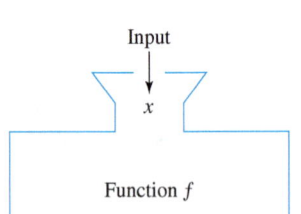

Function Machine

EXAMPLE 2 ▶ Domains and Ranges from Graphs

For each of the following graphs of functions, find the domain and range.

(a)

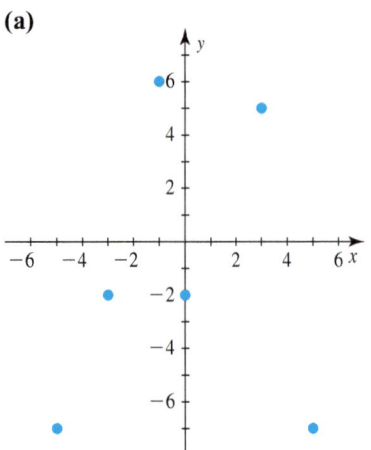

Figure 1.7

Table 1.5

x	y
−5	−7
−3	−2
−1	6
0	−2
3	5
5	−7

(b)

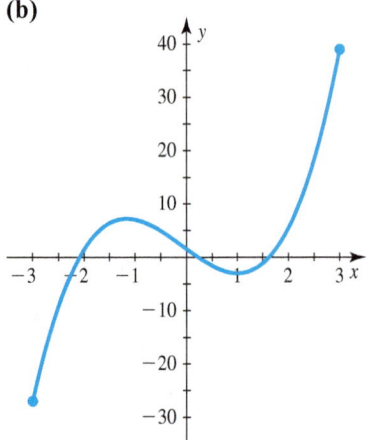

Figure 1.8

(c)

Figure 1.9

SOLUTION

a. The function defined by both Figure 1.7 and Table 1.5 contains the points whose ordered pairs are $(-5, -7), (-3, -2), (-1, 6), (0, -2), (3, 5), (5, -7)$. The domain is the set of all the x-coordinates of the points, $\{-5, -3, -1, 0, 3, 5\}$. The range is the set of all y-coordinates of the points, $\{-7, -2, 6, 5\}$. Note that the y-coordinates -2 and -7 appear twice in the ordered pairs but need to be listed only once in the range.

b. To find the domain of this function, look at all the inputs on the x-axis that correspond to points on the graph. The inputs go from -3 to 3, inclusive. (Because the points at the end of the curve are closed circles, the points are included.) Thus the domain can be written as the inequality $-3 \leq x \leq 3$ or the interval $[-3, 3]$. Similarly, to find the range of the function, look at all the outputs on the y-axis that correspond to points on the graph. The outputs go from -27 to 39, inclusive, so the range can be written as $-27 \leq y \leq 39$ or $[-27, 39]$.

c. The arrow on the left side of the graph indicates that the graph continues indefinitely in that direction, so the inputs will be all real numbers less than or equal to 4. Thus the domain is $x \leq 4$ or the interval $(-\infty, 4]$. The outputs are all real numbers greater than or equal to 0, so the range is $y \geq 0$ or the interval $[0, \infty)$.

Arrow diagrams that show how each individual input results in exactly one output can also represent functions. Each arrow in Figure 1.10(a) and in Figure 1.10(c) goes from an input to exactly one output, so each of these diagrams defines a function. On the other hand, the arrow diagram in Figure 1.10(b) does not define a function because one input, 8, goes to two different outputs, 6 and 9.

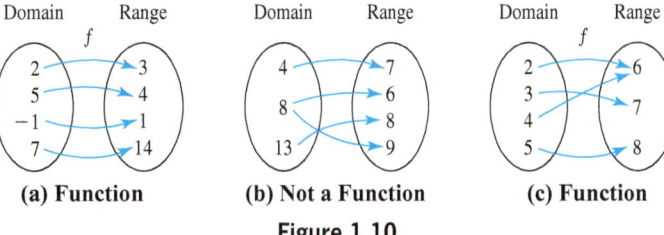

Figure 1.10

Tests for Functions

Functions play an important role in the solution of mathematical problems. Also, to graph a relationship using graphing calculators and computer software, it is usually necessary to express the association between the variables in the form of a function. It is therefore essential for you to recognize when a relationship is a function. Recall that a function is a rule or correspondence that determines exactly one output for each input.

EXAMPLE 3 ▶ Recognizing Functions

For each of the following, determine whether or not the indicated relationship represents a function. Explain your reasoning. For each function that is defined, give the domain and range.

a. Table 1.6 shows the number of female physicians F (in thousands) in the United States in the year x for selected years from 2015 and projected to 2040. Is F a function of x?

Table 1.6

Year	Female Physicians (thousands)	Year	Female Physicians (thousands)
2015	275.596	2030	374.397
2018	301.061	2035	390.328
2021	325.432	2040	405.175
2025	352.713		

(Source: U.S. Census Bureau)

b. The daily profit P (in dollars) from the sale of x pounds of candy, as shown in Figure 1.11. Is P a function of x?

c. The number of tons x of coal sold determined by the profit P that is made from the sale of the product, as shown in Table 1.7. Is x a function of P?

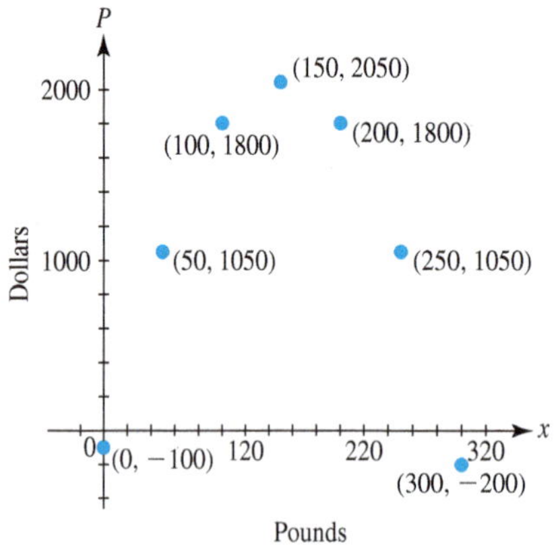

Figure 1.11 — scatter plot with points $(0, -100)$, $(50, 1050)$, $(100, 1800)$, $(150, 2050)$, $(200, 1800)$, $(250, 1050)$, $(300, -200)$; axes: Dollars (P) vs. Pounds (x).

Table 1.7

x (tons)	P ($)
0	−100
50	1050
100	1800
150	2050
200	1800
250	1050
300	−200

d. A is the amount in a person's checking account on a given day n. Is A a function of n?

SOLUTION

a. For each year x (input) listed in Table 1.6, only one value is given for the number of female physicians F (output), so Table 1.6 represents F as a function of x. The domain of this function is the set $\{2015, 2018, 2021, 2025, 2030, 2035, 2040\}$, and the range is the set $\{275.596, 301.061, 325.432, 352.713, 374.397, 390.328, 405.175\}$.

b. Each input x corresponds to only one daily profit P, so this scatter plot represents P as a function of x. The domain is $\{0, 50, 100, 150, 200, 250, 300\}$ pounds, and the range is $\{-200, -100, 1050, 1800, 2050\}$ dollars. For this function, x is the independent variable and P is the dependent variable.

c. The number of tons x of coal sold is not a function of the profit P that is made, because some values of P result in two values of x. For example, a profit of $480,000 corresponds to both 1000 tons of coal and 2000 tons of coal.

d. The amount in a person's checking account on a given day changes as a consequence of deposits, withdrawals, and/or fees. Thus, there is more than one output (amount in the account) for each input (day), and this relationship is not a function.

EXAMPLE 4 ▶ Functions

a. Does the equation $y^2 = 3x - 3$ define y as a function of x?

b. Does the equation $y = -x^2 + 4x$ define y as a function of x?

c. Does the graph in Figure 1.12 give the price of Stratus Building Supply stock as a function of the day for three months in 2015?

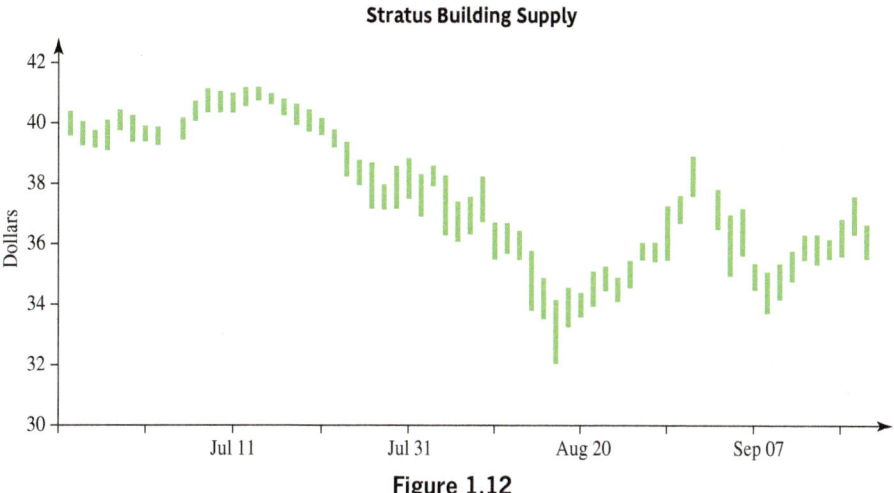

Figure 1.12

SOLUTION

a. This indicated relationship between x and y is not a function because there can be more than one output for each input. For instance, the rule $y^2 = 3x - 3$ determines both $y = 3$ and $y = -3$ for the input $x = 4$. Note that if we solve this equation for y, we get $y = \pm\sqrt{3x - 3}$, so two values of y will result for any value of $x > 1$. In general, if y raised to an even power is contained in an equation, y cannot be solved for uniquely and thus y cannot be a function of another variable.

b. Because each value of x results in exactly one value of y, this equation defines y as a function of x.

c. The graph in Figure 1.12 gives the stock price of Stratus Building Supply for each business day for ten months. The graph shows that the price of a share of stock during each day of these months is not a function. The vertical bar above each day shows that the stock has many prices between its daily high and low. Because of the fluctuation in price during each day of the month, the price of Stratus Building Supply during these months is not a function of the day.

Vertical Line Test

Another way to determine whether an equation defines a function is to inspect its graph. If y is a function of x, no two distinct points on the graph of $y = f(x)$ can have the same first coordinate. There are two points, $(4, 3)$ and $(4, -3)$, on the graph of $y^2 = 3x - 3$ shown in Figure 1.13, so the equation does not represent y as a function of x (as we concluded in Example 4). In general, no two points on its graph can lie on the same vertical line if the relationship is a function.

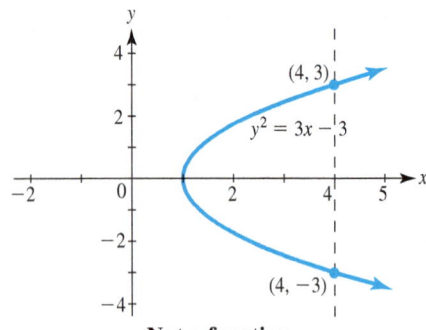

Not a function
Figure 1.13

Vertical Line Test

A set of points in a coordinate plane is the graph of a function if and only if no vertical line intersects the graph in more than one point.

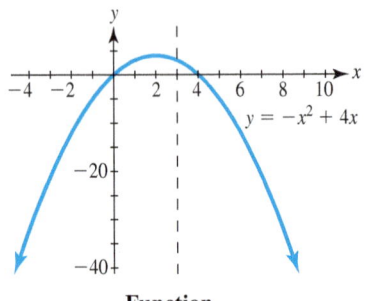

Function
Figure 1.14

When we look at the graph of $y = -x^2 + 4x$ in Figure 1.14, we can see that any vertical line will intersect the graph in at most one point for the portion of the graph that is visible. If we know that the graph will extend indefinitely in the same pattern, we can conclude that no vertical line will intersect the graph in two points and so the equation $y = -x^2 + 4x$ represents y as a function of x.

Function Notation

We can use the function notation $y = f(x)$, read "y equals f of x," to indicate that the variable y is a function of the variable x. For specific values of x, $f(x)$ represents the resulting outputs, or y-values. In particular, the point $(a, f(a))$ lies on the graph of $y = f(x)$ for any number a in the domain of the function. We can also say that $f(a)$ is $f(x)$ evaluated at $x = a$. Thus, if

$$f(x) = 4x^2 - 2x + 3$$

then

$$f(3) = 4(3)^2 - 2(3) + 3 = 33$$
$$f(-1) = 4(-1)^2 - 2(-1) + 3 = 9$$

This means that $(3, 33)$ and $(-1, 9)$ are points on the graph of $f(x) = 4x^2 - 2x + 3$. We can find function values using an equation, values from a table, or points on a graph.

EXAMPLE 5 ▶ U.S. Industrial Shipments

Table 1.8 shows the value (in billions of dollars) of U.S. industrial shipments for selected years from 2014 and projected to 2040.

Table 1.8

Year	Value ($ billions)	Year	Value ($ billions)
2014	7.17	2029	9.62
2017	7.78	2032	10.04
2020	8.35	2035	10.56
2023	8.84	2038	11.11
2026	9.26	2040	11.48

(*Source*: U.S. Department of Energy)

a. If S represents the value of U.S. industrial shipments in billions of dollars for year x, is S a function of x? If so, write it with function notation.

b. Find $f(2035)$ and explain what it means.

c. Find x if $f(x) = 8.84$. What does this mean?

SOLUTION

a. S is a function of x, because each input x is assigned one and only one output, S. So $S = f(x)$ represents the function.

b. In the table, the input 2035 corresponds to 10.56, so $f(2035) = 10.56$. This means that the value of U.S. industrial shipments is projected to be $10.56 billion in 2035.

c. In the table, the output 8.84 corresponds to 2023, so $f(2023) = 8.84$, and $x = 2023$. This means that the value of U.S. industrial shipments is projected to be $8.84 billion in 2023.

EXAMPLE 6 ▶ Function Notation

Figure 1.15 shows the graph of
$$f(x) = 2x^3 + 5x^2 - 28x - 15$$

a. Use the points shown on the graph to find $f(-2)$ and $f(4)$.

b. Use the equation to find $f(-2)$ and $f(4)$.

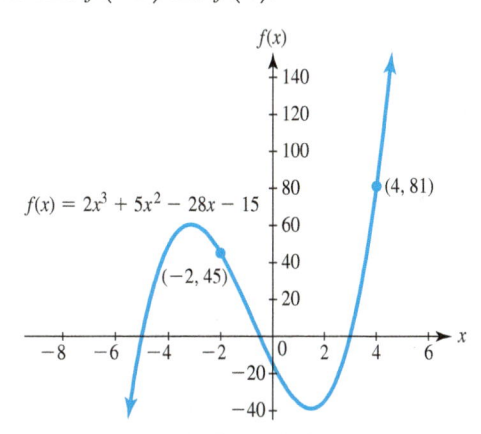

Figure 1.15

SOLUTION

a. By observing Figure 1.15, we see that the point $(-2, 45)$ is on the graph of $f(x) = 2x^3 + 5x^2 - 28x - 15$, so $f(-2) = 45$. We also see that the point $(4, 81)$ is on the graph, so $f(4) = 81$.

b. $f(-2) = 2(-2)^3 + 5(-2)^2 - 28(-2) - 15 = 2(-8) + 5(4) - 28(-2) - 15 = 45$
$f(4) = 2(4)^3 + 5(4)^2 - 28(4) - 15 = 2(64) + 5(16) - 28(4) - 15 = 81$

Note that the values found by substitution agree with the y-coordinates of the points on the graph.

EXAMPLE 7 ▶ Older Men in the Workforce

The points on the graph in Figure 1.16 give the number of older men in the workforce (in millions) as a function g of the year t for selected years from 1950 through the present and projected to 2050.

Figure 1.16

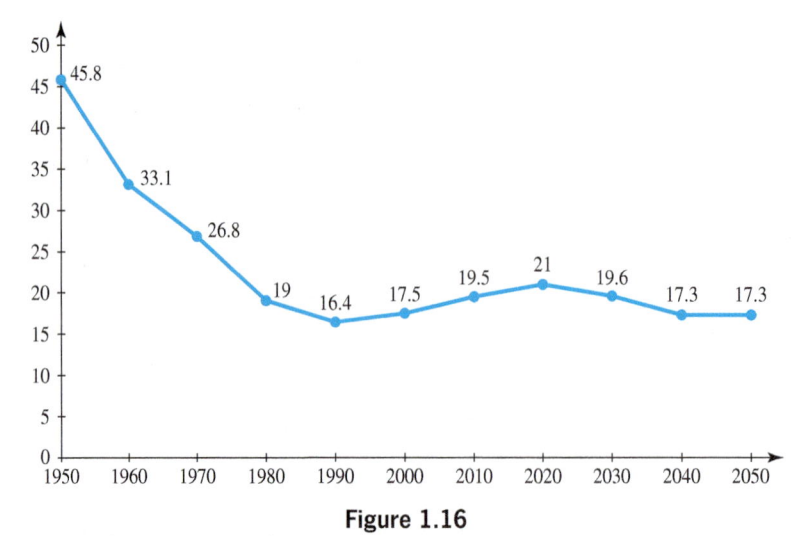

(*Source*: U.S. Department of Labor)

a. Find and interpret $g(2020)$.

b. What is the input t if the output is $g(t) = 19.6$ million older men?

c. What can be said about the number of older men in the workforce during the period 2020–2040?

d. What is the maximum number of older men in the workforce during the period shown on the graph?

SOLUTION

a. The point above $t = 2020$ has coordinates $(2020, 21)$, so $g(2020) = 21$. This means that there were 21 million older men in the workforce in 2020.

b. The point $(2030, 19.6)$ occurs on the graph, so $g(2030) = 19.6$ and $t = 2030$ is the input when the output is $g(t) = 19.6$.

c. The number of older men in the workforce is projected to decrease during the period 2020–2040.

d. The maximum number of older men in the workforce during the period was 45.8 million, in 1950.

Mathematical Models

The process of translating real-world information into a mathematical form so that it can be applied and then interpreted in the real-world setting is called **modeling**. As we most often use the term in this text, a **mathematical model** is a functional relationship (usually in the form of an equation) that includes not only the function rule but also descriptions of all involved variables and their units of measure. For example, the function $F = \frac{9}{5}C + 32$ was used to describe the relationship between temperature scales at the beginning of the section. The **model** that describes how to convert from one temperature scale to another must include the equation $\left(F = \frac{9}{5}C + 32\right)$ and a description of the variables (F is the temperature in degrees Fahrenheit and C is the temperature in degrees Celsius). A mathematical model can sometimes provide an exact description of a real situation (such as the Celsius/Fahrenheit model), but a model frequently provides only an approximate description of a real-world situation.

EXAMPLE 8 ▶ Blood Alcohol Percent

Table 1.9 gives the number of alcoholic drinks and resulting blood alcohol percent for a 90-pound woman. Figure 1.17 shows that the graph of the function $y = 0.05x$ lies on every point of the graph of the data in Table 1.9, so the function is a model of the data. This model can be used to find or accurately estimate the blood alcohol percent from the number of drinks or part of drinks between 0 and 9.

Table 1.9

Number of Drinks	0	1	2	3	4	5	6	7	8	9
Blood Alcohol Percent	0	0.05	0.10	0.15	0.20	0.25	0.30	0.35	0.40	0.45

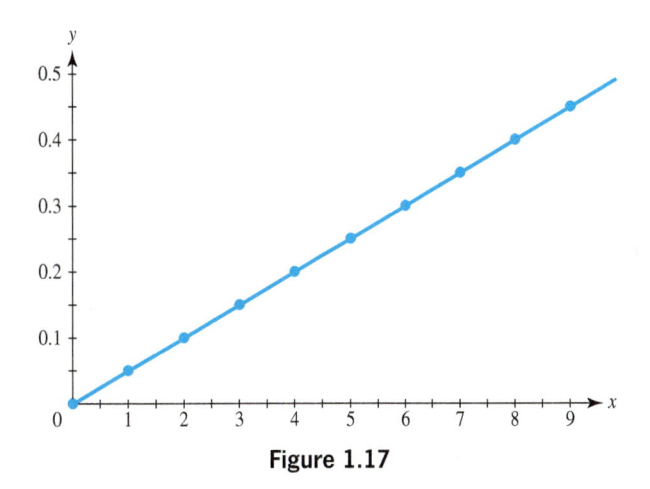

Figure 1.17

a. Use the model to estimate the blood alcohol level of a 90-pound woman who has consumed 1½ drinks.

b. Figure 1.17 shows the model graphed over the domain $0 \leq x \leq 9$. Why is it reasonable to use $x \geq 0$?

SOLUTION

a. We can find the blood alcohol percent after 1½ drinks by evaluating $y = 0.05x$ at $x = 1.5$, which gives $y = 0.075$. Because y is a function of x, we can use function notation to write the solution.

$$f(1.5) = 0.05(1.5) = 0.075 \text{ (percent)}$$

b. In this application, the number of drinks cannot be negative, so $x \geq 0$ is reasonable. The table could go beyond 9 drinks, but it would not be wise for a person to drink anything near 9.

Note that the model in Example 8 is an exact fit for the data points, but many models we find will be approximate fits to the data. Example 9 discusses a model that is not a perfect fit for the data, but that is useful nonetheless.

Aligning Data

When finding a model to fit a set of data, it is often easier to use aligned inputs rather than the actual data values. **Aligned inputs** are simply input values that have been converted to smaller numbers by subtracting the same number from each input. For instance, instead of using t as the actual year in the following example, it is more convenient to use an aligned input that is the number of years after 2010.

EXAMPLE 9 ▶ Medical 3-D Printing

Unlike ink printers, 3-D printers lay down layer after layer of plastic, liquefied metal, or living cells to create three-dimensional objects. Everything from hearing aids, hip replacements, knee replacements, and artificial arms to prescription pills can now be produced with 3-D printers, and 3-D–printed organs are on the horizon. Using data from 2014 and projected to 2024, the value of the medical 3-D printing market $V(t)$ (in millions of dollars) can be modeled (approximately) by the function $V(t) = 3.77t^2 + 24.4t + 105.6$, where t equals the number of years after 2010.

Years	3-D Medical Market Value	Years	3-D Medical Market Value
2014	250.3	2020	711.2
2015	325.4	2021	821.1
2016	394.4	2022	935.4
2017	471.9	2023	1062.5
2018	556.0	2024	1196.0
2019	629.3		

(*Source*: Statista)

a. What value of t represents 2025?

b. What is the projected value of the medical 3-D printing market in 2025?

c. Find $V(17)$ and interpret it in the context of this application.

SOLUTION

a. Because t is the number of years after 2010, $t = 2025 - 2010 = 15$ represents 2025.

b. Evaluating $V(t) = 3.77t^2 + 24.4t + 105.6$ at $t = 15$ gives

$$V(15) = 3.77(15)^2 + 24.4(15) + 105.6 = 1319.9$$

so the projected market value of the medical 3-D printing market in 2025 is approximately $1320 million (or approximately $1.320 billion).

c. $V(17) = 3.77(17)^2 + 24.4(17) + 105.6 = 1609.9$. This means that the projected value of the medical 3-D printing market is approximately $1610 million, or $1.610 billion, in 2027.

Note that in the solution to Example 9, we rounded to one decimal place because the original data were given to one decimal place.

SKILLS CHECK 1.1

Answers that are not seen can be found in the answer section at the back of the text.

Use the tables below in Exercises 1–6.

1. Table A gives y as a function of x, with $y = f(x)$.

   Table A

x	−9	−5	−7	6	12	17	20
y = f(x)	5	6	7	4	9	9	10

   a. Is −5 an input or an output of this function? Input
   b. Is $f(-5)$ an input or an output of this function? Output
   c. State the domain and range of this function.
      $D = \{-9, -5, -7, 6, 12, 17, 20\}; R = \{5, 6, 7, 4, 9, 10\}$
   d. Explain why this relationship describes y as a function of x. Each input gives exactly one output.

2. Table B gives y as a function of x, with $y = g(x)$.

   Table B

x	−4	−1	0	1	3	7	12
y = g(x)	5	7	3	15	8	9	10

   a. Is 0 an input or an output of this function? Input
   b. Is $g(7)$ an input or an output of this function? Output
   c. State the domain and range of this function.
      $D = \{-4, -1, 0, 1, 3, 7, 12\}; R = \{5, 7, 3, 15, 8, 9, 10\}$
   d. Explain why this relationship describes y as a function of x. Each input is assigned exactly one output.

3. Use Table A to find $y = f(-9)$ and $y = f(17)$. 5; 9

4. Use Table B to find $y = g(-4)$ and $y = g(3)$. 5; 8

5. Does Table A describe x as a function of y? Why or why not? No; input 9 for y gives two outputs for x.

6. Does Table B describe x as a function of y? Why or why not? Yes; each input is assigned exactly one output.

7. For each of the functions $y = f(x)$ described below, find $f(2)$.

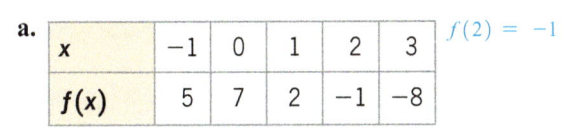

   a.
x	−1	0	1	2	3
f(x)	5	7	2	−1	−8

   $f(2) = -1$

   b. $y = 10 - 3x^2$ $f(2) = -2$

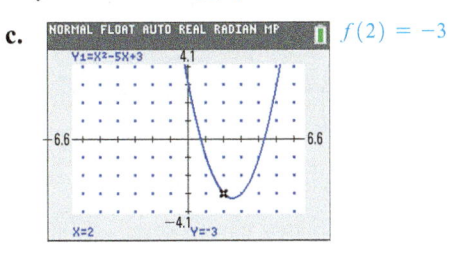

   c. [calculator screen: Y1=X²−5X+3, X=2, Y=−3] $f(2) = -3$

8. For each of the functions $y = f(x)$ described below, find $f(-1)$.

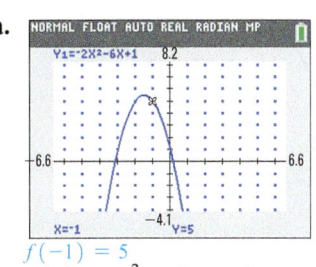

   a. [calculator screen: Y1=−2X²−6X+1, X=−1, Y=5] $f(-1) = 5$

   b.
x	f(x)
−3	12
−1	−8
0	5
3	16

   $f(-1) = -8$

   c. $f(x) = x^2 + 3x + 8$ $f(-1) = 6$

In Exercises 9 and 10, refer to the graph of the function $y = f(x)$ to complete the table.

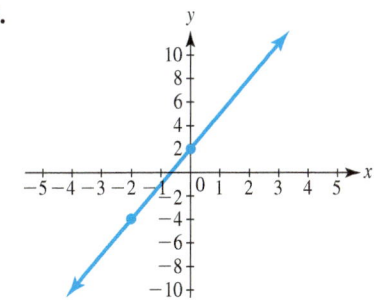

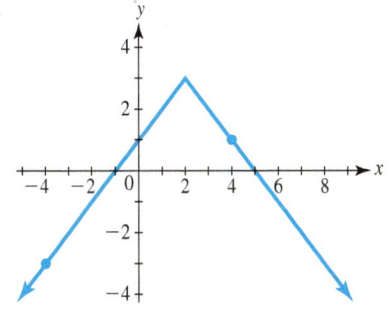

9.
x	y
0	2
−2	−4

10.
x	y
4	1
−4	−3

11. If $R(x) = 5x + 8$, find (a) $R(-3)$, (b) $R(-1)$, and (c) $R(2)$. (a) −7 (b) 3 (c) 18

12. If $C(s) = 16 - 2s^2$, find (a) $C(3)$, (b) $C(-2)$, and (c) $C(-1)$. (a) −2 (b) 8 (c) 14

13. Does the table below define y as a function of x? If so, give the domain and range of f. If not, state why not.

x	−1	0	1	2	3
y	5	7	2	−1	−8

    Yes; $D = \{-1, 0, 1, 2, 3\}; R = \{5, 7, 2, -1, -8\}$

14. Does the table below define y as a function of x? If so, give the domain and range of f. If not, state why not.
    No; input 2 is assigned two different outputs.

x	−3	2	0	3	2
y	12	−3	5	16	4

24 Chapter 1 Functions, Graphs, and Models; Linear Functions

Determine if each graph in Exercises 15–18 indicates that y is a function of x.

15. No

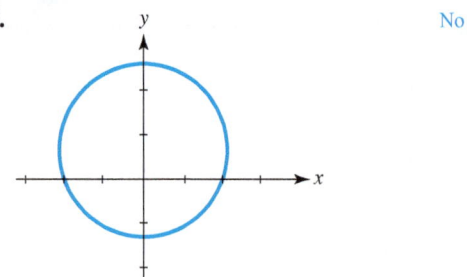

16. Yes

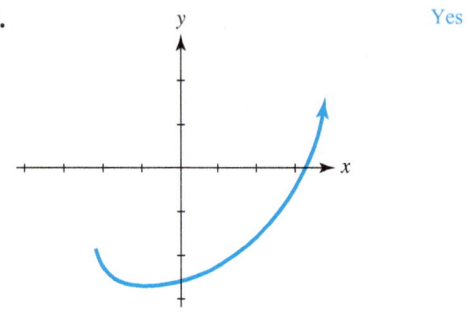

17. Yes

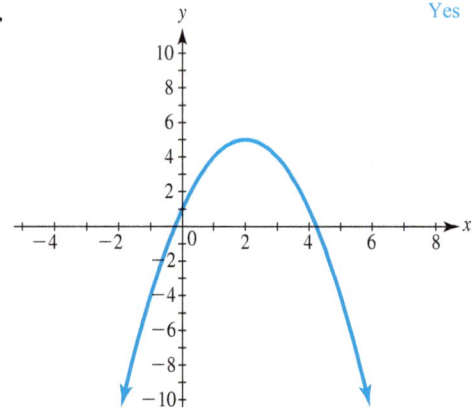

18. No

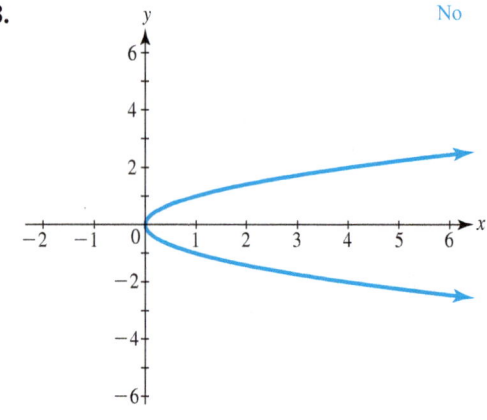

19. Determine whether the graph in the figure represents y as a function of x. Explain your reasoning.
No; one value of x, 3, gives two values of y.

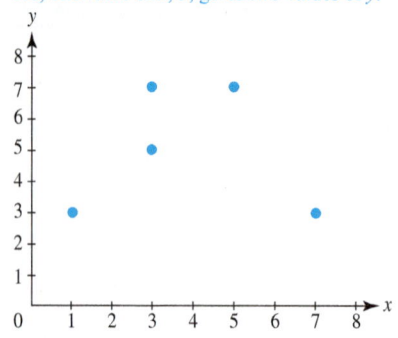

20. Determine whether the graph below represents y as a function of x. Explain your reasoning.
Yes; each input is assigned exactly one output.

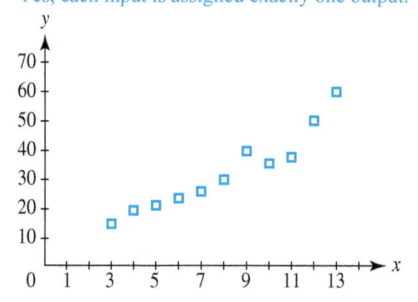

21. Which of the following sets of ordered pairs defines a function? b

  a. $\{(1, 6), (4, 12), (4, 8), (3, 3)\}$
  b. $\{(2, 4), (3, -2), (1, -2), (7, 7)\}$

22. Which of the following sets of ordered pairs defines a function? a

  a. $\{(1, 3), (-2, 4), (3, 5), (4, 3)\}$
  b. $\{(3, 4), (-2, 5), (4, 6), (3, 6)\}$

23. Which of the following arrow diagrams defines a function? b

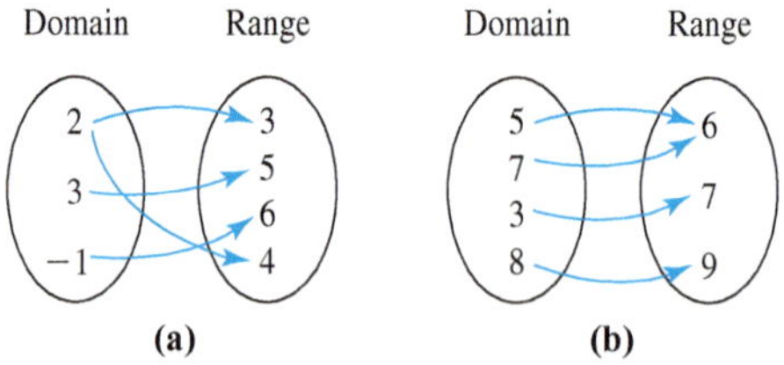

24. Which of the following arrow diagrams defines a function? a

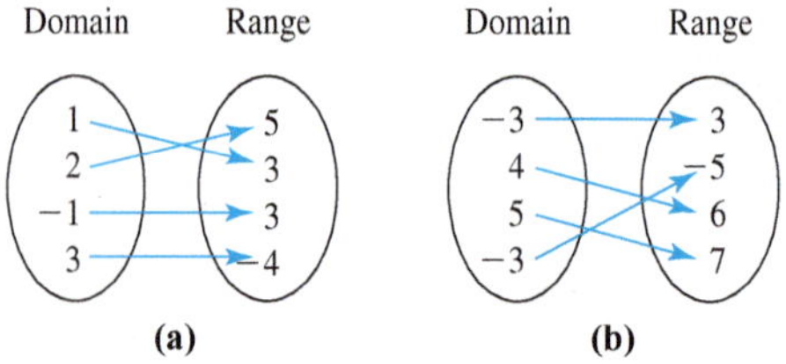

In Exercises 25–30, find the domain and range for the function shown in the graph.

25. 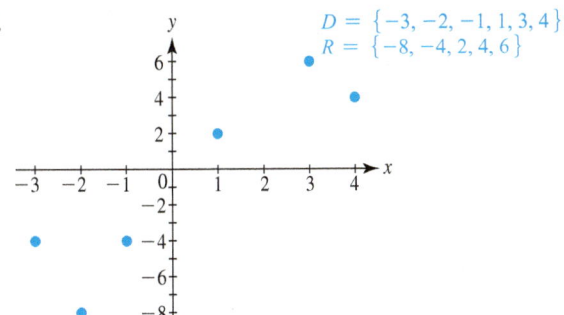 $D = \{-3, -2, -1, 1, 3, 4\}$; $R = \{-8, -4, 2, 4, 6\}$

26. 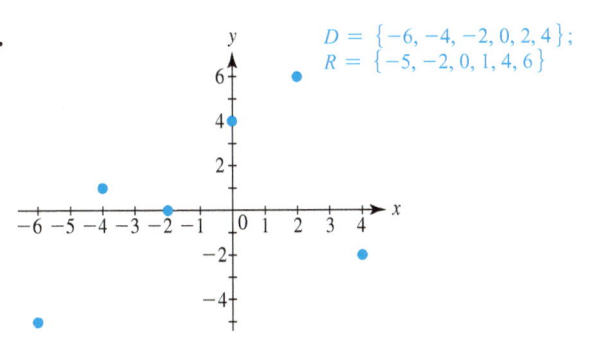 $D = \{-6, -4, -2, 0, 2, 4\}$; $R = \{-5, -2, 0, 1, 4, 6\}$

27. 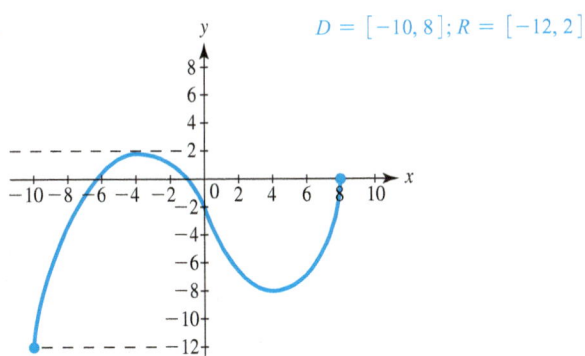 $D = [-10, 8]$; $R = [-12, 2]$

28. 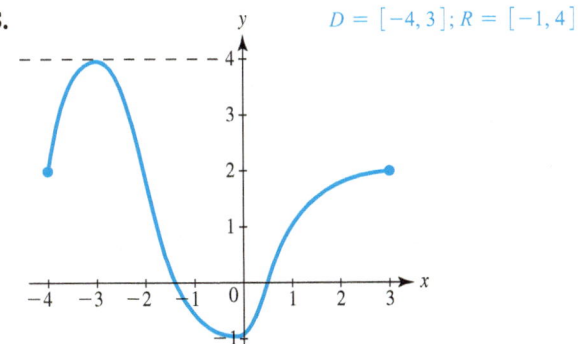 $D = [-4, 3]$; $R = [-1, 4]$

29. 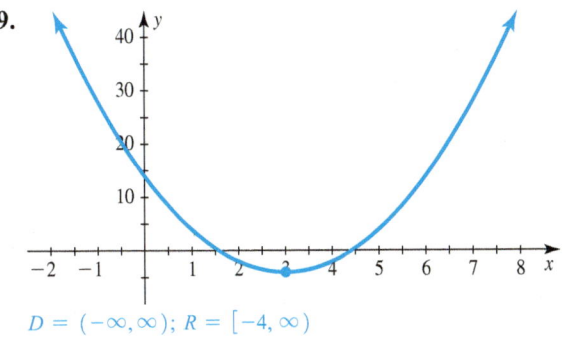 $D = (-\infty, \infty)$; $R = [-4, \infty)$

30. 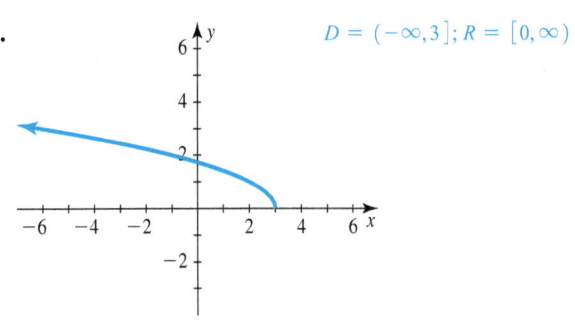 $D = (-\infty, 3]$; $R = [0, \infty)$

31. If a model has inputs aligned so that $x =$ the number of years after 2000, what inputs would be used to represent 2015 and 2022? 15; 22

32. If a set of data from the years 1990 to 2015 is given with a model, and $x =$ the number of years after 1990, what inputs would be used for the given years? 0 to 25

33. Does $x^2 + y^2 = 4$ describe y as a function of x? No

34. Does $x^2 + 2y = 9$ describe y as a function of x? Yes

35. The American Lung Association distributes a Conversion Chart that gives the following steps to convert Celsius temperatures to Fahrenheit temperatures:

  a. Multiply by 9. **b.** Divide by 5. **c.** Add 32.

  If F represents a Fahrenheit temperature and C represents the corresponding Celsius temperature, write an equation that gives the function that describes these steps. $F = \dfrac{9C}{5} + 32$

36. Write an equation to represent the function described by this statement: "The circumference C of a circle is found by multiplying 2π times the radius r." $C = 2\pi r$

37. Say F represents a Fahrenheit temperature and C represents the corresponding Celsius temperature. If $C = \dfrac{5}{9}(F - 32)$ is the function that converts Fahrenheit temperatures to Celsius temperatures for F between 32°F and 212°F, write the steps similar to those used in Exercise 35 to make this conversion.
a. Subtract 32.   b. Multiply by 5.   c. Divide by 9.

38. Write a verbal statement to represent the function $D = 3E^2 - 5$. *D is 3 times the square of E minus 5.*

EXERCISES 1.1

Answers that are not seen can be found in the answer section at the back of the text.

39. **Earnings** The figure shows the median lifetime earnings E in dollars determined by the highest level of educational attainment, in years of education Y. Is E a function of Y? Why?

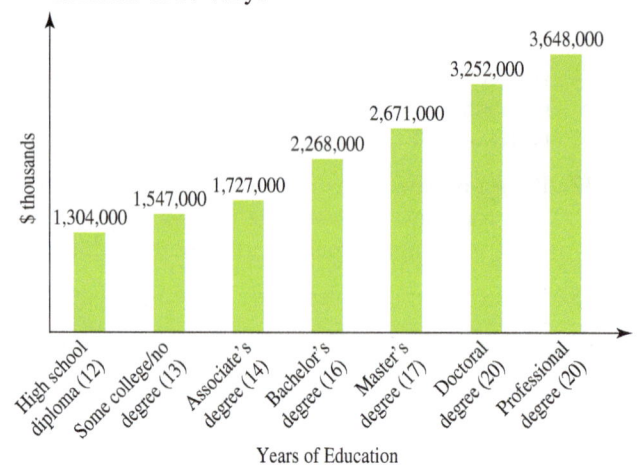

No; there are two different outputs for $Y = 20$.

40. **Obesity** The following table represents the percent p of Americans who are obese (that is, they have body mass index, or BMI, ≥ 30) for selected years y starting with 2000 and projected to 2030. Is the percent a function of the year? Why?
Yes; there is one percent p for each year Y.

Year	2000	2010	2015	2020	2025	2030
% Obese	22.1	30.9	34.5	37.4	39.9	42.2

(*Source:* American Journal of Preventive Medicine)

41. **Dental 3-D Printing** Dental 3-D printing is now being used to create bridges and other products in the dental field. The annual revenue from the sale of printed dental products R, in millions of dollars, for the year x for years from 2014 and projected to 2024 is shown in Table 1.10.

Table 1.10

Year, x	3-D Printer Revenue R ($ million)	Year, x	3-D Printer Revenue R ($ million)
2014	207.3	2020	475.1
2015	240.3	2021	497.0
2016	290.1	2022	515.7
2017	349.3	2023	532.2
2018	414.0	2024	544.8
2019	454.2		

(*Source:* Statista)

a. Is R a function of x? Why or why not?
Yes, because each input x is assigned exactly one output R.
b. If so, what are the domain and range of the function?

42. **Temperature** The figure below shows the graph of $T = 0.43m + 76.8$, which gives the temperature T (in degrees Fahrenheit) inside a concert hall m minutes after a 40-minute power outage during a summer rock concert. State what the variables m and T represent, and explain why T is a function of m.

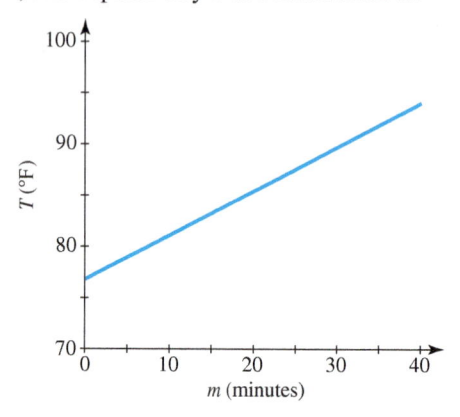

43. **Barcodes** A grocery scanner at Safeway connects the barcode number on a grocery item with the corresponding price.

a. Is the price a function of the barcode? Explain.
Yes; there is one price for each barcode.
b. Is the barcode a function of the price? Explain.
No; numerous items with the same price have different barcodes.

44. **Piano** A child's piano has 12 keys, each of which corresponds to a note.

a. Is the note sounded a function of the key pressed? If it is, how many elements are in the domain of the function? Yes; 12
b. Is the key pressed a function of the note desired? If it is, how many elements are in the range of the function? Yes; 12

45. **Depreciation** A business property valued at $300,000 is depreciated over 30 years by the straight-line method, so its value x years after the depreciation began is

$$V = 300{,}000 - 10{,}000x$$

Explain why the value of the property is a function of the number of years. There is one output for each input.

1.1 Functions and Models

46. *Seawater Pressure* In seawater, the pressure p is related to the depth d according to the model

$$p = \frac{18d + 496}{33}$$

where d is the depth in feet and p is in pounds per square inch. Is p a function of d? Why or why not? Yes; there is one output for each input.

47. *Weight* During the first two weeks in May, a man weighs himself daily (in pounds) and records the data in the table below.

   a. Does the table define weight as a function of the day in May? Yes

   b. What is the domain of this function? The days 1–14 of May

   c. What is the range? {171, 172, 173, 174, 175, 176, 177, 178}

   d. During what day(s) did he weigh most? May 1, May 3

   e. During what day(s) did he weigh least? May 14

   f. He claimed to be on a diet. What is the longest period of time during which his weight decreased? 3 days

May	1	2	3	4	5	6	7
Weight (lb)	178	177	178	177	176	176	175

May	8	9	10	11	12	13	14
Weight (lb)	176	175	174	173	173	172	171

48. *Test Scores*

   a. Is the average score on the final exam in a course a function of the average score on a placement test for the course, if the table below defines the relationship? No

   b. Is the average score on a placement test a function of the average score on the final exam in a course, if the table below defines the relationship? Yes

Average Score on Math Placement Test (%)	81	75	60	90	75
Average Score on Final Exam in Algebra Course (%)	86	70	58	95	81

49. *Car Financing* A couple wants to buy a $35,000 car and can borrow the money for the purchase at 8%, paying it off in 3, 4, or 5 years. The table below gives the monthly payment and total cost of the purchase (including the loan) for each of the payment plans.

t (years)	Monthly Payment ($)	Total Cost ($)
3	1096.78	39,484.08
4	854.46	41,014.08
5	709.68	42,580.80

(*Source:* Sky Financial Mortgage Tables)

Suppose that when the payment is over t years, $P(t)$ represents the monthly payment and $C(t)$ represents the total cost for the car and loan.

   a. Find $P(3)$ and write a sentence that explains its meaning. $1096.78; if the car is financed over 3 years, the payment is $1096.78.

   b. What is the total cost of the purchase if it is financed over 5 years? Write the answer using function notation. $C(5) = \$42,580.80$

   c. What is t if $C(t) = 41,014.08$? 4

   d. How much money will the couple save if they finance the car for 3 years rather than 5 years? $3096.72

50. *Mortgage* A couple can afford $800 per month to purchase a home. As indicated in the table, if they can get an interest rate of 7.5%, the number of years t that it will take to pay off the mortgage is a function of the dollar amount A of the mortgage for the home they purchase.

Amount A ($)	t (years)
40,000	5
69,000	10
89,000	15
103,000	20
120,000	30

(*Source:* Comprehensive Mortgage Tables [Publication No. 492], Financial Publishing Co.)

   a. If the couple wishes to finance $103,000, for how long must they make payments? Write this correspondence in function form if $t = f(A)$. 20 years; $f(103,000) = 20$

   b. What is $f(120,000)$? Write a sentence that explains its meaning. 30; if they borrow $120,000, they must make payments for 30 years.

   c. What is $f(3 \cdot 40,000)$? $f(120,000) = 30$

   d. What value of A makes $f(A) = 5$ true? 40,000

   e. Does $f(3 \cdot 40,000) = 3 \cdot f(40,000)$? Explain your reasoning. No; $30 \neq 3 \cdot 5$

51. *Women in the Workforce* From 1950 and projected to 2050, the number of women in the workforce increased. The points on the figure that follows give the number (in millions) of women in the workforce as a function f of the year for selected years.

   a. Approximately how many women were in the workforce in 2020? 79.3 million

   b. Estimate $f(2050)$ and write a sentence that explains its meaning. 91.5; Approximately 91.5 million women are projected to be in the workforce in 2050.

   c. What is the domain of this function if we consider only the indicated points? {1950, 1960, 1970, 1980, 1990, 2000, 2010, 2020, 2030, 2040, 2050}

d. How does the graph reflect the statement that the number of women in the workforce increases?
For each new indicated year, the number of women increases.

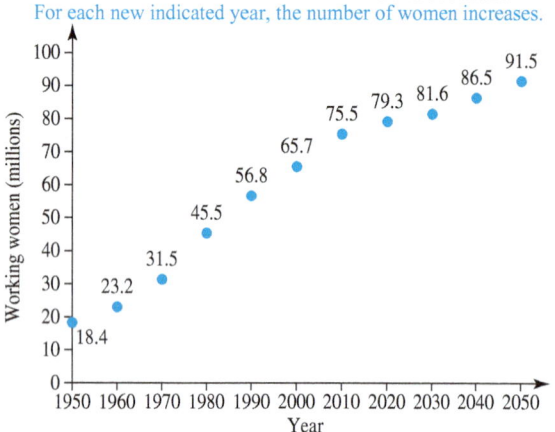

(Source: World Almanac)

52. Working Age The projected ratio of the working-age population (25- to 64-year-olds) to the elderly shown in the figure below defines the ratio as a function of the year shown. If this function is defined as $y = f(t)$ where t is the year, use the graph to answer the following:

a. What is the projected ratio of the working-age population to the elderly population in 2020?
approximately 3 to 1

b. Estimate $f(2005)$ and write a sentence that explains its meaning. 4; 4 is the projected ratio of the working-age population (25- to 64-year-olds) to the elderly in 2005.

c. What is the domain of this function?
{1995, 2000, 2005, 2010, 2015, 2020, 2025, 2030}

d. Is the projected ratio of the working-age population to the elderly increasing or decreasing over the domain shown in the figure? Decreasing

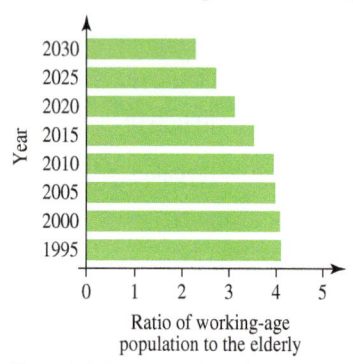

(Source: U.S. Department of Labor)

53. Life Insurance The following table gives the initial monthly premiums for males and females aged 50–54 as a function of the amount of insurance.

a. If the function $f(x)$ is the male's premium for insurance amount x, in dollars, find $f(25,000)$ and $f(75,000)$ in dollars. 14.15; 42.44

b. If the function $g(x)$ is the female's premium for insurance amount x, in dollars, find $g(50,000)$ and $g(100,000)$ in dollars. 12.83; 25.67

c. For what value of x is $f(x) = 28.29$? 50,000

d. Is $f(x)$ greater than $g(x)$ for all x-values? yes

Insurance Amount	Male Premium	Female Premium
$20,000	$11.32	$5.13
25,000	14.15	6.42
50,000	28.29	12.83
75,000	42.44	19.25
100,000	56.58	25.67

(Source: United State Insurance Company)

54. Population of Females Under 18 The graph below gives the population of females under the age of 18, in thousands, projected from 2020 to 2060, as a function f of the year t.

a. What is $f(2030)$? 39,244

b. Interpret that result. The population of females under the age of 18 is projected to be 39,244,000 in 2030.

c. Is this function increasing or decreasing? Increasing

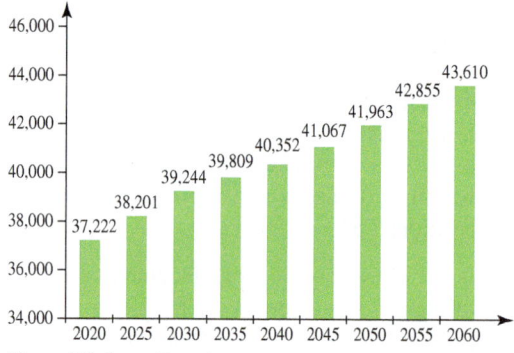

(Source: U.S. Census Bureau)

55. Alzheimer's Disease The figure gives the millions of U.S. citizens age 65 and older with Alzheimer's disease from 2000 and projected to 2050. If the number of millions of these U.S. citizens is the function $f(x)$, where x is the year, use the graph to answer the following.

a. How many U.S. citizens age 65 and older are expected to have Alzheimer's disease in 2020?
5.7 million

b. Find $f(2030)$ and explain its meaning.

c. In what year do we expect 11 million U.S. citizens age 65 and older to have Alzheimer's disease? 2040

d. Is this function increasing or decreasing? increasing

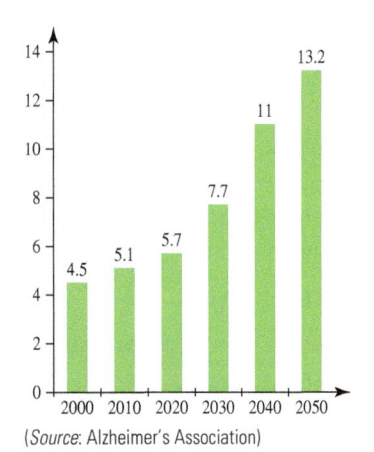

56. Employment in Manufacturing The table shows the total employment (number of jobs N, in millions) in manufacturing for selected years t from 2010 and projected to 2040.

   a. Is the number of jobs a function of the year? yes
   b. What is $f(2025)$ if $N = f(t)$? 12.9
   c. Write a sentence explaining $f(2025) = 12.9$.
   Total employment in manufacturing is projected to be 12.9 million in 2025.
   d. What is t if $f(t) = 11.8$? 2035

Year	Employment
2010	11.5
2015	12.4
2020	12.8
2025	12.9
2030	12.5
2035	11.8
2040	11.0

57. Foreign-Born Population The percent of the U.S. population that is foreign-born for selected years from 2020 and projected to 2060 is shown as a function of the given years in the figure. If this function is defined as $y = f(t)$, where t is the year and y is the percent, answer the following:

   a. What is the projected percent of the U.S. population that is foreign-born for 2030? 15.8%
   b. Find $f(2040)$ and write a sentence that explains its meaning. 17.1; The projected percent of the U.S. population that is foreign-born in 2040 is 17.1%.
   c. In what year is this population projected to be 18.2%? 2050
   d. What is the domain of this function? {2020, 2030, 2040, 2050, 2060}
   e. Is the projected percent of the U.S. population that is foreign-born increasing or decreasing over the domain of the function? increasing

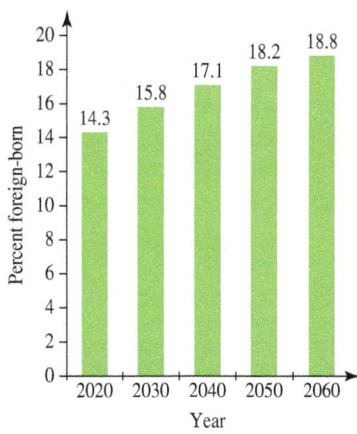

58. Social Security Funding Social Security benefits are funded by individuals who are currently employed. The following graph, based on known data and projections into the future, defines a function that gives the number of workers n supporting each retiree as a function of time t (given by the calendar year). Denote this function by $n = f(t)$.

   a. Find $f(1990)$ and explain its meaning.
   3.4; in 1990, there were 3.4 workers for each retiree.
   b. In what year will the number of workers supporting each retiree equal 2? 2030
   c. What does this function tell us about Social Security in the future? Funding is in jeopardy.

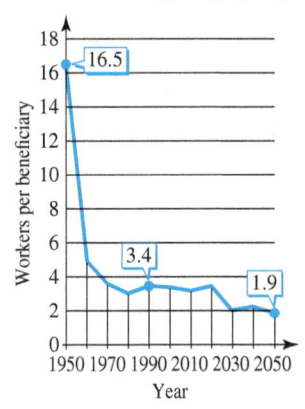

59. Revenue The revenue from the sale of specialty golf hats is given by the function $R(x) = 32x$ dollars, where x is the number of hats sold.

   a. What is $R(200)$? Interpret this result.
   6400; the revenue from the sale of 200 hats is $6400.
   b. What is the revenue from the sale of 2500 hats? Write this in function notation.
   $80,000; $R(2500) = 80,000$

60. Cost The cost of producing specialty golf hats is given by the function $C(x) = 4000 + 12x$, where x is the number of hats produced.

   a. What is $C(200)$? Interpret this result.
   6400; the cost of producing 200 hats is $6400.

b. What is the cost of producing 2500 hats? Write this in function notation. $C(2500) = \$34,000$

61. Utilities An electric utility company determines the monthly bill by charging 85.7 cents per kilowatt-hour (kWh) used plus a base charge of $19.35 per month. Thus, the monthly charge is given by the function

$$f(W) = 0.857W + 19.35 \text{ dollars}$$

where W is the number of kilowatt-hours.

a. Find $f(1000)$ and explain what it means.
876.35; the charge for 1000 kWh is $876.35.
b. What is the monthly charge if 1500 kWh are used? $1304.85

62. Profit The profit from the production and sale of iPod players is given by the function $P(x) = 450x - 0.1x^2 - 2000$, where x is the number of units produced and sold.

a. What is $P(500)$? Interpret this result.
198,000; the profit from the sale of 500 iPods is $198,000.
b. What is the profit from the production and sale of 4000 units? Write this in function notation.
$P(4000) = \$198,000$

63. Profit The daily profit from producing and selling Blue Chief bicycles is given by

$$P(x) = 32x - 0.1x^2 - 1000$$

where x is the number produced and sold and $P(x)$ is in dollars.

a. Find $P(100)$ and explain what it means. 1200; the daily profit for the sale of 100 bicycles is $1200.
b. Find the daily profit from producing and selling 160 bicycles. $1560

64. Projectiles Suppose a ball thrown into the air has its height (in feet) given by the function

$$h(t) = 6 + 96t - 16t^2$$

where t is the number of seconds after the ball is thrown.

a. Find $h(1)$ and explain what it means.
86; the height of the ball after 1 second is 86 ft.
b. Find the height of the ball 3 seconds after it is thrown. 150 ft
c. Test other values of $h(t)$ to decide if the ball eventually falls. When does the ball stop climbing?
After 3 sec

65. Test Reliability If a test that has reliability 0.7 has the number of questions increased by a factor n, the reliability R of the new test is given by

$$R(n) = \frac{0.7n}{0.3 + 0.7n}$$

a. What is the domain of the function defined by this equation? All real numbers except $-\frac{3}{7}$

b. If the application used requires that the size of the test be increased, what values of n make sense in the application? Positive values of n

66. Body-Heat Loss The description of body-heat loss due to convection involves a coefficient of convection K_c, which depends on wind speed s according to the equation $K_c = 4\sqrt{4s + 1}$.

a. Is K_c a function of s? Yes
b. What is the domain of the function defined by this equation? $s \geq -1/4$
c. What restrictions does the physical nature of the model put on the domain? $s \geq 0$

67. Cost–Benefit Suppose that the cost C (in dollars) of removing $p\%$ of the particulate pollution from the smokestack of a power plant is given by

$$C(p) = \frac{237{,}000p}{100 - p}$$

a. Use the fact that percent is measured between 0 and 100 and find the domain of this function.
$0 \leq p < 100$
b. Evaluate $C(60)$ and $C(90)$. 355,500; 2,133,000

68. Demand Suppose the number of units q of a product that is demanded by consumers is given as a function of p by

$$q = \frac{100}{\sqrt{2p + 1}}$$

where p is the price charged per unit.

a. What is the domain of the function defined by this equation? $\left(-\frac{1}{2}, \infty\right)$
b. What should the domain and range of this function be to make sense in the application?
Domain $p \geq 0$, range $q \geq 0$

69. Shipping Restrictions Some shipping restrictions say that in order for a box to be shipped, its length (longest side) plus its girth (distance around the box in the other two dimensions) must be equal to no more than 108 inches.

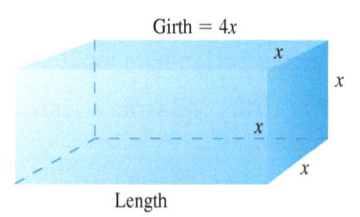

If the box with length plus girth equal to 108 inches has a square cross section that is x inches on each side, then the volume of the box is given by $V = x^2(108 - 4x)$ cubic inches.

a. Find $V(12)$ and $V(18)$. 8640; 11,664
b. What restrictions must be placed on x to satisfy the conditions of this model?
 $0 < x < 27$, so volume is positive and the box exists.
c. Create a table of function values to investigate the value of x that maximizes the volume. What are the dimensions of the box that has the maximum volume?

70. **Height of a Bullet** The height of a bullet shot into the air is given by $S(t) = -4.9t^2 + 98t + 2$, where t is the number of seconds after it is shot and $S(t)$ is in meters.
a. Find $S(0)$ and interpret it.
 2; the initial height of the bullet is 2 m when it is shot.
b. Find $S(9)$, $S(10)$, and $S(11)$. 487.1; 492; 487.1
c. What appears to be happening to the bullet at 10 seconds? Evaluate the function at some additional times near 10 seconds to confirm your conclusion. The bullet seems to reach a maximum height at 10 sec and then begins to fall.

1.2 Graphs of Functions

KEY OBJECTIVES
- Graph equations using the point-plotting method
- Graph equations using graphing calculators
- Graph equations using Excel spreadsheets
- Align inputs and scale outputs to model data
- Graph data points

SECTION PREVIEW Global EV/PHEV Sales

Using data from 2020 and projected to 2050, global sales of electric and plug-in hybrid electric vehicles (EV/PHEV) can be modeled by the function $y = 0.018x^2 + 2.32x - 19.25$, where y is the millions of vehicle sales and x is the number of years after 2010. (*Source*: International Energy Agency)

To graph this function, we can evaluate the function for selected x-values representing the years from 2020 through 2050 (that is, $x = 10$ through $x = 40$), plot the points representing these x- and y-values, and draw a smooth curve connecting the points. We will see that we can also use technology to graph this function, in which case the points determined above can be used to set a **viewing window** for the graph. (See Example 4.)

In this section, we will graph functions by point plotting and with technology. We will graph application functions on windows determined by the context of the applications, and we will align data so that smaller inputs can be used in models. We will also graph data points. ■

Graphs of Functions

If an equation defines y as a function of x, we can sketch the graph of the function by plotting enough points to determine the shape of the graph and then drawing a line or curve through the points. This is called the **point-plotting method** of sketching a graph.

EXAMPLE 1 ▶ Graphing an Equation by Plotting Points

a. Graph the equation $y = x^2$ by drawing a smooth curve through points determined by integer values of x between 0 and 3.

b. Graph the equation $y = x^2$ by drawing a smooth curve through points determined by integer values of x between -3 and 3.

SOLUTION

a. We use the values in Table 1.11 to determine the points. The graph of the function drawn through these points is shown in Figure 1.18(a).

Table 1.11

x	0	1	2	3
$y = x^2$	0	1	4	9
Points	(0, 0)	(1, 1)	(2, 4)	(3, 9)

b. We use Table 1.12 to find the additional points. The graph of the function drawn through these points is shown in Figure 1.18(b).

Table 1.12

x	−3	−2	−1	0	1	2	3
$y = x^2$	9	4	1	0	1	4	9
Points	(−3, 9)	(−2, 4)	(−1, 1)	(0, 0)	(1, 1)	(2, 4)	(3, 9)

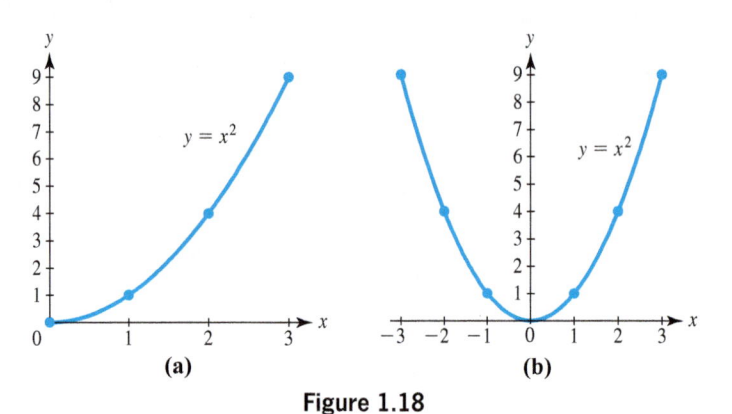

Figure 1.18

Although neither graph in Figure 1.18 shows all of the points satisfying the equation $y = x^2$, the graph in Figure 1.18(b) is a much better representation of the function (as you will learn later). When enough points are connected to determine the shape of the graph and the important parts of the graph and to suggest what the unseen parts of the graph look like, the graph is called **complete**.

Complete Graph

A graph is a complete graph if it shows the basic shape of the graph and important points on the graph (including points where the graph crosses the axes and points where the graph turns)* and suggests what the unseen portions of the graph will be.

EXAMPLE 2 ▶ Graphing a Complete Graph

Sketch the complete graph of the equation $f(x) = x^3 - 3x$, using the fact that the graph has at most two turning points.

SOLUTION

We use the values in Table 1.13 to determine some points. The graph of the function drawn through these points is shown in Figure 1.19.

---

*Points where graphs turn from rising to falling or from falling to rising are called turning points.

Table 1.13

x	−3	−2	−1	0	1	2	3
$y = x^3 - 3x$	−18	−2	2	0	−2	2	18
Points	(−3, −18)	(−2, −2)	(−1, 2)	(0, 0)	(1, −2)	(2, 2)	(3, 18)

Note that this graph has two turning points, so it is a complete graph.

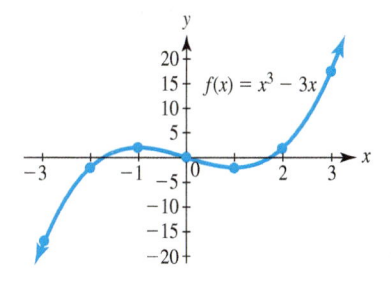

Figure 1.19

The ability to draw complete graphs, with or without the aid of technology, improves with experience and requires knowledge of the shape of the basic function(s) involved in the graph. As you learn more about types of functions, you will be better able to determine when a graph is complete.

Graphing with Technology

Obtaining a sufficient number of points to sketch a graph by hand can be time consuming. Computers and graphing calculators have graphing utilities that can be used to plot many points quickly, thereby producing the graph with minimal effort. However, graphing with technology involves much more than pushing a few buttons. Unless you know what to enter into the calculator or computer (that is, the domain of the function and sometimes the range), you cannot see or use the graph that is drawn. We suggest the following steps for graphing most functions with a graphing calculator.

Using a Graphing Calculator to Draw a Graph

1. Write the function with x representing the independent variable and y representing the dependent variable. Solve for y, if necessary.

2. Enter the function in the equation editor of the graphing utility. Use parentheses* as needed to ensure the mathematical correctness of the expression.

3. Activate the graph by pressing the (ZOOM) or (GRAPH) key. Most graphing utilities have several preset **viewing windows** under (ZOOM), including the **standard viewing window** that gives the graph of a function on a coordinate system in which the x-values range from −10 to 10 and the y-values range from −10 to 10.[†]

4. To see parts of the graph of a function other than those shown in a standard window, press (WINDOW) to set the x- and y-boundaries of the viewing window before pressing (GRAPH). Viewing window boundaries are discussed below and in the technology supplement. As you gain more knowledge of graphs of functions, determining viewing windows that give complete graphs will become less complicated.

Although the standard viewing window is a convenient window to use, it may not show the desired graph. To see parts of the graph of a function other than those that might be shown in a standard window, we change the x- and y-boundaries of the

---

*Parentheses should be placed around numerators and/or denominators of fractions, fractions that are multiplied by a variable, exponents consisting of more than one symbol, and in other places where the order of operations needs to be indicated.

[†]For more information on viewing windows and graphing, see Appendix A, page 689.

viewing window. The values that define the viewing window can be set manually under WINDOW or by using the GRAPH key. The boundaries of a viewing window are

x_{min}: the smallest value on the x-axis (the left boundary of the window)

y_{min}: the smallest value on the y-axis (the bottom boundary of the window)

x_{max}: the largest value on the x-axis (the right boundary of the window)

y_{max}: the largest value on the y-axis (the top boundary of the window)

x_{scl}: the distance between ticks on the x-axis (helps visually find x-intercepts and other points)

y_{scl}: the distance between ticks on the y-axis (helps visually find the y-intercept and other points)

When showing viewing window boundaries on calculator graphs in this text, we write them in the form

$$[x_{min}, x_{max}] \text{ by } [y_{min}, y_{max}]$$

Different viewing windows give different views of a graph. There are usually many different viewing windows that give complete graphs for a particular function, but some viewing windows do not show all the important parts of a graph.

EXAMPLE 3 ▶ Graphing a Complete Graph

Sketch the graph of $y = x^3 - 3x^2 - 13$

a. using the standard window.

b. using the window $x_{min} = -10$, $x_{max} = 10$, $y_{min} = -25$, $y_{max} = 10$.

Which graph gives a better view of the graph of the function?

SOLUTION

a. A graph of this function in the standard window appears to be a line (Figure 1.20(a)).

b. By setting the window with $x_{min} = -10$, $x_{max} = 10$, $y_{min} = -25$, $y_{max} = 10$, we obtain the graph in Figure 1.20(b). This window gives a better view of the graph of this equation. As you learn more about functions, you will see that the graph in Figure 1.20(b) is a complete graph of this function.

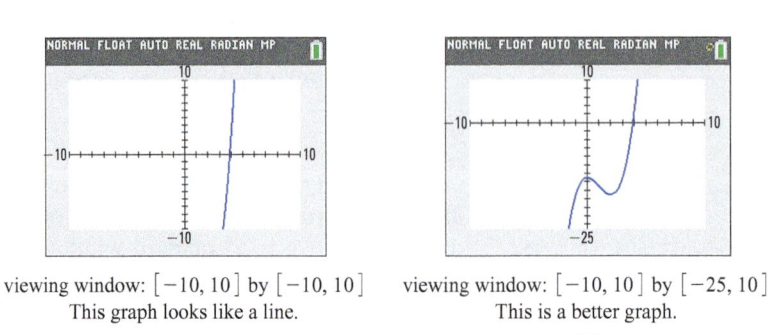

viewing window: $[-10, 10]$ by $[-10, 10]$
This graph looks like a line.
(a)

viewing window: $[-10, 10]$ by $[-25, 10]$
This is a better graph.
(b)

Figure 1.20

EXAMPLE 4 ▶ Global EV/PHEV Sales

Using data from 2020 and projected to 2050, global sales of electric and plug-in hybrid electric vehicles can be modeled by the function $y = 0.018x^2 + 2.32x - 19.25$, where y is millions of vehicle sales and x is the number of years after 2010. (*Source*: International Energy Agency)

a. Use a viewing window with x-values representing the years from 2020 through 2050 (that is, $x = 10$ through $x = 40$) and y representing the millions of vehicle sales from 0 through 120 to graph this function with a graphing calculator.

b. Find the Global EV/PHEV sales predicted for 2045 by this function.

c. Find the sales predicted for 2055. What does this tell us about sales of electric and plug-in hybrid electric vehicles over the period 2045 to 2055?

SOLUTION

a. Graphing the function with the viewing window using $x_{min} = 10$ and $x_{max} = 40$ for the x-values and $y_{min} = 0$ and $y_{max} = 120$ for the y-values gives the graph shown in Figure 1.21(a).

b. We evaluate $y = 0.018x^2 + 2.32x - 19.25$ at $x = 35$ by:

  1. Entering the function with the Y= key.

  2. Pressing **2ND** WINDOW (TBLSET), moving the cursor to Ask opposite Indpnt:, and pressing **ENTER**.

  3. Pressing **2ND** TABLE and entering 35 in the X column of Table. This gives output 84 in the Y_1 column, so the Global EV/PHEV Sales in 2045 is predicted to be 84 million (see Figure 1.21(b)).*

c. Evaluating $y = 0.018x^2 + 2.32x - 19.25$ at $x = 45$ gives approximately 121.6, so the Global EV/PHEV Sales in 2055 is predicted to be approximately 121.6 million. We can conclude that sales of these vehicles are expected to increase by approximately 50% over this 10-year period. (To see a graph that includes $x = 45$, increase the viewing window to $[0, 50]$ by $[0, 150]$.)

(a) (b)

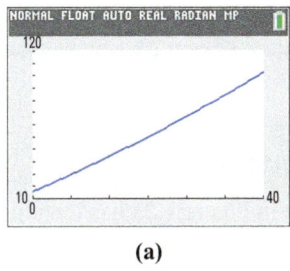

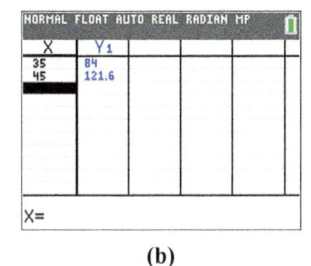

Figure 1.21

EXAMPLE 5 ▶ Cost–Benefit

Suppose that the cost C of removing $p\%$ of the pollution from drinking water is given by the model

$$C = \frac{5350p}{100 - p} \text{ dollars}$$

---

*For more details, see Appendix A, page 685.

a. Use the restriction on p to determine the limitations on the horizontal-axis values (which are the x-values on a calculator).

b. Graph the function on the viewing window $[0, 100]$ by $[0, 50{,}000]$. Why is it reasonable to graph this model on a viewing window with the limitation $C \geq 0$?

c. Find the point on the graph that corresponds to $p = 90$. Interpret the coordinates of this point.

SOLUTION

a. Because p represents the percent of pollution removed, it is limited to values from 0 to 100. However, $p = 100$ makes C undefined in this model, so p is restricted to $0 \leq p < 100$ for this model.

b. The graph of the function is shown in Figure 1.22(a), with x representing p and y representing C. The interval $[0, 100]$ contains all the possible values of p. The value of C is bounded below by 0 because C represents the cost of removing the pollution, which cannot be negative.

c. We can find (or estimate) the output of a function $y = f(x)$ at specific inputs with a graphing utility. We do this with TRACE or TABLE (accessed by **2ND** GRAPH). Using TRACE with the x-value 90 gives the point $(90, 48{,}150)$ (see Figure 1.22(b)). Because the values of p are represented by x-values and the values of C are represented by y, the coordinates of the point tell us that the cost of removing 90% of the pollution from the drinking water is \$48,150. Figure 1.22(c) shows the value of y for $x = 90$ and other values in a calculator table.

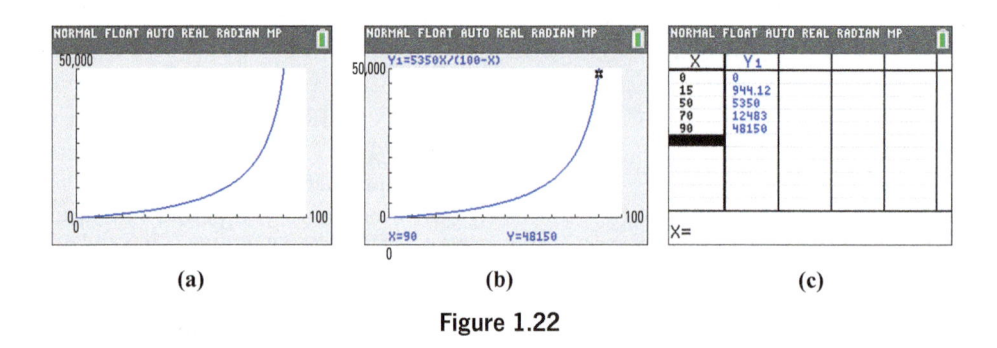

(a)　　　(b)　　　(c)

Figure 1.22

Spreadsheet ▶ SOLUTION We have discussed how to create graphs with a graphing calculator, and we can also use Excel spreadsheets to generate graphs. The table in Figure 1.23 has input data in the *cells* under column A. These cells have addresses identified by the column and row in which they lie. For example, 10 is in cell A3 and 40 is in cell A6. These cell addresses act like variables in an algebraic expression, so the inputs in column A represent values of p. These inputs can be entered manually. If the inputs follow a pattern as they do in Figure 1.23, those after the first two can be entered by selecting the first two, moving the mouse to the lower right corner (a thin plus sign will appear), and dragging the mouse down to the last cell where data are required. This process is called "filling down."

The output in cell B2 is 0. It is found by entering the formula

$$=5350*A2/(100 - A2)$$

in cell B2 and pressing ENTER; this gives the value of $C = \dfrac{5350p}{100 - p}$ at $p = 0$. To get the remaining outputs, move the mouse to the lower right corner of cell B2 and drag the mouse to the last cell where inputs are listed.

To create the graph, we highlight the two columns containing the inputs and outputs, select the insert tab, select the scatter plot icon in the Charts group, and then select the smooth curve icon. The graph will appear. Figure 1.23 shows the table of values and the graph of the functions.*

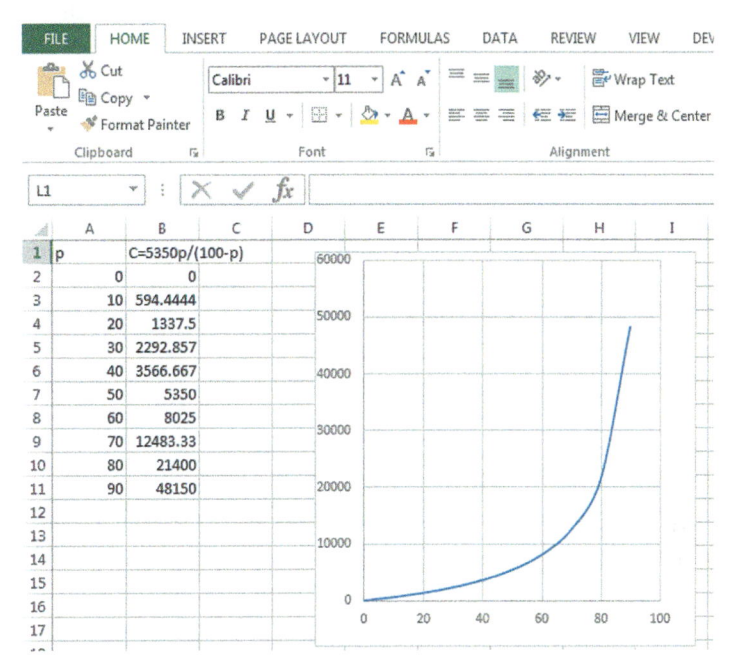

Figure 1.23

Determining Viewing Windows

Finding the function values (*y*-values) for selected inputs (*x*-values) can be useful when setting viewing windows for graphing utilities.

> **Technology Note**
>
> Once the input values for a viewing window have been selected, TRACE or TABLE (accessed by 2ND GRAPH) can be used to find enough output values to determine a *y*-view that gives a complete graph.†

EXAMPLE 6 ▶ Aging Workers

The millions of Americans who are working full time at selected ages from 20 to 65 can be modeled as a function of their age, *x*. A model for this is

$$y = -0.000362x^3 + 0.0401x^2 - 1.39x + 21.7$$

a. Use a viewing window with the *x*-values representing the ages from 20 to 65, and graph the function with a graphing calculator.

b. Use the model to estimate the number of Americans age 55 and age 64 who are working full time.

(*Source*: *Wall Street Journal*)

---

*For detailed steps needed to graph functions, see Appendix B, pages 702–705.
†It is occasionally necessary to make more than one attempt to find a window that gives a complete graph.

SOLUTION

a. Enter the equation in Y_1 and use the viewing window with $x_{min} = 20$ and $x_{max} = 65$ and with any values for y_{min} and y_{max}. Using TABLE with several values of x shows that we can get the desired graph using $y_{min} = 0, y_{max} = 10$. (See Figure 1.24(a).) Note that this viewing window can be written in interval notation as $[20, 65]$ by $[0, 10]$.* Figure 1.24(b) shows the suggested viewing window, containing x-scale and y-scale values that give appropriately spaced tick marks on the graph. Figure 1.24(c) shows the graph of the function.

b. Figure 1.24(d) shows the TABLE outputs for ages 55 and 64, as well as for age 20. It shows that the model estimates that 6.32 million Americans age 55 are working and 2.09 million Americans age 64 are working. We could also find the number working at these ages with TRACE on the graph.

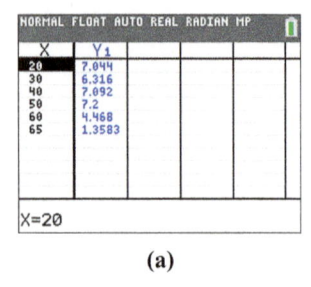

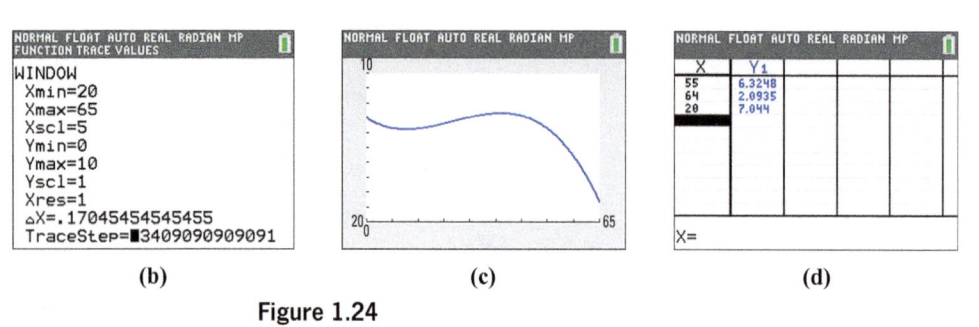

Figure 1.24

Spreadsheet ▶ SOLUTION

Excel can also be used to evaluate and graph the function $y = -0.000362x^3 + 0.0401x^2 - 1.39x + 21.7$, discussed in Example 6. A table of input and output values and the graph of the function are shown in Figure 1.25.

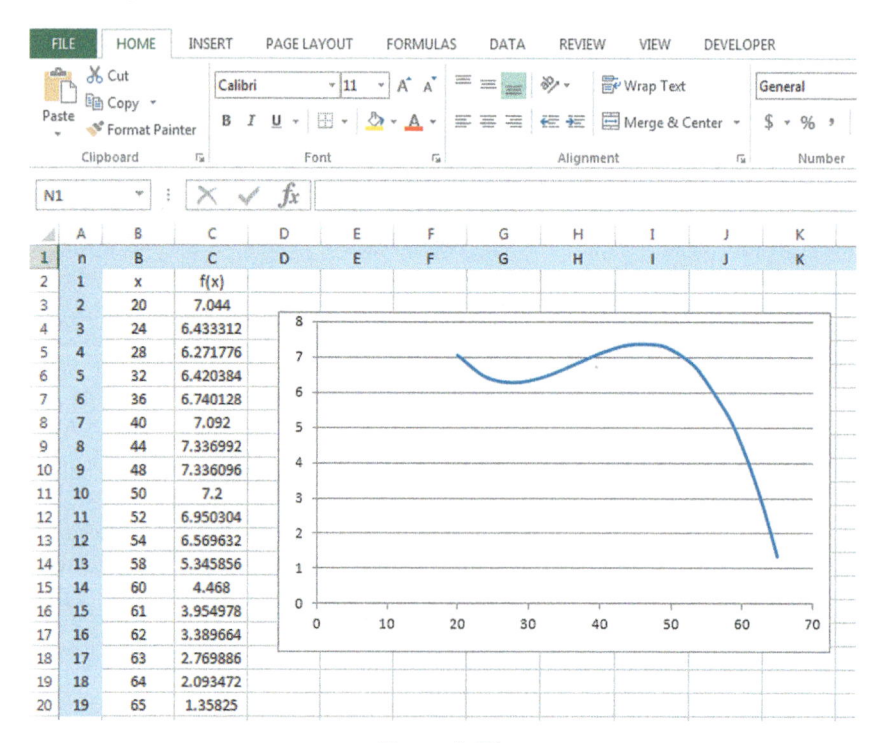

Figure 1.25

---

*Interval notation is discussed in the Algebra Toolbox, page 5.

Graphing Data Points

Graphing utilities can also be used to create lists of numbers and to create graphs of the data stored in the lists. To see how lists and scatter plots are created, consider the following example.

EXAMPLE 7 ▶ U.S. Diabetes

The number of U.S. adults with diabetes, in thousands, projected to 2050 is shown in Table 1.14.

Table 1.14

Year	Thousands
2015	37,300
2020	50,000
2025	59,500
2030	68,300
2035	76,200
2040	84,100
2045	91,700
2050	100,000

(*Source:* Centers for Disease Control and Prevention)

a. Align the data so that $x =$ the number of years after 2015, and enter these x-values in list L1 of your graphing calculator. Enter the number of millions of U.S. adults with diabetes in L2.

b. Use a graphing command* to create the scatter plot of these data points.

SOLUTION

a. If x represents the number of years after 2015, we must subtract 2015 from each year in the original table to obtain each corresponding x-value. The number of U.S. adults with diabetes was given in thousands in the original table, and because one million is 1000 times one thousand, we must divide each number by 1000 to obtain each y-value. Figure 1.26(a) shows the aligned inputs in L1 and the outputs from the data in L2.

b. The scatter plot is shown in Figure 1.26(b), with each entry in L1 represented by an x-coordinate of a point on the graph and the corresponding entry in L2 represented by the y-coordinate of that point on the graph. The window can be set automatically or manually to include x-values from L1 and y-values from L2.

(a)   (b)

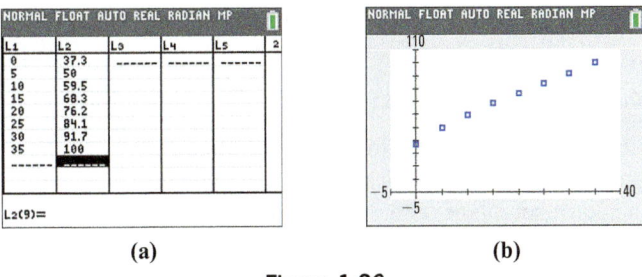

Figure 1.26

---

*Many calculators have a Stat Plot command. See Appendix A, page 683. Excel uses the points option under the Scatter icon in the Charts group, which is under the Insert tab. See Appendix B, page 707.

40 Chapter 1 Functions, Graphs, and Models; Linear Functions

Spreadsheet ► SOLUTION Figure 1.27 shows the aligned and scaled values from Table 1.14. To create a scatter plot of the data, we highlight the two columns containing the inputs and outputs, select the insert tab, select the scatter plot icon in the Charts group, and then select the scatter plot icon. The graph of the data will appear. (See Figure 1.27.)*

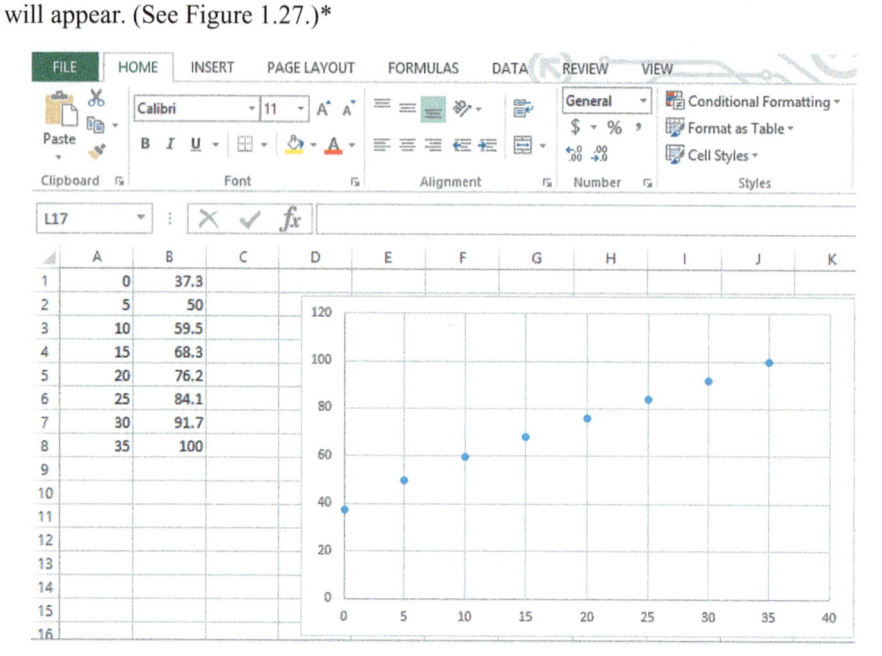

Figure 1.27

---

*See Appendix B, page 707–708, for detailed steps.

SKILLS CHECK 1.2

Answers that are not seen can be found in the answer section at the back of the text.

1. a. Complete the table of values for the function $y = x^3$, plot the values of x and y as points in a coordinate plane, and draw a smooth curve through the points.

x	−3	−2	−1	0	1	2	3
y	−27	−8	−1	0	1	8	27

b. Graph the function $y = x^3$ with a graphing calculator, using the viewing window $x_{min} = -4$, $x_{max} = 4$, $y_{min} = -30$, $y_{max} = 30$.

c. Compare the graphs in parts (a) and (b).
Graphs are the same.

2. a. Complete the table of values for the function $y = 2x^2 + 1$, plot the values of x and y as points in a coordinate plane, and draw a smooth curve through the points.

x	−3	−2	−1	0	1	2	3
y	19	9	3	1	3	9	19

b. Graph the function $y = 2x^2 + 1$ with a graphing calculator, using the viewing window $x_{min} = -5$, $x_{max} = 5$, $y_{min} = -1$, $y_{max} = 20$.

c. Compare the graphs in parts (a) and (b).
Graphs are the same.

In Exercises 3–8, make a table of values for each function and then plot the points to graph the function by hand.

3. $f(x) = 3x - 1$

4. $f(x) = 2x - 5$

5. $f(x) = \frac{1}{2}x^2$

6. $f(x) = 3x^2$

7. $f(x) = \frac{1}{x-2}$

8. $f(x) = \frac{x}{x+3}$

In Exercises 9–14, graph the functions with a graphing calculator using a standard viewing window. State whether the graph has a turning point in this window.

9. $y = x^2 - 5$ yes

10. $y = 4 - x^2$ yes

11. $y = x^3 - 3x^2$ yes **12.** $y = x^3 - 3x^2 + 4$ yes

13. $y = \dfrac{9}{x^2 + 1}$ yes **14.** $y = \dfrac{40}{x^2 + 4}$ yes

For Exercises 15–18, graph the given function with a graphing calculator using (a) the standard viewing window and (b) the specified window. Which window gives a better view of the graph of the function?

15. $y = x + 20$ with $x_{min} = -10$, $x_{max} = 10$, $y_{min} = -10$, $y_{max} = 30$

16. $y = x^3 - 3x + 13$ with $x_{min} = -5$, $x_{max} = 5$, $y_{min} = -10$, $y_{max} = 30$

17. $y = \dfrac{0.04(x - 0.1)}{x^2 + 300}$ on $[-20, 20]$ by $[-0.002, 0.002]$

18. $y = -x^2 + 20x - 20$ on $[-10, 20]$ by $[-20, 90]$

For Exercises 19–22, find an appropriate viewing window for the function, using the given x-values. Then graph the function. Find the coordinates of any turning points.

19. $y = x^2 + 50$, for x-values between -8 and 8.

20. $y = x^2 + 60x + 30$, for x-values between -60 and 0.

21. $y = x^3 + 3x^2 - 45x$, for x-values between -10 and 10.

22. $y = (x - 28)^3$, for x-values between 25 and 31.

23. Find a complete graph of $y = 10x^2 - 90x + 300$. (A complete graph of this function shows one turning point.)

24. Find a complete graph of $y = -x^2 + 34x - 120$. (A complete graph of this function shows one turning point.)

25. Use a calculator or a spreadsheet to find $S(t)$ for the values of t given in the following table in the next column.

t	S(t) = 5.2t − 10.5
12	51.9
16	72.7
28	135.1
43	213.1

26. Use a calculator or a spreadsheet to find $f(q)$ for the values of q given in the following table.

q	f(q) = 3q² − 5q + 8
−8	240
−5	108
24	1616
43	5340

27. Enter the data below into lists and graph the scatter plot of the data, using the window $[0, 110]$ by $[0, 550]$.

x	20	30	40	50	60	70	80	90	100
y	500	320	276	80	350	270	120	225	250

28. Enter the data below into lists and graph the scatter plot of the data, using the window $[-20, 120]$ by $[0, 80]$.

x	−15	−1	25	40	71	110	116
y	12	16	10	32	43	62	74

29. Use the table below in parts (a)–(c).

x	−3	−1	1	5	7	9	11
y	−42	−18	6	54	78	102	126

  a. Use a graphing utility to graph the points from the table.

  b. Use a graphing utility to graph the equation $f(x) = 12x - 6$ on the same set of axes as the data in part (a).

  c. Do the points fit on the graph of the equation? Do the respective values of x and y in the table satisfy the equation in part (b)? Yes; yes

30. Use the table below in parts (a)–(c).

x	1	3	6	8	10
y	−10	0	15	25	35

  a. Use a graphing utility to graph the points from the table.

  b. Use a graphing utility to graph the equation $f(x) = 5x - 15$ on the same set of axes as the data in part (a).

  c. Do the points fit on the graph of the equation? Do the respective values of x and y in the table satisfy the equation in part (b)? Yes; yes

31. Suppose $f(x) = x^2 - 5x$ million dollars are earned, where x is the number of years after 2000.

  a. What is $f(20)$? 300

  b. The answer to part (a) gives the number of millions of dollars earned for what year? 2020

32. Suppose $f(x) = 100x^2 - 5x$ thousand units are produced, where x is the number of years after 2000.

  a. What is $f(10)$? 9950

  b. How many units are produced in 2010, according to this function? 9,950,000

EXERCISES 1.2

Answers that are not seen can be found in the answer section at the back of the text.

In Exercises 33–38, plot the graphs by hand.

33. **Temperature Conversion** The function that converts temperature from Fahrenheit degree measurement F to Celsius degree measurement C is $C = \dfrac{5F - 160}{9}$. Graph this function for F from 32 to 212 degrees.

34. **Temperature Conversion** The function that converts temperature from Celsius degree measurement C to Fahrenheit degree measurement F is $F = \dfrac{9}{5}C + 32$. Graph this function for C from 0 to 100 degrees.

35. **Height of a Ball** If a ball is thrown into the air at 64 feet per second from the top of a 100-foot-tall building, its height can be modeled by the function $S = 100 + 64t - 16t^2$, where S is in feet and t is in seconds.

    a. Graph this function for t representing 0 to 6 seconds and S representing 0 to 200 feet.

    b. Find the height of the ball 1 second after it is thrown and 3 seconds after it is thrown. How can these values be equal? 148 ft and 148 ft; ball rising and falling

    c. Find the maximum height the ball will reach. 164 ft, in 2 seconds

36. **Depreciation** A business property valued at $600,000 is depreciated over 30 years by the straight-line method, so its value x years after the depreciation began is

    $$V = 600{,}000 - 20{,}000x$$

    a. Graph this function for x representing 0 to 30 years and v representing 0 to 600,000 dollars.

    b. What is the value 10 years after the depreciation is started? $400,000

37. **Cost** Suppose the cost of the production and sale of x Electra dishwashers is $C(x) = 15{,}000 + 100x + 0.1x^2$ dollars. Graph this function on a viewing window with x between 0 and 50.

38. **Revenue** Suppose the revenue from the sale of x coffee makers is given by $R(x) = 52x - 0.1x^2$. Graph this function on a viewing window with x between 0 and 100.

39. **Internet of Things** The cellular Internet of things (Iot) is the connection of devices that are connected to the Internet. The U.S. revenue from Iot for the years from 2014 and projected to 2025 can be modeled by the function $y = 1.31x^3 - 22.6x^2 + 210x - 164$, where x is the number of years after 2010 and y is the revenue in millions of dollars.
    (*Source*: Statista)

    a. What are the values of x that correspond to 2025 and 2030? 15 and 20

    b. Find the value of y when $x = 15$. Explain what this means. 2322.25; the U.S. revenue from Iot is projected to be approximately $2322 million in 2025.

    c. What is the projected U.S. revenue from Iot in 2030, according to this model? $5476 million

40. **Women in the Workforce** Using data starting with 1950 and projected to 2050, the number of women (in millions) in the workforce is given by the function $P = 0.79x + 20.86$, where x equals the number of years after 1950.

    a. Find the value of P when $x = 70$. Explain what this means. 76.16; the model predicts that there will be 76.16 million women in the workforce in 2020.

    b. What value of x represents 2030? Use the model to find the number of women in the workforce in 2030.
    (*Source*: U.S. Census Bureau) 80; 84,060,000

41. **Population of Americans 85 Years of Age and Older** Using data from 1910 to 2050, the number of thousands of Americans 85 years old and older can be modeled by $P = 1.458x^2 - 124.4x + 2516$, with $x = 0$ in 1900.

    a. Use a graphing utility to graph this function on the viewing window $[30, 165]$ by $[-1000, 25{,}000]$.

    b. Use the graphing utility to predict the population of Americans 85 years old and older in 2030.
    (*Source*: U.S. Census Bureau) 10,984,200

42. **Life Expectancy at 65** Using data for selected years from 1950 and projected to 2050, the function $y = 0.077x + 13.827$ gives the life expectancy at age 65, with y equal to the number of additional years of expected life at age 65 and x equal to the number of years after 1950.

    a. Graph the model on the viewing window $[0, 110]$ by $[0, 25]$.

    b. What time period does the viewing window in part (a) represent? From 1950 through 2060

    c. Use the graphing utility to find what the model predicts the life expectancy to be in 2038, rounded to one decimal place.
    (*Source*: Social Security Administration) 20.6 years

43. **Internet Usage** In 2015, 88% of U.S. residents used the Internet, up from 14% in 1995. Using data from 2000 and projected to 2025, the percent of U.S. residents who use the Internet can be modeled by the function $P(x) = -0.021x^2 + 1.71x + 68.18$, where x is the number of years after 2000.
    (*Source*: Nation and Statistics, Policy Exchange Analysis)

    a. $P(22)$ gives the value of P for what year? What is $P(22)$? 2022; 95.636%

b. What x_{min} and x_{max} should be used to set a window representing 2000 through 2025? 0 and 25

c. What is the maximum value that makes sense for y_{max} in the graph of this function? 100

d. Use the information above to graph the function with a graphing calculator.

44. Federal Tax per Capita The amount of federal tax per capita (per person) can be modeled by the function

$$S = 2.4807t^3 - 52.251t^2 + 528.68t + 5192.6$$

where t is the number of years after 2000.

a. What are the values of t that correspond to 2012, 2016, and 2020? 12, 16, 20

b. $S = f(18)$ gives the value of S for what year? What is $f(18)$? 2018; approximately 12,247

c. What x_{min} and x_{max} should be used to set a viewing window so that t represents 2000–2020?
(*Source*: Internal Revenue Service) 0 to 20

45. Student Loans Data for selected years from 2011 and projected to 2023 can be used to show that the balance of federal direct student loans y, in billions of dollars, is related to the number of years after 2010, x, by the function $y = 130.7x + 699.7$.

a. Graph the function for x-values corresponding 2010 to 2025.

b. Find the value of y when x is 13. 2398.8

c. What does this function predict the balance of federal direct student loans will be in 2029?
(*Source*: U.S. Office of Management and Budget) $3183 billion

46. Profit The profit from the production and sale of x laser printers is given by the function $P(x) = 200x - 0.01x^2 - 5000$, where x is the number of units produced and sold. Graph this function on a viewing window with x between 0 and 1000.

47. Profit The profit from the production and sale of x digital cameras is given by the function $P(x) = 1500x - 8000 - 0.01x^2$, where x is the number of units produced and sold. Graph this function on a viewing window with x between 0 and 500.

48. Drug Testing To test the body's ability to assimilate a certain drug, a pharmaceutical company administers a 200-mg dose and then monitors the rate of assimilation every half hour. The equation $R(t) = -11.6t^2 + 96.0t$ gives the rate of assimilation $R(t)$ in milligrams per hour as a function of the elapsed time t in hours.

a. Graph this function with a graphing calculator for t-values representing the hours from 0 through 3.

b. Does the amount of drug increase or decrease over this period? increases

49. Obesity Using data from 2000 and projected to 2030, the percent of Americans who are obese can be modeled by the function $y = 0.665x + 23.4$, where x is the number of years after 2000 and y is the percent of obese Americans.

a. What inputs correspond to the years 2000 through 2030? 0 to 30

b. What outputs correspond to the inputs for 2000 and 2030? 23.4; 43.35

c. Use the answers to parts (a) and (b) and the fact that the function increases to find an appropriate viewing window, and graph this function.
(*Source*: American Journal of Preventive Medicine)

50. Energy Use Energy use per dollar of GDP indexed to 1980 means that any dollar spent for energy use per year is viewed as a percent of the use per dollar of GDP in 1980. Using data for selected years from 1985 and projected to 2035, the function $y = -1.03x + 88.1$ models the energy use per dollar of GDP, as a percent, with x equal to the number of years after 1980.
(*Source*: U.S. Department of Energy)

a. What inputs correspond to the years from 1980 through 2035? 0 through 55

b. What outputs for y could be used to estimate the percent of energy use per dollar of GDP? 0 through 100

c. Based on your answers to (a) and (b), choose an appropriate window and graph the function on a graphing utility.

51. U.S. Population The projected population of the United States for selected years from 2000 to 2060 is shown in the table below, with the population given in millions.

a. According to this table, what should the U.S. population have been in 2010? 299.9 million, or 299,900,000

b. Create a new table with x representing the number of years after 2000 and y representing the number of millions.

c. Use a graphing utility to graph the data from the new table as a scatter plot.

Year	Population (millions)
2000	275.3
2010	299.9
2020	324.9
2030	351.1
2040	377.4
2050	403.7
2060	432.0

(*Source*: U.S. Census Bureau)

Chapter 1 Functions, Graphs, and Models; Linear Functions

52. *Total Energy Used for Industry* The table gives the total energy used for industry, in millions of British thermal units (BTUs), for the years from 2016 and projected to 2025.

  a. Graph the data from the table as a scatter plot, with y equaling energy used in millions of BTUs and x equal to the number of years after 2015.

  b. Graph the function $y = -0.0354x^2 + 0.959x + 29.8$ and the data points on the same axes.

  c. Does the function appear to be a good fit for the data? yes

Year	BTU (million)	Year	BTU (million)
2016	30.85	2021	34.18
2017	31.37	2022	34.77
2018	32.36	2023	35.33
2019	33.21	2024	35.64
2020	33.64	2025	35.74

(*Source*: U.S. Energy Information Administration)

53. *Men in the Workforce* The table gives the number of men (in millions) in the U.S. civilian workforce for selected years from 1950 and projected to 2050.

  a. Use a graphing utility to graph the data, with x equal to the number of years after 1950 and y equal to the number of millions of men in the workforce.

  b. Graph the equation $y = 0.587x + 43.145$ on the same axes as the scatter plot.

  c. Is the function a good visual fit for the data? yes

Year	Men (millions)	Year	Men (millions)
1950	43.8	2010	82.2
1960	46.4	2020	85.4
1970	51.2	2030	88.5
1980	61.5	2040	94.0
1990	69.0	2050	100.3
2000	75.2		

(*Source*: U.S. Bureau of Labor Statistics)

54. *Per Capita Expenditures for U.S. Health Care* The table shows the dollars spent per person for health care in the United States for selected years from 2002 and projected to 2024.

  a. Let x represent the number of years after 2000 and y represent the annual cost, and sketch the scatter plot of the data.

  b. Graph the function $y = 0.908x^3 - 25.32x^2 + 549x + 4544$ on the same axis as the scatter plot.

  c. Is the function a good visual fit for the data? yes

Year	Cost($)	Year	Cost($)
2002	5563	2014	9695
2004	6331	2016	10,527
2006	7091	2018	11,499
2008	7944	2020	12,741
2010	8428	2022	14,129
2012	8996	2024	15,618

(*Source*: U.S. Centers for Medicare and Medicaid Services)

55. *Crude Oil* The following table gives the U.S. crude oil production, in billions of barrels, for the years from 2010 projected to 2030.

  a. What does the table predict the crude oil production will be in 2026? 2.26 billion barrels

  b. Graph the data from this table as a scatter plot, with x equal to the number of years after 2010 and y equal to the number of billions of barrels of crude oil.

  c. Graph the function $y = -0.00107x^2 + 0.0369x + 1.95$ on the same axes as the scatter plot.

Year	Billions of Barrels
2010	1.94
2014	2.10
2018	2.16
2022	2.24
2026	2.26
2030	2.26

(*Source*: U.S. Energy Information Administration)

56. *Gross Domestic Product* The table gives the gross domestic product (GDP, the value of all goods and services) of the United States for selected years from 2005 to 2070.

  a. Create a scatter plot of the data, with y representing GDP in billions of dollars and x equal to the number of years after 2000.

  b. Graph the function $y = 117x^2 - 3792x + 45,330$ on the same axes as scatter plot.

  c. What does the function predict the GDP will be in 2046? $118,470 billion, or $118.47 trillion

Year	GDP ($ billions)	Year	GDP ($ billions)
2005	12,145	2040	79,680
2010	16,174	2045	103,444
2015	21,270	2050	133,925
2020	27,683	2055	173,175
2025	35,919	2060	224,044
2030	46,765	2065	290,042
2035	61,100	2070	375,219

(*Source*: Bureau of Economic Analysis)

1.3 Linear Functions

KEY OBJECTIVES

- Identify linear functions
- Find the intercepts and slopes of graphs of linear functions
- Graph linear functions
- Find the rate of change of a linear function
- Identify identity and constant functions
- Apply linear revenue, cost, and profit functions
- Find marginal revenue and marginal profit from linear revenue and linear profit functions

SECTION PREVIEW Hispanics in the United States

Using data and projections from 1990 through 2050, the percent of Hispanics in the U.S. population can be modeled by

$$H(x) = 0.224x + 9.01$$

with x equal to the number of years after 1990. The graph of this model is shown in Figure 1.28. (*Source*: U.S. Census Bureau)

Because the graph of this function is a line, it is called a linear function. The rate at which the Hispanic population is increasing is constant in this model, and it is equal to the slope of the line shown in Figure 1.28. (See Example 7.) In this section, we investigate linear functions and discuss slope, constant rates of change, intercepts, revenue, cost, and profit.

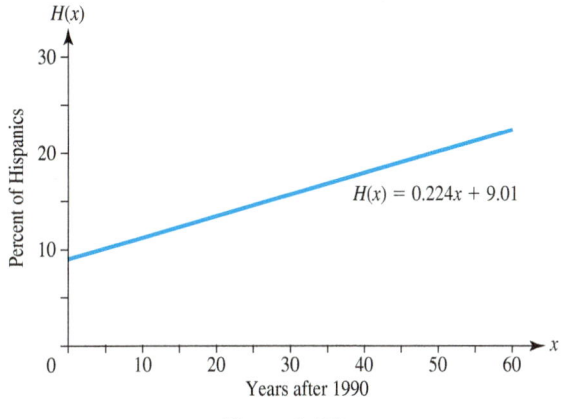

Figure 1.28

Linear Functions

A function whose graph is a line is a **linear function**.

> **Linear function**
>
> A linear function is a function that can be written in the form $f(x) = ax + b$, where a and b are constants.

If x and y are in separate terms in an equation and each appears to the first power (and not in a denominator), we can rewrite the equation relating them in the form

$$y = ax + b$$

for some constants a and b. So the original equation is a linear equation that represents a linear function. If no restrictions are stated or implied by the context of the problem situation and if its graph is not a horizontal line, both the domain and the range of a linear function consist of the set of all real numbers. Note that an equation of the form $x = d$, where d is a constant, is not a function; its graph is a vertical line.

Recall that in the form $y = ax + b$, a and b represent constants and the variables x and y can represent any variables, with x representing the independent (input) variable and y representing the dependent (output) variable. For example, the function $5q + p = 400$ can be written in the form $p = -5q + 400$, so we can say that p is a linear function of q.

46 Chapter 1 Functions, Graphs, and Models; Linear Functions

EXAMPLE 1 ▶ Linear Functions

Determine whether each equation represents a linear function. If so, give the domain and range.

a. $0 = 2t - s + 1$ **b.** $y = 5$ **c.** $xy = 2$

SOLUTION

a. The equation

$$0 = 2t - s + 1$$

does represent a linear function because each of the variables t and s appears to the first power and each is in a separate term. We can solve this equation for s, getting

$$s = 2t + 1$$

so s is a linear function of t. Because any real number can be multiplied by 2 and increased by 1 and the result is a real number, both the domain and the range consist of the set of all real numbers. (Note that we could also solve the equation for t so that t is a linear function of s.)

b. The equation $y = 5$ is in the form $y = ax + b$, where $a = 0$ and $b = 5$, so it represents a linear function. (It is in fact a constant function, which is a special linear function.) The domain is the set of all real numbers (because $y = 5$ regardless of what x we choose), and the range is the set containing 5.

c. The equation

$$xy = 2$$

does not represent a linear function because x and y are not in separate terms and the equation cannot be written in the form $y = ax + b$.

Intercepts

The points where a graph crosses or touches the x-axis and the y-axis are called the x-**intercepts** and y-**intercepts**, respectively, of the graph. For example, Figure 1.29 shows that the graph of the linear function $2x - 3y = 12$ crosses the x-axis at $(6, 0)$, so the x-intercept is $(6, 0)$. The graph crosses the y-axis at $(0, -4)$, so the y-intercept is $(0, -4)$. In this text, we will use the widely accepted convention that the x-coordinate of the x-intercept may also be called the x-intercept, and the y-coordinate of the y-intercept may also be called the y-intercept.* The procedure for finding intercepts is a direct result of these definitions.

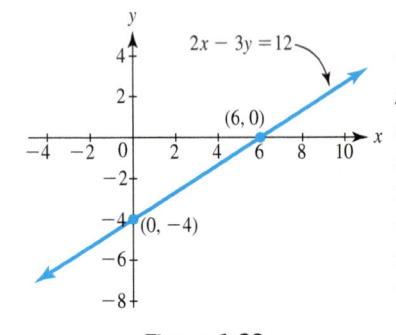

Figure 1.29

Finding Intercepts Algebraically

To find the y-intercept of a graph of $y = f(x)$, set $x = 0$ in the equation and solve for y. If the solution is b, we say the y-intercept is b and the graph intersects the y-axis at the point $(0, b)$.

To find the x-intercept(s) of the graph of $y = f(x)$, set $y = 0$ in the equation and solve for x. If the solution is a, we say the x-intercept is a and the graph intersects the x-axis at the point $(a, 0)$.

---

*We usually call the horizontal-axis intercept(s) the x-intercept(s) and the vertical-axis intercept(s) the y-intercept(s), but realize that symbols other than x and y can be used to represent the input and output. For example, if $p = f(q)$, the vertical intercept is called the p-intercept and the horizontal intercept is called the q-intercept.

Finding Intercepts Graphically

To find the intercept(s) of a graph of $y = f(x)$, first graph the function in a window that shows all intercepts.

To find the y-intercept, TRACE to $x = 0$ and the y-intercept will be displayed. To find the x-intercept(s) of the graph of $y = f(x)$, use the zero command under the CALC menu (accessed by 2ND TRACE).* The zero command gives you the x-intercept after asking you to choose a value of x to the left of the x-intercept, then a value of x to the right of the x-intercept, and then press ENTER.

EXAMPLE 2 ▶ Finding Intercepts Algebraically and Graphically

Find the x-intercept and the y-intercept of the graph of $2x - 3y = 12$

a. algebraically.

b. graphically.

SOLUTION

a. The y-intercept can be found by substituting 0 for x in the equation and solving for y.

$$2(0) - 3y = 12$$
$$-3y = 12$$
$$y = -4$$

Thus, the y-intercept is -4, and the graph crosses the y-axis at the point $(0, -4)$.

Similarly, the x-intercept can be found by substituting 0 for y in the equation and solving for x.

$$2x - 3(0) = 12$$
$$2x = 12$$
$$x = 6$$

Thus the x-intercept is 6, and the graph crosses the x-axis at the point $(6, 0)$.

b. To find the y-intercept graphically, first solve $2x - 3y = 12$ for y:

$$2x - 3y = 12$$
$$2x = 3y + 12 \quad \text{Add } 3y \text{ to both sides of the equation.}$$
$$2x - 12 = 3y \quad \text{Add 12 to both sides of the equation.}$$
$$\frac{2x - 12}{3} = y \quad \text{Divide both sides by 3.}$$
$$\frac{2}{3}x - 4 = y \quad \text{Simplify.}$$

Enter $y = \frac{2}{3}x - 4$ in the graphing utility and graph with the standard window (Figure 1.30(a)). To find the y-intercept, press TRACE and 0 (Figure 1.30(b)). The y-intercept is $(0, -4)$.

---

*For more details, see Appendix A, page 685.

To find the *x*-intercept, press **2ND** **TRACE** and 2:Zero. Enter a left bound (Figure 1.30(c)), right bound (Figure 1.30(d)), guess (Figure 1.30(e)), and press **ENTER** (Figure 1.30(f)). The zero, or the *x*-intercept, is $(6, 0)$.

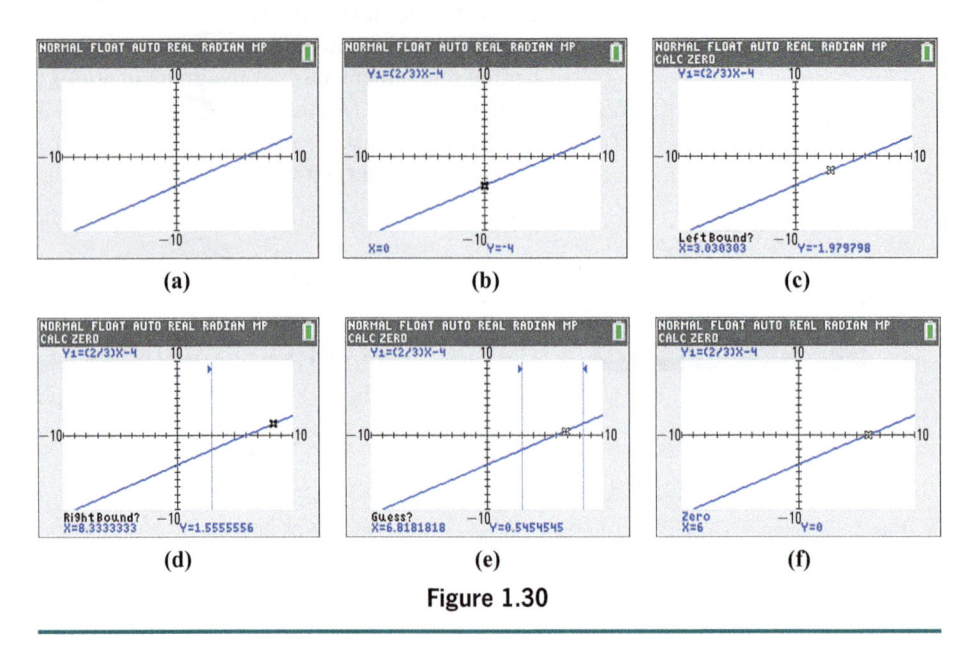

Figure 1.30

The graph of a linear function has one *y*-intercept and one *x*-intercept unless the graph is a horizontal line. The intercepts of the graph of a linear function are often easy to calculate. If the intercepts are distinct, then plotting these two points and connecting them with a line gives the graph.

Technology Note

When graphing with a graphing utility, finding or estimating the intercepts can help you set the viewing window for the graph of a linear equation.

EXAMPLE 3 ▶ Loan Balance

A business property is purchased with a promise to pay off a $60,000 loan plus the $16,500 interest on this loan by making 60 monthly payments of $1275. The amount of money, *y*, remaining to be paid on $76,500 (the loan plus interest) is reduced by $1275 each month. Although the amount of money remaining to be paid changes every month, it can be modeled by the linear function

$$y = 76{,}500 - 1275x$$

where *x* is the number of monthly payments made. We recognize that only integer values of *x* from 0 to 60 apply to this application.

a. Find the *x*-intercept and the *y*-intercept of the graph of this linear equation.

b. Interpret the intercepts in the context of this problem situation.

c. How should *x* and *y* be limited in this model so that they make sense in the application?

d. Use the intercepts and the results of part (c) to sketch the graph of the given equation.

SOLUTION

a. To find the *x*-intercept, set $y = 0$ and solve for *x*.

$$0 = 76{,}500 - 1275x$$
$$1275x = 76{,}500$$
$$x = \frac{76{,}500}{1275} = 60$$

Thus, 60 is the *x*-intercept.

To find the *y*-intercept, set $x = 0$ and solve for *y*.

$$y = 76{,}500 - 1275(0)$$
$$y = 76{,}500$$

Thus, 76,500 is the *y*-intercept.

b. The *x*-intercept corresponds to the number of months that must pass before the amount owed is $0. Therefore, a possible interpretation of the *x*-intercept is "The loan is paid off in 60 months." The *y*-intercept corresponds to the total (loan plus interest) that must be repaid 0 months after purchase—that is, when the purchase is made. Thus, the *y*-intercept tells us "A total of $76,500 must be repaid."

c. We know that the total time to pay the loan is 60 months. A value of *x* larger than 60 will result in a negative value of *y*, which makes no sense in the application, so the values of *x* vary from 0 to 60. The output, *y*, is the total amount owed at any time during the loan. The amount owed cannot be less than 0, and the value of the loan plus interest will be at its maximum, 76,500, when time is 0. Thus, the values of *y* vary from 0 to 76,500.

d. The graph intersects the horizontal axis at $(60, 0)$ and intersects the *y*-axis at $(0, 76{,}500)$, as indicated in Figure 1.31. Because we know that the graph of this equation is a line, we can simply connect the two points to obtain this first-quadrant graph.

Figure 1.31 shows the graph of the function on a viewing window determined by the context of the application.

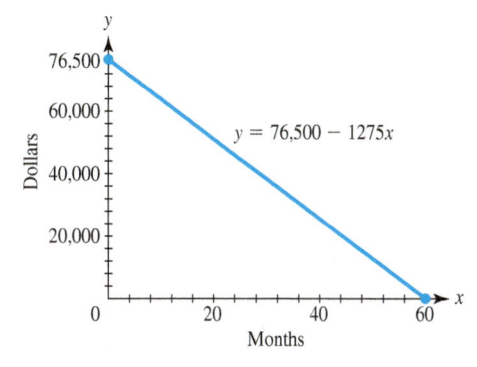

Figure 1.31

Slope of a Line

Consider two stairways: One goes from the park entrance to the shogun shrine at Nikko, Japan, and rises vertically 1 foot for every 1 foot of horizontal increase, and the second goes from the Tokyo subway to the street and rises vertically 20 centimeters for every 25 centimeters of horizontal increase. To see which set of stairs would be easier to climb, we can find the steepness, or **slope**, of each stairway. If a board is placed along the steps of each stairway, its slope is a measure of the incline of the stairway.

$$\text{shrine stairway steepness} = \frac{\text{vertical increase}}{\text{horizontal increase}} = \frac{1 \text{ foot}}{1 \text{ foot}} = 1$$

$$\text{subway stairway steepness} = \frac{\text{vertical increase}}{\text{horizontal increase}} = \frac{20 \text{ cm}}{25 \text{ cm}} = 0.8$$

The shrine stairway has a slope that is larger than that of the subway stairway, so it is steeper than the subway stairway. Thus, the subway steps would be easier to climb. In general, we define the slope of a line as follows.

Slope of a Line

The slope of a line is defined as

$$\text{slope} = \frac{\text{vertical change}}{\text{horizontal change}} = \frac{\text{rise}}{\text{run}}$$

The slope can be found by using any two points on the line (see Figure 1.32). If a nonvertical line passes through the two points, P_1 with coordinates (x_1, y_1) and P_2 with coordinates (x_2, y_2), its slope, denoted by m, is found by using

$$m = \frac{y_2 - y_1}{x_2 - x_1}$$

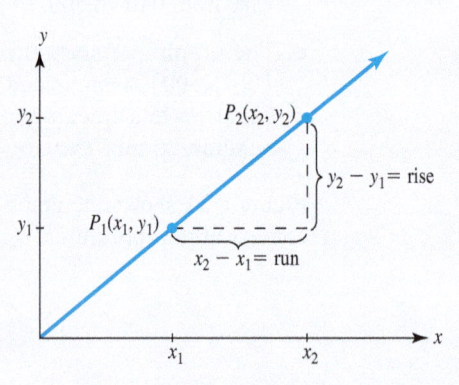

Figure 1.32

The slope of a vertical line is undefined because $x_2 - x_1 = 0$ and division by 0 is undefined.

The slope of any given nonvertical line is a constant. Thus, the same slope will result regardless of which two points on the line are used in its calculation.

EXAMPLE 4 ▶ Calculating the Slope of a Line

a. Find the slope of the line passing through the points $(-3, 2)$ and $(5, -4)$. What does the slope mean?

b. Find the slope of the line joining the *x*-intercept point and *y*-intercept point in the loan situation of Example 3.

SOLUTION

a. We choose one point as P_1 and the other point as P_2. Although it does not matter which point is chosen to be P_1, it is important to keep the correct order of the terms in the numerator and denominator of the slope formula. Letting $P_1 = (-3, 2)$ and $P_2 = (5, -4)$ and substituting in the slope formula gives

$$m = \frac{-4 - 2}{5 - (-3)} = \frac{-6}{8} = -\frac{3}{4}$$

Note that letting $P_1 = (5, -4)$ and $P_2 = (-3, 2)$ gives the same slope:

$$m = \frac{2 - (-4)}{-3 - 5} = \frac{6}{-8} = -\frac{3}{4}$$

A slope of $-\frac{3}{4}$ means that, from a given point on the line, by moving 3 units down and 4 units to the right or by moving 3 units up and 4 units to the left, we arrive at another point on the line.

b. Because $x = 60$ is the x-intercept in part (b) of Example 3, $(60, 0)$ is a point on the graph. The y-intercept is $y = 76{,}500$, so $(0, 76{,}500)$ is a point on the graph. Recall that x is measured in months and that y has units of dollars in the real-world setting (that is, the *context*) of Example 3. Substituting in the slope formula, we obtain

$$m = \frac{76{,}500 - 0}{0 - 60} = \frac{76{,}500}{-60} = -1275$$

This slope means that the amount owed decreases by \$1275 each month.

As Figure 1.33(a) to (d) indicates, the slope describes the direction of a line as well as the steepness.

The Relation Between Orientation of a Line and Its Slope

1. The slope is *positive* if the line *rises* upward toward the right.

2. The slope is *negative* if the line *falls* downward toward the right.

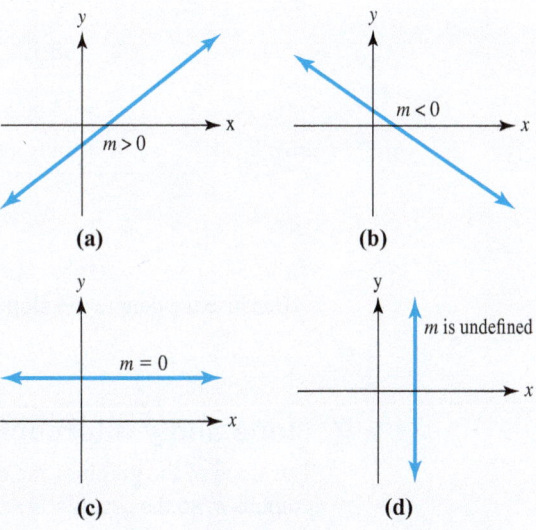

(a) $m > 0$

(b) $m < 0$

(c) $m = 0$

(d) m is undefined

Figure 1.33

3. The slope of a *horizontal line* is 0 because a horizontal line has a vertical change (rise) of 0 between any two points on the line.

4. The slope of a *vertical line* does not exist because a vertical line has a horizontal change (run) of 0 between any two points on the line.

Remembering how the orientation of a line is related to its slope can help you check that you have the correct order of the points in the slope formula. For instance, if you find that the slope of a line is positive and you see that the line falls downward from left to right when you observe its graph, you know that there is a mistake either in the slope calculation or in the graph that you are viewing.

Slope and y-Intercept of a Line

There is an important connection between the slope of the graph of a linear equation and its equation when written in the form $y = f(x)$. To investigate this connection, we can graph the equation

$$y = 3x + 2$$

by plotting points or by using a graphing utility. By creating a table for values of x equal to 0, 1, 2, 3, and 4 (see Table 1.15), we can see that each time x increases by 1, y increases by 3 (see Figure 1.34). Thus, the slope is

$$\frac{\text{change in } y}{\text{change in } x} = \frac{3}{1} = 3$$

From this table, we see that the y-intercept is 2, the same value as the constant term of the equation. Observe also that the slope of the graph of $y = 3x + 2$ is the same as the coefficient of x.

Table 1.15

x	y
0	2
1	5
2	8
3	11
4	14

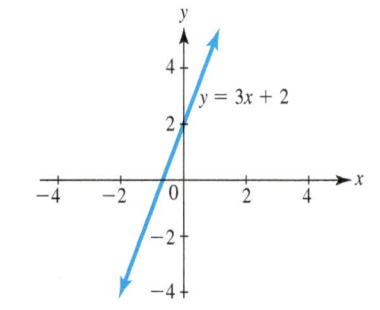

Figure 1.34

Because we denote the slope of a line by m, we have the following.

Slope and y-Intercept of a Line
The slope of the graph of the equation $y = mx + b$ is m and the y-intercept of the graph is b, so the graph crosses the y-axis at $(0, b)$.

Thus, when the equation of a linear function is written in the form $y = mx + b$ or $f(x) = mx + b$, we can "read" the values of the slope and y-intercept of its graph. If a function is linear but is not written in the form $y = ax + b$, the slope and y-intercept can be determined by solving the system for y to get that form.

1.3 Linear Functions   53

EXAMPLE 5 ▶ Finding the Slope and y-Intercept of a Line

Determine the slope and y-intercept of the linear function.

a. $y = 7x - 12$ **b.** $2x - 3y = 12$

SOLUTION

a. The equation is written in the form $y = ax + b$, so the slope is the coefficient of x and the y-intercept is the constant term; that is, $m = 7$ and $b = -12$.

b. The equation must be solved for y so that it is in the form $y = ax + b$.

$$2x - 3y = 12$$
$$2x = 12 + 3y \quad \text{Add 3y to both sides of the equation.}$$
$$2x - 12 = 3y \quad \text{Subtract 12 from both sides.}$$
$$\frac{2}{3}x - 4 = y \quad \text{Divide both sides by 3 and simplify.}$$

Thus, the slope of the line is $\frac{2}{3}$ and the y-intercept is -4.

EXAMPLE 6 ▶ Loan Balance

As we saw in Example 3, the amount of money y remaining to be paid on the loan of $60,000 with $16,500 interest is

$$y = 76,500 - 1275x$$

where x is the number of monthly payments that have been made.

a. What are the slope and y-intercept of the graph of this function?

b. How does the amount owed on the loan change as the number of months increases?

SOLUTION

a. Writing this equation in the form $y = mx + b$ gives $y = -1275x + 76,500$. The coefficient of x is -1275, so the slope is $m = -1275$. The constant term is 76,500, so the y-intercept is $b = 76,500$.

b. The slope of the line indicates that the amount owed decreases by $1275 each month. This can be verified from the graph of the function in Figure 1.35 and the sample inputs and outputs in Table 1.16.

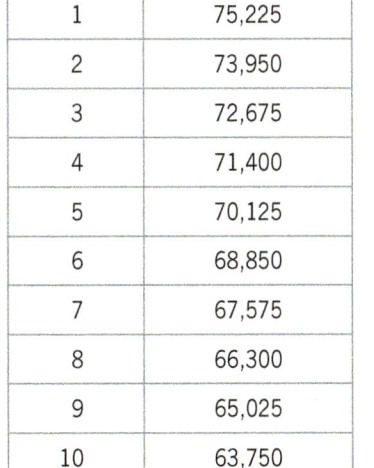

Table 1.16

Months (x)	Amount Owed (y)
1	75,225
2	73,950
3	72,675
4	71,400
5	70,125
6	68,850
7	67,575
8	66,300
9	65,025
10	63,750

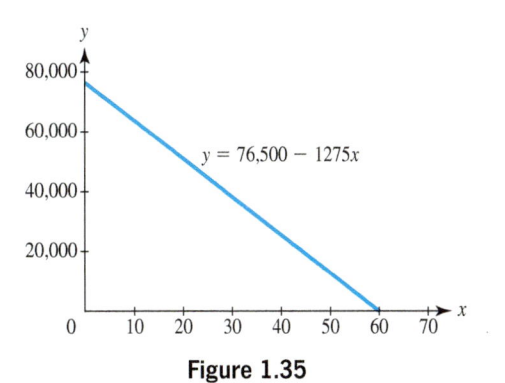

Figure 1.35

Constant Rate of Change

The function $y = 76,500 - 1275x$, whose graph is shown in Figure 1.35, gives the amount owed as a function of the months remaining on the loan. The coefficient of x, -1275, indicates that for each additional month, the value of y changes by -1275 (see Table 1.16). That is, the amount owed decreases at the **constant rate** of -1275 each month. Note that the **constant rate of change** of this linear function is equal to the **slope** of its graph. This is true for all linear functions.

> ### Constant Rate of Change
> The rate of change of the linear function $y = mx + b$ is the constant m, the slope of the graph of the function.

Note: The rate of change in an applied context should include appropriate units of measure. The rate of change describes by how much the output changes (increases or decreases) for every input unit.

EXAMPLE 7 ▶ Hispanics in the United States

Using data and projections from 1990 through 2050, the percent of Hispanics in the U.S. population can be modeled by

$$H(x) = 0.224x + 9.01$$

with x equal to the number of years after 1990. The graph of this model is shown in Figure 1.36. (*Source*: U.S. Census Bureau)

a. What is the slope of the graph of this function?

b. What does this slope tell us about the annual rate of change in the percent of Hispanics in the United States?

SOLUTION

a. The coefficient of x in the linear function $H(x) = 0.224x + 9.01$ is 0.224, so the slope of the line that is its graph is 0.224.

b. This slope tells us that the percent of Hispanics in the U.S. population is increasing by 0.224 each year since 1990. We say that the percent has increased by 0.224 *percentage point*.

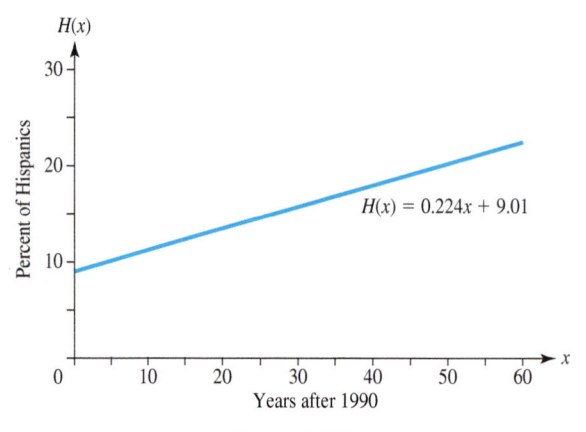

Figure 1.36

Revenue, Cost, and Profit

The **profit** that a company makes on its product is the difference between the amount received from sales (revenue) and the production and sales costs. If x units are produced and sold, we can write

$$P(x) = R(x) - C(x)$$

where

$P(x)$ = profit from sale of x units
$R(x)$ = total revenue from sale of x units
$C(x)$ = total cost of production and sale of x units

In general, **revenue** is found by using the equation

$$\text{revenue} = (\text{price per unit})(\text{number of units})$$

The **total cost** is composed of two parts: fixed costs and variable costs. **Fixed costs** (FC), such as depreciation, rent, and utilities, remain constant regardless of the number of units produced. **Variable costs** (VC) are those directly related to the number of units produced. Thus, the total cost, often simply called the cost, is found by using the equation

$$\text{cost} = \text{variable costs} + \text{fixed costs}$$

EXAMPLE 8 ▶ Cost, Revenue, and Profit

Suppose that a company manufactures 50-inch 3D plasma TVs and sells them for $1800 each. The costs incurred in the production and sale of the TVs are $400,000 plus $1000 for each TV produced and sold. Write the profit function for the production and sale of x TVs.

SOLUTION

The total revenue for x TVs is $1800x$, so the revenue function is $R(x) = 1800x$. The fixed costs are $400,000, so the total cost for x TVs is $1000x + 400{,}000$. Hence, $C(x) = 1000x + 400{,}000$. The profit function is given by $P(x) = R(x) - C(x)$, so

$$P(x) = 1800x - (1000x + 400{,}000)$$
$$P(x) = 800x - 400{,}000$$

Figures 1.37(a)–(c) show the graphs of the three linear functions: $R(x)$, $C(x)$, and $P(x)$.

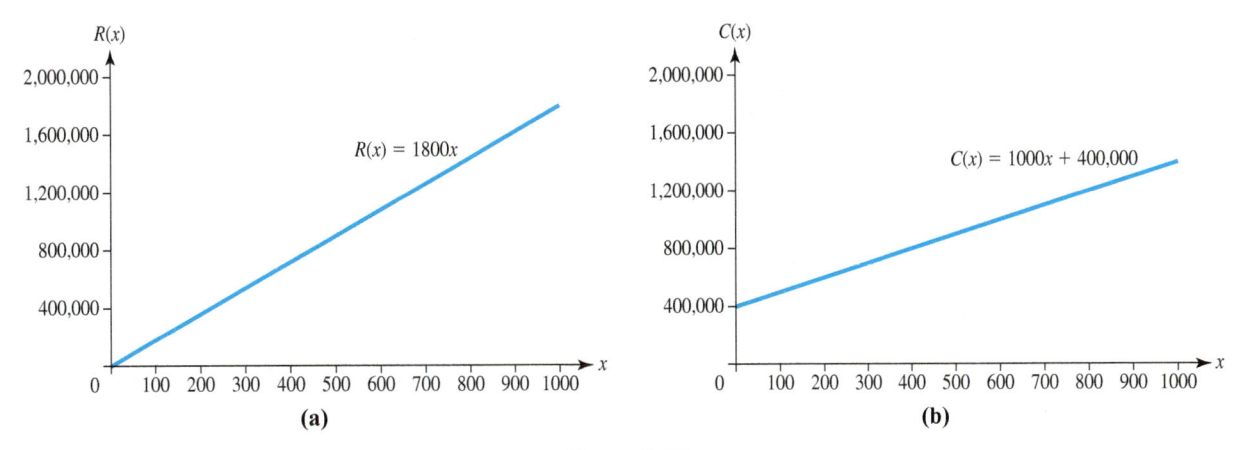

Figure 1.37

$P(x) = 800x - 400,000$

(c)

Figure 1.37

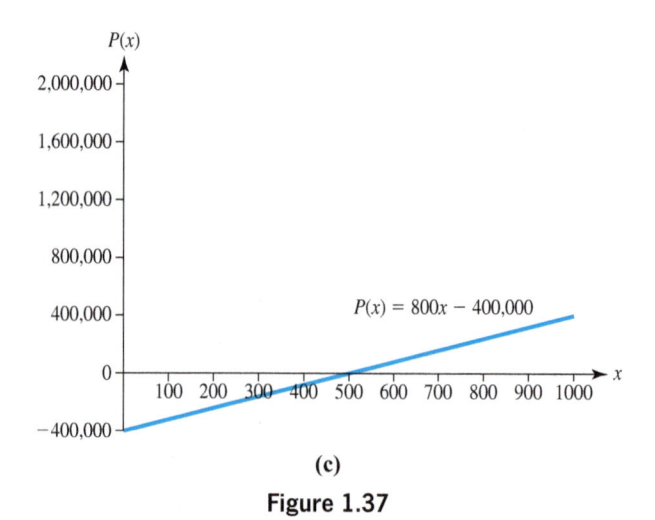

Marginal Cost, Revenue, and Profit

For total cost, total revenue, and profit functions* that are linear, the rates of change are called **marginal cost, marginal revenue**, and **marginal profit**, respectively. Suppose that the cost to produce and sell a product is $C(x) = 54.36x + 6790$ dollars, where x is the number of units produced and sold. This is a linear function, and its graph is a line with slope 54.36. Thus, the *rate of change* of this cost function, called the **marginal cost**, is $54.36 per unit produced and sold. This means that the production and sale of each additional unit will cost an additional $54.36.

EXAMPLE 9 ▶ Marginal Revenue and Marginal Profit

A company produces and sells a smartphone with revenue given by $R(x) = 89.50x$ dollars and cost given by $C(x) = 54.36x + 6790$ dollars, where x is the number of smartphones produced and sold.

a. What is the marginal revenue for this smartphone, and what does it mean?
b. Find the profit function.
c. What is the marginal profit for this smartphone, and what does it mean?

SOLUTION

a. The marginal revenue for this smartphone is the rate of change of the revenue function, which is the slope of its graph. Thus, the marginal revenue is $89.50 per unit sold. This means that the sale of each additional smartphone will result in additional revenue of $89.50.

b. To find the profit function, we subtract the cost function from the revenue function.
$$P(x) = 89.50x - (54.36x + 6790) = 35.14x - 6790$$

c. The marginal profit for this smartphone is the rate of change of the profit function, which is the slope of its graph. Thus, the marginal profit is $35.14 per smartphone sold. This means that the production and sale of each additional smartphone will result in an additional profit of $35.14.

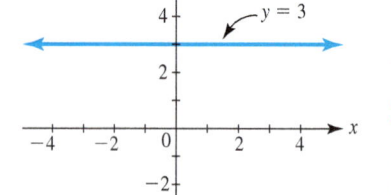

Special Linear Functions

A special linear function that has the form $y = 0x + b$, or $y = b$, where b is a real number, is called a **constant function**. The graph of the constant function $y = 3$ is shown in Figure 1.38(a). The temperature inside a sealed case containing an Egyptian

(a) **Constant Function**

*In this text, we frequently use "total cost" and "total revenue" interchangeably with "cost" and "revenue," respectively.

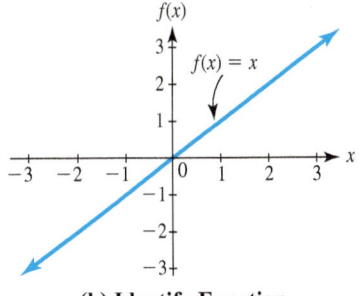

mummy in a museum is a constant function of time because the temperature inside the case never changes. Notice that even though the range of a constant function consists of a single value, the input of a constant function is any real number or any real number that makes sense in the context of an applied problem. Another special linear function is the **identity function**

$$y = 1x + 0, \quad \text{or} \quad y = x$$

which is a linear function of the form $y = mx + b$ with slope $m = 1$ and y-intercept $b = 0$. For the general identity function $f(x) = x$, the domain and range are each the set of all real numbers. A graph of the identity function f is shown in Figure 1.38(b).

(b) Identify Function
Figure 1.38

SKILLS CHECK 1.3

Answers that are not seen can be found in the answer section at the back of the text.

1. Which of the following functions are linear? b

   a. $y = 3x^2 + 2$ b. $3x + 2y = 12$

   c. $y = \dfrac{1}{x} + 2$

2. Is the graph in the figure below a function? No

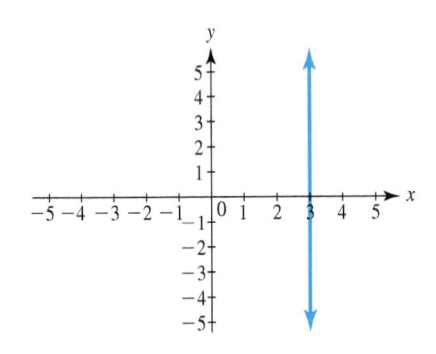

3. Find the slope of the line through $(4, 6)$ and $(28, -6)$. $-\dfrac{1}{2}$

4. Find the slope of the line through $(8, -10)$ and $(8, 4)$. Undefined

5. Find the slope of the line in the graph that follows. 2

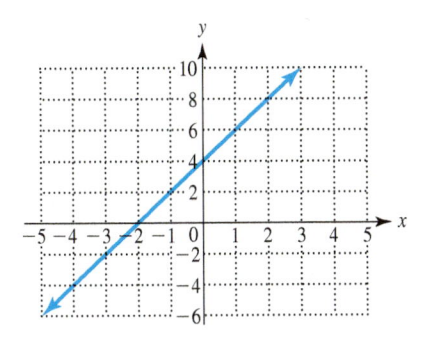

6. Find the slope of the line in the graph below. 0

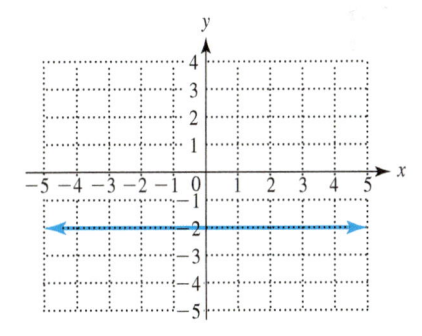

In Exercises 7–10, (a) find the x- and y-intercepts of the graph of the given equation, if they exist, and (b) graph the equation.

7. $5x - 3y = 15$
   a. x: 3; y: -5

8. $x + 5y = 17$
   a. x: 17; y: 3.4

9. $3y = 9 - 6x$ a. x: $\dfrac{3}{2}$; y: 3

10. $y = 9x$ a. x: 0; y: 0

11. If a line is horizontal, then its slope is _____. If a line is vertical, then its slope is _____. 0; undefined

12. Describe the line whose slope was determined in Exercise 4. Vertical line

For Exercises 13–14, determine whether the slope of the graph of the line is positive, negative, 0, or undefined.

13. a.   Positive

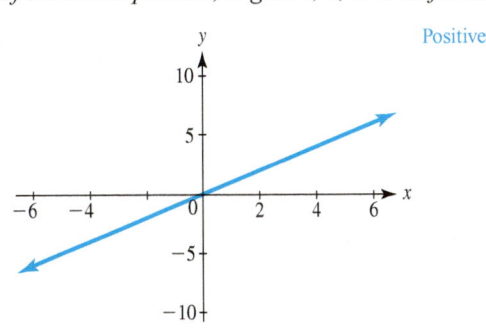

b.

Undefined

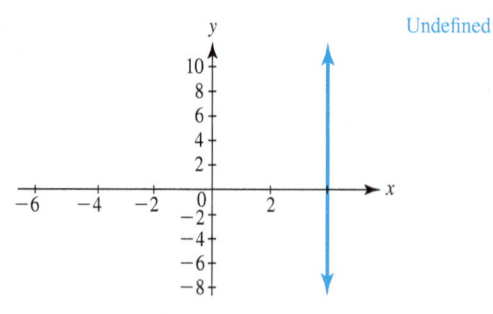

14. a.

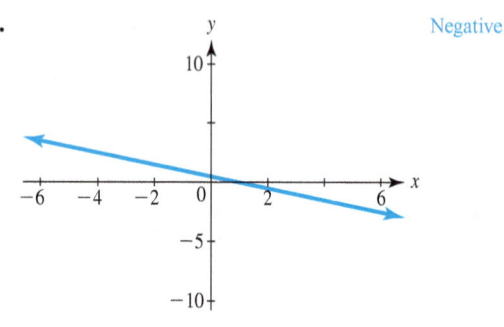

Negative

b.

0

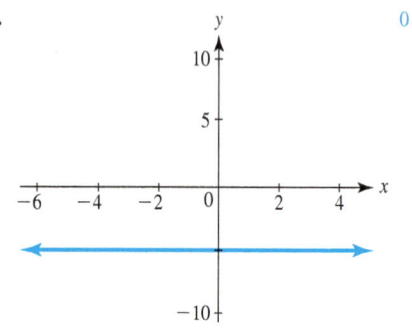

For Exercises 15–18, (a) give the slope of the line (if it exists) and the y-intercept (if it exists) and (b) graph the line.

15. $y = 4x + 8$
a. $m = 4; b = 8$

16. $3x + 2y = 7$
a. $m = -\frac{3}{2}; b = \frac{7}{2}$

17. $5y = 2$ a. $m = 0; b = \frac{2}{5}$

18. $x = 6$
a. No slope; no y-intercept

For each of the functions in Exercises 19–21, do the following:

a. Find the slope and y-intercept (if possible) of the graph of the function.

b. Determine whether the graph is rising or falling.

c. Graph each function on a window with the given x-range and a y-range that shows a complete graph.

19. $y = 4x + 5; [-5, 5]$ a. $m = 4; b = 5$ b. Rising

20. $y = 0.001x - 0.03; [-100, 100]$
a. $m = 0.001; b = -0.03$ b. Rising

21. $y = 50{,}000 - 100x; [0, 500]$
a. $m = -100; b = 50{,}000$ b. Falling

22. Rank the functions in Exercises 19–21 in order of increasing steepness. Exercise 20 is the least steep, followed by Exercise 19. Exercise 21 displays the greatest steepness.

For each of the functions in Exercises 23–26, find the rate of change.

23. $y = 4x - 3$ 4

24. $y = \frac{1}{3}x + 2$ $\frac{1}{3}$

25. $y = 300 - 15x$ -15

26. $y = 300x - 15$ 300

27. If a linear function has the points $(-1, 3)$ and $(4, -7)$ on its graph, what is the rate of change of the function? -2

28. If a linear function has the points $(2, 1)$ and $(6, 3)$ on its graph, what is the rate of change of the function? $\frac{1}{2}$

29. a. Does graph (i) or graph (ii) represent the identity function? ii

b. Does graph (i) or graph (ii) represent a constant function? i

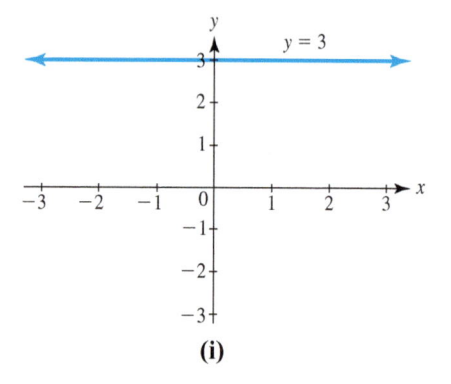

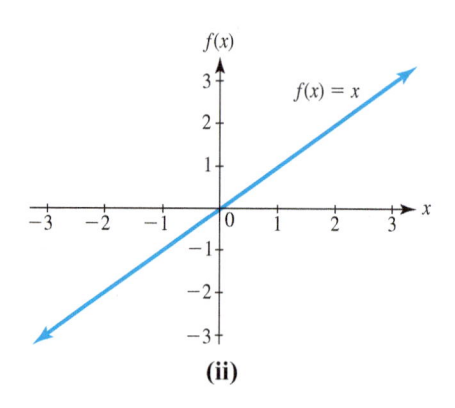

$y = 3$

(i)

$f(x) = x$

(ii)

30. What is the slope of the identity function? 1

31. a. What is the slope of the constant function $y = k$? 0

b. What is the rate of change of a constant function? 0

32. What is the rate of change of the identity function? 1

EXERCISES 1.3

Answers that are not seen can be found in the answer section at the back of the text.

33. **Lithium Demand** Because of the use of lithium batteries in all forms of applications, the demand for lithium is projected to steadily increase. Using data from 2017 and projected to 2025, the function $D = 22.71t + 183.6$ can be used to model the demand for lithium in metric tons, where t is the number of years after 2015.
   (*Source*: Statista)

   a. Is this a linear function? Why or why not?
      Yes; it has form $y = mx + b$.
   b. What is the slope of the graph of this function?
      22.71
   c. What is the D-intercept of the graph of this function?
      183.6

34. **Women in the Workforce** Using data and projections from 1950 through 2050 gives the number y (in thousands) of women in the workforce as the function $y = -0.005x^2 + 1.278x + 13.332$, where x is the number of years from 1950. Is this a linear function? Why or why not?
   (*Source*: U.S. Census Bureau, U.S. Dept. of Commerce)
   No; it cannot be written in the form $y = ax + b$.

35. **Life Expectancy at 65** Using data for selected years from 1950 and projected to 2050, the function $y = 0.077x + 13.827$ gives the life expectancy at age 65, with y equal to the number of additional years of expected life at age 65 and x equal to the number of years after 1950.

   a. Why is this a linear function?
      It is written in the form $y = mx + b$.
   b. Find and interpret the slope of the graph of this function. $m = 0.077$; the life expectancy is projected to increase
   (*Source*: Social Security Administration) by approximately 0.1 year per year.

36. **Student Loans** The balance of federal direct student loans, $S(t)$, in \$ billions, for selected years from 2011 and projected to 2023 can be modeled by $S = 130.7t + 699.7$, with t equal to the number of years after 2011 and S equal to the balance, in billions of dollars.

   a. What is the slope of the graph of this function?
      130.7
   b. Interpret the slope in the context of this problem.

Year	Balance ($ billions)	Year	Balance ($ billions)
2011	702	2019	1775
2013	940	2021	2000
2015	1220	2023	2250
2017	1500		

   (*Source*: U.S. Office of Management and Budget)
   The balance of federal direct student loans is projected to increase at a rate of \$130.7 billion per year.

37. **Depreciation** For tax purposes, a business depreciates its \$300,000 building using "straight-line depreciation" over 15 years by reducing its value by 1/15 of its original value each year. If we assume that the value of the building can be found or estimated at any point of time during this period, the value of the building can be modeled by the function $y = 300,000 - 20,000x$, where x is the number of years.

   a. Find and interpret the y-intercept of the graph of this function. 300,000; at the beginning of the 15 years, the building is worth \$300,000.
   b. Find and interpret the x-intercept of the graph of this function. 15; the value of the building is \$0 at the end of the 15 years.
   c. Use the intercepts to graph this function for values that give nonnegative values for x and y.

38. **Advertising** An advertising agency has determined that if it uses one of its advertising plans to introduce a new product in a city of 650,000 people, the rate of change in the number of people who are aware of the product is given by $R = 39,000 - 0.06x$, where x is the number of people who already are aware of the product.

   a. Find the R-intercept of the graph of this function.
      39,000
   b. Find and interpret the x-intercept of the graph of this equation. 650,000; the rate of change will be 0 when 650,000 people in the city already know about the product.
   c. Use the intercepts to graph this function.

39. **Advertising Impact** An advertising agency has found that when it promotes a new product in a city, the weekly rate of change R of the number of people who are aware of it x weeks after it is introduced is given by $R = 3500 - 70x$. Find the x-and R-intercepts and then graph the function on a viewing window that is meaningful in the application.

40. **Learning Rate** In a study using 50 foreign language vocabulary words, the learning rate R (in words per minute) was found to be related to the number of words already learned, x, according to

   $$10R + 4x = 200$$

   a. Find the R-intercept and the x-intercept of the graph of this equation. R: 20; x: 50
   b. Use the intercepts to graph the equation for nonnegative values of R and x.

Chapter 1 Functions, Graphs, and Models; Linear Functions

41. Life Insurance The monthly rates for a $100,000 life insurance policy for males aged 27–32 are shown in the table.

Age (years), x	27	28	29
Premium (dollars per month), y	11.81	11.81	11.81

Age (years), x	30	31	32
Premium (dollars per month), y	11.81	11.81	11.81

a. Could the data in this table be modeled by a constant function or the identity function? Constant function
b. Write an equation whose graph contains the data points in the table. $y = 11.81$
c. What is the slope of the graph of the function found in part (b)? 0
d. What is the rate of change of the data in the table? 0

42. Emigration The number of migrants per 1000 population from Spanish Caribbean and Latin American countries to the United States is projected in the following table.

2020	1.16	2045	1.16
2025	1.16	2050	1.16
2030	1.16	2055	1.16
2035	1.16	2060	1.16
2040	1.16		

(*Source*: U.S. Census Bureau)

a. Sketch the data as a scatter plot, with y equal to the given rate of emigration and x equal to the year.
b. Could the data be better modeled by a constant function or the identity function? constant
c. Write the equation of the function that fits the data points. $y = 1.16$
d. Sketch the graph of the function you found in part (c) on the same axes as the scatter plot.

43. Disposable Income Disposable income is the amount left after taxes have been paid and is one measure of health of the economy. The total disposable income for the United States, in billions of dollars, for selected years from 2010 and projected to 2040, can be modeled by $D(x) = 0.328x + 6.32$, where x is the number of years after 2000. (*Source*: U.S. Bureau of Economic Analysis)

a. What is the slope of the graph of this function? 0.328
b. Interpret this slope as a rate of change.
(*Source*: Bureau of Economic Analysis) From 2010 through 2040, the disposable income is projected to increase by $328 million per year.

44. Diabetes The figure below shows the projected percent of U.S. adults with diabetes for the years 2010 through 2050. What is the average rate of growth over this period of time? 0.465 percentage point per year

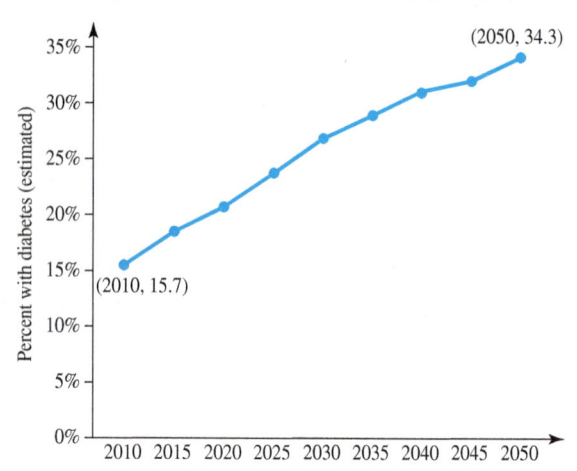

(*Source*: Centers for Disease Control and Prevention)

45. Women in the Workforce Using data from 1950 and projected to 2050, the number of women (in millions) in the workforce is given by the function $P = 0.79x + 20.86$, where x equals the number of years after 1950.
(*Source*: U.S. Census Bureau)

a. Find and interpret the P-intercept of the graph of this function. 20.86; approximately 21 million women were in the workforce in 1950.
b. What is the slope of the graph of this function? 0.79
c. Interpret the slope as a rate of change. The rate of change of the number of women in the workforce is 790 thousand per year.

46. World Population The world population from 2000 and projected to 2050, in billions, is given by the function $y = 0.063x + 6.191$, where x is the number of years after 2000.
(*Source*: U.S. Census Bureau)

a. What is the slope of the graph of this function? 0.063
b. Interpret this slope as a rate of change. The world population is projected to grow by 63 million per year during this period.

47. Black Population Using data and projections from 1990 through 2050, the percent of the U.S. population that is black can be modeled by $B(x) = 0.057x + 12.3$, where x is the number of years after 1990.

a. What is the slope of the graph of this function? 0.057
b. Interpret the slope as a rate of change.
(*Source*: U.S. Census Bureau) From 1990 to 2005, percent increased by 0.057 percentage point per year.

48. Seawater Pressure In seawater, the pressure p is related to the depth d according to the model $33p - 18d = 496$, where d is the depth in feet and p is in pounds per square inch.

a. What is the slope of the graph of this function? $\frac{6}{11}$
b. Interpret the slope as a rate of change.

49. **Crickets** The number of times per minute n that a cricket chirps can be modeled as a function of the Fahrenheit temperature T. The data can be approximated by the function

$$n = \frac{12T}{7} - \frac{52}{7}$$

   a. Is the rate of change of the number of chirps positive or negative? Positive

   b. What does this tell us about the relationship between temperature and the number of chirps? As the temperature increases, the number of chirps increases.

50. **EV/PHEV Sales** Using data from 2017 and projected to 2025, the percent of U.S. sales that are electric or plug-in hybrid electric vehicles can be modeled by the function $y = 0.770x + 1.271$, where x is the number of years after 2017. (*Source*: International Energy Agency)

   a. Find the slope and the y-intercept of the graph of this function. $m = 0.770$; $b = 1.271$

   b. What does the y-intercept of the graph represent in this context? The percent of U.S. sales that are electric or plug-in hybrid electric vehicles was approximately 1.271% in 2017.

   c. What does the slope of the graph represent?

51. **China Population** Using data from 2000 and projected to 2025, the population of China aged 15 years and older can be modeled by the function

$$y = 9.32x + 989$$

   where y is the population in millions and x is the number of years after 2000.
   (*Source*: World Health Organization)

   a. Find the slope and y-intercept of the graph of this function. $m = 9.32$; $b = 989$

   b. What interpretation could be given to the slope?

52. **Global Telecom Spending** Using data from 2018 through 2022, global telecom spending, in billions of dollars, can be modeled by the function $y = 14.6x + 1618.6$, where x is the number of years after 2015.
   (*Source*: Statista)

   a. According to the model, what is the rate of growth in spending during this period? $14.6 billion per year

   b. According to the model, what is the projected global telecom spending for 2025? $1764.6 billion

   c. What does the y-intercept tell us about global telecom spending? It was approximately 1618.6 billion in 2015.

53. **Depreciation** Suppose the cost of a business property is $1,920,000 and a company depreciates it with the straight-line method. Suppose V is the value of the property after x years and the line representing the value as a function of years passes through the points $(10, 1{,}310{,}000)$ and $(20, 700{,}000)$.

   a. What is the slope of the line through these points? $-61{,}000$

   b. What is the annual rate of change of the value of the property? $-$61,000 per year

54. **Men in the Workforce** The number of men in the workforce (in millions) for the years from 1950 to 2050 can be approximated by the linear model determined by connecting the points $(1950, 43.8)$ and $(2050, 100.3)$.

   a. Find the annual rate of change of the model whose graph is the line connecting these points. 0.565

   b. What does this tell us about men in the workforce? (*Source*: U.S. Census Bureau) The number of men in the workforce increased by 565,000 per year from 1950 to 2050.

55. **Profit** A company charting its profits notices that the relationship between the number of units sold x and the profit P is linear. If 300 units sold results in $4650 profit and 375 units sold results in $9000 profit, find the marginal profit, which is the rate of change of the profit. $58 per unit

56. **Cost** A company buys and retails baseball caps, and the total cost function is linear. The total cost for 200 caps is $2690, and the total cost for 500 caps is $3530. What is the marginal cost, which is the rate of change of the function? $2.80 per unit

57. **Marginal Cost** Suppose the monthly total cost for the manufacture of golf balls is $C(x) = 3450 + 0.56x$, where x is the number of balls produced each month.

   a. What is the slope of the graph of the total cost function? 0.56

   b. What is the marginal cost (rate of change of the cost function) for the product? $0.56 per ball

   c. What is the cost of each additional ball that is produced in a month? The cost will increase by $0.56 for each additional ball produced in a month.

58. **Marginal Cost** Suppose the monthly total cost for the manufacture of 19-inch television sets is $C(x) = 2546 + 98x$, where x is the number of TVs produced each month.

   a. What is the slope of the graph of the total cost function? 98

   b. What is the marginal cost for the product? $98 per unit

   c. Interpret the marginal cost for this product. Manufacturing one additional television increases the cost by $98.

59. **Marginal Revenue** Suppose the monthly total revenue for the sale of golf balls is $R(x) = 1.60x$, where x is the number of balls sold each month.

   a. What is the slope of the graph of the total revenue function? 1.60

b. What is the marginal revenue for the product?
$1.60 per ball
c. Interpret the marginal revenue for this product.
The revenue will increase by $1.60 for each additional ball sold in a month.

60. Marginal Revenue Suppose the monthly total revenue from the sale of 19-inch television sets is $R(x) = 198x$, where x is the number of TVs sold each month.

a. What is the slope of the graph of the total revenue function? 198

b. What is the marginal revenue for the product?
$198 per unit
c. Interpret the marginal revenue for this product.
Selling one additional television each month increases revenue by $198.

61. Profit The profit for a product is given by $P(x) = 19x - 5060$, where x is the number of units produced and sold. Find the marginal profit for the product.
$19 per unit

62. Profit The profit for a product is given by the function $P(x) = 939x - 12,207$, where x is the number of units produced and sold. Find the marginal profit for the product. $939 per unit

1.4 Equations of Lines

KEY OBJECTIVES

- Write equations of lines using the slope-intercept form and using the point-slope form
- Write equations of horizontal and vertical lines
- Write the equations of lines parallel or perpendicular to given lines
- Find the average rate of change over an interval for nonlinear functions
- Find the slope of the secant line between two points on a graph
- Find the difference quotient from $(x, f(x))$ to $(x + h, f(x + h))$
- Find average rates of change for approximately linear data

SECTION PREVIEW   Depreciation

For tax purposes, a business can depreciate its property and equipment over a number of years. One method of depreciation is called *straight-line depreciation* because it reduces the value of the property by an equal amount per year. For example, if an $800,000 building is depreciated by this method over 10 years, the value of the building is reduced by 1/10 of its original value, or $80,000 each year until its depreciated value is $0. The value of the building at the end of each year is shown in Figure 1.39. Clearly, the points representing the depreciated values lie on a line, with the rate of change of the building's value equal to the slope of the line. In Example 3, we will use the slope and one of the points on this line to write its equation.

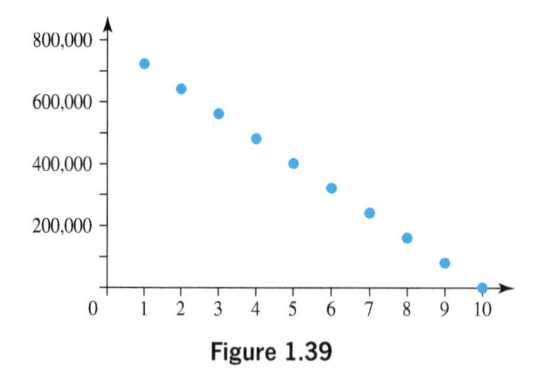

Figure 1.39

In this section, we see how to write the equation of a linear function from information about the line, such as the slope and a point on the line or two points on the line. We also discuss **average rates of change** for data that can be modeled by nonlinear functions, slopes of secant lines, and how to create linear models that approximate data that are nearly linear. ■

Writing Equations of Lines

Creating a linear model from data involves writing a linear equation that describes the mathematical situation. If we know two points on a line or the slope of the line and one point, we can write the equation of the line.

Recall that if a linear equation has the form $y = mx + b$, then the coefficient of x is the slope of the line that is the graph of the equation, and the constant b is the y-intercept of the line. Thus, if we know the slope and the y-intercept of a line, we can write the equation of this line.

Slope-Intercept Form of the Equation of a Line

The slope-intercept form of the equation of a line with slope m and y-intercept b is

$$y = mx + b$$

In an applied context, m is the rate of change and b is the initial value (when $x = 0$).

EXAMPLE 1 ▶ **Appliance Repair**

An appliance repairman charges $60 for a service call plus $25 per hour for each hour spent on the repair. Assuming his service call charges can be modeled by a linear function of the number of hours spent on the repair, write the equation of the function.

SOLUTION

Let x represent the number of hours spent on the appliance repair and let y be the service call charge in dollars. The slope of the line is the amount that the charge increases for every hour of work done, so the slope is 25. Because $60 is the basic charge before any time is spent on the repair, 60 is the y-intercept of the line. Substituting these values in the slope-intercept form of the equation gives the equation of the linear function modeling this situation. When the repairman works x hours on the service call, the charge is

$$y = 25x + 60 \text{ dollars}$$

We next consider a form of an equation of a line that can be written if we know the slope and a point. If the slope of a line is m, then the slope between a fixed point (x_1, y_1) and any other point (x, y) on the line is also m. That is,

$$m = \frac{y - y_1}{x - x_1}$$

Solving for $y - y_1$ (that is, multiplying both sides of this equation by $x - x_1$) gives the **point-slope form** of the equation of a line.

Point-Slope Form of the Equation of a Line

The equation of the line with slope m that passes through a known point (x_1, y_1) is

$$y - y_1 = m(x - x_1)$$

64 Chapter 1 Functions, Graphs, and Models; Linear Functions

EXAMPLE 2 ▶ Using a Point and Slope to Write an Equation of a Line

Write an equation for the line that passes through the point $(-1, 5)$ and has slope $\frac{3}{4}$.

SOLUTION

We are given $m = \frac{3}{4}$, $x_1 = -1$, and $y_1 = 5$. Substituting in the point-slope form, we obtain

$$y - 5 = \frac{3}{4}[x - (-1)]$$

$$y - 5 = \frac{3}{4}(x + 1)$$

$$y - 5 = \frac{3}{4}x + \frac{3}{4}$$

$$y = \frac{3}{4}x + \frac{23}{4}$$

Figure 1.40 shows the graph of this line.

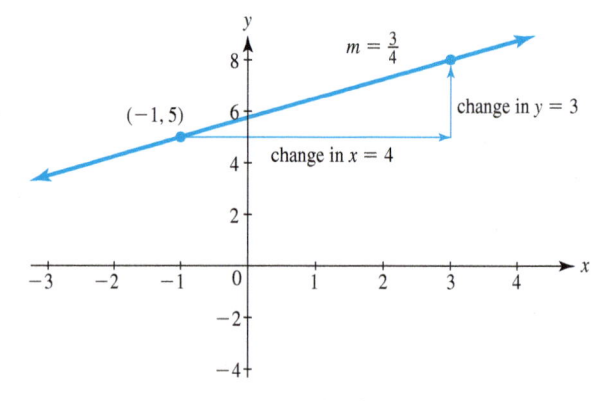

Figure 1.40

EXAMPLE 3 ▶ Depreciation

A business depreciates its $800,000 building using "straight-line depreciation" over 10 years, with the depreciated values (in dollars) at the end of each of the 10 years given in Table 1.17. In the Section Preview on page 62, we saw that the depreciated values of this building fit on a line.

a. Is the rate at which the property is depreciated a constant? What is it?

b. Write the equation of the function that models the depreciated values of the property as a function of the number of years.

Table 1.17

Year	1	2	3	4	5	6	7	8	9	10
Value	720,000	640,000	560,000	480,000	400,000	320,000	240,000	160,000	80,000	0

SOLUTION

a. The rate at which the property is depreciated is a constant. The depreciated value y decreases by 80,000 dollars each year, so the rate of change is $-80,000$ per year.

b. The rate of change is a constant, so the function is linear with its rate of change (and slope of its graph) equal to $-80,000$. Using this slope and any point on the graph gives the equation of the function, with x equal to the number of years and y equal to the corresponding depreciated value of the building. Using the point (5, 400,000) gives the equation

$$y - 400{,}000 = -80{,}000(x - 5), \quad \text{or} \quad y = -80{,}000x + 800{,}000$$

If we know two points on a line, we can find the slope of the line and use either of the points and this slope to write the equation of the line.

EXAMPLE 4 ▶ Using Two Points to Write an Equation of a Line

Write the equation of the line that passes through the points $(-1, 5)$ and $(2, 4)$.

SOLUTION

Because we know two points on the line, we can find the slope of the line.

$$m = \frac{4 - 5}{2 - (-1)} = \frac{-1}{3} = -\frac{1}{3}$$

We can now substitute one of the points and the slope in the point-slope form to write the equation. Using the point $(-1, 5)$ gives

$$y - 5 = -\frac{1}{3}[x - (-1)], \quad \text{or} \quad y - 5 = -\frac{1}{3}x - \frac{1}{3}, \quad \text{so} \quad y = -\frac{1}{3}x + \frac{14}{3}$$

Using the point (2, 4) gives

$$y - 4 = -\frac{1}{3}(x - 2)$$

$$y - 4 = -\frac{1}{3}x + \frac{2}{3}$$

$$y = -\frac{1}{3}x + \frac{14}{3}$$

Notice that the equations are the same, regardless of which of the two given points is used in the point-slope form.

If the rate of change of the outputs with respect to the inputs is a constant and we know two points that satisfy the conditions of the application, then we can write the equation of the linear function that models the application.

EXAMPLE 5 ▶ Illinois State Prison Population

The Illinois Department of Corrections projects that the state prison population will grow at a constant rate from 2020 until 2025, with 48 thousand inmates in 2020 and 52 thousand projected in 2025.

a. What is the rate of growth during this period?

b. Write the equation that models the number N of prisoners (in thousands) during this period, with x equal to the number of years.

c. Use the function to estimate the prison population in 2030.

66 Chapter 1 Functions, Graphs, and Models; Linear Functions

SOLUTION

a. The rate of growth is constant, so the data fit on a line connecting (2020, 48) and (2025, 52). The rate of change during this period is the slope of this line, given by

$$m = \frac{52 - 48}{2025 - 2020} = \frac{4}{5} = 0.8$$

Thus, the rate of growth of the Illinois state prison population during this period is $0.8(1000) = 800$ prisoners per year.

b. Substituting the slope in the point-slope form of the equation of a line (with either point) gives the equation of the line that contains the two points and thus models the application.

$$N - 48 = .8(x - 2020) \quad \text{or} \quad N = 0.8x - 1568$$

c. Evaluating the function at 2030 gives

$$N = 0.8(2030) - 1568 = 56$$

Thus, the prison population in Illinois is expected to be 56,000 in 2030.

EXAMPLE 6 ▶ Writing Equations of Horizontal and Vertical Lines

Write equations for the lines that pass through the point $(-1, 5)$ and have

a. slope 0. **b.** undefined slope.

SOLUTION

a. If $m = 0$, the point-slope form gives us the equation

$$y - 5 = 0[x - (-1)]$$
$$y = 5$$

Because the output is always the same value, the graph of this linear function is a horizontal line. See Figure 1.41(a).

b. Because m is undefined, we cannot use the point-slope form to write the equation of this line. Lines with undefined slope are vertical lines. Every point on the vertical line through $(-1, 5)$ has an x-coordinate of -1. Thus, the equation of the line is $x = -1$. Note that this equation does not represent a function. See Figure 1.41(b).

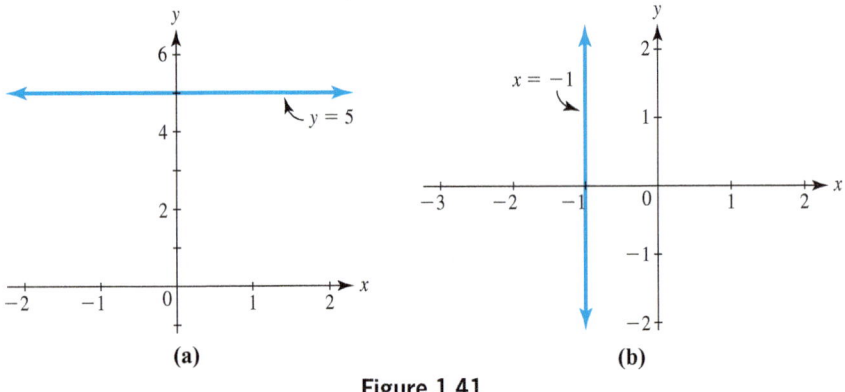

Figure 1.41

As we saw in Example 6, parts (a) and (b), there are special forms when the lines are horizontal or vertical.

Vertical and Horizontal Lines

A vertical line has the form $x = a$, where a is a constant and a is the x-coordinate of any point on the line.

A horizontal line has the form $y = b$, where b is a constant and b is the y-coordinate of any point on the line.

Parallel and Perpendicular Lines

Clearly horizontal lines are perpendicular to vertical lines. Two distinct nonvertical lines that have the same slope are **parallel**, and conversely. If a line has slope $m \neq 0$, any line **perpendicular** to it will have slope $-\dfrac{1}{m}$. That is, the slopes of perpendicular lines are negative reciprocals of each other if neither line is horizontal.

EXAMPLE 7 ▶ Parallel and Perpendicular Lines

Write the equation of the line through $(4, 5)$ and

a. parallel to the line with equation $3x + 2y = -1$.

b. perpendicular to the line with equation $3x + 2y = -1$.

SOLUTION

a. To find the slope of the line with equation $3x + 2y = -1$, we solve for y.

$$3x + 2y = -1$$
$$2y = -3x - 1$$
$$y = -\frac{3}{2}x - \frac{1}{2}$$

The line through $(4, 5)$ parallel to this line has the same slope, $-\dfrac{3}{2}$, and its equation is

$$y - 5 = -\frac{3}{2}(x - 4)$$
$$y = -\frac{3}{2}x + 11$$

b. The line through $(4, 5)$ perpendicular to $3x + 2y = -1$ has slope $\dfrac{2}{3}$, and its equation is

$$y - 5 = \frac{2}{3}(x - 4)$$
$$y = \frac{2}{3}x + \frac{7}{3}$$

In Example 7(b) we found that the equation of the line was

$$y = \frac{2}{3}x + \frac{7}{3}$$

By multiplying both sides of this equation by 3 and writing the new equation with x and y on the same side of the equation, we have a new form of the equation:

$$3y = 2x + 7, \quad \text{or} \quad 2x - 3y = -7$$

This equation is called the **general form** of the equation of the line.

General Form of the Equation of a Line

The general form of the equation of a line is $ax + by = c$, where a, b, and c are real numbers, with a and b not both equal to 0.

In summary, these are the forms we have discussed for the equation of a line.

Forms of Linear Equations

General form:	$ax + by = c$	where a, b, and c are real numbers, with a and b not both equal to 0.
Point-slope form:	$y - y_1 = m(x - x_1)$	where m is the slope of the line and (x_1, y_1) is a point on the line.
Slope-intercept form:	$y = mx + b$	where m is the slope of the line and b is the y-intercept.
Vertical line:	$x = a$	where a is a constant, and a is the x-coordinate of any point on the line. The slope is undefined.
Horizontal line:	$y = b$	where b is a constant, and b is the y-coordinate of any point on the line. The slope is 0.

Average Rate of Change

If the graph of a function is not a line, the function is nonlinear and the slope of a line joining two points on the curve may change as different points are chosen on the curve. The calculation and interpretation of the slope of a curve are topics you will study if you take a calculus course. However, there is a quantity called the **average rate of change** that can be applied to any function relating two variables.

In general, we can find the average rate of change of a function between two input values if we know how much the function outputs change between the two input values.

Average Rate of Change

The average rate of change of $f(x)$ with respect to x over the interval from $x = a$ to $x = b$ (where $a < b$) is calculated as

$$\text{average rate of change} = \frac{\text{change in } f(x) \text{ values}}{\text{corresponding change in } x \text{ values}} = \frac{f(b) - f(a)}{b - a}$$

How is the average rate of change over some interval of points related to the slope of a line connecting the points? For any function, the average rate of change between two points on its graph is the slope of the line joining the two points. Such a line is called a **secant line**.

EXAMPLE 8 ▶ Telecom Artificial Intelligence

The total worldwide software revenue for artificial intelligence used in telecommunications for the years from 2016 and projected to 2025 is shown in Figure 1.42(a). The revenue can be modeled by the function

$$R(t) = 170t^2 - 2356t + 8462$$

where t is the number of years after 2010 and $R(t)$ is in millions of dollars. A graph of $R(t)$ is shown in Figure 1.42(b). (*Source*: Tractica)

a. Find the average rate of change of $R(t)$ between $(6, R(6))$ and $(15, R(15))$ using the model.

b. Interpret the answer to part (a).

c. What is the relationship between the average rate of change and the slope of the secant line joining the points (6, 446) and (15, 11,372)?

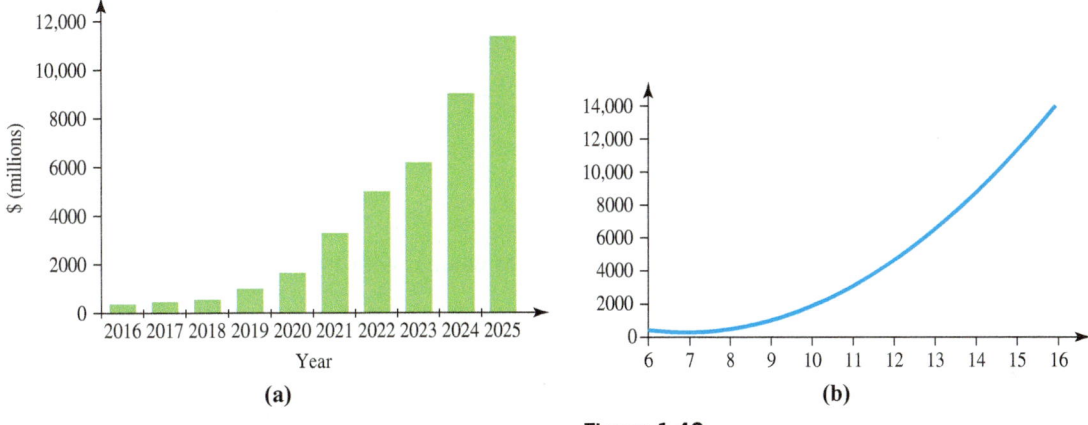

Figure 1.42

SOLUTION

a. Using the model to find the average rate of change of (t) between the two points gives

$$\frac{R(15) - R(6)}{15 - 6} = \frac{11{,}372 - 446}{9} = 1214$$

b. One possible interpretation is that the worldwide software revenue for artificial intelligence used in telecommunications is projected to increase by an average of $1214 million per year during the period from 2016 through 2025.

c. The slope of the secant line, shown in Figure 1.43, is numerically the same as the average rate of change of revenue between 2016 and 2025.

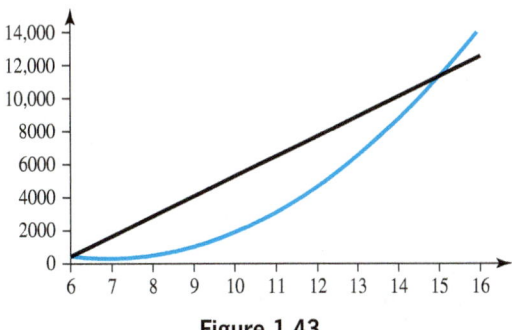

Figure 1.43

Difference Quotient

If we have points $(x, f(x))$ and $(x + h, f(x + h))$ on the graph of $y = f(x)$, we can find the slope of the secant line (and average rate of change) of the function from x to $x + h$, called the **difference quotient**, as follows.

Chapter 1 Functions, Graphs, and Models; Linear Functions

Difference Quotient

For $h \neq 0$, the slope of the secant line joining the points $(x, f(x))$ and $(x + h, f(x + h))$, and the average rate of change of the function $f(x)$ from x to $x + h$, is

$$\frac{f(x+h) - f(x)}{x+h-x} = \frac{f(x+h) - f(x)}{h}$$

EXAMPLE 9 ▶ Difference Quotient

For the function $f(x) = x^2 + 1$, whose graph is shown in Figure 1.44, find

a. $f(x + h)$.

b. $f(x + h) - f(x)$.

c. the average rate of change with the difference quotient $\dfrac{f(x+h) - f(x)}{h}$, for $h \neq 0$.

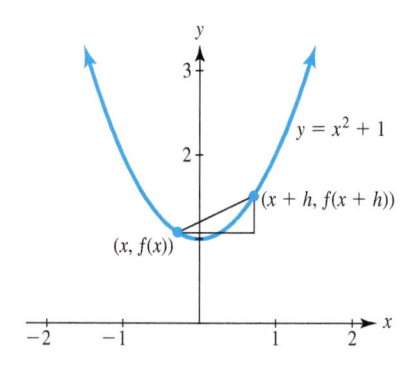

Figure 1.44

SOLUTION

a. $f(x + h)$ is found by substituting $x + h$ for x in $f(x) = x^2 + 1$:

$$f(x + h) = (x + h)^2 + 1 = (x^2 + 2xh + h^2) + 1$$

b. $f(x + h) - f(x) = (x^2 + 2xh + h^2) + 1 - (x^2 + 1)$
$$= x^2 + 2xh + h^2 + 1 - x^2 - 1 = 2xh + h^2$$

c. The slope of the secant line and the average rate of change of the function is

$$\frac{f(x+h) - f(x)}{h} = \frac{2xh + h^2}{h} = 2x + h, \text{ for } h \neq 0.$$

EXAMPLE 10 ▶ Average Velocity

Suppose a ball is shot straight upward at 64 feet per second. The equation that gives the height of the ball t seconds after it is shot is $f(t) = 64t - 16t^2$.

a. Find the average velocity (the average rate of change of the height per second) of this ball over the time interval from $t = 1$ to $t = 2$.

b. Find the difference quotient that can be used to find the average velocity of this ball over the time interval from $t = 1$ to $t = 1 + h$, for $h \neq 0$.

SOLUTION

a. The average velocity over the time interval from $t = 1$ to $t = 2$ is the change in the distance traveled divided by the difference in time.

$$V_{ave} = \frac{f(2) - f(1)}{2 - 1} = \frac{(128 - 64) - (64 - 16)}{1} = 16 \text{ feet per second}$$

b. To find the average velocity over the time interval from $t = 1$ to $t = 1 + h$, $h \neq 0$, we use the difference quotient.

$$V_{ave} = \frac{f(1 + h) - f(1)}{1 + h - 1} = \frac{[(64(1 + h) - 16(1 + h)^2)] - [(64 - 16)]}{h}$$

$$= \frac{64 + 64h - 16(1 + 2h + h^2) - 48}{h}$$

$$= \frac{64 + 64h - 16 - 32h - 16h^2 - 48}{h}$$

$$= \frac{32h - 16h^2}{h}$$

$$= \frac{h(32 - 16h)}{h}$$

$$= 32 - 16h, \text{ for } h \neq 0$$

Approximately Linear Data

Real-life data are rarely perfectly linear, but some sets of real data points lie sufficiently close to a line that two points can be used to create a linear function that models the data. Consider the following example.

EXAMPLE 11 ▶ Public School Enrollment

Table 1.18 and Figure 1.45 show the real and projected enrollment (in thousands) at U.S. public schools for selected years from 1980 through 2024. (*Source*: U.S. National Center for Education Statistics)

a. Create a scatter plot of the data. Does a line fit the data points exactly?

b. Find the average rate of change of the high school enrollment between 1980 and 2024.

c. Write the equation of the line determined by this rate of change and one of the two points.

Table 1.18

Year, x	Enrollment, y (thousands)
1980	41,651
2000	46,857
2010	49,373
2012	49,522
2022	51,485
2023	51,604
2024	52,113

(*Source*: U.S. National Center for Education Statistics)

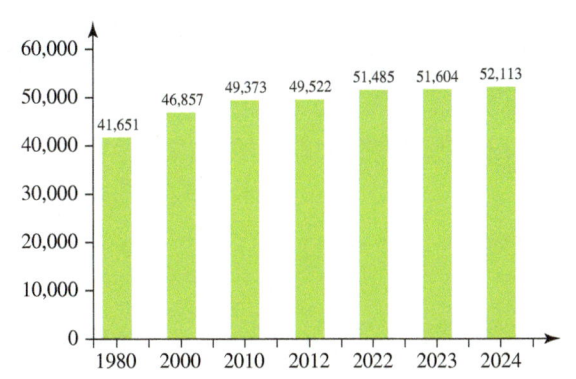

Figure 1.45

SOLUTION

a. A scatter plot is shown in Figure 1.46. The graph shows that the data do not fit a linear function exactly, but that a linear function could be used as an approximate model for the data.

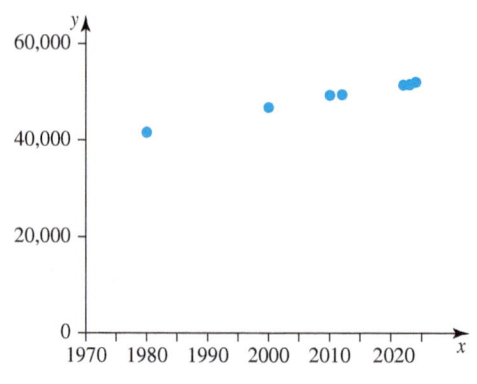

Figure 1.46

b. To find the average rate of change between 1980 and 2024, we use the points $(1980, 41{,}651)$ and $(2024, 52{,}113)$.

$$\frac{52{,}113 - 41{,}651}{2024 - 1980} = 237.8$$

This means that the public school enrollments have grown at an average rate of 237.8 thousand per year (because the outputs are in thousands).

c. Because the enrollment growth is approximately linear, the average rate of change can be used as the slope of the graph of a linear function describing the enrollments from 1980 through 2024. The graph connects the points $(1980, 41{,}651)$ and $(2024, 52{,}113)$, so we can use either of these two points to write the equation of the line. Using the point $(1980, 41{,}651)$ and $m = 237.8$ gives the equation

$$y - 41{,}651 = 237.8(x - 1980)$$
$$y - 41{,}651 = 237.8x - 470{,}844$$
$$y = 237.8x - 429{,}193$$

This equation is a model of the enrollment, in thousands, as a function of the year.

The equation in Example 11 approximates a model for the data, but it is not the best possible model for the data. In Section 2.2, we will see how technology can be used to find the linear function that is the best fit for a set of data of this type.

SKILLS CHECK 1.4

Answers that are not seen can be found in the answer section at the back of the text.

For Exercises 1–18, write the equation of the line with the given conditions.

1. Slope 4 and y-intercept $\frac{1}{2}$ $y = 4x + \frac{1}{2}$

2. Slope 5 and y-intercept $\frac{1}{3}$ $y = 5x + \frac{1}{3}$

3. Slope $\frac{1}{3}$ and y-intercept 3 $y = \frac{1}{3}x + 3$

4. Slope $-\frac{1}{2}$ and y-intercept -8 $y = -\frac{1}{2}x - 8$

5. Through the point $(4, -6)$ with slope $-\frac{3}{4}$ $y = \frac{-3}{4}x - 3$

6. Through the point $(-4, 3)$ with slope $-\frac{1}{2}$ $y = -\frac{1}{2}x + 1$

7. Vertical line, through the point $(9, -10)$ $x = 9$

8. Horizontal line, through the point $(9, -10)$ $y = -10$
9. Passing through $(-2, 1)$ and $(4, 7)$ $y = x + 3$
10. Passing through $(-1, 3)$ and $(2, 6)$ $y = x + 4$
11. Passing through $(5, 2)$ and $(-3, 2)$ $y = 2$
12. Passing through $(9, 2)$ and $(9, 5)$ $x = 9$
13. x-intercept -5 and y-intercept 4 $y = \frac{4}{5}x + 4$
14. x-intercept 4 and y-intercept -5 $y = \frac{5}{4}x - 5$
15. Passing through $(4, -6)$ and parallel to the line with equation $3x + y = 4$ $y = -3x + 6$
16. Passing through $(5, -3)$ and parallel to the line with equation $2x + y = -3$ $y = -2x + 7$
17. Passing through $(-3, 7)$ and perpendicular to the line with equation $2x + 3y = 7$ $y = \frac{3}{2}x + \frac{23}{2}$
18. Passing through $(-4, 5)$ and perpendicular to the line with equation $3x + 2y = -8$ $y = \frac{2}{3}x + \frac{23}{3}$

For Exercises 19–22, write the equation of the line whose graph is shown.

19. $y = 3x + 1$

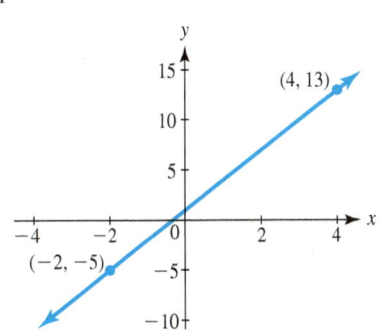

20. $y = -3x - 5$

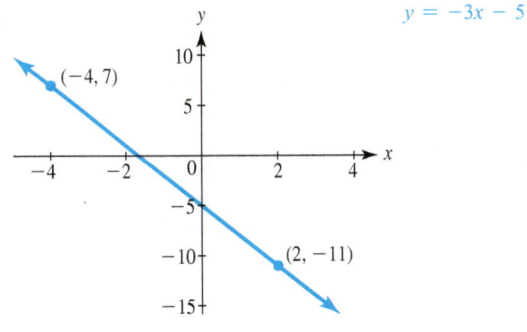

21. $y = -5$

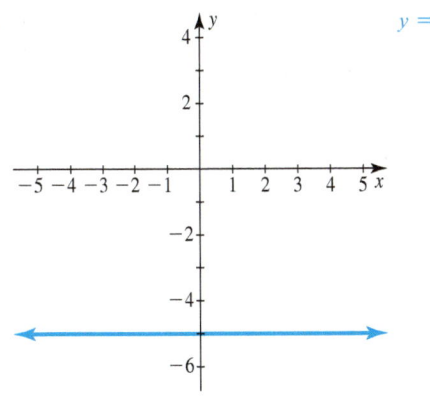

22. $x = 3$

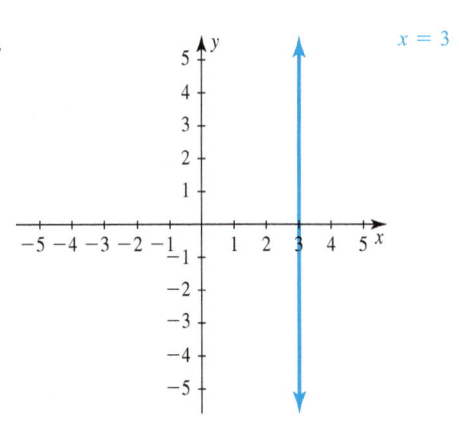

23. Write the equation of a function if its rate of change is -15 and its initial value is 12. $y = -15x + 12$
24. Write the equation of a function if its rate of change is -8 and its initial value is -7. $y = -8x - 7$
25. If a function has rate of change $\frac{2}{3}$ and $y = 9$ when $x = 3$, write its equation. $y = \frac{2}{3}x + 7$
26. If a function has rate of change $-\frac{1}{5}$ and $y = 12$ when $x = -2$, write its equation. $y = -\frac{1}{5}x + \frac{58}{5}$
27. For the function $y = x^2$, compute the average rate of change between $x = -1$ and $x = 2$. 1
28. For the function $y = x^3$, compute the average rate of change between $x = -1$ and $x = 2$. 3
29. For the function shown in the figure, find the average rate of change from $(-2, 7)$ to $(1, -2)$. -3

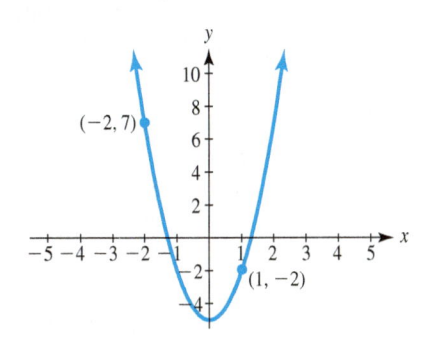

30. For the function shown in the figure, find the average rate of change from $(-1, 2)$ to $(2, -4)$. -2

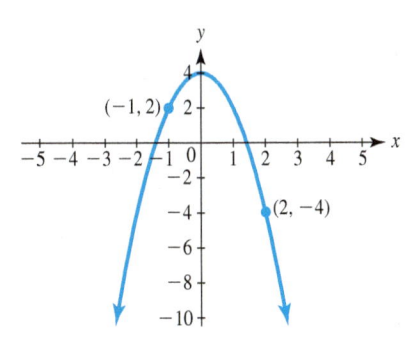

31. If one point on a line is $(1, -10)$ and the line's slope is -3, find the y-intercept. -7

32. If one point on a line is $(-2, 8)$ and the line's slope is $-\frac{3}{2}$, find the y-intercept. 5

For the functions given in Exercises 33–36, find
$$\frac{f(x+h) - f(x)}{h}, \text{ for } h \neq 0$$

33. $f(x) = 45 - 15x$ -15 **34.** $f(x) = 32x + 12$ 32

35. $f(x) = 2x^2 + 4$ $4x + 2h$ **36.** $f(x) = 3x^2 + 1$ $6x + 3h$

In Exercises 37–38, use the tables, which give a set of input values x and the corresponding outputs y that satisfy a function.

37. a. Would you say that a linear function could be used to model the data? Explain.
Yes; a scatter plot shows that data fit along a line.

b. Write the equation of the linear function that fits the data. $y = 3x + 555$

x	10	20	30	40	50
y	585	615	645	675	705

38. a. Verify that the values satisfy a linear function.
b. Write the equation of the linear function that fits the data. $y = \frac{3}{2}x - 2$

x	1	7	13	19
y	−0.5	8.5	17.5	26.5

38. a. The difference in y-coordinates is consistently 9, while the difference in x-coordinates is consistently 6.

EXERCISES 1.4

Answers that are not seen can be found in the answer section at the back of the text.

39. Utility Charges Palmetto Electric determines its monthly bills for residential customers by charging a base price of $12.00 plus an energy charge of 10.34 cents for each kilowatt-hour (kWh) used. Write an equation for the monthly charge y (in dollars) as a function of x, the number of kWh used.
$y = 12.00 + 0.1034x$ dollars

40. Phone Bills For interstate calls, AT&T charges 10 cents per minute plus a base charge of $2.99 each month. Write an equation for the monthly charge y as a function of the number of minutes of use.
$y = 0.10x + 2.99$ dollars

41. Depreciation A business uses straight-line depreciation to determine the value y of a piece of machinery over a 10-year period. Suppose the original value (when $t = 0$) is equal to $36,000 and the value is reduced by $3600 each year. Write the linear equation that models the value y of this machinery at the end of year t. $y = -3600t + 36{,}000$

42. Sleep Each day a young person should sleep 8 hours plus $\frac{1}{4}$ hour for each year the person is under 18 years of age.

a. Based on this information, how much sleep does a 10-year-old need? 10 hours

b. Based on this information, how much sleep does a 14-year-old need? 9 hours

c. Use the answers from parts (a) and (b) to write a linear equation relating hours of sleep y to age x, for $6 \leq x \leq 18$. $y = 12.5 - 0.25x$

d. Use your equation from part (c) to verify that an 18-year-old needs 8 hours of sleep.
$y = 12.5 - 0.25(18) = 8$

43. Heart Rate The desired heart rate for weight loss during exercise is a linear function of a person's age, with the desired rate for a 60-year-old person equal to 104 beats per minute and with the desired rate increasing by 6.5 beats per minute for each decrease of 10 years in age.
(*Source*: paramountfitness.com)

a. Write the equation of the function, with t equal to age and H equal to desired heart rate.
$H = -0.65t + 143$

b. What does the function give as the desired rate for a 40-year-old person? 117

44. Heart Rate The desired heart rate for cardiac health during exercise is a linear function of a person's age, with the desired rate for a 50-year-old person equal to 136 beats per minute and with the desired rate increasing by 8 beats per minute for each decrease of 10 years in age.
(*Source*: paramountfitness.com)

a. Write the equation of the function, with t equal to age and h equal to desired heart rate.
$h = -0.8t + 176$

b. What does the function give as the desired rate for a 75-year-old person? 116 beats per minute

45. Depreciation A business uses straight-line depreciation to determine the value y of an automobile over a 5-year period. Suppose the original value (when $t = 0$) is equal to $26,000 and the salvage value (when $t = 5$) is equal to $1000.

a. By how much has the automobile depreciated over the 5 years? $25,000

b. By how much is the value of the automobile reduced at the end of each of the 5 years? $5000

c. Write the linear equation that models the value s of this automobile at the end of year t.
$s = 26{,}000 - 5000t$

46. **Retirement** For Pennsylvania state employees for whom the average of the three best yearly salaries is $75,000, the retirement plan gives an annual pension of 2.5% of 75,000, multiplied by the number of years of service. Write the linear function that models the pension P in terms of the number of years of service y. (*Source*: Pennsylvania State Retirement Fund) $P = 1875y$

47. **Patrol Cars** The Beaufort County Sheriff's office assigns a patrol car to each of its deputy sheriffs, who keeps the car 24 hours a day. As the population of the county grew, the number of deputies and the number of patrol cars also grew. The data in the following table could describe how many patrol cars were necessary as the number of deputies increased. Write the function that models the relationship between the number of deputies and the number of patrol cars.

Number of Deputies	10	50	75	125	180	200	250
Number of Patrol Cars in Use	10	50	75	125	180	200	250

$y = x$; $x =$ deputies; $y =$ cars

48. **Life Insurance** The Life to 95 insurance plan provides insurance for persons from age 50 to 95 years of age. The monthly premiums (in dollars) for a $50,000 policy for females age 50 through 54 are shown in the table.

   a. Write the linear function that models the premium y for females in this age group as a function of their age x. What type of function is this? $y = 12.83$; constant function
   b. What is the rate of change of the premium for females in this age group? 0

Age	50	51	52	53	54
Premium ($)	12.83	12.83	12.83	12.83	12.83

(*Source*: United State Insurance Company)

49. **Profit** A company charting its profits notices that the relationship between the number of units sold, x, and the profit, P, is linear. If 300 units sold results in $4650 profit and 375 units sold results in $9000 profit, write the equation that models its profit. $P(x) = 58x - 12{,}750$; $x =$ number of units

50. **Cost** A company buys and retails baseball caps. The total cost function is linear, the total cost for 200 caps is $2680, and the total cost for 500 caps is $3580. Write the equation that models this cost function. $y = 3x + 2080$

51. **Depreciation** Suppose the cost of a business property is $1,920,000 and a company depreciates it with the straight-line method. If V is the value of the property after x years and the line representing the value as a function of years passes through the points $(10, 1{,}310{,}000)$ and $(20, 700{,}000)$, write the equation that gives the annual value of the property. $V = -61{,}000x + 1{,}920{,}000$

52. **Depreciation** Suppose the cost of a business property is $860,000 and a company depreciates it with the straight-line method. Suppose y is the value of the property after t years.

   a. What is the value at the beginning of the depreciation (when $t = 0$)? 860,000
   b. If the property is completely depreciated ($y = 0$) in 25 years, write the equation of the line representing the value as a function of years. $y = 860{,}000 - 34{,}400t$

53. **Social Security** Monthly social security benefits differ with the age a recipient decides to start receiving benefits. Recipients whose monthly benefits are $2000, if benefits start being paid at a full retirement age of 66, can have their benefits increase by $160 per month for each year after 66 they delay starting their benefits. Note that the monthly benefit does not increase if the starting age is after 70.

   a. Write the linear function that models the monthly benefit y as a function of age x at which benefits are started, for the starting ages from 66 to 70. $y = 160x - 8560$ for $66 \leq x \leq 70$
   b. What monthly benefit would these recipients receive if they started receiving benefits at age 70? (*Source*: Social Security Administration) 2640

54. **Consumer Price Index** The Social Security Administration predicts that goods and services that cost $100 in 1995 will cost $466 in 2035, so we say that the U.S. consumer price index (CPI) is 100 in 1995 and 466 in 2035.

   a. Write the equation of the linear function $y = f(x)$ that represents the CPI y as a function of the year x. $y = 9.15x - 18{,}154.25$
   b. Use the equation from part (a) to predict the CPI in 2040. 511.75

55. **Drinking and Driving** The following table gives the number of drinks and the resulting blood alcohol percent for a 100-lb woman legally considered to be driving under the influence (DUI). (One drink is equal to 1.25 oz of 80-proof liquor, 12 oz of regular beer, or 5 oz of table wine; many states have set 0.08% as the legal limit for driving under the influence.)

   a. The average rate of change of the blood alcohol percent with respect to the number of drinks is a constant. What is it? 0.045 percentage point per drink
   b. Use the rate of change and one point determined by a number of drinks and the resulting blood alcohol percent to write the equation of a linear model for this data. $y = 0.045x$

Number of Drinks	2	3	4	5	6	7
Blood Alcohol Percent	0.09	0.135	0.18	0.225	0.27	0.315

(*Source*: Pennsylvania Liquor Control Board)

76 Chapter 1 Functions, Graphs, and Models; Linear Functions

56. Blood Alcohol Percent The table below gives the number of drinks and the resulting blood alcohol percent for a 220-lb man.

  a. The rate of change in blood alcohol percent per drink for a 220-lb man is a constant. What is it? 0.017 percentage point per drink
  b. Write the equation of the function that models the blood alcohol percent as a function of the number of drinks. $f(x) = 0.017x$

Number of Drinks	0	1	2	3	4
Blood Alcohol Percent	0	0.017	0.034	0.051	0.068

Number of Drinks	5	6	7	8	9
Blood Alcohol Percent	0.085	0.102	0.119	0.136	0.153

(*Source*: Pennsylvania Liquor Control Board)

57. Men in the Workforce The number of men in the workforce (in millions) for selected decades from 1950 to 2050 is shown in the figure. The decade is defined by the year at the beginning of the decade, and $g(t)$ is defined by the average number of men (in millions) in the workforce during the decade (indicated by the point on the graph within the decade). The data can be approximated by the linear model determined by the line connecting (1950, 43.8) and (2050, 100.3).

  a. Write the equation of the line connecting these two points to find a linear model for the data. $g(t) = 0.565t - 1057.95$
  b. Does this line appear to be a reasonable fit to the data points? Yes
  c. How does the slope of this line compare with the average rate of change in the function during this period? They are the same.

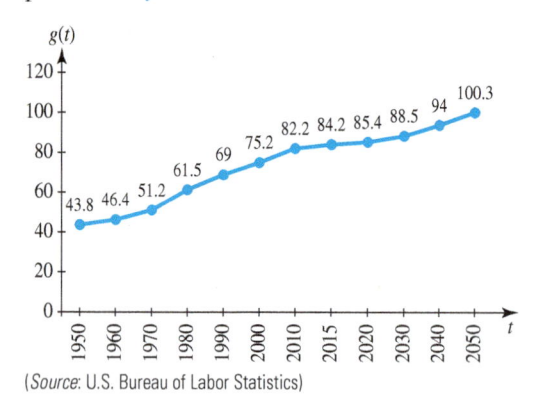

(*Source*: U.S. Bureau of Labor Statistics)

58. World Population The figure shows the real and projected total population w, in billions, of the world for selected years t from 1950 through 2050.

  a. Write the equation of the line connecting the points (1950, 2.556) and (2050, 9.346). $w = 0.0679t - 129.849$
  b. Does this line appear to be a reasonable fit to the data points? Yes
  c. Interpret the slope of this line as a rate of change of the world population. The population is projected to increase by 67.9 million each year.
  d. Use the equation from part (a) to predict what the world population will be in 2025. 7.649 billion

(*Source*: U.S. Census Bureau)

59. Disposable Income Disposable income is the amount left over after taxes have been paid, and it is one measure of the health of an economy. The table gives the total U.S. disposable income, in billions of dollars, for selected years from 2010 and projected to 2040.

  a. Is the line connecting the data points for 2010 and 2040 increasing or decreasing? What is its slope? increasing; 325.6
  b. Write the equation of the line connecting the data points for 2010 and 2040, with x equal to the number of years after 2010 and y equal to the disposable income, in billions of dollars. $y = 325.6x + 10,017$
  c. Do the data points for all the years in the table fit exactly on the graph of the equation in (b)? No

Year	Income	Year	Income
2010	10,017	2030	15,948
2015	11,120	2035	17,752
2020	12,655	2040	19,785
2025	14,259		

(*Source*: U.S. Energy Information Administration)

60. Obesity In 2000, 22.1% of Americans were obese, and the percent is projected to be 42.2% in 2030. Assuming that the growth in obesity is linear:

  a. Find the linear function that models this growth, with x representing the year and p representing the percent. $p = 0.67x - 1317.9$
  b. Use the function to predict what percent of Americans will be obese in 2020. 35.5%

(*Source*: American Journal of Preventive Medicine*)

61. *Women in the Workforce* The number of women in the workforce, based on data and projections from 1950 to 2050, can be modeled by a linear function. The number was 18.4 million in 1950 and is projected to be 81.6 million in 2030. Let x represent the number of years after 1950.

   a. What is the slope of the line through $(0, 18.4)$ and $(80, 81.6)$? 0.79

   b. What is the average rate of change in the number of women in the workforce during this time period? 0.79 million per year

   c. Use the slope from part (a) and the number of millions of women in the workforce in 1950 to write the equation of the line. $y = 0.79x + 18.4$
   (*Source*: U.S. Department of Labor)

62. *Worldwide 5G Shipments* According to Strategy Analytics, global shipments of 5G smartphones will grow from 11 million units in 2020 to 1.5 billion in 2025. Let t equal the number of years after 2020 and y equal the number of shipments, in millions.

   a. What is the slope of the line joining the points representing the number of shipments in 2020 and in 2025? 297.8

   b. What does this tell us about the rate of growth of global shipments of 5G smartphones during this period? The number of global shipments of 5G smartphones is increasing at an average rate of 297.8 million units per year during this period.

   c. Use the slope and one of the points to write the equation of the line whose slope was found in part (a). $y = 297.8t + 11$

63. *E-commerce in ASEAN Region* The Association of Southeast Asian Nations (ASEAN) is a regional grouping that promotes economic, political, and security cooperation among its 10 members: Brunei, Cambodia, Indonesia, Laos, Malaysia, Myanmar, the Philippines, Singapore, Thailand, and Vietnam. The graph shows the forecasted size of the e-commerce market in this region (in billions of dollars) for the years 2015 through 2025. Its size was 6 billion dollars in 2015 and is projected to be 88 billion dollars in 2025.

   a. What is the projected average rate of growth of the e-commerce market from 2015 to 2025? $8.2 billion per year

   b. What is the slope of the line connecting the two points representing the size of the e-commerce market in 2015 and 2025? 8.2

   c. Use the slope from part (b) and either point to write the equation of the secant line joining these two points on the curve. $y = 8.2x - 16{,}517$

   d. Does the secant line from part (c) give a good model for the data shown on the graph? No

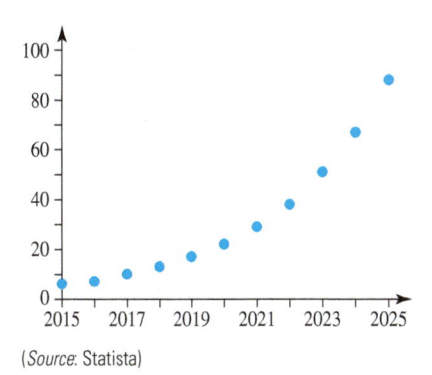

   (*Source*: Statista)

64. *Investment* The graph of the future value of an investment of $1000 for x years earning interest at a rate of 8% compounded continuously is shown in the following figure. The $1000 investment is worth approximately $1083 after 1 year and about $1492 after 5 years.

   a. What is the average rate of change in the future value over the 4-year period? 102.25

   b. Interpret this average rate of change.
   On average, value increases by $102.25 each year.

   c. What is the slope of the line connecting the points satisfying the conditions above? 102.25

   d. Write the equation of the secant line joining the two given points. $y = 102.25x + 980.75$

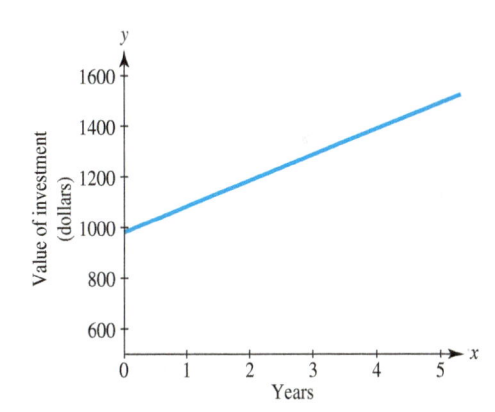

65. *Working Age* The scatter plot below projects the ratio of the working-age population to the elderly.

   a. Do the data appear to fit a linear function? No

   b. The data points shown in the scatter plot from 2010 to 2030 are projections made from a mathematical model. Do those projections appear to be made with a linear model? Explain.
   Yes, they appear to lie on a line.

   c. If the ratio is 3.9 in 2010 and projected to be 2.2 in 2030, what is the average annual rate of change of the data over this period of time? -0.085

   d. Write the equation of the line joining these two points. $y = -0.085x + 174.75$

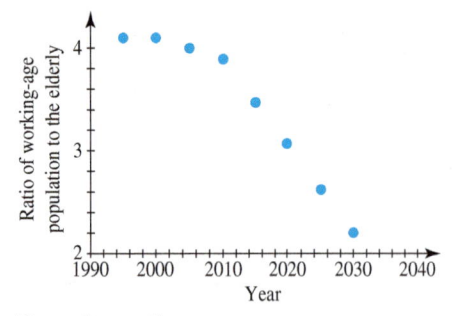

66. **Women in the Workforce** The number of women in the workforce (in millions) for selected years from 1950 to 2050 is shown in the following figure.

   a. Would the data in the scatter plot be modeled exactly by a linear function? Why or why not? No; the points in the scatter plot do not lie approximately on a line.
   b. The number of women in the workforce was 18.4 million in 1950 and 65.7 million in 2000. What is the average rate of change in the number of women in the workforce during this period? 0.964 million women per year (or 964,000 women per year)
   c. If the number of women in the workforce was 65.7 million in 2000 and is projected to be 91.5 million in 2050, what is the average rate of change in the number of women in the workforce during this period? 0.516 million women per year (or 516,000 women per year)
   d. Is it reasonable that these two average rates of change are different? How can you tell this from the graph? Yes; since the graph curves, the average rate of change is not constant.

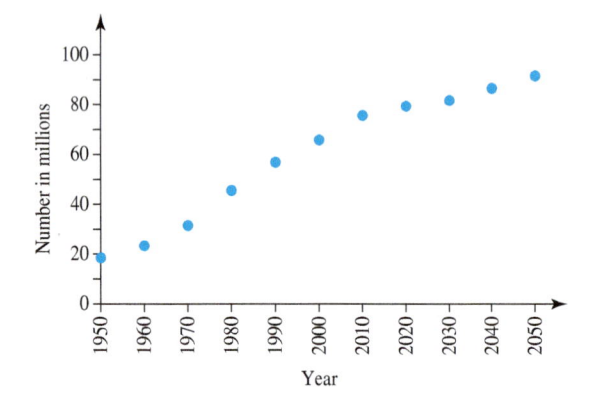

67. **U.S. Population** The number of white non-Hispanic individuals in the U.S. civilian non-institutional population 16 years and older was 153.1 million in 2000 and is projected to be 169.4 million in 2050. (*Source*: U.S. Census Bureau)

   a. Find the average annual rate of change in population during the period 2000–2050, with the appropriate units. 0.326 million per year
   b. Use the slope from part (a) and the population in 2000 to write the equation of the line associated with 2000 and 2050. $y = 0.326x - 498.9$
   c. What does this model project the population to be in 2020? 159.62 million

68. **Social Agency** A social agency provides emergency food and shelter to two groups of clients. The first group has x clients who need an average of $300 for emergencies, and the second group has y clients who need an average of $200 for emergencies. The agency has $100,000 to spend for these two groups.

   a. Write the equation that gives the number of clients who can be served in each group. $300x + 200y = 100{,}000$
   b. Find the y-intercept and the slope of the graph of this equation. Interpret each value. y-intercept 500, slope -1.5
   c. If 10 clients are added from the first group, what happens to the number served in the second group? Fifteen fewer clients can be served from the second group.

69. **Average Velocity** A ball is shot straight upward, with its height, in feet, after t seconds given by the function $f(t) = -16t^2 + 128t$. Find the average velocity of the ball from $t = 2$ to $t = 4$. 32 ft/sec

70. **Average Velocity** A ball is shot straight upward, with its height, in feet, after t seconds given by the function $f(t) = 32t - 16t^2$. Find the average velocity of the ball from $t = 2$ to $t = 3$. -48 ft/sec

71. **Velocity** A ball is shot straight upward, with its height, in feet, after t seconds given by the function $f(t) = -16t^2 + 128t$. Find the difference quotient to find the average velocity of the ball from $t = 2$ to $t = 2 + h$ seconds, with $h \neq 0$. $64 - 16h$

72. **Velocity** A ball is shot straight upward, with its height, in feet, after t seconds given by the function $f(t) = 112t - 16t^2$. Find the difference quotient to find the average velocity of the ball from $t = 3$ to $t = 3 + h$ seconds, with $h \neq 0$. $16 - 16h$

Preparing for CALCULUS

In Chapter 1, we **simplified expressions** by removing parentheses and combining like terms. These skills are often used in calculus. In Exercises 1–2, simplify the expressions.

1. $(x^2 - 3)(2) + (2x)(2x)$ $6x^2 - 6$
2. $(x^3 + 4)(5) + (5x)(3x^2)$ $20x^3 + 20$

We discussed function notation and its use in evaluating functions. Function notation is also used in calculus to evaluate the **derivative**, a new type of function. We can denote the derivative of the function $f(x)$ by $f'(x)$. Evaluate the derivatives at the given values in Exercises 3–5.

3. $f'(2)$ if $f'(x) = 3x^2 - 4x + 2$ 6
4. $f'(2)$ if $f'(x) = 4x^2 + 3x - 2$ 20
5. $f'(-1)$ if $f'(x) = (x^2 - 1)(6x) + (3x^2 + 1)(2x)$ -8

Evaluating the function $f(x + h)$ is also used as part the definition of the derivative. Simplify the expressions in Exercises 6–8.

6. $f(a + h) = 2(a + h) + 5$ $2a + 2h + 5$
7. $f(a + h) = 6(x + h) - 12$ $6x + 6h - 12$
8. $f(x + h) - f(x) = [3(x + h) + 1] - [3x + 1]$ $3h$

In Exercises 9–10, find and simplify $f(x + h)$ for the given functions.

9. $f(x) = 4x - 9$ $4x + 4h - 9$
10. $f(x) = -5x + 7$ $-5x - 5h + 7$

Another important calculation in calculus is the evaluation of the **difference quotient**. The difference quotient is $\dfrac{f(x + h) - f(x)}{h}$, where $h \neq 0$.

11. Find the difference quotient for $f(x) = 3x + 2$. 3
12. Find the difference quotient for $f(x) = -2x - 1$. -2

We also discussed **slope** of lines and **equations of lines**. Calculus can be used to find the slope and equation of the tangent to a curve. The slope of the tangent to the curve $y = f(x)$ at $x = a$ is $f'(a)$.

13. If $f'(x) = 4x^3 - 3x^2 + 1$, find the slope of the tangent at $x = 2$. 21
14. Find the slope of the tangent to $y = f(x)$ at $x = 2$ if $f'(x) = 8x^3 + 6x^2 - 5x - 10$. 68
15. If the slope of the line tangent to the graph of $f(x) = x^2$ at $(1, 1)$ is 2, write the equation of this line. $y = 2x - 1$
16. If the slope of the line tangent to the graph of $f(x) = x^3$ at $(2, 8)$ is 12, write the equation of this line. $y = 12x - 16$

chapter 1 SUMMARY

In this chapter, we studied the basic concept of a function and how to recognize functions with tables, graphs, and equations. We studied linear functions in particular—graphing, finding slope, model formation, and equation writing. We used graphing utilities to graph functions and to evaluate functions. We solved business and economics problems involving linear functions.

Key Concepts and Formulas

1.1 Functions and Models

Scatter plot of data	A scatter plot is a graph of pairs of values that represent real-world information collected in numerical form.
Function definition	A function is a rule or correspondence that determines exactly one output for each input. The function may be defined by a set of ordered pairs, a table, a graph, or an equation.
Domain and range	The set of possible inputs for a function is called its domain, and the set of possible outputs is called its range. In general, if the domain of a function is not specified, we assume that it includes all real numbers except • values that result in a denominator of 0 • values that result in an even root of a negative number
Independent variable and dependent variable	If x represents any element in the domain, then x is called the independent variable; if y represents an output of the function from an input x, then y is called the dependent variable.
Tests for functions	We can test for a function graphically, numerically, and analytically.
Vertical line test	A set of points in a coordinate plane is the graph of a function if and only if no vertical line intersects the graph in more than one point.
Function notation	We denote that y is a function of x using function notation when we write $y = f(x)$.
Modeling	The process of translating real-world information into a mathematical form so that it can be applied and then interpreted in the real-world setting is called modeling.
Mathematical model	A mathematical model is a functional relationship (usually in the form of an equation) that includes not only the function rule but also descriptions of all involved variables and their units of measure.

1.2 Graphs of Functions

Point-plotting method	The point-plotting method of sketching a graph involves sketching the graph of a function by plotting enough points to determine the shape of the graph and then drawing a smooth curve through the points.
Complete graph	A graph is a complete graph if it shows the basic shape of the graph and important points on the graph (including points where the graph crosses the axes and points where the graph turns) and suggests what the unseen portions of the graph will be.
Using a calculator to draw a graph	After writing the function with x representing the independent variable and y representing the dependent variable, enter the function in the equation editor of the graphing utility. Activate the graph with ZOOM or GRAPH. (Set the x- and y-boundaries of the viewing window before pressing GRAPH.)

Viewing windows	The values that define the viewing window can be set manually or by using the ZOOM keys. The boundaries of a viewing window are x_{min}: the smallest value on the x-axis (the left boundary of the window) x_{max}: the largest value on the x-axis (the right boundary of the window) y_{min}: the smallest value on the y-axis (the bottom boundary of the window) y_{max}: the largest value on the y-axis (the top boundary of the window)
Spreadsheets	Spreadsheets like Excel can be used to create accurate graphs, sometimes better looking than those created with graphing calculators.
Aligning data	Using aligned inputs (input values that have been shifted horizontally to smaller numbers), rather than the actual data values, results in smaller coefficients in models and less involved computations.
Evaluating functions with a calculator	We can find (or estimate) the output of a function $y = f(x)$ at specific inputs with a graphing calculator by using TRACE and moving the cursor to (or close to) the value of the independent variable.
Graphing data points	Graphing utilities can be used to create lists of numbers and to create graphs of the data stored in the lists.

1.3 Linear Functions

Linear functions	A linear function is a function of the form $f(x) = ax + b$, where a and b are constants. The graph of a linear function is a line.
Intercepts	A point (or the x-coordinate of the point) where a graph crosses or touches the horizontal axis is called an x-intercept, and a point (or the y-coordinate of the point) where a graph crosses or touches the vertical axis is called a y-intercept. To find the x-intercept(s) of the graph of an equation, set $y = 0$ in the equation and solve for x. To find the y-intercept(s), set $x = 0$ and solve for y.
Slope of a line	The slope of a line is a measure of its steepness and direction. The slope is defined as $$\text{slope} = \frac{\text{vertical change}}{\text{horizontal change}} = \frac{\text{rise}}{\text{run}}$$ If (x_1, y_1) and (x_2, y_2) are two points on a line, then the slope of the line is $$m = \frac{y_2 - y_1}{x_2 - x_1}$$
Slope and y-intercept of a line	The slope of the graph of the equation $y = mx + b$ is m, and the y-intercept of the graph is b. (This is called slope-intercept form.)
Constant rate of change	If a model is linear, the rate of change of the outputs with respect to the inputs will be constant, and the rate of change equals the slope of the line that is the graph of a linear function. Thus, we can determine whether a linear model fits a set of real data by determining whether the rate of change remains constant for the data.
Revenue, cost, and profit	If a company sells x units of a product for p dollars per unit, then the total revenue for this product can be modeled by the linear function $R(x) = px$. $$\text{Profit} = \text{Revenue} - \text{Cost}, \quad \text{or} \quad P(x) = R(x) - C(x)$$
Marginal cost, marginal revenue, and marginal profit	For total cost, revenue, and profit functions that are linear, the rates of change are called marginal cost, marginal revenue, and marginal profit, respectively.

Special linear functions

- **Constant function** — A special linear function that has the form $y = c$, where c is a real number, is called a constant function.

- **Identity function** — The identity function $y = x$ is a linear function of the form $y = mx + b$ with slope $m = 1$ and y-intercept $b = 0$.

1.4 Equations of Lines

Writing equations of lines

- **Slope-intercept form** — The slope-intercept form of the equation of a line with slope m and y-intercept b is

$$y = mx + b$$

- **Point-slope form** — The point-slope form of the equation of the line with slope m and passing through a known point (x_1, y_1) is

$$y - y_1 = m(x - x_1)$$

- **Horizontal line** — The equation of a horizontal line is $y = b$, where b is a constant.

- **Vertical line** — The equation of a vertical line is $x = a$, where a is a constant.

- **General form** — The general form of the equation of a line is $ax + by = c$, where a, b, and c are constants.

Parallel and perpendicular lines

Two distinct nonvertical lines are parallel if they have the same slope. Two nonvertical and nonhorizontal lines are perpendicular if their slopes are negative reciprocals.

Average rate of change

The average rate of change of a quantity over an interval describes how a change in the output of a function describing that quantity $f(x)$ is related to a change in the input x over that interval. The average rate of change of $f(x)$ with respect to x over the interval from $x = a$ to $x = b$ (where $a < b$) is calculated as

$$\text{average rate of change} = \frac{\text{change in } f(x) \text{ values}}{\text{corresponding change in } x \text{ values}} = \frac{f(b) - f(a)}{b - a}$$

Secant line

When a function is not linear, the average rate of change between two points is the slope of the line joining two points on the curve, which is called a secant line.

Difference quotient

The average rate of change of a function from a point $(x, f(x))$ to the point $(x + h, f(x + h))$ is the difference quotient $\dfrac{f(x + h) - f(x)}{h}$, for $h \neq 0$.

Approximately linear data

Some sets of real data lie sufficiently close to a line that two points can be used to create a linear function that models the data.

chapter 1 SKILLS CHECK

Answers that are not seen can be found in the answer section at the back of the text.

Use the values in the table below in Exercises 1–4.

x	−3	−1	1	3	5	7	9	11	13
y	9	6	3	0	−3	−6	−9	−12	−15

1. Explain why the relationship shown by the table describes y as a function of x.
   Each value of x is assigned exactly one value of y.

2. State the domain and range of the function.
   $D = \{-3, -1, 1, 3, 5, 7, 9, 11, 13\}$; $R = \{9, 6, 3, 0, -3, -6, -9, -12, -15\}$

3. If the function defined by the table is denoted by f, so that $y = f(x)$, what is $f(3)$? $f(3) = 0$

4. Do the outputs in this table indicate that a linear function fits the data? If so, write the equation of the line.
   Yes; $y = -\dfrac{3}{2}x + \dfrac{9}{2}$

5. If $C(s) = 16 - 2s^2$, find

   a. $C(3)$ -2 b. $C(-2)$ 8 c. $C(-1)$ 14

6. For each of the functions $y = f(x)$ described below, find $f(-3)$.

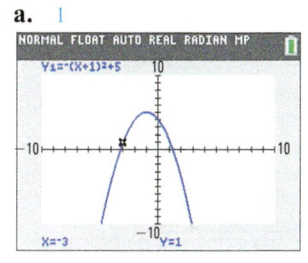

   a. 1

   b. -10

7. Graph the function $f(x) = -2x^3 + 5x$.

8. Graph the function $y = 3x^2$.

9. Graph $y = -10x^2 + 400x + 10$ on a standard window and on a window with $x_{min} = 0$, $x_{max} = 40$, $y_{min} = 0$, $y_{max} = 5000$. Which window gives a better view of the graph of the function?

10. Use a graphing utility to graph the points (x, y) from the table.

x	10	15	20	25	30	35	40
y	-8	-6	-3	0	5	8	10

11. Find the domain of each function.

   a. $y = \sqrt{2x - 8}$ $x \geq 4$
   b. $f(x) = \dfrac{x + 2}{x - 6}$ All real numbers except 6

12. One line passes through the points $(-12, 16)$ and $(-1, 38)$, and a second line has equation $2y + x = 23$. Are the lines parallel, perpendicular, or neither? Perpendicular

13. A line passes through $(-1, 4)$ and $(5, -3)$. Find the slope of a line parallel to this line and the slope of a line perpendicular to this line. $-\dfrac{7}{6}, \dfrac{6}{7}$

14. Find the slope of the line through $(-4, 6)$ and $(8, -16)$. $-\dfrac{11}{6}$

15. Given the equation $2x - 3y = 12$, (a) find the x- and y-intercepts of the graph and (b) graph the equation.
    a. $(0, -4)$ and $(6, 0)$

16. What is the slope of the graph of the function given in Exercise 15? $\dfrac{2}{3}$

17. Find the slope and y-intercept of the graph of $y = -6x + 3$. $m = -6, b = 3$

18. Find the rate of change of the function whose equation is given in Exercise 17. -6

19. Write the equation of the line that has slope $\dfrac{1}{3}$ and y-intercept 3. $y = \dfrac{1}{3}x + 3$

20. Write the equation of a line that has slope $\dfrac{-3}{4}$ and passes through $(4, -6)$. $y = -\dfrac{3}{4}x - 3$

21. Write the equation of the line that passes through $(-1, 3)$ and $(2, 6)$. $y = x + 4$

22. For the function $y = x^2$, compute the average rate of change between $x = 0$ and $x = 3$. 3

For the functions given in Exercises 23 and 24, find

   a. $f(x + h)$
   b. $f(x + h) - f(x)$
   c. $\dfrac{f(x + h) - f(x)}{h}$, for $h \neq 0$.

23. $f(x) = 5 - 4x$

   a. $5 - 4x - 4h$
   b. $-4h$
   c. -4

24. $f(x) = 10x - 50$

   a. $10x + 10h - 50$
   b. $10h$
   c. 10

chapter 1 REVIEW

Answers that are not seen can be found in the answer section at the back of the text.

25. **Artificial Intelligence** Enterprise AI refers to the use of artificial intelligence in enterprises to transform business operations and strategy. The table gives the revenue (in billions of dollars) from artificial intelligence for enterprise AI applications in North America for the years from 2016 with projections to 2025.

   a. Is the revenue R a function of the year y? Yes

   b. If $R = f(y)$, find $f(2024)$ and explain what it means. 11.395; the projected revenue from artificial intelligence for enterprise AI applications in North America in 2024 is $11.395 billion.

   c. What is y if $f(y) = 3.665$? 2021

Year	Revenue ($ billions)	Year	Revenue ($ billions)
2016	0.212	2021	3.665
2017	0.434	2022	5.633
2018	0.796	2023	8.237
2019	1.376	2024	11.395
2020	2.288	2025	14.896

(*Source:* Statista)

84   Chapter 1   Functions, Graphs, and Models; Linear Functions

26. *Artificial Intelligence* Graph the function defined by the table in Exercise 25 on the window $[2016, 2025]$ by $[0, 15]$.

27. *Artificial Intelligence* Find the average rate of change of the North American Enterprise AI revenue from 2016 to 2025, to 3 decimal places.
$1.632 billion per year

28. *Artificial Intelligence* Use your answer from Exercise 27 and the point representing the revenue in 2016 to write the equation of the line connecting them. $y = 1.632x - 3289.9$

When money is borrowed to purchase an automobile, the amount borrowed A determines the monthly payment P. In particular, if a dealership offers a 5-year loan at 2.9% interest, then the amount borrowed for the car determines the payment according to the following table. Use the table to define the function $P = f(A)$ in Exercises 29–31.

Amount Borrowed ($)	Monthly Payment ($)
10,000	179.25
15,000	268.87
20,000	358.49
25,000	448.11
30,000	537.73

(Source: Sky Financial)

29. *Car Loans*
   a. Explain why the monthly payment P is a linear function of the amount borrowed A. As each input value changes by 5000, the output value changes by 89.62.
   b. Find $f(25,000)$ and interpret its meaning. 448.11; the monthly payment to borrow $25,000 is $448.11.
   c. If $f(A) = 358.49$, what is A? $A = 20,000$

30. *Car Loans*
   a. What are the domain and range of the function $P = f(A)$ defined by the table? $D = \{10,000, 15,000, 20,000, 25,000, 30,000\}$; $R = \{179.25, 268.87, 358.49, 448.11, 537.73\}$
   b. Is this function $P = f(A)$ defined for a $12,000 loan? No

31. *Car Loans* The equation of the line that fits the data points in the table is $f(A) = 0.017924A + 0.01$.
   a. Use the equation to find $f(28,000)$ and explain what it means. $f(28,000) = 501.882$; if $28,000 is borrowed, the payment is $501.88.
   b. Can the function f be used to find the monthly payment for any dollar amount A of a loan if the interest rate and length of loan are unchanged? Yes

32. *Life Expectancy* The figure gives the number of years the average woman is estimated to live beyond age 65 during selected years between 1950 and 2030. Let x represent the year and let y represent the expected number of years a woman will live past age 65. Write $y = f(x)$ and answer the following questions:

   a. What is $f(1960)$ and what does it mean? 15.9; in 1960, the average woman was expected to live 15.9 years past 65.
   b. What is the life expectancy for the average woman in 2010? $65 + 19.4 = 84.4$ years
   c. In what year was the average woman expected to live 19 years past age 65? 1990

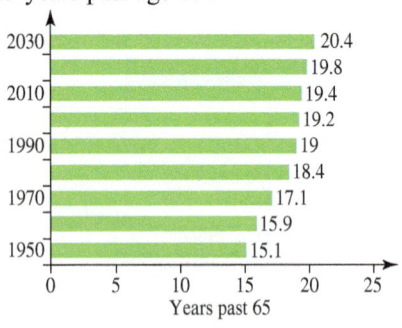

Women
(Source: National Center for Health Statistics)

33. *Life Expectancy* The figure gives the number of years the average man is estimated to live beyond age 65 during selected years between 1950 and 2030. Let x represent the year and let y represent the expected number of years a man will live past age 65. Write $y = g(x)$ and answer the following questions:

   a. What is $g(2020)$ and what does it mean? 16.9; a 65-year-old man in 2020 is expected to live 16.9 years past 65.
   b. What was the life expectancy for the average man in 1950? 77.8 years
   c. Write a function expression that indicates that the average man in 1990 has a life expectancy of 80 years. $g(1990) = 15$

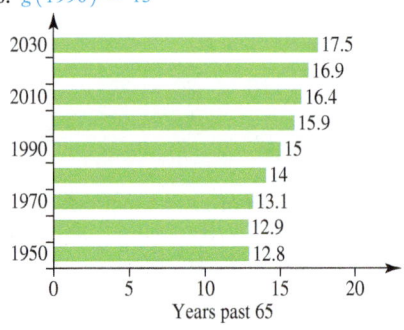

Men
(Source: National Center for Health Statistics)

34. *Civilian Workforce* The size of the U.S. civilian workforce for the years from 1950 and projected to 2050 can be modeled by the linear function $y = 1.296x - 2465$, where x is the year and y is the number in the civilian workforce (in millions). (Source: U.S. Bureau of Labor Statistics)

   a. Is this a linear function? yes
   b. Interpret the slope of this line as a rate of change. The number of civilian workers is projected to increase by approximately 1.296 million per year.
   c. What is the projected size of the civilian workforce in 2030? 165.88 million

35. *Heart Rate* A Paramount elliptical machine provides the following table, which gives the desired heart rate

during exercise for cardiac health based on the age of the person exercising.

a. Use the points corresponding to age 20 and age 90 to write the equation of the line connecting them. Let t represent age and h represent heart rate.
$h = -0.8t + 176$
b. Does every data point in the table fit the graph of this equation? Yes

c. Can the data in the table be expressed exactly as this linear function? Yes

Age	Heart Rate (beats per minute)	Age	Heart Rate (beats per minute)
20	160	60	128
30	152	70	120
40	144	80	112
50	136	90	104

(*Source*: paramount fitness.com)

36. *Crude Oil* The U.S. crude oil production, in billions of barrels, was 1.94 in the year 2010 and is projected to be 2.26 in 2030.

    a. Find the projected average rate of change in production from 2010 through 2030.
    0.016 billion barrels per year
    b. Use the rate of change from part (a) and the point corresponding to 2010 to write the equation of the line that joins the points (2010, 1.94) and (2030, 2.26), with x equal to the year 2010 and y equal to the number of billions of barrels of crude oil.
    (*Source*: U.S. Energy Information Administration)
    $y = 0.016x - 30.22$

37. *Fuel* The table below shows data for the number of gallons of gas purchased each day x of a certain week by the 150 taxis owned by the Inner City Transportation Taxi Company. Write the equation that models the data. $f(x) = 4500$

Days from First Day	0	1	2	3	4	5	6
Gas Used (gal)	4500	4500	4500	4500	4500	4500	4500

38. *Work Hours* The average weekly hours worked by employees at PriceLo Company are given by the data in the table below.

Month and Year	Average Weekly Hours	Month and Year	Average Weekly Hours
June 2016	33.8	Oct. 2016	33.8
July 2016	33.8	Nov. 2016	33.8
Aug. 2016	33.8	Dec. 2016	33.8
Sept. 2016	33.8	Jan. 2017	33.8
		Feb. 2017	33.8

a. Write the equation of a function that describes the average weekly hours using an input x equal to the number of months past May 2016. $f(x) = 33.8$

b. What type of function is this? Constant function

39. *Revenue* A company has revenue given by $R(x) = 564x$ dollars and total costs given by $C(x) = 40,000 + 64x$ dollars, where x is the number of units produced and sold.

    a. What is the revenue when 120 units are produced?
    $67,680
    b. What is the cost when 120 units are produced?
    $47,680
    c. What is the marginal cost and what is the marginal revenue for this product? MC: 64; MR: 564
    d. What is the slope of the graph of $C(x) = 40,000 + 64x$? $m = 64$
    e. Graph $R(x)$ and $C(x)$ on the same set of axes.

40. *Profit* A company has revenue given by $R(x) = 564x$ dollars and total cost given by $C(x) = 40,000 + 64x$ dollars, where x is the number of units produced and sold. The profit can be found by forming the function $P(x) = R(x) - C(x)$.

    a. Write the profit function. $P(x) = 500x - 40,000$
    b. Find the profit when 120 units are produced and sold. $20,000
    c. How many units give break-even? 80
    d. What is the marginal profit for this product? 500
    e. How is the marginal profit related to the marginal revenue and the marginal cost? Marginal revenue minus marginal cost

41. *Depreciation* A business property can be depreciated for tax purposes by using the formula $y + 3000x = 300,000$, where y is the value of the property x years after it was purchased.

    a. Find the y-intercept of the graph of this function. Interpret this value.
    $y = 300,000$; the initial value of the property is $300,000.
    b. Find the x-intercept. Interpret this value.
    $x = 100$; the value of the property after 100 years is zero dollars.

42. *Marginal Profit* A company has determined that its profit for a product can be described by a linear function. The profit from the production and sale of 150 units is $455, and the profit from 250 units is $895.

    a. What is the average rate of change of the profit for this product when between 150 and 250 units are sold? $4.40 per unit
    b. What is the slope of the graph of this profit function? $m = 4.4$
    c. Write the equation of the profit function for this product. $P(x) = 4.4x - 205$
    d. What is the marginal profit for this product?
    $4.40 per unit
    e. How many units give break-even for this product?
    Approximately 47

Group Activities ▶ EXTENDED APPLICATIONS

1. Body Mass Index

Obesity is a risk factor for the development of medical problems, including high blood pressure, high cholesterol, heart disease, and diabetes. Of course, how much a person can safely weigh depends on his or her height. One way of comparing weights that account for height is the *body mass index (BMI)*. The table below gives the BMI for a variety of heights and weights of people. Roche Pharmaceuticals states that a BMI of 30 or greater can create an increased risk of developing medical problems associated with obesity.

Describe how to assist a group of people in using the information. Some things you would want to include in your description follow.

1. How a person uses the table to determine his or her BMI.
2. How a person determines the weight that will put him or her at medical risk.
3. How a person whose weight or height is not in the table determine whether his or her BMI is 30? To answer this question, develop and explain a formula to find the weight that would give a BMI of 30 for a person of a given height:

   a. Pick the points from the table that correspond to a BMI of 30 and create a table of these heights and weights. Change the heights to inches to simplify the data.
   b. Create a scatter plot of the data.
   c. Using the two points with the smallest and largest heights, write a linear equation that models the data.
   d. Graph the linear equation from part (c) with the scatter plot and discuss the fit.
   e. Explain how to use the model to test for obesity.

Body Mass Index for Specified Height (ft/in.) and Weight (lb)

Height/Weight	120	130	140	150	160	170	180	190	200	210	220	230	240	250
5′0″	23	25	27	29	31	33	35	37	39	41	43	45	47	49
5′1″	23	25	27	28	30	32	34	36	38	40	42	44	45	47
5′2″	22	24	26	27	29	31	33	35	37	38	40	42	44	46
5′3″	21	23	25	27	28	30	32	34	36	37	39	41	43	44
5′4″	21	22	24	26	28	29	31	33	34	36	38	40	41	43
5′5″	20	22	23	25	27	28	30	32	33	35	37	38	40	42
5′6″	19	21	23	24	26	27	29	31	32	34	36	37	39	40
5′7″	19	20	23	24	25	27	28	30	31	33	35	36	38	39
5′8″	18	20	21	23	24	26	27	29	30	32	34	35	37	38
5′9″	18	19	21	22	24	25	27	28	30	31	33	34	36	37
5′10″	17	19	20	22	23	24	25	27	28	29	31	33	35	36
5′11″	17	18	20	21	22	24	25	27	28	29	31	32	34	35
6′0″	16	18	19	20	22	23	24	26	27	29	30	31	33	34
6′1″	16	17	19	20	21	22	24	25	26	28	29	30	32	33

(*Source*: Roche Pharmaceuticals)

2. Total Revenue, Total Cost, and Profit

The total revenue is the amount a company receives from the sales of its products. It can be found by multiplying the selling price per unit times the number of units sold. That is, the revenue is

$$R(x) = p \cdot x$$

where p is the price per unit and x is the number of units sold.

The total cost comprises two costs: the fixed costs and the variable costs. Fixed costs (FC), such as depreciation, rent, and utilities, remain constant regardless of the number of units produced. Variable costs (VC) are those directly related to the number of units produced. The variable cost is the cost per unit (c) times the number of units produced ($VC = c \cdot x$), and the fixed cost is constant ($FC = k$), so the total cost is found by using the equation

$$C(x) = c \cdot x + k$$

where c is the cost per unit and x is the number of units produced.

Also, as discussed in Section 1.3, the profit a company makes on x units of a product is the difference between the revenue and the cost from the production and sale of x units.

$$P(x) = R(x) - C(x)$$

Suppose a company manufactures MP3 players and sells them to retailers for $98 each. It has fixed costs of $262,500 related to the production of the MP3 players, and the cost per unit for production is $23.

1. What is the total revenue function?
2. What is the marginal revenue for this product?
3. What is the total cost function?
4. What is the marginal cost for this product? Is the marginal cost equal to the variable cost or the fixed cost?
5. What is the profit function for this product? What is the marginal profit?
6. What are the cost, revenue, and profit if 0 units are produced?
7. Graph the total cost and total revenue functions on the same axes and estimate where the graphs intersect.
8. Graph the profit function and estimate where the graph intersects the x-axis.
9. What do the points of intersection in Question 7 and Question 8 give?

2

Linear Models, Equations, and Inequalities

Data from the Social Security Administration can be used to show how the monthly Social Security benefits differ with the age at which a recipient decides to start receiving benefits. These data can be used to create linear equations that give the benefit as a function of the age of the recipient. We can also find when the manufacturing sector of China reaches and surpasses that of the United States, by finding linear models and solving them simultaneously. Market equilibrium occurs when the number of units of a product demanded equals the number of units supplied. In this chapter, we solve these and other problems by creating linear models, solving linear equations and inequalities, and solving systems of equations.

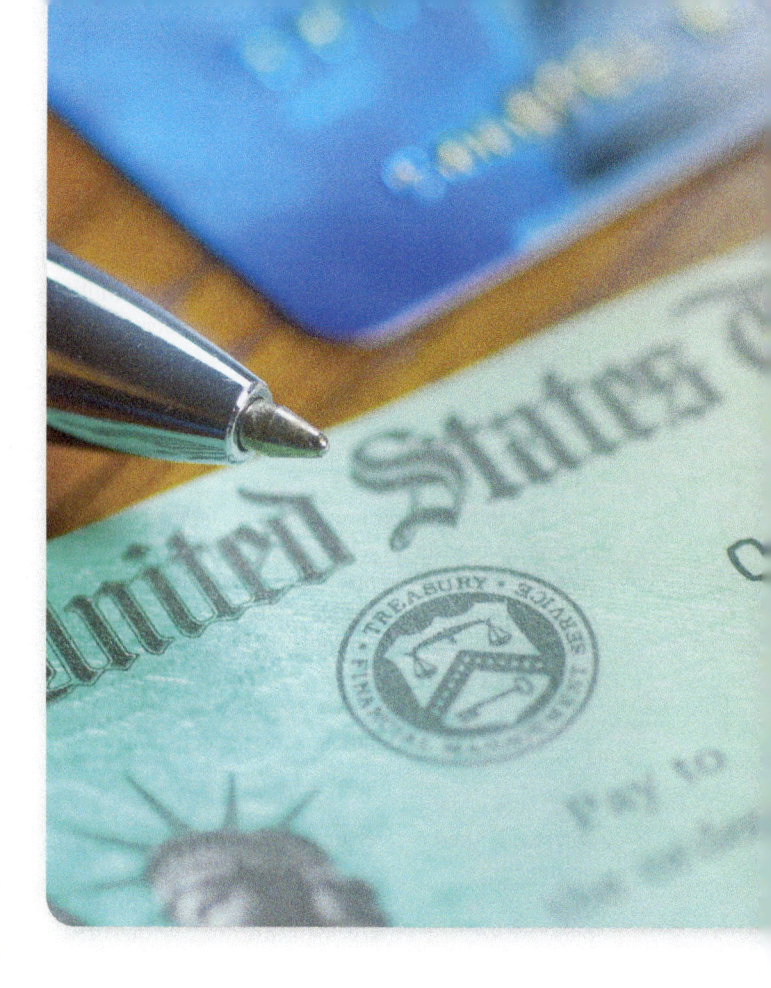

section	objectives	applications
2.1 Algebraic and Graphical Solution of Linear Equations	Find solutions algebraically; relate solutions, zeros, and x-intercepts; solve linear equations; find functional forms of equations in two variables; solve literal equations; apply direct variation	Digital ad spending, credit card debt, stock market, simple interest, blood alcohol percent
2.2 Fitting Lines to Data Points: Modeling Linear Functions	Fit lines to data points with linear regression; model with linear functions; apply linear models; evaluate goodness of fit	Big data market size, retirement, population of females, U.S. population
2.3 Systems of Linear Equations in Two Variables	Solve systems of linear equations in two variables with graphing, substitution, and elimination; find break-even and market equilibrium; find nonunique solutions if they exist	U.S. civilian workforce, break-even, market equilibrium, investments, medication
2.4 Solutions of Linear Inequalities	Solve linear inequalities with analytical and graphical methods; solve double inequalities	Profit, body temperature, apparent temperature, course grades, expected prison sentences

Algebra TOOLBOX

KEY OBJECTIVES

- Multiply and divide fractions
- Add and subtract fractions
- Determine if a linear equation is a conditional equation, an identity, or a contradiction
- Use properties of equations to solve linear equations
- Evaluate formulas for given values
- Use properties of inequalities to solve linear inequalities
- Rounding Function Coefficients and Constants

Multiplying and Dividing Fractions

To multiply two fractions, multiply their numerators and multiply their denominators. Often it is simpler to factor the numerator and denominator first, and then divide by common factors. To divide fractions, multiply by the reciprocal of the divisor. Recall that the dividend divided by the divisor yields the quotient.

EXAMPLE 1 ▶ Multiply Fractions

$$\frac{6}{7} \cdot \frac{21}{10} = \frac{6 \cdot 21}{7 \cdot 10} = \frac{2 \cdot 3 \cdot 3 \cdot 7}{7 \cdot 2 \cdot 5} = \frac{\cancel{2} \cdot 3 \cdot 3 \cdot \cancel{7}}{\cancel{7} \cdot \cancel{2} \cdot 5} = \frac{9}{5}$$

EXAMPLE 2 ▶ Divide Fractions

$$\left(-\frac{55}{18}\right) \div \frac{25}{4} = \left(-\frac{55}{18}\right) \cdot \frac{4}{25} = \frac{-5 \cdot 11 \cdot 2 \cdot 2}{2 \cdot 3 \cdot 3 \cdot 5 \cdot 5} = -\frac{\cancel{5} \cdot 11 \cdot \cancel{2} \cdot 2}{\cancel{2} \cdot 3 \cdot 3 \cdot \cancel{5} \cdot 5} = -\frac{22}{45}$$

Adding and Subtracting Fractions

To add or subtract fractions with *like* denominators, add or subtract the numerators, over their common denominator.

EXAMPLE 3 ▶ Subtract Fractions with a Common Denominator

$$\frac{72}{25} - \frac{64}{25} = \frac{72 - 64}{25} = \frac{8}{25}$$

To add or subtract fractions with *unlike* denominators, first find the least common denominator of the fractions. Then rewrite each fraction so that it has the common denominator, using the Multiplication Property of One. Finally, add or subtract their numerators, over their common denominators.

Recall that to find the *least common denominator (LCD)*, factor all the denominators. The LCD is the product of each different factor, used the maximum number of times it occurs in any one denominator. Equivalent fractions are found by multiplying the numerator and denominator by factors needed to obtain the LCD.

EXAMPLE 4 ▶ Add and Subtract Fractions with Unlike Denominators

$$\frac{2}{3} + \frac{7}{10} - \frac{15}{6} = \frac{2}{3} \cdot \frac{10}{10} + \frac{7}{10} \cdot \frac{3}{3} - \frac{15}{6} \cdot \frac{5}{5} = \frac{20}{30} + \frac{21}{30} - \frac{75}{30} = \frac{20 + 21 - 75}{30} = -\frac{34}{30}$$

Note: The denominators are factored as $3, 2 \cdot 5, 2 \cdot 3$. The LCD is $2 \cdot 3 \cdot 5 = 30$

Properties of Equations

In this chapter, we will solve equations. To solve an equation means to find the value(s) of the variable(s) that makes the equation a true statement. For example, the equation $3x = 6$ is true when $x = 2$, so 2 is a solution of this equation. Equations of this type are sometimes called **conditional equations** because they are true only for certain values of the variable. Equations that are true for all values of the variables for which both sides are defined are called **identities**. For example, the equation $7x - 4x = 5x - 2x$ is an identity. Equations that are not true for any value of the variable are called **contradictions**. For example, the equation $2(x + 3) = 2x - 1$ is a contradiction.

We frequently can solve an equation for a given variable by rewriting the equation in an equivalent form whose solution is easy to find. Two equations are **equivalent** if and only if they have the same solutions. The following operations give equivalent equations:

Properties of Equations

1. **Addition Property** Adding the same number to both sides of an equation gives an equivalent equation. For example, $x - 5 = 2$ is equivalent to $x - 5 + 5 = 2 + 5$, or to $x = 7$.

2. **Subtraction Property** Subtracting the same number from both sides of an equation gives an equivalent equation. For example, $z + 12 = -9$ is equivalent to $z + 12 - 12 = -9 - 12$, or to $z = -21$.

3. **Multiplication Property** Multiplying both sides of an equation by the same nonzero number gives an equivalent equation. For example, $\frac{y}{6} = 5$ is equivalent to $6\left(\frac{y}{6}\right) = 6(5)$, or to $y = 30$.

4. **Division Property** Dividing both sides of an equation by the same nonzero number gives an equivalent equation. For example, $17x = -34$ is equivalent to $\frac{17x}{17} = \frac{-34}{17}$, or to $x = -2$.

5. **Substitution Property** The equation formed by substituting one expression for an equal expression is equivalent to the original equation. For example, if $y = 3x$, then $x + y = 8$ is equivalent to $x + 3x = 8$, so $4x = 8$ and $x = 2$.

EXAMPLE 5 ▶ Properties of Equations

State the property (or properties) of equations that can be used to solve each of the following equations, and then use the property (or properties) to solve the equation.

a. $3x = 6$ **b.** $\frac{x}{5} = 12$ **c.** $3x - 5 = 17$ **d.** $4x - 5 = 7 + 2x$

SOLUTION

a. Division Property. Dividing both sides of the equation by 3 gives the solution to the equation.

$$3x = 6$$
$$\frac{3x}{3} = \frac{6}{3}$$
$$x = 2$$

Thus, $x = 2$ is the solution to the original equation.

b. Multiplication Property. Multiplying both sides of the equation by 5 gives the solution to the equation.

$$\frac{x}{5} = 12$$

$$5\left(\frac{x}{5}\right) = 5(12)$$

$$x = 60$$

Thus, $x = 60$ is the solution to the original equation.

c. Addition Property and Division Property. Adding 5 to both sides of the equation and dividing both sides by 3 gives an equivalent equation.

$$3x - 5 = 17$$

$$3x - 5 + 5 = 17 + 5$$

$$3x = 22$$

$$\frac{3x}{3} = \frac{22}{3}$$

$$x = \frac{22}{3}$$

Thus, $x = \frac{22}{3}$ is the solution to the original equation.

d. Addition Property, Subtraction Property, and Division Property. Adding 5 to both sides of the equation and subtracting $2x$ from both sides of the equation gives an equivalent equation.

$$4x - 5 = 7 + 2x$$

$$4x - 5 + 5 = 7 + 2x + 5$$

$$4x = 12 + 2x$$

$$4x - 2x = 12 + 2x - 2x$$

$$2x = 12$$

Dividing both sides by 2 gives the solution to the original equation.

$$\frac{2x}{2} = \frac{12}{2}$$

$$x = 6$$

Thus, $x = 6$ is the solution to the original equation.

EXAMPLE 6 ▶ Equations

Determine whether each equation is a conditional equation, an identity, or a contradiction.

a. $3(x - 5) = 2x - 7$ **b.** $5x - 6(x + 1) = -x - 9$

c. $-3(2x - 4) = 2x + 12 - 8x$

SOLUTION

a. To solve $3(x - 5) = 2x - 7$, we first use the Distributive Property, then the Addition Property.

$$3(x - 5) = 2x - 7$$
$$3x - 15 = 2x - 7$$
$$3x - 15 - 2x = 2x - 7 - 2x$$
$$x - 15 + 15 = -7 + 15$$
$$x = 8$$

Because this equation is true for only the value $x = 8$, it is a conditional equation.

b. To solve $5x - 6(x + 1) = -x - 9$, we first use the Distributive Property, then the Addition Property.

$$5x - 6(x + 1) = -x - 9$$
$$5x - 6x - 6 = -x - 9$$
$$-x - 6 = -x - 9$$
$$-x - 6 + x = -x - 9 + x$$
$$-6 = -9$$

Because $-6 = -9$ is false, the original equation is a contradiction.

c. To solve $-3(2x - 4) = 2x + 12 - 8x$, we first use the Distributive Property, then the Addition Property.

$$-3(2x - 4) = 2x + 12 - 8x$$
$$-6x + 12 = 12 - 6x$$
$$-6x + 12 + 6x = 12 - 6x + 6x$$
$$12 = 12$$

Because $12 = 12$ is a true statement, the equation is true for all real numbers and thus is an identity.

Evaluating Formulas for Given Values

In this chapter we will solve literal equations (formulas) for a specified variable. A formula contains one or more letters that represent constants or variables. To evaluate a formula, we simply substitute the values given to find a value that is requested.

EXAMPLE 7 ▶ Investment

The formula $A = P(1 + rt)$ gives the amount accumulated in an account if present value P is invested at annual interest rate r for t years. Solve for P if $A = 5000$, $r = 0.04$, and $t = 10$.

SOLUTION

Substituting 5000 for A, 0.04 for r, and 10 for t gives

$$5000 = P(1 + (0.04)(10))$$
$$5000 = 1.4P$$
$$\frac{5000}{1.4} = P$$
$$P = 3571.43$$

Properties of Inequalities

We will solve linear inequalities in this chapter. As with equations, we can find solutions to inequalities by finding equivalent inequalities from which the solutions can be easily seen. We use the following properties to reduce an inequality to a simple equivalent inequality.

Properties of Inequalities	Examples	
Substitution Property The inequality formed by substituting one expression for an equal expression is equivalent to the original inequality.	$7x - 6x \geq 8$ $x \geq 8$	$3(x - 1) < 8$ $3x - 3 < 8$
Addition and Subtraction Properties The inequality formed by adding the same quantity to (or subtracting the same quantity from) both sides of an inequality is equivalent to the original inequality.	$x - 6 < 12$ $x - 6 + 6 < 12 + 6$ $x < 18$	$3x > 13 + 2x$ $3x - 2x > 13 + 2x - 2x$ $x > 13$
Multiplication Property I The inequality formed by multiplying (or dividing) both sides of an inequality by the same *positive* quantity is equivalent to the original inequality.	$\frac{1}{3}x \leq 6$ $3\left(\frac{1}{3}x\right) \leq 3(6)$ $x \leq 18$	$3x > 6$ $\frac{3x}{3} > \frac{6}{3}$ $x > 2$
Multiplication Property II The inequality formed by multiplying (or dividing) both sides of an inequality by the same *negative* number and reversing the inequality symbol is equivalent to the original inequality.	$-x > 7$ $-1(-x) < -1(7)$ $x < -7$	$-4x \geq 12$ $\frac{-4x}{-4} \leq \frac{12}{-4}$ $x \leq -3$

Of course, these properties can be used in combination to solve an inequality. This means that the steps used to solve a linear inequality are the same as those used to solve a linear equation, except that the inequality symbol is reversed if both sides are multiplied (or divided) by a negative number.

EXAMPLE 8 ▶ Properties of Inequalities

State the property (or properties) of inequalities that can be used to solve each of the following inequalities, and then use the property (or properties) to solve the inequality.

a. $\dfrac{x}{2} \geq 13$ **b.** $-3x < 6$ **c.** $5 - \dfrac{x}{3} \geq -4$ **d.** $-2(x + 3) < 4$

SOLUTION

a. Multiplication Property I. Multiplying both sides by 2 gives the solution to the inequality.

$$\frac{x}{2} \geq 13$$

$$2\left(\frac{x}{2}\right) \geq 2(13)$$

$$x \geq 26$$

The solution to the inequality is $x \geq 26$.

b. Multiplication Property II. Dividing both sides by -3 and reversing the inequality symbol gives the solution to the inequality.

$$-3x < 6$$
$$\frac{-3x}{-3} > \frac{6}{-3}$$
$$x > -2$$

The solution to the inequality is $x > -2$.

c. Subtraction Property, Multiplication Property II. First we subtract 5 from both sides of the inequality.

$$5 - \frac{x}{3} \geq -4$$
$$5 - \frac{x}{3} - 5 \geq -4 - 5$$
$$-\frac{x}{3} \geq -9$$

Multiplying both sides by -3 and reversing the inequality symbol gives the solution to the inequality.

$$-\frac{x}{3} \geq -9$$
$$-3\left(-\frac{x}{3}\right) \leq -3(-9)$$
$$x \leq 27$$

The solution to the inequality is $x \leq 27$.

d. Substitution Property, Addition Property, Multiplication Property II. First we distribute -2 on the left side of the inequality.

$$-2(x + 3) < 4$$
$$-2x - 6 < 4$$

Adding 6 to both sides and then dividing both sides by -2 and reversing the inequality symbol gives the solution to the inequality.

$$-2x - 6 + 6 < 4 + 6$$
$$-2x < 10$$
$$\frac{-2x}{-2} > \frac{10}{-2}$$
$$x > -5$$

The solution to the inequality is $x > -5$.

Rounding Function Coefficients and Constants

In this chapter, we will create linear functions that approximately fit given data points. When we use technology to create the models, we usually have to round the coefficients and constants to report (write) the model. Unless we are instructed otherwise, it will be common practice to report the functions with three decimal places. Recall the following:

Tenths means 0.1; hundredths means 0.01, thousandths means 0.001, ten-thousandths means 0.0001, and so on.

Algebra Toolbox 95

In some cases, however, we will be asked to report the functions with three or four digits of accuracy. These **significant digits** carry meaning that contributes to the precision of the number. In general, significant digits are those from the first nonzero digit on the left to the last digit *after* the decimal point, if there is a decimal point. If there is no decimal point, any zeros to the right of the nonzero digits are not significant. Note that

1. All nonzero digits are significant.
2. All zeros between significant digits are significant.
3. All zeros that are both to the right of the decimal point and to the right of all nonzero significant digits are themselves significant.
4. Leading zeros are not significant.

Thus 0.0324, 56,300 and 30.0 all have three significant digits.

EXAMPLE 9 ▶ Significant Digits

Determine the number of significant digits in each of the following numbers.

a. 33,465.2   b. 0.004432   c. 5600   d. 2.0056   e. 22.370

SOLUTION

a. For the number 33,465.2, there are five nonzero digits to the left of the decimal point and one nonzero digit to the right of the decimal point, so there are six significant digits.

b. For the number 0.004432, there are three leading zeros, which are not significant, and four digits after those zeros, so there are four significant digits.

c. Since there is no decimal point, the zeros after the first two digits in the number 5600 are not significant, so there are two significant digits.

d. For the number 2.0056, there are three nonzero digits, and the zeros between nonzero digits are significant, so there are five significant digits.

e. For the number 22.370, the zero to the right of the decimal is significant, so there are five significant digits.

EXAMPLE 10 ▶ Rounding Function Coefficients and Constants

Rewrite the function $f(x) = 0.003457x^2 - 240.3669x + 360.71$ with the coefficients (and constant) rounded to:

a. three decimal places   b. three significant digits   c. four significant digits

SOLUTION

a. Rounded to three decimal places, the function $f(x) = 0.003457x^2 - 240.3669x + 360.71$ becomes $f(x) = 0.003x^2 - 240.367x + 360.710$.

b. The function $f(x) = 0.003457x^2 - 240.3669x + 360.71$ written with three significant digits is $f(x) = 0.00346x^2 - 240.0x + 361$.

c. The function $f(x) = 0.003457x^2 - 240.3669x + 360.71$ written with four significant digits is $f(x) = 0.003457x^2 - 240.4x + 360.7$.

Chapter 2 Linear Models, Equations, and Inequalities

Toolbox EXERCISES

In Exercises 1–6, multiply or divide the fractions, as indicated, and reduce to lowest terms.

1. $\dfrac{7}{5} \cdot \dfrac{10}{21}$ $\dfrac{2}{3}$
2. $\dfrac{15}{22} \cdot \dfrac{11}{9}$ $\dfrac{5}{6}$
3. $\dfrac{12}{13} \cdot \left(-\dfrac{65}{9}\right)$ $-\dfrac{20}{3}$
4. $\dfrac{11}{7} \div \dfrac{15}{7}$ $\dfrac{11}{15}$
5. $\dfrac{9}{16} \div \dfrac{63}{20}$ $\dfrac{5}{28}$
6. $\left(-\dfrac{405}{45}\right) \div \left(-\dfrac{90}{63}\right)$ $\dfrac{63}{10}$

In Exercises 7–14, add or subtract the fractions, as indicated, and simplify the result.

7. $\dfrac{5}{12} + \dfrac{11}{12}$ $\dfrac{4}{3}$
8. $-\dfrac{17}{5} + \dfrac{13}{5}$ $-\dfrac{4}{5}$
9. $\dfrac{7}{8} - \dfrac{15}{8}$ -1
10. $\dfrac{9}{20} + \dfrac{4}{15}$ $\dfrac{43}{60}$
11. $\dfrac{8}{21} - \dfrac{9}{28}$ $\dfrac{5}{84}$
12. $-\dfrac{43}{33} + \dfrac{17}{22}$ $-\dfrac{35}{66}$
13. $-\dfrac{13}{12} + \dfrac{11}{6} - \dfrac{49}{90}$ $\dfrac{37}{180}$
14. $\left(\dfrac{5}{7} - \dfrac{8}{21}\right) \div \left(-\dfrac{33}{6} + \dfrac{5}{14}\right)$ $-\dfrac{7}{108}$

In Exercises 15–22, state the property (or properties) of equations that can be used to solve each of the following equations; then use the property (or properties) to solve the equation.

15. $3x = 6$ Division Property; $x = 2$
16. $x - 7 = 11$ Addition Property; $x = 18$
17. $x + 3 = 8$ Subtraction Property; $x = 5$
18. $x - 5 = -2$ Addition Property; $x = 3$
19. $\dfrac{x}{3} = 6$ Multiplication Property; $x = 18$
20. $-5x = 10$ Division Property; $x = -2$
21. $2x + 8 = -12$ Subtraction Property and Division Property; $x = -10$
22. $\dfrac{x}{4} - 3 = 5$ Addition Property and Multiplication Property; $x = 32$

Solve the equations in Exercises 23–30.

23. $4x - 3 = 6 + x$ $x = 3$
24. $3x - 2 = 4 - 7x$ $x = \dfrac{3}{5}$
25. $\dfrac{3x}{4} = 12$ $x = 16$
26. $\dfrac{5x}{2} = -10$ $x = -4$
27. $3(x - 5) = -2x - 5$ $x = 2$
28. $-2(3x - 1) = 4x - 8$ $x = 1$
29. $2x - 7 = -4\left(4x - \dfrac{1}{2}\right)$ $x = \dfrac{1}{2}$
30. $-2(2x - 6) = 3\left(3x - \dfrac{1}{3}\right)$ $x = 1$

In Exercises 31–34, use the Substitution Property of Equations to solve the equation.

31. Solve for x if $y = 2x$ and $x + y = 12$ $x = 4$
32. Solve for x if $y = 4x$ and $x + y = 25$ $x = 5$
33. Solve for x if $y = 3x$ and $2x + 4y = 42$ $x = 3$
34. Solve for x if $y = 6x$ and $3x + 2y = 75$ $x = 5$

In Exercises 35–38, determine whether the equation is a conditional equation, an identity, or a contradiction.

35. $3x - 5x = 2x + 7$ Conditional
36. $3(x + 1) = 3x - 7$ Contradiction
37. $9x - 2(x - 5) = 3x + 10 + 4x$ Identity
38. $\dfrac{x}{2} - 5 = \dfrac{x}{4} + 2$ Conditional

In Exercises 39–44, solve each formula as requested.

39. Given the formula $A = lw$, if $l = 7$ and $w = 6$, find A. $A = 42$
40. Solve for h in the formula $A = \dfrac{1}{2}bh$ if $A = 320$ and $b = 40$. $h = 16$
41. If $V = \dfrac{1}{3}Bh$, solve for B if $V = 900$ and $h = 45$. $B = 60$
42. If $C = 2\pi r$, solve for C if $r = 9$. $C = 18\pi$
43. Solve for t in the formula $S = P + Prt$, if $P = 5000$, $r = 0.05$, and $S = 7000$. $t = 8$
44. Solve for C in the formula $5F - 9C = 160$, if $F = 75$. $C = 23.9$

In Exercises 45–52, solve the inequalities.

45. $5x + 1 > -5$ $x > -\dfrac{6}{5}$
46. $1 - 3x \geq 7$ $x \leq -2$
47. $\dfrac{x}{4} > -3$ $x > -12$
48. $\dfrac{x}{6} > -2$ $x > -12$
49. $\dfrac{x}{4} - 2 > 5x$ $x < \dfrac{-8}{19}$
50. $\dfrac{x}{2} + 3 > 6x$ $x < \dfrac{6}{11}$
51. $-3(x - 5) < -4$ $x > \dfrac{19}{3}$
52. $-\dfrac{1}{2}(x + 4) < 6$ $x > -16$

In Exercises 53–58, determine the number of significant digits in each of the following numbers.

53. 52,865.3 6

54. 0.003389 4

55. 4700 2

56. 4.0092 5

57. 58.610 5

58. 0.000041 2

59. Rewrite the function

$$f(x) = 0.007543x^2 - 632.5778x + 480.65$$

with the coefficients (and constant) rounded to:

a. three decimal places
   $f(x) = 0.008x^2 - 632.578x + 480.650$

b. three significant digits
   $f(x) = 0.00754x^2 - 633x + 481$

c. four significant digits
   $f(x) = 0.007543x^2 - 632.6x + 480.7$

2.1 Algebraic and Graphical Solution of Linear Equations

KEY OBJECTIVES

- Solve linear equations algebraically
- Solve real-world application problems
- Compare solutions of equations with zeros and x-intercepts of graphs of functions
- Solve linear equations graphically using the x-intercept and intersection methods
- Solve literal equations for a specified variable
- Solve direct variation problems

SECTION PREVIEW Digital Ad Spending

In 2017, worldwide digital ad spending reached $227 billion, while TV ad spending stood at $178 billion. The digital ad spending from 2017 and projected to 2023 is shown in Table 2.1. These data can be used to create the function $y = 48.953x - 613.505$, with x equal to the number of years after 2000 and y equal to worldwide digital spending in billions of dollars, as an approximate model for the data.

Table 2.1

Year	Spending ($ billions)
2017	226.561
2018	266.238
2019	311.398
2020	360.421
2021	411.919
2022	464.731
2023	517.620

(*Source:* Statista)

This function describes the upward trend in worldwide digital ad spending y as a function of x. To find the year when worldwide digital ad spending is predicted to reach $700 billion, we use algebraic or graphical methods to solve the linear equation

$$700 = 48.953x - 613.505$$

for x. (See Example 6.) In this section, we use additional algebraic methods and graphical methods to solve linear equations in one variable. ∎

Algebraic Solution of Linear Equations

We can use algebraic, graphical, or a combination of algebraic and graphical methods to solve linear equations. Sometimes it is more convenient to use algebraic solution methods rather than graphical solution methods, especially if an exact solution is desired. The steps used to solve a linear equation in one variable algebraically follow.

Chapter 2 Linear Models, Equations, and Inequalities

> ### Steps for Solving a Linear Equation in One Variable
> 1. If a linear equation contains fractions, multiply both sides of the equation by a number that will remove all denominators from the equation. If there are two or more fractions, use the least common denominator (LCD) of the fractions.
> 2. Perform any multiplications or divisions to remove any parentheses or other symbols of grouping.
> 3. Perform any additions or subtractions to get all terms containing the variable on one side and all other terms on the other side of the equation. Combine like terms.
> 4. Divide both sides of the equation by the coefficient of the variable.
> 5. Check the solution by substitution in the original equation. If a real-world solution is desired, check the algebraic solution for reasonableness in the real-world situation.

EXAMPLE 1 ▶ Algebraic Solutions

a. Solve for x: $\dfrac{2x-3}{4} = \dfrac{x}{3} + 1$

b. Solve for y: $\dfrac{y}{5} - \dfrac{1}{2}\left(\dfrac{y}{2}\right) = \dfrac{7}{20}$

c. Solve for x if $y = 0.72$: $y = 1.312x - 2.56$

SOLUTION

a. $\dfrac{2x-3}{4} = \dfrac{x}{3} + 1$

$12\left(\dfrac{2x-3}{4}\right) = 12\left(\dfrac{x}{3} + 1\right)$ Multiply both sides by the LCD, 12.

$3(2x-3) = 12\left(\dfrac{x}{3} + 1\right)$ Simplify the fraction $\dfrac{12}{4}$ on the left.

$6x - 9 = 4x + 12$ Remove parentheses using the distributive property.

$2x = 21$ Subtract $4x$ from both sides and add 9 to both sides.

$x = \dfrac{21}{2}$ Divide both sides by 2.

Check the result: $\dfrac{2\left(\frac{21}{2}\right) - 3}{4} \stackrel{?}{=} \dfrac{\frac{21}{2}}{3} + 1 \Rightarrow \dfrac{9}{2} = \dfrac{9}{2}$

b. $\dfrac{y}{5} - \dfrac{1}{2}\left(\dfrac{y}{2}\right) = \dfrac{7}{20}$

$\dfrac{y}{5} - \dfrac{y}{4} = \dfrac{7}{20}$ Remove parentheses first to find the LCD.

$20\left(\dfrac{y}{5} - \dfrac{y}{4}\right) = 20\left(\dfrac{7}{20}\right)$ Multiply both sides by the LCD, 20.

$4y - 5y = 7$ Remove parentheses using the distributive property.

$-y = 7$ Combine like terms.

$y = -7$ Multiply both sides by -1.

Check the result: $\dfrac{-7}{5} - \dfrac{1}{2}\left(\dfrac{-7}{2}\right) \stackrel{?}{=} \dfrac{7}{20} \Rightarrow \dfrac{-7}{5} + \dfrac{7}{4} \stackrel{?}{=} \dfrac{7}{20} \Rightarrow \dfrac{7}{20} = \dfrac{7}{20}$

2.1 Algebraic and Graphical Solution of Linear Equations

c. $y = 1.312x - 2.56$

$0.72 = 1.312x - 2.56$ Substitute 0.72 for y.

$3.28 = 1.312x$ Add 2.56 to both sides.

$x = 2.5$ Divide both sides by 1.312.

Check the result: $0.72 \stackrel{?}{=} 1.312(2.5) - 2.56 \Rightarrow 0.72 = 0.72$

As the next example illustrates, we solve an application problem that is set in a real-world context by using the same solution methods. However, you must remember to include units of measure with your answer and check that your answer makes sense in the problem situation.

EXAMPLE 2 ▶ Credit Card Debt

It is hard for some people to pay off credit card debt in a reasonable period of time because of high interest rates. The interest paid on a $10,000 debt over 3 years is approximated by

$$y = 175.393x - 116.287 \text{ dollars}$$

when the interest rate is $x\%$. What is the interest rate if the interest is $1637.60? (*Source*: Consumer Federation of America)

SOLUTION

To answer this question, we solve the linear equation

$$1637.60 = 175.393x - 116.287$$

$$1753.887 = 175.393x$$

$$x = 9.9998$$

Thus, if the interest rate is approximately 10%, the interest is $1637.60. (Note that if you check the approximate answer, you are checking only for the reasonableness of the estimate.)

EXAMPLE 3 ▶ Stock Market

For a period of time, a man is very successful speculating on an Internet stock, with its value growing to $100,000. However, the stock drops rapidly until its value has been reduced by 40%. What percent increase will have to occur before the latest value returns to $100,000?

SOLUTION

The value of the stock after the 40% loss is $100{,}000 - 0.40(100{,}000) = 60{,}000$. To find the percent p of increase that is necessary to return the value to 100,000, we solve

$$60{,}000 + 60{,}000p = 100{,}000$$

$$60{,}000p = 40{,}000$$

$$p = \frac{40{,}000}{60{,}000} = \frac{2}{3} = 66\frac{2}{3}\%$$

Thus, the stock value must increase by $66\frac{2}{3}\%$ to return to a value of $100,000.

Solutions, Zeros, and x-Intercepts

Because *x*-intercepts are *x*-values that make the output of the function equal to 0, these intercepts are also called **zeros** of the function.

> **Zero of a Function**
>
> Any number a for which $f(a) = 0$ is called a **zero** of the function $f(x)$. If a is real, a is an *x*-intercept of the graph of the function.

The zeros of the function are values that make the function equal to 0, so they are also solutions to the equation $f(x) = 0$.

> The following three concepts are numerically the same:
> The *x*-intercepts of the graph of $y = f(x)$
> The real zeros of the function f
> The real solutions to the equation $f(x) = 0$

The following example illustrates the relationships that exist among *x*-intercepts of the graph of a function, zeros of the function, and solutions to associated equations.

EXAMPLE 4 ▶ Relationships Among x-Intercepts, Zeros, and Solutions

For the function $f(x) = 13x - 39$, find

a. $f(3)$

b. The zero of $f(x) = 13x - 39$

c. The *x*-intercept of the graph of $y = 13x - 39$

d. The solution to the equation $13x - 39 = 0$

SOLUTION

a. Evaluate the output $f(3)$ by substituting the input 3 for *x* in the function:
$$f(3) = 13(3) - 39 = 0$$

b. Because $f(x) = 0$ when $x = 3$, we say that 3 is a zero of the function.

c. The *x*-intercept of a graph occurs at the value of *x* where $y = 0$, so the *x*-intercept is $x = 3$. The only *x*-intercept is $x = 3$ because the graph of $f(x) = 13x - 39$ is a line that crosses the *x*-axis at only one point (Figure 2.1).

d. The solution to the equation $13x - 39 = 0$ is $x = 3$ because $13x - 39 = 0$ gives $13x = 39$, or $x = 3$.

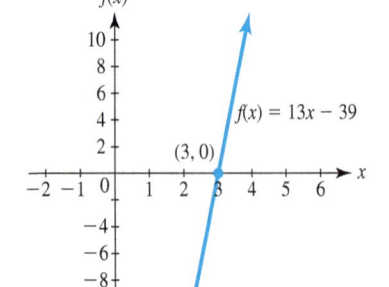

Figure 2.1

We can use TRACE or TABLE (accessed by 2ND GRAPH) on a graphing utility to check the reasonableness of a solution. The graph of $y = 13x - 39$ given in Figure 2.1 confirms that $x = 3$ is the *x*-intercept. We can also confirm that $f(3) = 0$ by using Excel. Figure 2.2 shows that when we enter 3 in cell A2 and the formula "=13 * A2 - 39" in cell B2, the function $f(x) = 13x - 39$ is evaluated at $x = 3$, with a result of 0.*

Figure 2.2

---

*See Appendix B, page 703.

Graphical Solution of Linear Equations

We can also view the graph or use TRACE on a graphing utility to obtain a quick estimate of an answer. TRACE may or may not provide the exact solution to an equation, but it will provide an estimate of the solution. If your calculator or computer has a graphical or numerical solver, the solver can be used to obtain the solution to the equation $f(x) = 0$. Recall that an x-intercept of the graph of $y = f(x)$, a real zero of $f(x)$, and the real solution to the equation $f(x) = 0$ are all different names for the same input value. If the graph of the linear function $y = f(x)$ is not a horizontal line, we can find the one solution to the linear equation $f(x) = 0$ as described below. We call this solution method the *x-intercept method*.

> ### Solving a Linear Equation Using the x-Intercept Method with Graphing Utilities
>
> 1. Rewrite the equation to be solved with 0 (and nothing else) on one side of the equation.
>
> 2. Enter the nonzero side of the equation found in the previous step in the equation editor of your graphing utility and graph the line in an appropriate viewing window. Be certain that you can see the line cross the horizontal axis on your graph.
>
> 3. Find the x-intercept by using the zero command after pressing 2ND TRACE. The x-intercept is the value of x that makes the equation equal to zero, so it is the solution to the equation. The value of x displayed by this method is sometimes a decimal approximation of the exact solution rather than the exact solution. *Note:* Using the Frac command after pressing MATH will often convert a decimal solution of a linear equation (approximated on the display) to the exact solution.*

EXAMPLE 5 ▶ Graphical Solution

Solve $\dfrac{2x-3}{4} = \dfrac{x}{3} + 1$ for x using the x-intercept method.

SOLUTION

To solve the equation using the x-intercept method, first rewrite the equation with 0 on one side.

$$0 = \frac{x}{3} + 1 - \frac{2x-3}{4}$$

Next, enter the right side of the equation,

$$\frac{x}{3} + 1 - \frac{2x-3}{4}$$

in the equation editor of your graphing utility as

$$y_1 = (x/3) + 1 - (2x-3)/4$$

and graph this function. Use parentheses as needed to preserve the order of operations. You should obtain a graph similar to the one seen in Figure 2.3, which was obtained using the viewing window $[-10, 15]$ by $[-4, 4]$. However, any graph in which you can see the x-intercept will do. Using "zero" gives the x-intercept of the graph, 10.5. (See Figure 2.3.) This is the value of x that makes $y = 0$, and thus the original equation true, so it is the solution to this linear equation.

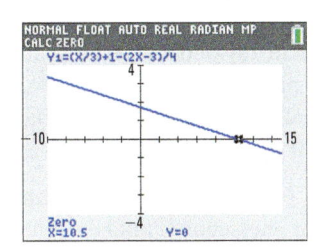

Figure 2.3

---

*See Appendix A, page 687.

You can determine if $x = 10.5 = \dfrac{21}{2}$, found graphically, is the exact answer by substituting this value into the original equation. Both sides of the original equation are equal when $x = 10.5$, so it is the exact solution.

We can also use the intersection method described below to solve linear equations:

> **Solving a Linear Equation Using the Intersection Method**
> 1. Enter the left side of the equation as y_1 and the right side as y_2. Graph both of these equations on a window that shows their point of intersection.
> 2. Find the point of intersection with the intersect command after pressing **2ND** **TRACE**.* This is the point where $y_1 = y_2$. The x-value of this point is the value of x that makes the two sides of the equation equal, so it is the solution to the original equation.

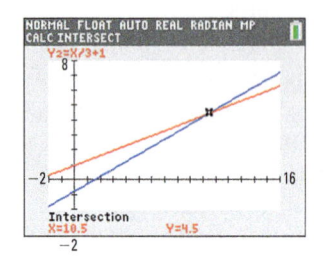

Figure 2.4

The equation in Example 5,

$$\dfrac{2x-3}{4} = \dfrac{x}{3} + 1$$

can be solved using the intersection method. Enter the left side as $y_1 = (2x-3)/4$ and the right side as $y_2 = (x/3) + 1$. Graphing these two functions gives the graphs in Figure 2.4. The point of intersection is found to be $(10.5, 4.5)$. So the solution is $x = 10.5$ (as we found in Example 5).

EXAMPLE 6 ▶ Digital Ad Spending

In 2017, worldwide digital ad spending reached $227 billion, while TV ad spending stood at $178 billion. The upward trend in digital ad spending from 2017 and projected to 2023 can be modeled by the function $y = 48.953x - 613.505$ billion dollars, where x is the number of years after 2000. To find the year when worldwide digital ad spending is predicted to reach $700 billion, write an equation and solve it by using (a) the x-intercept method and (b) the intersection method.
(*Source*: Statista)

a. To find x when $y = 700$, we solve the linear equation $700 = 48.953x - 613.505$ by using the x-intercept method as follows:

$0 = 48.953x - 613.505 - 700$ Rewrite the equation in a form with 0 on one side.

$0 = 48.953x - 1313.505$ Combine the constant terms.

Enter $y_1 = 48.953x - 1313.505$ and graph this equation in a window that shows the x-intercept (Figure 2.5). Using the zero command after pressing **2ND** **TRACE** gives an x-intercept of approximately 26.8 (Figure 2.5). Thus, worldwide digital ad spending is predicted to reach 300 billion during the year $2000 + 27$, or 2027.

b. To use the intersection method, enter $y_1 = 48.953x - 613.505$ and $y_2 = 700$ and graph these equations in a window that shows the point of intersection. Find the intersection of the lines with the intersect command found under **2ND** **TRACE**. Figure 2.6 shows the point of intersection, and again we see that x is approximately 26.8.

---

*See Appendix A, page 687.

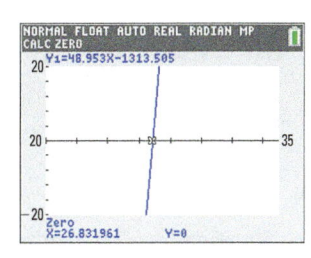

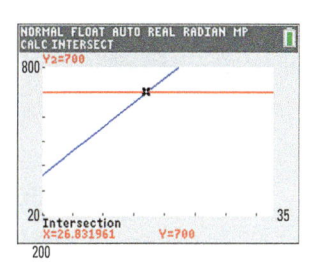

2.1 Algebraic and Graphical Solution of Linear Equations   103

Figure 2.5

Figure 2.6

Spreadsheet ▶ SOLUTION We can use the **Goal Seek** feature of Excel to find the *x*-intercept of the graph of a function—that is, the zeros of the function. To solve

$$120 = 0.55x - 2.886$$

we enter the nonzero side of $0 = 0.55x - 122.886$ in cell B2 as shown in Figure 2.7 and put any number in A2. We then select DATA, What-if-Analysis, and Goal Seek. In the Goal Seek menu, we select the target cell B2, set it to 0, and set the changing cell to A2. Figure 2.8 shows the solution in cell A2.*

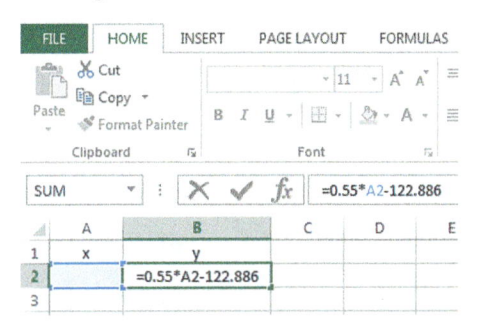

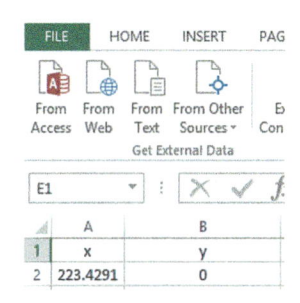

Figure 2.7

Figure 2.8

Literal Equations; Solving an Equation for a Specified Linear Variable

An equation that contains two or more letters that represent constants or variables is called a **literal equation**. Formulas are examples of literal equations. If one of two or more of the variables in an equation is present only to the first power, we can solve for that variable by treating the other variables as constants and using the same steps that we used to solve a linear equation in one variable. This is often useful because it is necessary to get equations in the proper form to enter them in a graphing utility.

EXAMPLE 7 ▶ Simple Interest

The formula for the future value of an investment of P dollars at simple interest rate r for t years is $A = P(1 + rt)$. Solve the formula for r, the interest rate.

SOLUTION

$\quad A = P(1 + rt)$

$\quad A = P + Prt$ Multiply to remove parentheses.

$A - P = Prt$ Get the term containing *r* by itself on one side of the equation.

$\quad r = \dfrac{A - P}{Pt}$ Divide both sides by *Pt*.

---

*For detailed steps, See Appendix B, page 709.

104 Chapter 2 Linear Models, Equations, and Inequalities

EXAMPLE 8 ▶ Solving an Equation for a Specified Variable

Solve the equation $2(2x - b) = \dfrac{5cx}{3}$ for x.

SOLUTION

We solve the equation for x by treating the other variables as constants:

$2(2x - b) = \dfrac{5cx}{3}$

$6(2x - b) = 5cx$ Clear the equation of fractions by multiplying by the LCD, 3.

$12x - 6b = 5cx$ Multiply to remove parentheses.

$12x - 5cx = 6b$ Get all terms containing x on one side and all other terms on the other side.

$x(12 - 5c) = 6b$ Factor x from the expression. The remaining factor is the coefficient of x.

$\dfrac{x(12 - 5c)}{(12 - 5c)} = \dfrac{6b}{12 - 5c}$ Divide both sides by the coefficient of x.

$x = \dfrac{6b}{12 - 5c}$

EXAMPLE 9 ▶ Writing an Equation in Functional Form and Graphing Equations

Solve each of the following equations for y so that y is expressed as a function of x. Then graph the equation on a graphing utility with a standard viewing window.

a. $2x - 3y = 12$ **b.** $x^2 + 4y = 4$

SOLUTION

a. Because y is to the first power in the equation, we solve the equation for y using linear equation solution methods:

$2x - 3y = 12$

$-3y = -2x + 12$ Isolate the term involving y by subtracting $2x$ from both sides.

$y = \dfrac{2x}{3} - 4$ Divide both sides by -3, the coefficient of y.

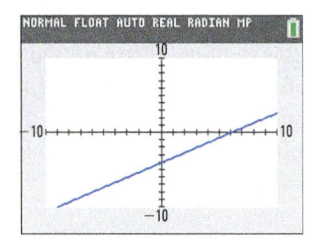

The graph of this equation is shown in Figure 2.9.

Figure 2.9

b. The variable y is to the first power in the equation

$x^2 + 4y = 4$

so we solve for y by using linear solution methods. (Note that to solve this equation for x would be more difficult; this method will be discussed in Chapter 3.)

$4y = -x^2 + 4$ Isolate the term containing y by subtracting x^2 from both sides.

$y = \dfrac{-x^2}{4} + 1$ Divide both sides by 4, the coefficient of y.

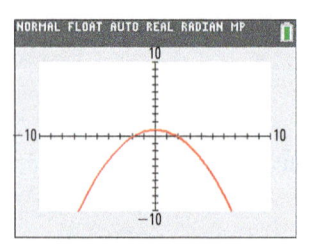

The graph of this equation is shown in Figure 2.10.

Figure 2.10

2.1 Algebraic and Graphical Solution of Linear Equations

Direct Variation

Often in mathematics we need to express relationships between quantities. One relationship that is frequently used in applied mathematics occurs when two quantities are proportional. Two variables x and y are proportional to each other (or vary directly) if their quotient is a constant. That is, y is **directly proportional** to x, or y **varies directly** with x, if y and x are related by

$$\frac{y}{x} = k \quad \text{or, equivalently,} \quad y = kx$$

where k is called the **constant of proportionality** or the **constant of variation**.

As we saw earlier, the revenue from the sale of x units at p dollars per unit is $R(x) = px$, so we can say that the revenue is directly proportional to the number of units, with the constant of proportionality equal to p.

We know that blood alcohol percent is dependent on the number of drinks consumed and that excess consumption can lead to impairment of the drinker's judgment and motor skills (and sometimes to a DUI arrest). The blood alcohol percent p of a 130-pound man is directly proportional to the number of drinks x, where a drink is defined as 1.25 oz of 80-proof liquor, 12 oz of regular beer, or 5 oz of table wine.

EXAMPLE 10 ▶ Blood Alcohol Percent

The blood alcohol percent of a 130-pound man is directly proportional to the number of drinks consumed, and 3 drinks give a blood alcohol percent of 0.087. Find the constant of proportionality and the blood alcohol percent resulting from 5 drinks.

SOLUTION

The equation representing blood alcohol percent is $0.087 = k(3)$. Solving for k in $0.087 = k(3)$ gives $k = 0.029$, so the constant of proportionality is 0.029. Five drinks would result in a blood alcohol percent of $5(0.029) = 0.145$.

SKILLS CHECK 2.1

In Exercises 1–12, solve the equations.

1. $5x - 14 = 23 + 7x$ $x = -\frac{37}{2}$
2. $3x - 2 = 7x - 24$ $x = \frac{11}{2}$
3. $3(x - 7) = 19 - x$ $x = 10$
4. $5(y - 6) = 18 - 2y$ $y = \frac{48}{7}$
5. $x - \frac{5}{6} = 3x + \frac{1}{4}$ $x = -\frac{13}{24}$
6. $3x - \frac{1}{3} = 5x + \frac{3}{4}$ $x = -\frac{13}{24}$
7. $\frac{5(x-3)}{6} - x = 1 - \frac{x}{9}$ $x = -63$
8. $\frac{4(y-2)}{5} - y = 6 - \frac{y}{3}$ $y = 57$
9. $5.92t = 1.78t - 4.14$ $t = -1$
10. $0.023x + 0.8 = 0.36x - 5.266$ $x = 18$
11. $\frac{3}{4} + \frac{1}{5}x - \frac{1}{3} = \frac{4}{5}x$ $x = \frac{25}{36}$
12. $\frac{2}{3}x - \frac{6}{5} = \frac{1}{2} + \frac{5}{6}x$ $x = -\frac{51}{5}$
13. $3(x - 1) + 5 = 4(x - 3) - 2(2x - 3)$ $x = -\frac{8}{3}$
14. $5x - (x - 2) + 7 = -(2x - 9) - 8(3x + 6)$ $x = -\frac{8}{5}$
15. $2[3(x - 10) - 3(2x + 1)] = 3[-(x - 6) - 5(2x - 4)]$ $x = \frac{16}{3}$
16. $\frac{3x}{5} - \frac{x-2}{6} = \frac{2x+3}{4} - 1$ $x = \frac{35}{4}$
17. $\frac{6-x}{4} - 2 - \frac{6-2x}{3} = -\left[\frac{5x-2}{3} - \frac{x}{5}\right]$ $x = \frac{190}{113}$
18. $12 - \frac{3-2x}{5} - \frac{4-x}{3} = -5\left[\frac{5x}{2} - \frac{2(3x+2)}{5}\right]$ $x = -\frac{26}{31}$

Chapter 2 Linear Models, Equations, and Inequalities

For Exercises 19–22, find (a) the solution to the equation $f(x) = 0$, (b) the x-intercept of the graph of $y = f(x)$, and (c) the zero of $f(x)$.

19. $f(x) = 32 + 1.6x$
    a. -20 b. -20 c. -20

20. $f(x) = 15x - 60$
    a. 4 b. 4 c. 4

21. $f(x) = \dfrac{3}{2}x - 6$
    a. 4 b. 4 c. 4

22. $f(x) = \dfrac{x - 5}{4}$
    a. 5 b. 5 c. 5

In Exercises 23 and 24, you are given a table showing input and output values for a given function $y_1 = f(x)$. Using this table, find (if possible) (a) the x-intercept of the graph of $y = f(x)$, (b) the y-intercept of the graph of $y = f(x)$, and (c) the solution to the equation $f(x) = 0$.

23. a. 2 b. -34 c. 2

24. a. -5 b. 17 c. -5

In Exercises 25 and 26, you are given the graph of a certain function $y = f(x)$ and the zero of that function. Using this graph, find (a) the x-intercept of the graph of $y = f(x)$ and (b) the solution to the equation $f(x) = 0$.

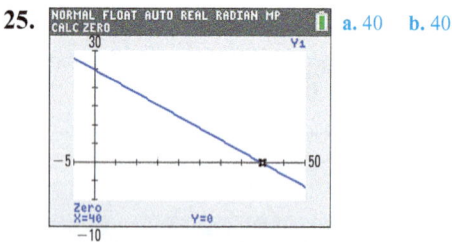

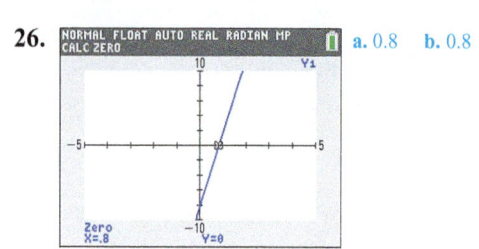

25. a. 40 b. 40

26. a. 0.8 b. 0.8

In Exercises 27 and 28, use the Y = screen to write the equation being solved if $Y_1 = Y_2$. Then use the table to solve the equation.

27. $2x - 5 = 3(x - 2)$; $x = 1$

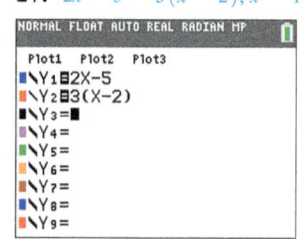

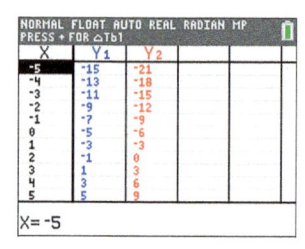

28. $-5(2x + 4) = 3x - 7$; $x = -1$

In Exercises 29–32, you are given the equation of a function. For each function, (a) find the zero of the function, (b) find the x-intercept of the graph of the function, and (c) solve the equation $f(x) = 0$.

29. $f(x) = 4x - 100$ a. 25 b. 25 c. 25

30. $f(x) = 6x - 120$ a. 20 b. 20 c. 20

31. $f(x) = 330 + 40x$ a. -8.25 b. -8.25 c. -8.25

32. $f(x) = 250 + 45x$ a. $-\dfrac{50}{9}$ b. $-\dfrac{50}{9}$ c. $-\dfrac{50}{9}$

In Exercises 33–40, solve the equations using graphical methods.

33. $14x - 24 = 27 - 3x$ $x = 3$

34. $3x - 8 = 15x + 4$ $x = -1$

35. $3(s - 8) = 5(s - 4) + 6$ $s = -5$

36. $5(2x + 1) + 5 = 5(x - 2)$ $x = -4$

37. $\dfrac{3t}{4} - 2 = \dfrac{5t - 1}{3} + 2$ $t = -4$

38. $4 - \dfrac{x}{6} = \dfrac{3(x - 2)}{4}$ $x = 6$

39. $\dfrac{t}{3} - \dfrac{1}{2} = \dfrac{t + 4}{9}$ $t = \dfrac{17}{4}$

40. $\dfrac{x - 5}{4} + x = \dfrac{x}{2} + \dfrac{1}{3}$ $x = \dfrac{19}{9}$

41. Solve $A = P(1 + rt)$ for r. $r = \dfrac{A - P}{Pt}$

42. Solve $V = \dfrac{1}{3}\pi r^2 h$ for h. $h = \dfrac{3V}{\pi r^2}$

43. Solve $5F - 9C = 160$ for F. $F = \dfrac{9}{5}C + 32$

44. Solve $4(a - 2x) = 5x + \dfrac{c}{3}$ for x. $x = \dfrac{c - 12a}{-39}$

45. Solve $\dfrac{P}{2} + A = 5m - 2n$ for n. $n = \dfrac{5m}{2} - \dfrac{P}{4} - \dfrac{A}{2}$

46. Solve $y - y_1 = m(x - x_1)$ for x. $x = \dfrac{y - y_1 + mx_1}{m}$

In Exercises 47–50, solve the equations for y and graph them with a standard window on a graphing utility.

47. $5x - 3y = 5$ $y = \dfrac{5x - 5}{3}$

48. $3x + 2y = 6$ $y = -\dfrac{3}{2}x + 3$

49. $x^2 + 2y = 6$ $y = \dfrac{6 - x^2}{2}$

50. $4x^2 + 2y = 8$ $y = -2x^2 + 4$

EXERCISES 2.1

51. Depreciation An $828,000 building is depreciated for tax purposes by its owner using the straight-line depreciation method. The value of the building y, after x months of use, is given by $y = 828{,}000 - 2300x$ dollars. After how many years will the value of the building be $690,000? *60 months, or 5 years*

52. Temperature Conversion The equation $5F - 9C = 160$ gives the relationship between Fahrenheit and Celsius temperature measurements. What Fahrenheit measure is equivalent to a Celsius measurement of $20°$? *68°F*

53. Investments The future value of a simple interest investment is given by $S = P(1 + rt)$. What principal P must be invested for $t = 5$ years at the simple interest rate $r = 10\%$ so that the future value grows to $9000? *$6000*

54. Temperature–Humidity Index The temperature–humidity index I is $I = t - 0.55(1 - h)(t - 58)$, where t is the air temperature in degrees Fahrenheit and h is the relative humidity expressed as a decimal. If the air temperature is $88°$, find the humidity h (as a percent) that gives an index value of 84.7. *80%*

55. PEV Vehicle Sales Plug-in electric vehicles and plug-in hybrid electric vehicles are referred to as PEVs. Using data from 2017 and projected to 2025, the percent of total vehicle sales that are PEVs can be modeled by the function $p = 0.770x - 0.270$, where x is in years after 2015. When is the percent of PEV sales projected to equal 8.2% of total vehicle sales? *In 2026 ($x = 11$)*

56. Game Show Question The following question was worth $32,000 on the game show *Who Wants to Be a Millionaire?*: "At what temperature are the Fahrenheit and Celsius temperature scales the same?" Answer this question. Recall that Fahrenheit and Celsius temperatures are related by $5F - 9C = 160$. *−40°*

57. Average Annual Wage Using data from 2012 and projected to 2050, the U.S. average annual wage can be modeled by the function $y = 3.94x + 26.40$, where x is the number of years after 2010 and y is in thousands of dollars. In what year does the model predict that the average annual wage will be 164.3 thousand dollars? (*Source*: Social Security Administration) *In 2045 ($x = 35$)*

58. Reading Score The average reading score on the National Assessment of Progress tests is given by $y = 0.155x + 255.37$, where x is the number of years after 1970. In what year would the average reading score be 259.4 if this model is accurate? *26 years from 1970, or 1996*

59. Lithium Demand The demand for lithium is projected to steadily increase because of the growth of production of electric vehicles. Using data from 2017 and projections to 2025, the function $D = 22.71t + 183.6$ can be used to model the demand for lithium in thousands of metric tons, where t is the number of years after 2015. In what year will the demand reach 365,200 metric tons, according to this model? (*Source*: Statista) *2023 ($x = 8$)*

60. Profit The profit from the production and sale of specialty golf hats is given by the function $P(x) = 20x - 4000$, where x is the number of hats produced and sold.

 a. Producing and selling how many units will give a profit of $8000? *600*

 b. How many units must be produced and sold to avoid a loss? *200*

61. Civilian Workforce The number of men in the U.S. civilian workforce for the years from 1950 and projected to 2050 can be modeled by the linear function $y = 0.587x + 43.1$, where x is the number of years after 1950 and y is the number of men in the civilian workforce (in millions) in year x. In what year does the model predict that there will be 100 million men in the civilian workforce? (*Source*: U.S. Bureau of Labor Statistics) *In 2047 ($x \approx 96.9$)*

62. U.S. Internet Users Using data from 2000 and projected to 2025, the percent p of the U.S. population that uses the Internet can be modeled by $p = 1.19x + 57.96$, where x is the number of years after 1990. When will the percent be 99.61%, according to this model? (*Source*: Nation and Statistics, Policy Exchange Analysis) *In 2025 ($x = 35$)*

63. U.S. Population The U.S. population can be modeled for the years 1960–2060 by the function $p = 2.6x + 177$, where p is in millions of people and x is in years from 1960. During what year does the model estimate the population to be 320,000,000? *$x = 55$, so 2015* (*Source*: www.census.gov/statab)

64. Disposable Income Disposable income is the amount left after taxes have been paid and is one measure of health of the economy. The total disposable income for the United States, in billions of dollars, for selected years from 2010 and projected to 2040 can be modeled by $D(x) = 0.328x + 6.32$, where x is the number of years after 2000. In what year does this model project the total disposable income for the United States to be $12.88 billion? *2020* (*Source*: U.S. Bureau of Economic Analysis)

65. Heart Rate The desired heart rate for weight loss for adults during exercise is given by the linear function $H = -0.65t + 143$, with t equal to the person's age and H equal to the heart rate in beats per minute.

   a. If a person's desired heart rate for weight loss is 130, what is the person's age? 20

   b. If a person's desired heart rate for weight loss is 104, what is the person's age? 60

   (*Source*: paramountfitness.com)

66. Population of Females Under 18 Using projected data from 2020 to 2060, the population of females under 18 in the United States is given by $y = 154.03x + 34{,}319$, with x equal to the number of years after 2000 and y equal to the projected population, in thousands. In what year will the population reach 38,169,750, according to this model? 2025
(*Source*: U.S. Census Bureau)

67. Hispanic Population Using data and projections from 1980 through 2050, the number (in millions) of Hispanics in the U.S. civilian non-institutional population is given by $y = 0.876x + 6.084$, where x is the number of years after 1980. During what year was the number of Hispanics 14.8 million, if this model is accurate? 1990
(*Source*: U.S. Census Bureau)

68. Internet Users The percent of the U.S. population with Internet access can be modeled by $y = 1.36x + 68.8$, with x equal to the number of years after 2000. When does this model indicate that the U.S. population with Internet access will reach 96%? 2020
(*Source*: Jupiter Media Metrix)

69. Life Expectancy at 65 Using data for selected years from 1950 and projected to 2050, the equation $y = 0.077x + 14$ gives the number of additional years of life expected at age 65, with x equal to the number of years after 1950. In what year does this model predict that the life expectancy at age 65 will reach 20 additional years? 2028
(*Source*: Social Security Administration)

70. Cell Phone Subscribers The number of cell phone subscribers (in billions) between 2014 and 2020 can be modeled by $y = 0.234x + 6.856$, where x is the number of years after 2010. In what year does this model indicate that there were 9.43 billion subscribers? 2021
(*Source*: Semiannual CTIA Wireless Survey)

71. Grades To earn an A in a course, a student must get an average score of at least 90 on five tests. If her first four test scores are 92, 86, 79, and 96, what score does she need on the last test to obtain a 90 average? 97

72. Grades To earn an A in a course, a student must get an average score of at least 90 on three tests and a final exam. If the final exam score is higher than the lowest score, then the lowest score is removed and the final exam score counts double. If the first three test scores are 86, 79, and 96, what is the lowest score the student can get on the last test and still obtain a 90 average? 89

73. Tobacco Judgment As part of the largest-to-date damage award in a jury trial in history, the penalty handed down against the tobacco industry included a $74 billion judgment against Philip Morris. If this amount was 94% of this company's 1999 revenue, how much was Philip Morris's 1999 revenue? $78.723 billion
(*Source*: Newsweek)

74. Tobacco Judgment As part of the largest-to-date damage award in a jury trial in history, the penalty handed down against the tobacco industry included a $36 billion judgment against R. J. Reynolds. If this amount was 479% of this company's 1999 revenue, how much was R. J. Reynolds' 1999 revenue?
(*Source*: Newsweek) Approximately $7.52 billion

75. Sales Commission A salesman earns $50,000 in commission in 1 year and then has his commission reduced by 20% the next year. What percent increase in commission over the second year will give him $50,000 in the third year? 25%

76. Salaries A woman earning $100,000 per year has her salary reduced by 5% because of a reduction in the company's market. A year later, she receives $104,500 in salary. What percent raise from the reduced salary does this represent? 10%

77. Sales Tax The total cost of a new automobile, including a 6% sales tax on the price of the automobile, is $29,998. How much of the total cost of this new automobile is sales tax? $1698

78. Wildlife Management In wildlife management, the capture–mark–recapture technique is used to estimate the population of certain types of fish or animals. To estimate the population, we enter information in the equation

$$\frac{\text{total in population}}{\text{total number marked}} = \frac{\text{total number in second capture}}{\text{number found marked in second capture}}$$

Suppose that 50 sharks are caught along a certain shoreline, marked, and then released. If a second capture of 50 sharks yields 20 sharks that have been marked, what is the resulting population estimate? 125

79. *Investment* The formula for the future value A of a simple interest investment is $A = P + Prt$, where P is the original investment, r is the annual interest rate, and t is the time in years. Solve this formula for t. $t = \dfrac{A - P}{Pr}$

80. *Investment* The formula for the future value A of a simple interest investment is $A = P + Prt$, where P is the original investment, r is the annual interest rate, and t is the time in years. Solve this formula for P. $P = \dfrac{A}{1 + rt}$

81. *Investment* If P dollars are invested for t years at simple interest rate r, the future value of the investment is $A = P + Prt$. If $2000 invested for 6 years gives a future value of $3200, what is the simple interest rate of this investment? 10%

82. *Investment* If an investment at 7% simple interest has a future value of $5888 in 12 years, what is the original investment? $3200

83. *Investment* The simple interest earned in 9 years is directly proportional to the interest rate r. If the interest is $920 when r is 12%, what is the amount of interest earned in 9 years at 8%? $613.33

84. *Investment* The interest earned at 9% simple interest is directly proportional to the number of years the money is invested. If the interest is $4903.65 in 5 years, in how many years will the interest earned at 9% be $7845.84? 8 years

85. *Circles* Does the circumference of a circle vary directly with the radius of the circle? If so, what is the constant of variation? Yes; 2π

86. *Calories* The amount of heat produced in the human body by burning protein is directly proportional to the amount of protein burned. If burning 1 gram of protein produces 32 calories of heat, how much protein should be burned to produce 180 calories? 5.625 g

87. *Body Mass Index* Body mass index (BMI) is a measure that helps determine obesity, with a BMI of 30 or greater indicating that the person is obese. A BMI table for heights in inches and weights in pounds is shown on page 86. BMI was originally defined in the metric system of measure, and the BMI is directly proportional to the weight of a person of a given height.

   a. If the BMI of a person who is 1.5 meters tall is 20 when the person weighs 45 kilograms, what is the constant of variation? 4/9

   b. If a woman of this height has a BMI of 32, what does she weigh? 72 kg

   (*Source*: Roche Pharmaceuticals)

88. *Land Cost* The cost of land in Savannah is directly proportional to the size of the land. If a 2500-square-foot piece of land costs $172,800, what is the cost of a piece of land that is 5500 square feet? $380,160

2.2 Fitting Lines to Data Points: Modeling Linear Functions

KEY OBJECTIVES

- Find exact linear models for data
- Determine whether a set of data can be modeled exactly or approximately
- Create scatter plots for sets of data
- Find approximate linear models for data
- Visually determine whether a linear model is a "good" fit for data
- Solve problems using linear models

SECTION PREVIEW Big Data Market Size

The term "big data" refers to the large amounts of information that are gathered and stored for eventual analysis. Data are collected in all types of formats, from numeric data in traditional databases to text documents, email, video, audio, stock ticker data, and financial transactions. Table 2.2 gives the global big data market size, based on revenue, from 2015 and projected through 2027. Figure 2.11 shows the scatter plot of the data, and it appears that a line would approximately fit these data points. We can determine the relationship between the big data revenue and the number of years from 2015 by creating a linear equation that gives the revenue as a function of the years. (See Example 3.) In this section, we will use graphing utilities to create linear models from data points and use the models to answer questions about the data.

Table 2.2

Year	Revenue ($ billions)	Year	Revenue ($ billions)
2015	22.6	2022	70
2016	28	2023	77
2017	35	2024	84
2018	42	2025	90
2019	49	2026	96
2020	56	2027	103
2021	64		

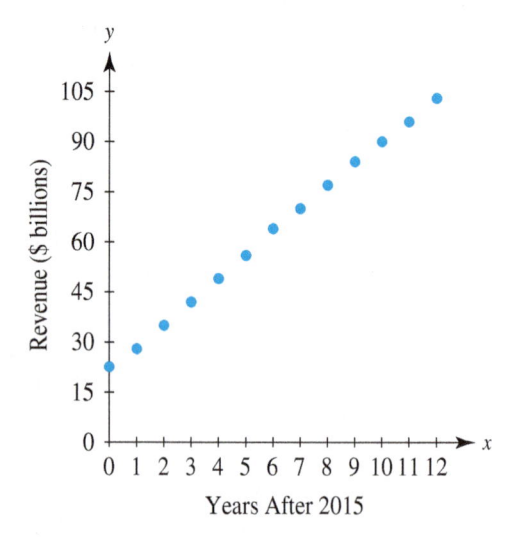

(*Source:* Statista)

Figure 2.11

Exact and Approximate Linear Models

We have seen that if data points fit exactly on a line, we can use two of the points to model the linear function (write the equation of the line). We can determine that the data points fit exactly on a line by determining that the changes in output values are equal for equal changes in the input values. In this case, we say that the inputs are **uniform** and the **first differences** are constant.

- If the first differences of data outputs are constant for uniform inputs, the rate of change is constant and a linear function can be found that fits the data exactly.
- If the first differences are "nearly constant," a linear function can be found that is an approximate fit for the data.

If the first differences of data outputs are constant for uniform inputs, we can use two of the points to write the linear equation that models the data. If the first differences of data outputs are constant for inputs differing by 1, this constant difference is the rate of change of the function, which is the slope of the line fitting the points exactly.

EXAMPLE 1 ▶ Retirement

Table 2.3 gives the annual retirement payment to a 62-year-old retiree with 21 or more years of service at Clarion State University as a function of the number of years of service and the first differences of the outputs.

Table 2.3

Year	21	22	23	24	25
Retirement Payment	40,950	42,900	44,850	46,800	48,750
First Differences		1950	1950	1950	1950

Because the first differences of the outputs of this function are constant (equal to 1950) for each unit change of the input (years), the rate of change is the constant 1950. Using this rate of change and a point gives the equation of the line that contains all the points.

Representing the annual retirement payment by y and the years of service by x and using the point (21, 40,950), we obtain the equation

$$y - 40{,}950 = 1950(x - 21)$$
$$y = 1950x$$

Note that the values in Table 2.3 represent points satisfying a **discrete function** (a function with a finite number of inputs), with each input representing the number of years of service. Although only points with integer inputs represent the annual retirement payments, we can model the application with the **continuous function** $y = 1950x$, whose graph is a line that passes through the 5 data points. Informally, a continuous function can be defined as a function whose graph can be drawn over its domain without lifting the pen from the paper.

In the context of this application, we must give a discrete interpretation to the model. This is because the only inputs of the function that make sense in this case are nonnegative integers representing the number of years of service. The graph of the discrete function defined in Table 2.3 is the scatter plot shown in Figure 2.12(a), and the continuous function that fits these data points, $y = 1950x$, is shown in Figure 2.12(b).

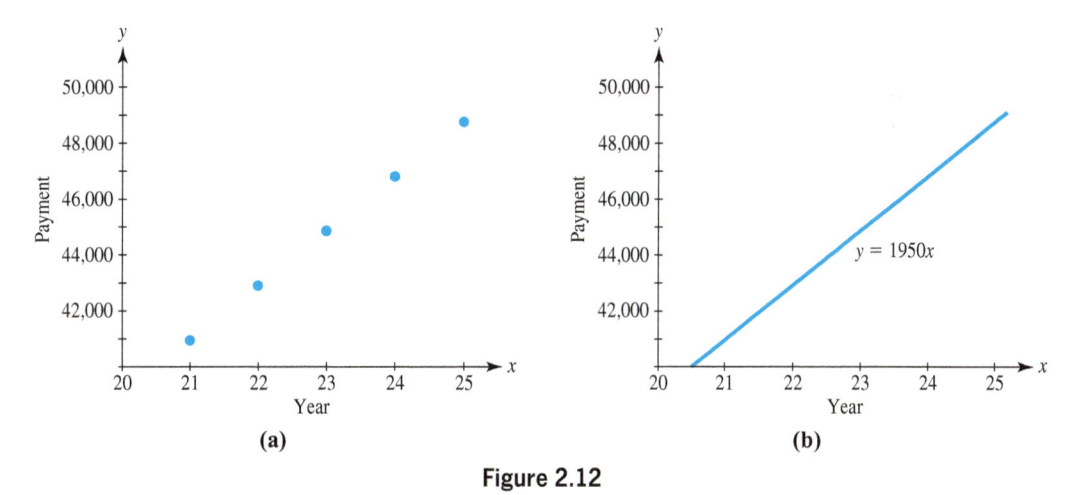

Figure 2.12

When a scatter plot of data can be approximately fitted by a line, we attempt to find the graph that visually gives the best fit for the data and find its equation. For now, we informally define the "best-fit" line as the one that appears to come closest to all the data points.

EXAMPLE 2 ▶ Population of Females Under 18

Table 2.4 gives the projected population, in millions, of females under the age of 18 projected for the years 2020 through 2060.

Table 2.4

Year	Population (millions)	Year	Population (millions)
2020	37.2	2045	41.0
2025	38.2	2050	42.0
2030	39.2	2055	42.9
2035	39.8	2060	43.6
2040	40.4		

(*Source*: U.S. Census Bureau)

a. Draw a scatter plot of the data with the x-value of each point representing the number of years after 2000 and the y-value representing the population of females under the age of 18, in millions.

b. Graph the equation $y = 0.1x + 36$ and the data points on the same axes. Is this a good fit for the data?

c. Use a piece of (uncooked!) spaghetti or a mechanical pencil "lead" to find a line through two of the data points that is a good visual fit for the points, and use the two points to write the equation of this line.

SOLUTION

a. By using x as the number of years after 2000, we have aligned the data with $x = 0$ representing 2000, with $x = 20$ representing 2020, and so forth. In the lists of a graphing calculator (or columns of an Excel spreadsheet), we enter the aligned input data representing the years in Table 2.4 and the output data representing the population of females under the age of 18 (in millions). Table 2.5 shows the lists containing the data. The window for this scatter plot can be set manually or with a command on the graphing utility such as ZoomStat, which is used to automatically set the window and display the graph. The graph of these points is shown in Figure 2.13.

b. The graph of the function $y = 0.1x + 36$ and the scatter plot of the data are shown in Figure 2.14. The line does not appear to be the best possible fit to the data points.

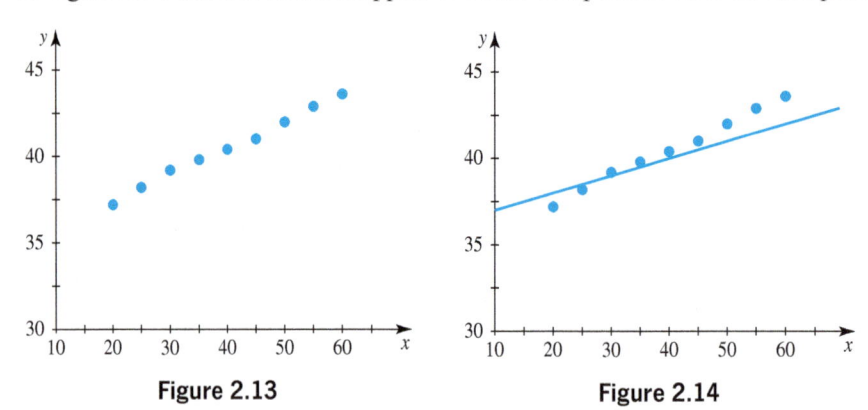

Table 2.5

L1	L2
20	37.2
25	38.2
30	39.2
35	39.8
40	40.4
45	41.0
50	42.0
55	42.9
60	43.6

Figure 2.13

Figure 2.14

c. Placing a piece of spaghetti or a pencil lead along the points shows a line that passes through the points $(25, 38.2)$ and $(60, 43.6)$ and appears to be a good fit (see Figure 2.15). The equation of this line is found as follows.

$$m = \frac{43.6 - 38.2}{60 - 25} \approx 0.154 \quad \text{so} \quad y - 43.6 = 0.154(x - 60)$$

$$\text{or} \quad y = 0.154x + 34.36$$

Comparing Figure 2.14 and Figure 2.16 shows that this function is a much better fit than $y = 0.1x + 36$ from part (b).

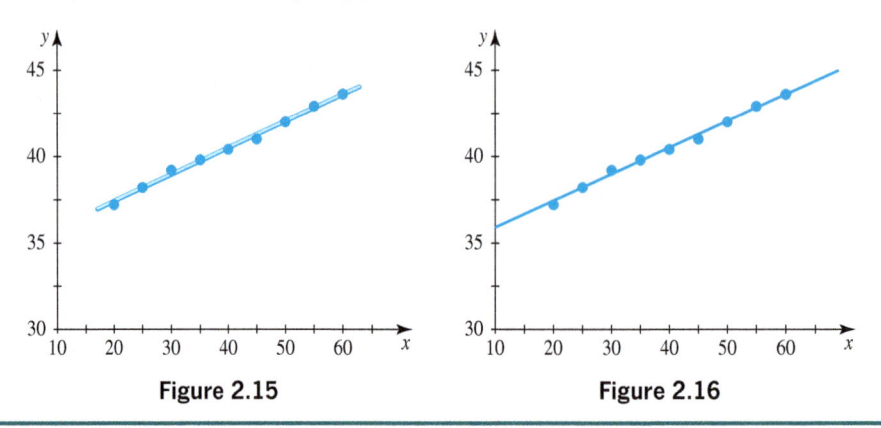

Figure 2.15

Figure 2.16

Is either of the lines in Figures 2.14 and 2.16 the best-fitting of all possible lines that could be drawn on the scatter plot? How do we determine the best-fit line? We now discuss the answers to these questions.

Fitting Lines to Data Points; Linear Regression

The points determined by the data in Table 2.4 on page 111 do not all lie on a line, but we can determine the equation of the line that is the "best fit" for these points by using a procedure called **linear regression**. This procedure defines the *best-fit line* as the line for which the sum of the squares of the vertical distances from the data points to the line is a minimum. For this reason, the linear regression procedure is also called the **least squares method**.

The vertical distance between a data point and the corresponding point on a line is simply the amount by which the line misses going through the point—that is, the difference in the *y*-values of the data point and the point on the line. If we call this difference in outputs d_i (where *i* takes on the values from 1 to *n* for *n* data points), the least squares method* requires that

$$d_1^2 + d_2^2 + d_3^2 + \cdots + d_n^2$$

be as small as possible. The line for which this sum of squared differences is as small as possible is called the *linear regression line* or *least squares line* and is the one that we consider to be the **best-fit line** for the data.

For an illustration of the linear regression process, consider again the two lines in Figures 2.14 and 2.16 and the data points in Table 2.4. Figures 2.17 and 2.18 indicate the vertical distances for each of these lines. To the right of each figure is the calculation of the sum of squared differences for the line.

$$d_1^2 + d_2^2 + d_3^2 + \cdots + d_9^2$$
$$= (37.2 - 38)^2 + (38.2 - 38.5)^2 + (39.2 - 39)^2$$
$$+ (39.8 - 39.5)^2 + (40.4 - 40.0)^2 + (41.0 - 40.5)^2$$
$$+ (42.0 - 41.0)^2 + (42.9 - 41.5)^2 + (43.6 - 42)^2$$
$$= 6.79$$

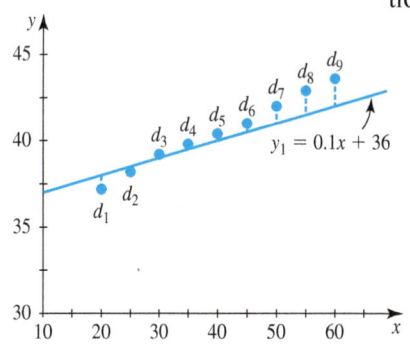

Figure 2.17

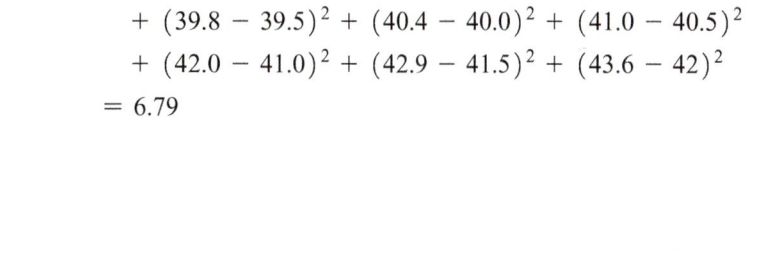

$$d_1^2 + d_2^2 + d_3^2 + \cdots + d_9^2$$
$$= (37.2 - 37.44)^2 + (38.2 - 38.21)^2 + (39.2 - 38.98)^2$$
$$+ (39.8 - 39.75)^2 + (40.4 - 40.52)^2 + (41.0 - 41.29)^2$$
$$+ (42.0 - 42.06)^2 + (42.9 - 42.83)^2 + (43.6 - 43.6)^2$$
$$= 0.2156$$

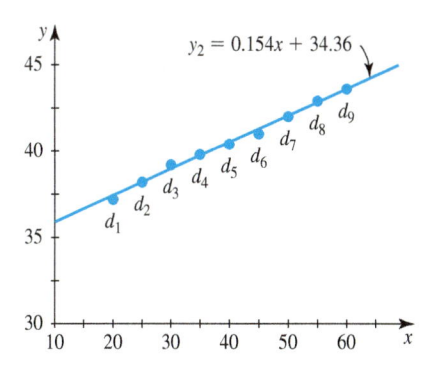

Figure 2.18

We see numerically from the sum of squared differences calculations, as well as visually from Figures 2.14 and 2.15, that the line

$$y_2 = 0.154x + 34.36$$

---

*The sum of the squared differences is often called SSE, the *sum of squared errors*. The development of the equations that lead to the minimum SSE is a calculus topic.

114   Chapter 2 Linear Models, Equations, and Inequalities

is a better model for the data than the line y_1. However, is this line the best-fit line for these data? Remember that the best-fit line must have the smallest sum of squared differences of *all* lines that can be drawn through the data! We will see that the line which is the best fit for these data points, with coefficients rounded to three decimal places, is

$$y = 0.155x + 34.278$$

The sum of squared differences for the regression line is approximately 0.198, which is slightly smaller than the sum 0.2156, which was found for $y_2 = 0.154x + 34.3$.

Development of the formulas that give the best-fit line for a set of data is beyond the scope of this text, but graphing calculators, computer programs, and spreadsheets have built-in formulas or programs that give the equation of the best-fit line. That is, we can use technology to find the best linear model for the data. The calculation of the sum of squared differences for the dental employees data was given only to illustrate what the best-fit line means. You will not be asked to calculate—nor do we use from this point on—the value of the sum of squared differences in this text.

We illustrate the use of technology to find the regression line by returning to the data of Example 2. To find the equation that gives the population of females under 18 as a function of the years after 2000, we use the following steps.

Modeling Data

1. Enter the data into lists of a graphing utility. Figure 2.19(a) shows the data from Example 2 entered in the lists of a TI-84 Plus C.

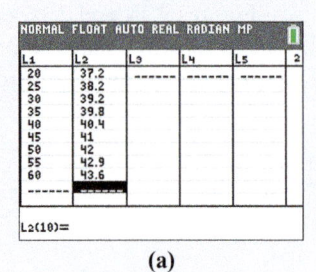

(a)

Figure 2.19

2. Create a scatter plot of the data to see whether a linear model is reasonable. The data should appear to follow a linear pattern with no distinct curvature. Figures 2.19(b) and (c) show the scatter plot of the data from Example 2 and the window that was used.

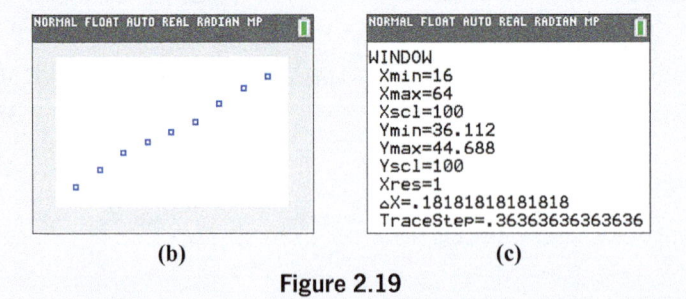

(b)   (c)

Figure 2.19

3. To use a graphing utility to obtain the linear equation that is the best fit for the data, press **STAT**, move to CALC, and select 4:LinReg. Then move to Store RegEQ:, press **VARS**, move to Y-VARS, and press **ENTER** four times. This will show the equation of the best-fitting line on the screen and on the Y= screen.*
Figure 2.19(d) shows the equation for Example 2, which can be approximated

---

*See Appendix A, page 690, for details.

by $y = 0.155x + 34.278$, and Figure 2.19(e) shows the equation on the Y= screen.

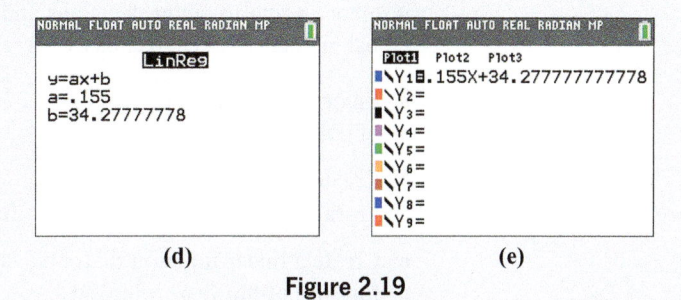

(d)

(e)

Figure 2.19

4. Graph the linear function (unrounded) and the data points on the same graph to see how well the function fits the data. The graph of the data and the best-fitting line is shown in Figure 2.19(f).

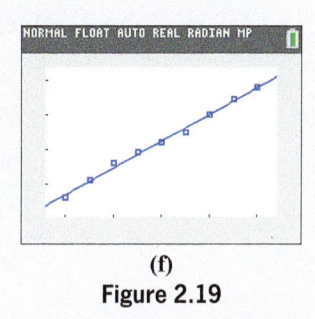

(f)

Figure 2.19

5. Report the function and/or numerical results in a way that makes sense in the context of the problem, with the appropriate units and with the variables identified. Unless otherwise indicated, report functions with coefficients rounded to three decimal places. Sometimes you will be asked to report a model with three or four significant digits. For example, the model above could be reported with three significant digits as $y = 0.155x + 34.3$. See the Toolbox, page 95, for a discussion of significant digits.

Figure 2.19(f) shows that

$$y = 0.155x + 34.278$$

is a good model of the population of females under 18, where y is in millions and x is the number of years after 2000.

The screens for the graphing utility that you use may vary slightly from those given in this text. Also, the regression line you obtain is dependent on your particular technology and may have some decimal places slightly different from those shown here.

It is important to be careful when rounding coefficients in equations and when rounding during calculations with the model. We will use the following guidelines in this text.

Technology Note: Rounding Guidelines

After a model for a data set has been found, it can be rounded for reporting purposes. However, do not use a rounded model in calculations, and do not round answers during the calculation process unless instructed to do so. When the model is used to find numerical answers, the answers should be rounded in a way that agrees with the context of the problem.

EXAMPLE 3 Big Data Market Size

Big data is a term that describes the large volume of data that can be analyzed for insights leading to better decisions and strategic business moves. Table 2.6 shows the global big data market size, based on revenue, from 2015 and projected through 2027.

a. Enter x, the number of years after 2015, in list L1 and enter y, the revenue in billions of dollars, in list L2.

b. Create the scatter plot of the data points in the lists, with x equal to the number of years after 2015 and y equal to the revenue in billions of dollars.

c. Create a linear function that models the big data revenue in billions of dollars as a function of the years after 2015.

d. Graph the linear function and the data points on the same graph and discuss how well the function models the data.

Table 2.6

Year	Revenue ($ billions)	Year	Revenue ($ billions)
2015	23	2022	70
2016	28	2023	77
2017	35	2024	84
2018	42	2025	90
2019	49	2026	96
2020	56	2027	103
2021	64		

(*Source:* Statista)

SOLUTION

a. If x represents the number of years after 2015, then we subtract 2015 from each year to obtain each corresponding x-value. Figure 2.20(a) shows the aligned inputs representing the number of years after 2015 in L1 and the outputs representing the big data revenue in billions of dollars in L2.

b. Figure 2.20(b) shows the scatter plot of the data using the window $[0, 12]$ by $[20, 105]$. The data points do not lie on any line perfectly, but the points on the scatter plot exhibit a nearly linear pattern, so a linear function could be used to model these data.

(a)   (b)

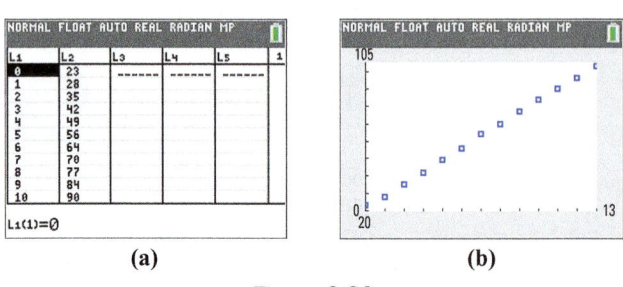

Figure 2.20

c. Using the instructions in step 3 on page 114 and/or in Appendix A, page 690, gives the result found in Figure 2.21(a). This model, reported with three significant digits, is

$$y = 6.79x + 22.1$$

Remember that a *model* gives not only the equation but also a description of the variables and their units of measure. The rounded model

$$y = 6.79x + 22.1$$

gives the big data revenue in billions of dollars as a function of the number of years after 2015.

d. The graph of the data and the (unrounded) model is shown in Figure 2.21(b). The graph of the function that models the data is an excellent visual fit to the data.*

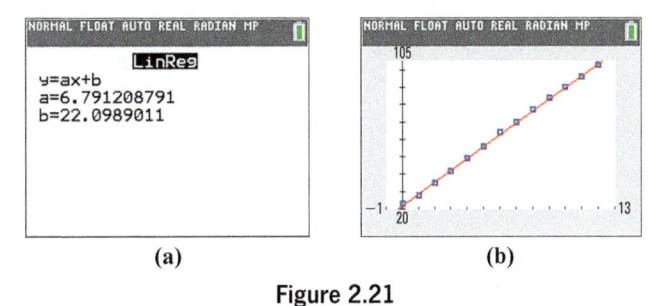

(a)   (b)

Figure 2.21

EXAMPLE 4 ▶ U.S. Population

The total U.S. population for selected years beginning in 1960 and projected to 2050 is shown in Table 2.7, with the population given in millions.

a. Align the data to represent the number of years after 1960, and draw a scatter plot of the data.

b. Create the linear equation that is the best fit for these data, where y is in millions and x is the number of years after 1960.

Table 2.7

Year	Population (millions)	Year	Population (millions)
1960	180.671	1995	263.044
1965	194.303	1998	270.561
1970	205.052	2000	281.422
1975	215.973	2003	294.043
1980	227.726	2025	358.030
1985	238.466	2050	408.695
1990	249.948		

(*Source*: U.S. Census Bureau)

---

*Even though we write rounded models found from data given in this text, all graphs and calculations use the unrounded model found by the graphing utility unless otherwise directed.

c. Graph the equation of the linear model on the same graph with the scatter plot and discuss how well the model fits the data.

d. Align the data to represent the years after 1950 and create the linear equation that is the best fit for the data, where y is in millions.

e. How do the x-values for a given year differ?

f. Use both unrounded models to estimate the population in 2000 and in 2010. Are the estimates equal?

SOLUTION

a. The aligned data have $x = 0$ representing 1960, $x = 5$ representing 1965, and so forth. Figure 2.22(a) shows the first eleven entries using the aligned data. The scatter plot of these data is shown in Figure 2.22(b).

(a)   (b)

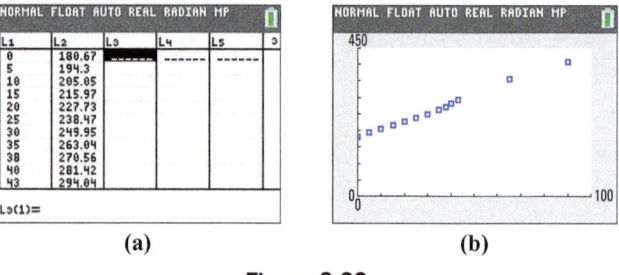

Figure 2.22

b. The equation of the best-fit line is found by using linear regression with a graphing calculator. With the decimals rounded to three places, the linear model for the U.S. population is

$$y = 2.607x + 177.195 \text{ million}$$

where x is the number of years after 1960.

c. Using the unrounded function in the equation editor (Figure 2.23(a)) and graphing it along with the scatter plot shows that the graph of the best-fit line is very close to the data points (Figure 2.23(b)). However, the points do not all fit the line because the U.S. population did not increase by exactly the same amount each year.

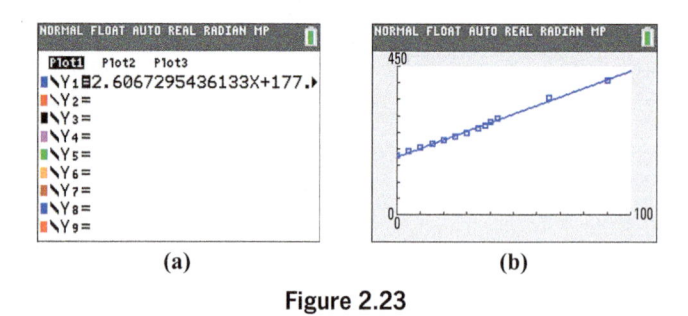

(a)   (b)

Figure 2.23

d. If we align the data to represent the years after 1950, then $x = 10$ corresponds to 1960, $x = 15$ corresponds to 1965, and so forth. Figure 2.24(a) shows the first eleven entries using the aligned data, and Figure 2.24(b) shows the scatter plot. The equation that best fits the data, found using linear regression with a calculator, is

$$y = 2.607x + 151.128$$

where x is the number of years from 1950 and y is in millions. Figure 2.24(c) shows the regression equation.

2.2 Fitting Lines to Data Points: Modeling Linear Functions 119

(a)   (b)   (c)

Figure 2.24

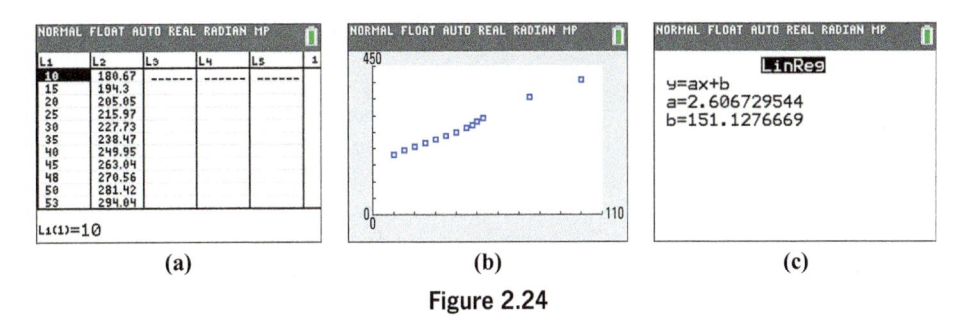

e. The x-values for a given year are 10 more with this model than with the first model.

f. Both models estimate the population to be 281.464 million in 2000 and to be 307.531 million in 2010. They are equal estimates. In fact, if you substitute $x - 10$ for x in the first model, you will get the second model.

Spreadsheet ▶ SOLUTION   Using Excel to find the linear function that is the best fit for the data involves several steps.

1. Enter the data in two columns.

2. Create a scatter plot of the data by highlighting the two columns containing the x- and y- values, select Insert > Scatter, and choose the Points Only Option to plot the points.

3. Right-click (with a Mac, use Control-click) on one of the data points and choose Add Trendline.

4. Select the function type (linear in this case) and click the box Display Equation on the chart at the bottom.

5. Close the dialog box to see the graph and the equation of the model. Figure 2.25 shows the best-fitting line and its equation for the data in Example 4, with x equal to the number of years after 1960 and y equal to the population in millions.*

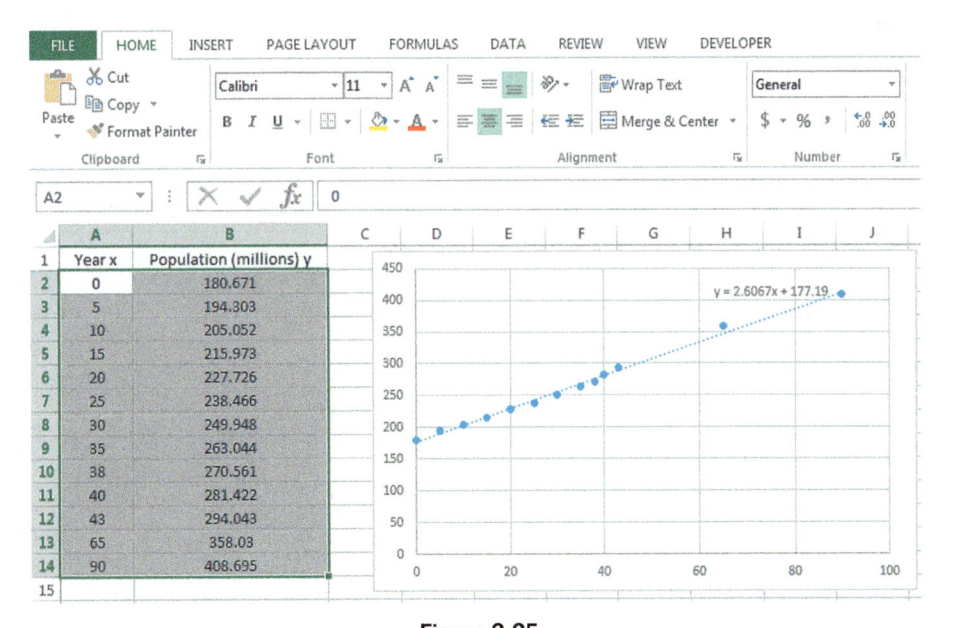

Figure 2.25

---

*See Appendix B, page 708 for details.

Applying Models

Because 2000 is a year between two given values in the table of Example 4, using the model to find the population in 2000 is called **interpolation**. When a model is evaluated for a *prediction* using input(s) outside the given data points, the process is called **extrapolation**.

A question arises whenever data that involve time are used. If we label the *x*-coordinate of a point on the input axis as 1999, what time during 1999 do we mean? Does 1999 refer to the beginning of the year, the middle, the end, or some other time? Because most data from which the functions in applications are derived represent end-of-year totals, we adopt the following convention when modeling, unless otherwise specified:

A point on the input axis indicating a time refers to the end of the time period.

For instance, the point representing the year 1999 means "at the end of 1999." Notice that this instant in time also represents the beginning of the year 2000. If, for example, data are aligned with *x* equal to the number of years after 1990, then any *x*-value greater than 9 and less than or equal to 10 represents some time in the year 2000. If a point represents anything other than the end of the period, this information will be clearly indicated. Also, if *a* and *b* are points in time, we use the phrases "from *a* to *b*" and "between *a* and *b*" to indicate the same time interval.

Goodness of Fit

Consider again Example 4, where we modeled the U.S. population for selected years. Looking at Table 2.8, which is a partial table of the data, we see that for uniform inputs the first differences of the outputs appear to be relatively close to the same constant, especially compared to the size of the population. (If these differences were closer to a constant, the fit would be better.)

Table 2.8

Uniform Inputs (year)	Outputs Population (millions)	First Differences in Output (difference in population)
1960	180.671	
1965	194.303	13.632
1970	205.052	10.749
1975	215.973	10.921
1980	227.726	11.753
1985	238.466	10.740
1990	249.948	11.482
1995	263.044	13.096
2000	281.422	18.378

So how "good" is the fit of the linear model $y = 2.6067295436133x + 151.1276669$, with *x* equal to the number of years after 1960, to the data in Example 4? Based on observation of the graph of the line and the data points on the same set of axes (see Figure 2.25), it is reasonable to say that the regression line provides a very good fit, but not an exact fit, to the data.

The goodness of fit of a line to a set of data points can be observed from the graph of the line and the data points on the same set of axes, and it can be measured if your

2.2 Fitting Lines to Data Points: Modeling Linear Functions

graphing utility computes the **correlation coefficient**.* The correlation coefficient is a number r, $-1 \leq r \leq 1$, that measures the strength of the linear relationship that exists between the two variables. The closer $|r|$ is to 1, the more closely the data points fit the linear regression line. (There is no linear relationship between the two variables if $r = 0$.) Positive values of r indicate that the output variable increases as the input variable increases, and negative values of r indicate that the output variable decreases as the input variable increases. But the *strength* of the relationship is indicated by how close $|r|$ is to 1. For the data of Example 4, computing the correlation coefficient gives $r = .997$ (Figure 2.26), which means that the linear relationship is strong and that the linear model is an excellent fit for the data. When a calculator feature or computer program is used to fit a linear model to data, the resulting equation can be considered the best linear fit for the data. As we will see later in this text, other mathematical models may be better fits to some sets of data, especially if the first differences of the outputs are not close to being constant.

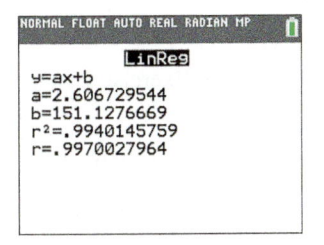

```
NORMAL FLOAT AUTO REAL RADIAN MP
        LinReg
y=ax+b
a=2.606729544
b=151.1276669
r²=.9940145759
r=.9970027964
```

Figure 2.26

*See Appendix A, page 690.

SKILLS CHECK 2.2

Answers that are not seen can be found in the answer section at the back of the text.

Report models to three decimal places unless otherwise specified. Use unrounded models to graph and calculate unless otherwise specified.

Discuss whether the data shown in the scatter plots in the figures for Exercises 1 and 2 should be modeled by a linear function.

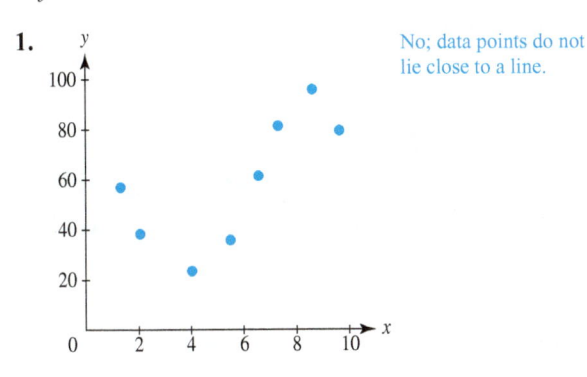

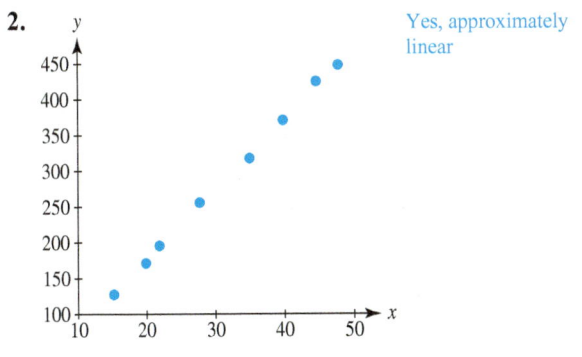

1. No; data points do not lie close to a line.

2. Yes, approximately linear

Discuss whether the data shown in the scatter plots in the figures for Exercises 3 and 4 should be modeled by a linear function exactly or approximately.

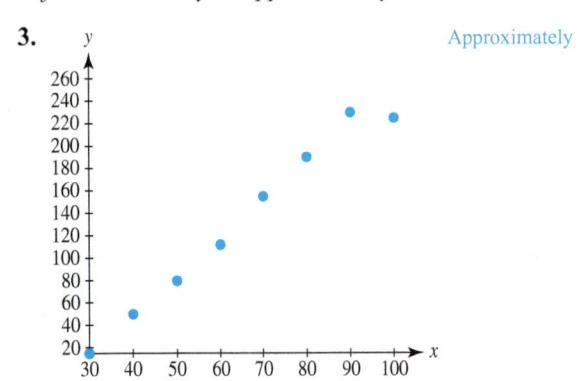

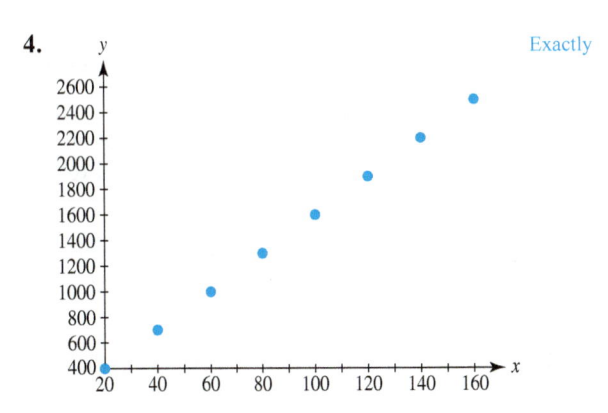

3. Approximately

4. Exactly

Create a scatter plot for each of the sets of data in Exercises 5 and 6.

5.

x	1	3	5	7	9
y	4	7	10	13	16

6.

x	1	2	3	5	7	9	12
y	1	3	6	1	9	2	6

7. Can the scatter plot in Exercise 5 be fit exactly or only approximately by a linear function? How do you know?
Exactly; first differences of the outputs are equal.
8. Can the scatter plot in Exercise 6 be fit exactly or only approximately by a linear function? How do you know?
Only approximately; no line will fit all points.
9. Find the linear function that is the best fit for the data in Exercise 5. $y = 1.5x + 2.5$

10. Find the linear function that is the best fit for the data in Exercise 6. $y = 0.282x + 2.428$

Use the data in the following table for Exercises 11–14.

x	5	8	11	14	17	20
y	7	14	20	28	36	43

11. Construct a scatter plot of the data in the table.

12. Determine whether the points plotted in Exercise 11 appear to lie near some line. Yes

13. Create a linear model for the data in the table.
$y = 2.419x - 5.571$
14. Use the function $y = f(x)$ created in Exercise 13 to evaluate $f(3)$ and $f(5)$. $f(3) \approx 1.7; f(5) \approx 6.5$

Use the data in the table for Exercises 15–18.

x	2	5	8	9	10	12	16
y	5	10	14	16	18	21	27

15. Construct a scatter plot of the data in the table.

16. Determine if the points plotted in Exercise 15 appear to lie near some line. Yes

17. Create a linear model for the data in the table.
$y = 1.577x + 1.892$
18. Use the rounded function $y = f(x)$ that was found in Exercise 17 to evaluate $f(3)$ and $f(5)$.
$f(3) \approx 6.6; f(5) \approx 9.8$
19. Determine which of the equations, $y = -2x + 8$ or $y = -1.5x + 8$, is the better fit for the data points $(0, 8), (1, 6), (2, 5), (3, 3)$. $y = -1.5x + 8$

20. Determine which of the equations, $y = 2.3x + 4$ or $y = 2.1x + 6$, is the better fit for the data points $(20, 50), (30, 73), (40, 96), (50, 119), (60, 142)$.
$y = 2.3x + 4$
21. Without graphing, determine which of the following data sets are exactly linear, approximately linear, or nonlinear.

a.
x	y
1	5
2	8
3	11
4	14
5	17

Exactly

b.
x	y
1	2
2	5
3	10
4	5
5	2

Nonlinear

c.
x	y
1	6
2	8
3	12
4	14
5	18

Approximately

22. Why can't first differences be used to tell if the following data are linear? $(1, 3), (4, 5), (5, 7), (7, 9)$
The difference between inputs is not constant.

EXERCISES 2.2

Answers that are not seen can be found in the answer section at the back of the text.
Report models to three decimal places unless otherwise specified. Use unrounded models to graph and calculate unless otherwise specified.

23. **Women in the Workforce** The number of women in the workforce for selected years from 1950 through 2050 is shown in the following figures.

   a. Do the points on the graph in the figure define the number of working women as a discrete or continuous function of the year? Discrete

   b. Does the graph of $y = W(x)$ shown in the figure define the number of working women as a discrete or continuous function of the year? Continuous

   c. Is $y = W(x)$ a reasonable model for the data points in the top graph? Is it a linear model?
   Yes; no

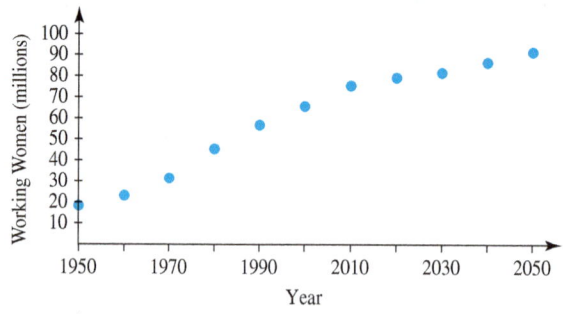

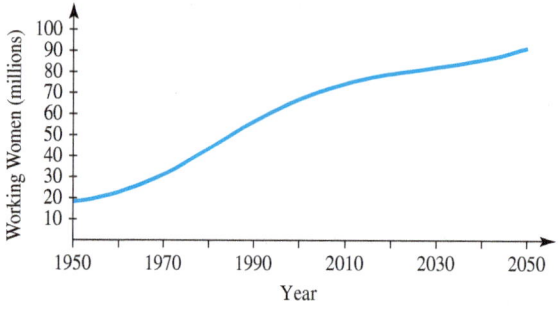

(*Source*: U.S. Census Bureau)

24. Worldwide Internet Users The data in the table show the percent of people in the world who are Internet users from 2014 and projected to 2025.

a. Create a scatter plot of the data, with *x* representing the number of years after 2010 and *p* representing the percent.

b. Use the scatter plot to determine whether the data can be fit exactly by a linear function. Not a perfect fit

Year	Percent	Year	Percent
2014	41.1	2018	51.1
2015	44.3	2019	53.0
2016	46.8	2020	54.6
2017	49.0	2025	61.0

(*Source*: eMarketer)

25. Social Security Benefits The figure shows how monthly Social Security benefits differ with the age a recipient decides to start receiving benefits. The example shown in this table is for recipients whose monthly benefit is $1000 if benefits start being paid at a full retirement age of 66. Note that benefits do not start being paid before age 62, and that the monthly benefit does not increase if the starting age is after 70.

a. Can the data shown be modeled exactly by a linear function? Explain. No; increases from one year to the next are not constant.

b. Can the data shown from age 66 to age 70 be modeled exactly by a linear function? Explain. Yes; increases from one year to the next are constant.

c. Write the linear function $S = f(x)$ that models the data shown from age 66 to age 70, where *x* is the age. $S = 80x - 4280$

Monthly benefit amounts differ based on the age you decide to start receiving benefits
This example assumes a benefit of $1,000 at a full retirement age of 66

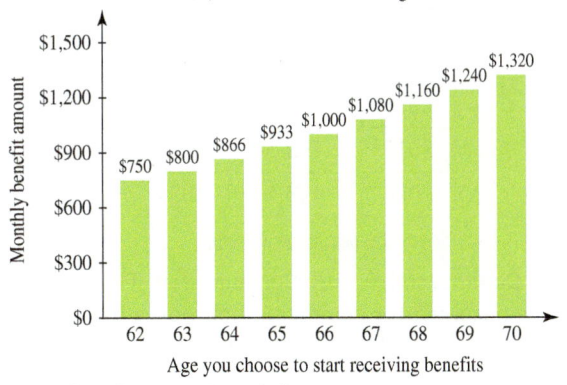

Bar chart showing monthly benefit amount by age: 62: $750, 63: $800, 64: $866, 65: $933, 66: $1,000, 67: $1,080, 68: $1,160, 69: $1,240, 70: $1,320

(*Source*: Social Security Administration)

26. Future Value of an Investment If $1000 is invested at 6% simple interest, the initial value and the future value *S* at the end of each of 5 years are shown in the table that follows.

a. Can a linear function model exactly the points from the table? Explain. Yes; the first differences are constant for uniform inputs.

b. If so, find a linear function $S = f(t)$ that models the points. $S = 60t + 1000$

c. Use the model to find the future value of this investment at the end of the 7th year. Is this an interpolation or an extrapolation from the data? $1420; extrapolation

d. Should this model be interpreted discretely or continuously? Discretely, because the interest payments occur only at the end of the year.

Year (t)	0	1	2	3	4	5
Future Value (S)	1000	1060	1120	1180	1240	1300

27. Drug Doses The table below shows the usual dosage for a certain prescription drug that combats bacterial infections for a person's weight.

Weight (lb)	Usual Dosage (mg)	Weight (lb)	Usual Dosage (mg)
88	40	154	70
99	45	165	75
110	50	176	80
121	55	187	85
132	60	198	90
143	65		

a. Can the data be modeled exactly by a linear function? Explain. Yes; the first differences are constant for uniform inputs.

b. If so, how many of the data points are needed to find the linear function that models the data exactly? Two

c. Find the linear function $D = f(w)$ that models the data exactly. $D = \frac{5w}{11}$

d. If the model is interpreted continuously, what dosage would be recommended for a 209-pound patient? 95 mg

28. Cardiac Health A Paramount elliptical machine provides the table below, which gives the desired heart rate during exercise for cardiac health based on the age of the person exercising on the elliptical machine.

a. Can the desired heart rate be expressed exactly as a linear function of age, based on these data? Yes

b. Write a linear function that models these data. $y = -0.8x + 176$

Age	Heart Rate (beats per minute)	Age	Heart Rate (beats per minute)
20	160	60	128
30	152	70	120
40	144	80	112
50	136	90	104

(*Source*: paramountfitness.com)

29. Disposable Income
Disposable income is the amount left after taxes have been paid and is one measure of health of the economy. The table gives the total disposable income for the United States, in billions of dollars, for selected years from 2010 and projected to 2040.

a. Can these data be fitted exactly by a linear model? No

b. Find a linear function that models the data, where x is the number of years after 2000 and y is the total disposable income in billions of dollars.
$y = 0.328x + 6.316$

c. What does the model predict the total disposable income will be in 2023? $13.850 billion

Year	Income ($ billions)
2010	10.017
2015	11.120
2020	12.655
2025	14.259
2030	15.948
2035	17.752
2040	19.785

30. Oil Imports
Crude oil and petroleum products are imported continuously by the United States. The table shows the billions of dollars of net expenditures for the imports of these products for several years.

a. Write a linear function that models these data, with x equal to the number of years after 2015 and y equal to the billions of dollars of expenditures. Report the model with two decimal places.
$y = 6.39x + 169.86$

b. Is the model an exact fit for the data? No

c. What does the model predict the billions of dollars of net expenditures will be for the imports in 2023? $220.98 billion

d. When does the model predict the billions of dollars of net expenditures for the imports will be $208.2 billion? 2021

Year	Expenditures ($ billions)
2018	191.0
2020	198.9
2022	214.5
2024	228.4

(Source: Energy Information Administration)

31. Annual Wage
The table gives the average annual wage of U.S. workers for selected years from 2012 projected through 2050.

a. Graph the data, with y equal to the average annual wage in thousands of dollars and x equal to the number of years after 2010.

b. Find a linear function that models the data, with y equal to the average annual wage in dollars and x equal to the number of years after 2010. Report the model with four significant digits.
$3.943x + 26.40$

c. Find the slope of the reported model, and interpret it in this context. $m = 3.943$; the average annual wage is expected to increase by $3943 per year.

Year	Wage ($ thousands)	Year	Wage ($ thousands)
2012	$44.6	2030	93.2
2014	48.6	2035	113.2
2016	53.3	2040	137.6
2018	58.7	2045	167.1
2020	63.7	2050	202.5
2025	76.8		

(Source: Social Security Administration)

32. Carbon Dioxide Emissions
For selected years from 2010 and projected to 2032, the millions of metric tons of carbon dioxide (CO_2) emissions from biomass energy combustion in the United States are shown in the following table.

a. Create a scatterplot of the data with x equal to the number of years after 2010 and y equal to the millions of metric tons of CO_2.

b. Create a linear function that models the data.
$y = 18.962x + 321.509$

c. Graph the data and the model on the same axes and comment on the fit of the model to the data.

Year	CO_2 Emissions	Year	CO_2 Emissions
2010	338.5	2022	556.2
2012	364.5	2024	590.9
2014	396.1	2026	629.7
2016	425.8	2028	663.1
2018	453.1	2030	701.1
2020	498.4	2032	743.7

(Source: U.S. Department of Energy)

33. Commercial Energy Consumption
The total energy consumed for commercial purposes from 2015 and projected to 2025 is shown in the table.

a. Create a scatter plot of the data, with y equal to the total energy consumption in quadrillion BTUs and x equal to the number of years after 2015.

b. Find a linear function that models the data in the table. $y = -0.017x + 17.847$

c. Graph the data and the model on the same axes. Is the model a good visual fit?

d. Use the model to find the projected total energy consumption for commercial purposes in 2029. 17.61 quadrillion BTUs

Year	Consumption (quadrillion BTUs)	Year	Consumption (quadrillion BTUs)
2016	17.82	2021	17.75
2017	17.83	2022	17.71
2018	17.78	2023	17.71
2019	17.78	2024	17.70
2020	17.77	2025	17.67

(*Source*: U.S. Energy Information Administration)

34. **Carbon Dioxide Emissions** The table gives the millions of metric tons of carbon dioxide emissions from biomass energy combustion in the United States for selected years from 2010 projected to 2032.

   a. Find a linear function that gives the millions of metric tons of carbon dioxide emissions, y, as function of x, the number of years after 2010.
   $y = 18.962x + 321.509$

   b. Graph the model on the same axes with the data. Is it a good fit for the data? Good fit

   c. What does the (unrounded) model predict the millions of metric tons of carbon dioxide emissions will be in 2027? 643.9

   d. When will the millions of metric tons of carbon dioxide emissions reach 776.6, according to the model? 2034

Year	CO$_2$ Emissions (millions of metric tons)
2010	338.5
2012	364.5
2014	396.1
2016	425.8
2018	453.1
2020	498.4
2022	556.2
2024	590.9
2026	629.7
2028	663.1
2030	701.1
2032	743.7

(*Source*: U.S. Department of Energy)

35. **U.S. Population over Age 16** The table gives the projections of the population, in millions, of U.S. residents over age 16 for selected years 2015 to 2060.

   a. Graph the data, with y equal to the population in millions and x equal to the number of years after 2010.

   b. Find a linear function that models the data, with y equal to the population in millions and x equal to the number of years after 2010. $y = 1.890x + 247.994$

   c. Graph the model and the data on the same axes and comment on the fit of the model to the data.

   d. What does the model predict that the population will be in 2042? 308.476 million

   e. In what year does the model predict that the population will be 336.827 million? 2057

Year	Population (millions)	Year	Population (millions)
2015	255.161	2040	306.634
2020	266.024	2045	315.152
2025	276.558	2050	323.314
2030	286.967	2055	331.770
2035	297.259	2060	340.868

36. **U.S. Population** The following table gives projections of the U.S. population from 2000 to 2100.

   a. Find a linear function that models the data, with x equal to the number of years after 2000 and $f(x)$ equal to the population in millions.
   $f(x) = 2.920x + 265.864$

   b. Find $f(65)$ and state what it means. 455.6; projected population is 455.6 million in 2065.

   c. What does this model predict the population to be in 2080? How does this compare with the value for 2080 in the table? 499.4 million; fairly close

Year	Population (millions)	Year	Population (millions)
2000	275.3	2060	432.0
2010	299.9	2070	463.6
2020	324.9	2080	497.8
2030	351.1	2090	533.6
2040	377.4	2100	571.0
2050	403.7		

(*Source*: www.census.gov/population/projections)

37. **U.S. National Consumption** The table gives the U.S. real consumption and disposable personal income (both in billions of dollars) for presidential election years from 2012 and projected to 2040.

   a. Find the linear function $C = f(y)$ that models these data, with y representing the disposable personal income and C representing the real consumption. Report the model with 3 significant digits.
   $C = 0.885y + 0.413$

   b. Graph the data and the linear model on the same axes, and discuss the fit.

c. Use the model to predict the U.S. real consumption when U.S. disposable personal income is $21 billion. $19 billion

Year	Income	Consumption
2012	10.2	9.6
2016	12.6	11.6
2020	14.2	12.9
2024	15.5	14.0
2028	17.1	15.4
2032	18.5	16.8
2036	20.1	18.3
2040	21.9	19.9

(*Source*: U.S. Energy Information Administration)

38. **Non-White Population** The graph gives the number of millions of individuals in the U.S. civilian noninstitutional population 16 years and older who are non-white or Hispanic for the years 2000 and projected to 2050.

   a. Create a scatter plot of the data with x equal to the number of years after 2000 y equal to the population 16 years and older who are non-white or Hispanic, in millions.

   b. Find the best linear function modeling y, the size of this population, as a function of x, the number of years after 2000. $y = 1.731x + 54.087$

   c. Use the model to predict that population in 2035. 114.7

   d. Graph the unrounded model on the same axes as the scatter plot of the data. Is the model a "good" fit? Yes.

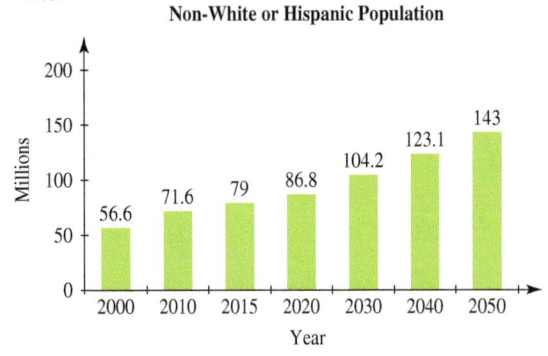

Non-White or Hispanic Population

(bar chart values: 56.6, 71.6, 79, 86.8, 104.2, 123.1, 143 for years 2000, 2010, 2015, 2020, 2030, 2040, 2050)

(*Source*: U.S. Census Bureau)

39. **Life Expectancy at 65** The graph gives the number of additional years of life expectancy at age 65 for selected years 1950 projected to 2050.

   a. Graph the data, with y equal to the number of additional years of life expectancy and x equal to the number of years after 1950.

   b. Find a linear function that models the data, with y equal to the life expectancy and x equal to the number of years after 1950. $y = 0.077x + 13.827$

   c. Graph the model and the data on the same axes, and comment on the fit of the model to the data. Good fit

   d. What does the model predict that the additional life expectancy will be in 2022? 19.4 years

   e. Graphically find the year that the unrounded model predicts the additional life expectancy will be 21 years. 2043

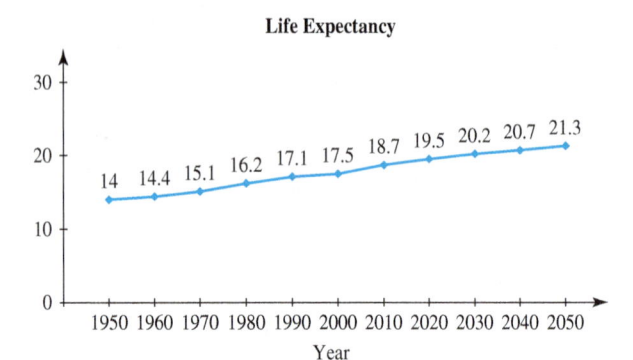

Life Expectancy

(data points: 14, 14.4, 15.1, 16.2, 17.1, 17.5, 18.7, 19.5, 20.2, 20.7, 21.3)

(*Source*: Social Security Administration)

40. **Civilian Labor Force** The table gives the size of the U.S. workforce, in millions, for selected years from 1950 and projected to 2050.

   a. Graph the data, with y equal to the number in the U.S. workforce, in millions, and x equal to the number of years after 1950.

   b. Find a linear function that models the data, with y equal to the number in millions and x equal to the number of years after 1950. $y = 1.377x + 64.068$

   c. Graph the data and the model on the same set of axes. Does the model fit the data exactly? No

   d. What does the model predict the size of the workforce to be in 2045? 194.9 million

   e. Graphically determine the year in which the workforce is projected to be 182.5 million according to the model. 2036

Year	Total Workforce (millions)	Year	Total Workforce (millions)
1950	62.2	2010	157.7
1960	69.6	2015	162.8
1970	82.7	2020	164.7
1980	107.0	2030	170.1
1990	125.8	2040	180.5
2000	140.9	2050	191.8

(*Source*: U.S. Bureau of Labor Statistics)

41. **Obesity** People who are severely obese (BMI ≥ 40) are at most risk for serious health problems, which are the most expensive to treat. The percent of

Americans who are severely obese from 2000 and projected to 2030 are shown in the table.

a. Find the linear function that models severe obesity, with x equal to the number of years after 2000 and y equal to the percent of severely obese Americans. $y = 0.297x + 2.043$

b. Graph the model and the data set on the same axes, and comment on the fit. Excellent fit

c. At what rate does the model indicate that severe obesity is growing during this period?
0.3 percentage point per year

Year	2000	2010	2015	2020	2025	2030
% Severely Obese	2.2	4.9	6.4	7.9	9.5	11.1

(Source: American Journal of Preventive Medicine.)

42. **Male Civilian Labor Force** The table gives the male civilian labor force, in millions, for selected years from 1950 projected to 2050.

a. Graph the data, with y equal to the number of men in the workforce, in millions, and x equal to the number of years after 1950.

b. Find a linear function that models the data, with y equal to the number of men in the workforce, in millions, and x equal to the number of years after 1950.
$y = 0.587x + 43.145$

c. Graph the model and the data on the same axes, and comment on the fit of the model to the data.

d. What does the model predict that the number of men in the workforce will be in 2051?
102.442 million

e. In what year does the model predict that the population will be 99.5 million? 2046

Year	Workforce (millions)	Year	Workforce (millions)
1950	43.8	2010	82.2
1960	46.4	2020	85.4
1970	51.2	2030	88.5
1980	61.5	2040	94.0
1990	69.0	2050	100.3
2000	75.2		

(Source: U.S. Bureau of Labor Statistics)

43. **Population of the World** The table gives the total population of the world from 1950 and projected to 2050.

a. Create a scatter plot of the data, with y equal to the world population in billions and x equal to the number of years after 1950.

b. Find a function that models the data, with y equal to the world population in billions and x equal to the number of years after 1950. Report the model with three significant digits. $y = 0.0715x + 2.42$

c. Graph the data points and the model on the same axes. Is the model a good fit?

d. What does the model estimate that the world population was in 2007? 6.5 billion

e. In what year does the model estimate the world population will be 10 billion? 2056

Year	Total World Population (billions)
1950	2.556
1960	3.039
1970	3.707
1980	4.454
1990	5.279
2000	6.083
2010	6.849
2020	7.585
2030	8.247
2040	8.850
2050	9.346

(Source: U.S. Census Bureau, International Database.)

44. **National Health Care** The following table shows the total national expenditures for health care (in billions of dollars) for selected years from 2002 and projected to 2024. (These data include expenditures for medical research and medical facilities construction.)

Year	Expenditures	Year	Expenditures
2002	1602	2014	3080
2004	1855	2016	3403
2006	2113	2018	3786
2008	2414	2020	4274
2010	2604	2022	4825
2012	2817	2024	5425

(Source: U.S. Centers for Medicare and Medicaid Services)

a. Find a linear function that models the data, with x equal to the number of years after 2000 and y equal to the expenditures for health in billions of dollars. Report the model with 3 significant digits.
$y = 163x + 1060$

b. What does the reported model predict the expenditures will be in 2030? $5950 billion

c. During what year does the reported model predict that the national expenditures for health care will be $7906 billion? 2042

128 Chapter 2 Linear Models, Equations, and Inequalities

45. Student Loans The table shows the total of federal direct student loans (in billions of dollars) for selected years from 2011 and projected to 2023.

a. Find the linear function that models the total loan debt y, in billions of dollars, as a function of the years after 2010. Report the model with three significant digits. $y = 131x + 569$

b. What does the reported model predict this debt to be in 2027? $2796 billion

c. When does the reported model predict that the debt will be $3189 billion? 2030 ($x = 20$)

Year	Debt ($ billions)
2011	702
2013	940
2015	1220
2017	1500
2019	1775
2021	2000
2023	2250

(*Source*: U.S. Office of Management and Budget)

2.3 Systems of Linear Equations in Two Variables

KEY OBJECTIVES

- Solve systems of linear equations graphically
- Solve systems of linear equations algebraically with the substitution method
- Solve systems of linear equations algebraically by elimination
- Model systems of equations to solve problems
- Determine whether a system of linear equations is inconsistent or dependent

SECTION PREVIEW U.S. Civilian Workforce

Using data from 1950 to 2050, the number of men in the workforce, in millions, can be modeled by the linear function $M(x) = 0.587x + 43.1$, and the number of women, in millions, can be modeled by $W(x) = 0.778x + 20.7$, where x equals the number of years after 1950. Figure 2.27 shows the graphs of these two linear functions for the years from 1950 to 2050. The lines in the figure eventually intersect, indicating that the number of women will eventually equal the number of men. To find when the number of women in the workforce will equal the number of men, we can solve the **system of equations** algebraically with substitution or elimination, or graphically by finding the point of intersection of the graphs of the two equations. The resulting solution is called the *simultaneous solution* of the two equations. (See Example 5.)

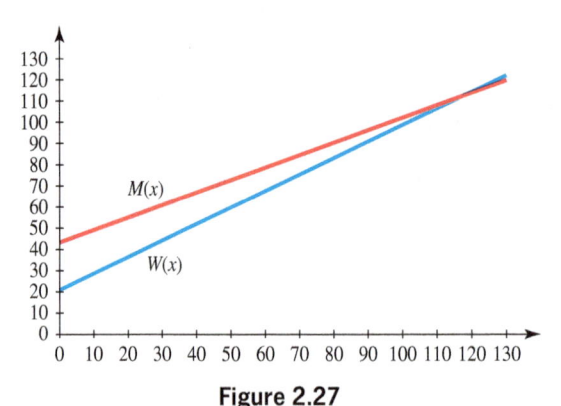

Figure 2.27
(*Source*: U.S. Bureau of Labor Statistics)

In this section, we solve systems of linear equations in two variables graphically, by substitution, and by the elimination method. ■

System of Equations

A **system of linear equations** is a collection of linear equations containing the same set of variables. A system of equations can have exactly one solution, no solution, or infinitely many solutions. A *solution to a system of equations* in two variables is an ordered pair that satisfies both equations in the system. We will solve systems of two equations in two variables by graphing, by substitution, and by the elimination method.

2.3 Systems of Linear Equations in Two Variables

Graphical Solution of Systems

In Section 2.1, we used the intersection method to solve a linear equation by first graphing functions representing the expressions on each side of the equation and then finding the intersection of these graphs. For example, to solve

$$3000x - 7200 = 5800x - 8600$$

we can graph

$$y_1 = 3000x - 7200 \quad \text{and} \quad y_2 = 5800x - 8600$$

and find the point of intersection to be $(0.5, -5700)$. The x-coordinate of the point of intersection of the lines is the value of x that satisfies the original equation, $3000x - 7200 = 5800x - 8600$. Thus, the solution to this equation is $x = 0.5$ (Figure 2.28). In this example, we were actually using a graphical method to solve a **system of two equations in two variables** denoted by

$$\begin{cases} y = 3000x - 7200 \\ y = 5800x - 8600 \end{cases}$$

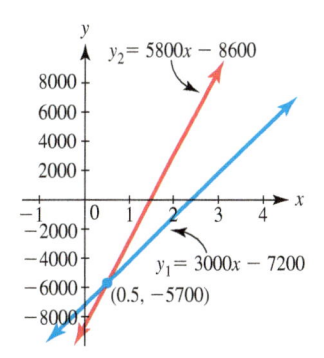

Figure 2.28

The coordinates of the point of intersection of the two graphs give the x- and y-values that satisfy both equations **simultaneously**, and these values are called the **solution** to the system. The following example uses the graphical method to solve a system of equations in two variables.*

EXAMPLE 1 ▶ Break-Even

A company is said to **break even** from the production and sale of a product if the total revenue equals the total cost—that is, if $R(x) = C(x)$. Because profit $P(x) = R(x) - C(x)$, we can also say that the company breaks even if the profit for the product is zero.

Suppose a company has its total revenue for a product given by $R = 5585x$ and its total cost given by $C = 61{,}740 + 440x$, where x is the number of thousands of tons of the product that are produced and sold per year. The company is said to break even when the total revenue equals the total cost—that is, when $R = C$. Find the number of thousands of tons of the product that gives break-even and how much the revenue and cost are at that level of production.

SOLUTION

We graph the revenue function as $y_1 = 5585x$ and the cost function as $y_2 = 61{,}740 + 440x$ (Figure 2.29(a)). We can find break-even with the **intersection method** on a calculator by graphing the two equations, $y_1 = 5585x$ and $y_2 = 61{,}740 + 440x$, on a window that contains the point of intersection and then finding the point of intersection, which is the point where the y-values are equal.* This point, which gives break-even, is $(12, 67{,}020)$ (Figure 2.29(b)).

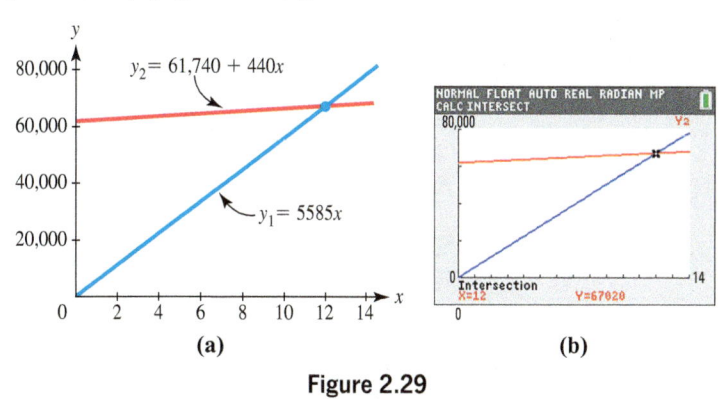

Figure 2.29

---

*See Appendix A, page 687, for details.

Thus, the company will break even on this product if 12 thousand tons of the product are sold, when both the cost and revenue equal $67,020.

It is frequently necessary to solve each equation for a variable so that the equation can be graphed with a graphing utility. It is also necessary to find a viewing window that contains the point of intersection. Consider the following example.

EXAMPLE 2 ▶ Solving a System of Linear Equations

Solve the system

$$\begin{cases} 3x - 4y = 21 \\ 2x + 5y = -9 \end{cases}$$

SOLUTION

To solve this system with a graphing utility, we first solve both equations for y:

$$3x - 4y = 21 \qquad\qquad 2x + 5y = -9$$
$$-4y = 21 - 3x \qquad\qquad 5y = -9 - 2x$$
$$y = \frac{21 - 3x}{-4} = \frac{3x - 21}{4} \qquad y = \frac{-9 - 2x}{5}$$

Graphing these equations with a window that contains the point of intersection (Figure 2.30(a)) and finding the point of intersection (Figure 2.30(b)) gives $x = 3$, $y = -3$, so the solution is $(3, -3)$.*

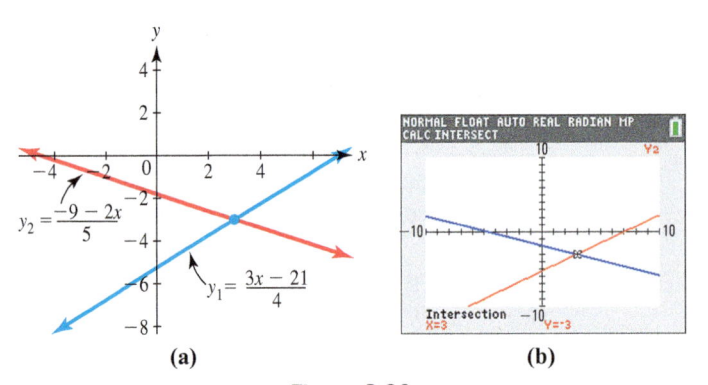

Figure 2.30

Solution by Substitution

Graphing with a graphing utility is not always the easiest method to use to solve a system of equations because the equations must be solved for y to be entered in the utility and an appropriate window must be found. A second solution method for a system of linear equations is the **substitution method**, where one equation is solved for a variable and that variable is replaced by the equivalent expression in the other equation.

The substitution method is illustrated by the following example, which finds market equilibrium. The quantity of a product that is demanded by consumers is called the **demand** for the product, and the quantity that is supplied is called the **supply**. In a free economy, both demand and supply are related to the price, and the price where the number of units demanded equals the number of units supplied is called the **equilibrium price**.

---
*For a discussion of solving systems of equations using Excel, see Appendix B, page 713.

EXAMPLE 3 ▶ Market Equilibrium

Suppose that the daily demand for a product is given by $p = 200 - 2q$, where q is the number of units demanded and p is the price per unit in dollars, and that the daily supply is given by $p = 60 + 5q$, where q is the number of units supplied and p is the price in dollars. If a price results in more units being supplied than demanded, we say there is a *surplus*, and if the price results in fewer units being supplied than demanded, we say there is a *shortfall*. **Market equilibrium** occurs when the supply quantity equals the demand quantity (and when the prices are equal)—that is, when q and p both satisfy the system

$$\begin{cases} p = 200 - 2q \\ p = 60 + 5q \end{cases}$$

a. If the price is $140, how many units are supplied and how many are demanded?

b. Does this price give a surplus or a shortfall of the product?

c. What price gives market equilibrium?

SOLUTION

a. If the price is $140, the number of units demanded satisfies $140 = 200 - 2q$, or $q = 30$, and the number of units supplied satisfies $140 = 60 + 5q$, or $q = 16$.

b. At this price, the quantity supplied is less than the quantity demanded, so a shortfall occurs.

c. Because market equilibrium occurs where q and p both satisfy the system

$$\begin{cases} p = 200 - 2q \\ p = 60 + 5q \end{cases}$$

we seek the solution to this system.

We can solve this system by substitution. Substituting $60 + 5q$ for p in the first equation gives

$$60 + 5q = 200 - 2q$$

Solving this equation gives

$$60 + 5q = 200 - 2q$$
$$7q = 140$$
$$q = 20$$

Thus, market equilibrium occurs when the number of units is 20, and the equilibrium price is

$$p = 200 - 2(20) = 60 + 5(20) = 160 \text{ dollars per unit}$$

The substitution in Example 3 was not difficult because both equations were solved for p. In general, we use the following steps to solve systems of two equations in two variables by substitution.

Solution of Systems of Equations by Substitution

1. Solve one of the equations for one of the variables in terms of the other variable.
2. Substitute the expression from step 1 into the other equation to give an equation in one variable.
3. Solve the linear equation for the variable.
4. Substitute this solution into the equation from step 1 or into one of the original equations and solve this equation for the second variable.
5. Check the solution in both original equations or check graphically.

EXAMPLE 4 ▶ Solution by Substitution

Solve the system $\begin{cases} 3x + 4y = 10 \\ 4x - 2y = 6 \end{cases}$ by substitution.

SOLUTION

To solve this system by substitution, we can solve either equation for either variable and substitute the resulting expression into the other equation. Solving the second equation for y gives

$$4x - 2y = 6$$
$$-2y = -4x + 6$$
$$y = 2x - 3$$

Substituting this expression for y in the first equation gives

$$3x + 4(2x - 3) = 10$$

Solving this equation gives

$$3x + 8x - 12 = 10$$
$$11x = 22$$
$$x = 2$$

Substituting $x = 2$ into $y = 2x - 3$ gives $y = 2(2) - 3 = 1$, so the solution to the system is $x = 2, y = 1$, or $(2, 1)$.

Checking shows that this solution satisfies both original equations.

EXAMPLE 5 ▶ U.S. Civilian Workforce

Using data from 1950 and projected to 2050, the number of women and men in the workforce (in millions) can be modeled by the linear functions

$$y = 0.778x + 20.7 \quad \text{and} \quad y = 0.587x + 43.1$$

respectively, where x is the number of years after 1950. Find the year when the number of women in the workforce is projected to equal the number of men. (*Source*: U.S. Bureau of Labor Statistics)

SOLUTION

The number of women in the workforce will equal the number of men when the y-values of the two models are equal, so we solve the following system by substitution.

$$\begin{cases} y = 0.778x + 20.7 \\ y = 0.587x + 43.1 \end{cases}$$

$$0.778x + 20.7 = 0.587x + 43.1$$
$$0.191x = 22.4$$
$$x \approx 117.3$$

This means that the number of women in the workforce will equal the number of men in the 118th year after 1950, in 2068. We can check the solution graphically by graphing both equations on a graphing calculator and using Intersect. (See Figure 2.31.)

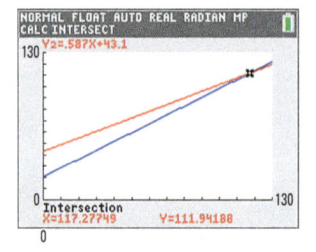

Figure 2.31

2.3 Systems of Linear Equations in Two Variables 133

Solution by Elimination

A second algebraic method, called the **elimination method**, is frequently an easier method to use to solve a system of linear equations. The elimination method is based on rewriting one or both of the equations in an equivalent form that allows us to eliminate one of the variables by adding or subtracting the equations.

> **Solving a System of Two Equations in Two Variables by Elimination**
>
> 1. If necessary, multiply one or both equations by a nonzero number that will make the coefficients of one of the variables in the equations equal, except perhaps for sign.
> 2. Add or subtract the equations to eliminate one of the variables.
> 3. Solve for the variable in the resulting equation.
> 4. Substitute the solution from step 3 into one of the original equations and solve for the second variable.
> 5. Check the solutions in the remaining original equation, or check graphically.

EXAMPLE 6 ▶ Solution by Elimination

Use the elimination method to solve the system

$$\begin{cases} 3x + 4y = 10 \\ 4x - 2y = 6 \end{cases}$$

and check the solution graphically.

SOLUTION

The goal is to convert one of the equations into an equivalent equation of a form such that addition of the two equations will eliminate one of the variables. Notice that the coefficient of y in the second equation, -2, is a factor of the coefficient of y in the first equation, 4. If we multiply both sides of the second equation by 2 and add the two equations, this will eliminate the y-variable.

$$\begin{cases} 3x + 4y = 10 & (1) \\ 4x - 2y = 6 & (2) \end{cases}$$

$$\begin{cases} 3x + 4y = 10 & (1) \\ 8x - 4y = 12 & (3) \end{cases}$$ Multiply 2 times Equation (2), getting equivalent Equation (3).

$$11x = 22$$ Add Equations (1) and (3) to eliminate y.
$$x = 2$$ Solve the new equation for x.

Substituting $x = 2$ in the first equation gives $3(2) + 4y = 10$, or $y = 1$. Thus, the solution to the system is $x = 2, y = 1$, or $(2, 1)$.

To check graphically, we solve both equations for y, getting $y_1 = \dfrac{10 - 3x}{4}$ and $y_2 = \dfrac{4x - 6}{2}$, then graph the equations and find the point of intersection to be $(2, 1)$ (Figure 2.32).

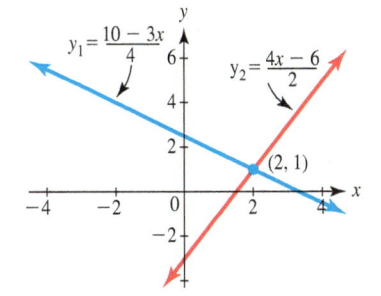

Figure 2.32

Modeling Systems of Linear Equations

Solving some real problems requires us to create two or more equations whose simultaneous solution is the solution to the problem. Consider the following examples.

EXAMPLE 7 ▶ Investments

An investor has $300,000 to invest, part at 12% and the remainder in a less risky investment at 7%. If her investment goal is to have an annual income of $27,000, how much should she put in each investment?

SOLUTION

If we denote the amount invested at 12% as x and the amount invested at 7% as y, the sum of the investments is $x + y$, so we have the equation

$$x + y = 300,000$$

The annual income from the 12% investment is $0.12x$, and the annual income from the 7% investment is $0.07y$. Thus, the desired annual income from the two investments is

$$0.12x + 0.07y = 27,000$$

We can write the given information as a system of equations:

$$\begin{cases} x + y = 300,000 \\ 0.12x + 0.07y = 27,000 \end{cases}$$

To solve this system, we multiply the first equation by -0.12 and add the two equations. This results in an equation with one variable:

$$\begin{cases} -0.12x - 0.12y = -36,000 \\ 0.12x + 0.07y = 27,000 \end{cases}$$

$$-0.05y = -9000$$

$$y = 180,000$$

Substituting 180,000 for y in the first original equation and solving for x gives $x = 120,000$. Thus, $120,000 should be invested at 12%, and $180,000 should be invested at 7%.

To check this solution, we see that the total investment is $120,000 + $180,000, which equals $300,000. The interest earned at 12% is $120,000(0.12) = $14,400, and the interest earned at 7% is $180,000(0.07) = $12,600. The total interest is $14,400 + $12,600, which equals $27,000. This agrees with the given information.

EXAMPLE 8 ▶ Medication

A nurse has two solutions that contain different concentrations of a certain medication. One is a 12% concentration, and the other is an 8% concentration. How many cubic centimeters (cc) of each should she mix together to obtain 20 cc of a 9% solution?

SOLUTION

We begin by denoting the amount of the first solution by x and the amount of the second solution by y. The total amount of solution is the sum of x and y, so

$$x + y = 20$$

The total medication in the combined solution is 9% of 20 cc, or $0.09(20) = 1.8$ cc, and the mixture is obtained by adding $0.12x$ and $0.08y$, so

$$0.12x + 0.08y = 1.8$$

We can use substitution to solve the system

$$\begin{cases} x + y = 20 \\ 0.12x + 0.08y = 1.8 \end{cases}$$

Substituting $20 - x$ for y in $0.12x + 0.08y = 1.8$ gives $0.12x + 0.08(20 - x) = 1.8$, and solving this equation gives

$$0.12x + 0.08(20 - x) = 1.8$$
$$0.12x + 1.6 - 0.08x = 1.8$$
$$0.04x = 0.2$$
$$x = 5$$

Thus, combining 5 cc of the first solution with $20 - 5 = 15$ cc of the second solution gives 20 cc of the 9% solution.

Dependent and Inconsistent Systems

The system of linear equations discussed in Example 2 has a unique solution, shown as the point of intersection of the graphs. It is possible that two equations in a system of linear equations in two variables describe the same line. When this happens, the equations are equivalent, and the values that satisfy one equation are also solutions to the other equation and to the system. Such a system is a **dependent system**. If a system contains two equations whose graphs are parallel lines, they have no point in common, and thus the system has no solution. Such a system of equations is **inconsistent**. Figure 2.33(a)–(c) represents these three situations: systems that have a unique solution, many solutions (dependent system), and no solution (inconsistent system), respectively. Note that the slopes of the lines are equal in Figure 2.33(b) and in Figure 2.33(c).

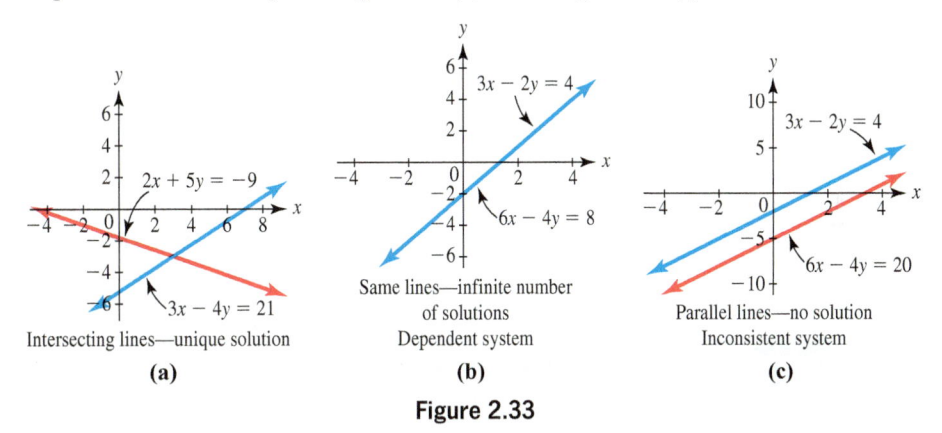

Intersecting lines—unique solution
(a)

Same lines—infinite number of solutions
Dependent system
(b)

Parallel lines—no solution
Inconsistent system
(c)

Figure 2.33

EXAMPLE 9 ▶ Systems with Nonunique Solutions

Use the elimination method to solve each of the following systems, if possible. Verify the solution graphically.

a. $\begin{cases} 2x - 3y = 4 \\ 6x - 9y = 12 \end{cases}$

b. $\begin{cases} 2x - 3y = 4 \\ 6x - 9y = 36 \end{cases}$

SOLUTION

a. To solve $\begin{cases} 2x - 3y = 4 \\ 6x - 9y = 12 \end{cases}$, we multiply the first equation by -3 and add the equations, getting

$$\begin{cases} -6x + 9y = -12 \\ 6x - 9y = 12 \end{cases}$$
$$0 = 0$$

This indicates that the graphs of the equations intersect when $0 = 0$, *which is always true*. Thus, any values of x and y that satisfy one of these equations also satisfy the other, and there are *infinitely many* solutions. Figure 2.34(a) shows that the graphs of the equations lie on the same line. Notice that the second equation is a multiple of the first, so the equations are equivalent. This system is *dependent*.

The infinitely many solutions all satisfy both of the two equations. That is, they are values of x and y that satisfy

$$2x - 3y = 4, \quad \text{or} \quad y = \frac{2}{3}x - \frac{4}{3}$$

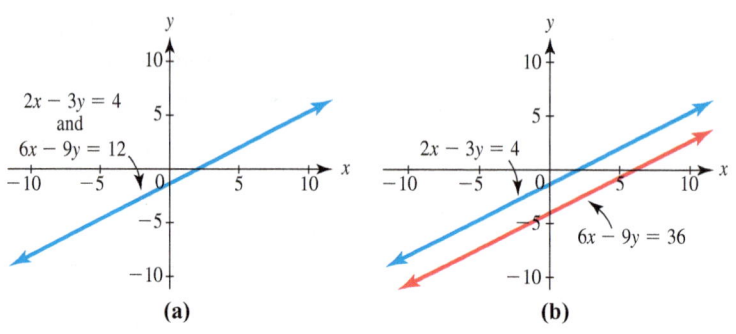

Figure 2.34

b. To solve $\begin{cases} 2x - 3y = 4 \\ 6x - 9y = 36 \end{cases}$, we multiply the first equation by -3 and add the equations, getting

$$\begin{cases} -6x + 9y = -12 \\ 6x - 9y = 36 \end{cases}$$
$$0 = 24$$

This indicates that the equations intersect when $0 = 24$, *which is never true*. Thus, no values of x and y satisfy both of the equations. Figure 2.34(b) shows that the graphs of the equations are parallel. This system is *inconsistent*.

EXAMPLE 10 ▶ Investment

Members of an investment club have set a goal of earning 15% on the money they invest in stocks. They are considering buying two stocks, for which the cost per share and the projected growth per share (both in dollars) are summarized in Table 2.9.

Table 2.9

	Utility	Technology
Cost/share	$30	$45
Growth/share	$4.50	$6.75

a. If they have $180,000 to invest, how many shares of each stock should they buy to meet their goal?

b. If they buy 1800 shares of the utility stock, how many shares of the technology stock should they buy to meet their goal?

2.3 Systems of Linear Equations in Two Variables

SOLUTION

a. The money available to invest in stocks is $180,000, so if x is the number of utility shares and y is the number of technology shares purchased, we have

$$30x + 45y = 180{,}000$$

A 15% return on their investment would be $0.15(180{,}000) = 27{,}000$ dollars, so we have

$$4.50x + 6.75y = 27{,}000$$

To find x and y, we solve the system

$$\begin{cases} 30x + 45y = 180{,}000 \\ 4.50x + 6.75y = 27{,}000 \end{cases}$$

Multiplying 4.5 times both sides of the first equation and -30 times both sides of the second equation gives

$$\begin{cases} 135x + 202.5y = 810{,}000 \\ -135x - 202.5y = -810{,}000 \end{cases}$$

Adding the equations gives $0 = 0$, so the system is dependent, with many solutions.

The number of shares of each stock that can be purchased satisfies both of the two original equations. In particular, it satisfies $30x + 45y = 180{,}000$, so

$$y = \frac{180{,}000 - 30x}{45}, \quad \text{or} \quad y = \frac{12{,}000 - 2x}{3}$$

with x between 0 and 6000 shares and y between 0 and 4000 shares (because neither x nor y can be negative).

b. Substituting 1800 for x in the equation gives $y = 2800$, so if they buy 1800 shares of the utility stock, they should buy 2800 shares of the technology stock to meet their goal.

SKILLS CHECK 2.3

Answers that are not seen can be found in the answer section at the back of the text.

Determine if each ordered pair is a solution of the system of equations given.

1. $\begin{cases} 2x + 3y = -1 \\ x - 4y = -6 \end{cases}$

   a. $(2, 1)$ No **b.** $(-2, 1)$ Yes

2. $\begin{cases} 4x - 2y = 7 \\ -2x + 2y = -4 \end{cases}$

   a. $\left(\dfrac{3}{2}, -\dfrac{1}{2}\right)$ Yes **b.** $\left(\dfrac{1}{2}, -\dfrac{3}{2}\right)$ No

3. What are the coordinates of the point of intersection of $y = 3x - 2$ and $y = 3 - 2x$? $(1, 1)$

4. Give the coordinates of the point of intersection of $3x + 2y = 5$ and $5x - 3y = 21$. $(3, -2)$

In Exercises 5–8, solve the systems of equations graphically.

5. $\begin{cases} y = 3x - 12 \\ y = 4x + 2 \end{cases}$ $x = -14, y = -54$

6. $\begin{cases} 2x - 4y = 6 \\ 3x + 5y = 20 \end{cases}$ $x = 5, y = 1$

7. $\begin{cases} 4x - 3y = -4 \\ 2x - 5y = -4 \end{cases}$ $x = -\dfrac{4}{7}, y = \dfrac{4}{7}$

8. $\begin{cases} 5x - 6y = 22 \\ 4x - 4y = 16 \end{cases}$ $x = 2, y = -2$

9. Does the system $\begin{cases} 2x + 5y = 6 \\ x + 2.5y = 3 \end{cases}$ have a unique solution, no solution, or many solutions? What does this mean graphically? Many solutions; the two equations have the same graph.

10. Does the system $\begin{cases} 6x + 4y = 3 \\ 3x + 2y = 3 \end{cases}$ have a unique solution, no solution, or many solutions? What does this mean graphically? No solution; the lines are parallel.

Chapter 2 Linear Models, Equations, and Inequalities

In Exercises 11–14, solve the systems of equations by substitution.

11. $\begin{cases} x = 5y + 12 \\ 3x + 4y = -2 \end{cases}$ $x = 2, y = -2$

12. $\begin{cases} 2x - 3y = 2 \\ y = 5x - 18 \end{cases}$ $x = 4, y = 2$

13. $\begin{cases} 2x - 3y = 5 \\ 5x + 4y = 1 \end{cases}$ $x = 1, y = -1$

14. $\begin{cases} 4x - 5y = -17 \\ 3x + 2y = -7 \end{cases}$ $x = -3, y = 1$

In Exercises 15–24, solve the systems of equations by elimination, if a solution exists.

15. $\begin{cases} x + 3y = 5 \\ 2x + 4y = 8 \end{cases}$ $x = 2, y = 1$

16. $\begin{cases} 4x - 3y = -13 \\ 5x + 6y = 13 \end{cases}$ $x = -1, y = 3$

17. $\begin{cases} 5x = 8 - 3y \\ 2x + 4y = 8 \end{cases}$ $x = \frac{4}{7}, y = \frac{12}{7}$

18. $\begin{cases} 3y = 5 - 3x \\ 2x + 4y = 8 \end{cases}$ $x = \frac{2}{3}, y = \frac{7}{3}$

19. $\begin{cases} 0.3x + 0.4y = 2.4 \\ 5x - 3y = 11 \end{cases}$ $x = 4, y = 3$

20. $\begin{cases} 8x - 4y = 0 \\ 0.5x + 0.3y = 2.2 \end{cases}$ $x = 2, y = 4$

21. $\begin{cases} 3x + 6y = 12 \\ 4y - 8 = -2x \end{cases}$ Dependent; many solutions

22. $\begin{cases} 6y - 12 = 4x \\ 10x - 15y = -30 \end{cases}$ Dependent; many solutions

23. $\begin{cases} 6x - 9y = 12 \\ 3x - 4.5y = -6 \end{cases}$ No solution

24. $\begin{cases} 4x - 8y = 5 \\ 6x - 12y = 10 \end{cases}$ No solution

In Exercises 25–34, solve the systems of equations by any convenient method, if a solution exists.

25. $\begin{cases} y = 3x - 2 \\ y = 5x - 6 \end{cases}$ $x = 2, y = 4$

26. $\begin{cases} y = 8x - 6 \\ y = 14x - 12 \end{cases}$ $x = 1, y = 2$

27. $\begin{cases} 4x + 6y = 4 \\ x = 4y + 8 \end{cases}$ $x = \frac{32}{11}, y = -\frac{14}{11}$

28. $\begin{cases} y = 4x - 5 \\ 3x - 4y = 7 \end{cases}$ $x = 1, y = -1$

29. $\begin{cases} 2x - 5y = 16 \\ 6x - 8y = 34 \end{cases}$ $x = 3, y = -2$

30. $\begin{cases} 4x - y = 4 \\ 6x + 3y = 15 \end{cases}$ $x = \frac{3}{2}, y = 2$

31. $\begin{cases} 3x = 7y - 1 \\ 4x = 11 - 3y \end{cases}$ $x = 2, y = 1$

32. $\begin{cases} 5x = 12 + 3y \\ -5y = 8 - 3x \end{cases}$ $x = \frac{9}{4}, y = -\frac{1}{4}$

33. $\begin{cases} 4x - 3y = 9 \\ 8x - 6y = 16 \end{cases}$ No solution

34. $\begin{cases} 5x - 4y = 8 \\ -15x + 12y = -12 \end{cases}$ No solution

EXERCISES 2.3

Answers that are not seen can be found in the answer section at the back of the text.

35. **Break-Even** A manufacturer of kitchen sinks has total revenue given by $R = 76.50x$ and has total cost given by $C = 2970 + 27x$, where x is the number of sinks produced and sold. Use graphical methods to find the number of units that gives break-even for the product. 60 units

36. **Break-Even** A jewelry maker has total revenue for her bracelets given by $R = 89.75x$ and incurs a total cost of $C = 23.50x + 1192.50$, where x is the number of bracelets produced and sold. Use graphical methods to find the number of units that gives break-even for the product. 18 units

37. **Break-Even** A manufacturer of reading lamps has total revenue given by $R = 15.80x$ and total cost given by $C = 8593.20 + 3.20x$, where x is the number of units produced and sold. Use a nongraphical method to find the number of units that gives break-even for this product. 682 units

38. **Break-Even** A manufacturer of automobile air conditioners has total revenue given by $R = 136.50x$ and total cost given by $C = 9661.60 + 43.60x$, where x is the number of units produced and sold. Use a nongraphical method to find the number of units that gives break-even for this product. 104

39. **Market Equilibrium** The demand for a brand of clock radio is given by $p + 2q = 320$, and the supply for these radios is given by $p - 8q = 20$, where p is the price and q is the number of clock radios. Solve the system containing these two equations to find (a) the price at which the quantity demanded equals the quantity supplied and (b) the equilibrium quantity. a. $260 b. 30

40. **Supply and Demand** A certain product has supply and demand functions given by $p = 5q + 20$ and $p = 128 - 4q$, respectively.

   a. If the price p is $60, how many units q are supplied and how many are demanded? 8 supplied, 17 demanded

   b. What price gives market equilibrium, and how many units are demanded and supplied at this price? $80; 12 units

41. Concerta and Ritalin Concerta and Ritalin are two different brands of a drug used to treat attention-deficit hyperactivity disorder (ADHD). After Ritalin had been on the market for a long time, Concerta was released.

  a. Use the fact that the market share for Concerta went from 2.4% to 10% in eleven weeks to write a linear function representing its market share (percent) as a function of the weeks after it was 2.4% (at $x = 0$). $y = 0.69x + 2.4$

  b. During the same time period, the market share for Ritalin went from 7.7% to 6.9%. Write a linear function representing its market share (percent) as a function of the weeks during this time.
  $y = -0.073x + 7.7$

  c. When during this period did the market share of Concerta reach that of Ritalin?
  (Source: Newsweek) In the 7th week ($x = 6.95$)

42. Market Equilibrium Wholesalers' willingness to sell laser printers is given by the supply function $p = 50.50 + 0.80q$, and retailers' willingness to buy the printers is given by $p = 400 - 0.70q$, where p is the price per printer in dollars and q is the number of printers. What price will give market equilibrium for the printers? $236.90

43. Population Demographics Using actual and projected data from 2000 through 2050, the number of millions of white non-Hispanics in the U.S. civilian non-institutional population 16 years and over can be modeled by $W(x) = 0.139x + 164$, and the millions in the remainder of this population can be modeled by $R(x) = 1.98x + 64.2$, where x is the number of years after 2000. If growth continues according to these models, in what year will the populations be equal?
(Source: U.S. Census Bureau) 2055 ($x = 54.2$)

44. Temperature conversion When U.S. residents travel to other countries, they may need to convert the reported Celsius temperatures to Fahrenheit temperatures so they know how to dress. The formula for converting from Celsius temperatures to Fahrenheit temperatures is $F = \frac{9}{5}C + 32$, but a useful approximation of it that is easy to remember and use is the "Tourist formula" $F = 2C + 30$.

  a. At what Fahrenheit temperature do these two formulas agree? $F = 50$ (and $C = 10$)

  b. As the temperature rises above the reading found in part (a), does the tourist formula overestimate or underestimate the Fahrenheit temperature? Why would you expect that from the formulas?
  Overestimate; its rate of change is greater

45. U.S. Population Using data and projections from 1980 through 2050, the percent of Hispanics in the U.S. civilian noninstitutional population is given by $y = 0.224x + 9.0$ and the percent of blacks is given by $y = 0.057x + 12.3$, where x is the number of years after 1990. During what year did the percent of Hispanics equal the percent of blacks in the United States?
$x = 19.8$, so 2010

46. Age at First Marriage and First Birth The graph shows the "Great Crossover" where the median age of women at first marriage and the median age at giving first birth are equal. After the year at this crossover point, many women are having children before getting married.

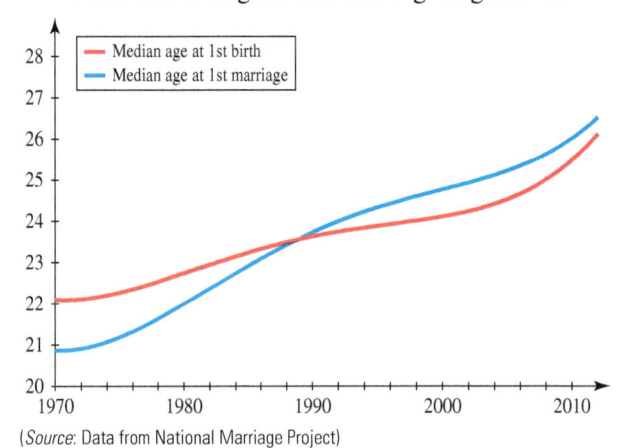

(Source: Data from National Marriage Project)

  a. Use the points (1970, 22.1) and (2010, 25.5) to find the equation of a line that models the median age at giving first birth as a function of the year.
  $y = 0.085x - 145.35$

  b. Use the points (1970, 20.9) and (2010, 26.1) to find the equation of a line that models the median age for women at first marriage as a function of the year. $y = 0.13x - 235.2$

  c. Use the models from parts (a) and (b) to approximate the year in which the median age at giving first birth was equal to the median age at first marriage. Does this approximation agree with the crossover point in the accompanying graph? 1997; no

47. China's Manufacturing The size of the manufacturing sector of China and that of the United States in this century can be modeled by the functions

$$y = 0.158x - 0.0949 \quad \text{and} \quad y = 0.037x + 1.477$$

respectively, with x representing the years after 2000 and y representing the sizes of the manufacturing sector in trillions of 2005 dollars. Find the year during which China reached the United States on its way to becoming the world's largest manufacturer.
(Source: HIS Global Insight) 2013 ($x = 12.99$)

48. Stock Prices The sum of the high and low prices of a share of stock in Johns, Inc., in 2012 is $83.50, and the difference between these two prices in 2012 is $21.88. Find the high and low prices. $52.69, $30.81

49. Pricing A concert promoter needs to make $84,000 from the sale of 2400 tickets. The promoter charges $30 for some tickets and $45 for the others.

  a. If there are x of the $30 tickets sold and y of the $45 tickets sold, write an equation that states that the total number of tickets sold is 2400. $x + y = 2400$

b. How much money is received from the sale of x tickets for $30 each? $30x$

c. How much money is received from the sale of y tickets for $45 each? $45y$

d. Write an equation that states that the total amount received from the sale is $84,000. $30x + 45y = 84,000$

e. Solve the equations simultaneously to find how many tickets of each type must be sold to yield the $84,000. 1600 $30 tickets, 800 $45 tickets

50. **Rental Income** A woman has $500,000 invested in two rental properties. One yields an annual return of 10% of her investment, and the other returns 12% per year on her investment. Her total annual return from the two investments is $53,000. Let x represent the amount of the 10% investment and let y represent the amount of the 12% investment.

   a. Write an equation that states that the sum of the investments is $500,000. $x + y = 500,000$

   b. What is the annual return on the 10% investment? $0.10x$

   c. What is the annual return on the 12% investment? $0.12y$

   d. Write an equation that states that the sum of the annual returns is $53,000. $0.10x + 0.12y = 53,000$

   e. Solve these two equations simultaneously to find how much is invested in each property. $350,000 in 10% property, $150,000 in 12% property

51. **Investment** One safe investment pays 8% per year, and a more risky investment pays 12% per year.

   a. How much must be invested in each account if an investor of $100,000 would like a return of $9000 per year? $75,000 at 8%, $25,000 at 12%

   b. Why might the investor use two accounts rather than put all the money in the 12% investment? 12% account is probably more risky; investor might lose money.

52. **Investment** A woman invests $52,000 in two different mutual funds, one that averages 10% per year and another that averages 14% per year. If her average annual return on the two mutual funds is $5720, how much did she invest in each fund? $39,000 in 10% fund, $13,000 in 14% fund

53. **Investment** Jake has $250,000 to invest. He chooses one money market fund that pays 6.6% and a mutual fund that has more risk but has averaged 8.6% per year. If his goal is to average 7% per year with minimal risk, how much should he invest in each fund? $200,000 at 6.6%, $50,000 at 8.6%

54. **Investment** Sue chooses one money market fund that pays 6.2% and a mutual fund that has more risk but has averaged 9.2% per year. If she has $300,000 to invest and her goal is to average 7.6% per year with minimal risk, how much should she invest in each fund? $160,000 at 6.2%, $140,000 at 9.2%

55. **Medication** A pharmacist wants to mix two solutions to obtain 100 cc of a solution that has an 8% concentration of a certain medicine. If one solution has a 10% concentration of the medicine and the second has a 5% concentration, how much of each of these solutions should she mix? 60 cc of 10% solution, 40 cc of 5% solution

56. **Medication** A pharmacist wants to mix two solutions to obtain 200 cc of a solution that has a 12% concentration of a certain medicine. If one solution has a 16% concentration of the medicine and the second has a 6% concentration, how much of each solution should she mix? 120 cc of 16% solution, 80 cc of 6% solution

57. **Nutrition** A glass of skim milk supplies 0.1 mg of iron and 8.5 g of protein. A quarter pound of lean meat provides 3.4 mg of iron and 22 g of protein. A person on a special diet is to have 7.1 mg of iron and 69.5 g of protein. How many glasses of skim milk and how many quarter-pound servings of meat will provide this? 3 glasses of milk, 2 servings of meat

58. **Nutrition** Each ounce of substance A supplies 6% of a nutrient a patient needs, and each ounce of substance B supplies 10% of the required nutrient. If the total number of ounces given to the patient was 14 and 100% of the nutrient was supplied, how many ounces of each substance was given? 10 oz of substance A, 4 oz of substance B

59. **Alcohol Use** According to the National Household Survey on Drug Abuse, the pattern of higher rates of current alcohol use, binge drinking, and heavy alcohol use among full-time college students may be decreasing but has been higher than use among others aged 18 to 22. The following equations represent the percents of young adults aged 18 to 22 who used alcohol in the past month, where x represents the number of years after 2000.

$$\text{Enrolled in college: } y = -0.282x + 19.553$$

$$\text{Not enrolled: } y = -0.086x + 13.643$$

Solve this system to estimate when the percent for those enrolled in college will equal that for those not enrolled. What will the percent be?
(*Source*: National Survey on Drug Abuse, U.S. Department of Health and Human Services) 2031; 11%

60. **Medication** A nurse has two solutions that contain different concentrations of a certain medication. One is a 30% concentration, and the other is a 15% concentration. How many cubic centimeters (cc) of each should she mix to obtain 45 cc of a 20% solution? 15 cc of 30% solution, 30 cc of 15% solution

61. **Supply and Demand** The table on the next page gives the quantity of graphing calculators demanded and the quantity supplied for selected prices.

   a. Find the linear equation that gives the price as a function of the quantity demanded. $p = -\frac{1}{2}q + 155$

   b. Find the linear equation that gives the price as a function of the quantity supplied. $p = \frac{1}{4}q + 50$

   c. Use these equations to find the market equilibrium price. $85

Price ($)	Quantity Demanded (thousands)	Quantity Supplied (thousands)
50	210	0
60	190	40
70	170	80
80	150	120
100	110	200

62. *Market Analysis* The supply function and the demand function for a product are linear and are determined by the table that follows. Create the supply and demand functions and find the price that gives market equilibrium. $p = \frac{1}{2}q; p = -\frac{1}{2}q + 600; \$300$

Supply Price	Function Quantity	Demand Price	Function Quantity
200	400	400	400
400	800	200	800
600	1200	0	1200

63. *Social Security* Persons scheduled to receive a $1000 monthly Social Security benefit at a full retirement age of 66 can begin drawing benefits at age 62 and receive $750 per month. At what age will these persons who started taking benefits at age 66 have the same in total benefits paid as those who started taking benefits at age 62? 78
(*Source*: Social Security Administration)

64. *Medication* Suppose combining x cubic centimeters (cc) of a 20% concentration of a medication and y cc of a 5% concentration of the medication gives $(x + y)$ cc of a 15.5% concentration. If 7 cc of the 20% concentration are added, by how much must the amount of 5% concentration be increased to keep the same concentration? 3 cc

65. *Social Agency* A social agency provides emergency food and shelter to two groups of clients. The first group has x clients who need an average of $300 for emergencies, and the second group has y clients who need an average of $200 for emergencies. The agency has $100,000 to spend for these two groups.

   a. Write an equation that describes the maximum number of clients who can be served with the $100,000. $300x + 200y = 100,000$

   b. If the first group has twice as many clients as the second group, how many clients are in each group if all the money is spent?
   250 in the first group, 125 in the second group

66. *Market Equilibrium* A retail chain will buy 800 televisions if the price is $350 each and 1200 if the price is $300. A wholesaler will supply 700 of these televisions at $280 each and 1400 at $385 each. Assuming that the supply and demand functions are linear, find the market equilibrium point and explain what it means. (1000, 325); if the price per unit is $325, both supply and demand will be 1000.

67. *Market Equilibrium* A retail chain will buy 900 cordless phones if the price is $10 each and 400 if the price is $60. A wholesaler will supply 700 phones at $30 each and 1400 at $50 each. Assuming that the supply and demand functions are linear, find the market equilibrium point and explain what it means.
700 units at $30; when the price is $30, the amount demanded equals the amount supplied equals 700.

2.4 Solutions of Linear Inequalities

KEY OBJECTIVES

- Solve linear inequalities algebraically
- Solve linear inequalities graphically with the intersection and x-intercept methods
- Solve double inequalities algebraically and graphically

SECTION PREVIEW Profit

For an electronic reading device, the respective weekly revenue and weekly cost are given by

$$R(x) = 400x \quad \text{and} \quad C(x) = 200x + 16{,}000$$

where x is the number of units produced and sold. For what levels of production will a profit result?

Profit will occur when revenue is greater than cost. So we find the level of production and sale x that gives a profit by solving the **linear inequality**

$$R(x) > C(x), \quad \text{or} \quad 400x > 200x + 16{,}000$$

(See Example 2.) In this section, we will solve linear inequalities of this type algebraically and graphically. ∎

Algebraic Solution of Linear Inequalities

An **inequality** is a statement that one quantity or expression is greater than, less than, greater than or equal to, or less than or equal to another.

Linear Inequality

A linear inequality (or first-degree inequality) in the variable x is an inequality that can be written in the form $ax + b > 0$, where $a \neq 0$. (The inequality symbol can be $>$, $\geq$, $<$, or $\leq$.)

The inequality $4x + 3 < 7x - 6$ is a linear inequality (or first-degree inequality) because the highest power of the variable (x) is 1. The values of x that satisfy the inequality form the solution set for the inequality. For example, 5 is in the solution set of this inequality because substituting 5 into the inequality gives

$$4 \cdot 5 + 3 < 7 \cdot 5 - 6, \quad \text{or} \quad 23 < 29$$

which is a true statement. On the other hand, 2 is not in the solution set because

$$4 \cdot 2 + 3 \not< 7 \cdot 2 - 6$$

Solving an inequality means finding its solution set. The solution to an inequality can be written as an inequality or in interval notation. The solution can also be represented by a graph on a real number line.

Two inequalities are *equivalent* if they have the same solution set.

We use the properties of inequalities discussed in the Algebra Toolbox to solve an inequality. In general, the steps used to solve a linear inequality are the same as those used to solve a linear equation, except that the inequality symbol is reversed if both sides are multiplied (or divided) by a negative number.

Steps for Solving a Linear Inequality Algebraically

1. If a linear inequality contains fractions with constant denominators, multiply both sides of the inequality by a positive number that will remove all denominators in the inequality. If there are two or more fractions, use the least common denominator (LCD) of the fractions.

2. Remove any parentheses by multiplication.

3. Perform any additions or subtractions to get all terms containing the variable on one side and all other terms on the other side of the inequality. Combine like terms.

4. Divide both sides of the inequality by the coefficient of the variable. *Reverse the inequality symbol if this number is negative.*

5. Check the solution by substitution or with a graphing utility. If a real-world solution is desired, check the algebraic solution for reasonableness in the real-world situation.

EXAMPLE 1 ▶ Solution of a Linear Inequality

Solve the inequality $3x - \dfrac{1}{3} \leq -4 + x$.

SOLUTION

To solve the inequality

$$3x - \frac{1}{3} \leq -4 + x$$

first multiply both sides by 3:

$$3\left(3x - \frac{1}{3}\right) \leq 3(-4 + x)$$

Removing parentheses gives

$$9x - 1 \leq -12 + 3x$$

Performing additions and subtractions to both sides to get the variables on one side and the constants on the other side gives

$$6x \leq -11$$

Dividing both sides by the coefficient of the variable gives

$$x \leq -\frac{11}{6}$$

The solution set contains all real numbers less than or equal to $-\frac{11}{6}$. The graph of the solution set $\left(-\infty, -\frac{11}{6}\right]$ is shown in Figure 2.35.

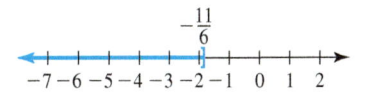

Figure 2.35

EXAMPLE 2 ▶ Profit

For an electronic reading device, the weekly revenue and weekly cost (in dollars) are given by

$$R(x) = 400x \quad \text{and} \quad C(x) = 200x + 16{,}000$$

respectively, where x is the number of units produced and sold. For what levels of production will a profit result?

SOLUTION

Profit will occur when revenue is greater than cost. So we find the level of production and sale that gives a profit by solving the linear inequality $R(x) > C(x)$, or $400x > 200x + 16{,}000$.

$$400x > 200x + 16{,}000$$
$$200x > 16{,}000 \quad \text{Subtract 200}x \text{ from both sides.}$$
$$x > 80 \quad \text{Divide both sides by 200.}$$

Thus, a profit occurs if more than 80 units are produced and sold.

EXAMPLE 3 ▶ Body Temperature

A child's health is at risk when his or her body temperature is 103°F or higher. What Celsius temperature reading would indicate that a child's health was at risk?

SOLUTION

A child's health is at risk if $F \geq 103$, and $F = \frac{9}{5}C + 32$, where F is the temperature in degrees Fahrenheit and C is the temperature in degrees Celsius. Substituting $\frac{9}{5}C + 32$ for F, we have

$$\frac{9}{5}C + 32 \geq 103$$

Now we solve the inequality for C:

$$\frac{9}{5}C + 32 \geq 103$$

$9C + 160 \geq 515$ Multiply both sides by 5 to clear fractions.

$9C \geq 355$ Subtract 160 from both sides.

$C \geq 39.\overline{4}$ Divide both sides by 9.

Thus, a child's health is at risk if his or her Celsius temperature is approximately 39.4° or higher.

Graphical Solution of Linear Inequalities

In Section 2.1, we used graphical methods to solve linear equations. In a similar manner, graphical methods can be used to solve linear inequalities. We will illustrate both the intersection of graphs method and the x-intercept method.*

Intersection Method

To solve an inequality by the intersection method, we use the following steps.

> ### Steps for Solving a Linear Inequality with the Intersection Method
>
> 1. Set the left side of the inequality equal to y_1, set the right side equal to y_2, and graph the equations using your graphing utility.
>
> 2. Choose a viewing window that contains the point of intersection and find the point of intersection, with x-coordinate a. This is the value of x where $y_1 = y_2$.
>
> 3. The values of x that satisfy the inequality represented by $y_1 < y_2$ are those values for which the graph of y_1 is below the graph of y_2. The values of x that satisfy the inequality represented by $y_1 > y_2$ are those values of x for which the graph of y_1 is above the graph of y_2.

To solve the inequality

$$5x + 2 < 2x + 6$$

by using the intersection method, let

$$y_1 = 5x + 2 \quad \text{and} \quad y_2 = 2x + 6$$

Graphing the equations (Figure 2.36) shows that their point of intersection occurs at $x = \frac{4}{3}$. Some graphing utilities show this answer in the form $x = 1.3333333$. The exact x-value $\left(x = \frac{4}{3}\right)$ can also be found by solving the equation $5x + 2 = 2x + 6$ algebraically. Figure 2.36 shows that the graph of y_1 is below the graph of y_2 when x is

---

*See Appendix A, page 687.

2.4 Solutions of Linear Inequalities 145

less than $\frac{4}{3}$. Thus, the solution to the inequality $5x + 2 < 2x + 6$ is $x < \frac{4}{3}$, which can be written in interval notation as $\left(-\infty, \frac{4}{3}\right)$.

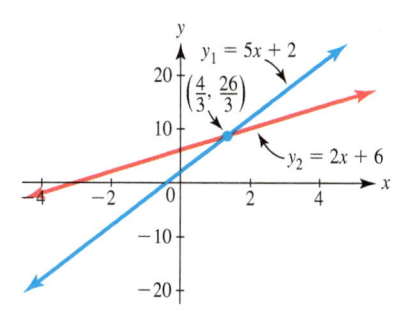

Figure 2.36

EXAMPLE 4 ▶ Intersection Method of Solution

Solve $\dfrac{5x + 2}{5} \geq \dfrac{4x - 7}{8}$ using the intersection of graphs method.

SOLUTION

Enter the left side of the inequality as $y_1 = (5x + 2)/5$, enter the right side of the inequality as $y_2 = (4x - 7)/8$, graph these lines, and find their point of intersection. As seen in Figure 2.37, the two lines intersect at the point where $x = -2.55$ and $y = -2.15$.

The solution to the inequality is the x-interval for which the graph of y_1 is above the graph of y_2, or the x-value for which the graph of y_1 intersects the graph of y_2. Figure 2.37 indicates that this is the interval to the right of and including the input value of the point of intersection of the two lines. Thus, the solution is $x \geq -2.55$, or $[-2.55, \infty)$.

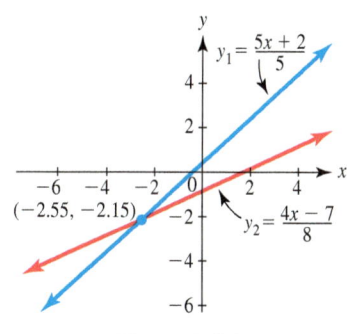

Figure 2.37

x-Intercept Method

To use the x-intercept method to solve a linear inequality, we use the following steps.

> ### Solving Linear Inequalities with the x-Intercept Method
> 1. Rewrite the inequality with all nonzero terms on one side of the inequality and combine like terms, getting $f(x) > 0, f(x) < 0, f(x) \leq 0,$ or $f(x) \geq 0$.
> 2. Graph the nonzero side of this inequality. (Any window in which the x-intercept can be clearly seen is appropriate.)
> 3. Find the x-intercept of the graph to find the solution to the equation $f(x) = 0$. (The exact solution can be found algebraically.)
> 4. Use the graph to determine where the inequality is satisfied.

To use the x-intercept method to solve the inequality $5x + 2 < 2x + 6$, we rewrite the inequality with all nonzero terms on one side of the inequality and combine like terms:

$$5x + 2 < 2x + 6$$

$$3x - 4 < 0 \qquad \text{Subtract } 2x \text{ and } 6 \text{ from both sides of the inequality.}$$

Graphing the nonzero side of this inequality as the linear function $f(x) = 3x - 4$ gives the graph in Figure 2.38. Finding the x-intercept of the graph (Figure 2.38) gives the solution to the equation $3x - 4 = 0$. The x-intercept (and zero of the function) is $x = 1.3333\cdots = \dfrac{4}{3}$.

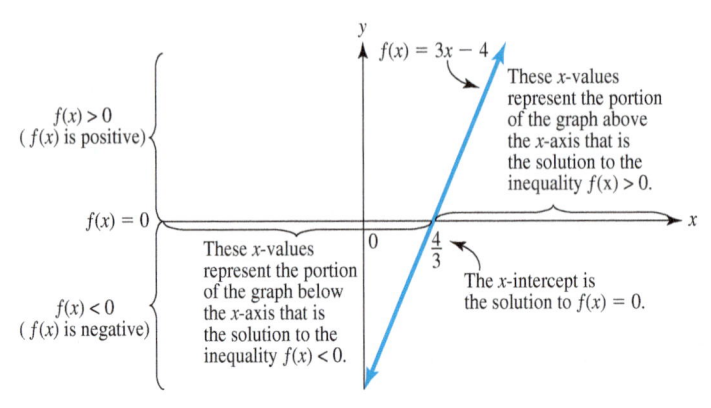

Figure 2.38

We now want to find where $f(x) = 3x - 4$ is less than 0. Notice that the portion of the graph *below* the x-axis gives $3x - 4 < 0$. Thus, the solution to $3x - 4 < 0$, and thus to $5x + 2 < 2x + 6$, is $x < \dfrac{4}{3}$, or $\left(-\infty, \dfrac{4}{3}\right)$.

EXAMPLE 5 ▶ Apparent Temperature

During a recent summer, Dallas, Texas, endured 29 consecutive days when the temperature was at least 100°F. On many of these days, the combination of heat and humidity made it feel even hotter than it was. When the temperature is 100°F, the apparent temperature A (or heat index) depends on the humidity h (expressed as a decimal) according to

$$A = 90.2 + 41.3h$$

For what humidity levels is the apparent temperature at least 110°F? (*Source*: W. Bosch and C. Cobb, "Temperature-Humidity Indices," *UMAP Journal*)

SOLUTION

If the apparent temperature is at least 110°F, the inequality to be solved is

$$A \geq 110, \quad \text{or} \quad 90.2 + 41.3h \geq 110$$

Rewriting this inequality with 0 on the right side gives $41.3h - 19.8 \geq 0$. Entering $y_1 = 41.3x - 19.8$ and graphing with a graphing utility gives the graph in Figure 2.39. The x-intercept of the graph is (approximately) 0.479.

The x-interval where the graph is on or above the x-axis is the solution that we seek, so the solution to the inequality is $[0.479, \infty)$. However, humidity is limited to 100%, so the solution is $0.479 \leq h \leq 1.00$, and we say that the apparent temperature is at least 110° when the humidity is between 47.9% and 100%, inclusive.

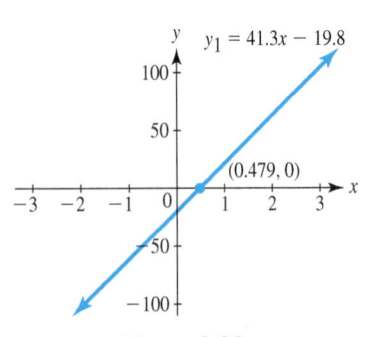

Figure 2.39

Double Inequalities

The inequality $0.479 \leq h \leq 1.00$ in Example 5 is a **double inequality**. A double inequality represents two inequalities connected by the word *and* or *or*. The inequality $0.479 \leq h \leq 1.00$ is a compact way of saying $0.479 \leq h$ and $h \leq 1.00$. Double inequalities can be solved algebraically or graphically, as illustrated in the following example. Note that any arithmetic operation is performed to *all three* parts of a double inequality.

EXAMPLE 6 ▶ Course Grades

A student has taken four tests and has earned grades of 90%, 88%, 93%, and 85%. If all five tests count the same, what grade must the student earn on the final test so that his course average is a B (that is, so his average is at least 80% and less than 90%)?

ALGEBRAIC SOLUTION

To receive a B, the final test score, represented by x, must satisfy

$$80 \leq \frac{90 + 88 + 93 + 85 + x}{5} < 90$$

Solving this inequality gives

$$80 \leq \frac{356 + x}{5} < 90$$

$400 \leq 356 + x < 450$ Multiply all three parts by 5.

$44 \leq x < 94$ Subtract 356 from all three parts.

Thus, he will receive a grade of B if his final test score is at least 44 but less than 94.

GRAPHICAL SOLUTION

To solve this inequality graphically, we assign the left side of the inequality to y_1, the middle to y_2, and the right side to y_3, and we graph these equations to obtain the graph in Figure 2.40.

$$y_1 = 80$$
$$y_2 = (356 + x)/5$$
$$y_3 = 90$$

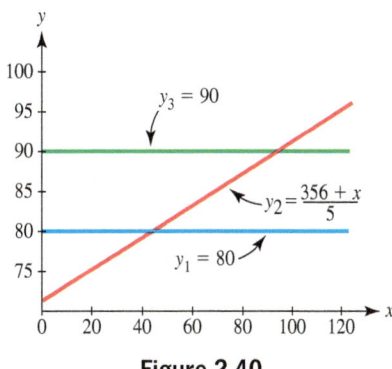

Figure 2.40

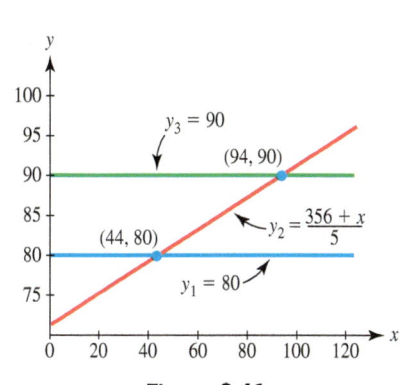

Figure 2.41

We seek the values of x where the graph of y_2 is above or on the graph of y_1 and below the graph of y_3. The left endpoint of this x-interval occurs at the point of intersection of y_2 and y_1, and the right endpoint of the interval occurs at the intersection of y_2 and y_3. These two points can be found using the intersection method. Figure 2.41 shows the points of intersection of these graphs.

The x-values of the points of intersection are 44 and 94, so the solution to $80 \leq \dfrac{356 + x}{5} < 90$ is $44 \leq x < 94$, which agrees with our algebraic solution.

EXAMPLE 7 ▶ Expected Prison Sentences

The mean (expected) time y served in prison for a serious crime can be approximated by a function of the mean sentence length x, with $y = 0.55x - 2.886$, where x and y are measured in months. According to this model, to how many months should a judge sentence a convicted criminal so that the criminal will serve between 37 and 78 months? (*Source*: National Center for Policy Analysis)

SOLUTION

We seek values of x that give y-values between 37 and 78, so we solve the inequality $37 \leq 0.55x - 2.886 \leq 78$ for x:

$$37 \leq 0.55x - 2.886 \leq 78$$
$$37 + 2.886 \leq 0.55x \leq 78 + 2.886$$
$$39.886 \leq 0.55x \leq 80.886$$
$$72.52 \leq x \leq 147.07$$

Thus, the judge could impose a sentence of 73 to 147 months if she wants the criminal to actually serve between 37 and 78 months.

SKILLS CHECK 2.4

Answers that are not seen can be found in the answer section at the back of the text.

In Exercises 1–12, solve the inequalities both algebraically and graphically. Draw a number line graph of each solution and give interval notation.

1. $6x - 1 \leq 11 + 2x$ $x \leq 3$
2. $2x + 6 < 4x + 5$ $x > \dfrac{1}{2}$
3. $4(3x - 2) \leq 5x - 9$ $x \leq -\dfrac{1}{7}$
4. $5(2x - 3) > 4x + 6$ $x > \dfrac{7}{2}$
5. $4x + 1 < -\dfrac{3}{5}x + 5$ $x < \dfrac{20}{23}$
6. $4x - \dfrac{1}{2} \leq -2 + \dfrac{x}{3}$ $x \leq -\dfrac{9}{22}$
7. $\dfrac{x - 5}{2} < \dfrac{18}{5}$ $x < \dfrac{61}{5}$
8. $\dfrac{x - 3}{4} < \dfrac{16}{3}$ $x < \dfrac{73}{3}$
9. $\dfrac{3(x - 6)}{2} \geq \dfrac{2x}{5} - 12$ $x \geq \dfrac{-30}{11}$
10. $\dfrac{2(x - 4)}{3} \geq \dfrac{3x}{5} - 8$ $x \geq -80$
11. $2.2x - 2.6 \geq 6 - 0.8x$ $x \geq 2.8\overline{6}$
12. $3.5x - 6.2 \leq 8 - 0.5x$ $x \leq 3.55$

In Exercises 13 and 14, solve graphically by the intersection method. Give the solution in interval notation.

13. $7x + 3 < 2x - 7$ $(-\infty, -2)$
14. $3x + 4 \leq 6x - 5$ $[3, \infty)$

In Exercises 15 and 16, solve graphically by the x-intercept method. Give the solution in interval notation.

15. $5(2x + 4) \geq 6(x - 2)$ $[-8, \infty)$

16. $-3(x - 4) < 2(3x - 1)$ $\left(\frac{14}{9}, \infty\right)$

17. The graphs of two linear functions f and g are shown in the following figure. (Domains are all real numbers.)

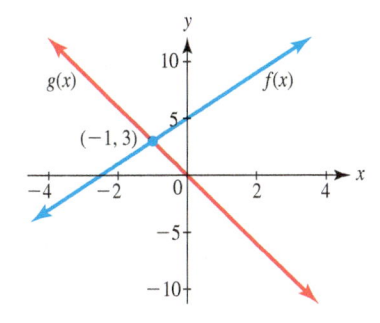

a. Solve the equation $f(x) = g(x)$. $x = -1$

b. Solve the inequality $f(x) < g(x)$. $(-\infty, -1)$

18. The graphs of three linear functions f, g, and h are shown in the following figure.

a. Solve the equation $f(x) = g(x)$. $x = 10$

b. Solve the inequality $h(x) \leq g(x)$. $(-\infty, 30]$

c. Solve the inequality $f(x) \leq g(x) \leq h(x)$. No solution

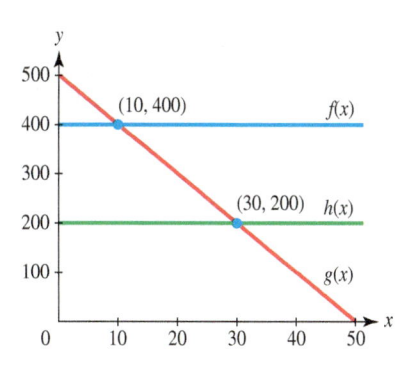

In Exercises 19–28, solve the double inequalities.

19. $17 \leq 3x - 5 < 31$ $\frac{22}{3} \leq x < 12$

20. $120 < 20x - 40 \leq 160$ $8 < x \leq 10$

21. $2x + 1 \geq 6$ and $2x + 1 \leq 21$ $\frac{5}{2} \leq x \leq 10$

22. $16x - 8 > 12$ and $16x - 8 < 32$ $\frac{5}{4} < x < \frac{5}{2}$

23. $3x + 1 < -7$ and $2x - 5 > 6$ No solution

24. $6x - 2 \leq -5$ or $3x + 4 > 9$ $x \leq -\frac{1}{2}$ or $x > \frac{5}{3}$

25. $\frac{3}{4}x - 2 \geq 6 - 2x$ or $\frac{2}{3}x - 1 \geq 2x - 2$ $x \geq \frac{32}{11}$ or $x \leq \frac{3}{4}$

26. $\frac{1}{2}x - 3 < 5x$ or $\frac{2}{5}x - 5 > 6x$ $x > -\frac{2}{3}$ or $x < -\frac{25}{28}$

27. $37.002 \leq 0.554x - 2.886 \leq 77.998$ $72 \leq x \leq 146$

28. $70 \leq \dfrac{60 + 88 + 73 + 65 + x}{5} < 80$ $64 \leq x < 114$

EXERCISES 2.4

Answers that are not seen can be found in the answer section at the back of the text.

29. *Depreciation* Suppose a business purchases equipment for $12,000 and depreciates it over 5 years with the straight-line method until it reaches its salvage value of $2000 (see the figure below). Assuming that the depreciation can be for any part of a year, do the following:

a. Write an equation that represents the depreciated value V as a function of the years t.
$V = 12,000 - 2000t$

b. Write an inequality that indicates that the depreciated value V of the equipment is less than $8000.
$12,000 - 2000t < 8000$

c. Write an inequality that describes the time t during which the depreciated value is at least half of the original value. $12,000 - 2000t \geq 6000$

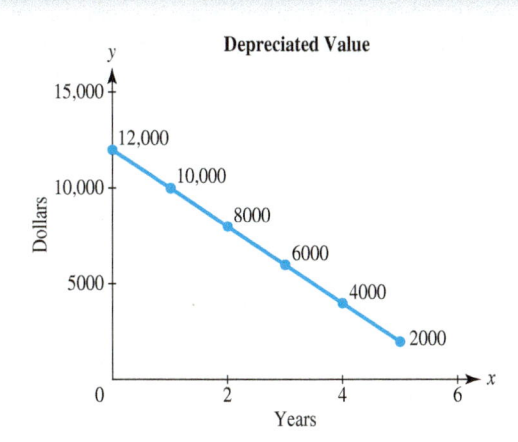

30. *Blood Alcohol Percent* The blood alcohol percent p of a 220-pound male is a function of the number of 12-oz beers consumed, and the percent at which a person is considered legally intoxicated (and guilty of DUI if driving) is 0.1% or higher (see the following figure).

a. Use an inequality to indicate the percent of alcohol in the blood when a person is considered legally intoxicated. $p \geq 0.1$

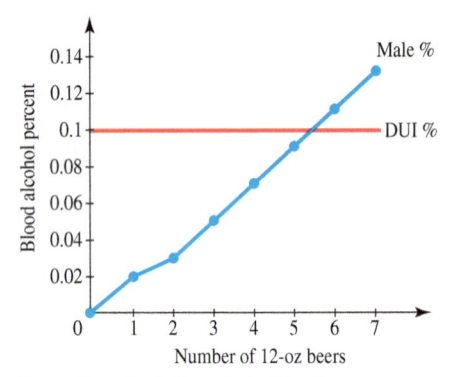

(Source: Pennsylvania Liquor Control Board)

b. If x is the number of beers consumed by a 220-pound male, write an inequality that gives the number of beers that will cause him to be legally intoxicated. $x \geq 6$

31. **Freezing** The equation $F = \frac{9}{5}C + 32$ gives the relationship between temperatures measured in degrees Celsius and degrees Fahrenheit. We know that a temperature at or below 32°F is "freezing." Use an inequality to represent the corresponding "freezing" Celsius temperature. $C \leq 0$

32. **Boiling** The equation $C = \frac{5}{9}(F - 32)$ gives the relationship between temperatures measured in degrees Celsius and degrees Fahrenheit. We know that a temperature at or above 100°C is "boiling." Use an inequality to represent the corresponding "boiling" Fahrenheit temperature. $F \geq 212$

33. **Job Selection** Deb Cook is given the choice of two positions, one paying $3100 per month and the other paying $2000 per month plus a 5% commission on all sales made during the month. What amount must she sell in a month for the second position to be more profitable? More than $22,000

34. **Stock Market** Susan Mason purchased 1000 shares of stock for $22 per share, and 3 months later the value had dropped by 20%. What is the minimum percent increase required for her to make a profit? At least 25%

35. **Grades** If Stan Cook has a course average score between 80 and 89, he will earn a grade of B in his algebra course. Suppose that he has four exam scores of 78, 69, 92, and 81 and that his teacher said the final exam score has twice the weight of each of the other exams. What range of scores on the final exam will result in Stan earning a grade of B? 80 through 100

36. **Grades** If John Deal has a course average score between 70 and 79, he will earn a grade of C in his algebra course. Suppose that he has three exam scores of 78, 62, and 82 and that his teacher said the final exam score has twice the weight of the other exams. What range of scores on the final exam will result in John earning a grade of C? 64 through 86.5

37. **China's Tobacco Smoking** Using data from 2000 and projected to 2025, the percent of the Chinese population over 15 who smoke tobacco daily can be modeled by the function $y = -0.209x + 25.9$, where x is the number of years after 2000. When does the model predict that the percent will be less than 19.63%? (Source: World Health Organization) After 2030

38. **SAT Scores** The College Board now reports SAT scores with a new scale, which has the new scale score y defined as a function of the old scale score x by the equation $y = 0.97x + 128.3829$. Suppose a college requires a new scale score greater than or equal to 1000 to admit a student. To determine what old score values would be equivalent to the new scores that would result in admission to this college, do the following:

a. Write an inequality to represent the problem, and solve it algebraically. $0.97x + 128.3829 \geq 1000; x \geq 899$

b. Solve the inequality from part (a) graphically to verify your result.

39. **Social Security** A person scheduled to receive a $2000 monthly Social Security benefit at a full retirement age of 66 can delay drawing benefits until age 70 and receive $2640 per month. At what age will the person who started taking benefits at age 70 have more in total benefits paid than if he or she had started taking benefits at age 66? After age 82 ½

40. **Break-Even** A large hardware store's monthly profit from the sale of PVC pipe can be described by the equation $P(x) = 6.45x - 9675$ dollars, where x is the number of feet of PVC pipe sold. What level of monthly sales is necessary to avoid a loss? At least 1500 feet

41. **HID Headlights** The high-intensity discharge (HID) headlights containing xenon gas have an expected life of 1500 hours. Because a complete system costs $1000, it is hoped that these lights will last for the life of the car. Suppose that the actual life of the lights could be 10% longer or shorter than the advertised expected life. Write an inequality that gives the range of life of these new lights. $1350 \leq x \leq 1650$ (Source: Automobile, July 2000)

42. **Prison Sentences** The mean time y spent in prison for a crime can be found from the mean sentence length x, using the equation $y = 0.554x - 2.886$, where x and y are measured in months. To how many months should a judge sentence a convicted criminal

if she wants the criminal to actually serve between 4 and 6 years? From 92 months to 135 months
(*Source*: Index of Leading Cultural Indicators)

43. **Average Annual Wage** The average annual wage of U.S. workers for the years from 2011 and projected to 2050 can be modeled by $y = 3943x + 26{,}398$, with y equal to the average annual wage and x equal to the number of years after 2010. When does this model predict that the average annual wage will be at least $93,429? In 2027 and after

44. **Doctorates** The number of new doctorates in mathematics employed in academic positions per year can be modeled by $y = 28.5x + 50.5$, where x is the number of years after 2000.

    a. If the model is accurate, algebraically determine the year in which the number of these doctorates employed was 250. 2007

    b. Use a graph to verify your answer to part (a).

    c. Use your graph to find when the number of these doctorates employed was below 250 per year.
    (*Source*: www.ams.org) Before 2007

45. **Home Appraisal** A home purchased in 1996 for $190,000 was appraised at $270,000 in 2000. Assuming the rate of increase in the value of the home is constant, do the following:

    a. Write an equation for the value of the home as a function of the number of years, x, after 1996.
    $y = 20{,}000x + 190{,}000$

    b. Assuming that the equation in part (a) remained accurate, write an inequality that gives the range of years (until the end of 2010) when the value of the home was greater than $400,000. $11 \leq x \leq 14$

    c. Does it seem reasonable that this model remained accurate until the end of 2010? No

46. **Car Sales Profit** A car dealer purchases 12 new cars for $32,500 each and sells 11 of them at a profit of 5.5%. For how much must he sell the remaining car to average a profit of at least 6% on the 12 cars?
    At least $36,238

47. **Electrical Components Profit** A company's daily profit from the production and sale of electrical components can be described by the equation $P(x) = 6.45x - 2000$ dollars, where x is the number of units produced and sold. What level of production and sales will give a daily profit of more than $10,900?
    More than 2000 units

48. **Profit** The yearly profit from the production and sale of Plumber's Helpers is $P(x) = -40{,}255 + 9.80x$ dollars, where x is the number of Plumber's Helpers produced and sold. What level of production and sales gives a yearly profit of more than $84,352?
    More than 12,715 units

49. **Global Insurance Premiums** Using actual and projected data from 2015 through 2025, the total worldwide insurance premiums by private payers can be modeled by the function $y = 0.145x + 1.127$, where y is in trillions of euros and x is in years after 2015. When will the premiums be more than 3012 billion euros?
    After 2028

50. **Break-Even** The yearly profit from the production and sale of Plumber's Helpers is $P(x) = -40{,}255 + 9.80x$ dollars, where x is the number of Plumber's Helpers produced and sold. What level of production and sales will result in a loss? Fewer than 4108 units

51. **Break-Even** A company produces a logic board for computers. The annual fixed cost for the board is $345,000, and the variable cost is $125 per board. If the logic board sells for $489, write an inequality that gives the number of logic boards that will give a profit for the product. $364x - 345{,}000 > 0$

52. **Temperature** The temperature T (in degrees Fahrenheit) inside a concert hall m minutes after a 40-minute power outage during a summer rock concert is given by $T = 0.43m + 76.8$. Write and solve an inequality that describes when the temperature in the hall is not more than 85°F.
    $0.43m + 76.8 \leq 85$; for the first 19 minutes

53. **Hispanic Population** Using data and projections from 1990 through 2050, the percent of Hispanics in the U.S. population is given by $H(x) = 0.224x + 9.0$, where x is the number of years after 1990. Find the years when the Hispanic population is projected to be at least 14.6% of the U.S. population. 2015 and after
    (*Source*: U.S. Census Bureau)

54. **Reading Tests** The average reading score of 17-year-olds on the National Assessment of Progress tests is given by $y = 0.155x + 244.37$ points, where x is the number of years after 1970. Assuming that this model were valid, write and solve an inequality that describes when the average 17-year-old's reading score on this test was between but not including 245 and 248. (Your answer should be interpreted discretely.)
    (*Source*: U.S. Department of Education)
    $245 < 0.155x + 244.37 < 248$; between 1974 and 1993

55. **National Health Care** Using data for selected years from 2002 and projected to 2024, the national expenditures (in billions of dollars) for health care in the United States can be modeled by $y = 190x + 579$, where x is the number of years after 2000. When does the model project that expenditures for national health care will exceed $6469 billion? After 2031

56. **National Health Care** The national expenditures for health care in the United States for selected years from 1995 with projections to 2020 are given by $y = 138.2x + 97.87$, with x equal to the number of years after 1990 and y equal to the expenditures for health care in billions of dollars. When does the model project that the expenditures for health care will exceed $4658.5 billion? After 2023
    (*Source*: U.S. Centers for Medicare and Medicaid Services)

Preparing for CALCULUS

In calculus, we often need to find where a derivative equals 0, so solving equations is an important skill. In Section 2.1, we solved **linear equations**. Recall that to solve a linear equation, we isolate the variable.

In Exercises 1–4, the derivative $f'(x)$ of a function is given. Find where $f'(x) = 0$.

1. $f'(x) = 3x - 6$ 2
2. $f'(x) = 8x - 4$ ½
3. $f'(x) = \frac{1}{2}x - 9$ 18
4. $f'(x) = \frac{1}{3}x + 4$ −12

Writing the equation defining the function in the form $y = f(x)$ usually makes it easier to find the derivative. Write each of the equations in Exercises 5–6 in this form by solving for y.

5. $8x^2 + 4y = 12$ $y = -2x^2 + 3$
6. $3x^2 - 2y = 6$ $y = \frac{3}{2}x^2 - 3$

To find where the derivative of $y = f'(x)$ is zero, we can find the x-intercepts of the graph of $y = f'(x)$. Find the (a) x-intercept(s) of the graphs and (b) real zero(s) of $f'(x)$ in Exercises 7–8.

7. $f'(x) = 32 + 1.6x$ a. −20   b. −20
8. $f'(x) = \frac{x-5}{4}$ a. 5   b. 5

9. The largest value of the expression $6 + 2x - x^2$ occurs at the x-value where $0 = 2 - 2x$. Find the x-value; then find the largest value of the expression.
   $x = 1; 7$

10. The smallest value of the expression $x^2 - 8x + 4$ occurs at the x-value where $0 = 2x - 8$. Find the x-value; then find the smallest value of the expression.
    $x = 4; -12$

In calculus, it is sometimes necessary to solve equations like those in the following exercises. Consider y' as a different variable than y, and solve each equation for y' in Exercises 11–14.

11. $2x + 2yy' = xy' + y$ $y' = \frac{y - 2x}{2y - x}$
12. $12y^2 y' - 2 = 3y' + 2x$ $y' = \frac{2x + 2}{12y^2 - 3}$
13. $3x^2 - (xy' + y) + 6yy' = 0$ $y' = \frac{3x^2 - y}{x - 6y}$
14. $2(x - y)(1 - y') = 2xyy' + y^2$ $y' = \frac{y^2 - 2x + 2y}{2y - 2x - 2xy}$

15. Solving a certain type of problem in calculus requires "splitting" an expression into two fractions. For example, it would require finding A and B so the expression

$$\frac{2x + 3}{(x - 1)(x + 2)}$$

can be written in the form

$$\frac{A}{x - 1} + \frac{B}{x + 2}$$

To do this, solve $\begin{cases} A + B = 2 \\ 2A - B = 3 \end{cases}$ $A = 5/3, B = 1/3$

16. In calculus, it is sometimes necessary to solve inequalities such as in the following.

Solve   a. $-1 < \frac{x}{2} < 1$ $-2 < x < 2$ b. $-1 < \frac{3x + 1}{2} < 1$ $-1 < x < \frac{1}{3}$

152

chapter 2 SUMMARY

In this chapter, we studied the solution of linear equations and systems of linear equations. We used graphing utilities to solve linear equations. We solved business and economics problems involving linear functions, solved application problems, solved linear inequalities, and used graphing utilities to model linear functions.

Key Concepts and Formulas

2.1 Algebraic and Graphical Solution of Linear Equations

Algebraic solution of linear equations	If a linear equation contains fractions, multiply both sides of the equation by a number that will remove all denominators in the equation. Next remove any parentheses or other symbols of grouping, and then perform any additions or subtractions to get all terms containing the variable on one side and all other terms on the other side of the equation. Combine like terms. Divide both sides of the equation by the coefficient of the variable. Check the solution by substitution in the original equation.
Solving real-world application problems	To solve an application problem that is set in a real-world context, use the same solution methods. However, remember to include units of measure with your answer and check that your answer makes sense in the problem situation.
Zero of a function	Any number a for which $f(a) = 0$ is called a zero of the function $f(x)$.
Solutions, zeros, and x-intercepts	If a is an x-intercept of the graph of a function f, then a is a real zero of the function f, and a is a real solution to the equation $f(x) = 0$.
Graphical solution of linear equations	
• x-intercept method	Rewrite the equation with 0 on one side, enter the nonzero side into the equation editor of a graphing calculator, and find the x-intercept of the graph. This is the solution to the equation.
• Intersection method	Enter the left side of the equation into y_1, enter the right side of the equation into y_2, and find the point of intersection. The x-coordinate of the point of intersection is the solution to the equation.
Literal equations; solving an equation for a specified linear variable	To solve an equation for one variable if two or more variables are in the equation and if that variable is to the first power in the equation, solve for that variable by treating the other variables as constants and using the same steps as are used to solve a linear equation in one variable.
Direct variation	Two variables x and y are directly proportional to each other if their quotient is a constant.

2.2 Fitting Lines to Data Points: Modeling Linear Functions

Fitting lines to data points	When real-world information is collected as numerical information called *data*, technology can be used to determine the pattern exhibited by the data (provided that a recognizable pattern exists). Such patterns can often be described by mathematical functions.
Constant first differences	If the first differences of data outputs are constant (for equally spaced inputs), a linear model can be found that fits the data exactly. If the first differences are "nearly constant," a linear model can be found that is an approximate fit for the data.

154 Chapter 2 Linear Models, Equations, and Inequalities

Linear regression	We can determine the equation of the line that is the best fit for a set of points by using a procedure called linear regression (or the least squares method), which defines the best-fit line as the line for which the sum of the squares of the vertical distances from the data points to the line is a minimum.
Modeling data	We can model a set of data by entering the data into a graphing utility, obtaining a scatter plot, and using the graphing utility to obtain the linear equation that is the best fit for the data. The equation and/or numerical results should be reported in a way that makes sense in the context of the problem, with the appropriate units and with the variables identified.
Discrete versus continuous	We use the term *discrete* to describe data or a function that is presented in the form of a table or a scatter plot. We use the term *continuous* to describe a function or graph when the inputs can be any real number or any real number between two specified values.
Applying models	Using a model to find an output for an input between two given data points is called *interpolation*. When a model is used to predict an output for an input outside the given data points, the process is called *extrapolation*.
Goodness of fit	The goodness of fit of a linear model can be observed from a graph of the model and the data points and/or measured with the correlation coefficient.

2.3 Systems of Linear Equations in Two Variables

System of equations	A system of linear equations is a set of equations in two or more variables. A solution of the system must satisfy every equation in the system.
Solving a system of linear equations in two variables	
• Graphing	Graph the equations and find their point of intersection.
• Substitution	Solve one of the equations for one variable and substitute that expression into the other equation, thus giving an equation in one variable.
• Elimination	Rewrite one or both equations in a form that allows us to eliminate one of the variables by adding or subtracting the equations.
Break-even analysis	A company is said to break even from the production and sale of a product if the total revenue equals the total cost—that is, if the profit for that product is zero.
Market equilibrium	*Market equilibrium* is said to occur when the quantity of a commodity demanded is equal to the quantity supplied. The price at this point is called the *equilibrium price*, and the quantity at this point is called the *equilibrium quantity*.
Dependent and inconsistent systems	
• Unique solution	Graphs are intersecting lines.
• No solution	Graphs are parallel lines; the system is *inconsistent*.
• Many solutions	Graphs are the same line; the system is *dependent*.
Modeling systems of equations	Solution of real problems sometimes requires us to create two or more equations whose simultaneous solution is the solution to the problem.

2.4 Solutions of Linear Inequalities

Linear inequality	A linear inequality (or first-degree inequality) is an inequality that can be written in the form $ax + b > 0$, where $a \neq 0$. (The inequality symbol can be $>$, $\geq$, $<$, or $\leq$.)

Algebraically solving linear inequalities	The steps used to solve a linear inequality are the same as those used to solve a linear equation, except that the inequality symbol is reversed if both sides are multiplied (or divided) by a negative number.
Graphical solution of linear inequalities	
• Intersection method	Set the left side of the inequality equal to y_1 and set the right side equal to y_2, graph the equations using a graphing utility, and find the x-coordinate of the point of intersection. The values of x that satisfy the inequality represented by $y_1 < y_2$ are those values for which the graph of y_1 is below the graph of y_2.
• x-intercept method	To use the x-intercept method to solve an inequality, rewrite the inequality with all nonzero terms on one side and zero on the other side of the inequality and combine like terms. Graph the nonzero side of this inequality and find the x-intercept of the graph. If the inequality to be solved is $f(x) > 0$, the solution will be the interval of x-values representing the portion of the graph above the x-axis. If the inequality to be solved is $f(x) < 0$, the solution will be the interval of x-values representing the portion of the graph below the x-axis.
Double inequalities	A double inequality represents two inequalities connected by the word *and* or *or*. Double inequalities can be solved algebraically or graphically. Any operation performed on a double inequality must be performed on *all three* parts.

chapter 2 SKILLS CHECK

Answers that are not seen can be found in the answer section at the back of the text.

In Exercises 1–6, solve the equation for x algebraically and graphically.

1. $3x + 22 = 8x - 12$ $x = \frac{34}{5}$

2. $2(x - 7) = 5(x + 3) - x$ $x = -14.5$

3. $\frac{3(x - 2)}{5} - x = 8 - \frac{x}{3}$ $x = -138$

4. $\frac{6x + 5}{2} = \frac{5(2 - x)}{3}$ $x = \frac{5}{28}$

5. $\frac{3x}{4} - \frac{1}{3} = 1 - \frac{2}{3}\left(x - \frac{1}{6}\right)$ $x = \frac{52}{51}$

6. $3.259x - 198.8546 = -3.8(8.625x + 4.917)$ $x = 5$

7. For the function $f(x) = 7x - 105$, (a) find the zero of the function, (b) find the x-intercept of the graph of the function, and (c) solve the equation $f(x) = 0$.
   a. 15 b. 15 c. 15

8. Solve $P(a - y) = 1 + \frac{m}{3}$ for y. $y = \frac{3 + m - 3Pa}{-3P}$

9. Solve $4x - 3y = 6$ for y and graph it on a graphing utility with a standard window. $y = \frac{4x - 6}{3}$

Use the table of data below in Exercises 10–13.

x	1	3	6	8	10
y	-9	-1	5	12	18

10. Create a scatter plot of the data.

11. Find the linear function that is the best fit for the data in the table. $y = 2.895x - 11.211$

12. Use a graphing utility to graph the function found in Exercise 11 on the same set of axes as the scatter plot in Exercise 10, with $x_{min} = 0$, $x_{max} = 12$, $y_{min} = -10$, and $y_{max} = 20$.

13. Do the data points in the table fit exactly on the graph of the function from Exercise 12? No

Solve the systems of linear equations in Exercises 14–19, if possible.

14. $\begin{cases} 3x + 2y = 0 \\ 2x - y = 7 \end{cases}$ $x = 2, y = -3$

15. $\begin{cases} 3x + 2y = -3 \\ 2x - 3y = 3 \end{cases}$ $x = \frac{-3}{13}, y = \frac{-15}{13}$

16. $\begin{cases} -4x + 2y = -14 \\ 2x - y = 7 \end{cases}$ Many solutions

17. $\begin{cases} -6x + 4y = 10 \\ 3x - 2y = 5 \end{cases}$ No solution

18. $\begin{cases} 2x + 3y = 9 \\ -x - y = -2 \end{cases}$ $x = -3, y = 5$

19. $\begin{cases} 2x + y = -3 \\ 4x - 2y = 10 \end{cases}$ $x = \frac{1}{2}, y = -4$

In Exercises 20–22, solve the inequalities both algebraically and graphically.

20. $3x + 8 < 4 - 2x$ $x < -\frac{4}{5}$

21. $3x - \frac{1}{2} \leq \frac{x}{5} + 2$ $x \leq \frac{25}{28}$

22. $18 \leq 2x + 6 < 42$ $6 \leq x < 18$

chapter 2 REVIEW

Answers that are not seen can be found in the answer section at the back of the text.

When money is borrowed to purchase an automobile, the amount borrowed A determines the monthly payment P. In particular, if a dealership offers a 5-year loan at 2.9% interest, then the amount borrowed for the car determines the payment according to the following table. Use the table to define the function $P = f(A)$ in Exercises 23–24.

Amount Borrowed ($)	Monthly Payment ($)
10,000	179.25
15,000	268.87
20,000	358.49
25,000	448.11
30,000	537.73

(*Source:* Sky Financial)

23. Car Loans
a. Are the first differences of the outputs in the table constant? Yes
b. Is there a line on which these data points fit exactly? Yes
c. Write the equation $P = f(A)$ of the line that fits the data points in the table, with coefficients rounded to three decimal places.
$P = f(A) = 0.018A + 0.010$

24. Car Loans
a. Use the rounded linear model found in part (c) of Exercise 23 to find $P = f(28,000)$ and explain what it means. 504.01; the predicted monthly payment on a car loan of $28,000 is $504.01.
b. Can the function f be used to find the monthly payment for any dollar amount A of a loan if the interest rate and length of loan are unchanged? Yes
c. Determine the amount of a loan that will keep the payment less than or equal to $500, using the unrounded model. $27,895

25. Crude Oil Using data for the years from 2010 and projected to 2030, the U.S. Crude Oil production is given by $y = 0.0154x + 1.85$ billions of barrels, with x equal to the number of years after 2010. In what year does the model predict the crude oil production will be 2.158 billion barrels? 2030
(*Source:* U.S. Energy Information Administration)

26. Fuel The table below shows data for the number of gallons of gas purchased each day of a certain week by the 250 taxis owned by the Inner City Transportation Taxi Company. Write the equation that models the data. $f(x) = 4500$

Days from First Day	0	1	2	3	4	5	6
Gas Used (gal)	4500	4500	4500	4500	4500	4500	4500

27. Marginal Profit A company has determined that its profit for a product can be described by a linear function. The profit from the production and sale of 150 units is $455, and the profit from 250 units is $895.
a. Write the equation of the profit function for this product. $P(x) = 4.4x - 205$
b. How many units must be produced and sold to make a profit on this product? At least 47 units

28. Job Selection A job candidate is given the choice of two positions, one paying $2100 per month and one paying $1000 per month plus a 5% commission on all sales made during the month.
a. How much (in dollars) must the employee sell in a month for the second position to pay as much as the first? $22,000
b. To be sure that the second position will pay more than the first, how much (in dollars) must the employee sell each month? More than $22,000 per month

29. Marketing A car dealer purchased 12 automobiles for $24,000 each. If she sells 8 of them at an average profit of 12%, for how much must she sell the remaining 4 to obtain an average profit of 10% on all 12? $25,440 each

30. Investment A retired couple has $420,000 to invest. They chose one relatively safe investment fund that has an annual yield of 6% and another riskier investment that has a 10% annual yield. How much should they invest in each fund to earn $30,000 per year? $300,000 in the safe account, $120,000 in the risky account

31. Life-Income Plan Pomona College offers a life-income plan whereby a donor can transfer cash or securities to the college, the college pays the donor a fixed income for life, and the remaining balance passes to Pomona College when the contract ends at the death of the donor. With the Pomona Plan, the donor can get monthly income based on the amount of the donation and the age of the donor. Suppose the annual rate of returns on the donation for selected years are as given in the table.

a. Write equation of the line connecting the points (70, 7.0) and (90, 13.0) from the table. Write the equation in the form $y = f(x)$, where y is the annual percent of return and where x is the age of the donor. $y = 0.3x - 14$

b. Graph the function and the data points on the same axes.

c. Do the points from the table fit the graph exactly? Yes

d. What does this function give as the annual rate of return on a donation at age 78? 9.4%

e. At what age could a donation be made to get a return with an annual rate of 12.4%? Age 88

Age at Donation	Annual Return Percent
70	7.0
75	8.5
80	10.0
85	11.5
90	13.0

(Source: pomonaplan.pomona.edu)

32. **Profit** A company has revenue given by $R(x) = 500x$ dollars and total costs given by $C(x) = 48,000 + 100x$ dollars, where x is the number of units produced and sold. How many units will give a profit? More than 120 units

33. **Profit** A company has revenue given by $R(x) = 564x$ dollars and total cost given by $C(x) = 40,000 + 64x$ dollars, where x is the number of units produced and sold. The profit can be found by forming the function $P(x) = R(x) - C(x)$.

   a. Write the profit function. $P(x) = 500x - 40,000$

   b. For what values of x is $P(x) > 0$? $x > 80$

   c. For how many units is there a profit? More than 80 units

34. **Depreciation** A business property can be depreciated for tax purposes by using the formula $y + 15,000x = 300,000$, where y is the value of the property x years after it was purchased.

   a. For what x-values is the property value below $150,000? $x > 10$

   b. After how many years is the property value below $150,000? 10 years after its purchase

35. **Worldwide Internet Users** The data in the table show the percent of people in the world who are Internet users from 2014 and projected to 2025.

   a. Find the linear equation that models these data, with x representing the number of years after 2010 and with p representing the percent. Report the model with 3 significant digits. $y = 1.78x + 35.9$

   b. Use the reported model to predict the percent of people in the world who are Internet users in 2030. 71.5%

   c. Use the reported model to project when the percent will be 80.4. 2035 ($x = 25$)

Year	Percent
2014	41.1
2015	44.3
2016	46.8
2017	49.0
2018	51.1
2019	53.0
2020	54.6
2025	61.0

(Source: eMarketer)

36. **Life Expectancy**
   a. Find a linear function $y = f(x)$ that models the data shown in the figure that follows, with x equal to the number of years after 1950 and y equal to the number of years the average 65-year-old woman is estimated to live beyond age 65. $y = 0.0638x + 15.702$

   b. Graph the data and the model on the same set of axes.

   c. Use the model to estimate $f(99)$ and explain what it means. 22; projected life span of a 65-year-old woman is 87 in 2049.

   d. Determine the time period (in years) for which the average 65-year-old woman can expect to live more than 84 years. 2002 and after

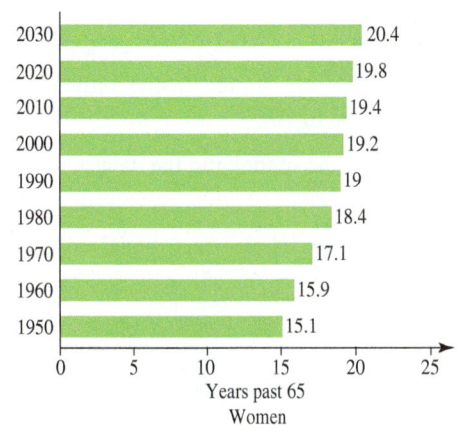

Year	Years past 65
2030	20.4
2020	19.8
2010	19.4
2000	19.2
1990	19
1980	18.4
1970	17.1
1960	15.9
1950	15.1

Women

(Source: National Center for Health Statistics)

37. **Life Expectancy**
   a. Find a linear function $y = g(x)$ that models the data shown in the figure, with x equal to the number of years after 1950 and y equal to the number of years the average 65-year-old man is estimated to live beyond age 65. $y = 0.0655x + 12.324$

   b. Graph the data and the model on the same set of axes.

c. Use the model to estimate $g(130)$ and explain what it means. 20.8; projected life span of a 65-year-old man is 86 in 2080.

d. In what year would the average 65-year-old man expect to live to age 90? 2144

e. Determine the time period for which the average man could expect to live less than 81 years. Before 2007

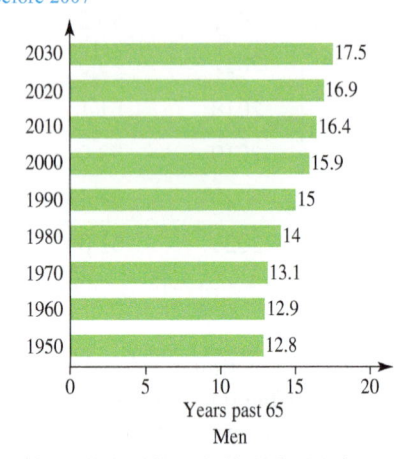

Years past 65	
2030	17.5
2020	16.9
2010	16.4
2000	15.9
1990	15
1980	14
1970	13.1
1960	12.9
1950	12.8

Men

(*Source*: National Center for Health Statistics)

38. *Social Security* The figure below shows how the monthly Social Security benefits differ with the age at which a recipient decides to start receiving benefits. The example shown in this table is for recipients whose monthly benefit is $1000 if benefits start being paid at a full retirement age of 66. Note that benefits do not start being paid before age 62 and that the monthly benefit does not increase if the starting age is after 70.

a. What is the linear function that best models the monthly benefit y as a function of age x at which benefits are started, using only the data points from 66 to 70? $y = 80x - 4280$ for $66 \le x \le 70$

b. Graph the model and the data points on the same axes.

c. Is this model an exact model or an approximate model? Exact

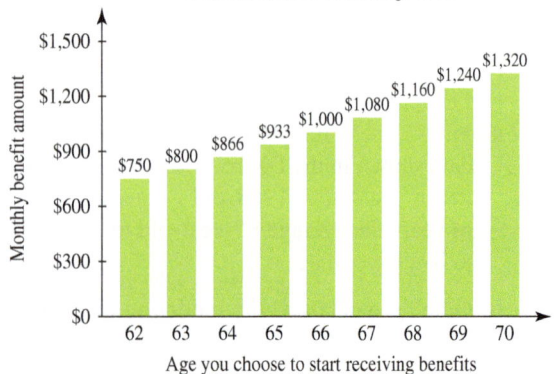

Monthly benefit amounts differ based on the age you decide to start receiving benefits
This example assumes a benefit of $1,000 at a full retirement age of 66

Age	Monthly benefit amount
62	$750
63	$800
64	$866
65	$933
66	$1,000
67	$1,080
68	$1,160
69	$1,240
70	$1,320

(*Source*: Social Security Administration)

39. *Social Security* The figure in Exercise 38 shows how the monthly Social Security benefits differ for recipients whose monthly benefit is $1000 if benefits start being paid at a full retirement age of 66 based on the age at which a recipient starts receiving the benefits. Note that benefits do not start being paid before age 62 and that the monthly benefit does not increase if the starting age is after 70.

a. What is the linear function that best models the monthly benefit y as a function of age x at which benefits are started, for the data points from 62 to 70? $y = 72.25x - 3751.94$ for $62 \le x \le 70$

b. Graph the model and the data points on the same axes.

c. Is this model an exact model or an approximate model? Approximate

(*Source*: Social Security Administration)

40. *Earnings per Share* The table below gives the earnings ($ thousands) per share (EPS) for ACS stock for the years 2020–2025.

a. Graph the data points to determine if a linear equation is a reasonable model for the data.
A linear equation is reasonable.

b. If it is reasonable, find the linear model that is the best fit for the data. Let x equal the number of years after 2005 and y equal the EPS, in thousands of dollars. $y = 0.209x - 2.283$

c. Graph the data points and the function on the same axes and discuss the goodness of fit.
The line seems to fit the data points very well.

Year	EPS ($ thousands)
2020	0.85
2021	1.05
2022	1.29
2023	1.46
2024	1.68
2025	1.90

41. *Marriage Rate* According to data from the *National Vital Statistics Report*, the marriage rate (the number of marriages per 1000 unmarried women) can be described by $132x + 1000y = 9570$, where x is the number of years after 1980. For what years does this model indicate that the rate will be above 6.25 per 1000? Will be below 3.63 per 1000?
2005 and before; 2025 and after

42. *Heart Rate* The table gives the desired heart rate during exercise for weight loss based on the age of the person exercising on an elliptical machine. (*Source*: paramountfitness.com)

a. Can the desired heart rate be expressed exactly as a linear function of age, based on these data? No

Age	Heart Rate (beats per minute)
20	130
30	123
40	117
50	110
60	104
70	97
80	91
90	84

b. Write a linear function that models the data above. $y = -0.652x + 142.881$

c. Can the desired heart rate be expressed exactly as a linear function of age, based on the data in the following table? If yes, what is the function?
Yes; $y = -0.65x + 143$

Age	Heart Rate
20	130
30	123.5
40	117
50	110.5
60	104
70	97.5
80	91
90	84.5

43. *Prison Sentences* The mean time y in prison for a crime can be found as a function of the mean sentence length x, using $y = 0.554x - 2.886$, where x and y are in months. If a judge sentences a convicted criminal to serve between 3 and 5 years, how many months would we expect the criminal to serve? (*Source*: Index of Leading Cultural Indicators)
Between 17 and 31 months

44. *Investment* A retired couple has $240,000 to invest. They chose one relatively safe investment fund that has an annual yield of 8% and another riskier investment that has a 12% annual yield. How much should they invest in each fund to earn $23,200 per year?
$140,000 at 8%, $100,000 at 12%

45. *Break-Even* A computer manufacturer has a new product with daily total revenue given by $R = 565x$ and daily total cost given by $C = 6000 + 325x$. How many units per day must be produced and sold to give break-even for the product? 25 units

46. *Medication* Medication A is given six times per day, and medication B is given twice per day. For a certain patient, the total intake of the two medications is limited to 25.2 mg per day. If the ratio of the dosage of medication A to the dosage of medication B is 2 to 3, how many milligrams are in each dosage?
A, 2.8 mg; B, 4.2 mg

47. *Market Equilibrium* The demand for a certain brand of women's shoes is given by $3q + p = 340$, and the supply of these shoes is given by $p - 4q = -220$, where p is the price in dollars and q is the number of pairs at price p. Solve the system containing these two equations to find the equilibrium price and the equilibrium quantity.
$p = \$100, q = 80$ pairs

48. *Market Analysis* Suppose that, for a certain product, the supply and demand functions are $p = \dfrac{q}{10} + 8$ and $10p + q = 1500$, respectively, where p is in dollars and q is in units. Find the equilibrium price and quantity. $p = \$79; q = 710$ units

49. *Pricing* A concert promoter needs to make $120,000 from the sale of 2600 tickets. The promoter charges $40 for some tickets and $60 for the others.

   a. If there are x of the $40 tickets and y of the $60 tickets, write an equation that states that the total number of the tickets sold is 2600. $x + y = 2600$

   b. How much money is made from the sale of x tickets for $40 each? $40x$

   c. How much money is made from the sale of y tickets for $60 each? $60y$

   d. Write an equation that states that the total amount made from the sale is $120,000. $40x + 60y = 120{,}000$

   e. Solve the equations simultaneously to find how many tickets of each type must be sold to yield the $120,000. 1800 $40 tickets, 800 $60 tickets

50. *Rental Income* A woman has $500,000 invested in two rental properties. One yields an annual return of 12% of her investment, and the other returns 15% per year on her investment. Her total annual return from the two investments is $64,500. If x represents the 12% investment and y represents the 15% investment, answer the following:

   a. Write an equation that states that the sum of the investments is $500,000. $x + y = 500{,}000$

   b. What is the annual return on the 12% investment? $0.12x$

   c. What is the annual return on the 15% investment? $0.15y$

   d. Write an equation that states that the sum of the annual returns is $64,500. $0.12x + 0.15y = 64{,}500$

   e. Solve these two equations simultaneously to find how much is invested in each property.
   $350,000 in the 12% property, $150,000 in the 15% property

Group Activities
▶ EXTENDED APPLICATIONS

1. Taxes

The table below gives the income tax due, $f(x)$, on each given taxable income, x.

1. What are the domain and the range of the function in the table?
2. Create a scatter plot of the data.
3. Do the points appear to lie on a line?
4. Do the inputs change by the same amount? Do the outputs change by the same amount?
5. Is the rate of change in tax per $1 of income constant? What is the rate of change?
6. Will a linear function fit the data points exactly?
7. Write a linear function $y = g(x)$ that fits the data points.
8. Verify that the linear model fits the data points by evaluating the linear function at $x = 63{,}900$ and $x = 64{,}100$ and comparing the resulting y-values with the income tax due for these taxable incomes.
9. Is the model a discrete or continuous function?
10. Can the model be used to find the tax due on any taxable income between $63,700 and $64,300?
11. What is the tax due on a taxable income of $64,150, according to this model?

U.S. Federal Taxes

Taxable Income ($)	Income Tax Due ($)
63,700	8779
63,800	8804
63,900	8829
64,000	8854
64,100	8879
64,200	8904
64,300	8929

(*Source*: U.S. Internal Revenue Service)

2. Research

Linear functions can be used to model many types of real data. Graphs displaying linear growth are frequently found in periodicals such as *Newsweek* and *Time*, in newspapers such as *USA Today* and the *Wall Street Journal*, and on numerous websites on the Internet. Tables of data can also be found in these sources, especially on federal and state government websites, such as www.census.gov. (This website is the source of the Florida resident population data used in the Chapter Review, for example.)

Your mission is to find a company sales record, a stock price, a biological growth pattern, or a sociological trend over a period of years (using at least four points) that is linear or "nearly linear," and then to create a linear function that is a model for the data.

A linear model will be a good fit for the data if

- the data are presented as a graph that is linear or "nearly linear."
- the data are presented in a table and the plot of the data points lies near some line.
- the data are presented in a table and the first differences of the outputs are nearly constant for equally spaced inputs.

After you have created the model, you should test the goodness of fit of the model to the data and discuss uses that you could make of the model.

Your completed project should include

a. A complete citation of the source of the data you are using.
b. An original copy or photocopy of the data being used.
c. A scatter plot of the data.
d. The equation that you have created.
e. A graph containing the scatter plot and the modeled equation.
f. A statement about how the model could be used to make estimations or predictions about the trend you are observing.

Some helpful hints:

1. If you decide to use a relation determined by a graph that you have found, read the graph very carefully to determine the data points or contact the source of the data to get the data from which the graph was drawn.
2. Align the independent variable by letting x represent the number of years after some convenient year and then enter the data into a graphing utility and create a scatter plot.
3. Use your graphing utility to create the equation of the function that is the best fit for the data. Graph this equation and the data points on the same axes to see if the equation is reasonable.

3

Quadratic, Piecewise-Defined, and Power Functions

Revenue and profit from the sale of products frequently cannot be modeled by linear functions because they increase at rates that are not constant. In this chapter, we use nonlinear functions, including quadratic and power functions, to model numerous applications in business, economics, and the life and social sciences.

sections	objectives	applications
3.1 Quadratic Functions; Parabolas	Graph quadratic functions; find vertices of parabolas; identify increasing and decreasing functions	Maximum revenue from sales, foreign-born population, height of a ball, minimizing cost
3.2 Solving Quadratic Equations	Solve equations by factoring; solve equations graphically; combine graphs and factoring methods; solve equations with the square root method and by completing the square; solve equations using the quadratic formula; find complex solutions; find discriminants	Worldwide 5G subscriptions, height of a ball, profit, hospital expenditures
3.3 Power and Root Functions	Graph and apply power and root functions; graph the reciprocal function; apply direct variation as a power	Foreign-born population, distance an object falls, wingspan of birds, Kepler's third law, production
3.4 Piecewise-Defined Functions and Absolute Value Functions	Graph and apply piecewise-defined and absolute value functions; solve absolute value equations; use the greatest integer function	Residential power costs, postage, wind-chill factor, service calls
3.5 Quadratic and Power Models	Model with quadratic functions; compare linear and quadratic models; model with power functions; compare power and quadratic models	Female physicians, artificial intelligence, the cloud, diabetes, purchasing power

Algebra TOOLBOX

KEY OBJECTIVES

- Simplify expressions involving integer and rational exponents
- Find the absolute values of numbers
- Simplify expressions involving radicals
- Convert rational exponents to radicals and vice versa
- Understand terminology related to polynomials
- Add and subtract monomials and polynomials
- Multiply monomials and polynomials
- Find special binomial products
- Divide polynomials by monomials
- Factor polynomial expressions completely
- Identify numbers as real, imaginary, or pure imaginary
- Determine values that make complex numbers equal

In this Toolbox, we discuss absolute value, integer and rational exponents, and radicals. We also discuss the multiplication of monomials and binomials, factoring, and complex numbers.

Integer Exponents

In this chapter and future ones, we will discuss functions and equations containing integer powers of variables. For example, we will discuss the function $y = x^{-1} = \dfrac{1}{x}$.

If a is a real number and n is a positive integer, then a^n represents a as a factor n times in a product

$$a^n = \underbrace{a \cdot a \cdot a \ldots \cdot a}_{n \text{ times}}$$

In a^n, a is called the base and n is called the exponent.

In particular, $a^2 = a \cdot a$ and $a^1 = a$. Note that, for positive integers m and n,

$$a^m \cdot a^n = \underbrace{a \cdot a \cdot a \ldots \cdot a}_{m \text{ times}} \underbrace{a \cdot a \ldots \cdot a}_{n \text{ times}} = \underbrace{a \cdot a \cdot a \ldots \cdot a}_{m+n \text{ times}} = a^{m+n}$$

and that, for $m \geq n$,

$$\dfrac{a^m}{a^n} = \dfrac{\overbrace{a \cdot a \cdot a \ldots \cdot a}^{m \text{ times}}}{\underbrace{a \cdot a \ldots \cdot a}_{n \text{ times}}} = \underbrace{a \cdot a \cdot a \ldots \cdot a}_{m-n \text{ times}} = a^{m-n} \quad \text{if } a \neq 0$$

These two important properties of exponents can be extended to all integers.

Properties of Exponents

For real numbers a and b and integers m and n,

1. $a^m \cdot a^n = a^{m+n}$ \quad (Product Property)
2. $\dfrac{a^m}{a^n} = a^{m-n}, a \neq 0$ \quad (Quotient Property)

We define an expression raised to zero and to a negative power as follows.

Zero and Negative Exponents

For $a \neq 0, b \neq 0$,

1. $a^0 = 1$
2. $a^{-1} = \dfrac{1}{a}$
3. $a^{-n} = \dfrac{1}{a^n}$
4. $\left(\dfrac{a}{b}\right)^{-n} = \left(\dfrac{b}{a}\right)^n$

Algebra Toolbox   163

> **EXAMPLE 1** ▶ **Zero and Negative Exponents**
>
> Simplify the following expressions by removing all zero and negative exponents, for nonzero a, b, and c.
>
> **a.** $(4c)^0$ **b.** $4c^0$ **c.** $(5b)^{-1}$ **d.** $5b^{-1}$ **e.** $\left(\dfrac{a}{b}\right)^{-3}$ **f.** $6a^{-3}$
>
> **SOLUTION**
>
> **a.** $(4c)^0 = 1$ **b.** $4c^0 = 4(1) = 4$ **c.** $(5b)^{-1} = \dfrac{1}{(5b)} = \dfrac{1}{5b}$
>
> **d.** $5b^{-1} = 5 \cdot \dfrac{1}{b} = \dfrac{5}{b}$ **e.** $\left(\dfrac{a}{b}\right)^{-3} = \left(\dfrac{b}{a}\right)^3 = \dfrac{b^3}{a^3}$ **f.** $6a^{-3} = 6 \cdot \dfrac{1}{a^3} = \dfrac{6}{a^3}$

Absolute Value

The distance the number a is from 0 on a number line is the **absolute value** of a, denoted by $|a|$. The absolute value of any nonzero number is positive, and the absolute value of 0 is 0. For example, the distance from 5 to 0 is 5, so $|5| = 5$, and the distance from -8 to 0 is 8, so $|-8| = 8$. Note that if a is a nonnegative number, then $|a| = a$, but if a is negative, then $|a|$ is the positive number $-a$. Formally, we say

$$|a| = \begin{cases} a & \text{if } a \geq 0 \\ -a & \text{if } a < 0 \end{cases}$$

For example, $|5| = 5$ and $|-5| = -(-5) = 5$.

Radicals

The number b is a square root of a number a if $b^2 = a$. The **principal** square root of a nonnegative number is its nonnegative square root. The symbol $\sqrt{a}$ represents the principal square root of a. The expression written under the radical symbol is called the **radicand**, and the root is called the **index**. The square root of a negative number is not a real number. The number c is the cube root of a number a if $c^3 = a$. The symbol $\sqrt[3]{a}$ represents the cube root of a. The cube root of a positive number is a positive number, and the cube root of a negative number is a negative number. In general, the (principal) nth root of a real number is defined as $\sqrt[n]{a} = b$ if $b^n = a$, and the even root of a negative number is not a real number.

> **EXAMPLE 2** ▶ **Radicals**
>
> **a.** $\sqrt[3]{64} = 4$ because $4^3 = 64$.
>
> **b.** $\sqrt[3]{-27} = -3$ because $(-3)^3 = -27$.
>
> **c.** $\sqrt{81} = 9$ because $9^2 = 81$.
>
> **d.** $\sqrt{-64}$ is not a real number because there is no real number that, when squared, gives -64.

We note that $\sqrt[n]{a^n} = |a|$, and if a is nonnegative, $\sqrt[n]{a^n} = a$.

The **Product Rule for Radicals** states that for nonnegative real numbers a and b, $\sqrt{a} \cdot \sqrt{b} = \sqrt{ab}$ and $\sqrt{ab} = \sqrt{a} \cdot \sqrt{b}$, and in general, $\sqrt[n]{a} \cdot \sqrt[n]{b} = \sqrt[n]{ab}$ and $\sqrt[n]{ab} = \sqrt[n]{a} \cdot \sqrt[n]{b}$, provided $\sqrt[n]{a}$ and $\sqrt[n]{b}$ are real. We can use this property to simplify some radicals. A radical is in **simplified form** when all of the following are true.

1. All possible operations are performed.
2. The radicand contains no factor (other than 1) that is a perfect power of the index.
3. There are no fractions in the radicand.
4. There are no radicals in a denominator.

To simplify a radical expression, look for factors of the radicand that are perfect squares if the radical is $\sqrt{}$, perfect cubes if the radical is $\sqrt[3]{}$, and so on. Rewrite the radicand using these perfect powers. Write the root of the perfect powers outside the radical.

EXAMPLE 3 ▶ Simplifying Radicals

Simplify the radicals, assuming the expressions are real and the variables represent nonnegative real numbers.

a. $\sqrt{20} = \sqrt{4 \cdot 5} = \sqrt{4} \cdot \sqrt{5} = 2\sqrt{5}$

b. $\sqrt[3]{54} = \sqrt[3]{27 \cdot 2} = \sqrt[3]{27} \cdot \sqrt[3]{2} = 3\sqrt[3]{2}$

c. $\sqrt{80x^3y^5z} = \sqrt{16 \cdot 5x^2xy^4yz} = \sqrt{16} \cdot \sqrt{x^2} \cdot \sqrt{y^4}\sqrt{5xyz} = 4xy^2\sqrt{5xyz}$

Rational Exponents and Radicals

In this chapter we will study functions involving a variable raised to a rational power (called **power functions**), and we will solve equations involving rational exponents and radicals. This may involve converting expressions involving radicals to expressions involving rational exponents, or vice versa.

Exponential expressions are defined for rational numbers in terms of radicals. Note that, for $a \geq 0$ and $b \geq 0$,

$$\sqrt{a} = b \text{ only if } a = b^2$$

Thus,

$$(\sqrt{a})^2 = b^2 = a, \text{ so } (\sqrt{a})^2 = a$$

We define $a^{1/2} = \sqrt{a}$, so $(a^{1/2})^2 = a$ for $a \geq 0$.

The following definitions show the connection between rational exponents and radicals.

Rational Exponents and Radicals

1. If a is a real number, variable, or algebraic expression and n is a positive integer $n \geq 2$, then

$$a^{1/n} = \sqrt[n]{a}$$

provided that $\sqrt[n]{a}$ exists.

2. If a is a real number and if m and n are integers containing no common factor with $n \geq 2$, then

$$a^{m/n} = \sqrt[n]{a^m} = \left(\sqrt[n]{a}\right)^m$$

provided that $\sqrt[n]{a}$ exists.

EXAMPLE 4 ▶ Rational Exponents

Write the following expressions with exponents rather than radicals.

a. $\sqrt[3]{x^2}$ b. $\sqrt[4]{x^3}$ c. $\sqrt{(3xy)^5}$ d. $3\sqrt{(xy)^5}$

SOLUTION

a. $\sqrt[3]{x^2} = x^{2/3}$ b. $\sqrt[4]{x^3} = x^{3/4}$ c. $\sqrt{(3xy)^5} = (3xy)^{5/2}$ d. $3(xy)^{5/2}$

EXAMPLE 5 ▶ Radical Notation

Write the following in radical form.

a. $y^{1/2}$ b. $(3x)^{3/7}$ c. $12x^{3/5}$

SOLUTION

a. $y^{1/2} = \sqrt{y}$ b. $(3x)^{3/7} = \sqrt[7]{(3x)^3} = \sqrt[7]{27x^3}$ c. $12x^{3/5} = 12\sqrt[5]{x^3}$

Polynomials

An algebraic expression containing a finite number of additions, subtractions, and multiplications of constants and nonnegative integer powers of variables is called a **polynomial**. When simplified, a polynomial cannot contain negative powers of variables, fractional powers of variables, variables in a denominator, or variables inside a radical. The expressions $5x - 2y$ and $7z^3 + 2y$ are polynomials, but $\dfrac{3x-5}{12+5y}$ and $3x^2 - 6\sqrt{x}$ are not polynomials. If the only variable in the polynomial is x, then the polynomial is called a **polynomial in x**. The general form of a polynomial in x is

$$a_n x^n + a_{n-1} x^{n-1} + \cdots + a_1 x + a_0$$

where a_0 and each coefficient $a_n, a_{n-1}, \ldots$ are real numbers and each exponent $n, n-1, \ldots$ is a positive integer.

For a polynomial in the single variable x, the power of x in each term is the **degree** of that term, with the degree of a constant term equal to 0. The term that has the highest power of x is called the **leading term** of the polynomial, the coefficient of this term is the **leading coefficient**, and the degree of this term is the **degree of the polynomial**. Thus, $5x^4 + 3x^2 - 6$ is a fourth-degree polynomial with leading coefficient 5. Polynomials with one term are called **monomials**, those with two terms are called **binomials**, and those with three terms are called **trinomials**. The right side of the equation $y = 4x + 3$ is a first-degree binomial, and the right side of $y = 6x^2 - 5x + 2$ is a second-degree trinomial.

EXAMPLE 6 ▶ Polynomials

For each polynomial, state the constant term, the leading coefficient, and the degree of the polynomial.

a. $5x^2 - 8x + 2x^4 - 3$ b. $5x^2 - 6x^3 + 3x^6 + 7$

SOLUTION

a. The constant term is -3; the term of highest degree is $2x^4$, so the leading coefficient is 2 and the degree of the polynomial is 4.

b. The constant term is 7; the term of highest degree is $3x^6$, so the leading coefficient is 3 and the degree of the polynomial is 6.

Operations with Polynomials

We add or subtract (**combine**) polynomials by combining the like terms. For example, the sum of the polynomials $(5sx - 2y + 7z^3)$ and $(2y + 5sx - 4z^3)$ is

$$(5sx - 2y + 7z^3) + (2y + 5sx - 4z^3) = 5sx - 2y + 7z^3 + 2y + 5sx - 4z^3$$
$$= 10sx + 3z^3$$

The difference of the polynomials $5sx - 2y + 7z^3$ and $2y + 5sx - 4z^3$ is

$$(5sx - 2y + 7z^3) - (2y + 5sx - 4z^3) = 5sx - 2y + 7z^3 - 2y - 5sx + 4z^3$$
$$= -4y + 11z^3$$

We multiply two monomials by multiplying the coefficients and adding the exponents of the respective variables that are in both monomials. For example,

$$(3x^3y^2)(4x^2y) = 3 \cdot 4 \cdot x^3 \cdot x^2 \cdot y^2 \cdot y = 3 \cdot 4x^{3+2}y^{2+1} = 12x^5y^3$$

We can multiply more than two monomials in the same manner.

We can use the **distributive property**,

$$a(b + c) = ab + ac$$

to multiply a monomial times a polynomial. For example,

$$x(3x + y) = x \cdot 3x + x \cdot y = 3x^2 + xy$$

We can extend the property $a(b + c) = ab + ac$ to multiply a monomial times any polynomial. For example,

$$3x(2x + xy + 6) = 3x \cdot 2x + 3x \cdot xy + 3x \cdot 6 = 6x^2 + 3x^2y + 18x$$

The product of two binomials can be found by using the distributive property as follows:

$$(a + b)(c + d) = a(c + d) + b(c + d) = ac + ad + bc + bd$$

Note that this product can be remembered as the sum of the products of the First, Outer, Inner, and Last terms of the binomials, and we use the word **FOIL** to denote this method.

$$(a + b)(c + d) = ac + ad + bc + bd$$

EXAMPLE 7 ▶ Multiplying Polynomials

Find the following products.

a. $(x - 4)(x - 5)$ **b.** $(2x - 3)(3x + 2)$

SOLUTION

a. $(x - 4)(x - 5) = x \cdot x + x(-5) + (-4)x + (-4)(-5)$
$$= x^2 - 5x - 4x + 20 = x^2 - 9x + 20$$

b. $(2x - 3)(3x + 2) = (2x)(3x) + (2x)2 + (-3)(3x) + (-3)2$
$$= 6x^2 + 4x - 9x - 6 = 6x^2 - 5x - 6$$

Certain products and powers involving binomials occur frequently, so the following special products should be remembered.

> **Special Binomial Products**
> 1. $(x + a)(x - a) = x^2 - a^2$ (difference of two squares)
> 2. $(x + a)^2 = x^2 + 2ax + a^2$ (perfect square trinomial)
> 3. $(x - a)^2 = x^2 - 2ax + a^2$ (perfect square trinomial)

EXAMPLE 8 ▶ Special Binomial Products

Find the following products by using the special binomial products formulas.

a. $(5x + 1)^2$ **b.** $(2x - 5)(2x + 5)$ **c.** $(3x - 4)^2$

SOLUTION

a. $(5x + 1)^2 = (5x)^2 + 2(5x)(1) + 1^2 = 25x^2 + 10x + 1$

b. $(2x - 5)(2x + 5) = (2x)^2 - 5^2 = 4x^2 - 25$

c. $(3x - 4)^2 = (3x)^2 - 2(3x)(4) + 4^2 = 9x^2 - 24x + 16$

Dividing Polynomials by Monomials

A polynomial can be divided by a monomial if the monomial can be divided into each term of the polynomial. In this case, the monomial is called a **factor** of the polynomial. For example, if x is not equal to 0, we can simplify the algebraic expression

$$\frac{3x^2 - 4x}{x} \quad \text{as} \quad \frac{3x^2}{x} - \frac{4x}{x} = 3x - 4$$

Factoring Polynomials

Factoring is the process of writing a number or an algebraic expression as the product of two or more numbers or expressions. For example, the distributive property justifies factoring of monomials from polynomials, as in

$$5x^2 - 10x = 5x(x - 2)$$

Factoring out the **greatest common factor** (gcf) from a polynomial is the first step in factoring.

EXAMPLE 9 ▶ Greatest Common Factor

Factor out the greatest common factor.

a. $4x^2y^3 - 18xy^4$ **b.** $3x(a - b) - 2y(a - b)$

SOLUTION

a. The gcf of 4 and 18 is 2. The gcf of x^2 and x is the lower power of x, which is x. The gcf of y^3 and y^4 is the lower power of y, which is y^3. Thus, the gcf of $4x^2y^3$ and $18xy^4$ is $2xy^3$. Factoring out the gcf gives

$$4x^2y^3 - 18xy^4 = 2xy^3(2x - 9y)$$

b. The gcf of $3x(a - b)$ and $2y(a - b)$ is the binomial $a - b$. Factoring out $a - b$ from each term gives

$$3x(a - b) - 2y(a - b) = (a - b)(3x - 2y)$$

By recognizing that a polynomial has the form of one of the special products given on the previous page, we can factor that polynomial.

EXAMPLE 10 ▶ Factoring Special Products

Use knowledge of binomial products to factor the following algebraic expressions.

a. $9x^2 - 25$ **b.** $4x^2 - 12x + 9$

SOLUTION

a. Both terms are squares, so the polynomial can be recognized as the **difference of two squares**. It will then factor as the product of the sum and the difference of the square roots of the terms (see Special Binomial Products, Formula 1).

$$9x^2 - 25 = (3x + 5)(3x - 5)$$

b. Recognizing that the second-degree term and the constant term are squares leads us to investigate whether $12x$ is twice the product of the square roots of these two terms (see Special Binomial Products, Formula 3). The answer is yes, so the polynomial is a **perfect square**, and it can be factored as follows:

$$4x^2 - 12x + 9 = (2x - 3)^2$$

This can be verified by expanding $(2x - 3)^2$.

The first step in factoring is to look for common factors. All applicable factoring techniques should be applied to factor a polynomial completely.

EXAMPLE 11 ▶ Factoring Completely

Factor the following polynomials completely.

a. $3x^2 - 33x + 72$ **b.** $6x^2 - x - 1$

SOLUTION

a. The number 3 can be factored from all three terms, giving

$$3x^2 - 33x + 72 = 3(x^2 - 11x + 24)$$

If the trinomial can be factored into the product of two binomials, the first term of each binomial must be x, and we seek two numbers whose product is 24 and whose sum is -11. Because -3 and -8 satisfy these requirements, we get

$$3(x^2 - 11x + 24) = 3(x - 3)(x - 8)$$

b. The four possible factorizations of $6x^2 - x - 1$ that give $6x^2$ as the product of the first terms and -1 as the product of the last terms follow:

$$(6x - 1)(x + 1) \qquad (6x + 1)(x - 1)$$
$$(2x - 1)(3x + 1) \qquad (2x + 1)(3x - 1)$$

The factorization that gives a product with middle term $-x$ is the correct factorization.

$$(2x - 1)(3x + 1) = 6x^2 - x - 1$$

Some polynomials, such as $6x^2 + 9x - 8x - 12$, can be factored by **grouping**. To do this, we factor out common factors from pairs of terms and then factor out a common binomial expression if it exists. For example,

$$6x^2 + 9x - 8x - 12 = 3x(2x + 3) - 4(2x + 3) = (2x + 3)(3x - 4)$$

Algebra Toolbox

When a second-degree trinomial can be factored but there are many possible factors to test to find the correct one, an alternative method of factoring can be used. The steps used to factor a trinomial using factoring by grouping techniques follow.

Factoring a Trinomial into the Product of Two Binomials Using Grouping

Steps	Example
To factor a quadratic trinomial in the variable x:	Factor $5x - 6 + 6x^2$:
1. Arrange the trinomial with the powers of x in descending order.	1. $6x^2 + 5x - 6$
2. Form the product of the second-degree term and the constant term (first and third terms).	2. $6x^2(-6) = -36x^2$
3. Determine whether there are two factors of the product in step 2 that will sum to the middle (first-degree) term. (If there are no such factors, the trinomial will not factor into two binomials.)	3. $-36x^2 = (-4x)(9x)$ and $-4x + 9x = 5x$
4. Rewrite the middle term from step 1 as a sum of the two factors from step 3.	4. $6x^2 + 5x - 6 = 6x^2 - 4x + 9x - 6$
5. Factor the four-term polynomial from step 4 by grouping.	5. $6x^2 - 4x + 9x - 6 = 2x(3x - 2) + 3(3x - 2) = (3x - 2)(2x + 3)$

EXAMPLE 12 ▶ Factor $10x^2 + 23x - 5$, using grouping.

SOLUTION

To use this method to factor $10x^2 + 23x - 5$, we

1. Note that the trinomial has powers of x in descending order. $10x^2 + 23x - 5$

2. Multiply the second-degree and constant terms. $10x^2(-5) = -50x^2$

3. Factor $-50x^2$ so the sum of the factors is $23x$. $-50x^2 = 25x(-2x)$, $25x + (-2x) = 23x$

4. Rewrite the middle term of the expression as a sum of the two factors from step 3. $10x^2 + 23x - 5 = 10x^2 + 25x - 2x - 5$

5. Factor by grouping.
$$10x^2 + 25x - 2x - 5$$
$$= 5x(2x + 5) - (2x + 5)$$
$$= (2x + 5)(5x - 1)$$

Complex Numbers

In this chapter, some equations do not have real number solutions but do have solutions that are complex numbers.

The numbers discussed up to this point are real numbers (either rational numbers such as $2, -3, \frac{5}{8}$, and $-\frac{2}{3}$ or irrational numbers such as $\sqrt{3}, \sqrt[3]{6}$, and π). But some equations do not have real solutions. For example, if

$$x^2 + 1 = 0, \quad \text{then} \quad x^2 = -1$$

but there is no real number that, when squared, will equal -1. However, we can denote one solution to this equation as the **imaginary unit** i, defined by

$$i = \sqrt{-1}$$

Note that $i^2 = \sqrt{-1} \cdot \sqrt{-1} = -1$.

The set of **complex numbers** is formed by adding real numbers and multiples of i.

> ### Complex Number
> The number $a + bi$, in which a and b are real numbers, is said to be a **complex number in standard form**. The a is the real part of the number, and bi is the imaginary part. If $b = 0$, the number $a + bi = a$ is a real number, and if $b \neq 0$, the number $a + bi$ is an **imaginary number**. If $a = 0$, bi is a **pure imaginary number**.

The complex number system includes the real numbers as well as the imaginary numbers (Figure 3.1). Examples of imaginary numbers are $3 + 2i$, $5 - 4i$, and $\sqrt{3} - \frac{1}{2}i$; examples of pure imaginary numbers are $-i$, $2i$, $12i$, $-4i$, $i\sqrt{3}$, and πi.

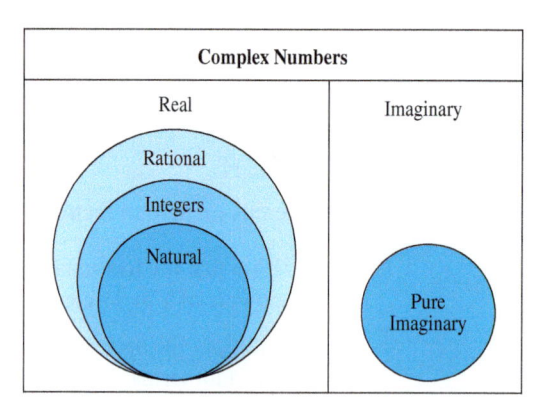

Figure 3.1

The complex number system is an extension of the real numbers that includes imaginary numbers. The term *imaginary number* seems to imply that the numbers do not exist, but in fact they do have important theoretical and technical applications. Complex numbers are used in the design of electrical circuits and airplanes, and they were used in the development of quantum physics. For example, one way of accounting for the amount as well as the phase of the current or voltage in an alternating current electrical circuit involves complex numbers. Special sets of complex numbers can be graphed on the **complex coordinate system** to create pictures called **fractal images**. (Figure 3.2 shows a fractal image called the Mandelbrot set, which can be generated with the complex number i.)

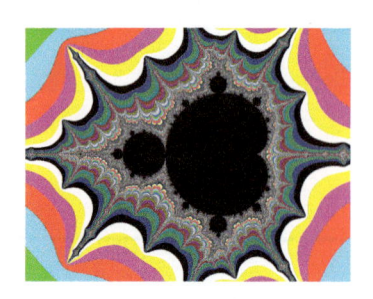

Figure 3.2

EXAMPLE 13 ▶ Simplifying Complex Numbers

Simplify the following numbers by writing them in the form a, bi, or $a + bi$.

a. $6 + \sqrt{-8}$ **b.** $\dfrac{4 - \sqrt{-6}}{2}$

SOLUTION

a. $6 + \sqrt{-8} = 6 + i\sqrt{8} = 6 + 2i\sqrt{2}$
because $\sqrt{8} = \sqrt{4 \cdot 2} = \sqrt{4} \cdot \sqrt{2} = 2\sqrt{2}$

b. $\dfrac{4 - \sqrt{-6}}{2} = \dfrac{4 - i\sqrt{6}}{2} = \dfrac{4}{2} - \dfrac{i\sqrt{6}}{2} = 2 - \dfrac{i\sqrt{6}}{2}$

EXAMPLE 14 ▶ Complex Numbers

Identify each number as one or more of the following: real, imaginary, pure imaginary.

a. $6 + 4i$ **b.** $3i - \sqrt{4}$ **c.** $3 - 2i^2$ **d.** $4 - \sqrt{-16}$ **e.** $\sqrt{-3}$

SOLUTION

a. Imaginary because it contains i.

b. Imaginary because it contains i.

c. Real because $3 - 2i^2 = 3 - 2(-1) = 3 + 2 = 5$.

d. Imaginary because it contains $\sqrt{-16} = 4i$, which gives $4 - 4i$.

e. Imaginary because it contains $\sqrt{-3} = i\sqrt{3}$, and pure imaginary because the real part is 0.

Toolbox EXERCISES

In Exercises 1–18, use the rules of exponents to simplify the following expressions and remove all zero and negative exponents. Assume that all variables are nonzero.

1. $\left(\dfrac{2}{3}\right)^{-2}$ $\dfrac{9}{4}$

2. $\left(\dfrac{3}{2}\right)^{-3}$ $\dfrac{8}{27}$

3. $10^{-2} \cdot 10^0$ $\dfrac{1}{100}$

4. $8^{-2} \cdot 8^0$ $\dfrac{1}{64}$

5. $(2^{-1})^3$ $\dfrac{1}{8}$

6. $(4^{-2})^2$ $\dfrac{1}{256}$

7. $a^5 \cdot a$ a^6

8. $y^{-5} \cdot y^2$ $\dfrac{1}{y^3}$

9. $\dfrac{x^8}{x^4}$ x^4

10. $\dfrac{a^5}{a^{-1}}$ a^6

11. $\dfrac{y^{-3}}{y^{-4}}$ y

12. $(x^4)^3$ x^{12}

13. $(2x^{-2}y)^{-4}$ $\dfrac{x^8}{16y^4}$

14. $\left(\dfrac{x^2}{y^3}\right)^5$ $\dfrac{x^{10}}{y^{15}}$

15. $(-32x^5)^{-2}$ $\dfrac{1}{1024x^{10}}$

16. $(4x^3y^{-1}z)^0$ 1

17. $\left(\dfrac{x^{-2}y}{z}\right)^{-3}$ $\dfrac{x^6 z^3}{y^3}$

18. $\left(\dfrac{a^{-2}b^{-1}c^{-4}}{a^4 b^{-3} c^0}\right)^{-3}$ $\dfrac{a^{18}c^{12}}{b^6}$

Find the absolute values in Exercises 19 and 20.

19. $|-6|$ 6

20. $|7 - 11|$ 4

In Exercises 21–24, find the roots, if they are real.

21. $\sqrt{121}$ 11

22. $\sqrt[6]{64}$ 2

23. $\sqrt[3]{-8}$ -2

24. $\sqrt{-16}$ not a real number

In Exercises 25–28, simplify the radicals, assuming the expressions are real and the variables represent nonnegative real numbers.

25. $\sqrt{48x^3}$ $4x\sqrt{3x}$

26. $\sqrt[3]{72a^3b^4}$ $2ab\sqrt[3]{9b}$

27. $\sqrt{8t^2s^5}$ $2ts^2\sqrt{2s}$

28. $\sqrt[3]{-128a^4b^5c^6}$ $-4abc^2\sqrt[3]{2ab^2}$

29. Write each of the following expressions in simplified exponential form.

  a. $\sqrt{x^3}$ $x^{3/2}$ **b.** $\sqrt[4]{x^3}$ $x^{3/4}$ **c.** $\sqrt[5]{x^3}$ $x^{3/5}$

  d. $\sqrt[6]{27y^9}$ $3^{1/2}y^{3/2}$ **e.** $27\sqrt[6]{y^9}$ $27y^{3/2}$

172 Chapter 3 Quadratic, Piecewise-Defined, and Power Functions

30. Write each of the following in radical form.
   a. $a^{3/4}$ $\sqrt[4]{a^3}$
   b. $-15x^{5/8}$ $-15\sqrt[8]{x^5}$
   c. $(-15x)^{5/8}$ $\sqrt[8]{(-15x)^5}$

31. Label each of the following as a monomial, binomial, or trinomial.
   a. $18z$ Monomial
   b. $-5y^2 + 4y - 7$ Trinomial
   c. $16ab^3cx$ Monomial
   d. $10 - 4x^2$ Binomial

32. State the degree of each term.
   a. $3x^2$ 2
   b. $6xy^3$ 4
   c. 16 0
   d. $-8k$ 1

33. State the degree and leading coefficient of each polynomial.
   a. $-5y^2 + 4y - 7$ 2; -5
   b. $4x^2 - 6x^4 + 3x - 2$ 4; -6
   c. $-x + 3$ 1; -1

34. Write the polynomial in descending order.
   $6x^2 - x^4 + 10 + 3x$ $-x^4 + 6x^2 + 3x + 10$

In Exercises 35–38, perform the indicated operation and simplify.

35. $(3y^3 - 6y + 1) + (7y^2 - 10y - 9)$ $3y^3 + 7y^2 - 16y - 8$
36. $(5p^4 + 6p^2 + 5) - (7p^4 + 8p^2 - 8p - 3)$ $-2p^4 - 2p^2 + 8p + 8$
37. Subtract $(-2c^3 - 2c^2 + 2)$ from $(2c^3 + 7c^2 - 8c - 1)$ and simplify. $4c^3 + 9c^2 - 8c - 3$
38. $-[(6v^3 + 7v^2 - 7) - (4v + 6v^2 - 2v^3)] + (3v^2 - 8v + 2)$ $-8v^3 + 2v^2 - 4v + 9$

In Exercises 39–47, multiply and simplify.

39. $(5x^3)(7x^2)$ $35x^5$
40. $(-3x^2y)(2xy^3)(4x^2y^2)$ $-24x^5y^6$
41. $(3mx)(2mx^2) - (4m^2x)x^2$ $2m^2x^3$
42. $ax^2(2x^2 + ax - ab)$ $2ax^4 + a^2x^3 - a^2bx^2$
43. $(4a + 5b - 6c)ac$ $4a^2c + 5abc - 6ac^2$
44. $(x - 4)(x + 3)$ $x^2 - x - 12$
45. $(3x + 2)(2x - 5)$ $6x^2 - 11x - 10$
46. $(1 - 2x^2)(2 - x^2)$ $2x^4 - 5x^2 + 2$
47. $(a - 2b)(a^2 - 3ab + b^2)$ $a^3 - 5a^2b + 7ab^2 - 2b^3$

In Exercises 48–53, find the special products.

48. $(6x - 5)^2$ $36x^2 - 60x + 25$
49. $(7s - 2t)(7s + 2t)$ $49s^2 - 4t^2$
50. $(8w + 2)^2$ $64w^2 + 32w + 4$
51. $(6x - y)^2$ $36x^2 - 12xy + y^2$
52. $(2y + 5z)^2$ $4y^2 + 20yz + 25z^2$
53. $(3x^2 + 5y)(3x^2 - 5y)$ $9x^4 - 25y^2$

In Exercises 54–73, factor each of the polynomials completely.

54. $3x^2 - 12x$ $3x(x - 4)$
55. $12x^5 - 24x^3$ $12x^3(x^2 - 2)$
56. $9x^2 - 25m^2$ $(3x - 5m)(3x + 5m)$
57. $x^2 - 8x + 15$ $(x - 3)(x - 5)$
58. $x^2 - 2x - 35$ $(x - 7)(x + 5)$
59. $3x^2 - 5x - 2$ $(x - 2)(3x + 1)$
60. $8x^2 - 22x + 5$ $(2x - 5)(4x - 1)$
61. $6n^2 + 18 + 39n$ $3(2n + 1)(n + 6)$
62. $9ab - 12ab^2 + 18b^2$ $3b(3a - 4ab + 6b)$
63. $8x^2y - 160x + 48x^2$ $8x(xy - 20 + 6x)$
64. $12y^3z + 4y^2x^2 - 8y^2z^3$ $4y^2(3yz + x^2 - 2z^3)$
65. $x^2(x^2 + 4)^2 + 2(x^2 + 4)^2$ $(x^2 + 2)(x^2 + 4)^2$
66. $x^2 + 8x + 16$ $(x + 4)^2$
67. $x^2 - 4x + 4$ $(x - 2)^2$
68. $4x^2 + 20x + 25$ $(2x + 5)^2$
69. $x^4 - 6x^2y + 9y^2$ $(x^2 - 3y)^2$
70. $5x^5 - 80x$ $5x(x^4 - 16) = 5x(x^2 + 4)(x + 2)(x - 2)$
71. $3y^4 + 9y^2 - 12y^2 - 36$ $3(y - 2)(y + 2)(y^2 + 3)$
72. $18p^2 + 12p - 3p - 2$ $(3p + 2)(6p - 1)$
73. $5x^2 - 10xy - 3x + 6y$ $(5x - 3)(x - 2y)$

74. Simplify $\dfrac{12x - 5x^2}{x}$ $12 - 5x$

75. Simplify $\dfrac{8x^2 + 2x}{2x}$ $4x + 1$

In Exercises 76 and 77, identify each number as one of the following: real, imaginary, pure imaginary.

76. a. $2 - i\sqrt{2}$ Imaginary
    b. $5i$ Pure imaginary
    c. $4 + 0i$ Real
    d. $2 - 5i^2$ Real

77. a. $3 + i\sqrt{5}$ Imaginary
    b. $3 + 0i$ Real
    c. $8i$ Pure imaginary
    d. $2i^2 - i$ Imaginary

In Exercises 78–80, find values for a and b that make the statement true.

78. $a + bi = 4$ $a = 4, b = 0$
79. $a + 3i = 15 - bi$ $a = 15, b = -3$
80. $a + bi = 2 + 4i$ $a = 2, b = 4$

3.1 Quadratic Functions; Parabolas

KEY OBJECTIVES

- Determine whether a function is quadratic
- Determine whether the graph of a quadratic function is a parabola that opens up or down
- Determine whether the vertex of the graph of a quadratic function is a maximum or a minimum
- Determine whether a function increases or decreases over a given interval
- Find the vertex of the graph of a quadratic function
- Graph a quadratic function
- Write the equation of a quadratic function given information about its graph
- Find the vertex form of the equation of a quadratic function

SECTION PREVIEW Revenue

When products are sold with variable discounts or with prices affected by supply and demand, revenue functions for these products may be nonlinear. Suppose the monthly revenue from the sale of Carlson 42-inch 3D televisions is given by the function

$$R(x) = -0.1x^2 + 600x \text{ dollars}$$

where x is the number of TVs sold. In this case, the revenue of this product is represented by a **second-degree function**, or **quadratic function**. A quadratic function is a function that can be written in the form

$$f(x) = ax^2 + bx + c$$

where a, b, and c are real numbers with $a \neq 0$.

The graph of the quadratic function $f(x) = ax^2 + bx + c$ is a **parabola** with a turning point called the **vertex**. Figure 3.3 shows the graph of the function $R(x) = -0.1x^2 + 600x$, which is a parabola that opens downward, and the vertex occurs where the function has its maximum value. Notice also that the graph of this revenue function is symmetric about a vertical line through the vertex. This vertical line is called the **axis of symmetry**.

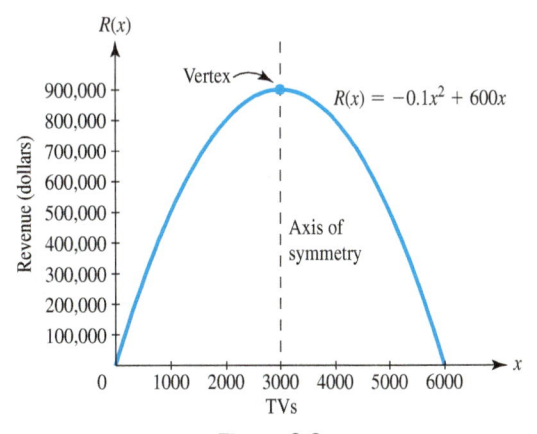

Figure 3.3

Knowing where the maximum value of $R(x)$ occurs can show the company how many units must be sold to obtain the largest revenue. Knowing the maximum output of $R(x)$ helps the company plan its sales campaign. (See Example 2.) In this section, we graph and apply quadratic functions. ■

Parabolas

The graph of every quadratic function has the distinctive shape known as a parabola. The graph of a quadratic function is determined by the location of the vertex and whether the parabola opens upward or downward.

Consider the basic quadratic function $y = x^2$. Each output y is obtained by squaring an input x (Table 3.1). The graph of $y = x^2$, shown in Figure 3.4, is a parabola that opens upward with the vertex (turning point) at the origin, $(0, 0)$.

Table 3.1

x	y
−4	16
−3	9
−2	4
−1	1
0	0
1	1
2	4
3	9
4	16

173

174 Chapter 3 Quadratic, Piecewise-Defined, and Power Functions

Observe that for $x > 0$, the graph of $y = x^2$ rises as it moves from left to right (that is, as the x-values increase), so the function $y = x^2$ is **increasing** for $x > 0$. For values of $x < 0$, the graph falls as it moves from left to right (as the x-values increase), so the function $y = x^2$ is **decreasing** for $x < 0$.

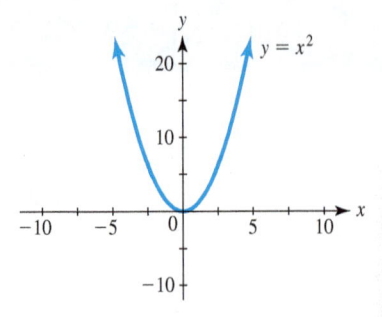

Figure 3.4

> ### Increasing and Decreasing Functions
> A function f is **increasing** on an interval if, for any x_1 and x_2 in the interval, when $x_2 > x_1$, it is true that $f(x_2) > f(x_1)$.
> A function f is **decreasing** on an interval if, for any x_1 and x_2 in the interval, when $x_2 > x_1$, it is true that $f(x_2) < f(x_1)$.

The quadratic function $y = -x^2$ has the form $y = ax^2$ with $a < 0$, and its graph is a parabola that opens downward with vertex at $(0, 0)$ (Figure 3.5(a)). The function $y = \frac{1}{2}x^2$ has the form $y = ax^2$ with $a > 0$, and its graph is a parabola opening upward (Figure 3.5(b)).

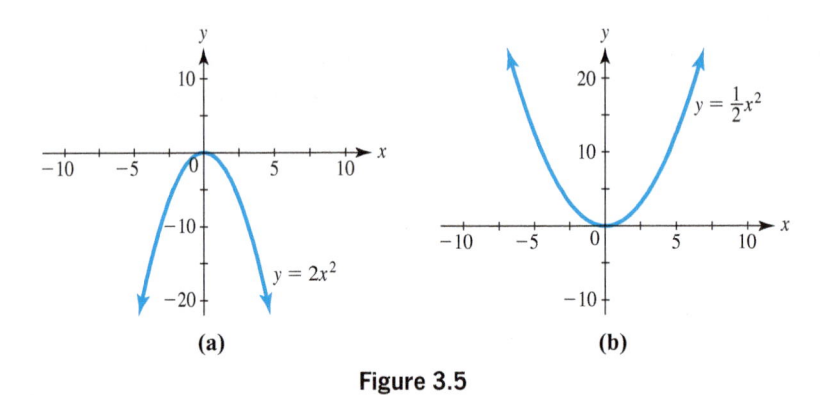

Figure 3.5

In general, the graph of a quadratic function of the form $y = ax^2$ is a parabola that opens upward (is **concave up**) if a is positive and opens downward (is **concave down**) if a is negative.* The vertex, which is the point where the parabola turns, is a **minimum point** if a is positive and is a **maximum point** if a is negative. The vertical line through the vertex is called the **axis of symmetry** because this line divides the graph into two halves that are reflections of each other (Figure 3.6(a)).

We can find the x-coordinate of the vertex of the graph of $y = ax^2 + bx + c$ by using the fact that the axis of symmetry of a parabola passes through the vertex. As Figure 3.6(b)

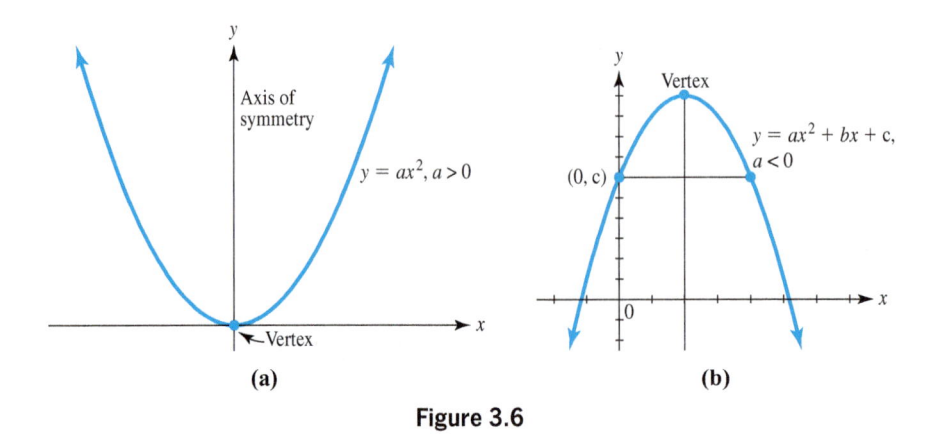

Figure 3.6

---

*A parabola that is concave up will appear to "hold water," and a parabola that is concave down will appear to "shed water."

shows, the y-intercept of the graph of $y = ax^2 + bx + c$ is $(0, c)$, and there is another point on the graph with y-coordinate c.

The x-coordinates of the points on this graph with y-coordinate c satisfy

$$c = ax^2 + bx + c$$

Solving this equation gives

$$0 = ax^2 + bx$$
$$0 = x(ax + b)$$
$$x = 0 \quad \text{or} \quad x = \frac{-b}{a}$$

The x-coordinate of the vertex is on the axis of symmetry, which is halfway from $x = 0$ to $x = \frac{-b}{a}$, so it is at $x = \frac{-b}{2a}$. The y-coordinate of the vertex can be found by evaluating the function at the x-coordinate of the vertex.

Graph of a Quadratic Function

The graph of the function

$$f(x) = ax^2 + bx + c$$

is a parabola that opens upward, and the vertex is a minimum, if $a > 0$. The parabola opens downward, and the vertex is a maximum, if $a < 0$.

The larger the value of $|a|$, the more narrow the parabola will be. Its vertex is at the point $\left(\frac{-b}{2a}, f\left(\frac{-b}{2a}\right)\right)$ (Figure 3.7).

The axis of symmetry of the parabola has equation $x = \frac{-b}{2a}$.

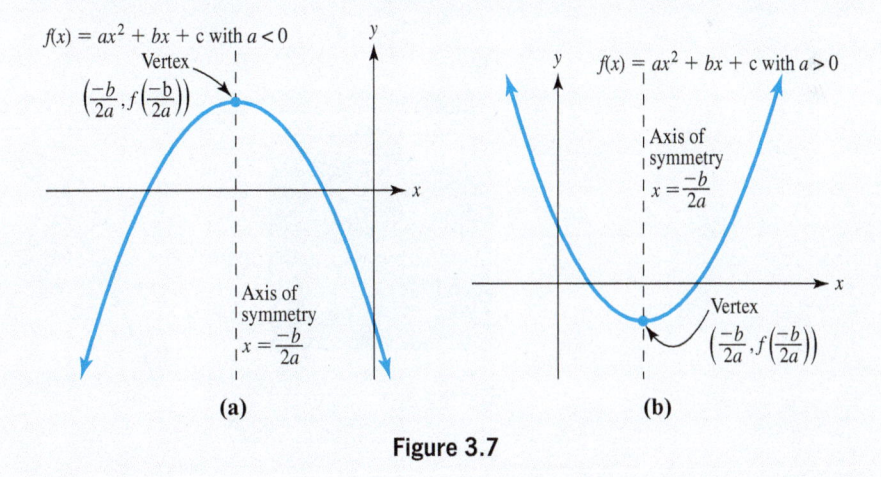

Figure 3.7

Observe that the graph of $y = x^2$ (Figure 3.4) is narrower than the graph of $y = \frac{1}{2}x^2$ (Figure 3.5(b)), and that 1 (the coefficient of x^2 in $y = x^2$) is larger than $\frac{1}{2}$ (the coefficient of x^2 in $y = \frac{1}{2}x^2$).

If we know the location of the vertex and the direction in which the parabola opens, we can make a good sketch of the graph by plotting just a few more points.

176 Chapter 3 Quadratic, Piecewise-Defined, and Power Functions

EXAMPLE 1 ▶ Graphing a Quadratic Function

Find the vertex and graph the quadratic function $f(x) = -2x^2 - 4x + 6$.

SOLUTION

Note that a, the coefficient of x^2, is -2, so the parabola opens downward. The x-coordinate of the vertex is $\dfrac{-b}{2a} = \dfrac{-(-4)}{2(-2)} = -1$, and the y-coordinate of the vertex is $f(-1) = -2(-1)^2 - 4(-1) + 6 = 8$. Thus, the vertex is $(-1, 8)$. The x-intercepts can be found by setting $f(x) = 0$ and solving for x:

$$-2x^2 - 4x + 6 = 0$$
$$-2(x^2 + 2x - 3) = 0$$
$$-2(x + 3)(x - 1) = 0$$
$$x = -3 \quad \text{or} \quad x = 1$$

The y-intercept is easily found by computing $f(0)$:

$$f(0) = -2(0)^2 - 4(0) + 6 = 6$$

The axis of symmetry is the vertical line $x = -1$. The graph is shown in Figure 3.8.

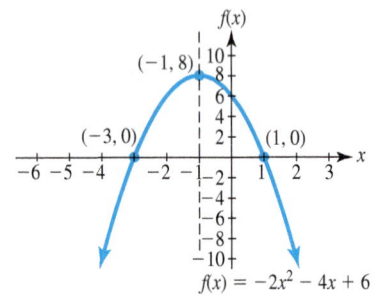

Figure 3.8

Spreadsheet ▶ SOLUTION

To graph the function $f(x) = -2x^2 - 4x + 6$ using Excel, first create a table containing values for x and $f(x)$. Then highlight the two columns containing these values, select insert, scatter plot, and the smooth curve option. Figure 3.9 shows the table of values and the graph.*

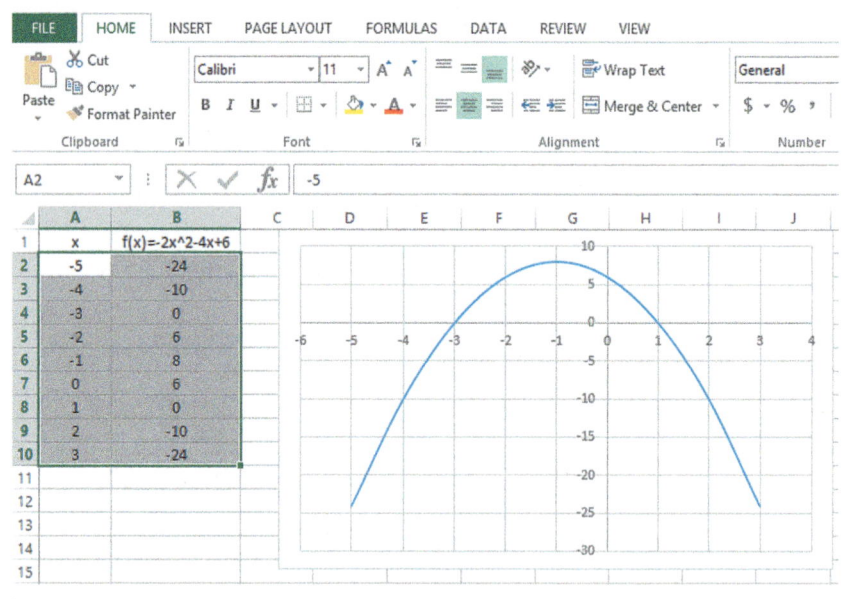

Figure 3.9

EXAMPLE 2 ▶ Maximizing Revenue

Suppose the monthly revenue from the sale of Carlson 42-inch 3D televisions is given by the function

$$R(x) = -0.1x^2 + 600x \text{ dollars}$$

where x is the number of televisions sold.

---

*For detailed steps, see Appendix B, page 704.

a. Find the vertex and the axis of symmetry of the graph of this function.

b. Determine if the vertex represents a maximum or minimum point.

c. Interpret the vertex in the context of the application.

d. Graph the function.

e. For what x-values is the function increasing? decreasing? What does this mean in the context of the application?

SOLUTION

a. The function is a quadratic function with $a = -0.1$, $b = 600$, and $c = 0$. The x-coordinate of the vertex is $\frac{-b}{2a} = \frac{-600}{2(-0.1)} = 3000$, and the axis of symmetry is the line $x = 3000$. The y-coordinate of the vertex is

$$R(3000) = -0.1(3000)^2 + 600(3000) = 900{,}000$$

so the vertex is $(3000, 900{,}000)$.

b. Because $a < 0$, the parabola opens downward, so the vertex is a maximum point.

c. The x-coordinate of the vertex gives the number of televisions that must be sold to maximize revenue, so selling 3000 sets will result in the maximum revenue. The y-coordinate of the vertex gives the maximum revenue, \$900,000.

d. Table 3.2 lists some values that satisfy $R(x) = -0.1x^2 + 600x$. The graph is shown in Figure 3.10.

Table 3.2

x	R(x)
0	0
1000	500,000
2000	800,000
3000	900,000
4000	800,000
5000	500,000
6000	0

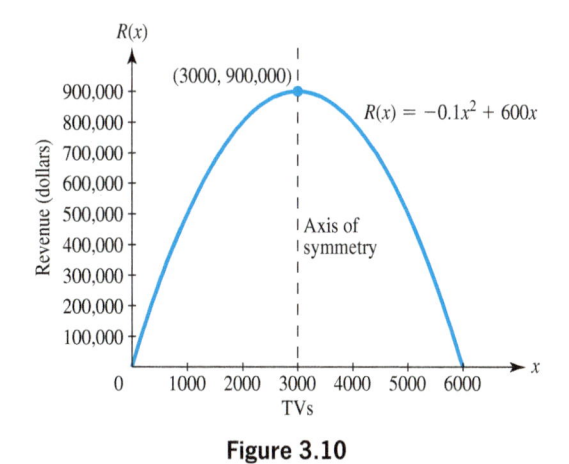

Figure 3.10

e. The function is increasing on the interval $(-\infty, 3000)$ and decreasing on $(3000, \infty)$. However, in the context of the application, negative inputs and outputs do not make sense, so we may say the revenue increases on $(0, 3000)$ and decreases on $(3000, 6000)$.

Even when using a graphing utility to graph a quadratic function, it is important to recognize that the graph is a parabola and to locate the vertex. Determining which way the parabola opens and the x-coordinate of the vertex is very useful in setting the viewing window so that a complete graph (that includes the vertex and the intercepts) is shown.

EXAMPLE 3 ▶ Foreign-Born Population

Using data from 1910 and projected to 2060, the percent of the U.S. population that is foreign-born can be modeled by the function

$$y = 0.0018x^2 - 0.216x + 13.0$$

where x is the number of years after 1910.

a. During what year does the model indicate that the percent of foreign-born population was a minimum?

b. What is the minimum percent?
(*Source*: U.S. Census Bureau)

SOLUTION

a. This equation is in the form $f(x) = ax^2 + bx + c$, so $a = 0.0018$. Because $a > 0$, the parabola opens upward, the vertex is a minimum, and the *x*-coordinate of the vertex is where the minimum percent occurs.

$$x = \frac{-b}{2a} = \frac{-(-0.216)}{2(0.0018)} = 60$$

So the percent of the U.S. population that was foreign born was a minimum in the 60th year after 1910, or in 1970.

b. The minimum for the model is found by evaluating the function at $x = 60$. The minimum, 6.52, can be found with TRACE or TABLE (accessed by 2ND GRAPH) or with direct evaluation (Figure 3.11). So the minimum percent is 6.52%.

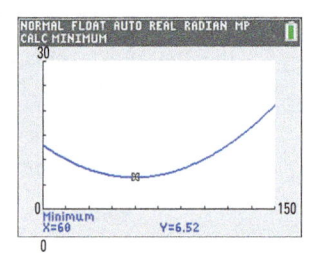

Figure 3.11

Suppose an object is shot or thrown into the air and then falls. If air resistance is ignored, the height in feet of the object after *t* seconds can be modeled by

$$S(t) = -16t^2 + v_0 t + h_0$$

where -16 ft/sec^2 is the acceleration due to gravity, v_0 ft/sec is the initial velocity (at $t = 0$ sec), and h_0 is the initial height in feet (at $t = 0$).

EXAMPLE 4 ▶ Height of a Ball

A ball is thrown upward at 64 feet per second from the top of an 80-foot-high building.

a. Write the quadratic function that models the height (in feet) of the ball as a function of the time *t* (in seconds).

b. Find the *t*-coordinate and *S*-coordinate of the vertex of the graph of this quadratic function.

c. Graph the model.

d. Explain the meaning of the coordinates of the vertex for this model.

SOLUTION

a. The model has the form $S(t) = -16t^2 + v_0 t + h_0$, where $v_0 = 64$ and $h_0 = 80$. Thus, the model is

$$S(t) = -16t^2 + 64t + 80 \text{ (feet)}$$

b. The height S is a function of the time t, and the t-coordinate of the vertex is

$$t = \frac{-b}{2a} = \frac{-64}{2(-16)} = 2$$

The S-coordinate of the vertex is the value of S at $t = 2$, so $S = -16(2)^2 + 64(2) + 80 = 144$ is the S-coordinate of the vertex. The vertex is $(2, 144)$.

c. The function is quadratic and the coefficient in the second-degree term is negative, so the graph is a parabola that opens down with vertex $(2, 144)$. To graph the function, we choose a window that includes the vertex $(2, 144)$ near the center top of the screen. Using the window $[0, 6]$ by $[-20, 150]$ gives the graph shown in Figure 3.12(a). Using CALC (accessed by **2ND** **TRACE**) maximum* verifies that the vertex is $(2, 144)$ (see Figure 3.12(b)).

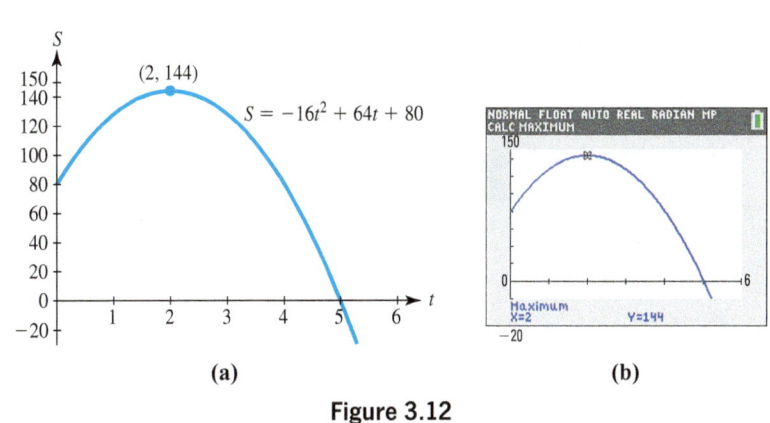

Figure 3.12

d. The graph is a parabola that opens down, so the vertex is the highest point on the graph and the function has its maximum there. The t-coordinate of the vertex, 2, is the time (in seconds) at which the ball reaches its maximum height, and the S-coordinate, 144, is the maximum height (in feet) that the ball reaches.

Vertex Form of a Quadratic Function

When a quadratic function is written in the form $f(x) = ax^2 + bx + c$, we can calculate the coordinates of the vertex. But if a quadratic function is written in the form

$$y = a(x - h)^2 + k$$

the vertex of the parabola is at (h, k) (Figure 3.13(a)). For example, the graph of $y = (x - 2)^2 + 3$ is a parabola opening upward with vertex $(2, 3)$ (Figure 3.13(b)).

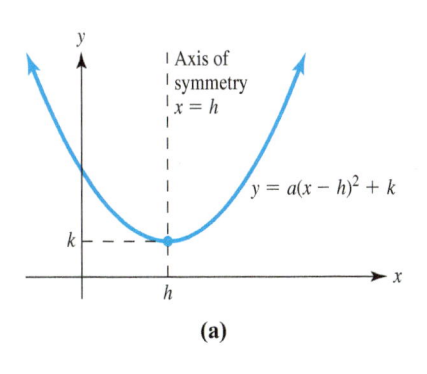

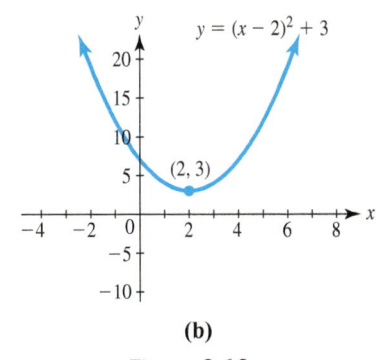

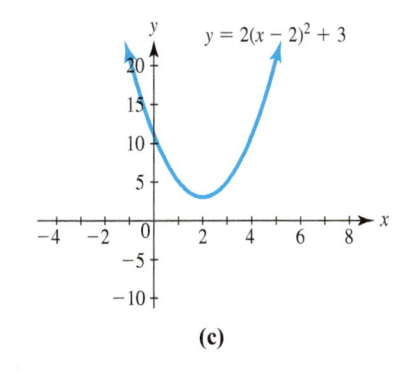

Figure 3.13

*For more details, see Appendix A, page 684.

Graph of a Quadratic Function

In general, the graph of the function

$$y = a(x - h)^2 + k$$

is a parabola with its vertex at the point (h, k).

- The parabola opens upward if $a > 0$, and the vertex is a minimum.
- The parabola opens downward if $a < 0$, and the vertex is a maximum.
- The axis of symmetry of the parabola has equation $x = h$.
- The a is the same as the leading coefficient in $y = ax^2 + bx + c$, so the larger the value of $|a|$, the narrower the parabola will be.

Note that the graph of $y = 2(x - 2)^2 + 3$ in Figure 3.13(c) is narrower than the graph of $y = (x - 2)^2 + 3$ in Figure 3.13(b).

EXAMPLE 5 ▶ Minimizing Cost

The cost for producing x Champions golf hats is given by the function

$$C(x) = 0.2(x - 40)^2 + 200 \text{ dollars}$$

a. Find the vertex of this function.

b. Is the vertex a maximum or minimum? Interpret the vertex in the context of the application.

c. Graph the function using a window that includes the vertex.

d. Describe what happens to the function between $x = 0$ and the x-coordinate of the vertex. What does this mean in the context of the application?

SOLUTION

a. This function is in the form $y = a(x - h)^2 + k$ with $h = 40$ and $k = 200$. Thus, the vertex of $C(x) = 0.2(x - 40)^2 + 200$ is $(40, 200)$.

b. Because $a = 0.2$, which is positive, the vertex is a minimum. This means that the cost of producing golf hats is at a minimum of $200 when 40 hats are produced.

c. We know that the vertex of the graph of this function is $(40, 200)$ and that it is a minimum, so we can choose a window with the $x = 40$ near the center of the screen and $y = 200$ near the bottom of the screen. The graph using the window $[0, 100]$ by $[-50, 1000]$ is shown in Figure 3.14.

d. For x-values between 0 and 40, the graph decreases. Thus, the cost of producing golf hats is decreasing until 40 hats are produced; after 40 hats are produced, the cost begins to increase.

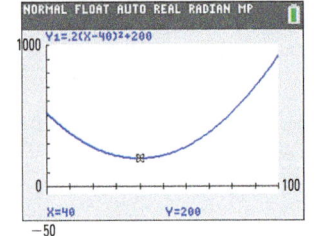

Figure 3.14

We can use the vertex form $y = a(x - h)^2 + k$ to write the equation of a quadratic function if we know the vertex and a point on its graph.

EXAMPLE 6 ▶ Profit

Right Sports Management had its monthly maximum profit, $450,000, when it produced and sold 5500 Waist Trimmers. Its fixed cost is $155,000. If the profit can be modeled by a quadratic function of x, the number of Waist Trimmers produced and sold each month, find this quadratic function $P(x)$.

SOLUTION

When 0 units are produced, the cost is $155,000 and the revenue is $0. Thus, the profit is $-\$155,000$ when 0 units are produced, and the y-intercept of the graph of the function is $(0, -155,000)$. The vertex of the graph of the quadratic function is $(5500, 450,000)$. Using these points gives

$$P(x) = a(x - 5500)^2 + 450,000$$

and

$$-155,000 = a(0 - 5500)^2 + 450,000$$

which gives

$$a = -0.02$$

Thus, the quadratic function that models the profit is $P(x) = -0.02(x - 5500)^2 + 450,000$, or $P(x) = -0.02x^2 + 220x - 155,000$, where $P(x)$ is in dollars and x is the number of units produced and sold.

EXAMPLE 7 ▶ Equation of a Quadratic Function

If the points in the table lie on a parabola, write the equation whose graph is the parabola.

x	−1	0	1	2
y	13	−2	−7	−2

SOLUTION

The x-values are a uniform distance apart. Because the points $(0, -2)$ and $(2, -2)$ both have a y-coordinate of -2, the symmetry of a parabola indicates that the vertex will be halfway between $x = 0$ and $x = 2$. Thus, the vertex of this parabola is at $(1, -7)$. (See Figure 3.15.)

The equation of the function is

$$y = a(x - 1)^2 - 7$$

The point $(2, -2)$ or any other point in the table besides $(1, -7)$ can be used to find a.

$$-2 = a(2 - 1)^2 - 7, \quad \text{or} \quad a = 5$$

Thus, the equation is

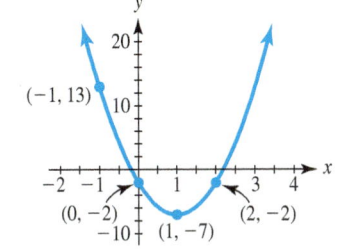

$$y = 5(x - 1)^2 - 7, \quad \text{or} \quad y = 5x^2 - 10x - 2$$

Figure 3.15

EXAMPLE 8 ▶ Vertex Form of a Quadratic Function

Write the vertex form of the equation of the quadratic function from the general form $y = 2x^2 - 8x + 5$ by finding the vertex and using the value of a.

SOLUTION

The vertex is at $x = \dfrac{-b}{2a} = \dfrac{-(-8)}{2(2)} = 2$, and $y = 2(2^2) - 8(2) + 5 = -3$. We know a is 2, because a is the same in both forms, so the equation has the form $y = a(x - 2)^2 - 3$. Thus,

$$y = 2(x - 2)^2 - 3$$

is the vertex form of the equation.

SKILLS CHECK 3.1

Answers that are not seen can be found in the answer section at the back of the text.

In Exercises 1–6, (a) determine whether the function is quadratic. If it is, (b) determine whether the graph is concave up or concave down. (c) Determine whether the vertex of the graph is a maximum point or a minimum point.

1. $y = 2x^2 - 8x + 6$
   a. Quadratic b. Up c. Minimum
2. $y = 4x - 3$ Not quadratic
3. $y = 2x^3 - 3x^2$
   Not quadratic
4. $f(x) = x^2 + 4x + 4$
   a. Quadratic b. Up c. Minimum
5. $g(x) = -5x^2 - 6x + 8$
   a. Quadratic b. Down c. Maximum
6. $h(x) = -2x^2 - 4x + 6$
   a. Quadratic b. Down c. Maximum

In Exercises 7–14, (a) graph each quadratic function on $[-10, 10]$ by $[-10, 10]$. (b) Does this window give a complete graph?

7. $y = 2x^2 - 8x + 6$
   b. Yes
8. $f(x) = x^2 + 4x + 4$
   b. Yes
9. $g(x) = -5x^2 - 6x + 8$ b. Yes
10. $h(x) = -2x^2 - 4x + 6$ b. Yes
11. $y = x^2 + 8x + 19$ b. Yes
12. $y = x^2 - 4x + 5$ b. Yes
13. $y = 0.01x^2 - 8x$
    b. No; the complete graph will be a parabola.
14. $y = 0.1x^2 + 8x + 2$
    b. No; the complete graph will be a parabola.
15. Write the equation of the quadratic function whose graph is shown. $y = (x - 2)^2 - 4$

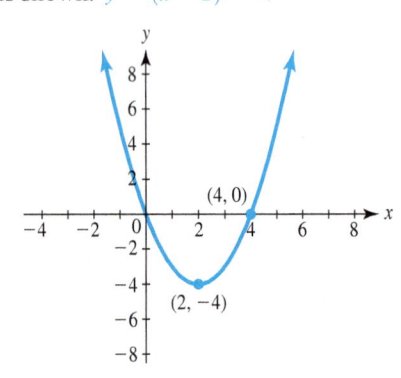

16. Write the equation of the quadratic function whose graph is shown. $y = -(x - 3)^2 + 5$

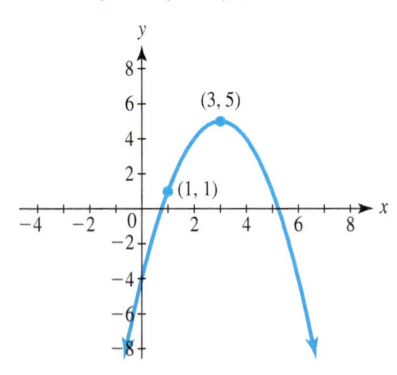

17. The two graphs shown have equations of the form $y = a(x - 2)^2 + 1$. Is the value of a larger for y_1 or y_2? y_1

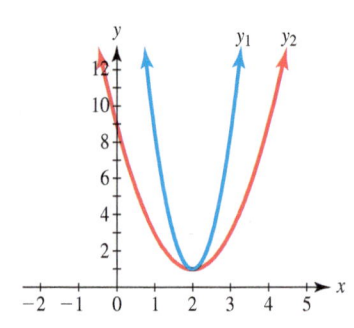

18. The two graphs shown have equations of the form $y = -a(x - 3)^2 + 5$. Is the value of $|a|$ larger for y_1 or y_2? y_2

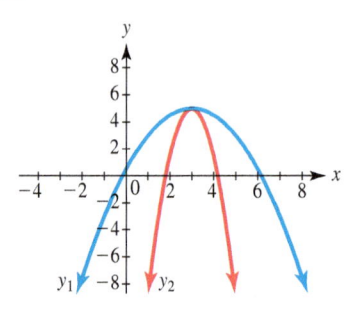

19. If the points in the table lie on a parabola, write the equation whose graph is the parabola.
    $y = -5x^2 + 10x + 8$

x	-1	1	3	5
y	-7	13	-7	-67

20. If the points in the table lie on a parabola, write the equation whose graph is the parabola.
    $y = 3x^2 + 12x - 3$

x	-6	-5	-4	-3	-2	-1
y	33	12	-3	-12	-15	-12

In Exercises 21–30, (a) give the coordinates of the vertex of the graph of each function. (b) Graph each function on a window that includes the vertex.

21. $y = (x - 1)^2 + 3$
    a. $(1, 3)$
22. $y = (x + 10)^2 - 6$
    a. $(-10, -6)$
23. $y = (x + 8)^2 + 8$
    a. $(-8, 8)$
24. $y = (x - 12)^2 + 1$
    a. $(12, 1)$
25. $f(x) = 2(x - 4)^2 - 6$ a. $(4, -6)$
26. $f(x) = -0.5(x - 2)^2 + 1$ a. $(2, 1)$
27. $y = 12x - 3x^2$
    a. $(2, 12)$
28. $y = 3x + 18x^2$ a. $\left(-\frac{1}{12}, -\frac{1}{8}\right)$
29. $y = 3x^2 + 18x - 3$
    a. $(-3, -30)$
30. $y = 5x^2 + 75x + 8$
    a. $(-7.5, -273.25)$

For Exercises 31–34, (a) find the x-coordinate of the vertex of the graph. (b) Set the viewing window so that the x-coordinate of the vertex is near the center of the window and the vertex is visible, and then graph the given equation. (c) State the coordinates of the vertex.

31. $y = 2x^2 - 40x + 10$ a. $x = 10$ c. $(10, -190)$
32. $y = -3x^2 - 66x + 12$ a. $x = -11$ c. $(-11, 375)$
33. $y = -0.2x^2 - 32x + 2$ a. $x = -80$ c. $(-80, 1282)$
34. $y = 0.3x^2 + 12x - 8$ a. $x = -20$ c. $(-20, -128)$

In Exercises 35–40, sketch complete graphs of the functions.

35. $y = x^2 + 24x + 144$
36. $y = x^2 - 36x + 324$
37. $y = -x^2 - 100x + 1600$
38. $y = -x^2 - 80x - 2000$
39. $y = 2x^2 + 10x - 600$
40. $y = 2x^2 - 75x - 450$

Use the graph of each function in Exercises 41–46 to estimate the x-intercepts.

41. $y = 2x^2 - 8x + 6$ $x = 1, x = 3$
42. $f(x) = x^2 + 4x + 4$ $x = -2$
43. $y = x^2 - x - 110$ $x = -10, x = 11$
44. $y = x^2 + 9x - 36$ $x = -12, x = 3$
45. $g(x) = -5x^2 - 6x + 8$ $x = -2, x = 0.8$
46. $h(x) = -2x^2 - 4x + 6$ $x = 1, x = -3$

EXERCISES 3.1

Answers that are not seen can be found in the answer section at the back of the text.

47. **Profit** The daily profit for a product is given by $P = 32x - 0.01x^2 - 1000$, where x is the number of units produced and sold.

    a. Graph this function for x between 0 and 3200.

    b. Describe what happens to the profit for this product when the number of units produced is between 1 and 1600. It increases.

    c. What happens to the profit after 1600 units are produced? Profit decreases.

48. **Profit** The daily profit for a product is given by $P = 420x - 0.1x^2 - 4100$ dollars, where x is the number of units produced and sold.

    a. Graph this function for x between 0 and 4200.

    b. Is the graph of the function concave up or down? The graph is concave down.

49. **Worldwide Internet Users** Using data from 2014 and projected to 2025, the percent of people in the world who are Internet users can be modeled by the function $y = -0.0921x^2 + 3.53x + 28.8$, with x representing the number of years after 2010 and p representing the percent.
    (Source: eMarketer)

    a. Graph the function for the years 2010 through 2040.

    b. Find the year in which the percent is maximized. 2030

    c. If the model is valid, what will happen to the percent of people in the world who are Internet users after 2030? It will decrease.

50. **World Population** A low-projection scenario of world population for the years 1995–2150 by the United Nations is given by the function $y = -0.36x^2 + 38.52x + 5822.86$, where x is the number of years after 1990 and the world population is measured in millions of people.

    a. Graph this function for $x = 0$ to $x = 120$.

    b. What would the world population have been in 2010 if the projections made using this model had been accurate? 6449.3 million, or 6.4493 billion
    (Source: World Population Prospects, United Nations)

51. **Global Biometrics Revenue** The annual revenue from the global biometrics market for the years from 2016 and projected to 2025 can be modeled by the function
    $$y = 0.143x^2 - 0.259x + 2.333$$
    with x equal to the number of years after 2015 and y equal to the revenue in billions of dollars.

    a. Graph this function for $x = 0$ to $x = 11$.

    b. Find the global biometrics revenue projected by this model for 2030. $30.623 billion

    c. Is the value in part (b) an interpolation or an extrapolation? Extrapolation
    (Source: Tractica)

52. **Flight of a Ball** If a ball is thrown upward at 96 feet per second from the top of a building that is 100 feet high, the height of the ball can be modeled by $S(t) = 100 + 96t - 16t^2$ feet, where t is the number of seconds after the ball is thrown.

    a. Describe the graph of the model.
       A parabola opening down

b. Find the *t*-coordinate and *S*-coordinate of the vertex of the graph of this quadratic function. $t = 3, S = 244$

c. Explain the meaning of the coordinates of the vertex for this model.
The ball reaches its maximum height of 244 feet in 3 seconds.

53. Flight of a Ball If a ball is thrown upward at 39.2 meters per second from the top of a building that is 30 meters high, the height of the ball can be modeled by $S(t) = 30 + 39.2t - 9.8t^2$ meters, where *t* is the number of seconds after the ball is thrown.

a. Find the *t*-coordinate and *S*-coordinate of the vertex of the graph of this quadratic function. $t = 2, S = 69.2$

b. Explain the meaning of the coordinates of the vertex for this function. Two seconds after the ball is thrown it reaches its maximum height of 69.2 meters.

c. Over what time interval is the function increasing? What does this mean in relation to the ball? Until $t = 2$ seconds; at 2 seconds, it reaches maximum height and then falls.

54. Photosynthesis The rate of photosynthesis *R* for a certain plant depends on the intensity of light *x*, in lumens, according to $R(x) = 270x - 90x^2$.

a. Sketch the graph of this function on a meaningful window.

b. Determine the intensity *x* that gives the maximum rate of photosynthesis. 1.5 lumens

55. Crude Oil The U.S. Crude Oil production, in billions of barrels, for the years from 2010 projected to 2030, can be modeled by $y = -0.001x^2 + 0.037x + 1.949$, with *x* equal to the number of years after 2010 and *y* equal to the number of billions of barrels of crude oil.

a. Find and interpret the vertex of the graph of this model. Vertex: (18.5, 2.291); the maximum number of barrels of crude oil projected to be produced during this period is 2.291 billion barrels during 2029.

b. What does the model predict the crude oil production will be in 2033? 2.271 billion barrels

c. Graph the function for the years 2010 to 2030.
(*Source*: U.S. Energy Information Administration)

56. Workers and Output The weekly output of graphing calculators is $Q(x) = 200x + 6x^2$. Graph this function for values of *x* and *Q* that make sense in this application, if *x* is the number of weeks, $x \leq 10$.

57. Profit The profit for a certain brand of MP3 player can be described by the function $P(x) = 40x - 3000 - 0.01x^2$ dollars, where *x* is the number of MP3 players produced and sold.

a. To maximize profit, how many MP3 players must be produced and sold? 2000

b. What is the maximum possible profit? $37,000

58. Profit The profit for Easy-Cut lawnmowers can be described by the function $P(x) = 840x - 75.6 - 0.4x^2$ dollars, where *x* is the number of mowers produced and sold.

a. To maximize profit, how many mowers must be produced and sold? 1050

b. What is the maximum possible profit? $440,924.40

59. Revenue The annual total revenue for Pilot V5 pens is given by $R(x) = 1500x - 0.02x^2$ dollars, where *x* is the number of pens sold.

a. To maximize the annual revenue, how many pens must be sold? 37,500

b. What is the maximum possible annual revenue? $28,125,000

60. Revenue The monthly total revenue for satellite radios is given by $R(x) = 300x - 0.01x^2$ dollars, where *x* is the number of radios sold.

a. To maximize the monthly revenue, how many radios must be sold? 15,000

b. What is the maximum possible monthly revenue? $2,250,000

61. Area If 200 feet of fence are used to enclose a rectangular pen, the resulting area of the pen is $A = x(100 - x)$, where *x* is the width of the pen.

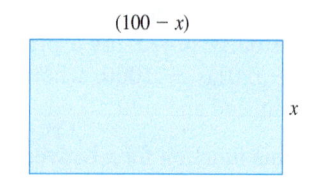

a. Is *A* a quadratic function of *x*? Yes

b. What is the maximum possible area of the pen?
$A = 100x - x^2$; maximum area of 2500 sq ft when *x* is 50 ft

62. Area If 25,000 feet of fence are used to enclose a rectangular field, the resulting area of the field is $A = (12,500 - x)x$, where *x* is the width of the pen. What is the maximum possible area of the pen? 39,062,500 sq ft

63. Obesity Obesity (BMI ≥ 30) increases the risk of diabetes, heart disease, and many other ailments. The percent of Americans who are obese, from 2000 projected to 2030, can be modeled by $y = -0.010x^2 + 0.971x + 22.1$, where *x* equals the number of years after 2000 and *y* is the percent of obese Americans.
(*Source*: American Journal of Preventive Medicine)

a. Graph the function for the years 2000 through 2030.

b. Find and interpret the vertex of the graph.

c. If the model remains valid, describe what happens to the percent of obese Americans after 2030.

64. U.S. Native Population The population of native-born citizens of the United States can be modeled by $y = -0.007x^2 + 1.722x + 270$, where *y* is in millions and *x* is the number of years after 2010.

a. Find the input and the output at the vertex of the graph of this function. (123, 375.903)

b. Interpret the results of part (b) if this model is valid though this period.
(*Source:* U.S. Census Bureau) The native-born U.S. population will reach a maximum of 375.9 million in 2133.

65. **U.S. Retail E-commerce Sales** Using data from 2016 and projected to 2022, the annual retail e-commerce sales can be modeled by

$$y = -1.56x^2 + 75.4x - 38.6$$

where x is the number of years after 2010 and y is the sales in millions of dollars.

a. Graph the function on the intervals $5 \leq x \leq 50$ and $300 \leq y \leq 1000$.

b. Find the coordinates of the vertex of the graph in part (a), rounded to three significant digits. (24.2, 872)

c. Explain the meaning of the coordinates of the vertex from part (b) in the context of the application.
(*Source:* FTI Consulting) The annual retail e-commerce sales will be maximized at approximately $872 million in 2035.

66. **Total Energy Used for Industry** The total energy used for industry, in millions of **British thermal units** (BTU), for the years from 2016 and projected to 2025 can be modeled by the function

$$y = -0.035x^2 + 0.959x + 29.8$$

where x is the number of years after 2015.

a. Find the vertex of the graph of this function. (14, 36.66)

b. Interpret the vertex in terms of the application.
(*Source:* U.S. Energy Information Administration) The model predicts that the total energy used for industry will be maximized at 36.66 million BTUs in 2029.

67. **Wind and Pollution** The amount of particulate pollution p in the air depends on the wind speed s, among other things, with the relationship between p and s approximated by $p = 25 - 0.01s^2$, where p is in ounces per cubic yard and s is in miles per hour.

a. Sketch the graph of this model with s on the horizontal axis and with nonnegative values of s and p.

b. Is the function increasing or decreasing on this domain? Decreasing

c. What is the p-intercept of the graph? 25

d. What does the p-intercept mean in the context of this application? When wind speed is 0 mph, amount of pollution is 25 oz per cubic yard.

68. **Drug Sensitivity** The sensitivity S to a drug is related to the dosage size x by $S = 1000x - x^2$.

a. Sketch the graph of this model using a domain and range with nonnegative x and S.

b. Is the function increasing or decreasing for x between 0 and 500? Increasing

c. What is the positive x-intercept of the graph? (1000, 0)

d. Why is this x-intercept important in the context of this application?

69. **Falling Object** A tennis ball is thrown downward into a swimming pool from the top of a tall hotel. The height of the ball from the pool is given by $D(t) = -16t^2 - 4t + 210$ feet, where t is the time, in seconds, after the ball is thrown. Graphically find the t-intercepts for this function. Interpret the value(s) that make sense in this problem context. 3.5, −3.75

70. **Break-Even** The profit for a product is given by $P = 1600 - 100x + x^2$, where x is the number of units produced and sold. Graphically find the x-intercepts of the graph of this function to find how many units will give break-even (that is, return a profit of zero). 20 or 80 units

71. **Flight of a Ball** A softball is hit with upward velocity 32 feet per second when $t = 0$, from a height of 3 feet.

a. Find the function that models the height of the ball as a function of time. $y = -16t^2 + 32t + 3$

b. Find the maximum height of the ball. 19 feet

72. **Flight of a Ball** A baseball is hit with upward velocity 48 feet per second when $t = 0$, from a height of 4 feet.

a. Find the function that models the height of the ball as a function of time. $y = -16t^2 + 48t + 4$

b. Find the maximum height of the ball and in how many seconds the ball will reach that height. 40 ft in 1.5 sec

73. **Apartment Rental** The owner of an apartment building can rent all 100 apartments if he charges $1200 per apartment per month, but the number of apartments rented is reduced by 2 for every $40 increase in the monthly rent.

a. Construct a table that gives the revenue if the rent charged is $1240, $1280, and $1320.

b. Does $R(x) = (1200 + 40x)(100 - 2x)$ model the revenue from these apartments if x represents the number of $40 increases? Yes

c. What monthly rent gives the maximum revenue for the apartments? $1600

74. **Rink Rental** The owner of a skating rink rents the rink for parties at $720 if 60 or fewer skaters attend, so the cost is $12 per person if 60 attend. For each 6 skaters above 60, she reduces the price per skater by $0.50.

a. Construct a table that gives the revenue if the number attending is 66, 72, and 78.

b. Does the function $R(x) = (60 + 6x)(12 - 0.5x)$ model the revenue from the party if x represents the number of increases of 6 people each? Yes

c. How many people should attend for the rink's revenue to be a maximum? 102

75. World Population A low-projection scenario for world population for 1995–2150 by the United Nations is given by the function $y = -0.36x^2 + 38.52x + 5822.86$, where x is the number of years after 1990 and the world population is measured in millions of people.

a. Find the input and output at the vertex of the graph of this model. $x = 53.5, y = 6853.3$

b. Interpret the values from part (a).
World population will be maximized at 6,853,300,000 in 2044.

c. For what years after 1995 does this model predict that the population will increase? Until 2044

(*Source: World Population Prospects*, United Nations)

3.2 Solving Quadratic Equations

KEY OBJECTIVES

- Solve quadratic equations using factoring
- Solve quadratic equations graphically using the x-intercept method and the intersection method
- Solve quadratic equations by combining graphical and factoring methods
- Solve quadratic equations using the square root method
- Solve quadratic equations by completing the square
- Solve quadratic equations using the quadratic formula
- Solve quadratic equations having complex solutions

SECTION PREVIEW Worldwide 5G Subscriptions

Using data for the years from 2020 and projected to 2025, the number of millions of 5G smartphone subscriptions worldwide for the years can be modeled by the function

$$S(t) = 54.5t^2 - 250t + 1.0$$

where t is the number of years after 2010. (*Source*: Business Wire)

To find the year after 2010 when the number of 5G subscriptions worldwide will equal 2951 billion (that is, 2.951 million), we solve the quadratic equation

$$2951 = 54.5t^2 - 250t + 1.0$$

The values of t that satisfy this equation are the solutions of the equation

$$54.5t^2 - 250t - 2950 = 0$$

and the zeros of the function

$$S(t) = 54.5t^2 - 250t - 2950$$

As with linear equations, we can solve quadratic equations by using algebraic, graphical, and numerical methods. (See Example 9.) The algebraic methods we will learn to use in this section include factoring methods, the square root method, completing the square, and using the quadratic formula. ■

Factoring Methods

An equation that can be written in the form $ax^2 + bx + c = 0$, with $a \neq 0$, is called a **quadratic equation**. Solutions to some quadratic equations can be found exactly by factoring; other quadratic equations require different types of solution methods to find or to approximate solutions.

Solution by factoring is based on the following property of real numbers.

Zero Product Property

If the product of two real numbers is 0, then at least one of them must be 0. That is, for real numbers a and b, if the product $ab = 0$, then either $a = 0$ or $b = 0$ or both a and b are equal to 0.

3.2 Solving Quadratic Equations

To use this property to solve a quadratic equation by factoring, we must first make sure that the equation is written in a form with zero on one side. If the resulting nonzero expression is factorable, we factor it and use the zero product property to convert the equation into two linear equations that are easily solved.* Before applying this technique to real-world applications, we consider the following example.

EXAMPLE 1 ▶ Solving a Quadratic Equation by Factoring

Solve the equation $3x^2 + 7x = 6$.

SOLUTION

We first subtract 6 from both sides of the equation to rewrite the equation with 0 on one side:

$$3x^2 + 7x - 6 = 0$$

To begin factoring the trinomial $3x^2 + 7x - 6$, we seek factors of $3x^2$ (that is, $3x$ and x) as the first terms of two binomials and factors of -6 as the last terms of the binomials. The factorization whose inner and outer products combine to $7x$ is $(3x - 2)(x + 3)$, so we have

$$(3x - 2)(x + 3) = 0$$

Using the zero product property gives

$$3x - 2 = 0, \quad \text{or} \quad x + 3 = 0$$

Solving these linear equations gives the two solutions to the original equation:

$$x = \frac{2}{3} \quad \text{or} \quad x = -3$$

EXAMPLE 2 ▶ Height of a Ball

The height above ground of a ball thrown upward at 64 feet per second from the top of an 80-foot-high building is modeled by $S(t) = 80 + 64t - 16t^2$ feet, where t is the number of seconds after the ball is thrown. How long will the ball be in the air?

SOLUTION

The ball will be in the air from $t = 0$ (with the height $S = 80$) until it reaches the ground ($S = 0$). Thus, we can find the time in the air by solving

$$0 = -16t^2 + 64t + 80$$

Because 16 is a factor of each of the terms, we can get a simpler but equivalent equation by dividing both sides of the equation by -16:

$$0 = t^2 - 4t - 5$$

This equation can be solved easily by factoring the right side:

$$0 = (t - 5)(t + 1)$$
$$0 = t - 5 \quad \text{or} \quad 0 = t + 1$$
$$t = 5 \quad \text{or} \quad t = -1$$

---

*For a review of factoring methods, see page 167 in the Algebra Toolbox.

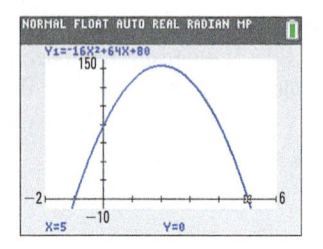

Figure 3.16

The time in the air starts at $t = 0$, so $t = -1$ has no meaning in this application. S also equals 0 at $t = 5$, which means that the ball is on the ground 5 seconds after it was thrown; that is, the ball is in the air 5 seconds. Figure 3.16 shows a graph of the function, which confirms that the height of the ball is 0 at $t = 5$.

Graphical Methods

In cases where factoring $f(x)$ to solve $f(x) = 0$ is difficult or impossible, graphing $y = f(x)$ can be helpful in finding the solution. Recall that if a is a real number, the following three statements are equivalent:

- a is a real solution to the equation $f(x) = 0$.
- a is a real zero of the function f.
- a is an x-intercept of the graph of $y = f(x)$.

It is important to remember that the above three statements are equivalent because connecting these concepts allows us to use different methods for solving equations. Sometimes graphical methods are the easiest way to find or approximate solutions to real data problems. If the x-intercepts of the graph of $y = f(x)$ are easily found, then graphical methods may be helpful in finding the solutions. Note that if the graph of $y = f(x)$ does not cross or touch the x-axis, there are no real solutions to the equation $f(x) = 0$.

We can also find solutions or decimal approximations of solutions to quadratic equations by using the intersection method with a graphing utility.

EXAMPLE 3 ▶ Profit

Consider the daily profit from the production and sale of x units of a product, given by

$$P(x) = -0.01x^2 + 20x - 500 \text{ dollars}$$

a. Use a graph to find the levels of production and sales that give a daily profit of $1400.

b. Is it possible for the profit to be greater than $1400?

SOLUTION

a. To find the level of production and sales, x, that gives a daily profit of 1400 dollars, we solve

$$1400 = -0.01x^2 + 20x - 500$$

To solve this equation by the intersection method, we graph

$$y_1 = -0.01x^2 + 20x - 500 \quad \text{and} \quad y_2 = 1400$$

To find the appropriate window for this graph, we note that the graph of the function $y_1 = -0.01x^2 + 20x - 500$ is a parabola with the vertex at

$$x = \frac{-b}{2a} = \frac{-20}{2(-0.01)} = 1000 \quad \text{and} \quad y = P(1000) = 9500$$

We use a viewing window containing this point, with $x = 1000$ near the center, to graph the function (Figure 3.17). Using the intersection method, we see that $(100, 1400)$ and $(1900, 1400)$ are the points of intersection of the graphs of the two functions. Thus, $x = 100$ and $x = 1900$ are solutions to the equation $1400 = -0.01x^2 + 20x - 500$, and the profit is $1400 when $x = 100$ units or $x = 1900$ units of the product are produced and sold.

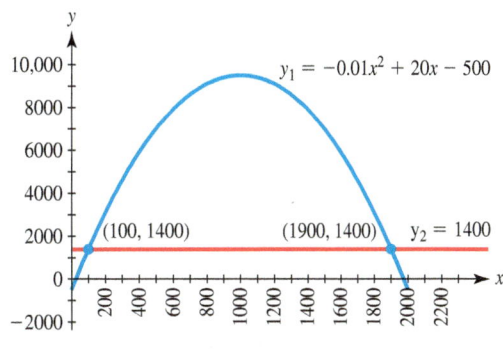

Figure 3.17

b. We can see from the graph in Figure 3.17 that the profit is more than $1400 for many values of x. Because the graph of this profit function is a parabola that opens down, the maximum profit occurs at the vertex of the graph. As we found in part (a), the vertex occurs at $x = 1000$, and the maximum possible profit is $P(1000) = \$9500$, which is more than $1400.

Combining Graphs and Factoring

If the factors of a quadratic function are not easily found, its graph may be helpful in finding the factors. Note once more the important correspondences that exist among solutions, zeros, and x-intercepts: the x-intercepts of the graph of $y = P(x)$ are the real solutions of the equation $0 = P(x)$ and the real zeros of $P(x)$. In addition, the factors of $P(x)$ are related to the zeros of $P(x)$. The relationship among the factors, solutions, and zeros is true for any polynomial function f and can be generalized by the following theorem.

Factor Theorem

The polynomial function f has a factor $(x - a)$ if and only if $f(a) = 0$. Thus, $(x - a)$ is a factor of $f(x)$ if and only if $x = a$ is a solution to $f(x) = 0$.

This means that we can verify our factorization of f and the real solutions to $0 = f(x)$ by graphing $y = f(x)$ and observing where the graph crosses the x-axis. We can also sometimes use our observation of the graph to assist us in the factorization of a quadratic function:

If one solution can be found exactly from the graph, it can be used to find one of the factors of the function. The second factor can then be found easily, leading to the second solution.

EXAMPLE 4 ▶ Graphing and Factoring Methods Combined

Solve $0 = 3x^2 - x - 10$ by using the following steps.

a. Graphically find one of the x-intercepts of $y = 3x^2 - x - 10$.

b. Algebraically verify that the zero found in part (a) is an exact solution to $0 = 3x^2 - x - 10$.

c. Use the method of factoring to find the other solution to $0 = 3x^2 - x - 10$.

190 Chapter 3 Quadratic, Piecewise-Defined, and Power Functions

SOLUTION

a. The vertex of the graph of $y = 3x^2 - x - 10$ is at $x = \dfrac{1}{6}$ and $y \approx -10.08$, and because the parabola opens up, we set a viewing window that includes this point near the bottom center of the window. Graphing the function and using the *x*-intercept method, we find that the graph crosses the *x*-axis at $x = 2$ (Figure 3.18). This means that 2 is a zero of $f(x) = 3x^2 - x - 10$.

(a)   (b)

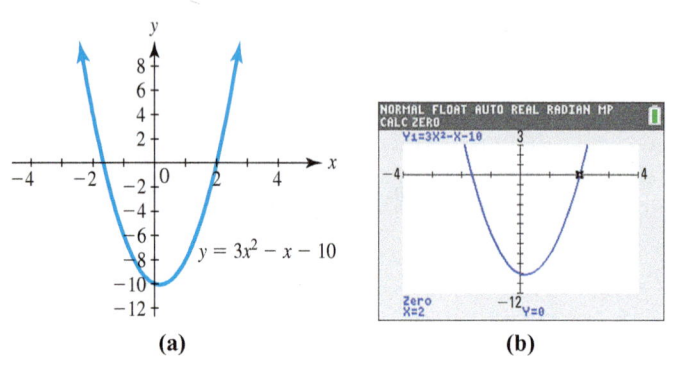

Figure 3.18

b. Because $x = 2$ was obtained graphically, it may be an approximation of the exact solution. To verify that it is exact, we show that $x = 2$ makes the equation $0 = 3x^2 - x - 10$ a true statement:

$$3(2)^2 - 2 - 10 = 12 - 2 - 10 = 0$$

c. Because $x = 2$ is a solution of $0 = 3x^2 - x - 10$, one factor of $3x^2 - x - 10$ is $(x - 2)$. Thus, we use $(x - 2)$ as one factor and seek a second binomial factor so that the product of the two binomials is $3x^2 - x - 10$:

$$3x^2 - x - 10 = 0$$
$$(x - 2)(\qquad) = 0$$
$$(x - 2)(3x + 5) = 0$$

The remaining factor is $3x + 5$, which we set equal to 0 to obtain the other solution:

$$3x = -5$$
$$x = -\dfrac{5}{3}$$

Thus, the two solutions to $0 = 3x^2 - x - 10$ are $x = 2$ and $x = -\dfrac{5}{3}$.

Graphical and Numerical Methods

When approximate solutions to quadratic equations are sufficient, graphical and/or numerical solution methods can be used. These methods of solving are illustrated in Example 5.

EXAMPLE 5 ▶ National Hospital Expenditures

In 2015, hospital spending was projected to increase 5.6% as a consequence of the continued effects of the Affordable Care Act insurance expansion, combined with the effect of faster economic growth. For 2016 through 2022, continued population aging and the impacts of improved economic conditions are expected to result in projected average annual growth of 6.4%. The function

$$E(x) = 2.171x^2 - 5.051x + 628.152$$

gives the national expenditures for hospitals, in billions of dollars, for the years 2006 and projected to 2022, where x equals the number of years after 2000.

a. Graph the function for the years 2006 to 2022.

b. Find the year after 2000 during which the national hospital expenditures are expected to reach $1,400,000,000,000, using a graphical method.

c. Verify the solution numerically.

(*Source*: Centers for Medicare & Medicaid Services)

SOLUTION

a. The graph of the function for the years 2006 to 2022 is shown in Figure 3.19(a). The inputs that represent these years are $x = 6$ to $x = 22$.

b. If expenditures are $1,400,000,000,000, the output for the function will be $E(x) = 1400$, so we solve the equation

$$2.171x^2 - 5.051x + 628.152 = 1400$$

We solve this equation with the intersection method. Figure 3.19(b) shows the graphs of $Y_1 = 2.171x^2 - 5.051x + 628.152$ and $Y_2 = 1400$ and the point where the graphs intersect.

The intersection point indicates that the hospital expenditures will reach 1400 billion dollars, or $1,400,000,000,000, when x is more than 20, so the expenditures reach $1400 billion in 2021.

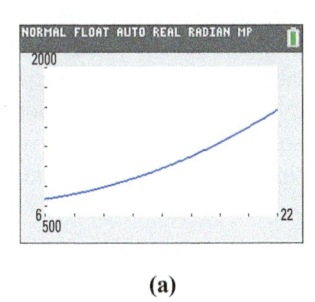

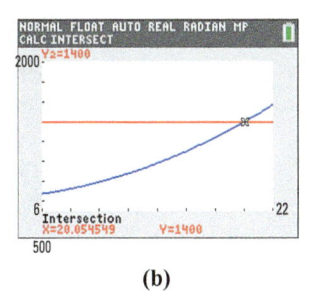

Figure 3.19

c. Figure 3.20 shows a table of inputs near 21 and the corresponding outputs. The table shows that the expenditures are close to $1400 billion in 2020 but do not actually reach $1400 billion until during 2021.

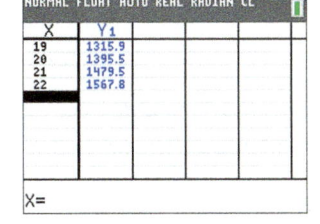

Figure 3.20

Spreadsheet ▶ SOLUTION To find the intersection of the graphs of $y = 2.171x^2 - 5.051x + 628.152$ and $y = 40$ with Excel, we enter their outputs in columns B and C, respectively, enter B − C in column D, and then use the Goal Seek command to find the x-value that makes the function in column D equal to 0, as we did with linear equations in Chapter 2. The x-value that makes the difference of these two functions equal to 0 is approximately 20.05 (see Figure 3.21).* This value agrees with the value found with a calculator in Figure 3.19(b). The graph of the two functions in Figure 3.21 confirms the solution.

---

*For details, see Appendix B, page 711.

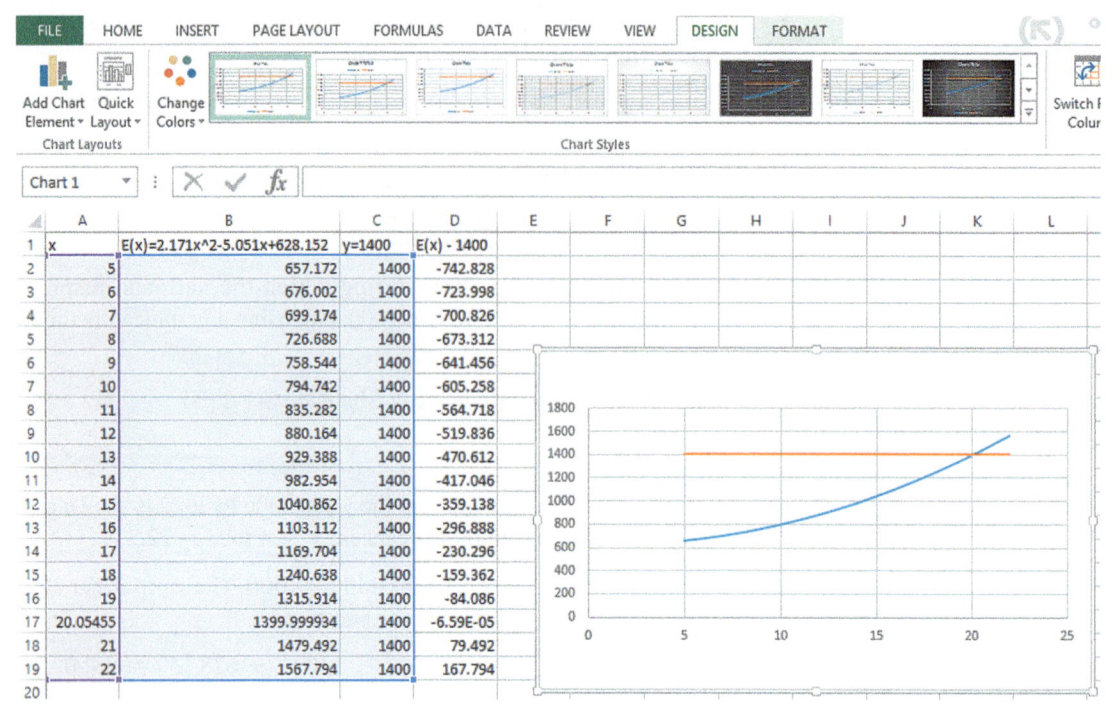

192   Chapter 3   Quadratic, Piecewise-Defined, and Power Functions

Figure 3.21

The Square Root Method

We have solved quadratic equations by factoring and by graphical methods. Another method can be used to solve quadratic equations that are written in a particular form. In general, we can find the solutions of quadratic equations of the form $x^2 = C$, where C is a constant, by taking the square root of both sides. For example, to solve $x^2 = 25$, we take the square root of both sides of the equation, getting $x = \pm 5$. Note that there are two solutions because $5^2 = 25$ and $(-5)^2 = 25$.

> ### Square Root Method
> The solutions of the quadratic equation $x^2 = C$ are $x = \pm\sqrt{C}$. Note that, when we take the square root of both sides, we use a $\pm$ symbol because there are both a positive and a negative value that, when squared, give C.

Note that this method can also be used to solve equations of the form $(ax + b)^2 = C$.

EXAMPLE 6 ▶ Square Root Solution Method

Solve the following equations by using the square root method.

a. $2x^2 - 16 = 0$

b. $(x - 6)^2 = 18$

SOLUTION

a. This equation can be written in the form $x^2 = C$, so the square root method can be used. We want to rewrite the equation so that the x^2-term is isolated and its coefficient is 1 and then take the square root of both sides.

$$2x^2 - 16 = 0$$
$$2x^2 = 16$$
$$x^2 = 8$$
$$x = \pm\sqrt{8} = \pm 2\sqrt{2}$$

The exact solutions to $2x^2 - 16 = 0$ are $x = 2\sqrt{2}$ and $x = -2\sqrt{2}$.

b. The left side of this equation is a square, so we can take the square root of both sides to find x.

$$(x - 6)^2 = 18$$
$$x - 6 = \pm\sqrt{18}$$
$$x = 6 \pm \sqrt{18}$$
$$= 6 \pm \sqrt{9}\sqrt{2}$$
$$= 6 \pm 3\sqrt{2}$$

Completing the Square

When we convert one side of a quadratic equation to a perfect binomial square and use the square root method to solve the equation, the method used is called **completing the square**.

EXAMPLE 7 ▶ Completing the Square

Solve the equation $x^2 - 12x + 7 = 0$ by completing the square.

SOLUTION

Because the left side of this equation is not a perfect square, we rewrite it in a form where we can more easily get a perfect square. We subtract 7 from both sides, getting

$$x^2 - 12x = -7$$

We complete the square on the left side of this equation by adding the appropriate number to both sides of the equation to make the left side a perfect square trinomial. Because $(x + a)^2 = x^2 + 2ax + a^2$, the constant term of a perfect square trinomial will be the *square of half the coefficient of x*. Using this rule with the equation $x^2 - 12x = -7$, we would need to take half the coefficient of x and add the square of this number to get a perfect square trinomial. Half of -12 is -6, so adding $(-6)^2 = 36$ to $x^2 - 12x$ gives the perfect square $x^2 - 12x + 36$. We also have to add 36 to the other side of the equation (to preserve the equality), giving

$$x^2 - 12x + 36 = -7 + 36$$

Factoring this perfect square trinomial gives

$$(x - 6)^2 = 29$$

We can now solve the equation with the square root method.

$$(x - 6)^2 = 29$$
$$x - 6 = \pm \sqrt{29}$$
$$x = 6 \pm \sqrt{29}$$

The Quadratic Formula

We can generalize the method of completing the square to derive a general formula that gives the solution to any quadratic equation. We use this method to find the general solution to $ax^2 + bx + c = 0, a \neq 0$.

$ax^2 + bx + c = 0$	Standard form
$ax^2 + bx = -c$	Subtract c from both sides.
$x^2 + \dfrac{b}{a}x = -\dfrac{c}{a}$	Divide both sides by a.

We would like to make the left side of the last equation a perfect square trinomial. Half the coefficient of x is $\dfrac{b}{2a}$, and squaring this gives $\dfrac{b^2}{4a^2}$. Hence, adding $\dfrac{b^2}{4a^2}$ to both sides of the equation gives a perfect square trinomial on the left side, and we can continue with the solution.

$x^2 + \dfrac{b}{a}x + \dfrac{b^2}{4a^2} = \dfrac{b^2}{4a^2} - \dfrac{c}{a}$	Add $\dfrac{b^2}{4a^2}$ to both sides of the equation.
$\left(x + \dfrac{b}{2a}\right)^2 = \dfrac{b^2 - 4ac}{4a^2}$	Factor the left side and combine the fractions on the right side.
$x + \dfrac{b}{2a} = \pm\sqrt{\dfrac{b^2 - 4ac}{4a^2}}$	Take the square root of both sides.
$x = -\dfrac{b}{2a} \pm \dfrac{\sqrt{b^2 - 4ac}}{2|a|}$	Subtract $\dfrac{b}{2a}$ from both sides and simplify.
$x = -\dfrac{b}{2a} \pm \dfrac{\sqrt{b^2 - 4ac}}{2a}$	$\pm 2|a| = \pm 2a$
$x = \dfrac{-b \pm \sqrt{b^2 - 4ac}}{2a}$	Combine the fractions.

The formula we have developed is called the **quadratic formula**.

Quadratic Formula

The solutions of the quadratic equation $ax^2 + bx + c = 0$ are given by the formula

$$x = \frac{-b \pm \sqrt{b^2 - 4ac}}{2a}$$

Note that a is the coefficient of x^2, b is the coefficient of x, and c is the constant term.

Because of the $\pm$ sign, the solutions can be written as

$$x = \frac{-b + \sqrt{b^2 - 4ac}}{2a} \quad \text{and} \quad x = \frac{-b - \sqrt{b^2 - 4ac}}{2a}$$

We can use the quadratic formula to solve all quadratic equations exactly, but it is especially useful for finding exact solutions to those equations for which factorization is difficult or impossible. For example, the solutions to $2.171x^2 - 5.051x + 628.152 = 1400$ were found approximately by graphical methods in Example 5. If we need to find the exact solutions, we could use the quadratic formula.

EXAMPLE 8 ▶ Solving Using the Quadratic Formula

Solve $6 - 3x^2 + 4x = 0$ using the quadratic formula.

SOLUTION

The equation $6 - 3x^2 + 4x = 0$ can be rewritten as $-3x^2 + 4x + 6 = 0$, so $a = -3$, $b = 4$, and $c = 6$. The two solutions to this equation are

$$x = \frac{-4 \pm \sqrt{4^2 - 4(-3)(6)}}{2(-3)} = \frac{-4 \pm \sqrt{88}}{-6} = \frac{-4 \pm 2\sqrt{22}}{-6} = \frac{2 \pm \sqrt{22}}{3}$$

Thus, the exact solutions are the irrational numbers $x = \dfrac{2 + \sqrt{22}}{3}$ and $x = \dfrac{2 - \sqrt{22}}{3}$.

Three-place decimal approximations for these solutions are $x \approx 2.230$ and $x \approx -0.897$.

Decimal approximations of irrational solutions found with the quadratic formula will often suffice as answers to an applied problem. The quadratic formula is especially useful when the coefficients of a quadratic equation are decimal values that make factorization impractical. This occurs in many applied problems, such as the one discussed at the beginning of this section. If the graph of the quadratic function $y = f(x)$ does not intersect the x-axis at "nice" values of x, the solutions may be irrational numbers, and using the quadratic formula allows us to find these solutions exactly.

EXAMPLE 9 ▶ Worldwide 5G Subscriptions

Using data for the years from 2020 and projected to 2025, the number of millions of 5G smartphone subscriptions worldwide for the years can be modeled by the function

$$S(t) = 54.5t^2 - 250t + 1.0$$

where t is the number of years after 2020. Find the year after 2020 when the number of 5G subscriptions worldwide will equal 2.951 billion. (*Source*: Business Wire)

SOLUTION

We seek to find when $S(t)$ is 2951 million, which equals 2.951 billion. We can do this by using algebraic, numerical, or graphical methods to solve the equation

$$2951 = 54.5t^2 - 250t + 1.0$$

We choose to solve this equation with the quadratic formula.
We first write the equation above with 0 on one side.

$$54.5t^2 - 250t - 2950 = 0$$

This gives $a = 54.5, b = -250$, and $c = -2950$. Substituting into the quadratic formula gives

$$t = \frac{-(-250) \pm \sqrt{(-250)^2 - 4(54.5)(-2950)}}{2(54.5)} = \frac{250 \pm 840}{109}$$

This gives $t = 10$ or $t \approx -5.4$

Thus, the model predicts that the number of 5G subscriptions worldwide will equal 2.951 billion in the 10th year after 2020, in 2030. (Note that the negative value for t will give 5.4 years before 2020.)

Figure 3.22(a) shows the outputs of

$$Y_1 = 54.5t^2 - 250t + 1.0$$

near and at $x = 10$, and Figure 3.22(b) shows the graph of the intersection of $Y_1 = 54.5t^2 - 250t + 1.0$ and $Y_2 = 2951$, which numerically and graphically confirms the conclusion we found algebraically.

(a)   (b)

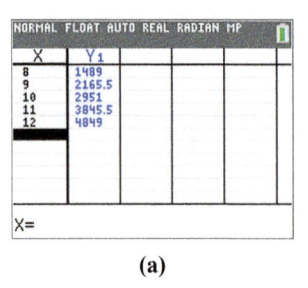

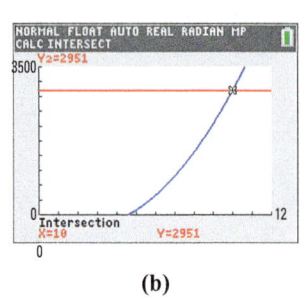

Figure 3.22

The Discriminant

We can also determine the type of solutions a quadratic equation has by looking at the expression $b^2 - 4ac$, which is called the **discriminant** of the quadratic equation $ax^2 + bx + c = 0$. The discriminant is the expression inside the radical in the quadratic formula $x = \dfrac{-b \pm \sqrt{b^2 - 4ac}}{2a}$, so it determines if the quantity inside the radical is positive, zero, or negative.

- If $b^2 - 4ac > 0$, there are two different real solutions.
- If $b^2 - 4ac = 0$, there is one real solution.
- If $b^2 - 4ac < 0$, there is no real solution.

For example, the equation $3x^2 + 4x + 2 = 0$ has no real solution because $4^2 - 4(3)(2) = -8 < 0$, and the equation $x^2 + 4x + 2 = 0$ has two different real solutions because $4^2 - 4(1)(2) = 8 > 0$.

Some quadratic equations have a solution that occurs twice. For example, the equation $x^2 - 6x + 9 = 0$ can be solved by factoring:

$$x^2 - 6x + 9 = 0$$
$$(x - 3)(x - 3) = 0$$
$$x = 3, x = 3$$

Because $x - 3$ is a factor twice, the solution $x = 3$ is said to have **multiplicity** 2. Note that the discriminant of this equation is $(-6)^2 - 4(1)(9) = 36 - 36 = 0$. When the discriminant equals 0, there will be only one solution to the quadratic equation, and it will have multiplicity 2.

Aids for Solving Quadratic Equations

The x-intercepts of the graph of a quadratic function $y = f(x)$ can be used to determine how to solve the quadratic equation $f(x) = 0$. Table 3.3 summarizes these ideas with suggested methods for solving and the graphical representations of the solutions.

Table 3.3 Connections Between Graphs of Quadratic Functions and Solution Methods

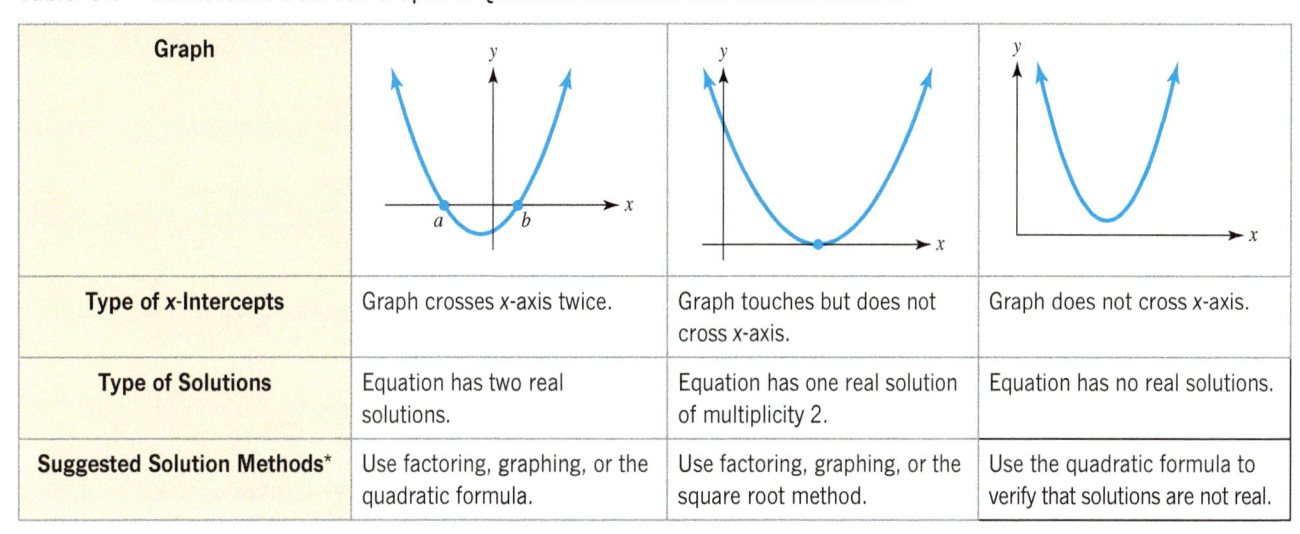

Graph			
Type of x-Intercepts	Graph crosses x-axis twice.	Graph touches but does not cross x-axis.	Graph does not cross x-axis.
Type of Solutions	Equation has two real solutions.	Equation has one real solution of multiplicity 2.	Equation has no real solutions.
Suggested Solution Methods*	Use factoring, graphing, or the quadratic formula.	Use factoring, graphing, or the square root method.	Use the quadratic formula to verify that solutions are not real.

Equations with Complex Solutions

Recall that the solutions of the quadratic equation $x^2 = C$ are $x = \pm \sqrt{C}$, so the solutions of the equation $x^2 = -a$ for $a > 0$ are

$$x = \pm\sqrt{-a} = \pm\sqrt{-1}\sqrt{a} = \pm i\sqrt{a}$$

EXAMPLE 10 ▶ Solution Using the Square Root Method

Solve the equations.

a. $x^2 = -9$ **b.** $3x^2 + 24 = 0$

SOLUTION

a. Taking the square root of both sides of the equation gives the solution of $x^2 = -9$:

$$x = \pm\sqrt{-9} = \pm\sqrt{-1}\sqrt{9} = \pm 3i$$

b. We solve $3x^2 + 24 = 0$ using the square root method, as follows:

$$3x^2 = -24$$
$$x^2 = -8$$
$$x = \pm\sqrt{-8} = \pm\sqrt{-1}\sqrt{4 \cdot 2} = \pm 2i\sqrt{2}$$

We used the square root method to solve the equations in Example 10 because neither equation contained a first-degree term (that is, a term containing x to the first power).

---

*Solution methods other than the suggested methods may also be successful.

Chapter 3 Quadratic, Piecewise-Defined, and Power Functions

We can also find complex solutions by using the quadratic formula.* Recall that the solutions of the quadratic equation $ax^2 + bx + c = 0$ are given by the formula

$$x = \frac{-b \pm \sqrt{b^2 - 4ac}}{2a}$$

EXAMPLE 11 ▶ Complex Solutions of Quadratic Equations

Solve the equations.

a. $x^2 - 3x + 5 = 0$ b. $3x^2 + 4x = -3$

SOLUTION

a. Using the quadratic formula, with $a = 1, b = -3,$ and $c = 5$, gives

$$x = \frac{-(-3) \pm \sqrt{(-3)^2 - 4(1)(5)}}{2(1)} = \frac{3 \pm \sqrt{-11}}{2} = \frac{3 \pm i\sqrt{11}}{2}$$

Note that the solutions can also be written in the form $\frac{3}{2} \pm \frac{\sqrt{11}}{2}i$.

Thus, the solutions are the complex numbers $\frac{3}{2} + \frac{\sqrt{11}}{2}i$ and $\frac{3}{2} - \frac{\sqrt{11}}{2}i$.

b. Writing $3x^2 + 4x = -3$ in the form $3x^2 + 4x + 3 = 0$ gives $a = 3, b = 4,$ and $c = 3$, so the solutions are

$$x = \frac{-4 \pm \sqrt{(4)^2 - 4(3)(3)}}{2(3)} = \frac{-4 \pm \sqrt{-20}}{6} = \frac{-4 \pm \sqrt{-1}\sqrt{4}\sqrt{5}}{6}$$

$$= \frac{-4 \pm 2i\sqrt{5}}{6} = \frac{2(-2 \pm i\sqrt{5})}{2 \cdot 3} = \frac{-2 \pm i\sqrt{5}}{3}$$

Thus,

$$x = -\frac{2}{3} + \frac{\sqrt{5}}{3}i \quad \text{and} \quad x = -\frac{2}{3} - \frac{\sqrt{5}}{3}i$$

---

*The solutions could also be found by completing the square. Recall that the quadratic formula was developed by completing the square on the quadratic equation $ax^2 + bx + c = 0$.

SKILLS CHECK 3.2

Answers that are not seen can be found in the answer section at the back of the text.

In Exercises 1–10, use factoring to solve the equations.

1. $x^2 - 3x - 10 = 0$
   $x = 5, x = -2$
2. $x^2 - 9x + 18 = 0$
   $x = 6, x = 3$
3. $x^2 - 11x + 24 = 0$
   $x = 8, x = 3$
4. $x^2 + 3x - 10 = 0$
   $x = -5, x = 2$
5. $2x^2 + 2x - 12 = 0$
   $x = -3, x = 2$
6. $2s^2 + s - 6 = 0$
   $s = 3/2, s = -2$
7. $0 = 2t^2 - 11t + 12$
   $t = 3/2, t = 4$
8. $6x^2 - 13x + 6 = 0$
   $x = 2/3, x = 3/2$
9. $6x^2 + 10x = 4$
   $x = -2, x = 1/3$
10. $10x^2 + 11x = 6$
    $x = -3/2, x = 2/5$

In Exercises 11–16, find the x-intercepts algebraically.

11. $f(x) = 3x^2 - 5x - 2 \quad -\frac{1}{3}, 2$
12. $f(x) = 5x^2 + 7x + 2 \quad -\frac{2}{5}, -1$
13. $f(x) = 4x^2 - 9 \quad \frac{3}{2}, -\frac{3}{2}$
14. $f(x) = 4x^2 + 20x + 25 \quad -\frac{5}{2}$
15. $f(x) = 3x^2 + 4x - 4 \quad \frac{2}{3}, -2$
16. $f(x) = 9x^2 - 1 \quad \frac{1}{3}, -\frac{1}{3}$

Use a graphing utility to find or to approximate the x-intercepts of the graph of each function in Exercises 17–22.

17. $y = x^2 + 7x + 10$
    $x = -2, x = -5$
18. $y = x^2 + 4x - 32$
    $x = -8, x = 4$

19. $y = 3x^2 - 8x + 4$ **20.** $y = 2x^2 + 8x - 10$
$x = 2/3, x = 2$ $x = -5, x = 1$
21. $y = 2x^2 + 7x - 4$ **22.** $y = 5x^2 - 17x + 6$
$x = 1/2, x = -4$ $x = 0.4, x = 3$

Use a graphing utility as an aid in factoring to solve the equations in Exercises 23–28.

23. $2w^2 - 5w - 3 = 0$ **24.** $3x^2 - 4x - 4 = 0$
$w = -1/2, w = 3$ $x = -2/3, x = 2$
25. $x^2 - 40x + 256 = 0$ **26.** $x^2 - 32x + 112 = 0$
$x = 8, x = 32$ $x = 28, x = 4$
27. $2s^2 - 70s = 1500$ **28.** $3s^2 - 130s = -1000$
$s = 50, s = -15$ $s = 10, s = \frac{100}{3}$

In Exercises 29–34, use the square root method to solve the quadratic equations.

29. $4x^2 - 9 = 0$ $x = \pm\frac{3}{2}$ **30.** $x^2 - 20 = 0$ $x = \pm 2\sqrt{5}$
31. $x^2 - 32 = 0$ **32.** $5x^2 - 25 = 0$
$x = \pm 4\sqrt{2}$ $x = \pm\sqrt{5}$
33. $(x - 5)^2 = 9$ $x = 2, 8$ **34.** $(2x + 1)^2 = 20$ $x = \frac{-1 \pm 2\sqrt{5}}{2}$

In Exercises 35–38, complete the square to solve the quadratic equations.

35. $x^2 - 4x - 9 = 0$ **36.** $x^2 - 6x + 1 = 0$
$x = 2 \pm \sqrt{13}$ $x = 3 \pm 2\sqrt{2}$
37. $x^2 - 3x + 2 = 0$ **38.** $2x^2 - 9x + 8 = 0$
$x = 1, x = 2$ $x = (9 \pm \sqrt{17})/4$

In Exercises 39–42, use the quadratic formula to solve the equations.

39. $x^2 - 5x + 2 = 0$ **40.** $3x^2 - 6x - 12 = 0$
$x = (5 \pm \sqrt{17})/2$ $x = 1 \pm \sqrt{5}$
41. $5x + 3x^2 = 8$ **42.** $3x^2 - 30x - 180 = 0$
$x = 1, x = -\frac{8}{3}$ $x = 5 \pm \sqrt{85}$

In Exercises 43–48, use a graphing utility to find or approximate solutions of the equations.

43. $2x^2 + 2x - 12 = 0$ **44.** $2x^2 + x - 6 = 0$
$x = -3, x = 2$ $x = -2, x = 1.5$
45. $0 = 6x^2 + 5x - 6$ **46.** $10x^2 = 22x - 4$
$x = 2/3 \approx 0.667, x = -3/2 = -1.5$ $x = 0.2, x = 2$
47. $4x + 2 = 6x^2 + 3x$ **48.** $(x - 3)(x + 2) = -4$
$x = -1/2 = -0.5, x = 2/3 \approx 0.667$ $x = 2, x = -1$

In Exercises 49–54, find the exact solutions to $f(x) = 0$ in the complex numbers and confirm that the solutions are not real by showing that the graph of $y = f(x)$ does not cross the x-axis.

49. $x^2 + 25 = 0$ **50.** $2x^2 + 40 = 0$
$x = \pm 5i$ $x = \pm 2i\sqrt{5}$
51. $(x - 1)^2 = -4$ **52.** $(2x + 1)^2 + 7 = 0$
$x = 1 \pm 2i$ $x = (-1 \pm i\sqrt{7})/2$
53. $x^2 + 4x + 8 = 0$ **54.** $x^2 - 5x + 7 = 0$
$x = -2 \pm 2i$ $x = (5 \pm i\sqrt{3})/2$

In Exercises 55–58, you are given the graphs of several functions of the form $f(x) = ax^2 + bx + c$ for different values of a, b, and c. For each function,

a. *Determine whether the discriminant is positive, negative, or zero.*
b. *Determine whether there are 0, 1, or 2 real solutions to $f(x) = 0$.*
c. *Solve the equation $f(x) = 0$, if possible.*

55.

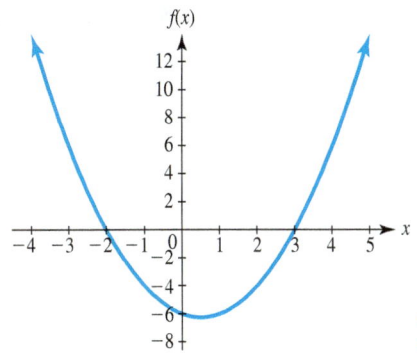

a. Positive   b. 2
c. $x = 3, x = -2$

56.

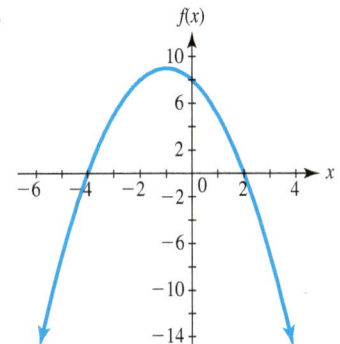

a. Positive   b. 2
c. $x = -4, x = 2$

57.

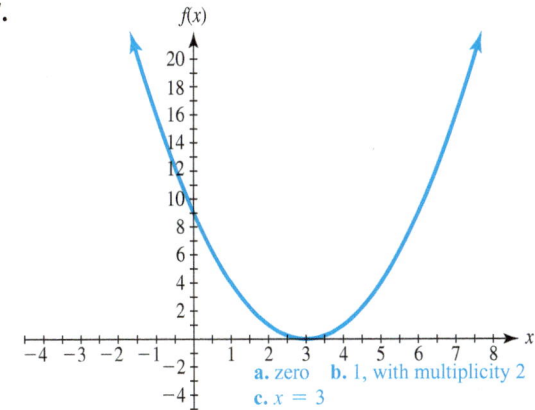

a. zero   b. 1, with multiplicity 2
c. $x = 3$

58.

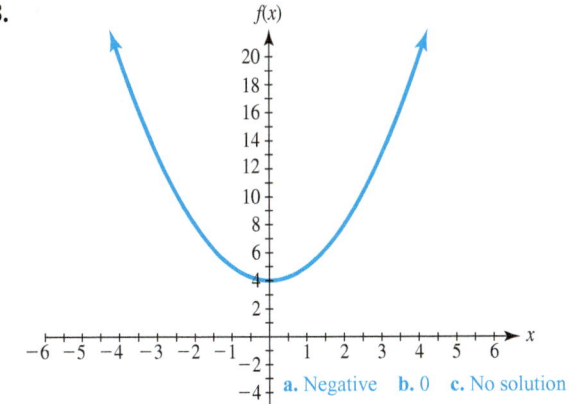

a. Negative   b. 0   c. No solution

For each function in Exercise 59–62,

a. *Calculate the discriminant.*
b. *Determine whether there are 0, 1, or 2 real solutions to $f(x) = 0$.*

59. $f(x) = 3x^2 - 5x - 2$ **60.** $f(x) = 2x^2 - x + 1$
49; two $-7; 0$
61. $f(x) = 4x^2 + 4x + 1$ **62.** $f(x) = 4x^2 + 25$
0; one, with multiplicity 2 $-400; 0$

EXERCISES 3.2

Answers that are not seen can be found in the answer section at the back of the text.

In Exercises 63–74, solve analytically and then check graphically.

63. **Flight of a Ball** If a ball is thrown upward at 96 feet per second from the top of a building that is 100 feet high, the height of the ball can be modeled by $S(t) = 100 + 96t - 16t^2$ feet, where t is the number of seconds after the ball is thrown. How long after the ball is thrown is the height 228 feet? $t = 2$ sec and 4 sec

64. **Falling Object** A tennis ball is thrown into a swimming pool from the top of a tall hotel. The height of the ball above the pool is modeled by $D(t) = -16t^2 - 4t + 200$ feet, where t is the time, in seconds, after the ball is thrown. How long after the ball is thrown is it 44 feet above the pool? 3 sec

65. **Break-Even** The profit for an electronic reader is given by $P(x) = -12x^2 + 1320x - 21{,}600$, where x is the number of readers produced and sold. How many readers give break-even (that is, give zero profit) for this product? 20 or 90

66. **Break-Even** The profit for Coffee Exchange coffee beans is given by $P(x) = -15x^2 + 180x - 405$ thousand dollars, where x is the number of tons of coffee beans produced and sold. How many tons give break-even (that is, give zero profit) for this product? 3 or 9

67. **Break-Even** The total revenue function for French door refrigerators is given by $R = 550x$ dollars, and the total cost function for this same product is given by $C = 10{,}000 + 30x + x^2$, where C is measured in dollars. For both functions, the input x is the number of refrigerators produced and sold.

   a. Form the profit function for the refrigerators from the two given functions. $P(x) = 520x - 10{,}000 - x^2$

   b. What is the profit when 18 refrigerators are produced and sold? −$964 (loss of $964)

   c. What is the profit when 32 refrigerators are produced and sold? $5616

   d. How many refrigerators must be sold to break even on this product? 20 or 500

68. **Break-Even** The total revenue function for a home theater system is given by $R = 266x$, and the total cost function for the system is $C = 2000 + 46x + 2x^2$, where R and C are each measured in dollars and x is the number of units produced and sold.

   a. Form the profit function for this product from the two given functions. $P(x) = -2x^2 + 220x - 2000$

   b. What is the profit when 55 systems are produced and sold? $4050

   c. How many systems must be sold to break even on this product? 10 or 100

69. **Wind and Pollution** The amount of particulate pollution p from a power plant in the air above the plant depends on the wind speed s, among other things, with the relationship between p and s approximated by $p = 25 - 0.01s^2$, with s in miles per hour.

   a. Find the value(s) of s that will make $p = 0$. $s = 50, s = -50$

   b. What does $p = 0$ mean in this application? There is no particulate pollution.

   c. What solution to $0 = 25 - 0.01s^2$ makes sense in the context of this application? $s = 50$; speed is positive.

70. **Velocity of Blood** Because of friction from the walls of an artery, the velocity of a blood corpuscle in an artery is greatest at the center of the artery and decreases as the distance r from the center increases. The velocity of the blood in the artery can be modeled by the function

$$v = k(R^2 - r^2)$$

where R is the radius of the artery and k is a constant that is determined by the pressure, viscosity of the blood, and the length of the artery. In the case where $k = 2$ and $R = 0.1$ centimeter, the velocity is $v = 2(0.01 - r^2)$ centimeters per second (cm/sec).

   a. What distance r would give a velocity of 0.02 cm/sec? $r = 0$

   b. What distance r would give a velocity of 0.015 cm/sec? $r = 0.05$

   c. What distance r would give a velocity of 0 cm/sec? Where is the blood corpuscle? $r = 0.1$; against the wall of the artery

71. **Drug Sensitivity** The sensitivity S to a drug is related to the dosage size x by $S = 100x - x^2$, where x is the dosage size in milliliters.

   a. What dosage(s) would give zero sensitivity? 0 mL or 100 mL

   b. Explain what your answer in part (a) might mean.

72. **Body-Heat Loss** The model for body-heat loss depends on the coefficient of convection K, which depends on wind speed s according to the equation $K^2 = 16s + 4$, where s is in miles per hour. Find the positive coefficient of convection when the wind speed is

   a. 20 mph. $K = 18$

b. 60 mph. $K = 2\sqrt{241}$

c. What is the change in K for a change in speed from 20 mph to 60 mph? 0.326 per unit increase in wind speed

73. Market Equilibrium Suppose that the demand for artificial Christmas trees is given by the function

$$p = 109.70 - 0.10q$$

and that the supply of these trees is given by

$$p = 0.01q^2 + 5.91$$

where p is the price of a tree in dollars and q is the quantity of trees that are demanded/supplied in hundreds. Find the price that gives the market equilibrium price and the number of trees that will be sold/bought at this price. $100 gives demand = supply = 97 trees.

74. Market Equilibrium The demand for diamond-studded watches is given by $p = 7000 - 2x$ dollars, and the supply of watches is given by $p = 0.01x^2 + 2x + 1000$ dollars, where x is the number of watches demanded and supplied when the price per watch is p dollars. Find the equilibrium quantity and the equilibrium price. 600; $5800

75. Global Internet Using data from 2014 and projected to 2021, the worldwide Internet penetration can be modeled by the function

$$y = -0.05x^2 + 2.65x + 30.9$$

where x is the number of years after 2010 and y is the percent of the worldwide population with access to the Internet.
(*Source*: Statista)

a. Write the equation that could be used to find x when $y = 66$, and then write the equation with one side set equal to 0 and a positive coefficient on the x^2-term.
$66 = -0.05x^2 + 2.65x + 30.9$; $0.05x^2 - 2.65x + 35.1 = 0$

b. One solution to the equation in part (a) is $x = 26$. Interpret this in terms of the application. The model projects that 66% of the worldwide population will have access to the Internet in 2036.

c. Use the solution in part (b) to write a factor of the nonzero side of the equation found in part (a) and then use division to find a second factor.
$x - 26$; $x - 27$

d. Use the second factor to find a second solution to $0.05x^2 - 2.65x + 35.1 = 0$. Interpret this in terms of the application. $x = 27$; The model projects that 66% of the worldwide population will also have access to the Internet in 2037.

76. Smartphone Users Using data from 2010 and projected to 2022, the number of smartphone users in the United States, in millions, can be modeled by the function $y = -1.17x^2 + 31.23x + 63.04$, where x is the number of years after 2010.
(*Source*: Statista)

a. Write the equation that could be used to find x when $y = 265$, and then write the equation with one side set equal to 0 and a positive coefficient on the x^2-term.
$265 = -1.17x^2 + 31.23x + 63.04$; $1.17x^2 - 31.23x + 201.96 = 0$

b. Verify that one solution to the equation in part (a) is $x = 11$. When does this estimate that the number of smartphone users in the United States was 265 million? $x = 11$ is a solution; in 2021

c. Use technology to find a second solution to $y = -1.17x^2 + 31.23x + 63.04 = 265$. Interpret this in terms of the application. $x = 15.69$; the model projects the number of smartphone users in the U.S. to be 265 million in 2026.

d. Does the model project that there will be more than 265 million smartphone users in the United States? Yes; it projects approximately 271 million in 2024.

77. Global Semiconductor and Sensor Market Using data from 2015 and projected to 2025, the value of the global semiconductor and sensor markets for the Internet of Things can be modeled by the function

$$y = 0.71x^2 + 1.2x + 29$$

where x is equal to the number of years after 2015 and y is equal to the market value in billions of dollars.
(*Source*: SEMI.ORG)

a. Use the quadratic formula to find the positive value of x that satisfies

$$337 = 0.71x^2 + 1.2x + 29 \quad x = 20$$

b. Interpret the solution to the equation in part (a) in terms of the application. The value of the global semiconductor and sensor markets for the Internet of Things is projected to be $337 billion in 2035.

c. Verify this value with technology. The solution checks.

78. Artificial Intelligence Enterprise AI refers to the use of artificial intelligence in enterprises to transform business operations and strategy. The revenue (in billions of dollars) from artificial intelligence for enterprise applications in North America for the years from 2016 through 2025 can be modeled by $y = 0.229x^2 - 0.483x + 0.543$, where x is the number of years after 2016.
(*Source*: Statista)

a. Use the quadratic formula to find when enterprise artificial intelligence revenue is projected to reach $74.035 billion, according to this model.
In 2035 ($x = 19$)

b. Use technology to verify your solution. The solution checks.

79. China's Labor Pool Using UN data from 1975 and projections to 2050, the number of millions of people age 15 to 59 in China can be modeled by $y = -0.224x^2 + 22x + 370.7$, with x equal to the number of years after 1970 and y equal to the number of millions of people in this labor pool. Use technology with the model to find when this population is expected to be equal to 903.1 million.
(*Source*: United Nations)
In 2014 and in 2025

80. **Alzheimer's Disease** Partially because of Americans living longer, the number with Alzheimer's disease and other dementia is projected to grow each year. The millions of U.S. citizens with Alzheimer's from 2000 and projected to 2050 can be modeled by $y = 0.002107x^2 + 0.09779x + 4.207$, with x equal to the number of years after 2000 and y equal to the number of millions of Americans with Alzheimer's disease. In what year does the model predict that 9.037 million Americans will have Alzheimer's disease or other dementia?

Year	2000	2010	2020	2030	2040	2050
Projected (millions affected)	4.0	5.9	6.8	8.7	11.8	14.3

(*Source*: National Academy on an Aging Society)
2030

81. **Lithium Demand** The demand for lithium is projected to steadily increase because of the growth in production of electric vehicles. Using data from 2017 and projections to 2025, the function $D = 0.676t^2 + 14.604t + 203.4$ can be used to model the demand for lithium in thousands of metric tons, where t is the number of years after 2015. Use graphical methods to find the year after 2015 in which the demand will reach 610,120 metric tons. 2031 ($x = 16$)
(*Source*: Statista)

82. **National Health Expenditures** The total national expenditures for health care (in billions of dollars) for selected years from 2002 and projected to 2024 can be modeled by the function $y = 4.61x^2 + 43.4x + 1620$, where x is the number of years after 2000. In what year after 2000 are the expenditures predicted by this model to be $7071 billion? 2030
(*Source*: U.S. Centers for Medicare and Medicaid Services)

83. **U.S. Households** The percent of total households that have married couples for selected years from 2000 and projected to 2050 can be modeled by the function

$$y = 0.00308x^2 - 0.460x + 60.2$$

with x representing the number of years after 2000 and y representing the percent. In what year will the percent fall to 45%? In 2050
(*Source*: Yi Zeng et al., *Household and Living Arrangement Projections*; Springer)

84. **Non-white Population** The number of millions of individuals in the U.S civilian non-institutional population 16 years and older who are non-white or Hispanic for the years 2000 and projected to 2050 can be modeled by $y = 0.007x^2 + 1.24x + 43.71$, with x equal to the number of years after 1990 and y equal to this population, in millions. Use technology to find the year after 1990 when this population is projected to be 104.51 million. 2030
(*Source*: U.S. Census Bureau)

85. **Retail Sales** November and December retail sales, excluding autos, for the years 2001–2010 can be modeled by the function $S(x) = -1.751x^2 + 38.167x + 388.997$ billion dollars, where x is the number of years after 2000.

   a. Graph the function for values of x representing 2001–2010.

   b. During what years does the model estimate the sales to be $550 billion?
   $x = 5.7187$ and $x = 16.07$, so in 2006 and 2017
   c. The recession in 2008 caused retail sales to drop. Does the model agree with the facts; that is, does it indicate that a maximum occurred in 2008?

(*Source*: U.S. Census Bureau)
No, the model indicates a maximum in 2011.

86. **Cell Phones** Using the CTIA Wireless Survey for 1985–2009, the number of U.S. cell phone subscribers (in millions) can be modeled by

$$y = 0.632x^2 - 2.651x + 1.209$$

where x is the number of years after 1985.

   a. Graphically find when the number of U.S. subscribers was 301,617,000. 2009

   b. When does the model estimate that the number of U.S. subscribers would reach 359,515,000? 2011

   c. What does the answer to (b) tell about this model?
   Number exceeds the U.S. population, so the model has become invalid.

87. **World Population** One projection of the world population by the United Nations (a low-projection scenario) is given in the table below. The data can be modeled by $y = -0.36x^2 + 38.52x + 5822.86$ million people, where x is the number of years after 1990. In what year after 1990 does this model predict the world population will first reach 6,702,000,000?
2023

Year	Projected Population (millions)
1995	5666
2000	6028
2025	7275
2050	7343
2075	6402
2100	5153
2125	4074
2150	3236

(*Source*: World Population Prospects, United Nations)

3.3 Power and Root Functions

KEY OBJECTIVES

- Evaluate and graph power functions
- Graph root functions and the reciprocal function
- Solve power and root equations
- Solve problems involving direct variation as the *n*th power

SECTION PREVIEW Foreign-Born Population

Table 3.4 gives the percent of the U.S. population that is foreign-born for selected years from 1970 and projected to 2060. We have modeled these data with a quadratic function, but we can also model data with a **power function**. We will later see that these data can be modeled by the function $y = 0.996x^{0.646}$, with x equal to the number of years after 1960 and y equal to the percent. In Example 2, we will evaluate, graph, and apply this power function. In this section, we will also evaluate and graph **root functions** and solve problems involving direct variation as an *n*th power. ∎

Power Functions

We define power functions as follows.

Table 3.4

Year	Percent	Year	Percent
1970	4.8	2020	14.3
1980	6.2	2030	15.8
1990	8.0	2040	17.1
2000	10.4	2050	18.2
2010	12.4	2060	18.8

(*Source:* U.S. Census Bureau)

Power Functions

A **power function** is a function of the form $y = ax^b$, where a and b are real numbers, $b \neq 0$.

Two simple power functions are $y = x^2$, whose graph is shown in Figure 3.23(a), and $y = x^3$, whose graph is shown in Figure 3.24(b). The function $y = x^2$ is also a basic quadratic function, called the **squaring function**. For positive values of x, it can represent the areas y (in square units) of squares that are x units on each side.

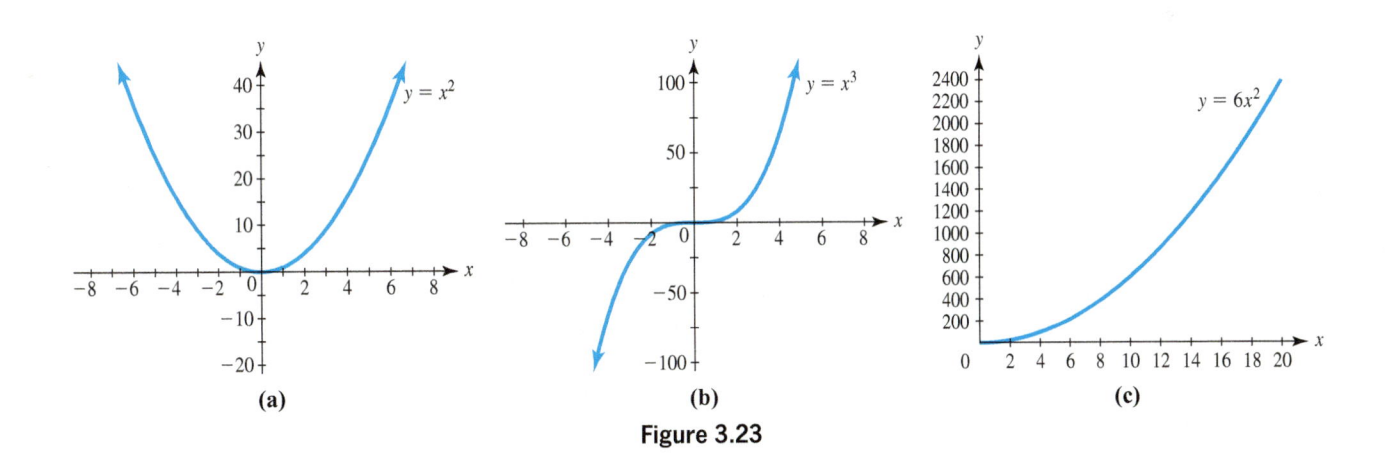

Figure 3.23

The function $y = x^3$ is called the **cubing function**. For positive values of x, it represents the volume (in cubic units) of cubes that are x units on each edge. Another power function is $y = 6x^2, x \geq 0$, which represents the surface area of a cube that is x units on an edge. (Figure 3.23(c))

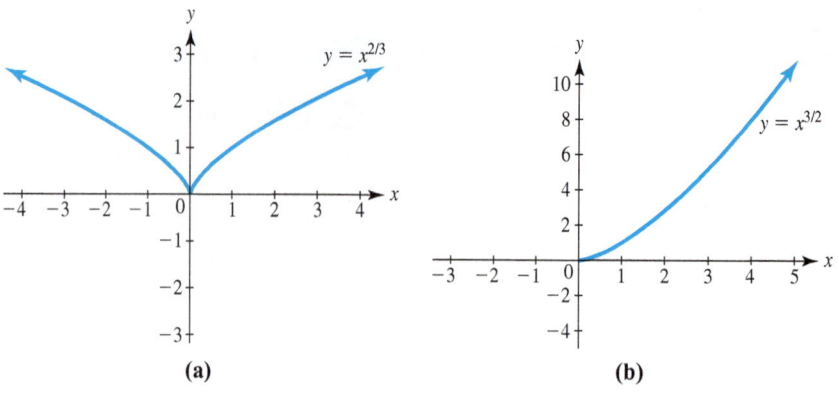

Figure 3.24

The graph of $y = x^{2/3}$ in Figure 3.24(a) shows that this function is defined for all real numbers. Notice that it decreases to the left of $x = 0$ and increases to the right of $x = 0$. The graph of $y = x^{3/2}$ in Figure 3.24(b) shows nothing to the left of $x = 0$. This is because $x^{3/2} = \sqrt{x^3}$, and the square root of any negative number is undefined.

Observe that the exponent in $y = x^{2/3}$ is a positive number less than 1, and that the function increases at a rate that is less than the rate of $y = x$ for values of x greater than 1. Because of this, we say that the graph is *concave down* in the first quadrant. We can see that the exponent in $y = x^{3/2}$ is a positive number greater than 1, and that the function increases at a rate that is greater than the rate of $y = x$ for values of x greater than 1. Because of this, we say that the graph is *concave up* in the first quadrant.

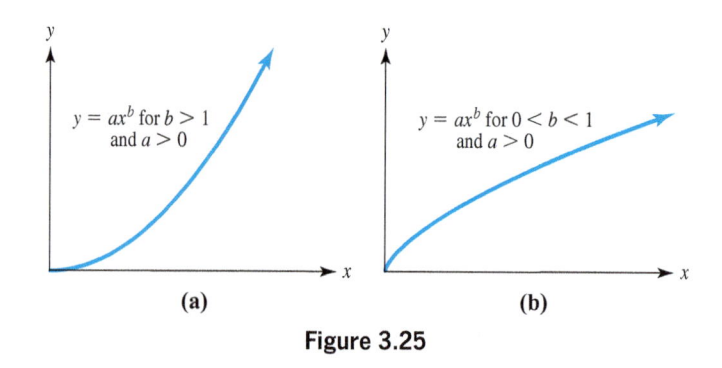

Figure 3.25

In general, if $a > 0$ and $x > 0$, the graph of $y = ax^b$ is concave up if $b > 1$ and concave down if $0 < b < 1$. Additionally, if a function is increasing and its graph is concave up on an interval, it is increasing at an increasing rate, and if it is increasing and its graph is concave down on an interval, it is increasing at a decreasing rate. Figures 3.25(a) and 3.25(b) show the first-quadrant portion of typical graphs of $y = ax^b$ for $a > 0$ and different values of b.

EXAMPLE 1 ▶ Distance an Object Falls

If an object is dropped from a 144-foot-tall building, the distance in feet that it travels after t seconds is given by the function $f(t) = 16t^2$ (neglecting air resistance).

a. What is the lower limit to the domain of this function in the context of the application?

b. Graph this function for $t = 0$ to $t = 6$.

c. Is the graph of the function concave up or concave down? How do you know?

d. How long does it take the object to hit the ground?

SOLUTION

a. Because time cannot be negative, we restrict the domain to $t \geq 0$.

b. The graph of the function is shown in Figure 3.26.

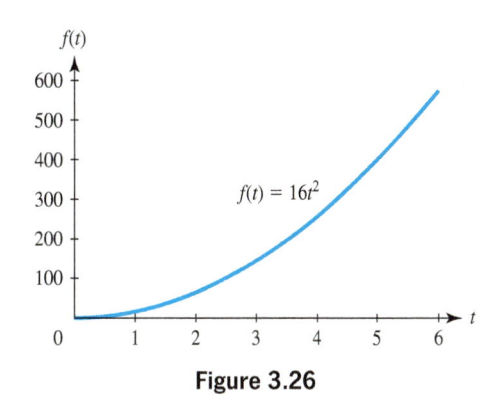

Figure 3.26

c. The graph is concave up, because the coefficient of t is positive and the power of t is greater than 1.

d. The object must travel 144 feet to reach the ground, so we solve the equation $16t^2 = 144$. We can do this algebraically or graphically. To solve algebraically, we use the square-root method:

$$16t^2 = 144$$
$$t^2 = 9 \quad \text{Divide both sides by 16.}$$
$$t = \pm 3 \quad \text{Take the square root of both sides, using a } \pm \text{ symbol.}$$

However, we are only interested in the nonnegative value for time, so it takes 3 seconds to hit the ground. To solve graphically, we enter $y_2 = 144$ in our graphing utility and find the intersection point (Figure 3.27).

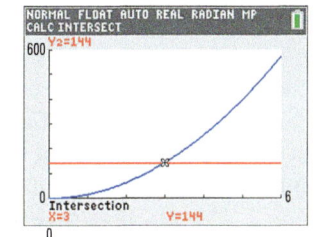

Figure 3.27

EXAMPLE 2 ▶ Foreign-Born Population

The percent of the U.S. population that is foreign-born for selected years from 1970 and projected to 2060 can be modeled by the function $y = 0.980x^{0.646}$, with x equal to the number of years after 1960 and y equal to the percent. (*Source*: U.S. Census Bureau)

a. Is the percent increasing or decreasing for the domain representing the period from 1970 through 2060?

b. Graph this function for values of x representing 1970 through 2060.

c. Is the graph of this function concave up or concave down?

d. What does the model project the percent of the U.S. population that is foreign-born to be in 2028?

SOLUTION

a. Because the exponent in the function is positive, the function is increasing.

b. The value of x that represents 1970 is $1970 - 1960 = 10$, and the value of x that represents 2060 is 100, so we graph the function over the interval $10 \leq x \leq 100$.

Evaluating $y = 0.980x^{0.646}$ at $x = 10$ and at $x = 100$ gives $y \approx 4.3$ and $y \approx 19.2$, so it is reasonable to use $0 \leq y \leq 20$ as the y-interval for the graph. This graph is found in Figure 3.28.

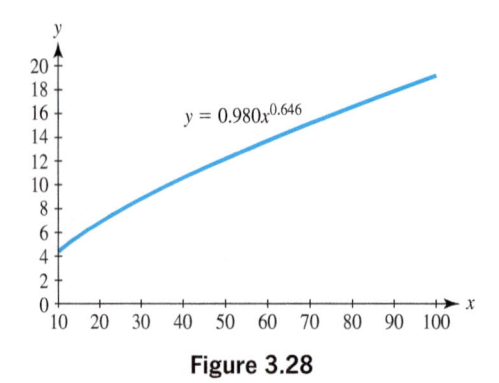

Figure 3.28

c. We can see that the graph of the function is concave down by observing its shape and by observing that the exponent of the function is a positive number less than 1.

d. To find what the percent of the U.S. population that is foreign-born will be in 2028, we evaluate the function at $x = 2028 - 1960 = 68$. This gives approximately 14.96, so the percent of the U.S. population that is foreign-born is projected to be 15%.

EXAMPLE 3 ▶ Wingspan of Birds

At 17 feet, the wingspan of a prehistoric bony-toothed bird may exceed that of any other bird that ever existed. The animal weighed about 64 pounds and soared the Chilean skies 5–10 million years ago. The wingspan of a bird can be estimated using the model

$$L = 2.43W^{0.3326}$$

where L is the wingspan in feet and W is the weight in pounds.
(*Source*: Discovery News)

a. Graph this function in the context of the application.

b. Use the function to compute the wingspan of a mute swan that weighs 40 pounds.

c. Use the function to approximate the weight of an albatross whose wingspan is 11 feet 4 inches.

d. Is the graph of this function concave up or concave down? What does this mean?

SOLUTION

a. It is appropriate to graph this model only in the first quadrant because both the weight and the wingspan must be nonnegative. The graph is shown in Figure 3.29(a).

b. To compute the wingspan of a mute swan that weighs 40 pounds, we evaluate the function with an input of 40, giving an output of $L = 2.43(40)^{0.3326} = 8.288$. Thus, the wingspan of the mute swan is approximately 8.3 feet.

c. If the wingspan of an albatross is 11 feet 4 inches, we graphically solve the equation $11.333 = 2.43W^{0.3326}$. Using the intersection method (Figure 3.29(b)) gives the weight of the albatross to be approximately 102.48 pounds.

d. The graph is concave down. This tells us that as the weight of the bird increases, the wingspan increases, but at a decreasing rate.

3.3 Power and Root Functions   **207**

(a) Graph showing $L = 2.43W^{0.3326}$ for $0 \leq W \leq 70$.

(b) Graph showing $L = 2.43W^{0.3326}$ intersecting $L = 11.333$ at $(102.48, 11.333)$.

Figure 3.29

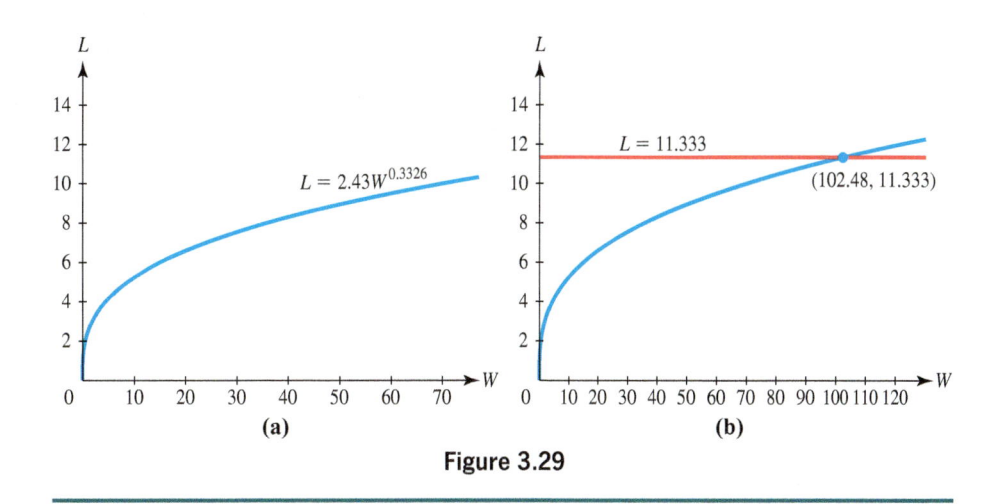

Note that the power in $L = 2.43W^{0.3326}$ is 0.3326, which can be written in the form $\dfrac{3326}{10{,}000}$ and is therefore a rational number.

EXAMPLE 4 ▶ Kepler's Third Law

Johannes Kepler discovered a simple relationship between the average distance of a planet from the Sun (called its semi-major axis, A, measured in astronomical units [AU]) and the amount of time it takes a planet to orbit the Sun once (called its orbital period, P, measured in years). For objects orbiting the Sun,

$$P = A^{3/2}$$

a. Graph this function for values $0 \leq A \leq 40$.

b. If Jupiter's semi-major axis is 5.2028 AU, find its orbital period.

c. Saturn has an orbital period of 29.458 years. What is its distance from the Sun?

SOLUTION

a. The graph of the function is shown in Figure 3.30(a).

b. Substituting 5.2028 for A, we get

$$P = (5.2028)^{3/2} \approx 11.867$$

So Mercury's orbital period is about 11.867 years.

c. To find Saturn's distance from the Sun, we solve $x^{3/2} = 29.458$. We do this graphically in Figure 3.30(b) and find the distance to be about 9.538 AU.

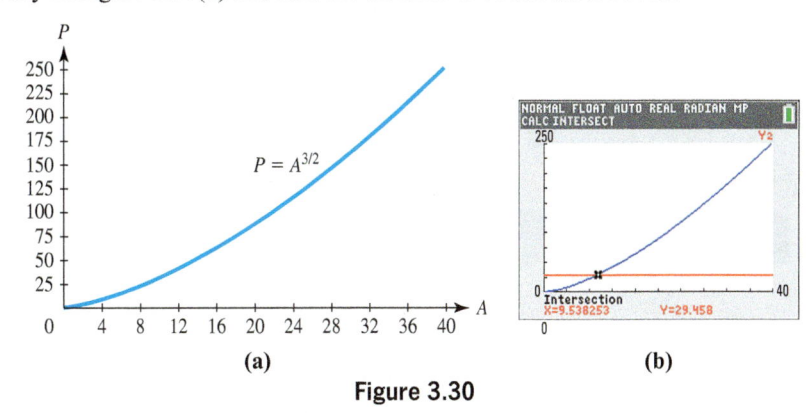

Figure 3.30

Power Functions with Negative Exponents

The power functions we have discussed so far have had positive exponents. What effect does a negative exponent have on a power function? Figures 3.31(a)–(d) show the graphs of some power functions with negative integer exponents.

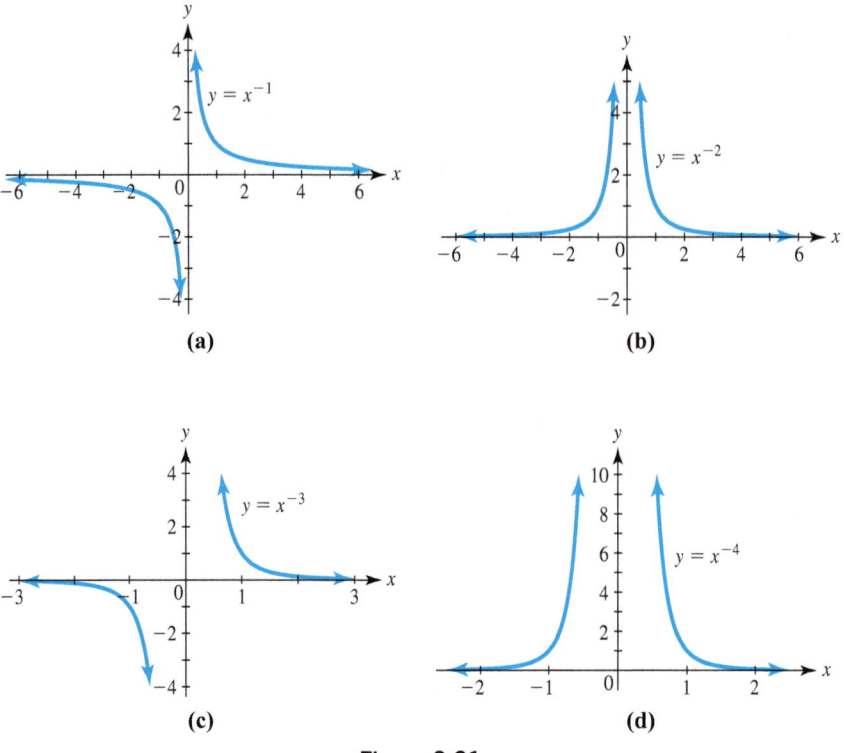

Figure 3.31

(a) $y = x^{-1}$

(b) $y = x^{-2}$

(c) $y = x^{-3}$

(d) $y = x^{-4}$

Note that, with each of these functions, x cannot equal 0, so 0 is not in the domain of these functions. The graphs show that as x gets close to 0, $|y|$ gets large and the graph approaches but does not touch the y-axis. We say that the y-axis is a **vertical asymptote** of these graphs. Similarly, the x-axis is a **horizontal asymptote**.

When the power of the function is odd, as in Figures 3.31(a) and (c), the graph resembles ⌐ or ¬. When the power of the function is even, as in Figures 3.31(b) and (d), the graph resembles ⊔ or ⊔.

The function $y = x^{-1} = \dfrac{1}{x}$, shown in Figure 3.31(a), is a special power function called the **reciprocal function.**

EXAMPLE 5 ▶ Power Function with Negative Exponent

For the function $f(x) = \dfrac{1}{x^7}$, complete the following.

a. Is this a power function?

b. If possible, evaluate $f(0), f(-2),$ and $f(1/2)$.

c. Give the domain and range of the function.

d. Graph the function on the interval $[-1.4, 1.4]$ by $[-10, 10]$.

e. Does $f(x)$ ever reach 0? How does this affect the graph?

SOLUTION

a. $f(x) = \dfrac{1}{x^7}$ is a power function because it can be written as $f(x) = x^{-7}$.

b. If $x = 0$, the function is undefined. The values of the function at $x = -2$ and $x = \frac{1}{2}$ are $f(-2) = \dfrac{1}{(-2)^7} \approx -0.0078$ and $f(1/2) = \dfrac{1}{(1/2)^7} \approx -128$.

c. The domain and range are both all real numbers except 0.

d. The graph is shown in Figure 3.32.

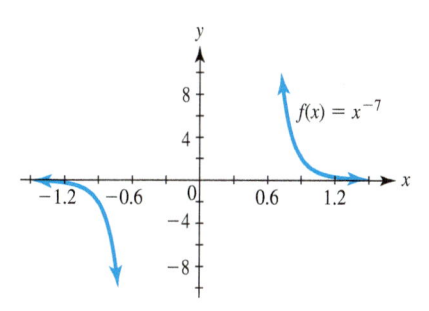

Figure 3.32

e. No, because $\dfrac{1}{x^7}$ can never equal 0. This means that the graph has a horizontal asymptote at $y = 0$.

Note that the values of $y = \dfrac{1}{x}$ get very small as $|x|$ gets large, and the graph approaches but does not touch the x-axis. So the x-axis is a **horizontal asymptote** for this graph.

Root Functions

Functions with rational powers can also be written with radicals. For example, $y = x^{1/3}$ can be written in the form $y = \sqrt[3]{x}$, and $y = x^{1/2}$ can be written in the form $y = \sqrt{x}$. Functions like $y = \sqrt[3]{x}$ and $y = \sqrt{x}$ are special power functions called **root functions**.

> **Root Functions**
>
> A root function is a function of the form $y = ax^{1/n}$, or $y = a\sqrt[n]{x}$, where n is an integer, $n \geq 2$.

The graphs of $y = \sqrt{x}$ and $y = \sqrt[3]{x}$ are shown in parts (a) and (b) of Figure 3.33. Note that the domain of $y = \sqrt{x}$ is $x \geq 0$ because $\sqrt{x}$ is undefined for negative values of x. Note also that these root functions can be written as power functions.

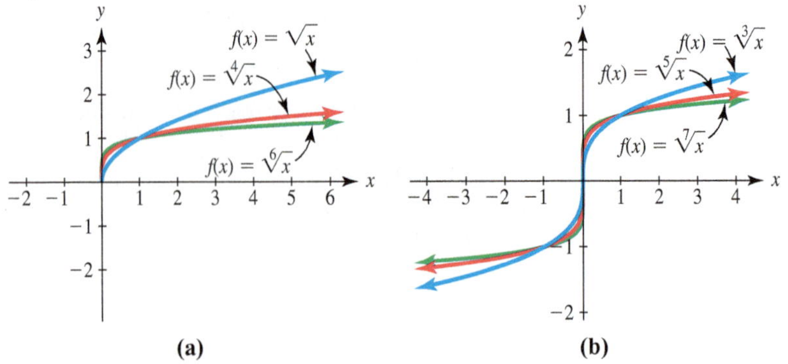

210   Chapter 3   Quadratic, Piecewise-Defined, and Power Functions

(a)

(b)

Figure 3.33

All root functions of the form $f(x) = \sqrt[n]{x}$, with n an even positive integer, have the same shape as shown in Figure 3.33(a); the domain and the range are both $[0, \infty)$. Root functions with n odd have the same shape as shown in Figure 3.33(b); the domain and range are both $(-\infty, \infty)$.

Solution of Power and Root Equations

If an equation has the form $k = ax^n$, where n is an integer, the solution (if it exists) can be found by rewriting the equation with x^n on one side and then taking the nth root of both sides of the equation. If n is even, you must take the $\pm$ nth root of both sides of the equation.

EXAMPLE 6 ▶ Power Equations

Solve the following equations.

a. $540 = 5x^3$ **b.** $4x^{-5} = 512$ **c.** $4(2x - 3)^4 = 2500$

SOLUTION

a. Dividing both sides of the equation by 3 and then taking the cube root of both sides gives the solution.

$$540 = 5x^3$$
$$108 = x^3$$
$$x = \sqrt[3]{108} = 3\sqrt[3]{4}.$$

b. The first step in solving for x is rewriting the equation with a positive power.

$$4x^{-5} = 512$$
$$\frac{4}{x^5} = 512$$
$$\frac{4}{512} = x^5$$
$$\frac{1}{128} = x^5$$
$$x = \sqrt[5]{\frac{1}{128}} = \frac{1}{\sqrt[5]{128}} = \frac{1}{2\sqrt[5]{4}}$$

c. If an equation has a term containing the variable raised to a power, we first solve for the term to the power, as follows.

$$4(2x - 3)^4 = 2500$$
$$(2x - 3)^4 = 625$$
$$2x - 3 = \pm\sqrt[4]{625}$$
$$2x - 3 = \pm 5$$
$$x = 4 \quad \text{or} \quad x = -1$$

An equation of the form $a\sqrt[n]{x} = k$, where n is a positive integer, is called a **root equation**. If an equation of this form can be solved, rewriting the equation with $\sqrt[n]{x}$ on one side of the equation and then raising both sides of the equation to the positive power n gives all *possible* solutions of the equation. But it is necessary to check all solutions in the original equation and eliminate those that do not check, because raising both sides of such an equation by a power may give a new equation that has *more* solutions than the original equation. For example, squaring both sides of $x = -3$ gives $x^2 = 9$. But $x^2 = 9$ has the 2 solutions $x = 3$ and $x = -3$, and $x = 3$ is not a solution to the original equation.

EXAMPLE 7 ▶ Solving Root Equations

Solve the following root equations.

a. $3\sqrt[3]{x} = 12$ **b.** $5\sqrt[4]{x} = -15$ **c.** $3\sqrt{2x - 1} = 9$

SOLUTION

a. As with power equations, we first solve the equation for the radicand containing the variable.

$$3\sqrt[3]{x} = 12$$
$$\sqrt[3]{x} = 4$$
$$x = 4^3 = 64$$

Checking the solution shows that 64 is the solution.

$$3\sqrt[3]{64} = 3(4) = 12$$

b. Checking the solution is important with this problem.

$$5\sqrt[4]{x} = -15$$
$$\sqrt[4]{x} = -3$$
$$x = 81$$

Checking the solution $x = 81$ in the original equation shows that 81 is not a solution. In fact, there is no solution to this equation.

$$5\sqrt[4]{81} = 5(3) = 15 \neq -15$$

c. If an algebraic expression is contained in the radicand of a root equation, the first step in solving the equation is rewriting the equation with the radicand on one side of the equation.

$$3\sqrt{2x - 1} = 9$$
$$\sqrt{2x - 1} = 3$$
$$2x - 1 = 9$$
$$2x = 10$$
$$x = 5$$

212 Chapter 3 Quadratic, Piecewise-Defined, and Power Functions

Checking shows that $x = 5$ is the solution to the equation.
$$3\sqrt{2(5) - 1} = 3\sqrt{9} = 3(3) = 9$$

Power equations and root equations can be solved with calculators and with spreadsheets in the same way that linear and quadratic equations are solved.

Direct Variation as a Power

In Section 2.1, we discussed direct variation. If a quantity y varies directly as a power of x, we say that y is directly proportional to the nth power of x.

> ### Direct Variation as the nth Power
> A quantity y varies directly as the nth power $(n > 0)$ of x if there is a constant k such that
> $$y = kx^n$$
> The number k is called the constant of variation or the constant of proportionality.

EXAMPLE 8 ▶ Production

In the production of an item, the number of units of one raw material required varies as the cube of the number of units of a second raw material that is required. Suppose 500 units of the first and 5 units of the second raw material are required to produce 100 units of the item. How many units of the first raw material are required if the number of units produced requires 10 units of the second raw material?

SOLUTION

If x is the number of units of the second raw material required and y is the number of units of the first raw material required, then y varies as the cube of x, or
$$y = kx^3$$

Because $y = 500$ when $x = 5$, we have
$$500 = k \cdot 5^3, \quad \text{or} \quad k = 4$$

Then $y = 4x^3$ and $y = 4(10^3) = 4000$. Thus, 4000 units of the first raw material are required if the number of units produced requires 10 units of the second raw material.

SKILLS CHECK 3.3

Answers that are not seen can be found in the answer section at the back of the text.

1. Which of the following are power functions?
   a. $y = 3x^4$
   b. $d = q^4$
   c. $f = 3^x$
   d. $g(x) = x^{1/4}$ a, b, d

2. Rewrite each of the following as power functions of the form $y = ax^b$.
   a. $y = \sqrt[3]{x}$ $y = x^{1/3}$
   b. $y = \dfrac{1}{x}$ $y = x^{-1}$
   c. $y = 3\sqrt[4]{x^5}$ $y = 3x^{5/4}$
   d. $y = \dfrac{1}{x^2}$ $y = x^{-2}$

3. Determine whether the function $y = 4x^3$ is increasing or decreasing for

   a. $x < 0$. Increasing b. $x > 0$. Increasing

4. Determine whether the function $y = -3x^4$ is increasing or decreasing for

   a. $x < 0$. Increasing b. $x > 0$. Decreasing

5. Complete the table of values for the function $y = x^{3.5}$, if possible. Round to one decimal place.

x	− 1	0	1	2	10
$y = x^{3.5}$					

   Undef, 0, 1, 11.3, 3162.3

6. Complete the table of values for the function $y = x^{-1.5}$, if possible. Round to two decimal places.

x	− 1	0	1	2	10
$y = x^{-1.5}$					

   Undef, undef, 1, 0.35, 0.03

7. For which value(s) of x in Exercise 5 is the function undefined? Explain. −1; The even root of a negative number is undefined.

8. For which value(s) of x in Exercise 6 is the function undefined? Explain. −1; The even root of a negative number is undefined; 0, Division by 0 is undefined.

For each of the functions in Exercises 9–12, determine if the function is concave up or concave down in the first quadrant.

9. $y = x^{1/2}$ Concave down 10. $y = x^{3/2}$ Concave up

11. $y = x^{1.4}$ Concave up 12. $y = x^{0.6}$ Concave down

13. Give an example of a power function that is increasing and concave down. Answers may vary.

14. Give an example of a power function that is decreasing and concave up. Answers may vary.

15. Graph the function $y = x^{3.5}$. What are the domain and range? $D: [0, \infty)$ $R: [0, \infty)$

16. Graph the function $y = x^{-1.5}$. What are the domain and range? Does this graph have any asymptotes? If so, where? $D: (0, \infty)$ $R: (0, \infty)$; yes, $x = 0$ and $y = 0$

17. Graph the function $y = 2\sqrt[3]{x}$. What are the domain and range? $D: (-\infty, \infty)$ $R: (-\infty, \infty)$

18. Graph the function $f(x) = -3\sqrt[4]{x}$. What are the domain and range? $D: [0, \infty)$ $R: (-\infty, 0]$

In Exercises 19–30, solve the equation and check for extraneous solutions.

19. $3x^3 = 120$ $2\sqrt[3]{5}$

20. $\frac{1}{8}x^{-5} = 384$ $\frac{1}{4\sqrt[5]{3}}$

21. $3(3x - 2)^4 = 768$ 2 or $-\frac{2}{3}$

22. $x^{2/3} - \frac{3}{4} = -\frac{1}{2}$ $\pm\frac{1}{8}$

23. $3\sqrt[4]{x} = 12$ 256

24. $5\sqrt[5]{x} = 10$ 32

25. $5\sqrt[4]{x} = -25$ No solution

26. $4\sqrt{2x - 3} = 36$ 42

27. $x^{1/4} = 4$ 256

28. $x^{1/3} = \frac{1}{4}$ 1/64

29. $x^{2/3} = 16$ 64

30. $x^{2/5} = 4$ 32

31. Suppose that S varies directly as the 2/3 power of T, and that $S = 64$ when $T = 64$. Find S when $T = 8$. 16

32. Suppose that y varies directly as the square root of x, and that $y = 16$ when $x = 4$. Find x when $y = 24$. 9

33. Let T be directly proportional to the 1/3 power of y and suppose that $T = 6$ when $y = 27$.

    a. Find the constant of proportionality. 2
    b. Find y when $T = 10$. 125

34. Let Q be directly proportional to the 3/4 power of s and suppose that $Q = 16$ when $s = 16$.

    a. Find the constant of proportionality. 2
    b. Find s when $Q = 54$. 81

EXERCISES 3.3

Answers that are not seen can be found in the answer section at the back of the text.

35. **Taxi Miles** The Inner City Taxi Company estimated, on the basis of collected data, that the number of taxi miles driven each day can be modeled by the function $Q = 489L^{0.6}$, when the company employs L drivers per day.

    a. Graph this function for $0 \le L \le 35$.
    b. How many taxi miles are driven each day if there are 32 drivers employed? 3912
    c. Does this model indicate that the number of taxi miles increases or decreases as the number of drivers increases? Is this reasonable? Increases; yes
    d. Is the graph of this function concave up or concave down? What does this mean? Concave down; it increases at a decreasing rate.

36. **Female Physicians** Representation of females in medicine continues to show steady increases. The number of female physicians can be modeled by $F(x) = 193x^{0.220}$,

where x is the number of years after 2010 and $F(x)$ is the number of female physicians in thousands.

a. What type of function is this? Power

b. What is $F(10)$? What does this mean? 320.3; in 2020, there were 320,300 female physicians.

c. How many female physicians will there be in 2027, according to the model? 359,962

(*Source*: U.S. Census Bureau Association)

37. *U.S. Households* The percent of total households that have married couples for selected years from 1982 and projected to 2050 can be modeled by $y = 66.164x^{-0.094}$, with x representing the number of years after 1980 and y representing the percent.
(*Source*: U.S. Census Bureau)

a. Is the function increasing or decreasing for the domain representing 1982 through 2050? Decreasing

b. Graph the function for $0 \le x \le 80$.

c. What is the predicted percent of total households that have married couples in 2042, according to this model? Approximately 44.9%

38. *U.S. Population* The U.S. population can be modeled by the function $y = 165.6x^{1.345}$, where y is in thousands and x is the number of years after 1800.

a. What was the population in 1960, according to this model? 152,617,000

b. Is the graph of this function concave up or concave down? What does this mean? Concave up; increasing at an increasing rate

c. Use numerical or graphical methods to find when the model estimates the population was 93,330,000. 1911

(*Source*: U.S. Census Bureau)

39. *Production Output* The monthly output of a product (in units) is given by $P = 1200x^{5/2}$, where x is the capital investment in thousands of dollars.

a. Graph this function for x from 0 to 10 and P from 0 to 200,000.

b. Is the graph concave up or concave down? What does this mean? Concave up; increasing at an increasing rate

40. *Purchasing Power* The purchasing power of a 2012 dollar for selected years from 2012 and projected to 2050 can be modeled by the function $y = 1.595x^{-0.343}$, where x is the number of years after 2010.

a. Graph this function, with x representing the years 2012 through 2050.

b. Is this function increasing or decreasing? Decreasing

c. What does the model predict the purchasing power of a 2012 dollar will be in 2050? $.45, or 45 cents

d. Does the model predict that the purchasing power of a 2012 dollar will ever be $0? No

41. *Personal Income* The total personal income in the United States (in billions of dollars) for selected years from 1960 and projected to 2024 can be modeled by $y = 2.15x^{2.12}$, where x equals the number of years after 1960.

a. Use the model to predict the total personal income in 2029. Approximately $17,013.7 billion

b. Graph this function for the years 1960 through 2030.

c. Use numerical or graphical methods to estimate when the total personal income will reach $18.620 trillion. In 2032

42. *Diabetes* The projected percent of the U.S. adult population with diabetes (diagnosed and undiagnosed) can be modeled by $y = 4.97x^{0.495}$, where x is the number of years after 2000.

a. Does this model indicate that the percent of the U.S. adult population with diabetes is projected to increase or decrease? Increase

b. What percent is projected for 2022? 23.0%

c. In what year does this model project the percent to be 17? 2012

(*Source*: Centers for Disease Control and Prevention)

43. *Artificial Intelligence* The revenue (in billions of dollars) from artificial intelligence for enterprise applications in North America for the years from 2016 through 2025 can be modeled by $y = 0.132x^{1.921}$, where x is the number of years after 2015.

a. Graph the function for the years from 2015 through 2030.

b. Is the function increasing or decreasing over this period of time? Increasing

c. Is the graph concave up or down? up

d. What is the revenue in 2037? Approximately $50.046 billion

44. *Worldwide Internet Users* Using data from 2014 and projected to 2025, the percent of people in the world who are Internet users can be modeled by the function $p(x) = 27.334\sqrt[10]{x^3}$, with x representing the number of years after 2010 and p representing the percent.
(*Source*: eMarketer)

a. Graph the function for the years 2014 through 2026.

b. Is the graph concave up or down? Down

c. If the model is valid, what is the percent of people in the world who will be Internet users in 2033? Approximately 70%

d. Use technology to determine when the percent of people in the world who will be Internet users is projected to be 59%. 2023

45. **Cloud Revenue** Many services are provided through the Cloud, and public vendor revenue is projected to grow rapidly. The global Cloud revenue from 2012 and projected to 2026, in billions of dollars, can be modeled by $f(x) = 7.53\sqrt{x^3}$, where x is equal to the number of years after 2012 and y is equal to the revenue in billions of dollars.

    a. Graph the model for values of x that represent 2012 through 2026.

    b. Is the graph of the function concave up or down? Up

    c. What does the model predict the revenue will be in 2032? $673.5 billion

    d. When does the model predict the revenue will be $941.25 billion? 2037

    (*Source:* Wikibon Server SAN & Cloud Research Projects)

46. **Harvesting** A farmer's main cash crop is tomatoes, and the tomato harvest begins in the month of May. The number of bushels of tomatoes harvested on the xth day of May is given by the equation
    $$y = 6\sqrt{(x+1)^3}.$$

    a. How many bushels did the farmer harvest on May 8? 162

    b. In what day of May does the farmer harvest 384 bushels of tomatoes? May 15

47. **Investing** If money is invested for 3 years with interest compounded annually, the future value of the investment varies directly as the cube of $1 + r$, where r is the annual interest rate. If the future value of the investment is $6298.56 when the interest rate is 8%, what rate gives a future value of $5955.08? 6%

48. **Investing** If money is invested for 4 years with interest compounded annually, the future value of the investment varies directly as the fourth power of $1 + r$, where r is the annual interest rate. If the future value of the investment is $17,569.20 when the interest rate is 10%, what rate gives a future value of $24,883.20? 20%

3.4 Piecewise-Defined Functions and Absolute Value Functions

KEY OBJECTIVES

- Evaluate and graph piecewise-defined functions
- Graph the greatest integer function
- Graph the absolute value function
- Solve absolute value equations

SECTION PREVIEW **Residential Power Costs**

The data in Table 3.5 give the rates that Georgia Power Company charges its residential customers for electricity during the months of June through September, excluding fuel adjustment costs and taxes. The monthly charges can be modeled by a **piecewise-defined function**, as we will see in Example 3. Piecewise-defined functions are used in applications in the life, social, and physical sciences, as well as in business, when there is not a single function that accurately represents the application.

A special piecewise-defined function is the **absolute value function**, which has many applications in mathematics. In this section, we evaluate, graph, and apply piecewise-defined and absolute value functions, and we will evaluate and graph the **greatest integer function**.

Table 3.5

Monthly Kilowatt-hours (kWh)	Monthly Charge
0 to 650	$0.05658 per kWh
More than 650, up to 1000	$36.78 plus $0.09398 per kWh above 650
More than 1000	$69.67 plus $0.09727 per kWh above 1000

Piecewise-Defined Functions

It is possible that a set of data cannot be modeled with a single equation. We can use a piecewise-defined function when there is not a single function that accurately represents the situation. A piecewise-defined function is so named because it is defined with different pieces for different parts of its domain rather than one equation.

EXAMPLE 1 ▶ Postage

The postage paid for a first-class letter "jumps" by 15 cents for each ounce after the first ounce, but does not increase until the weight increases by 1 ounce. Table 3.6 gives the 2019 postage for letters up to 3.5 ounces. Write a piecewise-defined function that models the price of postage and graph the function.

Table 3.6

Weight x (oz)	Postage y (cents)
$0 < x \leq 1$	55
$1 < x \leq 2$	70
$2 < x \leq 3$	85
$3 < x \leq 3.5$	100

SOLUTION

The function that models the price P of postage for x ounces, where x is between 0 and 3.5, is

$$P(x) = \begin{cases} 55 & \text{if } 0 < x \leq 1 \\ 70 & \text{if } 1 < x \leq 2 \\ 85 & \text{if } 2 < x \leq 3 \\ 100 & \text{if } 3 < x \leq 3.5 \end{cases}$$

The graph of this function (Figure 3.34) shows that each output is a constant with discontinuous "steps" at $x = 1, 2,$ and 3. This is a special piecewise-defined function, called a **step function**. Other piecewise-defined functions can be defined by polynomial functions over limited domains.

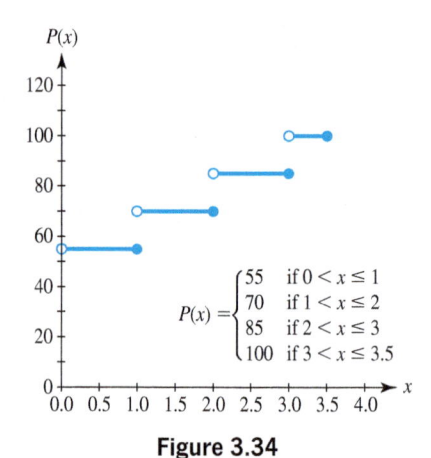

Figure 3.34

Note: An open circle indicates that the point is *not* on the graph. A closed circle indicates that the point *is* on the graph.

EXAMPLE 2 ▶ Piecewise-Defined Function

Graph the function

$$f(x) = \begin{cases} 5x + 2 & \text{if } 0 \leq x < 3 \\ x^3 & \text{if } 3 \leq x \leq 5 \end{cases}$$

Table 3.7(a)

x	0	1	2	2.99
f(x)	2	7	12	16.95

Table 3.7(b)

x	3	4	5
f(x)	27	64	125

SOLUTION

To graph this function, we can construct a table of values for each of the pieces. Table 3.7(a) gives outputs of the function for some sample inputs x on the interval $[0, 3)$; on this interval, $f(x)$ is defined by $f(x) = 5x + 2$. Table 3.7(b) gives outputs for some sample inputs on the interval $[3, 5]$; on this interval, $f(x)$ is defined by $f(x) = x^3$.

Plotting the points from Table 3.7(a) and connecting them with a smooth curve gives the graph of $y = f(x)$ on the x-interval $[0, 3)$ (Figure 3.35). The open circle on this piece of the graph indicates that this piece of the function is not defined for $x = 3$. Plotting the points from Table 3.7(b) and connecting them with a smooth curve gives the graph of $y = f(x)$ on the x-interval $[3, 5]$ (Figure 3.35). The closed circles on this piece of the graph indicate that this piece of the function is defined for $x = 3$ and $x = 5$. The graph of $y = f(x)$, shown in Figure 3.35, consists of these two pieces.

$$f(x) = \begin{cases} 5x + 2 & \text{if } 0 \leq x < 3 \\ x^3 & \text{if } 3 \leq x \leq 5 \end{cases}$$

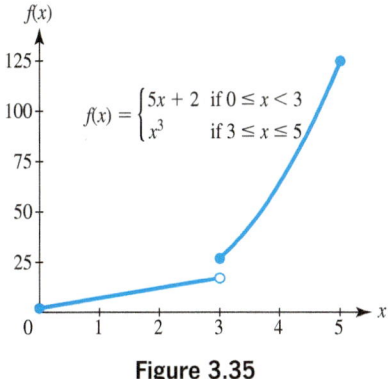

Figure 3.35

EXAMPLE 3 ▶ Residential Power Costs

Excluding fuel adjustment costs and taxes, the rates Georgia Power Company charges its residential customers for electricity during the months of June through September are shown in Table 3.8.

Table 3.8

Monthly Kilowatt-hours (kWh)	Monthly Charge
0 to 650	$0.05658 per kWh
More than 650, up to 1000	$36.78 plus $0.09398 per kWh above 650
More than 1000	$69.67 plus $0.09727 per kWh above 1000

a. Write the piecewise-defined function C that gives the monthly charge for residential electricity, with input x equal to the monthly number of kilowatt-hours.

b. Find $C(950)$ and explain what it means.

c. Find the charge for using 1560 kWh in a month.

d. Use a graph of the function to determine how many kilowatt-hours would be used if the monthly charge were $122.87.

SOLUTION

a. The monthly charge is given by the function

$$C(x) = \begin{cases} 0.05658x & \text{if } 0 \leq x \leq 650 \\ 36.78 + 0.09398(x - 650) & \text{if } 650 < x \leq 1000 \\ 69.67 + 0.09727(x - 1000) & \text{if } x > 1000 \end{cases}$$

where $C(x)$ is the charge in dollars for x kWh of electricity.

b. To evaluate $C(950)$, we must determine which "piece" defines the function when $x = 950$. Because 950 is between 650 and 1000, we use the "middle piece" of the function:

$$C(950) = 36.78 + 0.09398(950 - 650) = 64.974$$

Companies regularly round charges *up* to the next cent if any part of a cent is due. This means that if 950 kWh are used in a month, the bill is $64.98.

c. To find the charge for 1560 kWh, we evaluate $C(1560)$, using the "bottom piece" of the function because $1560 > 1000$. Evaluating gives

$$C(1560) = 69.27 + 0.09727(1560 - 1000) = 124.1412$$

so the charge for the month is $124.15.

d. The graph of the function is shown in Figure 3.36(a). We graph the function $y = 249.62$ and find the point of intersection of the two graphs. See Figure 3.36(b). Note that the intersection occurs with the third piece of the piecewise-defined function, where $x > 1000$. The intersection point shows that when 2850 kWh are used, the monthly cost is $249.62.

(a) (b)

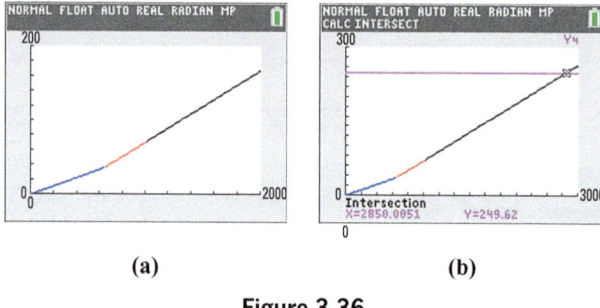

Figure 3.36

EXAMPLE 4 ▶ Wind-Chill Factor

Wind-chill factors are used to measure the effect of the combination of temperature and wind speed on human comfort by providing equivalent air temperatures with no wind blowing. One formula that gives the wind-chill factor for a 30°F temperature and a wind with velocity V in miles per hour is

$$W = \begin{cases} 30 & \text{if } 0 \leq V < 4 \\ 1.259V - 18.611\sqrt{V} + 62.255 & \text{if } 4 \leq V \leq 55.9 \\ -6.5 & \text{if } V > 55.9 \end{cases}$$

a. Find the wind-chill factor for the 30°F temperature if the wind is 40 mph.

b. Find the wind-chill factor for the 30°F temperature if the wind is 65 mph.

c. Graph this function for $0 \leq V \leq 70$.

d. What are the domain and range of the function graphed in (c)?
(*Source*: The National Weather Service)

SOLUTION

a. Because $V = 40$ is in the interval $4 \leq V \leq 55.9$, the wind-chill factor is

$$1.259(40) - 18.611\sqrt{40} + 62.255 \approx -5.091 \approx -5$$

This means that, if the wind is 40 mph, a temperature of 30°F would actually feel like −5°F.

b. Because $V = 65$ is in the interval $V > 55.9$, the wind-chill factor is −6.5. This means that, if the wind is 65 mph, a temperature of 30°F would actually feel like −6.5°F.

c. The graph is shown in Figure 3.37.

$$W = \begin{cases} 30 & \text{if } 0 \leq V < 4 \\ 1.259V - 18.611\sqrt{V} + 62.255 & \text{if } 4 \leq V \leq 55.9 \\ -6.5 & \text{if } V > 55.9 \end{cases}$$

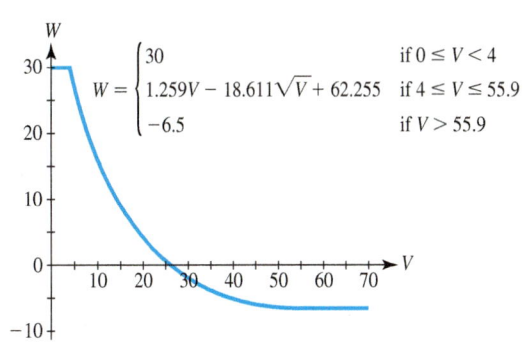

Figure 3.37

d. The domain is specified to be $0 \leq V \leq 70$, and the range, which is the set of outputs for W, is $-6.5 \leq W \leq 30$. That is, for wind speeds between 0 and 70 mph, the wind-chill factor is no lower than $-6.5°F$ and no higher than $30°F$.

EXAMPLE 5 ▶ Service Calls

The cost of weekend service calls by Airtech Services is shown by the graph in Figure 3.38. Write a piecewise-defined function that represents the cost for service during the first five hours, as defined by the graph.

SOLUTION

The graph indicates that there are three pieces to the function, and each piece is a constant function. For t-values between 0 and 1, $C = 330$; for t-values between 1 and 2, $C = 550$; and for t-values between 2 and 5, $C = 770$. The open and closed circles on the graph tell us where the endpoints of each interval of the domain are defined. Thus, the cost function, with C in dollars and t in hours, is

$$C = \begin{cases} 330 & \text{if } 0 < t \leq 1 \\ 550 & \text{if } 1 < t \leq 2 \\ 770 & \text{if } 2 < t \leq 5 \end{cases}$$

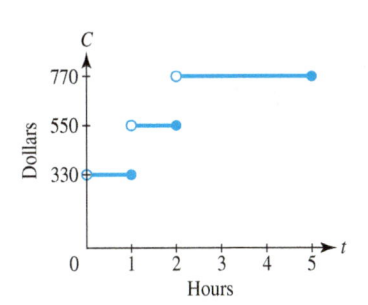

Figure 3.38

Greatest Integer Function

A special piecewise-defined function used in mathematics is the **greatest integer function**, denoted $f(x) = [\![x]\!]$. This function, also a step function, is defined as

$[\![x]\!]$ is the greatest integer less than or equal to x

Some examples of the greatest integer are

$$[\![5.8]\!] = 5, \quad [\![7]\!] = 7, \quad [\![-4.7]\!] = -5, \quad [\![-9]\!] = -9$$

The graph of $f(x) = [\![x]\!]$ is shown in Figure 3.39(a). Note that the domain of $f(x) = [\![x]\!]$ is $(-\infty, \infty)$ and the range is $\{\ldots, -3, -2, -1, 0, 1, 2, 3, \ldots\}$.

The graph $f(x) = [\![x]\!]$ using a graphing calculator is shown in Figure 3.39(b). The **TI-83** and the **TI-84** designate this **function** by using $Y1 = \text{int}(x)$ and it is found in the MATH NUM menu. Note that in some calculators you may need to set the mode to Dot so that the pieces of the graph are not connected.

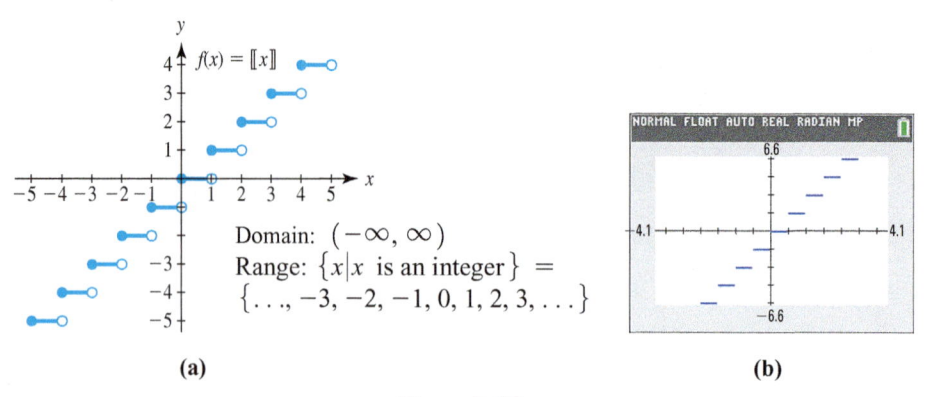

Figure 3.39

Absolute Value Function

We can construct a new function by combining two functions into a special piecewise-defined function. For example, we can write a function with the form

$$f(x) = \begin{cases} x & \text{if } x \geq 0 \\ -x & \text{if } x < 0 \end{cases}$$

This function is called the **absolute value function**, which is denoted by $f(x) = |x|$ and is derived from the definition of the absolute value of a number. Recall that the definition of the absolute value of a number is

$$|x| = \begin{cases} x & \text{if } x \geq 0 \\ -x & \text{if } x < 0 \end{cases}$$

To graph $f(x) = |x|$, we graph the portion of the line $y_1 = x$ for $x \geq 0$ (Figure 3.40(a)) and the portion of the line $y_2 = -x$ for $x < 0$ (Figure 3.40(b)). When these pieces are joined on the same graph (Figure 3.40(c)), we have the graph of $y = |x|$.

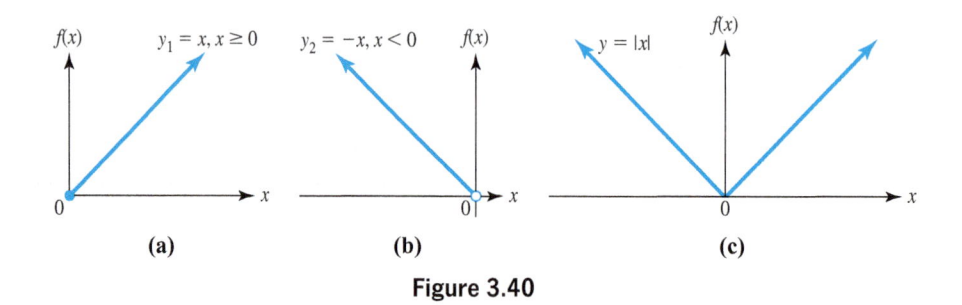

Figure 3.40

Note that the absolute value function $f(x) = |x|$ has domain $(-\infty, \infty)$ and range $[0, \infty)$. It is decreasing on the interval $(-\infty, 0)$ and increasing on the interval $(0, \infty)$.

Solving Absolute Value Equations

We now consider the solution of equations that contain absolute value symbols, called **absolute value equations**. Consider the equation $|x| = 5$. We know that $|5| = 5$ and $|-5| = 5$. Therefore, the solution to $|x| = 5$ is $x = 5$ or $x = -5$. Also, if $|x| = 0$, then x must be 0. Finally, because the absolute value of a number is never negative, we cannot solve $|x| = a$ if a is negative. We generalize this as follows.

3.4 Piecewise-Defined Functions and Absolute Value Functions

> **Absolute Value Equation**
>
> If $|x| = a$ and $a > 0$, then $x = a$ or $x = -a$.
> There is no solution to $|x| = a$ if $a < 0$; $|x| = 0$ has solution $x = 0$.

If one side of an equation is a function contained in an absolute value and the other side is a nonnegative constant, we can solve the equation by using the method above.

EXAMPLE 6 ▶ Absolute Value Equations

Solve the following equations.

a. $|x - 3| = 9$ **b.** $|2x - 4| = 8$ **c.** $|x^2 - 5x| = 6$

SOLUTION

a. If $|x - 3| = 9$, then $x - 3 = 9$ or $x - 3 = -9$. Thus, the solution is $x = 12$ or $x = -6$.

b. If $|2x - 4| = 8$, then

$$2x - 4 = 8 \quad \text{or} \quad 2x - 4 = -8$$
$$2x = 12 \quad | \quad 2x = -4$$
$$x = 6 \quad | \quad x = -2$$

Thus, the solution is $x = 6$ or $x = -2$.

c. If $|x^2 - 5x| = 6$, then

$$x^2 - 5x = 6 \quad \text{or} \quad x^2 - 5x = -6$$
$$x^2 - 5x - 6 = 0 \quad | \quad x^2 - 5x + 6 = 0$$
$$(x - 6)(x + 1) = 0 \quad | \quad (x - 3)(x - 2) = 0$$
$$x = 6 \quad \text{or} \quad x = -1 \quad | \quad x = 3 \quad \text{or} \quad x = 2$$

Thus, four values of x satisfy the equation. We can check these solutions by graphing or by substitution. Using the intersection method, we graph $y_1 = |x^2 - 5x|$ and $y_2 = 6$ and find the points of intersection. The solutions to the equation, $x = 6, -1, 3,$ and 2 found above, are also the x-values of the points of intersection (Figure 3.41).

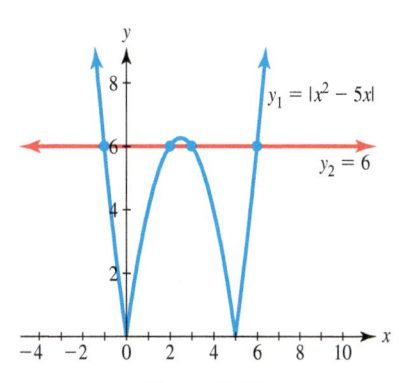

Figure 3.41

SKILLS CHECK 3.4

Answers that are not seen can be found in the answer section at the back of the text.

For each of the functions in Exercises 1–4, find the value of (a) $f(-1)$ and (b) $f(3)$, if possible.

1. $y = \begin{cases} 5 & \text{if } x \leq 1 \\ 6 & \text{if } x > 1 \end{cases}$
   a. 5 b. 6

2. $y = \begin{cases} -2 & \text{if } x < -1 \\ 4 & \text{if } x \geq -1 \end{cases}$
   a. 4 b. 4

3. $y = \begin{cases} x^2 - 1 & \text{if } x \leq 0 \\ x^3 + 2 & \text{if } x > 0 \end{cases}$ a. 0 b. 29

4. $y = \begin{cases} 3x + 1 & \text{if } x < 3 \\ x^2 & \text{if } x \geq 3 \end{cases}$ a. −2 b. 9

In Exercises 5–6, sketch the graph of each function using a window that gives a complete graph.

5. $y = \begin{cases} -1 & \text{if } x < 0 \\ 1 & \text{if } x \geq 0 \end{cases}$
6. $y = \begin{cases} 2 & \text{if } x \geq 2 \\ 6 & \text{if } x < 2 \end{cases}$

7. a. Graph the function $f(x) = \begin{cases} 5 & \text{if } 0 \leq x < 2 \\ 10 & \text{if } 2 \leq x < 4 \\ 15 & \text{if } 4 \leq x < 6 \\ 20 & \text{if } 6 \leq x < 8 \end{cases}$

   b. What type of function is this? Piecewise, step function

8. a. Graph the function
   $f(x) = \begin{cases} 100 & \text{if } 0 \leq x < 20 \\ 200 & \text{if } 20 \leq x < 40 \\ 300 & \text{if } 40 \leq x < 60 \\ 400 & \text{if } 60 \leq x < 80 \end{cases}$

   b. What type of function is this? Piecewise, step function

9. a. Graph $f(x) = \begin{cases} 4x - 3 & \text{if } x \leq 3 \\ x^2 & \text{if } x > 3 \end{cases}$
   b. Find $f(2)$ and $f(4)$. 5, 16
   c. State the domain of the function. All real numbers

10. a. Graph $f(x) = \begin{cases} 3 - x & \text{if } x \leq 2 \\ x^2 & \text{if } x > 2 \end{cases}$
    b. Find $f(2)$ and $f(3)$. 1, 9
    c. State the domain of the function. All real numbers

11. Recall the greatest integer function, where $[\![x]\!]$ equals the greatest integer that is less than or equal to x.
    a. If $f(x) = [\![x]\!]$, find $f(0.2)$, $f(3.8)$, $f(-2.6)$, and $f(5)$. 0, 3, −3, 5
    b. Graph $f(x) = [\![x]\!]$ for domain $[-5, 5]$.

12. Graph the following and describe how the graphs relate to the graph of $y = [\![x]\!]$:
    a. $f(x) = [\![x]\!] + 1.5$
    b. $g(x) = [\![x + 1.5]\!]$.

13. a. Graph $f(x) = |x|$.
    b. Find $f(-2)$ and $f(5)$. 2, 5
    c. State the domain of the function. All real numbers

14. a. Graph $f(x) = |x - 4|$.
    b. Find $f(-2)$ and $f(5)$. 6, 1
    c. State the domain of the function. All real numbers

15. Graph $f(x) = \begin{cases} x & \text{if } x \geq 0 \\ -x & \text{if } x < 0 \end{cases}$

16. Graph $f(x) = \begin{cases} x - 4 & \text{if } x \geq 4 \\ 4 - x & \text{if } x < 4 \end{cases}$

17. Graph $f(x) = \begin{cases} x & \text{if } x < 0 \\ -x & \text{if } x \geq 0 \end{cases}$

18. Compare the graph in Exercise 15 with the graph in Exercise 13(a). They are the same.

19. Compare the graph in Exercise 16 with the graph in Exercise 14(a). They are the same.

In Exercises 20–24, solve the equations and check graphically.

20. $|2x - 5| = 3$ $x = 4, x = 1$
21. $\left|x - \dfrac{1}{2}\right| = 3$ $x = \dfrac{7}{2}, x = -\dfrac{5}{2}$
22. $|x| = x^2 + 4x$ $x = 0, x = -5$
23. $|3x - 1| = 4x$ $x = \dfrac{1}{7}$
24. $|x - 5| = x^2 - 5x$ $x = 5, x = -1$

EXERCISES 3.4

Answers that are not seen can be found in the answer section at the back of the text.

25. **Postal Rates** The table that follows gives the 2019 postal rates as a function of the weight for media mail. Write a step function that gives the postage P as a function of the weight in pounds x for $0 < x \leq 5$.

Weight (lb)	Postal Rate ($)
$0 < x \leq 1$	2.75
$1 < x \leq 2$	3.27
$2 < x \leq 3$	3.79
$3 < x \leq 4$	4.31
$4 < x \leq 5$	4.83

    (*Source:* USPS)

26. **Electric Charges** For the nonextreme weather months, Palmetto Electric charges $7.10 plus 6.747 cents per kilowatt-hour (kWh) for customers using up to 1200 kWh, and charges $88.06 plus 5.788 cents per kWh above 1200 for customers using more than 1200 kWh.

    a. Write the function that gives the monthly charge in dollars as a function of the kilowatt-hours used.

    b. What is the monthly charge if 960 kWh are used? $71.87

    c. What is the monthly charge if 1580 kWh are used? $110.05

27. **Income Tax** The U.S. federal income tax owed by a married couple filing jointly for 2018 can be found from the following table.

    2018 Income Tax Brackets Filing Status: Married Filing Jointly

Rate	Taxable Income Bracket	Tax Owed
22%	$77,401 to $165,000	$8,907 plus 22% of the amount over $77,400
24%	$165,001 to $315,000	$28,179 plus 24% of the amount over $165,000
32%	$315,001 to $400,000	$64,179 plus 32% of the amount over $315,000
35%	$400,001 to $600,000	$91,379 plus 35% of the amount over $400,000

    a. Write the piecewise-defined function T with input x that models the federal tax dollars owed as a function of x, the taxable income dollars earned, with $77,400 < x \leq 600,000$.

    b. Use the function to find $T(300,000)$. $60,579

    c. Find the tax owed on a taxable income of $145,000. $23,779

    d. A friend tells Jack Waddell not to earn anything over $165,000 because it would raise his tax rate to 24% on all of his taxable income. Test this statement by finding the tax on $165,000 and $165,000 + $1. What do you conclude? The 24% rate applies only to the $1 over $165,000.

28. **First-Class Postage** The first-class postage charged for a letter in a large envelope is a function of its weight. The U.S. Postal Service uses the following table to describe the rates for 2015.

Weight Increment x (oz)	First-Class Postage $P(x)$
First ounce or fraction of an ounce	$1.00
Each additional ounce or fraction	$0.15

    (*Source:* pe.usps.gov/text)

    a. Convert this table to a piecewise-defined function that represents first-class postage for letters weighing up to 4 ounces, using x as the weight in ounces and P as the postage in cents.

    b. Find $P(1.2)$ and explain what it means. 115; the postage on a 1.2-oz letter in a large envelope is $1.15.

    c. Give the domain of P as it is defined above. $0 < x \leq 4$

    d. Find $P(2)$ and $P(2.01)$. 115; 130

    e. Find the postage for 2-ounce and 2.01-ounce letters in large envelopes. $1.15; $1.30

29. **Emigration to the United States** For the years from 2012 and projected to 2060, the rate of emigration to the United States from Spanish Caribbean and Latin American countries per thousand in their population can be modeled by the function

    $$f(x) = \begin{cases} 0.011x + 0.94 & \text{if } 2 \leq x < 20 \\ 1.16 & \text{if } 20 \leq x \leq 50 \end{cases}$$

    where x is the number of years after 2010.

    a. What does this model give as the rate of emigration per thousand population in 2025? In 2035? 1.105; 1.16

    b. Graph the function for $2 \leq x \leq 50$.

    (*Source*: U.S. Census Bureau)

30. **Wind Chill** The formula that gives the wind-chill factor for a 60°F temperature and a wind with velocity V in miles per hour is

    $$W = \begin{cases} 60 & \text{if } 0 \leq V < 4 \\ 0.644V - 9.518\sqrt{V} + 76.495 & \text{if } 4 \leq V \leq 55.9 \\ 41 & \text{if } V > 55.9 \end{cases}$$

224 Chapter 3 Quadratic, Piecewise-Defined, and Power Functions

a. Find the wind-chill factor for the 60° temperature if the wind is 20 mph. 47°

b. Find the wind-chill factor for the 60° temperature if the wind is 65 mph. 41°

c. Graph the function for $0 \leq V \leq 80$.

d. What are the domain and range of the function graphed in part (c)?
domain $0 \leq V \leq 80$; range $41 \leq W \leq 60$

31. *Fujita Scale* The Fujita Scale (F-Scale), or Fujita–Pearson Scale (FPP Scale), is a scale for rating tornado intensity, based primarily on the damage tornadoes inflict on human-built structures and vegetation. The scale is shown in the table.

Fujita scale

F0	40–72 mph	Light damage
F1	73–112 mph	Moderate damage
F2	113–157 mph	Considerable damage
F3	158–206 mph	Severe damage
F4	207–260 mph	Devastating damage
F5	261–318 mph	Incredible damage

a. Write a piecewise function $F(x)$ that models the F-value where x is the intensity of the tornado in mph.

b. Graph this function for $40 \leq x \leq 320$.

c. A typical tornado has winds of 110 mph or less, is approximately 250 feet across, and travels a mile or so before dissipating. What is the F-value for a typical tornado? 1

d. Evaluate $F(125)$. What does this mean in the context of the application?
2; A tornado with wind speed 125 mph has an F-Scale of F2.

32. *Library Mail* Library Mail is an inexpensive way for libraries, academic institutions, museums, nonprofits, and similar organizations to send items on loan to one another. These groups can mail up to 70 pounds of books, sound recordings, academic theses, and other related media at a time for one low cost. The USPS charges $2.61 for the first pound and an additional $0.49 for each additional pound up to 70 pounds.

a. Write a piecewise function $L(x)$ that models the charges for library mail, with x representing the weight in pounds.

b. Graph this function for $0 \leq x \leq 70$.

c. Evaluate $L(25)$. What does this mean in the context of the application? $14.37; Library mail that weighs 25 pounds will cost $14.37 to mail.

d. If it costs $30.54 for a library mail package, how much did the package weigh? 58 pounds

33. *First-Class Commercial Postage* The first-class postage for a commercial package is determined by its weight up to 16 ounces. The 2019 rates by the U.S. Postal Service for Zones 1 & 2 are shown in the table.

Weight Not Over (oz.)	Price ($)
1	2.66
2	2.66
3	2.66
4	2.66
5	3.18
6	3.18
7	3.18
8	3.18
9	3.82
10	3.82
11	3.82
12	3.82
13	4.94
14	4.94
15	4.94
15.999	4.94

a. Use this table to create a piecewise-defined function that represents first-class postage for commercial packages in Zones 1 & 2.

b. Graph the function using an appropriate window.

34. *Electricity Rates* The rates Georgia Power Company charges its residential customers, excluding fuel adjustments and taxes, for electricity during the months October–May are shown in the table.

Kilowatt-hours (kWh)	Monthly Charge
0 to 650	$0.05658 per kWh
More than 650, up to 1000	$36.78 plus $0.04853 per kWh over 650
More than 1000	$53.77 plus $0.04764 per kWh over 1000

(*Source*: Georgia Public Service Commission)

a. Write a piecewise-defined function C that gives the monthly charge for residential customers as a function of the number of kilowatt-hours used.

b. Find the charge for using 800 kWh. Round up to the nearest cent.
$44.06

c. Senior citizens over the age of 65 receive a discount of $18.00. How much would a senior citizen pay for using 1300 kWh?
$50.07

d. Graph the function.

e. A senior citizen received a bill for $74.31. How many kWh did she use that month?
1809 kWh

35. **Housing Starts** The construction industry was one of the hardest-hit by the recession of 2008, but housing starts began increasing again in 2010. The number of housing starts in thousands, during this time period can be modeled by three equations as follows, where x is the number of years after 2000:

For 2000–2006, $y = 1.3x^2 + 0.57x + 104$

For 2007–2009, $y = -31.6x + 318$

For 2010–2019, $y = 6.23x - 24.9$

(*Source*: U.S. Census Bureau)

a. Create a piecewise-defined function that represents the number of housing units, in thousands, from 2000–2019, with x equal to the number of years after 2000.

b. Use the function to find the number of housing starts in 2005, 2008, and 2019.
139,350; 65,200; 93,470

c. Explain the significance of the slope of the line in the second equation. The slope is negative, and this agrees with the fact that housing starts declined during this time period.

d. If the housing trend continues, how many new units will be started in 2025? About 131 thousand

3.5 Quadratic and Power Models

KEY OBJECTIVES

- Find the exact quadratic function that fits three points on a parabola
- Model data approximately using quadratic functions
- Model data using power functions
- Use first and second differences and visual comparison to determine whether a linear or quadratic function is the better fit to a set of data
- Determine whether a quadratic or power function gives the better fit to a given set of data

SECTION PREVIEW Female Physicians

Table 3.9 shows the number of female physicians (in thousands) in the United States for selected years from 2015 and projected to 2040. The scatter plot of the data, with x representing the number of years after 2010 and y representing the number of female physicians, in thousands, is shown in Figure 3.42. From the scatter plot we can see that a quadratic function or a power function might be a good fit to the data. In this section, we will use graphing calculators and Excel to create models with these functions by using steps similar to those that we used to model data with linear functions. In Example 6, we will show that a power function can be found that is an excellent fit for this data and we will see how well a quadratic function fits the data in Exercise 3.5.

Table 3.9

Year, x	Female Physicians (thousands)
2015	275.596
2018	301.061
2021	325.432
2025	352.713
2030	374.397
2035	390.328
2040	405.175

(*Source*: U.S. Census Bureau)

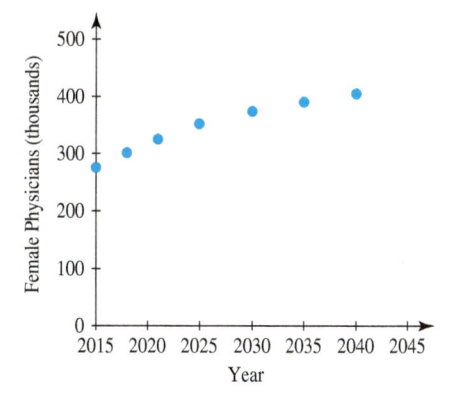

Figure 3.42

226 Chapter 3 Quadratic, Piecewise-Defined, and Power Functions

Modeling a Quadratic Function from Three Points on Its Graph

If we know three (or more) points that fit exactly on a parabola, we can find the quadratic function whose graph is the parabola.

EXAMPLE 1 ▶ Writing the Equation of a Quadratic Function

A parabola passes through the points $(0, 5)$, $(4, 13)$, and $(-2, 25)$. Write the equation of the quadratic function whose graph is this parabola.

SOLUTION

The general form of a quadratic function is $y = ax^2 + bx + c$. Because $(0, 5)$ is on the graph of the parabola, we can substitute 0 for x and 5 for y in $y = ax^2 + bx + c$, which gives $5 = a(0)^2 + b(0) + c$, so $c = 5$. Substituting the values of x and y for the other two points on the parabola and $c = 5$ in $y = ax^2 + bx + c$ gives

$$\begin{cases} 13 = a \cdot 4^2 + b \cdot 4 + 5 \\ 25 = a \cdot (-2)^2 + b \cdot (-2) + 5 \end{cases} \quad \text{or} \quad \begin{cases} 8 = 16a + 4b \\ 20 = 4a - 2b \end{cases}$$

Multiplying both sides of the bottom equation by 2 and adding the equations gives

$$\begin{cases} 8 = 16a + 4b \\ 40 = 8a - 4b \end{cases}$$
$$\overline{48 = 24a} \quad \Rightarrow a = 2$$

Substituting $a = 2$ in $8 = 16a + 4b$ gives $8 = 32 + 4b$, so $b = -6$. Thus, $a = 2, b = -6$, and $c = 5$, and the equation of this parabola is

$$y = 2x^2 - 6x + 5$$

EXAMPLE 2 ▶ Equation of a Quadratic Function

Find the equation of the quadratic function whose graph is a parabola containing the points $(-1, 9)$, $(2, 6)$, and $(3, 13)$.

SOLUTION

Using the three points $(-1, 9)$, $(2, 6)$, and $(3, 13)$, we substitute the values for x and y in the general equation $y = ax^2 + bx + c$, getting three equations.

$$\begin{cases} 9 = a(-1)^2 + b(-1) + c \\ 6 = a(2)^2 + b(2) + c \\ 13 = a(3)^2 + b(3) + c \end{cases} \quad \text{or} \quad \begin{cases} 9 = a - b + c \\ 6 = 4a + 2b + c \\ 13 = 9a + 3b + c \end{cases}$$

We use these three equations to solve for a, b, and c, using techniques similar to those used to solve two equations in two variables.*

Subtracting the third equation, $13 = 9a + 3b + c$, from each of the first and second equations gives a system of two equations in two variables.

$$\begin{cases} -4 = -8a - 4b \\ -7 = -5a - b \end{cases}$$

---

*We will discuss solution of systems of three equations in three variables further in Chapter 7.

Multiplying the second equation by -4 and adding the equations gives

$$\begin{cases} -4 = -8a - 4b \\ 28 = 20a + 4b \end{cases}$$
$$\overline{24 = 12a} \quad \Rightarrow a = 2$$

Substituting $a = 2$ in $-4 = -8a - 4b$ gives

$$-4 = -16 - 4b, \quad \text{or} \quad b = -3$$

and substituting $a = 2$ and $b = -3$ in the original first equation, $9 = a - b + c$, gives

$$9 = 2 - (-3) + c, \quad \text{or} \quad c = 4$$

Thus, $a = 2$, $b = -3$, and $c = 4$, and the quadratic function whose graph contains the points is

$$y = 2x^2 - 3x + 4$$

Modeling with Quadratic Functions

If the graph of a set of data has a pattern that approximates the shape of a parabola or part of a parabola, a quadratic function may be appropriate to model the data. In Example 3, we will use technology to model the dollar value of the U.S. mobile advertising market as a function of the number of years after 2000.

EXAMPLE 3 ▶ Artificial Intelligence

Enterprise AI refers to the use of artificial intelligence in enterprises to transform business operations and strategy. Table 3.10 gives the revenue (in billions of dollars) from artificial intelligence for enterprise applications in North America for the years 2016 through 2025.

Table 3.10

Year	Revenue ($ billions)	Year	Revenue ($ billions)
2016	0.212	2021	3.665
2017	0.434	2022	5.633
2018	0.796	2023	8.237
2019	1.376	2024	11.395
2020	2.288	2025	14.896

(*Source:* Statista)

a. Create a scatter plot of the data, with x equal to the number of years after 2015.

b. Find the quadratic function that is the best fit for the data, with x equal to the number of years after 2015 and y equal to the revenue in billions of dollars.

c. Graph the aligned data and model on the same axes. Is the model a good fit?

d. Use the model to find the enterprise AI revenue in 2026.

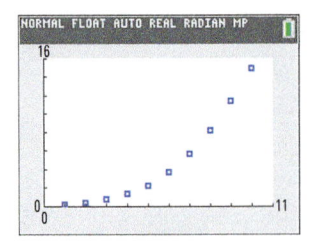

228 Chapter 3 Quadratic, Piecewise-Defined, and Power Functions

Figure 3.43

SOLUTION

a. Entering the aligned inputs 1 through 10 in one list of a graphing calculator and the corresponding outputs in a second list gives the scatter plot of the data shown in Figure 3.43. The shape looks as though it could be part of a parabola, so it is reasonable to find a quadratic function whose graph will at least approximately fit the data points.

b. Using quadratic regression with **STAT**, CALC, 5:QuadReg gives a quadratic function that models the data. The function, with coefficients rounded to three decimal places, is

$$y = 0.229x^2 - 0.941x + 1.255$$

c. The graphs of the aligned data points and the unrounded function on the same axes are shown in Figure 3.44(a). The model is an excellent fit for the data.

d. Recall that we use unrounded models in our calculations and graphs unless otherwise instructed. Using TABLE with the unrounded function shows that the enterprise AI revenue is approximately $18.6 billion during 2026, when $x = 11$. (See Figure 3.44(b).)

(a)   (b)

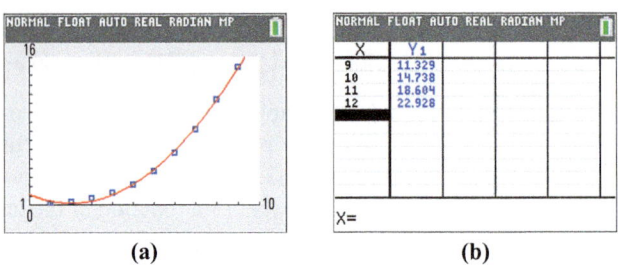

Figure 3.44

EXAMPLE 4 ▶ The Cloud

North American cloud computing revenues from 2008 to 2020 are shown in Table 3.11.

a. Create a scatter plot for the data with y equal to the number of billions of euros of revenue and x equal to the number of years after 2000, and determine the function types that can best be used to model the data.

b. Find the best-fitting quadratic model for the data.

Table 3.11

Year	Revenue (billions of euros)	Year	Revenue (billions of euros)
2008	7.800	2015	29.503
2009	9.553	2016	34.158
2010	11.665	2017	39.330
2011	14.568	2018	45.150
2012	17.831	2019	51.785
2013	21.393	2020	59.453
2014	25.270		

www.statista.com

c. Graph the aligned data and the function on the same axes. Does the model seem like a reasonable fit to the data?

d. Use the model to predict the revenue in 2019. Is this estimate close to the data projection?

e. In what year does the model predict the revenue will be 65.98 billion euros?

SOLUTION

a. The scatter plot, with x equal to the number of years after 2000 and y equal to the revenue in billions of euros, is shown in Figure 3.45(a). The shape appears as though it could be part of a parabola, so it is reasonable to find a quadratic function using the aligned data points.

b. Using quadratic regression on a graphing calculator gives the quadratic function that is the best fit for the data. The function, with the coefficients rounded to three decimal places, is

$$y = 0.231x^2 - 2.246x + 11.131$$

Note that we will use the unrounded model in graphing and performing calculations.

c. The scatter plot of the aligned data and the graph of the function are shown in Figure 3.45(b). The function appears to be an excellent fit to the data.

d. Evaluating the function at $x = 19$ gives $y = 51.966$, so the model predicts that the revenue will be 51.966 billion euros in 2019.

e. Finding the intersection of the model and the equation $y = 65.98$ gives $x = 21$ (see Figure 3.45(c)), so the model estimates that the revenue will be 65.98 billion euros in 2021.

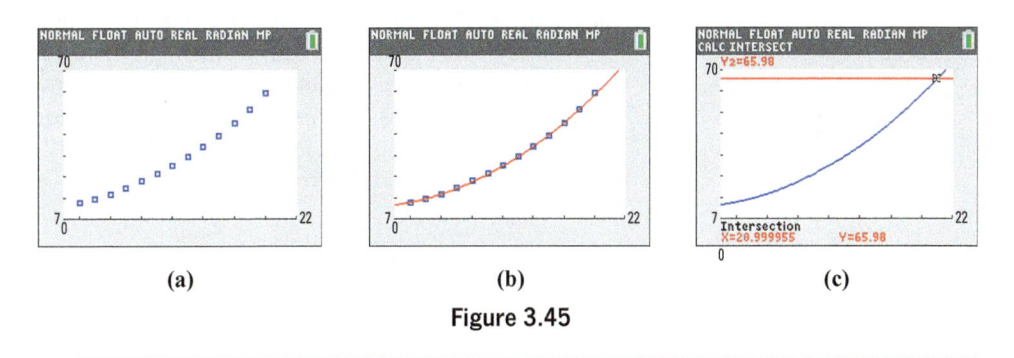

(a)   (b)   (c)

Figure 3.45

Spreadsheet ▶ SOLUTION   We can use graphing utilities, software programs, and spreadsheets to find the quadratic function that is the best fit for data. Figure 3.46 shows a partial Excel spreadsheet for the aligned data of Example 4. Selecting all the cells containing the data, getting the scatter plot of the data, selecting Add trendline, picking Polynomial with order 2, clicking Display Equation, and closing the dialog box gives the equation of the quadratic function that is the best fit for the data. Observe that this model is the same as the one found in Example 4. Figure 3.46 shown this equation and its graph, along with the scatter plot of the data.*

---

*See Appendix B, page 708, for details.

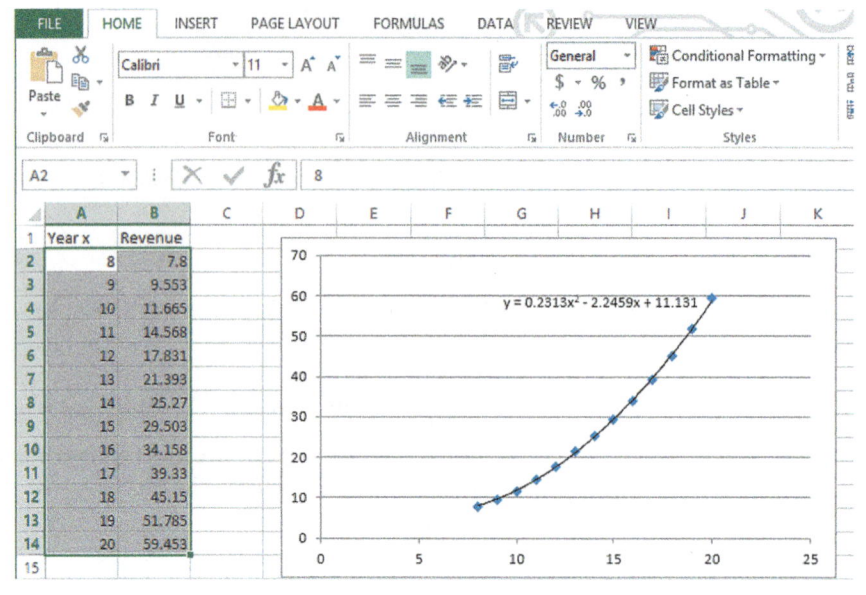

230   Chapter 3   Quadratic, Piecewise-Defined, and Power Functions

Figure 3.46

Comparison of Linear and Quadratic Models

Recall that when the changes in inputs are constant and the (first) differences of the outputs are constant or nearly constant, a linear model will give a good fit for the data. In a similar manner, we can compare the differences of the first differences, which are called the **second differences**. If the second differences are constant for equally spaced inputs, the data can be modeled exactly by a quadratic function.

Consider the data in Table 3.12, which gives the measured height y of a toy rocket x seconds after it has been shot into the air from the ground.

Figure 3.47 gives the first differences and second differences for the equally spaced inputs for the rocket height data in Table 3.12. Excel is especially useful for finding first and second differences.

Table 3.12   Height of a Rocket

Time x (seconds)	Height (meters)
1	68.6
2	117.6
3	147
4	156.8
5	147
6	117.6
7	68.6

Figure 3.47

The first differences are not constant, but each of the second differences is -19.6, which indicates that the data can be fit exactly by a quadratic function. We can find that the quadratic model for the height of the toy rocket is $y = 78.4x - 9.8x^2$ meters, where x is the time in seconds.

Modeling with Power Functions

We can model some experimental data by observing patterns. For example, by observing how the area of each of the (square) faces of a cube is found and that a cube has six faces, we can deduce that the surface area of a cube that is x units on each edge is

$$S = 6x^2 \text{ square units}$$

We can also measure and record the surface areas for cubes of different sizes to investigate the relationship between the edge length and the surface area for cubes. Table 3.13 contains selected measures of edges and the resulting surface areas.

Table 3.13

Edge Length x (units)	Surface Area of Cube (square units)
1	6
2	24
3	54
4	96
5	150

We can enter the lengths from the table as the independent (x) variable and the corresponding surface areas as the dependent (y) variable in a graphing utility, and then have the utility create the **power function** that is the best model for the data (Figure 3.48). This model also has the equation $y = 6x^2$ square units.

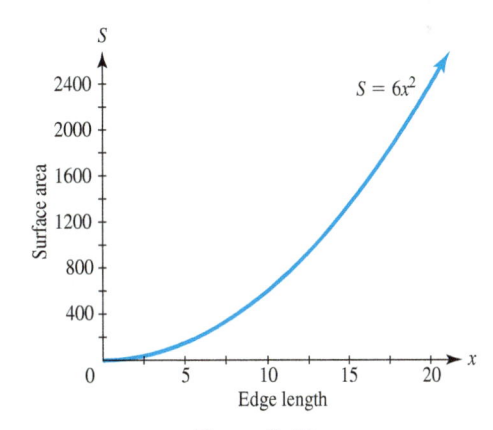

Figure 3.48

EXAMPLE 5 ▶ Diabetes

Figure 3.49 and Table 3.14 show that the percent of the U.S. adult population with diabetes (diagnosed and undiagnosed) is projected to grow rapidly in the future.

a. Find the power model that fits the data, with x equal to the number of years after 2000.

b. Use the model to predict the percent of U.S. adults with diabetes in 2015.

c. In what year does this model predict the percent to be 29.6?

232   Chapter 3   Quadratic, Piecewise-Defined, and Power Functions

Figure 3.49

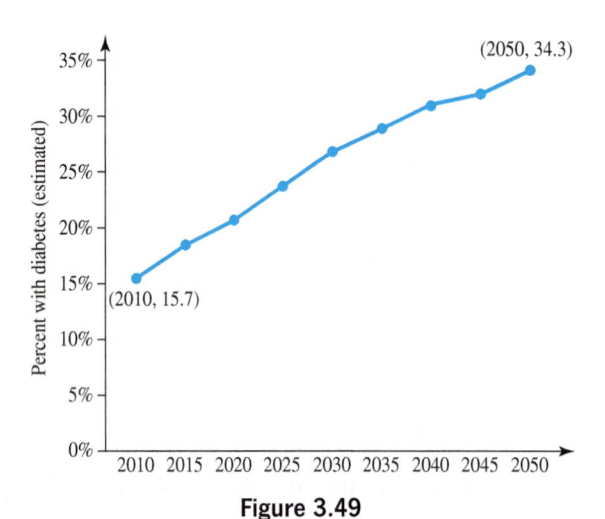

Table 3.14

Year	Percent	Year	Percent
2010	15.7	2035	29.0
2015	18.9	2040	31.4
2020	21.1	2045	32.1
2025	24.2	2050	34.3
2030	27.2		

(*Source*: Centers for Disease Control and Prevention)

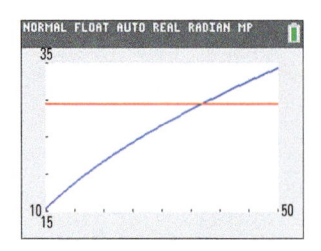

Figure 3.50

SOLUTION

a. The power model that fits the data is $f(x) = 4.947x^{0.495}$, with x equal to the number of years after 2000.

b. Evaluating the unrounded model at 15 gives the percent of U.S. adults with diabetes in 2015 to be

$$f(15) = 18.9\%$$

c. By intersecting the graphs of the unrounded model and $y = 29.6$ (Figure 3.50), we find the percent to be 29.6 when $x = 36.99$, or during 2037.

EXAMPLE 6 ▶ Female Physicians

Table 3.15 shows the number of female physicians, in thousands, in the United States for selected years from 2015 and projected to 2040. The scatter plot of the data, with x representing the number of years after 2010 and y representing the number of female physicians, in thousands, is shown in Figure 3.51.

a. Find the power function that models the data.

b. Graph the data and the model on the same axes.

c. What does the model predict the number of female physicians to be in 2045?

Table 3.15

Year, x	Female Physicians (thousands)
2015	275.596
2018	301.061
2021	325.432
2025	352.713
2030	374.397
2035	390.328
2040	405.175

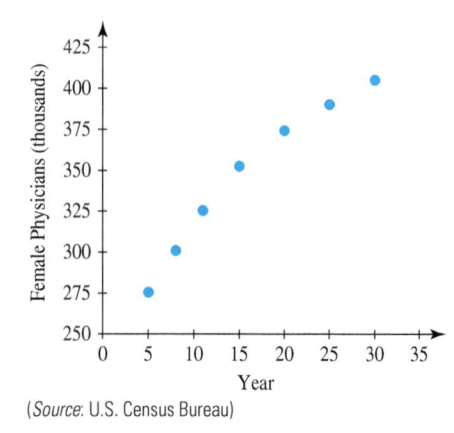

(*Source*: U.S. Census Bureau)

Figure 3.51

3.5 Quadratic and Power Models

SOLUTION

a. Using power regression with **STAT**, CALC, A:PwrReg on a graphing calculator gives the power function that models the data. The power function that is the best-fitting model for the data, reported with four significant digits, is

$$y = 192.5x^{0.2202}$$

where y is in thousands and x is the number of years after 2010.

b. The graphs of the data and the power function that models it are shown in Figure 3.52(a). The model is an excellent visual fit to the data.

c. Evaluating the unrounded model at $x = 35$ gives 421.091, so the model predicts that the number of female physicians will be 421,091 in 2045. (See Figure 3.52(b).)

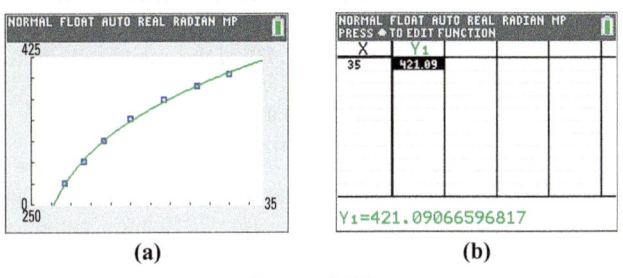

Figure 3.52

Spreadsheet ▶ SOLUTION

Power models can also be found with Excel. The Excel spreadsheet in Figure 3.53 shows the number of cohabiting households (in thousands) for 1960–2013, listed for selected years after 1950. Selecting the cells containing the data, getting the scatter plot of the data, selecting Add Trendline, and picking Power gives the equation of the power function that is the best fit for the data, along with the scatter plot and the graph of the best-fitting power function. The equation and graph of the function are shown in Figure 3.53.

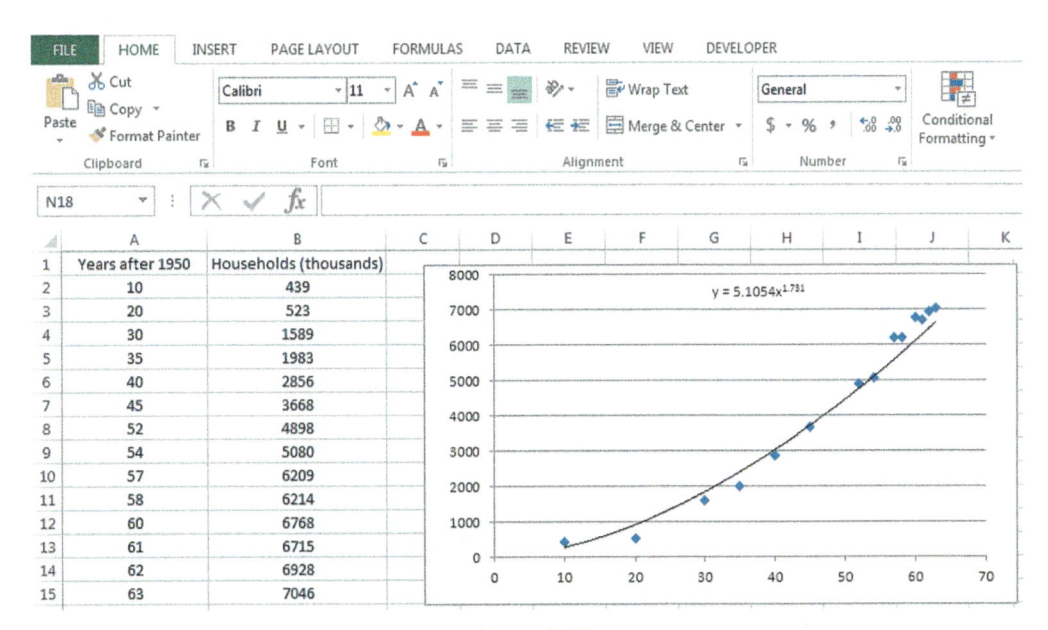

Years after 1950	Households (thousands)
10	439
20	523
30	1589
35	1983
40	2856
45	3668
52	4898
54	5080
57	6209
58	6214
60	6768
61	6715
62	6928
63	7046

$y = 5.1054x^{1.731}$

Figure 3.53

Comparison of Power and Quadratic Models

We found a power function that is a good fit for the data in Figure 3.53, but a linear function or a quadratic function may also be a good fit for the data. A quadratic function may fit data points even if there is no obvious turning point in the graph of the data points. If the data points appear to rise (or fall) more rapidly than a line, then a quadratic model or a power model may fit the data well. In some cases, it may be necessary to find both models to determine which is the better fit for the data.

EXAMPLE 7 ▶ Purchasing Power

Prices for goods and services traditionally rise over time, so that the value (purchasing power) of a dollar will decrease. Table 3.16 gives the purchasing power of a 2012 dollar for selected years from 2012 projected to 2050. The table indicates that a 2018 dollar purchased 87.7% of the goods and services that could be purchased for $1 in 2012.

a. Create a scatter plot of the data in Table 3.16, with x equal to the number of years after 2010.

b. Find the quadratic function that is the best fit for the data, with x equal to the number of years after 2010 and y equal to the purchasing power of a 2012 dollar.

c. Graph the aligned data and the quadratic model on the same axes. Is the model a good fit?

d. The shape of the scatter plot of the data shown in Figure 3.54 looks as though it could also be modeled by a power function. Find the power function that is the best fit for the data, with x equal to the number of years after 2010 and y equal to the purchasing power of a 2012 dollar. Report the model with three significant digits.

e. Graph the aligned data and the power model on the same axes. Is the model a good fit?

f. Use the better model to find the year in which the value of a 2012 dollar will fall to 60 cents.

Table 3.16

Year	Purchasing Power of $1	Year	Purchasing Power of $1
2012	1.00	2030	0.629
2014	0.962	2035	0.548
2016	0.921	2040	0.477
2018	0.877	2045	0.416
2020	0.829	2050	0.362
2025	0.722		

(*Source:* Social Security Administration)

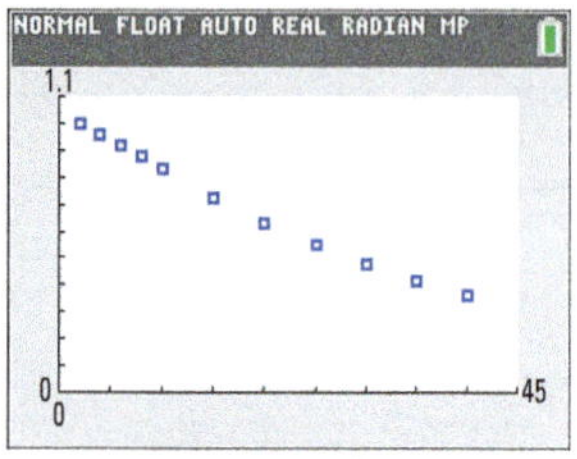

Figure 3.54

SOLUTION

a. The scatter plot of the data is shown in Figure 3.54. The shape looks as though it could be part of a parabola, so it is reasonable to find a quadratic function whose graph will approximately fit the data points.

b. Using quadratic regression gives a quadratic function that models the data. The function, with coefficients rounded to three significant digits, is

$$y = 0.000194x^2 - 0.0252x + 1.06$$

c. The graphs of the aligned data points and the unrounded quadratic function on the same axes are shown in Figure 3.55(a). The model is an excellent fit for the data.

d. Using power regression gives a power function that models the data. The function, with coefficients rounded to three significant digits, is

$$y = 1.59x^{-0.343}$$

e. The graph of the aligned data points and the unrounded power function on the same axes are shown in Figure 3.55(b). The model is not a good fit for the data.

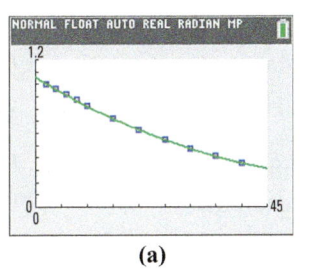

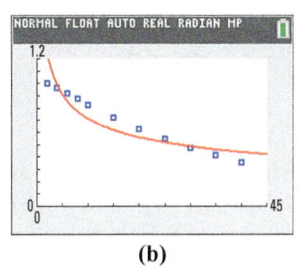

Figure 3.55

f. Recall that we use unrounded models in our calculations and graphs unless otherwise instructed. The quadratic model is the better fit, and we use TABLE with the unrounded quadratic function. The purchasing power falls to 0.615 at the end of 2031, falls to $0.60 during 2032, and is 0.598 by the end of 2032. (See Figure 3.56(b).)

We could also graphically find the intersection of the unrounded quadratic model and $y = 0.60$ to see when the purchasing power equals 0.60. (See Figure 3.56(b).)

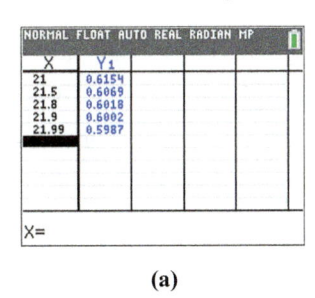

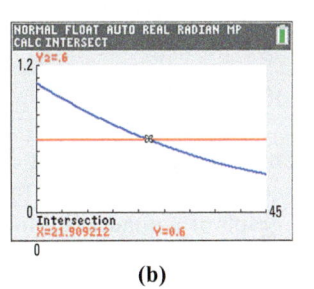

Figure 3.56

SKILLS CHECK 3.5

Answers that are not seen can be found in the answer section at the back of the text.

In Exercises 1–6, write the equation of the quadratic function whose graph is a parabola containing the given points.

1. $(0, 1)$, $(3, 10)$, and $(-2, 15)$ $y = 2x^2 - 3x + 1$

2. $(0, -3)$, $(4, 37)$, and $(-3, 30)$ $y = 3x^2 - 2x - 3$

3. $(6, 30)$, $(0, -3)$, and $(-3, 7.5)$ $y = x^2 - 0.5x - 3$

4. $(6, -22)$, $(-3, 23)$, and $(0, 2)$ $y = \frac{x^2}{3} - 6x + 2$

5. $(0, 6)$, $(2, \frac{22}{3})$, and $(-9, \frac{99}{2})$ $y = \frac{x^2}{2} - \frac{x}{3} + 6$

6. $(0, 7)$, $(2, 8.5)$, and $(-3, 12.25)$ $y = \frac{x^2}{2} - \frac{x}{4} + 7$

7. A ball is thrown upward from the top of a 48-foot-high building. The ball is 64 feet above ground level after 1 second, and it reaches ground level in 3 seconds. The height above ground is a quadratic function of the time after the ball is thrown. Write the equation of this function. $y = -16x^2 + 32x + 48$

8. A ball is dropped from the top of a 256-foot-high building. The ball is 192 feet above ground after 2 seconds, and it reaches ground level in 4 seconds. The height above ground is a quadratic function of the time after the ball is thrown. Write the equation of this function. $y = -16x^2 + 256$

9. Find the equation of the quadratic function whose graph is a parabola containing the points $(-1, 6)$, $(2, 3)$, and $(3, 10)$. $y = 2x^2 - 3x + 1$

236 Chapter 3 Quadratic, Piecewise-Defined, and Power Functions

10. Find the equation of the quadratic function whose graph is a parabola containing the points $(-2, -4)$, $(3, 1)$, and $(2, 4)$. $y = -x^2 + 2x + 4$

11. Find the quadratic function that models the data in the table below. $f(x) = 3x^2 - 2x$

x	-2	-1	0	1	2	3	4
y	16	5	0	1	8	21	40

x	5	6	7	8	9	10
y	65	96	133	176	225	280

12. The following table has the inputs, x, and the outputs for three functions, $f, g,$ and h. Use second differences to determine which function is exactly quadratic, which is approximately quadratic, and which is not quadratic. $f(x)$ is approximately quadratic; $g(x)$ is exactly quadratic; $h(x)$ is not quadratic.

x	f(x)	g(x)	h(x)
0	0	2	0
2	399	0.8	110
4	1601	1.2	300
6	3600	3.2	195
8	6402	6.8	230
10	9998	12	290

13. As you can verify, the following data points give constant second differences, but the points do not fit on the graph of a quadratic function. How can this be? The x-values are not equally spaced.

x	1	2	4	8	16	32	64
y	1	6	15	28	45	66	91

14. a. Make a scatter plot of the data in the table below.
    b. Does it appear that a quadratic model or a power model is the better fit for the data? Quadratic model

x	y
1	4
2	9
3	11
4	21
5	32
6	45

15. Find the quadratic function that is the best fit for $f(x)$ defined by the table in Exercise 12. $y = 99.9x^2 + 0.64x - 0.75$

16. Find the quadratic function that is the best fit for $g(x)$ defined by the table in Exercise 12. $y = 0.2x^2 - x + 2$

17. a. Find a power function that models the data in the table in Exercise 14. $y = 3.545x^{1.323}$
    b. Find a linear function that models the data. $y = 8.114x - 8.067$
    c. Visually determine which function is the better fit for the data. Power function

18. a. Make a scatter plot of the data in the table below.
    b. Does it appear that a linear model or a power model is the better fit for the data? Linear

x	y
3	4.7
5	8.6
7	13
9	17

19. a. Find a power function that models the data in the table in Exercise 18. $y = 1.292x^{1.178}$
    b. Find a linear function that models the data. $y = 2.065x - 1.565$
    c. Visually determine if each model is a good fit. They are both good fits.

20. Find the quadratic function that models the data in the tables that follow. $y = 3.2226x^2 - 7.2538x - 3.3772$

x	-2	-1	0	1	2	3
y	15	5	2	1	3	10

x	4	5	6	7	8
y	20	35	55	75	176

21. Find the power function that models the data in the table below. $y = 2.98x^{0.614}$

x	1	2	3	4	5	6	7	8
y	3	4.5	5.8	7	8	9	10	10.5

22. Is a power function or a quadratic function the better model for the data below? Both fit the data reasonably well.

x	0.5	1	2	3	4	5	6
y	1	7	17	32	49	70	90

EXERCISES 3.5

Answers that are not seen can be found in the answer section at the back of the text.
Calculate numerical results with the unrounded models, unless otherwise instructed. Report models to 3 decimal places unless otherwise stated.

23. **U.S. Households** The table gives the percent of total households that have married couples for selected years from 1982 and projected to 2050.

    a. Find the quadratic function that best models the data, with x representing the number of years after 1980 and y representing the percent.
    $y = 0.003x^2 - 0.460x + 60.207$

    b. Graph the function and the data points on the same axes. Comment on the fit.

    c. Use the model to find the predicted percent of total households that have married couples in 2037.
    Approximately 44%

Year	Percent	Year	Percent
1982	59	2020	46
1990	56	2030	45
2000	53	2040	44
2010	49	2050	43

    (*Source*: Yi Zeng et al., *Household and Living Arrangement Projections*, Springer)

24. **National Health Care** The following table shows the total national expenditures for health care, in billions of dollars, for selected years from 2002 and projected to 2024. (These data include expenditures for medical research and medical facilities construction.)

Year	Amount ($ billions)	Year	Amount ($ billions)
2002	1602	2014	3080
2004	1855	2016	3403
2006	2113	2018	3786
2008	2414	2020	4274
2010	2604	2022	4825
2012	2817	2024	5425

    (*Source*: U.S. Centers for Medicare and Medicaid Services)

    a. Use a scatter plot with x as the number of years after 2000 and y equal to the expenditures for health in billions of dollars to determine if a quadratic function would be a good model for the data.

    b. Find a quadratic function that models the data, with x equal to the number of years after 2000 and y equal to the expenditures for health in billions of dollars. Report the model with three significant digits.
    $y = 4.61x^2 + 43.4x + 1620$

    c. Graph the quadratic function and the data on the same axes to determine if the function is a good fit for the data.

    d. What does the unrounded model predict the expenditures will be in 2030?
    Approximately $7073 billion

25. **Internet of Things Devices** According to IHS, the Internet of Things (IoT) market had 15.4 billion devices connected in 2015, with the number of devices expected to double by 2025. The table shows the projected number of devices, in billions, from 2015 through 2025.
    (*Source*: Forbes.com)

    a. Create a scatter plot of the data in the table, with x equal to the number of years after 2015 and y equal to the billions of devices connected to the Internet, and determine if it is reasonable to fit a quadratic model to the data.

    b. Find the best-fitting quadratic model for the data, with x equal to the number of years after 2015 and y equal to the billions of devices connected to the Internet. $y = 0.590x^2 - 0.275x + 17.201$

    c. Graph the model on the same axes with the data, and discuss how well the function fits the data.

    d. What does the model predict that the number of devices will be in 2028? 113.27 billion

Year	Devices (Billions)	Year	Devices (Billions)
2015	15.41	2021	35.82
2016	17.68	2022	42.62
2017	20.35	2023	51.11
2018	23.14	2024	62.12
2019	26.66	2025	75.44
2020	30.73		

    (*Source*: Forbes.com)

26. **Obesity** Obesity (BMI $\geq$ 30) increases the risk of diabetes, heart disease, and many other ailments. The percent of Americans who are obese from 2000 projected to 2030 is shown in the table on the next page.

    a. Graph the data points to determine whether a quadratic function is a reasonable model for the data.

    b. If it is reasonable, find the quadratic function that models obesity, with x equal to the number of years after 2000 and y equal to the percent of obese Americans. Report the model with three significant digits.
    $y = -0.0102x^2 + 0.971x + 22.1$

c. Use the unrounded model to predict the percent with obesity in 2023. 39.1%

Year	2000	2010	2015	2020	2025	2030
% Obese	22.1	30.9	34.5	37.4	39.9	42.2

(Source: American Journal of Preventive Medicine)

27. Wind Chill The table gives the wind-chill temperature when the outside temperature is 20°F.

Wind (mph)	Wind Chill (°F)	Wind (mph)	Wind Chill (°F)
5	13	35	0
10	9	40	−1
15	6	45	−2
20	4	50	−3
25	3	55	−3
30	1	60	−4

(Source: National Weather Service)

a. Use x as the wind speed and create a quadratic model for the data. $y = 0.0052x^2 - 0.62x + 15.0$

b. At what wind speed does the model estimate that the wind-chill temperature will be −3°F? 50 mph

c. Do you think the model found in part (a) is valid for $x > 60$? Explain. No; with minimum near 59.4 mph, model predicts warmer temperatures for winds > 60 mph.

28. Foreign-Born Population The table gives the percent of the U.S. population that is foreign-born for selected years from 1970 and projected to 2060.

a. Create a scatter plot of the data, with x equal to the number of years after 1970 and with y equal to the percent.

b. Does it appear that the data could be modeled by a quadratic function? Yes

c. Find the quadratic function that is the best-fitting model for the data, with x equal to the number of years after 1970 and y equal to the percent. Report the model rounded to four decimal places.
$y = -0.0008x^2 + 0.2343x + 4.2245$

d. Graph the model on the same axes with the data, and comment on fit of the model to the data.

Year	Percent	Year	Percent
1970	4.8	2020	14.3
1980	6.2	2030	15.8
1990	8.0	2040	17.1
2000	10.4	2050	18.2
2010	12.4	2060	18.8

(Source: U.S. Census Bureau)

29. Aging Workers The table shows the millions of Americans who are working full time at selected ages from 20 to 62.

a. Find a quadratic model that gives the number (in millions) y of Americans working full time as a function of their age, x. Report the model with three significant digits.
$y = -0.00834x^2 + 0.690x - 6.81$

b. Graph the model and the data points on the same axes.

c. Use technology to find the age at which the model indicates that the number of workers is greatest. at age 42 ($x = 41.38$)

Age	Millions Working Full Time
20	3.62
27	6.30
32	6.33
37	6.96
42	7.07
47	7.52
52	7.10
57	5.58
62	3.54

(Source: Wall Street Journal, June 18, 2012)

30. U.S. Population The table gives the U.S. population, in millions, for selected years, with projections to 2050.

a. Create a scatter plot for the data in the table, with x equal to the number of years after 1960.

b. Use the scatter plot to determine the type of function that can be used to model the data, and create a function that best fits the data, with x equal to the number of years after 1960.
$y = 0.00138x^2 + 2.488x + 178.792$

Year	U.S. Population (millions)	Year	U.S. Population (millions)
1960	180.671	1995	263.044
1965	194.303	1998	270.561
1970	205.052	2000	281.422
1975	215.973	2003	294.043
1980	227.726	2025	358.030
1985	238.466	2050	408.695
1990	249.948		

(Source: U.S. Census Bureau)

31. Alzheimer's Disease Partially because of Americans living longer, the number with Alzheimer's disease and other dementia is projected to grow each year. The table below gives the millions of U.S. citizens with Alzheimer's from 2000 and projected to 2050.

Year	2000	2010	2020	2030	2040	2050
Number	4.0	5.9	6.8	8.7	11.8	14.3

(*Source*: National Academy on an Aging Society)

a. Create a scatter plot of the data, with x equal to the number of years after 2000 and with y equal to the number of millions of Americans with Alzheimer's disease or other dementia.

b. Find a quadratic function that models this data. Report the model with four significant digits.
$y = 0.002107x^2 + 0.09779x + 4.207$

c. Graph the data and the model on the same axes, and comment on the fit. Good fit

d. How many Americans does the model predict will have Alzheimer's disease or other dementia in 2056? 16.3 million

e. In what year after 2000 does the model predict that 10.2 million Americans will have Alzheimer's or other dementia? 2035

32. *Volume* The measured volume of a pyramid with each edge of the base equal to x units and with its altitude (height) equal to x units is given in the table below.

a. Determine whether the second differences of the outputs are constant. No

b. If the answer is yes, find the quadratic model that is the best fit for the data. Otherwise, find the power function that is the best fit. $y = \frac{1}{3}x^3$

Edge Length x (units)	Volume of Pyramid (cubic units)
1	1/3
2	8/3
3	9
4	64/3
5	125/3
6	72

33. *Telecom Artificial Intelligence* The table shows the worldwide software revenue for artificial intelligence used in telecommunications for the years from 2016 and projected to 2025.

a. Create a scatter plot for the data over the interval $6 \leq x \leq 16$, with x equal to the number of years after 2010 and y equal to the revenue in millions of dollars.

b. Find the quadratic function that models these data, with x equal to the number of years after 2010 and y equal to the revenue in millions of dollars. Report the model with four significant digits.
$y = 170.1x^2 - 2356x + 8463$

c. Graph the model on the same axes with the data, and discuss how well the function fits the data. Very good fit

d. What does the model predict that the worldwide software revenue for artificial intelligence used in telecommunication will be in 2024? Approximately $8818 million

e. Use the model to predict when the worldwide software revenue for artificial intelligence used in telecommunications will reach $55.871 billion. 2035

Year	Revenue ($ Million)	Year	Revenue ($ Million)
2016	363	2021	3273
2017	455	2022	5000
2018	545	2023	6182
2019	1000	2024	9000
2020	1636	2025	11,363

(*Source*: Tractica)

34. *World Population* One projection of the world population by the United Nations for selected years (a low-projection scenario) is given in the table below.

Year	Projected Population (millions)	Year	Projected Population (millions)
1995	5666	2075	6402
2000	6028	2100	5153
2025	7275	2125	4074
2050	7343	2150	3236

(*Source*: *World Population Prospects*, United Nations)

a. Find a quadratic function that fits the data, using the number of years after 1990 as the input.
$y = -0.360x^2 + 38.518x + 5822.864$

b. Find the positive x-intercept of this graph, to the nearest year. $x \approx 191$, or the year 2181

c. When can we be certain that this model no longer applies? After 2181

35. *China's Labor Pool* The table shows United Nations data from 1975 and projections to 2050 of the number of millions of people age 15 to 59 in China.

a. What is the quadratic function that best models this population as a function of the number of years after 1970? Let x represent the number of years after 1970 and y represent the number of millions of people in this labor pool.
$y = -0.224x^2 + 22.005x + 370.705$

b. Use technology with the model to find the maximum size of this population before it begins to shrink. 911 in 2020.

Year	Labor Pool (millions)	Year	Labor Pool (millions)
1975	490	2015	920
1980	560	2020	920
1985	650	2025	905
1990	730	2030	875
1995	760	2035	830
2000	800	2040	820
2005	875	2045	800
2010	910	2050	670

(Source: United Nations)

36. Personal Income The table shows the total personal income in the United States (in billions of dollars) for selected years from 1960 and projected to 2024.

a. These data can be modeled by a quadratic function. Write the equation of this function, with x equal to the number of years after 1960.
$y = 6.116x^2 - 49.273x + 620.318$

b. Graph the data and the model on the same axes. Is the model a good fit for the data?

c. Does the unrounded model overestimate or underestimate the total personal income given in the table for 2018? Underestimates

d. In what year does the model predict the total personal income will reach $25.548 trillion? 2028

Year	Income ($ billions)	Year	Income ($ billions)
1960	411.5	2008	12,100.7
1970	838.8	2014	14,728.6
1980	2307.9	2018	19,129.6
1990	4878.6	2024	22,685.1
2000	8429.7		

(Source: U.S. Bureau of Labor Statistics)

37. Consumer Price Index The table below gives the U.S. consumer price index (CPI) for selected years from 2012 and projected to 2050. With the reference year 2012, a 2020 CPI = 120.56 means goods and services that cost $100.00 in 2012 are expected to cost $120.56 in 2020.

a. Find the quadratic function that is the best fit for the data, with x as the number of years after 2010 and y as the CPI in dollars.
$y = 0.068x^2 + 1.746x + 96.183$

b. Graph the model and the data on the same axes and comment on the fit of the model to the data. Excellent fit

c. Use the model to predict the CPI in 2022. 126.98

d. According to the model, in what year will the CPI reach 216? 2041

Year	CPI	Year	CPI	Year	CPI
2012	100.00	2025	138.41	2040	209.44
2016	108.58	2030	158.90	2045	240.45
2020	120.56	2035	182.43	2050	276.05

(Source: Social Security Administration)

38. Global 3D Printing Market The table shows the market size of the global 3D printer market for selected years from 2013 and projected to 2025.

a. Using the input x equal to the number of years after 2010 and output y equal to the market size in billions of dollars, graph the aligned data points and the given equations below.

  I. $y = 0.692\sqrt{x^3}$

  II. $y = 0.34x^2 - 2.7x + 11$

b. Which function is the better model for the data? II, the quadratic function

Year	Market Size ($ billions)	Year	Market Size ($ billions)
2013	4.4	2019	14.5
2014	5.7	2020	17.5
2015	6.9	2021	21
2016	8.3	2022	28.5
2017	10	2023	32.8
2018	12	2025	49.1

(Source: Statista)

39. Mortgages The balance owed y on a $50,000 mortgage after x monthly payments is shown in the table that follows. Graph the data points with each of the equations below to determine which is the better model for the data, if x is the number of months that payments have been made. b

a. $y = 338,111.278x^{-0.676}$

b. $y = 4700\sqrt{110 - x}$

Monthly Payments	Balance Owed ($)
12	47,243
24	44,136
48	36,693
72	27,241
96	15,239
108	8074

40. Cloud Revenue Many services are provided through the Cloud, and public vendor revenue is projected to grow rapidly. The table shows the global Cloud revenue and projected revenue, in billions of dollars, from 2012 and projected to 2026.

   a. Create a scatter plot of the data, with x equal to the number of years after 2010 and y equal to the revenue in billions of dollars.

   b. Find the power model that is the best fit for the data, with x equal to the number of years after 2010 and y equal to the revenue in billions of dollars. $y = 7.531x^{1.554}$

   c. Graph the model on the same graph as the data points. Is this model an excellent fit? Not an excellent fit

   d. What does the model predict the revenue will be in 2030? $791.4 billion

Year	Revenue ($ billion)	Year	Revenue ($ billion)
2012	26	2020	298
2013	39	2021	345
2014	56	2022	387
2015	80	2023	422
2016	116	2024	451
2017	154	2025	474
2018	199	2026	493
2019	248		

(Source: Wikibon Server SAN & Cloud Research Projects)

41. Energy Use The following table shows the energy use per dollar of GDP, as a percent, for years from 2000 and projected to 2035. (GDP, gross domestic product, is the monetary value of all the finished goods and services produced within a country's borders in a specific time period, though GDP is usually calculated on an annual basis. It includes all of private and public consumption, government outlays, investments, and exports less imports that occur within a defined territory.) These data are indexed to 1980, which means that energy use for any year is viewed as a percent of the use per dollar of GDP in 1980.

   a. Find a quadratic function that models these data, with x equal to the number of years after 2000 and y equal to the energy use per dollar of GDP, as a percent of the energy use per dollar of GDP in 1980. Report the model with three significant digits.
   $y = 0.00785x^2 - 1.22x + 66.792$

   b. Graph the model and the data on the same axes.

   c. Use technology to find when this model predicts that energy use per dollar of GDP will reach a minimum. 2078

Energy Use per Dollar of GDP Indexed to 1980

Year	Percent	Year	Percent	Year	Percent
2000	67	2015	51	2030	37
2005	60	2020	45	2035	34
2010	56	2025	41		

(Source: U.S. Department of Energy)

42. Average Annual Wage The following table shows the U.S. average annual wage in thousands of dollars for selected years from 2014 and projected to 2050.

   a. Find a power function that models these data, with x equal to the number of years after 2000 and y equal to the average annual wage in thousands of dollars. Report the model with four significant digits. $y = 2.465x^{1.096}$

   b. What does the model predict that the average annual wage will be in 2043? $151.8 thousand

   c. Use technology to find when the model predicts that the average annual wage will be $159.6 thousand. 2045

Year	Average Annual Wage ($ thousands)
2014	48.6
2016	53.3
2018	58.7
2020	63.7
2025	76.8
2030	93.2
2035	113.2
2040	137.6
2045	167.1
2050	202.5

(Source: Social Security Administration)

43. Insurance Rates The following table gives the monthly insurance rates for a $100,000 life insurance policy for smokers 35–50 years of age.

   a. Create a scatter plot for the data.

   b. Does it appear that a quadratic function can be used to model the data? If so, find the best-fitting quadratic model. Yes; $y = 0.052x^2 - 3.234x + 66.656$

   c. Find the power model that is the best fit for the data. $y = 0.011x^{2.038}$

   d. Compare the two models by graphing each model on the same axes with the data points. Which model appears to be the better fit? Quadratic model

242 Chapter 3 Quadratic, Piecewise-Defined, and Power Functions

Age (yr)	Monthly Insurance Rate ($)	Age (yr)	Monthly Insurance Rate ($)
35	17.32	43	23.71
36	17.67	44	25.11
37	18.02	45	26.60
38	18.46	46	28.00
39	19.07	47	29.40
40	19.95	48	30.80
41	21.00	49	32.55
42	22.22	50	34.47

(*Source*: American General Life Insurance Company)

44. **Worldwide Internet Users** The data in the table show the percent of people in the world who are Internet users from 2014 and projected to 2025.

   a. Find a quadratic function that models these data, with x representing the number of years after 2010 and with p representing the percent. Report the model with three significant digits.
   $p = -0.0921x^2 + 3.53x + 28.8$

   b. Graph the model and the data on the same axes.

   c. Find a power function that models these data, with x representing the number of years after 2010 and with p representing the percent. Report the model with three significant digits.
   $p = 27.3x^{0.299}$

   d. Graph this model and the data on the same axes.

   e. Graph both models with the scatter plot for years after 2025 to determine which of these two models may not be useful for the years after 2020.
   Quadratic; percent decreases after 2020

Year	Percent
2014	41.1
2015	44.3
2016	46.8
2017	49.0
2018	51.1
2019	53.0
2020	54.6
2025	61.0

(*Source*: eMarketer)

45. **U.S. Gross Domestic Product** The table gives the U.S. gross domestic product (GDP) (in trillions of dollars) for selected years from 2005 and projected to 2070.

   a. Find the best-fitting quadratic model for the data, with x equal to number of years after 2000.
   $y = 0.117x^2 - 3.792x + 45.330$

   b. Find the power model that is the best fit for the data, with x equal to number of years after 2000. $y = 0.711x^{1.342}$

   c. Compare the two models by graphing each model on the same axes with the data points. Which model appears to be the better fit? Quadratic

Year	GDP ($ trillions)
2005	12.145
2010	16.174
2015	21.270
2020	27.683
2025	35.919
2030	46.765
2035	61.100
2040	79.680
2045	103.444
2050	133.925
2055	173.175
2060	224.044
2065	290.042
2070	375.219

(*Source*: U.S. Bureau of Economic Analysis)

46. **Auto Noise** The noise level of a Volvo S60 increases as the speed of the car increases. The table gives the noise, in decibels (db), at different speeds.

Speed (mph)	Noise Level (db)
10	42
30	57
50	64
70	66
100	71

(*Source*: Road & Track, 2011)

   a. Fit a power function model to the data.
   $y = 25.425x^{0.228}$

   b. Graph the data points and model on the same axes.

   c. Use the result from part (a) to estimate the noise level at 80 mph. 69 db

47. **Crude Oil** The following table gives the U.S. crude oil production, in billions of barrels, for the years from 2010 projected to 2030.

   a. Graph a scatter plot of the data, with x equal to the number of years after 2000 and y equal to the number of billions of barrels of crude oil.

b. Find the power function that models the data, with x equal to the number of years after 2000 and y equal to the number of billions of barrels of crude oil.
$y = 1.428x^{1.408}$

c. Find the quadratic function that models the data, with x equal to the number of years after 2000 and y equal to the number of billions of barrels of crude oil. $y = -0.001x^2 + 0.058x + 1.473$

d. Use the power model to predict the number of billions of barrels of crude oil in 2040. 2.4

e. Use the quadratic model to predict the number of billions of barrels of crude oil in 2040. 2.1

f. If the crude oil production begins to decrease after 2030, which model is the better fit for the data?
Quadratic model

Year	Billions of Barrels
2010	1.94
2014	2.10
2018	2.16
2022	2.24
2026	2.26
2030	2.26

(*Source*: U.S. Energy Information Administration)

48. **Total Energy Used for Industry** The table gives the total energy used for industry, in millions of BTUs, for the years from 2016 and projected to 2025.

a. Graph the data from the table as a scatter plot, with y equal to the energy used in millions of BTUs and x equal to the number of years after 2015.

b. Find the quadratic function that is the best fit to the data, with y equal to the energy used in millions of BTUs and x equal to the number of years after 2015. $y = -0.0354x^2 + 0.959x + 29.8$

c. Graph the quadratic function on the same axes with the data points.

d. Find the power function that is the best fit to the data, with y equal to the energy used in millions of BTUs and x equal to the number of years after 2015. $y = 30.3x^{0.0703}$

e. Graph the power function on the same axes with the data points.

f. Which function is the better visual fit for the data? Quadratic function

Year	BTUs (millions)
2016	30.85
2017	31.37
2018	32.36
2019	33.21
2020	33.64
2021	34.18
2022	34.77
2023	35.33
2024	35.64
2025	35.74

(*Source*: U.S. Energy Information Administration)

Preparing for CALCULUS

*An important calculation in calculus is the evaluation of the **difference quotient**, $\dfrac{f(x+h)-f(x)}{h}$, where $h \neq 0$, which can be found for **quadratic functions**.*

1. **a.** Find the difference quotient for $f(x) = x^2 + 2x$. $2x + h + 2$

   b. What does $\dfrac{f(x+h)-f(x)}{h}$ approach as h becomes very close to 0? $2x + 2$

2. **a.** Find the difference quotient for $f(x) = -2x^2 + 3x - 1$. $-4x - 2h + 3$

   b. What does $\dfrac{f(x+h)-f(x)}{h}$ approach as h becomes very close to 0? $-4x + 3$

*In calculus, we can find where the graph of the function $y = f(x)$ turns by finding where a derivative $f'(x)$ equals 0, so solving equations is an important skill. In Chapter 3 we solved **quadratic equations**. Recall that to solve a quadratic equation, we can use factoring, completing the square, the square root method, or the quadratic formula. In Exercises 3–6, the derivative $f'(x)$ of a function is given. Find where $f'(x) = 0$.*

3. $f'(x) = 3x^2 - 6x - 9$ $3, -1$

4. $f'(x) = 6x^2 - 7x + 2$ $\dfrac{2}{3}, \dfrac{1}{2}$

5. $f'(x) = x[3(x-2)^2] - (x-2)^3$ $2, -1$

6. $f'(x) = (x^2 - 1)3(x-5)^2 + (x-5)^3(2x)$ $5, \dfrac{5 + 2\sqrt{10}}{5}, \dfrac{5 - 2\sqrt{10}}{5}$

Equations containing y^2 occasionally must be solved for y. Sometimes this can be accomplished with the root method, and sometimes a method like the quadratic formula is necessary. Solve the equations in Exercises 7–8 for y.

7. $y^2 + 4x - 3 = 0$
   $y = \pm\sqrt{3 - 4x}$

8. $5x - y^2 = 12$
   $y = \pm\sqrt{5x - 12}$

To perform calculus operations on functions, it is frequently desirable or necessary to write the function in the form $y = cx^b$, where b is a constant. Write each of the functions in Exercises 9–12 in this form.

9. $y = 3\sqrt{x}$
   $3x^{1/2}$

10. $y = 6\sqrt[3]{x}$
    $6x^{1/3}$

11. $y = 2\sqrt[3]{x^2}$
    $2x^{2/3}$

12. $y = 4\sqrt{x^3}$
    $4x^{3/2}$

Write each of the functions in Exercises 13–16 without radicals by using fractional exponents.

13. $y = \sqrt[3]{x^2 + 1}$ $y = (x^2 + 1)^{1/3}$

14. $y = \sqrt[4]{x^3 - 2}$ $y = (x^3 - 2)^{1/4}$

15. $y = \sqrt{x} + \sqrt[3]{2x}$
    $x^{1/2} + (2x)^{1/3}$

16. $y = \sqrt[3]{(2x)^2} + \sqrt{(4x)^3}$
    $(2x)^{2/3} + (4x)^{3/2}$

*In calculus, **limits** are used to determine if a function is continuous. We can say that a function is continuous at all points in its domain if it can be drawn without lifting a pencil. For piecewise-defined functions, we are frequently interested in whether the two "pieces" fit together, for if they do, the function is continuous.*

17. Use graphs of the following functions to determine if they are continuous over the given domain.

    a. $\begin{cases} 9 - x^2 & \text{if } x < 3 \\ x - 3 & \text{if } x \geq 3 \end{cases}; 0 \leq x \leq 6$
    Continuous

    b. $\begin{cases} 2x^2 - 4 & \text{if } x < 2 \\ 5 & \text{if } x \geq 2 \end{cases}; -1 \leq x \leq 5$
    Not continuous

chapter 3 SUMMARY

In this chapter, we discussed in depth quadratic functions, including realistic applications that involve the vertex and x-intercepts of parabolas and solving quadratic equations. We then studied piecewise-defined functions, power functions, root functions, and other nonlinear functions. Real-world data are provided throughout the chapter, and we fit power and quadratic functions to some of these data.

Key Concepts and Formulas

3.1 Quadratic Functions; Parabolas

Quadratic function	Also called a *second-degree polynomial function*, this function can be written in the form $f(x) = ax^2 + bx + c$, where $a \neq 0$.
Parabola	A parabola is the graph of a quadratic function.
Vertex	The turning point on a parabola is the vertex.
Maximum point	If a quadratic function has $a < 0$, the vertex of the parabola is the maximum point, and the parabola opens downward.
Minimum point	If a quadratic function has $a > 0$, the vertex of the parabola is the minimum point, and the parabola opens upward.
Axis of symmetry	The axis of symmetry is the vertical line through the vertex of the parabola.
Forms of quadratic functions	$y = a(x-h)^2 + k$ Vertex at (h, k) Axis of symmetry is the line $x = h$ Parabola opens up if a is positive and down if a is negative.	$y = ax^2 + bx + c$ x-coordinate of vertex at $x = -\dfrac{b}{2a}$ Axis of symmetry is the line $x = -\dfrac{b}{2a}$ Parabola opens up if a is positive and down if a is negative.

3.2 Solving Quadratic Equations

Solving quadratic equations	An equation that can be written in the form $ax^2 + bx + c = 0$, $a \neq 0$, is called a *quadratic equation*.
Zero product property	For real numbers a and b, if the product $ab = 0$, then at least one of the numbers a or b must equal 0.
Solving by factoring	To solve a quadratic equation by factoring, write the equation in a form with 0 on one side. Then factor the nonzero side of the equation, if possible, and use the zero product property to convert the equation into two linear equations that are easily solved.
Solving and checking graphically	Solutions or decimal approximations of solutions to quadratic equations can be found by using TRACE, zero, or intersect with a graphing utility.
Factor Theorem	The factorization of a polynomial function $f(x)$ and the real solutions to $f(x) = 0$ can be verified by graphing $y = f(x)$ and observing where the graph crosses the x-axis. If $x = a$ is a solution to $f(x) = 0$, then $(x - a)$ is a factor of f.
Solving using the square root method	When a quadratic equation has the simplified form $x^2 = C$, the solutions are $x = \pm\sqrt{C}$. This method can also be used to solve equations of the form $(ax + b)^2 = C$.
Completing the square	A quadratic equation can be solved by converting one side to a perfect binomial square and taking the square root of both sides.
Solving using the quadratic formula	The solutions of the quadratic equation $ax^2 + bx + c = 0$, with $a \neq 0$, are given by the formula $$x = \frac{-b \pm \sqrt{b^2 - 4ac}}{2a}$$
Discriminant	The expression $b^2 - 4ac$ of the quadratic equation $y = ax^2 + bx + c$ determines the nature of the solutions of $ax^2 + bx + c = 0$.
Solutions, zeros, x-intercepts, and factors	If a is a real number, the following three statements are equivalent: • a is a real solution to the equation $f(x) = 0$. • a is a real zero of the function $f(x)$. • a is an x-intercept of the graph of $y = f(x)$.

3.3 Power and Root Functions

Power functions	A power function is a function of the form $y = ax^b$, where a and b are real numbers, $b \neq 0$.
Cubing function	A special power function, $y = x^3$, is called the cubing function.
Root functions	A root function is a function of the form $y = ax^{1/n}$, or $y = a\sqrt[n]{x}$, where n is an integer, $n \geq 2$.
Reciprocal function	This function is formed by the quotient of one and the identity function, $f(x) = \dfrac{1}{x}$, and can be written $f(x) = x^{-1}$.

Power and root equations	An equation of the form $k = ax^n$, where n is an integer, can be solved (if a solution exists) by rewriting the equation with x^n on one side and then taking the nth root of both sides of the equation. If n is even, you must take the $\pm$ nth root of both sides of the equation.
Direct variation as the nth power	If a quantity y varies directly as the nth power of x, then y is directly proportional to the nth power of x and $$y = kx^n$$ The number k is called the constant of variation or the constant of proportionality.

3.4 Piecewise-Defined Functions and Absolute Value Functions

Piecewise-defined functions	This is a function that is defined in pieces for different intervals of the domain. Piecewise-defined functions can be graphed by graphing their pieces on the same axes. A familiar example of a piecewise-defined function is the *absolute value function*.
Absolute value function	$$|x| = \begin{cases} x & \text{if } x \geq 0 \\ -x & \text{if } x < 0 \end{cases}$$
Absolute value equation	The solution of the absolute value equation $|x| = a$ is $x = a$ or $x = -a$ when $a \geq 0$. There is no solution to $|x| = a$ if $a < 0$.
Greatest integer function	A special piecewise-defined function is the **greatest integer function**, denoted $f(x) = [\![x]\!]$ and defined as $[\![x]\!]$ is the greatest integer less than or equal to x.

3.5 Quadratic and Power Models

Modeling a quadratic function from three points on its graph	If we know three (or more) points that fit exactly on a parabola, we can find the quadratic function whose graph is the parabola.
Quadratic modeling	The use of graphing utilities permits us to fit quadratic functions to data by using technology.
Second differences	If the second differences of data are constant for equally spaced inputs, a quadratic function is an exact fit for the data.
Power modeling	The use of graphing utilities permits us to fit power functions to nonlinear data by using technology.

chapter 3 SKILLS CHECK

Answers that are not seen can be found in the answer section at the back of the text.

In Exercises 1–8, (a) give the coordinates of the vertex of the graph of each quadratic function and (b) graph each function in a window that includes the vertex and all intercepts.

1. $y = (x - 5)^2 + 3$
   a. (5, 3)
2. $y = (x + 7)^2 - 2$
   a. (−7, −2)
3. $y = 3x^2 - 6x - 24$
   a. (1, −27)
4. $y = 2x^2 + 8x - 10$
   a. (−2, −18)
5. $y = -x^2 + 30x - 145$ a. (15, 80)
6. $y = -2x^2 + 120x - 2200$ a. (30, −400)
7. $y = x^2 - 0.1x - 59.998$ a. (0.05, −60.0005)
8. $y = x^2 + 0.4x - 99.96$ a. (−0.2, −100)

In Exercises 9 and 10, use factoring to solve the equations.

9. $x^2 - 5x + 4 = 0$
   $x = 4, x = 1$
10. $6x^2 + x - 2 = 0$
    $x = \frac{1}{2}, x = -\frac{2}{3}$

Use a graphing utility as an aid in factoring to solve the equations in Exercises 11 and 12.

11. $5x^2 - x - 4 = 0$ $x = -\frac{4}{5}, x = 1$

12. $3x^2 + 4x - 4 = 0$ $x = -2, x = \frac{2}{3}$

In Exercises 13 and 14, use the quadratic formula to solve the equations.

13. $x^2 - 4x + 3 = 0$ $x = 3, x = 1$

14. $4x^2 + 4x - 3 = 0$ $x = \frac{1}{2}, x = -\frac{3}{2}$

15. **a.** Use graphical and algebraic methods to find the x-intercepts of the graph of $f(x) = 3x^2 - 6x - 24$. $4, -2$

    b. Find the solutions to $f(x) = 0$ if $f(x) = 3x^2 - 6x - 24$. $4, -2$

16. **a.** Use graphical and algebraic methods to find the x-intercepts of the graph of $f(x) = 2x^2 + 8x - 10$. $-5, 1$

    b. Find the solutions to $f(x) = 0$ if $f(x) = 2x^2 + 8x - 10$. $-5, 1$

In Exercises 17 and 18, use the square root method to solve the equations.

17. $5x^2 - 20 = 0$ $2, -2$

18. $(x - 4)^2 = 25$ $9, -1$

In Exercises 19–22, find the exact solutions to the equations in the complex numbers.

19. $z^2 - 4z + 6 = 0$ $z = 2 \pm i\sqrt{2}$

20. $w^2 - 4w + 5 = 0$ $w = 2 \pm i$

21. $4x^2 - 5x + 3 = 0$ $x = \frac{5 \pm i\sqrt{23}}{8}$

22. $4x^2 + 2x + 1 = 0$ $x = \frac{-1 \pm i\sqrt{3}}{4}$

In Exercises 23–30, graph each function.

23. $f(x) = \begin{cases} 3x - 2 & \text{if } x < -1 \\ 4 - x^2 & \text{if } x \geq -1 \end{cases}$

24. $f(x) = \begin{cases} 4 - x & \text{if } x \leq 3 \\ x^2 - 5 & \text{if } x > 3 \end{cases}$

25. $f(x) = 2x^3$

26. $f(x) = x^{3/2}$

27. $f(x) = \sqrt{x - 4}$

28. $f(x) = \frac{1}{x} - 2$

29. $y = x^{4/5}$

30. $y = \sqrt[3]{x + 2}$

31. Determine whether the function $y = -3x^2$ is increasing or decreasing

    a. For $x < 0$. Increasing   **b.** For $x > 0$. Decreasing

32. For each of the functions, determine whether the function is concave up or concave down.

    a. $y = x^{5/4}$ Concave up   **b.** $y = x^{4/5}$ for $x > 0$ Concave down

In Exercises 33–36, solve the equation and check for extraneous solutions.

33. $4x^3 = 108$ $x = 3$

34. $\frac{1}{2}x^{-3} = -128$ $x = -\frac{1}{4\sqrt[3]{4}}$

35. $2p^{1/3} = \frac{1}{3}$ $p = \frac{1}{216}$

36. $2\sqrt[4]{x} = -16$ No solution

37. Solve $|3x - 6| = 24$. $x = 10, x = -6$

38. Solve $|2x + 3| = 13$. $x = 5, x = -8$

39. Find the equation of a quadratic function whose graph is a parabola passing through the points $(0, -2), (-2, 12)$, and $(3, 7)$. $y = 2x^2 - 3x - 2$

40. Find the equation of a quadratic function whose graph is a parabola passing through the points $(-2, -9), (2, 7)$, and $(4, -9)$. $y = -2x^2 + 4x + 7$

41. Find a power function that models the data below. $y = 3.545x^{1.323}$

x	1	2	3	4	5	6
y	4	9	11	21	32	45

42. Find a quadratic function that models the data below. $y = 1.043x^2 - 0.513x + 0.977$

x	1	3	4	6	8
y	2	8	15	37	63

43. Suppose that q varies directly as the 3/2 power of p and that $q = 16$ when $p = 4$. Find q when $p = 16$. 128

44. If $f(x) = \begin{cases} 3x - 2 & \text{if } -8 \leq x < 0 \\ x^2 - 4 & \text{if } 0 \leq x < 3, \\ -5 & \text{if } x \geq 3 \end{cases}$ find $f(-8), f(0),$ and $f(4)$. $-26, -4, -5$

chapter 3 REVIEW

Answers that are not seen can be found in the answer section at the back of the text.

45. **Maximizing Profit** The monthly profit from producing and selling x units of a product is given by the function $P(x) = -0.01x^2 + 62x - 12,000$.

    a. Producing and selling how many units will result in the maximum profit for this product? 3100

    b. What is the maximum possible profit for the product? $84,100

46. **Profit** The revenue from sales of x units of a product is given by $R(x) = 200x - 0.01x^2$, and the cost of producing and selling the product can be described by $C(x) = 38x + 0.01x^2 + 16,000$.

    a. Producing and selling how many units will give maximum profit? 4050

    b. What is the maximum possible profit from producing and selling the product? $312,050

47. Height of a Ball If a ball is thrown into the air at 64 feet per second from a height of 192 feet, its height (in feet) is given by $S(t) = 192 + 64t - 16t^2$, where t is in seconds.

  a. In how many seconds will the ball reach its maximum height? 2 sec

  b. What is the maximum possible height for the ball? 256 ft

48. Height of a Ball If a ball is thrown into the air at 29.4 meters per second from a height of 60 meters, its height (in meters) is given by $S(t) = 60 + 29.4t - 9.8t^2$, where t is in seconds.

  a. In how many seconds will the ball reach its maximum height? 1.5 sec

  b. What is the maximum possible height for the ball? 82.05 m

49. Home Range The home range of an animal is the region to which the animal confines its movements. The area, in hectares, of the home range of a meat-eating mammal can be modeled by the function $H(x) = 0.11x^{1.36}$, where x is the mass of the animal in grams. What would the home range be for a bobcat weighing 1.6 kg? 2506 ha
(*Source*: *The American Naturalist*)

50. Break-Even The profit for a product is given by $P = -3600 + 150x - x^2$, where x is the number of units produced and sold. How many units will give break-even (that is, return a profit of 0)? 30 or 120

51. Falling Ball If a ball is dropped from the top of a 400-foot-high building, its height S in feet is given by $S(t) = 400 - 16t^2$, where t is in seconds. In how many seconds will it hit the ground? 5 sec

52. Profit The profit from producing and selling x units of a product is given by the function $P(x) = -0.3x^2 + 1230x - 120,000$ dollars. Producing and selling how many units will result in a profit of $324,000 for this product? 400 or 3700

53. Millionaire's Tax Rate The effective tax rate for a head of household earning the equivalent of $1 million in non-investment income can be modeled by the function

$$T(x) = \begin{cases} 0.08x^2 - 2.64x + 22.35 & \text{if } 15 \leq x \leq 45 \\ -0.525x + 89.82 & \text{if } 45 < x \leq 110 \end{cases}$$

where x is the number of years after 1900.

  a. Graph the function for $15 \leq x \leq 110$.

  b. According to the model, what was the tax rate for a millionaire head of household in 2018? 27.87%

  c. In 2010, with President Bush's tax cuts in effect, the tax rate was 32.4%. Does the model agree with this rate? Yes; the model gives 32.07%.
(*Source*: The Tax Foundation)

54. Worldwide 5G Subscriptions The number of millions of 5G subscriptions worldwide for the years from 2020 and projected to 2025 can be modeled by the function $f(t) = 54.5t^2 + 251t + 1$, where t equals the number of years after 2020. Use the quadratic formula to find the year after 2020 when the projected subscriptions will be 7961 million. 2030 ($x = 10$)
(*Source*: Business Wire)

55. Internet of Things Devices The Internet of Things (IoT) market had 15.4 billion devices connected in 2015 and the number of devices is projected to double by 2025. According to IHS, the projected number of devices, in billions, can be modeled by

$$y = 0.590x^2 - 0.275x + 17.201$$

with x equal to the number of years after 2015 and y equal to the billions of devices connected to the Internet. In what year will the number of devices reach 98.8 billion, according to the model? 2027 ($x = 12$)

56. U.S. Households The percent of total households that have married couples for selected years from 1982 and projected to 2050 can be modeled by $y = 66.2x^{-0.094}$, with x representing the number of years after 1980 and y representing the percent. In what year after 1980 does the model predict that the percent of total households that have married couples will be 46.1%? 2027
(*Source*: U.S. Census Bureau)

57. Global Biometrics Revenue The annual revenue from the global biometrics market for the years from 2016 and projected to 2025 can be modeled by the function

$$y = 0.143x^2 - 0.259x + 2.333$$

with x equal to the number of years after 2015 and y equal to the revenue in billions of dollars. Use the quadratic formula to find the year after 2016 when the annual revenue is projected to be 23.133. 2028 ($x = 13$)
(*Source*: Tractica)

58. Telecom Artificial Intelligence The worldwide software revenue for artificial intelligence used in telecommunications, in millions of dollars, for the years from 2016 and projected to 2025 can be modeled by the function $y = 170x^2 - 2356x + 8463$, with x equal to the number of years after 2010. Use technology to predict when after 2016 the worldwide software revenue for artificial intelligence used in telecommunications will reach $29.343 billion. 2030
(*Source*: Tractica)

59. Emigration to the United States For the years from 2015 and projected to 2060, the rate of emigration to the United States from Spanish Caribbean and Latin American countries per thousand in their population can be modeled by the function

$$f(x) = \begin{cases} 0.011x + 0.94 & \text{if } 12 \leq x < 20 \\ 1.16 & \text{if } 20 \leq x \leq 50 \end{cases}$$

where x is the number of years after 2010.

a. Evaluate $f(22)$. What does this mean? to the U.S. from Spanish Caribbean and Latin American countries is expected to be 1.16 per thousand. 1.16; in 2032 the rate of emigration

b. According to the model, during what year is the rate of emigration expected to be 1.149 per thousand? 2029

60. *Income Tax* The U.S. federal income tax owed by a married couple filing jointly in 2018 for taxable incomes between $165,001 and $600,00 can be modeled by the piecewise-defined function

$$T(x) = \begin{cases} 28{,}179 + 0.24(x - 165{,}000) & \text{if } 165{,}001 \leq x \leq 315{,}000 \\ 64{,}179 + 0.32(x - 315{,}000) & \text{if } 315{,}001 \leq x \leq 400{,}000 \\ 91{,}379 + 0.35(x - 400{,}000) & \text{if } 400{,}001 \leq x \leq 600{,}000 \end{cases}$$

a. Use the function to find $T(360{,}000)$. What does this mean? 78,579; a couple with taxable income of $360,000 will pay $78,579 in taxes.

b. Find the tax owed on a taxable income of $430,000. $101,879

c. How much more is the tax owed in part (b) than the maximum tax owed for the bracket 315,001–400,000? $10,500

d. If a couple paid $58,179 in taxes, what was their taxable income? $290,000

61. *Insurance Premiums* The following table gives the monthly premiums required for a $250,000 term life insurance policy on a 35-year-old female nonsmoker for different guaranteed term periods.

a. Find a quadratic function that models the monthly premium as a function of the length of term for a 35-year-old female nonsmoking policyholder. Report your answer to five decimal places.
$y = 0.05143x^2 + 3.18286x + 65.40000$

b. Assuming that the domain of the function contains integer values between 10 years and 30 years, what term in years could a 35-year-old nonsmoking female purchase for $130 a month? 16 yr

Term Period (years)	Monthly Premium for 35-Year-Old Female ($)
10	103
15	125
20	145
25	183
30	205

(*Source*: Quotesmith.com)

62. *National Health Expenditures* Health spending is projected to grow at an average rate of 5.8% from 2012 through 2022, 1.0 percentage point faster than the expected average annual growth in the gross domestic product (GDP). The table gives the national health care expenditures, in billions of dollars, from 2006 and projected to 2022.

Year	Expenditures ($ billions)	Year	Expenditures ($ billions)
2006	2,163	2015	3,273
2007	2,298	2016	3,458
2008	2,407	2017	3,660
2009	2,501	2018	3,889
2010	2,600	2019	4,142
2011	2,701	2020	4,416
2012	2,807	2021	4,702
2013	2,915	2022	5,009
2014	3,093		

(*Source*: Centers for Medicare & Medicaid Services)

a. Create a scatter plot of the data, with x equal to the number of years after 2000 and y equal to the national health care expenditures in billions of dollars.

b. Choose an appropriate function to model the data, and write the equation of that model. Quadratic is most appropriate; $y = 7.897x^2 - 50.650x + 2267.916$

c. Graph your model and the scatter plot on the same axes.

d. Use the rounded model to find the average rate of change of health care expenditures from 2006 to 2012; from 2012 to 2022. Round to one decimal place.
91.5 billion dollars per year; 217.8 billion dollars per year

e. Use the rounded model to find the percent increase in health care expenditures from 2012 to 2013; from 2021 to 2022. Round to one decimal place. Is the statement that health spending is projected to grow at an average rate of 5.8 percent from 2012 through 2022 reasonable? 5.2%; 6.1%; yes

63. *Prescription Drug Expenditures* In 2012, prescription drug spending is estimated to have accounted for $260.8 billion of national health spending, a decline of 0.8%, compared to 2.9% growth in 2011. This decline was due to increased adoption of generic drugs as a number of popular brand-name drugs lost patent protection, increases in cost-sharing requirements, and lower spending on new medicines. The table gives expenditures for prescription drugs, in billions of dollars, for the years 2006 and projected to 2022.

a. Create a scatter plot of the data, with x equal to the number of years after 2000 and y equal to the prescription drug expenditures in billions of dollars.

b. For the data from 2006 to 2013, find a quadratic function to model the data with x equal to the number of years after 2000. Round to two decimal places. $y = -0.99x^2 + 24.27x + 114.11$

c. For the data from 2014 to 2022, find a quadratic function to model the data with x equal to the number of years after 2000. Round to two decimal places. $y = 0.93x^2 - 11.55x + 256.47$

d. Join the two functions from parts (b) and (c) to create a piecewise function that models the data from 2006 to 2022.

e. Use the model to estimate the prescription drug expenditures for 2010 and 2022.
$257.5 billion; $454.1 billion

Year	Expenditures ($ billions)	Year	Expenditures ($ billions)
2006	224.1	2015	294.9
2007	235.9	2016	311.6
2008	242.6	2017	330.7
2009	254.6	2018	350.6
2010	255.7	2019	372.7
2011	263.0	2020	397.9
2012	260.8	2021	425.5
2013	262.3	2022	455.0
2014	275.9		

(*Source*: Centers for Medicare & Medicaid Services)

64. **Personal Income** The table shows the total personal income in the United States (in billions of dollars) for selected years from 1960 and projected to 2024.

a. Find the best-fitting power function that will model the data, with x equal to the number of years after 1950 and y in billions of dollars. Report the model with three significant digits.
$y = 2.15(x^{2.12})$

b. Graph the data and the model on the same axes. Is the model an excellent fit for the data?
Not an excellent fit

c. Does the unrounded model overestimate or underestimate the total personal income given in the table for 2024? Underestimates

d. In what year does the model predict the total personal income will reach $21.971 trillion? 2028

Year	Income ($ billions)	Year	Income ($ billions)
1960	411.5	2008	12,100.7
1970	838.8	2014	14,728.6
1980	2307.9	2018	19,129.6
1990	4878.6	2024	22,685.1
2000	8429.7		

(*Source*: U.S. Bureau of Labor Statistics)

65. **Worldwide 5G Subscriptions** The table gives the number of millions of 5G subscriptions worldwide for the years from 2020 and projected to 2025.

a. Find the linear function that models $S(t)$, the millions of 5G subscriptions worldwide, as a function of t, the number of years after 2020. Report the model with three significant digits.
$S(t) = 523t - 180$

b. Find the quadratic function that models $S(t)$, the millions of 5G subscriptions worldwide, as a function of the number of years after 2020. Report the model with three significant digits.
$S(t) = 54.5t^2 + 251t + 1.07$

c. Graph each function on the same axes with the data to determine which function is the better model for the data. The quadratic function is better.

Year	Subscriptions (millions)	Year	Subscriptions (millions)
2020	40	2023	1310
2021	230	2024	1850
2022	720	2025	2610

(*Source*: Business Wire)

66. **Artificial Intelligence** The table gives the revenue (in billions of dollars) from artificial intelligence for enterprise applications in North America for the years from 2016 through 2025.

a. Find the power function that is the best fit for the data, with x equal to the number of years after 2015 and y equal to the corresponding revenue in billions of dollars.
$y = 0.132x^{1.921}$

b. Graph the aligned data and model on the same axes. Is the model an excellent fit?

c. Use the model to find the enterprise AI revenue in 2035.
$41.832 billion

Year	Revenue ($ billions)	Year	Revenue ($ billions)
2016	0.212	2021	3.665
2017	0.434	2022	5.633
2018	0.796	2023	8.237
2019	1.376	2024	11.395
2020	2.288	2025	14.896

(*Source*: Statista)

67. **Global Biometrics Revenue** The table gives the annual revenue from the global biometrics market for the years from 2016 and projected to 2025.

a. Create a scatter plot for the data, with x equal to the number of years after 2015 and y equal to the revenue in billions of dollars.

b. Find the quadratic function that models these data, with x equal to the number of years after 2015 and y equal to the revenue in billions of dollars.
$y = 0.143x^2 - 0.259x + 2.833$

c. Graph the model on the same axes with the data, and discuss how well the function fits the data.

d. What does the model predict that the global biometrics revenue will be in 2041?
$93.4 billion

e. In what year is the revenue $17.4 billion, according to the model?
In 2026

Year	Revenue ($ Billions)	Year	Revenue ($ Billions)
2016	2.4	2021	6.4
2017	2.9	2022	7.8
2018	3.6	2023	9.5
2019	4.5	2024	12.0
2020	5.3	2025	15.1

(Source: Tractica)

68. **Housing Starts** The numbers of housing units started, in thousands, for the years 2000–2019, are shown in the table. The construction industry was one of the hardest hit by the recession of 2008, but the table indicates that housing units began increasing again in 2010.

Year	Housing Units Started (thousands)	Year	Housing Units Started (thousands)
2000	104.0	2010	38.9
2001	106.4	2011	40.2
2002	110.4	2012	47.2
2003	117.8	2013	58.7
2004	124.5	2014	60.7
2005	142.9	2015	73.0
2006	153.0	2016	74.3
2007	95.0	2017	82.3
2008	70.8	2018	91.6
2009	31.9	2019	87.0

(Source: U.S. Census Bureau)

a. Find a quadratic model for the number of housing starts from 2000 to 2006, with x equal to the number of years after 2000.
$y = 1.299x^2 + 0.568x + 104.126$

b. Find a linear model for the number of housing starts from 2007 to 2009, with x equal to the number of years after 2000.
$y = -31.55x + 318.3$

c. Find a linear model for the number of housing starts from 2010 to 2019, with x equal to the number of years after 2000. $y = 6.23x - 24.9$

d. Using the equations from parts (a)–(c), create a piecewise-defined function that represents the number of housing units from 2000 to 2019, with x equal to the number of years after 2000.

e. If the housing trend continues, how many new units will be started in 2030? 162 thousand

f. According to the model, during what year will housing reach the actual number of housing starts for 2004? 2024

69. **Diabetes** As the following table shows, projections indicate that the percent of U.S. adults with diabetes could dramatically increase.

Year	Percent
2010	15.7
2015	18.9
2020	21.1
2025	24.2
2030	27.2
2035	29.0
2040	31.4
2045	32.1
2050	34.3

(Source: Centers for Disease Control and Prevention)

a. Find a quadratic model that fits the data in the table, with $x = 0$ in 2000.
$y = -0.00482x^2 + 0.754x + 8.512$

b. Use the model to predict the percent of U.S. adults with diabetes in 2022. 22.8%

c. In what year does this model predict the percent of U.S. adults with diabetes will be 30.2%?
$x = 38.0$, so in 2038

Group Activities
▶ EXTENDED APPLICATIONS

1. Research and Modeling

Graphs displaying linear and nonlinear growth are frequently found in periodicals such as *Newsweek* and *Time*, in newspapers such as *USA Today* and the *Wall Street Journal*, and on websites on the Internet. Tables of data can also be found in these sources, especially on federal and state government websites such as www.census.gov.

Your mission is to find a company sales, stock price, biological growth, or sociological trend over a period of years that is nonlinear, to determine which type of nonlinear function is the best fit for the data, and to find a nonlinear function that is a model for the data.

A quadratic model will be a good fit for the data under the following conditions:

- If the data are presented as a graph that resembles a parabola or part of a parabola.
- If the data are presented in a table and a plot of the data points shows that they lie near some parabola or part of a parabola.
- If the data are presented in a table and the second differences of the outputs are nearly constant for equally spaced inputs.

A power model may be a good fit if the shape of the graph or of the plot of the data points resembles part of a parabola but a parabola is not a good fit for the data.

After creating the model, you should test the goodness of fit of the model to the data and discuss uses that you could make of the model.

Your completed project should include

a. A complete citation of the source of the data that you are using.
b. An original copy or photocopy of the data being used.
c. A scatter plot of the data.
d. The equation that you have created.
e. A graph containing the scatter plot and the modeled equation.
f. A statement about how the model could be used to make estimations or predictions about the trend you are observing.

Some helpful hints:

1. If you decide to use a relation determined by a graph that you have found, read the graph very carefully to determine the data points or contact the source of the data to get the data from which the graph was drawn.
2. Align the independent variable by letting x represent the number of years after some convenient year and then enter the data into a graphing utility and create a scatter plot.
3. Use your graphing utility to create the equation of the function that is the best fit for the data. Graph this equation and the data points on the same axes to see if the equation is reasonable.

2. China's GDP vs. United States' GDP

I. It is widely predicted that China's economy, measured by Gross Domestic Product (GDP), will eventually overtake that of the United States. Table 3.17 shows the GDPs for China and the United States from 2005 and projected to 2040 under the assumption that China's GDP will grow at an average annual rate of 6.5% and that the United States' GDP will grow at an average annual rate of 2%. Investigate when China's GDP will reach that of the United States under these assumptions about the countries' growth rates by using the following steps.

1. a. Create a scatter plot for the data, with x equal to the number of years after 2000 and y equal to China's GDP in trillions of dollars. Would these data be better modeled by a linear function or a quadratic function?

Table 3.17

Year	China GDP ($ trillions)	U.S. GDP ($ trillions)	Year	China GDP ($ trillions)	U.S. GDP ($ trillions)
2005	2.3	13.1	2026	21.2	23.2
2008	4.6	14.7	2029	25.6	24.6
2011	7.5	15.5	2032	30.9	26.1
2014	10.5	17.4	2035	37.3	27.7
2017	12.0	19.4	2038	45.1	29.4
2020	14.5	20.6	2040	51.1	30.6
2023	17.5	21.8			

(*Source*: Bloomberg.com)

b. Find the quadratic function that models these data, with x equal to the number of years after 2000 and y equal to the GDP in trillions of dollars. Report the model with three significant digits.
c. Graph the model on the same axes with the data and discuss how well the function fits the data.
d. What does the model predict that the GDP will be in 2052?
e. When will it reach $51.8 trillion?

2. a. Create a scatter plot for the data, with x equal to the number of years after 2000 and y equal to the United States' GDP in trillions of dollars.
b. Find the linear function that models these data, with x equal to the number of years after 2010 and y equal to the GDP in trillions of dollars. Report the model with three significant digits.
c. Graph the model on the same axes with the data, and discuss how well the function fits the data.
d. What does the model predict that the GDP will be in 2048?
e. When will it reach $35.2 trillion?

3. a. Find the intersection of the graphs of the reported models for the GDPs of China and the United States to determine the year in which China's economy is projected to reach that of the United States under the assumptions above.
b. What is the projected GDP for the countries when they are equal?

II. Suppose recent trade negotiations and economic changes result in projections that China's GDP will grow at an average annual rate of 6.0% and that the United States' GDP will grow at an average annual

Table 3.18

Year	China GDP ($ trillions)	U.S. GDP ($ trillions)	Year	China GDP ($ trillions)	U.S. GDP ($ trillions)
2005	2.3	13.1	2026	20.3	24.2
2008	4.6	14.7	2029	24.2	26.1
2011	7.5	15.5	2032	28.8	28.1
2014	10.5	17.4	2035	34.3	30.2
2017	12.0	19.4	2038	40.8	32.6
2020	14.3	20.9	2040	45.9	34.2
2023	17.0	22.5			

rate of 2.5%. The projected GDPs for China and the United States under these assumptions are shown in Table 3.18. Find when China's GDP will reach that of the United States under these assumptions about the countries' growth rates by using the following steps.

4. Find the quadratic function that models the data for China, with x equal to the number of years after 2000 and y equal to the GDP in trillions of dollars. Report the model with three significant digits.

5. Find the quadratic function that models the data for the United States, with x equal to the number of years after 2000 and y equal to the GDP in trillions of dollars. Report the model with three significant digits.

6. a. Find the intersection of the graphs of the reported models for the GDPs of China and the United States to determine the year in which China's economy is projected to reach that of the United States under these new assumptions.
b. What is the projected GDP for the countries when they are equal?

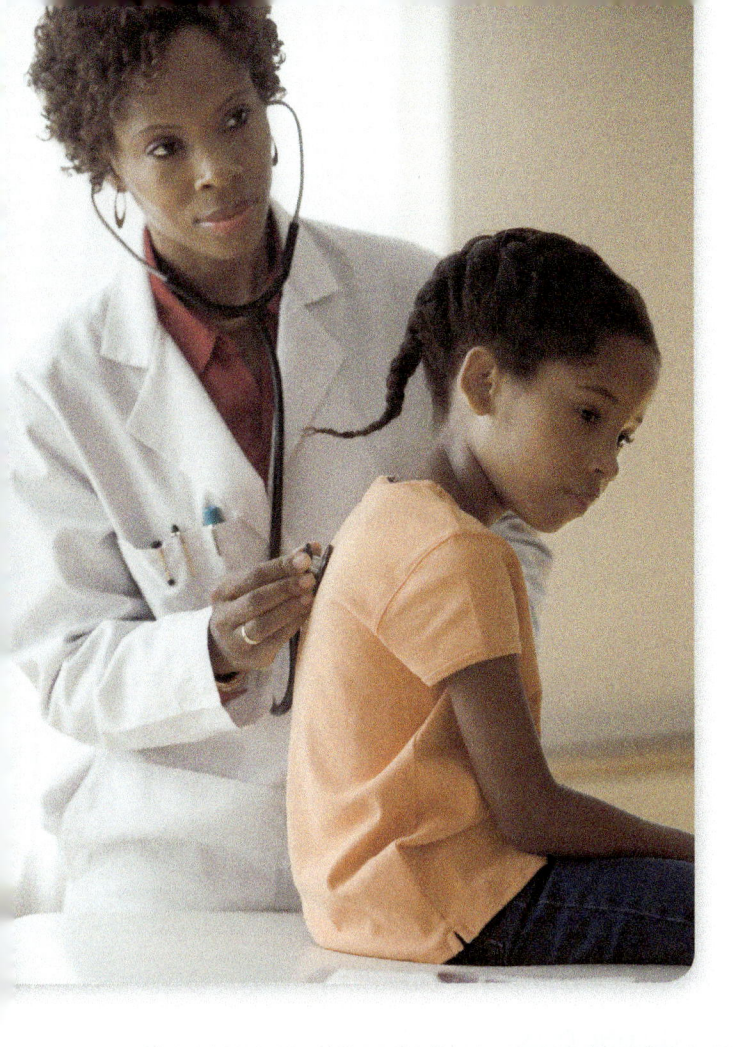

4

Additional Topics with Functions

A profit function can be formed by algebraically combining cost and revenue functions. An average cost function can be constructed by finding the quotient of two functions. Other new functions can be created using function composition and inverse functions. In this chapter, we study additional topics with functions to create functions and solve applied problems.

sections	objectives	applications
4.1 Transformations of Graphs and Symmetry	Find equations and graphs of vertically and horizontally shifted functions; stretch, compress, and reflect graphs; identify symmetry with respect to the x-axis, y-axis, and origin	Diabetes, profit, stimulus and response, pollution, velocity of blood
4.2 Combining Functions; Composite Functions	Add, subtract, multiply, and divide functions; find composition of functions	Profit, unconventional vehicle sales, average cost, canning orange juice
4.3 One-to-One and Inverse Functions	Identify inverse functions; find the inverse of a function; graph inverse functions; find inverse functions on limited domains	Loans, temperature measurement, investments, velocity of blood
4.4 Additional Equations and Inequalities	Solve radical equations, equations with rational powers, equations in quadratic form, and absolute value inequalities; solve quadratic inequalities analytically and graphically; solve power inequalities	Profit, height of a model rocket, obesity, investments

Algebra TOOLBOX

KEY OBJECTIVES

- Identify functions
- Find domains and ranges
- Determine if a function is increasing or decreasing over given intervals
- Determine over what intervals a function is increasing or decreasing

In this Toolbox, we create a library of functions studied so far, which sets the stage for the study of properties and operations with functions.

Linear Functions

$y = mx + b$

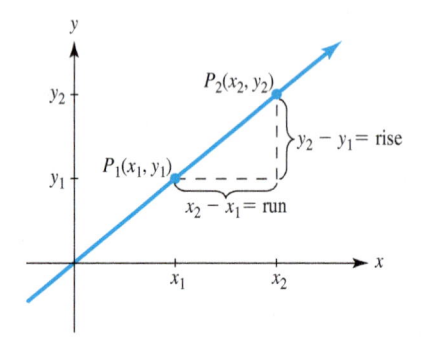

- m is the slope of the line, b is the y-intercept:

$$m = \frac{y_2 - y_1}{x_2 - x_1}$$

- The function is increasing if $m > 0$.
- The function is decreasing if $m < 0$.

Special Linear Functions

IDENTITY FUNCTION

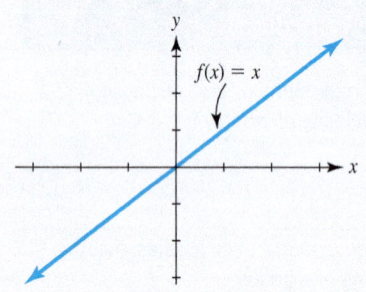

$f(x) = x$

Domain: $(-\infty, \infty)$
Range: $(-\infty, \infty)$

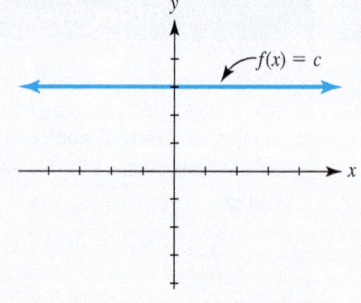

- Increases on its domain

CONSTANT FUNCTION

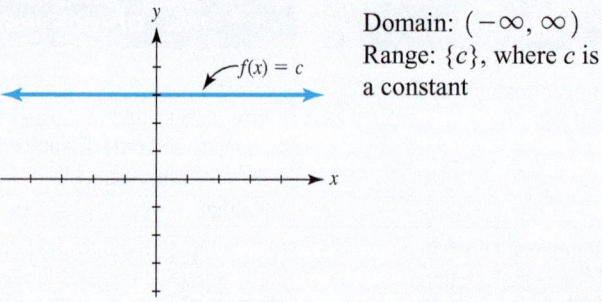

$f(x) = c$

Domain: $(-\infty, \infty)$
Range: $\{c\}$, where c is a constant

- Graph is a horizontal line.

Quadratic Functions

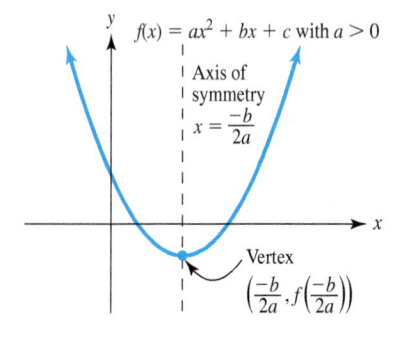

$f(x) = ax^2 + bx + c$ with $a > 0$

Axis of symmetry $x = \frac{-b}{2a}$

Vertex $\left(\frac{-b}{2a}, f\left(\frac{-b}{2a}\right)\right)$

$f(x) = ax^2 + bx + c$ with $a < 0$

Vertex $\left(\frac{-b}{2a}, f\left(\frac{-b}{2a}\right)\right)$

Axis of symmetry $x = \frac{-b}{2a}$

- Vertex is at the point $\left(\frac{-b}{2a}, f\left(\frac{-b}{2a}\right)\right)$.

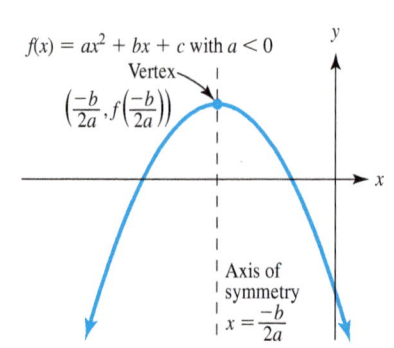

Piecewise-Defined Functions

$$f(x) = \begin{cases} -x^2 + 5 & \text{if } x < 1 \\ 3x - 4 & \text{if } x \geq 1 \end{cases}$$

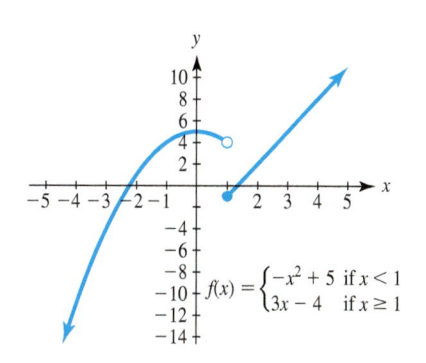

Special Piecewise-Defined Functions

ABSOLUTE VALUE FUNCTION

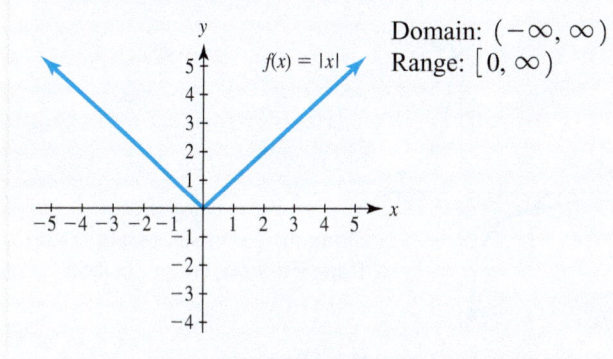

$f(x) = |x|$

Domain: $(-\infty, \infty)$
Range: $[0, \infty)$

- Decreases on $(-\infty, 0)$, increases on $(0, \infty)$

GREATEST INTEGER FUNCTION

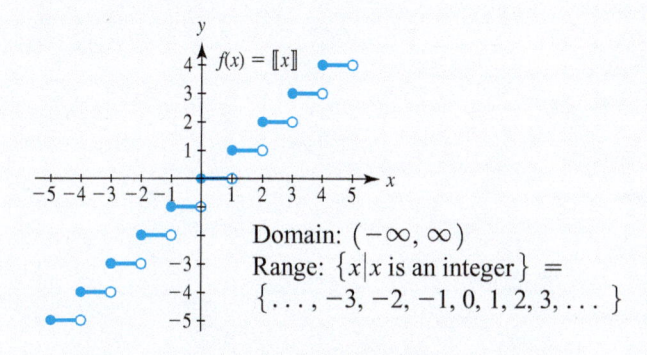

$f(x) = [\![x]\!]$

Domain: $(-\infty, \infty)$
Range: $\{x \mid x \text{ is an integer}\} = \{\ldots, -3, -2, -1, 0, 1, 2, 3, \ldots\}$

- $f(x) = [\![x]\!]$ is a special **step function**.

Power Functions

POWER FUNCTION, n EVEN

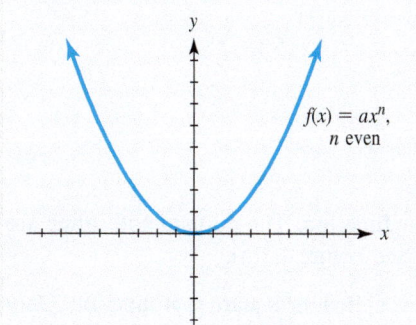

$f(x) = ax^n$, n even

- Decreases on $(-\infty, 0)$, increases on $(0, \infty)$ for $a > 0$
- Increases on $(-\infty, 0)$, decreases on $(0, \infty)$ for $a < 0$

POWER FUNCTION, n ODD

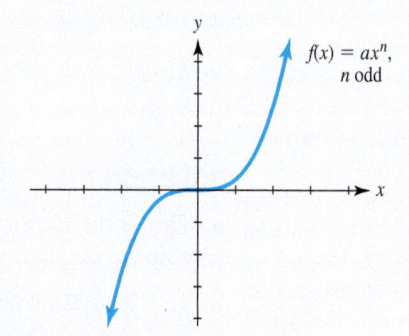

$f(x) = ax^n$, n odd

- Increases on $(-\infty, \infty)$ for $a > 0$
- Decreases on $(-\infty, \infty)$ for $a < 0$

Special Power Functions

ROOT FUNCTION, n EVEN

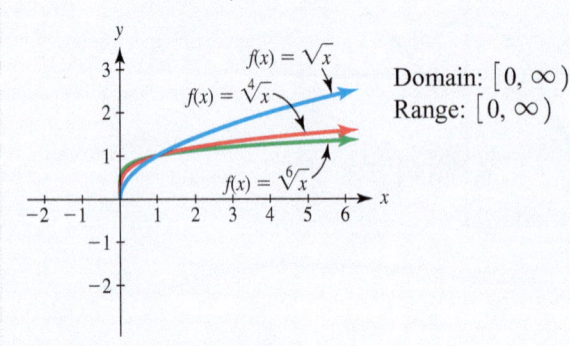

$f(x) = \sqrt{x}$
$f(x) = \sqrt[4]{x}$
$f(x) = \sqrt[6]{x}$

Domain: $[0, \infty)$
Range: $[0, \infty)$

- Increases on its domain

ROOT FUNCTION, n ODD

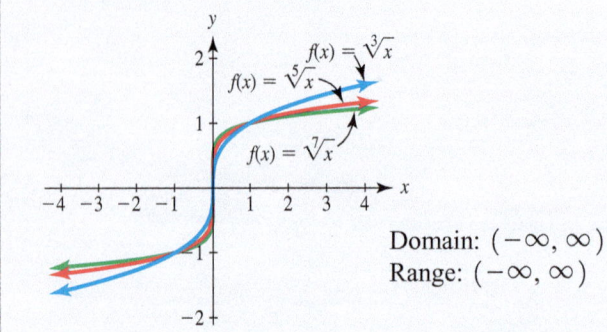

$f(x) = \sqrt[3]{x}$
$f(x) = \sqrt[5]{x}$
$f(x) = \sqrt[7]{x}$

Domain: $(-\infty, \infty)$
Range: $(-\infty, \infty)$

- Increases on its domain

RECIPROCAL FUNCTION

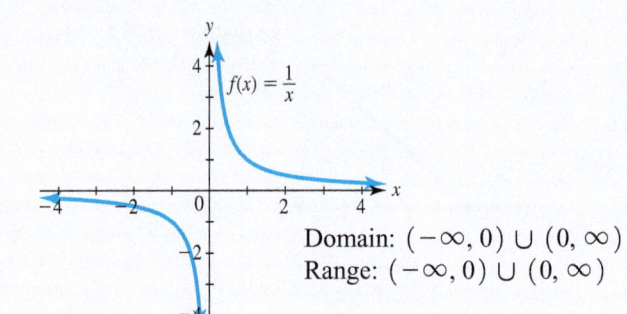

$f(x) = \dfrac{1}{x}$

Domain: $(-\infty, 0) \cup (0, \infty)$
Range: $(-\infty, 0) \cup (0, \infty)$

- Decreases on $(-\infty, 0)$ and $(0, \infty)$

EXAMPLE 1 ▶ **Special Functions**

Identify the type of each of the following functions from its equation. Then find the domain and range, and graph the function.

a. $f(x) = x$

b. $f(x) = \sqrt{x}$

c. $f(x) = 3$

SOLUTION

a. This is the identity function, a special linear function. The domain and range are both the set of all real numbers, $(-\infty, \infty)$. See Figure 4.1(a).

b. This is a special power function, $f(x) = x^{1/2}$, called the square root function. Only nonnegative values can be used for x, so the domain is $[0, \infty)$. The outputs will also be nonnegative values, so the range is $[0, \infty)$. See Figure 4.1(b).

c. This is a constant function, a special linear function. The domain is all real numbers, $(-\infty, \infty)$, and the range consists of the single value 3. See Figure 4.1(c).

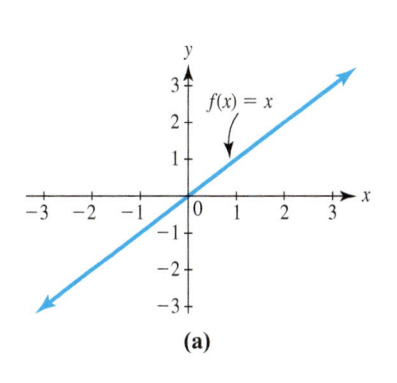

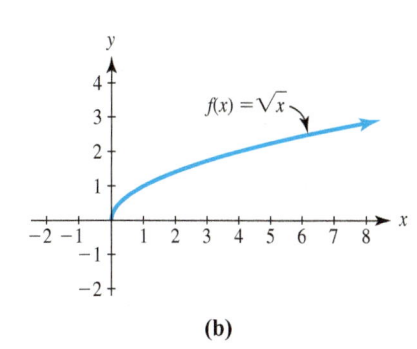

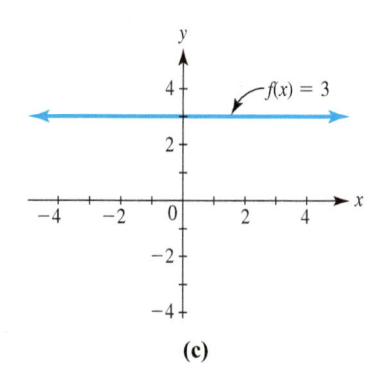

Figure 4.1

EXAMPLE 2 ▶ Special Functions

Identify the type of each of the following functions from its equation. Then find the domain and range, and graph the function.

a. $f(x) = x^3$
b. $f(x) = \sqrt[3]{x}$
c. $f(x) = \dfrac{5}{x^2}$

SOLUTION

a. This is a special power function, with power 3. It is called the cubing function, and both the domain and the range are the set of all real numbers, $(-\infty, \infty)$. See Figure 4.2(a).

b. This is a special power function, with power $1/3$. It is called the cube root function. Any real number input results in a real number output, so both the domain and the range are the set of all real numbers, $(-\infty, \infty)$. See Figure 4.2(b).

c. This function can be written as $f(x) = 5x^{-2}$, so it is a special power function, with power -2. If 0 is substituted for x, the result is undefined, so only nonzero real numbers can be used as inputs and only positive real numbers result as outputs. Thus, the domain is $(-\infty, 0) \cup (0, \infty)$, and the range is $(0, \infty)$. We will see in Chapter 6 that $f(x) = \dfrac{5}{x^2}$ is also called a **rational function**. See Figure 4.2(c).

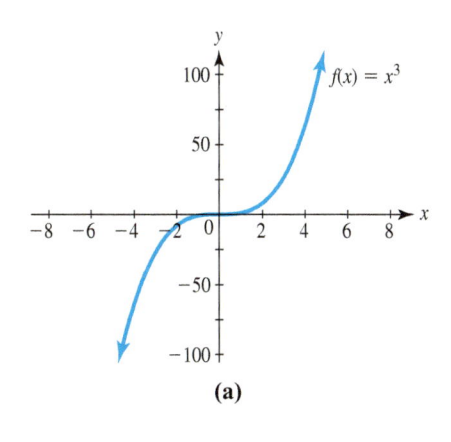

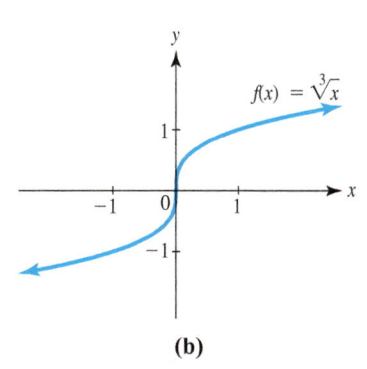

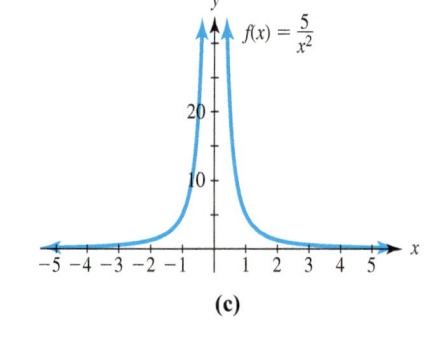

Figure 4.2

EXAMPLE 3 ▶ Functions

Identify the type of each of the following functions from its equation. Then find the domain and range, and graph the function.

a. $g(x) = 2x^2 - 4x + 3$ **b.** $f(x) = 3x - 18$ **c.** $f(x) = \begin{cases} 3x - 4 & \text{if } x < 2 \\ 8 - 2x & \text{if } x \geq 2 \end{cases}$

SOLUTION

a. This is a quadratic function. Any real number can be used as an input, and the resulting outputs are all real numbers greater than or equal to 1. Thus, the domain is $(-\infty, \infty)$, and the range is $[1, \infty)$. Its graph is a parabola that opens up. See Figure 4.3(a).

b. This is a linear function. Its graph is a line that rises as x increases, because its slope is positive. The domain and range are both $(-\infty, \infty)$. See Figure 4.3(b).

c. This is a piecewise-defined function, where each piece is a linear function. The domain is $(-\infty, \infty)$, and the range is $(-\infty, 4]$. See Figure 4.3(c).

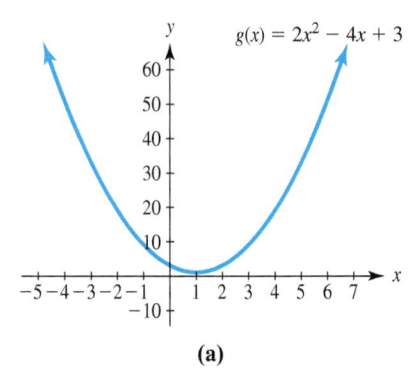

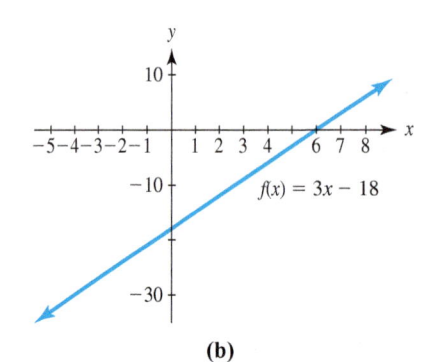

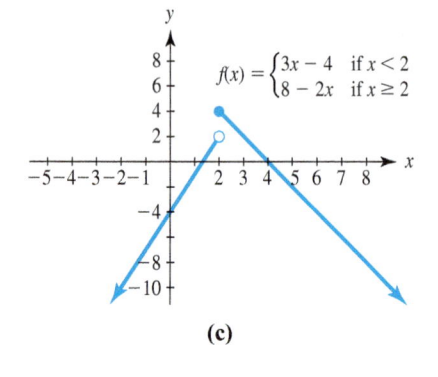

(a) (b) (c)

Figure 4.3

EXAMPLE 4 ▶ Functions

Determine whether each of the graphs represents a function.

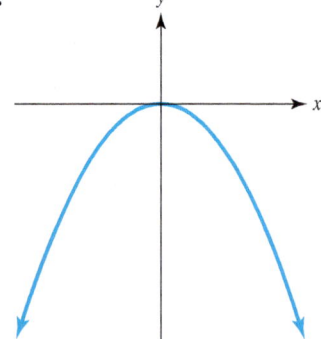

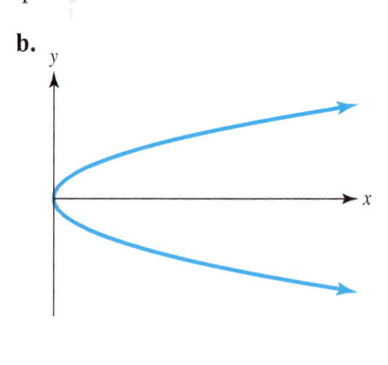

a. **b.**

SOLUTION

a. The vertical line test confirms that this graph represents a function. It is a parabola opening downward, so it is the graph of a quadratic function.

b. By the vertical line test, we see that this is not the graph of a function. This graph has the shape of a parabola and is called a **horizontal parabola**.

Toolbox EXERCISES

1. The domain of the reciprocal function is $(-\infty, 0) \cup (0, \infty)$, and its range is $(-\infty, 0) \cup (0, \infty)$.

2. The domain of the constant function $g(x) = k$ is $(-\infty, \infty)$, and its range is $\{k\}$.

3. The reciprocal function decreases on $(-\infty, 0) \cup (0, \infty)$.

4. The absolute value function increases on the interval $(0, \infty)$ and decreases on $(-\infty, 0)$.

5. The range of the squaring function is $[0, \infty)$.

6. The domain of the squaring function is $(-\infty, \infty)$.

In Exercises 7–12, determine whether the function is increasing or decreasing on the given interval.

7. $g(x) = \sqrt[5]{x}$; $(-\infty, \infty)$ — Increasing
8. $h(x) = \sqrt[4]{x}$; $[0, \infty)$ — Increasing
9. $f(x) = 5 - 0.8x$; $(-\infty, \infty)$ — Decreasing
10. $f(x) = \dfrac{-1}{x}$; $(-\infty, 0)$ — Increasing
11. $g(x) = -2x^{2/3}$; $(0, \infty)$ — Decreasing
12. $h(x) = 3x^{1/6}$; $[0, \infty)$ — Increasing

Identify the type of each of the following functions from its equation. Then graph the function.

13. $f(x) = x^3$ — Cubing
14. $f(x) = \sqrt{x}$ — Square root
15. $f(x) = 8\sqrt[6]{x}$ — Power function, with power $1/6$
16. $f(x) = \dfrac{-4}{x^3}$ — Power function, with power -3

Which of the following graphs represents a function?

17. Not a function

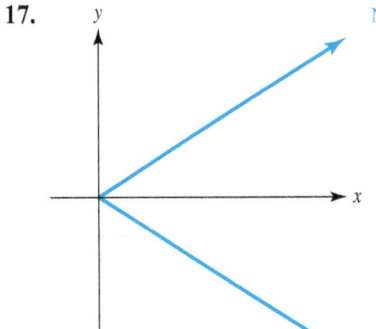

18. Not a function

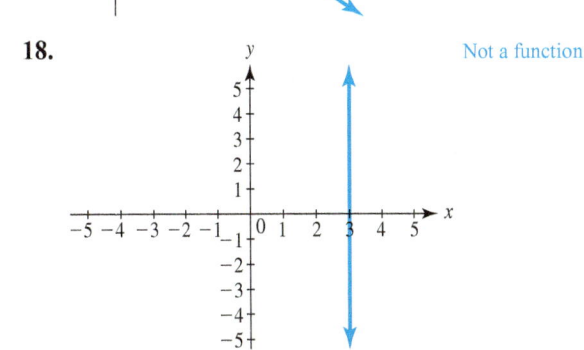

19. Function

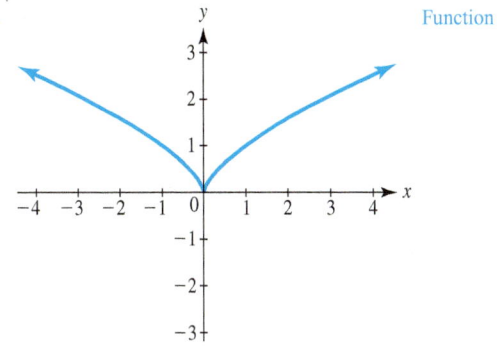

20. Function

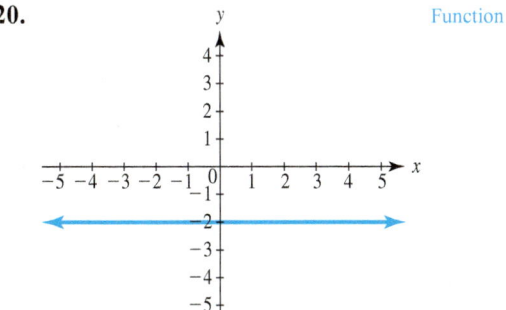

21. Function

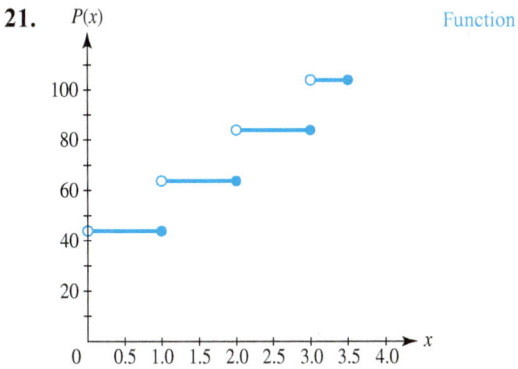

22. Function

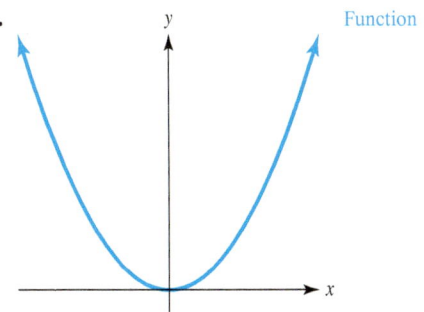

4.1 Transformations of Graphs and Symmetry

KEY OBJECTIVES

- Find equations and graphs of functions whose graphs have been vertically shifted, horizontally shifted, stretched, compressed, and reflected
- Determine if a graph is symmetric about the y-axis
- Determine if a graph is symmetric about the origin
- Determine if a graph is symmetric about the x-axis
- Determine if a function is even, odd, or neither

SECTION PREVIEW Diabetes

The projected percent of the U.S. adult population with diabetes (diagnosed and undiagnosed) for 2010 through 2050 can be modeled by the power function $f(x) = 4.947x^{0.495}$, with x equal to the number of years after 2000. We will see in Example 5 that we can write a new model for this population with x equal to the number of years after 2010 by replacing x in the old model by $x + 10$. This new model is called a shifted model of the original one.

In this section, we discuss shifting, stretching, compressing, and reflecting the graph of a function. This is useful in obtaining graphs of additional functions and also in finding windows in which to graph them. We also discuss symmetry of graphs. ■

Shifts of Graphs of Functions

Consider the quadratic function in the form $y = a(x - h)^2 + k$. Substituting h for x in this equation gives $y = k$, so the graph of this function contains the point (h, k). Because the graph is symmetric about the vertical line through the vertex, there is a second point on the graph with y-coordinate k, *unless* the point (h, k) is the vertex of the parabola (Figure 4.4). To see whether the second point exists or if (h, k) is the vertex, we solve

$$k = a(x - h)^2 + k$$

This gives

$$0 = a(x - h)^2$$
$$0 = (x - h)^2$$
$$0 = x - h$$
$$x = h$$

Thus, only one value of x corresponds to $y = k$, so the point (h, k) is the vertex of any parabola with an equation of the form $y = a(x - h)^2 + k$.* We can say that the vertex of $y = a(x - h)^2 + k$ has been **shifted** from $(0, 0)$ to the point (h, k), and this entire graph is in fact the graph of $y = ax^2$ shifted h units horizontally and k units vertically. We will see that graphs of other functions can be shifted similarly.

In Section 3.3, we studied absolute value functions. In the following example, we investigate shifts of the absolute value function.

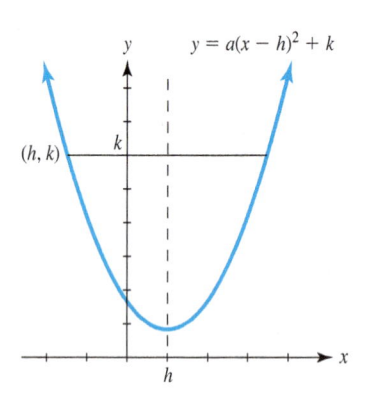

Figure 4.4

EXAMPLE 1 ▶ Shifts of Functions

Graph the functions $f(x) = |x|$ and $g(x) = |x| + 3$ on the same axes. What relationship do you notice between the two graphs?

SOLUTION

Table 4.1 shows inputs for x and outputs for both $f(x) = |x|$ and $g(x) = |x| + 3$. Observe that for a given input, each output value of $g(x) = |x| + 3$ is 3 more than the corresponding output value for $f(x) = |x|$. Thus, the y-coordinate of each point on the graph of $g(x) = |x| + 3$ is 3 more than the y-coordinate of the point on the graph of $f(x) = |x|$ with the same x-coordinate. We say that the graph of $g(x) = |x| + 3$ is the graph of $f(x) = |x|$ shifted up 3 units (Figure 4.5).

---

*This confirms use of the vertex form of a quadratic function in Section 3.1.

4.1 Transformations of Graphs and Symmetry

Table 4.1

x	f(x) = |x|	g(x) = |x| + 3
−3	3	6
−2	2	5
−1	1	4
0	0	3
1	1	4
2	2	5
3	3	6
4	4	7

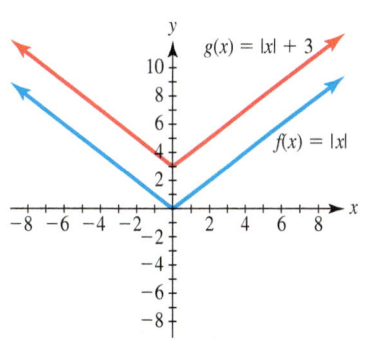

Figure 4.5

In Example 1, the equation of the shifted graph is $g(x) = |x| + 3 = f(x) + 3$. In general, we have the following.

> **Vertical Shifts of Graphs**
>
> If k is a positive real number:
>
> The graph of $g(x) = f(x) + k$ can be obtained by shifting the graph of $f(x)$ upward k units.
> The graph of $g(x) = f(x) - k$ can be obtained by shifting the graph of $f(x)$ downward k units.

EXAMPLE 2 ▶ Shifts of Functions

Graph the functions $f(x) = |x|$ and $g(x) = |x - 5|$ on the same axes. What relationship do you notice between the two graphs?

SOLUTION

Table 4.2 shows inputs for x and outputs for both $f(x) = |x|$ and $g(x) = |x - 5|$. Observe that the output values for both functions are equal *if* the input value for $g(x) = |x - 5|$ is 5 more than the corresponding input value for $f(x) = |x|$. Thus, the x-coordinate of each point on the graph of $g(x) = |x - 5|$ is 5 more than the x-coordinate of the point on the graph of $f(x) = |x|$ with the same y-coordinate. We say that the graph of $g(x) = |x - 5|$ is the graph of $f(x) = |x|$ shifted 5 units to the right (Figure 4.6).

Table 4.2

x	f(x) = |x|	x	g(x) = |x − 5|
−3	3	2	3
−2	2	3	2
−1	1	4	1
0	0	5	0
1	1	6	1
2	2	7	2
3	3	8	3
4	4	9	4

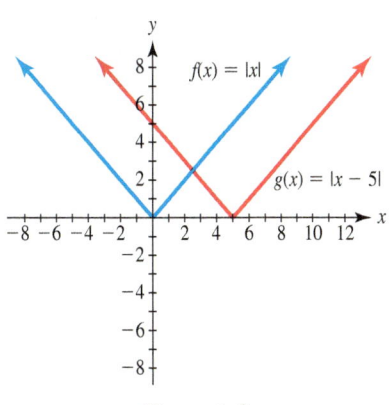

Figure 4.6

In Example 2, the equation of the shifted graph is $g(x) = |x - 5| = f(x - 5)$. In general, we have the following.

> **Horizontal Shifts of Graphs**
>
> If h is a positive real number:
>
> The graph of $g(x) = f(x - h)$ can be obtained by shifting the graph of $f(x)$ to the right h units.
>
> The graph of $g(x) = f(x + h)$ can be obtained by shifting the graph of $f(x)$ to the left h units.

EXAMPLE 3 ▶ Shifts of Functions

Graph the functions $f(x) = \sqrt{x}$ and $g(x) = \sqrt{x + 3} - 4$ on the same axes. What relationship do you notice between the two graphs?

SOLUTION

The graphs of $f(x) = \sqrt{x}$ and $g(x) = \sqrt{x + 3} - 4$ are shown in Figure 4.7. We see that the graph of $g(x) = \sqrt{x + 3} - 4$ can be obtained by shifting the graph of $f(x) = \sqrt{x}$ to the left 3 units and down 4 units. To verify this, consider a few points from each graph in Table 4.3.

Table 4.3

$f(x) = \sqrt{x}$		$g(x) = \sqrt{x + 3} - 4$	
x	$f(x) = \sqrt{x}$	x	$g(x) = \sqrt{x + 3} - 4$
0	0	-3	-4
1	1	-2	-3
2	$\sqrt{2}$	-1	$\sqrt{2} - 4$
3	$\sqrt{3}$	0	$\sqrt{3} - 4$
4	2	1	-2

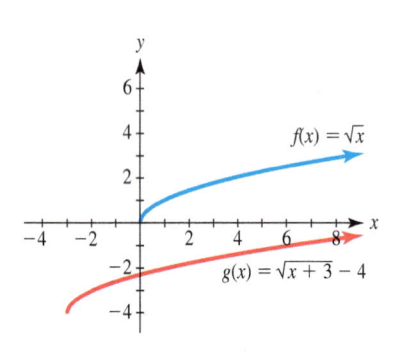

Figure 4.7

EXAMPLE 4 ▶ Profit

Right Sports Management sells elliptical machines, and the monthly profit from these machines can be modeled by

$$P(x) = -(x - 55)^2 + 4500$$

where x is the number of elliptical machines produced and sold.

a. Use the graph of $y = -x^2$ and the appropriate transformation to set the window and graph the profit function.

b. How many elliptical machines must be produced and sold to yield the maximum monthly profit? What is the maximum monthly profit?

4.1 Transformations of Graphs and Symmetry

(a)   (b)

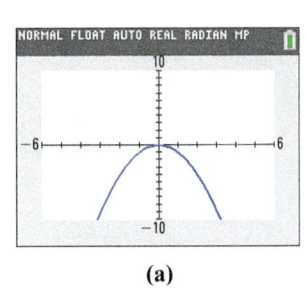

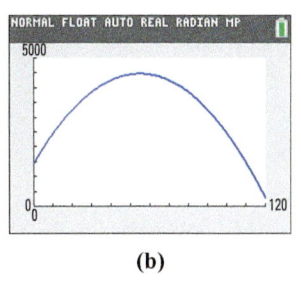

Figure 4.8

SOLUTION

a. Shifting the graph of $y = -x^2$ (Figure 4.8(a)) 55 units to the right and up 4500 units gives the graph of $P(x) = -(x - 55)^2 + 4500$ (Figure 4.8(b)). Note that we graph the function only in the first quadrant, because only nonnegative values make sense for this application.

b. The maximum value of the function occurs at the vertex, and the vertex of the graph is $(55, 4500)$. Thus, maximum monthly profit of $4500 is made when 55 elliptical machines are produced and sold.

EXAMPLE 5 ▶ Diabetes

The projected percent of the U.S. adult population with diabetes (diagnosed and undiagnosed) for 2010 through 2050 can be modeled by the power function $f(x) = 4.947x^{0.495}$, with x equal to the number of years after 2000. Find the power function that models this projection with x equal to the number of years after 2010.

SOLUTION

Because y represents the percent in both models, there is no vertical shift from one function to the other. To find the power model that will give the percent where x is the number of years after 2010 rather than the number of years after 2000, we rewrite the equation in a new form where $x = 0$ represents the same input as $x = 10$ does in the original model. This can be accomplished by replacing x in the original model with $x + 10$, giving the new model

$$f(x) = 4.947(x + 10)^{0.495}$$

where x is the number of years after 2010.

To see that these two models are equivalent, observe the inputs and outputs for selected years in Table 4.4 and the graphs in Figure 4.9.

Table 4.4

Original Model		New Model	
Years from 2000	Percent	Years from 2010	Percent
10	15.46	0	15.46
20	21.79	10	21.79
30	26.64	20	26.64
50	34.30	40	34.30

(*Source:* Centers for Disease Control and Prevention)

266 Chapter 4 Additional Topics with Functions

$f(x) = 4.947x^{0.495}$
Years after 2000
(a)

$f(x) = 4.947(x + 10)^{0.495}$
Years after 2010
(b)

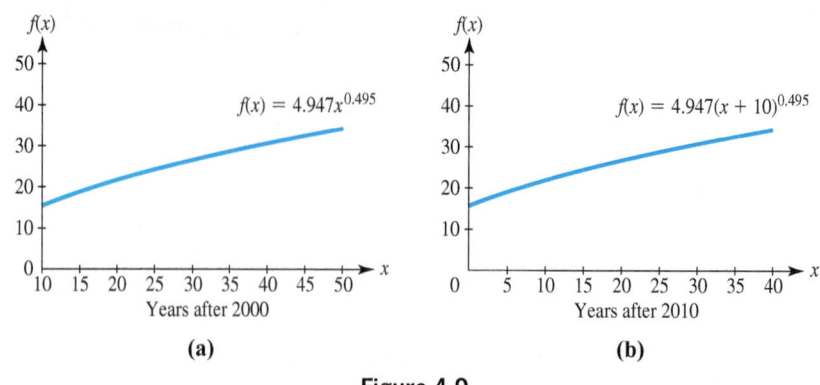

Figure 4.9

Stretching and Compressing Graphs

EXAMPLE 6 ▶ Stimulus-Response

One of the early results in psychology relates the magnitude of a stimulus x to the magnitude of the response y with the model $y = kx^2$, where k is an experimental constant. Compare the graphs of $y = kx^2$ for $k = 1, 2$, and $\frac{1}{2}$.

SOLUTION

Table 4.5 shows the values of y for $y = f(x) = x^2$, $y = 2f(x) = 2x^2$, and $y = \frac{1}{2}f(x) = \frac{1}{2}x^2$ for selected values of x. Observe that for these x-values, the y-values for $y = 2x^2$ are 2 times the y-values for $y = x^2$, and the points on the graph of $y = 2x^2$ have y-values that are 2 times the y-values on the graph of $y = x^2$ for equal x-values. We say that the graph of $y = 2x^2$ is a **vertical stretch** of $y = x^2$ by a factor of 2. (Compare Figure 4.10(a) and Figure 4.10(b).)

Observe also that for these x-values, the y-values for $y = \frac{1}{2}x^2$ are $\frac{1}{2}$ of the y-values for $y = x^2$, and the points on the graph of $y = \frac{1}{2}x^2$ have y-values that are $\frac{1}{2}$ of the y-values on the graph of $y = x^2$ for equal x-values. We say that the graph of $y = \frac{1}{2}x^2$ is a **vertical compression** of $y = x^2$ using a factor of $\frac{1}{2}$ (Figure 4.10(c)).

Table 4.5

x	0	1	2
f(x)	0	1	4

x	0	1	2
2f(x)	0	2	8

x	0	1	2
$\frac{1}{2}f(x)$	0	$\frac{1}{2}$	2

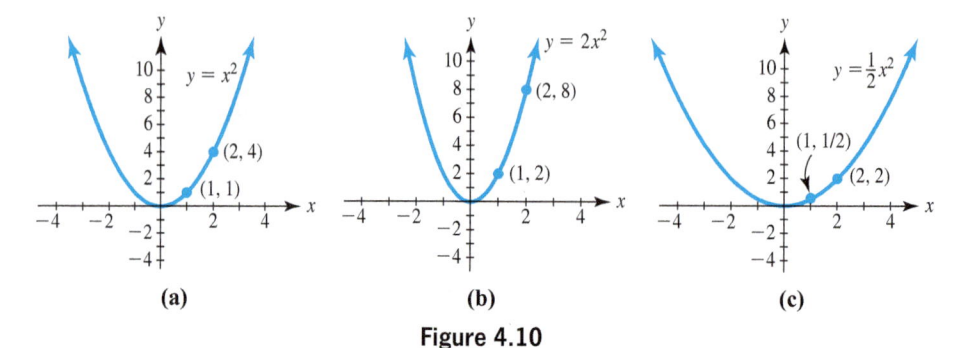

(a) (b) (c)

Figure 4.10

Stretching and Compressing Graphs

The graph of $y = af(x)$ is obtained by vertically stretching the graph of $f(x)$ using a factor of $|a|$ if $|a| > 1$ and vertically compressing the graph of $f(x)$ using a factor of $|a|$ if $0 < |a| < 1$.

Reflections of Graphs

In Figure 4.11(a), we see that the graph of $y = -x^2$ is a parabola that *opens down*. It can be obtained by reflecting the graph of $f(x) = x^2$, which *opens up*, across the x-axis. We can compare the y-coordinates of the graphs of these two functions by looking at Table 4.6 and Figure 4.11(b). Notice that for a given value of x, the y-coordinates of $y = x^2$ and $y = -x^2$ are negatives of each other.

Table 4.6

x	y = x²	y = -x²
-2	4	-4
-1	1	-1
0	0	0
1	1	-1
2	4	-4

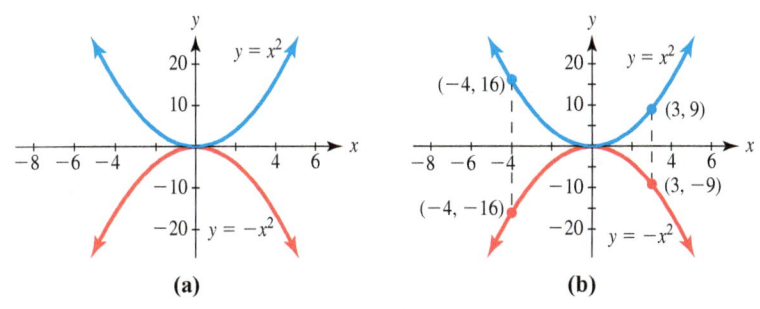

Figure 4.11

We can also see that the graph of $y = (-x)^3$ is a reflection of the graph of $y = x^3$ across the y-axis by looking at Table 4.7 and Figure 4.12. Notice that for a given value of y, the x-coordinates of $y = x^3$ and $y = (-x)^3$ are negatives of each other.

Table 4.7

x	y = x³	x	y = (-x)³
2	8	-2	8
1	1	-1	1
0	0	0	0
-1	-1	1	-1
-2	-8	2	-8

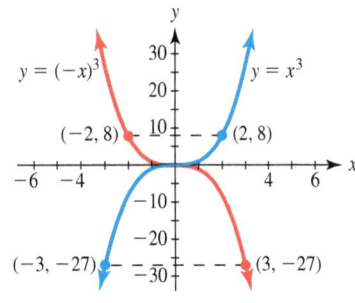

Figure 4.12

In general, we have the following:

Reflections of Graphs Across the Coordinate Axes

1. The graph of $y = -f(x)$ can be obtained by reflecting the graph of $y = f(x)$ across the x-axis.

2. The graph of $y = f(-x)$ can be obtained by reflecting the graph of $y = f(x)$ across the y-axis.

268   Chapter 4   Additional Topics with Functions

A summary of the transformations of a graph follows.

Graph Transformations

For a given function $y = f(x)$,

Vertical Shift	$y = f(x) + k$	Graph is shifted k units up if $k > 0$ and k units down if $k < 0$.
Horizontal Shift	$y = f(x - h)$	Graph is shifted h units right if $h > 0$ and h units left if $h < 0$.
Stretch/Compress	$y = af(x)$	Graph is vertically stretched using a factor of $|a|$ if $|a| > 1$. Graph is compressed using a factor of $|a|$ if $|a| < 1$.
Reflection	$y = -f(x)$	Graph is reflected across the x-axis.
	$y = f(-x)$	Graph is reflected across the y-axis.

EXAMPLE 7 ▶ Pollution

Suppose that for a certain city, the cost C of obtaining drinking water that contains $p\%$ impurities (by volume) is given by

$$C = \frac{120{,}000}{p} - 1200$$

a. Determine the domain of this function and graph the function without concern for the context of the problem.

b. Use knowledge of the context of the problem to graph the function in a window that applies to the application.

c. What is the cost of obtaining drinking water that contains 5% impurities?

SOLUTION

a. All values of p except $p = 0$ result in real values for the function. Thus, the domain of C is all real numbers except 0. The graph of this function is a transformation of the graph of $C = \frac{1}{p}$, stretched by a factor of 120,000, then shifted downward 1200 units. The viewing window should have its center near $p = 0$ (horizontally) and $C = -1200$ (vertically). We increase the vertical view to allow for the large stretching factor, using the viewing window $[-100, 100]$ by $[-20{,}000, 20{,}000]$. The graph, shown in Figure 4.13(a), has the same shape as the graph of $y = \frac{1}{x}$.

(a)   (b)

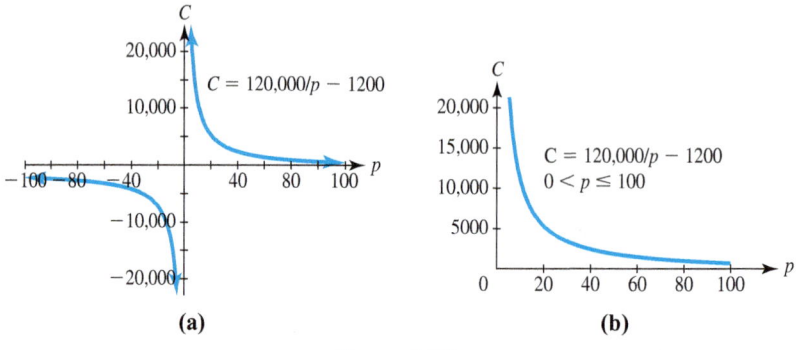

Figure 4.13

b. Because p represents the percent of impurities, the viewing window is set for values of p from 0 to 100. (Recall from part (a) that p cannot be 0.) The cost of reducing the impurities cannot be negative, so the vertical view is set from 0 to 20,000. The graph is shown in Figure 4.13(b).

c. To determine the cost of obtaining drinking water that contains 5% impurities, we substitute 5 for p in $C = \dfrac{120{,}000}{p} - 1200$, giving

$$C = \dfrac{120{,}000}{5} - 1200 = 22{,}800$$

Thus, the cost is $22,800 to obtain drinking water that contains 5% impurities.

EXAMPLE 8 ▶ Velocity of Blood

Because of friction from the walls of an artery, the velocity of blood is greatest at the center of the artery and decreases as the distance r from the center increases. The velocity of the blood in an artery can be modeled by the function

$$v = k(R^2 - r^2)$$

where R is the constant radius of the artery and k is a constant that is determined by the pressure, the viscosity of the blood, and the length of the artery. In the case where $k = 2$ and $R = 0.1$ centimeter, the velocity is

$$v = 2(0.01 - r^2) \text{ centimeter per second}$$

a. Graph this function and the functions $v = r^2$ and $v = -2r^2$ in the viewing window $[-0.1, 0.1]$ by $[-0.05, 0.05]$.

b. How is the function $v = 2(0.01 - r^2)$ related to the second-degree power function $v = r^2$?

SOLUTION

a. The graph of the function $v = 2(0.01 - r^2)$ is shown in Figure 4.14(a), the graph of $v = r^2$ is shown in Figure 4.14(b), and the graph of $v = -2r^2$ is shown in Figure 4.14(c).

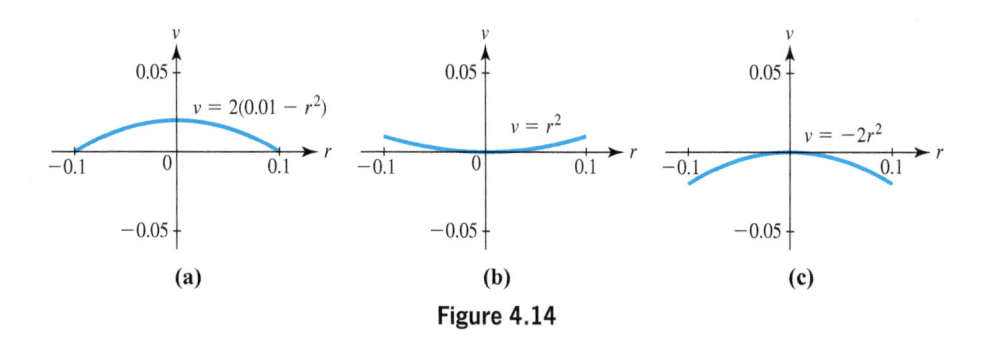

(a) (b) (c)

Figure 4.14

b. The function $v = 2(0.01 - r^2)$ can also be written in the form $v = -2r^2 + 0.02$, so it should be no surprise that the graph is a reflected about the r-axis, stretched, and shifted form of the graph of $v = r^2$. If the graph of $v = r^2$ is reflected about the r-axis, has each r^2-value doubled, and then is shifted upward 0.02 unit, the graph of $v = -2r^2 + 0.02 = 2(0.01 - r^2)$ results.

Symmetry; Even and Odd Functions

In Chapter 3, we observed that any parabola that is a graph of a quadratic function is symmetric about a vertical line called the axis of symmetry. This means that the two halves of the parabola are reflections of each other. In particular, the graph of $y = x^2$ in Figure 4.15 is symmetric (that is, is a reflection of itself) about the y-axis. For example, the point $(3, 9)$ is on the graph, and its mirror image across the y-axis is the point $(-3, 9)$. This suggests an algebraic way to determine whether the graph of an equation is symmetric with respect to the y-axis—that is, replace x with $-x$ and simplify. If the resulting equation is equivalent to the original equation, then the graph is symmetric with respect to the y-axis.

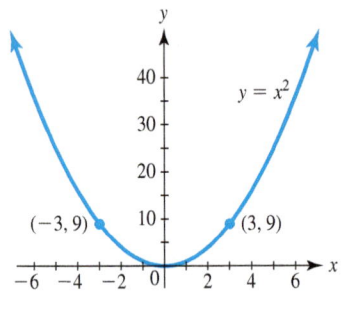

Figure 4.15

Symmetry with Respect to the y-axis

The graph of $y = f(x)$ is symmetric with respect to the y-axis if, for every point (x, y) on the graph, the point $(-x, y)$ is also on the graph—that is,

$$f(-x) = f(x)$$

for all x in the domain of f. Such a function is called an **even function**.

All power functions with even integer exponents will have graphs that are symmetric about the y-axis. This is why all functions satisfying this condition are called even functions, although functions other than power functions exist that are even functions.

A graph of $y = x^3$ is shown in Figure 4.16(a). Notice in Figure 4.16(b) that if we draw a line through the origin and any point on the graph, it will intersect the graph at a second point, and the distances from the two points to the origin will be equal.

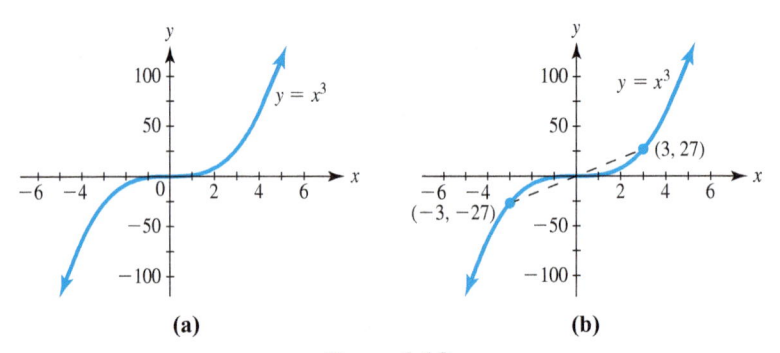

Figure 4.16

For example, the point $(3, 27)$ is on the graph of $y = x^3$, and the line through $(3, 27)$ and the origin intersects the graph at the point $(-3, -27)$, which is the same distance from the origin as $(3, 27)$. In general, the graph of a function is called symmetric with respect to the origin if for every point (x, y) on the graph, the point $(-x, -y)$ is also on the graph.

To determine algebraically whether the graph of an equation is symmetric about the origin, replace x with $-x$, replace y with $-y$, and simplify. If the resulting equation is equivalent to the original equation, then the graph is symmetric with respect to the origin.

Symmetry with Respect to the Origin

The graph of $y = f(x)$ is symmetric with respect to the origin if, for every point (x, y) on the graph, the point $(-x, -y)$ is also on the graph—that is,

$$f(-x) = -f(x)$$

for all x in the domain of f. Such a function is called an **odd function**.

The graphs of some equations can be symmetric with respect to the x-axis (see Figure 4.17). Such graphs do not represent y as a function of x (because they do not pass the vertical line test).

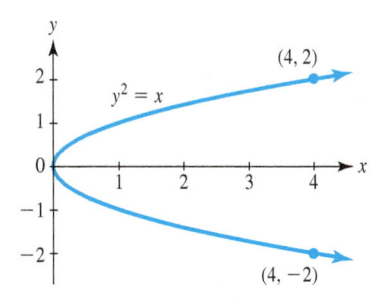

Figure 4.17

For example, we can graph the equation $x^2 + y^2 = 9$ by hand or with technology after solving it for y. Solving gives

$$y^2 = 9 - x^2$$
$$y = \pm\sqrt{9 - x^2}$$

Entering values for x in the equation gives points on the **circle**, and entering the two functions in a calculator, as in Figure 4.18(a), gives the graph. Note that the calculator graph will not look like a circle unless the calculator window is square. Figure 4.18(b) gives the graph with the standard window, Figure 4.18(c) gives the graph if ZOOM Square is used after ZOOM Standard, and Figure 4.18(d) gives the graph with ZOOM Decimal.* In general, if the equation contains one and only one even power of y, its graph will be symmetric about the x-axis.

(a)

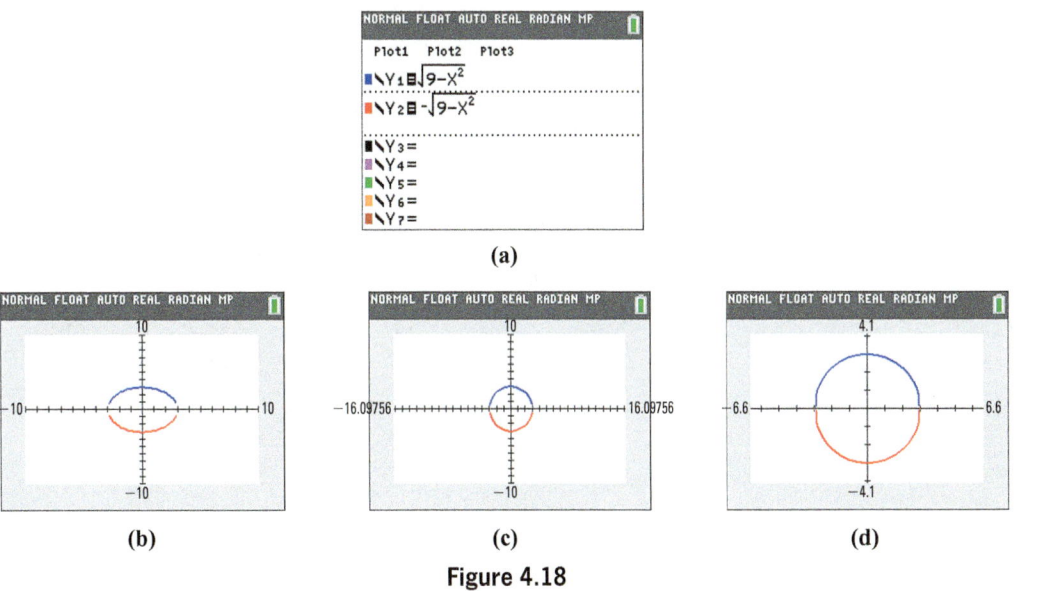

(b) (c) (d)

Figure 4.18

To determine algebraically whether the graph of an equation is symmetric about the x-axis, replace y with $-y$ and simplify. If the resulting equation is equivalent to the original equation, then the graph is symmetric with respect to the x-axis.

> ### Symmetry with Respect to the x-axis
> The graph of an equation is symmetric with respect to the x-axis if, for every point (x, y) on the graph, the point $(x, -y)$ is also on the graph.

---

*See Appendix A, page 683 for details.

272 Chapter 4 Additional Topics with Functions

EXAMPLE 9 ▶ Symmetry

Determine algebraically whether the graph of each equation is symmetric with respect to the *x*-axis, *y*-axis, or origin. Confirm your conclusion graphically.

a. $y = \dfrac{2x^2}{x^2 + 1}$ **b.** $y = x^3 - 3x$ **c.** $x^2 + y^2 = 16$

SOLUTION

a. To test the graph of $y = \dfrac{2x^2}{x^2 + 1}$ for symmetry with respect to the *x*-axis, we replace *y* with $-y$. Because the result,

$$-y = \dfrac{2x^2}{x^2 + 1}$$

is not equivalent to the original equation, the graph is not symmetric with respect to the *x*-axis. This equation gives *y* as a function of *x*.

To test for *y*-axis symmetry, we replace *x* with $-x$:

$$y = \dfrac{2(-x)^2}{(-x)^2 + 1} = \dfrac{2x^2}{x^2 + 1}$$

which is equivalent to the original equation. Thus, the graph is symmetric with respect to the *y*-axis, and *y* is an even function of *x*.

To test for symmetry with respect to the origin, we replace *x* with $-x$ and *y* with $-y$:

$$-y = \dfrac{2(-x)^2}{(-x)^2 + 1} = \dfrac{2x^2}{x^2 + 1}$$

which is not equivalent to the original equation. Thus, the graph is not symmetric with respect to the origin.

Figure 4.19 confirms the fact that the graph of $y = \dfrac{2x^2}{x^2 + 1}$ is symmetric with respect to the *y*-axis.

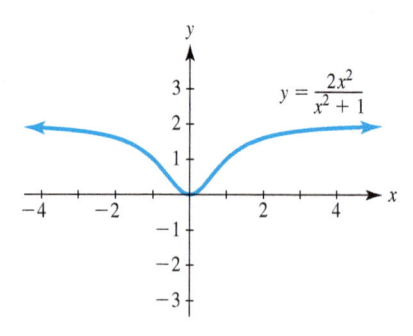

Figure 4.19

b. To test the graph of $y = x^3 - 3x$ for symmetry with respect to the *x*-axis, we replace *y* with $-y$. Because the result,

$$-y = x^3 - 3x$$

is not equivalent to the original equation, the graph is not symmetric with respect to the *x*-axis.

To test for *y*-axis symmetry, we replace *x* with $-x$:

$$y = (-x)^3 - 3(-x) = -x^3 + 3x$$

which is not equivalent to the original equation. Thus, the graph is not symmetric with respect to the y-axis.

To test for symmetry with respect to the origin, we replace x with $-x$ and y with $-y$:

$$-y = (-x)^3 - 3(-x) = -x^3 + 3x, \quad \text{or} \quad y = x^3 - 3x$$

which is equivalent to the original equation. Thus, the graph is symmetric with respect to the origin, and y is an odd function of x.

Figure 4.20 confirms this.

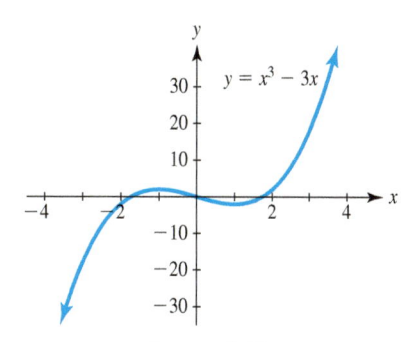

Figure 4.20

c. To test the graph of $x^2 + y^2 = 16$ for symmetry with respect to the x-axis, we replace y with $-y$:

$$x^2 + (-y)^2 = 16 \quad \text{or} \quad x^2 + y^2 = 16$$

This is equivalent to the original equation, so the graph is symmetric with respect to the x-axis. Thus, y is not a function of x in this equation.

To test for y-axis symmetry, we replace x with $-x$:

$$(-x)^2 + y^2 = 16 \quad \text{or} \quad x^2 + y^2 = 16$$

This is equivalent to the original equation, so the graph is symmetric with respect to the y-axis.

To test for symmetry with respect to the origin, we replace x with $-x$ and y with $-y$:

$$(-x)^2 + (-y)^2 = 16, \quad \text{or} \quad x^2 + y^2 = 16$$

which is equivalent to the original equation. Thus, the graph is symmetric with respect to the origin.

Solving for y gives the two functions shown in Figure 4.21(a). Choosing a window $[-7, 7]$ by $[-5, 5]$ and then selecting ZOOM 5:ZSquare gives the graph in Figure 4.21(b), which confirms the fact that $y = x^2 + y^2$ is symmetric with respect to the x-axis, the y-axis, and the origin.

(a)   (b)

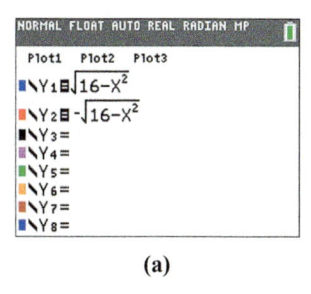

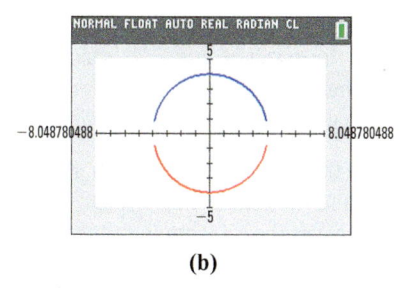

Figure 4.21

SKILLS CHECK 4.1

Answers that are not seen can be found in the answer section at the back of the text.

In Exercises 1–16, (a) sketch the graph of each pair of functions using a standard window, and (b) describe the transformations used to obtain the graph of the second function from the first function.

1. $y = x^3, y = x^3 + 5$ **b.** Vertical shift 5 units up
2. $y = x^2, y = x^2 + 3$ **b.** Vertical shift 3 units up
3. $y = \sqrt{x}, y = \sqrt{x} - 4$ **b.** Horizontal shift 4 units right
4. $y = x^2, y = (x + 2)^2$ **b.** Horizontal shift 2 units left
5. $y = \sqrt[3]{x}, y = \sqrt[3]{x + 2} - 1$
    b. Horizontal shift 2 units left, vertical shift 1 unit down
6. $y = x^3, y = (x - 5)^3 - 3$
    b. Horizontal shift 5 units right, vertical shift 3 units down
7. $y = |x|, y = |x - 2| + 1$
    b. Horizontal shift 2 units right, vertical shift 1 unit up
8. $y = |x|, y = |x + 3| - 4$
    b. Horizontal shift 3 units left, vertical shift 4 units down
9. $y = x^2, y = -x^2 + 5$
    b. Reflection across x-axis, vertical shift 5 units up
10. $y = \sqrt{x}, y = -\sqrt{x - 2}$
    b. Reflection across x-axis, horizontal shift 2 units right
11. $y = \dfrac{1}{x}, y = \dfrac{1}{x} - 3$ **b.** Vertical shift 3 units down
12. $y = \dfrac{1}{x}, y = \dfrac{2}{x - 1}$ **b.** Vertical stretch using a factor of 2, horizontal shift 1 unit right
13. $f(x) = x^2, g(x) = \dfrac{1}{3}x^2$ **b.** Vertical compression using a factor of 1/3
14. $f(x) = x^3, g(x) = 0.4x^3$
    b. Vertical compression using a factor of 0.4
15. $f(x) = |x|, g(x) = 3|x|$
    b. Vertical stretch using a factor of 3
16. $f(x) = \sqrt{x}, g(x) = 4\sqrt{x}$
    b. Vertical stretch using a factor of 4
17. How is the graph of $y = (x - 2)^2 + 3$ transformed from the graph of $y = x^2$? Shifted 2 units right and 3 units up
18. How is the graph of $y = (x + 4)^3 - 2$ transformed from the graph of $y = x^3$?
    Shifted 4 units left and 2 units down
19. Suppose the graph of $y = x^{3/2}$ is shifted to the left 4 units. What is the equation that gives the new graph? $y = (x + 4)^{3/2}$
20. Suppose the graph of $y = x^{3/2}$ is shifted down 5 units and to the right 4 units. What is the equation that gives the new graph? $y = (x - 4)^{3/2} - 5$
21. Suppose the graph of $y = x^{3/2}$ is stretched by a factor of 3 and then shifted up 5 units. What is the equation that gives the new graph? $y = 3x^{3/2} + 5$
22. Suppose the graph of $y = x^{2/3}$ is compressed by a factor of $\dfrac{1}{5}$ and then shifted right 6 units. What is the equation that gives the new graph? $y = \dfrac{1}{5}(x - 6)^{2/3}$
23. Suppose the graph of $f(x) = \sqrt{x}$ is shifted up 3 units and to the left 6 units. What is the equation of the new graph? Verify the result graphically.
24. Suppose the graph of $f(x) = x^2$ is shifted up 2 units and to the right 5 units. What is the equation of the new graph? Verify the result graphically.
25. Suppose the graph of $f(x) = \dfrac{1}{x}$ is stretched vertically by a factor of 4 and then shifted down 3 units. What is the equation of the new graph? Verify the result graphically.
26. Suppose the graph of $f(x) = x^3$ is compressed vertically using a factor of $\dfrac{1}{3}$ and then shifted to the left 5 units. What is the equation of the new graph? Verify the result graphically.
27. Suppose the graph of $f(x) = \sqrt{x}$ is reflected across the x-axis and shifted up 2 units. What is the equation of the new graph? Verify the result graphically.
28. Suppose the graph of $f(x) = \sqrt[3]{x}$ is reflected across the x-axis and shifted to the right 1 unit. What is the equation of the new graph? Verify the result graphically.

In Exercises 29–32, write the equation of the function $g(x)$ that is transformed from the given function $f(x)$ and whose graph is shown.

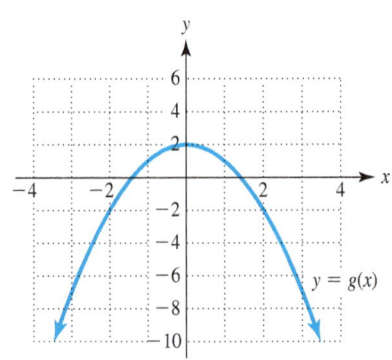

29. $f(x) = x^2 \quad g(x) = -x^2 + 2$

30. $f(x) = x^2$ $g(x) = -x^2 - 1$

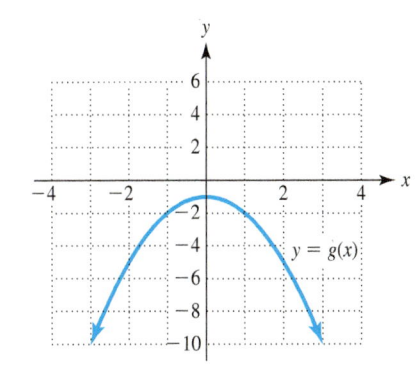

31. $f(x) = |x|$ $g(x) = |x + 3| - 2$

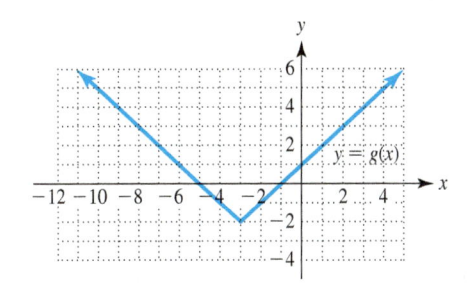

32. $f(x) = x^2$ $g(x) = (x - 4)^2 + 2$

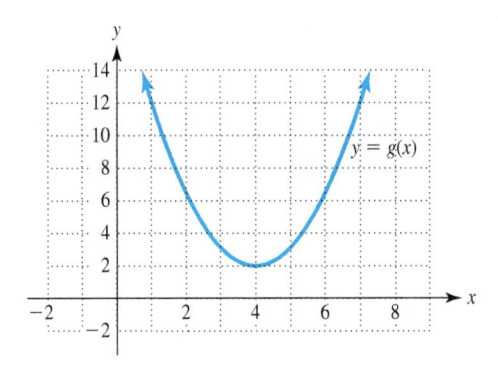

In Exercises 33–36, determine visually whether each of the graphs is symmetric about the x-axis, the y-axis, or neither. If the graph represents a function, determine whether the function is even, odd, or neither.

33. y-axis; even

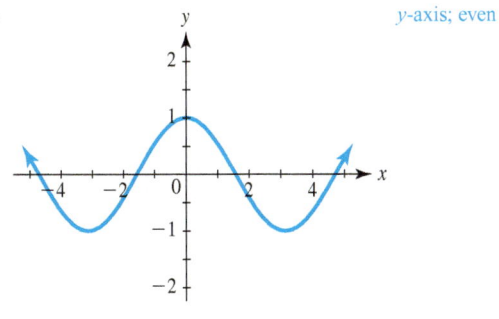

34. x-axis, y-axis; not a function

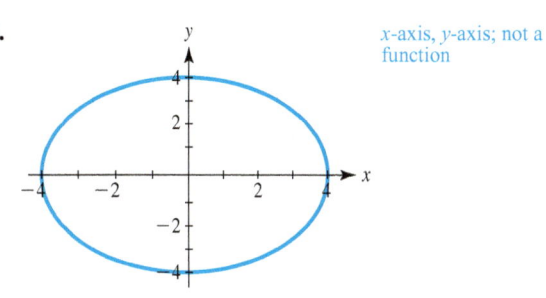

35. x axis; not a function

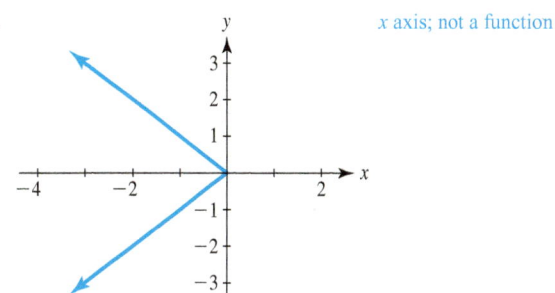

36. Neither; neither

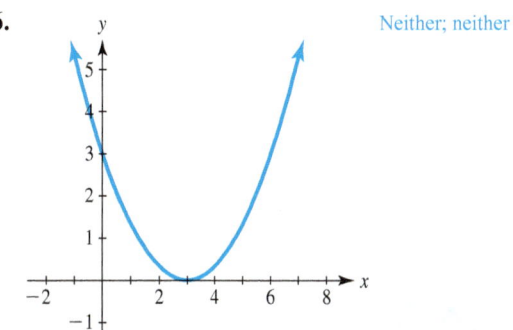

In Exercises 37 and 38, determine whether the graph of the equation in Y_1 is symmetric about the y-axis, based on the ordered pairs shown in each pair of tables.

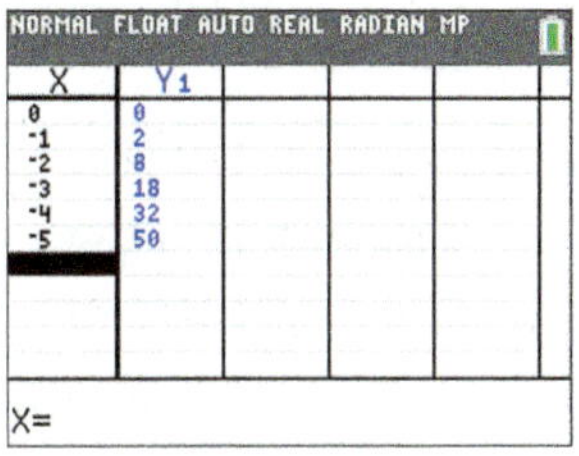

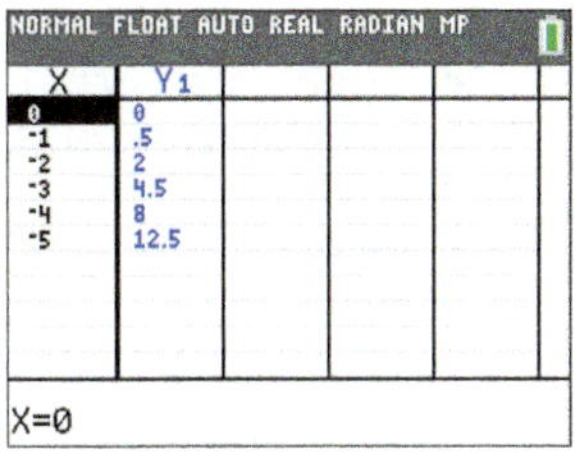

37. Yes

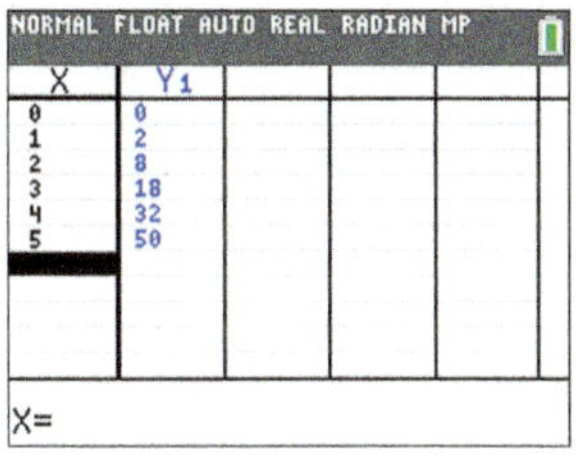

38. Yes

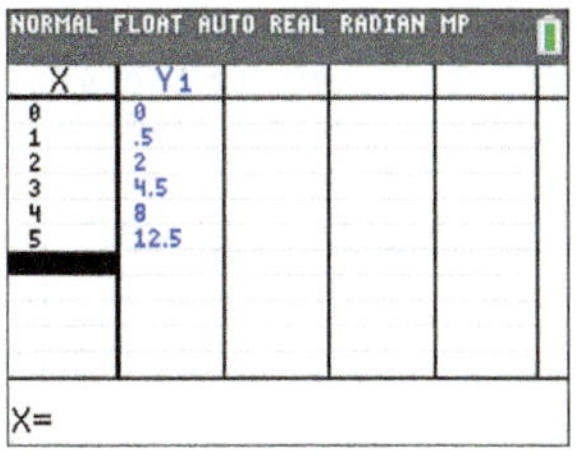

In Exercises 39–46, determine algebraically whether the graph of the given equation is symmetric with respect to the x-axis, the y-axis, and/or the origin. Confirm graphically.

39. $y = 2x^2 - 3$ y-axis

40. $y = -x^2 + 4$ y-axis

41. $y = x^3 - x$ Origin

42. $y = -x^3 + 5x$ Origin

43. $y = \dfrac{6}{x}$ Origin

44. $x = 3y^2$ x-axis

45. $x^2 + y^2 = 25$ x-axis, y-axis, origin

46. $x^2 - y^2 = 25$ y-axis, x-axis, origin

In Exercises 47–52, determine whether the function is even, odd, or neither.

47. $f(x) = |x| - 5$ Even

48. $f(x) = |x - 2|$ Neither

49. $g(x) = \sqrt{x^2 + 3}$ Even

50. $f(x) = \dfrac{1}{2}x^3 - x$ Odd

51. $g(x) = \dfrac{5}{x}$ Odd

52. $g(x) = 4x + x^2$ Neither

Determine whether each of the complete graphs in Exercises 53 and 54 represents a function that is even, odd, or neither.

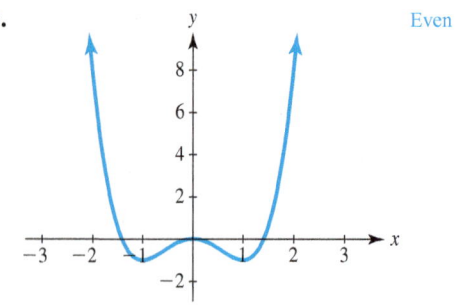

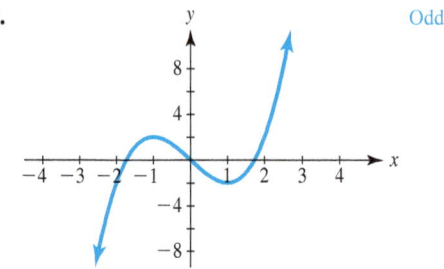

53. Even

54. Odd

55. Graph $(y + 1)^2 = x + 4$. Does this graph have any type of symmetry? Symmetric about the line $y = -1$

56. About what line is the function $y = 3 \pm \sqrt{x - 2}$ symmetric? $y = 3$

EXERCISES 4.1

Answers that are not seen can be found in the answer section at the back of the text.

57. **Medical Marijuana Revenue** Using data from 2016 and projected to 2025, the U.S. medical marijuana revenue, in billions of dollars, can be modeled by the function

    $M(x) = 0.037(x - 6)^2 + 0.652(x - 6) + 4.537$

    where x is the number of years after 2010. (*Source*: thecannabist.co)

    a. The graph of this function is a shifted graph of which basic function? $y = x^2$

    b. Find and interpret $M(24)$. $M(24) = 28.261$; the U.S. medical marijuana revenue is projected to be $28.261 billion in 2024.

    c. Graph $y = M(x)$ for x representing the years 2016 through 2025.

58. **Ballistics** Ballistic experts are able to identify the weapon that fired a certain bullet by studying the markings on the bullet. If a test is conducted by firing a bullet into a water tank, the distance that a bullet will travel is given by $s = 27 - (3 - 10t)^3$ inches for $0 \le t \le 0.3$, t in seconds.

    a. The graph of this function is a shifted graph of which basic function? $s = t^3$

    b. Graph this function for $0 \le t \le 0.3$.

    c. How far does the bullet travel during this time period? 27 in.

59. **Supply and Demand** The price per unit of a product is p, and the number of units of the product is denoted by q. The supply function for a product is given by $p = \dfrac{180 + q}{6}$, and the demand for the product is given by $p = \dfrac{30{,}000}{q} - 20$.

    a. Is the supply function a linear function or a shifted reciprocal function? Linear

    b. Is the demand function a shifted linear function or a shifted reciprocal function? Describe the transformations needed to obtain the specific function from the basic function. Reciprocal; vertical stretch using a factor of 30,000 and shift down 20 units

60. **Supply and Demand** The supply function for a commodity is given by $p = 58 + \dfrac{q}{2}$, and the demand function for this commodity is given by $p = \dfrac{2555}{q + 5}$.

a. Is the supply function a linear function or a shifted reciprocal function? Linear

b. Is the demand function a shifted linear function or a shifted reciprocal function? Describe the transformations needed to obtain the specific function from the basic function. Reciprocal; vertical stretch using a factor of 2555 and shift 5 units left

61. **Average Cost** The monthly average cost of producing x sofas is $\overline{C}(x) = \dfrac{30{,}000}{x} + 150$ dollars.

a. Will the average cost function decrease or increase as the number of units produced increases? Decrease

b. What transformations of the graph of the reciprocal function give the graph of this function? Stretch vertically by a factor of 30,000, then vertical shift 150 units up

62. **Pollution** The daily cost C (in dollars) of removing pollution from the smokestack of a coal-fired electric power plant is related to the percent of pollution p being removed according to the equation

$$C = \dfrac{10{,}500}{100 - p}$$

a. Describe the transformations needed to obtain this function from the function $C = \dfrac{1}{p}$.

b. Graph the function for $0 \le p < 100$.

c. What is the daily cost of removing 80% of the pollution? $525

63. **Population Growth** Suppose the population of a certain microorganism at time t (in minutes) is given by

$$P = -1000\left(\dfrac{1}{t + 10} - 1\right)$$

a. Describe the transformations needed to obtain this function from the function $f(t) = \dfrac{1}{t}$.

b. Graph this function for values of t representing 0 to 100 minutes.

64. **Mortgages** The balance owed y on a $50,000 mortgage after x monthly payments is shown in the table that follows. The function that models the data is

$$y = 4700\sqrt{110 - x}$$

a. Is this a shifted root function? Yes

b. What is the domain of the function in the context of this application? $(0, 110]$

c. Describe the transformations needed to obtain the graph from the graph of $y = \sqrt{x}$.

Monthly Payments	Balance Owed ($)
12	47,243
24	44,136
48	36,693
72	27,241
96	15,239
108	8074

65. **U.S. Personal Income** Data for the presidential years from 2012 and projected to 2040 indicates that U.S. disposable personal income (in billions of dollars) can be modeled by the function $y = 8.450x^{0.231}$, where x is the number of years after 2010. Rewrite the function if x equals the number of years after 2012. 65. $y = 8.450(x + 2)^{0.231}$
(Source: U.S. Energy Information Administration)

66. **Unconventional Vehicle Sales** The number of E85 flex fuel vehicles, in millions, projected to be sold in the United States can be modeled by the function $F(x) = 0.084x^{0.675}$, where x is the number of years after 2000. Convert the function so that x equals the number of years after 1990. $V(x) = 0.084(x - 10)^{0.675}$
(Source: www.eia.gov)

67. **Medical Marijuana Revenue** Using data from 2016 and projected to 2025, the U.S. medical marijuana revenue, in billions of dollars, can be modeled by the function

$$M(x) = 0.037(x - 6)^2 + 0.652(x - 6) + 4.537$$

where x is the number of years after 2010. Write the model $R(x)$ with x equal to the number of years after 2016. $y = 0.037x^2 + 0.652x + 4.537$
(Source: thecannabist.co)

68. **Cloud Revenue** The global Cloud revenue and projected revenue, in billions of dollars, for selected years 2012 and projected to 2026 can be modeled by

$$C(x) = 7.531x^{1.554}$$

with x equal to the number of years after 2010. Rewrite the model as $R(x)$ with x equal to the number of years after 2012. $R(x) = 7.531(x + 2)^{1.554}$
(Source: Wikibon Server SAN & Cloud Research Projects)

69. **U.S. National Consumption** The U.S. real consumption (in billions of dollars) for selected years from 2012 and projected to 2040 can be modeled by the function

$$S(t) = 0.008t^2 + 0.617t + 2.432$$

where x is the number of years after 2012.
(Source: U.S. Energy Administration)

a. Use the model to predict the real consumption in 2030. $16.13 billion

b. Write the model $C(t)$ with t equal to the number of years after 2010.
$C(t) = 0.008(t-2)^2 + 0.617(t-2) + 2.432$

c. With the model in part (b), what value of t should be used to find the real consumption in 2030? Does the value of $C(t)$ agree with the answer in part (a)? 20; yes

70. Cost-Benefit Suppose for a certain city the cost C of obtaining drinking water that contains $p\%$ impurities (by volume) is given by

$$C = \frac{120{,}000}{p} - 1200$$

a. What is the cost of drinking water that is 100% impure? $0

b. What is the cost of drinking water that is 50% impure? $1200

c. What transformations of the graph of the reciprocal function give the graph of this function?
Stretch using a factor of 120,000 and shift down 1200 units

71. Internet of Things Devices For the years from 2015 and projected through 2025, the number of devices connected to the Internet, in billions, can be modeled by the function

$$D(x) = 0.590x^2 - 0.275x + 17.201$$

where x equals the number of years after 2015.
(*Source*: Forbes.com)

a. What does the model predict that the number of devices will be in 2028? 113.34 billion

b. Rewrite the model as the function $T(x)$, with x equal to the number of years after 2010.
$T(x) = 0.590(x-5)^2 - 0.275(x-5) + 17.201$

c. What value of x should be used in $T(x)$ to predict what the number of devices will be in 2028? Does this x-value give a value of $T(x)$ that agrees with the answer to part (a)? 18; yes

72. Pollution The daily cost C (in dollars) of removing pollution from the smokestack of a coal-fired electric power plant is related to the percent of pollution p being removed according to the equation $100C - Cp = 10{,}500$.

a. Solve this equation for C to write the daily cost as a function of p, the percent of pollution removed. $C(p) = \dfrac{10{,}500}{100 - p}$

b. What is the daily cost of removing 50% of the pollution? $210

c. Why would this company probably resist removing 99% of the pollution? The daily cost of $10,500 is high.

4.2 Combining Functions; Composite Functions

KEY OBJECTIVES

- Find sums, differences, products, and quotients of two functions
- Form average cost functions
- Find the composition of two functions

SECTION PREVIEW Profit

If the daily total cost to produce x units of a product is

$$C(x) = 360 + 40x + 0.1x^2 \text{ thousand dollars}$$

and the daily revenue from the sale of x units of this product is

$$R(x) = 60x \text{ thousand dollars}$$

we can model the profit function as $P(x) = R(x) - C(x)$, giving

$$P(x) = 60x - (360 + 40x + 0.1x^2)$$

or

$$P(x) = -0.1x^2 + 20x - 360$$

In a manner similar to the one used to form this function, we can construct new functions by performing algebraic operations with two or more functions. For example, we can build an average cost function by finding the quotient of two functions. (See Example 4.) In addition to using arithmetic operations with functions, we can create new functions using function composition. ■

Operations with Functions

New functions that are the sum, difference, product, and quotient of two functions are defined as follows:

Operation	Formula	Example with $f(x) = \sqrt{x}$ and $g(x) = x^3$
Sum	$(f + g)(x) = f(x) + g(x)$	$(f + g)(x) = \sqrt{x} + x^3$
Difference	$(f - g)(x) = f(x) - g(x)$	$(f - g)(x) = \sqrt{x} - x^3$
Product	$(f \cdot g)(x) = f(x) \cdot g(x)$	$(f \cdot g)(x) = \sqrt{x} \cdot x^3 = x^3\sqrt{x}$
Quotient	$\left(\dfrac{f}{g}\right)(x) = \dfrac{f(x)}{g(x)} \quad (g(x) \neq 0)$	$\left(\dfrac{f}{g}\right)(x) = \dfrac{\sqrt{x}}{x^3} \quad (x \neq 0)$

The domain of the sum, difference, and product of f and g consists of all real numbers of the input variable for which f and g are defined. The domain of the quotient function $\dfrac{f}{g}$ consists of all real numbers for which f and g are defined and $g \neq 0$.

EXAMPLE 1 ▶ Operations with Functions

If $f(x) = x^3$ and $g(x) = x - 1$, find the following functions and give their domains.

a. $(f + g)(x)$ **b.** $(f - g)(x)$ **c.** $(f \cdot g)(x)$ **d.** $\left(\dfrac{f}{g}\right)(x)$

SOLUTION

a. $(f + g)(x) = f(x) + g(x) = x^3 + x - 1$; all real numbers

b. $(f - g)(x) = f(x) - g(x) = x^3 - (x - 1) = x^3 - x + 1$; all real numbers

c. $(f \cdot g)(x) = f(x) \cdot g(x) = x^3(x - 1) = x^4 - x^3$; all real numbers

d. $\left(\dfrac{f}{g}\right)(x) = \dfrac{f(x)}{g(x)} = \dfrac{x^3}{x - 1}$; all real numbers except 1

EXAMPLE 2 ▶ Revenue, Cost, and Profit

The demand for a certain electronic component is given by $p(x) = 1000 - 2x$. Producing and selling x components involves a monthly fixed cost of $1999 and a production cost of $4 for each component.

a. Write the equations that model total revenue and total cost as functions of the components produced and sold in a month.

b. Write the equation that models the profit as a function of the components produced and sold during a month.

c. Find the maximum possible monthly profit.

SOLUTION

a. The revenue for the components is given by the product of $p(x) = 1000 - 2x$ and x, the number of units sold.

$$R(x) = p(x) \cdot x = (1000 - 2x)x = 1000x - 2x^2 \text{ dollars}$$

The monthly total cost is the sum of the variable cost, 4x, and the fixed cost, 1999.

$$C(x) = 4x + 1999 \text{ dollars}$$

b. The monthly profit for the production and sale of the electronic components is the difference between the revenue and cost functions.

$$P(x) = R(x) - C(x) = (1000x - 2x^2) - (4x + 1999) = -2x^2 + 996x - 1999$$

c. The maximum monthly profit occurs where $x = \dfrac{-996}{2(-2)} = 249$. The maximum profit is $P(249) = 122{,}003$ dollars.

EXAMPLE 3 ▶ Unconventional Vehicle Sales

In December 2010, the U.S. Energy Information Administration (EIA) presented results projecting rapid growth in sales of unconventional vehicles through 2035. The EIA expects unconventional vehicles—vehicles using diesel or alternative fuels and/or hybrid electric systems—to account for over 40% of U.S. light-duty vehicles sold in 2035. Figure 4.22 shows the number of light-duty cars and trucks in each category sold, in millions, projected to 2035.

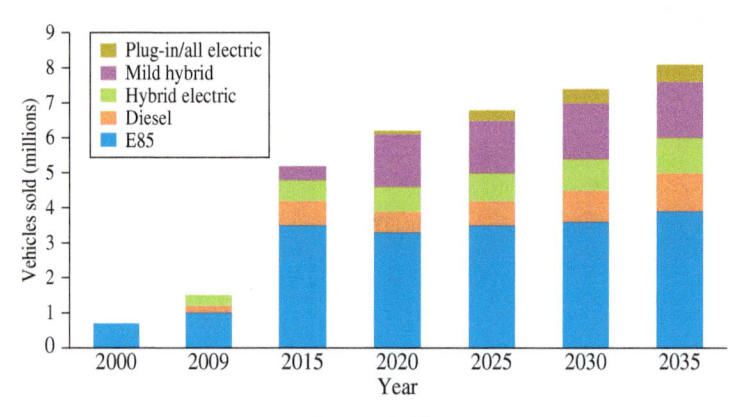

Figure 4.22

The functions that model the number of each type of vehicle sold, in millions, are

E85 flex fuel:	$F(x) = -0.0027x^2 + 0.193x + 0.439$
Diesel:	$D(x) = 0.028x + 0.077$
Hybrid electric:	$H(x) = 0.025x + 0.154$
Mild hybrid:	$M(x) = -0.006x^2 + 0.350x - 3.380$
Plug-in electric and all electric:	$E(x) = -0.001x^2 + 0.081x - 1.115$

where x is the number of years after 2000.

a. Add functions F and D to obtain a function that gives the total number of flex fuel and diesel vehicles sold.

b. Add functions H, M, and E to obtain a function that gives the total number of hybrid and electric vehicles sold.

c. Graph the functions from parts (a) and (b) together on the same axes for the years 2000–2035.

d. Graph the functions from parts (a) and (b) together on the same axes through 2050 ($x = 50$). Will the number of hybrid and electric vehicles exceed the number of flex fuel and diesel vehicles before 2050, according to the models? Based on this graph, do the models appear to be good predictors of unconventional vehicle sales after 2035?

e. Add the functions from parts (a) and (b) to obtain a function that gives the total number of unconventional vehicles sold. According to the EIA projection, there will be 8.1 million unconventional vehicles sold in 2035. Does the model agree with this projection?

SOLUTION

a. Adding functions F and D gives

$$F(x) + D(x) = -0.0027x^2 + 0.193x + 0.439 + 0.028x + 0.077$$
$$= -0.0027x^2 + 0.221x + 0.516$$

b. Adding functions H, M, and E gives

$$H(x) + M(x) + E(x) = 0.025x + 0.154 - 0.006x^2 + 0.350x - 3.380$$
$$-0.001x^2 + 0.081x - 1.115$$
$$= -0.007x^2 + 0.456x - 4.341$$

c. The graphs of the functions from parts (a) and (b) through $x = 35$ are shown in Figure 4.23.

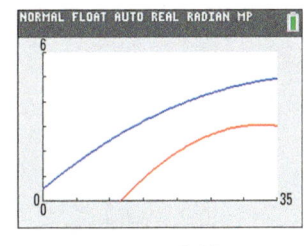

Figure 4.23

Figure 4.24

d. The graphs of the functions from parts (a) and (b) through $x = 50$ are shown in Figure 4.24. The graphs do not intersect in this time interval, so the number of hybrid and electric vehicles will not exceed the number of flex fuel and diesel vehicles before 2050. Based on these graphs, the models do not appear to be good predictors after 2035.

e. Adding the two functions from parts (a) and (b) gives

$$U(x) = -0.0027x^2 + 0.221x + 0.516 - 0.007x^2 + 0.456x - 4.341$$
$$= -0.0097x^2 + 0.677x - 3.825$$

Substituting $x = 35$ into function U gives

$$U(35) = -0.0097(35)^2 + 0.677(35) - 3.825 = 7.99$$

Thus, the model predicts $7.99 \approx 8.0$ million, which is close to the EIA projection.

Average Cost

A company's **average cost** per unit, when x units are produced, is the quotient of the function $C(x)$ (the total production cost) and the identity function $I(x) = x$ (the number of units produced). That is, the **average cost** function is

$$\overline{C}(x) = \frac{C(x)}{x}$$

EXAMPLE 4 ▶ Average Cost

Sunny's Greenhouse produces roses, and their total cost for the production of x hundred roses is

$$C(x) = 50x + 500$$

a. Form the average cost function.

b. For which input values is $\overline{C}$ defined? Give a real-world explanation of this answer.

c. Graph $\overline{C}$ for 0 to 5000 roses (50 units). What can you say about the average cost?

SOLUTION

a. The average cost function is the quotient of the cost function $C(x) = 50x + 500$ and the identity function $I(x) = x$.

$$\overline{C}(x) = \frac{50x + 500}{x}$$

b. The average cost function is defined for all real numbers such that $x > 0$ because producing negative units is not possible and the function is undefined for $x = 0$. (This is reasonable, because if nothing is produced, it does not make sense to discuss average cost per unit of product.)

c. The graph is shown in Figure 4.25. It can be seen from the graph that as the number of roses produced increases, the average cost decreases.

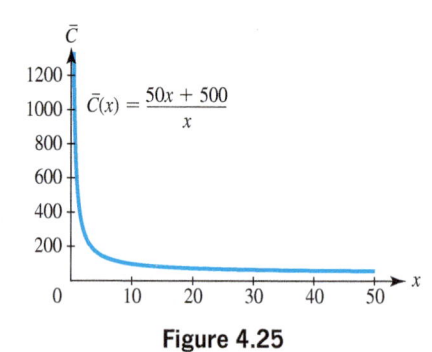

Figure 4.25

Composition of Functions

We have seen that we can convert a temperature of $x°$ Celsius to a Fahrenheit temperature by using the formula $F(x) = \frac{9}{5}x + 32$. There are also temperature scales in which 0° is absolute zero, which is set at $-273.15°C$. One of these, called the Kelvin temperature scale (K), uses the same degree size as Celsius until it reaches absolute zero, which is 0 K. To convert a Celsius temperature to a Kelvin temperature, we can add 273.15 to the Celsius temperature, and to convert a Kelvin temperature to a Celsius temperature, we can use the formula

$$C(x) = x - 273.15$$

To convert from x kelvins to a Fahrenheit temperature, we find $F(C(x))$ by substituting $C(x) = x - 273.15$ for x in $F(x) = \frac{9}{5}x + 32$, getting

$$F(C(x)) = F(x - 273.15) = \frac{9}{5}(x - 273.15) + 32 = \frac{9}{5}x - 459.67$$

4.2 Combining Functions; Composite Functions 283

That is, we can convert from the Kelvin scale to the Fahrenheit scale by using

$$F(K) = \frac{9}{5}K - 459.67$$

The process we have used to get a new function from these two functions is called **composition of functions**. The function $F(K(x))$ is called a **composite function**.

> **Composite Function**
>
> The composite function f of g is denoted by $f \circ g$ and defined by
>
> $$(f \circ g)(x) = f(g(x))$$
>
> The domain of $f \circ g$ is the subset of the domain of g for which $f \circ g$ is defined.
> The composite function $g \circ f$ is defined by $(g \circ f)(x) = g(f(x))$.
> The domain of $g \circ f$ is the subset of the domain of f for which $g \circ f$ is defined.

When computing the composite function $f \circ g$, keep in mind that the output of g becomes the input for f and that the rule for f is applied to this new input.

A composite function machine can be thought of as a machine within a machine. Figure 4.26 shows a composite "function machine" in which the input is denoted by x, the inside function is g, the outside function is f, the composite function rule is denoted by $f \circ g$, and the composite function output is symbolized by $f(g(x))$. Note that for most functions f and g, $f \circ g \neq g \circ f$.

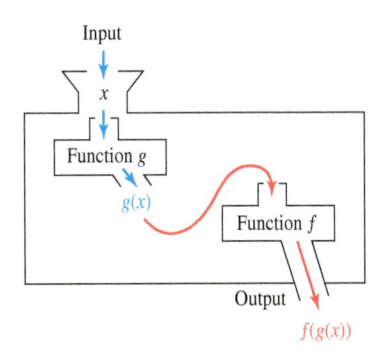

Figure 4.26

EXAMPLE 5 ▶ Function Composition and Orange Juice

There are two machines in a room; one of the machines squeezes oranges to make orange juice, and the other is a canning machine that puts the orange juice into cans. Thinking of these two processes as functions, we name the squeezing function g and the canning function f.

a. Describe the composite function $(f \circ g)(\text{orange}) = f(g(\text{orange}))$.

b. Describe the composite function $(g \circ f)(\text{orange}) = g(f(\text{orange}))$.

c. Which function, $f(g(\text{orange}))$, $g(f(\text{orange}))$, neither, or both, makes sense in context?

SOLUTION

a. The process by which the output $f(g(\text{orange}))$ is obtained is easiest to understand by thinking "from the inside out." Think of the input as an orange. The orange first goes into the *g* machine, so it is squeezed. The output of *g*, liquid orange juice, is then put into the *f* machine, which puts it into a can. The result is a can containing orange juice (Figure 4.27).

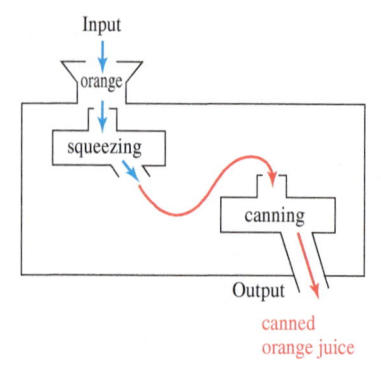

Figure 4.27

b. Again thinking from the inside out, the output $g(f(\text{orange}))$ is obtained by first putting an orange into the *f* machine, which puts it into a can. The output of *f*, the canned orange, is then put into the squeezing machine, *g*. The result is a compacted can, containing an orange (Figure 4.28).

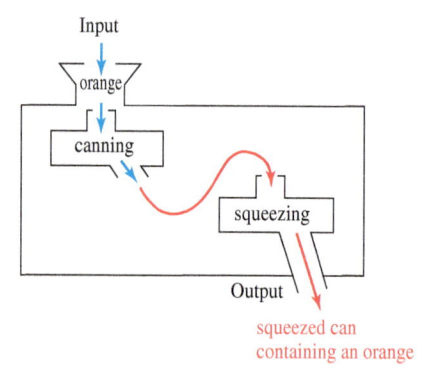

Figure 4.28

c. As can be seen from the results of parts (a) and (b), the order in which the functions appear in the composite function symbol does make a difference. The process that makes sense is the one in part (a), $f(g(\text{orange}))$.

EXAMPLE 6 ▶ Finding Composite Function Outputs

Find the following composite function outputs, using $f(x) = 2x - 5$, $g(x) = 6 - x^2$, and $h(x) = \dfrac{1}{x}$. Give the domain of each new function formed.

a. $(h \circ f)(x) = h(f(x))$ b. $(f \circ g)(x) = f(g(x))$ c. $(g \circ f)(x) = g(f(x))$

SOLUTION

a. First, the output of *f* becomes the input for *h*.

$$(h \circ f)(x) = h(f(x)) = h(2x - 5) = \frac{1}{2x - 5}$$

Next, the rule for h is applied to this input.

The domain of this function is all $x \neq \frac{5}{2}$ because $\frac{1}{2x-5}$ is undefined if $x = \frac{5}{2}$.

b. $(f \circ g)(x) = f(g(x)) = f(6 - x^2) = 2(6 - x^2) - 5 = 12 - 2x^2 - 5$
$= -2x^2 + 7$

The domain of this function is the set of all real numbers.

c. $(g \circ f)(x) = g(f(x)) = g(2x - 5) = 6 - (2x - 5)^2 = 6 - (4x^2 - 20x + 25)$
$= 6 - 4x^2 + 20x - 25 = -4x^2 + 20x - 19$

The domain of this function is the set of all real numbers.

Technology Note

If you have two functions input as Y_1 and Y_2, you can enter the sum, difference, product, quotient, or composition of the two functions in Y_3. These combinations of functions can then be graphed or evaluated for input values of x.

EXAMPLE 7 ▶ Combinations with Functions

If $f(x) = \sqrt{x - 5}$ and $g(x) = 2x^2 - 4$,

a. Graph $(f + g)(x)$ using the window $[-1, 10]$ by $[0, 200]$. What is the domain of $f + g$?

b. Graph $\left(\dfrac{f}{g}\right)(x)$ using the window $[5, 15]$ by $[0, 0.016]$. What is the domain of $\dfrac{f}{g}$?

c. Compute $(f \circ g)(-3)$ and $(g \circ f)(9)$.

SOLUTION

a. Functions f and g are entered as Y_1 and Y_2 in a graphing calculator and the sum of f and g is entered in Y_3, as shown in Figure 4.29(a). The graph of $(f + g)(x)$ is shown in Figure 4.29(b).

(a) (b)

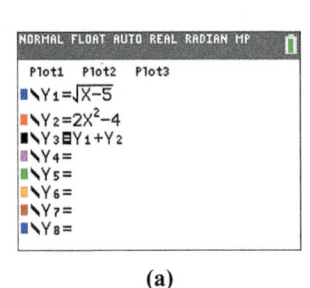

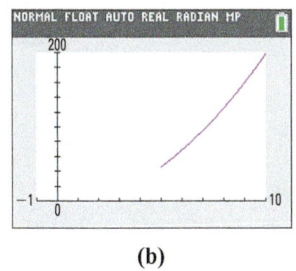

Figure 4.29

The domain of $(f + g)(x)$ is $x \geq 5$ because values of x less than 5 make $f(x) + g(x)$ undefined. This can also be seen in the graph in Figure 4.29(b).

b. The quotient of f and g is entered in Y_3, as shown in Figure 4.30(a). The graph of $\left(\dfrac{f}{g}\right)(x)$ is shown in Figure 4.30(b). The domain of $\left(\dfrac{f}{g}\right)(x)$ is $x \geq 5$ because values of x less than 5 make $\dfrac{f(x)}{g(x)}$ undefined. This can also be seen in the graph in Figure 4.30(b).

286　Chapter 4　Additional Topics with Functions

(a)

(b)

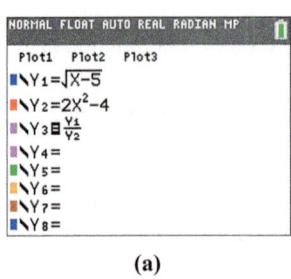

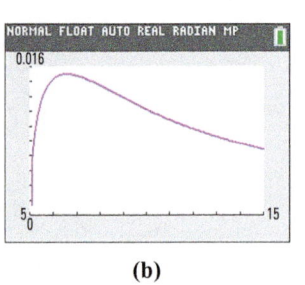

Figure 4.30

c. To compute $(f \circ g)(-3)$, we enter $Y_1(Y_2(-3))$, obtaining 3 (Figure 4.31(a)). To compute $(g \circ f)(9)$, we enter $Y_2(Y_1(9))$, obtaining 4 (Figure 4.31(b)).

(a)

(b)

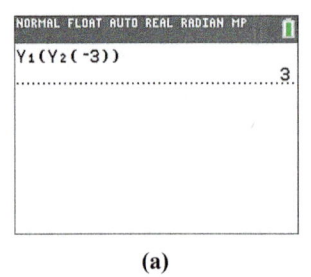

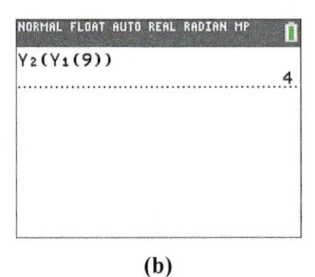

Figure 4.31

SKILLS CHECK 4.2

Answers that are not seen can be found in the answer section at the back of the text.

In Exercises 1–8, find the following: (a) $(f + g)(x)$, (b) $(f - g)(x)$, (c) $(f \cdot g)(x)$, (d) $\left(\dfrac{f}{g}\right)(x)$, and (e) the domain of $\dfrac{f}{g}$.

1. $f(x) = 3x - 5; g(x) = 4 - x$
2. $f(x) = 2x - 3; g(x) = 5 - x$
3. $f(x) = x^2 - 2x; g(x) = 1 + x$
4. $f(x) = 2x^2 - x; g(x) = 2x + 1$
5. $f(x) = \dfrac{1}{x}; g(x) = \dfrac{x + 1}{5}$
6. $f(x) = \dfrac{x - 2}{3}; g(x) = \dfrac{1}{x}$
7. $f(x) = \sqrt{x}; g(x) = 1 - x^2$
8. $f(x) = x^3; g(x) = \sqrt{x + 3}$

9. If $f(x) = x^2 - 5x$ and $g(x) = 6 - x^3$, evaluate
   a. $(f + g)(2)$ −8
   b. $(g - f)(-1)$ 1
   c. $(f \cdot g)(-2)$ 196
   d. $\left(\dfrac{g}{f}\right)(3)$ 3.5

10. If $f(x) = 4 - x^2$ and $g(x) = x^3 + x$, evaluate
    a. $(f + g)(1)$ 5
    b. $(f - g)(-2)$ 10
    c. $(f \cdot g)(-3)$ 150
    d. $\left(\dfrac{g}{f}\right)(2)$ Undefined

In Exercises 11–20, find (a) $(f \circ g)(x)$ and (b) $(g \circ f)(x)$.

11. $f(x) = 2x - 6; g(x) = 3x - 1$ a. $6x - 8$ b. $6x - 19$
12. $f(x) = 3x - 2; g(x) = 2x - 2$ a. $6x - 8$ b. $6x - 6$
13. $f(x) = x^2; g(x) = \dfrac{1}{x}$
14. $f(x) = x^3; g(x) = \dfrac{2}{x}$
15. $f(x) = \sqrt{x - 1}; g(x) = 2x - 7$

16. $f(x) = \sqrt{3-x}; g(x) = x - 5$

17. $f(x) = |x-3|; g(x) = 4x$ a. $|4x-3|$ b. $4|x-3|$

18. $f(x) = |4-x|; g(x) = 2x+1$
    a. $|3-2x|$ b. $2|4-x|+1$

19. $f(x) = \dfrac{3x+1}{2}; g(x) = \dfrac{2x-1}{3}$ a. x b. x

20. $f(x) = \sqrt[3]{x+1}; g(x) = x^3 + 1$ a. $\sqrt[3]{x^3+2}$ b. $x+2$

In Exercises 21 and 22, use $f(x)$ and $g(x)$ to evaluate each expression.

21. $f(x) = 2x^2; g(x) = \dfrac{x-5}{3}$

    a. $(f \circ g)(2)$ 2
    b. $(g \circ f)(-2)$ 1

22. $f(x) = (x-1)^2; g(x) = 3x - 1$

    a. $(f \circ g)(2)$ 16
    b. $(g \circ f)(-2)$ 26

In Exercises 23 and 24, use the following graphs of f and g to evaluate the functions.

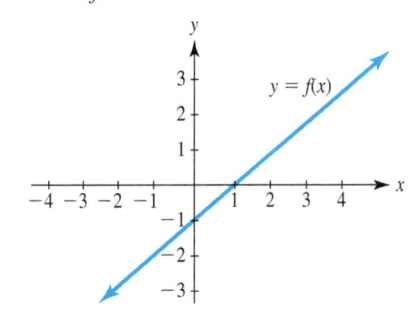

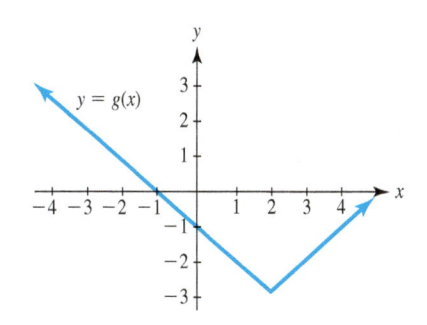

23. a. $(f+g)(2)$ −2 b. $(f \circ g)(-1)$ −1
    c. $\left(\dfrac{f}{g}\right)(4)$ −3 d. $(f \circ g)(1)$ −3
    e. $(g \circ f)(-2)$ 2

24. a. $(g - f)(-2)$ 4 b. $(f \circ g)(3)$ −3
    c. $\left(\dfrac{f}{g}\right)(0)$ 1 d. $(f \circ g)(-2)$ 0
    e. $(g \circ f)(2)$ −2

EXERCISES 4.2

Answers that are not seen can be found in the answer section at the back of the text.

25. **Profit** Suppose that the total weekly cost for the production and sale of x bicycles is $C(x) = 23x + 3420$ dollars and that the total revenue is given by $R(x) = 89x$ dollars, where x is the number of bicycles.

    a. Write the equation of the function that models the weekly profit from the production and sale of x bicycles. $P = 66x - 3420$

    b. What is the profit on the production and sale of 150 bicycles? $6480

    c. Write the function that gives the average profit per bicycle. $\overline{P}(x) = \dfrac{66x - 3420}{x}$

    d. What is the average profit per bicycle if 150 are produced and sold? $43.20

26. **Profit** Suppose that the total weekly cost for the production and sale of televisions is $C(x) = 189x + 5460$ and that the total revenue is given by $R(x) = 988x$, where x is the number of televisions and $C(x)$ and $R(x)$ are in dollars.

    a. Write the equation of the function that models the weekly profit from the production and sale of x televisions. $P = 799x - 5460$

    b. What is the profit on the production and sale of 80 television sets in a given week? $58,460

    c. Write the function that gives the average profit per television. $\overline{P}(x) = \dfrac{799x - 5460}{x}$

    d. What is the average profit per television if 80 are produced and sold? $730.75

27. **Revenue and Cost** The total revenue function for LED TVs is given by $R = 1050x$ dollars, and the total cost function for the TVs is $C = 10{,}000 + 30x + x^2$ dollars, where x is the number of TVs that are produced and sold.

    a. Which function is quadratic, and which is linear?
    Cost is quadratic; revenue is linear.

b. Form the profit function for the TVs from these two functions. $P = 1020x - x^2 - 10{,}000$

c. Is the profit function a linear function, a quadratic function, or neither of these? Quadratic

28. Revenue and Cost The total monthly revenue function for Easy-Ride golf carts is given by $R = 26{,}600x$ dollars, and the total monthly cost function for the carts is $C = 200{,}000 + 4600x + 2x^2$ dollars, where x is the number of golf carts that are produced and sold.

a. Which function is quadratic, and which is linear? Cost is quadratic; revenue is linear.

b. Form the profit function for the golf carts from these two functions. $P = -2x^2 + 22{,}000x - 200{,}000$

c. Is the profit function a linear function, a quadratic function, or neither of these? Quadratic

29. Revenue and Cost The total weekly revenue function for a certain digital camera is given by $R = 550x$ dollars, and the total weekly cost function for the cameras is $C = 10{,}000 + 30x + x^2$ dollars, where x is the number of cameras that are produced and sold.

a. Find the profit function. $P = 520x - x^2 - 10{,}000$

b. Find the number of cameras that gives maximum profit. 260

c. Find the maximum possible profit. $57,600

30. Revenue and Cost The total monthly revenue function for camcorders is given by $R = 6600x$ dollars, and the total monthly cost function for the camcorders is $C = 2000 + 4800x + 2x^2$ dollars, where x is the number of camcorders that are produced and sold.

a. Find the profit function. $P = -2x^2 + 1800x - 2000$

b. Find the number of camcorders that gives maximum profit. 450

c. Find the maximum possible profit. $403,000

31. Average Cost If the monthly total cost of producing 27-inch television sets is given by $C(x) = 50{,}000 + 105x$, where x is the number of sets produced per month, then the average cost per set is given by

$$\overline{C}(x) = \frac{50{,}000 + 105x}{x}$$

a. Explain how $C(x)$ and another function can be combined to obtain the average cost function.

b. What is the average cost per set if 3000 sets are produced? $121.67

32. Cost-Benefit Suppose that for a certain city the cost C of obtaining drinking water that contains $p\%$ impurities (by volume) is given by

$$C = \frac{120{,}000}{p} - 1200$$

a. This function can be considered as the difference of what two functions? $f(p) = \frac{120{,}000}{p}, g(p) = 1200$

b. What is the cost of drinking water that is 80% impure? $300

33. Printers The weekly total cost function for producing a laser printer is $C(x) = 3000 + 72x$, where x is the number of printers produced per week.

a. Form the weekly average cost function for this product. $\overline{C}(x) = \frac{3000 + 72x}{x}$

b. Find the average cost for the production of 100 printers. $102 per printer

34. Electronic Components The monthly cost of producing x electronic components is $C(x) = 2.15x + 2350$.

a. Find the monthly average cost function.

b. Find the average cost for the production of 100 components. $25.65

35. Football Tickets At a certain school, the number of student tickets sold for a home football game can be modeled by $S(p) = 62p + 8500$, where p is the winning percent of the home team. The number of nonstudent tickets sold for these home games is given by $N(p) = 0.5p^2 + 16p + 4400$.

a. Write an equation for the total number of tickets sold for a home football game at this school as a function of the winning percent p. $S(p) + N(p) = 0.5p^2 + 78p + 12{,}900$

b. What is the domain for the function in part (a) in this context? $0 \le p \le 100$

c. Assuming that the football stadium is filled to capacity when the team wins 90% of its home games, what is the capacity of the school's stadium? 23,970

36. T-Shirt Sales Let $T(c)$ be the number of T-shirts that are sold when the shirts have c colors and $P(c)$ be the price, in dollars, of a T-shirt that has c colors. Write a sentence explaining the meaning of the function $(T \cdot P)(c)$. Function represents number of T-shirts sold multiplied by price per shirt. Result is revenue for selling shirts with c colors.

37. Harvesting A farmer's main cash crop is tomatoes, and the tomato harvest begins in the month of May. The number of bushels of tomatoes harvested on the xth day of May is given by the equation $B(x) = 6(x + 1)^{3/2}$. The market price in dollars of 1 bushel of tomatoes on the xth day of May is given by the formula $P(x) = 8.5 - 0.12x$.

a. How many bushels did the farmer harvest on May 8? $B(8) = 162$

b. What was the market price of tomatoes on May 8? $P(8) = \$7.54$

c. How much was the farmer's tomato harvest worth on May 8? $(B \cdot P)(8) = \$1221.48$

d. Write a model for the worth W of the tomato harvest on the xth day of May. $W(x) = (B \cdot P)(x) = 6(x + 1)^{3/2}(8.5 - 0.12x)$

38. Total Cost If the fixed cost for producing a product is $3000 and the variable cost for the production is $3.30x^2$ dollars, where x is the number of units produced, form the total cost function $C(x)$.
$C(x) = 3000 + 3.30x^2$

39. Profit A manufacturer of satellite systems has monthly fixed costs of $32,000 and variable costs of $432 per system, and it sells the systems for $592 per unit.

   a. Write the function that models the profit P from the production and sale of x units of the system.
$P(x) = 160x - 32,000$

   b. What is the profit if 600 satellite systems are produced and sold in 1 month? $64,000

   c. At what rate does the profit grow as the number of units increases? $160 per unit

40. Profit A manufacturer of computers has monthly fixed costs of $87,500 and variable costs of $87 per computer, and it sells the computers for $295 per unit.

   a. Write the function that models the profit P from the production and sale of x computers.
$P(x) = 208x - 87,500$

   b. What is the profit if 700 computers are produced and sold in 1 month? $58,100

   c. What is the y-intercept of the graph of the profit function? What does it mean? $(0, -87,500)$

41. Tobacco Smoking in Brazil The table gives the percent of persons in Brazil aged 15 years and older who smoke daily for selected years from 2000 and projected to 2025.

Years	Men	Women
2000	22.8	14.1
2005	19.9	12.1
2010	17.6	10.3
2015	15.4	8.9
2020	13.6	7.7
2025	12.0	6.6

(*Source*: World Health Organization)

   a. Will adding the percent for men and the percent for women for each year create a new function that gives the total percent of persons in Brazil aged 15 years and older who smoke daily? no

   b. A function that approximately models the total percent of persons in Brazil aged 15 years and older who smoke daily is $p(t) = 0.005x^2 - 0.486x + 18.264$, where t is the number of years after 2000. Use the model to estimate the total percent in 2000 and in 2025. 18.3%; 9.3%

   c. Add the percent of men and the percent of women for the 2000 data and for the 2025 data in the table and divide each by 2. Do these answers agree with your answers in part (b)? 18.45%; 9.3%; they are close to agreeing.

42. Population of Children The table gives the estimated population (in thousands) of U.S. boys age 5 and under and U.S. girls age 5 and under in selected years. A function that models the population (in thousands) of U.S. boys age 5 and under t years after 2010 is $B(t) = -3.17t^2 + 119t + 10,253$, and a function that models the population (in thousands) of U.S. girls age 5 and under t years after 2010 is

$$G(t) = -3.02t^2 + 114t + 9802$$

   a. Find the equation of a function that models the total estimated population (in thousands) of all U.S children age 5 and under t years after 2010. $B(t) + G(t) = -6.19t^2 + 233t + 20,055$

   b. Use the result of part (a) to predict the total estimated population (in thousands) of all children age 5 and under in 2028. 22,243

Years	2015	2020	2025	2030
Boys	10,763	11,150	11,307	11,377
Girls	10,288	10,658	10,807	10,875

(*Source*: U.S. Census Bureau)

43. Function Composition Think of each of the following processes as a function designated by the indicated letter: f, placing in a styrofoam container; g, grinding. Describe each of the functions in parts (a) to (e), then answer part (f).

   a. $f(\text{meat})$ Meat put in container
   b. $g(\text{meat})$ Meat ground
   c. $(g \circ g)(\text{meat})$ Meat ground and ground again
   d. $f(g(\text{meat}))$ Meat ground and put in container
   e. $g(f(\text{meat}))$ Meat put in container, then both ground
   f. Which gives a sensible operation: the function in part (d) or the function in part (e)? d

44. Function Composition Think of each of the following processes as a function designated by the indicated letter: f, putting on a sock; g, taking off a sock. Describe each of the functions in parts (a) to (c).

   a. $f(\text{left foot})$ Sock placed on left foot
   b. $f(f(\text{left foot}))$ Sock placed on left foot, followed by second sock placed on left foot
   c. $(g \circ f)(\text{right foot})$ Sock placed on right foot and then removed

45. Shoe Sizes A woman's shoe that is size x in Japan is size $s(x)$ in the United States, where $s(x) = x - 17$. A woman's shoe that is size x in the United States is size $p(x)$ in Britain, where $p(x) = x - 1.5$. Find a function that will convert Japanese shoe size to British shoe size. $p(s(x)) = x - 18.5$
(*Source*: Kuru International Exchange Association)

46. Shoe Sizes A man's shoe that is size x in Britain is size $d(x)$ in the United States, where $d(x) = x + 0.5$. A man's shoe that is size x in the United States is size $t(x)$ in Continental size, where $t(x) = x + 34.5$. Find a function that will convert British shoe size to Continental shoe size. $t(d(x)) = x + 35$
(*Source*: Kuru International Exchange Association)

47. **Exchange Rates** On a certain date, each Japanese yen was worth 0.34954 Russian ruble and each Chilean peso was worth 0.171718 Japanese yen. Find the value of 100 Chilean pesos in Russian rubles at that time. Round the answer to two decimal places. 6.00 Russian rubles
(*Source*: x-rates.com)

48. **Exchange Rates** On a certain date, each euro was worth 1.3773 U.S. dollars and each Mexican peso was worth 0.06047 euro. Find the value of 100 Mexican pesos in U.S. dollars at that time. $8.33

49. **Facebook** If $f(x)$ represents the number of Facebook users x years after 2000 and $g(x)$ represents the number of MySpace users x years after 2000, what function gives the ratio of MySpace users to Facebook users x years after 2000? $\frac{g(x)}{f(x)}$

50. **Home Computers** If $f(x)$ represents the percent of American homes with computers and $g(x)$ represents the number of American homes, with x equal to the number of years after 1990, then what function represents the number of American homes with computers, with x equal to the number of years after 1990? $(f \cdot g)(x)$ or $(g \cdot f)(x)$

51. **Education** If the function $f(x)$ gives the number of female PhDs produced by American universities x years after 2005 and the function $g(x)$ gives the number of male PhDs produced by American universities x years after 2005, what function gives the total number of PhDs produced by American universities x years after 2005? $f(x) + g(x)$

52. **Wind Chill** If the air temperature is 25°F, the wind chill temperature C is given by $C = 59.914 - 2.35s - 20.14\sqrt{s}$, where s is the wind speed in miles per hour.

   a. State two functions whose difference gives this function. $f(s) = 59.914 - 2.35s$, $g(s) = 20.14\sqrt{s}$

   b. Graph this function for $3 \leq s \leq 12$.

   c. Is the function increasing or decreasing on this domain? Decreasing

53. **Discount Prices** Half-Price Books has a sale with an additional 20% off the regular (1/2) price of their books. What percent of the retail price is charged during this sale? 40%
(*Source*: Half-Price Books, Cleveland, Ohio)

54. **Average Cost** The monthly average cost of producing 42-inch plasma TVs is given by $\overline{C}(x) = \dfrac{100{,}000}{x} + 150$ dollars, where x is the number of sets produced.

   a. Graph this function for $x > 0$.

   b. Will the average cost function decrease or increase as the number of sets produced increases? Decrease

   c. What transformations of the graph of the reciprocal function give the graph of this function? Stretch using a factor of 100,000 and shift up 150 units

4.3 One-to-One and Inverse Functions

KEY OBJECTIVES

- Determine whether two functions are inverses
- Determine whether a function is one-to-one
- Find the inverse of a function
- Graph inverse functions
- Find inverse functions on limited domains

SECTION PREVIEW   Investment

If $1000 is invested for three years at interest rate x, compounded annually, the future value y of the investment is given by

$$y = f(x) = 1000(1 + x)^3$$

If we want a specific future value to be available in three years, and we want to know what interest rate is necessary to get this amount, we can find the inverse of the above function. (See Example 6.)

In this section, we determine whether two functions are inverses of each other and find the inverse of a function if it has one. We also solve applied problems by using inverse functions. ■

Inverse Functions

Suppose a property costing $153,000 is purchased with the promise of paying $1275 per month for 10 years. If all payments are made on time, the amount remaining to be paid when x monthly payments have been made is

$$f(x) = 153{,}000 - 1275x$$

and the function that can be used to find directly how many monthly payments have been made for any given amount x remaining to be paid is

$$g(x) = \frac{153{,}000 - x}{1275}$$

To see that one of these functions "undoes" what the other one does, see Table 4.8, which gives selected inputs and the resulting outputs for both functions. Observe that if we input the number 80 into $f(x)$, the output after it is operated on by f is 51,000, and if 51,000 is then operated on by g, the output is 80, which means that $g(f(80)) = 80$. In addition, if 51,000 is operated on by g, the output is 80, and if 80 is operated on by f, the output is 51,000, so $f(g(51{,}000)) = 51{,}000$.

Table 4.8

x	0	40	80	120
f(x)	153,000	102,000	51,000	0

x	153,000	102,000	51,000	0
g(x)	0	40	80	120

In fact, $g(f(x)) = x$ and $f(g(x)) = x$ for any input x. This means that both of these composite functions act as identity functions because their outputs are the same as their inputs. This happens because the second function performs the inverse operations of the first function. Because of this, we say that $g(x)$ and $f(x)$ are **inverse functions**.

Inverse Functions

Functions f and g for which $f(g(x)) = x$ for all x in the domain of g and $g(f(x)) = x$ for all x in the domain of f are called inverse functions. In this case, we denote g by f^{-1}, read as "f inverse."

EXAMPLE 1 ▶ Temperature Conversion

To convert from x degrees Celsius to y degrees Fahrenheit, we can use the function $y = f(x) = \frac{9}{5}x + 32$. To convert x degrees Fahrenheit to y degrees Celsius, we can use the function $y = g(x) = \frac{5x - 160}{9}$. Show that f and g are inverse functions.

SOLUTION

To show that these functions are inverse functions, we show that $f(g(x)) = x$ and $g(f(x)) = x$.

$$f(g(x)) = f\left(\frac{5x - 160}{9}\right) = \frac{9}{5}\left(\frac{5x - 160}{9}\right) + 32 = \frac{5x - 160}{5} + 32$$

$$= x - 32 + 32 = x$$

$$g(f(x)) = g\left(\frac{9}{5}x + 32\right) = \frac{5\left(\frac{9}{5}x + 32\right) - 160}{9} = \frac{9x + 160 - 160}{9} = \frac{9x}{9} = x$$

Thus the functions are inverse functions.

Consider a function that doubles each input. Its inverse takes half of each input. For example, the discrete function f with equation $f(x) = 2x$ and domain $\{1, 4, 5, 7\}$ has the range $\{2, 8, 10, 14\}$. The inverse of this function has the equation $f^{-1}(x) = \dfrac{x}{2}$.

The domain of the inverse function is the set of outputs of the original function, $\{2, 8, 10, 14\}$. The outputs of the inverse function form the set $\{1, 4, 5, 7\}$, which is the domain of the original function. Figure 4.32 illustrates the relationship between the domains and ranges of the function f and its inverse f^{-1}. In this case and in every case, the domain of the inverse function is the range of the original function, and the range of the inverse function is the domain of the original function.

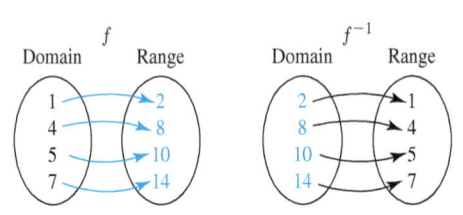

Figure 4.32

It is very important to note that the "-1" used in f^{-1} and g^{-1} is *not* an exponent, but rather a symbol used to denote the inverse of the function. The expression f^{-1} *always* refers to the inverse function of f and *never* to the reciprocal $\dfrac{1}{f}$ of f; that is,

$$f^{-1}(x) \neq \dfrac{1}{f(x)}.$$

One-to-One Functions

In general, we can show that a function has an inverse if it is a **one-to-one function**. We define a one-to-one function as follows.

One-to-One Function

A function f is a one-to-one function if each output of the function corresponds to exactly one input in the domain of the function. This means there is a one-to-one correspondence between the elements of the domain and the elements of the range.

This statement means that for a one-to-one function f, $f(a) \neq f(b)$ if $a \neq b$.

EXAMPLE 2 ▶ One-to-One Functions

Determine if each of the functions is a one-to-one function.

a. $f(x) = 3x^4$

b. $f(x) = x^3 - 1$

SOLUTION

a. Clearly $a = -2$ and $b = 2$ are different inputs, but they both give the same output for $f(x) = 3x^4$.

$$f(-2) = 3(-2)^4 = 48 \quad \text{and} \quad f(2) = 3(2)^4 = 48$$

Thus, $y = 3x^4$ is not a one-to-one function.

b. Suppose that $a \neq b$. Then $a^3 \neq b^3$ and $a^3 - 1 \neq b^3 - 1$, so if $f(x) = x^3 - 1$, $f(a) \neq f(b)$. This satisfies the condition $f(a) \neq f(b)$ if $a \neq b$, so the function $f(x) = x^3 - 1$ is one-to-one.

Recall that no vertical line can intersect the graph of a function in more than one point. The definition of a one-to-one function means that if a function is one-to-one, a horizontal line will intersect its graph in at most one point.

Horizontal Line Test

A function is one-to-one if no horizontal line can intersect the graph of the function in more than one point.

EXAMPLE 3 ▶ Horizontal Line Test

Determine if each of the functions is one-to-one by using the horizontal line test.

a. $y = 3x^4$ **b.** $y = x^3 - 1$

SOLUTION

a. We can see that $y = 3x^4$ is not a one-to-one function because we can observe that the graph of this function does not pass the horizontal line test (Figure 4.33(a)). Note that when $x = 2$, $y = 48$, and when $x = -2$, $y = 48$.

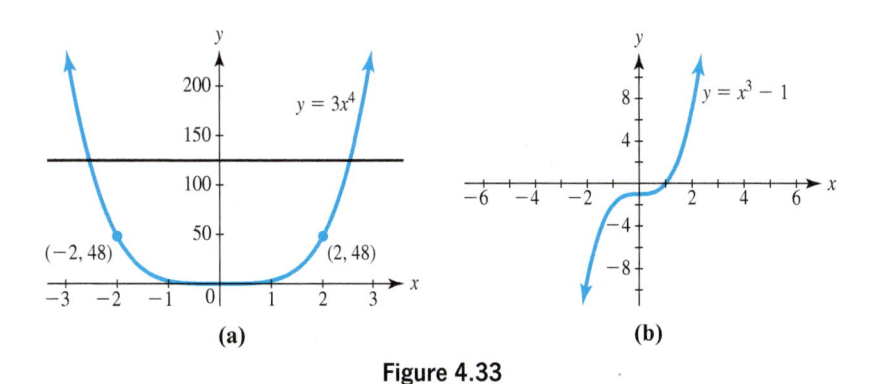

Figure 4.33

b. The graph in Figure 4.33(b) and the horizontal line test indicate that the function $y = x^3 - 1$ is one-to-one. That is, no horizontal line will intersect this graph in more than one point.

Inverse Functions

The functions f and g are inverse functions if, whenever the pair (a, b) satisfies $y = f(x)$, the pair (b, a) satisfies $y = g(x)$. Note that when this happens,

$$f(g(x)) = x \quad \text{and} \quad g(f(x)) = x$$

for all x in the domain of g and f, respectively. The domain of the function f is the range of its inverse g, and the domain of g is the range of f.

EXAMPLE 4 ▶ Inverse Functions

a. Determine whether $f(x) = x^3 - 1$ has an inverse function.

b. Verify that $g(x) = \sqrt[3]{x + 1}$ is the inverse function of $f(x) = x^3 - 1$.

c. Find the domain and range of each function.

SOLUTION

a. Because each output of the function $f(x) = x^3 - 1$ corresponds to exactly one input, the function is one-to-one. Thus, it has an inverse function.

b. To verify that the functions $f(x) = x^3 - 1$ and $g(x) = \sqrt[3]{x + 1}$ are inverse functions, we show that the compositions $f(g(x))$ and $g(f(x))$ are each equal to the identity function, x.

$$f(g(x)) = f(\sqrt[3]{x+1}) = (\sqrt[3]{x+1})^3 - 1 = (x+1) - 1 = x \quad \text{and}$$
$$g(f(x)) = g(x^3 - 1) = \sqrt[3]{(x^3 - 1) + 1} = \sqrt[3]{x^3} = x$$

Thus, we have verified that these functions are inverses.

c. The cube of any real number decreased by 1 is a real number, so the domain of f is the set of all real numbers. The cube root of any real number plus 1 is also a real number, so the domain of g is the set of all real numbers. The ranges of f and g are the set of real numbers.

By using the definition of inverse functions, we can find the equation for the inverse function of f by interchanging x and y in the equation $y = f(x)$ and solving the new equation for y.

Finding the Inverse of a Function

To find the inverse of the function f that is defined by the equation $y = f(x)$:

1. Rewrite the equation, replacing $f(x)$ with y.

2. Interchange x and y in the equation defining the function.

3. Solve the new equation for y. If this equation cannot be solved uniquely for y, the original function has no inverse function.

4. Replace y with $f^{-1}(x)$.

EXAMPLE 5 ▶ Finding an Inverse Function

a. Find the inverse function of $f(x) = \dfrac{2x - 1}{3}$.

b. Graph $f(x) = \dfrac{2x - 1}{3}$ and its inverse function on the same axes.

SOLUTION

a. Using the steps for finding the inverse of a function, we have

$$y = \dfrac{2x - 1}{3} \quad \text{Replace } f(x) \text{ with } y.$$

$$x = \dfrac{2y - 1}{3} \quad \text{Interchange } x \text{ and } y.$$

$$3x = 2y - 1 \quad \text{Solve for } y.$$

$$3x + 1 = 2y$$

$$\dfrac{3x + 1}{2} = y$$

$$f^{-1}(x) = \dfrac{3x + 1}{2} \quad \text{Replace } y \text{ with } f^{-1}(x).$$

b. The graphs of $f(x) = \dfrac{2x - 1}{3}$ and its inverse $f^{-1}(x) = \dfrac{3x + 1}{2}$ are shown in Figure 4.34.

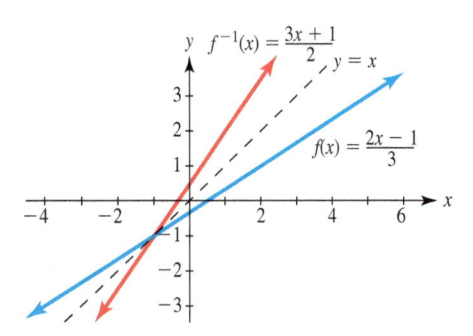

Figure 4.34

Notice that the graphs of $y = f(x)$ and $y = f^{-1}(x)$ in Figure 4.34 appear to be reflections of each other about the line $y = x$. This occurs because if you were to choose any point (a, b) on the graph of the function f and interchange the x- and y-coordinates, the new point (b, a) would be on the graph of the inverse function f^{-1}. This should make sense because the inverse function is formed by interchanging x and y in the equation defining the original function. In fact, this relationship occurs for every function and its inverse.

Graphs of Inverse Functions

The graphs of a function and its inverse are symmetric with respect to the line $y = x$.

We again illustrate the symmetry of graphs of inverse functions in Figure 4.35, which shows the inverse functions $f(x) = x^3 - 2$ and $f^{-1}(x) = \sqrt[3]{x + 2}$. Note that the point $(0, -2)$ is on the graph of $f(x) = x^3 - 2$ and the point $(-2, 0)$ is on the graph of $f^{-1}(x) = \sqrt[3]{x + 2}$.

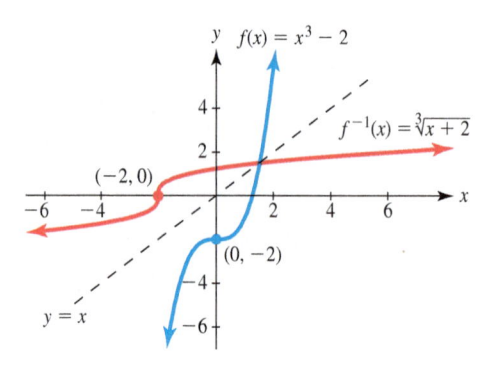

Figure 4.35

EXAMPLE 6 ▶ Investment

If $1000 is invested for 3 years at interest rate x, compounded annually, the future value y of the investment is given by

$$y = f(x) = 1000(1 + x)^3$$

a. To find the interest rate x that results in future value y, find the inverse function of

$$f(x) = 1000(1 + x)^3$$

b. Use the inverse function to find the interest rate that gives a future value of $1331 in 3 years.

SOLUTION

a. To find the inverse of $f(x) = 1000(1 + x)^3$, we proceed as follows.

1. Replace $f(x)$ with y. $y = 1000(1 + x)^3$
2. Interchange x and y. $x = 1000(1 + y)^3$
3. Solve this new equation for y.

$$\frac{x}{1000} = (1 + y)^3$$

$$\sqrt[3]{\frac{x}{1000}} = \sqrt[3]{(1 + y)^3}$$

$$\frac{\sqrt[3]{x}}{\sqrt[3]{1000}} = 1 + y$$

$$y = \frac{\sqrt[3]{x}}{10} - 1$$

4. Replace y with $f^{-1}(x)$.

$$f^{-1}(x) = \frac{\sqrt[3]{x}}{10} - 1$$

b. To find the interest rate that gives a future value of $1331 in 3 years, we evaluate

$$y = f^{-1}(x) = \frac{\sqrt[3]{x}}{10} - 1$$

at $x = 1331$, getting $y = f^{-1}(1331) = \frac{\sqrt[3]{1331}}{10} - 1 = \frac{11}{10} - 1 = 0.10 = 10\%$.

Inverse Functions on Limited Domains

As we have stated, a function cannot have an inverse function if it is not a one-to-one function. However, if there is a limited domain over which such a function is a one-to-one function, then it has an inverse function for this domain. Consider the function $f(x) = x^2$. The horizontal line test on the graph of this function (Figure 4.36(a)) indicates that this function is not a one-to-one function and that the function f does not have an inverse function. To see why, consider the attempt to find the inverse function:

$$y = x^2$$
$$x = y^2 \quad \text{Interchange } x \text{ and } y.$$
$$\pm\sqrt{x} = y \quad \text{Solve for } y.$$

This equation is not the inverse function because it is not a function (one value of x gives two values for y). Because we cannot solve $x = y^2$ uniquely for y, $f(x) = x^2$ does not have an inverse function.

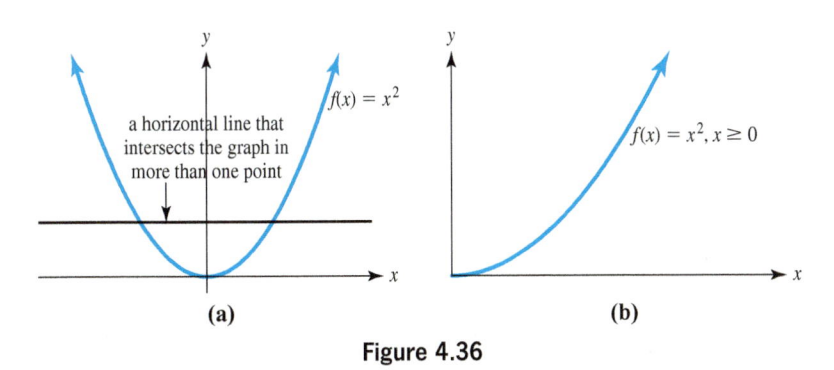

Figure 4.36

However, if we limit the domain of the function to $x \geq 0$, no horizontal line intersects the graph of $f(x) = x^2$ more than once, so the function is one-to-one on this limited domain (Figure 4.36(b)), and the function has an inverse. If we restrict the domain of the original function by requiring that $x \geq 0$, then the range of the inverse function is restricted to $y \geq 0$, and the equation $y = \sqrt{x}$, or $f^{-1}(x) = \sqrt{x}$, defines the inverse function. Figure 4.37 shows the graph of $f(x) = x^2$ for $x \geq 0$ and its inverse $f^{-1}(x) = \sqrt{x}$. Note that the graphs are symmetric about the line $y = x$.

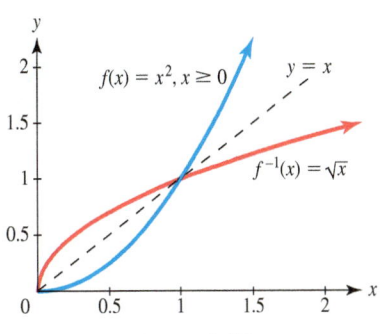

Figure 4.37

EXAMPLE 7 ▶ Velocity of Blood

Because of friction from the walls of an artery, the velocity of blood is greatest at the center of the artery and decreases as the distance x from the center increases. The (nonnegative) velocity in centimeters per second of a blood corpuscle in an artery can be modeled by the function

$$v = k(R^2 - x^2), k > 0, 0 \le x \le R$$

where R is the radius of the artery and k is a constant that is determined by the pressure, the viscosity of the blood, and the length of the artery. In the case where $k = 2$ and $R = 0.1$ centimeter, the velocity can be written as a function of the distance x from the center as

$$v(x) = 2(0.01 - x^2) \text{ centimeter/second}$$

a. Assuming the velocity is nonnegative, does the inverse of this function exist on a domain limited by the context of the application?

b. What is the inverse function?

c. What does the inverse function mean in the context of this application?

SOLUTION

a. Because x represents distance, it is nonnegative, and because $R = 0.1$ and x equals the distance of blood in an artery from the center of the artery, $0 \le x \le 0.1$. Hence, the function is one-to-one for $0 \le x \le 0.1$ and we can find its inverse.

b. We find the inverse function as follows.

$$f(x) = 2(0.01 - x^2)$$

$y = 2(0.01 - x^2)$ Replace $f(x)$ with y.

$x = 2(0.01 - y^2)$ Interchange y and x.

$x = 0.02 - 2y^2$ Solve for y.

$2y^2 = 0.02 - x$

$y^2 = \dfrac{0.02 - x}{2}$

$y = \sqrt{\dfrac{0.02 - x}{2}}$ Only nonnegative values of y are possible.

$f^{-1}(x) = \sqrt{\dfrac{0.02 - x}{2}}$ Replace y with $f^{-1}(x)$.

c. The inverse function gives the distance from the center of the artery as a function of the velocity of a blood corpuscle.

SKILLS CHECK 4.3

Answers that are not seen can be found in the answer section at the back of the text.

In Exercises 1 and 2, determine if the function f defined by the arrow diagram has an inverse. If it does, create an arrow diagram that defines the inverse. If it does not, explain why not.

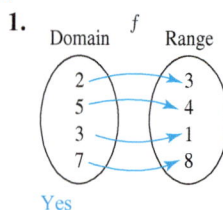

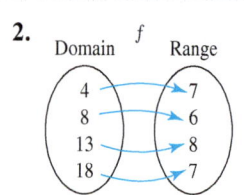

1.
Domain f Range
2 → 3
5 → 4
3 → 1
7 → 8

Yes

2.
Domain f Range
4 → 7
8 → 6
13 → 8
18 → 7

No; the function is not one-to-one.

In Exercises 3 and 4, determine whether the function f defined by the set of ordered pairs has an inverse. If it does, find the inverse.

3. $\{(5, 2), (4, 1), (3, 7), (6, 2)\}$ No inverse

4. $\{(2, 8), (3, 9), (4, 10), (5, 11)\}$
   Yes; $\{(8, 2), (9, 3), (10, 4), (11, 5)\}$

5. If $f(x) = 3x$ and $g(x) = \dfrac{x}{3}$,

   a. What are $f(g(x))$ and $g(f(x))$? x

   b. Are $f(x)$ and $g(x)$ inverse functions? Yes

6. If $f(x) = 4x - 1$ and $g(x) = \dfrac{x+1}{4}$,

   a. What are $f(g(x))$ and $g(f(x))$? x

   b. Are $f(x)$ and $g(x)$ inverse functions? Yes

7. If $f(x) = x^3 + 1$ and $g(x) = \sqrt[3]{x - 1}$, are $f(x)$ and $g(x)$ inverse functions? Yes

8. If $f(x) = (x - 2)^3$ and $g(x) = \sqrt[3]{x} + 2$, are $f(x)$ and $g(x)$ inverse functions? No

9. For the function f defined by $f(x) = 3x - 4$, complete the tables below for f and f^{-1}.

x	f(x)
−1	−7
0	−4
1	−1
2	2
3	5

x	f⁻¹(x)
−7	−1
−4	0
−1	1
2	2
5	3

10. For the function g defined by $g(x) = 2x^3 - 1$, complete the tables below for g and g^{-1}.

x	g(x)
−2	−17
−1	−3
0	−1
1	1
2	15

x	g⁻¹(x)
−17	−2
−3	−1
−1	0
1	1
15	2

In Exercises 11–14, determine whether the function is one-to-one.

11. $\{(1, 5), (2, 6), (3, 7), (4, 5)\}$ No

12. $\{(2, -4), (5, -8), (8, -12), (11, -16)\}$ Yes

13. $f(x) = (x - 3)^3$ Yes 14. $f(x) = \dfrac{1}{x}$ Yes

In Exercises 15 and 16, determine whether each graph is the graph of a one-to-one function.

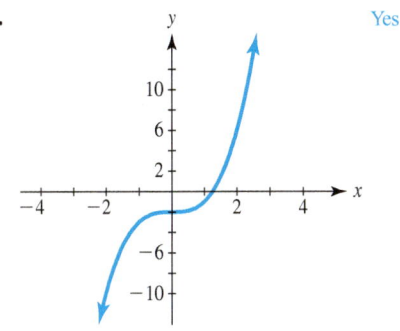

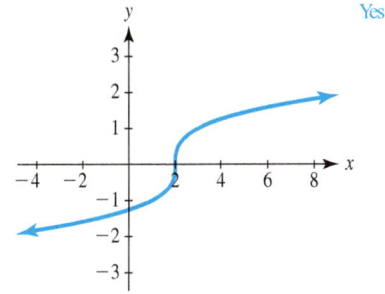

15. Yes

16. Yes

In Exercises 17 and 18, determine whether the function is one-to-one.

17. $y = -2x^4$ No   18. $y = \sqrt{x} + 3$ Yes

19. a. Write the inverse of $f(x) = 3x - 4$. $f^{-1}(x) = \dfrac{x+4}{3}$

    b. Do the values for f^{-1} in the table of Exercise 9 fit the equation for f^{-1}? Yes

20. a. Write the inverse of $g(x) = 2x^3 - 1$. $g^{-1}(x) = \sqrt[3]{\dfrac{x+1}{2}}$

    b. Do the values for g^{-1} in the table of Exercise 10 fit the equation for g^{-1}? Yes

21. If function h has an inverse and $h^{-1}(-2) = 3$, find $h(3)$. −2

22. Find the inverse of $f(x) = \dfrac{1}{x}$. $f^{-1}(x) = \dfrac{1}{x}$

23. Find the inverse of $g(x) = 4x + 1$. $g^{-1}(x) = \dfrac{x-1}{4}$

24. Find the inverse of $f(x) = 4x^2$ for $x \geq 0$. $f^{-1}(x) = \dfrac{\sqrt{x}}{2}$

25. Find the inverse of $g(x) = x^2 - 3$ for $x \geq 0$. $g^{-1}(x) = \sqrt{x+3}$

26. Graph $g(x) = \sqrt{x}$ and its inverse $g^{-1}(x)$ for $x \geq 0$ on the same axes.

27. Graph $g(x) = \sqrt[3]{x}$ and its inverse $g^{-1}(x)$ on the same axes.

28. $f(x) = (x - 2)^2$ and $g(x) = \sqrt{x} + 2$ are inverse functions for what values of x? $x \geq 2$

29. Is the function $f(x) = 2x^3 + 1$ a one-to-one function? Does it have an inverse? Yes; yes

30. Sketch the graph of $y = f^{-1}(x)$ on the axes with the graph of $y = f(x)$, shown below.

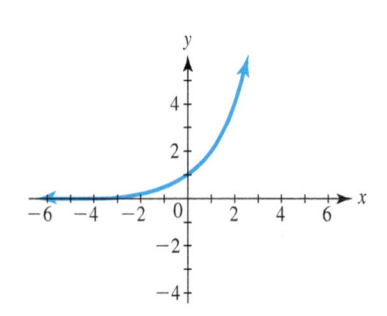

31. Sketch the graph of $y = f^{-1}(x)$ on the axes with the graph of $y = f(x)$, shown below.

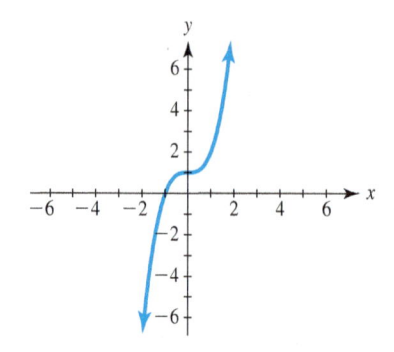

EXERCISES 4.3

Answers that are not seen can be found in the answer section at the back of the text.

32. Shoe Sizes If x is the size of a man's shoe in Britain, then $d(x) = x + 0.5$ is its size in the United States.

  a. Find the inverse of the function. $d^{-1}(x) = x - 0.5$

  b. Use the inverse function to find the British size of a shoe if it is U.S. size $8\frac{1}{2}$. 8

(*Source*: Kuru International Exchange Association)

33. Shoe Sizes If x is the size of a man's shoe in the United States, then $t(x) = x + 34.5$ is its Continental size.

  a. Find a function that will convert Continental shoe size to U.S. shoe size. $t^{-1}(x) = x - 34.5$

  b. Use the inverse function to find the U.S. size if the Continental size of a shoe is 43. $8\frac{1}{2}$

(*Source*: Kuru International Exchange Association)

34. Investments If x dollars are invested at 10% for 6 years, the future value of the investment is given by $S(x) = x + 0.6x$.

  a. Find the inverse of this function. $S^{-1}(x) = \dfrac{x}{1.6}$

  b. What do the outputs of the inverse function represent? The amount originally invested

  c. Use this function to find the amount of money that must be invested for 6 years at 10% to have a future value of $24,000. $15,000

35. Currency Conversion Suppose the function that converts from Canadian dollars to U.S. dollars is $f(x) = 1.0136x$, where x is the number of Canadian dollars and $f(x)$ is the number of U.S. dollars.

  a. Find the inverse function for f and interpret its meaning.
  $f^{-1}(x) = \dfrac{x}{1.0136}$; dividing U.S. dollars by 1.0136 gives Canadian dollars.

  b. Use f and f^{-1} to determine the money you will have if you take 500 U.S. dollars to Canada, convert them to Canadian dollars, don't spend any, and then convert them back to U.S. dollars. (Assume that there is no fee for conversion and the conversion rate remains the same.) $500

(*Source*: Expedia.com)

36. Apparent Temperature If the outside temperature is 90°F, the apparent temperature is given by $A(x) = 82.35 + 29.3x$, where x is the humidity written as a decimal. Find the inverse of this function and use it to find the percent humidity that will give an apparent temperature of 97° if the temperature is 90°F.
(*Source*: "Temperature-Humidity Indices," *The UMAP Journal*, Fall 1989) $A^{-1}(x) = \dfrac{x - 82.35}{29.3}$; 50%

37. Antidepressants The function that models the percent of children ages 0–19 taking antidepressants from 2004 to 2009 is $f(x) = -0.085x + 2.97$, where x is the number of years after 2000.

  a. Find the inverse of this function. What do the outputs of the inverse function represent? $f^{-1}(x) = \dfrac{2.97 - x}{0.085}$; years after 2000 when percentage taking antidepressants is x %

  b. Use the inverse function to find when the percentage is 2.3%. 2008

(*Source*: Medco Health Solutions)

38. Body-Heat Loss The model for body-heat loss depends on the coefficient of convection $K = f(x)$, which depends on wind speed x according to the equation $f(x) = 4\sqrt{4x + 1}$. $x \geq -\dfrac{1}{4}; f(x) \geq 0$

  a. What are the domain and range of this function without regard to the context of this application?

  b. Find the inverse of this function. $f^{-1}(x) = \dfrac{x^2 - 16}{64}$

  c. What are the domain and range of the inverse function? $x \geq 0; f^{-1}(x) \geq -\dfrac{1}{4}$

d. In the context of the application, what are the domain and range of the inverse function? $x \geq 4; f^{-1}(x) \geq 0$

39. **Algorithmic Relationship** For many species of fish, the weight W is a function of the length x, given by $W = kx^3$, where k is a constant depending on the species. Suppose $k = 0.002$, W is in pounds, and x is in inches, so the weight is $W(x) = 0.002x^3$.

   a. Find the inverse function of this function.
      $W^{-1}(x) = \sqrt[3]{500x}$
   b. What does the inverse function give?
      Given the weight, the inverse function calculates the length.
   c. Use the inverse function to find the length of a fish that weighs 2 pounds. 10 in.
   d. In the context of the application, what are the domain and range of the inverse function?
      $x > 0; W^{-1}(x) > 0$

40. **Decoding Messages** If we assign numbers to the letters of the alphabet as follows and assign 27 to a blank space, we can convert a message to a numerical sequence. We can "encode" a message by adding 3 to each number that represents a letter in a message.

A	B	C	D	E	F	G	H	I	J	K	L	M
1	2	3	4	5	6	7	8	9	10	11	12	13

N	O	P	Q	R	S	T	U	V	W	X	Y	Z
14	15	16	17	18	19	20	21	22	23	24	25	26

Thus, the message "Go for it" can be encoded by using the numbers to represent the letters and further encoded by using the function $C(x) = x + 3$. The coded message would be 10 18 30 9 18 21 30 12 23. Find the inverse of the function and use it to decode 23 11 8 30 21 8 4 15 30 23 11 12 17 10.
$C^{-1}(x) = x - 3$; the real thing

41. **Decoding Messages** Use the numerical representation from Exercise 40 and the inverse of the encoding function $C(x) = 3x + 2$ to decode 41 5 35 17 83 41 77 83 14 5 77. Make my day

42. **Social Security Numbers and Income Taxes** Consider the function that assigns each person who pays federal income tax his or her Social Security number. Is this a one-to-one function? Explain.
    Yes; each number identifies exactly one person.

43. **Checkbook Balance** Consider the function with the check number in your checkbook as input and the dollar amount of the check as output. Is this a one-to-one function? Explain.
    No; two checks could be written for the same amount.

44. **Volume of a Cube** The volume of a cube is $f(x) = x^3$ cubic inches, where x is the length of the edge of the cube in inches.

    a. Is this function one-to-one? Yes
    b. Find the inverse of this function. $f^{-1}(x) = \sqrt[3]{x}$

c. What are the domain and range of this inverse function in the context of the application?
   $x > 0; f^{-1}(x) > 0$
d. How could this inverse function be used?
   To find the edge length of a cube from its volume

45. **Volume of a Sphere** The volume of a sphere is $f(x) = \dfrac{4}{3}\pi x^3$ cubic inches, where x is the radius of the sphere in inches.

    a. Is this function one-to-one? Yes
    b. Find the inverse of this function. $f^{-1}(x) = \sqrt[3]{\dfrac{3x}{4\pi}}$
    c. What are the domain and range of the inverse function for this application? $x > 0; f^{-1}(x) > 0$
    d. How could this inverse function be used?
       To calculate the radius of a sphere from its volume
    e. What is the radius of a sphere if its volume is 65,450 cubic inches? 25 in.

46. **Surface Area** The surface area of a cube is $f(x) = 6x^2$ cm^2, where x is the length of the edge of the cube in centimeters.

    a. For what values of x does this model make sense? Is the model a one-to-one function for these values of x? $x > 0$; yes
    b. What is the inverse of this function on this interval? $f^{-1}(x) = \sqrt{\dfrac{x}{6}}$
    c. How could the inverse function be used?
       To calculate the length of the edge of a cube from its surface area

47. **Cloud Revenue.** The global cloud revenue and projected revenue, in billions of dollars, for selected years 2012 and projected to 2026 can be modeled by
    $$C^{-1}(x) = \left(\dfrac{2x}{15}\right)^{2/3}$$
    with x equal to the number of years after 2010. Find the equation of the inverse of function C. $C^{-1}(x) = \left(\dfrac{2x}{15}\right)^{2/3}$
    (*Source*: Wikibon Server SAN & Cloud Research Projects)

48. **Supply** The supply function for a product is $p(x) = \dfrac{1}{4}x^2 + 20$, where x is the number of thousands of units a manufacturer will supply if the price is $p(x)$ dollars.

    a. Is this function a one-to-one function? No
    b. What is the domain of this function in the context of the application? $x \geq 0$
    c. Is the function one-to-one for the domain in part (b)? Yes
    d. Find the inverse of this function and use it to find how many units the manufacturer is willing to supply if the price is $101. $p^{-1}(x) = 2\sqrt{x - 20}$; 18

49. **Illumination** The intensity of illumination of a light is a function of the distance from the light. For a

given light, the intensity is given by $I(x) = \dfrac{300,000}{x^2}$ candle-power, where x is the distance in feet from the light.

a. Is this function a one-to-one function if it is not limited by the context of the application? No

b. What is the domain of this function in the context of the application? $x > 0$; it is a distance.

c. Is the function one-to-one for the domain in part (b)? Yes

d. Find the inverse of this function on the domain from part (b) and use it to find the distance at which the intensity of the light is 75,000 candle-power. $I^{-1}(x) = \sqrt{\dfrac{300,000}{x}}$; 2 ft

50. **First-Class Postage** The postage charged for first-class mail is a function of its weight. The U.S. Postal Service uses the following table to describe the rates for 2015.

Weight Increment, x	Postal Rate
First ounce or fraction of an ounce	$0.55
Each additional ounce or fraction	$0.15

(Source: pe.usps.gov/text)

a. Convert this table to a piecewise-defined function $P(x)$ that represents postage for letters weighing more than 0 and no more than 3 ounces, using x as the weight in ounces and $P(x)$ as the postage in cents.

b. Does P have an inverse function? Why or why not? No; it is not a one-to-one function.

51. **Currency Conversion** Suppose the function that converts United Kingdom (U.K.) pounds to U.S. dollars is $f(x) = 1.6249x$, where x is the number of pounds and $f(x)$ is the number of U.S. dollars.

a. Find the inverse function for f and interpret its meaning. $f^{-1}(x) = 0.6154x$; converts U.S. dollars to U.K. pounds

b. Use f and f^{-1} to determine the money you will have if you take 1000 U.S. dollars to the United Kingdom, convert them to pounds, don't spend any, and then convert them back to U.S. dollars. (Assume that there is no fee for conversion and the conversion rate remains the same.) $1000
(Source: International Monetary Fund)

52. **Path of a Ball** If a ball is thrown into the air at a velocity of 96 feet per second from a building that is 256 feet high, the height of the ball after x seconds is $f(x) = 256 + 96x - 16x^2$ feet.

a. For how many seconds will the ball be in the air? 8 sec

b. Is this function a one-to-one function over this time interval? No

c. Give an interval over which the function is one-to-one. $0 \le x \le 3$ or $3 \le x \le 8$

d. Find the inverse of the function over the interval $0 \le x \le 3$. What does it give? $f^{-1}(x) = 3 - \dfrac{\sqrt{400-x}}{4}$; the number of seconds, from 0 sec to 3 sec, to attain height x

4.4 Additional Equations and Inequalities

KEY OBJECTIVES
- Solve radical equations
- Solve equations with rational powers
- Solve equations in quadratic form
- Solve quadratic inequalities
- Solve power inequalities
- Solve inequalities involving absolute value

SECTION PREVIEW Profit

The daily profit from the production and sale of x units of a product is given by

$$P(x) = -0.01x^2 + 20.25x - 500$$

Because a profit occurs when $P(x)$ is positive, there is a profit for those values of x that make $P(x) > 0$. Thus, the values of x that give a profit are the solutions to

$$-0.01x^2 + 20.25x - 500 > 0$$

4.4 Additional Equations and Inequalities

In this section, we solve quadratic inequalities by using both analytical and graphical methods. We also solve equations and inequalities involving radicals, rational powers, and absolute values. ∎

Radical Equations; Equations Involving Rational Powers

An equation containing a radical can frequently be converted to an equation that does not contain a radical by raising both sides of the equation to an appropriate power. For example, an equation containing a square root radical can usually be converted by squaring both sides of the equation. It is possible, however, that raising both sides of an equation to a power may produce an *extraneous* solution (a value that does not satisfy the original equation). For this reason, we must check solutions when using this technique. To solve an equation containing a radical, we use the following steps.

Solving Radical Equations

1. Isolate a single radical on one side of the equation.
2. Square both sides of the equation. (Or raise both sides to a power that is equal to the index of the radical.)
3. If a radical remains, repeat steps 1 and 2.
4. Solve the resulting equation.
5. All solutions must be checked in the original equation, and only those that satisfy the original equation are actual solutions.

EXAMPLE 1 ▶ Radical Equation

Solve $\sqrt{x + 5} + 1 = x$.

SOLUTION

To eliminate the radical, we isolate it on one side and square both sides:

$$\sqrt{x + 5} = x - 1$$
$$x + 5 = x^2 - 2x + 1$$
$$0 = x^2 - 3x - 4$$
$$0 = (x - 4)(x + 1)$$
$$x = 4 \quad \text{or} \quad x = -1$$

Checking these values shows that 4 is a solution and -1 is not.

$x = 4$	$x = -1$
$\sqrt{4 + 5} + 1 \stackrel{?}{=} 4$	$\sqrt{-1 + 5} + 1 \stackrel{?}{=} -1$
$\sqrt{9} + 1 \stackrel{?}{=} 4$	$\sqrt{4} + 1 \stackrel{?}{=} -1$
$3 + 1 \stackrel{?}{=} 4$	$2 + 1 \stackrel{?}{=} -1$
$4 = 4$	$3 \neq -1$

304 Chapter 4 Additional Topics with Functions

We can also check graphically by the intersection method. Graphing $y = \sqrt{x+5} + 1$ and $y = x$, we find the point of intersection to be $(4, 4)$ (Figure 4.38). Note that there is *not* a point of intersection at $x = -1$.

Thus, the solution to the equation $\sqrt{x+5} + 1 = x$ is $x = 4$.

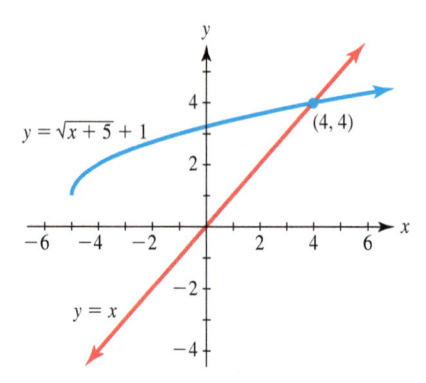

Figure 4.38

EXAMPLE 2 ▶ Radical Equation

Solve $\sqrt{4x - 8} - 1 = \sqrt{2x - 5}$.

SOLUTION

We begin by confirming that one of the radicals is isolated. Then we square both sides of the equation:

$$(\sqrt{4x - 8} - 1)^2 = (\sqrt{2x - 5})^2$$
$$4x - 8 - 2\sqrt{4x - 8} + 1 = 2x - 5$$
$$4x - 7 - 2\sqrt{4x - 8} = 2x - 5$$

Now we must isolate the remaining radical and square both sides of the equation again:

$$2x - 2 = 2\sqrt{4x - 8}$$
$$(2x - 2)^2 = (2\sqrt{4x - 8})^2$$
$$4x^2 - 8x + 4 = 4(4x - 8)$$
$$4x^2 - 8x + 4 = 16x - 32$$
$$4x^2 - 24x + 36 = 0$$
$$4(x^2 - 6x + 9) = 0$$
$$4(x - 3)^2 = 0$$
$$x = 3$$

We can check by substituting or graphing. Graphing $y = \sqrt{4x - 8} - 1$ and $y = \sqrt{2x - 5}$, we find the point of intersection at $x = 3$ (Figure 4.39). Thus, $x = 3$ is the solution.

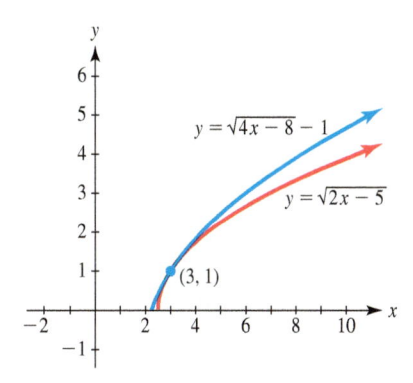

Figure 4.39

Equations Containing Rational Powers

Some equations containing rational powers can be solved by writing the equation as a radical equation.

EXAMPLE 3 ▶ **Equation with Rational Powers**

Solve the equation $(x - 3)^{2/3} - 4 = 0$.

SOLUTION

Rewriting the equation as a radical equation and with the radical isolated gives

$$(x - 3)^{2/3} = 4$$
$$\sqrt[3]{(x - 3)^2} = 4$$

Cubing both sides of the equation and solving gives

$$(x - 3)^2 = 64$$
$$\sqrt{(x - 3)^2} = \pm \sqrt{64}$$
$$x - 3 = \pm 8$$
$$x = 3 \pm 8$$
$$x = 11 \quad \text{or} \quad x = -5$$

Entering each solution in the original equation shows that each checks there.

$x = 11$	$x = -5$
$(11 - 3)^{2/3} - 4 \stackrel{?}{=} 0$	$(-5 - 3)^{2/3} - 4 \stackrel{?}{=} 0$
$8^{2/3} - 4 \stackrel{?}{=} 0$	$(-8)^{2/3} - 4 \stackrel{?}{=} 0$
$(\sqrt[3]{8})^2 - 4 \stackrel{?}{=} 0$	$(\sqrt[3]{-8})^2 - 4 \stackrel{?}{=} 0$
$2^2 - 4 \stackrel{?}{=} 0$	$(-2)^2 - 4 \stackrel{?}{=} 0$
$0 = 0$	$0 = 0$

Equations in Quadratic Form

Some equations that are not quadratic can be written as quadratic equations using substitution. These equations are said to be in **quadratic form**. For example, the equation $x^4 - 5x^2 - 36 = 0$ can be written as $(x^2)^2 - 5(x^2) - 36 = 0$ and then can be more easily solved if we substitute a simpler variable for x^2. If we let $u = x^2$, then the equation $x^4 - 5x^2 - 36 = 0$ can be written in the quadratic form $u^2 - 5u - 36 = 0$. In general, an equation in quadratic form will have the form $au^2 + bu + c = 0$, with u equal to some expression. We can then solve this quadratic equation for u, substitute the expression for u, and complete the solution, as Example 4 shows.

EXAMPLE 4 ▶ Solving an Equation in Quadratic Form

Solve the equation $x^4 - 5x^2 - 36 = 0$.

SOLUTION

The equation contains an expression to a power, x^2, that expression to a power, $(x^2)^2$, and a constant. If we let $u = x^2$, we can solve the equation as follows.

$$x^4 - 5x^2 - 36 = 0$$

$(x^2)^2 - 5(x^2) - 36 = 0$ Write the equation in quadratic form.

$u^2 - 5u - 36 = 0$ Substitute u for x^2.

$(u - 9)(u + 4) = 0$ Factor the quadratic equation.

$u - 9 = 0 \quad u + 4 = 0$ Use the zero-product property.

$u = 9 \quad u = -4$ Solve for u.

Now we have solved our new equation for u, but we must go back and solve the original equation for x. To do this, we substitute x^2 for u in the last two lines of the solution above:

$$u = 9 \qquad\qquad u = -4$$
$$x^2 = 9 \qquad\qquad x^2 = -4$$
$$x = \pm\sqrt{9} \qquad x = \pm\sqrt{-4}$$
$$x = 3, -3 \qquad x = 2i, -2i$$

We can check the solutions graphically. Figure 4.40 shows the graph of $y = x^4 - 5x^2 - 36$. Note that there are only two x-intercepts, 3 and -3. This is so because the solutions $2i$ and $-2i$ are imaginary and therefore are not real zeros of the function.

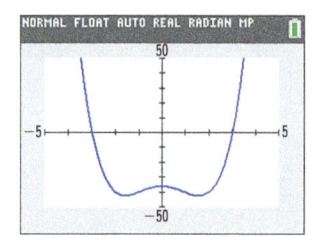

Figure 4.40

EXAMPLE 5 ▶ Solving an Equation in Quadratic Form

Solve the equation $x^{2/3} - x^{1/3} - 6 = 0$.

SOLUTION

Note that $x^{2/3} = (x^{1/3})^2$. The equation contains the expression $x^{1/3}$, that expression to the second power, $(x^{1/3})^2$, and a constant. If we let $u = x^{1/3}$, we can rewrite the equation as

$$x^{2/3} - x^{1/3} - 6 = 0$$

$(x^{1/3})^2 - (x^{1/3}) - 6 = 0$ Write the equation in quadratic form.

$u^2 - u - 6 = 0$ Substitute u for $x^{1/3}$.

$(u - 3)(u + 2) = 0$ Factor the quadratic equation.

$u = 3 \quad\quad u = -2$ Solve for u.

$x^{1/3} = 3 \quad\quad x^{1/3} = -2$ Substitute $x^{1/3}$ for u.

$x = 3^3 \quad\quad x = (-2)^3$ Cube both sides to eliminate the rational exponent.

$x = 27 \quad\quad x = -8$

Thus the solutions are 27 and -8, which can be verified graphically (Figure 4.41).

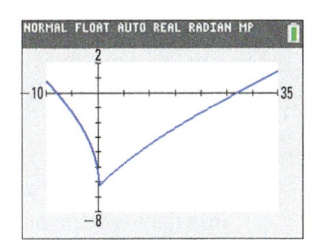

Figure 4.41

Quadratic Inequalities

A **quadratic inequality** is an inequality that can be written in the form

$$ax^2 + bx + c > 0$$

where a, b, and c are real numbers and $a \neq 0$ (or with $>$ replaced by $<, \geq,$ or $\leq$).

Algebraic Solution of Quadratic Inequalities

To solve a quadratic inequality $f(x) > 0$ or $f(x) < 0$ algebraically, we first need to find the zeros of $f(x)$. The zeros can be found by factoring or the quadratic formula. If the quadratic function has two real zeros, these two zeros divide the real number line into three intervals. Within each interval, the value of the quadratic function is either always positive or always negative, so we can use one value in each interval to test the function on the interval. We can use the following steps, which summarize these ideas, to solve quadratic inequalities.

> ### Solving a Quadratic Inequality Algebraically
>
> 1. Write an equivalent inequality with 0 on one side and with the function $f(x)$ on the other side.
> 2. Solve $f(x) = 0$.
> 3. Create a sign diagram that uses the solutions from step 2 to divide the number line into intervals. Pick a test value in each interval and determine whether $f(x)$ is positive or negative in that interval to create a sign diagram.*
> 4. Identify the intervals that satisfy the inequality in step 1. The values of x that define these intervals are solutions to the original inequality.

EXAMPLE 6 ▶ Solving Quadratic Inequalities

Solve the inequality $x^2 - 3x > 3 - 5x$.

SOLUTION

Rewriting the inequality with 0 on the right side of the inequality gives $f(x) > 0$ with $f(x) = x^2 + 2x - 3$:

$$x^2 + 2x - 3 > 0$$

Writing the equation $f(x) = 0$ and solving for x gives

$$x^2 + 2x - 3 = 0$$
$$(x + 3)(x - 1) = 0$$
$$x = -3 \quad \text{or} \quad x = 1$$

The two values of x, -3 and 1, divide the number line into three intervals: the numbers less than -3, the numbers between -3 and 1, and the numbers greater than 1. We need only find the signs of $(x + 3)$ and $(x - 1)$ in each interval and then find the sign of their product to find the solution to the original inequality. Testing a value in each interval determines the sign of each factor in each interval. (See the sign diagram in Figure 4.42.)

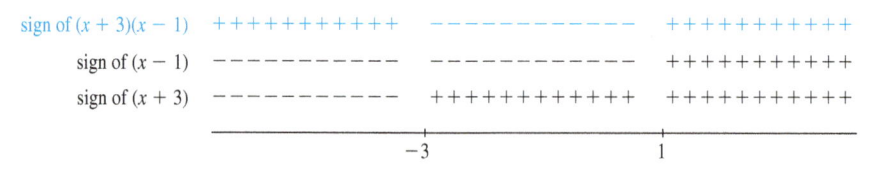

```
sign of (x + 3)(x − 1)   +++++++++++   -----------   +++++++++++
       sign of (x − 1)   -----------   -----------   +++++++++++
       sign of (x + 3)   -----------   +++++++++++   +++++++++++
                       ─────────────┼─────────────┼─────────────
                                   −3             1
```

Figure 4.42

The function $f(x) = (x + 3)(x - 1)$ is positive on the intervals $(-\infty, -3)$ and $(1, \infty)$, so the solution to $x^2 + 2x - 3 > 0$ and thus the original inequality is

$$x < -3 \quad \text{or} \quad x > 1$$

---

*The numerical feature of your graphing utility can be used to test the x-values.

Graphical Solution of Quadratic Inequalities

Recall that we solved a linear inequality $f(x) > 0$ (or $f(x) < 0$) graphically by graphing the related equation $f(x) = 0$ and observing where the graph is above (or below) the x-axis—that is, where $f(x)$ is positive (or negative).

Similarly, we can solve a quadratic inequality $f(x) > 0$ (or $f(x) < 0$) graphically by graphing the related equation $f(x) = 0$ and observing the x-values of the intervals where $f(x)$ is positive (or negative).

For example, we can graphically solve the inequality

$$x^2 - 3x - 4 \leq 0$$

by graphing

$$y = x^2 - 3x - 4$$

and observing the part of the graph that is below or on the x-axis. Using the x-intercept method, we find that $x = -1$ and $x = 4$ give $y = 0$, and we see that the graph is below or on the x-axis for values of x satisfying $-1 \leq x \leq 4$ (Figure 4.43). Thus, the solution to the inequality is $-1 \leq x \leq 4$.

Table 4.9 shows the possible graphs of quadratic functions and the solutions to the related inequalities. Note that if the graph of $y = f(x)$ lies entirely above the x-axis, the solution to $f(x) > 0$ is the set of all real numbers, and there is no solution to $f(x) < 0$. (The graph of $y = f(x)$ can also touch the x-axis in one point.)

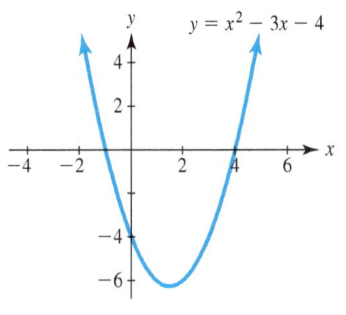

Figure 4.43

Table 4.9 Quadratic Inequalities

Orientation of Graph of $y = f(x)$	Inequality to Be Solved	Part of Graph That Satisfies the Inequality	Solution to Inequality
(upward parabola crossing x-axis at a, b)	$f(x) > 0$	where the graph is above the x-axis	$x < a$ or $x > b$
	$f(x) < 0$	where the graph is below the x-axis	$a < x < b$
(downward parabola crossing x-axis at a, b)	$f(x) > 0$	where the graph is above the x-axis	$a < x < b$
	$f(x) < 0$	where the graph is below the x-axis	$x < a$ or $x > b$
(upward parabola above x-axis)	$f(x) > 0$	the entire graph	all real numbers
	$f(x) < 0$	none of the graph	no solution
(downward parabola below x-axis)	$f(x) > 0$	none of the graph	no solution
	$f(x) < 0$	the entire graph	all real numbers

EXAMPLE 7 ▶ Height of a Model Rocket

A model rocket is projected straight upward from ground level according to the equation

$$h = -16t^2 + 192t, \; t \geq 0$$

where h is the height in feet and t is the time in seconds. During what time interval will the height of the rocket exceed 320 feet?

ALGEBRAIC SOLUTION

To find the time interval when the rocket is higher than 320 feet, we find the values of t for which

$$-16t^2 + 192t > 320$$

Getting 0 on the right side of the inequality, we have

$$-16t^2 + 192t - 320 > 0$$

First we determine when $-16t^2 + 192t - 320$ is equal to 0:

$$-16t^2 + 192t - 320 = 0$$
$$-16(t^2 - 12t + 20) = 0$$
$$-16(t - 2)(t - 10) = 0$$
$$t = 2 \quad \text{or} \quad t = 10$$

The values of $t = 2$ and $t = 10$ divide the number line into three intervals: the numbers less than 2, the numbers between 2 and 10, and the numbers greater than 10 (Figure 4.44). To solve the inequality, we find the sign of the product of -16, $(t - 2)$, and $(t - 10)$ in each interval. We do this by testing a value in each interval.

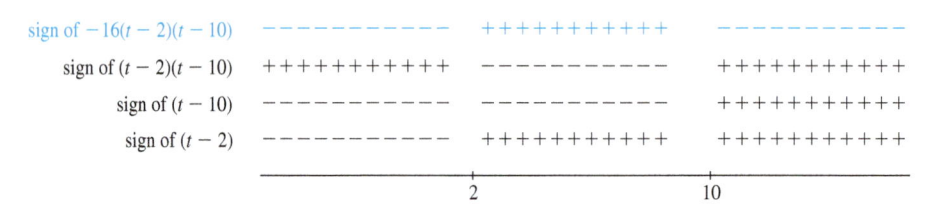

sign of $-16(t - 2)(t - 10)$	$----------$	$++++++++++$	$----------$
sign of $(t - 2)(t - 10)$	$++++++++++$	$----------$	$++++++++++$
sign of $(t - 10)$	$----------$	$----------$	$++++++++++$
sign of $(t - 2)$	$----------$	$++++++++++$	$++++++++++$
	2		10

Figure 4.44

This shows that the product $-16(t - 2)(t - 10)$ is positive only for $2 < t < 10$, so the solution to $-16t^2 + 192t - 320 > 0$ is $2 < t < 10$. This solution is also a solution to the original inequality, so the rocket exceeds 320 feet between 2 and 10 seconds after launch.

Note that if the inequality problem is applied, we must check that the solution makes sense in the context of the problem.

GRAPHICAL SOLUTION

To find the interval when the height exceeds 320 feet, we can also solve $-16t^2 + 192t > 320$ using a graphing utility.* To use the intercept method, we rewrite the inequality with 0 on the right side:

$$-16t^2 + 192t - 320 > 0$$

---

*In a graphing calculator, we would use the independent variable x instead of t.

4.4 Additional Equations and Inequalities 311

We enter $y = -16t^2 + 192t - 320$ and graph the function (Figure 4.45). The *t*-intercepts of the graph are $t = 2$ and $t = 10$.

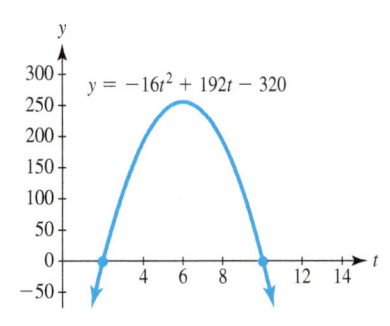

Figure 4.45

The solution to $-16t^2 + 192t - 320 > 0$ is the interval where the graph is above the *t*-axis, which is

$$2 < t < 10$$

Thus, the height of the rocket exceeds 320 feet between 2 and 10 seconds.

EXAMPLE 8 ▶ Obesity

Obesity (BMI ≥ 30) increases the risk of diabetes, heart disease, and many other ailments. The percent of Americans who are obese, from 2000 projected to 2030, can be modeled by $y = -0.0102x^2 + 0.971x + 22.1$, with *x* equal to the number of years after 2000 and *y* equal to the percent of obese Americans. For what years from 2000 to 2030 does this model estimate the percent to be at least 40%?

SOLUTION

To find the years from 2000 to 2030 when the percent of Americans who are obese is at least 40%, we solve the inequality

$$-0.0102x^2 + 0.971x + 22.1 \geq 40$$

We begin solving this inequality by graphing $y = -0.0102x^2 + 0.971x + 22.1$. We set the viewing window from $x = 0$ (2000) to $x = 30$ (2030). The graph of this function is shown in Figure 4.46(a).

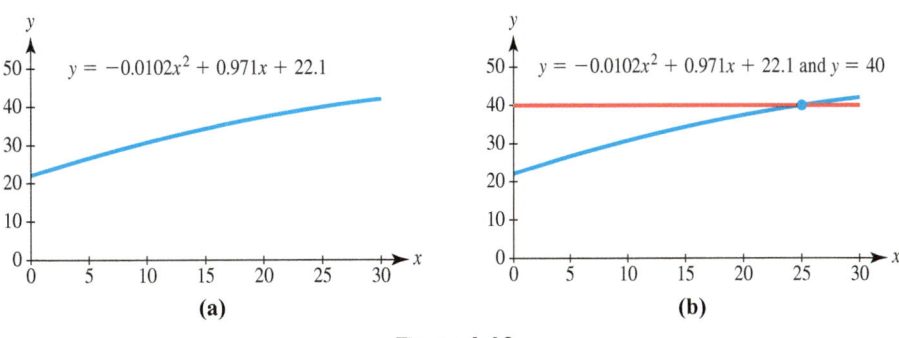

Figure 4.46

We seek the point(s) where the graph of $y = 40$ intersects the graph of $y = -0.0102x^2 + 0.971x + 22.1$. Graphing the line $y = 40$ and intersecting it with the graph of the quadratic function in the given window gives the point of intersection $(25, 40)$, as shown in Figure 4.46(b). This graph shows that the percent of Americans who are obese is at or above 40% for $25 \leq x \leq 30$. Because x represents the number of years after 2000, the percent of obese Americans is projected to be above 40% for the years 2025 through 2030.

Power Inequalities

To solve a power inequality, we will use a combination of analytical and graphical methods.

> **Power Inequalities**
>
> To solve a power inequality, first solve the related equation. Then use graphical methods to find the values of the variable that satisfy the inequality.

EXAMPLE 9 ▶ Investment

The future value of $3000 invested for 3 years at rate r, compounded annually, is given by $S = 3000(1 + r)^3$. What interest rate will give a future value of at least $3630?

SOLUTION

To solve this problem, we solve the inequality

$$3000(1 + r)^3 \geq 3630$$

We begin by solving the related equation using the root method.

$$3000(1 + r)^3 = 3630$$

$(1 + r)^3 = 1.21$ Divide both sides by 3000.

$1 + r = \sqrt[3]{1.21}$ Take the cube root $\left(\frac{1}{3} \text{ power}\right)$ of both sides.

$1 + r \approx 1.0656$

$r \approx 0.0656$

This tells us that the investment will have a future value of $3630 in 3 years if the interest rate is approximately 6.56%, and it helps us set the window to solve the inequality graphically. Graphing $y_1 = 3000(1 + r)^3$ and $y_2 = 3630$ on the same axes shows that $3000(1 + r)^3 \geq 3630$ if $r \geq 0.0656$ (Figure 4.47). Thus, the investment will result in at least $3630 if the interest rate is 6.56% or higher. Note that in the context of this problem, only values of r from 0 (0%) to 1 (100%) make sense.

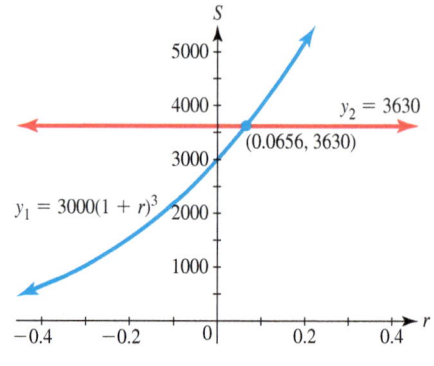

Figure 4.47

Inequalities Involving Absolute Values

Recall that if u is an algebraic expression, then for $a \geq 0$, $|u| = a$ means that $u = a$ or $u = -a$. We can represent inequalities involving absolute values as follows.

> For $a \geq 0$:
> $|u| < a$ means that $-a < u < a$.
> $|u| \leq a$ means that $-a \leq u \leq a$.
> $|u| > a$ means that $u < -a$ or $u > a$.
> $|u| \geq a$ means that $u \leq -a$ or $u \geq a$.

We can solve inequalities involving absolute values algebraically or graphically.

EXAMPLE 10 ▶ Solving Absolute Value Inequalities

Solve the following inequalities and verify the solutions graphically.

a. $|2x - 3| \leq 5$ **b.** $|3x + 4| - 5 > 0$

SOLUTION

a. This inequality is equivalent to $-5 \leq 2x - 3 \leq 5$. If we add 3 to all three parts of this inequality and then divide all parts by 2, we get the solution.

$$-5 \leq 2x - 3 \leq 5$$
$$-2 \leq 2x \leq 8$$
$$-1 \leq x \leq 4$$

Figure 4.48(a) shows that the graph of $y = |2x - 3|$ is on or below the graph of $y = 5$ for $-1 \leq x \leq 4$.

b. This inequality is equivalent to $|3x + 4| > 5$, which is equivalent to $3x + 4 < -5$ or $3x + 4 > 5$. We solve each of the inequalities as follows:

$$3x + 4 < -5 \quad \text{or} \quad 3x + 4 > 5$$
$$3x < -9 \quad\quad\quad 3x > 1$$
$$x < -3 \quad\quad\quad x > \frac{1}{3}$$

Figure 4.48(b) shows that the graph of $y = |3x + 4| - 5$ is above the x-axis for $x < -3$ or $x > \frac{1}{3}$.

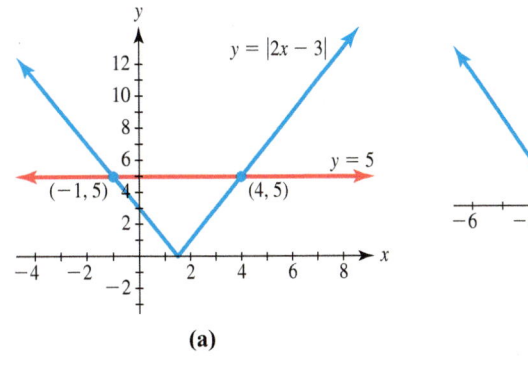

Figure 4.48

SKILLS CHECK 4.4

In Exercises 1–20, solve the equations algebraically and check graphically or by substitution.

1. $\sqrt{2x^2 - 1} - x = 0$ $x = 1$
2. $\sqrt{3x^2 + 4} - 2x = 0$ $x = 2$
3. $\sqrt[3]{x - 1} = -2$ $x = -7$
4. $\sqrt[3]{4 - x} = 3$ $x = -23$
5. $\sqrt{3x - 2} + 2 = x$ $x = 6$
6. $\sqrt{x - 2} + 2 = x$ $x = 3, x = 2$
7. $\sqrt[3]{4x + 5} = \sqrt[3]{x^2 - 7}$ $x = -2, x = 6$
8. $\sqrt{5x - 6} = \sqrt{x^2 - 2x}$ $x = 6$
9. $\sqrt{x} - 1 = \sqrt{x - 5}$ $x = 9$
10. $\sqrt{x - 10} = -\sqrt{x - 20}$ $x = 36$
11. $(x + 4)^{2/3} = 9$ $x = 23, x = -31$
12. $(x - 5)^{3/2} = 64$ $x = 21$
13. $4x^{5/2} - 8 = 0$ $2^{2/5}$ or $\sqrt[5]{4}$
14. $6x^{5/3} - 18 = 0$ $3^{3/5}$ or $\sqrt[5]{27}$
15. $x^4 - 5x^2 + 4 = 0$ $2, -2, 1, -1$
16. $x^4 - 11x^2 + 18 = 0$ $3, -3, \sqrt{2}, -\sqrt{2}$
17. $x^{-2} - x^{-1} - 30 = 0$ $\frac{1}{6}, -\frac{1}{5}$
18. $2x^{-2} + 3x^{-1} - 2 = 0$ $2, -\frac{1}{2}$
19. $2x^{2/3} + 5x^{1/3} - 12 = 0$ $\frac{27}{8}, -64$
20. $x^{2/5} - x^{1/5} - 6 = 0$ $-32, 243$

In Exercises 21–30, use algebraic methods to solve the inequalities.

21. $x^2 + 4x < 0$ $-4 < x < 0$
22. $x^2 - 25x < 0$ $0 < x < 25$
23. $9 - x^2 \geq 0$ $-3 \leq x \leq 3$
24. $x > x^2$ $0 < x < 1$
25. $-x^2 + 9x - 20 > 0$ $4 < x < 5$
26. $2x^2 - 8x < 0$ $0 < x < 4$
27. $2x^2 - 8x \geq 24$ $x \leq -2$ or $x \geq 6$
28. $t^2 + 17t \leq 8t - 14$ $-7 \leq t \leq -2$
29. $x^2 - 6x < 7$ $-1 < x < 7$
30. $4x^2 - 4x + 1 > 0$ $x < \frac{1}{2}$ or $x > \frac{1}{2}$

In Exercises 31–34, use graphical methods to solve the inequalities.

31. $2x^2 - 7x + 2 \geq 0$ $x \leq 0.314$ or $x \geq 3.186$
32. $w^2 - 5w + 4 > 0$ $w < 1$ or $w > 4$
33. $5x^2 \geq 2x + 6$ $x \leq -0.914$ or $x \geq 1.314$
34. $2x^2 \leq 5x + 6$ $-0.886 \leq x \leq 3.386$

In Exercises 35–42, solve the inequalities by using algebraic and graphical methods.

35. $(x + 1)^3 < 4$ $x < 0.587$
36. $(x - 2)^3 \geq -2$ $x \geq 0.74$
37. $(x - 3)^5 < 32$ $x < 5$
38. $(x + 5)^4 > 16$ $x < -7$ or $x > -3$
39. $|2x - 1| < 3$ $-1 < x < 2$
40. $|3x + 1| \leq 5$ $-2 \leq x \leq \frac{4}{3}$
41. $|x - 6| \geq 2$ $x \leq 4$ or $x \geq 8$
42. $|x + 8| > 7$ $x < -15$ or $x > -1$

In Exercises 43–46, you are given the graphs of several functions of the form $f(x) = ax^2 + bx + c$ for different values of a, b, and c. For each function, (a) solve $f(x) \geq 0$, and (b) solve $f(x) < 0$.

43.   a. $x \leq -2$ or $x \geq 3$ b. $-2 < x < 3$

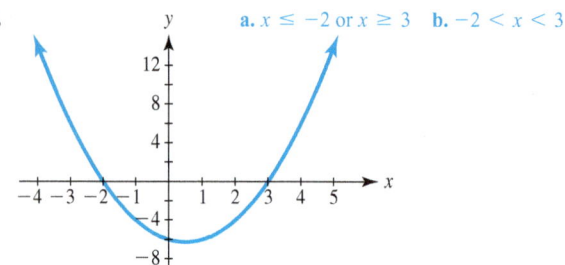

44.   a. $-4 \leq x \leq 2$ b. $x < -4$ or $x > 2$

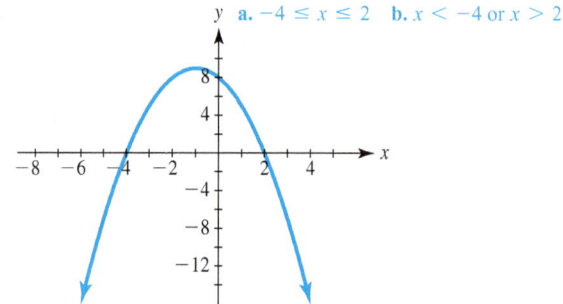

45.   a. No solution   b. All real numbers

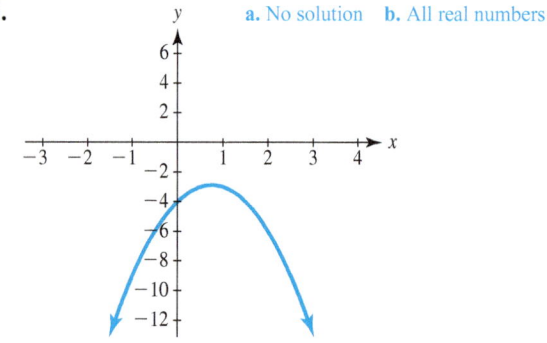

46.   a. All real numbers   b. No solution

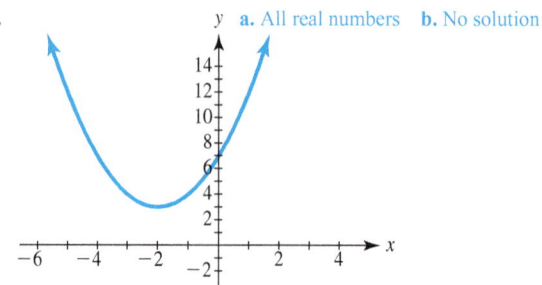

In Exercises 47 and 48, you are given the graphs of two functions $f(x)$ and $g(x)$. Solve $f(x) \leq g(x)$.

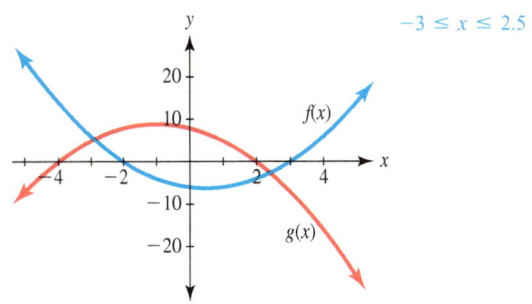

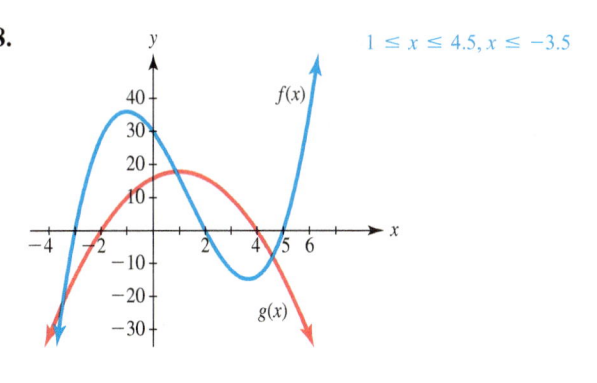

47. $-3 \leq x \leq 2.5$

48. $1 \leq x \leq 4.5, x \leq -3.5$

EXERCISES 4.4

Use algebraic and/or graphical methods to solve Exercises 49–58.

49. Profit The monthly profit from producing and selling x units of a product is given by

$$P(x) = -0.3x^2 + 1230x - 120{,}000$$

Producing and selling how many units will result in a profit for this product? $100 < x < 4000$ units

50. Profit The monthly profit from producing and selling x units of a product is given by

$$P(x) = -0.01x^2 + 62x - 12{,}000$$

Producing and selling how many units will result in a profit for this product? $200 < x < 6000$ units

51. Profit The revenue from sales of x units of a product is given by $R(x) = 200x - 0.01x^2$, and the cost of producing and selling the product is $C(x) = 38x + 0.01x^2 + 16{,}000$. Producing and selling how many units will result in a profit? $100 < x < 8000$ units

52. Profit The revenue from sales of x units of a product is given by $R(x) = 600x - 0.01x^2$, and the cost of producing and selling the product is $C(x) = 77x + 0.02x^2 + 52{,}000$. Producing and selling how many units will result in a profit? $100 < x < 17{,}333$ units

53. Projectiles Two projectiles are fired into the air over a lake, with the height of the first projectile given by $y = 100 + 130t - 16t^2$ and the height of the second projectile given by $y = -16t^2 + 180t$, where y is in feet and t is in seconds. Over what time interval, before the lower one hits the lake, is the second projectile above the first? $2 < t < 8.83$ sec

54. Projectile A rocket shot into the air has height $s = 128t - 16t^2$ feet, where t is the number of seconds after the rocket is shot. During what time after the rocket is shot is it at least 240 feet high? $3 \leq t \leq 5$ sec

55. United Kingdom Population Using data for selected years from 2000 and projected to 2100, the population of the United Kingdom (in millions) is approximated by

$$y = -0.002x^2 + 0.420x + 58.936$$

where x is the number of years after 2000.
(*Source*: Statista.com)

a. When in the 21st century does the model predict that the population will be at least 80.5 million? 2090 and after
b. If the model is valid in the 22nd century, when does it predict that the population of the United Kingdom will then fall below 80.5 million? after 2121

56. World Population The low long-range world population numbers and projections for the years 1995–2150 are given by the equation $y = -0.00036x^2 + 0.0385x + 5.823$, where x is the number of years after 1990 and y is in billions. During what years does this model estimate the population to be above 6 billion? 1995 to 2092
(*Source*: U.N. Department of Economic and Social Affairs)

57. Foreign-Born Population The percent of the U.S. population that is foreign-born for selected years from 1970 and projected to 2060 can be modeled by the function $y = -0.0008x^2 + 0.2343x + 4.225$, with x equal to the number of years after 1970. During what years in the 21st century does the model indicate that the percent is at least 13.94%?
(*Source*: U.S. Census Bureau) 2020 and after

58. Gross Domestic Product The U.S. gross domestic product (GDP) (in trillions of dollars) for selected years from 2005 and projected to 2070 can be modeled by $y = 0.116x^2 - 3.792x + 45.330$, where x is the number of years after 2000. During what years between 2005 and 2070 was the gross domestic product no more than $23.03 trillion? 2008 to 2025
(*Source*: Microsoft Encarta 98)

Use graphical and/or numerical methods to solve Exercises 59–64.

59. Wind Chill The wind chill temperature when the outside temperature is 20°F is given by $y = 0.0052x^2 - 0.62x + 15.0$, where x is the wind speed in mph. For what wind speeds between 0 mph and 90 mph is the wind chill temperature $-3°F$ or below? *50 to 69.23 mph*

60. China's Labor Pool Data from 1975 and projections to 2050 of the number of millions of people age 15 to 59 in China can be modeled by $y = -0.224x^2 + 22.005x + 370.705$, where x is the number of years after 1970 and y represents the number of millions of people in this labor pool. During what years from 1970 to 2050 is the number of people age 15 to 59 at least 903.38 million?
(*Source*: United Nations) *2014 through 2025 ($43.2 < x \leq 55$)*

61. Smartphone Users Using data from 2010 and projected to 2022, the number of smartphone users in the United States, in millions, can be modeled by the function $y = -1.17x^2 + 31.23x + 63.04$, where x is the number of years after 2010. In what years does the model project that there will be at least 263.2 million smartphone users in the United States?
(*Source*: Statista) *2021 through 2026*

62. Global Payment Market Data for selected years from 2015 and projected to 2023 indicate that the total global mobile payment market, in billions of dollars, can be modeled by $y = 79.2x^2 - 937.7x + 3298$, where x is the number of years after 2010. When does the model predict that the market will be at least $16.224 trillion? *2030*
(*Source*: Statista)

63. China vs. U.S. GDP Suppose that projections indicate that China's GDP will grow at an average annual rate of 5.0% and that the U.S. GDP will grow at an average annual rate of 3% over the period from 2005 through 2040. In this case, the respective GDPs of the countries can be modeled by

China: $y = 0.0178x^2 + 0.237x + 1.96$

U.S.: $y = 0.00953x^2 + 0.282x + 11.6$

where x is the number of years after 2000. Find when China's GDP will exceed that of the United States under these projections. *In 2037*
(*Source*: Bloomberg.com)

64. U.S. Retail E-commerce Sales Using data from 2016 and projected to 2022, the annual U.S. retail e-commerce sales can be modeled by

$$y = -1.56x^2 + 75.3x - 38.4$$

where x is the number of years after 2010 and y is the sales in millions of dollars. In what years does the model indicate that the sales will be at least $816.6 million? *2029 through 2040*
(*Source*: Statista)

65. Cloud Revenue Data from 2012 and projected to 2026 indicate that the global Cloud revenue, in billions of dollars, can be modeled by $y = 7.531x^{1.554}$, where x is the number of years after 2010. When after 2010 does the model indicate that the revenue will be less than or equal to $1105.6 billion?
(*Source*: Wikibon Server SAN & Cloud Research Projects) *2035 and before*

66. Voltage Required voltage for an electric oven is 220 volts, but it will function normally if the voltage varies from 220 by 10 volts.

  a. Write an absolute value inequality that gives the voltage x for which the oven will work normally. $|x - 220| \leq 10$
  b. Solve this inequality for x. $210 \leq x \leq 230$

67. Purchasing Power Inflation causes a decrease in the value of money used to purchase goods and services. The purchasing power of a 1983 dollar based on consumer prices for 1968–2010 can be modeled by the function $y = 34.394x^{-1.109}$, where x is the number of years after 1960. For what years through 2012 is the purchasing power of a 1983 dollar less than $1.00, according to the model? *1985 to 2012*

Preparing *for* CALCULUS

An equation of the form $y = (ax^k + b)^n$ can be thought of as the composition of two functions and can be written as $y = u^n$, where $u = ax^k + b$. This form is very useful in calculus.

1. If $y = \sqrt{x^2 + 1}$, then $y = u^?$, where $u = \underline{½; x^2 + 1}$.

2. If $y = (\underline{x^3 - x})^5$, then $y = (x^3 - x)^?$, where $u = \underline{5; x^3 - x}$.

3. If $f(x) = \underline{(x^5 - 4x)^6}$, then $f(u) = u^6$, where $u = x^5 - 4x$.

4. If $f(u) = u^{1/2}$ and $u = x^2 + 1$, then express $\frac{1}{2}u^{-1/2} \cdot 2x$ in terms of x alone, and simplify. $\frac{1}{2}(x^2 + 1)^{-1/2} \cdot 2x = \frac{x}{\sqrt{x^2 + 1}}$

5. If $f(u) = u^6$ and $u = x^4 + 5$, then express $6u^5 \cdot 4x^3$ in terms of x alone, and simplify. $24x^3(x^4 + 5)^5$

6. Write two functions whose composition gives $y = (x^3 - 6)^{3/2}$.
$y = u^{3/2}$ where $u = x^3 - 6$

Solving equations with rational powers is a skill that is useful in calculus.

7. For $x > 0$, the graph of the function $f(x) = 2x - 3x^{2/3}$ will have its lowest point where $0 = 2 - 2x^{-1/3}$. Find the value of x that is a solution to this equation. $x = 1$

8. The graph of the function $f(x) = \sqrt{(2x + 1)^3} - x$ will have its lowest point where $0 = 3\sqrt{2x + 1} - 1$. Find the value of x that is a solution to this equation. $-4/9$

9. The graph of the function $y = \frac{x^5}{5} - \frac{4}{3}x^3 + 3x$ has turning points for values of x that are solutions to $0 = x^4 - 4x^2 + 3$. Write this equation in quadratic form and solve it to find four turning points. $x = \pm 1, x = \pm\sqrt{3}$

10. In calculus, you may be asked to find the values of x that satisfy the inequality $\left|\frac{x}{2}\right| < 1$. Solve this inequality. $-2 < x < 2$

11. Solve $|x - 1| < 1$. $0 < x < 2$

12. Solve $\left|\frac{x - 1}{2}\right| < 1$. $-1 < x < 3$

chapter 4 SUMMARY

In this chapter, we presented the building blocks for function construction. We can construct new functions by horizontal or vertical transformation, stretching or compressing, or reflection. We can also construct new functions by combining functions algebraically or with function composition. Inverse functions can be used to "undo" the operations in a function. We finished the chapter by solving additional equations and inequalities.

Key Concepts and Formulas

4.1 Transformations of Graphs and Symmetry

Vertical shifts of graphs	If k is a positive real number, • The graph of $g(x) = f(x) + k$ can be obtained by shifting the graph of $f(x)$ upward k units. • The graph of $g(x) = f(x) - k$ can be obtained by shifting the graph of $f(x)$ downward k units.
Horizontal shifts of graphs	If h is a positive real number, • The graph of $g(x) = f(x - h)$ can be obtained by shifting the graph of $f(x)$ to the right h units. • The graph of $g(x) = f(x + h)$ can be obtained by shifting the graph of $f(x)$ to the left h units.
Stretching and compressing graphs	The graph of $y = af(x)$ is obtained by vertically stretching the graph of $f(x)$ using a factor of $\|a\|$ if $\|a\| > 1$ and by vertically compressing the graph of $f(x)$ using a factor of $\|a\|$ if $\|a\| < 1$.
Reflections of graphs across the coordinate axes	The graph of $y = -f(x)$ can be obtained by reflecting the graph of $y = f(x)$ across the x-axis. The graph of $y = f(-x)$ can be obtained by reflecting the graph of $y = f(x)$ across the y-axis.
Symmetry with respect to the y-axis	The graph of $y = f(x)$ is symmetric with respect to the y-axis if $f(-x) = f(x)$. Such a function is called an even function.
Symmetry with respect to the origin	The graph of $y = f(x)$ is symmetric with respect to the origin if $f(-x) = -f(x)$. Such a function is called an odd function.
Symmetry with respect to the x-axis	The graph of an equation is symmetric with respect to the x-axis if, for every point (x, y) on the graph, the point $(x, -y)$ is also on the graph.

4.2 Combining Functions; Composite Functions

Operations with functions	*Sum*: $(f + g)(x) = f(x) + g(x)$ *Difference*: $(f - g)(x) = f(x) - g(x)$ *Product*: $(f \cdot g)(x) = f(x) \cdot g(x)$ *Quotient*: $\left(\dfrac{f}{g}\right)(x) = \dfrac{f(x)}{g(x)}, g(x) \neq 0$

	The domain of the sum, difference, and product of f and g consists of all real numbers of the input variable for which f and g are defined. The domain of the quotient function consists of all real numbers for which f and g are defined and $g \neq 0$.
Average cost	The quotient of the cost function $C(x)$ and the identity function $I(x) = x$ gives the average cost function $\overline{C}(x) = \dfrac{C(x)}{x}$.
Composite functions	The notation for the composite function "f of g" is $(f \circ g)(x) = f(g(x))$. The domain of $f \circ g$ is the subset of the domain of g for which $f \circ g$ is defined.

4.3 One-to-One and Inverse Functions

Inverse functions	If $f(g(x)) = x$ and $g(f(x)) = x$ for all x in the domain of g and f, then f and g are inverse functions. In addition, the functions f and g are inverse functions if, whenever the pair (a, b) satisfies $y = f(x)$, the pair (b, a) satisfies $y = g(x)$. We denote g by f^{-1}, read as "f inverse."
One-to-one functions	For a function f to have an inverse, f must be one-to-one. A function f is one-to-one if each output of f corresponds to exactly one input of f.
Horizontal line test	If no horizontal line can intersect the graph of a function in more than one point, then the function is one-to-one.
Finding the inverse functions	To find the inverse of the function f that is defined by the equation $y = f(x)$: 1. Rewrite the equation with y replacing $f(x)$. 2. Interchange x and y in the equation defining the function. 3. Solve the new equation for y. If this equation cannot be solved uniquely for y, the function has no inverse. 4. Replace y by $f^{-1}(x)$.
Graphs of inverse functions	The graphs of a function and its inverse are symmetric with respect to the line $y = x$.
Inverse functions on limited domains	If a function is one-to-one on a limited domain, then it has an inverse function on that domain.

4.4 Additional Equations and Inequalities

Radical equations	An equation containing radicals can frequently be converted to an equation that does not contain radicals by raising both sides of the equation to a power that is equal to the index of the radical. The solutions must be checked when using this technique.
Equations with rational powers	Some equations with rational powers can be solved by writing the equation as a radical equation.
Equations in quadratic form	An equation of the form $au^2 + bu + c = 0$, where u is an algebraic expression, can be solved by substitution and a quadratic method.
Quadratic inequality	A quadratic inequality is an inequality that can be written in the form $ax^2 + bx + c > 0$, where a, b, c are real numbers and $a \neq 0$ (or with $>$ replaced by $<$, $\geq$, or $\leq$).
Solving a quadratic inequality algebraically	To solve a quadratic inequality $f(x) < 0$ or $f(x) > 0$ algebraically, solve $f(x) = 0$ and use the solutions to divide the number line into intervals. Pick a test value in each interval to determine whether $f(x)$ is positive or negative in that interval and identify the intervals that satisfy the original inequality.

320 Chapter 4 Additional Topics with Functions

Solving a quadratic inequality graphically	To solve a quadratic inequality $f(x) < 0$ (or $f(x) > 0$) graphically, graph the related equation $y = f(x)$ and observe the x-values of the intervals where the graph of $f(x)$ is below (or above) the x-axis.
Power inequalities	To solve a power inequality, first solve the related equation. Then use graphical methods to find the values of the variable that satisfy the inequality.
Absolute value inequalities	If $a > 0$, the solution to $|u| < a$ is $-a < u < a$. If $a > 0$, the solution to $|u| > a$ is $u < -a$ or $u > a$.

chapter 4 SKILLS CHECK

Answers that are not seen can be found in the answer section at the back of the text.

1. How is the graph of $g(x) = (x - 8)^2 + 7$ transformed from the graph of $f(x) = x^2$?
   Shift 8 units right and 7 units up
2. How is the graph of $g(x) = -2(x + 1)^3$ transformed from the graph of $f(x) = x^3$? Shift 1 unit left, vertical stretch using a factor of 2, reflect about x-axis
3. a. Graph the functions $f(x) = \sqrt{x}$ and $g(x) = \sqrt{x + 2} - 3$.

   b. How are the graphs related? Graph of $g(x)$ is graph of $f(x)$ shifted 2 units left and 3 units down.
4. What is the domain of the function $g(x) = \sqrt{x + 2} - 3$? $[-2, \infty)$

5. Suppose the graph of $f(x) = x^{1/3}$ is shifted up 4 units and to the right 6 units. What is the equation that gives the new graph? $y = (x - 6)^{1/3} + 4$

6. Suppose the graph of $f(x) = x^{1/3}$ is vertically stretched by a factor of 3 and then shifted down 5 units. What is the equation that gives the new graph? $y = 3x^{1/3} - 5$

For Exercises 7–9, match each graph with the correct equation.

a. $y = |x| + 2$ b. $y = x^3$

c. $y = \dfrac{3}{x - 1} + 1$ d. $y = \dfrac{-3}{x + 1} + 1$

e. $y = 2x^3$ f. $y = |x + 2|$

7. f

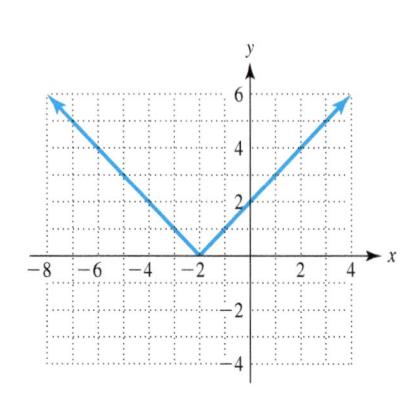

8. c

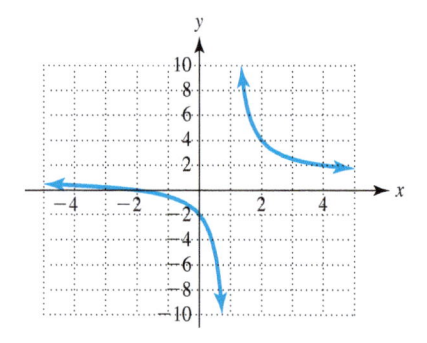

9. e

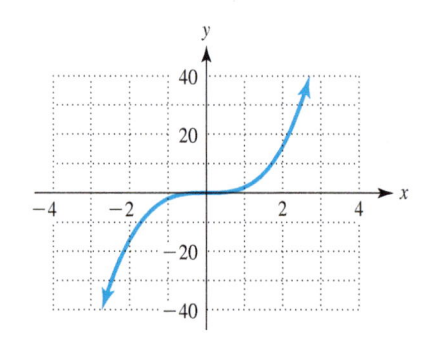

In Exercises 10 and 11, determine algebraically whether the graph of the function is symmetric with respect to the x-axis, y-axis, and/or origin. Confirm graphically.

10. $f(x) = x^3 - 4x$ 11. $f(x) = -x^2 + 5$
    Origin y-axis
12. Determine whether the function $f(x) = -\dfrac{2}{x}$ is even, odd, or neither. Odd

For Exercises 13–20, use the functions $f(x) = 3x^2 - 5x$, $g(x) = 6x - 4$, $h(x) = 5 - x^3$ to find the following.

13. $(f + g)(x)$ 14. $(h - g)(x)$
    $3x^2 + x - 4$ $9 - x^3 - 6x$

15. $(g \cdot f)(x)$ 16. $\left(\dfrac{h}{g}\right)(x)$ $\dfrac{5 - x^3}{6x - 4}, x \neq \dfrac{2}{3}$
    $18x^3 - 42x^2 + 20x$

17. $(f - g)(-2)$ 38
18. $(f \circ g)(x)$ $108x^2 - 174x + 68$
19. $(g \circ f)(x)$ $18x^2 - 30x - 4$
20. $(g \circ h)(-3)$ 188
21. For the functions $f(x) = 2x - 5$ and $g(x) = \dfrac{x + 5}{2}$,

    a. Find $f(g(x))$ and $g(f(x))$. $x; x$

    b. What is the relationship between $f(x)$ and $g(x)$? Inverse functions

22. Find the inverse of $f(x) = 3x - 2$. $f^{-1}(x) = \dfrac{x + 2}{3}$

23. Find the inverse of $g(x) = \sqrt[3]{x - 1}$. $g^{-1}(x) = x^3 + 1$

24. Graph $f(x) = (x + 1)^2$ and its inverse $f^{-1}(x)$ on the domain $[-1, 10]$.

25. Is the function $f(x) = \dfrac{x}{x - 1}$ a one-to-one function? Yes

26. Does the function $f(x) = x^2 - 3x$ have an inverse? No

27. Solve $\sqrt{4x^2 + 1} = 2x + 2$. $x = -\dfrac{3}{8}$

28. Solve $\sqrt{3x^2 - 8} + x = 0$. $x = -2$

29. Solve $4x - 5x^{1/2} + 1 = 0$. $\dfrac{1}{16}, 1$

30. Solve $(x - 6)^2 - 3(x - 6) - 10 = 0$. 4, 11

31. Solve the inequality $x^2 - 7x \le 18$. $[-2, 9]$

32. Solve the inequality $2x^2 + 5x \ge 3$. $x \le -3$ or $x \ge \dfrac{1}{2}$

33. Solve $|2x - 4| \le 8$. $-2 \le x \le 6$

34. Solve $|4x - 3| \ge 15$. $x \ge 4.5$ or $x \le -3$

35. Solve $(x - 4)^3 < 4096$. $x < 20$

36. Solve $(x + 2)^2 \ge 512$. $x \ge 16\sqrt{2} - 2$ or $x \le -16\sqrt{2} - 2$

chapter 4   REVIEW

Answers that are not seen can be found in the answer section at the back of the text.

37. **Ballistics** Ballistic experts are able to identify the weapon that fired a certain bullet by studying the markings on the bullet. If a test is conducted by firing a bullet into a bale of paper, the distance that the bullet will travel is given by $s = 64 - (4 - 10t)^3$ inches, for $0 \le t \le 0.4$, t in seconds.

    a. Graph this function for $0 \le t \le 0.4$.

    b. How far does the bullet travel during this time period? 64 in.

38. **Projectile** A toy rocket is fired into the air from the top of a building, with its height given by $S = -16(t - 4)^2 + 380$ feet, with t in seconds.

    a. In how many seconds will the rocket reach its maximum height? 4 sec

    b. What is the maximum possible height? 380 ft

    c. Describe the transformations needed to obtain the graph of $S(t)$ from the graph of $y = t^2$.

39. **Airline Traffic** The number of millions of passengers traveling on U.S. airlines can be modeled by $f(x) = -7.232x^2 + 95.117x + 441.138$, where x is the number of years after 2000.

    a. During what year does the model indicate that the number of passengers was at a maximum? 2007

    b. Determine the years during which the number of passengers was above 700 million. 2004 to 2009

40. **Pet Industry** U.S. pet industry expenditures can be modeled by $y = 0.037x^2 + 1.018x + 12.721$ million dollars, where x equals the number of years after 1990. During what years from 1990 through 2010 did Americans spend more than $40 million on their pets? 2007 to 2010

41. **Degrees Earned** Using data for selected years from 1980 and projected to 2025, the number of postsecondary degrees earned, in thousands (including associate's, bachelor's, master's, first professional, and doctoral degrees), can be modeled by $y = 0.225x^2 + 5.96x + 394$, where x is the number of years after 1980. If this model is accurate, during what years from 1980 on would the number of postsecondary degrees earned be less than or equal to 1,171,145. 1980 to 2027
    (Source: National Center for Education Statistics, *Digest of Education Statistics*)

42. **Average Cost** Suppose the total cost function for a product is determined to be $C(x) = 30x + 3150$, where x is the number of units produced and sold.

    a. Write the average cost function, which is the quotient of two functions. $\overline{C}(x) = \dfrac{30x + 3150}{x}$

    b. Graph this function for $0 < x \le 20$.

43. **Average Cost** The monthly average cost of producing x sofas is $\overline{C}(x) = \dfrac{50{,}000}{x} + 120$ dollars.

    a. Graph this function for $x > 0$.

    b. Will the average cost function decrease or increase as the number of units produced increases? Decrease

    c. What transformations of the graph of the reciprocal function give the graph of this function? Vertical stretch using a factor of 50,000 and shift 120 units up

44. Supply and Demand The price per unit of a product is p, and the number of units of the product is denoted by q. Suppose the supply function for a product is given by $p = \dfrac{180 + q}{6}$, and the demand for the product is given by $p = \dfrac{300}{q} - 20$.

  a. Which of these functions is a shifted reciprocal function? *Demand function*

  b. Graph the supply and demand functions on the same axes.

45. Supply and Demand The supply function for a commodity is given by $p = 58 + \dfrac{q}{2}$, and the demand function for this commodity is given by $p = \dfrac{2555}{q + 5}$.

  a. Which of these functions is a shifted reciprocal function? *Demand function*

  b. Graph the supply and demand functions on the same axes.

46. Personal Income The total personal income in the United States (in billions of dollars) for selected years from 1960 and projected to 2024 can be modeled by $P(x) = 6.12x^2 - 171x + 1725$, where x equals the number of years after 1950. Rewrite this model if x represents the number of years after 1960.
(*Source*: U.S. Bureau of Labor Statistics)
$P(x) = 6.12(x + 10)^2 - 171(x + 10) + 172$

47. Civilian Workforce Using data for selected years from 1950 and projected to 2050, the number of millions of men in the workforce can be modeled by $M(x) = 0.587x + 43.1$, where x is the number of years after 1950, and the number of millions of women in the workforce can be modeled by $W(M) = 1.33M - 37.1$, where M is the number of millions of men during this period.
(*Source*: U.S. Bureau of Labor)

  a. Find and interpret the meaning of $W(M(x))$. Write the coefficients of this function with three significant digits. $W(M(x)) = 0.781x + 20.2$

  b. Use the results of part (a) to find $W(M(80))$. Interpret this result. $W(M(80)) = 82.68$; the number of women in the workforce is projected to be 82.68 million in 2030.

48. Prison Sentences The mean time in prison y for certain crimes can be found as a function of the mean sentence length x, using $f(x) = 0.554x - 2.886$, where x and y are measured in months.
(*Source*: Index of Leading Cultural Indicators)

  a. Find the inverse of this function. $f^{-1}(x) = \dfrac{x + 2.886}{0.554}$

  b. Interpret the inverse function from part (a).
  Mean sentence length can be found as a function of mean time in prison.

49. Personal Income The total personal income in the United States (in billions of dollars) for selected years from 1960 and projected to 2024 can be modeled by $P(x) = \left(\dfrac{11}{5}\right)x^{11/5}$, where x equals the number of years after 1950.
(*Source*: U.S. Bureau of Labor Statistics)

  a. Find the equation of the inverse function of P. $P^{-1}(x) = \left(\dfrac{5x}{11}\right)^{5/11}$

  b. Find $P^{-1}(P(64))$. $P^{-1}(P(64)) = 64$

50. Total 3-D Printing Market The total market size of the global 3-D printer market for selected years from 2013 and projected to 2025 can be modeled by $M(x) = 336\sqrt{x^3}$, where x is the number of years after 2010 and y is the market size in millions of dollars. Rewrite this function as $P(x)$ with x representing the number of years after 2013.
(*Source*: Statista) $P(x) = 336\sqrt{(x + 3)^3}$

51. Profit The monthly profit from producing and selling x units of a product is given by

$$P(x) = -0.01x^2 + 62x - 12{,}000$$

Producing and selling how many units will result in profit for this product $(P(x) > 0)$?
More than 200 and less than 6000 units

52. The Cloud North American cloud computing revenues from 2008 and projected to 2020 can be modeled by $y = 0.231x^2 - 2.242x + 11.101$, where x equals the number of years after 2000 and y equals the revenue in billions of euros. For what years from 2008 to 2025 does this model estimate the North American cloud computing revenues to be at least 65.89 billion? *2021 through 2025*

Group Activities
▶ EXTENDED APPLICATIONS

1. Cost, Revenue, and Profit

The following table gives the weekly revenue and cost, respectively, for selected numbers of units of production and sale of a product by the Quest Manufacturing Company.

Number of Units	Revenue ($)	Number of Units	Cost ($)
100	6800	100	32,900
300	20,400	300	39,300
500	34,000	500	46,500
900	61,200	900	63,300
1400	95,200	1400	88,800
1800	122,400	1800	112,800
2500	170,000	2500	162,500

Provide the information requested and answer the questions.

1. Use technology to determine the functions that model revenue and cost functions for this product, using x as the number of units produced and sold.
2. a. Combine the revenue and cost functions with the correct operation to create the profit function for this product.
   b. Use the profit function to complete the following table:

x (units)	Profit, P(x) ($)
0	
100	
600	
1600	
2000	
2500	

3. Find the number of units of this product that must be produced and sold to break even.
4. Find the maximum possible profit and the number of units that gives the maximum profit.
5. a. Use operations with functions to create the average cost function for the product.
   b. Complete the following table.

x (units)	Average Cost, $\bar{C}(x)$ ($/unit)
1	
100	
300	
1400	
2000	
2500	

   c. Graph this function using the viewing window $[0, 2500]$ by $[0, 400]$.
6. Graph the average cost function using the viewing window $[0, 4000]$ by $[0, 100]$. Determine the number of units that should be produced to minimize the average cost and the minimum average cost.
7. Compare the number of units that produced the minimum average cost with the number of units that produced the maximum profit. Are they the same number of units? Discuss which of these values is more important to the manufacturer and why.

2. United States Population by Nativity

The figure shows the total U.S. population and the foreign-born population for the years from 2014 and projected to 2060.

1. Find the projected increase in the total U.S population from 2014 to 2060. Round to one decimal place.
2. Find the projected average rate of change per year in the total population. Round to one decimal place.
3. Use the data in the table to find the quadratic function that is the best fit to the total population data, with x equal to the number of years after 2010 and y equal to the total population in millions. Report the model with four decimal places.
4. What does the model predict the total population will be in 2050? In 2035?
5. Find the projected increase in the foreign-born population from 2014 to 2060. Round to one decimal place.
6. Find the projected average rate of change per year in the foreign-born population. Round to one decimal place.
7. Use the data in the table to find the quadratic function that is the best fit to the foreign-born population data, with x equal to the number of years after 2010 and y equal to the foreign-born population in millions. Report the model with four decimal places.
8. What does the model predict the foreign-born-population will be in 2050? In 2035?
9. Combine the functions that model the total U.S. population and the foreign-born population to find a function that models the native-born population in the United States.
10. What does the model predict the native-born population will be in 2050? In 2035?
11. Combine the functions that model the total U.S. population and the foreign-born population to find a function that models the percent that the foreign-born population is of the total population in the United States.
12. What percent is the projected foreign-born population of the projected total population in 2050? In 2035?

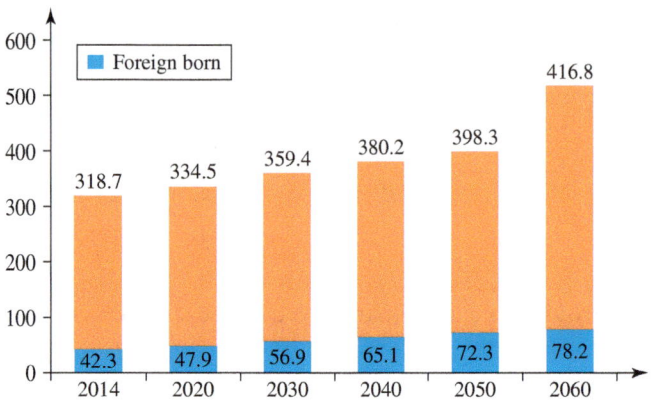

U.S. Population by Nativity: 2014 to 2060
(Population in millions)

Year	Total	Foreign born
2014	318.7	42.3
2020	334.5	47.9
2030	359.4	56.9
2040	380.2	65.1
2050	398.3	72.3
2060	416.8	78.2

(*Source*: U.S. Census Bureau)

5

Exponential and Logarithmic Functions

The intensities of earthquakes like the one in Japan in 2011 are measured with the Richter scale, which uses logarithmic functions. Logarithmic functions are also used in measuring loudness (in decibels) and stellar magnitude and in calculating pH values. Exponential functions are used in many real-world settings, such as the growth of investments, the growth of populations, sales decay, and radioactive decay.

sections	objectives	applications
5.1 Exponential Functions	Graph exponential functions; apply exponential growth and decay functions; compare transformations of graphs; graph exponential functions with base e	Growth of paramecia, inflation, personal income, sales decay, investments with continuous compounding, carbon-14 dating
5.2 Logarithmic Functions; Properties of Logarithms	Graph and evaluate logarithmic functions; convert equations to exponential and logarithmic forms; evaluate common and natural logarithms; simplify using logarithmic properties	Diabetes, pH scale, Richter scale, investments, decibel scale
5.3 Exponential and Logarithmic Equations	Solve exponential equations; solve logarithmic equations; convert logarithms using change of base; solve exponential and logarithmic inequalities	Carbon-14 dating, doubling time for investments, growth of investments, disposable income, global warming, sales decay
5.4 Exponential and Logarithmic Models	Model with exponential functions; use constant percent change to determine if data fit an exponential model; find logarithmic models; compare different models of data	Diabetes, insurance premiums, Consumer Price Index, sales decay, growth of bacteria, inflation, women in the labor force
5.5 Exponential Functions and Investing	Find the future value when interest is compounded periodically and when it is compounded continuously; find the present value of an investment; use graphing utilities to model investment data	Future value of an investment, present value of an investment, mutual fund growth
5.6 Annuities; Loan Repayment	Find the future value of an annuity; find the present value of an annuity; find the payments needed to amortize debt	Future value of an ordinary annuity, present value of an ordinary annuity, home mortgage
5.7 Logistic and Gompertz Functions	Graph and apply logistic growth functions; graph and apply logistic decay functions; graph and apply Gompertz functions	Women in the workforce, expected life span, uninsured Americans, deer populations, company growth

Algebra TOOLBOX

KEY OBJECTIVES

- Use properties of exponents with integers
- Use properties of exponents with real numbers
- Simplify exponential expressions
- Multiply and divide radicals
- Add and subtract radicals
- Rationalize denominators
- Write numbers in scientific notation
- Convert numbers in scientific notation to standard notation

Additional Properties of Exponents

In this Toolbox, we discuss properties of integer and real exponents and exponential expressions. These properties are useful in the discussion of exponential functions and of logarithmic functions, which are related to exponential functions. Integer exponents and rational exponents were discussed in Chapter 3, as well as the Product Property and Quotient Property.

For real numbers a and b and integers m and n,

1. $a^m \cdot a^n = a^{m+n}$ (Product Property)
2. $\dfrac{a^m}{a^n} = a^{m-n},\ a \neq 0$ (Quotient Property)

Additional properties of exponents, which can be developed using the properties above, follow:

For real numbers a and b and integers m and n,

3. $(ab)^m = a^m b^m$ (Power of a Product Property)
4. $\left(\dfrac{a}{b}\right)^m = \dfrac{a^m}{b^m}$, for $b \neq 0$ (Power of a Quotient Property)
5. $(a^m)^n = a^{mn}$ (Power of a Power Property)
6. $a^{-m} = \dfrac{1}{a^m}$, for $a \neq 0$

For example, we can prove Property 3 for positive integer m as follows:

$$(ab)^m = \underbrace{(ab)(ab)\cdots(ab)}_{m\text{ times}} = \underbrace{(a \cdot a \cdots a)}_{m\text{ times}}\underbrace{(b \cdot b \cdots b)}_{m\text{ times}} = a^m b^m$$

EXAMPLE 1 ▶ Using the Properties of Exponents

Use properties of exponents to simplify each of the following. Assume that denominators are nonzero.

a. $\dfrac{5^6}{5^4}$ b. $\dfrac{y^2}{y^5}$ c. $(3xy)^3$ d. $\left(\dfrac{y}{z}\right)^4$ e. $3^{15-2m} \cdot 3^{2m}$

SOLUTION

a. $\dfrac{5^6}{5^4} = 5^{6-4} = 5^2 = 25$ b. $\dfrac{y^2}{y^5} = y^{2-5} = y^{-3} = \dfrac{1}{y^3}$ c. $(3xy)^3 = 3^3 x^3 y^3 = 27 x^3 y^3$

d. $\left(\dfrac{y}{z}\right)^4 = \dfrac{y^4}{z^4}$ e. $3^{15-2m} \cdot 3^{2m} = 3^{(15-2m)+2m} = 3^{15-2m+2m} = 3^{15}$

We can also simplify and evaluate expressions involving powers of powers.

EXAMPLE 2 ▶ Powers of Powers

a. Simplify $(x^4)^5$. **b.** Evaluate $(2^3)^5$. **c.** Simplify $(x^2y^3)^5$. **d.** Evaluate 2^{3^2}.

SOLUTION

a. $(x^4)^5 = x^{4 \cdot 5} = x^{20}$

b. We can evaluate $(2^3)^5$ in two ways:

$$(2^3)^5 = 2^{15} = 32{,}768 \quad \text{or} \quad (2^3)^5 = (8)^5 = 32{,}768$$

c. $(x^2y^3)^5 = (x^2)^5(y^3)^5 = x^{10}y^{15}$ **d.** $2^{3^2} = 2^9 = 512$

EXAMPLE 3 ▶ Applying Exponent Properties

Compute the following products and quotients and write the answer with positive exponents.

a. $(-2x^{-2}y)(5x^{-2}y^{-3})$ **b.** $\dfrac{8xy^{-2}}{2x^4y^{-6}}$ **c.** $\dfrac{\dfrac{2x^{-1}y}{3a}}{\dfrac{6xy^{-2}}{5a}}$ **d.** $(4^0x^3y^{-2})^{-3}$

SOLUTION

a. $(-2x^{-2}y)(5x^{-2}y^{-3}) = -10x^{-2+(-2)}y^{1+(-3)} = -10x^{-4}y^{-2}$

$$= -10 \cdot \frac{1}{x^4} \cdot \frac{1}{y^2} = \frac{-10}{x^4y^2}$$

b. $\dfrac{8xy^{-2}}{2x^4y^{-6}} = 4x^{1-4}y^{-2-(-6)} = 4x^{-3}y^4 = 4 \cdot \dfrac{1}{x^3} \cdot y^4 = \dfrac{4y^4}{x^3}$

c. To perform this division, we invert and multiply.

$$\dfrac{\dfrac{2x^{-1}y}{3a}}{\dfrac{6xy^{-2}}{5a}} = \dfrac{2x^{-1}y}{3a} \cdot \dfrac{5a}{6xy^{-2}} = \dfrac{10ax^{-1}y}{18axy^{-2}} = \dfrac{5x^{-1-1}y^{1-(-2)}}{9} = \dfrac{5x^{-2}y^3}{9} = \dfrac{5y^3}{9x^2}$$

d. $(4^0x^3y^{-2})^{-3} = 4^0x^{-9}y^6 = 1 \cdot \dfrac{1}{x^9} \cdot y^6 = \dfrac{y^6}{x^9}$

Real Number Exponents

All the above properties also hold for rational exponents so long as no negative numbers in even powered roots result, and they also apply for all real numbers that give expressions that are real numbers.

EXAMPLE 4 ▶ Operations with Real Exponents

Perform the indicated operations.

a. $(x^{1/3})(x^{3/4})$ **b.** $\dfrac{y^{1/2}}{y^{2/5}}$ **c.** $(c^{1/3})^{3/4}$

SOLUTION

a. $(x^{1/3})(x^{3/4}) = x^{1/3+3/4} = x^{13/12}$ **b.** $\dfrac{y^{1/2}}{y^{2/5}} = y^{1/2-2/5} = y^{1/10}$

c. $(c^{1/3})^{3/4} = c^{1/3 \cdot 3/4} = c^{1/4}$

Multiplying Radicals

Recall that the Product Rule for Radicals states that for nonnegative real numbers a and b, $\sqrt{a} \cdot \sqrt{b} = \sqrt{ab}$, and in general, $\sqrt[n]{a} \cdot \sqrt[n]{b} = \sqrt[n]{ab}$, provided $\sqrt[n]{a}$ and $\sqrt[n]{b}$ are real. This allows us to multiply radicals with the same index.

EXAMPLE 5 ▶ **Using the Product Rule to Simplify Radicals**

Multiply and simplify the result, assuming the expressions are real and the variables represent nonnegative real numbers.

a. $\sqrt{50} \cdot \sqrt{5} = \sqrt{250} = \sqrt{25 \cdot 10} = 5\sqrt{10}$
b. $\sqrt{6x^3y} \cdot \sqrt{8x^5y} = \sqrt{48x^8y^2} = 4x^4y\sqrt{3}$

Dividing Radicals

The Quotient Rule for Radicals states that $\dfrac{\sqrt[n]{a}}{\sqrt[n]{b}} = \sqrt[n]{\dfrac{a}{b}}$, provided $\sqrt[n]{a}$ and $\sqrt[n]{b}$ are real and $b \neq 0$.

EXAMPLE 6 ▶ **Using the Quotient Rule to Simplify Radicals**

Divide and simplify the result, assuming the expressions are real and the variables represent positive real numbers.

a. $\dfrac{\sqrt{75}}{\sqrt{3}} = \sqrt{\dfrac{75}{3}} = \sqrt{25} = 5$ b. $\dfrac{\sqrt[3]{54x^5}}{\sqrt[3]{2x^2}} = \sqrt[3]{27x^3} = 3x$

Adding and Subtracting Radicals

We can add or subtract radicals if they are like radicals, that is, if they have the same index and the same radicand. We add (or subtract) like radicals by adding (or subtracting) their coefficients, as we add like terms in algebra. For example, $\sqrt{6} + 3\sqrt{6} = 4\sqrt{6}$. Before adding or subtracting radicals, be sure they are simplified.

EXAMPLE 7 ▶ **Add and Subtract Radicals**

Combine the following radicals, if possible.

a. $\sqrt[4]{2xy} + 8\sqrt[4]{2xy} = 9\sqrt[4]{2xy}$
b. $4a\sqrt{ab} - 6\sqrt{a^3b} = 4a\sqrt{ab} - 6a\sqrt{ab} = -2a\sqrt{ab}$

Rationalizing Denominators

Sometimes we wish to rewrite a radical expression so that there is no radical in the denominator. This process is called rationalizing the denominator, as we are changing the denominator from irrational to rational. If the denominator contains one term that is the square root of a positive number that is not a perfect square, we multiply the numerator and denominator by the smallest number that produces a perfect square in the denominator. (Multiplying the numerator and denominator by the same value gives an equivalent expression.)

EXAMPLE 8 ▶ Rationalizing the Denominator

Rationalize each denominator:

a. $\dfrac{12}{\sqrt{3}} = \dfrac{12}{\sqrt{3}} \cdot \dfrac{\sqrt{3}}{\sqrt{3}} = \dfrac{12\sqrt{3}}{3} = 4\sqrt{3}$

b. $\dfrac{15}{\sqrt{8}} = \dfrac{15}{\sqrt{8}} \cdot \dfrac{\sqrt{2}}{\sqrt{2}} = \dfrac{15\sqrt{2}}{\sqrt{16}} = \dfrac{15\sqrt{2}}{4}$

If the denominator contains two terms with at least one square root, we multiply the numerator and denominator by the **conjugate** of the denominator. The expressions $\sqrt{a} + \sqrt{b}$ and $\sqrt{a} - \sqrt{b}$ are conjugates of each other. Multiplying these two conjugates gives a real number, that is,

$$(\sqrt{a} + \sqrt{b})(\sqrt{a} - \sqrt{b}) = \sqrt{a^2} - \sqrt{b^2} = a - b.$$

EXAMPLE 9 ▶ Rationalizing a Denominator with Two Terms

Rationalize the denominator $\dfrac{5}{2 + \sqrt{7}}$.

SOLUTION

$\dfrac{5}{2 + \sqrt{7}} = \dfrac{5}{2 + \sqrt{7}} \cdot \dfrac{2 - \sqrt{7}}{2 - \sqrt{7}} = \dfrac{5(2 - \sqrt{7})}{(2 + \sqrt{7})(2 - \sqrt{7})} = \dfrac{5(2 - \sqrt{7})}{4 - 7}$

$= \dfrac{5(2 - \sqrt{7})}{-3} = -\dfrac{5(2 - \sqrt{7})}{3}$

Scientific Notation

Evaluating exponential and logarithmic functions in this chapter may result in outputs that are very large or very close to 0. **Scientific notation** is a convenient way to write very large (positive or negative) numbers or numbers close to 0. Numbers in scientific notation have the form

$$N \times 10^p \text{ (where } 1 \leq N < 10 \text{ and } p \text{ is an integer)}$$

For instance, the scientific notation form of 2,654,000 is 2.654×10^6. When evaluating functions or solving equations with a calculator, we may see an expression that looks like 5.122E-8 appear on the screen. This calculator expression represents 5.122×10^{-8}, which is scientific notation for $0.00000005122 \approx 0$. Figure 5.1 shows two numbers expressed in standard notation and in scientific notation on a calculator display.

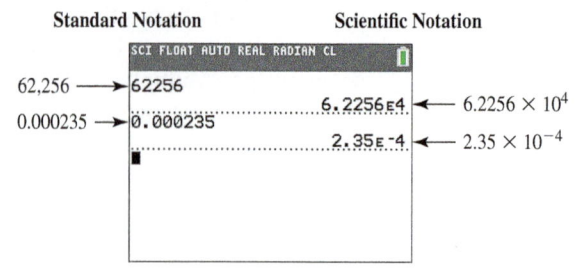

Figure 5.1

A number written in scientific notation can be converted to standard notation by multiplying (when the exponent on 10 is positive) or by dividing (when the exponent on 10 is negative). For example, we convert 7.983×10^5 to standard notation by multiplying 7.983 by $10^5 = 100{,}000$:

$$7.983 \times 10^5 = 7.983 \cdot 100{,}000 = 798{,}300$$

We convert 4.563×10^{-7} to standard notation by dividing 4.563 by $10^7 = 10{,}000{,}000$:

$$4.563 \times 10^{-7} = \frac{4.563}{10{,}000{,}000} = 0.0000004563$$

Multiplying two numbers in scientific notation involves adding the powers of 10, and dividing them involves subtracting the powers of 10.

EXAMPLE 10 ▶ Scientific Notation

Compute the following and write the answers in scientific notation.

a. $(7.983 \times 10^5)(4.563 \times 10^{-7})$

b. $(7.983 \times 10^5)/(4.563 \times 10^{-7})$

SOLUTION

a. $(7.983 \times 10^5)(4.563 \times 10^{-7}) = 36.426429 \times 10^{5+(-7)}$
$= 36.426429 \times 10^{-2} = (3.6426429 \times 10^1) \times 10^{-2} = 3.6426429 \times 10^{-1}$

The calculation using technology is shown in Figure 5.2(a).

b. $(7.983 \times 10^5)/(4.563 \times 10^{-7}) = 1.749506903 \times 10^{5-(-7)}$
$= 1.749506903 \times 10^{12}$. Figure 5.2(b) shows the calculation using technology.

(a) (b)

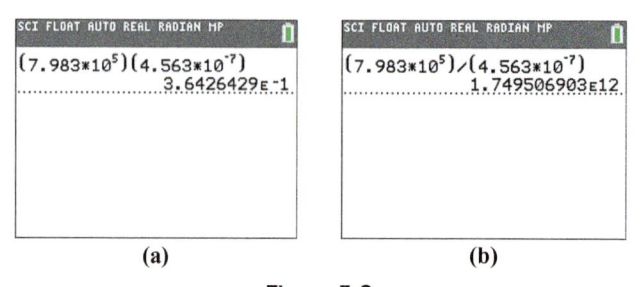

Figure 5.2

Toolbox EXERCISES

In Exercises 1 and 2, use the properties of exponents to simplify.

1. a. $x^4 \cdot x^3$ x^7 b. $\dfrac{x^{12}}{x^7}$ x^5 c. $(4ay)^4$ $256a^4y^4$ d. $\left(\dfrac{3}{z}\right)^4$ $\dfrac{81}{z^4}$

   e. $2^3 \cdot 2^2$ $2^5 = 32$ f. $(x^4)^2$ x^8

2. a. $y^5 \cdot y$ y^6 b. $\dfrac{w^{10}}{w^4}$ w^6 c. $(6bx)^3$ $216b^3x^3$ d. $\left(\dfrac{5z}{2}\right)^3$ $\dfrac{125z^3}{8}$

   e. $3^2 \cdot 3^3$ 3^5, or 243 f. $(2y^3)^4$ $16y^{12}$

In Exercises 3–18, use the properties of exponents to simplify the expressions and remove all zero and negative exponents. Assume that all variables are nonzero.

3. 10^{50} 10
4. 4^{2^2} 256
5. $x^{-4} \cdot x^{-3}$ $\dfrac{1}{x^7}$
6. $y^{-5} \cdot y^{-3}$ $\dfrac{1}{y^8}$
7. $(c^{-6})^3$ $\dfrac{1}{c^{18}}$
8. $(x^{-2})^4$ $\dfrac{1}{x^8}$
9. $\dfrac{a^{-4}}{a^{-5}}$ a
10. $\dfrac{b^{-6}}{b^{-8}}$ b^2
11. $(x^{-1/2})(x^{2/3})$ $x^{1/6}$
12. $(y^{-1/3})(y^{2/5})$ $y^{1/15}$
13. $(3a^{-3}b^2)(2a^2b^{-4})$ $\dfrac{6}{ab^2}$
14. $(4a^{-2}b^3)(-2a^4b^{-5})$ $\dfrac{-8a^2}{b^2}$
15. $\left(\dfrac{2x^{-3}}{x^2}\right)^{-2}$ $\dfrac{x^{10}}{4}$
16. $\left(\dfrac{3y^{-4}}{2y^2}\right)^{-3}$ $\dfrac{8y^{18}}{27}$
17. $\dfrac{28a^4b^{-3}}{-4a^6b^{-2}}$ $\dfrac{-7}{a^2b}$
18. $\dfrac{36x^5y^{-2}}{-6x^6y^{-4}}$ $\dfrac{-6y^2}{x}$

In Exercises 19–24, multiply or divide, as indicated, and simplify the result, assuming the expressions are real and the variables represent nonnegative real numbers.

19. $\sqrt{3} \cdot \sqrt{27}$ 9
20. $\sqrt{2a} \cdot \sqrt{6a}$ $2a\sqrt{3}$
21. $\sqrt{8xy^3z} \cdot \sqrt{4x^2y^2z}$ $4xy^2z\sqrt{2xy}$
22. $\sqrt[3]{2xy} \cdot \sqrt[3]{4x^2y}$ $2x\sqrt[3]{y^2}$
23. $\dfrac{\sqrt[3]{32}}{\sqrt[3]{4}}$ 2
24. $\dfrac{\sqrt{16a^3x}}{\sqrt{2ax}}$ $4a\sqrt{2}$

In Exercises 25–29, combine the radicals, if possible.

25. $2\sqrt{x} + 5\sqrt{x}$ $7\sqrt{x}$
26. $5\sqrt[3]{ab} - 2\sqrt[3]{ab}$ $3\sqrt[3]{ab}$
27. $3\sqrt{8} + 2\sqrt{2}$ $8\sqrt{2}$
28. $3\sqrt{27} + 6\sqrt{3} - \sqrt{12}$ $13\sqrt{3}$
29. $4x\sqrt[3]{xy} + \sqrt[3]{8x^4y}$ $6x\sqrt[3]{xy}$

In Exercises 30–34, rationalize the denominator and simplify.

30. $\dfrac{2}{\sqrt{10}}$ $\dfrac{\sqrt{10}}{5}$
31. $\dfrac{\sqrt{8}}{\sqrt{3}}$ $\dfrac{2\sqrt{6}}{3}$
32. $\dfrac{5}{\sqrt[3]{2}}$ $\dfrac{5\sqrt[3]{4}}{2}$
33. $\dfrac{9}{3+\sqrt{7}}$ $\dfrac{9(3-\sqrt{7})}{2}$
34. $\dfrac{6}{\sqrt{5}-\sqrt{3}}$ $3(\sqrt{5}+\sqrt{3})$

In Exercises 35–38, write the numbers in scientific notation.

35. $46{,}000{,}000$ 4.6×10^7
36. $862{,}000{,}000{,}000$ 8.62×10^{11}
37. 0.000094 9.4×10^{-5}
38. 0.00000278 2.78×10^{-6}

In Exercises 39–42, write the numbers in standard form.

39. 4.372×10^5 $437{,}200$
40. 7.91×10^6 $7{,}910{,}000$
41. 5.6294×10^{-4} 0.00056294
42. 6.3478×10^{-3} 0.0063478

In Exercises 43 and 44, multiply or divide, as indicated, and write the result in scientific notation.

43. $(6.250 \times 10^7)(5.933 \times 10^{-2})$ 3.708125×10^6
44. $\dfrac{2.961 \times 10^{-2}}{4.583 \times 10^{-4}}$ 6.460833515×10^1

In Exercises 45–54, use the properties of exponents to simplify the expressions. Assume that all variables are nonzero.

45. $x^{1/2} \cdot x^{5/6}$ $x^{4/3}$
46. $y^{2/5} \cdot y^{1/4}$ $y^{13/20}$
47. $(c^{2/3})^{5/2}$ $c^{5/3}$
48. $(x^{3/2})^{3/4}$ $x^{9/8}$
49. $\dfrac{x^{3/4}}{x^{1/2}}$ $x^{1/4}$
50. $\dfrac{y^{3/8}}{y^{1/4}}$ $y^{1/8}$
51. $\left(\dfrac{2b^{1/3}}{6c^{2/3}}\right)^3$ $\dfrac{b}{27c^2}$
52. $\left(\dfrac{9x^{2/3}}{16y^{5/6}}\right)^{3/2}$ $\dfrac{27x}{64y^{5/4}}$
53. $(x^{4/9}y^{2/3}z^{-1/3})^{3/2}$ $\dfrac{x^{2/3}y}{z^{1/2}}$
54. $\left(\dfrac{-8x^9y^{1/2}}{27z^{3/2}}\right)^{4/3}$ $\dfrac{16x^{12}y^{2/3}}{81z^2}$

5.1 Exponential Functions

KEY OBJECTIVES

- Graph and apply exponential functions
- Find horizontal asymptotes
- Graph and apply exponential growth functions
- Graph and apply exponential decay functions
- Compare transformations of graphs of exponential functions

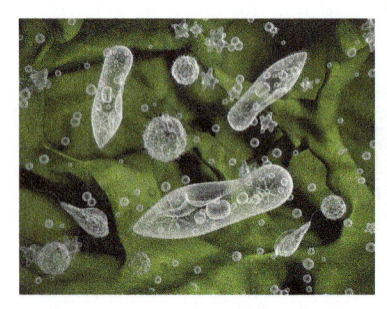

SECTION PREVIEW Paramecia

The primitive single-cell animal called a paramecium reproduces by splitting into two pieces (a process called binary fission), so the population doubles each time there is a split. If we assume that the population begins with 1 paramecium and doubles each hour, then there will be 2 paramecia after 1 hour, 4 paramecia after 2 hours, 8 after 3 hours, and so on. If we let y represent the number of paramecia in the population after x hours have passed, the points (x, y) that satisfy this function for the first 9 hours of growth are described by the data in Table 5.1 and the scatter plot in Figure 5.3(a). We can show that each of these points and the data points for $x = 10, 11$, and so forth lie on the graph of the function $y = 2^x$, with $x \geq 0$ (Figure 5.3(b)).

Functions like $y = 2^x$, which have a constant base raised to a variable power, are called exponential functions. We discuss exponential functions and their applications in this section. ■

Table 5.1

x (hours)	0	1	2	3	4	5	6	7	8	9
y (paramecia)	1	2	4	8	16	32	64	128	256	512

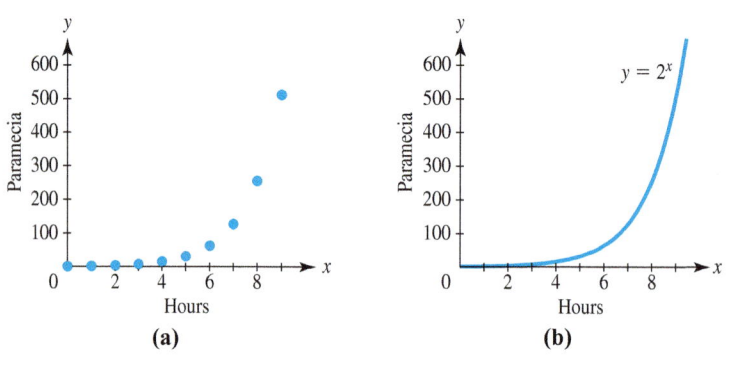

Figure 5.3

Exponential Functions

Recall that a linear function has a constant rate of change; that is, the outputs change by the same amount for each unit increase in the inputs. An exponential function has outputs that are *multiplied* by a fixed number for each unit increase in the inputs.

Note that in Example 1, the number y of paramecia is multiplied by 2 for each hour increase in time x, so the number is an exponential function of time.

EXAMPLE 1 ▶ Paramecia

As discussed in the Section Preview, the primitive single-cell animal called a paramecium reproduces by splitting into two pieces, so the population doubles each time there is a split. The number y of paramecia in the population after x hours have passed fits the graph of $y = 2^x$.

a. In the context of the paramecium application, what inputs can be used?

b. Is the function $y = 2^x$ discrete or continuous?

c. Graph the function $y = 2^x$, without regard to restricting inputs to those that make sense for the paramecium application.

d. What are the domain and range of this function?

SOLUTION

a. Not every point on the graph of $y = 2^x$ in Figure 5.3(b) describes a number of paramecia. For example, the point $(7.1, 2^{7.1})$ is on this graph, but $2^{7.1} \approx 137.187$ does not represent a number of paramecia because it is not a whole number. Since only whole number inputs will give whole number outputs, only whole number inputs can be used for this application.

b. Every point describing a number of paramecia is on the graph of $y = 2^x$, so the function $y = 2^x$ is a continuous model for this discrete paramecium growth function.

c. Some values satisfying the function $y = 2^x$ are shown in Table 5.2, and the graph of the function is shown in Figure 5.4.

d. The function that models the growth of paramecia was restricted because of the physical setting, but if the domain of the function $y = 2^x$ is not restricted, it is the set of real numbers. Note that the range of this function is the set of all positive real numbers because there is no power of 2 that results in a value of 0 or a negative number.

Table 5.2

x	y
−3	$2^{-3} = 0.125$
−1.5	$2^{-1.5} \approx 0.35$
−0.5	$2^{-0.5} \approx 0.71$
0	$2^0 = 1$
0.5	$2^{0.5} \approx 1.41$
1.5	$2^{1.5} \approx 2.83$
3	$2^3 = 8$
5	$2^5 = 32$

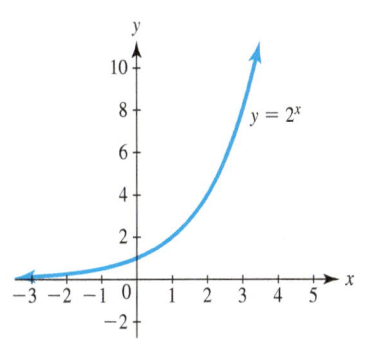

Figure 5.4

The graph in Figure 5.4 approaches but never touches the x-axis as x becomes more negative (approaching $-\infty$), so the x-axis is a **horizontal asymptote** for the graph. As previously mentioned, the function $y = 2^x$ is an example of a special class of functions called **exponential functions**. In general, we define an exponential function as follows.

Exponential Function

If b is a positive real number, $b \neq 1$, then the function $f(x) = b^x$ is an exponential function. The constant b is called the *base* of the function, and the variable x is the *exponent*.

If we graph another exponential function $y = b^x$ that has a base b greater than 1, the graph will have the same basic shape as the graph of $y = 2^x$. (See the graphs of $y = 1.5^x$ and $y = 12^x$ in Figure 5.5.)

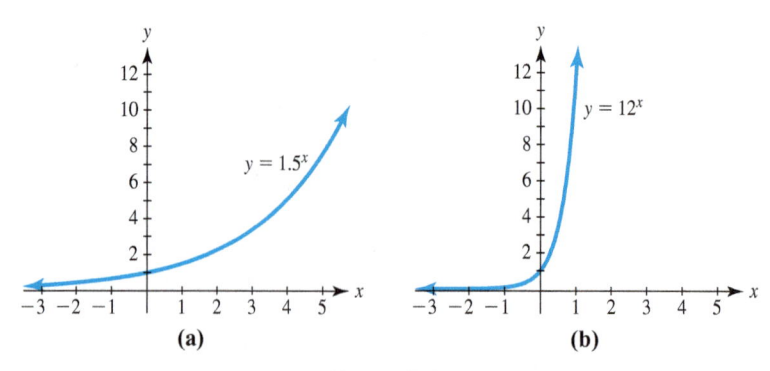

334 Chapter 5 Exponential and Logarithmic Functions

Figure 5.5

The exponential function $y = b^x$ with base b satisfying $0 < b < 1$ is a *decreasing* function. See the graph of $y = \left(\dfrac{1}{3}\right)^x$ in Figure 5.6(a). We can also write the function $y = \left(\dfrac{1}{3}\right)^x$ in the form $y = 3^{-x}$ because $3^{-x} = \dfrac{1}{3^x} = \left(\dfrac{1}{3}\right)^x$. (See Figure 5.6(b).) In general, a function of the form $y = b^{-x}$ with $b > 1$ can also be written as $y = c^x$ with $0 < c < 1$ and $c = \dfrac{1}{b}$, and their graphs are identical.

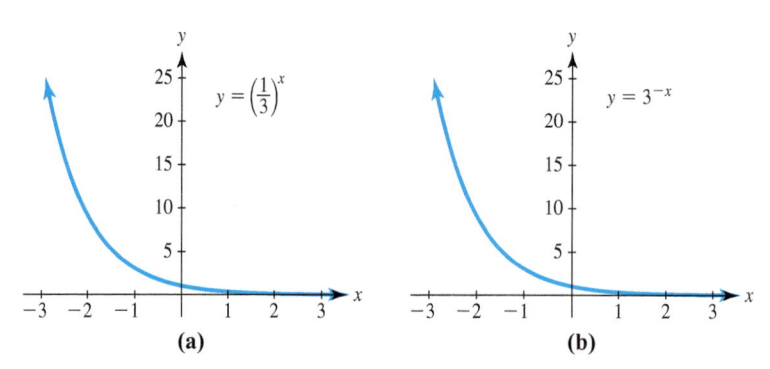

Figure 5.6

Graphs of Exponential Functions

Equation: $y = b^x$
x-intercept: none
y-intercept: $(0, 1)$
Domain: all real numbers
Range: all real numbers $y > 0$
Horizontal asymptote: *x*-axis
Shape: an increasing function if $b > 1$, $y = b^x$
a decreasing function if $y = b^x$ with $0 < b < 1$
or if $y = b^{-x}$ with $b > 1$

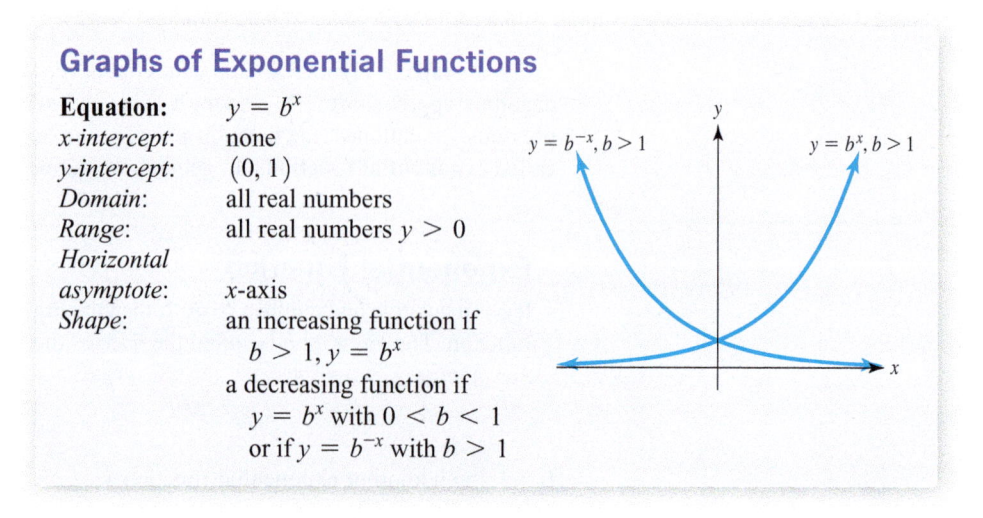

Transformations of Graphs of Exponential Functions

Graphs of exponential functions, like graphs of other types of functions, can be shifted, reflected, or stretched. Transformations, which were introduced in Chapter 4, can be applied to graphs of exponential functions.

Recall that graphs of functions can be shifted vertically or horizontally, reflected across either axis, and stretched or shrunk vertically. The following is a summary of the transformations of the graph of $f(x) = b^x$.

Vertical shift	$g(x) = b^x + k$	Graph is shifted k units up if k is positive, and $|k|$ units down if k is negative.
Horizontal shift	$g(x) = b^{x-h}$	Graph is shifted $|h|$ units right if h is positive, and $|h|$ units left if h is negative.
Reflection across the x-axis	$g(x) = -b^x$	Graph is reflected across the x-axis.
Reflection across the y-axis	$g(x) = b^{-x}$	Graph is reflected across the y-axis.
Vertical stretch/compress	$g(x) = a \cdot b^x$	Graph is vertically stretched if $|a| > 1$; graph is vertically compressed if $|a| < 1$.
Horizontal stretch/compress	$g(x) = b^{a \cdot x}$	Graph is horizontally stretched if $|a| < 1$; graph is horizontally compressed if $|a| > 1$.

EXAMPLE 2 ▶ Transformations of Graphs of Exponential Functions

Explain how the graph of each of the following functions compares with the graph of $y = 3^x$, and graph each function on the same axes as $y = 3^x$.

a. $y = 3^{x-4}$

b. $y = 2 + 3^{x-4}$

c. $y = 5(3^x)$

SOLUTION

a. The graph of $y = 3^{x-4}$ has the same shape as the graph of $y = 3^x$, but it is shifted 4 units to the right. The graph of $y = 3^x$ and the graph of $y = 3^{x-4}$ are shown in Figure 5.7(a).

b. The graph of $y = 2 + 3^{x-4}$ has the same shape as the graph of $y = 3^x$, but it is shifted 4 units to the right and 2 units up. The graph of $y = 2 + 3^{x-4}$ is shown in Figure 5.7(b). The horizontal asymptote is $y = 2$.

c. As we saw in Chapter 4, multiplication of a function by a constant greater than 1 stretches the graph of the function by a factor equal to that constant. Each of the y-values of $y = 5(3^x)$ is 5 times the corresponding y-value of $y = 3^x$. The graph of $y = 5(3^x)$ is shown in Figure 5.7(c).

336 Chapter 5 Exponential and Logarithmic Functions

Figure 5.7

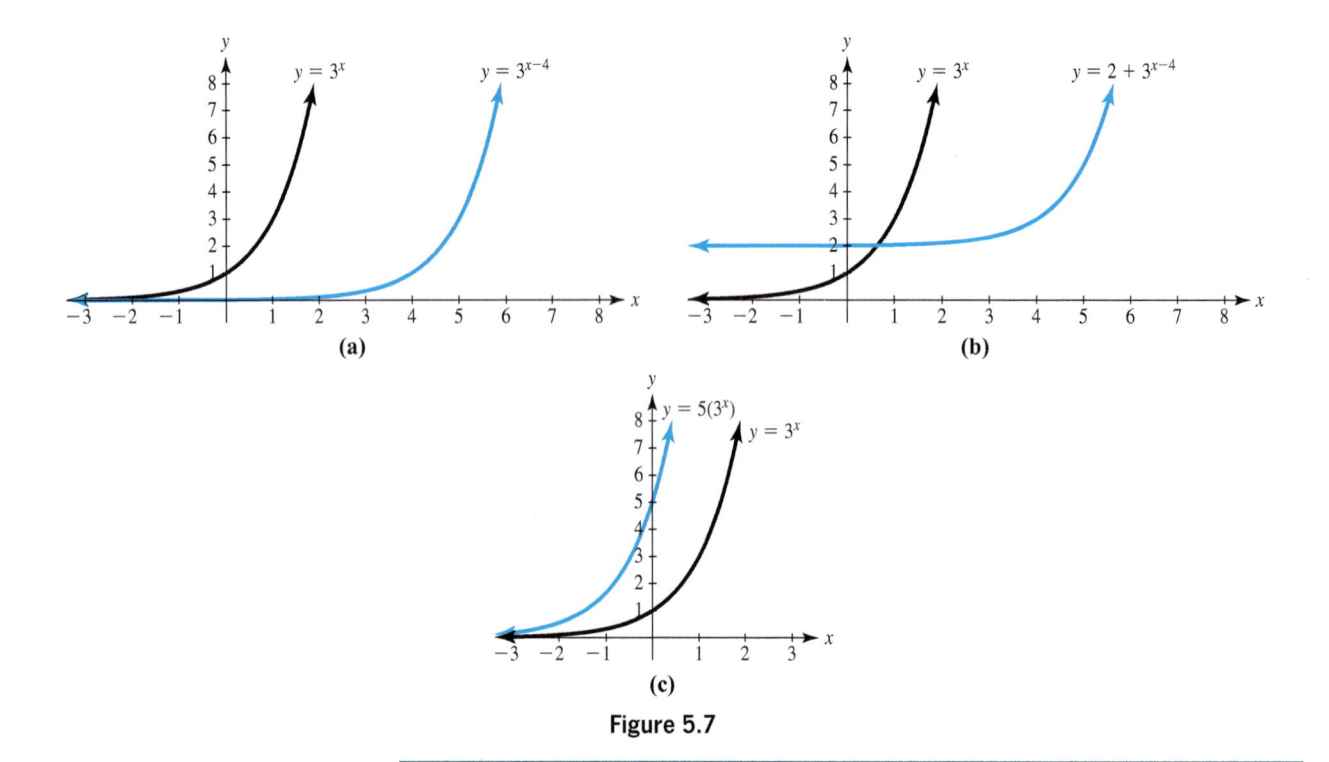

The graphs of exponential functions can also be horizontally stretched and compressed. In general, the graph of $f(x) = b^{ax}$ is the graph of $f(x) = b^x$ stretched horizontally if $|a| < 1$ and horizontally compressed if $|a| > 1$.

EXAMPLE 3 ▶ Horizontal Stretching and Compressing

Use a table of values from $x = -3$ to $x = 3$ to determine how the graphs of $y = 3^{2x}$ and $y = 3^{(1/2)x}$ are related to the graph of $y = 3^x$. Then graph each function on the same axes as $y = 3^x$.

SOLUTION

Table 5.3 shows the outputs of the functions $y_2 = 3^{2x}$ and $y_3 = 3^{(1/2)x}$ compared to those of $y_1 = 3^x$. It appears that the outputs of $y_2 = 3^{2x}$ are the squares of the outputs of $y_1 = 3^x$, and the outputs of $y_3 = 3^{(1/2)x}$ are the square roots of the outputs of $y_1 = 3^x$.

Table 5.3

x	$y_1 = 3^x$	$y_2 = 3^{2x}$	$y_3 = 3^{(1/2)x}$
−3	$\frac{1}{27}$	$\frac{1}{729}$	$\sqrt{\frac{1}{27}}$
−2	$\frac{1}{9}$	$\frac{1}{81}$	$\frac{1}{3}$
−1	$\frac{1}{3}$	$\frac{1}{9}$	$\sqrt{\frac{1}{3}}$
0	1	1	1
1	3	9	$\sqrt{3}$
2	9	81	3
3	27	729	$\sqrt{27}$

The graphs of each of these functions confirm our findings. The graph of $y = 3^{2x}$ is compressed horizontally from the graph of $y = 3^x$ (Figure 5.8(a)), and the graph of $y = 3^{(1/2)x}$ is stretched horizontally from the graph of $y = 3^x$ (Figure 5.8(b)).

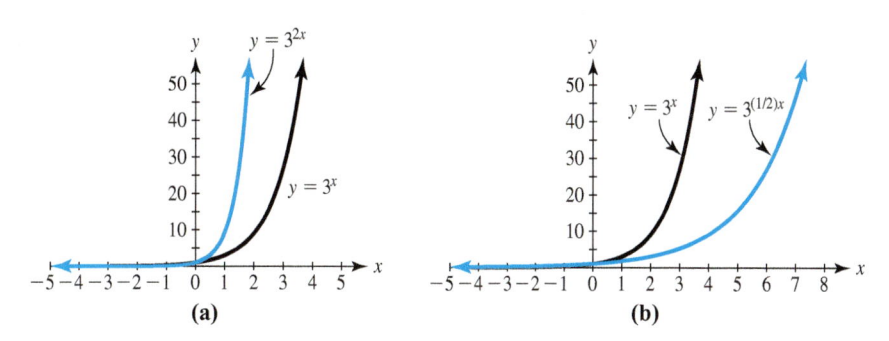

Figure 5.8

Exponential Growth Models

Functions of the form $y = b^x$ with $b > 1$ and, more generally, functions of the form $y = a(b^{kx})$ with $b > 1$, $a > 0$, and $k > 0$ are increasing and can be used to model growth.

Exponential Growth Models

Equation: $y = a(b^{kx})$, $b > 1$, $a > 0$, $k > 0$
x-intercept: none
y-intercept: $(0, a)$
Domain: all real numbers
Range: all real numbers $y > 0$
Horizontal asymptote: x-axis (the line $y = 0$)
Shape: increasing on domain and concave up

As we will see in Section 5.5, the future value S of an investment of P dollars invested for t years at interest rate r, compounded annually, is given by the exponential growth function

$$S = P(1 + r)^t$$

EXAMPLE 4 ▶ Inflation

Suppose that inflation is predicted to average 4% per year for each year from 2012 to 2025. This means that an item that costs $10,000 one year will cost $10,000(1.04) the next year and $10,000(1.04)(1.04) = $10,000(1.04^2)$ the following year.

a. Write the function that gives the cost of a $10,000 item t years after 2012.

b. Graph the growth model found in part (a) for $t = 0$ to $t = 13$.

c. If an item costs $10,000 in 2012, use the model to predict its cost in 2025.

SOLUTION

a. The function that gives the cost of a $10,000 item t years after 2012 is
$$y = 10{,}000(1.04^t)$$

b. The graph of $y = 10{,}000(1.04^t)$ is shown in Figure 5.9.

c. The year 2025 is 13 years from 2012, so the predicted cost of this item in 2025 is
$$y = 10{,}000(1.04^{13}) = 16{,}650.74 \text{ dollars}$$

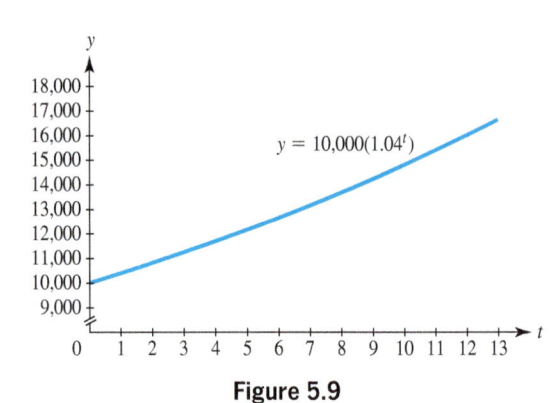

Figure 5.9

To graph exponential functions and find solutions involving exponential functions with a graphing calculator, we use the same steps as those used with other functions.

EXAMPLE 5 ▶ Personal Income

Total personal income in the United States (in billions of dollars) for selected years from 1960 and projected to 2018 can be modeled by
$$y = 492.4(1.07^x)$$
with x equal to the number of years after 1960.

a. What does the model predict the total U.S. personal income to be in 2014?

b. Graphically determine the year during which the model predicts that total U.S. personal income will reach $28.5 trillion.

SOLUTION

a. Evaluating the model at $x = 54$ gives total U.S. personal income in 2014 to be $19,013 billion, or $19.013 trillion.

b. Figure 5.10 shows the graph of the model and the graph of $y = 28{,}500$. Using Intersect gives $x = 59.98 \approx 60$, so the model predicts that total U.S. personal income will reach $28.5 trillion ($28,500 billion) in 2020.

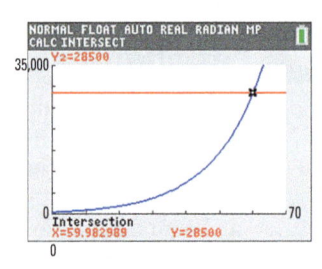

Figure 5.10

5.1 Exponential Functions 339

Spreadsheet ► SOLUTION Figure 5.11 shows the outputs of $y = 492.4(1.07^x)$ at several values of x, including $x = 60$. Using the data as we have to find other graphs, we get the graph of this exponential function, shown in Figure 5.11.*

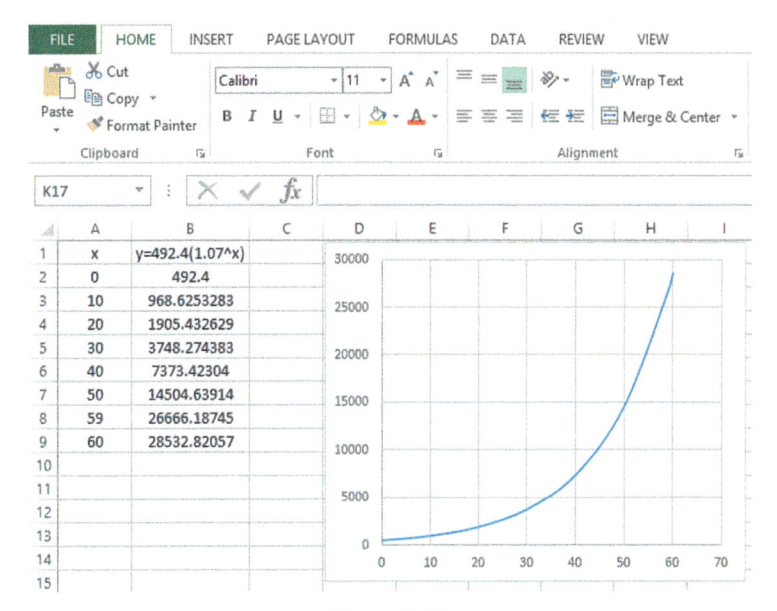

x	y=492.4(1.07^x)
0	492.4
10	968.6253283
20	1905.432629
30	3748.274383
40	7373.42304
50	14504.63914
59	26666.18745
60	28532.82057

Figure 5.11

Exponential Decay Models

Functions of the form $y = a(b^{kx})$ with $b > 1$, $a > 0$, and $k < 0$ or $y = a(b^{kx})$ with $0 < b < 1$, $a > 0$, and $k > 0$ are decreasing and can be used to model decay. For example, suppose a couple retires with a fixed income of $40,000 per year. If inflation averages 4% per year for each of the next 25 years, the value of what the couple can purchase with the $40,000 will decrease each year, with its *purchasing power* in the xth year after they retire given by the *exponential decay model*

$$P(x) = 40{,}000(0.96^x)$$

Using this model gives the value of what the $40,000 purchases in the 25th year after they retire to be

$$P(x) = 40{,}000(0.96^{25}) = 14{,}416 \text{ dollars}$$

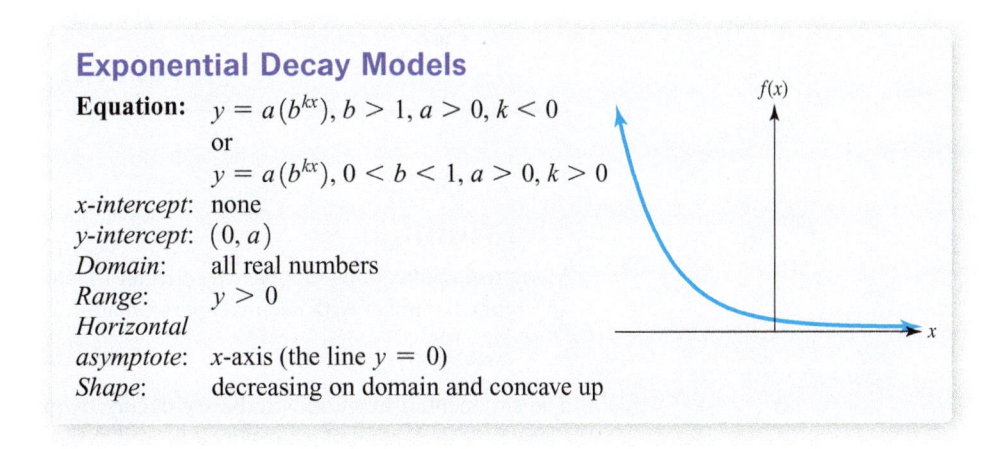

Exponential Decay Models

Equation: $y = a(b^{kx})$, $b > 1$, $a > 0$, $k < 0$
or
$y = a(b^{kx})$, $0 < b < 1$, $a > 0$, $k > 0$

x-intercept: none
y-intercept: $(0, a)$
Domain: all real numbers
Range: $y > 0$
Horizontal asymptote: *x*-axis (the line $y = 0$)
Shape: decreasing on domain and concave up

---

*For details, see Appendix B, page 704.

340   Chapter 5 Exponential and Logarithmic Functions

EXAMPLE 6 ▶ Sales Decay

It pays to advertise, and it is frequently true that weekly sales will drop rapidly for many products after an advertising campaign ends. This decline in sales is called *sales decay*. Suppose that the decay in the sales of a product is given by

$$S = 1000(2^{-0.5x}) \text{ dollars}$$

where x is the number of weeks after the end of a sales campaign. Use this function to answer the following.

a. What is the level of sales when the advertising campaign ends?

b. What is the level of sales 1 week after the end of the campaign?

c. Use a graph of the function to estimate the week in which sales equal $500.

d. According to this model, will sales ever fall to zero?

SOLUTION

a. The campaign ends when $x = 0$, so $S = 1000(2^{-0.5(0)}) = 1000(2^0) = 1000(1) = \$1000$.

b. At 1 week after the end of the campaign, $x = 1$. Thus, $S = 1000(2^{-0.5(1)}) = \$707.11$.

c. The graph of this sales decay function is shown in Figure 5.12(a). One way to find the x-value for which $S = 500$ is to graph $y_1 = 1000(2^{-0.5x})$ and $y_2 = 500$ and find the point of intersection of the two graphs. See Figure 5.12(b), which shows that $y = 500$ when $x = 2$. Thus, sales fall to half their original amount after 2 weeks.

d. The graph of this sales decay function approaches the positive x-axis as x gets large, but it never reaches the x-axis. Thus, sales will never reach a value of $0.

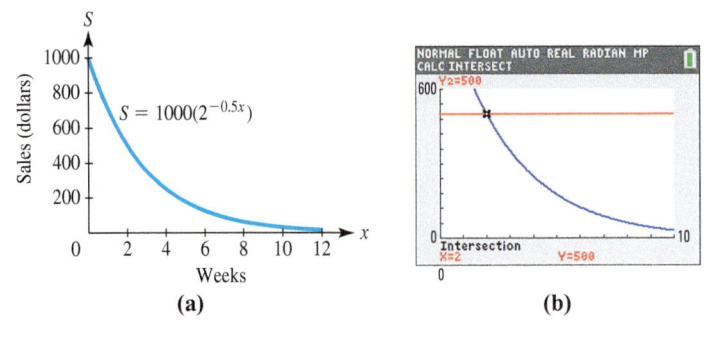

Figure 5.12

The Number e

Many real applications involve exponential functions with the base e. The number e is an irrational number with decimal approximation 2.718281828 (to nine decimal places).

$$e \approx 2.718281828$$

The exponential function with base e occurs frequently in biology and in finance. We discuss the derivation of e and how it is used in finance in Section 5.5. Because e is close in value to 3, the graph of the exponential function $y = e^x$ has the same basic shape as the graph of $y = 3^x$. (See Figure 5.13, which compares the graphs of $y = e^x$, $y = 3^x$, and $y = 2^x$.) The graph of $y = e^x$ can be found with a graphing calculator by using the e^x (**2ND** **LN**) key, and it can be found with Excel by using the formula =exp(cell#).

(a) $y = e^x$

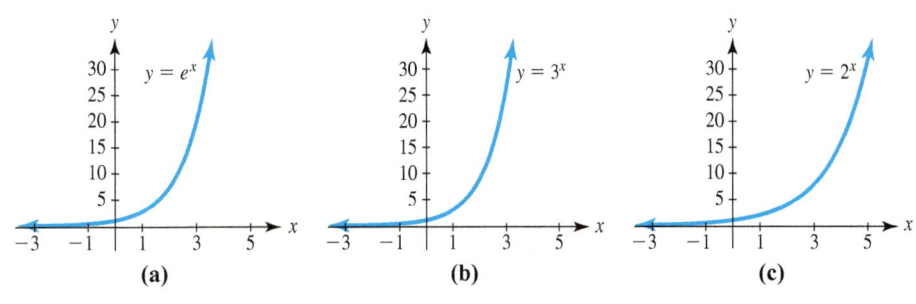

(b) $y = 3^x$

(c) $y = 2^x$

Figure 5.13

The growth in the value of an investment is another important application that can be modeled by an exponential function. As we prove in Section 5.5, the future value S of an investment of P dollars for t years at interest rate r, compounded continuously, is given by

$$S = Pe^{rt} \text{ dollars}$$

EXAMPLE 7 ▶ Growth of an Investment

If $10,000 is invested for 15 years at 10%, compounded continuously, what is the future value of the investment?

SOLUTION

The future value of this investment is

$$S = Pe^{rt} = 10{,}000e^{0.10(15)} = 10{,}000e^{1.5} = 44{,}816.89 \text{ dollars}$$

EXAMPLE 8 ▶ Carbon-14 Dating

During the life of an organism, the ratio of carbon-14 (C-14) atoms to carbon-12 (C-12) atoms remains about the same. When the organism dies, the number of C-14 atoms within its carcass begins to gradually decrease, but the number of C-12 atoms does not, so they can be measured to determine how many C-14 atoms were present at death. The number of C-14 atoms in an organism when it is discovered can then be compared to the number present at death to find how long ago the organism lived. The ratio of the present amount (y) of C-14 to the original amount (y_0) t years ago is given by

$$\frac{y}{y_0} = e^{-0.00012097t}$$

producing a function

$$y = f(t) = y_0 e^{-0.00012097t}$$

that gives the amount remaining after t years.
(*Source*: Thomas Jefferson National Accelerator Facility, http://www.jlab.org/)

a. If the original amount of carbon-14 present in an artifact is 100 grams, how much remains after 2000 years?

b. What percent of the original amount of carbon-14 remains after 4500 years?

SOLUTION

a. The initial amount y_0 is 100 grams and the time t is 2000 years, so we substitute these values in $f(t) = y_0 e^{-0.00012097t}$ and calculate:

$$f(2000) = 100e^{-0.00012097(2000)} \approx 78.51$$

Thus, the amount of carbon-14 present after 2000 years is approximately 78.5 grams.

b. If y_0 is the original amount, the amount that will remain after 4500 years is

$$f(4500) = y_0 e^{-0.00012097(4500)} = 0.58 y_0$$

Thus, approximately 58% of the original amount will remain after 4500 years. (Note that you do not need to know the original amount of carbon-14 in order to answer this question.)

SKILLS CHECK 5.1

Answers that are not seen can be found in the answer section at the back of the text.

1. Which of the following functions are exponential functions? c, d, e

   a. $y = 3x + 3$ b. $y = x^3$ c. $y = 3^x$
   d. $y = \left(\dfrac{2}{3}\right)^x$ e. $y = e^x$ f. $y = x^e$

2. Determine if each of the following functions models exponential growth or exponential decay.

   a. $y = 2^{0.1x}$ Growth   b. $y = 3^{-1.4x}$ Decay
   c. $y = 4e^{-5x}$ Decay   d. $y = 0.8^{3x}$ Decay

3. a. Graph the function $f(x) = e^x$ on the window $[-5, 5]$ by $[-10, 30]$.
   b. Find $f(1)$, $f(-1)$, and $f(4)$, rounded to three decimal places. 2.718; 0.368; 54.598
   c. What is the horizontal asymptote of the graph? x-axis
   d. What is the y-intercept? 1

4. a. Graph the function $f(x) = 5^x$ on the window $[-5, 5]$ by $[-10, 30]$.
   b. Find $f(1), f(3)$, and $f(-2)$. 5; 125; 0.04
   c. What is the horizontal asymptote of the graph? x-axis
   d. What is the y-intercept? 1

5. Graph the function $y = 3^x$ on $[-5, 5]$ by $[-10, 30]$.

6. Graph $y = 5e^{-x}$ on $[-4, 4]$ by $[-1, 20]$.

In Exercises 7–12, graph each function.

7. $y = 5^{-x}$ 8. $y = 3^{-2x}$
9. $y = 3^x + 5$ 10. $y = 2(1.5)^{-x}$
11. $y = 3^{(x-2)} - 4$ 12. $y = 3^{(x-1)} + 2$

In Exercises 13–18, match the graph with its equation.

13. $y = 4^{x+3}$ B

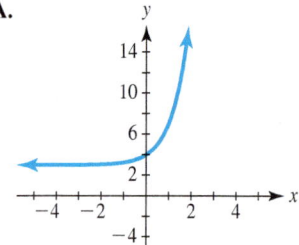

14. $y = 2 \cdot 4^x$ C

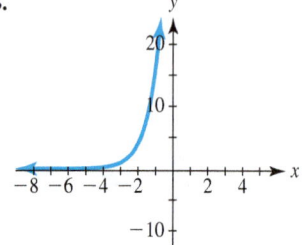

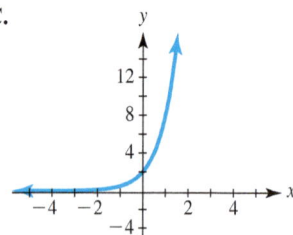

15. $y = 4^x + 3$ A

16. $y = 4^{-x}$ F

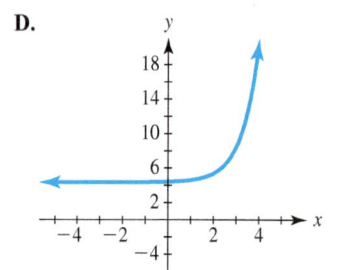

A.
B.
C.
D.

17. $y = -4^x$ E

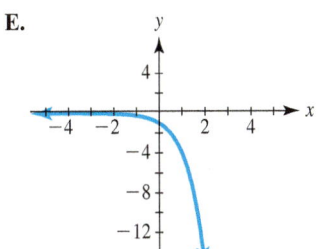

18. $y = 4^{x-2} + 4$ D

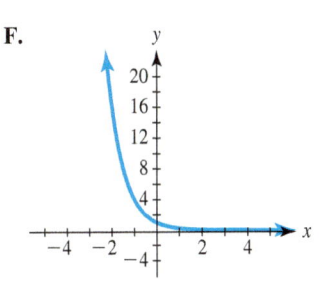

E.

F.

In Exercises 19–24, use your knowledge of transformations to compare the graph of the function with the graph of $f(x) = 4^x$. Then graph each function for $-4 \leq x \leq 4$.

19. $y = 4^x + 2$
20. $y = 4^{x-1}$
21. $y = 4^{-x}$
22. $y = -4^x$
23. $y = 3(4^x)$
24. $y = 3 \cdot 4^{(x-2)} - 3$

25. How does the graph of Exercise 24 compare with the graph of Exercise 23?

26. Which of the functions in Exercises 19–24 are increasing functions, and which are decreasing functions?
All are increasing except Exercises 21 and 22, which are decreasing.

In Exercises 27–34, use your knowledge of transformations to compare the graph of each function with the graph of $f(x) = e^x$. Then graph each function.

27. $y = e^{x-3}$ Horizontal shift, 3 units right
28. $y = e^x - 4$ Vertical shift, 4 units down
29. $y = \dfrac{1}{2}e^x$ Vertical compression with factor ½
30. $y = -e^x$ Reflection across the x-axis
31. $y = e^{-x}$ Reflection across the y-axis
32. $y = e^{x+2} + 1$ Horizontal shift to the left 2 units; vertical shift up 1 unit
33. $y = e^{3x}$ Horizontal compression; outputs are cubed
34. $y = e^{(1/3)x}$ Horizontal stretch; outputs are the cube roots

35. a. Graph $f(x) = 12e^{-0.2x}$ for $-5 \leq x \leq 15$.
    b. Find $f(10)$ and $f(-10)$. 1.624; 88.669
    c. Does this function represent growth or decay? Decay

36. If $y = 200(2^{-0.01x})$,
    a. Find the value of y (to two decimal places) when $x = 20$. 174.11
    b. Use graphical or numerical methods to determine the value of x that gives $y = 100$. 100

EXERCISES 5.1

Answers that are not seen can be found in the answer section at the back of the text.

37. **Sales Decay** At the end of an advertising campaign, weekly sales of smartphones declined according to the equation $y = 12{,}000(2^{-0.08x})$ dollars, where x is the number of weeks after the end of the campaign.
    a. Determine the sales at the end of the ad campaign. $12,000
    b. Determine the sales 6 weeks after the end of the campaign. $8603.73
    c. Does this model indicate that sales will eventually reach $0? No

38. **Sales Decay** At the end of an advertising campaign, weekly sales of camcorders declined according to the equation $y = 10{,}000(3^{-0.05x})$ dollars, where x is the number of weeks after the campaign ended.
    a. Determine the sales at the end of the ad campaign. $10,000
    b. Determine the sales 8 weeks after the end of the campaign. $6443.94
    c. How do we know, by inspecting the equation, that this function is decreasing? $b > 1, k < 0$

39. **Investment**
    a. Graph the function $S = 80{,}000(1.05^t)$, which gives the future value of $80,000 invested at 5%, compounded annually, for t years, $0 \leq t \leq 12$.
    b. Find the future value of $80,000 invested for 10 years at 5%, compounded annually. $130,311.57

40. **Investment**
    a. Graph the function $S = 56{,}000(1.09^t)$, which gives the future value of $56,000 invested at 9%, compounded annually, for t years, $0 \leq t \leq 15$.
    b. Find the future value of $56,000 invested for 13 years at 9%, compounded annually. $171,685.06

41. **Continuous Compounding** If $8000 is invested for t years at 8% interest, compounded continuously, the future value is given by $S = 8000e^{0.08t}$ dollars.
    a. Graph this function for $0 \leq t \leq 15$.
    b. Use the graph to estimate when the future value will be $20,000. In about 11.45 yr

c. Complete the following table.

t (year)	S ($)
10	17,804.33
20	39,624.26
22	46,449.50

42. Continuous Compounding If $35,000 is invested for t years at 9% interest, compounded continuously, the future value is given by $S = 35{,}000e^{0.09t}$ dollars.

  a. Graph this function for $0 \leq t \leq 8$.

  b. Use the graph to estimate when the future value will be $60,000. In about 6 yr

  c. Complete the following table.

t (year)	S ($)
10	86,086.11
15	135,009.89
22	253,496

43. Radioactive Decay The amount of radioactive isotope thorium-234 present at time t is given by $A(t) = 500e^{-0.02828t}$ grams, where t is the time in years that the isotope decays. The initial amount present is 500 grams.

  a. How many grams remain after 10 years? 376.84 g

  b. Graph this function for $0 \leq t \leq 100$.

  c. If the half-life is the time it takes for half of the initial amount to decay, use graphical methods to estimate the half-life of this isotope. About 24.5 yr

44. Radioactive Decay A breeder reactor converts stable uranium-238 into the isotope plutonium-239. The decay of this isotope is given by $A(t) = 100e^{-0.00002876t}$, where $A(t)$ is the amount of the isotope at time t (in years) and 100 grams is the original amount.

  a. How many grams remain after 100 years? 99.71 g

  b. Graph this function for $0 \leq t \leq 50{,}000$.

  c. The half-life is the time it takes for half of the initial amount to decay; use graphical methods to estimate the half-life of this isotope. 24,101 yr

45. Sales Decay At the end of an advertising campaign, weekly sales at an electronics store declined according to the equation $y = 2000(2^{-0.1x})$ dollars, where x is the number of weeks after the end of the campaign.

  a. Graph this function for $0 \leq x \leq 60$.

  b. Use the graph to find the weekly sales 10 weeks after the campaign ended. $1000

  c. Comment on "It pays to advertise" for this store. Sales declined drastically after the end of the ad campaign. (Answers may vary).

46. Sales Decay At the end of an advertising campaign, weekly retail sales of a product declined according to the equation $y = 40{,}000(3^{-0.1x})$ dollars, where x is the number of weeks after the campaign ended.

  a. Graph this function for $0 \leq x \leq 50$.

  b. Find the weekly sales 10 weeks after the campaign ended. $13,333

  c. Should the retailers consider another advertising campaign even if it costs $5000? Yes; sales are dropping rapidly and spending $5000 boosts sales to $40,000.

47. Purchasing Power The purchasing power (real value of money) decreases if inflation is present in the economy. For example, the purchasing power of R dollars after t years of 5% inflation is given by the model

$$P = R(0.95^t) \text{ dollars}$$

  a. What will be the purchasing power of $40,000 after 20 years of 5% inflation? $14,339.44

  b. How does this affect people planning to retire at age 50?
  (*Source*: *Viewpoints*, VALIC, 1993) Because of future inflation, they should plan to save money. (Answers may vary.)

48. Purchasing Power If a retired couple has a fixed income of $60,000 per year, the purchasing power (real value of money) after t years of 5% inflation is given by the equation $P = 60{,}000(0.95^t)$ dollars.

  a. What is the purchasing power of their income after 4 years? $48,870.38

  b. Test numerical values in this function to determine the years when their purchasing power will be less than $30,000. After 14 or more yr

49. Real Estate Inflation During a 5-year period of constant inflation, the value of a $100,000 property will increase according to the equation $v = 100{,}000e^{0.05t}$.

  a. What will be the value of this property in 4 years? $122,140.28

  b. Use a table or graph to estimate when this property will double in value. In about 14 yr

50. Inflation An antique table increases in value according to the function $v(x) = 850(1.04^x)$ dollars, where x is the number of years after 1990.

  a. How much was the table worth in 1990? $850

  b. If the pattern indicated by the function remains valid, what was the value of the table in 2005? $1530.80

  c. Use a table or graph to estimate the year when this table would reach double its 1990 value. 2008

51. Population The population in a certain city was 53,000 in 2000, and its future size is predicted to be $P(t) = 53{,}000e^{0.015t}$ people, where t is the number of years after 2000.

   a. Does this model indicate that the population is increasing or decreasing? Increasing

   b. Use this function to estimate the population of the city in 2005. 57,128

   c. Use this function to estimate the population of the city in 2010. 61,577

   d. What is the average rate of growth between 2000 and 2010? Approximately 858 people per year

52. Population The population in a certain city was 800,000 in 2003, and its future size is predicted to be $P = 800{,}000e^{-0.020t}$ people, where t is the number of years after 2003.

   a. Does this model indicate that the population is increasing or decreasing? Decreasing

   b. Use this model to estimate the population of the city in 2010. 695,487

   c. Use this model to predict the population of the city in 2020. 569,416

   d. What is the average rate of change in population between 2010 and 2020? Decreasing by 12,607 per year

53. Carbon-14 Dating An exponential decay function can be used to model the number of grams of a radioactive material that remain after a period of time. Carbon-14 decays over time, with the amount remaining after t years given by $y = 100e^{-0.00012097t}$ if 100 grams is the original amount.

   a. How much remains after 1000 years? About 88.6 g

   b. Use graphical methods to estimate the number of years until 10 grams of carbon-14 remain. About 19,034 yr

54. Drugs in the Bloodstream If a drug is injected into the bloodstream, the percent of the maximum dosage that is present at time t is given by

$$y = 100(1 - e^{-0.35(10-t)})$$

where t is in hours, with $0 \leq t \leq 10$.

   a. What percent of the drug is present after 2 hours? 93.92%

   b. Graph this function.

   c. When is the drug totally gone from the bloodstream? After 10 hours

55. Normal Curve The "curve" on which many students like to be graded is the bell-shaped normal curve. The equation $y = \dfrac{1}{\sqrt{2\pi}} e^{-(x-50)^2/2}$ describes the normal curve for a standardized test, where x is the test score before curving.

   a. Graph this function for x between 47 and 53 and for y between 0 and 0.5.

   b. The average score for the test is the score that gives the largest output y. Use the graph to find the average score. 50

56. IQ Measure The frequency of IQ measures x follows a bell-shaped normal curve with equation $y = \dfrac{1}{\sqrt{2\pi}} e^{-(x-100)^2/20}$. Graph this function for x between 80 and 120 and for y between 0 and 0.4.

5.2 Logarithmic Functions; Properties of Logarithms

KEY OBJECTIVES

- Graph and evaluate logarithmic functions
- Convert equations to logarithmic and exponential forms
- Evaluate and apply common logarithms
- Evaluate and apply natural logarithms
- Apply logarithmic properties

SECTION PREVIEW   Diabetes

Projections from 2010 to 2050 indicate that the fraction of U.S. adults with diabetes (diagnosed and undiagnosed) will eventually reach one-third of the adult population. The percent can be modeled by the logarithmic function

$$p(x) = -12.975 + 11.851 \ln x$$

where x is the number of years after 2000. To find the year when the percent reaches 33%, we can graph the equations $y_1 = -12.975 + 11.851 \ln x$ and $y_2 = 33$ on the same axes and find the point of intersection. (See Example 8.) (*Source*: Centers for Disease Control and Prevention)

In this section, we introduce logarithmic functions, discuss the relationship between exponential and logarithmic functions, and use the definition of logarithmic functions to rewrite exponential equations in a new form, called the logarithmic form of the equation. We also use logarithmic functions in real applications. ∎

Logarithmic Functions

Recall that we discussed inverse functions in Section 4.3. The inverse of a function is a second function that "undoes" what the original function does. For example, if the original function doubles each input, its inverse takes half of each input. If a function is defined by $y = f(x)$, the inverse of this function can be found by interchanging x and y in $y = f(x)$ and solving the new equation for y. As we saw in Section 4.3, a function has an inverse if it is a one-to-one function (that is, there is a one-to-one correspondence between the inputs and outputs defining the function).

Every exponential function of the form $y = b^x$, with $b > 0$ and $b \neq 1$, is one-to-one, so every exponential function of this form has an inverse function. The inverse function of

$$y = b^x$$

is found by interchanging x and y and solving the new equation for y. Interchanging x and y in $y = b^x$ gives

$$x = b^y$$

In this new function, y is the power to which we raise b to get the number x. To solve an expression like this for the exponent, we need new notation. We define y, the power to which we raise the base b to get the number x, as

$$y = \log_b x$$

This inverse function is called a **logarithmic function** with base b.

Logarithmic Function

For $x > 0$, $b > 0$, and $b \neq 1$, the logarithmic function with base b is

$$y = \log_b x$$

which is defined by $x = b^y$. That is, $\log_b x$ is the exponent to which we must raise the base b to get the number x.

This function is the inverse function of the exponential function $y = b^x$.

Note that, according to the definition of the logarithmic function, $y = \log_b x$ and $x = b^y$ are two different forms of the same equation. We call $y = \log_b x$ the *logarithmic form* of the equation and $x = b^y$ the *exponential form* of the equation. The number b is called the **base** in both $y = \log_b x$ and $x = b^y$, and y is the **logarithm** in $y = \log_b x$ and the **exponent** in $x = b^y$. Thus, a logarithm is an exponent.

EXAMPLE 1 ▶ **Converting from Exponential to Logarithmic Form**

Write each of the following exponential equations in logarithmic form.

a. $3^2 = 9$ **b.** $4^{-1} = \dfrac{1}{4}$

c. $5^1 = 5$ **d.** $x = 3^y$

SOLUTION

To write an exponential equation in its equivalent logarithmic form, remember that the base of the logarithm is the same as the base of the exponent in the exponential form. Also, the exponent in the exponential form is the value of the logarithm in the logarithmic form.

a. To write $3^2 = 9$ in logarithmic form, note that the base of the exponent is 3, so the base of the logarithm will also be 3. The exponent, 2, will be the value of the logarithm. Thus,

$$3^2 = 9 \quad \text{is equivalent to} \quad \log_3 9 = 2$$

We can refer to 9 as the **argument** of the logarithm.

b. To write $4^{-1} = \dfrac{1}{4}$ in logarithmic form, note that the base of the exponent is 4, so the base of the logarithm will also be 4. The exponent, -1, will be the value of the logarithm. Thus,

$$4^{-1} = \frac{1}{4} \quad \text{is equivalent to} \quad \log_4 \frac{1}{4} = -1$$

c. The base of the exponent in $5^1 = 5$ is 5, so the base of the logarithm will be 5, and the value of the logarithm will be 1. Thus,

$$5^1 = 5 \quad \text{is equivalent to} \quad \log_5 5 = 1$$

d. Similarly, the base of the exponent in $x = 3^y$ is 3, so the base of the logarithm will be 3, and the value of the logarithm will be y. Thus,

$$x = 3^y \quad \text{is equivalent to} \quad \log_3 x = y$$

EXAMPLE 2 ▶ Converting from Logarithmic to Exponential Form

Write each of the following logarithmic equations in exponential form.

a. $\log_2 16 = 4$ **b.** $\log_{10} 0.0001 = -4$ **c.** $\log_6 1 = 0$

SOLUTION

To write a logarithmic equation in its equivalent exponential form, remember that the base of the logarithm will be the base of the exponent in the exponential form. Also, the value of the logarithm will be the exponent in the exponential form. The argument of the logarithm will be the value of the exponential expression.

a. To write $\log_2 16 = 4$ in exponential form, note that the base of the exponent will be 2, the exponent will be 4, and the value of the exponential expression will be 16. Thus,

$$\log_2 16 = 4 \quad \text{is equivalent to} \quad 2^4 = 16$$

b. To write $\log_{10} 0.0001 = -4$ in exponential form, note that the base of the exponent will be 10, the exponent will be -4, and the value of the exponential expression will be 0.0001. Thus,

$$\log_{10} 0.0001 = -4 \quad \text{is equivalent to} \quad 10^{-4} = 0.0001$$

c. The base of the exponent in $\log_6 1 = 0$ will be 6, the exponent will be 0, and the value of the exponential expression will be 1. Thus,

$$\log_6 1 = 0 \quad \text{is equivalent to} \quad 6^0 = 1$$

We can sometimes more easily evaluate logarithmic functions for different inputs by changing the function from logarithmic form to exponential form.

EXAMPLE 3 ▶ Evaluating Logarithms

a. Write $y = \log_2 x$ in exponential form.

b. Use the exponential form from part (a) to find y when $x = 8$.

c. Evaluate $\log_2 8$.   **d.** Evaluate $\log_4 \dfrac{1}{16}$.   **e.** Evaluate $\log_{10} 0.001$.

SOLUTION

a. To rewrite $y = \log_2 x$ in exponential form, note that the base of the logarithm, 2, becomes the base of the exponential expression, and y equals the exponent in the equivalent exponential form. Thus, the exponential form is
$$2^y = x$$

b. To find y when x is 8, we solve $2^y = 8$. We can easily see that $y = 3$ because $2^3 = 8$.

c. To evaluate $\log_2 8$, we write $\log_2 8 = y$ in its equivalent exponential form $2^y = 8$. From part (b), $y = 3$. Thus, $\log_2 8 = 3$.

d. To evaluate $\log_4 \dfrac{1}{16}$, we write $\log_4 \dfrac{1}{16} = y$ in its equivalent exponential form $4^y = \dfrac{1}{16}$. Because $4^{-2} = \dfrac{1}{16}$, $y = -2$. Thus, $\log_4 \dfrac{1}{16} = -2$.

e. To evaluate $\log_{10} 0.001$, we write $\log_{10} 0.001 = y$ in its equivalent exponential form $10^y = 0.001$. Because $10^{-3} = 0.001$, $y = -3$. Thus, $\log_{10} 0.001 = -3$.

The relationship between the logarithmic and exponential forms of an equation is used in the following example.

EXAMPLE 4 ▶ Graphing a Logarithmic Function

Graph $y = \log_2 x$.

SOLUTION

We can graph $y = \log_2 x$ by graphing the equivalent equation $x = 2^y$. Table 5.4 gives the values (found by substituting values of y and calculating x), and the graph is shown in Figure 5.14.

Table 5.4

$x = 2^y$	y
$\dfrac{1}{8}$	-3
$\dfrac{1}{4}$	-2
$\dfrac{1}{2}$	-1
1	0
2	1
4	2
8	3

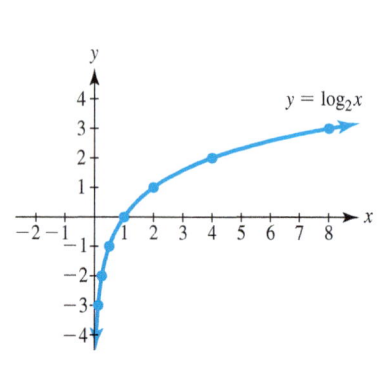

Figure 5.14

Figures 5.15(a) and (b) show graphs of the logarithmic function $y = \log_b x$ for a base $b > 1$ and a base $0 < b < 1$. Note that the graph of the function in Figure 5.15(a) is increasing and the graph of the function in Figure 5.15(b) is decreasing.

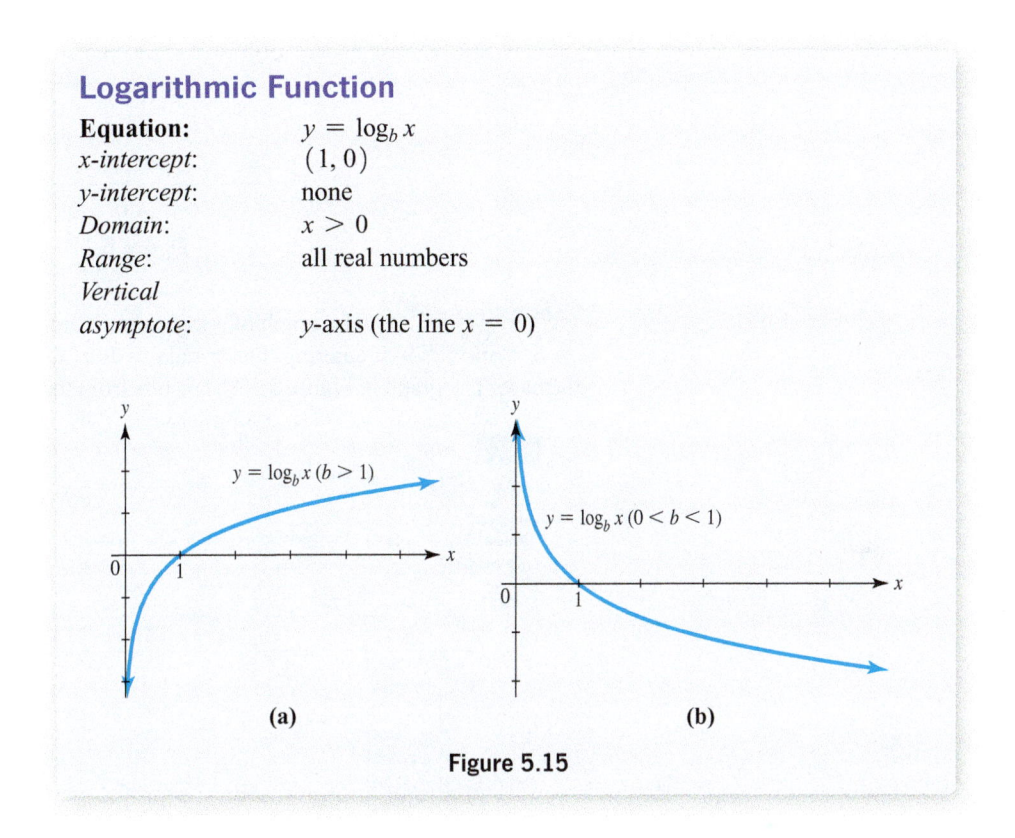

Logarithmic Function

Equation: $y = \log_b x$
x-intercept: $(1, 0)$
y-intercept: none
Domain: $x > 0$
Range: all real numbers
Vertical asymptote: y-axis (the line $x = 0$)

(a) $y = \log_b x \; (b > 1)$

(b) $y = \log_b x \; (0 < b < 1)$

Figure 5.15

Common Logarithms

We can use technology to evaluate logarithmic functions if the base of the logarithm is 10 or e. Because our number system uses base 10, logarithms with a base of 10 are called **common logarithms**, and, by convention, $\log_{10} x$ is simply denoted $\log x$. The graph of $y = \log x$ can be drawn by plotting points satisfying $x = 10^y$. Table 5.5 shows some points on the graph of $y = \log x$, using powers of 10 as x-values, and Figure 5.16 shows the graph.

Table 5.5

x	y = log x
1000	3
100	2
10	1
1	0
0.1	−1
0.01	−2
0.001	−3

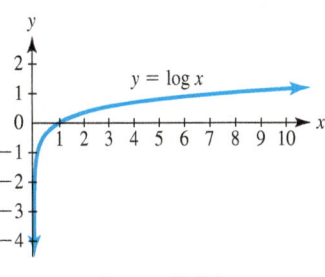

Figure 5.16

Chapter 5 Exponential and Logarithmic Functions

Graphing calculators can be used to evaluate and graph $y = \log x$, using the **LOG** key. Figure 5.17(a) shows the graph of $y = \log x$ on a graphing calculator.

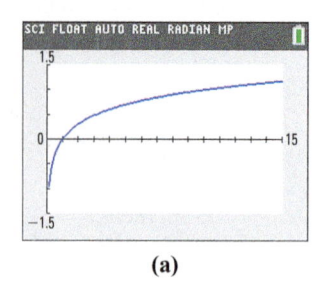

(a)

Figure 5.17

Spreadsheet ► SOLUTION The graph of a logarithmic function can be created with Excel by entering the formula $= \log(x, b)$ where x is a cell entry and b is the base. The graph in Figure 5.17(b) results from the Excel command $= \log(x, 10)$.*

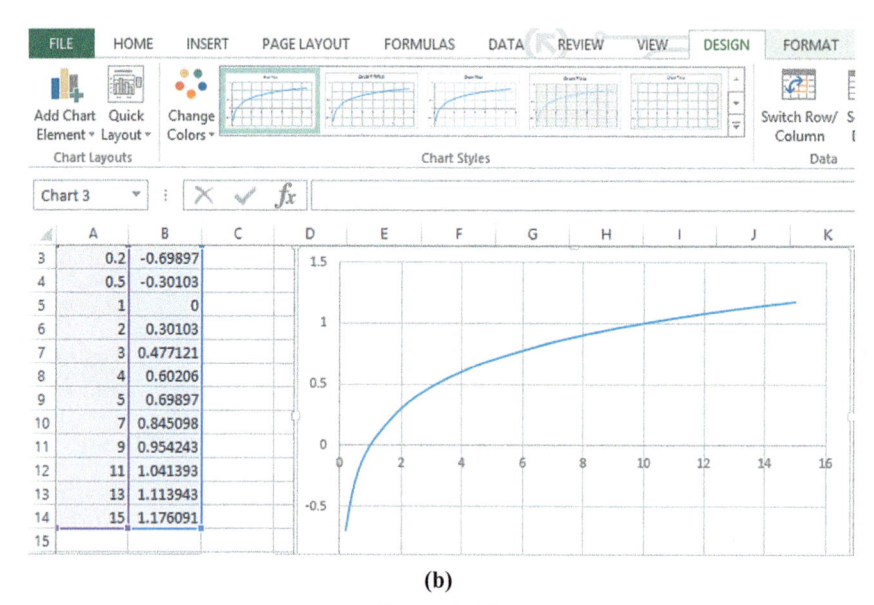

(b)

Figure 5.17

EXAMPLE 5 ▶ Evaluating Base 10 Logarithms

Using each of the indicated methods, find $f(10,000)$ if $f(x) = \log x$.

a. Write the equation $y = \log x$ in exponential form and find y when $x = 10,000$.

b. Use technology to evaluate $\log 10,000$.

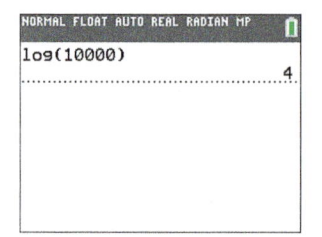

Figure 5.18

SOLUTION

a. The exponential form of $y = \log x = \log_{10} x$ is $10^y = x$. Substituting 10,000 for x in this equation gives $10^y = 10,000$. Because $10^4 = 10,000$, we see that $y = 4$. Thus, $f(10,000) = 4$.

b. The value of log 10,000 is shown to be 4 on the calculator screen in Figure 5.18, so $f(10,000) = 4$.

---

*For details, see Appendix B, page 704.

EXAMPLE 6 ▶ pH Scale

To simplify measurement of the acidity or basicity of a solution, the pH (hydrogen potential) scale was developed. The pH is given by the formula $pH = -\log[H^+]$, where $[H^+]$ is the concentration of hydrogen ions in moles per liter in the solution. The pH scale ranges from 0 to 14. A pH of 7 is neutral; a pH less than 7 is acidic; a pH greater than 7 is basic.

a. Using Table 5.6, find the pH level of bleach.

b. Using Table 5.6, find the pH level of lemon juice.

c. Lower levels of pH indicate a more acidic solution. How much more acidic is lemon juice than bleach?

Table 5.6

Concentration of Hydrogen Ions	Examples	Concentration of Hydrogen Ions	Examples
10^{-14}	Liquid drain cleaner	10^{-6}	Milk
10^{-13}	Bleach, oven cleaner	10^{-5}	Black coffee
10^{-12}	Soapy water	10^{-4}	Tomato juice
10^{-11}	Ammonia	10^{-3}	Orange juice
10^{-10}	Milk of magnesia	10^{-2}	Lemon juice
10^{-9}	Toothpaste	10^{-1}	Sulfuric acid
10^{-8}	Baking soda	1	Battery acid
10^{-7}	"Pure" water		

SOLUTION

a. The concentration of hydrogen ions for bleach is 10^{-13}. So the pH will be $-\log(10^{-13}) = 13$. See Figure 5.19(a).

(a)   (b)

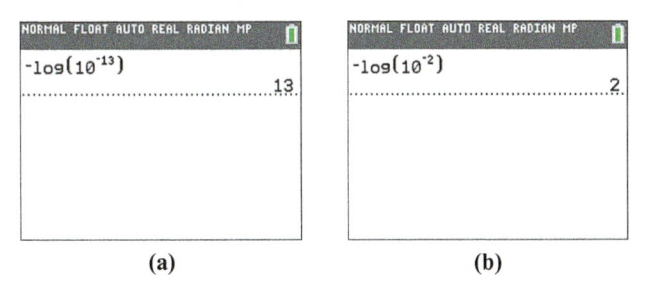

Figure 5.19

b. The concentration of hydrogen ions for lemon juice is 10^{-2}. So the pH will be $-\log(10^{-2}) = 2$. See Figure 5.19(b).

c. To find the ratio of the acidity of the two solutions, we divide the concentrations of hydrogen ions. Thus, the ratio is

$$\frac{10^{-2}}{10^{-13}} = \frac{10^{13}}{10^2} = 10^{11} = 100{,}000{,}000{,}000$$

indicating that lemon juice is 100,000,000,000 times as acidic as bleach.

EXAMPLE 7 ▶ Transformations of Graphs of Logarithmic Functions

Explain how the graph of each of the following functions compares with the graph of $y = \log x$, find the domain, and graph each function.

a. $y = \log(x + 3)$ **b.** $y = 4 + \log(x - 2)$ **c.** $y = \dfrac{1}{3}\log x$

SOLUTION

a. The graph of $y = \log(x + 3)$ has the same shape as the graph of $y = \log x$, but it is shifted 3 units to the left. Note that the vertical asymptote of the graph of $y = \log x$, $x = 0$, is also shifted 3 units left, so the vertical asymptote of $y = \log(x + 3)$ is $x = -3$. This means that, as the x-values approach -3, the y-values decrease without bound. See Figure 5.20(a). Because the graph has been shifted left 3 units, the domain is $(-3, \infty)$.

b. The graph of $y = 4 + \log(x - 2)$ has the same shape as the graph of $y = \log x$, but it is shifted 2 units to the right and 4 units up. The vertical asymptote is $x = 2$, and the graph is shown in Figure 5.20(b). The domain is $(2, \infty)$.

c. As we saw in Chapter 4, multiplication of a function by a constant less than 1 compresses the graph by a factor equal to that constant. So each of the y-values of $y = \dfrac{1}{3}\log x$ is $\dfrac{1}{3}$ times the corresponding y-value of $y = \log x$, and the graph is compressed. See Figure 5.20(c). The domain is $(0, \infty)$.

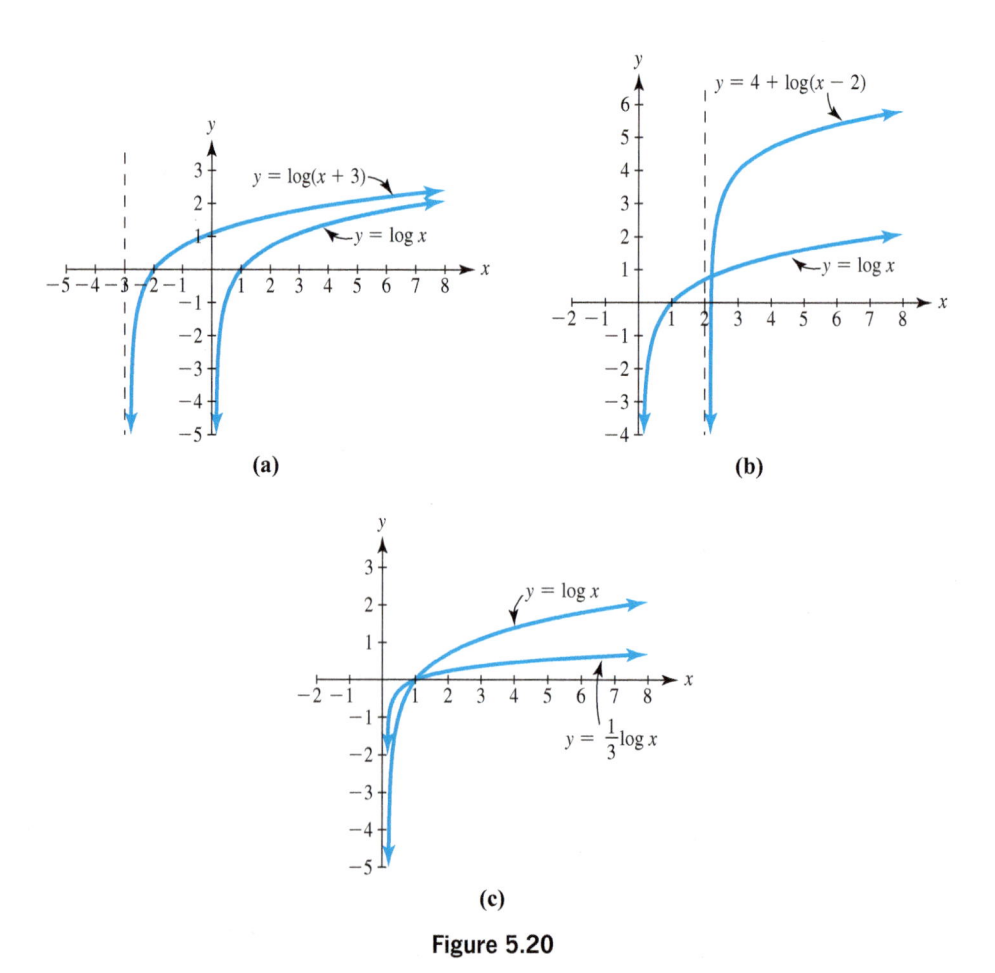

(a)

(b)

(c)

Figure 5.20

Natural Logarithms

As we have seen in the previous section, the number e is important in many business and life science applications. Likewise, logarithms to base e are important in many applications. Because they are used so frequently, especially in science, they are called **natural logarithms**, and we use the special notation $\ln x$ to denote $\log_e x$.

> ### Natural Logarithms
>
> The logarithmic function with base e is $y = \log_e x$, defined by $x = e^y$ for all positive numbers x and denoted by
>
> $$\ln x = \log_e x$$

Most graphing calculators and spreadsheets have the key or command **LN** to use in evaluating functions involving natural logarithms. Figure 5.21(a) shows the graph of $y = \ln x$, which is similar to the graph of $y = \log_3 x$ (Figure 5.21(b)) because the value of e, the base of $\ln x$, is close to 3.

(a) (b)

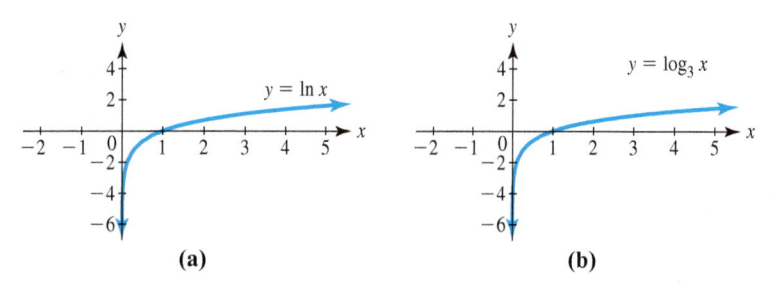

Figure 5.21

We can graph $y = \ln x$ with Excel using the command $=\ln(x)$.*

EXAMPLE 8 ▶ Diabetes

Projections from 2010 to 2050 indicate that the percent of U.S. adults with diabetes (diagnosed and undiagnosed) can be modeled by $p(x) = -12.975 + 11.851 \ln x$, where x is the number of years after 2000.

a. Graph this function.

b. Is the function increasing or decreasing? What does this mean in the context of the application?

c. What does this model predict the percent of U.S. adults with diabetes will be in 2022?

d. Use the graph to estimate the year in which this model predicts the percent will reach 33%.
(*Source*: Centers for Disease Control and Prevention)

SOLUTION

a. Using the window $0 \leq x \leq 60$ and $0 \leq y \leq 50$, the graph of $p(x) = -12.975 + 11.851 \ln x$ is as shown in Figure 5.22(a).

b. The function is increasing, which means the percent of U.S. adults with diabetes is predicted to increase from 2010 to 2050.

c. The year 2022 is 22 years after 2000, so the percent in 2022 is estimated to be

$$p(22) = -12.975 + 11.851 \ln 22 \approx 23.7$$

---

*For details, see Appendix B, page 704.

d. To find the x-value for which $y = p(x) = 33$, we graph $y_1 = -12.975 + 11.851 \ln x$ and $y_2 = 33$ on the same axes and find the point of intersection. This shows that the percent is 33% when $x \approx 48.4$, during 2049. (See Figure 5.22(b).)

(a)   (b)

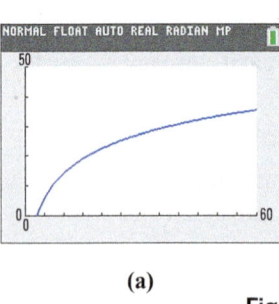

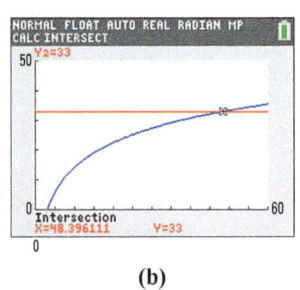

Figure 5.22

Logarithmic Properties

We have learned that logarithms are exponents and that a logarithmic function with base a is the inverse of an exponential function with base a. Thus, the properties of logarithms can be derived from the properties of exponents. These properties are useful in solving equations involving exponents and logarithms.

The basic properties of logarithms are easy to derive from the definition of a logarithm.

Basic Properties of Logarithms

For $b > 0$, $b \neq 1$,
1. $\log_b b = 1$
2. $\log_b 1 = 0$
3. $\log_b b^x = x$
4. $b^{\log_b x} = x$
5. For positive real numbers M and N, if $M = N$, then $\log_b M = \log_b N$.

These properties are easily justified by rewriting the logarithmic form as an exponential form or vice versa. In particular, $\log_b 1 = 0$ because $b^0 = 1$, $\log_b b = 1$ because $b^1 = b$, $\log_b b^x = x$ because $b^x = b^x$, and $b^{\log_b x} = x$ because $\log_b x = \log_b x$.

EXAMPLE 9 ▶ Logarithmic Properties

Use the basic properties of logarithms to simplify the following.

a. $\log_5 5^{10}$ **b.** $\log_4 4$ **c.** $\log_4 1$ **d.** $\log 10^7$ **e.** $\ln e^3$ **f.** $\log\left(\dfrac{1}{10^3}\right)$

SOLUTION

a. $\log_5 5^{10} = 10$ by Property 3   **b.** $\log_4 4 = 1$ by Property 1

c. $\log_4 1 = 0$ by Property 2   **d.** $\log 10^7 = \log_{10} 10^7 = 7$ by Property 3

e. $\ln e^3 = \log_e e^3 = 3$ by Property 3   **f.** $\log\left(\dfrac{1}{10^3}\right) = \log 10^{-3} = \log_{10} 10^{-3} = -3$

Other logarithmic properties are helpful in simplifying logarithmic expressions, which in turn help us solve some logarithmic equations. These properties are also directly related to the properties of exponents.

Additional Logarithmic Properties

For $b > 0$, $b \neq 1$, k a real number, and M and N positive real numbers,

6. $\log_b(MN) = \log_b M + \log_b N$ (Product Property)

7. $\log_b\left(\dfrac{M}{N}\right) = \log_b M - \log_b N$ (Quotient Property)

8. $\log_b M^k = k \log_b M$ (Power Property)

To prove Property 6, we let $x = \log_b M$ and $y = \log_b N$, so that $M = b^x$ and $N = b^y$. Then

$$\log_b(MN) = \log_b(b^x \cdot b^y) = \log_b(b^{x+y}) = x + y = \log_b M + \log_b N$$

To prove Property 8, we let $x = \log_b M$, so that $M = b^x$. Then

$$\log_b M^k = \log_b(b^x)^k = \log_b b^{kx} = kx = k \log_b M$$

EXAMPLE 10 ▶ Rewriting Logarithms

Rewrite each of the following expressions as a sum, difference, or product of logarithms, and simplify if possible.

a. $\log_4 5(x - 7)$ **b.** $\ln[e^2(e + 3)]$ **c.** $\log\left(\dfrac{x - 8}{x}\right)$

d. $\log_4 y^6$ **e.** $\ln\left(\dfrac{1}{x^5}\right)$

SOLUTION

a. By Property 6, $\log_4 5(x - 7) = \log_4 5 + \log_4(x - 7)$.

b. By Property 6, $\ln[e^2(e + 3)] = \ln e^2 + \ln(e + 3)$.
By Property 3, this equals $2 + \ln(e + 3)$.

c. By Property 7, $\log\left(\dfrac{x - 8}{x}\right) = \log(x - 8) - \log x$.

d. By Property 8, $\log_4 y^6 = 6 \log_4 y$.

e. By Property 7, Property 2, and Property 8,

$$\ln\left(\dfrac{1}{x^5}\right) = \ln 1 - \ln x^5 = 0 - 5 \ln x = -5 \ln x$$

Caution: There is no property of logarithms that allows us to rewrite a logarithm of a sum or a difference. That is why in Example 10(a) the expression $\log_4(x - 7)$ was *not* written as $\log_4 x - \log_4 7$.

EXAMPLE 11 ▶ Rewriting Logarithms

Rewrite each of the following expressions as a single logarithm.

a. $\log_3 x + 4 \log_3 y$ **b.** $\frac{1}{2} \log a - 3 \log b$ **c.** $\ln(5x) - 3 \ln z$

SOLUTION

a. $\log_3 x + 4 \log_3 y = \log_3 x + \log_3 y^4$ Use Logarithmic Property 8.

$ = \log_3 xy^4$ Use Logarithmic Property 6.

b. $\frac{1}{2} \log a - 3 \log b = \log a^{1/2} - \log b^3$ Use Logarithmic Property 8.

$\phantom{\frac{1}{2} \log a - 3 \log b} = \log\left(\frac{a^{1/2}}{b^3}\right)$ Use Logarithmic Property 7.

c. $\ln(5x) - 3 \ln z = \ln(5x) - \ln z^3$ Use Logarithmic Property 8.

$ = \ln\left(\frac{5x}{z^3}\right)$ Use Logarithmic Property 7.

EXAMPLE 12 ▶ Applying the Properties of Logarithms

a. Use logarithmic properties to estimate $\ln(5e)$ if $\ln 5 \approx 1.61$.

b. Find $\log_a\left(\frac{15}{8}\right)$ if $\log_a 15 = 2.71$ and $\log_a 8 = 2.079$.

c. Find $\log_a\left(\frac{15^2}{\sqrt{8}}\right)^3$ if $\log_a 15 = 2.71$ and $\log_a 8 = 2.079$.

SOLUTION

a. $\ln(5e) = \ln 5 + \ln e \approx 1.61 + 1 = 2.61$

b. $\log_a\left(\frac{15}{8}\right) = \log_a 15 - \log_a 8 = 2.71 - 2.079 = 0.631$

c. $\log_a\left(\frac{15^2}{\sqrt{8}}\right)^3 = 3 \log_a\left(\frac{15^2}{\sqrt{8}}\right) = 3[\log_a 15^2 - \log_a 8^{1/2}]$

$= 3\left[2 \log_a 15 - \frac{1}{2} \log_a 8\right] = 3\left[2(2.71) - \frac{1}{2}(2.079)\right] = 13.1415$

Richter Scale

Because the intensities of earthquakes are so large, the Richter scale was developed to provide a "scaled down" measuring system that permits earthquakes to be more easily measured and compared. The Richter scale gives the magnitude R of an earthquake using the formula

$$R = \log\left(\frac{I}{I_0}\right)$$

where I is the intensity of the earthquake and I_0 is a certain minimum intensity used for comparison.

Table 5.7 describes the typical effects of earthquakes of various magnitudes near the epicenter.

Table 5.7

Description	Richter Magnitudes	Earthquake Effects
Micro	Less than 2.0	Microearthquakes, not felt
Very minor	2.0–2.9	Generally not felt, but recorded
Minor	3.0–3.9	Often felt, but rarely causes damage
Light	4.0–4.9	Noticeable shaking of indoor items, rattling noises; significant damage unlikely
Moderate	5.0–5.9	Can cause damage to poorly constructed buildings over small regions
Strong	6.0–6.9	Can be destructive in areas up to about 100 miles across in populated areas
Major	7.0–7.9	Can cause serious damage over larger areas
Great	8.0–8.9	Can cause serious damage in areas several hundred miles across
Rarely, great	9.0–9.9	Devastating in areas several thousand miles across
Meteoric	10.0+	Never recorded

(*Source*: U.S. Geological Survey)

EXAMPLE 13 ▶ Richter Scale

a. If an earthquake has an intensity of 10,000 times I_0, what is the magnitude of the earthquake? Describe the effects of this earthquake.

b. Show that if the Richter scale reading of an earthquake is k, the intensity of the earthquake is $I = 10^k I_0$.

c. An earthquake that measured 9.0 on the Richter scale occurred in the Indian Ocean in December 2004, causing a devastating tsunami that killed thousands of people. Express the intensity of this earthquake in terms of I_0.

d. If an earthquake measures 7.0 on the Richter scale, give the intensity of this earthquake in terms of I_0. How much more intense is the earthquake in part (c) than the one with the Richter scale measurement of 7.0?

e. If one earthquake has intensity 320,000 times I_0 and a second has intensity 3,200,000 times I_0, what is the difference in their Richter scale measurements?

SOLUTION

a. If the intensity of the earthquake is $10,000 I_0$, we substitute $10,000 I_0$ for I in the equation $R = \log\left(\dfrac{I}{I_0}\right)$, getting $R = \log \dfrac{10,000 I_0}{I_0} = \log 10,000 = 4$. Thus, the magnitude of the earthquake is 4, signifying a light earthquake that would not likely cause significant damage.

b. If $R = k$, $k = \log\left(\dfrac{I}{I_0}\right)$, and the exponential form of this equation is $10^k = \dfrac{I}{I_0}$, so $I = 10^k I_0$.

c. The Richter scale reading for this earthquake is 9.0, so its intensity is

$$I = 10^{9.0} I_0 = 1{,}000{,}000{,}000 \, I_0$$

Thus, the earthquake had an intensity of 1 billion times I_0.

d. The Richter scale reading for this earthquake is 7.0, so its intensity is $I = 10^{7.0} I_0$. Thus, the ratio of intensities of the two earthquakes is

$$\dfrac{10^{9.0} I_0}{10^{7.0} I_0} = \dfrac{10^{9.0}}{10^{7.0}} = 10^2 = 100$$

Therefore, the Indian Ocean earthquake is 100 times as intense as the other earthquake.

e. Recall that the Richter scale measurement is given by $R = \log\left(\dfrac{I}{I_0}\right)$. The difference of the Richter scale measurements is

$$R_2 - R_1 = \log\dfrac{3{,}200{,}000\, I_0}{I_0} - \log\dfrac{320{,}000\, I_0}{I_0} = \log 3{,}200{,}000 - \log 320{,}000$$

Using Property 7 gives us

$$\log 3{,}200{,}000 - \log 320{,}000 = \log\dfrac{3{,}200{,}000}{320{,}000} = \log 10 = 1$$

Thus, the difference in the Richter scale measurements is 1.

Note from Example 13(d) that the Richter scale measurement of the Indian Ocean earthquake is 2 larger than the Richter scale measurement of the other earthquake, and that the Indian Ocean earthquake is 10^2 times as intense. In general, we have the following.

Richter Scale

1. If the intensity of an earthquake is I, its Richter scale measurement is

$$R = \log\dfrac{I}{I_0}$$

2. If the Richter scale reading of an earthquake is k, its intensity is

$$I = 10^k I_0$$

3. If the difference of the Richter scale measurements of two earthquakes is the positive number d, the intensity of the larger earthquake is 10^d times that of the smaller earthquake.

SKILLS CHECK 5.2

Answers that are not seen can be found in the answer section at the back of the text.

In Exercises 1–4, write the logarithmic equations in exponential form.

1. $y = \log_3 x$ $3^y = x$
2. $2y = \log_5 x$ $x = 5^{2y}$
3. $y = \ln(2x)$ $e^y = 2x$
4. $y = \log(-x)$ $10^y = -x$

In Exercises 5–8, write the exponential equations in logarithmic form.

5. $x = 4^y$ $\log_4 x = y$
6. $m = 3^p$ $p = \log_3 m$
7. $32 = 2^5$ $\log_2 32 = 5$
8. $9^{2x} = y$ $\log_9 y = 2x$

In Exercises 9 and 10, evaluate the logarithms, if possible. Round each answer to three decimal places.

9. a. $\log 7$ 0.845
   b. $\ln 86$ 4.454
   c. $\log 63{,}980$ 4.806

10. a. $\log 456$ 2.659
    b. $\log(-12)$ Undefined
    c. $\ln 10$ 2.303

In Exercises 11–13, find the value of the logarithms without using a calculator.

11. a. $\log_2 32$ 5
    b. $\log_9 81$ 2
    c. $\log_3 27$ 3
    d. $\log_4 64$ 3
    e. $\log_5 625$ 4

12. a. $\log_2 64$ 6
    b. $\log_9 27$ 3/2
    c. $\log_4 2$ 1/2
    d. $\ln(e^3)$ 3
    e. $\log 100$ 2

13. a. $\log_3 \dfrac{1}{27}$ -3
    b. $\ln 1$ 0
    c. $\ln e$ 1
    d. $\log 0.0001$ -4

14. Graph the functions by changing to exponential form.
    a. $y = \log_3 x$
    b. $y = \log_5 x$

In Exercises 15–18, graph the functions.

15. $y = 2 \ln x$
16. $y = 4 \log x$
17. $y = \log(x + 1) + 2$
18. $y = \ln(x - 2) - 3$

19. a. Write the inverse of $y = 4^x$ in logarithmic form.
    $y = \log_4 x$
    b. Graph $y = 4^x$ and its inverse and discuss the symmetry of their graphs.
    Graphs are symmetric about the line $y = x$.

20. a. Write the inverse of $y = 3^x$ in logarithmic form.
    $y = \log_3 x$
    b. Graph $y = 3^x$ and its inverse and discuss the symmetry of their graphs.
    Graphs are symmetric about the line $y = x$.

21. Write $\log_a a = x$ in exponential form and find x to evaluate $\log_a a$ for any $a > 0$, $a \neq 1$. $a^x = a; x = 1$

22. Write $\log_a 1 = x$ in exponential form and find x to evaluate $\log_a 1$ for any $a > 0$, $a \neq 1$. $a^x = 1; \log_a 1 = 0$

In Exercises 23–26, use the properties of logarithms to evaluate the expressions.

23. $\log 10^{14}$ 14
24. $\ln e^5$ 5
25. $10^{\log_{10} 12}$ 12
26. $6^{\log_6 25}$ 25

In Exercises 27–30, use $\log_a(20) = 1.4406$ and $\log_a(5) = 0.7740$ to evaluate each expression.

27. $\log_a(100)$ 2.2146
28. $\log_a(4)$ 0.6666
29. $\log_a 5^3$ 2.322
30. $\log_a \sqrt{20}$ 0.7203

In Exercises 31–34, rewrite each expression as a sum, difference, or product of logarithms, and simplify if possible.

31. $\ln \dfrac{3x - 2}{x + 1}$ $\ln(3x - 2) - \ln(x + 1)$

32. $\log[x^3(3x - 4)^5]$ $3 \log x + 5 \log(3x - 4)$

33. $\log_3 \dfrac{\sqrt[3]{4x + 1}}{4x^2}$

34. $\log_3 \dfrac{\sqrt[3]{3x - 1}}{5x^2}$

In Exercises 35–38, rewrite each expression as a single logarithm.

35. $3 \log_2 x + \log_2 y$ $\log_2 x^3 y$

36. $\log x - \dfrac{1}{3} \log y$ $\log \dfrac{x}{\sqrt[3]{y}}$

37. $4 \ln(2a) - \ln b$ $\ln \dfrac{(2a)^4}{b}$

38. $6 \ln(5y) + 2 \ln x$ $\ln(15{,}625 x^2 y^6)$

EXERCISES 5.2

Answers that are not seen can be found in the answer section at the back of the text.

39. **Life Span** On the basis of data and projections for the years 1920 through 2050, the expected life span of people in the United States can be described by the function $f(x) = 8.77 + 14.9 \ln x$ years, where x is the number of years from 1900 to the person's birth year.

   a. What does this model estimate the life span to the nearest year to be for people born in 1925? 57

   b. What does it predict the life span to the nearest year to be for people born in 2045? 83

   c. Explain why these numbers are so different. (*Source:* National Center for Health Statistics) Improved health care and better diet. (Answers may vary.)

40. **U.S. Population** Using data for selected years from 1990 and projected to 2050, the total U.S. population can be modeled by $y = -363.3 + 166.9 \ln x$, with x equal to the number of years after 1950 and y in millions. Use the model to predict the U.S. population in 2026 and in 2033. 359.5 million; 374.2 million

41. **Women in the Workforce** Using data from 1970 and projected to 2050, the number of millions of women in the civilian workforce can be modeled by $y = -35.700 + 27.063 \ln x$, where x is the number of years after 1960.

   a. What does this model predict the number of women in the workforce will be in 2025? 77.271 million

   b. Is this function increasing or decreasing? Increasing

   c. Graph this function from $x = 10$ to $x = 90$.

42. **Loan Repayment** The number of years t that it takes to pay off a $100,000 loan at 10% interest by making annual payments of R dollars is

   $$t = \frac{\ln R - \ln(R - 10{,}000)}{\ln 1.1}, R > 10{,}000$$

   If the annual payment is $16,274.54, in how many years will the loan be paid off? 10 yr

43. **Supply** Suppose that the supply of a product is given by $p = 20 + 6 \ln(2q + 1)$, where p is the price per unit and q is the number of units supplied. What price will give a supply of 5200 units? $75.50

44. **Demand** Suppose that the demand function for a product is $p = \dfrac{500}{\ln(q + 1)}$, where p is the price per unit and q is the number of units demanded. What price will give a demand for 6400 units? $57.05

45. **Female Workers** For the years 1970 to 2040, the percent of females over 15 years old in the workforce is given or projected by $y = 27.4 + 5.02 \ln x$, where x is the number of years from 1960.

   a. What does the model predict the percent to be in 2011? In 2015? 47.1%; 47.5%

   b. Is the percent of female workers increasing or decreasing? Increasing

46. **Obesity** The percent of Americans who are obese from 1990 projected through 2030 can be modeled by $y = -31.732 + 18.672 \ln x$, with x equal to the number of years after 1980.

   a. Use the model to predict the percent of Americans who will be obese in 2032. 42%

   b. Does this model predict that obesity will get better or worse in the near future? Worse (The percent will increase.)

47. **Doubling Time** If $4000 is invested in an account earning 10% annual interest, compounded continuously, then the number of years that it takes for the amount to grow to $8000 is $n = \dfrac{\ln 2}{0.10}$. Find the number of years. ≈6.9 yr

48. **Doubling Time** If $5400 is invested in an account earning 7% annual interest, compounded continuously, then the number of years that it takes for the amount to grow to $10,800 is $n = \dfrac{\ln 2}{0.07}$. Find the number of years. ≈ 9.9 yr

49. **Doubling Time** The number of quarters (a quarter equals 3 months) needed to double an investment when a lump sum is invested at 8%, compounded quarterly, is given by $n = \dfrac{\log 2}{0.0086}$. In how many years will the investment double? 35 quarters, or $8\frac{3}{4}$ yr

50. **Doubling Time** The number of periods needed to double an investment when a lump sum is invested at 12%, compounded semiannually, is $n = \dfrac{\log 2}{0.0253}$. How many years pass before the investment doubles in value? ≈6 yr

The number of years t it takes for an investment to double if it earns r percent (as a decimal), compounded annually, is $t = \dfrac{\ln 2}{\ln(1 + r)}$. *Use this formula in Exercises 51 and 52.*

51. **Investing** In how many years will an investment double if it is invested at 8%, compounded annually? 9 yr

52. *Investing* In how many years will an investment double if it is invested at 12.3%, compounded annually? 6 yr

53. *Earthquakes* If an earthquake has an intensity of 25,000 times I_0, what is the magnitude of the earthquake? 4.4

54. *Earthquakes*
    a. If an earthquake has an intensity of 250,000 times I_0, what is the magnitude of the earthquake? 5.4
    b. Compare the magnitudes of two earthquakes when the intensity of one is 10 times the intensity of the other. (See Exercise 53.)
    The magnitude of the stronger earthquake is 1 unit more.

55. *Earthquakes* The Richter scale measurement for the southern Sumatra earthquake of 2007 was 6.4. Find the intensity of this earthquake as a multiple of I_0. $10^{6.4}I_0 \approx 2{,}511{,}886 I_0$

56. *Earthquakes* The Richter scale measurement for the San Francisco earthquake that occurred in 1906 was 8.25. Find the intensity of this earthquake as a multiple of I_0. $10^{8.25}I_0 \approx 177{,}827{,}941 I_0$

57. *Earthquakes* The Richter scale reading for the San Francisco earthquake of 1989 was 7.1. Find the intensity I of this earthquake as a multiple of I_0. $10^{7.1}I_0 \approx 12{,}589{,}254 I_0$

58. *Earthquakes* On January 9, 2008, an earthquake in northern Algeria measuring 4.81 on the Richter scale killed 1 person, and on May 12, 2008, an earthquake in China measuring 7.9 on the Richter scale killed 67,180 people. How many times more intense was the earthquake in China than the one in northern Algeria? 1230 times as intense

59. *Earthquakes* The largest earthquake ever to strike San Francisco (in 1906) measured 8.25 on the Richter scale, and the second largest (in 1989) measured 7.1. How many times more intense was the 1906 earthquake than the 1989 earthquake?
    About 14.1 times as intense

60. *Earthquakes* The largest earthquake ever recorded in Japan occurred in 2011 and measured 9.0 on the Richter scale. An earthquake measuring 8.25 occurred in Japan in 1983. Calculate the ratio of the intensities of these earthquakes.
    Intensity of stronger is 5.62 times the intensity of other.

61. *Japan Tsunami* In May of 2008, an earthquake measuring 6.8 on the Richter scale struck near the east coast of Honshu, Japan. In March of 2011, an earthquake measuring 9.0 struck that same region, causing a devastating tsunami that killed thousands of people. The 2011 earthquake was how many times as intense as the one in 2008? 158.5 times as intense

Use the following information to answer the questions in Exercises 62–67. To quantify the intensity of sound, the decibel scale (named for Alexander Graham Bell) was developed. The formula for loudness L on the decibel scale is $L = 10 \log \dfrac{I}{I_0}$, where I_0 is the intensity of sound just below the threshold of hearing, which is approximately 10^{-16} watt per square centimeter.

62. *Decibel Scale* Find the decibel reading for a sound with intensity 20,000 times I_0. 43

63. *Decibel Scale* Compare the decibel readings of two sounds if the intensity of the first sound is 100 times the intensity of the second sound.
    Decibel reading of higher intensity sound is 20 more than the other.

64. *Decibel Scale* Use the exponential form of the function $L = 10 \log \dfrac{I}{I_0}$ to find the intensity of a sound if its decibel reading is 40. $10{,}000 I_0$

65. *Decibel Scale* Use the exponential form of $L = 10 \log \dfrac{I}{I_0}$ to find the intensity of a sound if its decibel reading is 140. (Sound at this level causes pain.) $10^{14} I_0$

66. *Decibel Scale* The intensity of a whisper is $115 I_0$, and the intensity of a busy street is $9{,}500{,}000 I_0$. How do their decibel levels compare? Differ by 49

67. *Decibel Scale* If the decibel reading of a sound that is painful is 140 and the decibel reading of rock music is 120, how much more intense is the painful sound than the rock music?
    Intensity of painful sound is 100 times the intensity of other.

To answer the questions in Exercises 68–72, use the fact that pH is given by the formula $\text{pH} = -\log[H^+]$.

68. *pH Levels* Find the pH value of beer for which $[H^+] = 0.0000631$. 4.2

69. pH Levels The pH value of eggs is 7.79. Find the hydrogen-ion concentration for eggs.
$[H^+] = 10^{-7.79} \approx 0.0000000162$

70. pH Levels The most common solutions have a pH range between 1 and 14. Use an exponential form of the pH formula to find the values of $[H^+]$ associated with these pH levels. $10^{-14} \leq [H^+] \leq 10^{-1}$

71. pH Levels Lower levels of pH indicate a more acidic solution. If the pH of ketchup is 3.9 and the pH of peanut butter is 6.3, how much more acidic is ketchup than peanut butter? 251.2 times as acidic

72. pH Levels Pure seawater in the middle of the ocean has a pH of 8.3. The pH of seawater in an aquarium must vary between 8 and 8.5; beyond these values, animals will experience certain physiological problems. If the pH of water in an aquarium is 8, how much more acidic is this water than pure seawater?
2 times as acidic

5.3 Exponential and Logarithmic Equations

KEY OBJECTIVES
- Solve an exponential equation by writing it in logarithmic form
- Convert logarithms using the change-of-base formula
- Solve an exponential equation by using properties of logarithms
- Solve logarithmic equations
- Solve exponential and logarithmic inequalities

SECTION PREVIEW Carbon-14 Dating

This is an artist's interpretation of the massive seabird *Pelagornis sandersi*, known from a 25-million-year-old fossil excavated in South Carolina. It had a wingspan of 24 feet.

Carbon-14 dating is one process by which scientists can tell the age of many fossils. To find the age of a fossil that originally contained 1000 grams of carbon-14 and now contains 1 gram, we solve the equation

$$1 = 1000e^{-0.00012097t}$$

An approximate solution of this equation can be found graphically, but sometimes a window that shows the solution is hard to find. If this is the case, it may be easier to solve an exponential equation like this one algebraically.

In this section, we consider two algebraic methods of solving exponential equations. The first method involves converting the equation to logarithmic form and then solving the logarithmic equation for the variable. The second method involves taking the logarithm of both sides of the equation and using the properties of logarithms to solve for the variable. ■

Solving Exponential Equations Using Logarithmic Forms

When we wish to solve an equation for a variable that is contained in an exponent, we can remove the variable from the exponent by converting the equation to its logarithmic form. The steps in this solution method follow.

> **Solving Exponential Equations Using Logarithmic Forms**
>
> To solve an exponential equation using logarithmic form:
>
> 1. Rewrite the equation with the term containing the exponent by itself on one side.
> 2. Divide both sides by the coefficient of the term containing the exponent.
> 3. Change the new equation to logarithmic form.
> 4. Solve for the variable.

This method is illustrated in the following example.

EXAMPLE 1 ▶ **Solving a Base 10 Exponential Equation**

a. Solve the equation $3000 = 150(10^{4t})$ for t by converting it to logarithmic form.

b. Solve the equation graphically to confirm the solution.

SOLUTION

a. 1. The term containing the exponent is by itself on one side of the equation.

   2. We divide both sides of the equation by 150:

   $$3000 = 150(10^{4t})$$
   $$20 = 10^{4t}$$

   3. We rewrite this equation in logarithmic form:

   $$4t = \log_{10} 20, \quad \text{or} \quad 4t = \log 20$$

   4. Solving for t gives the solution:

   $$t = \frac{\log 20}{4} \approx 0.32526$$

b. We solve the equation graphically by using the intersection method. Entering $y_1 = 3000$ and $y_2 = 150(10^{4t})$, we graph using the window $[0, 0.4]$ by $[0, 3500]$ and find the point of intersection to be about $(0.32526, 3000)$ (Figure 5.23). Thus, the solution to the equation $3000 = 150(10^{4t})$ is $t \approx 0.32526$, as we found in part (a).

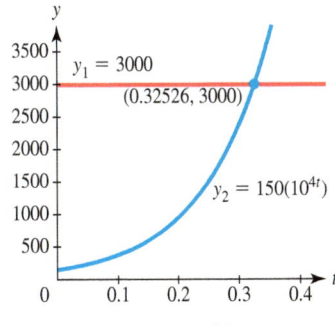

Figure 5.23

EXAMPLE 2 ▶ Doubling Time

a. Prove that the time it takes for an investment to double its value is $t = \dfrac{\ln 2}{r}$ if the interest rate is r, compounded continuously.

b. Suppose $2500 is invested in an account earning 6% annual interest, compounded continuously. How long will it take for the amount to grow to $5000?

SOLUTION

a. If P dollars are invested for t years at an annual interest rate r, compounded continuously, then the investment will grow to a future value S according to the equation

$$S = Pe^{rt}$$

and the investment will be doubled when $S = 2P$, giving

$$2P = Pe^{rt}, \quad \text{or} \quad 2 = e^{rt}$$

Converting this exponential equation to its equivalent logarithmic form gives

$$\log_e 2 = rt, \quad \text{or} \quad \ln 2 = rt$$

and solving this equation for t gives the doubling time,

$$t = \dfrac{\ln 2}{r}$$

b. We seek the time it takes to double the investment, and the doubling time is given by

$$t = \dfrac{\ln 2}{0.06} \approx 11.5525$$

Thus, it will take approximately 11.6 years for an investment of $2500 to grow to $5000 if it is invested at an annual rate of 6%, compounded continuously. We can confirm this solution graphically by finding the intersection of the graphs of $Y_1 = 2500e^{0.06x}$ and $Y_2 = 5000$, or numerically (see Figure 5.24). Note that this is the same time it would take any investment at 6%, compounded continuously, to double.

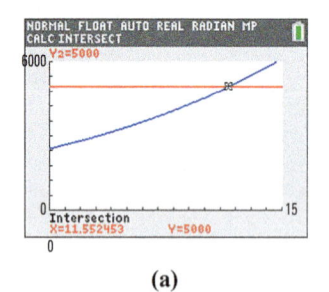

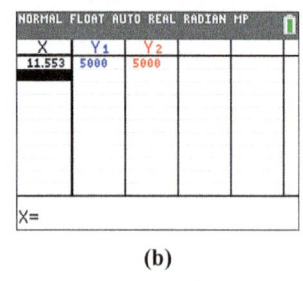

Figure 5.24

EXAMPLE 3 ▶ Carbon-14 Dating

Radioactive carbon-14 decays according to the equation

$$y = y_0 e^{-0.00012097t}$$

where y_0 is the original amount and y is the amount of carbon-14 at time t years. To find the age of a fossil if the original amount of carbon-14 was 1000 grams and the present amount is 1 gram, we solve the equation

$$1 = 1000 e^{-0.00012097t}$$

a. Find the age of the fossil by converting the equation to logarithmic form.

b. Find the age of the fossil using graphical methods.

SOLUTION

a. To write the equation so that the coefficient of the exponential term is 1, we divide both sides by 1000, obtaining

$$0.001 = e^{-0.00012097t}$$

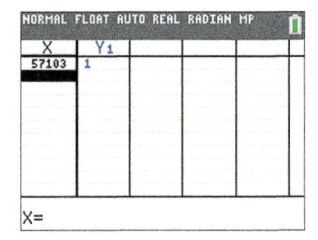

Writing this equation in logarithmic form gives the equation

$$-0.00012097t = \log_e 0.001, \quad \text{or} \quad -0.00012097t = \ln 0.001$$

Solving the equation for t gives

$$t = \frac{\ln 0.001}{-0.00012097} \approx 57{,}103$$

Thus, the age of the fossil is calculated to be about 57,103 years.

We can confirm that this is the solution to $1 = 1000e^{-0.00012097t}$ by using a table of values with $y_1 = 1000e^{-0.00012097t}$. The result of entering 57,103 for x is shown in Figure 5.25.

b. We can solve the equation graphically by using the intersection method. Entering

$$y_1 = 1000e^{-0.00012097t} \quad \text{and} \quad y_2 = 1$$

and investigating some y-values shows that we need very large x-values so that y is near 1 (see Figure 5.26(a)). Graphing using the window [0, 80000] by [−1, 3], we find the x-coordinate of the intersection point to be 57,103.044 (see Figure 5.26(b).) Thus, as we found in part (a), the age of the fossil is about 57,103 years.

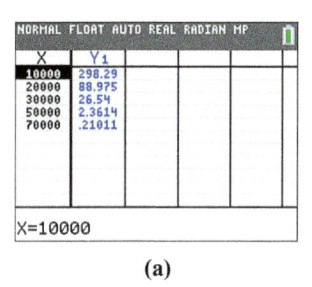

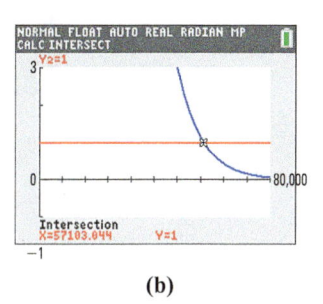

(a)   (b)

Figure 5.26

While finding this solution from a graph is easy, actually finding a window that contains the intersection point is not. It may be easier to solve the equation algebraically, as we did in part (a), and then use graphical or numerical methods to confirm the solution.

Change of Base

Most calculators and other graphing utilities can be used to evaluate logarithms and graph logarithmic functions if the base is 10 or e. Graphing logarithmic functions with other bases usually requires converting to exponential form to determine outputs and then graphing the functions by plotting points by hand. However, we can use a special formula called the **change-of-base formula** to rewrite logarithms so that the base is 10 or e. The formula for changing from base a to base b is developed below.

Suppose $y = \log_a x$. Then

$$a^y = x$$

$\log_b a^y = \log_b x$ Take logarithm, base b, of both sides (Property 5 of Logarithms).

$y \log_b a = \log_b x$ Use the Power Property of Logarithms.

$y = \dfrac{\log_b x}{\log_b a}$ Solve for y.

$\log_a x = \dfrac{\log_b x}{\log_b a}$ Substitute $\log_a x$ for y.

The general change-of-base formula is summarized on the next page.

366 Chapter 5 Exponential and Logarithmic Functions

> **Change-of-Base Formula**
>
> If $b > 0, b \neq 1, a > 0, a \neq 1$, and $x > 0$, then
>
> $$\log_a x = \frac{\log_b x}{\log_b a}$$
>
> In particular, for base 10 and base e,
>
> $$\log_a x = \frac{\log x}{\log a} \quad \text{and} \quad \log_a x = \frac{\ln x}{\ln a}$$

EXAMPLE 4 ▶ Applying the Change-of-Base Formula

a. Evaluate $\log_8 124$.

Graph the functions in parts (b) and (c) by changing each logarithm to a common logarithm and then by changing the logarithm to a natural logarithm.

b. $y = \log_3 x$ **c.** $y = \log_2(-3x)$

SOLUTION

a. The change-of-base formula (changing to base 10) can be used to evaluate this logarithm:

$$\log_8 124 = \frac{\log 124}{\log 8} = 2.318 \text{ approximately}$$

b. Changing $y = \log_3 x$ to base 10 gives $\log_3 x = \frac{\log x}{\log 3}$, so we graph $y = \frac{\log x}{\log 3}$ (Figure 5.27(a)). Changing to base e gives $\log_3 x = \frac{\ln x}{\ln 3}$, so we graph $y = \frac{\ln x}{\ln 3}$ (Figure 5.27(b)). Note that the graphs appear to be identical (as they should be because they both represent $y = \log_3 x$).

(a)   (b)

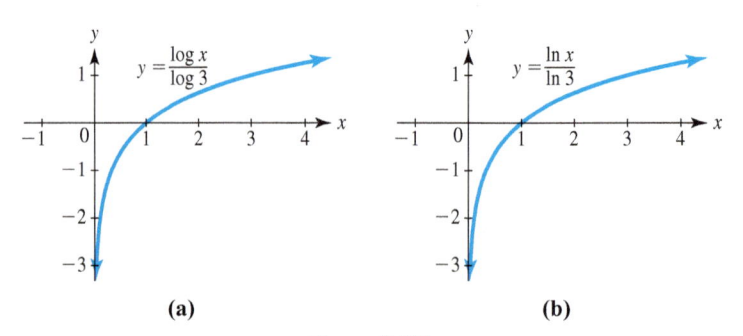

Figure 5.27

c. Changing $y = \log_2(-3x)$ to base 10 gives $\log_2(-3x) = \frac{\log(-3x)}{\log 2}$, so we graph $y = \frac{\log(-3x)}{\log 2}$ (Figure 5.28(a)). Changing to base e gives $\log_2(-3x) = \frac{\ln(-3x)}{\ln 2}$, so we graph $y = \frac{\ln(-3x)}{\ln 2}$ (Figure 5.28(b)). Note that the graphs are identical. Also notice that because logarithms are defined only for positive inputs, x must be negative so that $-3x$ is positive. Thus, the domain of $y = \log_2(-3x)$ is the interval $(-\infty, 0)$.

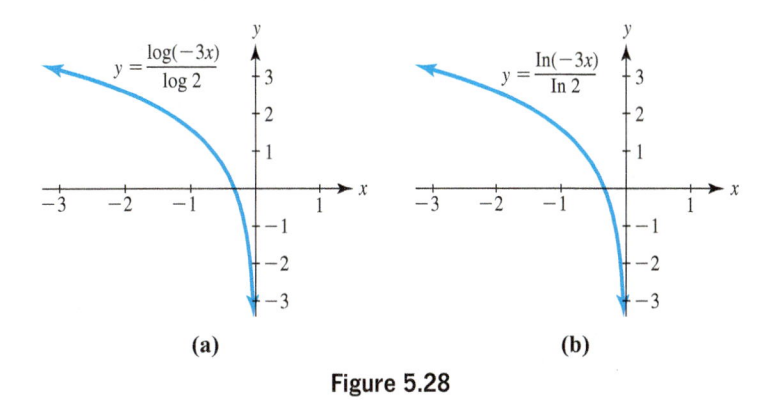

(a) $y = \dfrac{\log(-3x)}{\log 2}$

(b) $y = \dfrac{\ln(-3x)}{\ln 2}$

Figure 5.28

The change-of-base formula is also useful in solving exponential equations whose base is neither 10 nor e.

EXAMPLE 5 ▶ Investment

If $10,000 is invested for t years at 10%, compounded annually, the future value is given by

$$S = 10{,}000(1.10^t)$$

In how many years will the investment grow to $45,950?

SOLUTION

We seek to solve the equation

$$45{,}950 = 10{,}000(1.10^t)$$

Dividing both sides by 10,000 gives

$$4.5950 = 1.10^t$$

Writing the equation in logarithmic form solves the equation for t:

$$t = \log_{1.10} 4.5950$$

Using the change-of-base formula gives the number of years:

$$t = \log_{1.10} 4.5950 = \frac{\log 4.5950}{\log 1.10} = 16$$

Thus, $10,000 will grow to $45,950 in 16 years.

Solving Exponential Equations Using Logarithmic Properties

With the properties of logarithms at our disposal, we consider a second method for solving exponential equations, which may be easier to use. This method of solving an exponential equation involves taking the logarithm of both sides of the equation (Property 5 of Logarithms) and then using properties of logarithms to write the equation in a form that we can solve.

Solving Exponential Equations Using Logarithmic Properties

To solve an exponential equation using logarithmic properties:

1. Rewrite the equation with a base raised to a power on one side.
2. Take the logarithm, base e or 10, of both sides of the equation.
3. Use a logarithmic property to remove the variable from the exponent.
4. Solve for the variable.

This method is illustrated in the following example.

EXAMPLE 6 ▶ Solution of Exponential Equations

Solve the following exponential equations.

a. $4096 = 8^{2x}$ **b.** $6(4^{3x-2}) = 120$

SOLUTION

a. 1. This equation has the base 8 raised to a variable power on one side.

2. Taking the logarithm, base 10, of both sides of the equation $4096 = 8^{2x}$ gives

$$\log 4096 = \log 8^{2x}$$

3. Using the Power Property of Logarithms removes the variable x from the exponent:

$$\log 4096 = 2x \log 8$$

4. Solving for x gives the solution:

$$\frac{\log 4096}{2 \log 8} = x$$

$$x = 2$$

b. We first isolate the exponential expression on one side of the equation by dividing both sides by 6:

$$\frac{6(4^{3x-2})}{6} = \frac{120}{6}$$

$$4^{3x-2} = 20$$

Taking the natural logarithm of both sides leads to the solution:

$$\ln 4^{3x-2} = \ln 20$$

$$(3x - 2) \ln 4 = \ln 20$$

$$3x - 2 = \frac{\ln 20}{\ln 4}$$

$$x = \frac{1}{3}\left(\frac{\ln 20}{\ln 4} + 2\right) \approx 1.387$$

An alternative method of solving the equation is to write $4^{3x-2} = 20$ in logarithmic form:

$$\log_4 20 = 3x - 2$$

$$x = \frac{\log_4 20 + 2}{3}$$

The change-of-base formula can be used to compute this value and verify that this solution is the same as above:

$$x = \frac{\frac{\ln 20}{\ln 4} + 2}{3} \approx 1.387$$

Solution of Logarithmic Equations

Some logarithmic equations can be solved by converting to exponential form.

EXAMPLE 7 ▶ Disposable Income Per Capita

By using U.S. Energy Information Administration data for selected years from 2010 and projected to 2040, the U.S. real disposable income per capita (in thousands of dollars) can be modeled by

$$I(t) = 32.11(1.014)^t$$

In what year does the model predict that the disposable income will reach $39,000?

SOLUTION

We solve $39 = 32.11(1.014^t)$ as follows.

$$39 = 32.11(1.014^t)$$

$$\frac{39}{32.11} = 1.014^t$$

$$\ln \frac{39}{32.11} = t(\ln 1.014)$$

$$\frac{\ln \frac{39}{32.11}}{\ln 1.014} = t$$

$$t \approx 13.98$$

So the U.S. real disposable income will reach $39,000 in 2024, according to the model.

We can also solve this equation by graphing $Y_1 = 32.11(1.014^x)$ and $Y_2 = 39$ and finding the point of intersection. See Figure 5.29.

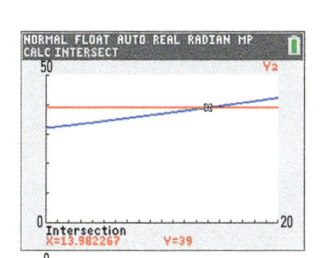

Figure 5.29

EXAMPLE 8 ▶ Solving a Logarithmic Equation

a. Solve $6 + 3 \ln x = 12$ by writing the equation in exponential form.

b. Solve the equation graphically.

SOLUTION

a. We first solve the equation for $\ln x$:

$$6 + 3 \ln x = 12$$

$$3 \ln x = 6$$

$$\ln x = 2$$

Writing $\log_e x = 2$ in exponential form gives

$$x = e^2$$

b. We solve $6 + 3 \ln x = 12$ graphically by the intersection method. Entering $y_1 = 6 + 3 \ln x$ and $y_2 = 12$, we graph using the window $[-2, 10]$ by $[-10, 15]$ and find the point of intersection to be about $(7.38906, 12)$ (Figure 5.30). Thus, the solution to the equation $6 + 3 \ln x = 12$ is $x \approx 7.38906$. Because $e^2 \approx 7.38906$, the solutions agree.

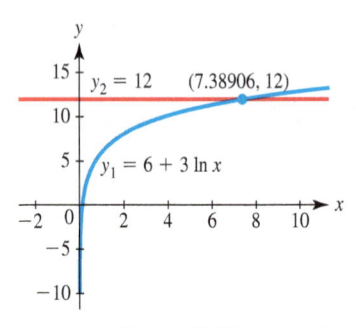

Figure 5.30

The properties of logarithms are frequently useful in solving logarithmic equations. Consider the following example.

EXAMPLE 9 ▶ Solving a Logarithmic Equation

Solve $\ln x + 3 = \ln(x + 4)$ by converting the equation to exponential form and then using algebraic methods.

SOLUTION

We first write the equation with the logarithmic expressions on one side:

$$\ln x + 3 = \ln(x + 4)$$
$$3 = \ln(x + 4) - \ln x$$

Using the Quotient Property of Logarithms gives

$$3 = \ln \frac{x + 4}{x}$$

Writing the equation in exponential form gives

$$e^3 = \frac{x + 4}{x}$$

We can now solve for x:

$$e^3 x = x + 4$$
$$e^3 x - x = 4$$
$$x(e^3 - 1) = 4$$
$$x = \frac{4}{e^3 - 1} \approx 0.21$$

EXAMPLE 10 ▶ Climate Change

In an effort to reduce climate change, it has been proposed that a tax be levied based on the emissions of carbon dioxide into the atmosphere. The cost–benefit equation $\ln(1 - P) = -0.0034 - 0.0053t$ estimates the relationship between the percent reduction of emissions of carbon dioxide P (as a decimal) and the tax t in dollars per ton of carbon.
(*Source*: W. Clime, *The Economics of Global Warming*)

a. Solve the equation for P, the estimated percent reduction in emissions.

b. Determine the estimated percent reduction in emissions if a tax of $100 per ton is levied.

SOLUTION

a. We solve $\ln(1 - P) = -0.0034 - 0.0053t$ for P by writing the equation in exponential form.

$$(1 - P) = e^{-0.0034 - 0.0053t}$$
$$P = 1 - e^{-0.0034 - 0.0053t}$$

b. Substituting 100 for t gives $P = 0.4134$, so a $100 tax per ton of carbon is estimated to reduce carbon dioxide emissions by 41.3%.

Some exponential and logarithmic equations are difficult or impossible to solve algebraically, and finding approximate solutions to real data problems is frequently easier with graphical methods.

Exponential and Logarithmic Inequalities

Inequalities involving exponential and logarithmic functions can be solved by solving the related equation algebraically and then investigating the inequality graphically. Consider the following example.

EXAMPLE 11 ▶ Sales Decay

After the end of an advertising campaign, the daily sales of Genapet fell rapidly, with daily sales given by $S = 3200e^{-0.08x}$ dollars, where x is the number of days from the end of the campaign. For how many days after the campaign ended were sales at least $1980?

SOLUTION

To solve this problem, we find the solution to the inequality $3200e^{-0.08x} \geq 1980$. We begin our solution by solving the related equation:

$$3200e^{-0.08x} = 1980$$
$$e^{-0.08x} = 0.61875 \quad \text{Divide both sides by 3200.}$$
$$\ln e^{-0.08x} = \ln 0.61875 \quad \text{Take the logarithm, base } e, \text{ of both sides.}$$
$$-0.08x \approx -0.4801 \quad \text{Use the Power Property of Logarithms.}$$
$$x \approx 6$$

We now investigate the inequality graphically.
 To solve this inequality graphically, we graph $y_1 = 3200e^{-0.08x}$ and $y_2 = 1980$ on the same axes, with nonnegative x-values, since x represents the number of days. The graph shows that $y_1 = 3200e^{-0.08x}$ is above $y_2 = 1980$ for $x < 6$ (Figure 5.31).

Thus, the daily revenue is at least $1980 for each of the first 6 days after the end of the advertising campaign.

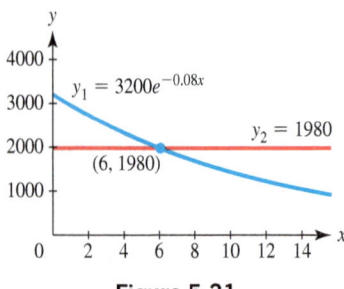

Figure 5.31

EXAMPLE 12 ▶ Cost–Benefit

The cost–benefit equation $\ln(1 - P) = -0.0034 - 0.0053t$ estimates the relationship between the percent reduction of emissions of carbon dioxide P (as a decimal) and the tax t in dollars per ton of carbon. What tax will give a reduction of at least 50%?

SOLUTION

To find the tax, we find the value of t that gives $P = 0.50$:

$$\ln(1 - 0.50) = -0.0034 - 0.0053t$$
$$-0.6931 = -0.0034 - 0.0053t$$
$$0.0053t = 0.6897$$
$$t = 130.14$$

In Example 10, we solved this cost–benefit for P, getting $P = 1 - e^{-0.0034 - 0.0053t}$. The graphs of this equation and $P = 0.50$ on the same axes are shown in Figure 5.32. The point of intersection of the graphs is $(130.14, 0.5)$, and the graph shows that the percent reduction of emissions is more than 50% if the tax is above $130.14 per ton of carbon.

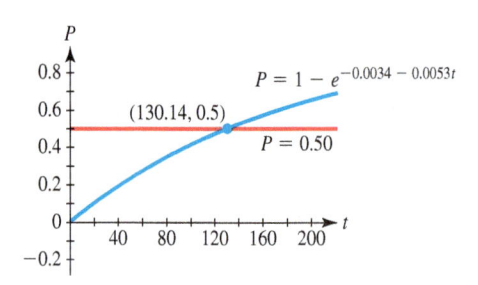

Figure 5.32

SKILLS CHECK 5.3

In Exercises 1–10, solve the equations algebraically and check graphically. Round to three decimal places.

1. $1600 = 10^x$ $x \approx 3.204$
2. $4600 = 10^x$ $x \approx 3.663$
3. $2500 = e^x$ $x \approx 7.824$
4. $54.6 = e^x$ $x \approx 4.000$
5. $8900 = e^{5x}$ $x \approx 1.819$
6. $2400 = 10^{8x}$ $x \approx 0.423$
7. $4000 = 200e^{8x}$ $x \approx 0.374$
8. $5200 = 13e^{12x}$ $x \approx 0.499$
9. $8000 = 500(10^x)$ $x \approx 1.204$
10. $9000 = 400(10^x)$ $x \approx 1.352$

In Exercises 11–14, use a change-of-base formula to evaluate each logarithm. Give your answers rounded to four decimal places.

11. $\log_6 18$ 1.6131
12. $\log_7 215$ 2.7600
13. $\log_8 \sqrt{2}$ 0.1667
14. $\log_4 \sqrt[3]{10}$ 0.5537

In Exercises 15–22, solve the equations.

15. $8^x = 1024$ $x = \frac{10}{3}$
16. $9^x = 2187$ $x = 3.5$
17. $2(5^{3x}) = 31{,}250$ $x = 2$
18. $2(6^{2x}) = 2592$ $x = 2$
19. $5^{x-2} = 11.18$ $x \approx 3.5$
20. $3^{x-4} = 140.3$ $x \approx 8.5$
21. $18{,}000 = 30(2^{12x})$ $x \approx 0.769$
22. $5880 = 21(2^{3x})$ $x \approx 2.710$

In Exercises 23–36, solve the logarithmic equations.

23. $\log_2 x = 3$ $x = 8$
24. $\log_4 x = -2$ $x = \frac{1}{16}$
25. $5 + 2\ln x = 8$ $x = e^{1.5} \approx 4.482$
26. $4 + 3\log x = 10$ $x = 100$
27. $5 + \ln(8x) = 23 - 2\ln x$ $x = \frac{e^6}{2} \approx 201.71$
28. $3\ln x + 8 = \ln(3x) + 12.18$ $x \approx 14$
29. $2\log x - 2 = \log(x - 25)$ $x = 50$
30. $\ln(x - 6) + 4 = \ln x + 3$ $x \approx 9.49$
31. $\log_3 x + \log_3 9 = 1$ $x = \frac{1}{3}$
32. $\log_2 x + \log_2(x - 6) = 4$ $x = 8$
33. $\log_2 x = \log_2 5 + 3$ $x = 40$
34. $\log_2 x = 3 - \log_2 2x$ $x = 2$
35. $\log 3x + \log 2x = \log 150$ $x = 5$
36. $\ln(x + 2) + \ln x = \ln(x + 12)$ $x = 3$

In Exercises 37–40, solve the inequalities.

37. $3^x < 243$ $x < 5$
38. $7^x \geq 2401$ $x \geq 4$
39. $5(2^x) \geq 2560$ $x \geq 9$
40. $15(4^x) \leq 15{,}360$ $x \leq 5$

EXERCISES 5.3

Answers that are not seen can be found in the answer section at the back of the text.

41. Supply The supply function for a certain size boat is given by $p = 340(2^q)$ boats, where p dollars is the price per boat and q is the quantity of boats supplied at that price. What quantity will be supplied if the price is $10,880 per boat? 5

42. Demand The demand function for a dining room table is given by $p = 4000(3^{-q})$ dollars per table, where p is the price and q is the quantity, in thousands of tables, demanded at that price. What quantity will be demanded if the price per table is $256.60? 2.5 thousand

43. Sales Decay After a television advertising campaign ended, the weekly sales of Korbel champagne fell rapidly. Weekly sales in a city were given by $S = 25{,}000 e^{-0.072x}$ dollars, where x is the number of weeks after the campaign ended.

  a. Write the logarithmic form of this function. $\ln(S/25{,}000) = -0.072x$
  b. Use the logarithmic form of this function to find the number of weeks after the end of the campaign before weekly sales fell to $16,230. 6 weeks

44. Sales Decay After a television advertising campaign ended, sales of Genapet fell rapidly, with daily sales given by $S = 3200 e^{-0.08x}$ dollars, where x is the number of days after the campaign ended.

  a. Write the logarithmic form of this function. $\ln(S/3200) = -0.08x$
  b. Use the logarithmic form of this function to find the number of days after the end of the campaign before daily sales fell to $2145. 5 days

45. Sales Decay After the end of an advertising campaign, the daily sales of Genapet fell rapidly, with daily sales given by $S = 3200 e^{-0.08x}$ dollars, where x is the number of days from the end of the campaign.

  a. What were daily sales when the campaign ended? $3200
  b. How many days passed after the campaign ended before daily sales were below half of what they were at the end of the campaign? 9 days

46. Sales Decay After the end of a television advertising campaign, the weekly sales of Korbel champagne fell rapidly, with weekly sales given by $S = 25{,}000 e^{-0.072x}$ dollars, where x is the number of weeks from the end of the campaign.

  a. What were weekly sales when the campaign ended? $25,000
  b. How many weeks passed after the campaign ended before weekly sales were below half of what they were at the end of the campaign? 10 weeks

47. **Super Bowl Ads** A minute ad during Super Bowl VII in 1973 cost $200,000. The price tag for a 30-second ad slot during the 2011 Super Bowl was $3 million. The cost of a 30-second slot of advertising for the Super Bowl can be modeled by $y = 0.0000966(1.101^x)$, where x is the number of years after 1900 and y is the cost in millions.

   a. According to the model, what was the cost of a 30-second Super Bowl ad in 2000? $1,457,837

   b. The cost of a 30-second ad during Super Bowl XLIX in 2015 was $4.5 million. Does the model underestimate or overestimate the cost of these ads?
   Overestimate; it estimates the cost for 2015 at $6.2 million

48. **Population Growth** The population in a certain city was 53,000 in 2000, and its future size was predicted to be $P = 53,000e^{0.015t}$ people, where t is the number of years after 2000. Determine algebraically when the population was predicted to reach 60,000, and verify your solution graphically. 2009

49. **Purchasing Power** The purchasing power (real value of money) decreases if inflation is present in the economy. For example, the purchasing power of $40,000 after t years of 5% inflation is given by the model

   $$P = 40,000e^{-0.05t} \text{ dollars}$$

   How long will it take for the value of a $40,000 pension to have a purchasing power of $20,000 under 5% inflation? 13.86 yr
   (*Source: Viewpoints,* VALIC)

50. **Purchasing Power** If a retired couple has a fixed income of $60,000 per year, the purchasing power of their income (adjusted value of the money) after t years of 5% inflation is given by the equation $P = 60,000e^{-0.05t}$. In how many years will the purchasing power of their income be half of their current income? ≈ 13.86 yr

51. **Real Estate Inflation** During a 5-year period of constant inflation, the value of a $100,000 property increases according to the equation $v = 100,000e^{0.03t}$ dollars. In how many years will the value of this building be double its current value? 23.11 yr

52. **Real Estate Inflation** During a 10-year period of constant inflation, the value of a $200,000 property is given by the equation $v = 200,000e^{0.05t}$ dollars. In how many years will the value of this building be $254,250? ≈ 4.8 yr

53. **Radioactive Decay** The amount of radioactive isotope thorium-234 present in a certain sample at time t is given by $A(t) = 500e^{-0.02828t}$ grams, where t years is the time since the initial amount was measured.

   a. Find the initial amount of the isotope present in the sample. 500 g

   b. Find the half-life of this isotope. That is, find the number of years until half of the original amount of the isotope remains. ≈ 24.51 yr

54. **Radioactive Decay** The amount of radioactive isotope thorium-234 present in a certain sample at time t is given by $A(t) = 500e^{-0.02828t}$ grams, where t years is the time since the initial amount was measured. How long will it take for the amount of isotope to equal 318 grams? ≈ 16 yr

55. **Drugs in the Bloodstream** The concentration of a drug in the bloodstream from the time the drug is injected until 8 hours later is given by

   $$y = 100(1 - e^{-0.312(8-t)}) \text{ percent, } t \text{ in hours,}$$

   where the drug is administered at time $t = 0$. In how many hours will the drug concentration be 79% of the initial dose? 3 hr

56. **Drugs in the Bloodstream** If a drug is injected into the bloodstream, the percent of the maximum dosage that is present at time t is given by $y = 100(1 - e^{-0.35(10-t)})$, where t is in hours, with $0 \le t \le 10$. In how many hours will the percent reach 65%? ≈ 7 hr

57. **International Air Travel** The number of passengers using international air traffic into and out of the United States was projected by the Federal Aviation Administration to grow at 4% per year from 2012 until 2032. If the number was 172 million in 2012, the equation of the exponential function that models this projection of the number of millions of passengers y is $y = 172(1.04^x)$, where x is the number of years after 2012. When does the model project that the number of passengers using international air traffic into and out of the United States will be 335 million? 17 years after 2012, in 2029

58. **Radioactive Decay** A breeder reactor converts stable uranium-238 into the isotope plutonium-239. The decay of this isotope is given by $A(t) = A_0 e^{-0.00002876t}$, where $A(t)$ is the amount of the isotope at time t (in years) and A_0 is the original amount. If the original amount is 100 pounds, find the half-life of this isotope. ≈ 24,101 yr

59. **Cost** Suppose the weekly cost for the production of x units of a product is given by $C(x) = 3452 + 50 \ln(x + 1)$ dollars. Use graphical methods to estimate the number of units produced if the total cost is $3556. 7 units

60. **Supply** Suppose the daily supply function for a product is $p = 31 + \ln(x + 2)$, where p is in dollars and x is the number of units supplied. Use graphical methods to estimate the number of units that will be supplied if the price is $35.70. Approximately 108 units

61. **Doubling Time** If P dollars are invested at an annual interest rate r, compounded annually for t years, the future value of the investment is given by $S = P(1.07)^t$. Find a formula for the number of years it will take to double the initial investment. $t = \dfrac{\ln 2}{\ln(1.07)}$

62. **Doubling Time** The future value of a lump sum P that is invested for n years at 10%, compounded annually, is $S = P(1.10)^n$. Show that the number of years it would take for this investment to double is $n = \log_{1.10} 2$.

63. **Investment** At the end of t years, the future value of an investment of $20,000 at 7%, compounded annually, is given by $S = 20{,}000(1 + 0.07)^t$. In how many years will the investment grow to $48,196.90? 13 yr

64. **Investment** At the end of t years, the future value of an investment of $30,000 at 9%, compounded annually, is given by $S = 30{,}000(1 + 0.09)^t$. In how many years will the investment grow to $129,829? 17 yr

65. **Investing** Find in how many years $40,000 invested at 10%, compounded annually, will grow to $64,420.40. 5 yr

66. **Investing** In how many years will $40,000 invested at 8%, compounded annually, grow to $86,357? 10 yr

67. **Life Span** Based on data from 1920 and projected to 2050, the expected life span of people in the United States can be described by the function $f(x) = 8.77 + 14.9 \ln x$ where x is the number of years from 1900.

   a. Use algebraic methods to find the birth year for which the expected life span reached 78 years. 2005

   b. Use graphical methods to determine the birth year for which the expected life span is 78 years. Does this agree with the solution in part (a)? 2005; yes
   (*Source*: National Center for Health Statistics)

68. **Supply** Suppose that the supply of a product is given by $p = 20 + 6\ln(2q + 1)$, where p is the price per unit and q is the number of units supplied. How many units will be supplied if the price per unit is $68.04? 1500 units

69. **Alzheimer's Disease** The millions of U.S. citizens with Alzheimer's disease from 2000 and projected to 2050 can be modeled by $y = 3.294(1.025^x)$, with x equal to the number of years after 1990 and y equal to the millions of Americans with Alzheimer's disease. In what year does the model predict the number of U.S. citizens with Alzheimer's disease will be 8 million? (*Source*: National Academy on an Aging Society) 2026

70. **Demand** Suppose that the demand function for a product is $p = \dfrac{500}{\ln(q+1)}$, where p is the price per unit and q is the number of units demanded. How many units will be demanded if the price is $61.71 per unit? 3301 units

71. **China's Shale-Natural Gas** The estimated annual production of shale-natural gas in China for the years 2013 through 2020, in billions of cubic feet, can be modeled by $y = 0.0131(1.725^x)$, with x equal to the number of years x after 2010. In what year does the model predict that the number of billions of cubic feet of shale-natural gas in China will be 1.77? 2019 (*Source*: Sanford C. Bernstein)

72. **Global Warming** In an effort to reduce global warming, it has been proposed that a tax be levied based on the emissions of carbon dioxide into the atmosphere. The cost–benefit equation $\ln(1 - P) = -0.0034 - 0.0053t$ estimates the relationship between the percent reduction of emissions of carbon dioxide P (as a decimal) and the tax t in dollars per ton of carbon dioxide.

   a. Solve the equation for t, giving t as a function of P. Graph the function. $t = \dfrac{-[\ln(1-P) + 0.0034]}{0.0053}$

   b. Use the equation in part (a) to find what tax will give a 30% reduction in emissions. $66.66 per ton of carbon
   (*Source*: W. Clime, *The Economics of Global Warming*)

73. **Rule of 72** The "Rule of 72" is a simplified way to determine how long an investment will take to double, given a fixed annual rate of interest. By dividing 72 by the annual interest rate, investors can get a rough estimate of how many years it will take for the initial investment to double. Algebraically we know that the time it takes an investment to double is $\dfrac{\ln 2}{r}$, when the interest is compounded continuously and r is written as a decimal.

   a. Complete the table to compare the exact time it takes for an investment to double to the "Rule of 72" time. (Round to two decimal places.)

Annual Interest Rate	Rule of 72 Years	Exact Years
2%	36	34.66
3%	24	23.10
4%	18	17.33
5%	14.4	13.86
6%	12	11.55
7%	10.29	9.90
8%	9	8.66
9%	8	7.70
10%	7.2	6.93
11%	6.55	6.30

   b. Compute the differences between the two sets of outputs. What conclusion can you reach about using the Rule of 72 estimate?
   As interest rate increases, estimate gets closer to actual value.

74. Doubling Time The number of quarters needed to double an investment when a lump sum is invested at 8%, compounded quarterly, is given by $n = \log_{1.02} 2$.

a. Use the change-of-base formula to find n. ≈35

b. In how many years will the investment double? 8.75 yr

75. Doubling Time The number of periods needed to double an investment when a lump sum is invested at 12%, compounded semiannually, is given by $n = \log_{1.06} 2$.

a. Use the change-of-base formula to find n. ≈11.9

b. How many years pass before the investment doubles in value? 6 yr

76. Annuities If $2000 is invested at the end of each year in an annuity that pays 5%, compounded annually, the number of years it takes for the future value to amount to $40,000 is given by $t = \log_{1.05} 2$. Use the change-of-base formula to find the number of years until the future value is $40,000. 15 yr

77. Annuities If $1000 is invested at the end of each year in an annuity that pays 8%, compounded annually, the number of years it takes for the future value to amount to $30,000 is given by $t = \log_{1.08} 3.4$. Use the change-of-base formula to find the number of years until the future value is $30,000. 16 yr

78. Deforestation One of the major causes of rain forest deforestation is agricultural and residential development. The number of hectares (2.47 acres) destroyed in a particular year t can be modeled by $y = -3.91435 + 2.62196 \ln t$, where $t = 0$ in 1950. When will more than 7 hectares be destroyed per year? In 2015 and after

79. Market Share Suppose that after a company introduces a new product, the number of months before its market share is x percent is given by

$$m = 20 \ln \frac{50}{50 - x}, x < 50$$

After how many months is the market share more than 45%, according to this model? 46 months

80. Drugs in the Bloodstream The concentration of a drug in the bloodstream from the time the drug is administered until 8 hours later is given by $y = 100(1 - e^{-0.312(8-t)})$ percent, where the drug is administered at time $t = 0$. For what time period is the amount of drug present more than 60%? First 5 hours

81. Carbon-14 Dating An exponential decay function can be used to model the number of grams of a radioactive material that remain after a period of time. Carbon-14 decays over time, with the amount remaining after t years given by $y = y_0 e^{-0.00012097t}$, where y_0 is the original amount. If the original amount of carbon-14 is 200 grams, find the number of years until 155.6 grams of carbon-14 remain. ≈2075 yr

82. Sales Decay After a television advertising campaign ended, the weekly sales of Korbel champagne fell rapidly. Weekly sales in a city were given by $S = 25,000e^{-0.072x}$ dollars, where x is the number of weeks after the campaign ended.

a. Use the logarithmic form of this function to find the number of weeks after the end of the campaign that passed before weekly sales fell below $16,230. 6 weeks

b. Check your solution by graphical or numerical methods.

83. Sales Decay After a television advertising campaign ended, the weekly sales of Turtledove bars fell rapidly. Weekly sales are given by $S = 600e^{-0.05x}$ thousand dollars, where x is the number of weeks after the campaign ended.

a. Use the logarithmic form of this function to find the number of weeks after the end of the campaign before weekly sales fell to $269.60. 16 weeks

b. Check your solution by graphical or numerical methods.

5.4 Exponential and Logarithmic Models

KEY OBJECTIVES

- Model data with exponential functions
- Use constant percent change to determine whether data fit an exponential model
- Compare quadratic and exponential models of data
- Model data with logarithmic functions

SECTION PREVIEW Diabetes

In Example 8 of Section 5.2, we solved problems about diabetes by using the fact that the percent of U.S. adults with diabetes (diagnosed and undiagnosed) can be modeled by the logarithmic function

$$p(x) = -12.975 + 11.851 \ln x$$

where x is the number of years after 2000. In this section, we will use projections from the Centers for Disease Control and Prevention to create this logarithmic model using technology. (See Example 5.)

Many sources provide data that can be modeled by exponential growth and decay functions, and technology can be used to find exponential functions that model data. In this section, we model real data with exponential functions and determine when this type of model is appropriate. We also create exponential functions that model phenomena characterized by constant percent change, and we model real data with logarithmic functions when appropriate. ■

Modeling with Exponential Functions

Exponential functions can be used to model real data if the data exhibit rapid growth or decay. When the scatter plot of data shows a very rapid increase or decrease, it is possible that an exponential function can be used to model the data. Consider the following examples.

EXAMPLE 1 ▶ Insurance Premiums

The monthly premiums for $250,000 in term-life insurance over a 10-year term period increase with the age of the men purchasing the insurance. The monthly premiums for nonsmoking males are shown in Table 5.8.

Table 5.8

Age (years)	Monthly Premium for 10-Year Term Insurance ($)	Age (years)	Monthly Premium for 10-Year Term Insurance ($)
35	123	60	783
40	148	65	1330
45	225	70	2448
50	338	75	4400
55	500		

(*Source*: Quotesmith.com)

a. Graph the data in the table with x as age and y in dollars.

b. Create an exponential function that models these premiums as a function of age.

c. Graph the data and the exponential function that models the data on the same axes.

SOLUTION

a. The scatter plot of the data is shown in Figure 5.33(a). The plot shows that the outputs rise rapidly as the inputs increase.

378 Chapter 5 Exponential and Logarithmic Functions

b. Using technology gives the exponential model for the monthly premium, rounded to four decimal places, as

$$y = 4.0389(1.0946^x) \text{ dollars}$$

where x is the age in years.

c. Figure 5.33(b) shows a scatter plot of the data and a graph of the (unrounded) exponential equation used to model it. The model appears to be a good, but not perfect, fit for the data.

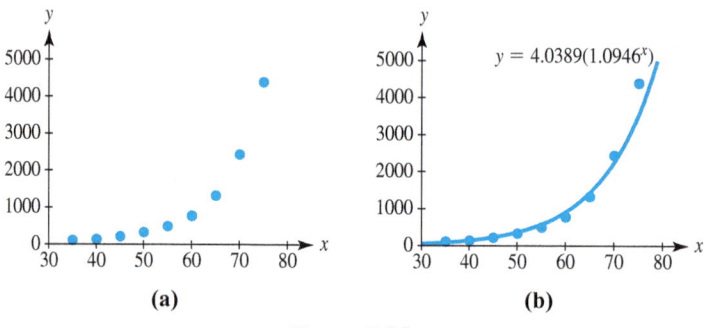

Figure 5.33

EXAMPLE 2 ▶ Consumer Price Index

The Social Security Administration makes projections about the consumer price index (CPI) to understand the effects of inflation on Social Security benefits and to plan for cost-of-living increases. Table 5.9 gives the CPI for selected years from 1995 and projected to 2070, with the reference year 1995. The CPI of 566.94 for 2040 in the table means that goods and services that cost $100 in 1995 are projected to cost $566.94 in 2040.

a. Graph the data, with x equal to the number of years after 1990.

b. Find the exponential function that models the data, with x equal to the years after 1990 and y equal to the percent.

c. Graph the data and the function on the same axes.

d. What does the unrounded model predict the CPI will be in 2030?

Table 5.9

Year	CPI	Year	CPI
1995	100.00	2035	465.98
2000	118.21	2040	566.94
2005	143.67	2045	689.77
2010	174.80	2050	839.21
2015	212.67	2055	1021.02
2020	258.74	2060	1242.23
2025	314.80	2065	1511.36
2030	383.00	2070	1838.81

(*Source:* Social Security Administration)

5.4 Exponential and Logarithmic Models

SOLUTION

a. The graph of the data is shown in Figure 5.34(a).

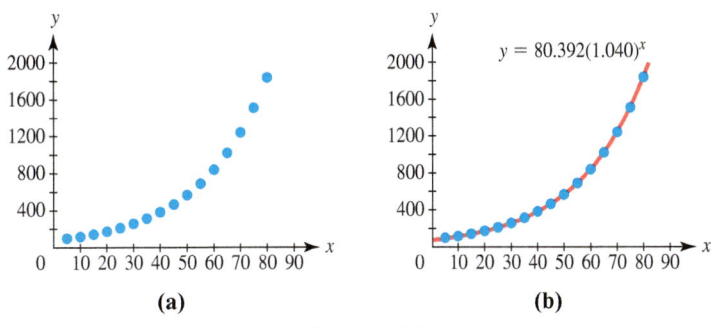

Figure 5.34

b. The exponential function that models the data, rounded to three decimal places, is

$$y = 80.392(1.040^x)$$

c. The graphs of the data and the exponential model are shown in Figure 5.34(b). The model is a near-perfect fit for the data.

d. Evaluating the unrounded function at $x = 40$ for the year 2030 gives 383.87. So the consumer price index is projected by the model to be 383.87 in 2030.

Spreadsheet ▶ SOLUTION We can use software programs and spreadsheets to find the exponential function that is the best fit for the data. Figure 5.35 shows the Excel spreadsheet for the aligned data of Example 2. Highlight the two data columns, select the Insert tab, then the Scatter option under the Charts group, then the points only option to get the scatter plot of the data. To get the graph and equation of the model, right click on one of the data points, select Add Trendline, then select Exponential and Display Equation on Chart, and close the box.*

Years after 1990	CPI
5	100
10	118.21
15	143.67
20	174.8
25	212.67
30	258.74
35	314.8
40	383
45	465.98
50	566.94
55	689.77
60	839.21
65	1021.02
70	1242.23
75	1511.36
80	1838.81

Trendline equation: $y = 80.392e^{0.0391x}$

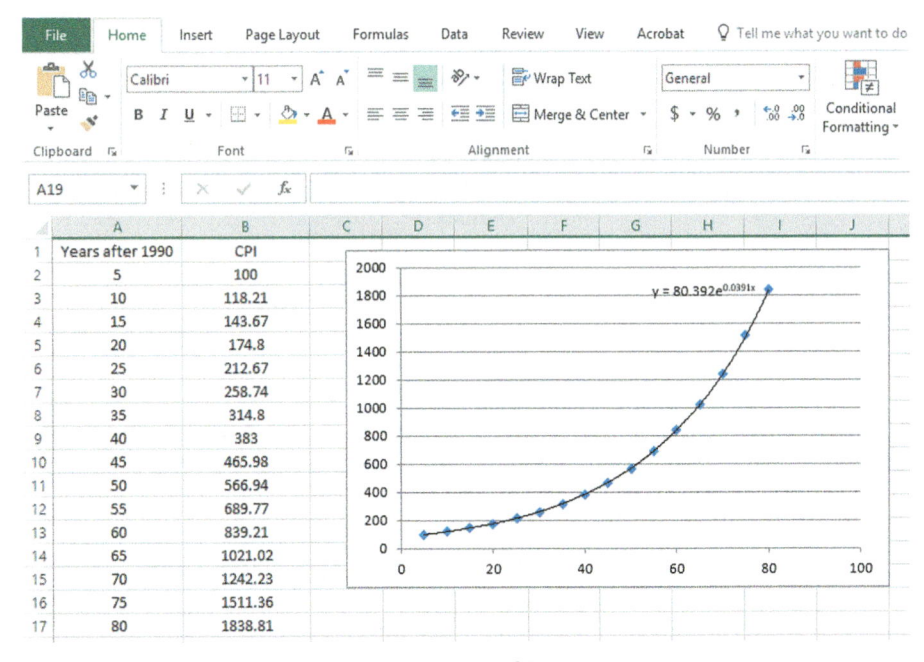

Figure 5.35

---

*See Appendix B, page 708, for details.

The equation created by Excel

$$y = 80.382e^{0.0391x}$$

is in a different form than the model found in Example 2, which is

$$y = 80.392(1.040)^x$$

However, they are equivalent because $e^{0.0391} \approx 1.040$, which means that

$$y = 80.382e^{0.0391x} = 80.392(1.040)^x$$

Finding Rates of Change in Exponential Growth and Decay Models

Recall that a set of data can be modeled exactly by a linear function if the first differences in outputs are constant for equally spaced inputs.

How can we determine whether a set of data can be modeled by an exponential function? To investigate this, we look at Table 5.10, which gives the growth of paramecia for each hour up to 9 hours.

Table 5.10

x (hours)	0	1	2	3	4	5	6	7	8	9
y (paramecia)	1	2	4	8	16	32	64	128	256	512

If we look at the first differences in the output, we see that they are not constant (Table 5.11). But if we calculate the percent change of the outputs for equally spaced inputs, we see that they are constant. For example, from hour 2 to hour 3, the population grew by 4 units; this represents a 100% increase from the hour-2 population. From hour 3 to hour 4, the population grew by 8, which is a 100% increase over the hour-3 population. In fact, this population increases by 100% each hour. This means that the population 1 hour from now will be the present population plus 100% of the present population.

Table 5.11

Outputs	1	2	4	8	16	32	64	128	256	512
First Differences		1	2	4	8	16	32	64	128	256
Percent Change		100	100	100	100	100	100	100	100	100

Because the percent change in the outputs is constant for equally spaced inputs in this example, an exponential model ($y = 2^x$) fits the data perfectly.

Constant Percent Changes

If the percent change of the outputs of a set of data is constant for equally spaced inputs, an exponential function will be a perfect fit for the data.

If the percent change of the outputs is approximately constant for equally spaced inputs, an exponential function will be an approximate fit for the data.

EXAMPLE 3 ▶ Sales Decay

Suppose a company develops a product that is released with great expectations and extensive advertising, but sales suffer because of bad word of mouth from dissatisfied customers.

a. Use the monthly sales data shown in Table 5.12 to determine the percent change for each of the months given.

Table 5.12

Month	1	2	3	4	5	6	7	8
Sales (thousands of $)	780	608	475	370	289	225	176	137

b. Find the exponential function that models the data.

c. Graph the data and the model on the same axes.

SOLUTION

a. The inputs are equally spaced; the differences of outputs and the percent changes are shown in Table 5.13.

Table 5.13

Outputs	780		608		475		370		289		225		176		137
First Differences		−172		−133		−105		−81		−64		−49		−39	
Percent Change		−22.1		−21.9		−22.1		−21.9		−22.1		−21.8		−22.2	

The percent change of the outputs is approximately −22%. This means that the sales 1 month from now will be approximately 22% less than the sales now.

b. Because the percent change is nearly constant, an exponential function should fit these data well. Technology gives the model

$$y = 999.781(0.780^x)$$

where x is the month and y is the sales in thousands of dollars.

c. The graphs of the data and the (unrounded) model are shown in Figure 5.36.

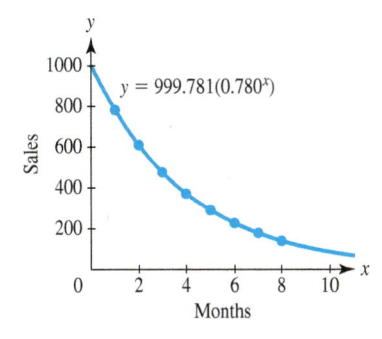

Figure 5.36

Exponential Models

Most graphing utilities give the best-fitting exponential model in the form

$$y = a \cdot b^x$$

It can be shown that the constant percent change of this function is

$$(b - 1) \cdot 100\%$$

Thus, if $b > 1$, the function is increasing (growing), and if $0 < b < 1$, the function is decreasing (decaying).

For the function $y = 2^x$, $b = 2$, so the constant percent change is $r = (2 - 1)100\% = 100\%$ (which we found in Table 5.11). For the function $y = 1000(0.78^x)$, the base is $b = 0.78$, and the percent change is $r = (0.78 - 1) \cdot 100\% = -22\%$.

Because $a \cdot b^0 = a \cdot 1 = a$, the value of $y = a \cdot b^x$ is a when $x = 0$, so a is the y-intercept of the graph of the function. Because $y = a$ when $x = 0$, we say that the **initial value** of the function is a. If the constant percent change (as a decimal) is $r = b - 1$, then $b = 1 + r$, and we can write $y = a \cdot b^x$ as

$$y = a(1 + r)^x$$

Exponential Model

If a set of data has initial value a and a constant percent change r (written as a decimal) for equally spaced inputs x, the data can be modeled by the exponential function

$$y = a(1 + r)^x$$

for exponential growth and by the exponential function

$$y = a(1 - r)^x$$

for exponential decay.

To illustrate this, suppose we know that the present population of a city is 100,000 and that it will grow at a constant percent of 10% per year. Then we know that the future population can be modeled by an exponential function with initial value $a = 100{,}000$ and constant percent change $r = 10\% = 0.10$ per year. That is, the population can be modeled by

$$y = 100{,}000(1 + 0.10)^x = 100{,}000(1.10)^x$$

where x is the number of years from the present.

EXAMPLE 4 ▶ Inflation

Suppose inflation averages 4% per year for each year from 2000 to 2010. This means that an item that costs $1 one year will cost $1.04 one year later. In the second year, the $1.04 cost will increase by a factor of 1.04, to $(1.04)(1.04) = 1.04^2$.

a. Write an expression that gives the cost t years after 2000 of an item costing $1 in 2000.

b. Write an exponential function that models the cost of an item t years from 2000 if its cost was $100 in 2000.

c. Use the model to find the cost of the item from part (b) in 2010.

SOLUTION

a. The cost of an item is $1 in 2000, and the cost increases at a constant percent of $4\% = 0.04$ per year, so the cost after t years will be

$$1(1 + 0.04)^t = 1.04^t \text{ dollars}$$

b. The inflation rate is $4\% = 0.04$. Thus, if an item costs $100 in 2000, the function that gives the cost after t years is

$$f(t) = 100(1 + 0.04)^t = 100(1.04^t) \text{ dollars}$$

c. The year 2010 is 10 years after 2000, so the cost of an item costing $100 in 2000 is

$$f(10) = 100(1.04^{10}) = 148.02 \text{ dollars}$$

Exponential Models of Growth or Decay

Recall that functions of the form $y = a(b^{kx})$, with $b > 1$, $a > 0$, and $k > 0$, can be used to model exponential growth, and if $b > 1$, $a > 0$, and $k < 0$, they can be used to model exponential decay. Many natural phenomena follow the law $A(t) = A_0 e^{kt}$, where k is the **rate of growth**. If we know that an amount follows this model, and we know two points, then we can solve for the rate of growth and write a function that models the data.

EXAMPLE 5 ▶ Growth of Bacteria

A biologist researching a newly discovered species of bacteria puts 50 bacteria in a growth medium at time $t = 0$ hours. Five hours later, she finds that the bacteria have grown to 550. Assuming exponential growth according to $A(t) = A_0 e^{kt}$:

a. Find the rate of growth and the function that models the growth of the bacteria.

b. Use the equation to predict the number of bacteria present after 24 hours.

SOLUTION

a. The initial amount is $A_0 = 50$, and the $A(5) = 550$. Substituting in the equation $A(t) = A_0 e^{kt}$, we have

$$550 = 50e^{5k}$$
$$11 = e^{5k}$$

To solve for k, we take the natural logarithm of both sides:

$$\ln 11 = \ln e^{5k}$$
$$\ln 11 = 5k$$
$$k = \frac{\ln 11}{5} \approx 0.479579$$

Thus, the rate of growth is $k = 0.48$ (rounded to two decimal places). The equation modeling the growth is

$$A(t) = 50e^{0.48t}$$

b. To predict the number of bacteria present after 24 hours, we substitute 24 for t:

$$A(24) = 50e^{0.48(24)} \approx 5{,}035{,}498 \text{ bacteria}$$

Comparison of Models

Sometimes it is hard to determine what type of model is the best fit for a set of data. When a scatter plot exhibits a single curvature without a visible high or low point, it is sometimes difficult to determine whether a power, quadratic, or exponential function should be used to model the data. Suppose after graphing the data points we are unsure which of these models is most appropriate. We can find the three models, graph them on the same axes as the data points, and inspect the graphs to see which is the best visual fit.*

Consider the exponential model for insurance premiums as a function of age that we found in Example 1. The curvature indicated by the scatter plot suggests that either a power, quadratic, or exponential model may fit the data. To compare the goodness of fit of the models, we find each model and compare its graph with the scatter plot (Figure 5.37).

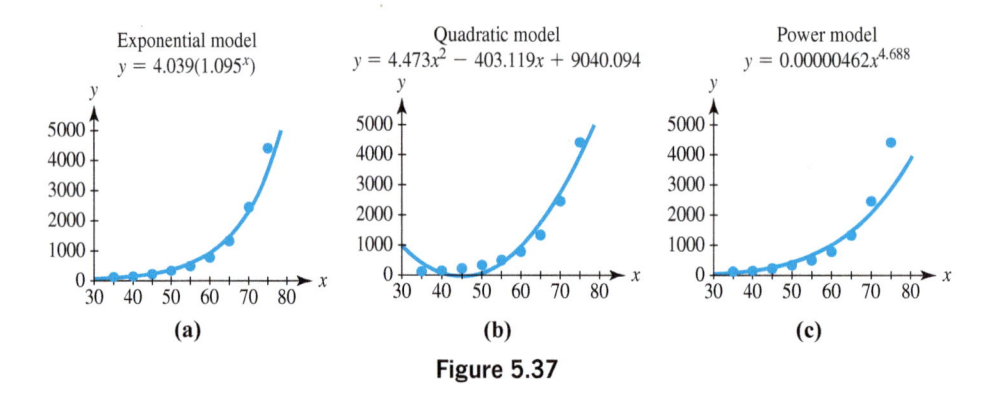

Exponential model
$y = 4.039(1.095^x)$
(a)

Quadratic model
$y = 4.473x^2 - 403.119x + 9040.094$
(b)

Power model
$y = 0.00000462x^{4.688}$
(c)

Figure 5.37

The exponential function appears to provide a better fit than the quadratic function because the data points seem to be approaching the x-axis as the input becomes closer to zero from the right. (Recall that the x-axis is a horizontal asymptote for the basic exponential function.) The exponential function also appears to be a better fit than the power function, especially for larger values of x.

When fitting exponential functions to data using technology, it is important to know that problems arise if the input values are large. For instance, when the input values are years (like 2000 and 2010), a technology-determined model may appear to be of the form $y = 0 \cdot b^x$, a function that certainly does not make sense as a model for exponential data. At other times when large inputs are used, your calculator or computer software may give an error message and not even produce an equation. To avoid these problems, it is helpful to align the inputs by converting years to the number of years after a specified year (for example, inputting years from 1990 reduces the size of the inputs).

Technology Note

When using technology to fit an exponential model to data, you should align the inputs to reasonably small values.

---

*Statistical measures of goodness of fit exist, but caution must be used in applying these measures. For example, the correlation coefficient r found for linear fits to data is different from the coefficient of determination R^2 found for quadratic fits to data.

Logarithmic Models

As with other functions we have studied, we can create models involving logarithms. Data that exhibit an initial rapid increase and then have a slow rate of growth can often be described by the function

$$f(x) = a + b \ln x \quad (\text{for } b > 0)$$

Note that the parameter a is the vertical shift of the graph of $y = \ln x$ and the parameter b affects how much the graph of $y = \ln x$ is stretched. If $b < 0$, the graph of $f(x) = a + b \ln x$ decreases rather than increases. Most graphing utilities will create logarithmic models in the form $y = a + b \ln x$.

EXAMPLE 6 ▶ Diabetes

As Table 5.14 shows, projections indicate that the percent of U.S. adults with diabetes could dramatically increase.

a. Find a logarithmic model that fits the data in Table 5.14, with $x = 0$ in 2000.

b. Use the reported model to predict the percent of U.S. adults with diabetes in 2027.

c. In what year does this model predict the percent to be 26.9%?

Table 5.14

Year	Percent	Year	Percent	Year	Percent
2010	15.7	2025	24.2	2040	31.4
2015	18.9	2030	27.2	2045	32.1
2020	21.1	2035	29.0	2050	34.3

(*Source:* Centers for Disease Control and Prevention)

SOLUTION

a. Entering the aligned input data (number of years after 2000) as the x-values and the percents as the y-values in a graphing calculator, and pressing STAT, CALC, and 9:LnReg give the logarithmic function that models the data. This function, rounded to three decimal places, is

$$y = -12.975 + 11.851 \ln x$$

where x is the number of years after 2000. The graphs of the aligned data and the function are shown in Figure 5.38.

b. Evaluating the reported model at $x = 27$ gives $y = -12.975 + 11.851 \ln 27 \approx$ 26.1 percent in 2027.

c. Setting $y = 26.9$ gives

$$26.9 = -12.975 + 11.851 \ln x$$

$$39.875 = 11.851 \ln x$$

$$3.3647 = \ln x$$

$$x = e^{3.3647} = 28.9$$

We could also solve graphically by intersecting the graphs of $y_1 = -12.975 + 11.851 \ln x$ and $y_2 = 26.9$, as shown in Figure 5.39. Thus, the percent reaches 27% in 2029.

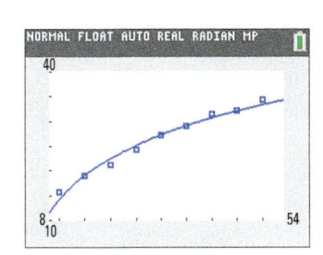

Figure 5.38

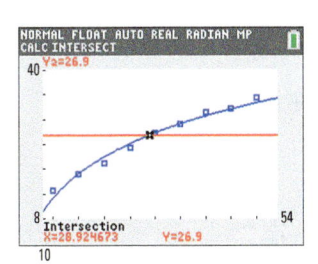

Figure 5.39

Note that the input data in Example 6 were not aligned as the number of years after 2010 because the first aligned input value would be 0 and the logarithm of 0 does not exist.

Technology Note

When using technology to fit a logarithmic model to data, you must align the data so that all input values are positive.

EXAMPLE 7 ▶ Women in the Labor Force

Table 5.15 contains the total percent of the labor force that is female for selected years from 1940 and projected to 2030.

Table 5.15

Year	Percent Female	Year	Percent Female
1940	24.3	1990	45.2
1950	29.6	2000	46.6
1960	33.4	2010	47.9
1970	38.1	2020	48.1
1980	42.5	2030	48.0

(*Source*: U.S. Bureau of Labor Statistics)

a. Find a logarithmic function that models these data. Align the input to be the number of years after 1900.

b. Graph the equation and the aligned data points, with x representing the years from 1930 through 2040. Comment on how the model fits the data.

c. Assuming that the model is valid in 2024, use it to estimate the percent of women in the labor force in 2024.

SOLUTION

a. A logarithmic function that models these data is

$$f(x) = -54.775 + 21.729 \ln x$$

where x is the number of years after 1900.

b. The scatter plot of the data and the graph of the logarithmic function that models the data are shown in Figure 5.40. The model appears to be a reasonably good fit for the data.

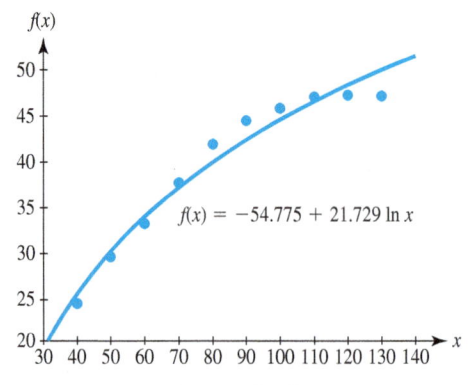

Figure 5.40

c. The percent of women in the labor force in 2024 can be estimated by evaluating $f(124) = 50\%$.

5.4 Exponential and Logarithmic Models

Spreadsheet ▶ SOLUTION

We can use software programs and spreadsheets to find the logarithmic function that is the best fit for data. Figure 5.41 shows the Excel spreadsheet for the data of Example 7. Selecting the cells containing the data, getting the scatter plot of the data, and selecting Logarithmic Trendline gives the equation of the logarithmic function that is the best fit for the data, along with the scatter plot and the graph of the best-fitting function. Checking Display equation gives the equation on the graph, shown in Figure 5.41. The model found here agrees with that found in Example 7.*

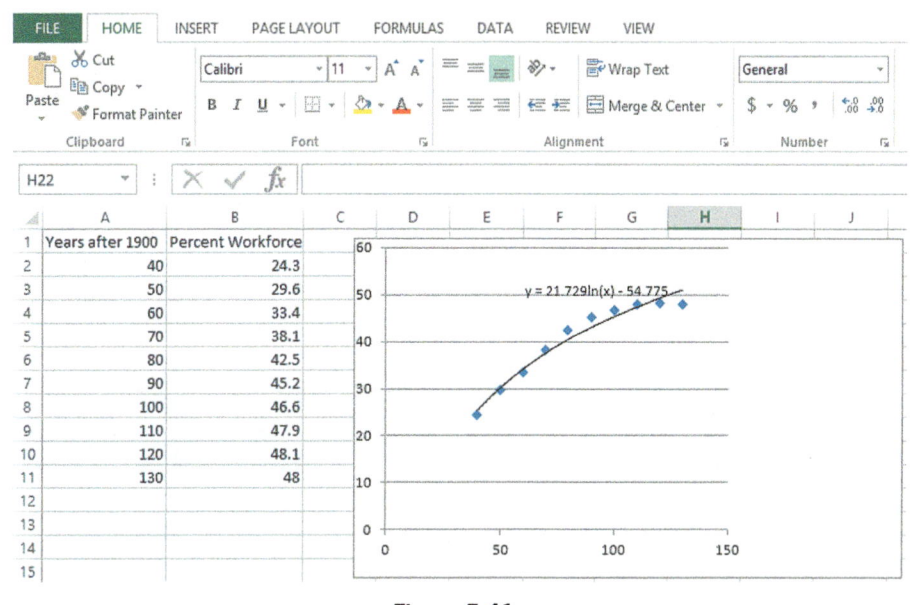

Years after 1900	Percent Workforce
40	24.3
50	29.6
60	33.4
70	38.1
80	42.5
90	45.2
100	46.6
110	47.9
120	48.1
130	48

$y = 21.729\ln(x) - 54.775$

Figure 5.41

Exponents, Logarithms, and Linear Regression

Now that we have knowledge of the properties of logarithms and how exponential functions are related to logarithmic functions, we can show that linear regression can be used to create exponential models that fit data. (The development of linear regression is discussed in Section 2.2.)

Logarithmic Property 3 states that $\log_b b^x = x$ and, in particular, that $\ln e^x = x$, so if we have data that can be approximated by an exponential function, we can convert the data to a linear form by taking the logarithm, base e, of the outputs. We can use linear regression to find the linear function that is the best fit for the converted data and then use the linear function as the exponent of e, which gives the exponential function that is the best fit for the original data. Consider Table 5.16, which gives the number y of paramecia in a population after x hours. We showed in Section 5.1 that the population can be modeled by $y = 2^x$.

The third row of Table 5.16 has the logarithms, base e, of the numbers (y-values) in the second row rounded to four decimal places, and the relationship between x and

Table 5.16

x (hours)	0	1	2	3	4	5	6	7	8	9
y (paramecia)	1	2	4	8	16	32	64	128	256	512
ln y	0	0.6931	1.3863	2.0794	2.7726	3.4657	4.1589	4.8520	5.5452	6.2383

*See Appendix B, page 708, for details.

ln y is linear. The first differences of the ln y values are constant, with the difference approximately equal to 0.6931. Using linear regression on the x and ln y values (to ten decimal places) gives the equation of the linear model:*

$$\ln y = 0.6931471806x$$

Because we seek the equation solved for y, we can write this equation in its exponential form, getting

$$y = e^{0.6931471806x}$$

To show that this model for the data is equal to $y = 2^x$, which we found in Section 5.1, we use properties of exponents and note that $e^{0.6931471806} \approx 2$:

$$y = e^{0.6931471806x} = (e^{0.6931471806})^x \approx 2^x$$

Fortunately, most graphing utilities have combined these steps to give the exponential model directly from the data.

---

*This equation could be found directly, in the form $\ln y = ax + b$, where $b = 0$ and a equals the constant difference. This is because the exponential function is a perfect fit for the data in this application.

SKILLS CHECK 5.4

Answers that are not seen can be found in the answer section at the back of the text.

1. Find the exponential function that models the data in the table below. $y = 2(3^x)$

x	−2	−1	0	1	2	3	4	5
y	2/9	2/3	2	6	18	54	162	486

2. The following table has input x and output $f(x)$. Test the percent change of the outputs to determine if the function is exactly exponential, approximately exponential, or not exponential. Exactly exponential

x	1	2	3	4	5	6
$f(x)$	4	16	64	256	1024	4096

3. The following table has input x and output $g(x)$. Test the percent change of the outputs to determine if the function is exactly exponential, approximately exponential, or not exponential. Not exponential

x	1	2	3	4	5	6
$g(x)$	2.5	6	8.5	10	8	6

4. The following table has input x and output $h(x)$. Test the percent change of the outputs to determine if the function is exactly exponential, approximately exponential, or not exponential. Approximately exponential

x	1	2	3	4	5	6
$h(x)$	1.5	2.25	3.8	5	11	17

5. Find the exponential function that is the best fit for $f(x)$ defined by the table in Exercise 2. $f(x) = 4^x$

6. Find the exponential function that is the best fit for $h(x)$ defined by the table in Exercise 4.
$h(x) = 0.859(1.633^x)$

7. a. Make a scatter plot of the data in the table below.
   b. Does it appear that a linear model or an exponential model is the better fit for the data? Linear model

x	1	2	3	4	5	6
y	2	3.1	4.3	5.4	6.5	7.6

8. Find the exponential function that models the data in the table below. Round the model with three-decimal-place accuracy. $y = 152.201(5.251^x)$

x	−2	−1	0	1	2	3
y	5	30	150	1000	4000	20,000

9. Compare the first differences and the percent change of the outputs to determine if the data in the table below should be modeled by a linear or an exponential function. Exponential function

x	1	2	3	4	5
y	2	6	14	34	81

10. Use a scatter plot to determine if a linear or exponential function is the better fit for the data in Exercise 9. Exponential function

11. Find the linear *or* exponential function that is the better fit for the data in Exercise 9. Round the model to three-decimal-place accuracy. $y = 0.876(2.494^x)$

12. Find the logarithmic function that models the data in the table below. Round the model to two-decimal-place accuracy. $y = 3.00 \ln x + 2.00$

x	1	2	3	4	5	6	7
y	2	4.08	5.3	6.16	6.83	7.38	7.84

13. a. Make a scatter plot of the data in the table below.
    b. Does it appear that a linear model or a logarithmic model is the better fit for the data?
       Logarithmic model

x	1	3	5	7	9
y	−2	1	3	4	5

14. a. Find a logarithmic function that models the data in the table in Exercise 13. Round the model to three-decimal-place accuracy. $y = 3.183 \ln x - 2.161$
    b. Find a linear function that models the data. $y = 0.850x - 2.050$
    c. Visually determine which model is the better fit for the data. Logarithmic model

15. a. Make a scatter plot of the data in the table below.
    b. Find a power function that models the data. Round to three decimal places. $y = 3.671x^{0.505}$
    c. Find a quadratic function that models the data. Round to three decimal places.
       $y = -0.125x^2 + 1.886x + 1.960$
    d. Find a logarithmic function that models the data. Round to three decimal places. $y = 3.468 + 2.917 \ln x$

x	1	2	3	4	5	6
y	3.5	5.5	6.8	7.2	8	9

16. Let $y = f(x)$ represent the power model found in Exercise 15(b), $y = g(x)$ represent the quadratic model found in Exercise 15(c), and $y = h(x)$ represent the logarithmic model found in Exercise 15(d). Graph each function on the same axes as the scatter plot, using the window $[0, 12]$ by $[0, 15]$. Which model appears to be the best fit?
    The logarithmic model, $y = h(x)$, is a slightly better fit.

EXERCISES 5.4

Answers that are not seen can be found in the answer section at the back of the text.
Report models accurate to three decimal places unless otherwise specified. Use the unrounded function to calculate and to graph the function.

Use the exponential form $y = a(1 + r)^x$ to model the information in Exercises 17–20.

17. **Inflation** Suppose that the retail price of an automobile is $30,000 in 2000 and that it increases at 4% per year.
    a. Write the equation of the exponential function that models the retail price of the automobile t years after 2000. $y = 30,000(1.04^t)$
    b. Use the model to predict the retail price of the automobile in 2015. $54,028

18. **Population** Suppose that the population of a city is 190,000 in 2000 and that it grows at 3% per year.
    a. Write the equation of the exponential function that models the annual growth. $y = 190,000(1.03^x)$
    b. Use the model to find the population of this city in 2010. 255,344

19. **Sales Decay** At the end of an advertising campaign, weekly sales amounted to $20,000. They then decreased by 2% each week after the end of the campaign.
    a. Write the equation of the exponential function that models the weekly sales. $y = 20,000(0.98^x)$
    b. Find the sales 5 weeks after the end of the advertising campaign. $18,078

20. **Inflation** The average price of a house in a certain city was $220,000 in 2008, and it increases at 3% per year.
    a. Write the equation of the exponential function that models the average price of a house t years after 2008. $220,000(1.03^t)$
    b. Use the model to predict the average price of a house in 2013. $255,040.30

21. **Personal Income** The table shows the total personal income in the United States (in billions of dollars) for selected years from 1960 and projected to 2024.
    a. Find the best-fitting exponential function that will model the data, with x equal to the number of years after 1960 and y in billions of dollars. Report the model with four significant digits.
       $y = 533.6(1.065^x)$
    b. If the model is valid in 2030, what is the projected total personal income then?
       Approximately $43,822 billion

c. In what year does the reported model predict the total personal income will reach $44.147 trillion? In 2031

Year	Income ($ billions)
1960	411.5
1970	838.8
1980	2307.9
1990	4878.6
2000	8429.7
2008	12,100.7
2014	14,728.6
2018	19,129.6
2024	22,685.1

(*Source*: U.S. Bureau of Labor Statistics)

22. Severe Obesity Severely obese people (BMI ≥ 40) are most at risk for serious health problems, and are the most expensive to treat. The table gives the percent of the Americans who are severely obese for the years from 1990 and projected to 2030.

a. Find an exponential function that models the data, with x equal to the number of years after 1990 and y equal to the percent. $y = 1.045(1.067^x)$

b. Use the reported model to predict what percent of Americans will be severely obese in 2032. 15.9%

Year	% Severely Obese
1990	0.8
2000	2.2
2010	4.9
2015	6.4
2020	7.9
2025	9.5
2030	11.1

(*Source:* American Journal of Preventative Medicine)

23. China's Shale Natural Gas The estimated annual production of shale natural gas in China for the years 2013 through 2020, in millions of cubic feet, is shown in the table.

a. Find the equation of the exponential function that models these estimates of the number of millions of cubic feet of shale natural gas in China y as a function of the number of years x after 2010. $y = 11.682(1.748^x)$

b. Graph the model and the data on the same axes and discuss the fit. Good fit

c. What does the model estimate that the number of millions of cubic feet of shale natural gas in China will be in 2021? 5438 million cubic feet

d. When does the model estimate that the number of millions of cubic feet of shale natural gas in China will be 9506? 2022

Year	Natural Gas (millions of cubic feet)
2013	50
2014	120
2015	200
2016	400
2017	600
2018	1000
2019	1700
2020	2900

(*Source:* Sanford C. Bernstein)

24. Insurance Premiums The table below gives the annual premiums required for a $250,000 20-year term-life insurance policy on female nonsmokers of different ages.

a. Find an exponential function that models the monthly premium as a function of the age of the female nonsmoking policyholder.
$y = 4.304(1.096^x)$

b. Find the quadratic function that is the best fit for the data. $y = 6.182x^2 - 565.948x + 12,810.482$

c. Graph each function on the same axes with the data points to determine visually which model is the better fit for the data. Use the window $[30, 80]$ by $[-10, 6800]$. Exponential function

Age	Monthly Premium for a 20-Year Policy ($)
35	145
40	185
45	253
50	363
55	550
60	845
65	1593
70	2970
75	5820

(*Source*: Quotesmith.com)

25. Average Annual Wage
The following table shows the U.S. average annual wage, in thousands of dollars, for selected years from 1980 and projected to 2050.

a. Find an exponential function that models these data, with x equal to the number of years after 1980 and y equal to the average annual wage in thousands of dollars. $y = 13.410(1.040^x)$

b. Graph the model and the data on the same axes.

c. What does the model predict that the average annual wage will be in 2040? Is this interpolation or extrapolation? 137.2 thousand dollars

d. When does the model predict that the average annual wage will reach $225,000? 2053 ($x = 72.75$)

Year	Average Annual Wage ($ thousands)
1980	12.5
1990	21.0
2000	32.2
2010	41.7
2012	44.6
2014	48.6
2016	53.3
2018	58.7
2020	63.7
2025	76.8
2030	93.2
2035	113.2
2040	137.6
2045	167.1
2050	202.5

(*Source*: Social Security Administration)

26. Alzheimer's Disease
Partially because people in the United States are living longer, the number with Alzheimer's disease and other forms of dementia is projected to grow each year. The table below gives the number of millions of U.S. citizens with Alzheimer's from 2000 and projected to 2050.

Year	2000	2010	2020	2030	2040	2050
Number	4.0	5.9	6.8	8.7	11.8	14.3

(*Source*: National Academy on an Aging Society)

a. Create a scatterplot of the data, with x equal to the number of years after 1990 and y equal to the number of millions of U.S. citizens with Alzheimer's disease.

b. Find an exponential function that models these data, with x equal to the number of years after 1990 and y equal to the millions of U.S. citizens with Alzheimer's disease. $y = 3.294(1.025^x)$

c. Find a power function that models these data, with x equal to the number of years after 1990 and y equal to the millions of U.S. citizens with Alzheimer's disease. $y = 0.759x^{0.688}$

d. Graph each function on the same axes as the data points, and determine which model better fits the data. Exponential model

27. Life Span
The table below gives the life expectancy for people in the United States for the birth years from 1920 and projected to 2050.

a. Find the logarithmic function that models the data, with x equal to 0 in 1900. $y = 8.774 + 14.907 \ln x$

b. Find the quadratic function that is the best fit for the data, with $x = 0$ in 1900. Round the quadratic coefficient to four decimal places.
$y = -0.0012x^2 + 0.4185x + 47.9448$

c. Graph each of these functions on the same axes with the data points.

d. Evaluate both models for the birth year 2022.
Logarithmic: 80.4; quadratic: 81.0

e. If the model is valid through the year 2100, which model is a better predictor of the life expectancy? logarithmic

Birth Year	Life Expectancy	Birth Year	Life Expectancy
1920	54.1	1990	75.4
1930	59.7	2000	76.8
1940	62.9	2010	78.7
1950	68.2	2020	80.2
1960	69.7	2030	81.7
1970	70.8	2040	83.0
1980	73.7	2050	84.4

(*Source*: National Center for Health Statistics)

28. Obesity
Obesity (BMI ≥ 30) is a serious problem in the United States and is expected to get worse. Being overweight increases the risk of diabetes, heart disease, and many other ailments. The table gives the percent of Americans who are obese from 1990 projected through 2030.

a. Find a logarithmic function that models the percent, y, as a function of the number of years after 1980, x. $y = -31.732 + 18.672 \ln x$

b. Graph the model and the data on the same set of axes, and comment on the fit of the model to the data. Good fit

c. Use the model to predict what percent of Americans will be obese in 2027. 40.2%

d. Graphically find the year in which the percent obese is projected to reach 38. 2022

Year	1990	2000	2010	2015	2020	2025	2030
% Obese	12.7	22.1	30.9	34.5	37.4	39.9	42.2

(Source: American Journal of Preventive Medicine, Vol 42, (June 2012), 563–570. ajpmonline.org)

29. Rate of Growth of Population The table gives the annual rates of growth of the U.S. population for selected years from 1970 projected to 2049.

  a. Find a logarithmic function that models the rate, y, as a function of the number of years after 1960.
  $y = 3.885 - 0.733 \ln x$

  b. Find an exponential function that models the rate, y, as a function of the number of years after 1960.
  $y = 2.572(0.982^x)$

  c. Graph each model on the same axes with the data to determine which model is the better fit to the data. The exponential model is better.

Year	Annual Growth Rate
1970	2.07
1980	1.69
1990	1.56
2000	1.25
2010	1.10
2020	0.90
2030	0.76
2040	0.62
2049	0.46

30. Diabetes The table shows the number of U.S. adults with diabetes (in millions), projected from 2015 through 2050.

Year	Millions of Adults with Diabetes
2015	37.3
2020	50.0
2025	59.5
2030	68.3
2035	77.2
2040	84.1
2045	91.7
2050	100.0

  a. Create a scatter plot of the data, with x equal to the number of years after 2010.

  b. Find a logarithmic function to model the data, with x equal to the number of years after 2010 and y equal to the number of millions of U.S. adults with diabetes. $y = -16.249 + 29.731 \ln x$

  c. When does the model predict that the number of millions of U.S. adults with diabetes will be 60.01? 2023 ($x = 13$)

(Source: Centers for Disease Control and Prevention)

31. Vehicle Temperature The temperature inside a vehicle can rise almost 20°F in just 10 minutes and almost 30°F in 20 minutes. The vehicle can quickly reach a temperature that puts a pet at risk of serious illness and even death. The estimated interior air temperature in a car after various numbers of minutes is shown in the table.

Estimated Vehicle Interior Air Temperature vs. Elapsed Time

Elapsed time	Outside Air Temperature (°F)					
	70	75	80	85	90	95
0 minutes	70	75	80	85	90	95
10 minutes	89	94	99	104	109	114
20 minutes	99	104	109	114	119	124
30 minutes	104	109	114	119	124	129
40 minutes	108	113	118	123	128	133
50 minutes	111	116	121	126	131	136
60 minutes	113	118	123	128	133	138
>1 hour	115	120	125	130	135	140

(Source: Courtesy Jan Null, CCM; Department of Geosciences, San Francisco State University)

  a. Create a scatter plot of the vehicle interior air temperature as a function of the number of minutes in the first hour, when the outside air temperature is 80°F.

  b. Omit the first data point and find a logarithmic function that models the data. (Why do we need to omit the first data point?) $y = 68.371 + 13.424 \ln x$; the input cannot be 0 for a logarithmic function

  c. Graph the model on the same axes as the data and comment on the fit.

  d. What does the model predict that the interior air temperature will be after 45 minutes? 119.5°F

32. Drone Services Revenue Commercial drone-enabled services revenue is growing rapidly and is projected to reach $8.7 billion annually by 2025. The table on the next page gives its annual revenue, in billions, for the years from 2015 through 2025.

  a. Create a scatter plot of the data, with x equal to the number of years after 2015 and y equal to the revenue in billions of dollars.

b. Find the best-fitting exponential function that models the data described in part (a).
$y = 0.109(1.557^x)$
c. Graph the data and the function from part (b) on the same axes, and comment on the fit of the model to the data. Good fit
d. Use the reported model to predict the commercial drone-enabled services revenue in 2027. $22.1 billion

Year	Revenue ($ billions)
2015	0.1
2016	0.2
2017	0.3
2018	0.4
2019	0.6
2020	0.8
2021	1.5
2022	2.5
2023	4.2
2024	6.4
2025	8.7

(Source: Tractica)

33. **Sexually Active Girls** The percents of girls age x or younger who have been sexually active are given in the table below.

   a. Create a logarithmic function that models the data, using an input equal to the age of the girls.
   $y = -681.976 + 251.829 \ln x$
   b. Use the model to estimate the percent of girls age 17 or younger who have been sexually active. 31.5%
   c. Find the quadratic function that is the best fit for the data. $y = 0.627x^2 - 7.400x - 26.675$
   d. Graph each of these functions on the same axes with the data points to determine which function is the better model for the data. Quadratic function

Age	Cumulative Percent Sexually Active Girls	Cumulative Percent Sexually Active Boys
15	5.4	16.6
16	12.6	28.7
17	27.1	47.9
18	44.0	64.0
19	62.9	77.6
20	73.6	83.0

(Source: "National Longitudinal Survey of Youth," Risking the Future, Washington D.C.: National Academy Press)

34. **Sexually Active Boys** The percents of boys age x or younger who have been sexually active are given in the table in Exercise 33.

   a. Create a logarithmic function that models the data, using an input equal to the age of the boys.
   $y = 246.612 \ln x - 651.703$
   b. Use the model to estimate the percent of boys age 17 or younger who have been sexually active. 47.0%
   c. Compare the percents that are sexually active for the two genders (see Exercise 33). What do you conclude? More males than females are sexually active at given ages.

35. **Fuel Economy** The lifetime gasoline use of light-duty vehicles is a function of the fuel economy, as shown in the figure below.

   a. Should the data be modeled by an exponential growth or an exponential decay function? Exponential decay
   b. Find an exponential function to model the data.
   $y = 16,278.587(0.979^x)$

Light-Duty Vehicle Fuel Consumption

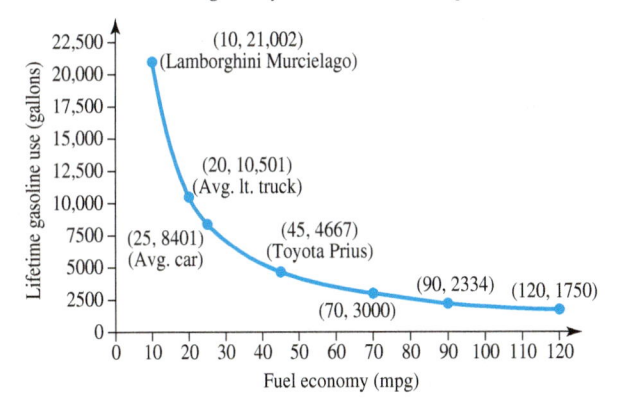

(10, 21,002) (Lamborghini Murcielago)
(20, 10,501) (Avg. lt. truck)
(25, 8401) (Avg. car)
(45, 4667) (Toyota Prius)
(70, 3000)
(90, 2334)
(120, 1750)

Lifetime gasoline use (gallons) vs. Fuel economy (mpg)

(Source: U.S. Department of Energy, Energy Efficiency, and Renewable Energy)

   c. Find a power function to model the data.
   $y = 210,022.816x^{-1}$
   d. Which function is the better fit to the data? Power function
   e. If a vehicle had a fuel economy of 100 mpg, what would its lifetime gasoline use be, according to the power model? 2100 gal

36. **E-Commerce Sales** E-commerce is taking a bigger slice of the overall retail sales and is growing far faster than total retail sales. U.S. retail sales amounted to $504.6 billion in 2018 and are projected to be $735.4 billion in 2023. The table gives the online retail e-commerce sales in billions of dollars for the years from 2016 and projected to 2023.

   a. Find the logarithmic function that models these data, with x equal to the number of years after 2010 and y equal to the retail sales in billions of dollars. Report the model with coefficients accurate to four significant digits.
   $y = -498.0 + 482.2 \ln x$

b. Graph the equation and the aligned data points on the same axes. Comment on how well the model fits the data.

c. Assuming the model is valid in 2025, use it to predict the online retail sales in 2025. $807.8 billion

Years	Sales ($ billions)
2016	360.3
2017	446.8
2018	504.6
2019	560.7
2020	613
2021	660
2022	700.6
2023	735.4

(*Source:* Statista)

37. *Growth of a Colony of Bees* The population of a colony of bees grows according to the equation $P(t) = P_0 e^{kt}$.

a. If there are 100 bees initially and there are 130 after 1 day, what is the size of the colony after 3 days? About 220 bees

b. How long will it be until the population of bees is 1200? About 9.5 days

38. *Decomposition of Chlorine in a Pool* Under certain water conditions, the free chlorine in a swimming pool decomposes according to the equation $C(t) = C_0 e^{kt}$. Initially, the amount of free chlorine in a pool was 2.8 parts per million (ppm). Eighteen hours later, the amount was 2.1 ppm.

a. What is the rate of decay of the amount of free chlorine in the pool? -0.016 ppm per hour

b. When the chlorine level reaches 1.0 ppm, the pool must be shocked. How long will it be before the pool must be shocked? About 64.4 hours

39. *Mexico Population* The population of Mexico is growing rapidly and will soon reach that of Japan. Mexico's population in 1960 was 38.17 million, and in 2019 its population was 132.1 million. If the population grows according to the equation $P(t) = P_0 e^{kt}$, then

a. find the growth rate and write an equation to model the population growth. $k = 0.021$ million per year; $P(t) = 38.17 e^{0.021t}$, where t is the number of years after 1960.

b. find the year during which the population doubled that of the population in 1960. 1993

5.5 Exponential Functions and Investing

KEY OBJECTIVES

- Find future value of investments when interest is compounded k times per year
- Find future value of investments when interest is compounded continuously
- Find the present value of an investment
- Use graphing utilities to model investment data

SECTION PREVIEW Investments

If $1000 is invested in an account that earns 6% interest, with the interest added to the account at the end of each year, the interest is said to be **compounded annually**. The amount to which an investment grows over a period of time is called its **future value**. In this section, we use general formulas to find the future value of money invested when the interest is compounded at regular intervals of time and when it is compounded continuously. We also find the lump sum that must be invested to have it grow to a specified amount in the future. This is called the **present value** of the investment. ∎

Compound Interest

In Section 5.4, we found that if a set of data has initial value a and a constant percent change r for equally spaced inputs x, the data can be modeled by the exponential function

$$y = a(1 + r)^x$$

5.5 Exponential Functions and Investing

Thus, if an investment of $1000 earns 6% interest, compounded annually, the future value at the end of t years will be

$$S = \$1000(1.06)^t \text{ dollars}$$

EXAMPLE 1 ▶ Future Value of an Account

To numerically and graphically view how $1000 grows at 6% interest, compounded annually, do the following:

a. Construct a table that gives the future values $S = 1000(1 + 0.06)^t$ for $t = 0, 1, 2, 3, 4,$ and 5 years after the $1000 is invested.

b. Graph $S = 1000(1 + 0.06)^t$ for the values of t given in part (a).

c. Graph the function $S = 1000(1 + 0.06)^t$ as a continuous function of t for $0 \le t \le 5$.

d. Is there another graph that more accurately represents the amount that would be in the account at any time during the first 5 years that the money is invested?

Table 5.17

t (years)	Future Value, S ($)
0	1000.00
1	1060.00
2	1123.60
3	1191.02
4	1262.48
5	1338.23

SOLUTION

a. Substitute the values of t into the equation $S = 1000(1 + 0.06)^t$. Because the output units for S are dollars, we round to the nearest cent to obtain the values in Table 5.17.

b. Figure 5.42 shows the scatter plot of the data in Table 5.17.

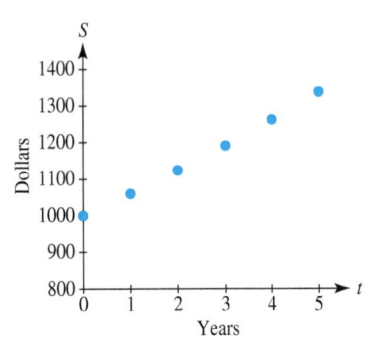

Figure 5.42

c. Figure 5.43(a) shows a continuous graph of the function $S = 1000(1 + 0.06)^t$ for $0 \le t \le 5$.

d. The scatter plot in Figure 5.42 shows the amount in the account at the end of each year, but not at any other time during the 5-year period. The continuous graph in Figure 5.43(a) shows the amount continually increasing, but interest is paid only at

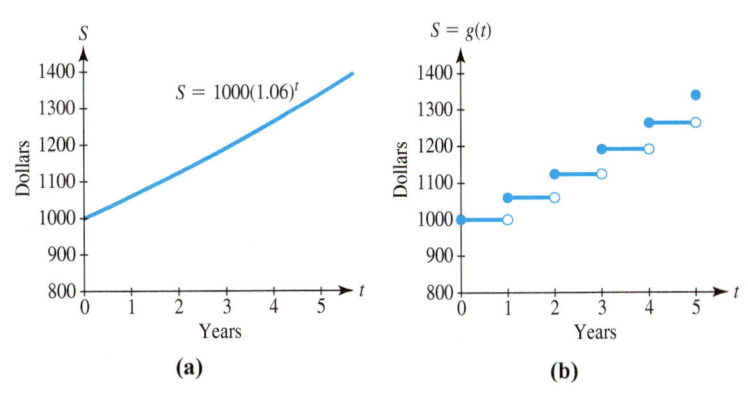

Figure 5.43

the end of each year and the amount in the account is constant at the previous level until more interest is added. So the graph in Figure 5.43(b) is a more accurate graph of the amount that would be in this account because it shows the amount remaining constant until the end of each year, when interest is added.

The function g graphed in Figure 5.43(b) is the step function defined by

$$g(t) = \begin{cases} 1000.00 & \text{if } 0 \leq t < 1 \\ 1060.00 & \text{if } 1 \leq t < 2 \\ 1123.60 & \text{if } 2 \leq t < 3 \\ 1191.02 & \text{if } 3 \leq t < 4 \\ 1262.48 & \text{if } 4 \leq t < 5 \\ 1338.23 & \text{if } t = 5 \end{cases} \quad \text{dollars in } t \text{ years}$$

This step function is rather cumbersome to work with, so we will find the future value by using the graph in Figure 5.43(a) and the function $S = 1000(1 + 0.06)^t$ *interpreted discretely*. That is, we draw the graph of this function as it appears in Figure 5.43(a) and evaluate the formula for S with the understanding that the only inputs that make sense in this investment example are nonnegative integers. In general, we have the following.

Future Value of an Investment with Annual Compounding

If P dollars are invested at an interest rate r per year, compounded annually, the future value S at the end of t years is

$$S = P(1 + r)^t$$

The **annual interest rate** r is also called the **nominal interest rate**, or simply the **rate**. Remember that the interest rate is usually stated as a percent and that it is converted to a decimal when computing future values of investments.

If the interest is compounded more than once per year, then the additional compounding will result in a larger future value. For example, if the investment is compounded twice per year (semiannually), there will be twice as many compounding periods, with the interest rate in each period equal to

$$r\left(\frac{1}{2}\right) = \frac{r}{2}$$

Thus, the future value is found by doubling the number of periods and halving the interest rate if the compounding is done twice per year. In general, we use the following model, interpreted discretely, to find the amount of money that results when P dollars are invested and earn compound interest.

Future Value of an Investment with Periodic Compounding

If P dollars are invested for t years at the annual interest rate r, where the interest is compounded k times per year, then the interest rate per period is $\frac{r}{k}$, the number of compounding periods is kt, and the future value that results is given by

$$S = P\left(1 + \frac{r}{k}\right)^{kt} \text{ dollars}$$

5.5 Exponential Functions and Investing

For example, if $1000 is placed in an account that earns 6% interest per year, with the interest added to the account at the end of every month (compounded monthly), the future value of this investment in 5 years is given by

$$S = 1000\left(1 + \frac{0.06}{12}\right)^{12(5)} = \$1348.85$$

EXAMPLE 2 ▶ Future Value

Find the future value of $10,000 placed in an account earning 8% per year for 10 years, if the interest is compounded

a. Annually

b. Daily

c. How much more is earned by compounding daily?

SOLUTION

a. Substituting $P = 10{,}000$, $r = 0.08$, $k = 1$, and $t = 10$ in $S = P\left(1 + \frac{r}{k}\right)^{kt}$ gives

$$S = 10{,}000\left(1 + \frac{0.08}{1}\right)^{1 \cdot 10} = S = 10{,}000(1.08)^{10} = 21{,}589.25$$

Thus the future value is $21,589.25.

b. Substituting $P = 10{,}000$, $r = 0.08$, $k = 365$ and $t = 10$ in $S = P\left(1 + \frac{r}{k}\right)^{kt}$ gives

$$S = 10{,}000\left(1 + \frac{0.08}{365}\right)^{365 \cdot 10} = 10{,}000\left(1 + \frac{0.08}{365}\right)^{3650} = 22{,}253.46$$

Thus the future value is $22,253.46.

c. Daily compounding gives $22,253.46 − $21,589.25 = $664.21 more than annual compounding.

Graphing calculators and spreadsheets such as Excel have built-in features that can be used to find future values and solve other financial problems. We can find the future value of investments such as those in Example 2, (called lump-sum investments) with a graphing calculator as follows:

Under the **APPS** menu, select Finance and then TMV Solver.

a. Set N = the total number of compounding periods.

b. Set I% = the annual percentage rate.

c. Set PV = the lump-sum investment (with a negative sign because the lump sum is *leaving* the investor and going into the investment).

d. Set PMT = 0 because no additional money is being invested.

e. Set both P/Y (payments per year) and C/Y (compounding periods per year) equal to the number of compounding periods per year.

f. Highlight END, move the cursor to FV, and press **ALPHA** **ENTER** to get the future value.

Figure 5.44(a) and (b) show the data entered and the future value for part (a) of Example 2, and Figure 5.44(c) shows the solution for part (b).* The solutions agree with those in Example 2.

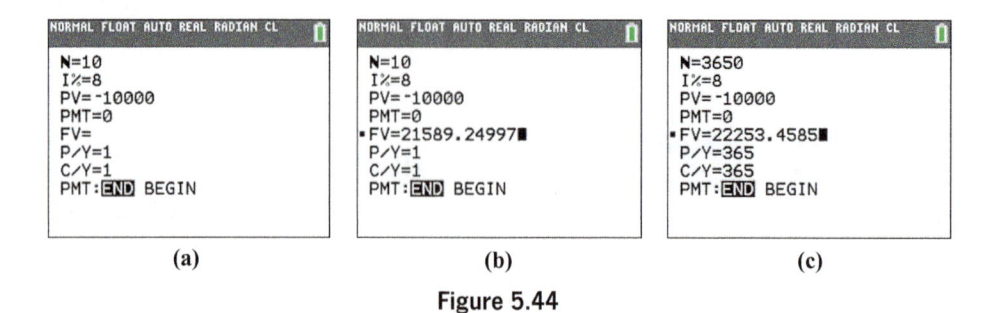

(a)　　　　　　　　　(b)　　　　　　　　　(c)

Figure 5.44

Spreadsheet ▶ SOLUTION　We can also use Excel to find the future value of a lump-sum investment by using the formula

$$= \text{fv}(F2, B2, C2, A2, 0), \text{ where}$$

Cell F2 has the periodic interest rate as a decimal.
Cell B2 has the total number of periods.
Cell C2 has the payment amount, 0.
Cell A2 has the lump-sum investment amount P.

Figure 5.45 shows the input for part (b) of Example 2 and the future value of $10,000 invested at 8% per year for 10 years, compounded daily, in cell B4.†

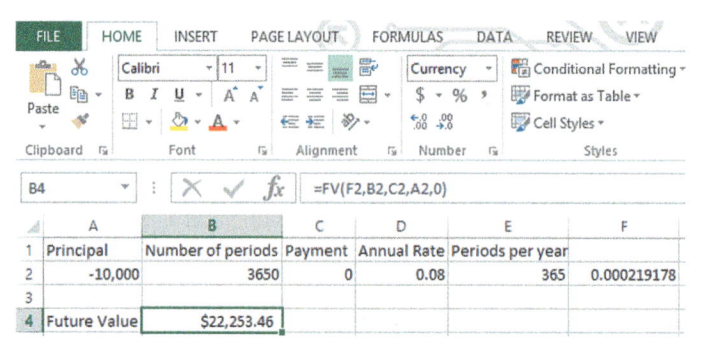

Figure 5.45

Continuous Compounding and the Number e

In Section 5.1, we stated that if P dollars are invested for t years at interest rate r, compounded continuously, the future value of the investment is

$$S = Pe^{rt}$$

To see how the number e becomes part of this formula, we consider the future value of $1 invested at an annual rate of 100%, compounded for 1 year with different compounding periods. If we denote the number of periods per year by k, the model that gives the future value is

$$S = \left(1 + \frac{1}{k}\right)^k \text{ dollars}$$

---

*For details, see Appendix A, page 693.
†For details, see Appendix B, page 711.

Table 5.18 shows the future value of this investment for different compounding periods.

Table 5.18

Type of Compounding	Number of Compounding Periods per Year	Future Value ($)
Annually	1	$\left(1 + \frac{1}{1}\right)^1 = 2$
Quarterly	4	$\left(1 + \frac{1}{4}\right)^4 = 2.44140625$
Monthly	12	$\left(1 + \frac{1}{12}\right)^{12} \approx 2.61303529022$
Daily	365	$\left(1 + \frac{1}{365}\right)^{365} \approx 2.71456748202$
Hourly	8760	$\left(1 + \frac{1}{8760}\right)^{8760} \approx 2.71812669063$
Each minute	525,600	$\left(1 + \frac{1}{525,600}\right)^{525,600} \approx 2.7182792154$
x times per year	x	$\left(1 + \frac{1}{x}\right)^x$

As Table 5.18 indicates, the future value increases (but not very rapidly) as the number of compounding periods during the year increases. As x gets very large, the future value approaches the number e, which is 2.718281828 to nine decimal places. In general, the outputs that result from larger and larger inputs in the function

$$f(x) = \left(1 + \frac{1}{x}\right)^x$$

approach the number e. (In calculus, e is defined as the *limit* of $\left(1 + \frac{1}{x}\right)^x$ as x approaches ∞.)

This definition of e permits us to create a new model for the future value of an investment when interest is compounded continuously. We can let $m = \frac{k}{r}$ to rewrite the formula

$$S = P\left(1 + \frac{r}{k}\right)^{kt} \quad \text{as} \quad S = P\left(1 + \frac{1}{m}\right)^{mrt}, \quad \text{or} \quad S = P\left[\left(1 + \frac{1}{m}\right)^m\right]^{rt}$$

Now as the compounding periods k increase without bound, $m = \frac{k}{r}$ increases without bound and

$$\left(1 + \frac{1}{m}\right)^m$$

approaches e, so $S = P\left[\left(1 + \frac{1}{m}\right)^m\right]^{rt}$ approaches Pe^{rt}.

> **Future Value of an Investment with Continuous Compounding**
>
> If P dollars are invested for t years at an annual interest rate r, compounded continuously, then the future value S is given by
>
> $$S = Pe^{rt} \text{ dollars}$$

For a given principal and interest rate, the function $S = Pe^{rt}$ is a continuous function whose domain consists of real numbers greater than or equal to 0. Unlike the other compound interest functions, this one is not discretely interpreted because of the continuous compounding.

EXAMPLE 3 ▶ Future Value and Continuous Compounding

a. What is the future value of $2650 invested for 8 years at 12%, compounded continuously?

b. How much interest will be earned on this investment?

SOLUTION

a. The future value of this investment is $S = 2650e^{0.12(8)} = 6921.00$.

b. The interest earned on this investment is the future value minus the original investment:

$$\$6921 - \$2650 = \$4271$$

EXAMPLE 4 ▶ Continuous Versus Annual Compounding of Interest

a. For each of 9 years, compare the future value of an investment of $1000 at 8%, compounded annually, and of $1000 at 8%, compounded continuously.

b. Graph the functions for annual compounding and for continuous compounding for $t = 30$ years on the same axes.

c. What conclusion can be made regarding compounding annually and compounding continuously?

SOLUTION

a. The future value of $1000, compounded annually, is given by the function $S = 1000(1 + 0.08)^t$, and the future value of $1000, compounded continuously, is given by $S = 1000e^{0.08t}$. By entering these formulas in an Excel spreadsheet and evaluating them for each of 9 years (see Figure 5.46), we can compare the future value for each of these 9 years. At the end of the 9 years, we see that compounding continuously results in $2054.43 − $1999.00 = $55.43 more than compounding annually yields.

b. The graphs are shown in Figure 5.46.

c. The value of the investment increases more under continuous compounding than under annual compounding.

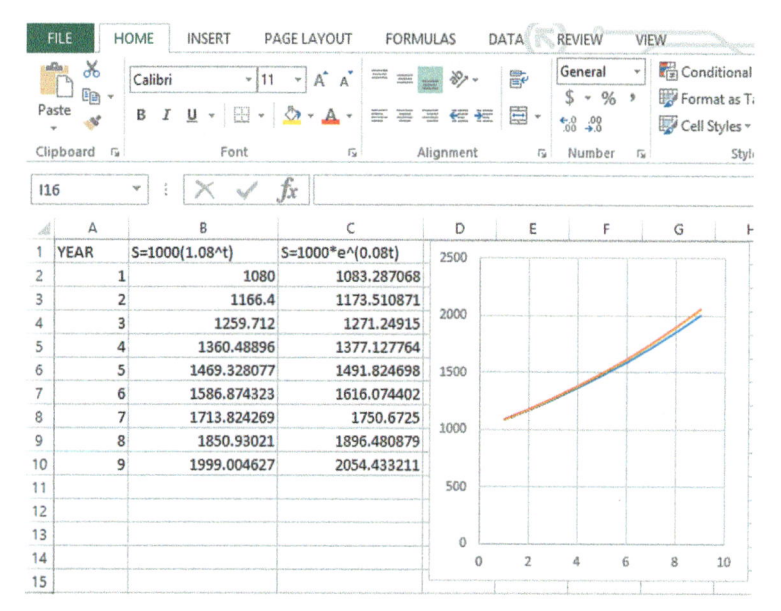

Figure 5.46

Present Value of an Investment

We can write the formula for the future value S when a lump sum P is invested for $n = kt$ periods at interest rate $i = \dfrac{r}{k}$ per compounding period as

$$S = P(1 + i)^n$$

We sometimes need to know the lump-sum investment P that will give an amount S in the future. This is called the **present value** P, and it can be found by solving $S = P(1 + i)^n$ for P. Solving for P gives the present value as follows.

> **Present Value**
>
> The lump sum that will give future value S in n compounding periods at rate i per period is the present value
>
> $$P = \frac{S}{(1 + i)^n} \quad \text{or, equivalently,} \quad P = S(1 + i)^{-n}$$

EXAMPLE 5 ▶ Present Value

What lump sum must be invested at 10%, compounded semiannually, for the investment to grow to $15,000 in 7 years?

SOLUTION

We have the future value $S = 15{,}000$, $i = \dfrac{0.10}{2} = 0.05$, and $n = 2 \cdot 7 = 14$, so the present value of this investment is

$$P = \frac{15{,}000}{(1 + 0.05)^{14}} = 7576.02 \text{ dollars}$$

We can find the present value of the lump-sum investment in Example 5 with a graphing calculator by using the same steps used to find future value, with the following changes:

Set FV = the desired future value = 15000
Highlight END, move the cursor to PV, and press **ALPHA** **ENTER** to get the present value.*

Figures 5.47(a) and (b) show the data entered and the present value for Example 5. The present value shown agrees with that computed in Example 5. The present value is negative because it is the investment that *leaves* the investor.

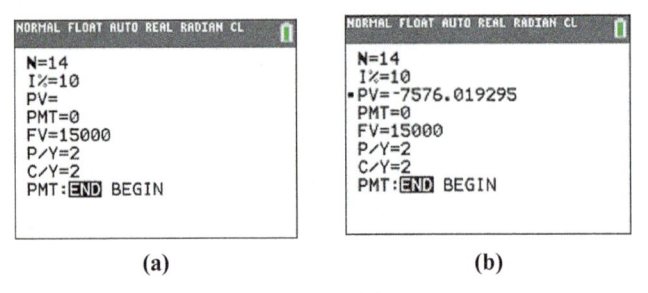

Figure 5.47

Spreadsheet SOLUTION We can use Excel to find the present value (investment) that gives the given future value by using the formula

$$= \text{pv}(F2, B2, C2, A2, 0), \text{ where}$$

Cell F2 has the periodic interest rate as a decimal = 0.05.
Cell B2 has the total number of periods = 14.
Cell C2 has the payment amount = 0.
Cell A2 has the future value.

Figure 5.48 shows inputs for the investment in Example 5, which gives a future value of $15,000, with the lump-sum inverstment (present value) in cell B4.† The negative value reflects the fact that $7576.02 is the necessary investment.

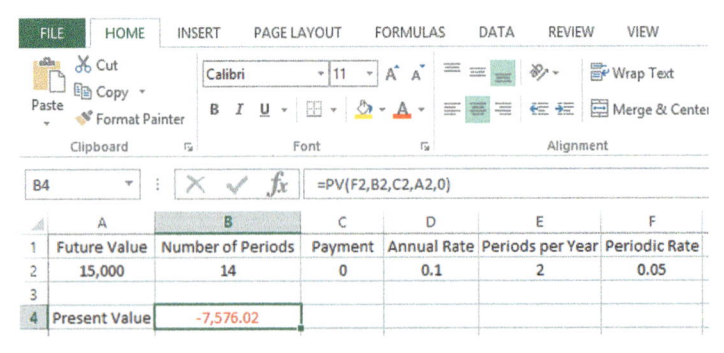

Figure 5.48

---

*For details, see Appendix A, page 698.
†For details, see Appendix B, page 712.

Investment Models

We have developed and used investment formulas. We can also use graphing utilities to model actual investment data. Consider the following example.

EXAMPLE 6 ▶ Mutual Fund Growth

The data in Table 5.19 give the annual return on an investment of $1.00 made on March 31, 1990, in the AIM Value Fund, Class A Shares. The values reflect reinvestment of all distributions and changes in net asset value but exclude sales charges.

Table 5.19

Time (years)	Average Annual Total Return ($)
1	1.23
3	2.24
5	3.11
10	6.82

(*Source:* FundStation)

a. Use an exponential function to model these data.

b. Use the model to find what the fund would amount to on March 31, 2001, if $10,000 was invested March 31, 1990, and the fund continued to follow this model after 2000.

c. Is it likely that this fund continued to grow at this rate in 2002?

SOLUTION

a. Using technology to create an exponential model for the data gives

$$A(x) = 1.1609(1.2004)^x$$

Thus, $1.00 invested in this fund grows to $A(x) = 1.1609(1.2004)^x$ dollars x years after March 31, 1990.

b. March 31, 2001, is 11 years after March 31, 1990, so using the equation from part (a) with $x = 11$ gives $A(11) = 1.1609(1.2004)^{11}$ as the future value of an investment of $1.00. Thus, the future value of an investment of $10,000 in 11 years is

$$\$10{,}000[1.1609(1.2004)^{11}] = \$86{,}572.64$$

c. No. Most stocks and mutual funds decreased in value in 2002 because of terrorist attacks in 2001.

SKILLS CHECK 5.5

Answers that are not seen can be found in the answer section at the back of the text.

Evaluate the expressions in Exercises 1–10. Write approximate answers rounded to two decimal places.

1. $15{,}000e^{0.06(20)}$ 49,801.75
2. $8000e^{0.05(10)}$ 13,189.77
3. $3000(1.06^x)$ for $x = 30$ 17,230.47
4. $20{,}000(1.07^x)$ for $x = 20$ 77,393.69
5. $12{,}000\left(1 + \dfrac{0.10}{k}\right)^{kn}$ for $k = 4$ and $n = 8$ 26,445.08
6. $23{,}000\left(1 + \dfrac{0.08}{k}\right)^{kn}$ for $k = 12$ and $n = 20$ 113,316.46
7. $P\left(1 + \dfrac{r}{k}\right)^{kn}$ for $P = 3000$, $r = 8\%$, $k = 2$, $n = 18$ 12,311.80
8. $P\left(1 + \dfrac{r}{k}\right)^{kn}$ for $P = 8000$, $r = 12\%$, $k = 12$, $n = 8$ 20,794.18
9. $300\left(\dfrac{1.02^n - 1}{0.02}\right)$ for $n = 240$ 1,723,331.03
10. $2000\left(\dfrac{1.10^n - 1}{0.10}\right)$ for $n = 12$ 42,768.57

11. Find $g(2.5)$, $g(3)$, and $g(3.5)$ if

$$g(x) = \begin{cases} 1000.00 & \text{if } 0 \leq x < 1 \\ 1060.00 & \text{if } 1 \leq x < 2 \\ 1123.60 & \text{if } 2 \leq x < 3 \\ 1191.00 & \text{if } 3 \leq x < 4 \\ 1262.50 & \text{if } 4 \leq x < 5 \\ 1338.20 & \text{if } x = 5 \end{cases}$$

1123.60; 1191.00; 1191.00

12. Find $f(2)$, $f(1.99)$, and $f(2.1)$ if

$$f(x) = \begin{cases} 100 & \text{if } 0 \leq x \leq 1 \\ 200 & \text{if } 1 < x < 2 \\ 300 & \text{if } 2 \leq x \leq 3 \end{cases}$$

300; 200; 300

13. Solve $S = P\left(1 + \dfrac{r}{k}\right)^{kn}$ for P. $P = S\left(1 + \dfrac{r}{k}\right)^{-kn}$

14. Solve $S = P(1 + i)^n$ for P. $P = S(1 + i)^{-n}$

EXERCISES 5.5

Answers that are not seen can be found in the answer section at the back of the text.

15. **Future Value** If $8800 is invested for x years at 8% interest, compounded annually, find the future value that results in

    a. 8 years. $16,288.19   b. 30 years. $88,551.38

16. **Investments** Suppose $6400 is invested for x years at 7% interest, compounded annually. Find the future value of this investment at the end of

    a. 10 years. $12,589.77   b. 30 years. $48,718.43

17. **Future Value** If $3300 is invested for x years at 10% interest, compounded annually, the future value that results is $S = 3300(1.10)^x$ dollars.

    a. Graph the function for $x = 0$ to $x = 8$.

    b. Use the graph to estimate when the money in the account will double. 8 yr

18. **Investments** If $5500 is invested for x years at 12% interest, compounded quarterly, the future value that results is $S = 5500(1.03)^{4x}$ dollars.

    a. Graph this function for $0 \leq x \leq 8$.

    b. Use the graph to estimate when the money in the account will double. 6 yr

19. **Future Value** If $10,000 is invested at 12% interest, compounded quarterly, find the future value in 10 years. $32,620.38

20. **Future Value** If $8800 is invested at 6% interest, compounded semiannually, find the future value in 10 years. $15,893.78

21. **Future Value** An amount of $10,000 is invested at 12% interest, compounded daily.

    a. Find the future value in 10 years. $33,194.62

    b. How does this future value compare with the future value in Exercise 19? Why are they different?
    More; compounded more frequently

22. **Future Value** A total of $8800 is invested at 6% interest, compounded daily.

    a. Find the future value in 10 years. $16,033.85

    b. How does this future value compare with the future value in Exercise 20? Why are they different?
    More; compounded more frequently

23. **Compound Interest** If $10,000 is invested at 12% interest, compounded monthly, find the interest earned in 15 years. $49,958.02

24. **Compound Interest** If $20,000 is invested at 8% interest, compounded quarterly, find the interest earned in 25 years. $124,892.92

25. **Continuous Compounding** Suppose $10,000 is invested for t years at 6% interest, compounded continuously. Give the future value at the end of

    a. 12 years. $20,544.33

    b. 18 years. $29,446.80

26. **Continuous Compounding** If $42,000 is invested for t years at 7% interest, compounded continuously, find the future value in

    a. 10 years. $84,577.61

    b. 20 years. $170,318.40

27. **Continuous Versus Annual Compounding**

    a. If $10,000 is invested at 6%, compounded annually, find the future value in 18 years. $28,543.39

    b. How does this compare with the result from Exercise 25(b)?
    Continuous compounding yields $903.41 more.

28. **Continuous Versus Annual Compounding**

    a. If $42,000 is invested at 7%, compounded annually, find the future value in 20 years. $162,526.75

    b. How does this compare with the result from Exercise 26(b)?
    Continuous compounding yields $7791.65 more.

29. **Doubling Time** Use a spreadsheet, a table, or a graph to estimate how long it takes for an amount to double if it is invested at 10% interest

    a. Compounded annually. ≈7.27 yr

    b. Compounded continuously. ≈6.93 yr

30. **Doubling Time** Use a spreadsheet, a table, or a graph to estimate how long it takes for an amount to double if it is invested at 6% interest

    a. Compounded annually. ≈11.9 yr

    b. Compounded continuously. ≈11.55 yr

31. **Future Value** Suppose $2000 is invested in an account paying 5% interest, compounded annually. What is the future value of this investment

    a. After 8 years? $2954.91

    b. After 18 years? $4813.24

32. **Future Value** If $12,000 is invested in an account that pays 8% interest, compounded quarterly, find the future value of this investment

    a. After 2 quarters. $12,484.80

    b. After 10 years. $26,496.48

33. **Investments** Suppose $3000 is invested in an account that pays 6% interest, compounded monthly. What is the future value of this investment after 12 years? $6152.25

34. **Investments** If $9000 is invested in an account that pays 8% interest, compounded quarterly, find the future value of this investment

    a. After 0.5 year. $9363.60

    b. After 15 years. $29,529.28

35. **Doubling Time** If money is invested at 10% interest, compounded quarterly, the future value of the investment doubles approximately every 7 years.

    a. Use this information to complete the table below for an investment of $1000 at 10% interest, compounded quarterly.

Years	0	7	14	21	28
Future Value ($)	1000	2000	4000	8000	16,000

    b. Create an exponential function, rounded to three decimal places, that models the discrete function defined by the table. $y = 1000(1.104)^x$

    c. Because the interest is compounded quarterly, this model must be interpreted discretely. Use the rounded function to find the value of the investment in 5 years and in $10\frac{1}{2}$ years. $1640.01; $2826.02

36. **Doubling Time** If money is invested at 11.6% interest, compounded monthly, the future value of the investment doubles approximately every 6 years.

    a. Use this information to complete the table below for an investment of $1000 at 11.6% interest, compounded monthly.

Years	0	6	12	18	24
Future Value ($)	1000	2000	4000	8000	16,000

    b. Create an exponential function, rounded to three decimal places, that models the discrete function defined by the table. $S = 1000(1.122)^t$

    c. Because the interest is compounded monthly, this model must be interpreted discretely. Use the rounded model to find the value of the investment in 2 months, in 4 years, and in $12\frac{1}{2}$ years. $1019.37; $1584.79; $4216.10

37. **Present Value** What lump sum must be invested at 10%, compounded monthly, for the investment to grow to $65,000 in 8 years? $29,303.36

38. **Present Value** What lump sum must be invested at 8%, compounded quarterly, for the investment to grow to $30,000 in 12 years? $11,596.13

39. **Present Value** What lump-sum investment will grow to $10,000 in 10 years if it is invested at 6%, compounded annually? $5583.95

40. **Present Value** What lump-sum investment will grow to $30,000 if it is invested for 15 years at 7%, compounded annually? $10,873.38

41. **College Tuition** New parents want to put a lump sum into a money market fund to provide $30,000 in 18 years to help pay for college tuition for their child. If the fund averages 10% per year, compounded monthly, how much should they invest? $4996.09

42. **Trust Fund** Grandparents decide to put a lump sum of money into a trust fund on their granddaughter's 10th birthday so that she will have $1,000,000 on her 60th birthday. If the fund pays 11%, compounded monthly, how much money must they put in the account? $4190.46

43. **College Tuition** The Toshes need to have $80,000 in 12 years for their son's college tuition. What amount must they invest to meet this goal if they can invest money at 10%, compounded monthly? $24,215.65

44. **Retirement** Hennie and Bob inherit $100,000 and plan to invest part of it for 25 years at 10%, compounded monthly. If they want it to grow to $1 million for their retirement, how much should they invest? $82,939.75

45. **Investment** At the end of t years, the future value of an investment of $10,000 in an account that pays 8%, compounded monthly, is

    $$S = 10{,}000\left(1 + \frac{0.08}{12}\right)^{12t} \text{ dollars}$$

    Assuming no withdrawals or additional deposits, how long will it take for the investment to amount to $40,000? $t \approx 17.39$; 17 yr, 5 mo

406　Chapter 5　Exponential and Logarithmic Functions

46. *Investment* At the end of t years, the future value of an investment of $25,000 in an account that pays 12%, compounded quarterly, is

$$S = 25{,}000\left(1 + \frac{0.12}{4}\right)^{4t} \text{ dollars}$$

In how many years will the investment amount to $60,000? $t \approx 7.4$; 7 yr, 5 mo

47. *Investment* At the end of t years, the future value of an investment of $38,500 in an account that pays 8%, compounded monthly, is

$$S = 38{,}500\left(1 + \frac{0.08}{12}\right)^{12t} \text{ dollars}$$

For what time period (in years) will the future value of the investment be more than $100,230? 12 yr and after

48. *Investment* At the end of t years, the future value of an investment of $12,000 in an account that pays 8%, compounded monthly, is

$$S = 12{,}000\left(1 + \frac{0.08}{12}\right)^{12t} \text{ dollars}$$

Assuming no withdrawals or additional deposits, for what time period (in years) will the future value of the investment be less than $48,000? First 17 years, 4 months

49. *Interest Rate* Suppose that $15,000 is invested at continuous rate r and grows to $140,900 in 28 years. Solve the equation $140{,}900 = 15{,}000(e^{28r})$ to find the interest rate r. 8%

50. *Interest Rate* Suppose that $10,000 is invested at continuous rate r and grows to $17,160 in 9 years. Find the interest rate r. 6%

51. *Interest Rate* Suppose that $20,000 is invested at continuous rate r and grows to $81,104 in 28 years. Find the interest rate r. 5%

52. *Investing* If an investment of P dollars earns r percent (as a decimal), compounded m times per year, its future value in t years is $A = P\left(1 + \dfrac{r}{m}\right)^{mt}$. Prove that the number of years it takes for this investment to double is $t = \dfrac{\ln 2}{m \ln(1 + r/m)}$.

5.6　Annuities; Loan Repayment

KEY OBJECTIVES
- Find the future value of an ordinary annuity
- Find the present value of an ordinary annuity
- Find the payments needed to amortize a loan

SECTION PREVIEW　Annuities

To prepare for future retirement, people frequently invest a fixed amount of money at the end of each month into an account that pays interest that is compounded monthly. Such an investment plan or any other characterized by regular payments is called an annuity. In this section, we develop a model to find future values of annuities.

A person planning retirement may also want to know how much money is needed to provide regular payments to him or her during retirement, and a couple may want to know what lump sum to put in an investment to pay future college expenses for their child. The amount of money they need to invest to receive a series of payments in the future is the present value of an annuity.

We can also use the present value concept to determine the payments that are necessary to repay a loan with equal payments made on a regular schedule, which is called amortization. The formulas used to determine the future value and the present value of annuities and the payments necessary to repay loans are applications of exponential functions. ∎

Future Value of an Annuity

An **annuity** is a financial plan characterized by regular payments. We can view an annuity as a savings plan where regular payments are made to an account, and we can use an exponential function to determine what the future value of the account will be. One type of annuity, called an **ordinary annuity**, is a financial plan where equal payments are contributed at the end of each period to an account that pays a fixed rate of interest compounded at the same time as payments are made. For example, suppose $1000 is invested at the end of each year for 5 years in an account that pays interest at 10%, compounded annually. To find the future value of this annuity, we can think of it as the sum of 5 investments of $1000, one that draws interest for 4 years (from the end of the first year to the end of the fifth year), one that draws interest for 3 years, and so on (Table 5.20).

Table 5.20

Investment	Future Value
Invested at end of 1st year	$1000(1.10)^4$
Invested at end of 2nd year	$1000(1.10)^3$
Invested at end of 3rd year	$1000(1.10)^2$
Invested at end of 4th year	$1000(1.10)^1$
Invested at end of 5th year	1000

Note that the last investment is at the end of the fifth year, so it earned no interest. The future value of the annuity is the sum of the separate future values. It is

$$S = 1000 + 1000(1.10)^1 + 1000(1.10)^2 + 1000(1.10)^3 + 1000(1.10)^4$$
$$= 6105.10 \text{ dollars}$$

We can find the value of an ordinary annuity like the one above with the following formula.

Future Value of an Ordinary Annuity

If R dollars are contributed at the end of each period for n periods into an annuity that pays interest at rate i at the end of the period, the future value of the annuity is

$$S = R\left(\frac{(1+i)^n - 1}{i}\right) \text{ dollars}$$

Note that the interest rate used in the above formula is the rate per period, not the rate for a year. We can use this formula to find the future value of an annuity in which $1000 is invested at the end of each year for 5 years into an account that pays interest at 10%, compounded annually. This future value, which we found earlier without the formula, is

$$S = R\left(\frac{(1+i)^n - 1}{i}\right) = 1000\left(\frac{(1+0.10)^5 - 1}{0.10}\right) = 6105.10 \text{ dollars}$$

EXAMPLE 1 ▶ **Future Value of an Ordinary Annuity**

Find the 5-year future value of an ordinary annuity with a contribution of $500 per quarter into an account that pays 8% per year, compounded quarterly.

SOLUTION

The payments and interest compounding occur quarterly, so the interest rate per period is $\frac{0.08}{4} = 0.02$ and the number of compounding periods is $4(5) = 20$. Substituting the information into the formula for the future value of an annuity gives

$$S = 500\left(\frac{(1+0.02)^{20} - 1}{0.02}\right) = 12{,}148.68$$

Thus, the future value of this investment is $12,148.68.

We can find the future value of the ordinary annuity in Example 1 with a graphing calculator as follows:

Under the **APPS** menu, select Finance and then TMV Solver.
a. Set N = the total number of compounding periods = 5*4 = 20.
b. Set I% = the annual percentage rate = 8.
c. Set PV = lump-sum investment = 0.
d. Set PMT = quarterly payment = −500.
e. Set both P/Y (payments per year) and C/Y (compounding periods per year) equal to the number of compounding periods per year = 4.
f. Highlight END, move the cursor to FV, and press **ALPHA** **ENTER** to get the future value.

Figures 5.49(a) and (b) show the data entered and the future value found in Example 1.*

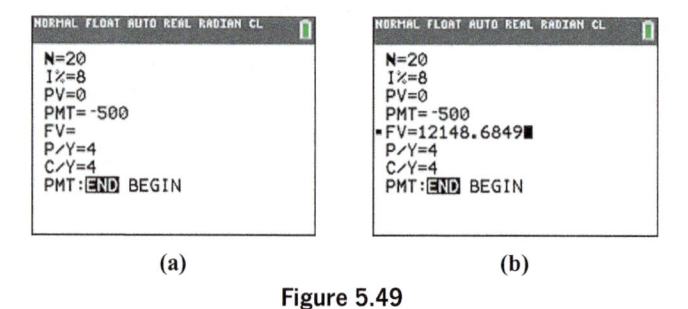

(a) (b)

Figure 5.49

Spreadsheet ▶ SOLUTION To find the future value of the ordinary annuity of Example 1 with Excel, we use the formula

$$= \text{fv}(F2, B2, C2, A2, 0)$$

with the periodic rate in cell B4, and where

Cell F2 has the periodic interest rate as a decimal = 0.02.
Cell B2 has the total number of periods = 20.
Cell C2 has the payment amount = −500.
Cell A2 has the future value = 0.

Figure 5.50 shows the input for the investment of Example 1, with the future value in cell B4.†

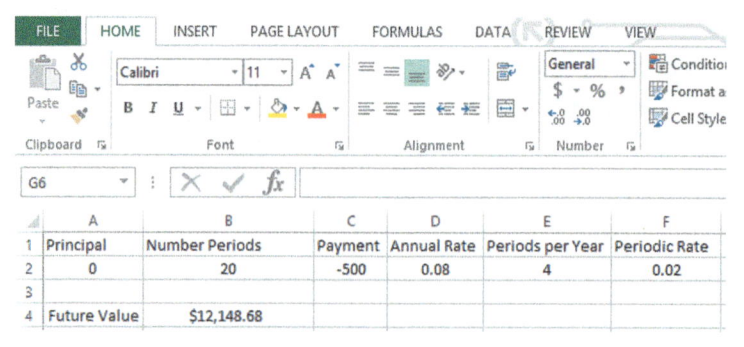

	A	B	C	D	E	F
1	Principal	Number Periods	Payment	Annual Rate	Periods per Year	Periodic Rate
2	0	20	-500	0.08	4	0.02
3						
4	Future Value	$12,148.68				

Figure 5.50

---

*For details, see Appendix A, page 698.
†For details, see Appendix B, page 712.

EXAMPLE 2 ▶ Future Value of an Annuity

Harry deposits $200 at the end of each month into an account that pays 12% interest per year, compounded monthly. Find the future value for every 4-month period, for up to 36 months.

SOLUTION

To find the future value of this annuity, we note that the interest rate per period (month) is $\dfrac{12\%}{12} = 0.01$, so the future value at the end of n months is given by the function

$$S = 200\left(\dfrac{1.01^n - 1}{0.01}\right) \text{ dollars}$$

This model for the future value of an ordinary annuity is a continuous function with discrete interpretation because the future value changes only at the end of each period. A graph of the function and the table of future values for every 4 months up to 36 months are shown in Figure 5.51 and Table 5.21, respectively. Keep in mind that the graph should be interpreted discretely; that is, only positive integer inputs and corresponding outputs make sense.

Table 5.21

Number of Months	Value of Annuity ($)	Number of Months	Value of Annuity ($)
4	812.08	24	5394.69
8	1657.13	28	6425.82
12	2536.50	32	7498.81
16	3451.57	36	8615.38
20	4403.80		

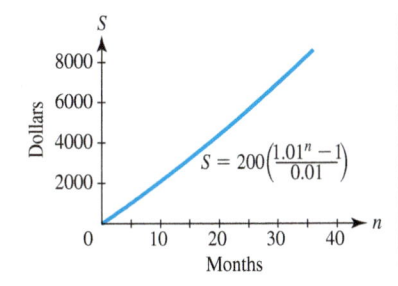

Figure 5.51

Present Value of an Annuity

Many retirees purchase annuities that give them regular payments through a period of years. Suppose a lump sum of money is invested to return payments of R dollars at the end of each of n periods, with the investment earning interest at rate i per period, after which $0 will remain in the account. This lump sum is the **present value** of this ordinary annuity, and it is equal to the sum of the present values of each of the future payments. Table 5.22 shows the present value of each payment of this annuity.

Table 5.22

Present	Present Value of This Payment ($)
Paid at end of 1st period	$R(1 + i)^{-1}$
Paid at end of 2nd period	$R(1 + i)^{-2}$
Paid at end of 3rd period	$R(1 + i)^{-3}$
⋮	⋮
Paid at end of $(n - 1)$st period	$R(1 + i)^{-(n-1)}$
Paid at end of nth period	$R(1 + i)^{-n}$

We must have enough money in the account to provide the present value of each of these payments, so the present value of the annuity is

$$A = R(1+i)^{-1} + R(1+i)^{-2} + R(1+i)^{-3} + \cdots + R(1+i)^{-(n-1)} + R(1+i)^{-n}$$

This sum can be written in the form

$$A = R\left(\frac{1 - (1+i)^{-n}}{i}\right)$$

Present Value of an Ordinary Annuity

If a payment of R dollars is to be made at the end of each period for n periods from an account that earns interest at a rate of i per period, then the account is an **ordinary annuity**, and the **present value** is

$$A = R\left(\frac{1 - (1+i)^{-n}}{i}\right)$$

EXAMPLE 3 ▶ Present Value of an Annuity

Suppose a retiring couple wants to establish an annuity that will provide $2000 at the end of each month for 20 years. If the annuity earns 6%, compounded monthly, how much must the couple put in the account to establish the annuity?

SOLUTION

We seek the present value of an annuity that pays $2000 at the end of each month for $12(20) = 240$ months, with interest at $\frac{6\%}{12} = \frac{0.06}{12} = 0.005$ per month. The present value is

$$A = 2000\left(\frac{1 - (1+0.005)^{-240}}{0.005}\right) = 279{,}161.54 \text{ dollars}$$

Thus, the couple can receive a payment of $2000 at the end of each month for 20 years if they put a lump sum of $279,161.54 in an annuity. (Note that the sum of the money they will receive over the 20 years is $2000 \cdot 20 \cdot 12 = \$480{,}000$.)

We can find the present value of the annuity in Example 3 with a graphing calculator by using the same steps used to find future value, with the inputs for PV and FV changed.

Under the **APPS** menu, select Finance and then TMV Solver.
a. Set N = the total number of compounding periods = 240.
b. Set I% = the annual percentage rate = 6.
c. Set FV = lump-sum investment = 0.
d. Set PMT = quarterly payment = 2000.
e. Set both P/Y (payments per year) and C/Y (compounding periods per year) equal to the number of compounding periods per year = 12.
f. Highlight END, move the cursor to PV, and press **ALPHA** **ENTER** to get the present value.

Figures 5.52(a) and (b) show the data entered and the present value for the investment in Example 3.*

---

*For details, see Appendix A, page 695.

5.6 Annuities; Loan Repayment   411

```
NORMAL FLOAT AUTO REAL RADIAN MP
N=240
I%=6
PV=0
PMT=2000
FV=0
P/Y=12
C/Y=12
PMT:END BEGIN
```
(a)

```
NORMAL FLOAT AUTO REAL RADIAN MP
N=240
I%=6
•PV=-279161.5434
PMT=2000
FV=0
P/Y=12
C/Y=12
PMT:END BEGIN
```
(b)

Figure 5.52

Spreadsheet ► SOLUTION

To find the present value of the ordinary annuity of Example 3 with Excel, we use the formula

$$= \text{pv}(F2, B2, C2, A2, 0)$$

with the periodic rate in cell B4, where

Cell F2 has the periodic interest rate as a decimal = 0.005.
Cell B2 has the total number of periods = 240.
Cell C2 has the payment amount = 2000.
Cell A2 has the future value = 0.

Figure 5.53 shows the input for the investment in Example 3, with the present value in cell B4.* The parentheses indicate a negative value, because the money *leaves* the couple to start the annuity.

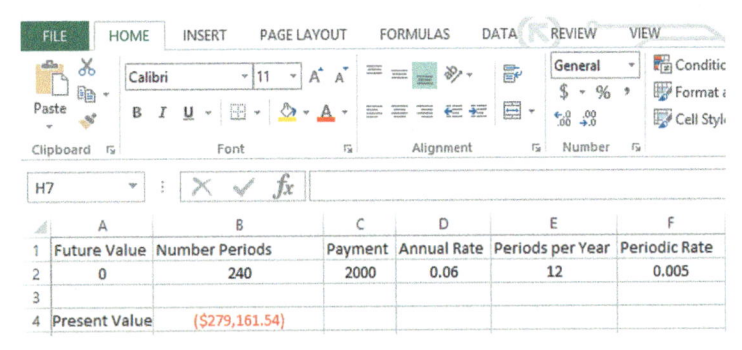

	A	B	C	D	E	F
1	Future Value	Number Periods	Payment	Annual Rate	Periods per Year	Periodic Rate
2	0	240	2000	0.06	12	0.005
3						
4	Present Value	($279,161.54)				

Figure 5.53

EXAMPLE 4 ▶ Home Mortgage

A couple who wants to purchase a home has $30,000 for a down payment and wants to make monthly payments of $2200. If the interest rate for a 25-year mortgage is 6% per year on the unpaid balance, what is the price of a house they can buy?

SOLUTION

The amount of money that they can pay for the house is the sum of the down payment and the present value of the payments that they can afford. The present value of the $12(25) = 300$ monthly payments of $2200 with interest at $\frac{6\%}{12} = \frac{0.06}{12} = 0.005$ per month is

$$A = 2200\left(\frac{1 - (1 + 0.005)^{-300}}{0.005}\right) = 341{,}455.10$$

They can buy a house costing $341,455.10 + $30,000 = $371,455.10.

---

*For details, see Appendix B, page 713.

Loan Repayment

When money is borrowed, the borrower must repay the total amount that was borrowed (the debt) plus interest on that debt. Most loans (including those for houses, for cars, and for other consumer goods) require regular payments on the debt plus payment of interest on the unpaid balance of the loan. That is, loans are paid off by a series of partial payments with interest charged on the unpaid balance at the end of each period. The stated interest rate (the **nominal rate**) is the *annual rate*. The rate per period is the nominal rate divided by the number of payment periods per year.

There are two popular repayment plans for these loans. One plan applies an equal amount to the debt each payment period plus the interest for the period, which is the interest on the unpaid balance. For example, a loan of $240,000 for 10 years at 12% could be repaid with 120 monthly payments of $2000 plus 1% $\left(\frac{1}{12} \text{ of } 12\%\right)$ of the unpaid balance each month. When this payment method is used, the payments will decrease as the unpaid balance decreases. A few sample payments are shown in Table 5.23.

Table 5.23

Payment Number (month)	Unpaid Balance ($)	Interest on the Unpaid Balance ($)	Balance Reduction ($)	Monthly Payment ($)	New Balance ($)
1	240,000	2400	2000	4400	238,000
2	238,000	2380	2000	4380	236,000
⋮					
51	140,000	1400	2000	3400	138,000
⋮					
101	40,000	400	2000	2400	38,000
⋮					
118	6000	60	2000	2060	4000

A loan of this type can also be repaid by making all payments (including the payment on the balance and the interest) of equal size. The process of repaying the loan in this way is called **amortization**. If a bank makes a loan of this type, it uses an amortization formula that gives the size of the equal payments. This formula is developed by considering the loan as an annuity purchased from the borrower by the bank, with the annuity paying a fixed return to the bank each payment period. The lump sum that the bank gives to the borrower (the principal of the loan) is the present value of this annuity, and each payment that the bank receives from the borrower is a payment from this annuity. To find the size of these equal payments, we solve the formula for the present value of an annuity, $A = R\left(\frac{1 - (1 + i)^{-n}}{i}\right)$, for R. This gives the following formula.

Amortization Formula

If a debt of A dollars, with interest at a rate of i per period, is amortized by n equal periodic payments made at the end of each period, then the size of each payment is

$$R = A\left(\frac{i}{1 - (1 + i)^{-n}}\right)$$

We can use this formula to find the *equal* monthly payments that will amortize the loan of $240,000 for 10 years at 12%. The interest rate per month is 0.01, and there are 120 periods.

$$R = A\left(\frac{i}{1-(1+i)^{-n}}\right) = 240{,}000\left(\frac{0.01}{1-(1+0.01)^{-120}}\right) \approx 3443.303$$

Thus, repaying this loan requires 120 equal payments of $3443.31. If we compare payment plans, we see that the first payment in Table 5.23 is nearly $1000 more than the equal amortization payment in Table 5.24 and that every payment after the 101st is $1000 less than the amortization payment. The advantage of the amortization method of repaying the loan is that the borrower can budget better by knowing what the payment will be each month. Of course, the balance of the loan will not be reduced as fast with this payment method, because most of the monthly payment will be needed to pay interest in the first few months. Table 5.24 shows the interest paid getting smaller, and balance reduction size increasing, as the payments increase.

Table 5.24

Payment Number	Unpaid Balance	Interest	Monthly Payment	Balance Reduction	New Balance
1	240,000	2400	3443.31	1043.31	238,956.69
2	238,956.69	2389.57	3443.31	1053.74	237,902.95
⋮					
51	172,745.38	1727.45	3443.31	1715.88	171,029.52
⋮					
101	63,136.30	631.36	3443.31	2811.95	60,324.35

EXAMPLE 5 ▶ Home Mortgage

A couple who wants to purchase a home with a price of $230,000 has $50,000 for a down payment. If they can get a 25-year mortgage at 9% per year on the unpaid balance,

a. What will be their equal monthly payments?

b. What is the total amount they will pay before they own the house outright?

c. How much interest will they pay?

SOLUTION

a. The amount of money that they must borrow is $230,000 − $50,000 = $180,000. The number of monthly payments is 12(25) = 300, and the interest rate is $\frac{9\%}{12} = \frac{0.09}{12} = 0.0075$ per month. The monthly payment is

$$R = 180{,}000\left(\frac{0.0075}{1-(1.0075)^{-300}}\right) \approx 1510.553$$

so the required payment would be $1510.56.

b. The amount that they must pay before owning the house is the down payment plus the total of the 300 payments:

$$\$50{,}000 + 300(\$1510.56) = \$503{,}168$$

c. The total interest is the total amount paid minus the price paid for the house. The interest is

$$\$503{,}168 - \$230{,}000 = \$273{,}168$$

We can find the equal payments needed to amortize the loan in Example 5 by using the same steps as those for finding the future value or present value of an ordinary annuity, with all the values entered except for the payment in PMT; moving the cursor to PMT and pressing ALPHA ENTER will give the payment amount.*

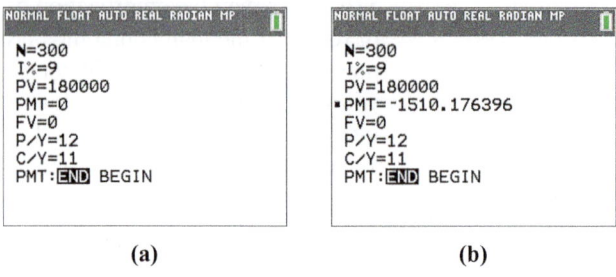

```
NORMAL FLOAT AUTO REAL RADIAN MP
N=300
I%=9
PV=180000
PMT=0
FV=0
P/Y=12
C/Y=11
PMT:END BEGIN
```
(a)

```
NORMAL FLOAT AUTO REAL RADIAN MP
N=300
I%=9
PV=180000
•PMT=-1510.176396
FV=0
P/Y=12
C/Y=11
PMT:END BEGIN
```
(b)

Figure 5.54

Excel can also be used to find the equal payments needed to amortize the loan. As with future value and present value calculations, the given data are entered appropriately; the payment amount is found using the formula Pmt(F2,B2,A2,C2,0). See Appendix B, page 712, for detailed steps.

When the interest paid is based on the periodic interest rate on the unpaid balance of the loan, the nominal rate is also called the **annual percentage rate (APR)**. You will occasionally see an advertisement for a loan with a stated interest rate and a different, larger APR. This can happen if a lending institution charges fees (sometimes called points) in addition to the stated interest rate. Federal law states that all these charges must be included in computing the APR, which is the true interest rate that is being charged on the loan.

*See Appendix A, page 695.

SKILLS CHECK 5.6

Give answers to two decimal places.

1. Solve $S = P(1 + i)^n$ for P with positive exponent n. $P = \dfrac{S}{(1+i)^n}$

2. Solve $S = P(1 + i)^n$ for P, with no denominator. $P = S(1+i)^{-n}$

3. Solve $Ai = R[1 - (1 + i)^{-n}]$ for A. $A = R\left(\dfrac{1-(1+i)^{-n}}{i}\right)$

4. Evaluate $2000\left(\dfrac{1 - (1 + 0.01)^{-240}}{0.01}\right)$. 181,638.83

5. Solve $A = R\left(\dfrac{1 - (1 + i)^{-n}}{i}\right)$ for R. $R = A\left(\dfrac{i}{1-(1+i)^{-n}}\right)$

6. Evaluate $240{,}000\left(\dfrac{0.01}{1 - (1 + 0.01)^{-120}}\right)$. 3443.30

EXERCISES 5.6

7. **IRA** Anne Wright decides to invest $4000 in an IRA CD at the end of each year for 10 years. If she makes these payments and the certificates all pay 6%, compounded annually, how much will she have at the end of the 10 years? $52,723.18

8. **Future Value** Find the future value of an annuity of $5000 paid at the end of each year for 20 years, if interest is earned at 9%, compounded annually. $255,800.60

9. **Down Payment** To start a new business, Beth deposits $1000 at the end of each 6-month period in an account

that pays 8%, compounded semiannually. How much will she have at the end of 8 years? $21,824.53

10. *Future Value* Find the future value of an annuity of $2600 paid at the end of each 3-month period for 5 years, if interest is earned at 6%, compounded quarterly. $60,121.53

11. *Retirement* Mr. Lawrence invests $600 at the end of each month in an account that pays 7%, compounded monthly. How much will be in the account in 25 years? $486,043.02

12. *Down Payment* To accumulate money for the down payment on a house, the Kings deposit $800 at the end of each month into an account paying 7%, compounded monthly. How much will they have at the end of 5 years? $57,274.32

13. *Dean's List* Parents agree to invest $1000 (at 10%, compounded semiannually) for their daughter on the December 31 or June 30 following each semester that she makes the Dean's List during her 4 years in college. If she makes the Dean's List in each of the 8 semesters, how much will the parents have to give her when she graduates? $9549.11

14. *College Tuition* To help pay for tuition, parents deposited $1000 into an account on each of their child's birthdays, starting with the first birthday. If the account pays 6%, compounded annually, how much will they have for tuition on her 19th birthday? $33,759.99

15. *Annuities* Find the present value of an annuity that will pay $1000 at the end of each year for 10 years if the interest rate is 7%, compounded annually. $7023.58

16. *Annuities* Find the present value of an annuity that will pay $500 at the end of each year for 20 years if the interest rate is 9%, compounded annually. $4564.27

17. *Lottery Winnings* The winner of a "million dollar" lottery is to receive $50,000 plus $50,000 at the end of each year for 19 years or the present value of this annuity in cash. How much cash would she receive if money is worth 8%, compounded annually? $530,179.96

18. *College Tuition* A couple wants to establish a fund that will provide $3000 for tuition at the end of each 6-month period for 4 years. If a lump sum can be placed in an account that pays 8%, compounded semiannually, what lump sum is required? $20,198.23

19. *Insurance Payment* A man is disabled in an accident and wants to receive an insurance payment that will provide him with $3000 at the end of each month for 30 years. If the payment can be placed in an account that pays 9%, compounded monthly, what size payment should he seek? $372,845.60

20. *Auto Leasing* A woman wants to lease rather than buy a car but does not want to make monthly payments. A dealer has the car she wants, which leases for $400 at the end of each month for 48 months. If money is worth 8%, compounded monthly, what lump sum should she offer the dealer to keep the car for 48 months? $16,384.77

21. *Business Sale* A man can sell his Thrifty Electronics business for $800,000 cash or for $100,000 plus $122,000 at the end of each year for 9 years.

    a. Find the present value of the annuity that is offered if money is worth 10%, compounded annually. $702,600.91

    b. If he takes the $800,000, spends $100,000 of it, and invests the rest in a 9-year annuity at 10%, compounded annually, what size annuity payment will he receive at the end of each year? $121,548.38

    c. Which is better, taking the $100,000 and the annuity or taking the cash settlement? Discuss the advantages of your choice. $100,000 plus the annuity has a larger present value than $800,000.

22. *Sale of a Practice* A physician can sell her practice for $1,200,000 cash or for $200,000 plus $250,000 at the end of each year for 5 years.

    a. Find the present value of the annuity that is offered if money is worth 7%, compounded annually. $1,025,049.36

    b. If she takes the $1,200,000, spends $200,000 of it, and invests the rest in a 5-year annuity at 7%, compounded annually, what size annuity payment will she receive at the end of each year? $243,890.69

    c. Which is better, taking the $200,000 and the annuity or taking the cash settlement? Discuss the advantages of your choice. $200,000 plus the annuity has a larger present value than $1,200,000.

23. *Home Mortgage* A couple wants to buy a house and can afford to pay $1600 per month.

    a. If they can get a loan for 30 years with interest at 9% per year on the unpaid balance and make monthly payments, how much can they pay for a house? $198,850.99

    b. What is the total amount paid over the life of the loan? $576,000

    c. What is the total interest paid on the loan? $377,149.01

24. *Auto Loan* A man wants to buy a car and can afford to pay $400 per month.

    a. If he can get a loan for 48 months with interest at 12% per year on the unpaid balance and make monthly payments, how much can he pay for a car? $15,189.58

    b. What is the total amount paid over the life of the loan? $19,200

    c. What is the total interest paid on the loan? $4010.42

25. *Loan Repayment* A loan of $10,000 is to be amortized with quarterly payments over 4 years. If the interest on the loan is 8% per year, paid on the unpaid balance,

a. What is the interest rate charged each quarter on the unpaid balance? 2%

b. How many payments are made to repay the loan? 16

c. What payment is required quarterly to amortize the loan? $736.50

26. Loan Repayment A loan of $36,000 is to be amortized with monthly payments over 6 years. If the interest on the loan is 6% per year, paid on the unpaid balance,

a. What is the interest rate charged each month on the unpaid balance? 0.5%

b. How many payments are made to repay the loan? 72

c. What payment is required each month to amortize the loan? $596.62

27. Home Mortgage A couple who wants to purchase a home with a price of $350,000 has $100,000 for a down payment. If they can get a 30-year mortgage at 6% per year on the unpaid balance,

a. What will be their monthly payments? $1498.88

b. What is the total amount they will pay before they own the house outright? $639,596.80

c. How much interest will they pay over the life of the loan? $289,596.80

28. Business Loan Business partners want to purchase a restaurant that costs $750,000. They have $300,000 for a down payment, and they can get a 25-year business loan for the remaining funds at 8% per year on the unpaid balance, with quarterly payments.

a. What will be the payments? $10,441.23

b. What is the total amount that they will pay over the 25-year period? $1,344,123

c. How much interest will they pay over the life of the loan? $594,123

5.7 Logistic and Gompertz Functions

KEY OBJECTIVES

- Graph and apply logistic functions
- Find limiting values of logistic functions
- Model with logistic functions
- Graph and apply Gompertz functions

SECTION PREVIEW Women in the Workforce

Figure 5.55 shows the total number of women in the workforce (in millions) for selected years from 1950 and projected to 2050. The S-shaped graph shows that the number of women in the workforce grew rapidly during the 20th century and that it is projected to grow more slowly in the first half of the 21st century. Very seldom does exponential growth continue indefinitely, so functions with this type of graph better represent many types of sales and organizational growth. Two functions that are characterized by rapid growth followed by leveling off are **logistic functions** and **Gompertz functions**. We consider applications of these functions in this section.

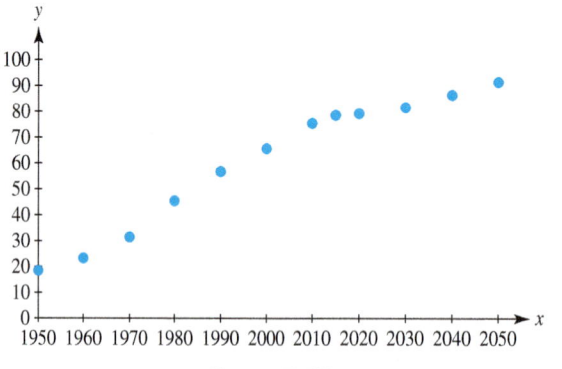

Figure 5.55

Logistic Functions

When growth begins slowly, then increases at a rapid rate, and finally slows over time to a rate that is almost zero, the amount (or number) present at any given time frequently fits on an S-shaped curve. Growth of business organizations and the spread of a virus or

5.7 Logistic and Gompertz Functions

disease sometimes occur according to this pattern. For example, the total number y of people on a college campus infected by a virus can be modeled by

$$y = \frac{10{,}000}{1 + 9999e^{-0.99t}}$$

where t is the number of days after an infected student arrives on campus. Table 5.25 shows the number infected on selected days. Note that the number infected grows rapidly and then levels off near 10,000. The graph of this function is shown in Figure 5.56.

Table 5.25

Days	1	3	5	7	9	11	13	15	17
Number Infected	3	19	139	928	4255	8429	9749	9965	9995

A function that can be used to model growth of this type is called a **logistic function**; its graph is similar to the one shown in Figure 5.56.

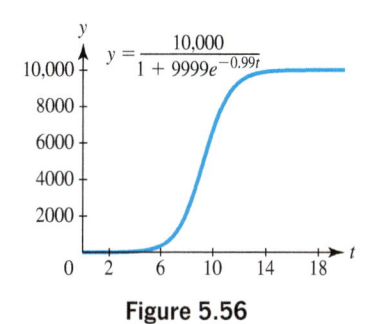

Figure 5.56

Logistic Function

For real numbers a, b, and c, the function

$$f(x) = \frac{c}{1 + ae^{-bx}}$$

is a logistic function.

If $a > 0$, a logistic function increases when $b > 0$ and decreases when $b < 0$.

The parameter c is often called the **limiting value**,* or **upper limit**, because the line $y = c$ is a horizontal asymptote for the logistic function. As seen in Figure 5.57, the line $y = 0$ is also a horizontal asymptote. Logistic growth begins at a rate that is near zero, rapidly increases in an exponential pattern, and then slows to a rate that is again near zero.

Logistic Growth Function

Equation: $y = \dfrac{c}{1 + ae^{-bx}} \quad (a > 0, b > 0)$

x-intercept: none

y-intercept: $\left(0, \dfrac{c}{1 + a}\right)$

Horizontal asymptotes: $y = 0, y = c$

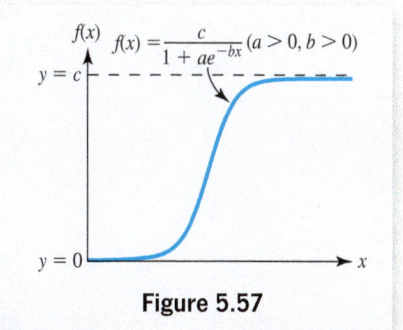

Figure 5.57

Even though the graph of an increasing logistic function has the elongated S-shape indicated in Figure 5.57, it may be the case that we have data showing only a portion of the S.

---

*The logistic function fit by most technologies is a best-fit (least squares) logistic equation. There may therefore be data values that are larger than the value of c. To avoid confusion, we talk about the limiting value of the logistic function or the upper limit in the problem context.

EXAMPLE 1 ▶ Women in the Workforce

Table 5.26 gives the total number of women in the workforce (in millions) for selected years from 1950 and projected to 2050.

a. Create a scatter plot of the data, with x equal to the number of years after 1940 and y equal to the number of women in the workforce, in millions.

b. Find the logistic function that models these data, with x and y defined as in (a).

c. Graph the data and the model on the same axes, and comment on the fit.

d. What will be the number of women in the workforce in 2032, according to the model?

e. According to the model, what is the upper limit on the number of women in the workforce?

Table 5.26

Year	Women in Workforce (millions)
1950	18.4
1960	23.2
1970	31.5
1980	45.5
1990	56.8
2000	65.7
2010	75.5
2015	78.6
2020	79.3
2030	81.6
2040	86.5
2050	91.5

(*Source:* U.S. Bureau of Labor Statistics)

SOLUTION

a. A scatter plot of the data, with x equal to the number of years after 1940 and y equal to the number of women in the workforce, in millions, is shown in Figure 5.58(a).

b. The logistic function that models the data is $f(x) = \dfrac{92.227}{1 + 7.537e^{(-0.049x)}}$, rounded to three decimal places. (See Figure 5.58(b).)

c. The graphs of the data and the logistic model are shown in Figure 5.58(c). The fit is visually very good.

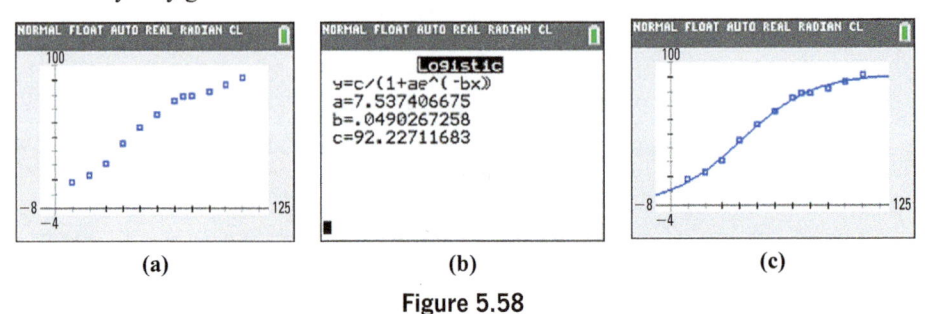

(a) (b) (c)

Figure 5.58

d. According to the model, the number of women in the workforce in 2032 will be $f(92) = 85.17$ million. See Figure 5.59.

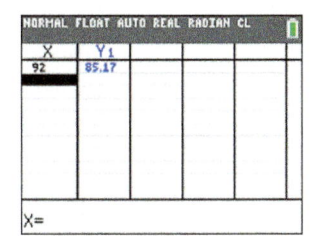

Figure 5.59

e. The value of c in the logistic function is 92.227, so this means the upper limit on the number of women in the workforce is 92.227 million according to this model. The model will probably become invalid eventually.

A logistic function can decrease; this behavior occurs when $b < 0$. Demand functions for products often follow such a pattern. Decreasing logistic functions can also be used to represent decay over time. Figure 5.60 shows a decreasing logistic function.

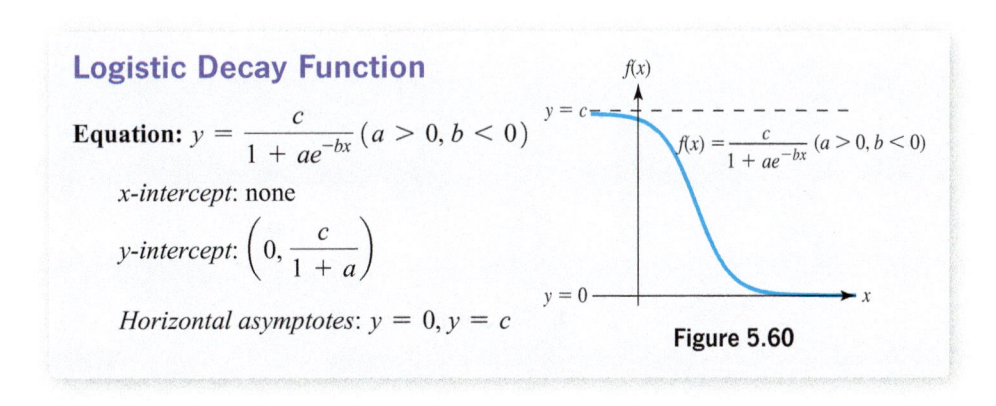

Logistic Decay Function

Equation: $y = \dfrac{c}{1 + ae^{-bx}}$ $(a > 0, b < 0)$

x-intercept: none

y-intercept: $\left(0, \dfrac{c}{1 + a}\right)$

Horizontal asymptotes: $y = 0, y = c$

Figure 5.60

The properties of logarithms can also be used to solve for a variable in an exponent in a logistic function.

EXAMPLE 2 ▶ Expected Life Span

Using data from 1920 and projected to 2050, the expected life span at birth of people born in the United States can be modeled by the equation

$$y = \dfrac{87.423}{1 + 0.842e^{-0.019x}}$$

where x is the number of years from 1900.

Use this model to estimate the birth year after 1900 that gives an expected life span of 82 years with

a. Graphical methods.

b. Algebraic methods.

SOLUTION

a. To solve the equation

$$82 = \frac{87.423}{1 + 0.842e^{-0.019x}}$$

with the *x*-intercept method, we solve

$$0 = \frac{87.423}{1 + 0.842e^{-0.019x}} - 82$$

Graphing the function

$$y = \frac{87.423}{1 + 0.842e^{-0.019x}} - 82$$

on the window [100, 150] by [−4, 4] shows a point where the graph crosses the *x*-axis (Figure 5.61(a)). Finding the *x*-intercept gives the approximate solution

$$x \approx 133.9$$

Since $x = 133.9$ represents the number of years from 1900, we conclude that people born in 2034 have an expected life span of 82 years.

We can also use the intersection method to solve the equation graphically. To do this, we graph the equations $y_1 = \dfrac{87.423}{1 + 0.842e^{-0.019x}}$ and $y_2 = 82$ and find the point of intersection of the graphs (Figure 5.61(b)).

(a) (b)

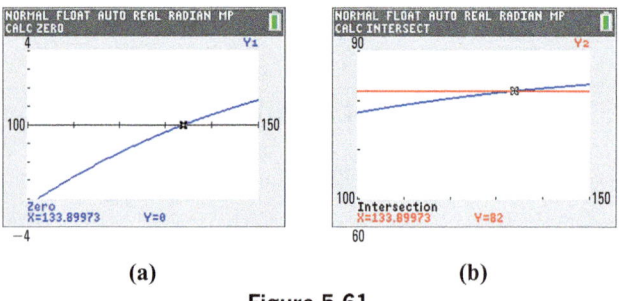

Figure 5.61

b. We can also solve the equation

$$82 = \frac{87.423}{1 + 0.842e^{-0.019x}}$$

algebraically. To do this, we first multiply both sides by the denominator and then solve for $e^{-0.019x}$:

$$82(1 + 0.842e^{-0.019x}) = 87.423$$
$$82 + 69.044e^{-0.019x} = 87.423$$
$$69.044e^{-0.019x} = 5.423$$
$$e^{-0.019x} = \frac{5.423}{69.044}$$

Taking the natural logarithm of both sides of the equation and using a logarithmic property are the next steps in the solution:

$$\ln e^{-0.019x} = \ln \frac{5.423}{69.044}$$
$$-0.019x = \ln \frac{5.423}{69.044}$$
$$x = \frac{\ln \dfrac{5.423}{69.044}}{-0.019} \approx 133.9$$

5.7 Logistic and Gompertz Functions   421

This solution is the same solution that was found (more easily) by using the graphical method in Figure 5.61(a); 2034 is the birth year for people with an expected life span of 82 years.

EXAMPLE 3 ▶ Uninsured Americans

The projected annual decrease in the number of uninsured people (in millions) under the Affordable Health Care law from 2014 and projected to 2022 is shown in Table 5.27.

a. Create a scatter plot of the data. Does a logistic function seem appropriate to model the data?

b. Find a logistic function that models the data, with y in millions and x in years after 2012.

c. Graph the model and the data on the same axes and comment on the fit.

d. What does this model project the decrease in the number of uninsured people to be in 2020?

e. What is the upper limit on the annual decrease in the number of uninsured Americans, according to the model?

Table 5.27

Years	Annual Decrease (millions)
2014	15
2015	20
2016	26
2017	28
2018	29
2019	29
2020	29
2021	31
2022	31

(Source: Wall Street Journal, July 25, 2012)

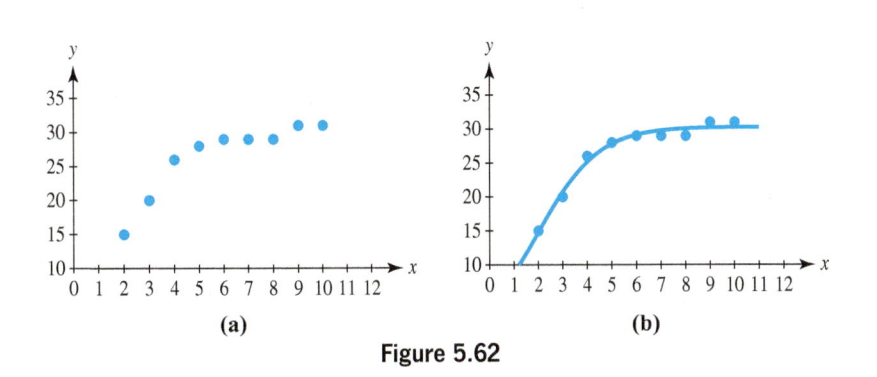

SOLUTION

a. The scatter plot of the data is shown in Figure 5.62(a), with y representing the annual decrease of uninsured people, in millions, and x representing the number of years after 2012. Because the scatter plot resembles an S-shape, with initial rapid growth and a slowing over time, a logistic function seems appropriate to model the data.

b. Technology gives the equation of the best-fitting logistic function as

$$y = \frac{30.298}{1 + 5.436e^{-0.822x}}$$

c. The graphs of the model and the data are shown in Figure 5.62(b). The fit appears to be excellent.

d. The model projects that the decrease in the number of uninsured people will be 30.1 million in 2020.

e. The parameter c in the logistic function is 30.298, which means that the upper limit on the annual decrease of uninsured Americans is 30.298 million. However, the actual data show that the annual decrease is expected to exceed 30.298 million in 2021 and 2022.

Figure 5.62

Gompertz Functions

Another type of function that models rapid growth that eventually levels off is called a **Gompertz function**. This type of function can also be used to describe human growth and development and the growth of organizations in a limited environment. These functions have equations of the form

$$N = Ca^{R^t}$$

where t represents the time, $R\,(0 < R < 1)$ is the expected rate of growth of the population, C is the maximum possible number of individuals, and a represents the proportion of C present when $t = 0$.

For example, the equation

$$N = 3000(0.2)^{0.6^t}$$

could be used to predict the number of employees t years after the opening of a new facility. Here the maximum number of employees C would be 3000, the proportion of 3000 present when $t = 0$ is $a = 0.2$, and R is 0.6. The graph of this function is shown in Figure 5.63. Observe that the initial number of employees, at $t = 0$, is

$$N = 3000(0.2)^{0.6^0} = 3000(0.2)^1 = 3000(0.2) = 600$$

To verify that the maximum possible number of employees is 3000, observe the following. Because $0.6 < 1$, higher powers of t make 0.6^t smaller, with 0.6^t approaching 0 as t approaches ∞. Thus, $N \to 3000(0.2)^0 = 3000(1) = 3000$ as $t \to \infty$.

Figure 5.63

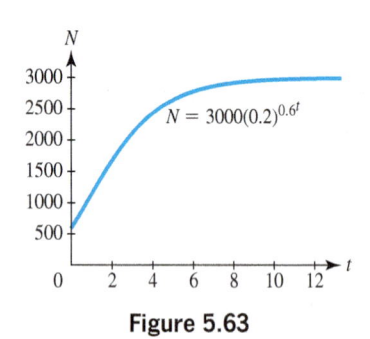

EXAMPLE 4 ▶ Deer Population

The Gompertz equation

$$N = 1000(0.06)^{0.2^t}$$

predicts the size of a deer herd on a small island t decades from now.

a. What is the number of deer on the island now $(t = 0)$?

b. How many deer are predicted to be on the island 1 decade from now $(t = 1)$?

c. Graph the function.

d. What is the maximum number of deer predicted by this model?

e. In how many years will there be 840 deer on the island?

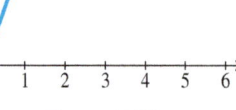

Figure 5.64

SOLUTION

a. If $t = 0$, $N = 1000(0.06)^{0.2^0} = 1000(0.06)^1 = 60$.

b. If $t = 1$, $N = 1000(0.06)^{0.2^1} = 1000(0.06)^{0.2} = 570$ (approximately).

c. The graph is shown in Figure 5.64.

d. According to the model, the maximum number of deer is predicted to be 1000. Evaluating N as t gets large, we see the values of N approach, but never reach, 1000. We say that $N = 1000$ is an asymptote for this graph.

e. To find the number of years from now when there will be 840 deer on the island, we graph $y = 840$ with $y = 1000(0.06)^{0.2^x}$ and find the point of intersection to be $(1.728, 840)$. Thus, after about 17 years there will be 840 deer on the island. (See Figure 5.65.)

Figure 5.65

EXAMPLE 5 ▶ Company Growth

A new dotcom company starts with 3 owners and 5 employees but tells investors that it will grow rapidly, with the total number of people in the company given by the model

$$N = 2000(0.004)^{0.5^t}$$

where t is the number of years from the present. Use graphical methods to determine the year in which the number of employees is predicted to be 1000.

SOLUTION

If the company has 1000 employees plus the 3 owners, the total number in the company will be 1003; so we seek to solve the equation

$$1003 = 2000(0.004)^{0.5^t}$$

Using the equations $y_1 = 1003$ and $y_2 = 2000(0.004)^{0.5^t}$ and the intersection method gives $t = 3$ (Figure 5.66). Because t represents the number of years, the owners predict that they will have 1000 employees in 3 years.

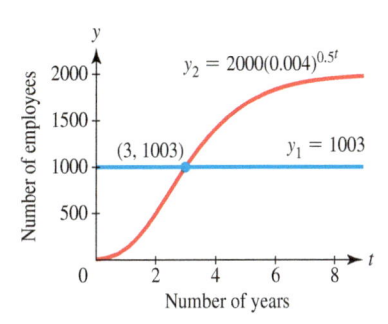

Figure 5.66

SKILLS CHECK 5.7

Answers that are not seen can be found in the answer section at the back of the text.
Give approximate answers to two decimal places.

1. Evaluate $\dfrac{79.514}{1 + 0.835e^{-0.0298(80)}}$. 73.83

2. If $y = \dfrac{79.514}{1 + 0.835e^{-0.0298x}}$ find

    a. y when $x = 10$. $y \approx 49.09$

    b. y when $x = 50$. $y \approx 66.92$

3. Evaluate $1000(0.06)^{0.2^t}$ for $t = 4$ and for $t = 6$.
   995.51; 999.82

4. Evaluate $2000(0.004)^{0.5^t}$ for $t = 5$ and for $t = 10$.
   ≈ 1683.04; ≈ 1989.24

5. a. Graph the function $f(x) = \dfrac{100}{1 + 3e^{-x}}$ for $x = 0$ to $x = 15$.

    b. Find $f(0)$ and $f(10)$. 25; 99.99

    c. Is this function increasing or decreasing? Increasing

    d. What is the limiting value of this function? 100

6. a. Graph $f(t) = \dfrac{1000}{1 + 9e^{-0.9t}}$ for $t = 0$ to $t = 15$.

    b. Find $f(2)$ and $f(5)$. 401.98; 909.11

    c. What is the limiting value of this function? 1000

7. a. Graph $y = 100(0.05)^{0.3^x}$ for $x = 0$ to $x = 10$.

    b. What is the initial value of this function (the y-value when $x = 0$)? 5

    c. What is the limiting value of this function? 100

8. a. Graph $N = 2000(0.004)^{0.5^t}$ for $t = 0$ to $t = 10$.

    b. What is the initial value of this function (the y-value when $x = 0$)? 8

    c. What is the limiting value of this function? 2000

EXERCISES 5.7

Answers that are not seen can be found in the answer section at the back of the text.

9. **Spread of a Disease** The spread of a highly contagious virus in a high school can be described by the logistic function

$$y = \frac{5000}{1 + 999e^{-0.8x}}$$

where x is the number of days after the virus is identified in the school and y is the total number of people who are infected by the virus.

a. Graph the function for $0 \leq x \leq 15$.

b. How many students had the virus when it was first discovered? 5

c. What is the upper limit of the number infected by the virus during this period? 5000

10. **Population Growth** Suppose that the size of the population of an island is given by

$$p(t) = \frac{98}{1 + 4e^{-0.1t}} \text{ thousand people}$$

where t is the number of years after 1988.

a. Graph this function for $0 \leq x \leq 30$.

b. Find and interpret $p(10)$.
  ≈ 39.652; 1998 population is approximately 39,652.
c. Find and interpret $p(100)$.
  ≈ 97.982; 2088 population is approximately 97,982.
d. What appears to be an upper limit for the size of this population? 98,000

11. **Sexually Active Boys** The percent of boys between ages 15 and 20 who have been sexually active at some time (the cumulative percent) can be modeled by the logistic function

$$y = \frac{89.786}{1 + 4.6531e^{-0.8256x}}$$

where x is the number of years after age 15.

a. Graph the function for $0 \leq x \leq 5$.

b. What does the model estimate the cumulative percent to be for boys whose age is 16? 29.56%

c. What cumulative percent does the model estimate for boys of age 21, if it is valid after age 20? 86.93%

d. What is the limiting value implied by this model?
(*Source*: "National Longitudinal Survey of Youth," riskingthefuture.com) 89.786%

12. **Sexually Active Girls** The percent of girls between ages 15 and 20 who have been sexually active at some time (the cumulative percent) can be modeled by the logistic function

$$y = \frac{83.84}{1 + 13.9233e^{-0.9248x}}$$

where x is the number of years after age 15.

a. Graph the function for $0 \leq x \leq 5$.

b. What does the model estimate the cumulative percent to be for girls of age 16? 12.85%

c. What cumulative percent does the model estimate for girls of age 20? 73.76%

d. What is the upper limit implied by the given logistic model? 83.84%
(*Source*: "National Longitudinal Survey of Youth," riskingthefuture.com)

13. **Spread of a Rumor** The number of people in a small town who are reached by a rumor about the mayor and an intern is given by

$$N = \frac{10{,}000}{1 + 100e^{-0.8t}}$$

where t is the number of days after the rumor begins.

a. How many people will have heard the rumor by the end of the first day? Approximately 218

b. How many will have heard the rumor by the end of the fourth day? 1970

c. Use graphical or numerical methods to find the day on which 7300 people in town have heard the rumor. Seventh day

14. **Sexually Active Girls** The cumulative percent of sexually active girls ages 15 to 20 is given in the table below.

Age (years)	Cumulative Percent Sexually Active
15	5.4
16	12.6
17	27.1
18	44.0
19	62.9
20	73.6

(*Source*: "National Longitudinal Survey of Youth," riskingthefuture.com)

a. A logarithmic function was used to model these data in Exercises 5.4, problem 33. Find the logistic function that models the data, with x equal to the number of years after age 15. $y = \dfrac{83.84}{1 + 13.9233e^{-0.9248x}}$

b. Does this model agree with the model used in Exercise 12? Yes

c. Find the linear function that is the best fit for the data. $y = 14.537x + 1.257$

d. Graph each function on the same axes with the data points to determine which model appears to be the better fit for the data. Logistic model

15. **Sexually Active Boys** The cumulative percent of sexually active boys ages 15 to 20 is given in the table.

    a. A logarithmic function was used to model this data in Exercises 5.4, problem 34. Create the logistic function that models the data, with x equal to the number of years after age 15. $y = \dfrac{89.786}{1 + 4.6531e^{-0.8256x}}$

    b. Does this agree with the model used in Exercise 11? Yes

    c. Find the linear function that is the best fit for the data. $y = 14.137x + 17.624$

    d. Determine which model appears to be the better fit for the data and graph this function on the same axes with the data points. Logistic model

Age (years)	Cumulative Percent Sexually Active
15	16.6
16	28.7
17	47.9
18	64.0
19	77.6
20	83.0

(Source: "National Longitudinal Survey of Youth," riskingthefuture.com)

16. **Population Aged 20–64** The population of U.S. citizens aged 20–64, in millions, is shown in the table near the top of the next column.

    a. Find a logistic function that is the best fit for the data, with x equal to the number of years after 1940. Report the best model as $f(x)$ with 3 significant digits. $f(x) = \dfrac{250}{1 + 2.53e^{-0.0280x}}$

    b. Graph the data and the model on the same axes and comment on the fit.

    c. Use the model to predict the population in 2023. 199.9 million

    d. What does the model predict will be the maximum number in this population? 250 million

e. Is it realistic to assume this population will be limited to the number found in part (d) forever, or to assume this model ceases to be valid eventually? This model ceases to be valid eventually.

Year	Population (millions)	Year	Population (millions)
1950	92.8	2010	189.1
1960	99.8	2015	194.6
1970	113.0	2020	198.2
1980	134.0	2030	202.8
1990	152.7	2040	213.3
2000	169.8	2050	224.3

(Source: Social Security Administration)

17. **Global EV/PHEV Sales** Using data from 2020 and projected to 2050, global sales of electric and plug-in hybrid electric vehicles can be modeled by the function

$$y = \dfrac{113}{1 + 63.0e^{-0.155x}}$$

where y is the millions of vehicle sales and x is the number of years after 2010. (Source: International Energy Agency)

a. Find the Global EV/PHEV sales predicted for 2035 by this function. 48.97 million

b. What is the maximum number of Global EV/PHEV sales, according to this model? 113 million

18. **World Population** The total population of the world, by decade, from 1950 and projected to 2050 is shown in the table below.

a. Find the logistic function that is the best fit for the data, with x equal to the number of years after 1940 and y equal to the world population, in billions. $y = \dfrac{11.355}{1 + 4.726e^{-0.028x}}$

b. Graph the model and the data on the same axes, and comment on the fit. Excellent fit

c. What does the model predict the world population will be in 2028? 8.1 billion

Year	Total World Population	Year	Total World Population
1950	2,556,000,053	2010	6,848,932,929
1960	3,039,451,023	2020	7,584,821,144
1970	3,706,618,163	2030	8,246,619,341
1980	4,453,831,714	2040	8,850,045,889
1990	5,278,639,789	2050	9,346,399,468
2000	6,082,966,429		

(Source: U.S. Census Bureau, International Database)

19. Life Span The expected life span of people in the United States for certain birth years is given in the table.

Birth Year	Life Expectancy	Birth Year	Life Expectancy
1920	54.1	1990	75.4
1930	59.7	2000	76.8
1940	62.9	2010	78.7
1950	68.2	2020	80.2
1960	69.7	2030	81.7
1970	70.8	2040	83.0
1980	73.7	2050	84.4

(*Source*: National Center for Health Statistics)

a. Find a logistic function to model the data, with x equal to how many years after 1900 the birth year is. $y = \dfrac{87.423}{1 + 0.842e^{-0.019x}}$

b. Estimate the expected life span of a person born in the United States in 1955 and a person born in 2025. 67.5 years; 81.1 years

c. Find an upper limit for a person's expected life span in the United States, according to this model. 87.4 years

20. Worldwide Internet Penetration The table shows the percent of people in the world who are Internet users during selected years from 2014 and projected to 2025.

a. Find the logistic function that models these data, with x representing the number of years after 2010 and p representing the percent. Report the model with three significant digits.
$y = 66.0/(1 + 1.21e^{-0.178x})$

b. Use the reported model to predict the percent of people in the world who are Internet users in 2030. 63.8%

c. Use the reported model to project when the percent will reach 65. In 2035

Year	Percent
2014	41.1
2015	44.3
2016	46.8
2017	49.0
2018	51.1
2019	53.0
2020	54.6
2025	61.0

(*Source*: eMarketer)

21. Organizational Growth A new community college predicts that its student body will grow rapidly at first and then begin to level off according to the Gompertz curve with equation $N = 10{,}000(0.4)^{0.2^t}$ students, where t is the number of years after the college opens.

a. What does this model predict the number of students to be when the college opens? 4000

b. How many students are predicted to attend the college after 4 years? Approximately 9985

c. Graph the equation for $0 \leq t \leq 10$ and estimate an upper limit on the number of students at the college. 10,000

22. Organizational Growth A new technology company started with 6 employees and predicted that its number of employees would grow rapidly at first and then begin to level off according to the Gompertz function

$$N = 150(0.04)^{0.5^t}$$

where t is the number of years after the company started.

a. What does this model predict the number of employees to be in 8 years? Approximately 148

b. Graph the function for $0 \leq t \leq 10$ and estimate the maximum predicted number of employees. 150

23. Sales Growth The president of a company predicts that sales will increase rapidly after a new product is brought to the market and that the number of units sold monthly can be modeled by $N = 40{,}000(0.2)^{0.4^t}$, where t represents the number of months after the product is introduced.

a. How many units will be sold by the end of the first month? 21,012

b. Graph the function for $0 \leq t \leq 10$.

c. What is the predicted upper limit on sales? 40,000

24. Company Growth Because of a new research grant, the number of employees in a firm is expected to grow, with the number of employees modeled by

$N = 1600(0.6)^{0.2^t}$, where t is the number of years after the grant was received.

a. How many employees did the company have when the grant was received? 960

b. How many employees did the company have at the end of 3 years after the grant was received? Approximately 1593

c. What is the expected upper limit on the number of employees? 1600

d. Graph the function.

25. **Company Growth** Suppose that the number of employees in a new company is expected to grow, with the number of employees modeled by $N = 1000(0.01)^{0.5^t}$, where t is the number of years after the company was formed.

   a. How many employees did the company have when it started? 10

   b. How many employees did the company have at the end of 1 year? 100

   c. What is the expected upper limit on the number of employees? 1000

   d. Use graphical or numerical methods to find the year in which 930 people are employed.
   6 yr after the company was formed

26. **Sales Growth** The president of a company predicts that sales will increase rapidly after a new advertising campaign, with the number of units sold weekly modeled by $N = 8000(0.1)^{0.3^t}$, where t represents the number of weeks after the advertising campaign begins.

   a. How many units per week will be sold at the beginning of the campaign? 800

   b. How many units will be sold at the end of the first week? Approximately 4009

   c. What is the expected upper limit on the number of units sold per week? 8000

   d. Use graphical or numerical methods to find the first week in which 6500 units will be sold.
   Second week

27. **Spread of Disease** An employee brings a contagious disease to an office with 150 employees. The number of employees infected by the disease t days after the employees are first exposed to it is given by

$$N = \frac{100}{1 + 79e^{-0.9t}}$$

Use graphical or numerical methods to find the number of days until 99 employees have been infected.
10 days

28. **Advertisement** The number of people in a community of 15,360 who are reached by a particular advertisement t weeks after it begins is given by

$$N(t) = \frac{14{,}000}{1 + 100e^{-0.6t}}$$

Use graphical or numerical methods to find the number of weeks until at least half of the community is reached by this advertisement. 8 weeks

29. **Population Growth** A pair of deer are introduced on a small island, and the population grows until the food supply and natural enemies of the deer on the island limit the population. If the number of deer is

$$N = \frac{180}{1 + 89e^{-0.5554t}}$$

where t is the number of years after the deer are introduced, how long does it take for the deer population to reach 150? 11 yr

30. **Spread of Disease** A student brings a contagious disease to an elementary school of 1200 students. If the number of students infected by the disease t days after the students are first exposed to it is given by

$$N = \frac{800}{1 + 799e^{-0.9t}}$$

use numerical or graphical methods to find in how many days at least 500 students will be infected.
8 days

Preparing for CALCULUS

In Exercises 1 and 2, functions involving exponential functions and their derivatives are given. To see where turns occur in the graphs of the functions, set the derivatives equal to 0 and solve for x.

1. $y = x^2 e^x$; $y' = x^2 e^x + e^x(2x)$ $x = 0, x = -2$

2. $y = \dfrac{x^3}{e^{x^2}}$; $y' = \dfrac{e^{x^2}(3x^2) - x^3 e^{x^2}(2x)}{(e^{x^2})^2}$ $x = 0, x = \sqrt{3/2}, x = -\sqrt{\dfrac{3}{2}}$

3. The number e is developed in calculus by showing that $(1 + x)^{1/x}$ approaches e as x approaches 0. Use your calculator to evaluate $(1 + x)^{1/x}$ to five decimal places for the following values of x.

x	1	0.1	0.01	0.001	0.0001	0.00001
$(1+x)^{1/x}$	a	b	c	d	e	f

a. 2 b. 2.59374 c. 2.70481 d. 2.71692 e. 2.71815 f. 2.71827

4. Compute e^1 on your calculator and compare the results in (a) through (f) with it.
$e \approx 2.71828$; the values in (a)–(f) approach the approximate value of e in the calculator.

Use the graph of $y = \ln x$ to answer the questions in 5 and 6.

5. As x approaches values closer and closer to 0, what values does $\ln x$ assume?
y approaches $-\infty$

6. As x approaches increasing larger values, without bound, what values does $\ln x$ assume? y approaches ∞

Calculus operations on expressions involving logarithms can frequently be performed more easily if the properties of logarithms are used. In Exercises 7–12, use log properties to write each expression as the sum or difference of two logarithmic expressions.

7. $\log\left(\dfrac{x}{x+1}\right)$ $\log x - \log(x+1)$

8. $\log\left(\dfrac{a}{x}\right)$ $\log a - \log x$

9. $\ln(ax^2)$ $\ln a + 2\ln x$

10. $\ln\left[(x^2+1)(4x^3+5)\right]$ $\ln(x^2+1) + \ln(4x^3+5)$

11. $\ln\left(\dfrac{x^2 - 4x + 3}{x^2 + 9}\right)$ $\ln(x^2 - 4x + 3) - \ln(x^2 + 9)$

12. $\ln\sqrt[4]{\dfrac{(x+3)^2}{x^3 - 4}}$ $\dfrac{1}{4}[\ln(x+3)^2 - \ln(x^3 - 4)]$

Derivatives of functions involving logarithms frequently require simplification. Simplify the following derivative expressions.

13. $\dfrac{1}{3} \cdot \dfrac{1}{2x^6 - 3x + 2}(12x^5 - 3)$ $\dfrac{(4x^5 - 1)}{2x^6 - 3x + 2}$

14. $\dfrac{(2x+1)\frac{1}{2x+1}(2) - [\ln(2x+1)]2}{(2x+1)^2}$ $\dfrac{2 - 2\ln(2x+1)}{(2x+1)^2}$

428

chapter 5 | SUMMARY

In this chapter, we discussed exponential and logarithmic functions and their applications. We explored the fact that logarithmic and exponential functions are inverses of each other. Many applications in today's world involve exponential and logarithmic functions, including the pH of substances, stellar magnitude, intensity of earthquakes, the growth of bacteria, the decay of radioactive isotopes, and compound interest. Logarithmic and exponential equations are solved by using both algebraic and graphical solution methods.

Key Concepts and Formulas

5.1 Exponential Functions

Exponential function	If a and b are real numbers, $b > 0$ and $b \neq 1$, then the function $f(x) = a(b^x)$ is an exponential function.
Exponential growth	Whenever the base b of $f(x) = a(b^x)$ is greater than 1 and $a > 0$, the exponential function is increasing and can be used to model exponential growth. The graph of an exponential growth model resembles the graph at the right. The x-axis is a horizontal asymptote, with the graph approaching $-\infty$ on the left.
Exponential decay	Whenever the base b of $f(x) = a(b^{kx})$ is greater than 1 and $a > 0$ and $k < 0$, the exponential function is decreasing and can be used to model exponential decay. The graph of an exponential decay model resembles the graph at the right. The x-axis is a horizontal asymptote, with the graph approaching $+\infty$ on the right. The function $f(x) = a(c^{kx})$, $a > 0, 0 < c < 1, k > 0$, is also an exponential decay function.
Number e	Many real applications involve exponential functions with the base e. The number e is an irrational number with decimal approximation 2.718281828 (to nine decimal places).

5.2 Logarithmic Functions; Properties of Logarithms

Logarithmic function	For $x > 0, a > 0$, and $a \neq 1$, the logarithmic function with base a is $y = \log_a x$, which is defined by $x = a^y$. The graph of $y = \log_a x$ for the base $a > 0$ is shown at the right. The y-axis is a vertical asymptote for the graph.
Logarithmic and exponential forms	These forms are equivalent: $$y = \log_a x \quad \text{and} \quad x = a^y$$

Common logarithms	Logarithms with a base of 10; $\log_{10} x$ is denoted $\log x$.
Natural logarithms	Logarithms with a base of e; $\log_e x$ is denoted $\ln x$.
Properties of logarithms	For $b > 0$, $b \neq 1$, k a real number, and M and N positive real numbers, Property 1. $\log_b b = 1$ Property 2. $\log_b 1 = 0$ Property 3. $\log_b b^x = x$ Property 4. $b^{\log_b x} = x$ Property 5. If $M = N$, then $\log_b M = \log_b N$ Property 6. $\log_b(MN) = \log_b M + \log_b N$ Property 7. $\log_b\left(\dfrac{M}{N}\right) = \log_b M - \log_b N$ Property 8. $\log_b M^k = k \log_b M$

5.3 Exponential and Logarithmic Equations

Solving exponential equations	Many exponential equations can be solved by converting to logarithmic form; that is, $x = a^y$ is equivalent to $y = \log_a x$.
Change of base	If $a > 0$, $a \neq 1$, $b > 0$, $b \neq 1$, and $x > 0$, then $\log_b x = \dfrac{\log_a x}{\log_a b}$.
Solving exponential equations	Some exponential equations can be solved by taking the logarithm of both sides and using logarithmic properties.
Solving logarithmic equations	Many logarithmic equations can be solved by converting to exponential form; that is, $y = \log_a x$ is equivalent to $x = a^y$. Sometimes properties of logarithms can be used to rewrite the equation in a form that can be more easily solved.
Exponential and logarithmic inequalities	Solving inequalities involving exponential and logarithmic functions involves rewriting and solving the related equation. Then graphical methods are used to find the values of the variable that satisfy the inequality.

5.4 Exponential and Logarithmic Models

Exponential models	Many sources provide data that can be modeled by exponential growth and decay functions according to the model $y = ab^x$ for real numbers a and b, $b > 0$ and $b \neq 1$.
Constant percent change	If the percent change of the outputs of a set of data is constant for equally spaced inputs, an exponential function will be a perfect fit for the data. If the percent change of the outputs is approximately constant for equally spaced inputs, an exponential function will be an approximate fit for the data.
Exponential model	If a set of data has initial value a and has a constant percent change r, the data can be modeled by the exponential function $$y = a(1 + r)^x$$
Comparison of models	To determine which type of model is the best fit for a set of data, we can look at the first differences and second differences of the outputs and the percent change of the outputs.

Logarithmic models	Models can be created involving $\ln x$. Data that exhibit an initial rapid increase and then have a slow rate of growth can often be described by the function $f(x) = a + b \ln x$ for $b > 0$.

5.5 Exponential Functions and Investing

Future value with annual compounding	If P dollars are invested for n years at an interest rate r, compounded annually, then the future value of P is given by $S = P(1 + r)^n$ dollars.
Future value of an investment	If P dollars are invested for t years at the annual interest rate r, where the interest is compounded k times per year, then the interest rate per period is $\dfrac{r}{k}$, the number of compounding periods is kt, and the future value that results is given by $S = P\left(1 + \dfrac{r}{k}\right)^{kt}$ dollars.
Future value with continuous compounding	If P dollars are invested for t years at an annual interest rate r, compounded continuously, then the future value S is given by $S = Pe^{rt}$ dollars.
Present value of an investment	The lump sum P that will give future value S if invested for n periods at interest rate i per compounding period is $$P = \frac{S}{(1+i)^n}, \quad \text{or} \quad P = S(1+i)^{-n}$$

5.6 Annuities; Loan Repayment

Future value of an ordinary annuity	If R dollars are contributed at the end of each period for n periods into an annuity that pays interest at rate i at the end of the period, the future value of the annuity is $$S = R\left(\frac{(1+i)^n - 1}{i}\right)$$
Present value of an ordinary annuity	If a payment of R dollars is to be made at the end of each period for n periods from an account that earns interest at a rate of i per period, then the account is an ordinary annuity, and the present value is $$A = R\left(\frac{1 - (1+i)^{-n}}{i}\right)$$
Amortization formula	If a debt of A dollars, with interest at a rate of i per period, is amortized by n equal periodic payments made at the end of each period, then the size of each payment is $$R = A\left(\frac{i}{1 - (1+i)^{-n}}\right)$$

5.7 Logistic and Gompertz Functions

Logistic function	For real numbers a, b, and c, the function $$f(x) = \frac{c}{1 + ae^{-bx}}$$ is a logistic function. A logistic function increases when $a > 0$ and $b > 0$. The graph of a logistic function resembles the graph at right.

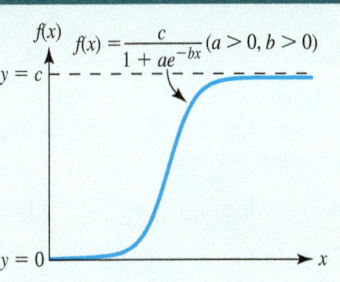

Logistic modeling	If data points indicate that growth begins slowly, then increases very rapidly, and finally slows over time to a rate that is almost zero, the data may fit the logistic model $$f(x) = \frac{c}{1 + ae^{-bx}}$$
Gompertz function	For real numbers C, a, and R ($0 < R < 1$), the function $N = Ca^{R^t}$ is a Gompertz function. A Gompertz function increases rapidly, then levels off as it approaches a limiting value. The graph of a Gompertz function resembles the graph at right.

chapter 5 SKILLS CHECK

Answers that are not seen can be found in the answer section at the back of the text.

1. **a.** Graph the function $f(x) = 4e^{-0.3x}$.
   b. Find $f(-10)$ and $f(10)$. ≈80.342; ≈0.19915

2. Is $f(x) = 4e^{-0.3x}$ an increasing or a decreasing function? Decreasing

3. Graph $f(x) = 3^x$ on $[-10, 10]$ by $[-10, 20]$.

4. Graph $y = 3^{(x-1)} + 4$ on $[-10, 10]$ by $[0, 15]$.

5. How does the graph of Exercise 4 compare with the graph of Exercise 3? Graph of $y = 3^{(x-1)} + 4$ is graph of $f(x) = 3^x$ shifted right 1 unit and up 4 units.

6. Is $y = 3^{(x-1)} + 4$ an increasing or a decreasing function? Increasing

7. **a.** If $y = 1000(2^{-0.1x})$, find the value of y when $x = 10$. $y = 500$
   b. Use graphical or numerical methods to determine the value of x that gives $y = 250$ if $y = 1000(2^{-0.1x})$. $x = 20$

In Exercises 8 and 9, write the exponential equations in logarithmic form.

8. $x = 6^y$ $y = \log_6 x$

9. $y = 7^{3x}$ $3x = \log_7 y$

In Exercises 10–12, write the logarithmic equations in exponential form.

10. $y = \log_4 x$ $x = 4^y$

11. $y = \log(x)$ $x = 10^y$

12. $y = \ln x$ $x = e^y$

13. Write the inverse of $y = 4^x$ in logarithmic form.
    $y = \log_4 x$

In Exercises 14–16, evaluate the logarithms, if possible. Give approximate solutions to four decimal places.

14. $\log 22$ 1.3424

15. $\ln 56$ 4.0254

16. $\log 10$ 1

In Exercises 17–19, find the value of the logarithms without using a calculator.

17. $\log_2 16$ 4

18. $\ln(e^4)$ 4

19. $\log 0.001$ −3

In Exercises 20 and 21, use a change-of-base formula to evaluate each of the logarithms. Give approximate solutions to four decimal places.

20. $\log_3 54$ 3.6309

21. $\log_8 56$ 1.9358

In Exercises 22 and 23, graph the functions with technology.

22. $y = \ln(x - 3)$

23. $y = \log_3 x$

In Exercises 24–27, solve the exponential equations. Give approximate solutions to four decimal places.

24. $340 = e^x$ 5.8289

25. $1500 = 300e^{8x}$ 0.2012

26. $9200 = 23(2^{3x})$ 2.8813

27. $4(3^x) = 36$ 2

28. Rewrite $\ln \dfrac{(2x - 5)^3}{x - 3}$ as the sum, difference, or product of logarithms, and simplify if possible.
    $3\ln(2x - 5) - \ln(x - 3)$

29. Rewrite $6 \log_4 x - 2 \log_4 y$ as a single logarithm. $\log_4(x^6/y^2)$

30. Determine whether a linear or an exponential function is the better fit for the data in the table. Then find the equation of the function with three-decimal-place accuracy. Exponential function; $y = 0.810(2.470^x)$

x	1	2	3	4	5
y	2	5	12	30	75

31. Evaluate $P\left(1 + \dfrac{r}{k}\right)^{kn}$ for $P = 1000$, $r = 8\%$, $k = 12$, $n = 20$, to two decimal places. 4926.80

32. Evaluate $1000\left(\dfrac{1 - 1.03^{-240+n}}{0.03}\right)$ for $n = 120$, to two decimal places. 32,373.02

33. a. Graph $f(t) = \dfrac{2000}{1 + 8e^{-0.8t}}$ for $t = 0$ to $t = 10$.

    b. Find $f(0)$ and $f(8)$, to two decimal places. 222.22; 1973.76

    c. What is the limiting value of this function? 2000

34. a. Graph $y = 500(0.1^{0.2^x})$ for $x = 0$ to $x = 10$.

    b. What is the initial value (the value of y when $x = 0$) of this function? 50

    c. What is the limiting value of this function? 500

chapter 5 REVIEW

Answers that are not seen can be found in the answer section at the back of the text.

35. **Average Annual Wage** The average annual wage of U.S. workers for the years from 2011 and projected to 2050 can be modeled by $y = 42.12(1.040^x)$ with y equal to the average annual wage in thousands and x equal to the number of years after 2010. (*Source*: Social Security Administration)

    a. Is this function a model of exponential growth or exponential decay? Explain. Growth; exponent is positive

    b. Graph the function for values of x that represent 2010 through 2050.

    c. When does this model predict that the average annual wage will be at least $96,000? After 2031

36. **Sales Decay** At the end of an advertising campaign, the daily sales (in dollars) declined, with daily sales given by the equation $y = 2000(2^{-0.1x})$, where x is the number of weeks after the end of the campaign. According to this model, what will sales be 4 weeks after the end of the ad campaign? $1515.72

37. **Sales Decay** At the end of an advertising campaign, the weekly sales (in dollars) declined, with weekly sales given by the equation $y = 2000(2^{-0.1x})$, where x is the number of weeks after the end of the campaign. In how many weeks will sales be half of the sales at the end of the ad campaign? 10 weeks

38. **Earthquakes**

    a. If an earthquake has an intensity of 1000 times I_0, what is the magnitude of the earthquake? 3 on Richter scale

    b. An earthquake that measured 6.5 on the Richter scale occurred in Pakistan in February 1991. Express this intensity in terms of I_0. $3,162,278\, I_0$

39. **Earthquakes** On January 23, 2001, an earthquake registering 7.9 hit western India, killing more than 15,000 people. On the same day, an earthquake registering 4.8 hit the United States, killing no one. How much more intense was the Indian earthquake than the American earthquake? Approximately 1259 times as intense

40. **Investments** The number of years needed for an investment of $10,000 to grow to $30,000 when money is invested at 12%, compounded annually, is given by

$$t = \log_{1.12}\dfrac{30{,}000}{10{,}000} = \log_{1.12} 3$$

In how many years will this occur? 10 yr

41. **Dental 3-D Printing** Dental 3-D printing is now being used to create bridges and other products in the dental field. The annual 3-D dental printer revenue R, in millions of dollars, for the year x for years from 2014 and projected to 2024 is shown in the table.

    a. Find a logistic function that models the data, with x equal to the number of years after 2010 and R equal to the revenue in millions of dollars. Report the model with three significant digits. $y = \dfrac{563}{1 + 9.01e^{-0.390x}}$

    b. According to the reported model, what will be the annual dental printer revenue in 2036? Approximately 562.8 million

    c. If the reported model is valid beyond 2036, what does it indicate that the maximum annual 3-D dental printer revenue will be? About 563 million

Year, x	3-D Printer Revenue R ($ million)	Year, x	3-D Printer Revenue R ($ million)
2014	207.3	2020	475.1
2015	240.3	2021	497.0
2016	290.1	2022	515.7
2017	349.3	2023	532.2
2018	414.0	2024	544.8
2019	454.2		

(Source: Statista)

42. Investments If $1000 is invested at 10%, compounded quarterly, the future value of the investment is given by $S = 1000(2^{x/7})$, where x is the number of years after the investment is made.

a. Use the logarithmic form of this function to solve the equation for x. $x = 7 \log_2 \frac{S}{1000}$

b. Find when the future value of the investment is $19,504. 30 yr

43. Hearing High-pitched tones that young people can hear at 30 decibels must be produced at 90 decibels for elderly people to hear them. How many times louder must the tones be for the elderly to hear them? (Source: Discover) 1 million times as loud.

44. Mobile Home Sales A company that buys and sells used mobile homes estimates its cost to be given by $C(x) = 2x + 50$ thousand dollars when x mobile homes are purchased. The same company estimates that its revenue from the sale of x mobile homes is given by $R(x) = 10(1.26^x)$ thousand dollars.

a. Combine C and R into a single function that gives the profit for the company when x used mobile homes are bought and sold.
$P(x) = 10(1.26^x) - 2x - 50$

b. Use a graph to find how many mobile homes the company must sell for revenue to be at least $30,000 more than cost. 10

45. Sales Decay At the end of an advertising campaign, the sales (in dollars) declined, with weekly sales given by the equation $y = 40,000(3^{-0.1x})$, where x is the number of weeks after the end of the campaign. In how many weeks will sales be less than half of the sales at the end of the ad campaign? 7 weeks

46. Carbon-14 Dating An exponential decay function can be used to model the number of atoms of a radioactive material that remain after a period of time. Carbon-14 decays over time, with the amount remaining after t years given by

$$y = y_0 e^{-0.00012097t}$$

where y_0 is the original amount.

a. If a sample of carbon-14 weighs 100 grams originally, how many grams will be present in 5000 years? 54.62 g

b. If a sample of wood at an archaeological site contains 36% as much carbon-14 as living wood, determine when the wood was cut. 8445 years ago

47. Purchasing Power If a retired couple has a fixed income of $60,000 per year, the purchasing power of their income (adjusted value of the money) after t years of 5% inflation is given by the equation $P = 60,000e^{-0.05t}$. In how many years will the purchasing power of their income fall below half of their current income? 14 yr

48. Investments If $2000 is invested at 8%, compounded continuously, the future value of the investment after t years is given by $S = 2000e^{0.08t}$ dollars. What is the future value of this investment in 10 years? Approximately $4451.08

49. Investments If $3300 is invested for x years at 10%, compounded annually, the future value that will result is $S = 3300(1.10)^x$ dollars. In how many years will the investment result in $13,784.92? 15 yr

50. Health Care Spending In 2011, national health care spending was $2.7 trillion, and it was projected to grow at an average annual rate of 5.7% for 2011 through 2021.

a. Write an exponential function that models the national health care spending, y, in trillions, as a function of the number of years after 2011, x.
$y = 2.7(1.057^x)$

b. What is the projected national health care spending for 2021, according to this model? $4.7 trillion

51. International Air Travel The number of passengers using international air traffic into and out of the United States was projected by the Federal Aviation Administration to grow at 4% per year from 2012 until 2032.

a. If the number was 172 million in 2012, write the equation of the exponential function that models this projection of the number of millions of passengers y as a function of the number of years x after 2012. $y = 172(1.04^x)$

b. What does this model project that the number of passengers using international air traffic into and out of the United States will be in 2032? 376.9 million

c. When does the model project that the number of passengers using international air traffic into and out of the United States will be 254.6 million? 2022

52. Energy Use Energy use per dollar of GDP is indexed to 1980, which means that energy use for any year is viewed as a percent of the use per dollar of GDP in 1980. The following data show the energy use per dollar of GDP, as a percent, for selected years from 1990 and projected to 2035.

a. Find an exponential function that models these data, with x equal to the number of years after 1990 and y equal to the energy use per dollar of GDP, as a percent. Is this function increasing or decreasing? $y = 81.041(0.981^x)$; decreasing

b. Graph the model and the data on the same axes.

c. What does this model predict the energy use will be in 2032? 36% of GDP in 1980

Energy Use per Dollar of 1980 GDP

Year	Percent	Year	Percent
1990	79	2015	51
1995	75	2020	45
2000	67	2025	41
2005	60	2030	37
2010	56	2035	34

(*Source*: U.S. Department of Energy)

53. **White Population** The table gives the number of millions of white, non-Hispanic individuals 16 years and older in the U.S. civilian non-institutional population.

a. Find the logarithmic function that is the best fit for the data, with y in millions of Americans and x equal to the number of years after 1990.
$y = 133.240 + 9.373 \ln x$

b. Graph the model and the data on the same axes.

c. Use the model to predict the number of this group in the U.S. civilian non-institutional population in 2019. 164.8 million

Year	White Non-Hispanic
2000	153.1
2010	162.1
2015	164.6
2020	166.3
2030	168.8
2040	169.7
2050	169.4

(*Source*: U.S. Census Bureau)

54. **Corvette Acceleration** The following table shows the times that it takes a Corvette to reach speeds from 0 mph to 100 mph, in increments of 10 mph after 30 mph. Find a logarithmic function that gives the speed as a function of the time. $y = 4.337 + 40.890 \ln x$

Time (sec)	Speed (mph)	Time (sec)	Speed (mph)
1.7	30	5.3	70
2.4	40	6.5	80
3.3	50	7.9	90
4.1	60	9.5	100

55. **Diabetes** The table gives the number of U.S. adults with diabetes (in millions) from 2010 and projected to 2050.

a. Find the logistic function that best models these data, with x equal to the number of years after 2000 and y equal to the number of millions of adults with diabetes. Report the model with three significant digits. $y = \dfrac{120}{1 + 5.26e^{-0.0639x}}$

b. Use the model to estimate the number of U.S. adults who will have diabetes in 2028. 63.9 million

c. According to the model, what will be the maximum number of U.S. adults to have diabetes? 120 million

d. In what year does the model predict that the number of U.S. adults with diabetes will reach 80 million? 2037

Year	Number (millions)	Year	Number (millions)
2010	32.2	2035	76.2
2015	37.3	2040	84.1
2020	50.0	2045	91.7
2025	59.5	2050	100.0
2030	68.3		

(*Source*: Centers for Disease Control and Prevention)

56. **Internet of Things Devices** The table on the next page shows the projected number of devices, in billions, connected to the Internet from 2015 through 2025.

a. Create a scatter plot of the data in the table, with x equal to the number of years after 2015 and y equal to the billions of devices connected to the Internet. 80

b. Find the best-fitting exponential model for the data, with x equal to the number of years after 2015 and y equal to the billions of devices connected to the Internet. $y = 14.697(1.170^x)$

c. Graph the model on the same axes with the data and discuss how well the function fits the data.

d. What does the model predict that the number of devices will be in 2029? 132 billion

436 Chapter 5 Exponential and Logarithmic Functions

Year	Devices (billions)	Year	Devices (billions)
2015	15.41	2021	35.82
2016	17.68	2022	42.62
2017	20.35	2023	51.11
2018	23.14	2024	62.12
2019	26.66	2025	75.44
2020	30.73		

(Source: Forbes.com)

57. **Female Physicians** The table shows the percent of physicians who are female in the United States for selected years from 1900 and projected to 2040.

   a. Find the logistic function that models the data, with x equal to the number of years after 1900 and y equal to the percent of female physicians. Report the model with three significant digits. $y = \dfrac{50.4}{1 + 333e^{-0.0579x}}$

   b. Use the model to predict the percent of physicians who will be female in 2050. Approximately 48%

   c. What does the model predict the maximum percent of female physicians will be? Approximately 50% (50.4)

Year, x	Female Physicians (percent)	Year, x	Female Physicians (percent)
1900	5	2021	40
1950	6	2025	42
1990	17	2030	43
2000	23	2035	44
2015	36	2040	44
2018	38		

(Source: U.S. Census Bureau)

58. **Investments** Find the 7-year future value of an investment of $20,000 placed into an account that pays 6%, compounded annually. $30,072.61

59. **Annuities** At the end of each quarter, $1000 is placed into an account that pays 12%, compounded quarterly. What is the future value of this annuity in 6 years? $34,426.47

60. **Annuities** Find the 10-year future value of an ordinary annuity with a contribution of $1500 at the end of each month, placed into an account that pays 8%, compounded monthly. $274,419.05

61. **Present Value** Find the present value of an annuity that will pay $2000 at the end of each month for 15 years if the interest rate is 8%, compounded monthly. $209,281.18

62. **Present Value** Find the present value of an annuity that will pay $500 at the end of each 6-month period for 12 years if the interest rate is 10%, compounded semiannually. $6899.32

63. **Loan Amortization** A debt of $2000 with interest at 12%, compounded monthly, is amortized by equal monthly payments for 36 months. What is the size of each payment? $66.43

64. **Loan Amortization** A debt of $120,000 with interest at 6%, compounded monthly, is amortized by equal monthly payments for 25 years. What is the size of each monthly payment? $773.16

65. **Medicaid** The projected annual increases in the number of people receiving Medicaid (in millions) from 2013 and projected to 2022 are shown in the table.

   a. Find a logistic function that models the data, with y in millions and x in years after 2012.

   b. Graph the model and the data on the same axes, and comment on the fit. Good fit

   c. What does this model predict that the maximum annual increase in the number of people receiving Medicaid will be? Does this number agree with the projections given in the table? 10.641 million; This number is close to, but less than, the projections given in the table.

Years	Annual Increase (millions)	Years	Annual Increase (millions)
2013	1	2018	11
2014	7	2019	11
2015	9	2020	11
2016	10	2021	11
2017	10	2022	11

(Source: Wall Street Journal, July 25, 2012)

66. **Global EV/PHEV Sales** Using data from 2020 and projected to 2050, global sales of electric and plug-in hybrid electric vehicles can be modeled by the function

$$y = \dfrac{113}{1 + 63.0e^{-0.155x}}$$

where y is the millions of vehicle sales and x is the number of years after 2010.
(Source: International Energy Agency)

   a. Find the global EV/PHEV sales predicted for 2035 by this function. 48.9 million

   b. What is the maximum number of global EV/PHEV sales, according to this model? 113 million

67. Spread of Disease A student brings a contagious disease to an elementary school of 2000 students. The number of students infected by the disease is given by

$$n = \frac{1400}{1 + 200e^{-0.5x}}$$

where x is the number of days after the student brings the disease.

a. How many students will be infected in 14 days? Approximately 1184
b. How many days will it take for 1312 students to be infected? 16 days

68. Organizational Growth The president of a new campus of a university predicts that the student body will grow rapidly after the campus is open, with the number of students at the beginning of year t given by

$$N = 4000 \, (0.06^{0.4^{t-1}})$$

a. How many students does this model predict for the beginning of the second year $(t = 2)$? 1298

b. How many students are predicted for the beginning of the tenth year? 3997

c. What is the limit on the number of students that can attend this campus, according to the model? 4000

69. Sales Growth The number of units of a new product that were sold each month after the product was introduced is given by

$$N = 18{,}000 \, (0.03^{0.4^t})$$

where t is the number of months.

a. How many units were sold 10 months after the product was introduced? Approximately 17,993

b. What is the limit on sales if this model is accurate? 18,000

70. Investing In how many years will $50,000 invested at continuous rate 7% grow to $202,760? 20 years

71. Interest Rate Suppose that $30,000 is invested at continuous rate r and grows to $121,656 in 28 years. Find the interest rate r. 5%

72. Interest Rate Suppose that $25,000 is invested at continuous rate r and grows to $45,552.97 in 15 years. Find the interest rate r. 4%

Group Activities
▶ **EXTENDED APPLICATIONS**

1. Chain Letters

Suppose you receive a chain letter asking you to send $1 to the person at the top of the list of six names, then to add your name to the bottom of the list, and finally to send the revised letter to six people. The promise is that when the letters have been sent to each of the people above you on the list, you will be at the top of the list and you will receive a large amount of money. To investigate whether this is worth the dollar that you are asked to spend, create the requested models and answer the following questions.

1. Suppose each of the six people on the original list sent six letters. How much money would the person on the top of the list receive?
2. Consider the person who is second on the original list. How much money would this person receive?
3. Complete the partial table of amounts that will be sent to the person on the top of the list during each "cycle" of the chain letter.

Cycle Number (after original six names)	Money Sent to Person on Top of List ($)
1	6 × 6 = 36
2	6 × 36 = 216
3	
4	
5	

4. If you were going to start such a chain letter (don't; it's illegal), would you put your name at the top of the first 6 names or as number 5?
5. Find the best quadratic, power, and exponential models for the data in the table. Which of these function types gives the best model for the data?
6. Use the exponential function found in part (5) to determine the amount of money the person who was at the bottom of the original list would receive if all people contacted sent the $1 and mailed the letter to 6 additional people.
7. How many people will have to respond to the chain letter for the sixth person on the original list to receive all the money that was promised?
8. If you receive the letter in its tenth cycle, how many other people have been contacted, along with you, assuming that everyone who receives the letter cooperates with its suggestions?
9. Who remains in the United States for you to send your letter to, if no one sends a letter to someone who has already received it?
10. Why do you think the federal government has made it illegal to send chain letters in the U.S. mail?

2. Modeling

Your mission is to find real-world data for company sales, a stock price, biological growth, or some sociological trend over a period of years that fit an exponential, logistic, or logarithmic function. To do this, you can look at a statistical graph called a histogram, which describes the situation, or at a graph or table describing it. Once you have found data consisting of numerical values, you are to find the equation that is the best exponential, logistic, or logarithmic fit for the data and then write the model that describes the relationship. Some helpful steps for this process are given in the instructions below.

1. Look in newspapers, in periodicals, on the Internet, or in statistical abstracts for a graph or table of data containing at least six data points. The data must have numerical values for the independent and dependent variables.
2. If you decide to use a relation determined by a graph that you have found, read the graph very carefully to determine the data points or (for full credit) contact the source of the data to obtain the data from which the graph was drawn.
3. If you have a table of values for different years, create a graph of the data to determine the type of function that will be the best fit for the data. To do this, first align the independent variable by letting the input represent the number of years after some convenient year and then enter the data into your graphing calculator and draw a scatter plot. Check to see if the points on the scatter plot lie near some curve that could be described by an exponential, logistic, or logarithmic function. If not, save the data for possible later use and renew your search. You can also test to

see if the percent change of the outputs is nearly constant for equally spaced inputs. If so, the data can be modeled by an exponential function.

4. Use your calculator to create the equation of the function that is the best fit for the data. Graph this equation and the data points on the same axes to see if the equation is reasonable.

5. Write some statements about the data that you have been working with. For example, describe how the quantity is increasing or decreasing over periods of years, why it is changing as it is, or when this model is no longer appropriate and why.

6. Your completed project should include the fol-lowing:

   a. A proper bibliographical reference for the source of your data.
   b. An original copy or photocopy of the data being used.
   c. A scatter plot of the data and reasons why you chose the model that you did to fit it.
   d. The equation that you have created, labeled with appropriate units of measure and variable descriptions.
   e. A graph of the scatter plot and the model on the same axes.
   f. One or more statements about how you think the model you have created could actually be used. Include any restrictions that should be placed on the model.

6

Higher-Degree Polynomial and Rational Functions

The amount of money spent by global tourists can be modeled by the function

$$y = 0.153x^3 - 2.914x^2 + 36.854x + 253.208$$

where y represents the amount spent by global travelers (in billions of dollars) and x represents the number of years after 1990. If the total cost of producing x units of a product is $C(x) = 100 + 30x + 0.01x^2$, the function

$$\overline{C}(x) = \frac{100 + 30x + 0.01x^2}{x}$$

models the average cost per unit. These are examples of polynomial and rational functions, which we will discuss in this chapter.

sections	objectives	applications
6.1 Higher-Degree Polynomial Functions	Identify graphs of higher-degree polynomial functions; graph cubic functions and quartic functions; find local extrema and absolute extrema	Energy from crude oil, older workers
6.2 Modeling with Cubic and Quartic Functions	Model data with cubic functions; model data with quartic functions	Aging workers, diabetes, Supplemental Nutrition Assistance Program, international travel
6.3 Solution of Polynomial Equations	Solve polynomial equations by factoring and the root method; estimate solutions with technology	Women in the workforce, maximizing volume, photosynthesis, cost, future value of an investment, China's labor pool
6.4 Polynomial Equations Continued; Fundamental Theorem of Algebra	Divide polynomials with synthetic division; solve cubic equations; combine graphical and algebraic methods to solve polynomial equations; solve quartic equations; use the Fundamental Theorem of Algebra to find the number of complex solutions of a polynomial equation; solve polynomial equations with complex solutions	Break-even
6.5 Rational Functions and Rational Equations	Graph rational functions; solve rational equations algebraically and graphically	Average cost of production, cost–benefit, advertising and sales, illumination, body mass index, volume of a pyramid
6.6 Polynomial and Rational Inequalities	Solve polynomial inequalities; solve rational inequalities	Average cost, box construction

Algebra TOOLBOX

KEY OBJECTIVES

- Identify polynomials
- Factor higher-degree polynomials
- Simplify rational expressions
- Multiply rational expressions
- Divide rational expressions
- Add rational expressions
- Subtract rational expressions
- Simplify complex fractions
- Divide polynomials using long division
- Perform operations with complex numbers

Polynomials

Recall from Chapter 1 that an expression containing a finite number of additions, subtractions, and multiplications of constants and positive integer powers of variables is called a **polynomial**. The general form of a polynomial in x is

$$a_n x^n + a_{n-1} x^{n-1} + \cdots + a_1 x + a_0$$

where a_0 and each coefficient of x are real numbers and each power of x is a positive integer. If $a_n \neq 0$, n is the highest power of x, a_n is called the **leading coefficient**, and n is the **degree** of the polynomial. Thus, $5x^4 + 3x^2 - 6$ is a fourth-degree (quartic) polynomial with leading coefficient 5.

EXAMPLE 1 ▶ Polynomials

What are the degree and the leading coefficient of each of the following polynomials?

a. $4x^2 + 5x^3 - 17$ **b.** $30 - 6x^5 + 3x^2 - 7$

SOLUTION

a. This is a third-degree polynomial. The leading coefficient is 5, the coefficient of the highest-degree term.

b. This is a fifth-degree polynomial. The leading coefficient is -6, the coefficient of the highest-degree term.

Factoring Higher-Degree Polynomials

Some higher-degree polynomials can be factored by expressing them as the product of a monomial and another polynomial, which sometimes can be factored to complete the factorization.

EXAMPLE 2 ▶ Factoring

Factor: **a.** $3x^3 - 21x^2 + 36x$ **b.** $-6x^4 - 10x^3 + 4x^2$

SOLUTION

a. $3x^3 - 21x^2 + 36x = 3x(x^2 - 7x + 12) = 3x(x-3)(x-4)$

b. $-6x^4 - 10x^3 + 4x^2 = -2x^2(3x^2 + 5x - 2) = -2x^2(3x-1)(x+2)$

Some polynomials can be written in *quadratic form* by using substitution. Then quadratic factoring methods can be used to begin the factorization.

EXAMPLE 3 ▶ **Factoring Polynomials in Quadratic Form**

Factor: **a.** $x^4 - 6x^2 + 8$ **b.** $x^4 - 18x^2 + 81$

SOLUTION

a. The expression $x^4 - 6x^2 + 8$ is not a quadratic polynomial, but substituting u for x^2 converts it into the quadratic expression $u^2 - 6u + 8$, so we say that the original expression is in *quadratic form*. We factor it using quadratic methods, as follows:

$$u^2 - 6u + 8 = (u - 2)(u - 4)$$

Replacing u with x^2 gives

$$(x^2 - 2)(x^2 - 4)$$

Completing the factorization by factoring $x^2 - 4$ gives

$$x^4 - 6x^2 + 8 = (x^2 - 2)(x - 2)(x + 2)$$

b. Substituting u for x^2 converts $x^4 - 18x^2 + 81$ into the quadratic expression $u^2 - 18u + 81$. Factoring this expression gives

$$u^2 - 18u + 81 = (u - 9)(u - 9)$$

Replacing u with x^2 and completing the factorization gives

$$x^4 - 18x^2 + 81 = (x^2 - 9)(x^2 - 9)$$
$$= (x - 3)(x + 3)(x - 3)(x + 3) = (x - 3)^2(x + 3)^2$$

Rational Expressions

An expression that is the quotient of two nonzero polynomials is called a **rational expression**. For example, $\dfrac{3x + 1}{x^2 - 2}$ is a rational expression. A rational expression is undefined when the denominator equals 0.

> ### Fundamental Principle of Rational Expressions
> The **Fundamental Principle of Rational Expressions** provides the means to simplify them.
>
> $$\frac{ac}{bc} = \frac{a}{b} \quad \text{for } b \neq 0, c \neq 0$$

We simplify a rational expression by factoring the numerator and denominator and dividing both the numerator and denominator by any common factors. We will assume that all rational expressions are defined for those values of the variables that do not make any denominator 0.

EXAMPLE 4 ▶ **Rational Expressions**

Simplify the rational expression.

a. $\dfrac{2x^2 - 8}{x + 2}$ **b.** $\dfrac{3x}{3x + 6}$ **c.** $\dfrac{3x^2 - 14x + 8}{x^2 - 16}$

Algebra Toolbox 443

SOLUTION

a. $\dfrac{2x^2 - 8}{x + 2} = \dfrac{2(x^2 - 4)}{x + 2} = \dfrac{2\cancel{(x+2)}(x-2)}{\cancel{(x+2)}} = 2(x-2)$

b. Note that we cannot divide both numerator and denominator by $3x$, because $3x$ is not a factor of $3x + 6$. Instead, we factor the denominator.

$$\dfrac{3x}{3x + 6} = \dfrac{\cancel{3}x}{\cancel{3}(x + 2)} = \dfrac{x}{x + 2}$$

c. $\dfrac{3x^2 - 14x + 8}{x^2 - 16} = \dfrac{(3x - 2)\cancel{(x-4)}}{(x + 4)\cancel{(x-4)}} = \dfrac{3x - 2}{x + 4}$

Multiplying and Dividing Rational Expressions

To multiply rational expressions, we write the product of the numerators divided by the product of the denominators and then simplify. We may also simplify prior to multiplying; this is usually easier than multiplying first.

EXAMPLE 5 ▶ **Products of Rational Expressions**

Multiply and simplify.

a. $\dfrac{4x^2}{5y^2} \cdot \dfrac{25y}{12x}$ b. $\dfrac{6 - 3x}{2x + 4} \cdot \dfrac{4x - 20}{2 - x}$

SOLUTION

a. $\dfrac{4x^2}{5y^2} \cdot \dfrac{25y}{12x} = \dfrac{4x^2 \cdot 25y}{5y^2 \cdot 12x} = \left(\dfrac{4 \cdot 25}{5 \cdot 12}\right)\left(\dfrac{x^2}{x}\right)\left(\dfrac{y}{y^2}\right) = \left(\dfrac{5}{3}\right)\left(\dfrac{x}{1}\right)\left(\dfrac{1}{y}\right) = \dfrac{5x}{3y}$

b. $\dfrac{6 - 3x}{2x + 4} \cdot \dfrac{4x - 20}{2 - x} = \dfrac{(6 - 3x)(4x - 20)}{(2x + 4)(2 - x)} = \dfrac{3\cancel{(2-x)} \cdot \overset{2}{\cancel{4}}(x - 5)}{\cancel{2}(x + 2) \cdot \cancel{(2-x)}} = \dfrac{6x - 30}{x + 2}$

To divide rational expressions, we multiply by the reciprocal of the divisor.

EXAMPLE 6 ▶ **Quotients of Rational Expressions**

Divide and simplify.

a. $\dfrac{a^2 b}{c} \div \dfrac{ab^3}{c^2}$ b. $\dfrac{x^2 + 7x + 12}{2 - x} \div \dfrac{x^2 - 9}{x^2 - x - 2}$

SOLUTION

a. $\dfrac{a^2 b}{c} \div \dfrac{ab^3}{c^2} = \dfrac{a^2 b}{c} \cdot \dfrac{c^2}{ab^3} = \left(\dfrac{a^2}{a}\right)\left(\dfrac{b}{b^3}\right)\left(\dfrac{c^2}{c}\right) = \dfrac{ac}{b^2}$

b. $\dfrac{x^2 + 7x + 12}{2 - x} \div \dfrac{x^2 - 9}{x^2 - x - 2} = \dfrac{x^2 + 7x + 12}{2 - x} \cdot \dfrac{x^2 - x - 2}{x^2 - 9}$

$= \dfrac{\cancel{(x+3)}(x + 4)\cancel{(x-2)}(x + 1)}{-\cancel{(x-2)}(x - 3)\cancel{(x+3)}} = \dfrac{(x + 4)(x + 1)}{-(x - 3)} = \dfrac{x^2 + 5x + 4}{3 - x}$

Adding and Subtracting Rational Expressions

We can add or subtract two fractions with the same denominator by adding or subtracting the numerators over their common denominator. If the denominators are not the same, we can write each fraction as an equivalent fraction with the common denominator. Our work is easier if we use the **least common denominator (LCD)**. The LCD is the lowest-degree polynomial that all denominators will divide into. For example, if several denominators are $3x$, $6x^2y$, and $9y^3$, the lowest-degree polynomial that all three denominators will divide into is $18x^2y^3$. We can find the LCD as follows.

> **Finding the Least Common Denominator of a Set of Fractions**
> 1. Completely factor all the denominators.
> 2. The LCD is the product of each factor used the maximum number of times it occurs in any one denominator.

EXAMPLE 7 ▶ Least Common Denominator

Find the LCD of the fractions $\dfrac{3x}{x^2 - 4x - 5}$ and $\dfrac{5x - 1}{x^2 - 2x - 3}$.

SOLUTION

The factored denominators are $(x - 5)(x + 1)$ and $(x - 3)(x + 1)$. The factors $(x - 5)$, $(x + 1)$, and $(x - 3)$ each occur a maximum of one time in any one denominator. Thus, the LCD is $(x - 5)(x + 1)(x - 3)$.

The procedure for adding or subtracting rational expressions follows.

> **Adding or Subtracting Rational Expressions**
> 1. Find the LCD of all the rational expressions.
> 2. Write the equivalent of each rational expression with the LCD as its denominator.
> 3. Combine the like terms in the numerators and write this expression in the numerator with the LCD in the denominator.
> 4. Simplify the rational expression, if possible.

EXAMPLE 8 ▶ Adding Rational Expressions

Add $\dfrac{3}{x^2y} + \dfrac{5}{xy^2}$.

SOLUTION

The factor x occurs a maximum of two times in any denominator and the factor y occurs a maximum of two times in any denominator, so the LCD is x^2y^2. We rewrite the fractions with the common denominator by multiplying the numerator and denominator of the first fraction by y and multiplying the numerator and denominator of the second fraction by x, as follows:

$$\frac{3}{x^2y} + \frac{5}{xy^2} = \frac{3 \cdot y}{x^2y \cdot y} + \frac{5 \cdot x}{xy^2 \cdot x} = \frac{3y}{x^2y^2} + \frac{5x}{x^2y^2}$$

We can now add the fractions by adding the numerators over the common denominator:

$$\frac{3y}{x^2y^2} + \frac{5x}{x^2y^2} = \frac{3y + 5x}{x^2y^2}$$

This result cannot be simplified because there is no common factor of the numerator and denominator.

EXAMPLE 9 ▶ Subtracting Rational Expressions

Subtract $\dfrac{x + 1}{x^2 + x - 6} - \dfrac{2x - 1}{x - 2}$.

SOLUTION

We factor the first denominator to get the LCD:

$$x^2 + x - 6 = (x + 3)(x - 2)$$

so the LCD is $(x + 3)(x - 2)$. We rewrite the second fraction so that it has the common denominator, subtract the second numerator, then combine the like terms of the numerators.

$$\begin{aligned}
\frac{x + 1}{x^2 + x - 6} - \frac{2x - 1}{x - 2} &= \frac{x + 1}{(x - 2)(x + 3)} - \frac{(2x - 1)(x + 3)}{(x - 2)(x + 3)} \\
&= \frac{x + 1}{(x - 2)(x + 3)} - \frac{2x^2 + 5x - 3}{(x - 2)(x + 3)} \\
&= \frac{x + 1 - (2x^2 + 5x - 3)}{(x - 2)(x + 3)} \\
&= \frac{x + 1 - 2x^2 - 5x + 3}{(x - 2)(x + 3)} \\
&= \frac{-2x^2 - 4x + 4}{(x - 2)(x + 3)}
\end{aligned}$$

This result cannot be simplified.

Simplifying Complex Fractions

A complex fraction has a fraction in the numerator, in the denominator, or in both the numerator and the denominator. To simplify a complex fraction, find the LCD of all fractions contained in the numerator and denominator of the complex fraction. Multiply both the numerator and the denominator of the complex fraction by the LCD. (By the distributive property, this means that every term of the complex fraction will be multiplied by the LCD.) Simplify each term of the complex fraction either by dividing out common factors or by multiplying. Factor the resulting numerator and denominator, and simplify if possible.

EXAMPLE 10 ▶ **Simplifying a Complex Fraction—Method 1**

Simplify $\dfrac{\dfrac{1}{3} + \dfrac{4}{x}}{3 - \dfrac{1}{xy}}$.

SOLUTION

To simplify $\dfrac{\dfrac{1}{3} + \dfrac{4}{x}}{3 - \dfrac{1}{xy}}$, note that the LCD of the numerator and denominator of the complex fraction is $3xy$. We multiply the numerator and denominator (and each term) of the complex fraction by $3xy$.

$$\dfrac{\dfrac{1}{3} + \dfrac{4}{x}}{3 - \dfrac{1}{xy}} = \dfrac{\left[\dfrac{1}{3} + \dfrac{4}{x}\right] \cdot 3xy}{\left[3 - \dfrac{1}{xy}\right] \cdot 3xy}$$

$$= \dfrac{\dfrac{1}{3} \cdot 3xy + \dfrac{4}{x} \cdot 3xy}{3 \cdot 3xy - \dfrac{1}{xy} \cdot 3xy} = \dfrac{xy + 12y}{9xy - 3} = \dfrac{y(x + 12)}{3(3xy - 1)}$$

Another way to simplify a complex fraction involves treating the fraction as a division. Rewrite the complex fraction with the main fraction bar written as a division symbol. If more than one fraction appears in the dividend (the numerator of the complex fraction) or divisor (the denominator of the complex fraction), find the LCD for each. Write the equivalent of each fraction with the desired LCD as its denominator. Add or subtract the fractions in the numerator and denominator so that the numerator contains a single fraction and the denominator contains a single fraction. Divide the fractions by inverting the divisor and multiplying. Write the indicated product of the numerators and the indicated product of the denominators. Simplify the fraction if possible.

EXAMPLE 11 ▶ **Simplifying a Complex Fraction—Method 2**

Simplify $\dfrac{\dfrac{1}{3} + \dfrac{4}{x}}{3 - \dfrac{1}{xy}}$.

SOLUTION

To simplify $\dfrac{\dfrac{1}{3} + \dfrac{4}{x}}{3 - \dfrac{1}{xy}}$, we first rewrite the fraction as a division:

$$\frac{\frac{1}{3}+\frac{4}{x}}{3-\frac{1}{xy}} = \left[\frac{1}{3}+\frac{4}{x}\right] \div \left[3-\frac{1}{xy}\right]$$

Next we get the LCD of the dividend and the LCD of the divisor and combine the fractions in the resulting dividend and divisor.

$$= \left[\frac{1}{3}\cdot\frac{x}{x}+\frac{4}{x}\cdot\frac{3}{3}\right] \div \left[3\cdot\frac{xy}{xy}-\frac{1}{xy}\right] = \left[\frac{x}{3x}+\frac{12}{3x}\right] \div \left[\frac{3xy}{xy}-\frac{1}{xy}\right]$$

$$= \frac{x+12}{3x} \div \frac{3xy-1}{xy}$$

We then invert the divisor, multiply, and simplify the result.

$$= \frac{x+12}{3x} \cdot \frac{xy}{3xy-1} = \frac{(x+12)xy}{3x(3xy-1)} = \frac{y(x+12)}{3(3xy-1)}$$

Division of Polynomials

When the divisor has a degree less than the dividend, the division of one polynomial by another is done in a manner similar to long division in arithmetic. The degree of the quotient polynomial will be less than the degree of the dividend polynomial, and the degree of the remainder will be less than the degree of the divisor. (If the remainder is 0, the divisor is a factor of the dividend.) The procedure follows.

Long Division of Polynomials

1. Write both the divisor and the dividend in descending order of a variable. Include missing terms with a 0 coefficient in the dividend.

    Divide $(4x^3 + 4x^2 + 5)$ by $(2x^2 + 1)$.

    $2x^2 + 1 \overline{\smash{)}4x^3 + 4x^2 + 0x + 5}$

2. Divide the highest power of the divisor into the highest power of the dividend, and write this partial quotient above the dividend. Multiply the partial quotient times the divisor, write the product under the dividend, subtract, and bring down the next term, getting a new dividend.

    $\begin{array}{r} 2x \\ 2x^2 + 1 \overline{\smash{)}4x^3 + 4x^2 + 0x + 5} \\ \underline{4x^3 + 2x } \\ 4x^2 - 2x + 5 \end{array}$

3. Repeat until the degree of the new dividend is less than the degree of the divisor. If a nonzero expression remains, it is the remainder from the division.

    $\begin{array}{r} 2x + 2 \\ 2x^2 + 1 \overline{\smash{)}4x^3 + 4x^2 + 0x + 5} \\ \underline{4x^3 + 2x } \\ 4x^2 - 2x + 5 \\ \underline{4x^2 + 2} \\ -2x + 3 \end{array}$

Thus, the quotient is $2x + 2$, with remainder $-2x + 3$.

EXAMPLE 12 ▶ Division of Polynomials

Perform the indicated division: $\dfrac{x^4 + 6x^3 - 4x + 2}{x + 2}$.

SOLUTION

Begin by placing $0x^2$ in the dividend.

$$
\begin{array}{r}
x^3 + 4x^2 - 8x + 12 \\
x + 2 \overline{) x^4 + 6x^3 + 0x^2 - 4x + 2} \\
\underline{x^4 + 2x^3} \\
4x^3 + 0x^2 \\
\underline{4x^3 + 8x^2} \\
-8x^2 - 4x \\
\underline{-8x^2 - 16x} \\
12x + 2 \\
\underline{12x + 24} \\
-22
\end{array}
$$

Thus, the quotient is $x^3 + 4x^2 - 8x + 12$, with remainder -22. We can write the result as

$$\dfrac{x^4 + 6x^3 - 4x + 2}{x + 2} = x^3 + 4x^2 - 8x + 12 - \dfrac{22}{x + 2}$$

Operations with Complex Numbers

Recall that complex numbers have the form $a + bi$, where a and b are real numbers and i is called the imaginary unit with $i^2 = -1$ or $i = \sqrt{-1}$.

Adding and Subtracting Complex Numbers

To add (or subtract) two complex numbers, add (or subtract) their real parts and add (or subtract) their imaginary parts.

$$(a + bi) + (c + di) = (a + c) + (b + d)i$$
$$(a + bi) - (c + di) = (a - c) + (b - d)i$$

EXAMPLE 13 ▶ Add Complex Numbers

Add $(5 - 4i) + (-6 - 3i)$.

SOLUTION

To add $(5 - 4i) + (-6 - 3i)$, we add the real parts and add the imaginary parts.

$$(5 - 4i) + (-6 - 3i) = (5 + (-6)) + (-4i + (-3i)) = -1 - 7i$$

EXAMPLE 14 ▶ **Subtract Complex Numbers**

Subtract $(-2 + 6i) - (-4 - 8i)$.

SOLUTION

To subtract, we first use the distributive property to change subtraction to addition. Then we add the real parts and add the imaginary parts.

$$(-2 + 6i) - (-4 - 8i) = (-2 + 6i) + (4 + 8i) = (-2 + 4) + (6i + 8i) = 2 + 14i$$

Multiplying and Dividing Complex Numbers

To multiply two complex numbers, we use binomial multiplication and $i^2 = -1$.

$$(a + bi)(c + di) = ac + adi + bci + bdi^2 = ac + (ad + bc)i + bd(-1)$$
$$= ac - bd + (ad + bc)i$$

To divide two complex numbers:

1. If the denominator contains a complex number with only an imaginary part, multiply the numerator and denominator by i.

2. If the denominator contains a complex number with both real and imaginary parts, multiply the numerator and denominator by the conjugate of the denominator.

$$\frac{a + bi}{c + di} = \frac{a + bi}{c + di} \cdot \frac{c - di}{c - di} = \frac{ac - adi + bci - bdi^2}{c^2 - cdi + cdi - di^2}$$

$$= \frac{ac - bd(i^2) + (bc - ad)i}{c^2 + d^2}$$

$$= \frac{ac + bd}{c^2 + d^2} + \frac{(bc - ad)i}{c^2 + d^2}$$

EXAMPLE 15 ▶ **Multiply Complex Numbers**

Multiply $(9 - 2i)(5 + i)$.

SOLUTION

$$(9 - 2i)(5 + i) = 45 + 9i - 10i - 2i^2 = 45 - i - 2(-1) = 45 - i + 2 = 47 - i$$

EXAMPLE 16 ▶ **Divide Complex Numbers**

Divide $\dfrac{2 + 7i}{2i}$.

SOLUTION

$$\frac{2 + 7i}{2i} = \frac{2 + 7i}{2i} \cdot \frac{i}{i} = \frac{2i + 7i^2}{2i^2} = \frac{2i + 7(-1)}{2(-1)} = \frac{-7 + 2i}{-2} = \frac{7}{2} - i$$

EXAMPLE 17 ▶ **Divide Complex Numbers**

Divide $\dfrac{3 - 7i}{3 - i}$.

Chapter 6 Higher-Degree Polynomial and Rational Functions

SOLUTION

$$\frac{3-7i}{3-i} = \frac{3-7i}{3-i} \cdot \frac{3+i}{3+i} = \frac{9+3i-21i-7i^2}{9-i^2} = \frac{9-18i-7(i^2)}{9-(i^2)} = \frac{9-18i+7}{9+1}$$

$$= \frac{16-18i}{10} = \frac{8}{5} - \frac{9}{5}i$$

Note that operations with complex numbers can be calculated on a graphing calculator.

Figure 6.1(a) shows solutions to Examples 13–15. Figure 6.1(b) shows solutions to Examples 16 and 17.

(a)   (b)

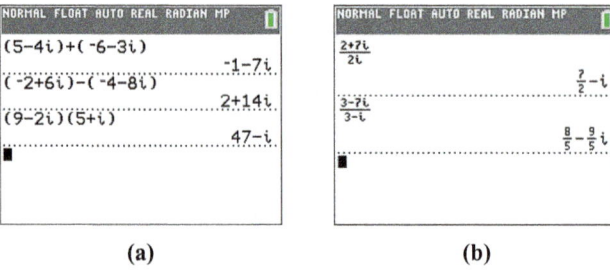

Figure 6.1

Toolbox EXERCISES

For Exercises 1–4, (a) give the degree of the polynomial and (b) give the leading coefficient.

1. $3x^4 - 5x^2 + \frac{2}{3}$
   a. Fourth b. 3
2. $5x^3 - 4x + 7$
   a. Third b. 5
3. $7x^2 - 14x^5 + 16$
   a. Fifth b. -14
4. $2y^5 + 7y - 8y^6$
   a. Sixth b. -8

In Exercises 5–10, factor the polynomials completely.

5. $4x^3 - 8x^2 - 140x$
   $4x(x-7)(x+5)$
6. $4x^2 + 7x^3 - 2x^4$
   $-x^2(2x+1)(x-4)$
7. $x^4 - 13x^2 + 36$
   $(x-3)(x+3)(x-2)(x+2)$
8. $x^4 - 21x^2 + 80$
   $(x-4)(x+4)(x^2-5)$
9. $2x^4 - 8x^2 + 8$
   $2(x^2-2)^2$
10. $3x^5 - 24x^3 + 48x$
    $3x(x-2)^2(x+2)^2$

In Exercises 11–16, simplify each rational expression.

11. $\frac{x-3y}{3x-9y}$ $\frac{1}{3}$
12. $\frac{x^2-9}{4x+12}$ $\frac{x-3}{4}$
13. $\frac{2y^3 - 2y}{y^2 - y}$ $2y+2$
14. $\frac{4x^3 - 3x}{x^2 - x}$ $\frac{4x^2-3}{x-1}$
15. $\frac{x^2 - 6x + 8}{x^2 - 16}$ $\frac{x-2}{x+4}$
16. $\frac{3x^2 - 7x - 6}{x^2 - 4x + 3}$ $\frac{3x+2}{x-1}$

In Exercises 17–33, perform the indicated operations and simplify.

17. $\frac{6x^3}{8y^3} \cdot \frac{16x}{9y^2} \cdot \frac{15y^4}{x^3}$ $\frac{20x}{y}$

18. $\frac{x-3}{x^3} \cdot \frac{x^2 - 4x}{x^2 - 7x + 12}$ $\frac{1}{x^2}$

19. $(x^2 - 4) \cdot \frac{2x-3}{x+2}$ $2x^2 - 7x + 6$

20. $(x^2 - x - 6) \div \frac{9-x^2}{x^2 + 3x}$ $-x(x+2)$

21. $\frac{4x+4}{x-4} \div \frac{8x^2 + 8x}{x^2 - 6x + 8}$ $\frac{x-2}{2x}$

22. $\frac{6x^2}{4x^2y - 12xy} \div \frac{3x^2 + 12x}{x^2 + x - 12}$ $\frac{1}{2y}$

23. $\frac{x^2+x}{x^2-5x+6} \cdot \frac{x^2-2x-3}{2x+4} \div \frac{x^3-3x^2}{4-x^2}$ $\frac{-(x+1)^2}{2x(x-3)}$

24. $\frac{6x-2}{3xy} + \frac{3x+2}{3xy}$ $\frac{3}{y}$

25. $\frac{2x+3}{x^2-1} + \frac{4x+3}{x^2-1}$ $\frac{6}{x-1}$

26. $3 + \frac{1}{x^2} - \frac{2}{x^3}$ $\frac{3x^3+x-2}{x^3}$

27. $\frac{5}{x} - \frac{x-2}{x^2} + \frac{4}{x^3}$ $\frac{4x^2+2x+4}{x^3}$

28. $\frac{a}{a^2-2a} - \frac{a-2}{a^2}$ $\frac{4a-4}{a^3-2a^2}$

29. $\frac{5x}{x^4-16} + \frac{8x}{x+2}$ $\frac{8x^4-16x^3+32x^2-59x}{x^4-16}$

30. $\dfrac{x-1}{x+1} - \dfrac{2}{x^2+x} \quad \dfrac{x-2}{x}$

31. $1 + \dfrac{1}{x-2} - \dfrac{2}{x^2} \quad \dfrac{x^3 - x^2 - 2x + 4}{x^2(x-2)}$

32. $\dfrac{x-7}{x^2 - 9x + 20} + \dfrac{x+2}{x^2 - 5x + 4} \quad \dfrac{2x^2 - 11x - 3}{(x-5)(x-4)(x-1)}$

33. $\dfrac{2x+1}{4x-2} + \dfrac{5}{2x} - \dfrac{x+1}{2x^2 - x} \quad \dfrac{2x^2 + 9x - 7}{4x^2 - 2x}$

In Exercises 34–37, simplify the complex fraction.

34. $\dfrac{\dfrac{1}{x} + \dfrac{1}{y}}{\dfrac{1}{x} - \dfrac{1}{y}} \quad \dfrac{y+x}{y-x}$

35. $\dfrac{\dfrac{5}{2y} + \dfrac{3}{y}}{\dfrac{1}{4} + \dfrac{1}{3y}} \quad \dfrac{66}{3y+4}$

36. $\dfrac{2 - \dfrac{1}{x}}{2x - \dfrac{3x}{x+1}} \quad \dfrac{x+1}{x^2}$

37. $\dfrac{1 - \dfrac{2}{x-2}}{x - 6 + \dfrac{10}{x+1}} \quad \dfrac{x+1}{(x-2)(x-1)}$

In Exercises 38–41, perform the long division.

38. $(x^5 + x^3 - 1) \div (x + 1) \quad x^4 - x^3 + 2x^2 - 2x + 2, R(-3)$

39. $(a^4 + 3a^3 + 2a^2) \div (a + 2) \quad a^3 + a^2$

40. $(3x^5 - x^4 + 5x - 1) \div (x^2 - 2) \quad 3x^3 - x^2 + 6x - 2, R(17x - 5)$

41. $\dfrac{3x^4 + 2x^2 + 1}{3x^2 - 1} \quad x^2 + 1, R2$

In Exercises 42–44, add or subtract, as indicated, and simplify.

42. $(8 + 2i) + (-3 - 4i) \quad 5 - 2i$

43. $(-3 - 12i) - (9 + 6i) \quad -12 - 18i$

44. $(4 + i) + (2 - \sqrt{-4}) - (17 + 8i) \quad -11 - 9i$

In Exercises 45–52, perform the indicated operations and simplify.

45. $(3 + 4i)(5 - 2i) \quad 23 + 14i$

46. $\dfrac{1 + 3i}{5 + 2i} \quad \dfrac{11}{29} + \dfrac{13}{29}i$

47. $i^{17}(3i + 2) \quad -3 + 2i$

48. $\left(\dfrac{3 + \sqrt{-16}}{2}\right)^2 \quad -\dfrac{7}{4} + 6i$

49. $\dfrac{3 + 7i}{4i} \quad \dfrac{7}{4} - \dfrac{3}{4}i$

50. $\dfrac{2 - \sqrt{-8}}{\sqrt{2} - \sqrt{-4}} \quad \sqrt{2}$

51. $\dfrac{(3 - i)(7 + 2i)}{4 - 3i} \quad \dfrac{19}{5} + \dfrac{13}{5}i$

52. $\sqrt{-4}\sqrt{-3}\sqrt{-24} \quad -12\sqrt{2}i$

6.1 Higher-Degree Polynomial Functions

KEY OBJECTIVES

- Identify the graphs of higher-degree polynomial functions
- Graph cubic functions
- Graph quartic functions
- Find local minima, local maxima, absolute minima, and absolute maxima of polynomial functions

SECTION PREVIEW Energy from Crude Oil

U.S. crude oil production continues to grow as a result of the further development of tight oil resources. Table 6.1 shows the total energy supply from crude oil products, in quadrillion British thermal units (BTUs), for selected years from 2010 and projected to 2040.

Table 6.1

Year	Quadrillion BTUs	Year	Quadrillion BTUs
2010	11.6	2030	13.5
2015	15.6	2035	13.4
2020	16.0	2040	13.1
2025	14.5		

(*Source*: U.S. Energy Information Administration)

The data can be used to find a **polynomial** function

$$C(x) = -0.0000752x^4 + 0.00589x^3 - 0.156x^2 + 1.48x + 11.6$$

that models the U.S. crude oil production as a function of the number of years after 2010.

452   Chapter 6   Higher-Degree Polynomial and Rational Functions

In Example 4, we will graph this fourth-degree function and use the graph and technology to find the year after 2010 in which the total energy supply from crude oil products is a maximum and the year before 2045 in which it is a minimum.

In Chapters 2–4, we solved many real problems by modeling data with linear functions and with quadratic functions. (Recall that $ax + b$ is a first-degree polynomial and $ax^2 + bx + c$ is a second-degree polynomial.) Both of these functions are types of **polynomial functions**. **Higher-degree polynomial functions** are functions with degree higher than 2. Examples of higher-degree polynomial functions are

$$y = x^3 - 16x^2 - 2x \qquad f(x) = 3x^4 - x^3 + 2 \qquad y = 2x^5 - x^3$$

In this section, we graph higher-degree polynomial functions and use technology to find local maxima and minima. ∎

Cubic Functions

Using data from 1986 to 2010, the number of international visitors to the United States can be modeled by

$$y = 0.010x^3 - 0.575x^2 + 10.814x - 20.752$$

where y is in millions and x is the number of years after 1980. This function is a third-degree function, or **cubic function**, because the highest-degree term in the function has degree 3. The graph of this function is shown in Figure 6.2.

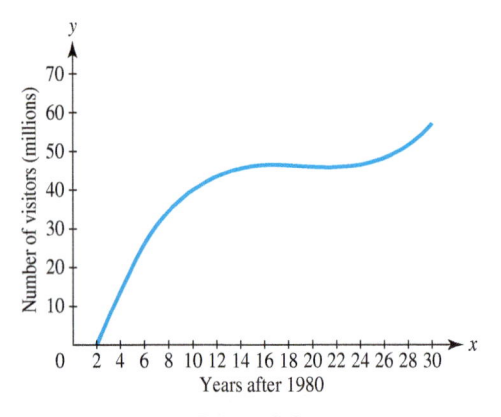

Figure 6.2

A **cubic function** in the variable x has the form

$$f(x) = ax^3 + bx^2 + cx + d \qquad (a \neq 0)$$

The basic cubic function $f(x) = x^3$ was discussed in Section 3.3. The graph of this function, shown in Figure 6.3(a), is one of the possible shapes of the graph of a cubic function. The graphs of two other cubic functions are shown in Figures 6.3(b) and (c).

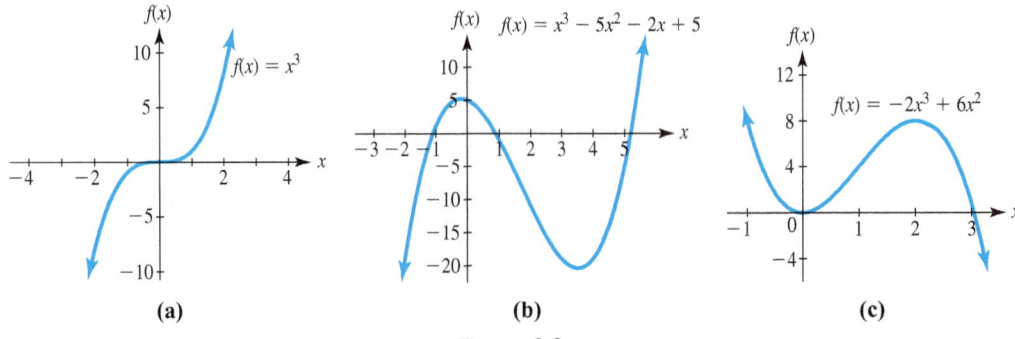

(a)   (b)   (c)

Figure 6.3

The graph of a cubic function will have 0 or 2 turning points, as shown in Figures 6.2 and 6.3. Also, it is characteristic of a cubic polynomial function to have a graph that has one of the end behaviors shown in these figures. This end behavior can be described as "one end opening up and one end opening down." The specific end behavior is determined by the leading coefficient (the coefficient of the third-degree term); the curve opens up on the right if the leading coefficient is positive, and it opens down on the right if the leading coefficient is negative.

Notice that the graph of the cubic function shown in Figure 6.3(a) has one x-intercept, the graph of the cubic function shown in Figure 6.3(b) has three x-intercepts, and the graph of the cubic function in Figure 6.3(c) has two x-intercepts.

> In general, the graph of a polynomial function of degree n has at most n x-intercepts.

Recall that the graph of a quadratic (second-degree) function is a parabola with 1 turning point, called a vertex, and that the vertex is a maximum point or a minimum point. Notice that in each of the graphs of the cubic (third-degree) functions shown in Figures 6.3(b) and (c), there are 2 points where the function changes from increasing to decreasing or from decreasing to increasing. (Recall that a curve is increasing on an interval if the curve rises as it moves from left to right on that interval.) These *turning points* are called the **local extrema points** of the graph of the function. In particular, a point where the curve changes from decreasing to increasing is called a **local minimum point**, and a point where the curve changes from increasing to decreasing is called a **local maximum point**. For example, in Figure 6.3(c), the point $(0, 0)$ is a local minimum point, and the point $(2, 8)$ is a local maximum point. Note that at the point $(0, 0)$ the function changes from decreasing to increasing, and at the point $(2, 8)$ the function changes from increasing to decreasing. Observe that the graph in Figure 6.3(a) has no local extrema points. In general, a cubic equation will have 0 or 2 local extrema points.

The highest point on the graph over an interval is called the **absolute maximum point** on the interval, and the lowest point on the graph over an interval is called the **absolute minimum point** on that interval. If the graph in Figure 6.3(c) were limited to the x-interval $[-0.5, 3.2]$, the absolute maximum point on this interval would be $(2, 8)$, and the absolute minimum point on this interval would be $(3.2, -4.096)$.

EXAMPLE 1 ▶ Cubic Graph

a. Using an appropriate window, graph $y = x^3 - 27x$.

b. Find the local maximum and local minimum, if possible.

c. Where is the absolute maximum of this function on the interval $[0, 3]$?

SOLUTION

a. This function is a third-degree (cubic) function, so it has 0 or 2 local extrema. We want to graph the function in a window that allows us to see any turning points that exist and recognize the end behavior of the function. Using the window $[-40, 30]$ by $[-100, 100]$, we get the graph shown in Figure 6.4(a). This graph has a shape like Figure 6.3(b); it has three x-intercepts and 2 local extrema, so it is a complete graph. A different viewing window can be used to provide a more detailed view of the turning points of the graph (Figure 6.4(b)).

454 Chapter 6 Higher-Degree Polynomial and Rational Functions

(a)

(b)

Figure 6.4

(a)

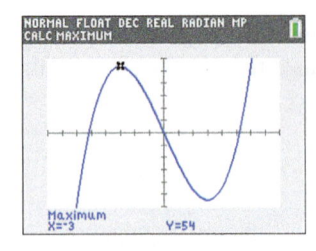

(b)

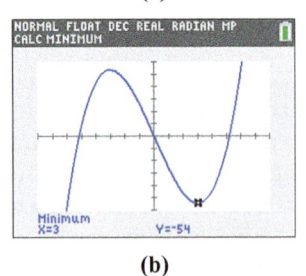

Figure 6.5

b. To find the local maximum of the function graphed in Figure 6.4:

1. Under the CALC menu (**2ND** **TRACE**), choose 4: maximum.

2. Respond to the prompts, by choosing a point to the left and right of the suspected local maximum, and then pressing **ENTER** after each response and after Guess.*

These steps give a local maximum at $(-3, 54)$. See Figure 6.5(a).

Using the same steps while choosing 3:minimum gives a local minimum at $(3, -54)$. See Figure 6.5(b).

c. The highest point on the graph on the interval $[0, 3]$ is $(0, 0)$, so the absolute maximum is $y = 0$ at $x = 0$.

EXAMPLE 2 ▶ Older Workers

Table 6.2 shows the number of millions of older persons in the workforce for the years from 1970 and projected to 2050. These data can be modeled by the cubic function $y = -0.0000597x^3 + 0.00934x^2 - 0.308x + 5.63$, where x is the number of years after 1960 and y is in millions.

Table 6.2

Years	Older Workforce (millions)
1970	3.22
1980	3.05
1990	3.45
2000	4.20
2010	5.44
2020	8.24
2030	10.1
2040	9.60
2050	10.2

(*Source*: U.S. Bureau of Labor Statistics)

---

*For detailed steps, see Appendix A, page 686.

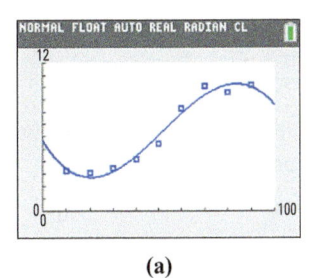

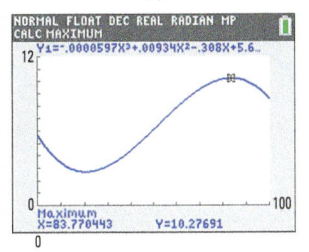

a. Graph the data points and the function on the same axes, using the window $[0, 100]$ by $[0, 12]$. Comment on the fit.

b. Use the model to predict the number of older workers in 2025.

c. Use technology and the model to find the year in the period from 1960 to 2060 during which the number of older workers is expected to reach its maximum.

d. A local minimum occurs at $(20.5, 2.73)$. What does the function do on the interval $[20.5, 83.8]$?

SOLUTION

a. Figure 6.6(a) shows a graph of the model and the data on the same axes. The model is a good fit.

b. Evaluating the function at $x = 65$ gives 8.68, so the model predicts that the number of older persons in the workforce will be 8.68 million in 2025.

c. Using the "maximum" feature under the CALC menu (see Figure 6.6(b)) shows that the local maximum is at approximately $(83.8, 10.3)$. This indicates that the model predicts that the number of older persons in the workforce will reach its maximum, 10.3 million, during 2044.

d. The function increases over this interval.

Quartic Functions

A polynomial function of the form $f(x) = ax^4 + bx^3 + cx^2 + dx + e, a \neq 0$, is a fourth-degree, or **quartic**, function. The basic quartic function $f(x) = x^4$ has the graph shown in Figure 6.7(a). Notice that this graph resembles a parabola, although it is not actually a parabola (recall the shape of a parabola, which was discussed in Section 3.1). The graph in Figure 6.7(a) is one of the possible shapes of the graph of a quartic function. The graphs of two additional quartic functions are shown in Figures 6.7(b) and (c).

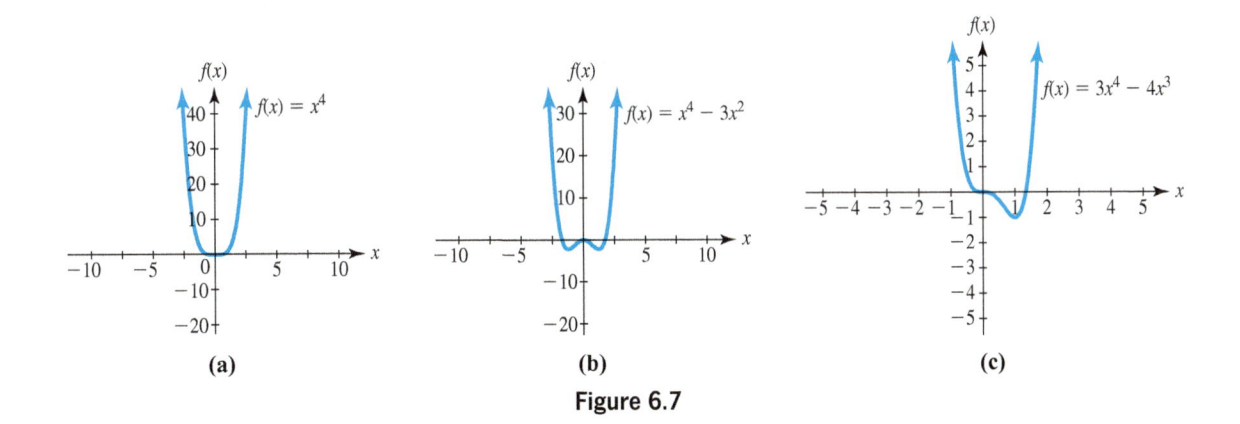

Notice that in the graph of the quartic function shown in Figure 6.7(b) there are 3 turning points and that in the graphs shown in Figures 6.7(a) and 6.7(c) there is 1 turning point each. The end behavior of a quartic function can be described as "both ends opening up" (if the leading coefficient is positive) or "both ends opening down" (if the leading coefficient is negative). Variations in the curvature may occur for different quartic functions, but these are the only possible end behaviors.

Table 6.3 summarizes the information about turning points and end behavior for graphs of cubic and quartic functions as well as for linear and quadratic functions.

Chapter 6 Higher-Degree Polynomial and Rational Functions

Table 6.3

Function	Possible Graphs	Degree	Number of Turning Points	End Behavior — Positive leading coefficient	End Behavior — Negative leading coefficient	Degree: Even or Odd
Linear	Positive slope / Negative slope	1	0			odd
Quadratic	Positive leading coefficient / Negative leading coefficient	2	1			even
Cubic — Positive leading coefficient / Negative leading coefficient		3	2 or 0			odd
Quartic — Positive leading coefficient / Negative leading coefficient		4	3 or 1			even

Table 6.3 helps us observe the following.

Polynomial Graphs

1. The graph of a polynomial function of degree n has at most $n - 1$ turning points.
2. The graph of a polynomial function of degree n has at most n x-intercepts.
3. The end behavior of the graph of a polynomial function with odd degree can be described as "one end opening up and one end opening down."
4. The end behavior of the graph of a polynomial function with even degree can be described as "both ends opening up" or "both ends opening down."

EXAMPLE 3 Graphs of Polynomial Functions

Figures 6.8(a)–(d) show the complete graphs of several polynomial functions. For each function, determine

i. the number of x-intercepts.

ii. the number of turning points.

iii. whether the leading coefficient is positive or negative.

iv. whether the degree of the polynomial is even or odd.

v. the minimum possible degree.

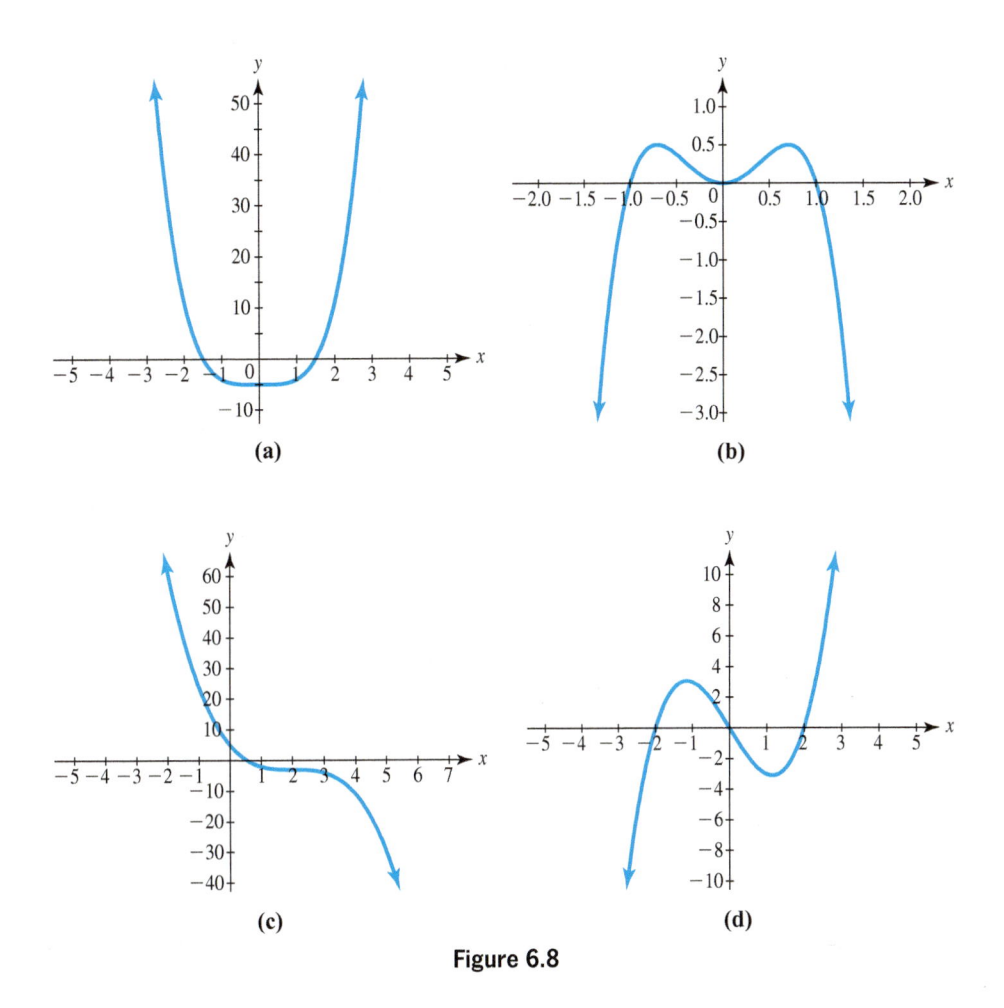

Figure 6.8

SOLUTION

a. The graph clearly shows two x-intercepts and 1 turning point. Because the end behavior is "both ends opening up," the degree of the polynomial function is even and the leading coefficient is positive. Because there is only 1 turning point, the function could be quadratic or possibly quartic, so the minimum degree of the polynomial is 2.

b. The graph shows 3 turning points and three x-intercepts. The end behavior is "both ends opening down," so the degree of the polynomial function is even and the leading coefficient is negative. The minimum possible degree is 4 because there is more than 1 turning point, and thus the function cannot be quadratic.

c. The graph shows one x-intercept and 0 turning points. The degree of the function is odd because the end behavior is "one end up and one end down." The leading coefficient is negative because the left end is up. Because the graph is not a line, the minimum degree is 3.

d. The graph shows three x-intercepts and 2 turning points. The end behavior is "one end up and one end down," with the right end opening up, so the function has odd degree and the leading coefficient is positive. Because the graph has three x-intercepts, the minimum degree is 3.

EXAMPLE 4 ▶ Energy from Crude Oil

Using data for selected years from 2010 and projected to 2040, the total energy supply from crude oil products, in quadrillion BTUs, can be modeled by the function

$$C(x) = -0.0000752x^4 + 0.00589x^3 - 0.156x^2 + 1.48x + 11.6$$

with x equal to the number of years after 2010. (*Source*: U.S. Energy Information Administration)

a. Graph the function for values of x from 0 to 35.

b. Use the graph and technology to approximate the year after 2010 in which the total energy supply from crude oil products is a maximum and the year before 2045 in which it is a minimum.

c. Is it likely that this model can be used to project the total energy supply from crude oil products for long after 2045?

SOLUTION

a. The graph is shown in Figure 6.9(a), with two local maximums and one local minimum.

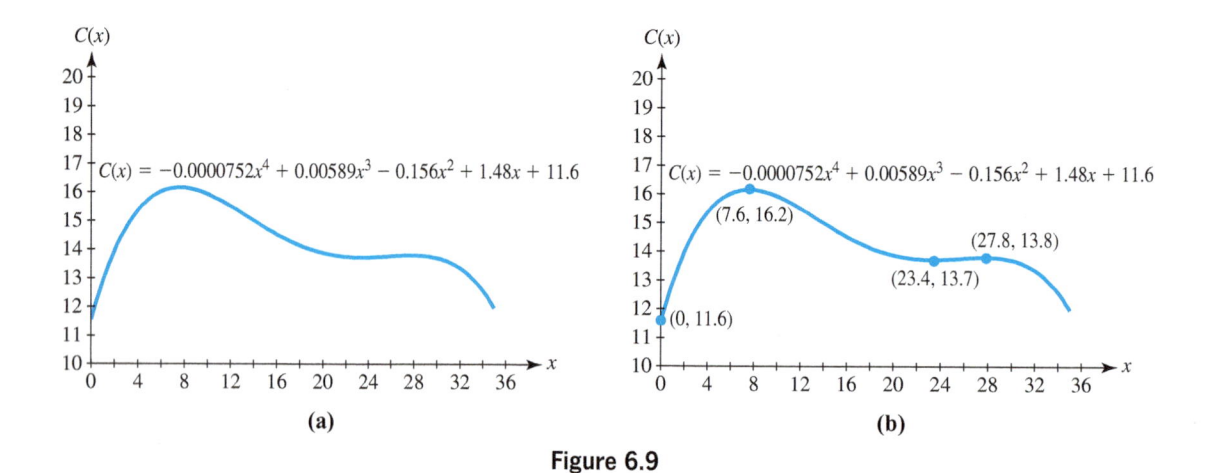

Figure 6.9

b. The graph in Figure 6.9(b) shows that the absolute maximum occurs at $x \approx 7.6$, $C(x) \approx 16.2$, so it estimates that the maximum energy supply from crude oil products, 16.2 quadrillion BTUs, occurred in 2018. The graph also shows that the absolute minimum occurs at $x = 0$, where $C(x) \approx 11.6$, so the model estimates that the minimum energy supply from crude oil products for this period, 11.6 quadrillion BTUs, occurred in 2010.

c. No. The graph of the model shows all three of its possible turning points, so it will continue to fall after 2045 and the model would eventually show a negative energy supply, which is not possible. Thus the model ceases to be valid at some point.

SKILLS CHECK 6.1

Answers that are not seen can be found in the answer section at the back of the text.

1. Graph the function $h(x) = 3x^3 + 5x^2 - x - 10$ on the windows given in parts (a) and (b). Which window gives a complete graph? Window (b)

    a. $[-5, 5]$ by $[-5, 5]$ b. $[-5, 5]$ by $[-20, 20]$

2. Graph the function $f(x) = 2x^3 - 3x^2 - 6x$ on the windows given in parts (a) and (b). Which window gives a complete graph? Window (b)

    a. $[-5, 5]$ by $[-5, 5]$

    b. $[-10, 10]$ by $[-10, 10]$

3. Graph the function $g(x) = 3x^4 - 12x^2$ on the windows given in parts (a) and (b). Which window gives a complete graph? Window (b)

    a. $[-5, 5]$ by $[-5, 5]$ b. $[-3, 3]$ by $[-12, 10]$

4. Graph the function $g(x) = 3x^4 - 4x^2 + 10$ on the windows given in parts (a) and (b). Which window gives a complete graph? Window (b)

    a. $[-10, 10]$ by $[-10, 10]$

    b. $[-5, 5]$ by $[-20, 20]$

For Exercises 5–10, use the given graph of the polynomial function to (a) estimate the x-intercept(s), (b) state whether the leading coefficient is positive or negative, and (c) determine whether the polynomial function is cubic or quartic.

5. a. $-2, 1, 2$
   b. Positive
   c. Cubic

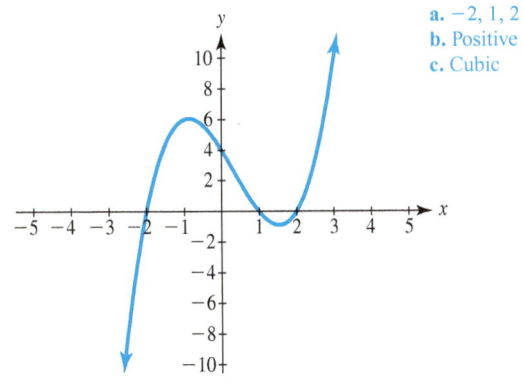

6. a. $-1, 2, 3$
   b. Negative
   c. Quartic

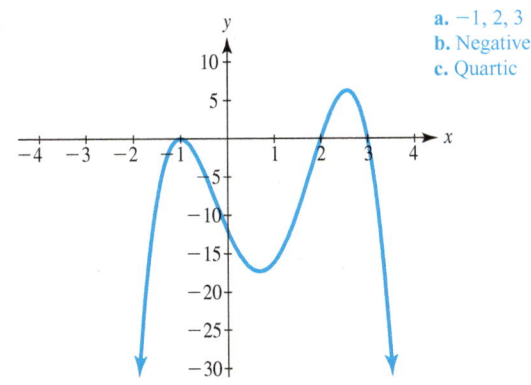

7. a. $-1, 1, 5$
   b. Negative
   c. Cubic

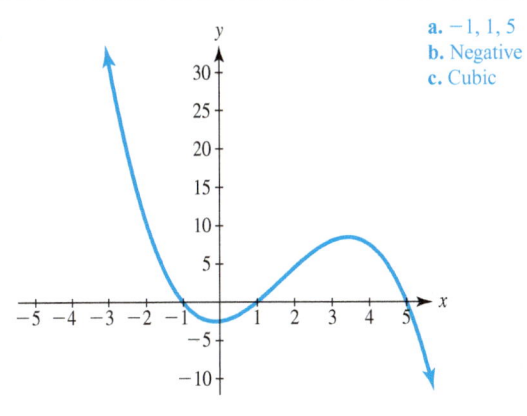

8. a. $-1, 2, 5$
   b. Positive
   c. Quartic

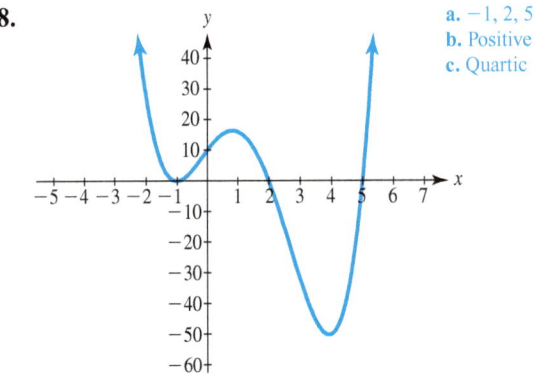

9. a. $-1.5, 1.5$
   b. Positive
   c. Quartic

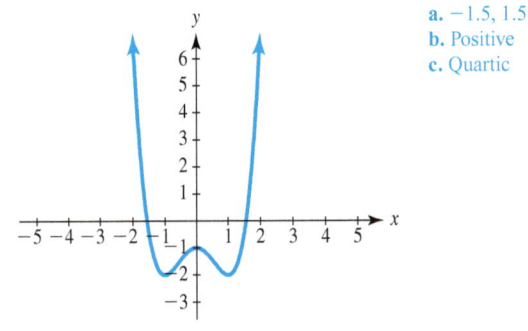

10. a. $-2, 3$
    b. Negative
    c. Quartic

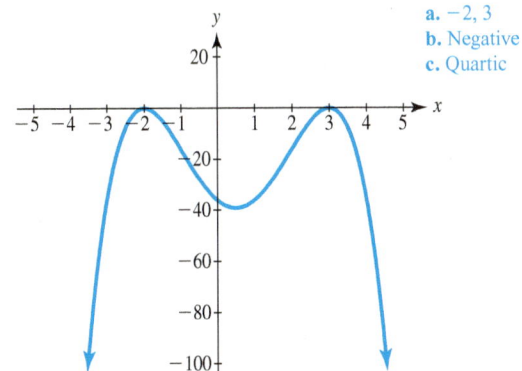

For Exercises 11–16, match the polynomial function with its graph (below or on the next page).

11. $y = 2x^3 + 3x^2 - 23x - 12$ C

460 Chapter 6 Higher-Degree Polynomial and Rational Functions

12. $y = -2x^3 - 8x^2 + 0.5x + 2$ A
13. $y = -6x^3 + 5x^2 + 17x - 6$ E
14. $y = x^4 - 0.5x^3 - 14.5x^2 + 17x + 12$ B
15. $y = x^4 + 3$ F
16. $y = -x^4 + 16x^2$ D

A.

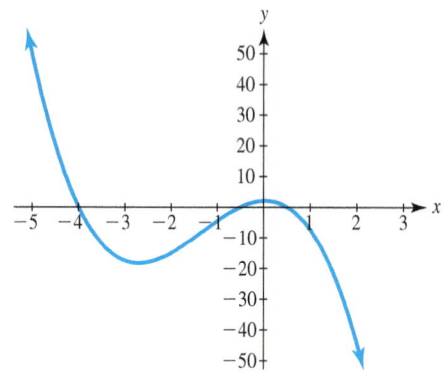

B.

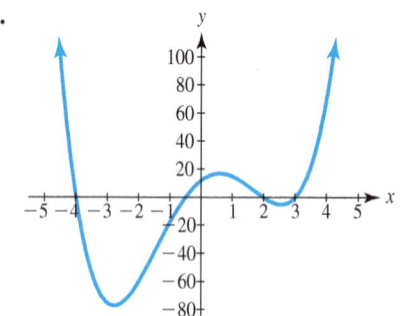

C.

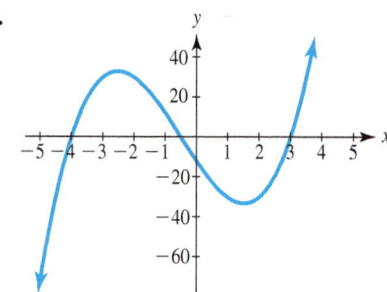

D.

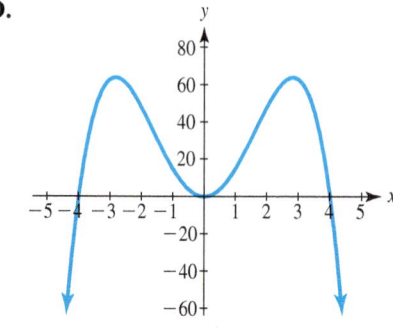

E.

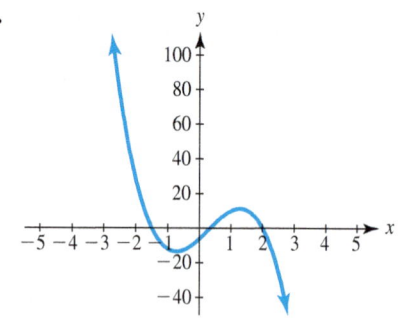

F.

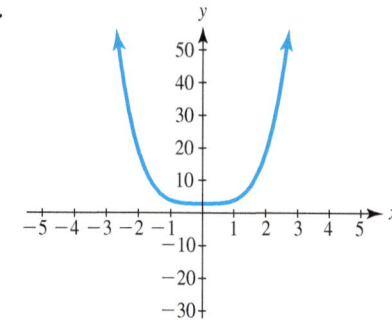

For Exercises 17–20, use the equation of the polynomial function to (a) state the degree and the leading coefficient and (b) describe the end behavior of the graph of the function. (c) Support your answer by graphing the function.

17. $f(x) = 2x^3 - x$

18. $g(x) = 0.3x^4 - 6x^2 + 17x$

19. $f(x) = -2(x - 1)(x^2 - 4)$

20. $g(x) = -3(x - 3)^2(x - 1)^2$

21. a. Graph the function $y = x^3 - 3x^2 - x + 3$ using the window $[-10, 10]$ by $[-10, 10]$.

    b. Is the graph complete? Yes; three x-intercepts show, along with the y-intercept.

22. a. Graph $y = x^3 + 6x^2 - 4x$ using the window $[-10, 10]$ by $[-10, 10]$.

    b. Is the graph complete? No; one turning point does not show.

23. Graph $y = 25x - x^3$ using

    a. the window $[-10, 10]$ by $[-10, 10]$.

    b. a window that shows 2 turning points.

24. Graph the function $y = x^3 - 16x$ using

    a. the window $[-10, 10]$ by $[-10, 10]$.

    b. a window that shows 2 turning points.

25. Graph the function $y = x^4 - 4x^3 + 4x^2$ using

    a. the window $[-10, 10]$ by $[-10, 10]$.

    b. the window $[-4, 4]$ by $[-4, 4]$.

    c. Which window gives a more detailed view of the graph near the turning points? Window (b)

26. a. Graph $y = x^4 - 4x^2$ using the window $[-10, 10]$ by $[-10, 10]$.

    b. Is the graph complete? Yes

27. a. Graph $y = x^4 - 4x^2 - 12$ on $[-8, 8]$ by $[-20, 10]$.

    b. How many turning points does the graph have? 3

    c. Could the graph of this function have more turning points? No

28. a. Graph $y = x^4 + 6x^2$ on $[-10, 10]$ by $[-10, 50]$.

    b. Is it possible for the graph of this function to have only 1 turning point? Yes; the graph of a fourth-degree polynomial function can have 3 or 1 turning points.

29. Sketch a graph of any cubic polynomial function that has a negative leading coefficient and one x-intercept.

30. Sketch a graph of any polynomial function that has degree 3, a positive leading coefficient, and three x-intercepts.

31. Sketch a graph of any polynomial function that has degree 4, a positive leading coefficient, and two x-intercepts.

32. Sketch a graph of any cubic polynomial function that has a negative leading coefficient and three x-intercepts.

33. a. Graph $y = x^3 + 4x^2 + 5$ on a window that shows a local maximum and a local minimum.

    b. A local maximum occurs at what point? $(-2.67, 14.48)$

    c. A local minimum occurs at what point? $(0, 5)$

34. a. Graph $y = x^4 - 8x^2$ on a window that shows 2 local minima and 1 local maximum.

    b. A local maximum occurs at what point? $(0, 0)$

    c. The local minima occur at what points? $(-2, -16)$ and $(2, -16)$

35. Use technology to find a local maximum and 2 local minima of the graph of the function $y = x^4 - 4x^3 + 4x^2$. Maximum: $(1, 1)$; minima: $(0, 0)$, $(2, 0)$

36. a. Graph the function $y = -x^3 - x^2 + 9x$ using the window $[-5, 5]$ by $[-15, 10]$.

    b. Graph the function on an interval with $x \geq 0$ and with $y \geq 0$.

    c. The graph of what other type of function resembles the *piece* of the graph of $y = -x^3 - x^2 + 9x$ shown in part (b)? A second-degree (quadratic) function

EXERCISES 6.1

Answers that are not seen can be found in the answer section at the back of the text.

37. **Daily Revenue** The daily revenue in dollars from the sale of a product is given by $R = -0.1x^3 + 11x^2 - 100x$, where x is the number of units sold.

    a. Graph this function on the window $[-100, 100]$ by $[-5000, 25,000]$. How many turning points do you see? 2 turning points

    b. Because x represents the number of units sold, what restriction should be placed on x in the context of the problem? What restriction should be placed on R? Both x and R need to be nonnegative.

    c. Using your result from part (b), graph the function on a new window that makes sense for the problem.

    d. What does this function give as the revenue if 50 units are produced? $10,000

38. **Weekly Revenue** A firm has total weekly revenue in dollars for its product given by $R(x) = 2000x + 30x^2 - 0.3x^3$, where x is the number of units sold.

    a. Graph this function on the window $[-100, 150]$ by $[-30,000, 220,000]$.

    b. Because x represents the number of units sold, what restrictions should be placed on x in the context of the problem? What restrictions should be placed on y? Both x and y need to be nonnegative.

    c. Using your result from part (b), graph the function on a new window that makes sense for the problem.

    d. What does this function give as the revenue if 60 units are produced? $163,200

39. **Daily Revenue** The daily revenue in dollars from the sale of a product is given by $R = 600x - 0.1x^3 + 4x^2$, where x is the number of units sold.

    a. Graph this function on the window $[0, 100]$ by $[0, 30,000]$.

    b. Use the graph of this function to estimate the number of units that will give the maximum daily revenue and the maximum possible daily revenue. 60 units; $28,800

    c. Find a window that will show a complete graph—that is, will show all of the turning points and intercepts. Graph the function using this window.

d. Does the graph in part (a) or the graph in part (c) better represent the revenue from the sale of *x* units of a product? Part (a)

e. Over what interval of *x*-values is the revenue increasing, if $x \geq 0$? $0 < x < 60$

40. Weekly Revenue A firm has total weekly revenue in dollars for its product given by $R(x) = 2800x - 8x^2 - x^3$, where *x* is the number of units sold.

a. Graph this function on the window $[0, 50]$ by $[0, 51,000]$.

b. Use technology to find the maximum possible revenue and the number of units that gives the maximum revenue. $50,176; 28 units

c. Find a window that will show a complete graph—that is, will show all of the turning points and intercepts. Graph the function using this window.

d. Does the graph in part (a) or the graph in part (c) better represent the revenue from the sale of *x* units of a product? Part (a)

e. Over what interval of *x*-values is the revenue increasing, if $x \geq 0$? $0 < x < 28$

41. Investment If $2000 is invested for 3 years at rate *r*, compounded annually, the future value of this investment is given by $S = 2000(1 + r)^3$, where *r* is the rate written as a decimal.

a. Complete the following table to see how increasing the interest rate affects the future value of this investment.

Rate, r	Future Value, S ($)
0.00	2000.00
0.05	2315.25
0.10	2662.00
0.15	3041.75
0.20	3456.00

b. Graph this function for $0 \leq r \leq 0.24$.

c. Use the table and/or graph to compare the future value if $r = 10\%$ and if $r = 20\%$. How much more money is earned at 20%? $794

d. Which interest rate, 10% or 20%, is more realistic for an investment? 10% is much more likely than 20%.

42. Investment The future value of $10,000 invested for 5 years at rate *r*, compounded annually, is given by $S = 10,000(1 + r)^5$, where *r* is the rate written as a decimal.

a. Complete the following table to see how increasing the interest rate affects the future value of this investment.

Rate, r	Future Value, S ($)
0.00	10,000.00
0.05	12,762.82
0.07	14,025.52
0.12	17,623.42
0.18	22,877.58

b. Graph this function for $0 \leq r \leq 0.24$.

c. Use the graph to compare the future value if $r = 10\%$ and if $r = 24\%$. How much more money is earned at 24%? $13,211.15

d. Is it more likely that you can get an investment paying 10% or 24%? 10% is more likely.

43. Industrial Shipments The value of U.S. industrial shipments for selected years from 2014 and projected to 2040 can be modeled by the function

$$y = 0.000185x^3 - 0.00968x^2 + 0.305x + 6.08$$

where *x* is the number of years after 2010 and *y* is the billions of dollars of U.S. industrial shipments. (*Source:* U.S. Department of Energy)

a. Graph the function for values representing 2010 through 2040.

b. Does the value of the U.S. industrial shipments ever decrease over this period of time? No

c. Use the model to predict the U.S. industrial shipments in 2040. $11.513 billion

44. U.S. Population Using data from 1980 and projected to 2050, the U.S. population, in millions, can be modeled by $y = -0.000233x^3 + 0.0186x^2 + 2.32x + 235$, with *x* equal to the number of years after 1980.

a. Graph the function on an interval representing 1980 through 2050.

b. What does the model predict the population will be in 2022? Round to the nearest million. 348 million

c. What is the maximum predicted population over this period of time, according to the model? 408.6 million (*Source*: Social Security Administration)

45. Drug Levels in Blood A new drug is tested to see how a 300-mg capsule is released into the bloodstream. Volunteers are given a capsule, blood samples are drawn every half-hour, and the average number of milligrams of the drug in the bloodstream is calculated for the sample. The cubic function that models these data is $N(t) = 24.1t^3 - 165t^2 + 241t + 167$.

a. Graph the model for the time period $t = 0$ through $t = 3.5$ hours.

b. Use technology to find the maximum number of milligrams of the drug in the bloodstream during this period. Approximately 268 mg at approximately 0.9 hour

c. Use technology to find the minimum number of milligrams of the drug in the bloodstream during this period. Approximately 22.5 mg at 3.5 hours

46. *Aging Workers* The number of millions of Americans who are working full time at selected ages from 27 to 62 can be modeled by the cubic function $f(x) = -0.000362x^3 + 0.0401x^2 - 1.39x + 21.7$, where x is their age.
(*Source: Wall Street Journal*, June 18, 2012)

a. Use the model to predict the number of Americans age 50 who are working full time. 7.2 million

b. Use technology to find the local maximum on the interval $[27, 62]$. What does this point indicate?

c. Find the local minimum on the interval $[27, 62]$. What does this indicate?

47. *Risky Drivers* Both inexperienced young drivers and very old drivers are involved in far more accidents than other age groups. The quartic function that models the police-reported crashes for different age groups is $y = 0.00001828x^4 - 0.003925x^3 + 0.3031x^2 - 9.907x + 118.2$, where x is the age of the driver and y equals the number of crashes per million miles traveled.

a. Graph the function for the ages 15 to 86.

b. What does the model estimate the number of crashes per million miles to be for 22-year-old drivers? 9.4

c. At what age among older drivers does the model predict the number of crashes will reach 8 per million miles? 83 ($x = 83.7$)

d. At what age among younger drivers does the model predict the number of crashes will be the minimum number? 34 ($x = 34.6$)
(*Source: Consumer Reports*, October 2012)

48. *Women in the Workforce* The number of women in the workforce for selected years from 1950 and projected to 2050 can be modeled by

$$y = 0.00358x^4 - 0.783x^3 + 50.1x^2 - 24.8x + 18,600$$

with y in thousands and x equal to the number of years after 1950.
(*Source*: U.S. Bureau of Labor Statistics)

a. Graph this function on an interval representing 1950 through 2050.

b. How many women are projected by this model to be in the workforce in 2030? 82,996.8 thousand, or 82,996,800

c. What does the model predict the maximum number of women in the workforce will be during this period? When? 92,120,000; in 2050

d. Does the number of women in the workforce ever decrease during this period, according to the model? No

49. *Profit* The weekly revenue for a product is given by $R(x) = 120x - 0.015x^2$, and the weekly cost is $C(x) = 10,000 + 60x - 0.03x^2 + 0.00001x^3$, where x is the number of units produced and sold.

a. How many units will give maximum profit? 2000

b. What is the maximum possible profit? $90,000

50. *Profit* The annual revenue for a product is given by $R(x) = 60,000x - 50x^2$, and the annual cost is $C(x) = 800 + 100x^2 + x^3$, where x is the number of thousands of units produced and sold.

a. How many units will give maximum profit? 100,000

b. What is the maximum possible profit? $3,499,200

6.2 Modeling with Cubic and Quartic Functions

KEY OBJECTIVES
- Model and apply data with cubic functions
- Model and apply data with quartic functions

SECTION PREVIEW   Aging Workers

It is reasonable to assume that the numbers of Americans of differing ages working full time will vary considerably. Table 6.4 shows the number of millions of Americans who are working full time at selected ages from 27 to 62. The scatter plot of the data in Figure 6.10 shows that neither a line nor a parabola fits the data well, and it appears that a cubic function would be a better fit.

Table 6.4

Age	Millions Working Full Time
27	6.30
32	6.33
37	6.96
42	7.07
47	7.52
52	7.10
57	5.58
62	3.54

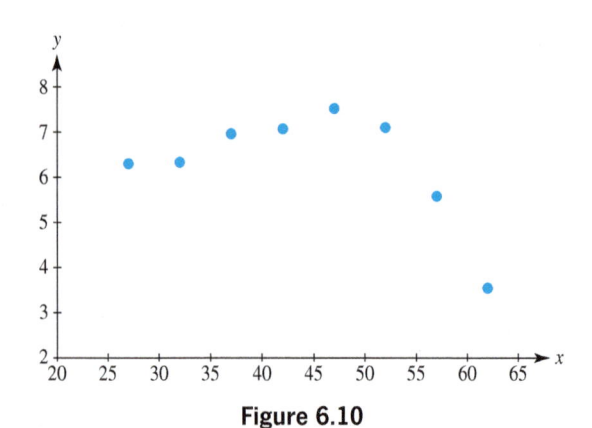

(Source: Wall Street Journal)

Figure 6.10

The use of graphing calculators and spreadsheets enables us to find cubic and quartic functions that model nonlinear data. In this section, we will model and apply these functions. ■

Modeling with Cubic Functions

The scatter plot in Figure 6.10 indicates that the data may fit close to one of the possible shapes of the graph of a cubic function, so it is possible to find a cubic function that models the data.

EXAMPLE 1 ▶ Aging Workers

Use the data in Table 6.4 to do the following:

a. Find a cubic model that gives the number of Americans working full time, y (in millions), as a function of their age, x. Report the model with three significant digits in the coefficients.

b. Graph the unrounded model from part (a) on the same axes with the data, and comment on the fit.

c. Use the model to estimate the number of Americans working at age 59.

SOLUTION

a. A cubic function that models the number of Americans working full time, y (in millions), as a function of their age, x, is

$$f(x) = -0.000362x^3 + 0.0401x^2 - 1.39x + 21.7$$

b. Figure 6.11 shows a graph of the model and the data on the same axes. The model is an excellent visual fit.

c. The number of Americans age 59 who are working is estimated from the model to be $y = f(59) \approx 5$ million.

6.2 Modeling with Cubic and Quartic Functions 465

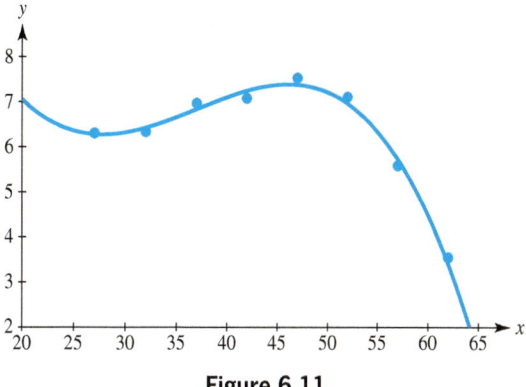

Figure 6.11

Spreadsheet ▶ SOLUTION

We can use graphing calculators, software programs, and spreadsheets to find the polynomial function that is the best fit for a set of data. Figure 6.12 includes an Excel spreadsheet for the data of Example 1. Selecting the cells containing the data, getting the scatter plot of the data, and selecting Polynomial Trendline with the order (degree) of the polynomial gives the equation of the polynomial function that is the best fit for the data, along with the scatter plot and the graph of the best-fitting curve. Figure 6.12 shows an Excel worksheet with the data from Example 1 and the graph of the cubic function that fits the data.*

Age	Millions
27	6.3
32	6.33
37	6.96
42	7.07
47	7.52
52	7.1
57	5.58
62	3.54

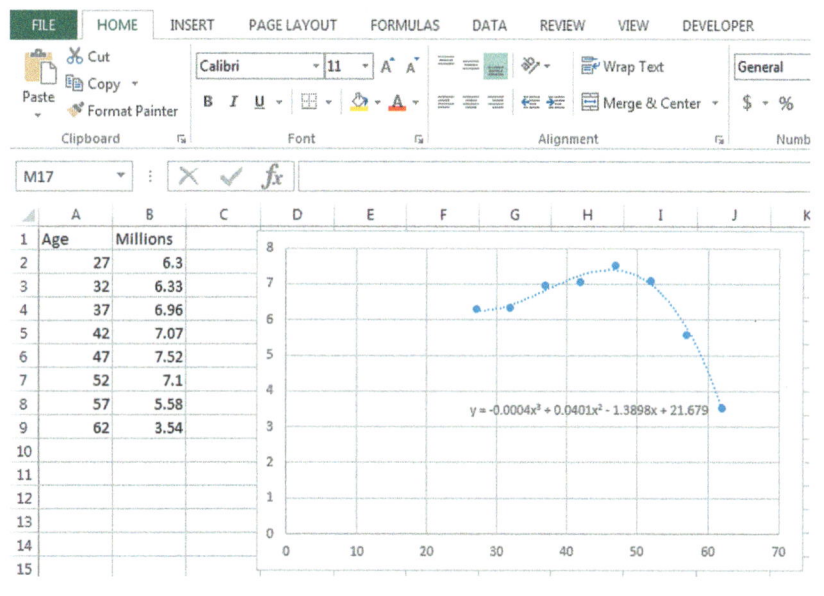

$y = -0.0004x^3 + 0.0401x^2 - 1.3898x + 21.679$

Figure 6.12

Sometimes a set of data points can be modeled reasonably well by both a linear function and a cubic function. Consider the example on the next page.

---

* For detailed steps, see Appendix B, page 708.

EXAMPLE 2　Diabetes

The number of U.S. adults with diabetes is projected to increase rapidly in the future. The projected numbers for selected years, in millions, are shown in Table 6.5.

a. Create a scatter plot of the data, with x equal to the number of years after 2015 and y equal to the number of millions of U.S. adults with diabetes.

b. Create a linear function that models the data. Report the model with three significant digits.

c. Create a cubic function that models the data. Report the model with three significant digits.

d. Graph the data on the same axes as each model, and determine which is the better fit to the data.

Table 6.5

Year	Millions	Year	Millions
2015	37.3	2035	76.2
2020	50.0	2040	84.1
2025	59.5	2045	91.7
2030	68.3	2050	100.0

(*Source:* Centers for Disease Control and Prevention)

SOLUTION

a. The scatter plot of the data is shown in Figure 6.13(a).

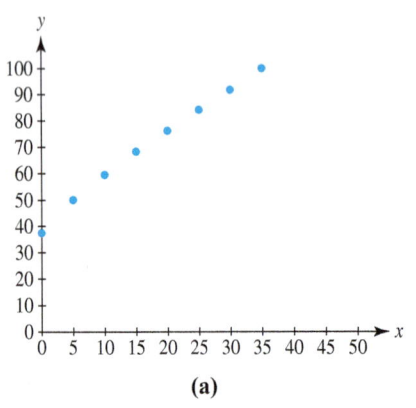

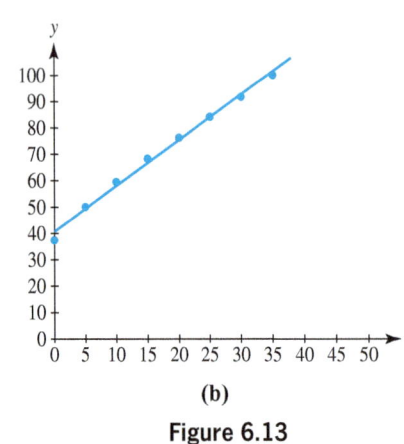

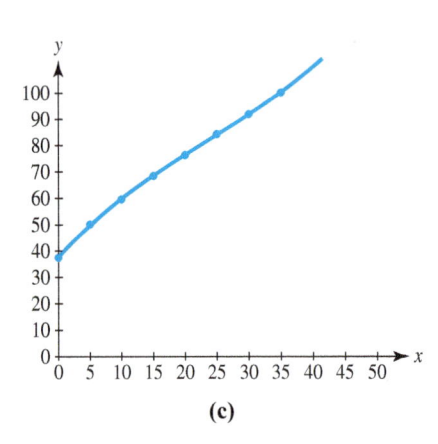

(a)　(b)　(c)

Figure 6.13

b. The linear function that is the best fit for the data is
$$y = 1.74x + 40.5$$

c. The cubic function that is the best fit for the data is
$$y = 0.000697x^3 - 0.0486x^2 + 2.64x + 37.5$$

d. The graphs of the data on the same axes as the linear function and the cubic function are shown in Figure 6.13(b) and Figure 6.13(c). The fit of the linear function to the data is excellent, but the cubic function is nearly an exact fit to the data, so the cubic function is the better fit.

Modeling with Quartic Functions

If a scatter plot of data indicates that the data may fit close to one of the possible shapes that are graphs of quartic functions, it is possible to find a quartic function that models the data. Consider the following example.

EXAMPLE 3 ▶ Supplemental Nutrition Assistance Program

The Supplemental Nutrition Assistance Program (SNAP) caseloads grew significantly between 2007 and 2011 as the recession and lagging economic recovery led more low-income households to qualify and apply for help. The Congressional Budget Office (CBO) expects that, as the economy improves, the number of participants will fall by about 2–4% each year over the next decade: from 46.5 million in fiscal year 2014 to 46.0 million in 2015, 44.3 million in 2016, and 32.8 million by 2025. The CBO forecasts that by 2025, the share of the population receiving SNAP will return to close to 2007 levels (about 9%). Table 6.6 shows the number of SNAP participants, in thousands, from 1990 to 2015 and projected to 2025.

Table 6.6

Year	SNAP Participants (thousands)	Year	SNAP Participants (thousands)
1990	20,049	2004	23,811
1991	22,625	2005	25,628
1992	25,407	2006	26,549
1993	26,987	2007	26,316
1994	27,474	2008	28,223
1995	26,619	2009	33,490
1996	25,543	2010	40,302
1997	22,858	2011	44,709
1998	19,791	2012	46,609
1999	18,183	2013	47,636
2000	17,194	2014	46,536
2001	17,318	2015	46,000
2002	19,096	2016	44,300
2003	21,250	2025	32,800

(*Source*: Center on Budget and Policy Priorities)

a. Create a scatter plot of the data, with $x = 0$ in 1990 and y equal to the number of thousands of SNAP participants, to determine whether the data can be modeled by a quartic function.

b. Find the quartic function that models the data, with $x = 0$ in 1990. Round the model to four decimal places.

c. Graph the unrounded model and the scatter plot of the data on the same axes.

d. Use the unrounded model to determine whether the CBO's projection for SNAP participants in 2025 is accurate.

SOLUTION

a. The scatter plot of the data is shown in Figure 6.14(a). The scatter plot indicates that a function whose graph has end behavior such that both ends open down would be a good fit. Thus, a quartic model is appropriate.

b. The quartic function that models the data is

$$y = -0.4445x^4 + 24.3876x^3 - 333.3264x^2 + 1086.8474x + 23441.2847$$

where x is the number of years after 1990.

c. The graph of the model and the scatter plot of the data are shown in Figure 6.14(b).

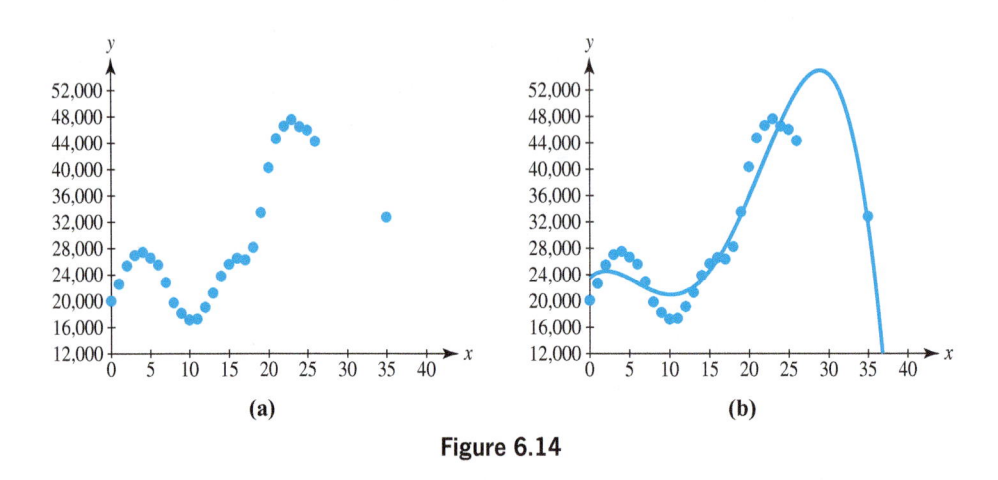

(a)

(b)

Figure 6.14

d. The year 2025 corresponds to an x-value of 35, and when $x = 35$, $y = 31,731$. This number is close to the number projected by the CBO: 32,800.

Model Comparisons

Even when a scatter plot does not show all the characteristics of the graph of a quartic function, a quartic function may be a good fit for the data. Consider the following example, which shows a comparison of a cubic and a quartic model for a set of data. It also shows that many decimal places are sometimes required in reporting a model so that the leading coefficient is not written as zero.

EXAMPLE 4 ▶ International Visitors to the United States

The numbers of inbound international visitors to the United States from 2000 and projected to 2022 are shown in Table 6.7 and Figure 6.15.

a. Create a scatter plot for the data, using the number of years after 2000 as the input x and the number of millions of visitors as the output y.

b. Find a cubic function to model the data and graph the function on the same axes with the data points. Report the model by rounding coefficients to four decimal places.

c. Find a quartic function to model the data and graph the function on the same axes with the data points. Report the model by rounding coefficients to four decimal places.

d. Use the graphs of the unrounded functions in parts (b) and (c) to determine which function is the better fit for the data.

6.2 Modeling with Cubic and Quartic Functions

Table 6.7

Year	Number of Visitors (millions)	Year	Number of Visitors (millions)
2000	51.2	2012	66.7
2001	46.9	2013	70
2002	43.6	2014	75
2003	41.2	2015	77.5
2004	46.1	2016	75.9
2005	49.2	2017	75.1
2006	51	2018	78
2007	56.1	2019	80.9
2008	58	2020	83.4
2009	55.1	2021	86.2
2010	60	2022	89
2011	62.8		

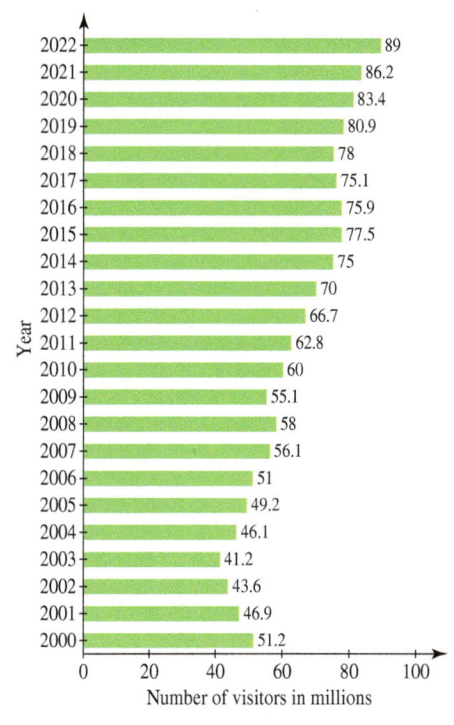

(*Source*: statista.com)

Figure 6.15

SOLUTION

a. The scatter plot of the data is shown in Figure 6.16(a).

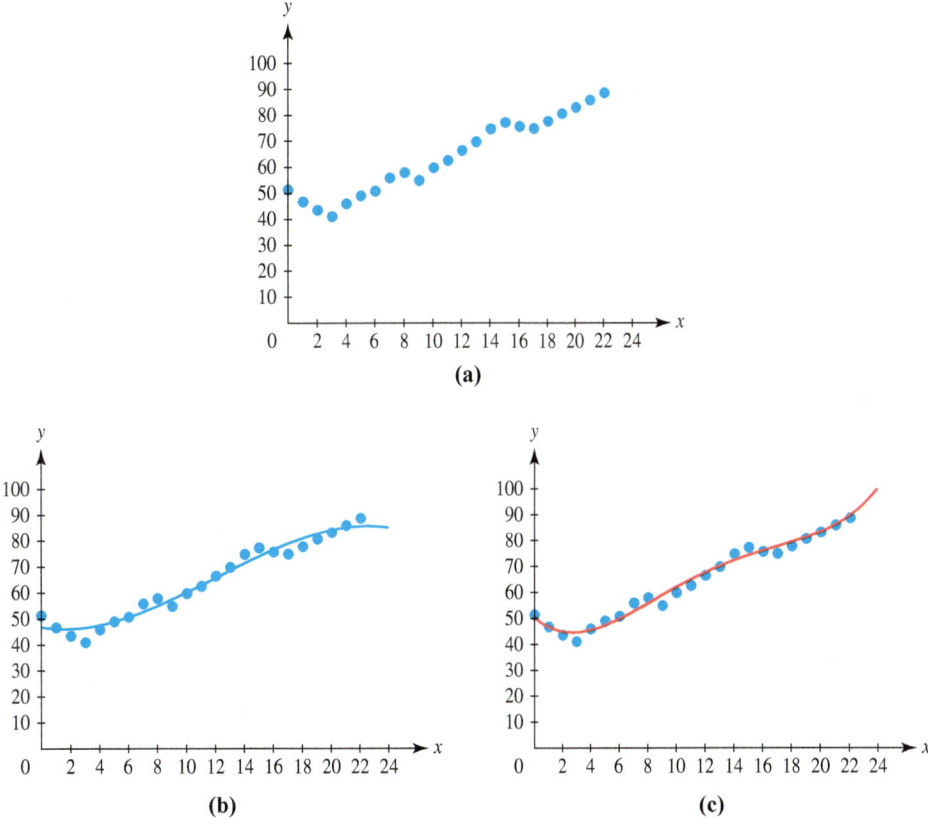

Figure 6.16

b. A cubic function that models the data is

$$y = -0.0086x^3 + 0.3111x^2 - 0.9179x + 46.9352$$

with $x = 0$ representing 2000. Figure 6.16(b) shows the graph of the unrounded model on the same axes with the data points.

c. A quartic function that models the data is

$$y = 0.0013x^4 - 0.0657x^3 + 1.1083x^2 - 4.6218x + 50.1939$$

with $x = 0$ representing 2000. Figure 6.16(c) shows the graph of the unrounded model on the same axes with the data points.

d. The quartic model seems to be a slightly better fit.

Third and Fourth Differences

Recall that for equally spaced inputs, the first differences of outputs are constant for data modeled by linear functions, and the second differences are constant for data modeled by quadratic functions. In a like manner, the third differences of outputs are constant for data modeled by cubic functions, and the fourth differences are constant for data modeled by quartic functions, if the inputs are equally spaced.

EXAMPLE 5 ▶ Model Comparisons

Table 6.8 has a set of equally spaced inputs, x, in column 1 and three sets of outputs in columns 2, 3, and 4. Use third and/or fourth differences with each set of outputs to determine if each set of data can be modeled by a cubic or quartic function. Then find the function that is the best fit for each set of outputs and the input, using

a. Output Set I   b. Output Set II   c. Output Set III

Table 6.8

Inputs, x	Output Set I	Output Set II	Output Set III
1	3	4	−16
2	14	4	−13
3	47	18	1
4	114	88	33
5	227	280	87
6	398	684	173

SOLUTION

a. The inputs (x-values in column 1) are equally spaced. The third differences for Output Set I data in column 2 follow.

Outputs	3		14		47		114		227		398
First Differences		11		33		67		113		171	
Second Differences			22		34		46		58		
Third Differences				12		12		12			

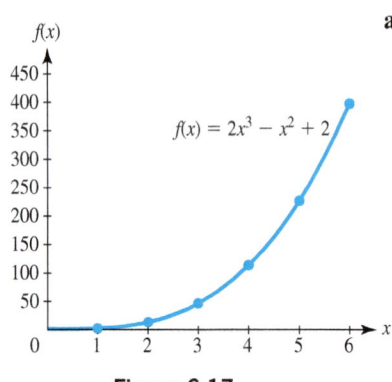

The third differences are constant, so the data points fit exactly on the graph of a cubic function. The function that models the data is $f(x) = 2x^3 - x^2 + 2$. Figure 6.17 shows the graph of this function for $0 \leq x \leq 6$.

Figure 6.17

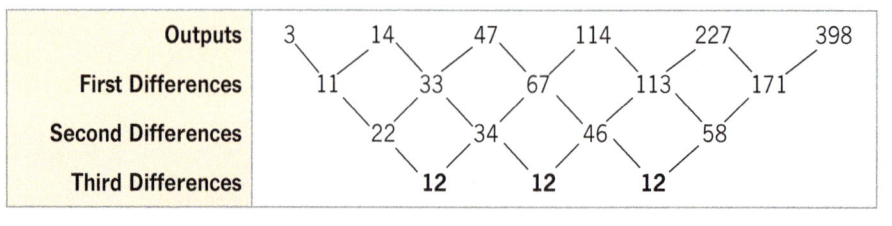

b. The third and fourth differences for Output Set II data in column 3 follow.

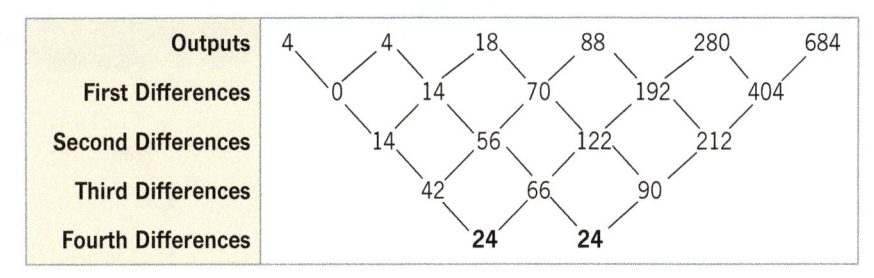

Outputs	4	4	18	88	280	684
First Differences		0	14	70	192	404
Second Differences			14	56	122	212
Third Differences				42	66	90
Fourth Differences					24	24

The fourth differences are constant, so the data points fit exactly on the graph of a quartic function. The function that models the data is $g(x) = x^4 - 3x^3 + 6x$ (see Figure 6.18).

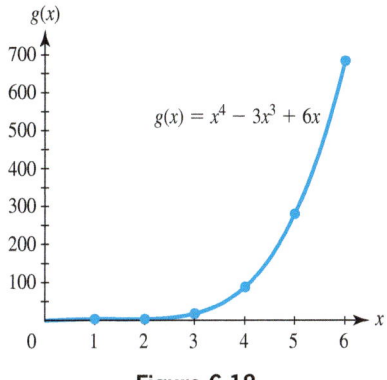

Figure 6.18

c. The third differences for Output Set III data in column 4 follow.

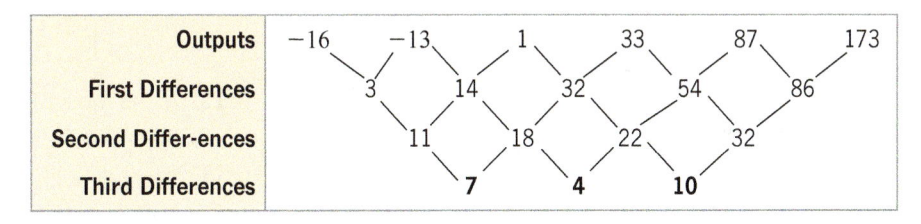

Outputs	−16	−13	1	33	87	173
First Differences		3	14	32	54	86
Second Differ-ences			11	18	22	32
Third Differences				7	4	10

The third differences are not constant, but are relatively close, so the data points can be approximately fitted by a cubic function. The cubic function that is the best fit for the data is $y = h(x) = 1.083x^3 - 1.107x^2 - 1.048x - 15$. Figure 6.19 shows that the graph is a good fit for the data points.

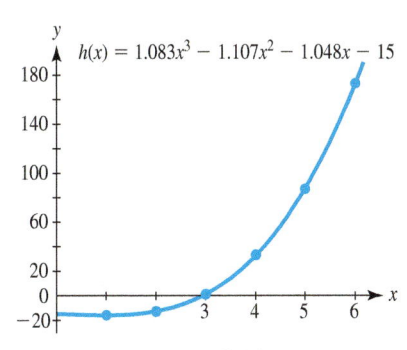

Figure 6.19

Note that the third (or fourth) differences will not be constant unless the data points fit on the graph of the cubic (or quartic) function exactly, and real data points will rarely fit on the graph of a function exactly. As with other models, statistical measures exist to measure the goodness of fit of the model to the data.

SKILLS CHECK 6.2

Answers that are not seen can be found in the answer section at the back of the text.

1. Find the cubic function that models the data in the table below. $y = x^3 - 2x^2$

x	−2	−1	0	1	2	3	4
y	−16	−3	0	−1	0	9	32

2. Find the cubic function that is the best fit for the data in the table below. $y = 2.972x^3 - 3.36x^2 - 3.7x + 1$

x	0	5	10	15
y	1	270	2600	9220

3. Find the quartic function that models the data in the table below. $y = x^4 - 4x^2$

x	−2	−1	0	1	2	3	4
y	0	−3	0	−3	0	45	192

4. Find the quartic function that is the best fit for the data in the table below. $y = 1.5x^4 - 1.75x^2$

x	−2	−1	0	1	4
y	17	−0.25	0	−0.25	356

5. a. Make a scatter plot of the data in the table below.

x	1	2	3	4	5	6
y	−2	−1	0	4	8	16

   b. Does it appear that a linear model or a cubic model is the better fit for the data? Cubic

6. a. Make a scatter plot of the data in the table below.

x	−2	−1	0	1	2	3
y	−8	7	1	2	9	57

   b. Does it appear that a cubic model or a quartic model is the better fit for the data? Cubic

7. a. Find a cubic function that models the data in the table in Exercise 5.
   $y = 0.102x^3 - 0.230x^2 + 0.811x - 2.667$
   b. Find a linear function that models the data.
   $y = 3.457x - 7.933$
   c. Visually determine which model is the better fit for the data. Cubic

8. a. Find a cubic function that models the data in the table in Exercise 6.
   $y = 2.843x^3 - 0.389x^2 - 6.612x + 3.079$
   b. Find a quartic function that models the data.
   $y = 0.146x^4 + 2.551x^3 - 1.160x^2 - 5.696x + 3.579$
   c. Visually determine which model is the better fit for the data. Perhaps quartic model is slightly better.

9. a. Find the cubic function that is the best fit for the data in the table below.
   $y = 35x^3 - 333.667x^2 + 920.762x - 677.714$

x	1	2	3	4	5	6	7
y	−2	0	4	16	54	192	1500

   b. Find the quartic function that is the best fit for the data in the table.
   $y = 12.515x^4 - 165.242x^3 + 748x^2 - 1324.814x + 738.286$

10. a. Graph each of the functions found in Exercise 9 on the same axes with the data in the table.
    b. Does it appear that a cubic model or a quartic model is the better fit for the data? Quartic

11. Find the quartic function that is the best fit for the data in the table below. $y = x^4 - 4x^2 - 3x + 1$

x	−3	−2	−1	0	1	2	3
y	55	7	1	1	−5	−5	37

12. Is the model found in Exercise 11 an exact fit to the data? Yes

13. The following table has the inputs, x, and the outputs for two functions, f and g. Use third differences to determine whether a cubic function exactly fits the data with input x and output $f(x)$. Not exactly

x	0	1	2	3	4	5
f(x)	0	1	5	24	60	110
g(x)	0	0.5	4	13.5	32	62.5

14. Use third differences with the data in the table in Exercise 13 to determine whether a cubic function exactly fits the data with input x and output $g(x)$. Exactly

15. Find the cubic function that is the best fit for $f(x)$ defined by the table in Exercise 13.
    $y = 0.565x^3 + 2.425x^2 - 4.251x + 0.556$

16. Find the cubic function that is the best fit for $g(x)$ defined by the table in Exercise 13. $y = 0.5x^3$

EXERCISES 6.2

Answers that are not seen can be found in the answer section at the back of the text.

Use unrounded models for graphing and calculations unless otherwise stated. Report models with coefficients to three decimal places unless otherwise instructed.

17. **National Health Care** The following table shows the total national expenditures for health care (in billions of dollars) for selected years from 2002 and projected to 2024. (These data include expenditures for medical research and medical facilities construction.)

Year	Amount	Year	Amount
2002	1602	2014	3080
2004	1855	2016	3403
2006	2113	2018	3786
2008	2414	2020	4274
2010	2604	2022	4825
2012	2817	2024	5425

(*Source*: U.S. Centers for Medicare and Medicaid Services)

a. Find a cubic function that models the data, with x equal to the number of years after 2000 and y equal to the expenditures for health in billions of dollars. Report the model with 4 significant digits.
$y = 0.3326x^3 - 8.358x^2 + 183.8x + 1256$

b. Graph the cubic function and the data on the same axes to determine if the function is a good fit for the data.

c. What does the unrounded model predict the expenditures will be in 2032? Approximately $9476 billion

18. **Social Security Beneficiaries** The table that follows gives the numbers of Social Security beneficiaries (in millions) for selected years from 1950 to 2000, with projections through 2030.

a. Find the cubic equation that models the data, with x equal to the number of years after 1950 and $B(x)$ equal to the number of millions of beneficiaries. Report the model with four decimal places.
$y = 0.0002x^3 - 0.0264x^2 + 1.6019x + 2.1990$

Year	Beneficiaries (millions)	Year	Beneficiaries (millions)
1950	2.9	2000	44.8
1960	14.3	2010	53.3*
1970	25.2	2020	68.8*
1980	35.1	2030	82.7*
1990	39.5		

(*Source*: 2000 Social Security Trustees Report) *Projected.

b. Graph the data points and the model on the same axes.

c. Is the model a good fit for the data? Yes

19. **Population of the World** The table gives the total population of the world from 1950 and projected to 2050.

a. Create a scatter plot of the data, with y equal to the world population in billions and x equal to the number of years after 1940.

b. Find a cubic function that models the data, with y equal to the world population in billions and x equal to the number of years after 1940. Report the model with three significant digits.
$y = -0.00000486x^3 + 0.000825x^2 + 0.0336x + 2.11$

c. Graph the data points and the model on the same axes. Is the model a good fit? Good fit

d. What does the model predict the world population will be in 2022? 7.737

e. In what year does the model estimate the world population will reach 8 billion? 2026

f. Is the model also a good predictor of the population after 2050? Why or why not?

Year	Total World Population (billions)	Year	Total World Population (billions)
1950	2.556	2010	6.849
1960	3.039	2020	7.585
1970	3.707	2030	8.247
1980	4.454	2040	8.850
1990	5.279	2050	9.346
2000	6.083		

(*Source*: U.S. Census Bureau, International Database)

20. **Income by Age** The median income for workers age 20 to 62 is shown in the table on the next page.

a. Find a cubic function that models the data with x equal to age and y equal to median salary in thousands.
$y = 0.000864x^3 - 0.128x^2 + 6.606x - 62.650$

b. Graph the function on the same axes as the scatter plot.

c. At what age does the model estimate the median income to be $56,520? 57

474 Chapter 6 Higher-Degree Polynomial and Rational Functions

Age	Median Income ($)	Age	Median Income ($)
20	25,404	47	54,260
27	37,896	52	54,405
32	45,770	57	56,427
37	50,740	62	59,222
42	51,613		

(*Source: Wall Street Journal*, June 18, 2012)

21. **Energy from Crude Oil** The table shows the total energy supply from crude oil products, in quadrillion BTUs, for selected years from 2010 and projected to 2040.

   a. Find the quartic function that is the best model for the data, with x equal to the number of years after 2010; let $C(x)$ equal the number of quadrillion BTUs of energy. Report the model with three significant digits.
   $C(x) = -0.0000752x^4 + 0.00589x^3 - 0.156x^2 + 1.48x + 11.6$

   b. Graph the model and the aligned data on the same axes and comment on the fit of the model to the data.

   c. What does the model predict the total energy supply from crude oil products will be in 2042?
   12.7 quadrillion BTUs

Year	Quadrillion BTUs	Year	Quadrillion BTUs
2010	11.6	2030	13.5
2015	15.6	2035	13.4
2020	16.0	2040	13.1
2025	14.5		

(*Source*: U.S. Energy Information Administration)

22. **Energy from Crude Oil** The table shows the total energy supply from crude oil products, in quadrillion BTUs, for selected years from 2010 and projected to 2040.

   a. Find the cubic function that is the best model for the data, with x equal to the number of years after 2010 and $C(x)$ equal to the number of quadrillion BTUs of energy. Report the model with three significant digits. $y = 0.00138x^3 - 0.0730x^2 + 1.00x + 11.8$

   b. Graph the model and the aligned data on the same axes.

   c. What does the model predict the total energy supply from crude oil products will be in 2043?
   14.9 quadrillion BTUs

   d. Compare the graph in part (b) with the graph in Exercise 21(b) to determine whether the cubic or the quartic model is the better fit for the data.
   quartic model

Year	Quadrillion BTUs	Year	Quadrillion BTUs
2010	11.6	2030	13.5
2015	15.6	2035	13.4
2020	16.0	2040	13.1
2025	14.5		

(*Source*: U.S. Energy Information Administration)

23. **China's Labor Pool** The following table shows UN data from 1975 and projections to 2050 of the size of China's labor pool (ages 15 – 59).

   a. Find the cubic function that best models this population as a function of the number of years after 1970. Let x represent the number of years after 1970 and y represent the number of millions of people in this labor pool. Report the model with coefficients rounded to three significant digits.
   $y = -0.000487x^3 - 0.162x^2 + 19.8x + 388$

   b. Graph the model and the data on the same axes, and comment on the fit.

   c. Use the model to predict the size of the labor pool in 2022. 913 million

   d. According to the model, when is the number in the labor pool projected to be at a maximum? What is the maximum? 2020; 914 million

Year	Labor Pool (millions)	Year	Labor Pool (millions)
1975	490	2015	920
1980	560	2020	920
1985	650	2025	905
1990	730	2030	875
1995	760	2035	830
2000	800	2040	820
2005	875	2045	800
2010	910	2050	670

(*Source*: United Nations; *Wall Street Journal*, January 19, 2013)

24. **Consumer Price Index** The table gives the consumer price index (CPI) for selected years from 1995 and projected to 2070, with the reference year 1995. The CPI of 566.94 for 2040 in the table means that goods and services that cost $100 in 1995 are projected to cost $566.94 in 2040.

   a. Graph the data, with x equal to the number of years after 1990.

   b. Find the cubic function that models the data, with x equal to the years after 1990 and y equal to the CPI.
   $y = 0.00481x^3 - 0.227x^2 + 9.660x + 43.707$. The model is a near-perfect fit for the data.

c. Graph the aligned data and the function on the same axes and comment on the fit of the model to the data.

d. What does the model predict the CPI will be in 2037? 494.66

Year	CPI	Year	CPI
1995	100.00	2035	465.98
2000	118.21	2040	566.94
2005	143.67	2045	689.77
2010	174.80	2050	839.21
2015	212.67	2055	1021.02
2020	258.74	2060	1242.23
2025	314.80	2065	1511.36
2030	383.00	2070	1838.81

(*Source*: Social Security Administration)

25. **Older Men in the Workforce** The table below gives the percent of men 65 years or older in the workforce for selected years from 1920 and projected to 2030.

Year	Percent
1920	55.6
1930	54.0
1940	41.8
1950	45.8
1960	33.1
1970	21.8
1980	19.0
1990	16.3
2000	17.7
2010	22.6
2020	27.2
2030	27.6

(*Source*: U.S. Bureau of the Census)

a. With $x = 0$ representing 1900, find the cubic function that models the data.
$y = 0.000078x^3 - 0.01069x^2 - 0.1818x + 64.6848$

b. Use the model to determine when the percent of older men in the workforce reached its minimum. 2000

26. **World Cell Phone Subscribership** The table shows the number of billions of world cell phone subscriberships from 1990 with projections to 2020.

Year	Subscriberships (billions)	Year	Subscriberships (billions)
1990	0.011	2006	2.76
1991	0.016	2007	3.30
1992	0.023	2008	4.10
1993	0.034	2009	4.72
1994	0.056	2010	5.40
1995	0.091	2011	6.05
1996	0.145	2012	6.65
1997	0.215	2013	7.19
1998	0.318	2014	7.65
1999	0.490	2015	8.03
2000	0.738	2016	8.35
2001	0.961	2017	8.60
2002	1.16	2018	8.79
2003	1.42	2019	8.95
2004	1.76	2020	9.07
2005	2.22		

(*Source*: International Telecommunications Union)

a. Find a cubic function that models the world cell phone subscriberships, y, in billions, as a function of the number of years after 1990, x.
$f(x) = -0.000841x^3 + 0.0481x^2 - 0.390x + 0.581$

b. Graph the function on the window $[0, 30]$ by $[0, 10]$.

c. What does the model give as the number of subscriberships in 2024? 9.83 billion

27. **U.S. Population** The table gives the U.S. population, in millions, for selected years from 1980 and projected to 2050.

Year	Millions	Year	Millions
1980	235.1	2020	342.9
1990	259.6	2030	369.5
2000	287.9	2040	390.8
2010	315.2	2050	409.1
2015	328.4		

(*Source*: Social Security Administration)

a. Create a scatter plot for the data in the table. Use x as the number of years after 1980 and y as the number of millions of Americans.

b. Find a cubic function that models the data with x equal to the number of years after 1980. Report the model with three significant digits.
$y = -0.000233x^3 + 0.0186x^2 + 2.33x + 235$

c. Graph the model and the data on the same axes, and comment on the fit. Excellent fit

d. Change the view of the graph by increasing the x-max to 140. Does the model seem appropriate for this increased time period?

28. Industrial Shipments The following table gives the value (in billions of dollars) of U.S. industrial shipments for selected years from 2014 and projected to 2040.

a. Find a cubic function that models these data, with x as the number of years after 2010 and y as the billions of dollars of U.S. industrial shipments. Report your model with three significant digit coefficients.
$y = 0.000184x^3 - 0.00968x^2 + 0.305x + 6.083$

b. Graph the model and the aligned data on the same axes and comment on the fit of the model to the data.

c. Use the reported model to predict U.S. industrial shipments in 2045. $12.81 billion

Year	Value ($ billions)	Year	Value ($ billions)
2014	7.17	2029	9.62
2017	7.78	2032	10.04
2020	8.35	2035	10.56
2023	8.84	2038	11.11
2026	9.26	2040	11.48

(Source: U.S. Department of Energy)

29. Internet of Things Devices According to IHS, the Internet of Things (IoT) market had 15.4 billion devices connected in 2015 with the number of devices expected to double by 2025. The table shows the projected number of devices, in billions, from 2015 through 2025.
(Source: Forbes.com)

a. Find the best-fitting cubic model for the data, with x equal to the number of years after 2015 and y equal to the billions of devices connected to the Internet. $y = 0.055x^3 - 0.230x^2 + 2.850x + 15.235$

b. Graph the model on the same axes with the data, and discuss how well the function fits the data.

c. What does the model predict that the number of devices will be in 2030? 190.65 billion

d. When does the model predict that the number of devices will reach 100 billion? In 2027

Year	Devices (Billions)	Year	Devices (Billions)
2015	15.41	2021	35.82
2016	17.68	2022	42.62
2017	20.35	2023	51.11
2018	23.14	2024	62.12
2019	26.66	2025	75.44
2020	30.73		

(Source: Forbes.com)

30. Sulfur Emissions The table shows the millions of short tons of sulfur emissions from electricity generation for selected years from 2000 and projected to 2035.

a. Find the cubic function that can be used to model these emissions, with x equal to the number of years after 2000 and $E(x)$ equal to the millions of short tons of sulfur emissions.
$E(x) = 0.0000347x^3 + 0.00858x^2 - 0.566x + 11.8$

b. Graph the model and the aligned data on the same axes, and comment on the fit.

c. What does the model project the amount of sulfur emissions to be in 2040? 5.1 million short tons

Year	Emissions (millions of tons)
2000	11.4
2005	10.2
2008	7.6
2015	4.7
2020	4.2
2025	3.8
2030	3.7
2035	3.8

31. Women in the Workforce The following table gives the number of women age 16 years and older (in millions) in the U.S. civilian workforce for selected years from 1950 and projected to 2050.

a. Use x as the number of years past 1950 to create a cubic model using these data. Report the model with three significant digits.
$y = -0.0000700x^3 + 0.00567x^2 + 0.863x + 16.0$

b. During what year does the model indicate that the number of women in the workforce will reach its maximum? 2046

Year	Women in the Workforce (millions)	Year	Women in the Workforce (millions)
1950	18.4	2010	75.5
1960	23.2	2015	78.6
1970	31.5	2020	79.2
1980	45.5	2030	81.6
1990	56.8	2040	86.5
2000	65.6	2050	91.5

(*Source*: Bureau of Labor Statistics)

32. Employment in Manufacturing The table shows the total employment (number of jobs N, in millions) in manufacturing for selected years t from 2010 and projected to 2040.

   a. Find the cubic function that is the best fit for the data, with t equal to the number of years after 2010 and N equal to the total employment in millions. Report the model with three significant digits. $N = 0.0000889t^3 - 0.0111t^2 + 0.237t + 11.5$

   b. Graph the model and the aligned data on the same axes and comment on the fit of the model to the data.

   c. What does the model predict the total manufacturing employment will be in 2032?
   Approximately 12.3 million

Year	Employment (millions)	Year	Employment (millions)
2010	11.5	2030	12.5
2015	12.4	2035	11.8
2020	12.8	2040	11.0
2025	12.9		

(*Source*: U.S. Department of Energy)

33. Women in the Workforce The following table gives the number of women in the workforce (in millions) for selected years from 1950 and projected to 2050. We saw in Exercise 31 that the data can be modeled by the cubic function $y = -0.0000700x^3 + 0.00567x^2 + 0.863x + 16.0$, with y in millions and x equal to the number of years after 1950.

   a. Graph this cubic model on the same axes with the aligned data.

   b. Find the quartic function that models these data, with y in millions and x equal to the number of years after 1950.
   $y = 0.00000358x^4 - 0.000783x^3 + 0.0501x^2 - 0.0248x + 18.6$

   c. Graph this quartic model on the same axes with the aligned data.

   d. Which model is a better fit for the data?
   The quartic model

Year	Women in Workforce (millions)	Year	Women in Workforce (millions)
1950	18.4	2010	75.5
1960	23.2	2015	78.6
1970	31.5	2020	79.2
1980	45.5	2030	81.6
1990	56.8	2040	86.5
2000	65.6	2050	91.5

(*Source*: U.S. Bureau of Labor Statistics)

34. Personal Income The table shows the total personal income in the United States (in billions of dollars) for selected years from 1960 and projected to 2024.

   a. Find the best-fitting cubic function that will model the data, with x equal to the number of years after 1960 and y in billions of dollars.
   $y = 0.034x^3 + 2.837x^2 + 30.328x + 345.869$

   b. Graph the data and the model on the same axes. Is the model an excellent fit for the data?

   c. What does the unrounded model predict the total personal income will be in 2030? $28,043.9 billion

Year	Income ($ billions)	Year	Income ($ billions)
1960	411.5	2008	12,100.7
1970	838.8	2014	14,728.6
1980	2307.9	2018	19,129.6
1990	4878.6	2024	22,685.1
2000	8429.7		

(*Source*: U.S. Bureau of Labor Statistics)

35. Carbon Dioxide Emissions For selected years from 2010 and projected to 2032, the table on the next page shows the number of millions of metric tons of carbon dioxide (CO_2) emissions from biomass energy combustion in the United States.

   a. Create a scatter plot of the data with x equal to the number of years after 2000 and y equal to the millions of metric tons of CO_2.

   b. Create a cubic function that models the data, with x equal to the number of years after 2000. Report the model with four significant digits.
   $y = -0.02189x^3 + 1.511x^2 - 13.67x + 347.4$

   c. Graph the data and the model on the same axes, and comment on the fit of the model to the data.

   d. Use the model to predict the CO_2 emissions in 2026. 628.4 million metric tons

e. Use technology to estimate the year in which CO_2 emissions are projected to be 781 million metric tons with the model. 2035

Year	CO_2 Emissions (millions of metric tons)	Year	CO_2 Emissions (millions of metric tons)
2010	338.5	2022	556.2
2012	364.5	2024	590.2
2014	396.1	2026	629.7
2016	425.8	2028	663.1
2018	453.1	2030	701.1
2020	498.4	2032	743.7

(*Source*: U.S. Department of Energy)

36. **Gross Domestic Product** The table gives the U.S. gross domestic product (GDP), in trillions of dollars, for selected years from 2005 and projected to 2070.

   a. Create a scatter plot of the data, with y representing GDP in billions of dollars and x representing the number of years after 2000.

   b. Find the quadratic function that best fits the data, with x equal to the number of years after 2000. Report the model with three significant digits. $y = 0.117x^2 - 3.79x + 45.3$

   c. Find the cubic function that best fits the data, with x equal to the number of years after 2000. Report the model with three significant digits. $y = 0.00188x^3 - 0.0954x^2 + 2.79x - 2.70$

   d. Graph each of these functions on the same axes as the data points, and visually determine which model is the better fit for the data. Cubic model

Year	GDP ($ trillions)	Year	GDP ($ trillions)
2005	12.145	2040	79.680
2010	16.174	2045	103.444
2015	21.270	2050	133.925
2020	27.683	2055	173.175
2025	35.919	2060	224.044
2030	46.765	2065	290.042
2035	61.100	2070	375.219

(*Source*: Bureau of Economic Analysis)

37. **Cloud Revenue** Many services are provided through the Cloud, and public vendor revenue is projected to grow rapidly. The table shows revenue, in billions of dollars, from 2012 and projected to 2026.

   a. Create a scatter plot of the data, with x equal to the number of years after 2010 and y equal to the revenue in billions of dollars.

   b. Find the cubic model that is the best fit for the data, with x equal to the number of years after 2010 and y equal to the revenue in billions of dollars. Report the model with three significant digits. $y = -0.295x^3 + 8.18x^2 - 28.2x + 54.7$

   c. Graph the model on the same graph as the data points, and discuss the fit of the model to the points.

   d. What does the model predict the revenue will be in 2027? $490.9 billion

   e. What happens to this function after 2027 that likely means it is no longer a valid model? The function decreases rapidly.

Year	Revenue ($ billion)	Year	Revenue ($ billion)
2012	26	2020	298
2013	39	2021	345
2014	56	2022	387
2015	80	2023	422
2016	116	2024	451
2017	154	2025	474
2018	199	2026	493
2019	248		

(*Source*: Wikibon Server SAN & Cloud Research Projects)

38. **Population of Children** The table gives the estimated population (in thousands) of U.S. children age 5 and under in selected years from 1985 and projected to 2050.

   a. Find the equation of a cubic function that models the estimated population (in thousands) of U.S children age 5 and under x years after 1980. Report the model with three significant digits. $y = 0.000169x^3 - 0.0224x^2 + 1.07x + 57.3$

   b. Graph this model on the same axes with the aligned data and comment on the fit of the model to the data.

   c. Use the result of part (a) to predict the total estimated population (in thousands) of children age 5 and under in 2038. 77.1 thousand

Years	Population (thousands)	Years	Population (thousands)
1985	62.6	2020	74.1
1990	64.2	2025	75.0
1995	69.5	2030	76.3
2000	72.4	2035	77.4
2005	73.5	2040	78.2
2010	74.1	2045	78.9
2015	73.6	2050	79.9

(*Source*: Childstats.gov)

6.3 Solution of Polynomial Equations

KEY OBJECTIVES

- Solve polynomial equations using factoring, factoring by grouping, and the root method
- Find factors, zeros, x-intercepts, and solutions
- Estimate solutions with technology
- Solve polynomial equations using the intersection method and the x-intercept method

SECTION PREVIEW Women in the Workforce

The number of women in the workforce (in millions) for selected years from 1950 and projected to 2050 can be modeled by

$$y = 0.00358x^4 - 0.783x^3 + 50.1x^2 - 24.8x + 18{,}600$$

with y in thousands and x equal to the number of years after 1950. The graph of this model, in Figure 6.20, shows that the number of women in the workforce has consistently increased over this period of time. In Example 8 we will find when the number of women in the workforce is projected to reach 92,120 thousand by solving the equation $0.00358x^4 - 0.783x^3 + 50.1x^2 - 24.8x + 18{,}600 = 92{,}120$.
(*Source*: U.S. Bureau of Labor Statistics)

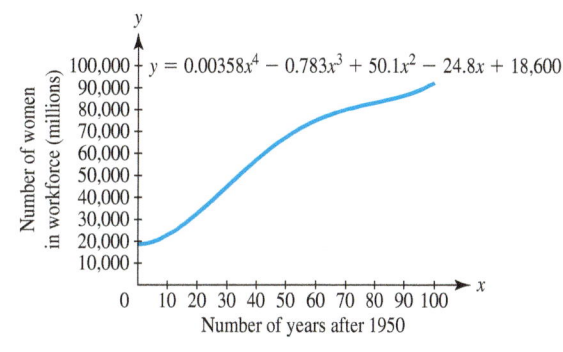

Figure 6.20

The technology of graphing utilities allows us to find approximate solutions of equations such as this one.

Although many equations derived from real data require technology to find or to approximate solutions, some higher-degree equations can be solved by factoring, and some can be solved by the root method. In this section, we discuss these methods for solving polynomial equations; in the next section, we discuss additional methods of solving higher-degree polynomial equations. ■

Solving Polynomial Equations by Factoring

Some polynomial equations can be solved by factoring and using the zero-product property, much like the method used to solve quadratic equations described in Chapter 3. From our discussion in Chapter 3, we know that the following statements regarding a function f are equivalent:

- $(x - a)$ is a factor of $f(x)$.
- a is a real zero of the function f.
- a is a real solution to the equation $f(x) = 0$.
- a is an x-intercept of the graph of $y = f(x)$.
- The graph crosses the x-axis at the point $(a, 0)$. Note that if the graph of $y = f(x)$ touches, but does not cross, the x-axis at $(a, 0)$, then $(x - a)^2$ is a factor of $f(x)$, and $(x - a)$ is a factor with even multiplicity.

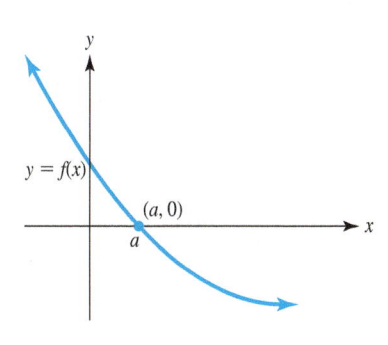

479

EXAMPLE 1 ▶ Factors, Zeros, Intercepts, and Solutions

If $P(x) = (x - 1)(x + 2)(x + 5)$, find

a. The zeros of $P(x)$

b. The solutions of $P(x) = 0$

c. The x-intercepts of the graph of $y = P(x)$

SOLUTION

a. The factors of $P(x)$ are $(x - 1)$, $(x + 2)$, and $(x + 5)$; thus, the zeros of $P(x)$ are $x = 1$, $x = -2$, and $x = -5$.

b. The solutions to $P(x) = 0$ are the zeros of $P(x)$: $x = 1$, $x = -2$, and $x = -5$.

c. The x-intercepts of the graph of $y = P(x)$ are 1, -2, and -5 (Figure 6.21).

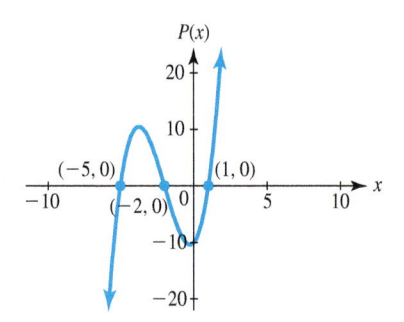

Figure 6.21

EXAMPLE 2 ▶ Finding the Maximum Volume

A box is to be formed by cutting a square of x inches per side from each corner of a square piece of cardboard that is 24 inches on each side and folding up the sides. This will give a box whose height is x inches, with the length of each side of the bottom $2x$ less than the original length of the cardboard (see Figure 6.22). Using $V = lwh$, the volume of the box is given by

$$V = (24 - 2x)(24 - 2x)x$$

where x is the length of the side of the square that is cut out.

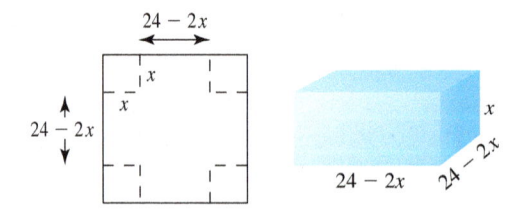

Figure 6.22

a. Use the factors of V to find the values of x that give volume 0.

b. Graph the function that gives the volume as a function of the side of the square that is cut out over an x-interval that includes the x-values found in part (a).

c. What input values make sense for this problem (that is, actually result in a box)?

6.3 Solution of Polynomial Equations 481

d. Graph the function using the input values that make sense for the problem (found in part (c)).

e. Use technology to determine the size of the square that should be cut out to give the maximum volume.

SOLUTION

a. We use the zero-product property to solve the equation $0 = (24 - 2x)(24 - 2x)x$.

$$0 = (24 - 2x)(24 - 2x)x$$
$$x = 0 \quad \text{or} \quad 24 - 2x = 0$$
$$-2x = -24$$
$$x = 0 \quad \text{or} \quad x = 12$$

b. A graph of $V(x)$ from $x = -1$ to $x = 16$ is shown in Figure 6.23(a).

c. The box can exist only if each of the dimensions is positive, resulting in a volume that is positive. Thus, $24 - 2x > 0$, or $x < 12$, so the box can exist only if the value of x is greater than 0 and less than 12. Note that if squares 12 inches on a side are removed, no material remains to make a box, and no squares larger than 12 inches on a side can be removed from all four corners.

d. The graph of $V(x)$ over the x-interval $0 < x < 12$ is shown in Figure 6.23(b).

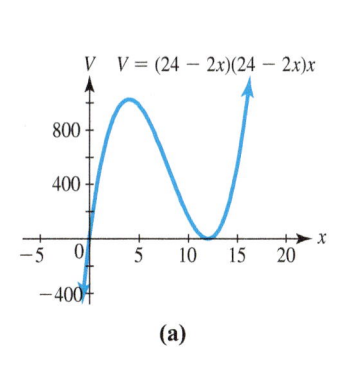

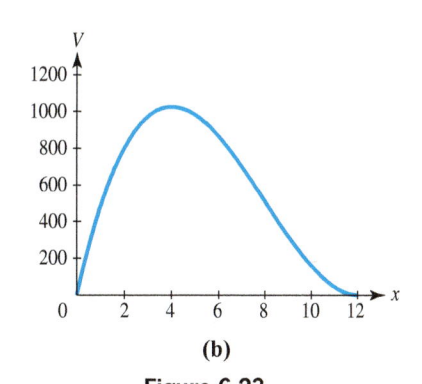

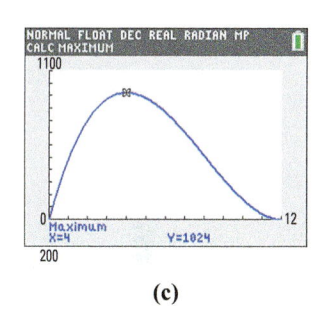

(a)   (b)   (c)

Figure 6.23

e. By observing the graph, we see that the function has a maximum value, which gives the maximum volume over the interval $0 < x < 12$. Using the "maximum" feature on a graphing utility, we see that the maximum occurs at or near $x = 4$ (Figure 6.23(c)). Cutting a square with side 4 inches from each corner and folding up the sides gives a box that is 16 inches by 16 inches by 4 inches. The volume of this box is $16 \cdot 16 \cdot 4 = 1024$ cubic inches, and testing values near 4 shows that this is the maximum possible volume.

EXAMPLE 3 ▶ Photosynthesis

The amount y of photosynthesis that takes place in a certain plant depends on the intensity x of the light (in lumens) present, according to the function $y = 120x^2 - 20x^3$. The model is valid only for nonnegative x-values that produce a positive amount of photosynthesis.

a. Graph this function with a viewing window large enough to see 2 turning points.

b. Use factoring to find the x-intercepts of the graph.

482 Chapter 6 Higher-Degree Polynomial and Rational Functions

c. What nonnegative values of x will give positive photosynthesis?

d. Graph the function with a viewing window that contains only the values of x found in part (c) and nonnegative values of y.

SOLUTION

a. The graph in Figure 6.24 shows the function with 2 turning points.

b. We can find the x-intercepts of the graph by solving $120x^2 - 20x^3 = 0$ for x. Factoring gives

$$20x^2(6 - x) = 0$$
$$20x^2 = 0 \quad \text{or} \quad 6 - x = 0$$
$$x = 0 \quad \text{or} \quad x = 6$$

Thus, the x-intercepts are 0 and 6.

c. From the graph of $y = 120x^2 - 20x^3$ in Figure 6.24, we see that positive output values result when the input values are between 0 and 6. Thus, the nonnegative x-values that produce positive amounts of photosynthesis are $0 < x < 6$.

d. Graphing the function in the viewing window $[0, 6]$ by $[0, 700]$ will show the region that is appropriate for the model (see Figure 6.25).

Figure 6.24

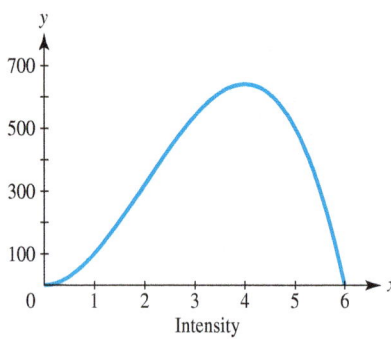

Figure 6.25

Solution Using Factoring by Grouping

Some higher-degree equations can be solved by using the method of factoring by grouping.* Consider the following examples.

EXAMPLE 4 ▶ Cost

The total cost of producing a product is given by the function

$$C(x) = x^3 - 12x^2 + 3x + 9 \text{ thousand dollars}$$

where x is the number of hundreds of units produced. How many units must be produced to give a total cost of $45,000?

SOLUTION

Because $C(x)$ is the total cost in thousands of dollars, a total cost of $45,000 is represented by $C(x) = 45$. So we seek to solve

$$45 = x^3 - 12x^2 + 3x + 9, \quad \text{or} \quad 0 = x^3 - 12x^2 + 3x - 36$$

---

*Factoring by grouping was discussed in the Chapter 3 Algebra Toolbox.

This equation appears to have a form that permits factoring by grouping. The solution steps follow.

$0 = (x^3 - 12x^2) + (3x - 36)$ — Separate the terms into two groups, each having a common factor.

$0 = x^2(x - 12) + 3(x - 12)$ — Factor x^2 from the first group, then 3 from the second group.

$0 = (x - 12)(x^2 + 3)$ — Factor the common factor $(x - 12)$ from each of the two terms.

$x - 12 = 0$ or $x^2 + 3 = 0$ — Set each factor equal to 0 and solve for x.

$x = 12$ or $x^2 = -3$

$x = 12$

Because no real number x can satisfy $x^2 = -3$, the only solution is 12 hundred units. Thus, producing 1200 units gives a total cost of $45,000.

The Root Method

We solved quadratic equations of the form $x^2 = C$ in Chapter 3 by taking the square root of both sides and using a $\pm$ symbol to indicate that we get both the positive and the negative root: $x = \pm\sqrt{C}$. In the same manner, we can solve $x^3 = C$ by taking the cube root of both sides. There will not be a $\pm$ sign because there is only one real cube root of a number. In general, there is only one real nth root if n is odd, and there are zero or two real roots if n is even and $C \geq 0$.

Root Method

The real solutions of the equation $x^n = C$ are found by taking the nth root of both sides:

$x = \sqrt[n]{C}$ if n is odd   and $x = \pm\sqrt[n]{C}$ if n is even and $C \geq 0$

EXAMPLE 5 ▶ Root Method

Solve the following equations.

a. $x^3 = 125$ **b.** $5x^4 = 80$ **c.** $4x^2 = 18$

SOLUTION

a. Taking the cube root of both sides gives

$x^3 = 125$
$x = \sqrt[3]{125} = 5$

No $\pm$ sign is needed because the root is odd.

b. We divide both sides by 5 to isolate the x^4 term:

$5x^4 = 80$
$x^4 = 16$

484 Chapter 6 Higher-Degree Polynomial and Rational Functions

Then we take the fourth root of both sides, using a $\pm$ sign (because the root is even):

$$x = \pm \sqrt[4]{16} = \pm 2$$

The solutions are 2 and -2.

c. We divide both sides by 4 to isolate the x^2 term:

$$4x^2 = 18$$

$$x^2 = \frac{9}{2}$$

Then we take the square root of both sides, using a $\pm$ sign (because the root is even):

$$x = \pm \sqrt{\frac{9}{2}} = \pm \frac{3}{\sqrt{2}}$$

The *exact* solutions are $\pm \dfrac{3}{\sqrt{2}}$; the solutions are *approximately* 2.121 and -2.121.

EXAMPLE 6 ▶ Future Value of an Investment

The future value of $10,000 invested for 4 years at interest rate r, compounded annually, is given by $S = 10{,}000(1 + r)^4$. Find the rate r, as a percent, for which the future value is $14,641.

SOLUTION

We seek to solve $14{,}641 = 10{,}000(1 + r)^4$ for r. We can rewrite this equation with $(1 + r)^4$ on one side by dividing both sides by 10,000:

$$14{,}641 = 10{,}000(1 + r)^4$$

$$1.4641 = (1 + r)^4$$

We can now remove the exponent by taking the fourth root of both sides of the equation, using a $\pm$ sign:

$$\pm \sqrt[4]{1.4641} = 1 + r$$

$$\pm 1.1 = 1 + r$$

Solving for r gives

$$1.1 = 1 + r \quad \text{or} \quad -1.1 = 1 + r$$

$$0.1 = r \quad \text{or} \quad -2.1 = r$$

The negative value cannot represent an interest rate, so the interest rate that gives this future value is

$$r = 0.1 = 10\%$$

The graphs of $y_1 = 10{,}000(1 + x)^4$ and $y_2 = 14{,}641$ shown in Figure 6.26 support the conclusion that $r = 0.10$ gives the future value 14,641.*

---

*The graph of the quartic function $y = 10{,}000(1 + x)^4$ is actually curved, but within this small window it appears to be nearly linear.

6.3 Solution of Polynomial Equations   485

Figure 6.26

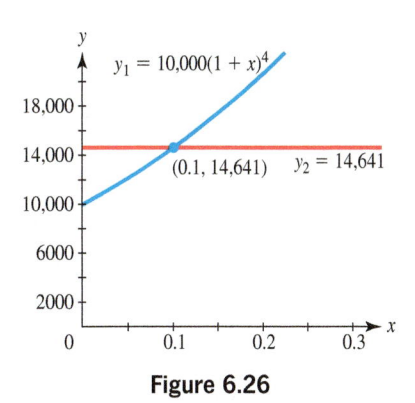

Estimating Solutions with Technology

Although we can solve some higher-degree equations by using factoring, many real problems do not have "nice" solutions, and factoring is not an appropriate solution method. Thus, many equations resulting from real data applications require graphical or numerical methods of solution.

EXAMPLE 7 ▶ China's Labor Pool

Using United Nations data from 1975 with projections to 2050, the size of China's labor pool (age 15 to 59) can be modeled by the function

$$y = -0.000487x^3 - 0.162x^2 + 19.8x + 388$$

where x is the number of years after 1970 and y represents the number of millions of people in this labor pool. Use this model to estimate graphically the years after 1970 when the population is and will be 900 million. (*Source*: United Nations; *Wall Street Journal*)

SOLUTION

To find the years when China's labor pool is or will be 900 million, we solve

$$900 = -0.000487x^3 - 0.162x^2 + 19.8x + 388$$

To solve this equation graphically, we can use either the intersection method or the x-intercept method. To use the intersection method, we graph

$$y_1 = -0.000487x^3 - 0.162x^2 + 19.8x + 388 \quad \text{and} \quad y_2 = 900$$

for $x \geq 0$ and find the x-coordinates of the points of intersection.

Figure 6.27(a) shows that the graph of $y_1 = -0.000487x^3 - 0.162x^2 + 19.8x + 388$ crosses the graph of $y_2 = 900$ in two places. Using 5: intersect under the CALC menu gives the x-coordinates of the two points of intersection. From these graphs, the model estimates that China's labor pool reached 900 million in 2013 ($x = 42.6$) and will fall back to 900 million in 2027 ($x = 57$).

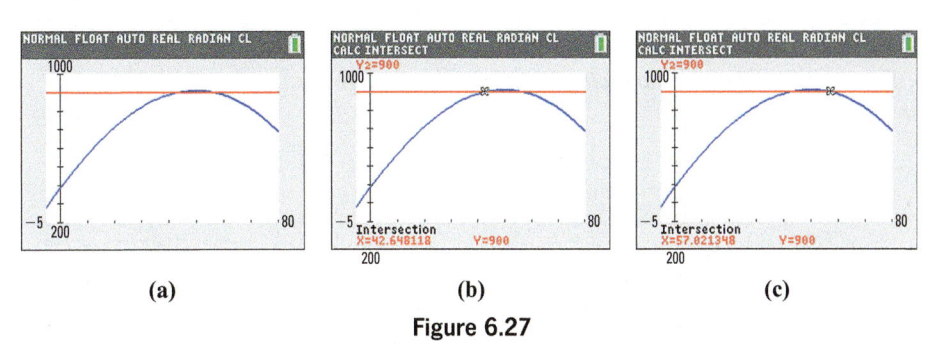

(a)　　　　　(b)　　　　　(c)

Figure 6.27

EXAMPLE 8 ▶ Women in the Workforce

The number of women in the workforce for selected years from 1950 and projected to 2050 can be modeled by

$$y = 0.00358x^4 - 0.783x^3 + 50.1x^2 - 24.8x + 18{,}600$$

with y in thousands and x equal to the number of years after 1950. If the model remains accurate, when is the number of women in the workforce projected to reach 92.12 million? (*Source*: U.S. Bureau of Labor Statistics)

SOLUTION

To answer this question, we solve $0.00358x^4 - 0.783x^3 + 50.1x^2 - 24.8x + 18{,}600 = 92{,}120$ (because 92.12 million is 92,120 thousand). To solve this equation with the intersection method, we graph

$$y_1 = 0.00358x^4 - 0.783x^3 + 50.1x^2 - 24.8x + 18{,}600 \text{ and } y_2 = 92{,}120$$

and then find the point of intersection (Figure 6.28). Because $1950 + 100 = 2050$, the number of women in the workforce is projected to reach 92.12 million in 2050.

Figure 6.28

SKILLS CHECK 6.3

Answers that are not seen can be found in the answer section at the back of the text.

In Exercises 1–4, solve the polynomial equation.

1. $(2x - 3)(x + 1)(x - 6) = 0$ $\frac{3}{2}, -1, 6$
2. $(3x + 1)(2x - 1)(x + 5) = 0$ $-\frac{1}{3}, \frac{1}{2}, -5$
3. $(x + 1)^2(x - 4)(2x - 5) = 0$ $-1, 4, \frac{5}{2}$
4. $(2x + 3)^2(5 - x)^2 = 0$ $-\frac{3}{2}, 5$

In Exercises 5–10, solve the polynomial equations by factoring, and check the solutions graphically.

5. $x^3 - 16x = 0$ $0, 4, -4$
6. $2x^3 - 8x = 0$ $0, -2, 2$
7. $x^4 - 4x^3 + 4x^2 = 0$ $0, 2$
8. $x^4 - 6x^3 + 9x^2 = 0$ $0, 3$
9. $4x^3 - 4x = 0$ $0, 1, -1$
10. $x^4 - 3x^3 + 2x^2 = 0$ $0, 1, 2$

In Exercises 11–14, use factoring by grouping to solve the equations.

11. $x^3 - 4x^2 - 9x + 36 = 0$ $4, 3, -3$
12. $x^3 + 5x^2 - 4x - 20 = 0$ $-5, -2, 2$
13. $3x^3 - 4x^2 - 12x + 16 = 0$ $2, -2, \frac{4}{3}$
14. $4x^3 + 8x^2 - 36x - 72 = 0$ $-2, -3, 3$

In Exercises 15–18, solve the polynomial equations by using the root method, and check the solutions graphically.

15. $2x^3 - 16 = 0$ 2
16. $3x^3 - 81 = 0$ 3
17. $\frac{1}{2}x^4 - 8 = 0$ $2, -2$
18. $2x^4 - 162 = 0$ $-3, 3$

In Exercises 19–24, use factoring and the root method to solve the polynomial equations.

19. $4x^4 - 8x^2 = 0$ $0, \sqrt{2}, -\sqrt{2}$
20. $3x^4 - 24x^2 = 0$ $0, 2\sqrt{2}, -2\sqrt{2}$
21. $0.5x^3 - 12.5x = 0$ $0, 5, -5$
22. $0.2x^3 - 24x = 0$ $0, 2\sqrt{30}, -2\sqrt{30}$
23. $x^4 - 6x^2 + 9 = 0$ $\sqrt{3}, -\sqrt{3}$
24. $x^4 - 10x^2 + 25 = 0$ $\sqrt{5}, -\sqrt{5}$

In Exercises 25–30, use the graph of the polynomial function $f(x)$ to (a) solve $f(x) = 0$, and (b) find the factorization of $f(x)$.

25. $f(x) = x^3 - 2x^2 - 11x + 12$

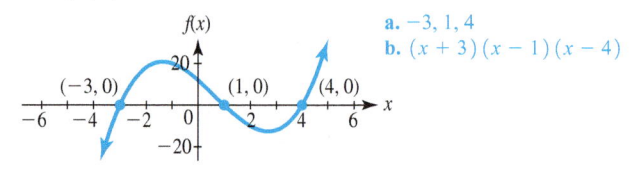

   a. $-3, 1, 4$
   b. $(x + 3)(x - 1)(x - 4)$

26. $f(x) = -x^3 + 6.5x^2 + 13x - 8$

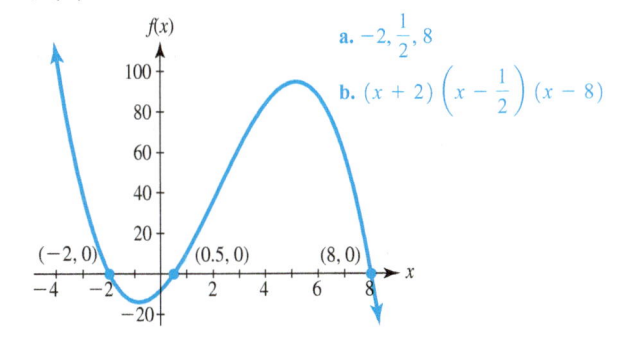

a. $-2, \frac{1}{2}, 8$
b. $(x + 2)\left(x - \frac{1}{2}\right)(x - 8)$

27. $y = -2x^4 + 6x^3 + 6x^2 - 14x - 12$

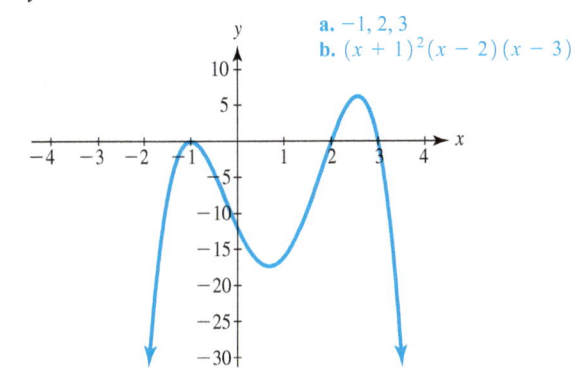

a. $-1, 2, 3$
b. $(x + 1)^2(x - 2)(x - 3)$

28. $y = x^4 - 5x^3 - 3x^2 + 13x + 10$

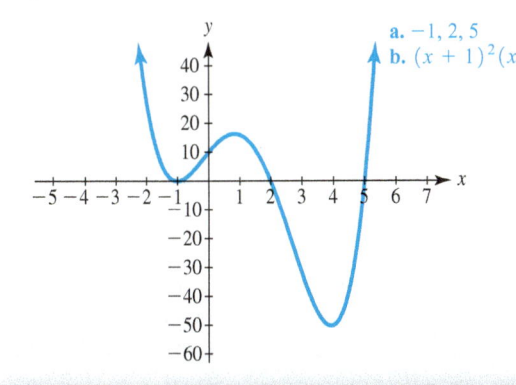

a. $-1, 2, 5$
b. $(x + 1)^2(x - 2)(x - 5)$

29. $y = -0.5x^3 + 2.5x^2 + 0.5x - 2.5$

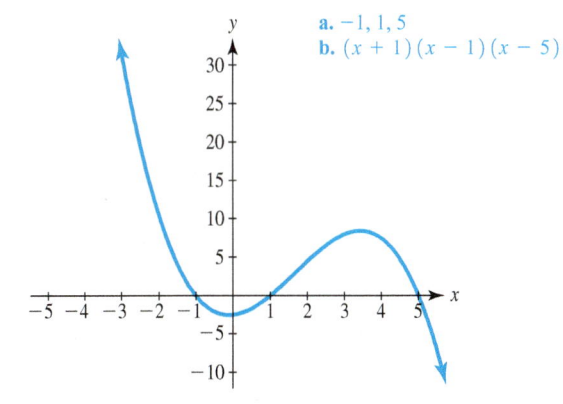

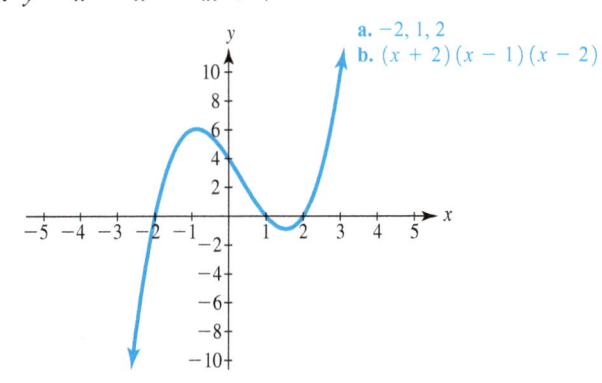

a. $-1, 1, 5$
b. $(x + 1)(x - 1)(x - 5)$

30. $y = x^3 - x^2 - 4x + 4$

a. $-2, 1, 2$
b. $(x + 2)(x - 1)(x - 2)$

In Exercises 31 and 32, find the solutions graphically.

31. $4x^3 - 15x^2 - 31x + 30 = 0$ $-2, 5, \frac{3}{4}$

32. $2x^3 - 15x^2 - 62x + 120 = 0$ $-4, 1.5, 10$

EXERCISES 6.3

Answers that are not seen can be found in the answer section at the back of the text.

33. **Revenue** The revenue from the sale of a product is given by the function $R = 400x - x^3$.

    a. Use factoring to find the numbers of units that must be sold to give zero revenue. 0, 20

    b. Does the graph of the revenue function verify this solution? Yes

34. **Revenue** The revenue from the sale of a product is given by the function $R = 12{,}000x - 0.003x^3$.

    a. Use factoring and the root method to find the numbers of units that must be sold to give zero revenue. 0, 2000

    b. Does the graph of the revenue function verify this solution? Yes

35. **Revenue** The price for a product is given by $p = 100{,}000 - 0.1x^2$, where x is the number of units sold, so the revenue function for the product is $R = px = (100{,}000 - 0.1x^2)x$.

    a. Find the numbers of units that must be sold to give zero revenue. 0, 1000

    b. Does the graph of the revenue function verify this solution? Yes

36. **Revenue** If the price from the sale of x units of a product is given by the function $p = 100x - x^2$, the revenue function for this product is given by $R = px = (100x - x^2)x$.

a. Find the numbers of units that must be sold to give zero revenue. 0, 100

b. Does the graph of the revenue function verify this solution? Yes

37. *Future Value* The future value of $2000 invested for 3 years at rate r, compounded annually, is given by $S = 2000(1 + r)^3$.

   a. Complete the table below to determine the future value of $2000 at certain interest rates.

Rate	Future Value
4%	$2249.73
5%	2315.25
7.25%	2467.30
10.5%	2698.47

   b. Graph this function on the window $[0, 0.24]$ by $[0, 5000]$.

   c. Use the root method to find the rate r, as a percent, for which the future value is $2662. 10%

   d. Use the root method to find the rate r, as a percent, for which the future value is $3456. 20%

38. *Future Value* The future value of $5000 invested for 4 years at rate r, compounded annually, is given by $S = 5000(1 + r)^4$.

   a. Graph this function on the window $[0, 0.24]$ by $[0, 12{,}000]$.

   b. Use the root method to find the rate r, as a percent, for which the future value is $10,368. 20%

   c. What rate, as a percent, gives $2320.50 in interest on this investment? 10%

39. *Constructing a Box* A box can be formed by cutting a square out of each corner of a piece of tin and folding the sides up. Suppose the piece of tin is 18 inches by 18 inches and each side of the square that is cut out has length x.

   a. Write an expression for the height of the box that is constructed. x in.

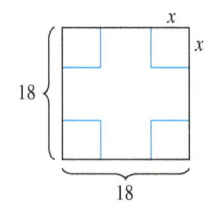

   b. Write an expression for the dimensions of the base of the box that is constructed. $18 - 2x$ in. by $18 - 2x$ in.

   c. Use the formula $V = lwh$ to find an equation that represents the volume of the box.
   $V = (18 - 2x)(18 - 2x)x = 324x - 72x^2 + 4x^3$

d. Use the equation that you constructed to find the values of x that make $V = 0$. $x = 0, x = 9$

e. For which of these values of x does a box exist if squares of length x are cut out and the sides are folded up? Neither

40. *Constructing a Box* A box can be formed by cutting a square out of each corner of a piece of cardboard and folding the sides up. If the piece of cardboard is 12 inches by 12 inches and each side of the square that is cut out has length x, the function that gives the volume of the box is $V = 144x - 48x^2 + 4x^3$.

   a. Find the values of x that make $V = 0$. $x = 0, x = 6$

   b. For each of the values of x in part (a), discuss what happens to the box if squares of length x are cut out.

   c. For what values of x does a box exist? $0 < x < 6$

   d. Graph the function that gives the volume as a function of the side of the square that is cut out, using an x-interval that makes sense for the problem (see part (c)).

41. *Profit* The profit function for a product is given by $P(x) = -x^3 + 2x^2 + 400x - 400$, where x is the number of units produced and sold and P is in hundreds of dollars. Use factoring by grouping to find the numbers of units that will give a profit of $40,000. 2, 20

42. *Cost* The total cost function for a product is given by $C(x) = 3x^3 - 6x^2 - 300x + 1800$, where x is the number of units produced and C is the cost in hundreds of dollars. Use factoring by grouping to find the numbers of units that will give a total cost of $120,000. 2, 10

43. *Ballistics* Ballistics experts are able to identify the weapon that fired a certain bullet by studying the markings on the bullet after it is fired. They test the rifling (the grooves in the bullet as it travels down the barrel of the gun) by comparing it to that of a second bullet fired into a bale of paper. The speed, s, in centimeters per second, that the bullet travels through the paper is given by $s = 30(3 - 10t)^3$, where t is the time after the bullet strikes the bale and $t \leq 0.3$ second.

   a. Complete the following table to find the speed for given values of t.

t	0	0.1	0.2	0.3
s (cm/sec)	810	240	30	0

   b. Use the root method to find the number of seconds to give $s = 0$. Does this agree with the data in the table? 0.3 sec; yes

44. *National Health Care* Using data for selected years from 2002 and projected to 2024, the total national expenditures for health care (in billions of dollars) can be modeled by the function

$$y = 0.3326x^3 - 8.358x^2 + 183.8x + 1256$$

where x is the number of years after 2000. When does the model project that the expenditures will be $8228 billion? In 2030
(*Source*: U.S. Centers for Medicare and Medicaid Services)

45. *Civilian Labor Force* The size of the civilian labor force (in millions) for selected years from 1950 and projected to 2050 can be modeled by

$$y = -0.0001x^3 + 0.0088x^2 + 1.43x + 57.9$$

where x is the number of years after 1950.

   a. What does this model project the civilian labor force to be in 2015? 160.6 million

   b. In what year does the model project the civilian labor force to be 163 million? 2017

46. *Internet of Things Devices* Using data from 2015 and projected to 2025, the projected number devices, in billions, can be modeled by $y = 0.055x^3 - 0.230x^2 + 2.850x + 15.235$, where x is the number of years after 2015. When does the model predict that the number of devices will reach 100 billion? In 2027
(*Source*: Forbes.com)

47. *College Enrollment* Using data for selected years from 1990 and projected to 2024, the total enrollment in post-secondary degree institutions, in thousands, can be modeled by the function

$$y = 0.0494x^4 - 3.74x^3 + 89.8x^2 + 339x + 13{,}958$$

where x is the number of years after 1990. When does this model project that enrollment will be 58,302,000? In 2030

48. *Foreign-Born Population* Using data for selected years from 1970 and projected to 2060, the percent of the U.S. population that is foreign-born can be modeled by the function

$$y = -0.0000172x^3 + 0.00156x^2 + 0.155x + 4.66$$

with x equal to the number of years after 1970. When does the model project that the percent of the U.S. population that is foreign-born equals 16.6%? In 2035
(*Source*: U.S. Census Bureau)

49. *Gross Domestic Product* The U.S. gross domestic product (GDP), in trillions of dollars, for selected years from 2005 and projected to 2070 can be modeled by

$$y = 0.00188x^3 - 0.0954x^2 + 2.79x - 2.70,$$

with x equal to the number of years after 2000. In what year does the model indicate the GDP will be $85.48 trillion? 2042
(*Source*: Bureau of Economic Analysis)

50. *Employment in Manufacturing* Using data from 2010 and projected to 2040, the total employment (number of jobs N, in millions) in manufacturing can be modeled by

$$N = 0.00889t^3 - 0.0111t^2 + 0.237t + 11.5$$

where t equals the number of years after 2010. When after 2010 does the model predict that the total employment in manufacturing will be 83.72 million? In 2030
(*Source*: U.S. Department of Energy)

6.4 Polynomial Equations Continued; Fundamental Theorem of Algebra

KEY OBJECTIVES

- Divide polynomials with synthetic division
- Solve cubic equations using division
- Combine graphical and algebraic methods to solve polynomial equations
- Solve quartic equations
- Use the Fundamental Theorem of Algebra to find the number of complex solutions of a polynomial equation
- Find the complex solutions of polynomial equations

SECTION PREVIEW Break-Even

Suppose the profit function for a product is given by

$$P(x) = -x^3 + 98x^2 - 700x - 1800 \text{ dollars}$$

where x is the number of units produced and sold. To find the number of units that gives break-even for the product, we must solve $P(x) = 0$. Solving this equation by factoring would be difficult. Finding all the zeros of $P(x)$ by using a graphical method may also be difficult without knowing an appropriate viewing window. In general, if we seek to find the exact solutions to a polynomial equation, a number of steps are involved if the polynomial is not easily factored.

In this section, we investigate methods of finding additional solutions to polynomial equations after one or more solutions have been found. Specifically, if we know that a is one of the solutions to a polynomial equation, we know that $(x - a)$ is a factor of $P(x)$, and we can divide $P(x)$ by $(x - a)$ to find a second factor of $P(x)$ that may give additional solutions. ∎

Division of Polynomials; Synthetic Division

Suppose $f(x)$ is a cubic function and we know that a is a solution of $f(x) = 0$. We can write $(x - a)$ as a factor of $f(x)$, and if we divide this factor into the cubic function, the quotient will be a quadratic factor of $f(x)$. If there are additional real solutions to $f(x) = 0$, this quadratic factor can be used to find the remaining solutions.

Division of a polynomial by a binomial was discussed in the Algebra Toolbox for this chapter. When we divide a polynomial by a binomial of the form

$$x - a$$

the division process can be simplified by using a technique called **synthetic division**. Note in the division of $x^4 + 6x^3 + 5x^2 - 4x + 2$ by $x + 2$ on the left below, many terms are recopied. Because the divisor is linear, the quotient has degree one less than the degree of the dividend, and the degree decreases by one in each remaining term. This is true in general, so we can simplify the division process by omitting all variables and writing only the coefficients. If we rewrite the work without the terms that were recopied in the division and without the x's, the process looks like the one on the right below.

$$
\begin{array}{r}
x^3 + 4x^2 - 3x + 2 \\
x + 2 \overline{)x^4 + 6x^3 + 5x^2 - 4x + 2} \\
\underline{x^4 + 2x^3} \\
4x^3 + 5x^2 \\
\underline{4x^3 + 8x^2} \\
-3x^2 - 4x \\
\underline{-3x^2 - 6x} \\
2x + 2 \\
\underline{2x + 4} \\
-2
\end{array}
\qquad
\begin{array}{r}
1 + 4 - 3 + 2 \\
+2 \overline{)1 + 6 + 5 - 4 + 2} \\
\underline{2} \\
4 \\
\underline{8} \\
-3 \\
\underline{-6} \\
2 \\
\underline{4} \\
-2
\end{array}
$$

Look at the process on the right above and observe that the coefficient of the first term of the dividend and the differences found in each step are identical to the coefficients of the quotient; the last difference, -2, is the remainder. Because of this, we can compress the division process as follows:

$$
\begin{array}{r}
+2 \overline{)1 + 6 + 5 - 4 + 2} \\
\underline{2 + 8 - 6 + 4} \\
1 + 4 - 3 + 2 - 2
\end{array}
$$

Note that the last line in this process gives the coefficients of the quotient, with the last number equal to the remainder. Changing the sign in the divisor allows us to use addition instead of subtraction in the process, as follows:

$$
\begin{array}{r}
-2 \overline{)1 + 6 + 5 - 4 + 2} \\
\underline{-2 - 8 + 6 - 4} \\
1 + 4 - 3 + 2 - 2
\end{array}
$$

The bottom line of this process gives the coefficients of the quotient, and the degree of the quotient is one less than the degree of the dividend, so the quotient is

$$x^3 + 4x^2 - 3x + 2, \quad \text{with remainder } -2$$

just as it was in the original division.

The synthetic division process uses the following steps.

6.4 Polynomial Equations Continued; Fundamental Theorem of Algebra 491

Synthetic Division Steps to Divide a Polynomial by $x - a$

To divide a polynomial by $x - a$:

Divide $2x^3 - 9x - 27$ by $x - 3$.

1. Arrange the coefficients in descending powers of x, with a 0 for any missing power. Place a from $x - a$ to the left of the coefficients.

   $3 \overline{) 2 + 0 - 9 - 27}$

2. Bring down the first coefficient to the third line. Multiply the last number in the third line by a and write the product in the second line under the next term.

   $3 \overline{) 2 + 0 - 9 - 27}$
   6
   2

3. Add the last number in the second line to the number above it in the first line. Continue this process until all numbers in the first line are used.

   $3 \overline{) 2 + 0 - 9 - 27}$
   $6 + 18 + 27$
   $2 + 6 + 9 + 0$

4. The third line represents the coefficients of the quotient, with the last number the remainder. The quotient is a polynomial of degree one less than the dividend.

   The quotient is
   $2x^2 + 6x + 9$

   If the remainder is 0, $x - a$ is a factor of the polynomial, and the polynomial can be written as the product of the divisor $x - a$ and the quotient.

   The remainder is 0, so $x - 3$ is a factor of $2x^3 - 9x - 27$.
   $2x^3 - 9x - 27$
   $= (x - 3)(2x^2 + 6x + 9)$

Using Synthetic Division to Solve Cubic Equations

If we know one solution of a cubic equation $P(x) = 0$, we can divide by the corresponding factor using synthetic division to obtain a second quadratic factor that can be used to find any additional solutions of the polynomial.

EXAMPLE 1 ▶ Solving an Equation with Synthetic Division

If $x = -2$ is a solution of $x^3 + 8x^2 + 21x + 18 = 0$, find the remaining solutions.

SOLUTION

If -2 is a solution of the equation $x^3 + 8x^2 + 21x + 18 = 0$, then $x + 2$ is a factor of

$$f(x) = x^3 + 8x^2 + 21x + 18$$

Synthetically dividing $f(x)$ by $x + 2$ gives

$$-2 \overline{) 1 + 8 + 21 + 18}$$
$$-2 - 12 - 18$$
$$1 + 6 + 9 + 0$$

Thus, the quotient is $x^2 + 6x + 9$, with remainder 0, so $x + 2$ is a factor of $x^3 + 8x^2 + 21x + 18$ and

$$x^3 + 8x^2 + 21x + 18 = (x + 2)(x^2 + 6x + 9)$$

Because $x^2 + 6x + 9 = (x + 3)(x + 3)$, we have

$$x^3 + 8x^2 + 21x + 18 = 0$$
$$(x + 2)(x^2 + 6x + 9) = 0$$
$$(x + 2)(x + 3)(x + 3) = 0$$

and the solutions to the equation are $x = -2$, $x = -3$, and $x = -3$.

Because one of the solutions of this equation, -3, occurs *twice*, we say that it is a double solution, or a solution of **multiplicity** 2.

To check the solutions, we can graph the function $f(x) = x^3 + 8x^2 + 21x + 18$ and observe x-intercepts at -2 and -3 (Figure 6.29). Observe that the graph touches but does not cross the x-axis at $x = -3$, where a *double* solution to the equation (and zero of the function) occurs.

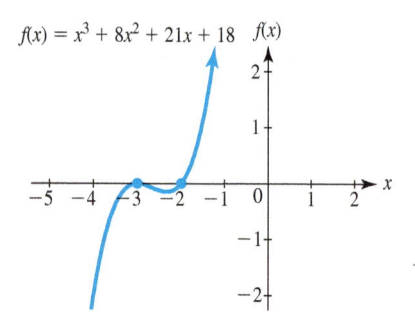

Figure 6.29

A graph touches but does not cross the x-axis at a zero of even multiplicity, and it crosses the x-axis at a zero of odd multiplicity.

Graphs and Solutions

Sometimes a combination of graphical methods and factoring can be used to solve a polynomial equation. If we are seeking exact solutions to a cubic equation and one solution can be found graphically, we can use that solution and division to find the remaining quadratic factor and to find the remaining two solutions. Consider the following example.

EXAMPLE 2 ▶ Combining Graphical and Algebraic Methods

Solve the equation $x^3 + 9x^2 - 610x + 600 = 0$.

SOLUTION

If we graph $P(x) = x^3 + 9x^2 - 610x + 600$ in a standard window, shown in Figure 6.30, we see that the graph appears to cross the x-axis at $x = 1$, so one solution to $P(x) = 0$ may be $x = 1$. (Notice that this window does not give a complete graph of the cubic polynomial function.)

To verify that $x = 1$ is a solution to $P(x) = 0$, we use synthetic division to divide $x^3 + 9x^2 - 610x + 600$ by $x - 1$.

$$\begin{array}{r|rrrr} 1) & 1 + & 9 - & 610 + & 600 \\ & & 1 + & 10 - & 600 \\ \hline & 1 + & 10 - & 600 + & 0 \end{array}$$

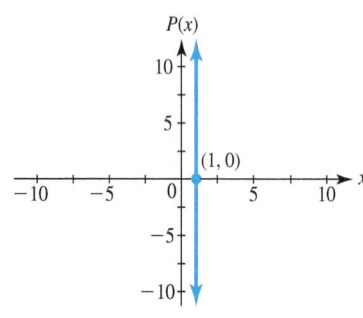

Figure 6.30

6.4 Polynomial Equations Continued; Fundamental Theorem of Algebra   **493**

The fact that the division results in a 0 remainder verifies that 1 is a solution to the equation $P(x) = 0$. The quotient is $x^2 + 10x - 600$, so the remaining solutions to $P(x) = 0$ can be found by factoring or by using the quadratic formula to solve $x^2 + 10x - 600 = 0$.

$$x^3 + 9x^2 - 610x + 600 = 0$$
$$(x - 1)(x^2 + 10x - 600) = 0$$
$$x - 1 = 0 \quad \text{or} \quad x^2 + 10x - 600 = 0$$
$$(x + 30)(x - 20) = 0$$
$$x = 1 \quad \text{or} \quad x = -30 \quad \text{or} \quad x = 20$$

Graphing the function $P(x) = x^3 + 9x^2 - 610x + 600$ using a viewing window that includes x-values of -30, 1, and 20 verifies that -30, 1, and 20 are zeros of the function and thus solutions to the equation $x^3 + 9x^2 - 610x + 600 = 0$ (Figure 6.31).

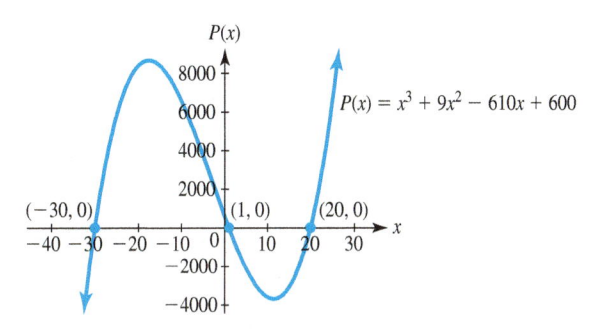

Figure 6.31

EXAMPLE 3 ▶ Break-Even

The weekly profit for a product is $P(x) = -0.1x^3 + 11x^2 - 80x - 2000$ thousand dollars, where x is the number of thousands of units produced and sold. To find the number of units that gives break-even,

a. Graph the function using a window representing up to 50 thousand units and find one x-intercept of the graph.

b. Use synthetic division to find a quadratic factor of $P(x)$.

c. Find all of the zeros of $P(x)$.

d. Determine the levels of production that give break-even.

SOLUTION

a. Break-even occurs where the profit is 0, so we seek the solution to

$$0 = -0.1x^3 + 11x^2 - 80x - 2000$$

for $x \geq 0$ (because x represents the number of units). Because $x = 50$ represents 50 thousand units and we seek an x-value where $P(x) = 0$, we use the viewing window [0, 50] by [-2500, 8000]. The graph of

$$P(x) = -0.1x^3 + 11x^2 - 80x - 2000$$

is shown in Figure 6.32. (Note that this is not a complete graph of the polynomial function.)

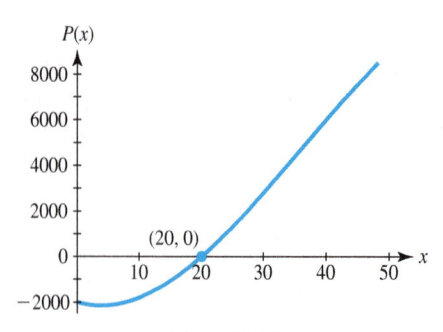

Figure 6.32

An x-intercept of this graph appears to be $x = 20$ (see Figure 6.32).

b. To verify that $P(x) = 0$ at $x = 20$ and to determine whether any other x-values give break-even, we divide

$$P(x) = -0.1x^3 + 11x^2 - 80x - 2000$$

by the factor $x - 20$:

$$\begin{array}{r} 20\overline{)\,-0.1 + 11 - 80 - 2000} \\ \underline{-2 + 180 + 2000} \\ -0.1 + 9 + 100 + 0 \end{array}$$

Thus, $P(x) = 0$ at $x = 20$ and the quotient is the quadratic factor $-0.1x^2 + 9x + 100$, so

$$-0.1x^3 + 11x^2 - 80x - 2000 = (x - 20)(-0.1x^2 + 9x + 100)$$

c. The solution to $P(x) = 0$ follows.

$$-0.1x^3 + 11x^2 - 80x - 2000 = 0$$
$$(x - 20)(-0.1x^2 + 9x + 100) = 0$$
$$(x - 20)[-0.1(x^2 - 90x - 1000)] = 0$$
$$-0.1(x - 20)(x - 100)(x + 10) = 0$$
$$x - 20 = 0 \quad \text{or} \quad x - 100 = 0 \quad \text{or} \quad x + 10 = 0$$
$$x = 20 \quad \text{or} \quad x = 100 \quad \text{or} \quad x = -10$$

Thus, the zeros are 20, 100, and -10. Figure 6.33 shows the graph of $P(x) = -0.1x^3 + 11x^2 - 80x - 2000$ on a window that shows all three x-intercepts.

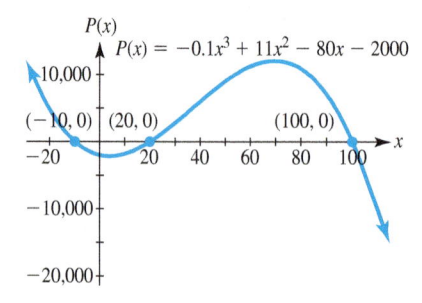

Figure 6.33

d. The solutions indicate that break-even occurs at $x = 20$ and at $x = 100$. The value $x = -10$ also makes $P(x) = 0$, but a negative number of units does not make sense in the context of this problem. Thus, producing and selling 20,000 or 100,000 units gives break-even.

6.4 Polynomial Equations Continued; Fundamental Theorem of Algebra

Combining graphical methods and synthetic division, as we did in Example 3, is especially useful in finding exact solutions to polynomial equations. If some of the solutions are irrational or nonreal solutions, the quadratic formula can be used rather than factoring. Remember that the quadratic formula can always be used to solve a quadratic equation, and factoring cannot always be used.

Rational Solutions Test

The Rational Solutions Test provides information about the rational solutions of a polynomial equation with integer coefficients.

Rational Solutions Test

The rational solutions of the polynomial equation

$$a_n x^n + a_{n-1} x^{n-1} + \cdots + a_1 x + a_0 = 0$$

with integer coefficients must be of the form $\frac{p}{q}$, where p is a factor of the constant term a_0 and q is a factor of a_n, the leading coefficient.

Recall that if a is a solution to $f(x) = 0$, then a is also a zero of the function $f(x)$. The Rational Solutions Test provides a list of *possible* rational zeros of a polynomial function. For a polynomial function with integer coefficients, if there is a rational zero, then it will be in this list of possible rational zeros. Keep in mind that there may be *no* rational zeros of a polynomial function. Graphing a polynomial function $f(x)$ will give us a better sense of the location of the zeros, and we can also use our list of possible zeros to find an appropriate viewing window for the function.

For example, the rational solutions of $x^3 + 8x^2 + 21x + 18 = 0$ must be numbers that are factors of 18 divided by factors of 1. Thus, the possible rational solutions are $\pm 1, \pm 18, \pm 2, \pm 9, \pm 3$, and ± 6. A viewing window containing x-values from -18 to 18 will show all possible rational solutions. Note that the solutions to this equation, found in Example 1 to be -2 and -3, are contained in this list.

Many cubic and quartic equations can be solved by using the following steps.

Solving Cubic and Quartic Equations of the Form $f(x) = 0$

1. Determine the possible rational solutions of $f(x) = 0$.

2. Graph $y = f(x)$ to see if any of the values from step 1 are x-intercepts. Those values that are x-intercepts are rational solutions to $f(x) = 0$.

3. Find the factors associated with the x-intercepts from step 2.

4. Use synthetic division to divide $f(x)$ by the factors from step 3 to confirm the graphical solutions and find additional factors. Continue until a quadratic factor remains.

5. Use factoring or the quadratic formula to find the solutions associated with the quadratic factor. These solutions are also solutions to $f(x) = 0$.

Example 4 shows how the Rational Solutions Test is used when the leading coefficient of the polynomial is not 1.

EXAMPLE 4 ▶ Solving a Quartic Equation

Solve the equation $2x^4 + 10x^3 + 13x^2 - x - 6 = 0$.

SOLUTION

The rational solutions of this equation must be factors of the constant -6 divided by factors of 2, the leading coefficient. Thus, the possible rational solutions are

$$\pm 1, \pm 2, \pm 3, \pm 6 \quad \text{and} \quad \pm\frac{1}{2}, \pm\frac{2}{2}, \pm\frac{3}{2}, \pm\frac{6}{2}*$$

All of these possible rational solutions are between -6 and 6. Using a window with x-values from -4 to 4, Figure 6.34 shows that the graph of

$$y = P(x) = 2x^4 + 10x^3 + 13x^2 - x - 6$$

appears to cross the x-axis at $x = -1$ and at $x = -2$, so two of the solutions to $y = 0$ appear to be $x = -1$ and $x = -2$. We will use synthetic division to confirm that these values are solutions to the equation and to find the two remaining solutions. (We know there are two additional solutions, which may be irrational, because the graph crosses the x-axis at two other points.)

We can confirm that $x = -1$ is a solution and find the remaining factor of $P(x)$ by using synthetic division to divide $2x^4 + 10x^3 + 13x^2 - x - 6$ by $x + 1$.

$$\begin{array}{r} -1\overline{)2 + 10 + 13 - 1 - 6} \\ \underline{-2 - 8 - 5 + 6} \\ 2 + 8 + 5 - 6 + 0 \end{array}$$

Thus, we have $2x^4 + 10x^3 + 13x^2 - x - 6 = (x + 1)(2x^3 + 8x^2 + 5x - 6)$.

We can also confirm that $x = -2$ is a solution of $P(x)$ by showing that $x + 2$ is a factor of $2x^3 + 8x^2 + 5x - 6$ and thus a factor of $P(x)$. Dividing $x + 2$ into $2x^3 + 8x^2 + 5x - 6$ using synthetic division gives a quadratic factor of $P(x)$:

$$\begin{array}{r} -2\overline{)2 + 8 + 5 - 6} \\ \underline{-4 - 8 + 6} \\ 2 + 4 - 3 + 0 \end{array}$$

So the quotient is the quadratic factor $2x^2 + 4x - 3$, and $P(x)$ has the following factorization:

$$2x^4 + 10x^3 + 13x^2 - x - 6 = (x + 1)(x + 2)(2x^2 + 4x - 3)$$

Thus, $P(x) = 0$ is equivalent to $(x + 1)(x + 2)(2x^2 + 4x - 3) = 0$, so the solutions can be found by solving $x + 1 = 0$, $x + 2 = 0$, and $2x^2 + 4x - 3 = 0$.

The two remaining x-intercepts of the graph of $y = 2x^2 + 4x - 3$ do not appear to cross the x-axis at integer values, and $2x^2 + 4x - 3$ does not appear to be factorable. Using the quadratic formula is the obvious method to solve $2x^2 + 4x - 3 = 0$ to find the two remaining solutions:

$$x = \frac{-4 \pm \sqrt{16 - 4(2)(-3)}}{2(2)} = \frac{-4 \pm \sqrt{40}}{4} = \frac{-2 \pm \sqrt{10}}{2}$$

So the solutions are $-1, -2, \dfrac{-2 + \sqrt{10}}{2}$, and $\dfrac{-2 - \sqrt{10}}{2}$. Note that two of these solutions are rational and two are irrational, approximated by 0.58114 and -2.58114. The two irrational solutions can also be verified graphically (Figure 6.35).

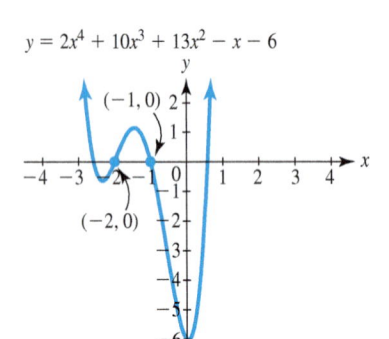

---

*Some of these possible solutions are duplicates.

$$y = 2x^4 + 10x^3 + 13x^2 - x - 6$$

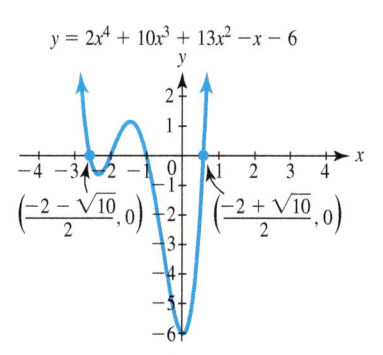

$$\left(\frac{-2 - \sqrt{10}}{2}, 0\right) \quad \left(\frac{-2 + \sqrt{10}}{2}, 0\right)$$

Figure 6.35

Note that some graphing utilities and spreadsheets have commands that can be used to find or to approximate solutions to polynomial equations.

Fundamental Theorem of Algebra

As we saw in Section 3.2, not all quadratic equations have real solutions. However, every quadratic equation has complex solutions. If the complex number a is a solution to $P(x) = 0$, then a is a complex zero of P. The **Fundamental Theorem of Algebra** states that every polynomial function has a complex zero.

Fundamental Theorem of Algebra
If $f(x)$ is a polynomial function of degree $n \geq 1$, then f has at least one complex zero.

As a result of this theorem, it is possible to prove the following.

Complex Zeros
Every polynomial function $f(x)$ of degree $n \geq 1$ has exactly n complex zeros. Some of these zeros may be imaginary, and some may be repeated. Nonreal zeros occur in conjugate pairs, $a + bi$ and $a - bi$.

EXAMPLE 5 ▶ **Solution in the Complex Number System**

Find the complex solutions to $x^3 + 2x^2 - 3 = 0$.

SOLUTION

Because the function is cubic, there are three solutions, and because any nonreal solutions occur in conjugate pairs, at least one of the solutions must be real. To estimate that solution, we graph

$$y = x^3 + 2x^2 - 3$$

on a window that includes the possible rational solutions, which are ± 1 and ± 3 (see Figure 6.36).

The graph appears to cross the x-axis at 1, so we (synthetically) divide $x^3 + 2x^2 - 3$ by $x - 1$:

$$\begin{array}{r} 1\overline{)1 + 2 + 0 - 3} \\ 1 + 3 + 3 \\ \hline 1 + 3 + 3 + 0 \end{array}$$

The division confirms that 1 is a solution to $x^3 + 2x^2 - 3 = 0$ and shows that

$$x^2 + 3x + 3$$

is a factor of $x^3 + 2x^2 - 3$. Thus, any remaining solutions to the equation satisfy

$$x^2 + 3x + 3 = 0$$

Using the quadratic formula gives the remaining solutions:

$$x = \frac{-3 \pm \sqrt{3^2 - 4(1)(3)}}{2(1)} = \frac{-3 \pm \sqrt{-3}}{2} = \frac{-3 \pm i\sqrt{3}}{2}$$

Thus, the solutions are $1, -\frac{3}{2} + \frac{\sqrt{3}}{2}i$, and $-\frac{3}{2} - \frac{\sqrt{3}}{2}i$. Because there can only be three solutions to a cubic equation, we have all of the solutions.

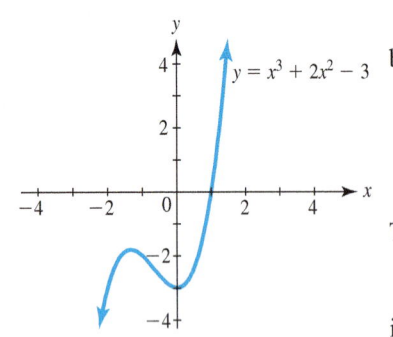

Figure 6.36

SKILLS CHECK 6.4

In Exercises 1–4, use synthetic division to find the quotient and the remainder.

1. $(x^4 - 4x^3 + 3x + 10) \div (x - 3)$ $x^3 - x^2 - 3x - 6 - \frac{8}{x-3}$
2. $(x^4 + 2x^3 - 3x^2 + 1) \div (x + 4)$ $x^3 - 2x^2 + 5x - 20 + \frac{81}{x+4}$
3. $(2x^4 - 3x^3 + x - 7) \div (x - 1)$ $2x^3 - x^2 - x - \frac{7}{x-1}$
4. $(x^4 - 1) \div (x + 1)$ $x^3 - x^2 + x - 1$

In Exercises 5 and 6, determine whether the given constant is a solution to the given polynomial equation.

5. $2x^4 - 4x^3 + 3x + 18 = 0; 3$ No
6. $x^4 + 3x^3 - 10x^2 + 8x + 40 = 0; -5$ Yes

In Exercises 7 and 8, determine whether the second polynomial is a factor of the first polynomial.

7. $(-x^4 - 9x^2 + 3x); (x + 3)$ No
8. $(2x^4 + 5x^3 - 6x - 4); (x + 2)$ Yes

In Exercises 9–12, one or more solutions of a polynomial equation are given. Use synthetic division to find any remaining solutions.

9. $-x^3 + x^2 + x - 1 = 0; -1$ 1 (a double solution)
10. $x^3 + 4x^2 - x - 4 = 0; 1$ $-1, -4$
11. $x^4 + 2x^3 - 21x^2 - 22x = -40; -5, 1$ $-2, 4$
12. $2x^4 - 17x^3 + 51x^2 - 63x + 27 = 0; 3, 1$ $\frac{3}{2}$

In Exercises 13–16, find one solution graphically and then find the remaining solutions using synthetic division.

13. $x^3 + 3x^2 - 18x - 40 = 0$ $4, -2, -5$
14. $x^3 - 3x^2 - 9x - 5 = 0$ $5, -1$ (a double solution)
15. $3x^3 + 2x^2 - 7x + 2 = 0$ $-2, 1, \frac{1}{3}$
16. $4x^3 + x^2 - 27x + 18 = 0$ $-3, 2, \frac{3}{4}$

In Exercises 17–20, determine all possible rational solutions of the polynomial equation.

17. $x^3 - 6x^2 + 5x + 12 = 0$ $\pm 1, \pm 2, \pm 3, \pm 4, \pm 6, \pm 12$
18. $4x^3 + 3x^2 - 9x + 2 = 0$ $\pm 1, \pm 2, \pm \frac{1}{2}, \pm \frac{1}{4}$
19. $9x^3 + 18x^2 + 5x - 4 = 0$ $\pm 1, \pm 2, \pm 4, \pm \frac{1}{3}, \pm \frac{2}{3}, \pm \frac{4}{3}, \pm \frac{1}{9}, \pm \frac{2}{9}, \pm \frac{4}{9}$
20. $6x^4 - x^3 - 42x^2 - 29x + 6 = 0$ $\pm 1, \pm 2, \pm 3, \pm 6, \pm \frac{1}{2}, \pm \frac{3}{2}, \pm \frac{1}{3}, \pm \frac{2}{3}, \pm \frac{1}{6}$

In Exercises 21–24, find all rational zeros of the polynomial function.

21. $f(x) = x^3 - 6x^2 + 5x + 12$ $-1, 3, 4$
22. $f(x) = 4x^3 + 3x^2 - 9x + 2$ $1, \frac{1}{4}, -2$

23. $g(x) = 9x^3 + 18x^2 + 5x - 4$ $\frac{1}{3}, -1, -\frac{4}{3}$

24. $g(x) = 6x^3 + 19x^2 - 19x + 4$ $\frac{1}{3}, \frac{1}{2}, -4$

Solve each of the equations in Exercises 25–30 exactly in the complex number system. (Find one solution graphically and then use the quadratic formula.)

25. $x^3 = 10x - 7x^2$ $x = 0, x = \frac{-7 \pm \sqrt{89}}{2}$

26. $t^3 - 2t^2 + 3t = 0$ $t = 0, t = 1 \pm i\sqrt{2}$

27. $w^3 - 5w^2 + 6w - 2 = 0$ $w = 1, w = 2 \pm \sqrt{2}$

28. $2w^3 + 3w^2 + 3w + 2 = 0$ $w = -1, w = \frac{-1 \pm i\sqrt{15}}{4}$

29. $z^3 - 8 = 0$ $z = 2, z = -1 \pm i\sqrt{3}$

30. $x^3 + 1 = 0$ $x = -1, x = \frac{1 \pm i\sqrt{3}}{2}$

EXERCISES 6.4

Answers that are not seen can be found in the answer section at the back of the text.

In Exercises 31–36, use synthetic division and factoring to solve the problems.

31. **Break-Even** The profit function, in dollars, for a product is given by $P(x) = -0.2x^3 + 66x^2 - 1600x - 60,000$, where x is the number of units produced and sold. If break-even occurs when 50 units are produced and sold,

   a. Use synthetic division to find a quadratic factor of $P(x)$. $-0.2x^2 + 56x + 1200$

   b. Use factoring to find a number of units other than 50 that gives break-even for the product, and verify your answer graphically. 300 units

32. **Break-Even** The profit function, in dollars, for a product is given by $P(x) = -x^3 + 98x^2 - 700x - 1800$, where x is the number of units produced and sold. If break-even occurs when 10 units are produced and sold,

   a. Use synthetic division to find a quadratic factor of $P(x)$. $-x^2 + 88x + 180$

   b. Find a number of units other than 10 that gives break-even for the product, and verify your answer graphically. 90 units

33. **Break-Even** The weekly profit for a product, in thousand of dollars, is $P(x) = -0.1x^3 + 50.7x^2 - 349.2x - 400$, where x is the number of thousands of units produced and sold. To find the number of units that gives break-even,

   a. Graph the function on the window $[-8, 10]$ by $[-60, 100]$.

   b. Graphically find one x-intercept of the graph. 8

   c. Use synthetic division to find a quadratic factor of $P(x)$. $-0.1x^2 + 49.9x + 50$

   d. Find all of the zeros of $P(x)$.
   $x = 8, x = 500, x = -1$

   e. Determine the levels of production and sale that give break-even. 8 units, 500 units

34. **Break-Even** The weekly profit for a product, in thousands of dollars, is $P(x) = -0.1x^3 + 10.9x^2 - 97.9x - 108.9$, where x is the number of thousands of units produced and sold. To find the number of units that gives break-even,

   a. Graph the function on the window $[-10, 20]$ by $[-100, 100]$.

   b. Graphically find one x-intercept of the graph. 11

   c. Use synthetic division to find a quadratic factor of $P(x)$. $-0.1x^2 + 9.8x + 9.9$

   d. Find all of the zeros of $P(x)$. $x = 11, x = -1, x = 99$

   e. Determine the levels of production and sale that give break-even. 11 units, 99 units

35. **Revenue** The revenue from the sale of a product is given by $R = 1810x - 81x^2 - x^3$. If the sale of 9 units gives a total revenue of $9000, use synthetic division to find another number of units that will give $9000 in revenue. 10 units

36. **Revenue** The revenue from the sale of a product is given by $R = 250x - 5x^2 - x^3$. If the sale of 5 units gives a total revenue of $1000, use synthetic division to find another number of units that will give $1000 in revenue. 10 units

37. **Older Workers** Using data from 1970 and projected to 2050, the number of older workers, in thousands, can be modeled by the function

$$y = -0.06x^3 + 9.34x^2 - 308x + 5630$$

where x is the number of years after 1970.
(*Source*: U.S. Bureau of Labor Statistics)

This model indicates that the number of older workers was 3424 thousand in 1980. To find any other years that the number was 3424 thousand:

   a. Set the function equal to 3424 and write the resulting equation with 0 on one side.
   $-0.06x^3 + 9.34x^2 - 308x + 2206 = 0$

b. Find the factor that represents 1980; then use it and synthetic division to find a quadratic factor of the nonzero side of the equation in (b).
$x - 10; -0.06x^2 + 8.74x - 220.6$

c. Use the quadratic formula to find any other solutions to this equation. $x \approx 32.48, x \approx 113.18$

d. Use this information to find the years after 1970 in which the number of older workers was or is projected to be 3424 thousand, if this model is accurate. 1980, 2003, 2084

38. **Employment in Manufacturing** The total employment (number of workers y, in thousands) in manufacturing for selected years t from 2010 and projected to 2040 can be modeled by the function

$$y = 0.089x^3 - 11x^2 + 237x + 11{,}500$$

where x is the number of years after 2010.
(Source: U.S. Department of Energy)

This model indicates that the number of workers in manufacturing was 12,859 thousand in 2020. To find any other years that the number was 12,859 thousand:

a. Set the function equal to 12,859 and write the resulting equation with 0 on one side.
$0.089x^3 - 11x^2 + 237x - 1359 = 0$

b. Find the factor that represents 2020; then use it and synthetic division to find a quadratic factor of the nonzero side of the equation in part (a).
$x - 10; 0.089x^2 - 10.11x + 135.9$

c. Use the quadratic formula to find any other solutions to this equation. $x \approx 15.6, x \approx 98$

d. Use this information to find the years after 1970 in which the number of workers in manufacturing was or is projected to be 12,859 thousand, if this model is accurate beyond 2040. 2020, 2026, 2108

39. **Foreign-Born Population** Using data for selected years from 1970 and projected to 2060, the percent of the U.S. population that is foreign-born can be modeled by the function

$$y = -0.000017x^3 + 0.00156x^2 + 0.156x + 4.601$$

with x equal to the number of years after 1970. This model indicates that the percent will be 6.3% in 1980. To find any other years that the percent is 6.3%:

a. Set the function equal to 6.3 and write the resulting equation with 0 on one side.
$-0.000017x^3 + 0.00156x^2 + 0.156x - 1.699 = 0$

b. Find the factor that represents 1980; then use it and synthetic division to find a quadratic factor of the nonzero side of the equation in part (a).
$x - 10; -0.000017x^2 + 0.00139x + 0.1699$

c. Use the quadratic formula to find any other solutions to this equation. $x \approx -67.1, x \approx 148.9$

d. Use this information to find the years after 1970 in which the percent of the U.S. population that is foreign-born was or is projected to be 6.3% if this model is accurate beyond 2060. 1980, 2119

40. **Aging Workers** The number of Americans working full time at various ages, in thousands, can be modeled by the function

$$y = -0.36x^3 + 40x^2 - 1390x + 21{,}700$$

where x is the age.
(Source: Wall Street Journal)

This model indicates that the number of workers at age 20 is 7020 thousand. To find any other ages at which the number is 7020 thousand:

a. Set the function equal to 7020 and write the resulting equation with 0 on one side.
$-0.36x^3 + 40x^2 - 1390x + 14{,}680 = 0$

b. Find the factor that represents age 20; then use it and synthetic division to find a quadratic factor of the nonzero side of the equation in part (a).
$x - 20; -0.36x^2 + 32.8x - 734$

c. Use the quadratic formula to find any other solutions to this equation. $x \approx 39.52, x \approx 51.59$

d. Use this information to find the ages at which the number of workers is estimated to be 7020 thousand by this model. 20, 39, 51

41. **Drug Levels in Blood** A new drug is tested to see how a 300-mg capsule is released into the bloodstream. Volunteers are given a capsule, blood samples are drawn every half-hour, and the average number of milligrams of the drug in the bloodstream is calculated for the sample. The cubic function that models these data is $N(t) = 24.1t^3 - 165t^2 + 241t + 167$, where t is in hours. This model indicates that the drug level is 249.26 mg one-half hour after taking the drug.

a. Use this fact and the quadratic formula to find other times after taking the drug when the model indicates that the drug level is 249.26 mg.
Approximately 1.4 and approximately 5 hours

b. Do all of the times found in part (a) sound reasonable? Why? No; it's not reasonable for the drug level to increase after it has started to decrease.

c. How long after taking the drug does this model fail to be accurate? After 3.7 hours, when the level begins to increase again.

42. **Sulfur Emissions** The thousands of short tons of sulfur emissions from electricity generation for selected years from 2000 and projected to 2035 can be modeled by

$$E(x) = 0.0347x^3 + 8.58x^2 - 566x + 11{,}800$$

where x is equal to the number of years after 2000.

This model indicates that 11,800 thousand short tons of sulfur were emitted in 2000. Use this fact and the quadratic formula to find any other years in the 21st century that the sulfur emissions from electricity generation are projected to be 11,800 thousand short tons. In 2055 ($x \approx 54.1, x \approx 301.4$)

6.5 Rational Functions and Rational Equations

KEY OBJECTIVES

- Graph rational functions
- Find vertical asymptotes, horizontal asymptotes, slant asymptotes, and missing points on graphs of rational functions
- Solve rational equations algebraically and graphically
- Solve inverse variation problems

SECTION PREVIEW   Average Cost

The function that gives the daily average cost (in hundreds of dollars) for the production of Stanley golf carts is formed by dividing the total cost function for the production of the golf carts

$$C(x) = 25 + 13x + x^2$$

by x, the number of golf carts produced. Thus, the average cost per golf cart is given by

$$\overline{C}(x) = \frac{25 + 13x + x^2}{x} \quad (x > 0)$$

where x is the number of golf carts produced. The graph of this function with x restricted to positive values is shown in Figure 6.37(a). The graph of the function

$$\overline{C}(x) = \frac{25 + 13x + x^2}{x}$$

with x not restricted is shown in Figure 6.37(b). (See Example 3.)

This function, as well as any other function that is formed by taking the quotient of two polynomials, is called a **rational function**. In this section, we discuss the characteristics of the graphs of rational functions.

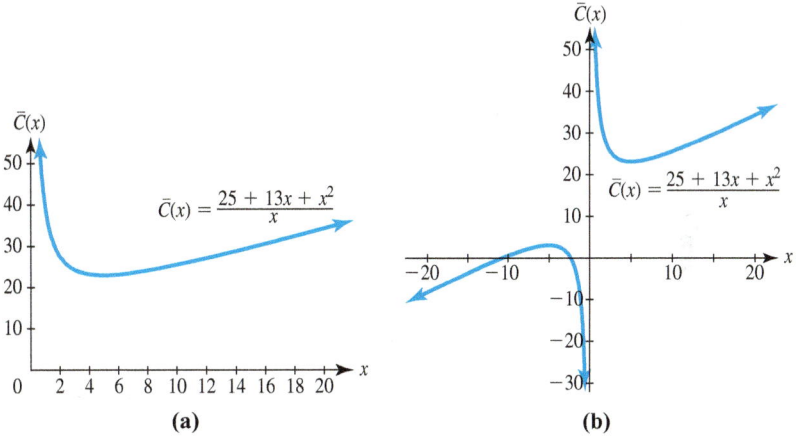

Figure 6.37

Graphs of Rational Functions

A rational function is defined as follows.

Rational Function

The function f is a rational function if

$$f(x) = \frac{P(x)}{Q(x)}$$

where $P(x)$ and $Q(x)$ are polynomials and $Q(x) \neq 0$.

Figure 6.38

In Section 3.3, we discussed the simple rational function $y = \dfrac{1}{x}$, whose graph is shown in Figure 6.38. We now expand this discussion.

Because division by 0 is not possible, those real values of x for which $Q(x) = 0$ are not in the domain of the rational function $f(x) = \dfrac{P(x)}{Q(x)}$. Observe that the graph of $y = \dfrac{1}{x}$ approaches, but does not touch, the y-axis. Thus, the y-axis is a **vertical asymptote**. The graph of a rational function frequently has vertical asymptotes.

To find where the vertical asymptote(s) occur, set the denominator equal to 0 and solve for x. If any of these values of x do not also make the numerator equal to 0, then a vertical asymptote occurs at those values.

Vertical Asymptote

A vertical asymptote occurs in the graph of $f(x) = \dfrac{P(x)}{Q(x)}$ at those values of x where $Q(x) = 0$ and $P(x) \neq 0$—that is, at values of x where the denominator equals 0 but the numerator does not equal 0.

EXAMPLE 1 ▶ Rational Function

Find the vertical asymptote and sketch the graph of $y = \dfrac{x+1}{(x+2)^2}$.

SOLUTION

Setting the denominator equal to 0, taking the square root of both sides of the equation, and solving gives

$$(x+2)^2 = 0$$
$$(x+2) = 0$$
$$x = -2$$

The value $x = -2$ makes the denominator of the function equal to 0 and does not make the numerator equal to 0, so a vertical asymptote occurs at $x = -2$. The graph of the function $y = \dfrac{x+1}{(x+2)^2}$ is shown in Figure 6.39.

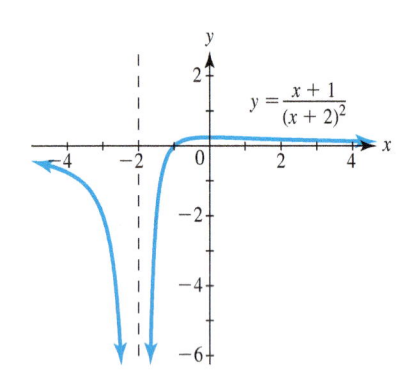

Figure 6.39

EXAMPLE 2 ▶ Cost–Benefit

Suppose that for specified values of p, the function

$$C(p) = \dfrac{800p}{100 - p}$$

can be used to model the cost of removing $p\%$ of the particulate pollution from the exhaust gases at an industrial site.

    a. Graph this function on the window $[-100, 200]$ by $[-4000, 4000]$.

    b. Does the graph of this function have a vertical asymptote on this window? Where?

    c. For what values of p does this function serve as a model for the cost of removing particulate pollution?

    d. Use the information determined in part (c) to graph the model.

    e. What does the part of the graph near the vertical asymptote tell us about the cost of removing particulate pollution?

6.5 Rational Functions and Rational Equations 503

SOLUTION

a. The graph is shown in Figure 6.40(a).

b. The denominator is equal to 0 at $p = 100$, and the numerator does not equal 0 at $p = 100$, so a vertical asymptote occurs at $p = 100$.

c. Because p represents the percent of pollution, p must be limited to values from 0 to 100. The domain cannot include 100, so the p-interval is $[0, 100)$.

d. The graph of the function on the window [0, 100] by $[-1000, 10,000]$ is shown in Figure 6.40(b).

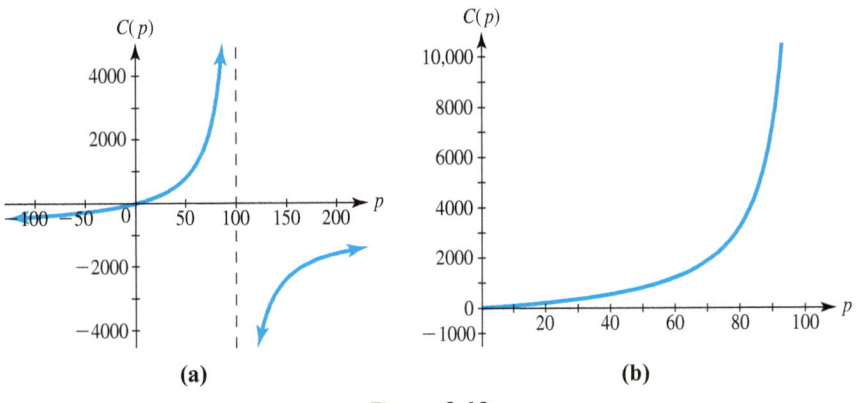

Figure 6.40

e. From Figure 6.40(b), we see that the graph is increasing rapidly as it approaches the vertical asymptote at $p = 100$, which tells us that the cost of removing pollution gets extremely high as the amount of pollution removed approaches 100%. It is impossible to remove 100% of the particulate pollution.

We see in Figure 6.41 that the graph of

$$y = \frac{3}{x + 1}$$

approaches the x-axis asymptotically on the left and on the right. Because the x-axis is a horizontal line, we say that the x-axis is a **horizontal asymptote** for this graph.

We can study the graph of a rational function to determine if the curve approaches a horizontal asymptote. If the graph approaches the horizontal line $y = a$ as $|x|$ gets very large, the graph has a horizontal asymptote at $y = a$. We can denote this as follows.

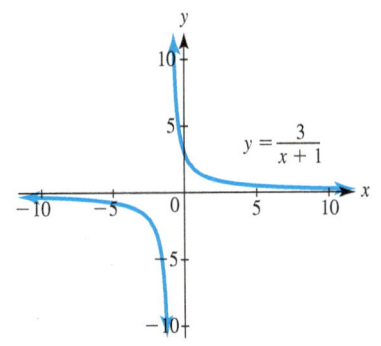

Figure 6.41

Horizontal Asymptote

If y approaches a as x approaches $+\infty$ or as x approaches $-\infty$, the graph of $y = f(x)$ has a horizontal asymptote at $y = a$.

We can also compare the degrees of the numerator and denominator of the rational function to determine if the curve approaches a horizontal asymptote, and to find the horizontal asymptote if it exists.

Determining the Horizontal Asymptotes of a Rational Function

Consider the rational function

$$f(x) = \frac{P(x)}{Q(x)} = \frac{a_n x^n + \cdots + a_1 x + a_0}{b_m x^m + \cdots + b_1 x + b_0}, \quad a_n \neq 0 \text{ and } b_m \neq 0$$

1. If $n < m$ (that is, if the degree of the numerator is less than the degree of the denominator), a horizontal asymptote occurs at $y = 0$ (the x-axis).

2. If $n = m$ (that is, if the degree of the numerator is equal to the degree of the denominator), a horizontal asymptote occurs at $y = \dfrac{a_n}{b_m}$. (This is the ratio of the leading coefficients.)

3. If $n > m$ (that is, if the degree of the numerator is greater than the degree of the denominator), there is no horizontal asymptote.

In the function $y = \dfrac{3}{x+1}$, the degree of the numerator is less than the degree of the denominator, so a horizontal asymptote occurs at $y = 0$, the x-axis (Figure 6.41).

In the function $C(p) = \dfrac{800p}{100 - p}$ of Example 2, the degrees of the numerator and the denominator are equal, so the graph has a horizontal asymptote at $y = C(p) = \dfrac{800}{-1} = -800$. Observing the graph of the function, shown in Figure 6.40(a) on the previous page, we see that this horizontal asymptote is reasonable for the graph of $C(p) = \dfrac{800p}{100 - p}$.

Although it is impossible for a graph of any function to cross a vertical asymptote (because the function is undefined at that value of x), the curve *may* cross a horizontal asymptote because the horizontal asymptote describes the end behavior (as x approaches ∞ or x approaches $-\infty$) of a graph. Figure 6.42 shows the graph of $y = \dfrac{6 - 5x}{x^2}$; it crosses the x-axis but approaches the line $y = 0$ (the x-axis) as x approaches ∞ and as x approaches $-\infty$.

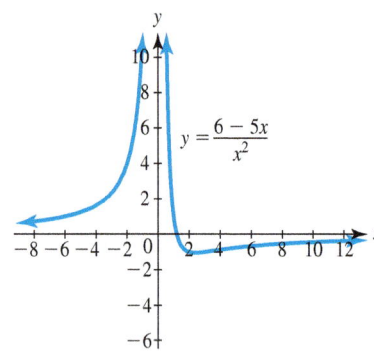

Figure 6.42

EXAMPLE 3 ▶ Average Cost

The function $\overline{C}(x) = \dfrac{25 + 13x + x^2}{x}$ represents the daily average cost (in hundreds of dollars) for the production of Stanley golf carts, with x equal to the number of golf carts produced.

a. Graph the function on the window $[-20, 20]$ by $[-30, 50]$.

b. Does the graph in part (a) have a horizontal asymptote?

c. Graph the function on the window $[0, 20]$ by $[0, 50]$.

d. Does the graph of the function using the window in part (a) or part (b) better model the average cost function? Why?

e. Use technology to find the minimum daily average cost and the number of golf carts that gives the minimum daily average cost.

6.5 Rational Functions and Rational Equations

SOLUTION

a. The graph on the window $[-20, 20]$ by $[-30, 50]$ is shown in Figure 6.43(a). Note that the graph has a vertical asymptote at $x = 0$ (the $\overline{C}(x)$-axis).

b. The graph in part (a) does not have a horizontal asymptote, because the degree of the numerator is greater than the degree of the denominator.

c. The graph on the window $[0, 20]$ by $[0, 50]$ is shown in Figure 6.43(b).

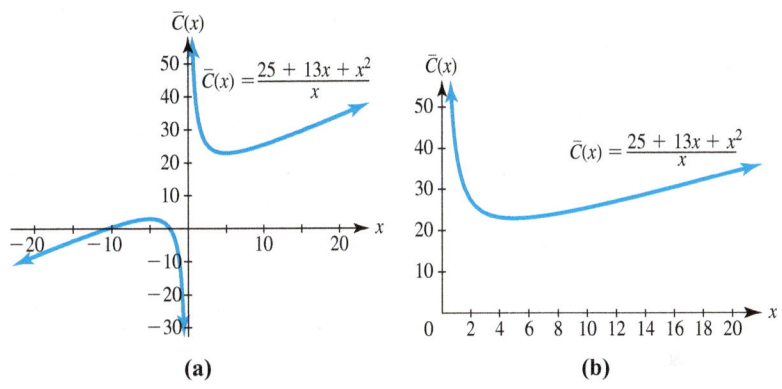

Figure 6.43

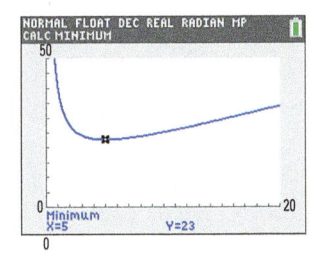

Figure 6.44

d. Because the number of golf carts produced cannot be negative, the window used in part (b) gives a better representation of the graph of the average cost function.

e. It appears that the graph reaches a low point somewhere between $x = 4$ and $x = 6$. By using technology, we can find the minimum point. Figure 6.44 shows that the minimum value of y is 23 at $x = 5$. Thus, the minimum daily average cost is $2300 per golf cart when 5 golf carts are produced.

Slant Asymptotes and Missing Points

The average cost function from Example 3,

$$\overline{C}(x) = \frac{25 + 13x + x^2}{x}$$

does not have a horizontal asymptote because the degree of the numerator is greater than the degree of the denominator. However, if the degree of the numerator of a rational function is *one* more than the degree of the denominator, its graph approaches a **slant asymptote**, which we can find by using the following method:

1. Divide the numerator by the denominator, getting a linear function plus a rational expression with a constant numerator (the remainder).

2. Observe that the rational expression found in step 1 above will approach 0 as x approaches ∞ or $-\infty$, so the graph of the original function will have a slant asymptote, which is the graph of the linear function.

The average cost function $\overline{C}(x) = \dfrac{25 + 13x + x^2}{x}$ can be written as

$$\overline{C}(x) = \frac{25}{x} + 13 + x$$

so its graph approaches the graph of the slant asymptote $y = 13 + x$ (Figure 6.45).

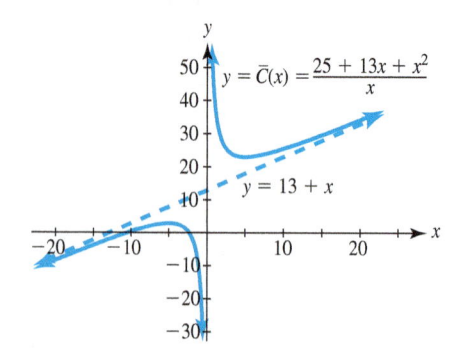

Figure 6.45

EXAMPLE 4 ▶ Slant Asymptote

Find the equations of the vertical, horizontal, and slant asymptotes (if any) for the graph of the function

$$f(x) = \frac{x^2 + 4}{x - 2}$$

Verify your results graphically.

SOLUTION

The denominator of the rational function is equal to 0 at $x = 2$, and the numerator does not equal 0 at $x = 2$, so there is a vertical asymptote at $x = 2$. The degree of the numerator of the rational function, 2, is greater than the degree of the denominator, 1, so there is no horizontal asymptote. Because the degree of the numerator is exactly one more than the degree of the denominator, there is a slant asymptote. To find the equation of the slant asymptote, we synthetically divide the numerator by the denominator:

$$\begin{array}{r} 2\overline{)1 \quad 0 \quad 4} \\ \underline{2 \quad 4} \\ 1 \quad 2 \quad 8 \end{array}$$

Thus, when $x^2 + 4$ is divided by $x - 2$, the quotient is $x + 2$ and the remainder is 8. So the equation of the slant asymptote is $y = x + 2$. Figure 6.46 shows the graph of $f(x) = \dfrac{x^2 + 4}{x - 2}$ and its asymptotes.

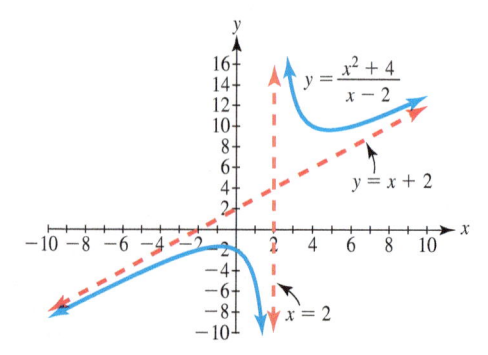

Figure 6.46

There is another type of rational function that is undefined at a value of x but does not have a vertical asymptote at that value. Note that if there is a value a such that $Q(a) = 0$ and $P(a) = 0$, then the function

$$f(x) = \frac{P(x)}{Q(x)}$$

is undefined at $x = a$, but its graph may have a "hole" in it at $x = a$ rather than an asymptote. Figure 6.47 shows the graph of such a function.

$$y = \frac{x^2 - 4}{x - 2}$$

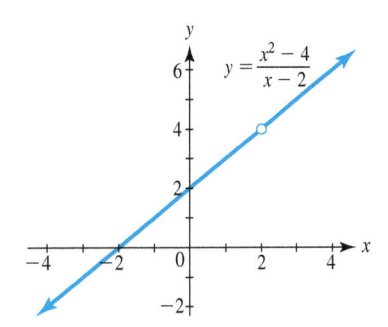

Figure 6.47

Observe that the function is undefined at $x = 2$, but the graph does not have an asymptote at this value of x. Notice that for all values of x except 2,

$$y = \frac{x^2 - 4}{x - 2} = \frac{(x + 2)(x - 2)}{x - 2} = x + 2$$

Thus, the graph of this function looks like the graph of $y = x + 2$, except that it has a missing point (hole) at $x = 2$ (Figure 6.47).

Algebraic and Graphical Solution of Rational Equations

To solve an equation involving rational expressions algebraically, we multiply both sides of the equation by the least common denominator (LCD) of the fractions in the equation. This will result in a linear or polynomial equation that can be solved by using the methods discussed earlier in the text. For these types of equations it is essential that all solutions be checked in the original equation, because some solutions to the resulting polynomial equation may not be solutions to the original equation. In particular, some solutions to the polynomial equation may result in a zero in the denominator of the original equation and thus cannot be solutions to the equation. Such solutions are called **extraneous solutions**.

The steps used to solve a rational equation follow.

Solving a Rational Equation Algebraically

To solve a rational equation algebraically:

1. Multiply both sides of the equation by the LCD of the fractions in the equation.

2. Solve the resulting polynomial equation for the variable.

3. Check each solution in the original equation. Some solutions to the polynomial equation may not be solutions to the original rational equation (these are called extraneous solutions).

EXAMPLE 5 ▶ Solving a Rational Equation

Solve the equation $x^2 + \dfrac{x}{x-1} = x + \dfrac{x^3}{x-1}$ for x.

SOLUTION

We first multiply both sides of the equation by $x - 1$, the LCD of the fractions in the equation, and then we solve the resulting polynomial equation:

$$x^2 + \frac{x}{x-1} = x + \frac{x^3}{x-1}$$

$$(x-1)\left(x^2 + \frac{x}{x-1}\right) = (x-1)\left(x + \frac{x^3}{x-1}\right)$$

$$(x-1)x^2 + (x-1)\frac{x}{x-1} = (x-1)x + (x-1)\frac{x^3}{x-1}$$

$$x^3 - x^2 + x = x^2 - x + x^3$$

$$0 = 2x^2 - 2x$$

$$0 = 2x(x-1)$$

$$x = 0 \quad \text{or} \quad x = 1$$

We must check these answers in the original equation.
Check $x = 0$:

$$0^2 + \frac{0}{0-1} = 0 + \frac{0^3}{0-1} \quad \text{checks}$$

Check $x = 1$:

$$1^2 + \frac{1}{1-1} = 1 + \frac{1^3}{1-1}, \quad \text{or} \quad 1 + \frac{1}{0} = 1 + \frac{1}{0}$$

But $1 + \dfrac{1}{0}$ cannot be evaluated, so $x = 1$ does not check in the original equation. This means that 1 is an extraneous solution of the equation. Thus, the only solution is $x = 0$.

As with polynomial equations derived from real data, some equations involving rational expressions may require technology to find or to approximate their solutions. The following example shows how applied problems involving rational functions can be solved algebraically and graphically.

EXAMPLE 6 ▶ Advertising and Sales

Monthly sales y (in thousands of dollars) for Yang products are related to monthly advertising expenses x (in thousands of dollars) according to the function

$$y = \frac{300x}{15 + x}$$

Determine the amount of money that must be spent on advertising to generate $100,000 in sales

a. Algebraically

b. Graphically

SOLUTION

a. To solve the equation $100 = \dfrac{300x}{15 + x}$ algebraically, we multiply both sides of the equation by $15 + x$, getting

$$100(15 + x) = 300x$$

Solving this equation gives

$$1500 + 100x = 300x$$
$$1500 = 200x$$
$$x = 7.5$$

Checking this solution in the original equation gives

$$100 = \dfrac{300(7.5)}{15 + 7.5}, \quad \text{or} \quad 100 = \dfrac{2250}{22.5}$$

so the solution checks. Thus, spending 7.5 thousand dollars ($7500) per month on advertising results in monthly sales of $100,000.

b. To solve the equation graphically, we graph the rational functions $y_1 = \dfrac{300x}{15 + x}$ and $y_2 = 100$ on the same axes on an interval with $x \geq 0$ (Figure 6.48). We then find the point of intersection (7.5, 100), which gives the same solution as in part (a).

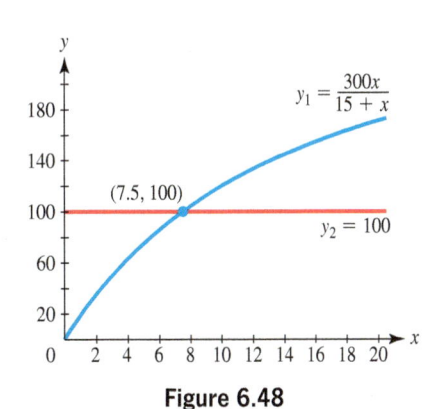

Figure 6.48

Inverse Variation

Recall from Section 2.1 that two variables are directly proportional (or vary directly) if their quotient is constant. When two variables x and y are **inversely proportional** (or vary inversely), an increase in one variable results in a decrease in the other. That is, y is inversely proportional to x (or y varies inversely as x) if there exists a nonzero number k such that

$$y = \dfrac{k}{x}, \quad \text{or} \quad y = kx^{-1}$$

Also, y is inversely proportional to the nth power of x if there exists a nonzero number k such that

$$y = \dfrac{k}{x^n}, \quad \text{or} \quad y = kx^{-n}$$

We can also say that y varies inversely as the nth power of x.

EXAMPLE 7 ▶ Illumination

The illumination produced by a light varies inversely as the square of the distance from the source of the light. If the illumination 30 feet from a light source is 60 candela, what is the illumination 20 feet from the source?

SOLUTION

If L represents the illumination and d represents the distance, the relation is

$$L = \frac{k}{d^2}$$

Substituting for L and d and solving for k gives

$$60 = \frac{k}{30^2}, \quad \text{or} \quad k = 54{,}000$$

Thus, the relation is $L = \dfrac{54{,}000}{d^2}$, and when $d = 20$ feet,

$$L = \frac{54{,}000}{20^2} = 135 \text{ candela}$$

So the illumination 20 feet from the source of the light is 135 candela.

Combined Variation

In combined variation, direct and inverse variation occur together. For example, suppose y varies directly as x and inversely as w; then we would say $y = \dfrac{kx}{w}$, where k is the constant of proportionality.

EXAMPLE 8 ▶ Body Mass Index

Body mass index (BMI) is a measure of body fat based on height and weight that applies to adult men and women. The formula $\text{BMI} = 703 \cdot \dfrac{w}{h^2}$, where w is weight in pounds and h is height in inches, means that an adult's BMI varies directly as the weight in pounds and inversely as the square of the height in inches. In adults, normal values for the BMI are between 20 and 25. Values below 20 indicate that a person is underweight, while values above 30 indicate that a person is obese.

a. What is the BMI for a person who is 190 pounds and 5 feet 4 inches tall? Would this person be considered obese?

b. If a person is 6 feet tall, what is the maximum weight for that person to be considered not obese?

SOLUTION

a. To calculate the person's BMI, we first change 5 feet 4 inches to $(5 \times 12) + 4$ inches $= 64$ inches. Then $\text{BMI} = 703 \cdot \dfrac{190}{64^2} = 32.6$. This value is over 30, so the person would be considered obese.

b. For a person who is 6 feet tall to be considered not obese, his or her BMI should be less than or equal to 30. So we solve

$$703 \cdot \frac{w}{72^2} \leq 30$$

$$703w \leq 155520$$

$$w \leq 221.2$$

Thus, a person who is 6 feet tall would need to weigh 221 pounds or less to be considered not obese.

Joint Variation

With joint variation, a variable varies directly as the product of two or more other variables. For example, Isaac Newton's formula for gravitation, $F = G\frac{m_1 m_2}{r^2}$, states that the force of gravitation F between two bodies varies jointly as the product of their masses m_1 and m_2, and inversely as the square of the distance r between them, where G is the gravitational constant, 6.67408×10^{-11} m³ kg⁻¹ s⁻².

EXAMPLE 9 ▶ Volume of a Pyramid

The volume of a pyramid varies jointly as its height and the area of its base. A pyramid with a height of 12 feet and a base with area of 23 square feet has a volume of 92 cubic feet. Find the volume of a pyramid with a height of 17 feet and a base with an area of 27 square feet.

SOLUTION

If the volume of a pyramid varies jointly as its height and the area of its base, then $V = k \cdot h \cdot b$, where h is the height and b is the area of the base. Substituting the given values in this equation gives

$$92 = k \cdot 12 \cdot 23$$

So $k = \frac{1}{3}$ and the equation is $V = \frac{1}{3} \cdot h \cdot b$

A pyramid with a height of 17 feet and base with an area of 27 square feet has volume

$$V = \frac{1}{3} \cdot 17 \cdot 27 = 153 \text{ cubic feet}$$

SKILLS CHECK 6.5

Answers that are not seen can be found in the answer section at the back of the text.

Give the equations of any (a) vertical and (b) horizontal asymptotes for the graphs of the rational functions $y = f(x)$ in Exercises 1–6.

1. $f(x) = \dfrac{3}{x - 5}$ **a.** $x = 5$ **b.** $y = 0$

2. $f(x) = \dfrac{7}{x - 4}$ **a.** $x = 4$ **b.** $y = 0$

3. $f(x) = \dfrac{x - 4}{5 - 2x}$ **a.** $x = \dfrac{5}{2}$ **b.** $y = -\dfrac{1}{2}$

4. $f(x) = \dfrac{2x - 5}{3 - x}$ **a.** $x = 3$ **b.** $y = -2$

5. $f(x) = \dfrac{x^3 + 4}{x^2 - 1}$ **a.** $x = 1, x = -1$ **b.** None

512   Chapter 6   Higher-Degree Polynomial and Rational Functions

6. $f(x) = \dfrac{x^2 + 6}{x^2 + 3}$ a. None   b. $y = 1$

7. Which of the following functions has a graph that does not have a vertical asymptote? Why?

   a. $f(x) = \dfrac{x - 3}{x^2 - 4}$ b. $f(x) = \dfrac{3}{x^2 - 4x}$

   c. $f(x) = \dfrac{3x^2 - 6}{x^2 + 6}$

   d. $f(x) = \dfrac{x + 4}{(x - 6)(x + 2)}$

   c; no value of x makes the denominator 0.

8. Which of the following functions has a graph that does not have a horizontal asymptote? Why?

   a. $f(x) = \dfrac{2x + 3}{x - 4}$ b. $f(x) = \dfrac{5x}{x^2 - 16}$

   c. $f(x) = \dfrac{x^3}{3x^2 + 2}$

   d. $f(x) = \dfrac{5}{(x + 2)(x - 3)}$

   c; degree of numerator > degree of denominator

In Exercises 9–14, match each function with its graph.

9. $y = \dfrac{3}{x - 1}$ E

10. $y = \dfrac{x + 1}{x - 2}$ B

11. $y = \dfrac{x - 3}{x^2 - 2x - 8}$ F

12. $y = \dfrac{3}{x^2 - 4}$ A

13. $y = \dfrac{-3x^2 + 2x}{x - 3}$ C

14. $y = \dfrac{2x^2 - 3}{x - 4}$ D

A.

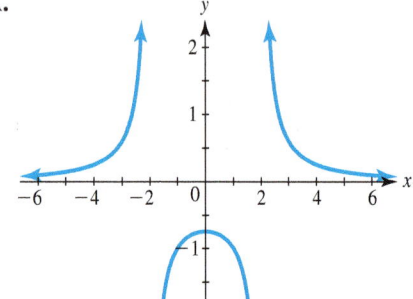

B.

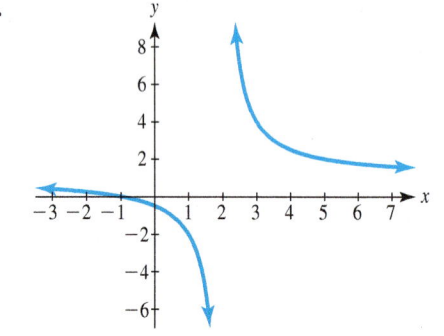

C.

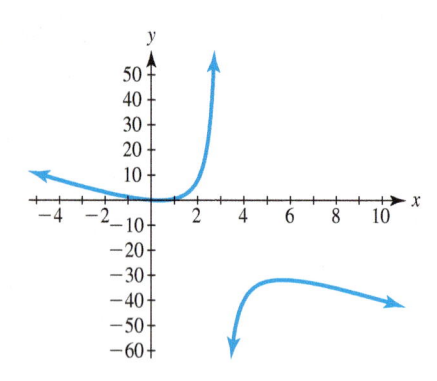

D.

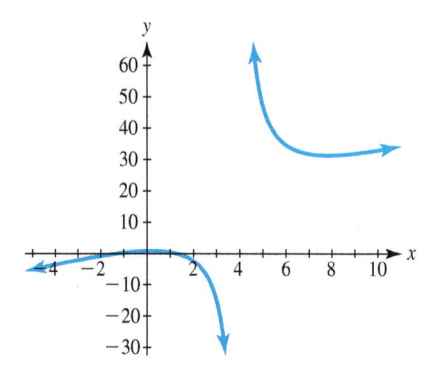

E.

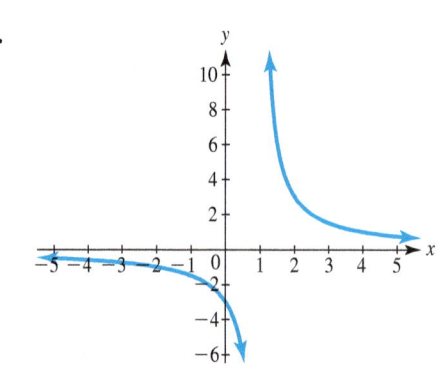

F.

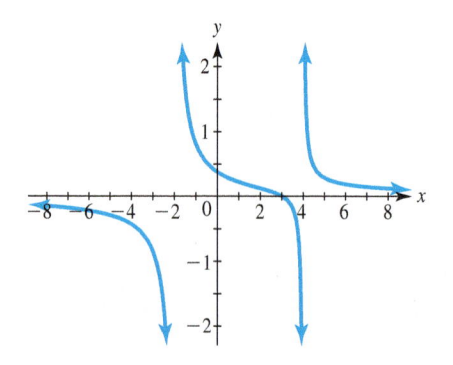

In Exercises 15–18, (a) find the horizontal asymptotes, (b) find the vertical asymptotes, and (c) sketch a graph of the function.

15. $f(x) = \dfrac{x + 1}{x - 2}$
    a. $y = 1$ b. $x = 2$

16. $f(x) = \dfrac{5x}{x - 3}$
    a. $y = 5$ b. $x = 3$

17. $f(x) = \dfrac{5 + 2x}{1 - x^2}$
    a. $y = 0$ b. $x = -1, x = 1$

18. $f(x) = \dfrac{2x^2 + 1}{2 - x}$
    a. None   b. $x = 2$

19. Graph the function $y = \dfrac{x^2 - 9}{x - 3}$. What happens at $x = 3$? A hole in the graph

20. Graph the function $y = \dfrac{x^2 - 16}{x + 4}$. What happens at $x = -4$? A hole in the graph

In Exercises 21–24, find the equations of any vertical, horizontal, and slant asymptotes. Then graph the function.

21. $f(x) = \dfrac{x^2 - 4}{x + 1}$ Vertical: $x = -1$; horizontal: none; slant: $y = x - 1$

22. $f(x) = \dfrac{x^2 - x - 6}{x + 3}$ Vertical: $x = -3$; horizontal: none; slant: $y = x - 4$

23. $f(x) = \dfrac{x^2 - x + 2}{x - 1}$ Vertical: $x = 1$; horizontal: none; slant: $y = x$

24. $f(x) = \dfrac{x^3 - 1}{x^2 - 9}$ Vertical: $x = 3, x = -3$; horizontal: none; slant: $y = x$

In Exercises 25–28, use graphical methods to find any turning points of the graph of the function.

25. $f(x) = \dfrac{x^2}{x - 1}$ 26. $f(x) = \dfrac{(x-1)^2}{x}$
    $(0, 0)$ and $(2, 4)$ $(-1, -4)$ and $(1, 0)$

27. $g(x) = \dfrac{x^2 + 4}{x}$ 28. $g(x) = \dfrac{x}{x^2 + 1}$
    $(-2, -4)$ and $(2, 4)$ $(-1, -0.5)$ and $(1, 0.5)$

29. a. Graph $y = \dfrac{1 - x^2}{x - 2}$ on the window $[-10, 10]$ by $[-10, 10]$.

    b. Use the graph to find y when $x = 1$ and when $x = 3$. $y = 0, y = -8$

    c. Use the graph to find the value(s) of x that give $y = -7.5$. $x = 4, x = 3.5$

    d. Use algebraic methods to solve $-7.5 = \dfrac{1 - x^2}{x - 2}$. $x = 4, x = 3.5$

30. a. Graph $y = \dfrac{2 + 4x}{x^2 + 1}$ on the window $[-10, 10]$ by $[-5, 5]$.

    b. Use the graph to find y when $x = -3$ and when $x = 3$. $y = -1, y = 1.4$

    c. Use the graph to find the value(s) of x that give $y = 2$. $x = 0, x = 2$

    d. Use algebraic methods to solve $2 = \dfrac{2 + 4x}{x^2 + 1}$. $x = 0, x = 2$

31. a. Graph $y = \dfrac{3 - 2x}{x}$ on the window $[-10, 10]$ by $[-10, 10]$.

    b. Use the graph to find y when $x = -3$ and when $x = 3$. $y = -3, y = -1$

    c. Use the graph to find the value(s) of x that give $y = -5$. $x = -1$

    d. Use algebraic methods to solve $-5 = \dfrac{3 - 2x}{x}$. $x = -1$

32. a. Graph $y = \dfrac{x^2}{(x + 1)^2}$ on the window $[-10, 10]$ by $[-6, 6]$.

    b. Use the graph to find y when $x = 0$ and when $x = -2$. $y = 0, y = 4$

    c. Use the graph to find the value(s) of x that give $y = \dfrac{1}{4}$. $x = -\dfrac{1}{3}, x = 1$

    d. Use algebraic methods to solve $\dfrac{1}{4} = \dfrac{x^2}{(x + 1)^2}$. $x = -\dfrac{1}{3}, x = 1$

33. Use algebraic methods to solve
    $\dfrac{x^2 + 1}{x - 1} + x = 2 + \dfrac{2}{x - 1}$. $x = \dfrac{1}{2}$

34. Use algebraic methods to solve
    $\dfrac{x}{x - 2} - x = 1 + \dfrac{2}{x - 2}$. $x = 0$

35. If y varies inversely as the 4th power of x and $y = 5$ when $x = -1$, what is y when $x = 0.5$? 80

36. If S varies inversely as the square root of T and $S = 4$ when $T = 4$, what is S when $T = 16$? 2

EXERCISES 6.5

Answers that are not seen can be found in the answer section at the back of the text.

37. **Average Cost** The average cost per unit for the production of a certain brand of DVD players is given by

$$\overline{C} = \dfrac{400 + 50x + 0.01x^2}{x}$$

where x is the number of units produced.

a. What is the average cost per unit when 500 units are produced? $55.80

b. What is the average cost per unit when 60 units are produced? $57.27

c. What is the average cost per unit when 100 units are produced? $55

d. Is it reasonable to say that the average cost continues to fall as the number of units produced rises? No; for example, at 600 units the average cost is $56.67 per unit.

38. **Average Cost** The average cost per set for the production of a certain brand of television sets is given by

$$\overline{C}(x) = \frac{1000 + 30x + 0.1x^2}{x}$$

where x is the number of units produced.

a. What is the average cost per set when 30 sets are produced? $66.33

b. What is the average cost per set when 300 sets are produced? $63.33

c. What happens to the function when $x = 0$? What does this tell you about the average cost when 0 units are produced? Undefined

39. **Advertising and Sales** The monthly sales volume y (in thousands of dollars) of a product is related to monthly advertising expenditures x (in thousands of dollars) according to the equation

$$y = \frac{400x}{x + 20}$$

a. What monthly sales will result if $5000 is spent monthly on advertising? $80,000

b. What value of x makes the denominator of this function 0? Will this ever happen in the context of this problem? −20; no

40. **Productivity** As an 8-hour day shift progresses, the rate at which workers produce picture frames, in units per hour, changes according to the equation

$$f(t) = \frac{100(t^2 + 3t)}{(t^2 + 3t + 12)^2} \quad (0 \le t \le 8)$$

where t is the number of hours after the beginning of the shift.

a. Graph this function using the window $[-10, 10]$ by $[-5, 5]$.

b. Graph this function using the window $[0, 8]$ by $[0, 5]$.

c. Does the graph in part (a) or part (b) better represent the rate of change of productivity function? Why? Part (b); it displays the function over its domain of $0 \le t \le 8$.

d. Is the rate of productivity higher near lunch ($t = 4$) or near quitting time ($t = 8$)? Near lunch

41. **Average Cost** The average cost per set for the production of television sets is given by

$$\overline{C}(x) = \frac{1000 + 30x + 0.1x^2}{x}$$

where x is the number of hundreds of units produced.

a. Graph this function using the window $[-10, 10]$ by $[-200, 300]$.

b. Graph this function using the window $[0, 50]$ by $[0, 300]$.

c. Which window makes sense for this application? Window in part (b)

d. Use the graph with the window $[0, 250]$ by $[0, 300]$ to find the minimum average cost and the number of units that gives the minimum average cost. $50; 10,000 units

42. **Average Cost** The average cost per unit for the production of a certain brand of DVD players is given by

$$\overline{C}(x) = \frac{400 + 50x + 0.01x^2}{x}$$

where x is the number of hundreds of units produced.

a. Graph this function using the window $[-100, 100]$ by $[-100, 400]$.

b. Graph this function using the window $[0, 300]$ by $[0, 400]$.

c. Which window is more appropriate for this problem? Window in part (b)

d. Use the graph to find the minimum average cost and the number of units that gives the minimum average cost. $54; 20,000 units

43. **Population** Suppose the number of employees of a start-up company is given by

$$f(t) = \frac{30 + 40t}{5 + 2t}$$

where t is the number of months after the company is organized.

a. Graph this function using the window $[0, 20]$ by $[0, 20]$.

b. Use the graph to find $f(0)$. What does this represent? 6; employees when the company starts

c. Use the graph to find $f(12)$. What does this represent? If $f(t)$ represents a number of individuals, how should you report your answer? 17.586; employees after 12 months; approximately 18 employees

44. **Drug Concentration** Suppose the concentration of a drug (as a percent) in a patient's bloodstream t hours after injection is given by

$$C(t) = \frac{200t}{2t^2 + 32}$$

a. Graph the function using the window $[0, 20]$ by $[0, 20]$.

b. What is the drug concentration 1 hour after injection? 5 hours? 5.88%; 12.20%

c. What is the highest percent concentration? In how many hours will it occur? 12.5%; in 4 hours

d. Describe how the end behavior of the graph of this function relates to the drug concentration.

45. **Cost–Benefit** Suppose the cost C of removing $p\%$ of the impurities from the waste water in a manufacturing process is given by

$$C(p) = \frac{3600p}{100 - p}$$

  a. Where does the graph of this function have a vertical asymptote? $p = 100$

  b. What does this tell us about removing the impurities from this process?
  It is impossible to remove 100% of the impurities.

46. **Cost–Benefit** The percent p of particulate pollution that can be removed from the smokestacks of an industrial plant by spending C dollars is given by

$$p = \frac{100C}{8300 + C}$$

  a. Find the percent of pollution that could be removed if spending were allowed to increase without bound. It approaches 100%.

  b. Can 100% of the pollution be removed? Explain.
  No; one can't spend an infinite amount of money.

47. **Sales Volume** Suppose the weekly sales volume (in thousands of units) for a product is given by

$$V = \frac{640}{(p + 2)^2}$$

  where p is the price in dollars per unit.

  a. Graph this function on the p-interval $[-10, 10]$. Does the graph of this function have a vertical asymptote on this interval? Where? Yes; $p = -2$

  b. Complete the table below.

Price/Unit ($)	5	20	50	100	200	500
Weekly Sales Volume						

  c. What values of p give a weekly sales volume that makes sense? Does the graph of this function have a vertical asymptote on this interval?
  Nonnegative values of p; no

  d. What is the horizontal asymptote of the graph of the weekly sales volume? Explain what this means. $V = 0$; weekly sales approach 0 as price increases.

48. **Population** Suppose the number of employees of a start-up company is given by

$$N = \frac{30 + 40t}{5 + 2t}$$

  where t is the number of months after the company is organized.

  a. Does the graph of this function have a vertical asymptote on $[-10, 10]$? Where? Yes; $t = -2.5$

  b. Does the graph of this function have a vertical asymptote for values of t in this application? No

  c. Does the graph of this function have a horizontal asymptote? Where? Yes; $N = 20$

  d. What is the maximum number of employees predicted by this model? It approaches 20.

49. **Demand** The quantity of a product demanded by consumers is defined by the function

$$p = \frac{100{,}000}{(q + 1)^2}$$

  where p is the price and q is the quantity demanded.

  a. Graph this function using the window $[0, 100]$ by $[0, 1000]$.

  b. What is the horizontal asymptote that this graph approaches? $p = 0$

  c. Use the graph to explain what happens to quantity demanded as price becomes lower.
  As the price drops, the quantity demanded increases.

50. **Advertising and Sales** Weekly sales y (in hundreds of dollars) are related to weekly advertising expenses x (in hundreds of dollars) according to the equation

$$y = \frac{800x}{20 + 5x}$$

  a. Graph this function using the window $[0, 100]$ by $[0, 300]$.

  b. Find the horizontal asymptote for this function. $y = 160$

  c. Complete the table below.

Weekly Expenses ($)	0	50	100	200	300	500
Weekly Sales						

  d. Use your results from parts (a) and (b) to determine the maximum weekly sales even if an unlimited amount of money is spent on advertising. $16,000

51. **Sales and Training** During the first 3 months of employment, the monthly sales S (in thousands of dollars) for an average new salesperson depend on the number of hours of training x, according to

$$S = \frac{40}{x} + \frac{x}{4} + 10 \quad \text{for} \quad x \geq 4$$

  a. Combine the terms of this function over a common denominator to create a rational function.
  $S = (x^2 + 40x + 160)/4x$

  b. How many hours of training should result in monthly sales of $21,000? 4 hours or 40 hours

52. **Sales and Training** The average monthly sales volume (in thousands of dollars) for a company depends

on the number of hours of training x of its sales staff, according to

$$S(x) = \frac{20}{x} + 40 + \frac{x}{2} \quad \text{for} \quad 4 \leq x \leq 120$$

a. Graph this function.

b. How many hours of training will give average monthly sales of $51,000? 20 hours

53. *Worker Productivity* Suppose the average time (in hours) that a new production team takes to assemble 1 unit of a product is given by

$$H = \frac{5 + 3t}{2t + 1}$$

where t is the number of days of training for the team.

a. Graph this function using the window $[0, 20]$ by $[0, 8]$.

b. What is the horizontal asymptote? What does this mean in this application? $H = \frac{3}{2}$

c. Use a graphical or numerical solution method to find the number of days of training necessary to reduce the production time to 1.6 hours. 17 days

54. *Advertising and Sales* Weekly sales y (in hundreds of dollars) are related to weekly advertising expenses x (in hundreds of dollars) according to the equation

$$y = \frac{800x}{20 + 5x}$$

Use graphical or numerical methods to find the amount of weekly advertising expenses that will result in weekly sales of $14,000, according to this model. $2800

55. *Per Capita Expenditures for U.S. Health Care* The table shows the dollars spent per person for health care in the United States for selected years from 2002 and projected to 2024. These data can be modeled by the rational function

$$y = \frac{4.6(x + 10)^2 - 49(x + 10) + 1600}{0.0026x + 0.28}$$

where x is the number of years after 2000 and y is the per capita expenditures for health care.

a. Graph the function on a window representing the years from 2000 through 2025.

b. What is the predicted per capita expenditure for health care in 2033? $21,865.50

c. In what year does the model predict that the per capita expenditure for health care is $16,000? 2025

Year	Cost ($)	Year	Cost ($)
2002	5563	2014	9695
2004	6331	2016	10,527
2006	7091	2018	11,499
2008	7944	2020	12,741
2010	8428	2022	14,129
2012	8996	2024	15,618

(*Source*: U.S. Centers for Medicare and Medicaid Services)

56. *Fences* Suppose that a rectangular field is to have an area of 51,200 square feet and that it needs to be enclosed by fence on three of its four sides (see the figure). What is the minimum length of fence needed? What dimensions should this field have to minimize the amount of fence needed? To solve this problem,

a. Write an equation that describes the area of the proposed field, with x representing the length and y representing the width of the field. $xy = 51,200$

b. Write an equation that gives the length of fence L as a function of the dimensions of the field, remembering that only three sides are fenced in. $L = x + 2y$

c. Use the equation from part (a) to write the length L as a function of one of the dimensions.

d. Graph the function L on a window that applies to the context of the problem: $[0, 400]$ by $[0, 1000]$.

e. Use graphical or numerical methods to find the minimum length of fence needed and the dimensions required to give this length. $L = 640$ feet; $x = 320$, $y = 160$

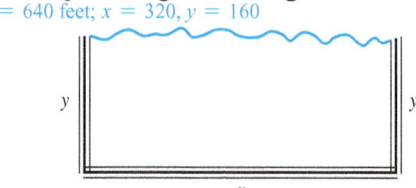

57. *Females in the Workforce* For selected years from 1950 and projected to 2050, the following table shows the percent of total U.S. workers who were female.

Year	% Female	Year	% Female
1950	29.6	2010	47.9
1960	33.4	2015	48.3
1970	38.1	2020	48.1
1980	42.5	2030	48.0
1990	45.2	2040	47.9
2000	46.6	2050	47.7

(*Source*: U.S. Bureau of Labor Statistics)

Assume these data can be modeled with the function

$$p(t) = \frac{78.6t + 2090}{1.38t + 64.1}$$

where t is the number of years after 1950.

a. What does this model project the percent to be in 2030? Does this agree with the data? 48.0%; yes

b. What is the maximum possible percent of women in the workforce, according to this model? 57%

c. In what year does the model indicate that the percent will reach 47.4%? 2022

58. **Wind Chill** If x is the wind speed in miles per hour and is greater than or equal to 5, then the wind chill (in degrees Fahrenheit) for an air temperature of 0°F can be approximated by the function

$$f(x) = \frac{289.173 - 58.5731x}{x + 1}, \quad x \geq 5$$

a. Ignoring the restriction $x \geq 5$, does $f(x)$ have a vertical asymptote? If so, what is it? Yes; $x = -1$

b. Does $f(x)$ have a vertical asymptote within its domain? No

c. Does $f(x)$ have a horizontal asymptote? If so, what is it? Yes; $y = -58.5731$

59. **Investing** The present value that will give a future value S in 3 years with interest compounded annually varies inversely as the cube of $1 + r$, where r is the annual interest rate. If the present value of $8396.19 gives a future value of $10,000, what would be the present value of $16,500?

60. **Investing** The present value that will give a future value S in 4 years with interest compounded annually varies inversely as the fourth power of $1 + r$, where r is the annual interest rate. If the present value of $1525.79 gives a future value of $2000, what would be the present value of $8000?

61. **Electrical Resistance** The electrical resistance of a wire varies directly as its length in feet and inversely as the square of its diameter in inches. If the resistance of 640 feet of copper wire 1/4 inch in diameter is 16 ohms, find the resistance of 320 feet of the same type of copper wire 1/2 inch in diameter. 2 ohms

62. **Volume of Gas** The volume of gas varies directly as the temperature and inversely as the pressure. If the volume is 230 cubic centimeters when the temperature is 300°K and the pressure is 20 pounds per square centimeter, what is the volume when the temperature is 270°K and the pressure is 30 pounds per square centimeter? 138 cm³

63. **Centrifugal Force** The centrifugal force of an object moving in a circle varies jointly with the radius of the circular path and the mass of the object and inversely as the square of the time it takes to move about one full circle. A 6-gram object moving in a circle with a radius of 75 centimeters at a rate of 1 revolution every 3 seconds has a centrifugal force of 5000 dynes. Find the centrifugal force of a 14-gram object moving in a circle with radius 125 centimeters at a rate of 1 revolution every 2 seconds. 43,750 dynes

64. **Machine Assembly** The number of hours h that it takes m men to assemble x machines varies directly as the number of machines and inversely as the number of men. If four men can assemble 12 machines in four hours, how many men are needed to assemble 36 machines in eight hours? 6 men

6.6 Polynomial and Rational Inequalities

KEY OBJECTIVES

- Solve polynomial inequalities using graphical and algebraic solution methods
- Solve rational inequalities using graphical and algebraic solution methods

SECTION PREVIEW Average Cost

The average cost per set for the production of 42-inch plasma televisions is given by

$$\overline{C}(x) = \frac{5000 + 80x + x^2}{x}$$

where x is the number of hundreds of units produced. To find the number of televisions that must be produced to keep the average cost to at most $590 per TV, we solve the rational inequality

$$\frac{5000 + 80x + x^2}{x} \leq 590$$

(See Example 2.) In this section, we use algebraic and graphical methods similar to those used in Section 4.4 to solve inequalities involving polynomial functions and rational functions. ■

Polynomial Inequalities

The quadratic inequalities studied in Section 4.4 are examples of polynomial inequalities. Other examples of polynomial inequalities are

$$3x^3 + 3x^2 - 4x \geq 2, \qquad x - 3 < 8x^4, \qquad x \geq (x - 2)^3$$

We can solve the inequality $144x - 48x^2 + 4x^3 > 0$ with a graphing utility. The graph of the function $y = 144x - 48x^2 + 4x^3$ is shown in Figure 6.49.

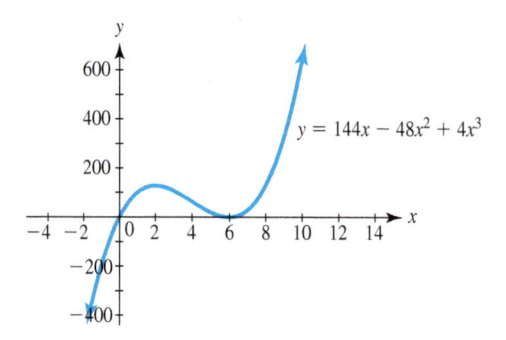

Figure 6.49

We can determine that the x-intercepts of the graph are $x = 0$ and $x = 6$. The graph touches but does not cross the x-axis at $x = 6$, and the graph of the function is above the x-axis for values of $x > 0$ and $x \neq 6$. The solution to $144x - 48x^2 + 4x^3 > 0$ is the set of all x that give positive values for this function: $0 < x < 6$ and $x > 6$, or x in the intervals $(0, 6)$ and $(6, \infty)$.

The steps used to solve polynomial inequalities algebraically are similar to those used to solve quadratic inequalities.

Solving Polynomial Inequalities

To solve a polynomial inequality algebraically:

1. Write an equivalent inequality with 0 on one side and with the function $f(x)$ on the other side.

2. Solve $f(x) = 0$.

3. Create a sign diagram that uses the solutions from step 2 to divide the number line into intervals. Pick a test value in each interval to determine whether $f(x)$ is positive or negative in that interval.*

4. Identify the intervals that satisfy the inequality in step 1. The values of x that define these intervals are the solutions to the original inequality.

---

*The numerical feature of your graphing utility can be used to test the x-values.

6.6 Polynomial and Rational Inequalities

We illustrate the use of this method in the following example.

EXAMPLE 1 ▶ Constructing a Box

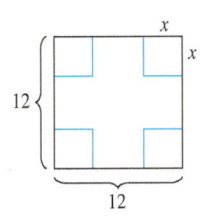

A box can be formed by cutting a square out of each corner of a piece of tin and folding the sides up. If the piece of tin is 12 inches by 12 inches and each side of the square that is cut out has length x, the function that gives the volume of the box is $V(x) = 144x - 48x^2 + 4x^3$.

a. Use factoring and then find the values of x that give positive values for $V(x)$.

b. Which of the values of x that give positive values for $V(x)$ result in a box?

SOLUTION

a. We seek those values of x that give positive values for $V(x)$—that is, for which

$$144x - 48x^2 + 4x^3 > 0$$

Writing the related equation $V(x) = 0$ and solving for x gives

$$144x - 48x^2 + 4x^3 = 0$$
$$4x(x^2 - 12x + 36) = 0$$
$$4x(x - 6)(x - 6) = 0$$
$$x = 0 \quad \text{or} \quad x = 6$$

These two values of x, 0 and 6, divide the number line into three intervals. Testing a value in each of the intervals (Figure 6.50(a)) determines the intervals where $V(x) > 0$ and thus determines the solutions to the inequality. The values of x that give positive values for V satisfy $0 < x < 6$ and $x > 6$. We found the same solutions to this inequality using graphical methods (see Figure 6.49) on the previous page.

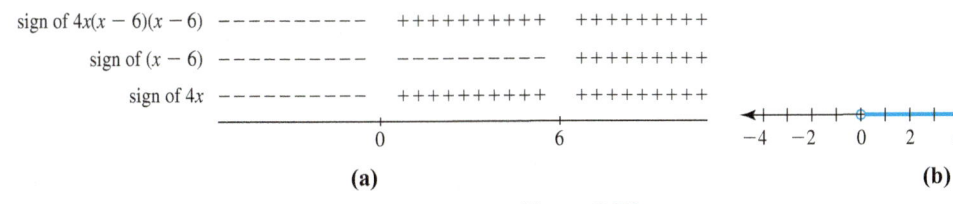

Figure 6.50

The interval notation for this solution set is $(0, 6) \cup (6, \infty)$, and the graph of the solution set is shown in Figure 6.50(b).

b. The box is formed by cutting 4 squares of length x inches from a piece of tin that is 12 inches by 12 inches, so it is impossible to cut squares longer than 6 inches. Thus, the solution $x > 6$ does not apply to the physical building of the box, and so the values of x that result in a box satisfy $0 < x < 6$.

Rational Inequalities

To solve a **rational inequality** using an algebraic method, we first get 0 on the right side of the inequality; then, if necessary, we get a common denominator and combine the rational expressions on the left side. We find the numbers that make the numerator of the rational expression equal to zero and those that make the denominator equal to zero. These numbers are used to create a sign diagram much like the one we used when solving polynomial inequalities. Note that we should avoid multiplying both sides of an inequality by any term containing a variable, because we cannot easily determine when the term is positive or negative.

EXAMPLE 2 ▶ Average Cost

The average cost per set for the production of 42-inch plasma televisions is given by

$$\overline{C}(x) = \frac{5000 + 80x + x^2}{x}$$

where x is the number of hundreds of units produced. Find the number of televisions that must be produced to keep the average cost to at most $590 per TV.

SOLUTION

To find the number of televisions that must be produced to keep the average cost to at most $590 per TV, we solve the rational inequality

$$\frac{5000 + 80x + x^2}{x} \leq 590$$

To solve this inequality with algebraic methods, we rewrite the inequality with 0 on the right side of the inequality, getting

$$\frac{5000 + 80x + x^2}{x} - 590 \leq 0$$

Combining the terms over a common denominator and factoring gives

$$\frac{5000 + 80x + x^2 - 590x}{x} \leq 0$$

$$\frac{x^2 - 510x + 5000}{x} \leq 0$$

$$\frac{(x - 10)(x - 500)}{x} \leq 0$$

The values of x that make the numerator equal to 0 are 10 and 500, and the value $x = 0$ makes the denominator equal to 0. We cannot have a negative number of units, so the average cost is defined only for values of $x > 0$. The values 0, 10, and 500 divide a number line into three intervals. Next we find the sign of $f(x) = \frac{(x - 10)(x - 500)}{x}$ in each interval. Testing a value in each interval determines which interval(s) satisfy the inequality. We do this by using the sign diagram in Figure 6.51.

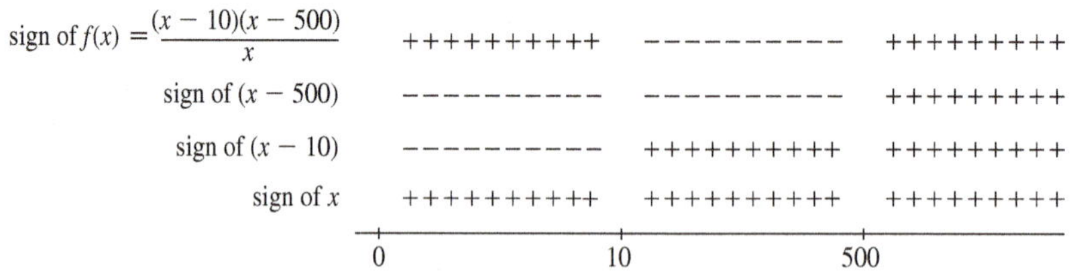

sign of $f(x) = \frac{(x - 10)(x - 500)}{x}$ ++++++++++   ----------   ++++++++++

sign of $(x - 500)$ ----------   ----------   ++++++++++

sign of $(x - 10)$ ----------   ++++++++++   ++++++++++

sign of x ++++++++++   ++++++++++   ++++++++++

          0   10   500

Figure 6.51

The function $f(x)$ is negative on $10 < x < 500$, so the solution to

$$\frac{5000 + 80x + x^2}{x} - 590 \leq 0$$

and thus to the original inequality, is $10 \leq x \leq 500$. Because x represents the number of hundreds, between 1000 and 50,000 televisions must be produced to keep the average cost to at most $590 per TV.

6.6 Polynomial and Rational Inequalities

We can also use graphical methods to solve rational inequalities. For example, we can solve the inequality

$$\frac{5000 + 80x + x^2}{x} \leq 590$$

for $x > 0$ by graphing

$$y_1 = \frac{5000 + 80x + x^2}{x} \quad \text{and} \quad y_2 = 590$$

on the same axes. We are only interested in the graph for positive values of x, so we restrict our graph to the first quadrant. The points of intersection can be found by the intersection method (Figure 6.52). The inequality is satisfied by the x-interval for which the graph of y_1 is below (or on) the graph of y_2—that is, for $10 \leq x \leq 500$. Recall that this is the solution that we found using algebraic methods in Example 2.

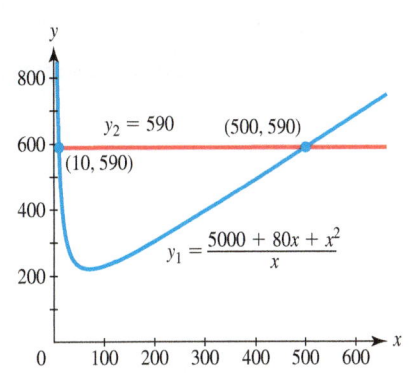

Figure 6.52

SKILLS CHECK 6.6

Answers that are not seen can be found in the answer section at the back of the text.

In Exercises 1–16, use algebraic and/or graphical methods to solve the inequalities.

1. $16x^2 - x^4 \geq 0$ $-4 \leq x \leq 4$
2. $x^4 - 4x^2 \leq 0$ $-2 \leq x \leq 2$
3. $2x^3 - x^4 < 0$ $x < 0$ or $x > 2$
4. $3x^3 \geq x^4$ $0 \leq x \leq 3$
5. $(x - 1)(x - 3)(x + 1) \geq 0$ $-1 \leq x \leq 1$ or $x \geq 3$
6. $(x - 3)^2(x + 1) < 0$ $x < -1$
7. $\dfrac{4 - 2x}{x} > 2$ $0 < x < 1$
8. $\dfrac{x - 3}{x + 1} \geq 3$ $-3 \leq x < -1$
9. $\dfrac{x}{2} + \dfrac{x - 2}{x + 1} \leq 1$ $x \leq -3$ or $-1 < x \leq 2$
10. $\dfrac{x}{x - 1} \leq 2x + \dfrac{1}{x - 1}$ $x \geq 0.5$ and $x \neq 1$
11. $(x - 1)^3 > 27$ $x > 4$
12. $(2x + 3)^3 \leq 8$ $x \leq -\dfrac{1}{2}$
13. $(x - 1)^3 < 64$ $x < 5$
14. $(x + 4)^3 - 125 \geq 0$ $x \geq 1$
15. $-x^3 - 10x^2 - 25x \leq 0$ $x = -5, x \geq 0$
16. $x^3 + 10x^2 + 25x < 0$ $x < 0$ and $x \neq -5$

For Exercises 17–20, use the graph of $y = f(x)$ to solve the requested inequality.

17.   **a.** Solve $f(x) < 0$. $x < -3$ or $0 < x < 2$
     b. Solve $f(x) \geq 0$. $-3 \leq x \leq 0$ or $x \geq 2$

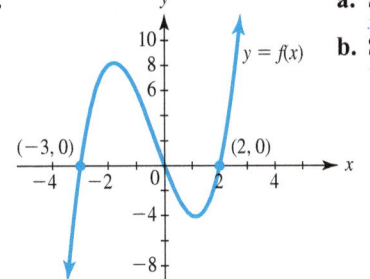

18.   **a.** Solve $f(x) < 0$. $x < 3$
     b. Solve $f(x) \geq 0$. $x \geq 3$

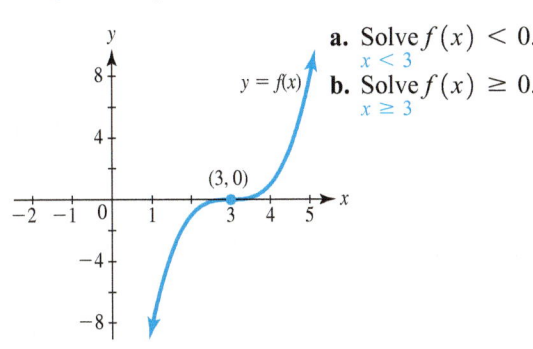

522 Chapter 6 Higher-Degree Polynomial and Rational Functions

19. 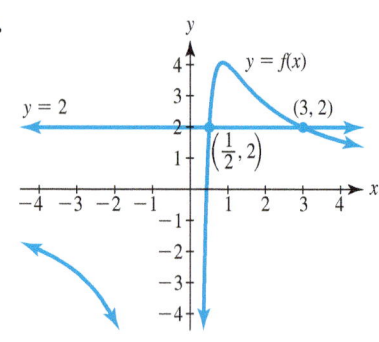 Solve $f(x) \geq 2$.
$\frac{1}{2} \leq x \leq 3$

20. 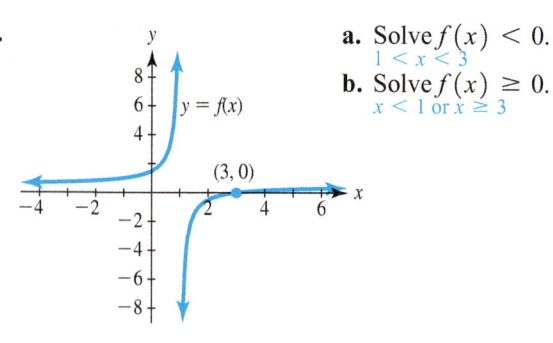
a. Solve $f(x) < 0$.
$1 < x < 3$
b. Solve $f(x) \geq 0$.
$x < 1$ or $x \geq 3$

EXERCISES 6.6

Combine factoring with graphical and/or numerical methods to solve Exercises 21–26.

21. Revenue The revenue from the sale of a product is given by the function $R = 400x - x^3$. Selling how many units will give positive revenue?
$0 < x < 20$; more than 0 and less than 20 units

22. Revenue The price for a product is given by $p = 1000 - 0.1x^2$, where x is the number of units sold.

a. Form the revenue function for the product.
$R(x) = 1000x - 0.1x^3$
b. Selling how many units gives positive revenue?
$0 < x < 100$; more than 0 and less than 100 units

23. Constructing a Box A box can be formed by cutting a square out of each corner of a piece of cardboard and folding the sides up. If the piece of cardboard is 36 cm by 36 cm and each side of the square that is cut out has length x cm, the function that gives the volume of the box is $V = 1296x - 144x^2 + 4x^3$.

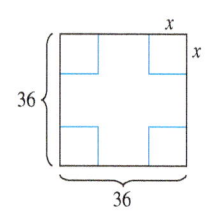

a. Find the values of x that make $V > 0$.
$0 < x < 18$ or $x > 18$
b. What size squares can be cut out to construct a box? More than 0 and less than 18 cm

24. Constructing a Box A box can be formed by cutting a square out of each corner of a piece of cardboard and folding the sides up. If the piece of cardboard is 12 inches by 16 inches and each side of the square that is cut out has length x inches, the function that gives the volume of the box is $V = 192x - 56x^2 + 4x^3$. What size squares can be cut out to construct a box?
$0 < x < 6$; more than 0 and less than 6 inches

25. Cost The total cost function for a product is given by $C(x) = 3x^3 - 6x^2 - 300x + 1800$, where x is the number of units produced and C is the cost in hundreds of dollars. Use factoring by grouping and then find the number of units that will give a total cost of at least $120,000. Verify your conclusion with a graphing utility.
$0 \leq x \leq 2$ or $x \geq 10$; between 0 and 2 units or at least 10 units

26. Profit The profit function for a product is given by $P(x) = -x^3 + 2x^2 + 400x - 400$, where x is the number of units produced and sold and P is in hundreds of dollars. Use factoring by grouping to find the number of units that will give a profit of at least $40,000. Verify your conclusion with a graphing utility.
$2 \leq x \leq 20$; between 2 and 20 units, inclusive

27. Advertising and Sales The monthly sales volume y (in thousands of dollars) is related to monthly advertising expenditures x (in thousands of dollars) according to the equation

$$y = \frac{400x}{x + 20}$$

Spending how much money on advertising will result in sales of at least $200,000 per month?
$x \geq 20$; at least $20,000

28. Average Cost The average cost per set for the production of a portable stereo system is given by

$$\overline{C} = \frac{100 + 30x + 0.1x^2}{x}$$

where x is the number of hundreds of units produced. What number of units can be produced while keeping the average cost to at most $41?
$10 \leq x \leq 100$; between 1000 and 10,000 units, inclusive

29. Future Value The future value of $2000 invested for 3 years at rate r, compounded annually, is given by $S = 2000(1 + r)^3$. Find the rate r that gives a future value from $2662 to $3456, inclusive.
$0.10 \leq r \leq 0.20$; between 10% and 20%

30. *Future Value* The future value of $5000 invested for 4 years at rate r, compounded annually, is given by $S = 5000(1 + r)^4$. Find the rate r that gives a future value of at least $6553.98. $r \geq 0.07$; at least 7%

31. *Revenue* The revenue from the sale of a product is given by the function $R = 4000x - 0.1x^3$ dollars. Use graphical or numerical methods to determine how many units should be sold to give a revenue of at least $39,990. $10 \leq x \leq 194$; between 10 and 194 units

32. *Revenue* The price for a product is given by $p = 1000 - 0.1x^2$, where x is the number of units sold.

  a. Form the revenue function for the product.
  $R(x) = 1000x - 0.1x^3$
  b. Use graphical or numerical methods to determine how many units should be sold to give a revenue of at most $37,500.
  Between 0 and 50 units, inclusive, or more than 65 units

33. *Supply and Demand* Suppose the supply function for a product is given by $6p - q = 180$ and the demand is given by $(p + 20)q = 30,000$. Over what meaningful price interval does supply exceed demand?
$p > 80$; above $80 per unit

34. *Drug Concentration* A pharmaceutical company claims that the concentration of a drug in a patient's bloodstream will be at least 10% for 8 hours. Suppose clinical tests show that the concentration of a drug (as percent) t hours after injection is given by

$$C(t) = \frac{200t}{2t^2 + 32}$$

During what time period is the concentration at least 10%? Is the company's claim supported by the evidence? Between 2 and 8 hours, inclusive; no

35. *Population* Suppose the number of employees of a start-up company is given by

$$f(t) = \frac{30 + 40t}{5 + 2t}$$

where t is the number of months after the company is organized.

  a. For what values of t is $f(t) \leq 18$? $0 \leq t \leq 15$
  b. During what months is the number of employees below 18? First 15 months

Preparing for CALCULUS

To take the derivative of a function or integrate a function in calculus, it is sometimes necessary or useful to rewrite a rational function as a sum of terms containing powers of x. Use division to do this for the following functions. Use negative exponents where necessary so no variables are in the denominator.

1. $f(x) = \dfrac{x^3 + 5x^2 - 4x}{x}$
   $f(x) = x^2 + 5x - 4$

2. $f(x) = \dfrac{4x^3 - 2x^2 + 8x}{2x}$
   $f(x) = 2x^2 - x + 4$

3. $g(x) = \dfrac{3x^3 + 9x^2 - 12x - 6}{3x}$
   $g(x) = x^2 + 3x - 4 - 2x^{-1}$

4. $g(x) = \dfrac{x^3 + 9x^2 - 10x + 4}{5}$
   $g(x) = \dfrac{1}{5}x^2 + \dfrac{9}{5}x - 2 + \dfrac{4}{5}x^{-1}$

5. $f(t) = \dfrac{2t^5 - 12t^4 - 10t^2 + 8t}{2t^2}$
   $f(t) = t^3 - 6t^2 - 5 + 4t^{-1}$

6. $f(t) = \dfrac{24t^6 + t^4 - 6t^2 + 12t}{6t^2}$
   $f(t) = 4t^4 + \dfrac{1}{6}t^2 - 1 + 2t^{-1}$

In Exercises 7 and 8, divide by a term containing the variable to remove the term from the denominator.

7. $f(t) = \dfrac{(t+6)^5 + 3(t+6)^4 - 6(t+6)^2}{(t+6)^2}$
   $f(t) = (t+6)^3 + 3(t+6)^2 - 6$

8. $g(t) = \dfrac{(t+3)^6 - 5(t+3)^4 + 6(t+3)^2 + 2}{(t+3)^3}$
   $g(t) = (t+3)^3 - 5(t+3) + 6(t+3)^{-1} + 2(t+3)^{-3}$

To take the derivative of a function or integrate a function in calculus, sometimes it is necessary or convenient to use polynomial division or synthetic division to rewrite a rational function without any variables in the denominator. Do this for the following functions.

9. $f(x) = \dfrac{x^3 + 5x^2 - 8x - 12}{x + 1}$
   $f(x) = x^2 + 4x - 12$

10. $f(x) = \dfrac{2x^3 + x^2 - 8x - 4}{x - 2}$
    $f(x) = 2x^2 + 5x + 2$

11. $f(x) = \dfrac{2x^3 + 4x^2 - 6x + 2}{x - 1}$
    $f(x) = 2x^2 + 6x + 2(x-1)^{-1}$

12. $f(x) = \dfrac{5x^3 + 18x^2 + 7x - 10}{x + 3}$
    $f(x) = 5x^2 + 3x - 2 - 4(x+3)^{-1}$

The derivatives of rational functions can look quite complicated. In Exercises 13–16, a function is given with its derivative. In these exercises, find where the derivative is zero, which may give a turning point for the graph of the function.

13. $f(x) = \dfrac{x+1}{(x+2)^2}$ $f'(x) = \dfrac{(x+2)^2 - 2(x+1)(x+2)}{(x+2)^4}$ $x = 0$

14. $f(x) = \dfrac{x^2 - 9}{x^2 - 4}$ $f'(x) = \dfrac{(x^2 - 4)(2x) - (x^2 - 9)(2x)}{(x^2 - 4)^2}$ $x = 0$

15. $g(x) = \dfrac{(3x+6)^2}{x - 2}$ $g'(x) = \dfrac{(x-2)2(3x+6)(3) - (3x+6)^2(1)}{(x-2)^2}$
    $x = -2, x = 6$

16. $g(x) = \dfrac{(4x+8)^2}{x - 1}$ $g'(x) = \dfrac{(x-1)2(4x+8)(4) - (4x+8)^2(1)}{(x-1)^2}$
    $x = -2, x = 4$

chapter 6 SUMMARY

In this chapter, we studied higher-degree polynomial functions, their graphs, and the solution of polynomial equations by factoring, by the root method, by graphical methods, and by synthetic division. We also studied rational functions, their graphs, vertical asymptotes, horizontal asymptotes, slant asymptotes, and the solution of equations and inequalities containing rational functions, as well as inverse, combined, and joint variation.

Key Concepts and Formulas

6.1 Higher-Degree Polynomial Functions

Polynomial function	A polynomial function of degree n with independent variable x is a function in the form $P(x) = a_n x^n + a_{n-1} x^{n-1} + \cdots + a_1 x + a_0$ with $a_n \neq 0$. Each a_i represents a real number and each n is a positive integer.
Cubic function	A polynomial function of the form $f(x) = ax^3 + bx^2 + cx + d$ with $a \neq 0$.
Quartic function	A polynomial function of the form $f(x) = ax^4 + bx^3 + cx^2 + dx + e$ with $a \neq 0$.
Extrema points	Turning points on the graph of a polynomial function: local minimum, local maximum, absolute minimum, absolute maximum.
Graphs of cubic functions	End behavior: one end opening up and one end opening down. Turning points: 0 or 2
Graphs of quartic functions	End behavior: Both ends opening up or both ends opening down. Turning points: 1 or 3

6.2 Modeling with Cubic and Quartic Functions

Cubic models	If a scatter plot of data indicates that the data may fit close to one of the possible shapes of the graph of a cubic function, it is possible to find a cubic function that is an approximate model for the data.
Quartic models	If a scatter plot of data indicates that the data may fit close to one of the possible shapes of the graph of a quartic function, it is possible to find a quartic function that is an approximate model for the data.
Third differences	For equally spaced inputs, the third differences are constant for data fitting cubic functions exactly.
Fourth differences	For equally spaced inputs, the fourth differences are constant for data fitting quartic functions exactly.

6.3 Solution of Polynomial Equations

Solving polynomial equations	
By factoring	Some polynomial equations can be solved by factoring and using the zero-product property.
By the root method	The real solutions of the equation $x^n = C$ are $x = \pm\sqrt[n]{C}$ if n is even and $C \geq 0$, or $x = \sqrt[n]{C}$ if n is odd.
Factors, zeros, intercepts, and solutions	If $(x - a)$ is a factor of $f(x)$, then a is an x-intercept of the graph of $y = f(x)$, a is a solution to the equation $f(x) = 0$, and a is a zero of the function f.
Estimating solutions	Solving equations derived from real data requires graphical or numerical methods using technology.

6.4 Polynomial Equations Continued; Fundamental Theorem of Algebra

Division of polynomials	If $f(x)$ is a function and we know that $x = a$ is a solution of $f(x) = 0$, then we can write $x - a$ as a factor of $f(x)$. If we divide this factor into the function, the quotient will be another factor of $f(x)$.
Synthetic division	To divide a polynomial by a binomial of the form $x - a$ (or $x + a$), we can use synthetic division to simplify the division process by omitting the variables and writing only the coefficients.
Solving cubic equations	If we know one exact solution of a cubic equation $P(x) = 0$, we can divide by the corresponding factor using synthetic division, obtaining a second quadratic factor that can be used to find any additional solutions of the equation.
Combining graphical and algebraic methods	To find exact solutions to a polynomial equation when one solution can be found graphically, we can use that solution and synthetic division to find a second factor.
Rational Solutions Test	The rational solutions of the polynomial equation $a_n x^n + a_{n-1}x^{n-1} + \cdots + a_1 x + a_0 = 0$ with integer coefficients must be of the form $\frac{p}{q}$, where p is a factor of the constant term a_0 and q is a factor of the leading coefficient a_n.
Fundamental Theorem of Algebra	If $f(x)$ is a polynomial function of degree $n \geq 1$, then f has at least one complex zero.
Complex solutions of polynomial equations	Every polynomial function $f(x)$ of degree $n \geq 1$ has exactly n complex zeros. Some of these zeros may be imaginary, and some may be repeated.

6.5 Rational Functions and Rational Equations

Rational function	The function f is a rational function if $f(x) = \frac{P(x)}{Q(x)}$, where $P(x)$ and $Q(x)$ are polynomials and $Q(x) \neq 0$.
Graphs of rational functions	The graph of a rational function may have vertical asymptotes and/or a horizontal asymptote. The graph may have a "hole" at $x = a$ rather than a vertical asymptote.
Vertical asymptotes	A vertical asymptote occurs in the graph of $f(x) = \frac{P(x)}{Q(x)}$ at those values of x where $Q(x) = 0$ and $P(x) \neq 0$, that is, where the denominator equals 0 but the numerator does not equal 0.

Horizontal asymptotes	Consider the rational function $$f(x) = \frac{P(x)}{Q(x)} = \frac{a_n x^n + \cdots + a_1 x + a_0}{b_m x^m + \cdots + b_1 x + b_0}$$ 1. If $n < m$, a horizontal asymptote occurs at $y = 0$ (the x-axis). 2. If $n = m$, a horizontal asymptote occurs at $y = \dfrac{a_n}{b_m}$. 3. If $n > m$, there is no horizontal asymptote.
Slant asymptote	If the degree of the numerator of a rational function is *one* more than the degree of the denominator, its graph approaches a slant asymptote. The equation of the slant asymptote can be found by dividing the numerator by the denominator, getting a linear function plus a rational expression with a constant numerator. The graph of the original function will have a slant asymptote, which is the graph of the linear function.
Missing point (hole)	If there is a value a such that $Q(a) = 0$ and $P(a) = 0$, then the graph of the function is undefined at $x = a$ but may have a "hole" at $x = a$ rather than an asymptote.
Solution of rational equations	
Algebraically	Multiplying both sides of the equation by the LCD of any fractions in the equation converts the equation to a polynomial equation that can be solved. All solutions must be checked in the original equation because some may be extraneous.
Graphically	Solutions to rational equations can be found or approximated by graphical methods.
Inverse variation	Two variables are inversely proportional if an increase in one results in a decrease in the other. That is, y is inversely proportional to x if $y = \dfrac{k}{x}$.
Combined variation	In combined variation, direct and inverse variation occur together. For example, suppose y varies directly as x and inversely as w; then we would say $y = \dfrac{kx}{w}$.
Joint variation	With joint variation, a variable varies directly as the product of two or more other variables. Isaac Newton's formula for gravitation, $F = G\dfrac{m_1 m_2}{r^2}$, is an example of joint variation.

6.6 Polynomial and Rational Inequalities

Polynomial inequalities	The steps used to solve polynomial inequalities graphically are similar to those used to solve quadratic inequalities. To solve a polynomial inequality algebraically, solve the corresponding polynomial equation and use a sign diagram to complete the solution.
Rational inequalities	To solve a rational inequality, get 0 on the right side of the inequality; then, if necessary, get a common denominator and combine the rational expressions on the left side. Find the numbers that make the numerator of the rational expression equal to 0 and those that make the denominator equal to 0. Use these numbers to create a sign diagram and complete the solution. Graphical methods can be used to find or to verify the solutions.

chapter 6 SKILLS CHECK

Answers that are not seen can be found in the answer section at the back of the text.

1. What is the degree of the polynomial $5x^4 - 2x^3 + 4x^2 + 5$? 4

2. What is the type of the polynomial function $y = 5x^4 - 2x^3 + 4x^2 + 5$? Quartic

3. Graph $y = -4x^3 + 4x^2 + 1$ using the window $[-10, 10]$ by $[-10, 10]$. Is the graph complete on this window? Yes

4. Graph the function $y = x^4 - 4x^2 - 20$ using
   a. The window $[-8, 8]$ by $[-8, 8]$.
   b. A window that gives a complete graph.

5. a. Graph $y = x^3 - 3x^2 - 4$ using a window that shows a local maximum and local minimum.
   b. A local maximum occurs at what point? $(0, -4)$
   c. A local minimum occurs at what point? $(2, -8)$

6. a. Graph $y = x^3 + 11x^2 - 16x + 40$ using the window $[-10, 10]$ by $[0, 100]$.
   b. Is the graph complete? No
   c. Sketch a complete graph of this function.
   d. Find the points where a local maximum and a local minimum occur. Maximum: $(-8, 360)$; minimum: $(0.667, 34.5)$

7. Solve $x^3 - 16x = 0$. $0, -4, 4$

8. Solve $2x^4 - 8x^2 = 0$. $0, 2, -2$

9. Solve $x^4 - x^3 - 20x^2 = 0$. $0, 5, -4$

10. Solve $0 = x^3 - 15x^2 + 56x$ by factoring. $0, 7, 8$

11. Use factoring by grouping to solve $4x^3 - 20x^2 - 4x + 20 = 0$. $-1, 1, 5$

12. Solve $12x^3 - 9x^2 - 48x + 36 = 0$ using factoring by grouping. $\frac{3}{4}, 2, -2$

13. Use technology to find the solutions to $x^4 - 3x^3 - 3x^2 + 7x + 6 = 0$. $-1, 2, 3$

14. Use technology to solve $6x^3 - 59x^2 - 161x + 60 = 0$. $12, -2.5, \frac{1}{3}$

15. Use the root method to solve $(x - 4)^3 = 8$. 6

16. Use the root method to solve $5(x - 3)^4 = 80$. $5, 1$

17. Use synthetic division to divide $4x^4 - 3x^3 + 2x - 8$ by $x - 2$. $4x^3 + 5x^2 + 10x + 22$ R36

18. Use synthetic division to solve $2x^3 + 5x^2 - 11x + 4 = 0$, given that $x = 1$ is one solution. $1, -4, \frac{1}{2}$

19. Find one solution of $3x^3 - x^2 - 12x + 4 = 0$ graphically, and use synthetic division to find any additional solutions. $2, \frac{1}{3}, -2$

20. Find one solution of $2x^3 + 5x^2 - 4x - 3 = 0$ graphically, and use synthetic division to find any additional solutions. $1, -3, -\frac{1}{2}$

21. For the function $y = \dfrac{1 - x^2}{x + 2}$,
    a. Find the x-intercepts and y-intercept if they exist.
    b. Find any horizontal asymptotes and vertical asymptotes that exist. Vertical asymptote: $x = -2$; no horizontal asymptote
    c. Find any slant asymptote that exists. $y = 2 - x$
    d. Sketch the graph of the function.

22. For the function $y = \dfrac{3x - 2}{x - 3}$,
    a. Find the x-intercepts and y-intercept if they exist.
    b. Find any horizontal asymptotes and vertical asymptotes that exist. Horizontal asymptote: $y = 3$; vertical asymptote: $x = 3$
    c. Sketch the graph of the function.

23. Use a graph and technology to find any local maxima and/or minima of the function $y = \dfrac{x^2}{x - 4}$. Maximum: $(0, 0)$; minimum: $(8, 16)$

24. Use a graph and technology to find any local maxima and/or minima of the function $y = \dfrac{x^2 + 3}{1 - x}$. Maximum: $(3, -6)$; minimum: $(-1, 2)$

25. a. Graph $y = \dfrac{1 + 2x^2}{x + 2}$ using the window $[-10, 10]$ by $[-30, 10]$.
    b. Use the graph to find y when $x = 1$ and when $x = 3$. $y = 1, y = 3.8$
    c. Use the graph to find the value(s) of x that give $y = \dfrac{9}{4}$. $x = 2, x = -0.875$
    d. Use an algebraic method to solve $\dfrac{9}{4} = \dfrac{1 + 2x^2}{x + 2}$. $2, -\frac{7}{8}$

26. Solve $x^4 - 13x^2 + 36 = 0$. $-3, 3, -2, 2$

In Exercises 27 and 28, find the exact solutions to $f(x) = 0$ in the complex number system.

27. $x^3 + x^2 + 2x - 4 = 0$ $1, -1 \pm i\sqrt{3}$

28. $4x^3 + 10x^2 + 5x + 2 = 0$ $-2, -\dfrac{1}{4} \pm \dfrac{\sqrt{3}}{4}i$

29. Solve $x^3 - 5x^2 \geq 0$. $x = 0, x \geq 5$

30. Use a graphing utility to solve $x^3 - 5x^2 + 2x + 8 \geq 0$. $x \geq 4$ or $-1 \leq x \leq 2$

31. Solve $2 < \dfrac{4x - 6}{x}$. $x > 3$ or $x < 0$

32. Solve $\dfrac{5x - 10}{x + 1} \geq 20$. $-2 \leq x < -1$

chapter 6 REVIEW

Answers that are not seen can be found in the answer section at the back of the text.

33. **Revenue** The monthly revenue for a product is given by $R = -0.1x^3 + 15x^2 - 25x$, where x is the number of thousands of units sold.

    a. Graph this function using the window $[-100, 200]$ by $[-2000, 60{,}000]$.

    b. Graph the function on a window that makes sense for the problem—that is, with nonnegative x and R.

    c. What is the revenue when 50,000 units are produced? $23,750

34. **Revenue** The monthly revenue for a product is given by $R = -0.1x^3 + 13.5x^2 - 150x$, where x is the number of thousands of units sold. Use technology to find the number of units that gives the maximum possible revenue. 84,051 units

35. **Investment** If $5000 is invested for 6 years at interest rate r (as a decimal), compounded annually, the future value of the investment is given by $S = 5000(1 + r)^6$ dollars.

    a. Find the future value of this investment for selected interest rates by completing the following table.

Rate, r	Future Value, S ($)
0.01	5307.60
0.05	6700.48
0.10	8857.81
0.15	11,565.30

    b. Graph this function for $0 \leq r \leq 0.20$.

    c. Compute the future value if the rate is 10% and 20%. How much more money is earned at 20%? $6072.11

36. **Photosynthesis** The amount y of photosynthesis that takes place in a certain plant depends on the intensity x of the light present, according to the function $y = 120x^2 - 20x^3$. The model is valid only for x-values that are nonnegative and that produce a positive amount of photosynthesis. Use graphical or numerical methods to find what intensity allows the maximum amount of photosynthesis. Intensity 4

37. **Population of Children** The estimated population (in thousands) of U.S. children age 5 and under in selected years from 1985 and projected to 2050 can be modeled by the function

    $$y = 0.000169x^3 - 0.0224x^2 + 1.07x + 57.3$$

    where x is the number of years after 1980.

    a. Graph the function on a window representing 1980 through 2050.

    b. What does the model predict the population of U.S. children age 5 and under will be in 2040? 77.364 thousand, or 77,364

    c. When does the model project the estimated population will be 75,925?

    (*Source*: Childstats.gov) 2030

38. **Older Persons in the Workforce** The table gives the number of millions of older Americans in the workforce for selected years from 1970 and projected to 2050.

    a. Create a scatter plot of the data on the window $[0, 100]$ by $[0, 12]$, with x equal to the number of years after 1960 and y equal to the number of millions of older workers.

    b. Find a cubic model for these data, with x equal to the number of years after 1960 and y equal to the number of older persons in the workforce, in millions. Report the model with three significant digits. $y = -0.0000596x^3 + 0.00933x^2 - 0.308x + 5.63$

    c. Graph the model and the data on the window representing 1960 through 2060.

    d. According to the model, how many older Americans will be in the workforce in 2060? 8.578 million

Year	Workforce (millions)	Year	Workforce (millions)
1970	3.22	2015	6.99
1980	3.05	2020	8.24
1990	3.45	2030	10.10
2000	4.20	2040	9.60
2010	5.44	2050	10.20

    (*Source*: U.S. Bureau of Labor Statistics)

39. Foreign-Born Population The table gives the percent of the U.S. population that is foreign-born for selected years from 1970 and projected to 2060.

   a. Create a scatter plot of the data, with x equal to the number of years after 1970 and y equal to the percent.

   b. Does it appear that the data could be modeled by a cubic function? Yes

   c. Find the cubic function that is the best-fitting model for the data, with x equal to the number of years after 1970 and y equal to the percent. Report the model with three significant digits.
   $y = -0.0000172x^3 + 0.00156x^2 + 0.155x + 4.66$

   d. Graph the model on the same axes with the data, and comment on fit of the model to the data.

Year	Percent	Year	Percent
1970	4.8	2020	14.3
1980	6.2	2030	15.8
1990	8.0	2040	17.1
2000	10.4	2050	18.2
2010	12.4	2060	18.8

(*Source*: U.S. Census Bureau)

40. Investment The future value of an investment of $8000 at interest rate r (as a decimal), compounded annually, for 3 years is given by $S = 8000(1 + r)^3$. Use the root method to find the rate r that gives a future value of $9261. 0.05

41. Constructing a Box A box can be formed by cutting a square out of each corner of a piece of tin and folding the sides up. If the piece of tin is 18 inches by 18 inches and each side of the square that is cut out has length x, the function that gives the volume of the box is $V = 324x - 72x^2 + 4x^3$.

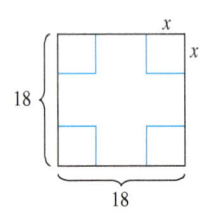

   a. Use factoring to find the values of x that make $V = 0$. 0, 9

   b. For these values of x, discuss what happens to the box if squares of length x are cut out.
   No sides if $x = 0$; no box material left if $x = 9$

   c. What values of x are reasonable for the squares that can be cut out to make a box? $0 < x < 9$

42. Break-Even The daily profit in dollars for a product is given by $P(x) = -0.2x^3 + 20.5x^2 - 48.8x - 120$ where x is the number of hundreds of units produced. To find the number of units that gives break-even,

   a. Graph the function on the window $[0, 40]$ by $[-15, 20]$, and graphically find an x-intercept of the graph.

   b. Use synthetic division to find a quadratic factor of $P(x)$. $-0.2x^2 + 19.7x + 30$

   c. Find all of the zeros of $P(x)$. 4, −1.5, 100

   d. Determine the levels of production that give break-even. 400 units, 10,000 units

43. Industrial Shipments The value (in millions of dollars) of U.S. industrial shipments for selected years from 2014 and projected to 2040 can be modeled by the function $y = 0.185x^3 - 9.68x^2 + 305x + 6083$, with x as the number of years after 2010. When does this model predict that the value of U.S. industrial shipments will be $11,516 million? In 2040
(*Source*: U.S. Department of Energy)

44. Drugs in the Bloodstream The concentration of a drug in a patient's bloodstream is given by

$$C = \frac{0.3t}{t^2 + 1}$$

where t is the number of hours after the drug was ingested and C is in mg/cm^3.

   a. What is the horizontal asymptote for the graph of this function? $C = 0$

   b. What does this say about the concentration of the drug?
   As t increases without bound, concentration approaches 0.

   c. Use technology to find the maximum concentration of the drug, and when it occurs, for $0 \le t \le 4$.
   0.15 when $t = 1$

45. Average Cost The average cost for production of a product is given by

$$\overline{C}(x) = \frac{50x + 5600}{x}$$

where x is the number of units produced.

   a. Find $\overline{C}(0)$, if it exists. What does this tell us about average cost? $\overline{C}(0)$ does not exist. If no units are produced, an average cost per unit cannot be calculated.

   b. Find the horizontal asymptote for the graph of this function. What does this tell us about the average cost of this product? $\overline{C}(x) = 50$; as number of units increases without bound, average cost approaches $50.

   c. Does this function increase or decrease for $x > 0$?
   Decrease

46. Average Cost The average cost for production of a product is given by

$$\overline{C}(x) = \frac{30x^2 + 12,000}{x}$$

where x is the number of units produced.

a. Graph the function for $x > 0$.

b. Use technology to find the number of units that gives the minimum average cost. What is the minimum average cost? 20 units; $1200

47. **Demand** The quantity of a certain product that is demanded is related to the price per unit by the equation $p = \dfrac{30,000 - 20q}{q}$, where p is the price per unit. Graph the function on a window that fits the context of the problem, with $0 \leq q \leq 1000$, with q on the horizontal axis.

48. **Printing** A printer has a contract to run 20,000 posters for a state fair. He can use any number of plates from 1 to 8 to run the posters; each stamp of the press will produce as many posters as there are plates. The cost of printing all the posters is given by $C(x) = 200 + 20x + \dfrac{180}{x}$, where x is the number of plates.

   a. Combine the terms of this function over a common denominator to create a rational function.
   $C = (200x + 20x^2 + 180)/x$

   b. Use numerical or graphical methods to find the number of plates that will make the cost $336. 5

   c. How many plates should the printer create to produce the posters at the minimum cost? 3

49. **Global E-Commerce Sales** Using data from 2015 through 2020, the fraction of total global retail sales that are e-commerce sales can be modeled by the function

$$y = \dfrac{0.502x - 1.073}{1.394x + 13.740}$$

where x is the number of years after 2010.

   a. Graph the function on a window that represents the years 2015 through 2030.

   b. What fraction of total global retail sales are e-commerce sales in 2035, according to the model? 0.2362

   c. Use technology to find the year in which the model predicts that the fraction of total global retail sales that are e-commerce sales is 0.2544. 2041

50. **Sales and Training** During the first month of employment, the monthly sales S (in thousands of dollars) for an average new salesperson depend on the number of hours of training x, according to

$$S = \dfrac{40}{x} + \dfrac{x}{4} + 10 \text{ for } x \geq 4$$

How many hours of training should result in monthly sales greater than $23,300? 50 or more hours

51. **Risky Drivers** The following table gives the police-reported crashes per million miles traveled for different age groups.

   a. Create a scatter plot of the data, with x equal to the age of the drivers and y equal to the number of crashes per million miles travel.

   b. Find the quartic function that is the best fit for the data, where x is the age of the drivers and y equals the number of crashes per million miles traveled. Report the model with four significant digits.
   $y = 0.00001828x^4 - 0.003925x^3 + 0.3031x^2 - 9.907x + 118.2$

   c. Graph the function and the data on the same set of axes, for $x = 15$ to $x = 85$.

   d. At what ages is the number of crashes less than 3, according to the model?
   From 29 to 43, and from 65 to 77

Age	Crashes	Age	Crashes
16	32	45	3.5
17	21	50	3.5
18	10	55	3
19	9	60	2
20	9.5	65	2.5
25	8	70	3
30	5	75	4
35	3.5	80	5
40	3	85	8

(Source: Consumer Reports, October 2012)

52. **Revenue** The revenue from the sale of x units of a product is given by the function $R = 1200x - 0.003x^3$. Selling how many units will give a revenue of at least $59,625? At least 50 units but no more than 605 units

53. **Average Cost** The average cost per set for the production of television sets is given by

$$\overline{C} = \dfrac{100 + 30x + 0.1x^2}{x}$$

where x is the number of hundreds of units produced. For how many units is the average cost at most $37?
Between 2000 and 5000 units, inclusive

54. **Cost–Benefit** The percent p of particulate pollution that can be removed from the smokestacks of an industrial plant by spending C dollars per week is given by $p = \dfrac{100C}{9600 + C}$. How much would it cost to remove at least 30.1% of the particulate pollution?
$4134 or more

55. Wind Resistance Wind resistance varies jointly as an object's surface area and velocity. If an object traveling at 40 miles per hour with a surface area of 25 square feet experiences a wind resistance of 225 newtons, how fast must a car with 40 square feet of surface area travel to experience a wind resistance of 270 newtons? 30 mph

Group Activities
▶ EXTENDED APPLICATIONS

1. Global Climate Change

1. Suppose the annual cost C (in hundreds of dollars) of removing $p\%$ of the particulate pollution from the smokestack of a power plant is given by

$$C(p) = \frac{242{,}000}{100 - p} - 2420$$

   a. What is the domain of the function described by this equation?
   b. Taking into account the meaning of the variables, give the domain and range of the function.

2. Find $C(60)$ and $C(80)$ and interpret $C(80)$.
3. Find the asymptotes of the function. What does the vertical asymptote tell you about the cost of removing pollution?
4. Find and interpret the y-intercept.
5. Suppose the annual fine for removing less than 80% of the particulate pollution from the smokestack is $700,000. If you were the company's chief financial officer (CFO), would you advise the company to remove the pollution or pay the fine? Why?
6. If the company has already paid to remove 60% of the particulate pollution, would you advise the company to remove another 20% or pay the fine?
7. If you were a lawmaker, to what amount would you raise the fine to encourage power plants to comply with the pollution regulations?
8. Compare the costs of removing 80% and 90% of the pollution from the smokestack. Do you think the benefit of removing 90% of particulate pollution rather than 80% is worth the cost?

2. Printing

A printer has a contract to print 100,000 invitations for a political candidate. He can run the invitations by using any number of metal printing plates from 1 to 20 on his press. Each stamp of the press will produce as many invitations as there are plates. For example, if he prepares 10 plates, each impression of his press will make 10 invitations. Preparing each plate costs $8, and he can make 1000 impressions per plate per hour. If it costs $128 per hour to run the press, how many metal printing plates should the printer use to minimize the cost of printing the 100,000 invitations?

A. To understand the problem:
   1. Determine how many hours it would take to print the 100,000 invitations with 10 plates.
   2. How much would it cost to print the invitations with 10 plates?
   3. Determine how much it would cost to prepare 10 printing plates.
   4. What is the total cost to make the 10 plates and print 100,000 invitations?

B. To find the number of plates that will minimize the total cost of printing the invitations:
   1. Find the cost of preparing x plates.
   2. Find the number of invitations that can be printed with one impression of x plates.
   3. Find the number of invitations that can be made per hour with x plates and the number of hours it would take to print 100,000 invitations.
   4. Write a function of x that gives the total cost of producing the invitations.
   5. Use technology to graph the cost function and to estimate the number of plates that will give the minimum cost of producing 100,000 invitations.
   6. Use numerical methods to find the number of plates that will give the minimum cost and the minimum possible cost of producing 100,000 invitations.

7

Systems of Equations and Matrices

To improve the highway infrastructure, departments of transportation can use systems of equations to analyze traffic flow. Solving systems of equations is also useful in finding break-even and market equilibrium. We can represent systems with matrices and use matrix reduction to solve the systems and related problems. We can also use matrices to make comparisons of data, to solve pricing problems and advertising problems, and to encode and decode messages.

sections	objectives	applications
7.1 Systems of Linear Equations in Three Variables	Solve systems of linear equations using left-to-right elimination; find solutions of dependent systems; determine that a system is inconsistent	Manufacturing, purchasing
7.2 Matrix Solution of Systems of Linear Equations	Represent systems of equations with matrices; solve systems with matrices; solve systems of equations with Gauss–Jordan elimination; find nonunique solutions of dependent systems	Transportation, investment
7.3 Matrix Operations	Add and subtract two matrices; multiply a matrix by a constant; multiply two matrices	Life expectancy, pricing, advertising, manufacturing
7.4 Inverse Matrices; Matrix Equations	Find the inverse of a matrix; find inverses with technology; encode and decode messages; solve matrix equations; solve matrix equations with technology	Credit card security, encoding messages, decoding messages, manufacturing
7.5 Determinants and Cramer's Rule	Find determinants of matrices; solve systems of equations using Cramer's Rule; use Cramer's Rule with inconsistent and dependent systems	Business properties
7.6 Systems of Nonlinear Equations	Solve systems of nonlinear equations algebraically and graphically	Market equilibrium, constructing a box

533

Algebra TOOLBOX

KEY OBJECTIVES

- Solve problems involving proportions
- Determine whether sets of triples are proportional
- Determine whether a pair of equations in three variables represent the same plane, parallel planes, or neither
- Determine whether given values of three variables satisfy a system of three equations in the three variables

Proportion

Often in mathematics we need to express relationships between quantities. One relationship that is frequently used in applied mathematics occurs when two quantities are proportional. Two variables x and y are proportional to each other (or vary directly) if their quotient is constant. That is, y is **directly proportional** (or simply **proportional**) to x if y and x are related by the equation

$$\frac{y}{x} = k \quad \text{or, equivalently,} \quad y = kx$$

where $k \neq 0$ is called the **constant of proportionality**.

EXAMPLE 1 ▶ Circles

Is the circumference of a circle proportional to the radius of the circle?

SOLUTION

The circumference C of a circle is proportional to the radius r of the circle because $C = 2\pi r$. In this case, C and r are the variables, and 2π is the constant of proportionality. We could write an equivalent equation,

$$\frac{C}{r} = 2\pi$$

which says that the quotient of the circumference divided by the radius is constant, and that constant is 2π.

Proportionality

One pair of numbers is said to be **proportional** to another pair of numbers if the ratio of one pair is equal to the ratio of the other pair. In general, the pair a, b is proportional to the pair c, d if and only if

$$\frac{a}{b} = \frac{c}{d} \quad (b \neq 0, d \neq 0)$$

This proportion can be written $a:b$ as $c:d$.

If $b \neq 0, d \neq 0$, we can verify that $\frac{a}{b} = \frac{c}{d}$ if $ad = bc$—that is, if the cross-products are equal.

EXAMPLE 2 ▶ Proportional Pairs

Is the pair 4, 3 proportional to the pair 12, 9?

SOLUTION

It is true that $\frac{4}{3} = \frac{12}{9}$ because we can reduce $\frac{12}{9}$ to $\frac{4}{3}$; or we can show that $\frac{4}{3} = \frac{12}{9}$ is true by showing that the cross-products are equal: $4 \cdot 9 = 3 \cdot 12$. Thus, the pair 4, 3 is proportional to 12, 9.

Proportional Triples

For two sets of triples a, b, c and d, e, f, if there exists some number k such that multiplying each number in the first set of triples by k results in the second set of triples, then the two sets of triples are proportional. That is, a, b, c is proportional to d, e, f if there exists a number k such that $ak = d$, $bk = e$, and $ck = f$.

EXAMPLE 3 ▶ Proportional Triples

Determine whether the sets of triples are proportional.

a. 2, 3, 5 and 4, 6, 10   **b.** 4, 6, 10 and 6, 9, 12

SOLUTION

a. If the sets of triples are proportional, then there is a number k such that $2k = 4$, $3k = 6$, and $5k = 10$. Because $k = 2$ is the solution to all three equations, we conclude that the sets of triples are proportional.

b. If the sets of triples are proportional, then there is a number k such that $4k = 6$, $6k = 9$, and $10k = 12$. Because $k = 1.5$ is the solution to $4k = 6$ but not a solution to $10k = 12$, the sets of triples are not proportional.

Linear Equations in Three Variables

If $a, b, c,$ and d represent nonzero constants, then

$$ax + by + cz = d$$

is a first-degree equation in three variables. We also call these equations linear equations in three variables, but unlike linear equations in two variables, their graphs are not lines. When equations of this form are graphed in a three-dimensional coordinate system, their graphs are planes. Similar to the graphs of linear equations in two variables, the graphs of two linear equations in three variables will be parallel planes if the respective coefficients of $x, y,$ and z are proportional and the constants are not in the same proportion. If the respective coefficients of $x, y,$ and z in two equations are not proportional, then the two planes will intersect along a line. And if the two equations are equivalent, the equations represent the same plane.

EXAMPLE 4 ▶ Planes

Determine whether the planes represented by the following pairs of equations represent the same plane, parallel planes, or neither.

a. $3x + 4y + 5z = 1$ and $6x + 8y + 10z = 2$

b. $2x - 5y + 3z = 4$ and $-6x + 10y - 6z = -12$

c. $3x - 6y + 9z = 12$ and $x - 2y + 3z = 26$

SOLUTION

a. If we multiply both sides of the first equation by 2, the second equation results, so the equations are equivalent, and thus the graphs of both equations are the same plane.

b. The respective coefficients of the variables are not in proportion, so the graphs are not the same plane and are not parallel planes, and thus the planes representing these equations must intersect in a line.

c. The respective coefficients of the variables are in proportion and the constants are not in the same proportion, so the graphs are parallel planes.

Three different planes may intersect in a single point (Figure 7.1), may intersect in a line (as in a paddle wheel; see Figure 7.2), or may not have a common intersection (Figure 7.3). Thus, three linear equations in three variables may have a unique solution, infinitely many solutions, or no solution. For example, the solution of the system

$$\begin{cases} 3x + 2y + z = 6 \\ x - y - z = 0 \\ x + y - z = 4 \end{cases}$$

is the single point with coordinates $x = 1$, $y = 2$, and $z = -1$, because these three values satisfy all three equations, and these are the only values that satisfy them.

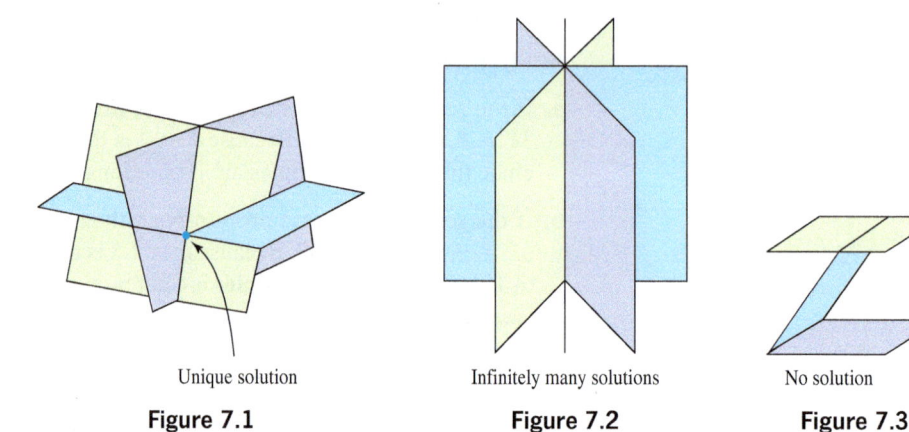

Unique solution
Figure 7.1

Infinitely many solutions
Figure 7.2

No solution
Figure 7.3

EXAMPLE 5 ▶ Solutions of Systems of Equations

Determine whether the given values of x, y, and z satisfy the system of equations.

a. $\begin{cases} 2x + y + z = -3 \\ x - y - 2z = -4; \quad x = -2, y = 0, z = 1 \\ x + y + 3z = 1 \end{cases}$

b. $\begin{cases} x + y + z = 4 \\ x - y - z = 2; \quad x = 3, y = -1, z = 2 \\ 2x + y - 2z = 3 \end{cases}$

SOLUTION

a. Substituting the values of the variables into each equation gives

$$2(-2) + 0 + 1 = -4 + 1 = -3 \quad \checkmark$$
$$-2 - 0 - 2(1) = -2 - 2 = -4 \quad \checkmark$$
$$-2 + 0 + 3(1) = -2 + 3 = 1 \quad \checkmark$$

The values of x, y, and z satisfy all three equations, so we conclude that $x = -2$, $y = 0$, $z = 1$ is a solution to the system.

b. Substituting the values of the variables into each equation gives

$$3 + (-1) + 2 = 2 + 2 = 4 \quad \checkmark$$
$$3 - (-1) - 2 = 4 - 2 = 2 \quad \checkmark$$
$$2(3) + (-1) - 2(2) = 5 - 4 = 1 \neq 3$$

Although the values of x, y, and z satisfy the first two equations, they do not satisfy the third equation, so we conclude that $x = 3$, $y = -1$, $z = 2$ is not a solution to the system.

Toolbox EXERCISES

1. If y is proportional to x, and $y = 12x$, what is the constant of proportionality? 12

2. If x is proportional to y, and $y = 10x$, what is the constant of proportionality? $\frac{1}{10}$

3. Is the pair of numbers 5, 6 proportional to the pair 10, 18? No

4. Is the pair of numbers 8, 3 proportional to the pair 24, 9? Yes

5. Is the pair of numbers 4, 6 proportional to the pair 2, 3? Yes

6. Is the pair of numbers 5.1, 2.3 proportional to the pair 51, 23? Yes

7. Suppose y is proportional to x, and when $x = 2$, $y = 18$. Find y when $x = 11$. 99

8. Suppose x is proportional to y, and when $x = 3$, $y = 18$. Find y when $x = 13$. 78

In Exercises 9–12, determine whether the pairs of triples are proportional.

9. 8, 6, 2 and 20, 12.5, 5 No

10. 4.1, 6.8, 9.3 and 8.2, 13.6, 18.6 Yes

11. $-1, 6, 4$ and $-3.4, 10.2, 13.6$ No

12. 5, -12, 16 and 2, -3, 4 No

In Exercises 13–16, determine whether the given values of x, y, and z satisfy the system of equations.

13. $\begin{cases} x - 2y + z = 8 \\ 2x - y + 2z = 10; \quad x = 1, y = -2, z = 3 \quad \text{No} \\ 3x - 2y + z = 5 \end{cases}$

14. $\begin{cases} x - 2y + 2z = -9 \\ 2x - 3y + z = -14; \quad x = -1, y = 4, z = 0 \\ x + 2y - 3z = 7 \quad \text{Yes} \end{cases}$

15. $\begin{cases} x + y - z = 4 \\ 2x + 3y - z = 5; \quad x = 5, y = -2, z = -1 \\ 3x - 2y + 5z = 14 \quad \text{Yes} \end{cases}$

16. $\begin{cases} x - 2y - 3z = 2 \\ 2x + 3y + 3z = 19; \quad x = 2, y = 3, z = -2 \\ -x - 2y + 2z = -12 \quad \text{No} \end{cases}$

17. Determine whether the planes represented by these equations represent the same plane, parallel planes, or neither: Parallel planes

    $x + 3y - 7z = 12$ and $2x + 6y - 14z = 18$

18. Determine whether the planes represented by these equations represent the same plane, parallel planes, or neither: Same plane

    $x - 3y - 2z = 15$ and $2x - 6y - 4z = 30$

19. Determine whether the planes represented by these equations represent the same plane, parallel planes, or neither: Neither

    $2x - y - z = 2$ and $2x + y - 4z = 8$

20. Determine whether the planes represented by these equations represent the same plane, parallel planes, or neither: Same plane

    $3x - y + 2z = 9$ and $9x - 3y + 6z = 27$

7.1 Systems of Linear Equations in Three Variables

KEY OBJECTIVES

- Solve systems of linear equations in three variables using left-to-right elimination
- Find solutions of dependent systems
- Determine when a system of equations is inconsistent

SECTION PREVIEW Manufacturing

A manufacturer of furniture has three models of chairs: Anderson, Blake, and Colonial. The numbers of hours required for framing, upholstery, and finishing for each type of chair are given in Table 7.1. The company has 1500 hours per week for framing, 2100 hours for upholstery, and 850 hours for finishing.

Finding how many of each type of chair can be produced under these conditions involves solving a system with three linear equations in three variables. It is not possible to solve this equation graphically, and new algebraic methods are needed. In this section, we discuss how to solve a system of equations by using the left-to-right elimination method.

Table 7.1

	Anderson	Blake	Colonial
Framing	2	3	1
Upholstery	1	2	3
Finishing	1	2	1/2

Systems in Three Variables

In Section 2.3, we used the elimination method to reduce a system of two linear equations in two variables to one equation in one variable. If a system of three linear equations in three variables has a unique solution, we can reduce the system to an equivalent system where at least one of the equations contains only one variable. We can then easily solve that equation and use the resulting value to find the values of the remaining variables. For example, we will show later that the system

$$\begin{cases} x + y + z = 2 \\ x - y - z = 0 \\ x + y - z = 4 \end{cases}$$

can be reduced to the equivalent system

$$\begin{cases} x + y + z = 2 & (1) \\ y + z = 1 & (2) \\ z = -1 & (3) \end{cases}$$

Clearly, Equation (3) in the reduced system has solution $z = -1$. Using this value of z in Equation (2) gives

$$y + (-1) = 1, \quad \text{or} \quad y = 2$$

Using $z = -1$ and $y = 2$ in Equation (1) gives

$$x + 2 + (-1) = 2, \quad \text{or} \quad x = 1$$

Thus, the solution to the reduced system is $x = 1, y = 2, z = -1$. The process of using the value of the third variable to work back to the values of the other variables is called **back substitution**.

We say that the reduced system is equivalent to the original system because both systems have the same solutions. We verify this by showing that these values of $x, y,$ and z satisfy the equations of the original system.

$$\begin{cases} 1 + 2 - 1 = 2 \\ 1 - 2 + 1 = 0 \\ 1 + 2 + 1 = 4 \end{cases}$$

Left-to-Right Elimination

The reduced system of equations that we discussed above is in a special form called the **echelon form**, which, in general, looks like

$$\begin{cases} x + ay + bz = d \\ y + cz = e \\ z = f \end{cases}$$

where $a, b, c, d, e,$ and f are constants. Notice that in this form, each equation begins with a new (leading) variable with coefficient 1. Any system of linear equations with a unique solution can be reduced to this form. By using the following operations, we can be assured that the new system will be equivalent to the original system.

Operations on a System of Equations

The following operations result in an equivalent system.

1. Interchange any two equations.
2. Multiply both sides of any equation by the same nonzero number.
3. Multiply any equation by a number, add the result to a second equation, and then replace the second equation with the sum.

The following process, which uses these operations to reduce a system of linear equations, is called the **left-to-right elimination method**.

Left-to-Right Elimination Method of Solving Systems of Linear Equations in Three Variables x, y, and z

1. If necessary, interchange two equations or use multiplication to make the coefficient of x in the first equation a 1.
2. Add a multiple of the first equation to each of the following equations so that the coefficients of x in the second and third equations become 0.
3. Multiply (or divide) both sides of the second equation by a number that makes the coefficient of y in the second equation equal to 1.
4. Add a multiple of the (new) second equation to the (new) third equation so that the coefficient of y in the newest third equation becomes 0.
5. Multiply (or divide) both sides of the third equation by a number that makes the coefficient of z in the third equation equal to 1. This gives the solution for z in the system of equations.
6. Use the solution for z to solve for y in the second equation. Then substitute values for y and z to solve for x in the first equation. (This is called back substitution.)

EXAMPLE 1 ▶ Solution by Elimination

Solve the system

$$\begin{cases} 2x - 3y + z = -1 & (1) \\ x - y + 2z = -3 & (2) \\ 3x + y - z = 9 & (3) \end{cases}$$

SOLUTION

1. To write the system with x-coefficient 1 in the first equation, interchange Equations (1) and (2).

$$\begin{cases} x - y + 2z = -3 \\ 2x - 3y + z = -1 \\ 3x + y - z = 9 \end{cases}$$

2. To eliminate x in the second equation, multiply the (new) first equation by -2 and add it to the second equation. To eliminate x in the third equation, multiply the (new) first equation by -3 and add it to the third equation. This gives the equivalent system

$$\begin{cases} x - y + 2z = -3 \\ -y - 3z = 5 \\ 4y - 7z = 18 \end{cases}$$

3. To get 1 as the coefficient of y in the second equation, multiply the (new) second equation by -1.

$$\begin{cases} x - y + 2z = -3 \\ \phantom{x -{}} y + 3z = -5 \\ \phantom{x -{}} 4y - 7z = 18 \end{cases}$$

4. To eliminate y from the (new) third equation, multiply the (new) second equation by -4 and add it to the third equation.

$$\begin{cases} x - y + 2z = -3 \\ \phantom{x -{}} y + 3z = -5 \\ \phantom{x - y +{}} -19z = 38 \end{cases}$$

5. To solve for z, divide both sides of the (new) third equation by -19.

$$\begin{cases} x - y + 2z = -3 \\ \phantom{x -{}} y + 3z = -5 \\ z = -2 \end{cases}$$

This gives the solution $z = -2$.

6. Substituting -2 for z in the second equation and solving for y gives $y + 3(-2) = -5$, so $y = 1$. Substituting -2 for z and 1 for y in the first equation and solving for x gives $x - 1 + 2(-2) = -3$, so $x = 2$, and the solution is $(2, 1, -2)$.

The process used in step 6 is called *back substitution* because we use the solution for the third variable to work back to the solutions of the first and second variables. Checking in the original system verifies that the solution satisfies all the equations:

$$\begin{cases} 2(2) - 3(1) + (-2) = -1 \\ 2 - 1 + 2(-2) = -3 \\ 3(2) + 1 - (-2) = 9 \end{cases}$$

Modeling Systems of Equations

Solution of real problems sometimes requires us to create three or more equations whose simultaneous solution is the solution to the problem.

EXAMPLE 2 ▶ Manufacturing

A manufacturer of furniture has three models of chairs: Anderson, Blake, and Colonial. The numbers of hours required for framing, upholstery, and finishing for each type of chair are given in Table 7.2. The company has 1500 hours per week for framing, 2100 hours for upholstery, and 850 hours for finishing. How many of each type of chair can be produced under these conditions?

Table 7.2

	Anderson	Blake	Colonial
Framing	2	3	1
Upholstery	1	2	3
Finishing	1	2	$\frac{1}{2}$

SOLUTION

If we represent the number of units of the Anderson model by x, the number of units of the Blake model by y, and the number of units of the Colonial model by z, then we have the following equations:

$$\begin{aligned} \text{Framing:} & \quad 2x + 3y + z = 1500 \\ \text{Upholstery:} & \quad x + 2y + 3z = 2100 \\ \text{Finishing:} & \quad x + 2y + \tfrac{1}{2}z = 850 \end{aligned}$$

We seek the values of x, y, and z that satisfy all three equations above. That is, we seek the solution to the system

$$\begin{cases} 2x + 3y + z = 1500 & (1) \\ x + 2y + 3z = 2100 & (2) \\ x + 2y + \tfrac{1}{2}z = 850 & (3) \end{cases}$$

We use the following steps to solve the system by left-to-right elimination. The notation that we use to represent each step is shown beside the system. We use this notation in subsequent examples.

To write the system with x-coefficient 1 in the first equation, interchange Equations (1) and (2).

$$\begin{cases} x + 2y + 3z = 2100 & \\ 2x + 3y + z = 1500 & \text{Eq1} \leftrightarrow \text{Eq2} \\ x + 2y + 0.5z = 850 & \end{cases}$$

To eliminate x in the second equation, multiply the first equation by -2 and add it to the second equation. To eliminate x in the third equation, multiply the first equation by -1 and add it to the third equation.

$$\begin{cases} x + 2y + 3z = 2100 & \\ -y - 5z = -2700 & (-2)\text{Eq1} + \text{Eq2} \rightarrow \text{Eq2} \\ -2.5z = -1250 & (-1)\text{Eq1} + \text{Eq3} \rightarrow \text{Eq3} \end{cases}$$

Observe that we can solve Equation (3) for z, getting $z = -1250/(-2.5) = 500$. Substituting $z = 500$ in Equation (2) gives $-y - 5(500) = -2700$, or $y = 200$. Substituting $z = 500$ and $y = 200$ in Equation (1) gives $x + 2(200) + 3(500) = 2100$, or $x = 200$.

Thus 200 Anderson, 200 Blake, and 500 Colonial chairs should be made under the given conditions.

Nonunique Solutions

It is not always possible to reduce a system of three equations in three variables to a system in which the third equation contains one variable. For example, the system

$$\begin{cases} x + 2y + z = 2 & (1) \\ x - y - 2z = -4 & (2) \\ 2x + y - z = 4 & (3) \end{cases}$$

reduces to

$$\begin{cases} x + 2y + z = 2 \\ y + z = 2 \\ 0 + 0 = 6 \end{cases}$$

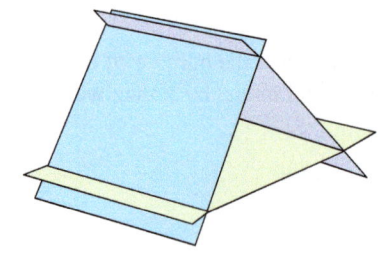

542 Chapter 7 Systems of Equations and Matrices

The third equation in the reduced system is $0 = 6$, which is impossible. Thus, this system and the original system have no solution. Such a system is called an **inconsistent system**.

When a system of three linear equations in three variables has no solution, the graphs of the equations will not have a common intersection, as shown in Figure 7.4.

The system

$$\begin{cases} x + 2y + z = 2 & (1) \\ x - y - 2z = -4 & (2) \\ 2x + y - z = -2 & (3) \end{cases}$$

No solution

Figure 7.4

looks similar but is slightly different from the inconsistent system just discussed. Reducing this system with left-to-right elimination results in the system

$$\begin{cases} x + 2y + z = 2 \\ y + z = 2 \\ 0 + 0 = 0 \end{cases}$$

The third equation in this system is $0 = 0$, and we will see that the system has infinitely many solutions. Notice that the z is not a leading variable in any of the equations, so we can solve for the other variables in terms of z. Back substitution gives

$$y = 2 - z$$
$$x = 2 - 2(2 - z) - z = -2 + z$$

Thus, the solutions have the form $x = -2 + z$, $y = 2 - z$, and $z =$ any number.

By selecting different values of z, we can find some particular solutions of this system.

For example,

$$z = 1 \quad \text{gives} \quad x = -1, \quad y = 1, \quad z = 1$$
$$z = 2 \quad \text{gives} \quad x = 0, \quad y = 0, \quad z = 2$$
$$z = 5 \quad \text{gives} \quad x = 3, \quad y = -3, \quad z = 5$$

A system of this type, with infinitely many solutions, is called a **dependent system**. When a system of three equations in three variables has infinitely many solutions, the graphs of the equations may be planes that intersect in a line, as in the paddle wheel shown in Figure 7.5.

Infinitely many solutions

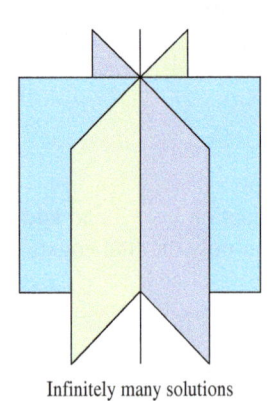

Figure 7.5

EXAMPLE 3 ▶

Find the solution to each of the following systems, if it exists.

a. $\begin{cases} x - y + 4z = -5 \\ 3x + z = 0 \\ -x + y - 4z = 20 \end{cases}$ **b.** $\begin{cases} -x + y + 2z = 0 \\ x + 2y + z = 6 \\ -2x - y + z = -6 \end{cases}$

SOLUTION

a. The system

$$\begin{cases} x - y + 4z = -5 & (1) \\ 3x + z = 0 & (2) \\ -x + y - 4z = 20 & (3) \end{cases}$$

reduces as follows:

$$\begin{cases} x - y + 4z = -5 & \\ 3y - 11z = 15 & (-3)\text{Eq1} + \text{Eq2} \to \text{Eq2} \\ 0 = 15 & \text{Eq1} + \text{Eq3} \to \text{Eq3} \end{cases}$$

7.1 Systems of Linear Equations in Three Variables

The third equation in the reduced system is $0 = 15$, which is impossible. Thus, the system is inconsistent and has no solution.

b. The system

$$\begin{cases} -x + y + 2z = 0 & (1) \\ x + 2y + z = 6 & (2) \\ -2x - y + z = -6 & (3) \end{cases}$$

reduces as follows:

$$\begin{cases} x - y - 2z = 0 & (-1)\text{Eq1} \rightarrow \text{Eq1} \\ x + 2y + z = 6 \\ -2x - y + z = -6 \end{cases}$$

$$\begin{cases} x - y - 2z = 0 \\ 3y + 3z = 6 & (-1)\text{Eq1} + \text{Eq2} \rightarrow \text{Eq2} \\ -3y - 3z = -6 & (2)\text{Eq1} + \text{Eq3} \rightarrow \text{Eq3} \end{cases}$$

$$\begin{cases} x - y - 2z = 0 \\ y + z = 2 & (1/3)\text{Eq2} \rightarrow \text{Eq2} \\ -3y - 3z = -6 \end{cases}$$

$$\begin{cases} x - y - 2z = 0 \\ y + z = 2 \\ 0 = 0 & (3)\text{Eq2} + \text{Eq3} \rightarrow \text{Eq3} \end{cases}$$

Because the third equation in the reduced system is $0 = 0$, the system has infinitely many solutions. We let z be any number and use back substitution to solve the other variables in terms of z.

$$y = 2 - z$$
$$x = 2 - z + 2z = 2 + z$$

Thus, the solution has the form

$$x = 2 + z$$
$$y = 2 - z$$
$$z = \text{any number}$$

If a system of equations has fewer equations than variables, the system must be dependent or inconsistent. Consider the following example.

EXAMPLE 4 ▶ Purchasing

A young man wins $200,000 and (foolishly) decides to buy a new car for each of the seven days of the week. He wants to choose from cars that are priced at $40,000, $30,000, and $25,000. How many cars of each price can he buy with the $200,000?

SOLUTION

If he buys x cars costing $40,000 each, y cars costing $30,000 each, and z cars costing $25,000 each, then

$$x + y + z = 7 \quad \text{and} \quad 40{,}000x + 30{,}000y + 25{,}000z = 200{,}000$$

Writing these two equations as a system of equations gives

$$\begin{cases} x + y + z = 7 & (1) \\ 40{,}000x + 30{,}000y + 25{,}000z = 200{,}000 & (2) \end{cases}$$

Because there are three variables and only two equations, the solution cannot be unique. The solution follows.

$$\begin{cases} x + y + z = 7 \\ -10{,}000y - 15{,}000z = -80{,}000 \end{cases} \quad -40{,}000\,\text{Eq1} + \text{Eq2} \to \text{Eq2}$$

$$\begin{cases} x + y + z = 7 \\ y + 1.5z = 8 \end{cases} \quad -0.0001\,\text{Eq2} \to \text{Eq2}$$

In the solution to this system, we see that z can be any value and that x and y depend on z.

$$y = 8 - 1.5z$$
$$x = -1 + 0.5z$$

In the context of this application, the solutions must be nonnegative integers (none of x, y, or z can be negative), so z must be at least 2 (or x will be negative) and z cannot be more than 5 (or y will be negative). In addition, z must be an even number (to get integer solutions). Thus, the only possible selections are

$$z = 2, x = 0, y = 5 \quad \text{or} \quad z = 4, x = 1, y = 2$$

Thus, the man can buy two $25,000 cars, zero $40,000 cars, and five $30,000 cars, or he can buy four $25,000 cars, one $40,000 car, and two $30,000 cars. So even though there are an infinite number of solutions to this system of equations, there are only two solutions in the context of this application.

SKILLS CHECK 7.1

Answers that are not seen can be found in the answer section at the back of the text.

In Exercises 1–4, solve the systems of equations.

1. $\begin{cases} x + 2y - z = 3 \\ y + 3z = 11 \\ z = 3 \end{cases}$ $x = 2, y = 2, z = 3$

2. $\begin{cases} x - 4y - 3z = 3 \\ y - 2z = 11 \\ z = 4 \end{cases}$ $x = 91, y = 19, z = 4$

3. $\begin{cases} x + 2y - z = 6 \\ y + 3z = 3 \\ z = -2 \end{cases}$ $x = -14, y = 9, z = -2$

4. $\begin{cases} x + 2y - z = 22 \\ y + 3z = 21 \\ z = -5 \end{cases}$ $x = -55, y = 36, z = -5$

In Exercises 5–16, use left-to-right elimination to solve the systems of equations.

5. $\begin{cases} x - y - 4z = 0 \\ y + 2z = 4 \\ 3y + 7z = 22 \end{cases}$ $x = 24, y = -16, z = 10$

6. $\begin{cases} x + 4y - 11z = 33 \\ y - 3z = 11 \\ 2y + 7z = -4 \end{cases}$ $x = -9, y = 5, z = -2$

7. $\begin{cases} x - 2y + 3z = 0 \\ y - 2z = -1 \\ y + 5z = 6 \end{cases}$ $x = -1, y = 1, z = 1$

8. $\begin{cases} x - 2y + 3z = -10 \\ y - 2z = 7 \\ y - 3z = 6 \end{cases}$ $x = 5, y = 9, z = 1$

9. $\begin{cases} x + 2y - 2z = 0 \\ x - y + 4z = 3 \\ x + 2y + 2z = 3 \end{cases}$ $x = 1/2, y = 1/2, z = 3/4$

10. $\begin{cases} x + 4y - 14z = 22 \\ x + 5y + z = -2 \\ x + 4y - z = 9 \end{cases}$ $x = 44, y = -9, z = -1$

11. $\begin{cases} x + 3y - z = 0 \\ x - 2y + z = 8 \\ x - 6y + 2z = 6 \end{cases}$ $x = 2, y = 4, z = 14$

12. $\begin{cases} x - 5y - 2z = 7 \\ x - 3y + 4z = 21 \\ x - 5y + 2z = 19 \end{cases}$ $x = 3, y = -2, z = 3$

13. $\begin{cases} 2x + 4y - 14z = 0 \\ 3x + 5y + z = 19 \\ x + 4y - z = 12 \end{cases}$ $x = 1, y = 3, z = 1$

14. $\begin{cases} 3x - 6y + 6z = 24 \\ 2x - 6y + 2z = 6 \\ x - 3y + 2z = 4 \end{cases}$ $x = 14, y = 4, z = 1$

15. $\begin{cases} 3x - 4y + 6z = 10 \\ 2x - 4y - 5z = -14 \\ x + 2y - 3z = 0 \end{cases}$ $x = 2, y = 2, z = 2$

16. $\begin{cases} x - 3y + 4z = 7 \\ 2x + 2y - 3z = -3 \\ x - 3y + z = -2 \end{cases}$ $x = 1, y = 2, z = 3$

22. $\begin{cases} x + 5y - 2z = 5 \\ x + 3y - 2z = 23 \\ x + 4y - 2z = 14 \end{cases}$ Infinitely many solutions of the form $x = 2z + 50, y = -9, z = z$

23. $\begin{cases} 3x - 4y - 6z = 10 \\ 2x - 4y - 5z = -14 \\ x - z = 0 \end{cases}$ Inconsistent system, no solution

The systems in Exercises 17–26 do not have unique solutions. Solve each system, if possible.

17. $\begin{cases} x - 3y + z = -2 \\ y - 2z = -4 \end{cases}$

18. $\begin{cases} x + 2y + 2z = 4 \\ y + 3z = 6 \end{cases}$

19. $\begin{cases} x - y + 2z = 4 \\ y + 8z = 16 \end{cases}$

20. $\begin{cases} x - 3y - 2z = 3 \\ y + 4z = 5 \\ 0 = 4 \end{cases}$ Inconsistent system, no solution

21. $\begin{cases} x - y + 2z = -1 \\ 2x + 2y + 3z = -3 \\ x + 3y + z = -2 \end{cases}$ Infinitely many solutions of the form $x = -\frac{7}{4}z - \frac{5}{4}, y = \frac{1}{4}z - \frac{1}{4}, z = z$

24. $\begin{cases} 3x - 9y + 4z = 24 \\ 2x - 6y + 2z = 6 \\ x - 3y + 2z = 20 \end{cases}$ Inconsistent system, no solution

25. $\begin{cases} 2x + 3y - 14z = 16 \\ 4x + 5y - 30z = 34 \\ x + y - 8z = 9 \end{cases}$ Infinitely many solutions of the form $x = 10z + 11, y = -2z - 2, z = z$

26. $\begin{cases} 2x + 4y - 4z = 6 \\ x + 5y - 3z = 3 \\ 3x + 3y - 5z = 9 \end{cases}$ Infinitely many solutions of the form $x = \frac{4}{3}z + 3, y = \frac{1}{3}z, z = z$

EXERCISES 7.1

Answers that are not seen can be found in the answer section at the back of the text.

27. **Rental Cars** A car rental agency rents compact, midsize, and luxury cars. Its goal is to purchase 60 cars with a total of $1,400,000 and to earn a daily rental of $2200 from all the cars. The compact cars cost $15,000 and earn $30 per day in rental, the midsize cars cost $25,000 and earn $40 per day, and the luxury cars cost $45,000 and earn $50 per day. To find the number of each type of car the agency should purchase, solve the system of equations

$$\begin{cases} x + y + z = 60 \\ 15{,}000x + 25{,}000y + 45{,}000z = 1{,}400{,}000 \\ 30x + 40y + 50z = 2200 \end{cases}$$

where x, y, and z are the numbers of compact cars, midsize cars, and luxury cars, respectively.
30 compact, 20 midsize, 10 luxury

28. **Ticket Pricing** A theater owner wants to divide an 1800-seat theater into three sections, with tickets costing $20, $35, and $50, depending on the section. He wants to have twice as many $20 tickets as the sum of the other kinds of tickets, and he wants to earn $48,000 from a full house. To find how many seats he should have in each section, solve the system of equations

$$\begin{cases} x + y + z = 1800 \\ x = 2(y + z) \\ 20x + 35y + 50z = 48{,}000 \end{cases}$$

where x, y, and z are the numbers of $20 tickets, $35 tickets, and $50 tickets, respectively.
1200 $20 tickets, 400 $35 tickets, 200 $50 tickets

29. **Pricing** A concert promoter needs to make $120,000 from the sale of 2600 tickets. The promoter charges $40 for some tickets and $60 for the others.

   a. If there are x of the $40 tickets and y of the $60 tickets, write an equation that states that the total number of the tickets sold is 2600. $x + y = 2600$

   b. How much money is made from the sale of x tickets for $40 each? $40x$

   c. How much money is made from the sale of y tickets for $60 each? $60y$

   d. Write an equation that states that the total amount made from the sale is $120,000. $40x + 60y = 120{,}000$

   e. Solve the equations simultaneously to find how many tickets of each type must be sold to yield the $120,000. 1800 $40 tickets, 800 $60 tickets

30. **Investment** A trust account manager has $500,000 to be invested in three different accounts. The accounts pay 8%, 10%, and 14%, respectively, and the goal is to earn $49,000, with the amount invested at 8% equal to the sum of the other two investments. To accomplish this, assume that x dollars are invested at 8%, y dollars are invested at 10%, and z dollars are invested at 14%.

   a. Write an equation that describes the sum of money in the three investments. $x + y + z = 500{,}000$

   b. Write an equation that describes the total amount of money earned by the three investments.
   $0.08x + 0.10y + 0.14z = 49{,}000$

c. Write an equation that describes the relationship among the three investments. $x = y + z$

d. Solve the system of equations to find how much should be invested in each account to satisfy the conditions. $250,000 at 8%, $150,000 at 10%, $100,000 at 14%

31. Investment A man has $400,000 invested in three rental properties. One property earns 7.5% per year on the investment, the second earns 8%, and the third earns 9%. The total annual earnings from the three properties is $33,700, and the amount invested at 9% equals the sum of the first two investments. Let x equal the investment at 7.5%, y equal the investment at 8%, and z represent the investment at 9%.

a. Write an equation that represents the sum of the three investments. $x + y + z = 400,000$

b. Write an equation that states that the sum of the returns from all three investments is $33,700. $0.075x + 0.08y + 0.09z = 33,700$

c. Write an equation that states that the amount invested at 9% equals the sum of the other two investments. $z = x + y$

d. Solve the system of equations to find how much is invested in each property. $60,000 at 7.5%, $140,000 at 8%, $200,000 at 9%

32. Loans A bank loans $285,000 to a development company to purchase three business properties. One of the properties costs $45,000 more than another, and the third costs twice the sum of these two properties.

a. Write an equation that represents the total loaned as the sum of the costs of the three properties, if the costs are x, y, and z, respectively. $x + y + z = 285,000$

b. Write an equation that states that one cost is $45,000 more than another. $x = y + 45,000$

c. Write an equation that states that the third cost is equal to twice the sum of the other two. $z = 2(x + y)$

d. Solve the system of equations to find the cost of each property. $70,000 for 1st, $25,000 for 2nd, $190,000 for 3rd

33. Transportation Ace Trucking Company has an order for delivery of three products, A, B, and C. The table below gives the volume in cubic feet, the weight in pounds, and the value for insurance in dollars for a unit of each of the products. If the carrier can carry 8000 cubic feet and 12,400 pounds and is insured for $52,600, how many units of each product can be carried? A: 100 units; B: 120 units; C: 80 units

	Product A	Product B	Product C
Volume (cu ft)	24	20	40
Weight (lb)	40	30	60
Value ($)	150	180	200

34. Nutrition The following table gives the calories, fat, and carbohydrates per ounce for three brands of cereal, as well as the total amount of calories, fat, and carbohydrates required for a special diet.

	Calories	Fat	Carbohydrates
All Bran	50	0	22.0
Sugar Frosted Flakes	108	0.1	25.5
Natural Mixed Grain	127	5.5	18.0
Total Required	393	5.7	91.0

To find the number of ounces of each brand that should be combined to give the required totals of calories, fat, and carbohydrates, let x represent the number of ounces of All Bran, y represent the number of ounces of Sugar Frosted Flakes, and z represent the number of ounces of Natural Mixed Grain.

a. Use the information in the table to write a system of three equations.

b. Solve the system to find the amount of each brand of cereal that will give the necessary nutrition. 1 oz All Bran, 2 oz Frosted Flakes, 1 oz Natural Mixed Grain

Exercises 35–40 have nonunique solutions.

35. Investment A trust account manager has $500,000 to invest in three different accounts. The accounts pay 8%, 10%, and 14%, and the goal is to earn $49,000. To solve this problem, assume that x dollars are invested at 8%, y dollars are invested at 10%, and z dollars are invested at 14%. Then x, y, and z must satisfy the equations $x + y + z = 500,000$ and $0.08x + 0.10y + 0.14z = 49,000$.

a. Explain the meaning of each of these equations. Sum of investments is $500,000; sum of interest earned is $49,000.

b. Solve the systems containing these two equations. Find x and y in terms of z. $x = 50,000 + 2z$, $y = 450,000 - 3z$

c. What limits must there be on z so that all investment values are nonnegative? $0 \leq z \leq 150,000$

36. Loans A bank gives three loans totaling $200,000 to a development company for the purchase of three business properties. The largest loan is $45,000 more than the sum of the other two. Represent the amount of money in each loan as x, y, and z, respectively.

a. Write a system of two equations in three variables to represent the problem.

b. Does the system have a unique solution? No

c. Find the amount of the largest loan. $122,500

d. What is the relationship between the remaining two loans? Their sum is $77,500.

37. Manufacturing A company manufactures three types of air conditioners: the Acclaim, the Bestfrig, and the

Cool King. The hours needed for assembly, testing, and packing for each product are given in the table below. The numbers of hours that are available daily are 300 for assembly, 120 for testing, and 210 for packing. Can all of the available hours be used to produce these three types of air conditioners? If so, how many of each type can be produced?
Inconsistent system; can't use all hours

	Assembly	Testing	Packing
Acclaim	5	2	1.4
Bestfrig	4	1.4	1.2
Cool King	4.5	1.7	1.3

38. *Transportation* Hohman Trucking Company has an order for delivery of three products, A, B, and C. The following table gives the volume in cubic feet, the weight in pounds, and the value for insurance in dollars per unit for each of the products. If one of the company's trucks can carry 4000 cubic feet and 6000 pounds and is insured to carry $24,450, how many units of each product can be carried on the truck?

	Product A	Product B	Product C
Volume (cu ft)	24	20	50
Weight (lb)	36	30	75
Value ($)	150	180	120

39. *Nutrition* A psychologist, studying the effects of good nutrition on the behavior of rabbits, feeds one group a combination of three foods: I, II, and III. Each of these foods contains three additives: A, B, and C. The following table gives the percent of each additive that is present in each food. If the diet being used requires 6.88 g per day of A, 6.72 g of B, and 6.8 g of C, find the number of grams of each food that should be used each day.

	Food I	Food II	Food III
Additive A (%)	12	14	8
Additive B (%)	8	6	16
Additive C (%)	10	10	12

40. *Investment* A woman has $1,400,000 to purchase three investments. One investment earns 8%, a second earns 7.5%, and the third earns 10%.

a. What investment strategies will give annual earnings of $120,000 from these three investments?

b. What one investment strategy will minimize her risk and return $120,000 if the 10% investment has the most risk?
$400,000 at 10%, $0 at 7.5%, $1,000,000 at 8%

7.2 Matrix Solution of Systems of Linear Equations

KEY OBJECTIVES

- Represent systems of equations with matrices
- Find dimensions of matrices
- Identify square matrices
- Identify an identity matrix
- Form an augmented matrix
- Identify a coefficient matrix
- Reduce a matrix with row operations
- Reduce a matrix to its row-echelon form
- Reduce a matrix to its reduced row-echelon form
- Solve systems of equations using the Gauss–Jordan elimination method

SECTION PREVIEW Transportation

Ace Trucking Company has an order for delivery of three products: A, B, and C. Table 7.3 gives the volume in cubic feet, the weight in pounds, and the value for insurance in dollars per unit of each of the products. If the carrier can carry 30,000 cubic feet and 62,000 pounds and is insured for $276,000, how many units of each product can be carried?

Table 7.3

	Product A	Product B	Product C
Volume (cu ft)	25	22	30
Weight (lb)	25	38	70
Value ($)	150	180	300

If the numbers of units of products A, B, and C are x, y, and z, respectively, the system of linear equations that satisfies these conditions is

$$\begin{cases} 25x + 22y + 30z = 30{,}000 & \text{Volume} \\ 25x + 38y + 70z = 62{,}000 & \text{Weight} \\ 150x + 180y + 300z = 276{,}000 & \text{Value} \end{cases}$$

As we will see in Example 5, there is a solution, but not a unique solution. In this section, we use matrices to solve systems with unique solutions and with infinitely many solutions, and we see how to determine when a system has no solution. ∎

Matrix Representation of Systems of Equations

Recall from Section 7.1 that in solving a system of three linear equations in three variables, our goal in each step was to use elimination to remove each variable, in turn, from the equations below it in the system. This process can be simplified if we represent the system in a new way without including the symbols for the variables. We can do this by creating a rectangular array containing the coefficients and constants from the system. This array, called a **matrix** (the plural is **matrices**), uses each row (horizontal) to represent an equation and each column (vertical) to represent a variable. For example, the system of equations that was solved in Example 1 of Section 7.1,

$$\begin{cases} 2x - 3y + z = -1 \\ x - y + 2z = -3 \\ 3x + y - z = 9 \end{cases}$$

can be represented in a matrix as follows:

$$\begin{bmatrix} 2 & -3 & 1 & | & -1 \\ 1 & -1 & 2 & | & -3 \\ 3 & 1 & -1 & | & 9 \end{bmatrix}$$

In this matrix, the first column contains the coefficient of x for each of the equations, the second column contains the coefficients of y, and the third column contains the coefficients of z. The column to the right of the vertical line, containing the constants of the equations, is called the **augment** of the matrix, and a matrix containing an augment is called an **augmented matrix**. The matrix that contains only the coefficients of the variables is called a **coefficient matrix**. The coefficient matrix for this system is

$$\begin{bmatrix} 2 & -3 & 1 \\ 1 & -1 & 2 \\ 3 & 1 & -1 \end{bmatrix}$$

The augmented matrix above has 3 rows and 4 columns, so its **dimension** is 3×4 (three by four). The coefficient matrix above is a 3×3 matrix. Matrices like this one are called **square matrices**. A square matrix that has 1s down its diagonal and 0s everywhere else, like matrix I below, is called an **identity matrix**. Matrix I is a 3×3 identity matrix.

$$I = \begin{bmatrix} 1 & 0 & 0 \\ 0 & 1 & 0 \\ 0 & 0 & 1 \end{bmatrix}$$

Echelon Forms of Matrices; Solving Systems with Matrices

The matrix that represents the system of equations

$$\begin{cases} x - y + 2z = -3 \\ y + 3z = -5 \\ z = -2 \end{cases} \text{ is } \begin{bmatrix} 1 & -1 & 2 & | & -3 \\ 0 & 1 & 3 & | & -5 \\ 0 & 0 & 1 & | & -2 \end{bmatrix}$$

The coefficient part of this augmented matrix has all 1s on its diagonal and all 0s below its diagonal.

7.2 Matrix Solution of Systems of Linear Equations

Any augmented matrix that has 1s or 0s on the diagonal of its coefficient part and 0's below the diagonal is said to be in **row-echelon form**.

If the rows of an augmented matrix represent the equations of a system of linear equations, we can perform **row operations** on the matrices that are equivalent to operations that we performed on equations to solve a system of equations. These operations are as follows.

Matrix Row Operations

Row Operation	Corresponding Equation Operation
1. Interchange two rows of the matrix.	1. Interchange two equations.
2. Multiply a row by any nonzero constant.	2. Multiply an equation by any nonzero constant.
3. Add a multiple of one row to another row.	3. Add a multiple of one equation to another.

When a new matrix results from one or more of these row operations performed on a matrix, the new matrix is **equivalent** to the original matrix.

EXAMPLE 1 ▶ Matrix Solution of Linear Systems

Use matrix row operations to solve the system of equations

$$\begin{cases} 2x - 3y + z = -1 & (1) \\ x - y + 2z = -3 & (2) \\ 3x + y - z = 9 & (3) \end{cases}$$

SOLUTION

We begin by writing the augmented matrix that represents the system.

$$\begin{bmatrix} 2 & -3 & 1 & | & -1 \\ 1 & -1 & 2 & | & -3 \\ 3 & 1 & -1 & | & 9 \end{bmatrix}$$

To get a first row that has 1 in the first column, we interchange row 1 and row 2. (Compare this to the equivalent system of equations, which is to the right.)

$$\xrightarrow{R_1 \leftrightarrow R_2} \begin{bmatrix} 1 & -1 & 2 & | & -3 \\ 2 & -3 & 1 & | & -1 \\ 3 & 1 & -1 & | & 9 \end{bmatrix} \quad \begin{cases} x - y + 2z = -3 \\ 2x - 3y + z = -1 \\ 3x + y - z = 9 \end{cases}$$

To get 0 as the first entry in the second row, we multiply the first row by -2 and add it to the second row. To get 0 as the first entry in the third row, we multiply the first row by -3 and add it to the third row. (Compare this to the equivalent system of equations, which is to the right.)

$$\xrightarrow[-3R_1 + R_3 \to R_3]{-2R_1 + R_2 \to R_2} \begin{bmatrix} 1 & -1 & 2 & | & -3 \\ 0 & -1 & -3 & | & 5 \\ 0 & 4 & -7 & | & 18 \end{bmatrix} \quad \begin{cases} x - y + 2z = -3 \\ -y - 3z = 5 \\ 4y - 7z = 18 \end{cases}$$

To get 1 as the second entry in the second row, we multiply the second row by -1. (Compare this to the equivalent system of equations, which is to the right.)

Chapter 7 Systems of Equations and Matrices

$$-1R_2 \to R_2 \quad \begin{bmatrix} 1 & -1 & 2 & | & -3 \\ 0 & 1 & 3 & | & -5 \\ 0 & 4 & -7 & | & 18 \end{bmatrix} \quad \begin{cases} x - y + 2z = -3 \\ y + 3z = -5 \\ 4y - 7z = 18 \end{cases}$$

To get 0 as the second entry in the third row, we multiply the new second row by -4 and add it to the third row. (Compare this to the equivalent system of equations, which is to the right.)

$$-4R_2 + R_3 \to R_3 \quad \begin{bmatrix} 1 & -1 & 2 & | & -3 \\ 0 & 1 & 3 & | & -5 \\ 0 & 0 & -19 & | & 38 \end{bmatrix} \quad \begin{cases} x - y + 2z = -3 \\ y + 3z = -5 \\ -19z = 38 \end{cases}$$

To get 1 as the third entry in the third row, we multiply the third row by $-\dfrac{1}{19}$. (Compare this to the equivalent system of equations, which is to the right.)

$$(-1/19)R_3 \to R_3 \quad \begin{bmatrix} 1 & -1 & 2 & | & -3 \\ 0 & 1 & 3 & | & -5 \\ 0 & 0 & 1 & | & -2 \end{bmatrix} \quad \begin{cases} x - y + 2z = -3 \\ y + 3z = -5 \\ z = -2 \end{cases}$$

The matrix is now in row-echelon form. The equivalent system can be solved by back substitution, giving the solution $(2, 1, -2)$, the same solution that was found in Example 1 of Section 7.1.

Gauss–Jordan Elimination

One method that can be used to solve a system of n equations in n variables is called the **Gauss–Jordan elimination method**. To use this method, we attempt to reduce the $n \times n$ coefficient matrix to one that contains 1s on its diagonal and 0s everywhere else.

> The augmented matrix representing n equations in n variables is said to be in **reduced row-echelon form** if it has 1s or 0s on the diagonal of its coefficient part and 0s everywhere else.

When the original augmented matrix represents a system of n equations in n variables with a unique solution, this method will reduce the coefficient matrix of the augmented matrix to an identity matrix. When this happens, we can easily find the solution to the system, as we will see below.

Rather than use back substitution to complete the solution to the system of Example 1, we can apply several additional row operations to reduce the matrix to reduced row-echelon form. That is, we can reduce the matrix

$$\begin{bmatrix} 1 & -1 & 2 & | & -3 \\ 0 & 1 & 3 & | & -5 \\ 0 & 0 & 1 & | & -2 \end{bmatrix}$$

to reduced row-echelon form with the following steps:

1. Add row 2 to row 1:

$$R_2 + R_1 \to R_1 \quad \begin{bmatrix} 1 & 0 & 5 & | & -8 \\ 0 & 1 & 3 & | & -5 \\ 0 & 0 & 1 & | & -2 \end{bmatrix}$$

2. Add (-5) times row 3 to row 1 and (-3) times row 3 to row 2:

$$\xrightarrow[-3R_3 + R_2 \to R_2]{-5R_3 + R_1 \to R_1} \begin{bmatrix} 1 & 0 & 0 & | & 2 \\ 0 & 1 & 0 & | & 1 \\ 0 & 0 & 1 & | & -2 \end{bmatrix}$$

3. The coefficient matrix part of this reduced row-echelon matrix is the identity matrix, so the solutions to the system can be easily "read" from the reduced augmented matrix. The rows of this reduced row-echelon matrix translate into the equations

$$\begin{cases} x + 0y + 0z = 2 \\ 0x + y + 0z = 1 \\ 0x + 0y + z = -2 \end{cases} \text{ or } \begin{cases} x = 2 \\ y = 1 \\ z = -2 \end{cases}$$

This solution agrees with the solution to the system found in Example 1.

EXAMPLE 2 ▶ Investment

The Trust Department of Century Bank divided a $150,000 investment among three mutual funds with different levels of risk and return. The Potus Fund returns 10% per year, the Stong Fund returns 8%, and the Franklin Fund returns 7%. If the annual return from the combined investments is $12,900 and if the investment in the Potus Fund has $20,000 less than the sum of the investments in the other two funds, how much is invested in each fund?

SOLUTION

If we represent the amount invested in the Potus Fund by x, the amount invested in the Stong Fund by y, and the amount invested in the Franklin Fund by z, the equations that represent this situation are

$$x + y + z = 150{,}000$$
$$0.10x + 0.08y + 0.07z = 12{,}900$$
$$x = y + z - 20{,}000$$

We can write these equations in a system, create an augmented matrix, and reduce the matrix representing the system to reduced row-echelon form.

$$\begin{cases} x + y + z = 150{,}000 \\ 0.10x + 0.08y + 0.07z = 12{,}900 \\ x - y - z = -20{,}000 \end{cases}$$

$$\begin{bmatrix} 1 & 1 & 1 & | & 150{,}000 \\ 0.10 & 0.08 & 0.07 & | & 12{,}900 \\ 1 & -1 & -1 & | & -20{,}000 \end{bmatrix} \xrightarrow[-R_1 + R_3 \to R_3]{-0.10R_1 + R_2 \to R_2} \begin{bmatrix} 1 & 1 & 1 & | & 150{,}000 \\ 0 & -0.02 & -0.03 & | & -2{,}100 \\ 0 & -2 & -2 & | & -170{,}000 \end{bmatrix}$$

$$\xrightarrow{-50R_2 \to R_2} \begin{bmatrix} 1 & 1 & 1 & | & 150{,}000 \\ 0 & 1 & 1.5 & | & 105{,}000 \\ 0 & -2 & -2 & | & -170{,}000 \end{bmatrix} \xrightarrow[2R_2 + R_3 \to R_3]{-R_2 + R_1 \to R_1} \begin{bmatrix} 1 & 0 & -0.5 & | & 45{,}000 \\ 0 & 1 & 1.5 & | & 105{,}000 \\ 0 & 0 & 1 & | & 40{,}000 \end{bmatrix}$$

$$\xrightarrow[-1.5R_3 + R_2 \to R_2]{0.5R_3 + R_1 \to R_1} \begin{bmatrix} 1 & 0 & 0 & | & 65{,}000 \\ 0 & 1 & 0 & | & 45{,}000 \\ 0 & 0 & 1 & | & 40{,}000 \end{bmatrix}$$

From the reduced row-echelon matrix, we see that

$$x = 65{,}000, \quad y = 45{,}000, \quad \text{and} \quad z = 40{,}000$$

Thus, the investments are $65,000 in the Potus Fund, $45,000 in the Stong Fund, and $40,000 in the Franklin Fund.

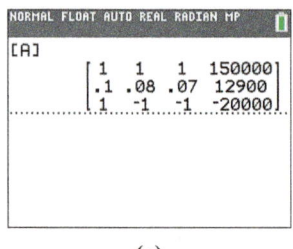

Solution with Technology

Graphing calculators are very useful in reducing an augmented matrix to solve a system of linear equations. To solve a system of linear equations with a graphing calculator, we use the following steps.

1. Enter the coefficients and constants in a matrix: Select MATRIX (accessed by **2ND** **x⁻¹**), move to EDIT, select a matrix name, enter the number of rows and columns of the matrix, and enter the coefficients of x, y, and z and the constants into columns 1, 2, 3, and 4, respectively. See Figure 7.6(a), which shows the augmented matrix for the system of equations in Example 2.*

2. Reduce the augmented matrix: Select MATRIX, move to MATH, choose "B:rref(" on the display, and select MATRIX and [A] under NAMES (getting rref([A]) on the screen). Pressing **ENTER** gives the reduced matrix, from which the solution can be read. See Figure 7.6(b), which shows the same solution matrix as the one found in Example 2.†

Solving systems of linear equations in more than three variables can also be done using Gauss–Jordan elimination. Because of the increased number of steps involved, technology is especially useful in solving larger systems.††

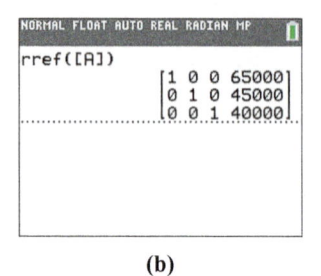

Figure 7.6

EXAMPLE 3 ▶ Four Equations in Four Variables

Solve the system

$$\begin{cases} x + y + z + w = 3 \\ x - 2y + z - 4w = -5 \\ x - z + w = 0 \\ y + z + w = 2 \end{cases}$$

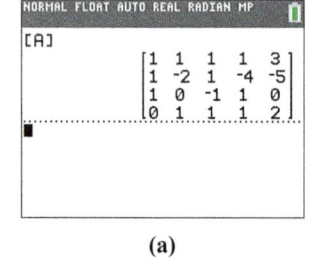

SOLUTION

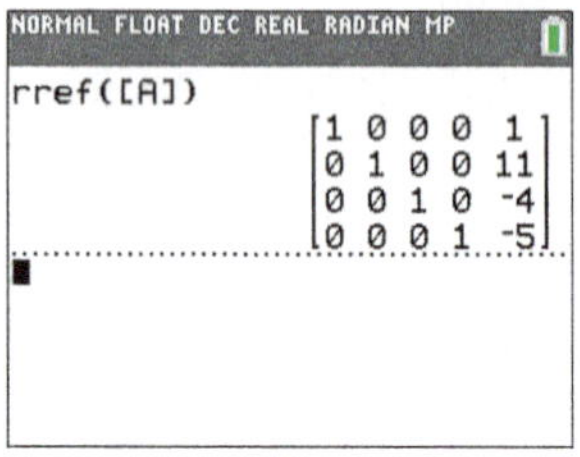

This system can be represented by the augmented matrix

$$\begin{bmatrix} 1 & 1 & 1 & 1 & | & 3 \\ 1 & -2 & 1 & -4 & | & -5 \\ 1 & 0 & -1 & 1 & | & 0 \\ 0 & 1 & 1 & 1 & | & 2 \end{bmatrix}$$

We can enter this augmented matrix into a graphing calculator (Figure 7.7(a)) and reduce the matrix to its reduced row-echelon form (Figure 7.7(b)). The solution to the system can be "read" from the reduced matrix:

$$x = 1, \quad y = 11, \quad z = -4, \quad w = -5, \quad \text{or} \quad (1, 11, -4, -5)$$

Figure 7.7

Nonunique Solution

If we represent a system of three equations in three variables by an augmented matrix and the coefficient matrix reduces to a 3×3 identity matrix, then the system has a unique solution. If the coefficient matrix representing a system of three equations in three variables does not reduce to an identity matrix, then either there is no solution to the system or there are an infinite number of solutions.

If a row of the reduced row-echelon coefficient matrix associated with a system contains all 0s and the augment of that row contains a nonzero number, the system has no solution and is an **inconsistent system**.

---

*For detailed steps, see Appendix A, page 695.
†For detailed steps, see Appendix A, page 697.
††For details of this solution method, see Appendix A, page 697.

If a row of the reduced 3 × 3 row-echelon coefficient matrix associated with a system contains all 0s and the augment of that row also contains 0, then there are infinitely many solutions. If any column of the reduced matrix does not contain a leading 1, that variable can assume any value, and the variables corresponding to columns containing leading 1s will have values dependent on that variable. (There may be more than one variable that can assume any value, with other variables depending on them.) This is a **dependent system**, and it has infinitely many solutions.

Of course, any system with two equations in three variables cannot have a unique solution.

A system with fewer equations than variables has either infinitely many solutions or no solutions.

Dependent Systems

The following example illustrates the solution of a dependent system.

EXAMPLE 4 ▶ A Dependent System

Solve the system

$$\begin{cases} 5x + 10y + 12z = 6{,}140 \\ 10x + 18y + 30z = 13{,}400 \\ 300x + 480y + 1080z = 435{,}600 \end{cases}$$

SOLUTION

The augmented matrix that represents the system is

$$\begin{bmatrix} 5 & 10 & 12 & | & 6{,}140 \\ 10 & 18 & 30 & | & 13{,}400 \\ 300 & 480 & 1080 & | & 435{,}600 \end{bmatrix}$$

The procedure follows:

$$\begin{bmatrix} 5 & 10 & 12 & | & 6{,}140 \\ 10 & 18 & 30 & | & 13{,}400 \\ 300 & 480 & 1080 & | & 435{,}600 \end{bmatrix} \xrightarrow{(1/5)R_1 \to R_1} \begin{bmatrix} 1 & 2 & 2.4 & | & 1{,}228 \\ 10 & 18 & 30 & | & 13{,}400 \\ 300 & 480 & 1080 & | & 435{,}600 \end{bmatrix}$$

$$\xrightarrow[\substack{-10R_1 + R_2 \to R_2 \\ -300R_1 + R_3 \to R_3}]{} \begin{bmatrix} 1 & 2 & 2.4 & | & 1{,}228 \\ 0 & -2 & 6 & | & 1{,}120 \\ 0 & -120 & 360 & | & 67{,}200 \end{bmatrix}$$

$$\xrightarrow{(-1/2)R_2 \to R_2} \begin{bmatrix} 1 & 2 & 2.4 & | & 1{,}228 \\ 0 & 1 & -3 & | & -560 \\ 0 & -120 & 360 & | & 67{,}200 \end{bmatrix}$$

$$\xrightarrow[\substack{120R_2 + R_3 \to R_3 \\ -2R_2 + R_1 \to R_1}]{} \begin{bmatrix} 1 & 0 & 8.4 & | & 2348 \\ 0 & 1 & -3 & | & -560 \\ 0 & 0 & 0 & | & 0 \end{bmatrix}$$

Note that the third row contains only 0s and there is no leading 1 in the "z" column of the reduced matrix, and that z occurs in the equations that correspond to the first two rows of the matrix. Thus, we can solve for x in terms of z in the first equation and for y in terms of z in the second equation. These solutions are

$$x = -8.4z + 2348$$
$$y = 3z - 560$$
$$z = \text{any real number}$$

Different values of z give different solutions to the system. Two sample solutions are $z = 0, y = -560, x = 2348$ and $z = 10, y = -530, x = 2264$. If $z = a$, the solution is $(-8.4a + 2348, 3a - 560, a)$.

Notice that the row-reduction technique is the same for systems that have nonunique solutions as it is for systems that have unique solutions. To solve a system of n linear equations in n variables, we attempt to reduce the coefficient matrix to the identity matrix. If we succeed, the system has a unique solution. If we are unable to reduce the coefficient matrix to the identity matrix, either there is no solution or the system is dependent and we can solve it in terms of one or more of the variables.

Figure 7.8(a) shows the original augmented matrix for the system of Example 4 on a calculator screen, and Figure 7.8(b) shows the reduced row-echelon form that is equivalent to the original matrix. This is the same as the reduced matrix found in Example 4 and yields the same solution.

(a)   (b)

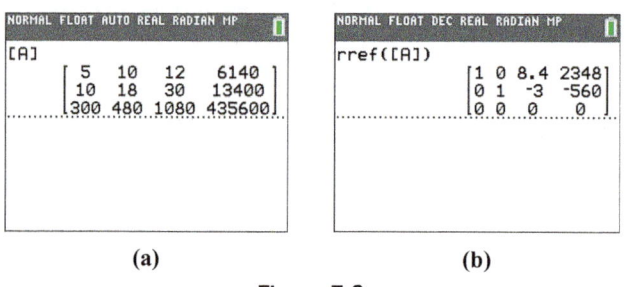

Figure 7.8

EXAMPLE 5 ▶ Transportation

Ace Trucking Company has an order for delivery of three products: A, B, and C. Table 7.4 gives the volume in cubic feet, the weight in pounds, and the value for insurance in dollars per unit of each of the products. If the company can carry 30,000 cubic feet and 62,000 pounds and is insured for $276,000, how many units of each product can be carried if the shipment of product C is as large as possible?

Table 7.4

	Product A	Product B	Product C
Volume (cu ft)	25	22	30
Weight (lb)	25	38	70
Value ($)	150	180	300

SOLUTION

If we represent the number of units of product A by x, the number of units of product B by y, and the number of units of product C by z, then we can write a system of equations to represent the problem.

$$\begin{cases} 25x + 22y + 30z = 30{,}000 & \text{Volume} \\ 25x + 38y + 70z = 62{,}000 & \text{Weight} \\ 150x + 180y + 300z = 276{,}000 & \text{Value} \end{cases}$$

7.2 Matrix Solution of Systems of Linear Equations

The Gauss–Jordan elimination method gives

$$\begin{bmatrix} 25 & 22 & 30 & | & 30{,}000 \\ 25 & 38 & 70 & | & 62{,}000 \\ 150 & 180 & 300 & | & 276{,}000 \end{bmatrix} \xrightarrow{(1/25)R_1 \to R_1} \begin{bmatrix} 1 & \frac{22}{25} & \frac{6}{5} & | & 1{,}200 \\ 25 & 38 & 70 & | & 62{,}000 \\ 150 & 180 & 300 & | & 276{,}000 \end{bmatrix}$$

$$\xrightarrow[-150R_1 + R_3 \to R_3]{-25R_1 + R_2 \to R_2} \begin{bmatrix} 1 & \frac{22}{25} & \frac{6}{5} & | & 1{,}200 \\ 0 & 16 & 40 & | & 32{,}000 \\ 0 & 48 & 120 & | & 96{,}000 \end{bmatrix}$$

$$\xrightarrow{(1/16)R_2 \to R_2} \begin{bmatrix} 1 & \frac{22}{25} & \frac{6}{5} & | & 1{,}200 \\ 0 & 1 & \frac{5}{2} & | & 2{,}000 \\ 0 & 48 & 120 & | & 96{,}000 \end{bmatrix}$$

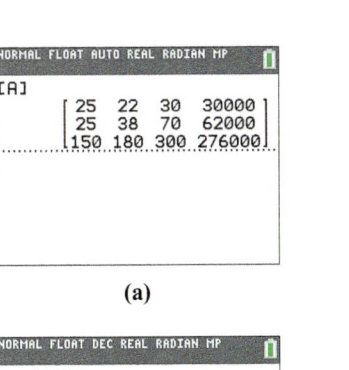

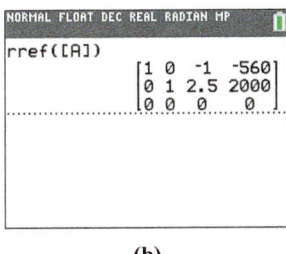

$$\xrightarrow[-48R_2 + R_3 \to R_3]{(-22/25)R_2 + R_1 \to R_1} \begin{bmatrix} 1 & 0 & -1 & | & -560 \\ 0 & 1 & \frac{5}{2} & | & 2000 \\ 0 & 0 & 0 & | & 0 \end{bmatrix}$$

Reducing this augmented matrix by algebraic methods (above) is quite time consuming. Much time and effort can be saved if we use technology to obtain the reduced row-echelon form (Figure 7.9).

The solution to this system is $x = -560 + z$, $y = 2000 - 2.5z$, with the values of z limited so that all values are nonnegative integers. Because x must be nonnegative, z must be at least 560, and because y must be a nonnegative integer, z must be an even number that is no more than 800. This also means that the size of the shipments of products A and B is limited by the size of the shipments of product C, because the limits on z put limits on x and y. We can write the solution as follows.

Product A: $x = -560 + z$
Product B: $y = 2000 - 2.5z$
Product C: $560 \le z \le 800$ (z is even)

The maximum shipment of product C is 800 units, making the shipment of products A and B equal to 240 units and 0 units, respectively.

(a)

(b)

Figure 7.9

Inconsistent Systems

The following example shows how we can use matrices to determine that a system has no solution.

EXAMPLE 6 ▶ An Inconsistent System

Solve the following system of equations if a solution exists.

$$\begin{cases} 5x + 10y + 12z = 6{,}140 \\ 10x + 18y + 30z = 13{,}400 \\ 300x + 480y + 1080z = 214{,}800 \end{cases}$$

SOLUTION

The augmented matrix for this system is $\begin{bmatrix} 5 & 10 & 12 & | & 6{,}140 \\ 10 & 18 & 30 & | & 13{,}400 \\ 300 & 480 & 1080 & | & 214{,}800 \end{bmatrix}$.

We reduce this matrix as follows.

$\begin{bmatrix} 5 & 10 & 12 & | & 6{,}140 \\ 10 & 18 & 30 & | & 13{,}400 \\ 300 & 480 & 1080 & | & 214{,}800 \end{bmatrix} \xrightarrow{(1/5)R_1 \to R_1} \begin{bmatrix} 1 & 2 & 2.4 & | & 1{,}228 \\ 10 & 18 & 30 & | & 13{,}400 \\ 300 & 480 & 1080 & | & 214{,}800 \end{bmatrix}$

$\xrightarrow[-300R_1 + R_3 \to R_3]{-10R_1 + R_2 \to R_2} \begin{bmatrix} 1 & 2 & 2.4 & | & 1{,}228 \\ 0 & -2 & 6 & | & 1{,}120 \\ 0 & -120 & 360 & | & -153{,}600 \end{bmatrix}$

$\xrightarrow{(-1/2)R_2 \to R_2} \begin{bmatrix} 1 & 2 & 2.4 & | & 1{,}228 \\ 0 & 1 & -3 & | & -560 \\ 0 & -120 & 360 & | & -153{,}600 \end{bmatrix}$

$\xrightarrow[120R_2 + R_3 \to R_3]{-2R_2 + R_1 \to R_1} \begin{bmatrix} 1 & 0 & 8.4 & | & 2{,}348 \\ 0 & 1 & -3 & | & -560 \\ 0 & 0 & 0 & | & -220{,}800 \end{bmatrix}$

This matrix represents the reduced system that is equivalent to the original system.

$$\begin{cases} x + 0y + 8.4z = 2{,}348 \\ 0x + y - 3z = -560 \\ 0x + 0y + 0z = -220{,}800 \end{cases}$$

The third equation is $0 = -220{,}800$, which is impossible. Thus, the system has no solution.

Figure 7.10(a) shows the augmented matrix and Figure 7.10(b) gives the resulting reduced row-echelon form for the system of equations in Example 6.

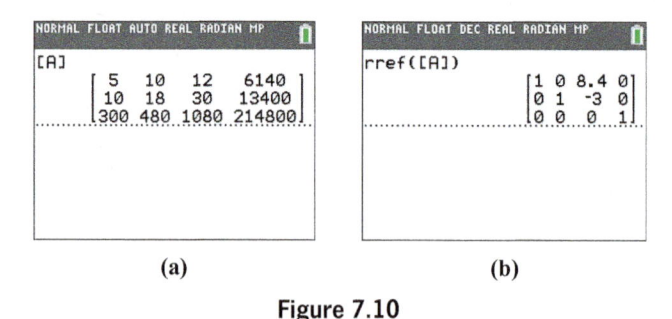

(a)   (b)

Figure 7.10

Finally, we consider systems with more equations than variables. Such systems of equations may have zero, one, or many solutions, and they are solved in the same manner as other systems.

1. If the reduced augmented matrix contains a row of 0s in the coefficient matrix with a nonzero number in the augment, the system has no solution.
2. If the coefficient matrix in the reduced augmented matrix contains an identity matrix and all remaining rows of the reduced augmented matrix contain all 0s, there is a unique solution to the system.
3. Otherwise, the system has many solutions.

SKILLS CHECK 7.2

Answers that are not seen can be found in the answer section at the back of the text.

In Exercises 1–4, write the augmented matrix associated with the given system.

1. $\begin{cases} x + y - z = 4 \\ x - 2y - z = -2 \\ 2x + 2y + z = 11 \end{cases}$

2. $\begin{cases} 3x - 4y + 6z = 10 \\ 2x - 4y - 5z = -14 \\ x + 2y - 3z = 0 \end{cases}$

3. $\begin{cases} 5x - 3y + 2z = 12 \\ 3x + 6y - 9z = 4 \\ 2x + 3y - 4z = 9 \end{cases}$

4. $\begin{cases} x - 3y + 4z = 7 \\ 2x + 2y - 3z = -3 \\ x - 3y + z = -2 \end{cases}$

In Exercises 5–14, the matrix associated with the solution to a system of linear equations in x, y, and z is given. Write the solution to the system, if it exists.

5. $\begin{bmatrix} 1 & 0 & 0 & | & -1 \\ 0 & 1 & 0 & | & 4 \\ 0 & 0 & 1 & | & -2 \end{bmatrix}$ $x = -1, y = 4, z = -2$

6. $\begin{bmatrix} 1 & 0 & 0 & | & -2 \\ 0 & 1 & 0 & | & 4 \\ 0 & 0 & 1 & | & 8 \end{bmatrix}$ $x = -2, y = 4, z = 8$

7. $\begin{bmatrix} 1 & 1 & -1 & | & 4 \\ 1 & -2 & -1 & | & -2 \\ 2 & 2 & 1 & | & 11 \end{bmatrix}$ $x = 3, y = 2, z = 1$

8. $\begin{bmatrix} 1 & -1 & 3 & | & 4 \\ 2 & 3 & -1 & | & 11 \\ 4 & 2 & -4 & | & 12 \end{bmatrix}$ $x = 3, y = 2, z = 1$

9. $\begin{bmatrix} 2 & -3 & 4 & | & 13 \\ 1 & -2 & 1 & | & 3 \\ 2 & -3 & 1 & | & 4 \end{bmatrix}$ $x = 2, y = 1, z = 3$

10. $\begin{bmatrix} 3 & 1 & 2 & | & 1 \\ 2 & 3 & -4 & | & -20 \\ 2 & 4 & 8 & | & 14 \end{bmatrix}$ $x = -1, y = -2, z = 3$

11. $\begin{bmatrix} 1 & 0 & 4 & | & 3 \\ 0 & 1 & 2 & | & 2 \\ 0 & 0 & 0 & | & 1 \end{bmatrix}$ Inconsistent system, no solution

12. $\begin{bmatrix} 1 & 0 & 2 & | & 1 \\ 0 & 1 & 3 & | & 5 \\ 0 & 0 & 0 & | & 0 \end{bmatrix}$

13. $\begin{bmatrix} 1 & 0 & 3 & | & 2 \\ 0 & 1 & -5 & | & 5 \\ 0 & 0 & 0 & | & 0 \end{bmatrix}$

14. $\begin{bmatrix} 1 & 0 & -1 & | & 3 \\ 0 & 1 & 2 & | & -2 \\ 0 & 0 & 0 & | & 0 \end{bmatrix}$

Solve the systems in Exercises 15–22.

15. $\begin{cases} x + y - z = 0 \\ x - 2y - z = 6 \\ 2x + 2y + z = 3 \end{cases}$ $x = 3, y = -2, z = 1$

16. $\begin{cases} x - 2y + z = -5 \\ 2x - y + 2z = 6 \\ 3x + 2y - z = 1 \end{cases}$ $x = -1, y = \dfrac{16}{3}, z = \dfrac{20}{3}$

17. $\begin{cases} 3x - 2y + 5z = 15 \\ x - 2y - 2z = -1 \\ 2x - 2y = 0 \end{cases}$ $x = -\dfrac{25}{3}, y = -\dfrac{25}{3}, z = \dfrac{14}{3}$

18. $\begin{cases} x + 3y + 5z = 5 \\ 2x + 4y + 3z = 9 \\ 2x + 3y + z = 1 \end{cases}$ $x = -24, y = 18, z = -5$

19. $\begin{cases} 2x + 3y + 4z = 5 \\ 6x + 7y + 8z = 9 \\ 2x + y + z = 1 \end{cases}$ $x = 0, y = -1, z = 2$

20. $\begin{cases} 4x + 3y + 8z = 1 \\ 2x + 3y + 8z = 5 \\ 2x + 5y + 5z = 6 \end{cases}$ $x = -2, y = \dfrac{7}{5}, z = \dfrac{3}{5}$

21. $\begin{cases} x - y + z - w = -2 \\ 2x + 4z + w = 5 \\ 2x - 3y + z = -5 \\ y + 2z + 20w = 4 \end{cases}$ $x = 40, y = 22, z = -19, w = 1$

22. $\begin{cases} x - 2y + z - 3w = 10 \\ 2x - 3y + 4z + w = 12 \\ 2x - 3y + z - 4w = 7 \\ x - y + z + w = 4 \end{cases}$ $x = -57, y = -70, z = -25, w = 16$

In Exercises 23–32, find the solutions, if any exist, to the systems.

23. $\begin{cases} -2x + 3y + 2z = 13 \\ -2x - 2y + 3z = 0 \\ 4x + y + 4z = 11 \end{cases}$ $x = 0, y = 3, z = 2$

24. $\begin{cases} 2x + 3y + 4z = 5 \\ x + y + z = 1 \\ 6x + 7y + 8z = 9 \end{cases}$

25. $\begin{cases} 2x + 5y + 6z = 6 \\ 3x - 2y + 2z = 4 \\ 5x + 3y + 8z = 10 \end{cases}$

26. $\begin{cases} -x + 5y - 3z = 10 \\ 3x + 7y + 2z = 5 \\ 4x + 12y - z = 15 \end{cases}$ $x = 0, y = \dfrac{35}{31}, z = -\dfrac{45}{31}$

27. $\begin{cases} -x - 5y + 3z = -2 \\ 3x + 7y + 2z = 5 \\ 4x + 12y - z = 7 \end{cases}$ Infinitely many solutions of the form $x = \dfrac{11}{8} - \dfrac{31z}{8}, y = \dfrac{1}{8} + \dfrac{11z}{8}, z = z$

28. $\begin{cases} x - 3y + 2z = 12 \\ 2x - 6y + z = 7 \end{cases}$

29. $\begin{cases} 2x - 3y + 2z = 5 \\ 4x + y - 3z = 6 \end{cases}$

30. $\begin{cases} 3x + 2y - z = 4 \\ 2x - 3y + z = 3 \end{cases}$

31. $\begin{cases} x - 3z - 3w = -2 \\ x + y + z + 3w = 2 \\ 2x + y - 2z - 2w = 0 \\ 3x + 2y - z + w = 2 \end{cases}$ Infinitely many solutions of the form $x = -2 + 3z, y = 4 - 4z, z = z, w = 0$

32. $\begin{cases} x + y + z + 5w = 10 \\ x + 2y + z + 6w = 16 \\ x + y + 2z + 7w = 11 \\ 2x + 3y + 3z + 13w = 27 \end{cases}$ Infinitely many solutions of the form $x = 3 - 2w, y = 6 - w, z = 1 - 2w, w = w$

EXERCISES 7.2

Answers that are not seen can be found in the answer section at the back of the text.

33. **Ticket Pricing** A theater owner wants to divide a 3600-seat theater into three sections, with tickets costing $40, $70, and $100, depending on the section. He wants to have twice as many $40 tickets as the sum of the other types of tickets, and he wants to earn $192,000 from a full house. Find how many seats he should have in each section.
2400 $40 seats, 800 $70 seats, 400 $100 seats

34. **Rental Cars** A car rental agency rents compact, midsize, and luxury cars. Its goal is to purchase 90 cars with a total of $2,270,000 and to earn a daily rental of $3150 from all the cars. The compact cars cost $18,000 each and earn $25 per day in rental, the midsize cars cost $25,000 each and earn $35 per day, and the luxury cars cost $40,000 each and earn $55 per day. Find the number of each type of car the agency should purchase to meet its goal.
40 compact, 30 midsize, 20 luxury

35. **Testing** A professor wants to create a test that has 15 true-false questions, 10 multiple-choice questions, and 5 essay questions. She wants the test to be worth 100 points, with each multiple-choice question worth twice as many points as a true-false question and with each essay question equal to three times the number of points of a true-false question.

    a. Write a system of equations to represent this problem, with x, y, and z equal to the number of points for a true-false, multiple-choice, and essay question, respectively.

    b. How many points should be assigned to each type of problem? Note that only positive integer answers are useful.
    2 points for T-F, 4 points for MC, 6 points for essay

36. **Testing** A professor wants to create a test that has true-false questions, multiple-choice questions, and essay questions. She wants the test to have 35 questions, with twice as many multiple-choice questions as essay questions and with twice as many true-false questions as multiple-choice questions.

    a. Write a system of equations to represent this problem.

    b. How many of each type question should the professor put on the test? Note that only positive integer answers are useful. 20 T-F, 10 MC, and 5 essay

37. **Investment** A company offers three mutual fund plans for its employees. Plan I consists of 4 blocks of common stock and 2 municipal bonds. Plan II consists of 8 blocks of common stock, 4 municipal bonds, and 6 blocks of preferred stock. Plan III consists of 14 blocks of common stock, 6 municipal bonds, and 6 blocks of preferred stock. If an employee wants to combine these plans so that she has 42 blocks of common stock, 20 municipal bonds, and 18 blocks of preferred stock, how many units of each plan does she need?
3 units of I, 2 units of II, 1 unit of III

38. **Manufacturing** A manufacturer of swing sets has three models, Deluxe, Premium, and Ultimate, which must be painted, assembled, and packaged for shipping. The following table gives the number of hours required for each of these operations for each type of swing set. If the manufacturer has 55 hours available per day for painting, 75 hours for assembly, and 40 hours for packaging, how many of each type swing set can be produced each day?
25 Deluxe, no Premium, 25 Ultimate

	Deluxe	Premium	Ultimate
Painting	0.8	1	1.4
Assembly	1	1.5	2
Packaging	0.6	0.75	1

39. **Nutrition** A psychologist, studying the effects of good nutrition on the behavior of rabbits, feeds one group a combination of three foods: I, II, and III. Each of these foods contains three additives: A, B, and C. The table below gives the percent of each additive that is present in each food. If the diet being used requires 3.74 grams per day of A, 2.04 grams of B, and 1.35 grams of C, find the number of grams of each food that should be used each day.
5 g of I, 6 g of II, 8 g of III

	Food I	Food II	Food III
Additive A	12%	15%	28%
Additive B	8%	6%	16%
Additive C	15%	2%	6%

40. **Manufacturing** To expand its manufacturing capacity, Krug Industries borrowed $440,000, part at 6%, part at 8%, and part at 10%. The sum of the money borrowed at 6% and 8% was three times that borrowed at 10%. The loan was repaid in full at the end of 5 years and the annual interest paid was $34,400. How much was borrowed at each rate?
$150,000 at 6%, $180,000 at 8%, $110,000 at 10%

Some of the following exercises have nonunique solutions.

41. **Social Services** A social agency is charged with providing services to three types of clients: A, B, and C. A total of 500 clients are to be served, with $300,000 available for counseling and $200,000 available for emergency food and shelter. Type A clients require an average of $400 for counseling and $600 for emergencies. Type B clients require an average of

$1000 for counseling and $400 for emergencies. Type C clients require an average of $600 for counseling and $200 for emergencies. How many of each type of client can be served?
200 type A clients, 100 type B clients, 200 type C clients

42. **Investment** A man has $235,000 invested in three rental properties. One property earns 7.5% per year on the investment, a second earns 10%, and the third earns 8%. The annual earnings from the properties total $18,000.

   a. Write a system of two equations to represent the problem, with x, y, and z representing the 7.5%, 10%, and 8% investments, respectively.

   b. Solve this system. $x = 220{,}000 - 0.8z$, $y = 15{,}000 - 0.2z$, z = any number, subject to x, y, z positive integers

   c. If $60,000 is invested at 8%, how much is invested in each of the other properties?
   $172,000 at 7.5%, $3000 at 10%

43. **Investment** A brokerage house offers three stock portfolios for its clients. Portfolio I consists of 10 blocks of common stock, 2 municipal bonds, and 3 blocks of preferred stock. Portfolio II consists of 12 blocks of common stock, 8 municipal bonds, and 5 blocks of preferred stock. Portfolio III consists of 10 blocks of common stock, 6 municipal bonds, and 4 blocks of preferred stock. A client wants to combine these portfolios so that she has 180 blocks of common stock, 140 municipal bonds, and 110 blocks of preferred stock. Can she do this? To answer this question, let x equal the number of units of portfolio I, y equal the number of units of portfolio II, and z equal the number of units of portfolio III, so that the equation $10x + 12y + 10z = 180$ represents the total number of blocks of common stock.

   a. Write the remaining two equations to create a system of three equations. $2x + 8y + 6z = 140$, $3x + 5y + 4z = 110$

   b. Solve the system of equations, if possible. Not possible

44. **Investment** A company offers three mutual fund plans for its employees. Plan I consists of 14 blocks of common stock, 4 municipal bonds, and 6 blocks of preferred stock. Plan II consists of 4 blocks of common stock and 2 municipal bonds. Plan III consists of 18 blocks of common stock, 6 municipal bonds, and 6 blocks of preferred stock. Suppose an employee wants to combine these plans so that she has 58 blocks of common stock, 20 municipal bonds, and 18 blocks of preferred stock. How many units of each plan does she need if she wants 1 unit of Plan III?
   2 units of Plan I, 3 units of Plan II, 1 unit of plan III

45. **Purchasing** A young man wins $100,000 and decides to buy four new cars. He wants to choose from cars that are priced at $40,000, $30,000, and $20,000 and spend all of the money.

   a. Write a system of two equations in three variables to represent the problem.

   b. Can this system have a unique solution? No

   c. Solve the system. $x = z - 2$, $y = 6 - 2z$, $2 \leq z \leq 3$

   d. Use the context of the problem to find how many cars of each price he can buy with the $100,000.

46. **Investment** A trust account manager has $400,000 to invest in three different accounts. The accounts pay 8%, 10%, and 12%, respectively, and the goal is to earn $42,400 with minimum risk. To solve this problem, assume that x dollars are invested at 8%, y dollars are invested at 10%, and z dollars are invested at 12%.

   a. Write a system of two equations in three variables to represent the problem.

   b. How much can be invested in each account with the largest possible amount invested at 8%?
   $260,000 at 12%, $0 at 10%, $140,000 at 8%

47. **Investments** An investor wants to place a total of $500,000 in two or more of three investments with different degrees of risk. One investment pays 4%, one pays 8%, and one pays 10%, and the investor wants to earn $35,000 from the investments.

   a. Write a system of two equations to represent the problem, representing the amount in the 4% investment by x, the amount in the 8% investment by y, and the amount in the 10% investment by z.

   b. Solve the system.

   c. If the investor wants to put $225,000 in the 8% investment, how much should she put in each of the other investments?

48. **Investing** An investor wants to place a total of $700,000 in two or more of three investments with different degrees of risk. One investment pays 4%, one pays 8%, and one pays 12%, and the investor wants to earn $42,000 from the investments.

   a. Write a system of two equations to represent the problem, representing the amount in the 4% investment by x, the amount in the 8% investment by y, and the amount in the 12% investment by z.

   b. Solve the system.

   c. If the investor wants to put the maximum amount in the 4% investment because of lower risk, how much should he put in each investment?

49. **Traffic Flow** In an analysis of traffic, a certain city estimates the traffic flow as illustrated in the figure on the next page, where the arrows indicate the flow of the traffic. If x_1 represents the number of cars traveling from intersection A to intersection B, x_2 represents the number of cars traveling from intersection B to intersection C, and so on, we can formulate equations based on the principle that the number of vehicles entering the intersection equals the number leaving it.

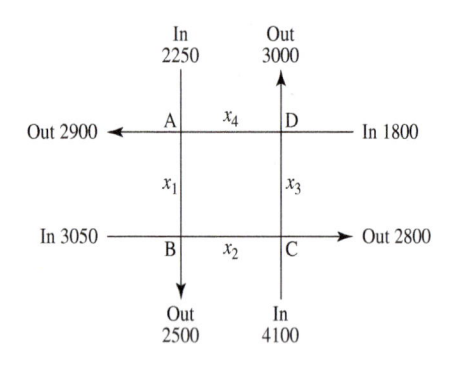

560　Chapter 7　Systems of Equations and Matrices

The arrows indicate that $3050 + x_1$ cars are entering intersection B and that $2500 + x_2$ cars are leaving intersection B. Thus, the equation that describes the number of cars entering and leaving intersection B is $3050 + x_1 = 2500 + x_2$. The equations that describe the numbers of cars entering and leaving all intersections are

B:　$3050 + x_1 = 2500 + x_2$
C:　$4100 + x_2 = x_3 + 2800$
D:　$x_3 + 1800 = 3000 + x_4$
A:　$x_4 + 2250 = x_1 + 2900$

Solve the system of these four equations to find how traffic between the other intersections is related to the traffic from intersection D to intersection A.

50. *Traffic Flow* In an analysis of traffic, a retirement community estimates the traffic flow on its "town square" at 6 P.M. to be as illustrated in the figure. If x_1 illustrates the number of cars moving from intersection A to intersection B, x_2 represents the number of cars traveling from intersection B to intersection C, and so on, we can formulate equations based on the principle that the number of vehicles entering the intersection equals the number leaving it. For example, the equation that represents the traffic through A is $x_4 + 470 = x_1 + 340$.

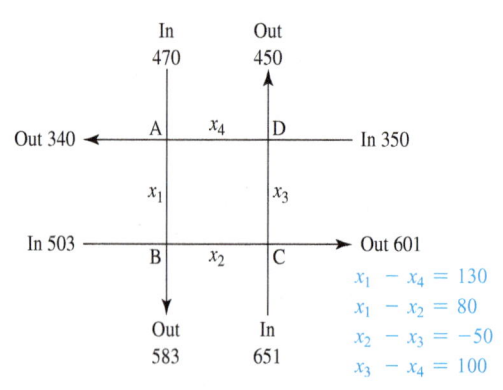

$x_1 - x_4 = 130$
$x_1 - x_2 = 80$
$x_2 - x_3 = -50$
$x_3 - x_4 = 100$

a. Formulate an equation for the traffic at each of the four intersections.

b. Solve the system of these four equations to find how traffic between the other intersections is related to the traffic from intersection D to intersection A.

51. *Irrigation* An irrigation system allows water to flow in the pattern shown in the figure below. Water flows into the system at A and exits at B, C, and D, with amounts shown. If x_1 represents the number of gallons of water moving from A to B, x_2 represents the number of gallons moving from A to C, and so on, we can formulate equations using the fact that at each point the amount of water entering the system equals the amount exiting. For example, the equation that represents the water flow through C is $x_2 = x_3 + 200,000$.

a. Formulate an equation for the water flow at each of the other three points. At A, $400,000 = x_1 + x_2$; at B, $x_1 = x_4 + 100,000$; at D, $x_3 + x_4 = 100,000$

b. Solve the system of these four equations.

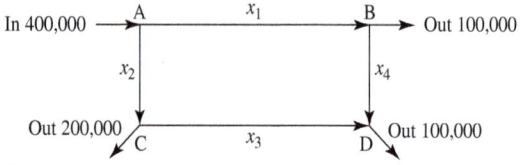

$x_1 = 100,000 + x_4$, $x_2 = 300,000 - x_4$, $x_3 = 100,000 - x_4$, where x_4 is the number of gallons flowing from B to D; $x_4 \le 100,000$

7.3　Matrix Operations

KEY OBJECTIVES
- Add two matrices
- Subtract two matrices
- Multiply a matrix by a constant
- Multiply two matrices

SECTION PREVIEW　Expected Life Span

Table 7.5 gives the years of life expected at birth for male and female blacks and whites born in the United States for selected years from 2020 and projected to 2060. To determine what the data tell us about the relationships among race, sex, and life expectancy, we can make a matrix W containing the information for whites and a matrix B for blacks, and use these matrices to find additional information. For example, we can find a matrix D that shows how many more years whites in each category are expected to live than blacks.

Table 7.5 Life Expectancy

	Whites		Blacks	
Year	Males	Females	Males	Females
2020	78.6	82.9	74.0	79.8
2030	80.2	84.1	76.1	81.3
2040	81.7	85.2	78.0	82.8
2050	83.2	86.4	79.8	84.1
2060	84.5	87.4	81.4	85.3

(*Source*: National Center for Health Statistics)

We used matrices to solve systems of equations in the last section. In this section, we consider the operations of addition, subtraction, and multiplication of matrices, and we use these operations in applications like the one above. ■

Addition and Subtraction of Matrices

If two matrices have the same numbers of rows and columns (the same dimensions), we can add them by adding the corresponding entries of the two matrices.

> **Matrix Addition**
>
> The sum of two matrices with the same dimensions is the matrix that is formed by adding the corresponding entries of the two matrices. Addition is not defined if the matrices do not have the same number of rows and the same number of columns.

EXAMPLE 1 ▶ Matrix Addition

Find the following sums of matrices.

a. $\begin{bmatrix} a & b \\ c & d \end{bmatrix} + \begin{bmatrix} w & y \\ x & z \end{bmatrix}$

b. $\begin{bmatrix} 1 & 3 & -8 & 0 \\ 3 & 6 & 1 & -3 \\ -4 & 5 & 3 & 2 \end{bmatrix} + \begin{bmatrix} -1 & 4 & 2 & 4 \\ 5 & -2 & 4 & 2 \\ -5 & 1 & 4 & -1 \end{bmatrix}$

c. $A + B$ if $A = \begin{bmatrix} 2 & -3 \\ -1 & 4 \end{bmatrix}$ and $B = \begin{bmatrix} -2 & 3 \\ 1 & -4 \end{bmatrix}$

SOLUTION

a. $\begin{bmatrix} a & b \\ c & d \end{bmatrix} + \begin{bmatrix} w & y \\ x & z \end{bmatrix} = \begin{bmatrix} a+w & b+y \\ c+x & d+z \end{bmatrix}$

b. $\begin{bmatrix} 1 & 3 & -8 & 0 \\ 3 & 6 & 1 & -3 \\ -4 & 5 & 3 & 2 \end{bmatrix} + \begin{bmatrix} -1 & 4 & 2 & 4 \\ 5 & -2 & 4 & 2 \\ -5 & 1 & 4 & -1 \end{bmatrix}$

$= \begin{bmatrix} 1+(-1) & 3+4 & -8+2 & 0+4 \\ 3+5 & 6+(-2) & 1+4 & -3+2 \\ -4+(-5) & 5+1 & 3+4 & 2+(-1) \end{bmatrix} = \begin{bmatrix} 0 & 7 & -6 & 4 \\ 8 & 4 & 5 & -1 \\ -9 & 6 & 7 & 1 \end{bmatrix}$

c. $A + B = \begin{bmatrix} 2 & -3 \\ -1 & 4 \end{bmatrix} + \begin{bmatrix} -2 & 3 \\ 1 & -4 \end{bmatrix} = \begin{bmatrix} 0 & 0 \\ 0 & 0 \end{bmatrix}$

562 Chapter 7 Systems of Equations and Matrices

The matrix that is the sum in Example 1(c) is called a **zero matrix** because each of its elements is 0. Matrix B in Example 1(c) is called the **negative** of matrix A, denoted $-A$, because the sum of matrices A and B is a zero matrix. Similarly, matrix A is the negative of matrix B and can be denoted by $-B$.

> **Matrix Subtraction**
>
> If matrices M and N have the same dimension, the difference $M - N$ is found by subtracting the elements of N from the corresponding elements of M. This difference can also be defined as
>
> $$M - N = M + (-N)$$

EXAMPLE 2 ▶ Matrix Subtraction

Complete the following matrix operations.

a. $\begin{bmatrix} 2 & -1 & 3 \\ 5 & -4 & 2 \\ 1 & 4 & -2 \end{bmatrix} - \begin{bmatrix} 1 & 2 & 4 \\ 5 & -2 & 3 \\ 6 & 2 & -3 \end{bmatrix}$

b. $\begin{bmatrix} 1 & 4 \\ 3 & 6 \\ -2 & 1 \end{bmatrix} - \begin{bmatrix} -5 & 2 \\ 3 & 5 \\ 12 & 3 \end{bmatrix} + \begin{bmatrix} 2 & 3 \\ -4 & -8 \\ 2 & 0 \end{bmatrix}$

SOLUTION

a. $\begin{bmatrix} 2 & -1 & 3 \\ 5 & -4 & 2 \\ 1 & 4 & -2 \end{bmatrix} - \begin{bmatrix} 1 & 2 & 4 \\ 5 & -2 & 3 \\ 6 & 2 & -3 \end{bmatrix}$

$= \begin{bmatrix} 2-1 & -1-2 & 3-4 \\ 5-5 & -4-(-2) & 2-3 \\ 1-6 & 4-2 & -2-(-3) \end{bmatrix} = \begin{bmatrix} 1 & -3 & -1 \\ 0 & -2 & -1 \\ -5 & 2 & 1 \end{bmatrix}$

b. $\begin{bmatrix} 1 & 4 \\ 3 & 6 \\ -2 & 1 \end{bmatrix} - \begin{bmatrix} -5 & 2 \\ 3 & 5 \\ 12 & 3 \end{bmatrix} + \begin{bmatrix} 2 & 3 \\ -4 & -8 \\ 2 & 0 \end{bmatrix}$

$= \begin{bmatrix} 6 & 2 \\ 0 & 1 \\ -14 & -2 \end{bmatrix} + \begin{bmatrix} 2 & 3 \\ -4 & -8 \\ 2 & 0 \end{bmatrix} = \begin{bmatrix} 8 & 5 \\ -4 & -7 \\ -12 & -2 \end{bmatrix}$

We can use technology to perform the operations of addition and subtraction of matrices, such as those in Examples 1 and 2. Figure 7.11 shows the computations for Example 2(b).*

---

*For more details, see Appendix A, page 714.

7.3 Matrix Operations 563

[A] $\begin{bmatrix} 1 & 4 \\ 3 & 6 \\ -2 & 1 \end{bmatrix}$	[B] $\begin{bmatrix} -5 & 2 \\ 3 & 5 \\ 12 & 3 \end{bmatrix}$	[C] $\begin{bmatrix} 2 & 3 \\ -4 & -8 \\ 2 & 0 \end{bmatrix}$	[A]-[B]+[C] $\begin{bmatrix} 8 & 5 \\ -4 & -7 \\ -12 & -2 \end{bmatrix}$
(a)	(b)	(c)	(d)

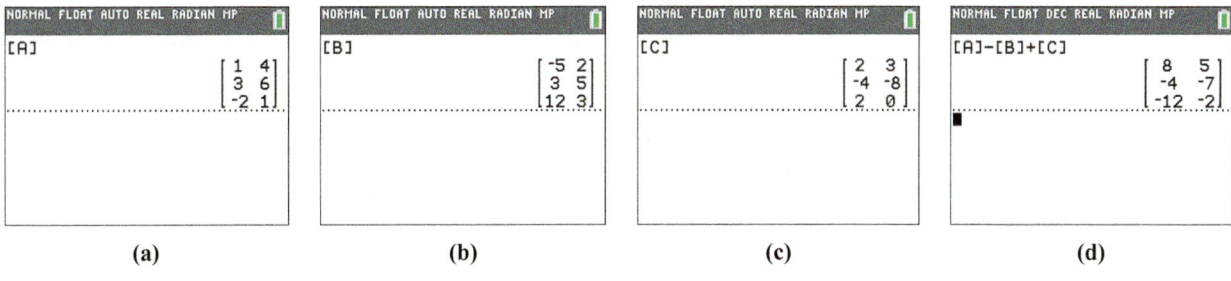

Figure 7.11

EXAMPLE 3 ▶ Life Expectancy

Table 7.6 gives the years of life expected at birth for male and female blacks and whites born in the United States for selected years from 2020 and projected to 2060.

a. Make a matrix W containing the life expectancy data for whites and a matrix B for blacks.

b. Use these matrices to find matrix $D = W - B$, which represents the difference between white and black life expectancy.

c. What does this tell us about race and life expectancy?

Table 7.6

	Whites		Blacks	
Year	Males	Females	Males	Females
2020	78.6	82.9	74.0	79.8
2030	80.2	84.1	76.1	81.3
2040	81.7	85.2	78.0	82.8
2050	83.2	86.4	79.8	84.1
2060	84.5	87.4	81.4	85.3

(*Source*: National Center for Health Statistics)

SOLUTION

a. $W = \begin{bmatrix} 78.6 & 82.9 \\ 80.2 & 84.1 \\ 81.7 & 85.2 \\ 83.2 & 86.4 \\ 84.5 & 87.4 \end{bmatrix}$ $B = \begin{bmatrix} 74.0 & 79.8 \\ 76.1 & 81.3 \\ 78.0 & 82.8 \\ 79.8 & 84.1 \\ 81.4 & 85.3 \end{bmatrix}$ **b.** $D = W - B = \begin{bmatrix} 4.6 & 3.1 \\ 4.1 & 2.8 \\ 3.7 & 2.4 \\ 3.4 & 2.3 \\ 3.1 & 2.1 \end{bmatrix}$

c. Because each element in the difference matrix D is positive, we conclude that the life expectancy for whites is longer than that for blacks for all birth years and for both sexes.

Spreadsheet ▶ SOLUTION We can use spreadsheets as well as calculators to perform operations with matrices. Figure 7.12 shows an Excel spreadsheet with W, B, and D from Example 3. The matrix W is entered in cells that we denote as B1:C5, and the matrix B is entered in cells B7:C11. The first entry of the difference matrix $D = W - B$ is found by entering the formula "$= B1 - B7$" in cell $B13$ and pressing ENTER. Using Fill Across to C17 and then Fill Down to copy the new row B13:C13 to B17:C17 gives the matrix containing the differences.*

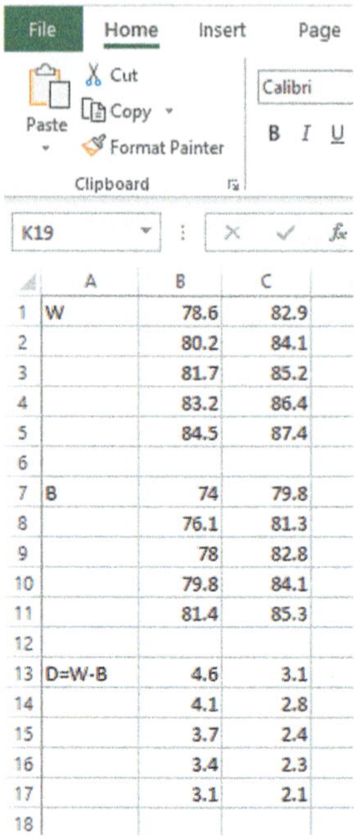

	A	B	C
1	W	78.6	82.9
2		80.2	84.1
3		81.7	85.2
4		83.2	86.4
5		84.5	87.4
6			
7	B	74	79.8
8		76.1	81.3
9		78	82.8
10		79.8	84.1
11		81.4	85.3
12			
13	D=W-B	4.6	3.1
14		4.1	2.8
15		3.7	2.4
16		3.4	2.3
17		3.1	2.1
18			

Figure 7.12

Multiplication of a Matrix by a Number

As in operations with real numbers, we can multiply a matrix by a positive integer to find the sum of repeated additions. For example, if

$$A = \begin{bmatrix} 1 & -2 & 4 \\ 6 & 3 & -3 \end{bmatrix}$$

then we can find $2A$ in two ways:

$$2A = A + A = \begin{bmatrix} 1 & -2 & 4 \\ 6 & 3 & -3 \end{bmatrix} + \begin{bmatrix} 1 & -2 & 4 \\ 6 & 3 & -3 \end{bmatrix} = \begin{bmatrix} 2 & -4 & 8 \\ 12 & 6 & -6 \end{bmatrix}$$

and

$$2A = \begin{bmatrix} 2 \cdot 1 & 2(-2) & 2 \cdot 4 \\ 2 \cdot 6 & 2 \cdot 3 & 2(-3) \end{bmatrix} = \begin{bmatrix} 2 & -4 & 8 \\ 12 & 6 & -6 \end{bmatrix}$$

Product of a Number and a Matrix

Multiplying a matrix A by a real number c results in a matrix in which each entry of matrix A is multiplied by the number c.

---

*For details, see Appendix B, page 715.

In general, we define multiplication of a matrix by a real number as follows.

For example, if $A = \begin{bmatrix} a & b & c & d \\ e & f & g & h \end{bmatrix}$, then $nA = \begin{bmatrix} na & nb & nc & nd \\ ne & nf & ng & nh \end{bmatrix}$.

Note that $-A = (-1)A = \begin{bmatrix} -a & -b & -c & -d \\ -e & -f & -g & -h \end{bmatrix}$.

EXAMPLE 4 ▶ Price Increases

Table 7.7 contains the purchase prices and delivery costs (per unit) for plywood, siding, and 2 × 4 lumber. If the supplier announces a 5% increase in all of these prices and in delivery costs, find the matrix that gives the new prices and costs.

Table 7.7

	Plywood	Siding	2 × 4's
Purchase Price ($)	32	23	2.60
Delivery Cost ($)	3	1	0.60

SOLUTION

The matrix that represents the original prices and costs is

$$\begin{bmatrix} 32 & 23 & 2.60 \\ 3 & 1 & 0.60 \end{bmatrix}$$

To find the prices and costs after a 5% increase, we multiply the matrix by 1.05 (100% of the old prices plus the 5% increase).

$$1.05 \begin{bmatrix} 32 & 23 & 2.60 \\ 3 & 1 & 0.60 \end{bmatrix} = \begin{bmatrix} 33.60 & 24.15 & 2.73 \\ 3.15 & 1.05 & 0.63 \end{bmatrix}$$

Matrix Multiplication

Suppose Circuitown made a special purchase for one of its stores, consisting of 22 televisions, 15 washers, and 12 dryers. If the value of each television is $550, each washer is $435, and each dryer is $325, then the value of this purchase is

$$550 \cdot 22 + 435 \cdot 15 + 325 \cdot 12 = 22{,}525 \text{ dollars}$$

If we write the value of each item in a 1 × 3 *row matrix*

$$A = \begin{bmatrix} 550 & 435 & 325 \end{bmatrix}$$

and the number of each of the items in the special purchase in a 3 × 1 *column matrix*

$$B = \begin{bmatrix} 22 \\ 15 \\ 12 \end{bmatrix}$$

then the value of the special purchase can be represented by the **matrix product**

$$AB = \begin{bmatrix} 550 & 435 & 325 \end{bmatrix} \begin{bmatrix} 22 \\ 15 \\ 12 \end{bmatrix}$$

$$= \begin{bmatrix} 550 \cdot 22 + 435 \cdot 15 + 325 \cdot 12 \end{bmatrix} = \begin{bmatrix} 22{,}525 \end{bmatrix}$$

In general, we have the following.

Product of a Row Matrix and a Column Matrix

The product of a $1 \times n$ row matrix and an $n \times 1$ column matrix is a 1×1 matrix given by

$$[a_1 \quad a_2 \quad \cdots \quad a_n] \begin{bmatrix} b_1 \\ b_2 \\ \vdots \\ b_n \end{bmatrix} = [a_1 b_1 + a_2 b_2 + \cdots + a_n b_n].$$

We can expand the multiplication to larger matrices. Suppose Circuitown has a second store and purchases 28 televisions, 21 washers, and 26 dryers for it. Rather than writing two matrices to represent the two stores, we can use a two-column matrix C to represent the purchases for the two stores.

$$C = \begin{matrix} & \text{Store I} & \text{Store II} \\ & \begin{bmatrix} 22 & 28 \\ 15 & 21 \\ 12 & 26 \end{bmatrix} & \begin{matrix} \text{TVs} \\ \text{Washers} \\ \text{Dryers} \end{matrix} \end{matrix}$$

If these products have the same values as given previously, we can find the value of the purchases for each store by multiplying the row matrix times each of the column matrices. The value of the store I purchases is found by multiplying the row matrix A times the first column of matrix C, and the value of the store II purchases is found by multiplying matrix A times the second column of matrix C. The result is

$$AC = [550 \quad 435 \quad 325] \begin{bmatrix} 22 & 28 \\ 15 & 21 \\ 12 & 26 \end{bmatrix}$$

$$= [550 \cdot 22 + 435 \cdot 15 + 325 \cdot 12 \quad 550 \cdot 28 + 435 \cdot 21 + 325 \cdot 26]$$

$$= [22{,}525 \quad 32{,}985].$$

This indicates that the value of the store I purchase is \$22,525 (the value found before) and the value of the store II purchase is \$32,985.

The matrix AC is the product of the 1×3 matrix A and the 3×2 matrix C. This product is a 1×2 matrix. In general, the product of an $m \times n$ matrix and an $n \times k$ matrix is an $m \times k$ matrix, and the product is undefined if the number of columns in the first matrix does not equal the number of rows in the second matrix.

In general, we can define the product of two matrices by defining how each element of the product is formed.

Product of Two Matrices

The product of an $m \times n$ matrix A and an $n \times k$ matrix B is the $m \times k$ matrix $C = AB$. The element in the ith row and jth column of matrix C has the form

$$c_{ij} = [a_{i1} \quad a_{i2} \quad \cdots \quad a_{in}] \begin{bmatrix} b_{1j} \\ b_{2j} \\ \vdots \\ b_{nj} \end{bmatrix} = [a_{i1}b_{1j} + a_{i2}b_{2j} + \cdots + a_{in}b_{nj}].$$

We illustrate below the product AB, with each of the c_{ij} elements found as shown in the box on the previous page.

$$C = AB = \begin{bmatrix} a_{11} & a_{12} & \cdots & a_{1n} \\ a_{21} & a_{22} & \cdots & a_{2n} \\ \vdots & \vdots & & \vdots \\ a_{i1} & a_{i2} & \cdots & a_{in} \\ \vdots & \vdots & & \vdots \\ a_{m1} & a_{m2} & \cdots & a_{mn} \end{bmatrix} \begin{bmatrix} b_{11} & b_{12} & \cdots & b_{1j} & \cdots & b_{1k} \\ b_{21} & b_{22} & \cdots & b_{2j} & \cdots & b_{2k} \\ \vdots & \vdots & & \vdots & & \vdots \\ b_{n1} & b_{n2} & \cdots & b_{nj} & \cdots & b_{nk} \end{bmatrix} = \begin{bmatrix} c_{11} & c_{12} & \cdots & c_{1j} & \cdots & c_{1k} \\ c_{21} & c_{22} & \cdots & c_{2j} & \cdots & c_{2k} \\ \vdots & \vdots & & \vdots & & \vdots \\ c_{i1} & c_{i2} & \cdots & c_{ij} & \cdots & c_{ik} \\ \vdots & \vdots & & \vdots & & \vdots \\ c_{m1} & c_{m2} & \cdots & c_{mj} & \cdots & c_{mk} \end{bmatrix}$$

$$m \times n \qquad\qquad n \times k \qquad\qquad m \times k$$

EXAMPLE 5 ▶ Matrix Product

Compute the products AB and BA for the matrices $A = \begin{bmatrix} 1 & 2 \\ 3 & 4 \end{bmatrix}$ and $B = \begin{bmatrix} a & b \\ c & d \end{bmatrix}$.

SOLUTION

$$AB = \begin{bmatrix} 1 & 2 \\ 3 & 4 \end{bmatrix} \begin{bmatrix} a & b \\ c & d \end{bmatrix} = \begin{bmatrix} 1a + 2c & 1b + 2d \\ 3a + 4c & 3b + 4d \end{bmatrix}$$

$$BA = \begin{bmatrix} a & b \\ c & d \end{bmatrix} \begin{bmatrix} 1 & 2 \\ 3 & 4 \end{bmatrix} = \begin{bmatrix} 1a + 3b & 2a + 4b \\ 1c + 3d & 2c + 4d \end{bmatrix}$$

Note that in Example 5 the product AB is quite different from the product BA. That is, $AB \neq BA$. For some matrices, but not all, $AB \neq BA$. We indicate this by saying that *matrix multiplication is not commutative.*

EXAMPLE 6 ▶ Advertising

A business plans to use three methods of advertising—newspapers, radio, and cable TV—in each of its two markets, I and II. The cost per ad type in each market (in thousands of dollars) is given by matrix A. The business has three target groups: teenagers, single women, and men aged 35 to 50. Matrix B gives the number of ads per week directed at each of these groups.

$$A = \begin{bmatrix} 12 & 10 \\ 10 & 8 \\ 5 & 9 \end{bmatrix} \begin{matrix} \text{Paper} \\ \text{Radio} \\ \text{TV} \end{matrix} \qquad B = \begin{bmatrix} 3 & 12 & 15 \\ 5 & 16 & 10 \\ 10 & 11 & 6 \end{bmatrix} \begin{matrix} \text{Teens} \\ \text{Single females} \\ \text{Men 35–50} \end{matrix}$$

with column headers Mkt I, Mkt II for A and Paper, Radio, TV for B.

a. Does AB or BA give the matrix that represents the cost of ads for each target group in each market?

b. Find this matrix.

c. For what group of people is the most money spent on advertising?

SOLUTION

a. Multiplying matrix B times matrix A gives the total cost of ads for each target group in each market. Note that the product AB is undefined, because multiplying a 3×2 matrix times a 3×3 matrix is not possible.

b. $BA = \begin{bmatrix} 3 & 12 & 15 \\ 5 & 16 & 10 \\ 10 & 11 & 6 \end{bmatrix} \begin{bmatrix} 12 & 10 \\ 10 & 8 \\ 5 & 9 \end{bmatrix}$

$= \begin{bmatrix} 36 + 120 + 75 & 30 + 96 + 135 \\ 60 + 160 + 50 & 50 + 128 + 90 \\ 120 + 110 + 30 & 100 + 88 + 54 \end{bmatrix} = \begin{bmatrix} 231 & 261 \\ 270 & 268 \\ 260 & 242 \end{bmatrix}$

The columns of this matrix represent the markets and the rows represent the target groups.

$\begin{matrix} & \text{Mkt I} & \text{Mkt II} \\ \begin{bmatrix} 231 & 261 \\ 270 & 268 \\ 260 & 242 \end{bmatrix} & \begin{matrix} \text{Teens} \\ \text{Single females} \\ \text{Men 35–50} \end{matrix} \end{matrix}$

c. The largest amount is spent on single females, $270,000 in market I and $268,000 in market II.

Multiplication with Technology

We can multiply two matrices by using technology. We can find the product of matrix B times matrix A on a graphing utility by entering [B] * [A] (or [B][A]) and pressing **ENTER**.* Figure 7.13 shows the product BA from Example 6.

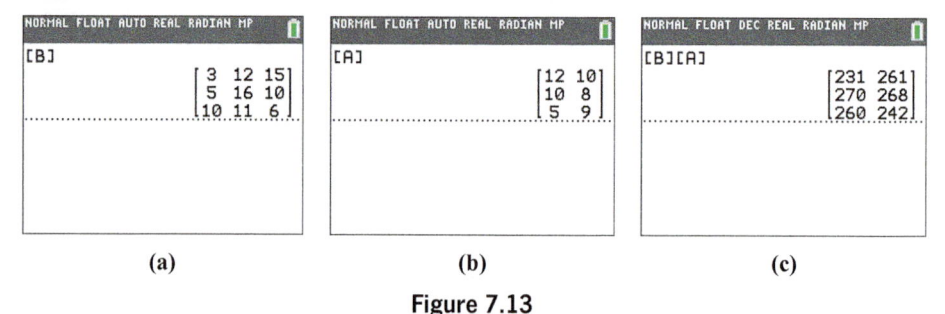

(a)   (b)   (c)

Figure 7.13

EXAMPLE 7 ▶ Multiplying with Technology

Use technology to compute BA and AB if $A = \begin{bmatrix} 1 & 2 \\ 0 & -1 \\ 3 & -2 \end{bmatrix}$ and $B = \begin{bmatrix} 2 & 4 & -1 \\ 3 & -2 & 1 \\ 2 & 0 & 2 \\ 1 & -3 & 0 \end{bmatrix}$.

SOLUTION

Figure 7.14 shows displays with matrix A, matrix B, and the product BA.

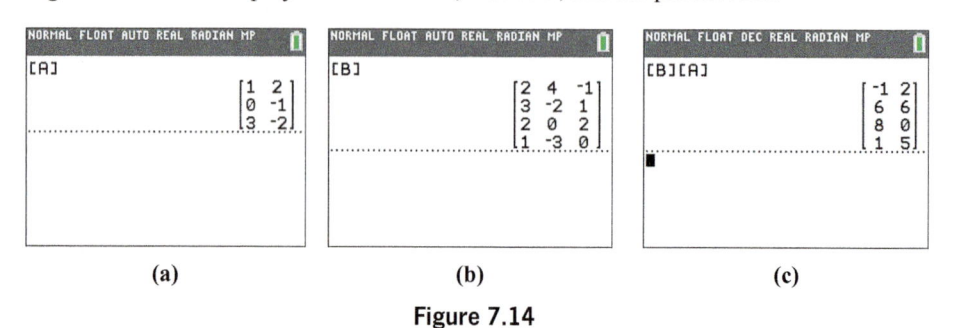

(a)   (b)   (c)

Figure 7.14

---

*For more details, see Appendix A, page 696.

Figure 7.15 shows that the matrix product AB does not exist because the dimensions do not match (that is, the number of columns of A does not equal the number of rows of B).

(a)   (b)

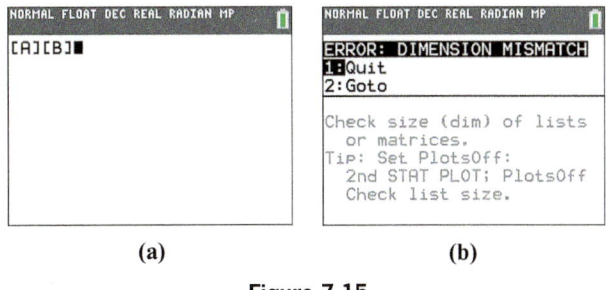

Figure 7.15

Spreadsheet ▶ SOLUTION Like graphing calculators and software programs, spreadsheets can be used to find products of matrices. Consider the following example, which is solved with Excel.

EXAMPLE 8 ▶ Manufacturing

A furniture company manufactures two products, A and B, which are constructed using steel, plastic, and fabric. The number of units of each raw material that is required for each product is given in Table 7.8.

Table 7.8

	Steel	Plastic	Fabric
Product A	2	3	8
Product B	3	1	10

Because of transportation costs to the company's two plants, the unit costs for some of the raw materials differ. Table 7.9 gives the unit costs for each of the raw materials at the two plants. Create two matrices from the information in the two tables and use matrix multiplication with a spreadsheet to find the cost of manufacturing each product at each plant.

Table 7.9

	Plant I	Plant II
Steel	$15	$16
Plastic	$11	$10
Fabric	$ 6	$ 7

SOLUTION

We represent the number of units of raw materials for each product by matrix P and the costs of the raw materials at each plant by matrix C. The cost of manufacturing each product at each plant is given by the matrix product PC. The spreadsheet shown in Figure 7.16 gives matrix P, matrix C, and the matrix product PC.

The product *PC* is found by selecting cells B8:C9, entering "=mmult(", then selecting the cells B1:D2, typing a comma, selecting the cells B4:C6, and closing the parentheses, and then holding the CTRL and SHIFT keys down and pressing ENTER.*

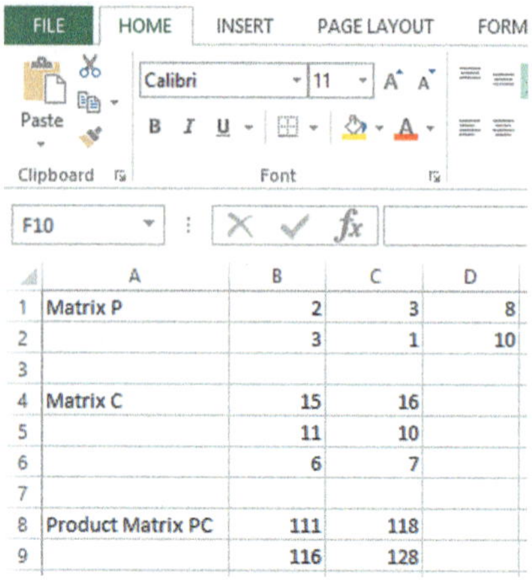

	A	B	C	D
1	Matrix P	2	3	8
2		3	1	10
3				
4	Matrix C	15	16	
5		11	10	
6		6	7	
7				
8	Product Matrix PC	111	118	
9		116	128	

Figure 7.16

The rows of matrix *PC* represent Product A and Product B, respectively, and the columns represent Plant I and Plant II. The entries give the cost of each product at each plant.

$$\begin{matrix} & \text{Plant I} & \text{Plant II} \\ \begin{bmatrix} 111 & 118 \\ 116 & 128 \end{bmatrix} & \text{Product A} \\ & \text{Product B} \end{matrix}$$

Recall that a square ($n \times n$) matrix with 1s on the diagonal and 0s elsewhere is an identity matrix. For any $n \times k$ matrix A and the $n \times n$ matrix I, $IA = A$, and for any $m \times n$ matrix B and the $n \times n$ matrix I, $BI = B$. If matrix C is an $n \times n$ matrix, then for the $n \times n$ matrix I, $IC = C$ and $CI = C$, so the product of an identity matrix and another square matrix is *commutative*. (Recall that multiplication of matrices, in general, is not commutative.)

EXAMPLE 9 ▶ The Identity Matrix

a. Write the 2×2 identity matrix I.

b. For the matrix $A = \begin{bmatrix} 3 & -2 \\ -1 & 5 \end{bmatrix}$, find IA and AI.

SOLUTION

a. $\begin{bmatrix} 1 & 0 \\ 0 & 1 \end{bmatrix}$

b. $\begin{bmatrix} 1 & 0 \\ 0 & 1 \end{bmatrix}\begin{bmatrix} 3 & -2 \\ -1 & 5 \end{bmatrix} = \begin{bmatrix} 3 & -2 \\ -1 & 5 \end{bmatrix}$ and $\begin{bmatrix} 3 & -2 \\ -1 & 5 \end{bmatrix}\begin{bmatrix} 1 & 0 \\ 0 & 1 \end{bmatrix} = \begin{bmatrix} 3 & -2 \\ -1 & 5 \end{bmatrix}$

Thus, $IA = A$ and $AI = A$.

---

*For detailed steps, see Appendix B, page 715.

SKILLS CHECK 7.3

Answers that are not seen can be found in the answer section at the back of the text.

Use the following matrices for Exercises 1–12.

$$A = \begin{bmatrix} 1 & 3 & -2 \\ 3 & 1 & 4 \\ -5 & 3 & 6 \end{bmatrix} \quad B = \begin{bmatrix} 2 & 1 & -1 \\ 3 & 2 & 4 \end{bmatrix}$$

$$C = \begin{bmatrix} 1 & 3 \\ 2 & 1 \\ 3 & -1 \end{bmatrix} \quad D = \begin{bmatrix} 2 & 3 & 1 \\ 3 & 4 & -1 \\ 2 & 5 & 1 \end{bmatrix}$$

$$E = \begin{bmatrix} 9 & 2 & -7 \\ -5 & 0 & 5 \\ 7 & -4 & -1 \end{bmatrix} \quad F = \begin{bmatrix} 2 & 1 & 3 \\ 4 & 0 & 1 \end{bmatrix}$$

1. Which pairs of the matrices can be added?
   A and D, A and E, D and E, B and F
2. Use letters to represent the matrix products that are defined.
3. Find the sum of A and D if it is defined.
4. a. Find $D + E$ and $E + D$.
   b. Are the sums equal? Yes
5. Find $3A$.
6. Find $-4F$. $\begin{bmatrix} -8 & -4 & -12 \\ -16 & 0 & -4 \end{bmatrix}$
7. Find $2D - 4A$.
8. Find $2B - 4F$. $\begin{bmatrix} -4 & -2 & -14 \\ -10 & 4 & 4 \end{bmatrix}$
9. a. Find AD and DA if these products exist.
   b. Are these products equal? No
   c. Do these products have the same dimension? Yes
10. a. Find BC and CB if these products exist.
    b. Are these products equal? No
    c. Do these products have the same dimension? No
11. a. Compute DE and ED.
    b. What is the name of $\frac{1}{10}DE$? 3×3 identity matrix
12. If I is a 3×3 identity matrix, find ID and DI.
    Both are D.
13. Compute the sum of $A = \begin{bmatrix} 1 & 5 \\ 3 & 2 \end{bmatrix}$ and
    $B = \begin{bmatrix} 2a & 3b \\ -c & -2d \end{bmatrix}$. $\begin{bmatrix} 1+2a & 5+3b \\ 3-c & 2-2d \end{bmatrix}$
14. Compute the difference $A - B$ if $A = \begin{bmatrix} a & b \\ c & d \\ f & g \end{bmatrix}$ and
    $B = \begin{bmatrix} 1 & 2 \\ 3 & 4 \\ 5 & 6 \end{bmatrix}$. $\begin{bmatrix} a-1 & b-2 \\ c-3 & d-4 \\ f-5 & g-6 \end{bmatrix}$
15. Compute $3A - 2B$ if $A = \begin{bmatrix} a & b \\ c & d \end{bmatrix}$ and $B = \begin{bmatrix} 1 & 2 \\ 3 & 4 \end{bmatrix}$.

16. Compute $2A - 3B$ if $A = \begin{bmatrix} 1 & -3 & 2 \\ 2 & 2 & -1 \\ 3 & 4 & 2 \end{bmatrix}$ and
    $B = \begin{bmatrix} 2 & 2 & 2 \\ 3 & -2 & -1 \\ 1 & 1 & 2 \end{bmatrix}$. $\begin{bmatrix} -4 & -12 & -2 \\ -5 & 10 & 1 \\ 3 & 5 & -2 \end{bmatrix}$

17. If an $m \times n$ matrix A is multiplied by an $n \times k$ matrix B, what is the dimension of the matrix that is the product AB? $m \times k$

18. If A and B are any two matrices, does $AB = BA$ always, sometimes, or never? Sometimes

19. If A is a 2×3 matrix and B is a 4×2 matrix:
    a. Which product is defined, AB or BA? BA
    b. What is the dimension of the product that is defined? 4×3

20. If C is a 3×4 matrix and D is a 4×3 matrix, what are the dimensions of CD and DC?
    CD is 3×3; DC is 4×4.

21. Find AB and BA if $A = \begin{bmatrix} a & b & c \\ d & e & f \end{bmatrix}$ and $B = \begin{bmatrix} 1 & 2 \\ 3 & 4 \\ 5 & 6 \end{bmatrix}$.

22. Find EF and FE if $E = \begin{bmatrix} a & b \\ c & d \end{bmatrix}$ and $F = \begin{bmatrix} e & f \\ g & h \end{bmatrix}$.

23. If $A = \begin{bmatrix} 1 & 5 \\ 3 & 2 \end{bmatrix}$ and $B = \begin{bmatrix} 2 & 3 \\ -1 & -2 \end{bmatrix}$, compute AB and BA, if possible. $AB = \begin{bmatrix} -3 & -7 \\ 4 & 5 \end{bmatrix}$; $BA = \begin{bmatrix} 11 & 16 \\ -7 & -9 \end{bmatrix}$

24. If $A = \begin{bmatrix} 1 & 4 \\ 3 & -1 \\ -2 & 2 \end{bmatrix}$ and $B = \begin{bmatrix} 4 & 2 & 2 \\ -1 & 3 & 1 \end{bmatrix}$, compute AB and BA, if possible.

25. If $A = \begin{bmatrix} 1 & -1 & 2 \\ 3 & 4 & 4 \end{bmatrix}$ and $B = \begin{bmatrix} 3 & 1 \\ 1 & 3 \\ -2 & 1 \end{bmatrix}$, compute AB and BA, if possible.

26. Suppose $A = \begin{bmatrix} 1 & -\frac{1}{2} & -\frac{1}{4} \\ -\frac{1}{2} & 0 & \frac{1}{2} \\ 0 & \frac{1}{2} & -\frac{1}{4} \end{bmatrix}$ and
    $B = \begin{bmatrix} 2 & 2 & 2 \\ 1 & 2 & 3 \\ 2 & 4 & 2 \end{bmatrix}$.
    a. Compute AB and BA, if possible.
    b. Are the products equal? Yes

EXERCISES 7.3

Answers that are not seen can be found in the answer section at the back of the text.

27. Future Demographics The tables below show important demographics for China, Bangladesh, and the Philippines for 2012 and projected for 2062.

a. Create matrix B for the 2012 table and matrix A for the 2062 table.

b. Find a matrix C that shows the changes of demographics from 2012 to 2062 for these countries.

c. Which negative entries in matrix C definitely indicate a positive change for the countries?

2012

	China	Bangladesh	The Philippines
Population (millions)	1341.3	148.7	93.3
Life Expectancy	73.8	69.4	69.2
Fertility Rate*	1.56	2.16	3.05
Infant Mortality Rate**	19.6	41.8	20.9

2062

	China	Bangladesh	The Philippines
Population (millions)	1211.5	192.4	165.5
Life Expectancy	80.8	79.0	78.4
Fertility Rate*	1.88	1.68	1.95
Infant Mortality Rate**	8.3	12.5	10.2

(Source: Discover Almanac)

28. Population The following tables give U.S. Census Bureau data and projections of the population by selected age groups and gender in the United States, in thousands.

a. Matrix M gives the data for males, and matrix F gives the data for females. Find a matrix $D = M - F$ and discuss what it means.

b. For what years and ages is the female population projected to be greater than the male population?

Males

	Under Age 5	5–13	14–17	18–64	Over Age 64
2015	10,763	18,784	8541	99,232	21,041
2020	11,150	19,309	8478	100,904	24,970
2025	11,307	20,210	8472	102,004	29,204
2030	11,377	20,649	9078	103,510	32,709

Females

	Under Age 5	5–13	14–17	18–64	Over Age 64
2015	10,288	17,988	8153	99,918	26,654
2020	10,658	18,640	8104	100,863	30,999
2025	10,807	19,301	8093	101,162	35,848
2030	10,875	19,717	8653	101,839	40,066

29. Cellular IoT Revenue The tables give the cellular Internet of Things market revenue from 2G, 3G, 4G, and 5G technologies in the United States for the years 2014 through 2019 and the projections for the years 2020 through 2025, in millions of dollars. (Source: Statista)

a. Create matrix A from the table for the years 2014–2019 and matrix B for the years 2020–2025.

b. Find a matrix C that gives the changes in each of the categories for the years 2014–2019 to the years that are 6 years later.

c. What trends do the entries in matrix C reveal? Revenues from 2G and 3G technologies will decrease while revenues from 4G and 5G technologies will increase.

Years 2014–2019

Years	2G	3G	4G	5G
2014	210	140	50	0
2015	240	200	30	0
2016	260	270	60	0
2017	230	300	80	0
2018	210	360	100	0
2019	210	330	120	0

---

*The fertility rate is the number of children an average woman will produce in her lifetime.

**Infant mortality is the number of deaths per thousand births.

Years 2020–2025

Years	2G	3G	4G	5G
2020	180	350	170	0
2021	160	340	210	0
2022	140	330	260	90
2023	120	330	300	160
2024	100	310	360	290
2025	40	270	390	560

30. *Population by Nativity* The tables give the total U.S. population by age group and the foreign-born population by age group (both in thousands) for selected years from 2014 and projected through 2060.
(*Source*: U.S. Census Bureau)

  a. Create a matrix A that represents the total U.S. population by age group for these years and a matrix B that represents the foreign-born population by age group for these years.

  b. If the difference of the total population and the foreign-born population equals the native population in the United States, find a matrix C representing this population.

  c. What can we conclude about the projections of the U.S. native population from matrix C?
The U.S. native population is projected to increase in all age groups.

A. Total U.S. Population

Age	2020	2060
<18	74,128	82,304
18–44	120,073	136,310
45–64	83,861	100,013
65+	56,441	98,164

B. Foreign-Born Population

Age	2020	2060
<18	2445	3254
18–44	20,704	25,169
45–64	16,665	24,520
65+	8079	25,288

31. *Energy Consumption* The table shows the U.S. annual consumption data and projections (in quadrillion BTUs) for various energy types (with liquid fuels and petroleum abbreviated as LF&P) in selected years.

  a. Create a matrix C containing this information.

  b. Find the matrix M that provides the average monthly consumption of these energy types for these years. Round the entries to two decimal places.

Energy Consumption

	2015	2020	2025
LF&P	37.04	37.54	36.87
Natural Gas	25.86	26.77	27.28
Other	34.82	36.73	38.19

(*Source*: U.S. Energy Information Administration)

32. *Insurance Premiums* The table gives one company's monthly life insurance premiums for females by age groups, in dollars. Suppose that because of the effect of obesity and diabetes on life span, the company decides to increase all its premiums by 10% in all age groups.

  a. Create a matrix A that represents the data in the table.

  b. Find a matrix B that represents the new premiums for its female customers.

Monthly Insurance Premiums for Females

Age	$20,000 Policy	$25,000 Policy	$50,000 Policy	$100,000 Policy
50–54	5.13	6.42	12.83	25.67
55–59	6.53	8.17	16.33	32.67
60–64	8.63	10.79	21.58	43.17
65–69	13.77	17.21	34.42	68.83
70–74	22.75	28.44	56.88	113.75

33. *Advertising* A political candidate plans to use three methods of advertising: newspapers, radio, and cable TV. The cost per ad (in thousands of dollars) for each type of media is given by matrix A. Matrix B shows the number of ads per month in these three media that are targeted to single people, to married males aged 35 to 55, and to married females over 65 years of age. Find the matrix that gives the cost of ads for each target group.

$$A = \begin{bmatrix} 12 \\ 15 \\ 5 \end{bmatrix} \begin{matrix} \text{TV} \\ \text{Radio} \\ \text{Papers} \end{matrix}$$

$$B = \begin{bmatrix} 30 & 45 & 35 \\ 25 & 32 & 40 \\ 22 & 12 & 30 \end{bmatrix} \begin{matrix} \text{Singles} \\ \text{Males 35–55} \\ \text{Females 65+} \end{matrix}$$

34. Cost Men and women in a church choir wear choir robes in the sizes shown in matrix A. Matrix B contains the prices (in dollars) of new robes and hoods according to size.

$$A = \begin{bmatrix} 10 & 24 \\ 22 & 10 \\ 33 & 3 \end{bmatrix} \begin{matrix} \text{Small} \\ \text{Medium} \\ \text{Large} \end{matrix}$$

with columns labeled Men, Women.

$$B = \begin{bmatrix} 45 & 50 & 55 \\ 20 & 20 & 20 \end{bmatrix} \begin{matrix} \text{Robes} \\ \text{Hoods} \end{matrix}$$

with columns labeled S, M, L.

a. Find the product BA and label the rows and columns to show what each row represents.

b. What is the cost of the robes for all the men? For all the women? $3365; $1745

35. Manufacturing Two departments, A and B, of a firm need differing amounts of steel, wood, and plastic. The following table gives the amounts of the products needed by the departments.

	Steel	Wood	Plastic
Department A	60	40	20
Department B	40	20	40

These three products are supplied by two suppliers, DeTuris and Marriott, with the unit prices (in dollars) given in the following table.

	DeTuris	Marriott
Steel	600	560
Wood	300	200
Plastic	300	400

a. Use matrix multiplication to determine how much these orders will cost each department at each of the two suppliers.

b. From which supplier should each department make its purchase? Dept. A from Marriott; dept. B from DeTuris

36. Manufacturing A furniture manufacturer produces three styles of chairs, with the number of units of each type of raw material needed for each style given in the table below.

	Wood	Nylon	Velvet	Springs
Style A	5	20	0	10
Style B	10	9	0	0
Style C	5	10	10	10

The cost in dollars per unit for each of the raw materials is given in the table below.

Wood	15
Nylon	12
Velvet	14
Springs	30

Create two matrices to represent the data and use matrix multiplication to find the price of manufacturing each style of chair.

37. Politics In a midwestern state, it is determined that 90% of all Republicans vote for Republican candidates and the remainder for Democratic candidates, while 80% of all Democrats vote for Democratic candidates and the remainder for Republican candidates. The percent of each party predicted to win the next election is given by

$$\begin{bmatrix} R \\ D \end{bmatrix} = \begin{bmatrix} 0.90 & 0.20 \\ 0.10 & 0.80 \end{bmatrix} \begin{bmatrix} a \\ b \end{bmatrix}$$

where a is the percent of Republicans and b is the percent of Democrats who won the last election. If 50% of those winning the election last time were Republicans and 50% were Democrats, what is the percent of each party that is predicted to win the next election? 55% Republican; 45% Democrat

38. Competition Two phone companies compete for customers in the southeastern region of a state. Company X retains $\frac{3}{5}$ of its customers and loses $\frac{2}{5}$ of its customers to company Y; company Y retains $\frac{2}{3}$ of its customers and loses $\frac{1}{3}$ of its customers to company X. If we represent the fraction of the market held last year by

$$\begin{bmatrix} a \\ b \end{bmatrix}$$

where a is the number of customers that company X had and b is the number of customers that company Y had, then the fraction that each company will have this year can be found from

$$\begin{bmatrix} x \\ y \end{bmatrix} = \begin{bmatrix} 3/5 & 1/3 \\ 2/5 & 2/3 \end{bmatrix} \begin{bmatrix} a \\ b \end{bmatrix}$$

If company X had 120,000 customers and company Y had 90,000 customers last year, how many customers did each have this year? Company X: 102,000 customers; Company Y: 108,000 customers

39. Wages The graph on the next page gives the median weekly earnings for male and female full-time wage and salary workers in selected ethnic groups in 2019.

a. Create a 2 × 4 matrix W containing the data.

b. Suppose that the median weekly earnings for 2025 for these groups are projected to increase over the 2019 earnings by 6% for men and by 8% for women. Use matrix multiplication by a 2 × 2 matrix to find the median weekly earnings for male and female workers in these selected ethnic groups in 2025.

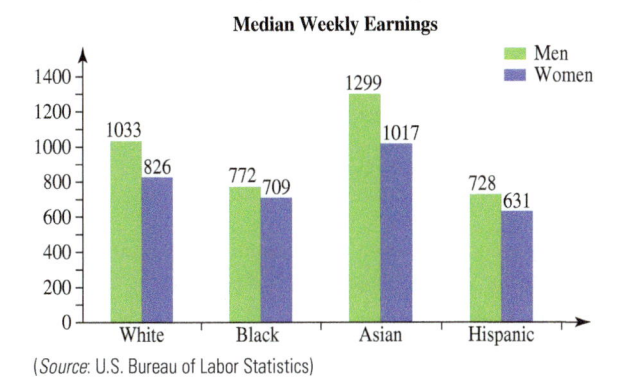

Median Weekly Earnings

Men: White 1033, Black 772, Asian 1299, Hispanic 728
Women: White 826, Black 709, Asian 1017, Hispanic 631

(*Source*: U.S. Bureau of Labor Statistics)

40. Global Tobacco Smoking The table gives the prevalence of tobacco smoking (%) by global region and by gender in 2010 and 2025.

a. Create a matrix A containing the data for 2010 and a matrix B containing the data for 2025.

b. Use these two matrices to find the changes from 2010 to 2025 in each of the categories.

c. In which of the categories has the prevalence increased? AFRO males, EMRO males

Prevalence of Current Smoking (%)

REGION	2010 Male	2010 Female	2025 Male	2025 Female
AFRO	23.2	2.5	34.7	1.6
AMRO	24.1	14.2	16.3	8.6
EMRO	35.1	3.1	45.3	2.5
EURO	40.3	19.9	31.3	15.9
SEARO	33.1	2.9	27.5	1.2
WPRO	49.4	3.6	43.3	2.4

(*Source*: World Health Organization)

7.4 Inverse Matrices; Matrix Equations

KEY OBJECTIVES
- Find the inverse of a matrix
- Find matrix inverses with technology
- Encode and decode messages
- Solve matrix equations
- Solve matrix equations with technology

SECTION PREVIEW Encryption

Security of credit card numbers on the Internet depends on encryption of the data. Encryption involves providing a way for the sender to encode a message so that the message is not apparent and a way for the receiver to decode the message so that it can be read. Throughout history, different military units have used encoding and decoding systems of varying sophistication. In this section, we find the inverse of a matrix, use it to decode messages, and use it to solve matrix equations that have unique solutions. ■

Inverse Matrices

If the product of matrices A and B is an identity matrix, I, we say that B is the inverse of A (and A is the inverse of B). B is called the **inverse matrix** of A and is denoted A^{-1}.

> **Inverse Matrices**
>
> Two square matrices, A and B, are called **inverses** of each other if
>
> $$AB = I \quad \text{and} \quad BA = I$$
>
> where I is the identity matrix. We denote this by $B = A^{-1}$ and $A = B^{-1}$.

EXAMPLE 1 ▶ Inverse Matrices

Show that A and B are inverse matrices if

$$A = \begin{bmatrix} 1 & -0.6 & -0.2 \\ 0 & 0.4 & -0.2 \\ -1 & 0.4 & 0.8 \end{bmatrix} \quad \text{and} \quad B = \begin{bmatrix} 2 & 2 & 1 \\ 1 & 3 & 1 \\ 2 & 1 & 2 \end{bmatrix}$$

576 Chapter 7 Systems of Equations and Matrices

SOLUTION

$$AB = \begin{bmatrix} 1 & -0.6 & -0.2 \\ 0 & 0.4 & -0.2 \\ -1 & 0.4 & 0.8 \end{bmatrix} \begin{bmatrix} 2 & 2 & 1 \\ 1 & 3 & 1 \\ 2 & 1 & 2 \end{bmatrix}$$

$$= \begin{bmatrix} 2 - 0.6 - 0.4 & 2 - 1.8 - 0.2 & 1 - 0.6 - 0.4 \\ 0 + 0.4 - 0.4 & 0 + 1.2 - 0.2 & 0 + 0.4 - 0.4 \\ -2 + 0.4 + 1.6 & -2 + 1.2 + 0.8 & -1 + 0.4 + 1.6 \end{bmatrix}$$

$$= \begin{bmatrix} 1 & 0 & 0 \\ 0 & 1 & 0 \\ 0 & 0 & 1 \end{bmatrix}$$

The product AB is the 3×3 identity matrix; we can use technology (Figure 7.17) to see that the product BA is also the identity matrix. Thus, A and B are inverse matrices.

(a)

(b)

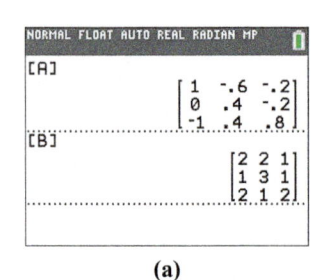

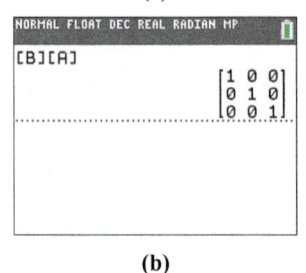

Figure 7.17

We have used elementary row operations on augmented matrices to solve systems of equations. We can also find the inverse of a matrix A, if it exists, by using elementary row operations. Note that if a matrix is not square, then it does not have an inverse, and that not all square matrices have inverses.

Finding the Inverse of a Square Matrix

1. Write the matrix with the same dimension identity matrix in its augment, getting a matrix of the form $[A|I]$.

2. Use elementary row operations on $[A|I]$ to attempt to transform A into an identity matrix, giving a new matrix of the form $[I|B]$. The matrix B is the inverse of A.

3. If A does not have an inverse, the reduction process will yield a row of zeros in the left half (representing the original matrix) of the augmented matrix.

EXAMPLE 2 ▶ Finding an Inverse Matrix

Find the inverse of $A = \begin{bmatrix} 2 & 2 \\ 2 & 1 \end{bmatrix}$.

SOLUTION

Creating the matrix $[A|I]$ and performing the operations to convert A to I gives

$$\begin{bmatrix} 2 & 2 & | & 1 & 0 \\ 2 & 1 & | & 0 & 1 \end{bmatrix} \xrightarrow{(1/2)R_1 \to R_1} \begin{bmatrix} 1 & 1 & | & \frac{1}{2} & 0 \\ 2 & 1 & | & 0 & 1 \end{bmatrix}$$

$$\xrightarrow{-2R_1 + R_2 \to R_2} \begin{bmatrix} 1 & 1 & | & \frac{1}{2} & 0 \\ 0 & -1 & | & -1 & 1 \end{bmatrix}$$

$$\xrightarrow{-R_2 \to R_2} \begin{bmatrix} 1 & 1 & | & \frac{1}{2} & 0 \\ 0 & 1 & | & 1 & -1 \end{bmatrix} \xrightarrow{-R_2 + R_1 \to R_1} \begin{bmatrix} 1 & 0 & | & -\frac{1}{2} & 1 \\ 0 & 1 & | & 1 & -1 \end{bmatrix}$$

Thus, the inverse matrix of A is $A^{-1} = \begin{bmatrix} -\frac{1}{2} & 1 \\ 1 & -1 \end{bmatrix}$. We can verify this by observing that $\begin{bmatrix} 2 & 2 \\ 2 & 1 \end{bmatrix} \begin{bmatrix} -\frac{1}{2} & 1 \\ 1 & -1 \end{bmatrix} = \begin{bmatrix} 1 & 0 \\ 0 & 1 \end{bmatrix}$.

EXAMPLE 3 ▶ **Inverse of a 3 × 3 Matrix**

Find the inverse of $A = \begin{bmatrix} -2 & 1 & 2 \\ 1 & 0 & -1 \\ 4 & -2 & -3 \end{bmatrix}$.

SOLUTION

Creating the matrix $[A \mid I]$ and performing the operations to convert A to I gives

$$\begin{bmatrix} -2 & 1 & 2 & | & 1 & 0 & 0 \\ 1 & 0 & -1 & | & 0 & 1 & 0 \\ 4 & -2 & -3 & | & 0 & 0 & 1 \end{bmatrix} \xrightarrow{R_1 \leftrightarrow R_2} \begin{bmatrix} 1 & 0 & -1 & | & 0 & 1 & 0 \\ -2 & 1 & 2 & | & 1 & 0 & 0 \\ 4 & -2 & -3 & | & 0 & 0 & 1 \end{bmatrix}$$

$$\xrightarrow[-4R_1 + R_3 \to R_3]{2R_1 + R_2 \to R_2} \begin{bmatrix} 1 & 0 & -1 & | & 0 & 1 & 0 \\ 0 & 1 & 0 & | & 1 & 2 & 0 \\ 0 & -2 & 1 & | & 0 & -4 & 1 \end{bmatrix}$$

$$\xrightarrow{2R_2 + R_3 \to R_3} \begin{bmatrix} 1 & 0 & -1 & | & 0 & 1 & 0 \\ 0 & 1 & 0 & | & 1 & 2 & 0 \\ 0 & 0 & 1 & | & 2 & 0 & 1 \end{bmatrix}$$

$$\xrightarrow{R_3 + R_1 \to R_1} \begin{bmatrix} 1 & 0 & 0 & | & 2 & 1 & 1 \\ 0 & 1 & 0 & | & 1 & 2 & 0 \\ 0 & 0 & 1 & | & 2 & 0 & 1 \end{bmatrix}$$

Thus, we have the inverse of A.

$$A^{-1} = \begin{bmatrix} 2 & 1 & 1 \\ 1 & 2 & 0 \\ 2 & 0 & 1 \end{bmatrix}$$

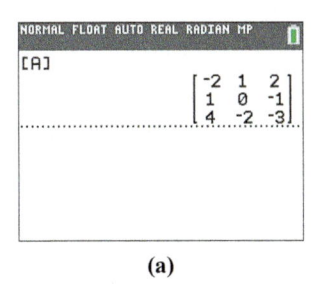

Inverses and Technology

Computer software, spreadsheets, and calculators can be used to find the inverse of a matrix. We will see that if the inverse of a square matrix exists, it can be found easily with a calculator.* Figure 7.18(a) shows matrix A from Example 3, and Figure 7.18(b) shows the inverse of A found with a graphing calculator. We can confirm that these matrices are inverses by computing AA^{-1} (Figure 7.19).

(a)

(b)

Figure 7.18

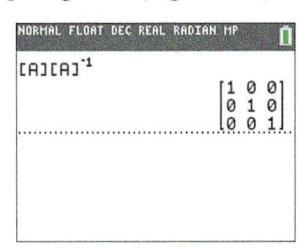

Figure 7.19

*For more details. see Appendix A. page 697.

578 Chapter 7 Systems of Equations and Matrices

Spreadsheet ▶ SOLUTION The inverse of matrix A is found with Excel by entering the matrix in cells B1:D3, selecting cells B5:D7 to contain the inverse, entering "=minverse(", and then selecting the cells B1:D3, closing the parentheses, and holding the CTRL and SHIFT keys down while pressing ENTER. Figure 7.20 shows the spreadsheet with matrix A and its inverse.*

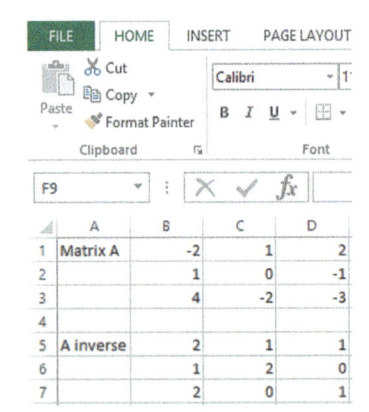

	A	B	C	D
1	Matrix A	-2	1	2
2		1	0	-1
3		4	-2	-3
4				
5	A inverse	2	1	1
6		1	2	0
7		2	0	1

Figure 7.20

EXAMPLE 4 ▶ Does the Inverse Exist?

Find the inverse of $A = \begin{bmatrix} 1 & 2 & -2 \\ 2 & 0 & 2 \\ 6 & 4 & 0 \end{bmatrix}$, if it exists.

SOLUTION

Attempting to find the inverse of matrix A with technology results in an error statement, indicating that the inverse does not exist (Figure 7.21).

Recall that if A does not have an inverse, the reduction process using elementary row operations will yield a row of zeros in the left half of the augmented matrix.†

$$\begin{bmatrix} 1 & 2 & -2 & | & 1 & 0 & 0 \\ 2 & 0 & 2 & | & 0 & 1 & 0 \\ 6 & 4 & 0 & | & 0 & 0 & 1 \end{bmatrix} \xrightarrow{\begin{array}{c}(-2)R_1 + R_2 \to R_2 \\ (-6)R_1 + R_3 \to R_3\end{array}} \begin{bmatrix} 1 & 2 & -2 & | & 1 & 0 & 0 \\ 0 & -4 & 6 & | & -2 & 1 & 0 \\ 0 & -8 & 12 & | & -6 & 0 & 1 \end{bmatrix}$$

$$\xrightarrow{(-1/4)R_2 \to R_2} \begin{bmatrix} 1 & 2 & -2 & | & 1 & 0 & 0 \\ 0 & 1 & -\frac{3}{2} & | & \frac{1}{2} & -\frac{1}{4} & 0 \\ 0 & -8 & 12 & | & -6 & 0 & 1 \end{bmatrix}$$

$$\xrightarrow{(8)R_2 + R_3 \to R_3} \begin{bmatrix} 1 & 2 & -2 & | & 1 & 0 & 0 \\ 0 & 1 & -\frac{3}{2} & | & \frac{1}{2} & -\frac{1}{4} & 0 \\ 0 & 0 & 0 & | & -2 & -2 & 1 \end{bmatrix}$$

Figure 7.21

We see that the left half of the bottom row of the reduced matrix contains all 0s, so it is not possible to reduce the original matrix to the identity matrix, and thus the matrix A does not have an inverse.

Encoding and Decoding Messages

In sending messages during military maneuvers, in business transactions, and in sending secure data on the Internet, encoding (or encryption) of messages is important. Suppose we want to encode the message "Cheer up." The following simple code could be used to change letters of the alphabet to the numbers 1 to 26, respectively, with the number 27 representing a blank space.

---

*For detailed steps, see Appendix B, page 716.
†The methods used by technology will sometimes yield approximations for numbers that should be zeros.

a	b	c	d	e	f	g	h	i	j	k	l	m	n	o	p	q	r	s	t	u	v	w	x	y	z	
1	2	3	4	5	6	7	8	9	10	11	12	13	14	15	16	17	18	19	20	21	22	23	24	25	26	27

Then "Cheer up" can be represented by the numbers

$$3\ 8\ 5\ 5\ 18\ 27\ 21\ 16$$

To further encode the message, we put these numbers in pairs and then create a 2×1 matrix for each pair of numbers. Next we choose an *encoding matrix*, like

$$A = \begin{bmatrix} 3 & -2 \\ -1 & 1 \end{bmatrix}$$

and multiply each pair of numbers (in order) by the encoding matrix, as follows.

$$\begin{bmatrix} 3 & -2 \\ -1 & 1 \end{bmatrix}\begin{bmatrix} 3 \\ 8 \end{bmatrix} = \begin{bmatrix} -7 \\ 5 \end{bmatrix} \quad \begin{bmatrix} 3 & -2 \\ -1 & 1 \end{bmatrix}\begin{bmatrix} 5 \\ 5 \end{bmatrix} = \begin{bmatrix} 5 \\ 0 \end{bmatrix}$$

$$\begin{bmatrix} 3 & -2 \\ -1 & 1 \end{bmatrix}\begin{bmatrix} 18 \\ 27 \end{bmatrix} = \begin{bmatrix} 0 \\ 9 \end{bmatrix} \quad \begin{bmatrix} 3 & -2 \\ -1 & 1 \end{bmatrix}\begin{bmatrix} 21 \\ 16 \end{bmatrix} = \begin{bmatrix} 31 \\ -5 \end{bmatrix}$$

Because multiplying a 2×2 matrix by a 2×1 matrix gives a 2×1 matrix, the resulting products are pairs of numbers. Combining the pairs of numbers gives the encoded numerical message

$$-7\ 5\ 5\ 0\ 0\ 9\ 31\ -5$$

Note that another encoding matrix could be used rather than the one used above. We can also encode a message by putting triples of numbers in 3×1 matrices and multiplying each 3×1 matrix by a 3×3 encoding matrix.

EXAMPLE 5 ▶ Encoding Messages

Use the encoding matrix $A = \begin{bmatrix} 1 & -3 & 2 \\ 2 & -2 & 2 \\ 3 & -1 & 1 \end{bmatrix}$ to encode the message "Meet me for lunch."

SOLUTION

Converting the letters of the message to triples of numbers gives

$$13\ 5\ 5 \quad 20\ 27\ 13 \quad 5\ 27\ 6 \quad 15\ 18\ 27 \quad 12\ 21\ 14 \quad 3\ 8\ 27$$

with 27 used to complete the last triple.

Placing these triples of numbers in 3×1 matrices and multiplying by matrix A gives

$$\begin{bmatrix} 1 & -3 & 2 \\ 2 & -2 & 2 \\ 3 & -1 & 1 \end{bmatrix}\begin{bmatrix} 13 \\ 5 \\ 5 \end{bmatrix} = \begin{bmatrix} 8 \\ 26 \\ 39 \end{bmatrix} \quad \begin{bmatrix} 1 & -3 & 2 \\ 2 & -2 & 2 \\ 3 & -1 & 1 \end{bmatrix}\begin{bmatrix} 20 \\ 27 \\ 13 \end{bmatrix} = \begin{bmatrix} -35 \\ 12 \\ 46 \end{bmatrix}$$

$$\begin{bmatrix} 1 & -3 & 2 \\ 2 & -2 & 2 \\ 3 & -1 & 1 \end{bmatrix}\begin{bmatrix} 5 \\ 27 \\ 6 \end{bmatrix} = \begin{bmatrix} -64 \\ -32 \\ -6 \end{bmatrix} \quad \begin{bmatrix} 1 & -3 & 2 \\ 2 & -2 & 2 \\ 3 & -1 & 1 \end{bmatrix}\begin{bmatrix} 15 \\ 18 \\ 27 \end{bmatrix} = \begin{bmatrix} 15 \\ 48 \\ 54 \end{bmatrix}$$

$$\begin{bmatrix} 1 & -3 & 2 \\ 2 & -2 & 2 \\ 3 & -1 & 1 \end{bmatrix}\begin{bmatrix} 12 \\ 21 \\ 14 \end{bmatrix} = \begin{bmatrix} -23 \\ 10 \\ 29 \end{bmatrix} \quad \begin{bmatrix} 1 & -3 & 2 \\ 2 & -2 & 2 \\ 3 & -1 & 1 \end{bmatrix}\begin{bmatrix} 3 \\ 8 \\ 27 \end{bmatrix} = \begin{bmatrix} 33 \\ 44 \\ 28 \end{bmatrix}$$

The products are triples of numbers. Combining the triples of numbers gives the encoded message

$$8\ 26\ 39\ -35\ 12\ 46\ -64\ -32\ -6\ 15\ 48\ 54\ -23\ 10\ 29\ 33\ 44\ 28$$

We can also find the encoded triples of numbers for the message of Example 5 by writing the original triples of numbers as columns in one matrix and then multiplying this matrix by the encoding matrix. The columns of the product will be the encoded triples of numbers.

$$\begin{bmatrix} 1 & -3 & 2 \\ 2 & -2 & 2 \\ 3 & -1 & 1 \end{bmatrix} \begin{bmatrix} 13 & 20 & 5 & 15 & 12 & 3 \\ 5 & 27 & 27 & 18 & 21 & 8 \\ 5 & 13 & 6 & 27 & 14 & 27 \end{bmatrix} = \begin{bmatrix} 8 & -35 & -64 & 15 & -23 & 33 \\ 26 & 12 & -32 & 48 & 10 & 44 \\ 39 & 46 & -6 & 54 & 29 & 28 \end{bmatrix}$$

Observe that the columns of this product have the same triples, respectively, as the individual products in Example 5. Figure 7.22 shows a calculator display of the product.

Sending an encoded message is of little value if the recipient is not able to decode it. If the message is encoded with a matrix, then the message can be decoded with the inverse of the matrix.

Recall that the message "Cheer up" was represented by the numbers 3 8 5 5 18 27 21 16 and encoded to -7 5 5 0 0 9 31 -5 with the encoding matrix

$$A = \begin{bmatrix} 3 & -2 \\ -1 & 1 \end{bmatrix}$$

To decode the encoded message, we multiply each pair of numbers in the encoded message on the left by the inverse of matrix A.

$$A^{-1} = \begin{bmatrix} 1 & 2 \\ 1 & 3 \end{bmatrix}$$

Multiplying the pairs of the coded message by A^{-1} gives us back the original message.

$$\begin{bmatrix} 1 & 2 \\ 1 & 3 \end{bmatrix} \begin{bmatrix} -7 \\ 5 \end{bmatrix} = \begin{bmatrix} 3 \\ 8 \end{bmatrix} \quad \begin{bmatrix} 1 & 2 \\ 1 & 3 \end{bmatrix} \begin{bmatrix} 5 \\ 0 \end{bmatrix} = \begin{bmatrix} 5 \\ 5 \end{bmatrix}$$

$$\begin{bmatrix} 1 & 2 \\ 1 & 3 \end{bmatrix} \begin{bmatrix} 0 \\ 9 \end{bmatrix} = \begin{bmatrix} 18 \\ 27 \end{bmatrix} \quad \begin{bmatrix} 1 & 2 \\ 1 & 3 \end{bmatrix} \begin{bmatrix} 31 \\ -5 \end{bmatrix} = \begin{bmatrix} 21 \\ 16 \end{bmatrix}$$

Thus, the decoded numbers are 3 8 5 5 18 27 21 16, which correspond to the letters in the message "Cheer up."

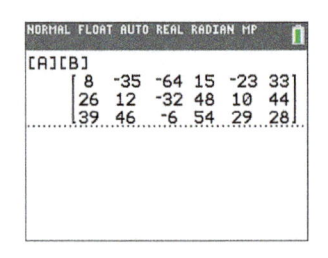

Figure 7.22

EXAMPLE 6 ▶ Decoding Messages

Suppose we encoded messages using a simple code that changed letters of the alphabet to numbers, with the number 27 representing a blank space, and the encoding matrix

$$A = \begin{bmatrix} 1 & 2 & -1 \\ 2 & 1 & 2 \\ 3 & 2 & -3 \end{bmatrix}$$

Decode the following coded message:

18 41 34 61 93 75 39 77 9 34 62 46 3 62 -23

7.4 Inverse Matrices; Matrix Equations 581

SOLUTION

To find the numbers representing the message, we place triples of numbers from the coded message into columns of a 3 × 5 matrix C and multiply that matrix by A^{-1}. The inverse of A is

$$A^{-1} = \begin{bmatrix} \frac{-7}{16} & \frac{1}{4} & \frac{5}{16} \\ \frac{3}{4} & 0 & \frac{-1}{4} \\ \frac{1}{16} & \frac{1}{4} & \frac{-3}{16} \end{bmatrix}$$

Multiplying by A^{-1} on the left gives

$$A^{-1} \begin{bmatrix} 18 & 61 & 39 & 34 & 3 \\ 41 & 93 & 77 & 62 & 62 \\ 34 & 75 & 9 & 46 & -23 \end{bmatrix} = \begin{bmatrix} \frac{-7}{16} & \frac{1}{4} & \frac{5}{16} \\ \frac{3}{4} & 0 & \frac{-1}{4} \\ \frac{1}{16} & \frac{1}{4} & \frac{-3}{16} \end{bmatrix} \begin{bmatrix} 18 & 61 & 39 & 34 & 3 \\ 41 & 93 & 77 & 62 & 62 \\ 34 & 75 & 9 & 46 & -23 \end{bmatrix}$$

$$= \begin{bmatrix} 13 & 20 & 5 & 15 & 7 \\ 5 & 27 & 27 & 14 & 8 \\ 5 & 13 & 20 & 9 & 20 \end{bmatrix}$$

If we use technology to perform this multiplication, we do not need to display the elements of A^{-1} in the computation. We enter $[A]$ and $[C]$ and compute $[A]^{-1}[C]$ (Figure 7.23).

Reading down the columns of the product gives the numbers representing the message:

$$13\ 5\ 5\ 20\ 27\ 13\ 5\ 27\ 20\ 15\ 14\ 9\ 7\ 8\ 20$$

The message is "Meet me tonight."

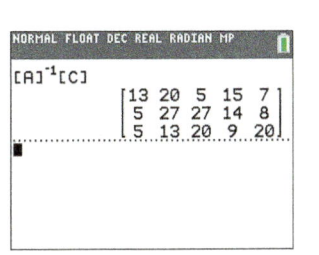

Figure 7.23

Matrix Equations

The system of equations

$$\begin{cases} -2x + y + 2z = 5 \\ x - z = 2 \\ 4x - 2y - 3z = 4 \end{cases}$$

can be written as the matrix equation

$$\begin{bmatrix} -2x + y + 2z \\ x - z \\ 4x - 2y - 3z \end{bmatrix} = \begin{bmatrix} 5 \\ 2 \\ 4 \end{bmatrix}$$

Because

$$\begin{bmatrix} -2 & 1 & 2 \\ 1 & 0 & -1 \\ 4 & -2 & -3 \end{bmatrix} \begin{bmatrix} x \\ y \\ z \end{bmatrix} = \begin{bmatrix} -2x + y + 2z \\ x - z \\ 4x - 2y - 3z \end{bmatrix}$$

we can write the system of equations as the matrix equation in the form

$$\begin{bmatrix} -2 & 1 & 2 \\ 1 & 0 & -1 \\ 4 & -2 & -3 \end{bmatrix} \begin{bmatrix} x \\ y \\ z \end{bmatrix} = \begin{bmatrix} 5 \\ 2 \\ 4 \end{bmatrix}$$

Note that the 3 × 3 matrix on the left side of the matrix equation is the coefficient matrix for the original system. We will call the coefficient matrix A, the matrix containing the variables X, and the matrix containing the constants C, giving the form

$$AX = C$$

If A^{-1} exists and we multiply both sides of this equation on the left by A^{-1}, the product is as follows:*

$$A^{-1}AX = A^{-1}C$$

Because $A^{-1}A = I$ and because multiplication of matrix X by an identity matrix gives matrix X, we have

$$IX = A^{-1}C \quad \text{or} \quad X = A^{-1}C$$

In general, multiplying both sides of the matrix equation $AX = C$ on the left by A^{-1} gives the solution to the system that the matrix equation represents, if the solution is unique.

EXAMPLE 7 ▶ Solution of Matrix Equations with Inverses

Solve the system

$$\begin{cases} -2x + y + 2z = 5 \\ x \qquad - z = 2 \\ 4x - 2y - 3z = 4 \end{cases}$$

by writing a matrix equation and using an inverse matrix.

SOLUTION

We can write this system of equations as $AX = C$, where

$$A = \begin{bmatrix} -2 & 1 & 2 \\ 1 & 0 & -1 \\ 4 & -2 & -3 \end{bmatrix}, \quad X = \begin{bmatrix} x \\ y \\ z \end{bmatrix} \quad \text{and} \quad C = \begin{bmatrix} 5 \\ 2 \\ 4 \end{bmatrix}$$

That is,

$$\begin{bmatrix} -2 & 1 & 2 \\ 1 & 0 & -1 \\ 4 & -2 & -3 \end{bmatrix} \begin{bmatrix} x \\ y \\ z \end{bmatrix} = \begin{bmatrix} 5 \\ 2 \\ 4 \end{bmatrix}$$

As we saw in Example 3, the inverse of matrix A is

$$A^{-1} = \begin{bmatrix} 2 & 1 & 1 \\ 1 & 2 & 0 \\ 2 & 0 & 1 \end{bmatrix}$$

Thus, we can solve the system by multiplying both sides of the matrix equation on the left by A^{-1} as follows:

$$A^{-1}A \begin{bmatrix} x \\ y \\ z \end{bmatrix} = A^{-1} \begin{bmatrix} 5 \\ 2 \\ 4 \end{bmatrix}$$

$$I \begin{bmatrix} x \\ y \\ z \end{bmatrix} = \begin{bmatrix} 2 & 1 & 1 \\ 1 & 2 & 0 \\ 2 & 0 & 1 \end{bmatrix} \begin{bmatrix} 5 \\ 2 \\ 4 \end{bmatrix}$$

---

*Recall that multiplication on the right may give a different product than multiplication on the left.

$$\begin{bmatrix} x \\ y \\ z \end{bmatrix} = \begin{bmatrix} 16 \\ 9 \\ 14 \end{bmatrix}$$

Thus, we have the solution to the system, $x = 16$, $y = 9$, and $z = 14$, or $(16, 9, 14)$.

Matrix Equations and Technology

We can multiply both sides of the matrix equation $AX = C$ by the inverse of A using technology and get the solution to the system that the matrix equation represents. If we use technology to solve a system of linear equations, it is not necessary to display the inverse of the coefficient matrix.

EXAMPLE 8 ▶ Manufacturing

Sharper Technology Company manufactures three types of calculators: a business calculator, a scientific calculator, and a graphing calculator. The production requirements are given in Table 7.10. If during each month the company has 134,000 circuit components, 56,000 hours for assembly, and 14,000 cases, how many of each type of calculator can it produce each month?

Table 7.10

	Business Calculator	Scientific Calculator	Graphing Calculator
Circuit Components	5	7	12
Assembly Time (hours)	2	3	5
Cases	1	1	1

SOLUTION

If the company produces x business calculators, y scientific calculators, and z graphing calculators, then the problem can be represented using the following system of equations.

$$\begin{cases} 5x + 7y + 12z = 134{,}000 \\ 2x + 3y + 5z = 56{,}000 \\ x + y + z = 14{,}000 \end{cases}$$

This system can be represented by the matrix equation $AX = C$, where A is the coefficient matrix and C is the constant matrix.

$$\begin{bmatrix} 5 & 7 & 12 \\ 2 & 3 & 5 \\ 1 & 1 & 1 \end{bmatrix} \begin{bmatrix} x \\ y \\ z \end{bmatrix} = \begin{bmatrix} 134{,}000 \\ 56{,}000 \\ 14{,}000 \end{bmatrix}$$

Multiplying both sides of this equation by the inverse of the coefficient matrix gives the solution (Figure 7.24):

$$\begin{bmatrix} x \\ y \\ z \end{bmatrix} = A^{-1} \begin{bmatrix} 134{,}000 \\ 56{,}000 \\ 14{,}000 \end{bmatrix} = \begin{bmatrix} 2000 \\ 4000 \\ 8000 \end{bmatrix}$$

(a)

(b)

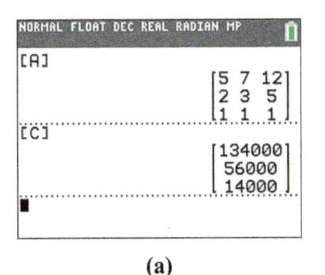

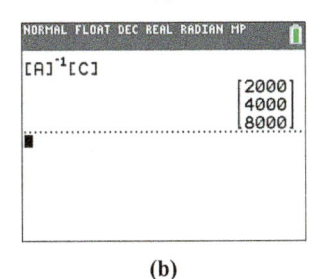

Figure 7.24

This shows that the company can produce 2000 business calculators, 4000 scientific calculators, and 8000 graphing calculators each month. (Recall that we could also solve this system of linear equations by using elementary row operations with an augmented matrix, as discussed in Section 7.2.)

Spreadsheet ▸ SOLUTION We can also use inverse matrices with Excel to solve a system of linear equations if a unique solution exists. We can solve such a system by finding the inverse of the coefficient matrix and multiplying this inverse times the matrix containing the constants. The spreadsheet in Figure 7.25 shows the solution of Example 8 found with Excel.

We can find this solution as follows:

1. Enter the coefficient matrix A in cells B1:D3, compute the inverse of A in B5:D7, enter matrix C containing the constants in B9:B11, and select cells b13:B15.

2. Type "=mmult(" in the formula bar, select B5:D7, type a comma, select B9:B11, and close the parentheses.

3. Hold the CTRL and SHIFT keys down and press ENTER, getting the solution in B13:B15. The inverse is shown in Figure 7.25.*

	A	B	C	D
1	Matrix A	5	7	12
2		2	3	5
3		1	1	1
4				
5	Inverse of A	2	-5	1
6		-3	7	1
7		1	-2	-1
8				
9	Matrix C	134,000		
10		56,000		
11		14,000		
12				
13	Solution X	2,000		
14		4,000		
15		8,000		

Formula bar: {=MMULT(B5:D7,B9:B11)}

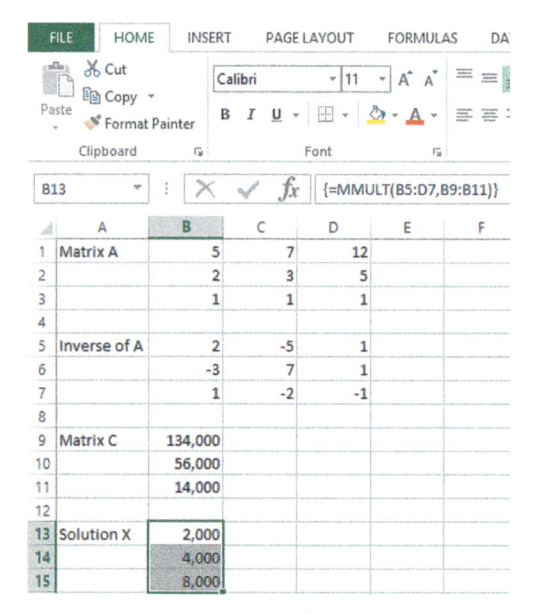

Figure 7.25

We can use the inverse of the coefficient matrix to solve a system of linear equations only if the system has a unique solution. Otherwise, the inverse of the coefficient matrix will not exist, and an attempt to use technology to solve the system with an inverse will give an error message.

---

*For detailed steps, see Appendix B, page 716.

SKILLS CHECK 7.4

Answers that are not seen can be found in the answer section at the back of the text.

1. Suppose $A = \begin{bmatrix} 3 & 1 \\ 4 & 2 \end{bmatrix}$ and $B = \begin{bmatrix} 1 & -0.5 \\ -2 & 1.5 \end{bmatrix}$.

   a. Compute AB and BA, if possible.

   b. What is the relationship between A and B?
   They are inverses.

2. Suppose $A = \begin{bmatrix} 1 & -\frac{1}{2} & -\frac{1}{4} \\ -\frac{1}{2} & 0 & \frac{1}{2} \\ 0 & \frac{1}{2} & -\frac{1}{4} \end{bmatrix}$ and
   $B = \begin{bmatrix} 2 & 2 & 2 \\ 1 & 2 & 3 \\ 2 & 4 & 2 \end{bmatrix}$.

   a. Compute AB and BA, if possible.

   b. What is the relationship between A and B?
   They are inverses.

3. Show that A and B are inverse matrices if
$$A = \begin{bmatrix} 1 & -1 & 1 \\ 2 & -1 & 0 \\ -2 & 2 & -1 \end{bmatrix} \text{ and } B = \begin{bmatrix} 1 & 1 & 1 \\ 2 & 1 & 2 \\ 2 & 0 & 1 \end{bmatrix}.$$

4. Show that C and D are inverse matrices if
$$C = \begin{bmatrix} -\frac{1}{2} & \frac{1}{2} & 1 \\ \frac{1}{2} & \frac{1}{2} & 0 \\ -\frac{1}{2} & \frac{3}{2} & 1 \end{bmatrix} \text{ and } D = \begin{bmatrix} 1 & 2 & -1 \\ -1 & 0 & 1 \\ 2 & 1 & -1 \end{bmatrix}.$$

In Exercises 5–14, find the inverse of the given matrix.

5. $A = \begin{bmatrix} 1 & 3 \\ 2 & 7 \end{bmatrix}$

6. $A = \begin{bmatrix} 2 & 4 \\ 2 & 5 \end{bmatrix}$

7. $A = \begin{bmatrix} 2 & -2 & 2 \\ 2 & 1 & 2 \\ 2 & 0 & 1 \end{bmatrix}$ $A^{-1} = \begin{bmatrix} -1/6 & -1/3 & 1 \\ -1/3 & 1/3 & 0 \\ 1/3 & 2/3 & -1 \end{bmatrix}$

8. $A = \begin{bmatrix} \frac{1}{2} & -1 & 1 \\ 1 & -2 & 0 \\ 1 & 2 & -1 \end{bmatrix}$

9. $A = \begin{bmatrix} 1 & 1 & -1 \\ -1 & 0 & 1 \\ 1 & 1 & 2 \end{bmatrix}$

10. $C = \begin{bmatrix} 0 & 1 & -1 \\ 1 & 0 & 0 \\ 0 & 0 & 1 \end{bmatrix}$

11. $A = \begin{bmatrix} 2 & 3 & 1 \\ 3 & 4 & -1 \\ 2 & 5 & 1 \end{bmatrix}$

12. $B = \begin{bmatrix} 9 & 2 & -7 \\ -5 & 0 & 5 \\ 7 & -4 & -1 \end{bmatrix}$ $B^{-1} = \begin{bmatrix} 0.2 & 0.3 & 0.1 \\ 0.3 & 0.4 & -0.1 \\ 0.2 & 0.5 & 0.1 \end{bmatrix}$

13. $C = \begin{bmatrix} 1 & 0 & -1 & -1 \\ 0 & 1 & -1 & -1 \\ 0 & 0 & 1 & 0 \\ 0 & 0 & 0 & 1 \end{bmatrix}$ $C^{-1} = \begin{bmatrix} 1 & 0 & 1 & 1 \\ 0 & 1 & 1 & 1 \\ 0 & 0 & 1 & 0 \\ 0 & 0 & 0 & 1 \end{bmatrix}$

14. $A = \begin{bmatrix} -1 & 0 & 1 & 1 \\ 0 & 1 & 1 & 1 \\ 1 & -1 & 0 & 0 \\ 0 & 0 & 0 & 1 \end{bmatrix}$ $A^{-1} = \begin{bmatrix} -0.5 & 0.5 & 0.5 & 0 \\ -0.5 & 0.5 & -0.5 & 0 \\ 0.5 & 0.5 & 0.5 & -1 \\ 0 & 0 & 0 & 1 \end{bmatrix}$

15. If $A^{-1} = \begin{bmatrix} 1 & 2 \\ 4 & 3 \end{bmatrix}$, solve $AX = \begin{bmatrix} 2 \\ 4 \end{bmatrix}$ for X. $X = \begin{bmatrix} 10 \\ 20 \end{bmatrix}$

16. If $A^{-1} = \begin{bmatrix} 1 & 2 & -1 \\ 0 & 2 & 1 \\ 2 & 0 & -2 \end{bmatrix}$, solve $AX = \begin{bmatrix} 1 \\ 2 \\ -2 \end{bmatrix}$ for X.

17. Solve the matrix equation
$$\begin{bmatrix} -1 & 1 & 0 \\ -2 & 3 & -2 \\ 2 & -2 & 1 \end{bmatrix} \begin{bmatrix} x \\ y \\ z \end{bmatrix} = \begin{bmatrix} 3 \\ 5 \\ 8 \end{bmatrix}$$
$x = 24, y = 27, z = 14$

18. Solve the matrix equation
$$\begin{bmatrix} -2 & 1 & 0 \\ -1 & 3 & -2 \\ 2 & -2 & 1 \end{bmatrix} \begin{bmatrix} x \\ y \\ z \end{bmatrix} = \begin{bmatrix} 2 \\ 4 \\ 8 \end{bmatrix}$$
$x = 22, y = 46, z = 56$

In Exercises 19–24, use an inverse matrix to find the solution to the systems.

19. $\begin{cases} 4x - 3y + z = 2 \\ -6x + 5y - 2z = -3 \\ x - y + z = 1 \end{cases}$ $x = 1, y = 1, z = 1$

20. $\begin{cases} 5x + 3y + z = 12 \\ 4x + 3y + 2z = 9 \\ x + y + 2z = 2 \end{cases}$ $x = 3, y = -1, z = 0$

21. $\begin{cases} 2x + y + z = 4 \\ x + 4y + 2z = 4 \\ 2x + y + 2z = 3 \end{cases}$ $x = 2, y = 1, z = -1$

22. $\begin{cases} x + 2y - z = 2 \\ -x + z = 4 \\ 2x + y - z = 6 \end{cases}$ $x = 7, y = 3, z = 11$

23. $\begin{cases} x_1 + x_3 + x_4 = 90 \\ x_2 + x_3 + x_4 = 72 \\ 2x_1 + 5x_2 + x_3 = 108 \\ 3x_2 + x_4 = 144 \end{cases}$
$x_1 = 34, x_2 = 16, x_3 = -40, x_4 = 96$

24. $\begin{cases} 2x_1 + x_2 + 3x_3 + 4x_4 = 1 \\ x_1 + x_2 + x_3 - x_4 = 2 \\ x_1 + x_2 + x_3 + x_4 = 10 \\ 2x_1 - x_2 + 3x_3 - x_4 = 5 \end{cases}$
$x_1 = 57, x_2 = -12, x_3 = -39, x_4 = 4$

EXERCISES 7.4

Answers that are not seen can be found in the answer section at the back of the text.

25. **Competition** Two phone companies compete for customers in the southeastern region of a state. Each year, company X retains $\frac{3}{5}$ of its customers and loses $\frac{2}{5}$ of its customers to company Y, while company Y retains $\frac{2}{3}$ of its customers and loses $\frac{1}{3}$ of its customers to company X. If we represent the fraction of the market held last year by

$$\begin{bmatrix} a \\ b \end{bmatrix}$$

where a is the number that company X had last year and b is the number that company Y had last year, then the number that each company will have this year can be found by

$$\begin{bmatrix} A \\ B \end{bmatrix} = \begin{bmatrix} \frac{3}{5} & \frac{1}{3} \\ \frac{2}{5} & \frac{2}{3} \end{bmatrix} \begin{bmatrix} a \\ b \end{bmatrix}$$

If company X has $A = 150{,}000$ customers and company Y has $B = 120{,}000$ customers this year, how many customers did each have last year?
Company X: 225,000; company Y: 45,000

26. **Competition** A satellite company and a cable company compete for customers in a city. The satellite company retains $\frac{2}{3}$ of its customers and loses $\frac{1}{3}$ of its customers to the cable company, while the cable company retains $\frac{3}{5}$ of its customers and loses $\frac{2}{5}$ of its customers to the satellite company. If we represent the fraction of the market held last year by

$$\begin{bmatrix} a \\ b \end{bmatrix}$$

where a is the number of customers that the satellite company had last year and b is the number that the cable company had last year, then the number that each company will have this year can be found by

$$\begin{bmatrix} A \\ B \end{bmatrix} = \begin{bmatrix} \frac{2}{3} & \frac{2}{5} \\ \frac{1}{3} & \frac{3}{5} \end{bmatrix} \begin{bmatrix} a \\ b \end{bmatrix}$$

If the satellite company has $A = 90{,}000$ customers and the cable company has $B = 85{,}000$ customers this year, how many customers did each have last year?
Satellite: 75,000; cable: 100,000

27. **Politics** In a midwestern state, it is observed that 90% of all Republicans vote for the Republican candidate for governor and the remainder vote for the Democratic candidate, while 80% of all Democrats vote for the Democratic candidate and the remaining Democrats vote for the Republican candidate. The percent voting for each party in the next election is given by

$$\begin{bmatrix} R \\ D \end{bmatrix} = \begin{bmatrix} 0.90 & 0.20 \\ 0.10 & 0.80 \end{bmatrix} \begin{bmatrix} r \\ d \end{bmatrix}$$

where r and d are the respective percents voting for each party in the last election. If 55% of the votes in this election were for a Republican and 45% were for a Democrat, what are the percents for each party candidate in the last election?
50% Republican, 50% Democrat

28. **Politics** Suppose that in a certain city, the Democratic, Republican, and Consumer parties always nominate candidates for mayor. The percent of people voting for each party candidate in the next election depends on the percent voting for that party in the previous election. The percent voting for each party in the next election is given by

$$\begin{bmatrix} D \\ R \\ C \end{bmatrix} = \begin{bmatrix} 50 & 40 & 30 \\ 40 & 50 & 30 \\ 10 & 10 & 40 \end{bmatrix} \begin{bmatrix} d \\ r \\ c \end{bmatrix}$$

where d, r, and c are the respective percents (in decimals) voting for each party in the last election. If the percent voting for a Democrat in this election is $D = 42\%$, the percent voting for a Republican in this election is $R = 42\%$, and the percent voting for the Consumer party candidate in this election is $C = 16\%$, what were the respective percents of people voting for these parties in the last election?
Democrat: 40%; Republican: 40%; Consumer: 20%

29. **Loans** A bank gives three loans totaling $400,000 to a development company for the purchase of three business properties. The largest loan is $100,000 more than the sum of the other two, and the smallest loan is one-half of the next larger loan. Represent the

amount of money in each loan as x, y, and z, respectively.

a. Write a system of three equations in three variables to represent the problem.

b. Use a matrix equation to solve the system and find the amount of each loan. $50,000; $100,000; $250,000

30. **Transportation** Ross Freightline has an order to deliver two products, A and B, to a store. The table gives the volume and weight for one unit of each of the two products.

	Product A	Product B
Volume (cu ft)	40	65
Weight (lb)	320	360

If the truck that can deliver the products can carry 5200 cubic feet and 31,360 pounds, how many units of each product can the truck carry?
26 units of product A, 64 units of product B

31. **Investment** An investor has $400,000 in three accounts, paying 6%, 8%, and 10%, respectively. If she has twice as much invested at 8% as she has at 6%, how much does she have invested in each account if she earns a total of $36,000 in interest?
$50,000 at 6%, $100,000 at 8%, $250,000 at 10%

32. **Pricing** A theater has 1000 seats divided into orchestra, main, and balcony. The orchestra seats cost $80, the main seats cost $50, and the balcony seats cost $40. If all the seats are sold, the revenue is $50,000. If all the orchestra and balcony seats are sold and $\frac{3}{4}$ of the main seats are sold, the revenue is $42,500. How many of each type of seat does the theater have? 100 orchestra, 600 main, 300 balcony

33. **Venture Capital** Suppose a bank draws its venture-capital funds annually from three sources of income: business loans, auto loans, and home mortgages. One spreadsheet shows the income from each of these sources for the years 2023, 2024, and 2025, and the second spreadsheet shows the venture capital for these years. If the bank uses a fixed percent of its income from each of the business loans, find the percent of income from each of these loans. 47% from business loans, 27% from auto loans, 32% from home loans

	A	B	C	D
1		Income from Loans ($millions)		
2	Years	Business	Auto	Home
3	2023	532	58	682
4	2024	562	62	695
5	2025	578	69	722

	A	B
1	Years	Capital (millions)
2	2023	483.94
3	2024	503.28
4	2025	521.33

34. **Bookcases** A company produces three types of bookcases: 4-shelf metal, 6-shelf metal, and 4-shelf wooden. The 4-shelf metal bookcase requires 2 hours for fabrication, 1 hour to paint, and $\frac{1}{2}$ hour to package for shipping. The 6-shelf bookcase requires 3 hours for fabrication, 1.5 hours to paint, and $\frac{1}{2}$ hour to package. The wooden bookcase requires 3 hours to fabricate, 2 hours to paint, and $\frac{1}{2}$ hour to package. The daily amount of time available is 124 hours for fabrication, 68 hours for painting, and 24 hours for packaging. How many of each type of bookcase can be produced each day?
20 metal 4-shelf, 16 metal 6-shelf, 12 wooden

Encoding Messages *We have encoded messages by assigning the numbers 1 to 26 to the letters a to z of the alphabet, respectively, and assigning 27 to a blank space.*

a b c d e f g h i j k l m n o p
1 2 3 4 5 6 7 8 9 10 11 12 13 14 15 16

q r s t u v w x y z blank
17 18 19 20 21 22 23 24 25 26 27

To further encode the messages, we can use an encoding matrix A to convert these numbers into new pairs or triples of numbers. In Exercises 35–38, use the given matrix A to encode the message.

35. **Encoding Messages**
    a. Convert "Just do it" from letters to numbers.
    10, 21, 19, 20, 27, 4, 15, 27, 9, 20
    b. Multiply $A = \begin{bmatrix} 4 & 4 \\ 1 & 2 \end{bmatrix}$ times 2×1 matrices created with columns containing pairs of numbers from part (a) to encode the message "Just do it" into pairs of coded numbers.
    124, 52; 156, 59; 124, 35; 168, 69; 116, 49

36. **Encoding Messages**
    a. Convert "Call home" from letters to numbers.
    3, 1, 12, 12, 27, 8, 15, 13, 5, 27
    b. Multiply $A = \begin{bmatrix} 2 & 3 \\ 2 & 2 \end{bmatrix}$ times 2×1 matrices created with columns containing pairs of numbers from part (a) to encode the message "Call home" into pairs of coded numbers.
    9, 8; 60, 48; 78, 70; 69, 56; 91, 64

37. **Encoding Messages**

    Use the matrix $\begin{bmatrix} 4 & 4 & 4 \\ 1 & 2 & 3 \\ 2 & 4 & 2 \end{bmatrix}$ to encode the message "Neatness counts" into triples of numbers.
    80, 27, 50; 156, 63, 106; 260, 138, 168; 156, 96, 108; 212, 111, 146

38. Encoding Messages

Use the matrix $\begin{bmatrix} 4 & 4 & 4 \\ 1 & 2 & 3 \\ 2 & 4 & 2 \end{bmatrix}$ to encode the message "Meet for lunch" into triples of numbers.
92, 38, 56; 212, 92, 160; 240, 132, 156; 188, 96, 136; 152, 100, 92

Decoding Messages We have encoded messages by assigning the numbers 1 to 26 to the letters a to z of the alphabet, respectively, and assigning 27 to a blank space. We can decode messages of this type by finding the inverse of the encoding matrix and multiplying it times the coded message. Use A^{-1} and the conversion table below to decode the messages in Exercises 39–44.

a	b	c	d	e	f	g	h	i	j	k	l	m	n	o	p
1	2	3	4	5	6	7	8	9	10	11	12	13	14	15	16

q	r	s	t	u	v	w	x	y	z	blank
17	18	19	20	21	22	23	24	25	26	27

39. Decoding Messages The encoding matrix is $A = \begin{bmatrix} 3 & -1 \\ -2 & 1 \end{bmatrix}$, and the encoded message is 51, −29, 55, −35, 76, −49, −15, 16, 11, 1. Vote early

40. Decoding Messages The encoding matrix is $A = \begin{bmatrix} 2 & 3 \\ 2 & 4 \end{bmatrix}$, and the encoded message is 59, 74, 72, 78, 84, 102, 81, 90, 121, 148. Go for it

41. Decoding Messages The encoding matrix is $A = \begin{bmatrix} -1 & 0 & 1 \\ -1 & 1 & 0 \\ 3 & -1 & -1 \end{bmatrix}$, and the encoded message is 1, −4, 16, 21, 23, −40, 3, 6, 6, −26, −14, 67, −9, 0, 23, 9, 1, 8. Mind your manners

42. Decoding Messages The encoding matrix is $A = \begin{bmatrix} 1 & 1 & 0 \\ 1 & 3 & 0 \\ 5 & 3 & 3 \end{bmatrix}$, and the encoded message is 16, 34, 128, 32, 86, 145, 32, 86, 109, 29, 33, 195, 6, 8, 61. Give me a break

43. Decoding Messages Use the encoding matrix $A = \begin{bmatrix} 1 & 2 & -1 \\ 2 & 1 & -3 \\ -1 & 4 & 2 \end{bmatrix}$ to decode the message 29, −1, 75, −19, −66, 50, 46, 41, 47, 3, −38, 65. Monday night

44. Decoding Messages Use the encoding matrix $A = \begin{bmatrix} 2 & 3 & -2 \\ 3 & 2 & -4 \\ -2 & 5 & 3 \end{bmatrix}$ to decode the message 20, −33, 142, 74, 51, 107, 67, 87, −26, 22, −22, 99. Arrive early

45. Coded message
   a. Choose a partner and create a message to send to the partner. Change the message from letters to numbers with the code

a	b	c	d	e	f	g	h	i	j	k	l	m	n	o	p
1	2	3	4	5	6	7	8	9	10	11	12	13	14	15	16

q	r	s	t	u	v	w	x	y	z	blank
17	18	19	20	21	22	23	24	25	26	27

   b. Encode your message with encoding matrix
   $A = \begin{bmatrix} 1 & 2 & 1 \\ 2 & 1 & -3 \\ -1 & 4 & 2 \end{bmatrix}$

   c. Exchange coded messages.

   d. Decode the message that you receive from your partner, and report your partner's message as you have decoded it along with your original message.

7.5 Determinants and Cramer's Rule

KEY OBJECTIVES
- Find determinants of matrices
- Solve systems of equations with Cramer's Rule
- Use Cramer's Rule with inconsistent and dependent systems

SECTION PREVIEW Business Properties

Suppose a bank lends $285,000 to a development company to purchase three business properties, where one of the properties costs $45,000 more than another, and the third costs twice the sum of these two properties. To find the cost of each property, we can solve the system of linear equations defining this problem using one of several methods. In this section, we consider the use of determinants to solve certain systems of equations in two variables and three variables. After we have defined determinants, we will use them to solve systems of this type with Cramer's Rule. (See Example 5.)

The Determinant of a 2 × 2 Matrix

Every square matrix has a number associated with it called the **determinant.** The determinant for a 2 × 2 matrix is defined as follows:

> The determinant of the matrix
> $$\begin{bmatrix} a & b \\ c & d \end{bmatrix} \text{ is } D = \det\begin{bmatrix} a & b \\ c & d \end{bmatrix} = \begin{vmatrix} a & b \\ c & d \end{vmatrix} = ad - bc$$

Note that the difference of the two diagonal products gives the determinant.

$$\begin{vmatrix} a & b \\ c & d \end{vmatrix} = ad - bc$$

EXAMPLE 1 ▶ Evaluating the Determinant of a 2 × 2 Matrix

Evaluate the determinants of the matrices

a. $\begin{bmatrix} 2 & 4 \\ -1 & -4 \end{bmatrix}$ b. $\begin{bmatrix} -3 & 2 \\ -1 & 4 \end{bmatrix}$

SOLUTION

a. $\begin{vmatrix} 2 & 4 \\ -1 & -4 \end{vmatrix} = (2)(-4) - (4)(-1) = -8 + 4 = -4$

b. $\begin{vmatrix} -3 & 2 \\ -1 & 4 \end{vmatrix} = (-3)(4) - (2)(-1) = -12 + 2 = -10$

Cramer's Rule

Determinants can be used to solve systems of equations if the number of variables equals the number of equations. Let's solve the system

$$\begin{cases} ax + by = u \\ cx + dy = w \end{cases}$$

using the elimination method. Assuming $d \neq 0$ and $b \neq 0$,

$\begin{cases} ax + by = u \\ cx + dy = w \end{cases} \rightarrow \begin{cases} adx + bdy = du \quad \text{Multiply by } d. \\ -bcx - bdy = -bw \quad \text{Multiply by } -b. \end{cases}$

$adx - bcx = du - bw$ Add these equations.

$(ad - bc)x = du - bw$ Factor out x.

$x = \dfrac{du - bw}{ad - bc}$ Divide by the coefficient of x.

Note that, by the definition of determinants, $\begin{vmatrix} a & b \\ c & d \end{vmatrix} = ad - bc$ and $\begin{vmatrix} u & b \\ w & d \end{vmatrix} = du - bw$. Thus, the solution x can be expressed as the quotient of two determinants:

$$x = \frac{\begin{vmatrix} u & b \\ w & d \end{vmatrix}}{\begin{vmatrix} a & b \\ c & d \end{vmatrix}}.$$

Similarly, we could use the elimination method to solve the original system for y as the quotient of two determinants.

$$\begin{cases} ax + by = u \\ cx + dy = w \end{cases} \rightarrow \begin{cases} -acx - bcy = -cu \\ acx + ady = aw \end{cases} \quad \text{Multiply by } -c.$$
$$\text{Multiply by } a.$$

$$-bcy + ady = -cu + aw \qquad \text{Add these equations.}$$
$$(-bc + ad)y = -cu + aw \qquad \text{Factor out } y.$$
$$y = \frac{aw - cu}{ad - bc} \qquad \text{Divide by the coefficient of } y.$$

Because $\begin{vmatrix} a & b \\ c & d \end{vmatrix} = ad - bc$ and $\begin{vmatrix} a & u \\ c & w \end{vmatrix} = aw - cu$, y can be expressed as the quotient of two determinants:

$$y = \frac{\begin{vmatrix} a & u \\ c & w \end{vmatrix}}{\begin{vmatrix} a & b \\ c & d \end{vmatrix}}.$$

These solutions can be summarized in the form below, which is called **Cramer's Rule**.

Solving a System of Two Equations in Two Variables Using Cramer's Rule

If $\begin{cases} ax + by = u \\ cx + dy = w \end{cases}$, then $x = \dfrac{\begin{vmatrix} u & b \\ w & d \end{vmatrix}}{\begin{vmatrix} a & b \\ c & d \end{vmatrix}}$ and $y = \dfrac{\begin{vmatrix} a & u \\ c & w \end{vmatrix}}{\begin{vmatrix} a & b \\ c & d \end{vmatrix}}$,

provided $D = \begin{vmatrix} a & b \\ c & d \end{vmatrix} = ad - bc \neq 0$.

Look carefully at the pattern when using Cramer's Rule. The denominator of the solution for both x and y above is the determinant for the coefficients of the variables,

$$D = \begin{vmatrix} a & b \\ c & d \end{vmatrix}.$$

In the solution for x, the numerator is the determinant formed by replacing the entries in the first column (the coefficients of x) by the constants on the right side of the equation. We write this as $D_x = \begin{vmatrix} u & b \\ w & d \end{vmatrix}$.

In the solution for y, the numerator is the determinant formed by replacing the entries in the second column (the coefficients of y) by the constants on the right side of the equation. We write this as $D_y = \begin{vmatrix} a & u \\ c & w \end{vmatrix}$.

Thus, Cramer's Rule states that if $D \neq 0$, then

$$x = \frac{D_x}{D} \quad \text{and} \quad y = \frac{D_y}{D}$$

EXAMPLE 2 ▶ Using Cramer's Rule to Solve a Linear System

Use Cramer's Rule to solve the system

$$\begin{cases} x - 2y = 5 \\ 5x - y = -2 \end{cases}$$

SOLUTION

We need to evaluate three determinants.
D, the determinant of the coefficients of the variables:

$$D = \begin{vmatrix} 1 & -2 \\ 5 & -1 \end{vmatrix} = (1)(-1) - (-2)(5) = -1 + 10 = 9$$

D_x, the determinant obtained by replacing the coefficients of x with the constants on the right side of the equations:

$$D_x = \begin{vmatrix} 5 & -2 \\ -2 & -1 \end{vmatrix} = (5)(-1) - (-2)(-2) = -5 - 4 = -9$$

D_y, the determinant obtained by replacing the coefficients of y with the constants on the right side of the equations:

$$D_y = \begin{vmatrix} 1 & 5 \\ 5 & -2 \end{vmatrix} = (1)(-2) - (5)(5) = -2 - 25 = -27$$

Then, by Cramer's Rule,

$$x = \frac{D_x}{D} = \frac{-9}{9} = -1 \quad \text{and} \quad y = \frac{D_y}{D} = \frac{-27}{9} = -3$$

Thus the solution to the system is $(-1, -3)$. The solution can be checked by substituting these values into the original equations.

Remember, if the determinant of the coefficients of the variables, D, equals 0, a solution cannot be found with Cramer's Rule.

The Determinant of a 3 × 3 Matrix

The determinant of a 3 × 3 matrix is defined as follows:

$$\begin{vmatrix} a_1 & b_1 & c_1 \\ a_2 & b_2 & c_2 \\ a_3 & b_3 & c_3 \end{vmatrix} = a_1 b_2 c_3 + b_1 c_2 a_3 + c_1 a_2 b_3 - a_3 b_2 c_1 - b_3 c_2 a_1 - c_3 a_2 b_1$$

We can rearrange the terms and factor this expression to obtain

$$a_1 b_2 c_3 - b_3 c_2 a_1 - c_3 a_2 b_1 + c_1 a_2 b_3 + b_1 c_2 a_3 - a_3 b_2 c_1$$
$$= a_1 (b_2 c_3 - b_3 c_2) - a_2 (b_1 c_3 - b_3 c_1) + a_3 (b_1 c_2 - b_2 c_1)$$
$$= a_1 \begin{vmatrix} b_2 & c_2 \\ b_3 & c_3 \end{vmatrix} - a_2 \begin{vmatrix} b_1 & c_1 \\ b_3 & c_3 \end{vmatrix} + a_3 \begin{vmatrix} b_1 & c_1 \\ b_2 & c_2 \end{vmatrix}$$

Thus, the determinant of a 3 × 3 determinant is defined by

$$\begin{vmatrix} a_1 & b_1 & c_1 \\ a_2 & b_2 & c_2 \\ a_3 & b_3 & c_3 \end{vmatrix} = a_1 \begin{vmatrix} b_2 & c_2 \\ b_3 & c_3 \end{vmatrix} - a_2 \begin{vmatrix} b_1 & c_1 \\ b_3 & c_3 \end{vmatrix} + a_3 \begin{vmatrix} b_1 & c_1 \\ b_2 & c_2 \end{vmatrix}$$

Note that each of the three terms in the definition contains a numerical factor that is an element from the first column of the 3 × 3 determinant, and a second-order determinant obtained by crossing out the row and column containing the numerical factor. That is,

To obtain $a_1 \begin{vmatrix} b_2 & c_2 \\ b_3 & c_3 \end{vmatrix}$, remove the first row and first column of $\begin{vmatrix} a_1 & b_1 & c_1 \\ a_2 & b_2 & c_2 \\ a_3 & b_3 & c_3 \end{vmatrix}$ and multiply by a_1.

To obtain $a_2 \begin{vmatrix} b_1 & c_1 \\ b_3 & c_3 \end{vmatrix}$, remove the second row and first column of $\begin{vmatrix} a_1 & b_1 & c_1 \\ a_2 & b_2 & c_2 \\ a_3 & b_3 & c_3 \end{vmatrix}$ and multiply by a_2.

To obtain $a_3 \begin{vmatrix} b_1 & c_1 \\ b_2 & c_2 \end{vmatrix}$, remove the third row and first column of $\begin{vmatrix} a_1 & b_1 & c_1 \\ a_2 & b_2 & c_2 \\ a_3 & b_3 & c_3 \end{vmatrix}$ and multiply by a_3.

The 2 × 2 determinants in the definition above are called **minors** of the 3 × 3 determinant.

Note that the signs alternate on the expressions involving the different minors, with a positive sign on the first expression, which corresponds to the element in the first row and first column.

EXAMPLE 3 ▶ Evaluating a 3 × 3 Determinant

Find the determinant of $\begin{bmatrix} -1 & 2 & -3 \\ 3 & 0 & -2 \\ 1 & 2 & -2 \end{bmatrix}$.

SOLUTION

For the first minor, we cross out row 1 and column 1: $\begin{vmatrix} -1 & 2 & -3 \\ 3 & 0 & -2 \\ 1 & 2 & -2 \end{vmatrix}$

For the second minor, we cross out row 2 and column 1: $\begin{vmatrix} -1 & 2 & -3 \\ 3 & 0 & -2 \\ 1 & 2 & -2 \end{vmatrix}$

For the third minor, we cross out row 3 and column 1: $\begin{vmatrix} -1 & 2 & -3 \\ 3 & 0 & -2 \\ 1 & 2 & -2 \end{vmatrix}$

To evaluate the determinant of $\begin{vmatrix} -1 & 2 & -3 \\ 3 & 0 & -2 \\ 1 & 2 & -2 \end{vmatrix}$, we multiply -1 by the first minor, multiply 3 by the second minor, and multiply 1 by the third minor. Then we add the first and third products and subtract the second product.

$$\begin{vmatrix} -1 & 2 & -3 \\ 3 & 0 & -2 \\ 1 & 2 & -2 \end{vmatrix} = (-1)\begin{vmatrix} 0 & -2 \\ 2 & -2 \end{vmatrix} - 3\begin{vmatrix} 2 & -3 \\ 2 & -2 \end{vmatrix} + 1\begin{vmatrix} 2 & -3 \\ 0 & -2 \end{vmatrix}$$

$$= (-1)((0)(-2) - (2)(-2)) - 3((2)(-2) - (2)(-3)) +$$
$$1((2)(-2) - (0)(-3))$$
$$= (-1)(0 + 4) - 3(-4 + 6) + 1(-4 - 0)$$
$$= (-1)(4) - 3(2) + 1(-4)$$
$$= -4 - 6 - 4$$
$$= -14$$

A graphing utility can be used to evaluate the determinant of a matrix.

> **Using a Graphing Calculator to Evaluate a Determinant**
> 1. Enter the matrix and name it A.
> 2. Select "det(" under MATRIX, MATH.
> 3. Enter MATRIX, A, and a right parenthesis, and press **ENTER**.
>
> **Figure 7.26**

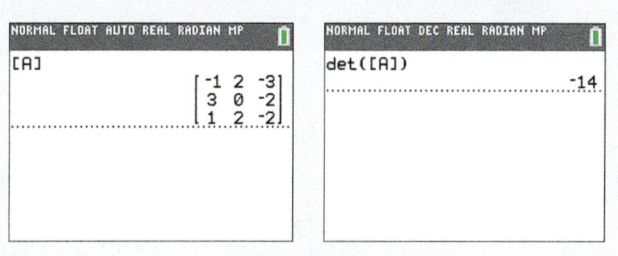

The determinant of $\begin{bmatrix} -1 & 2 & -3 \\ 3 & 0 & -2 \\ 1 & 2 & -2 \end{bmatrix}$, from Example 3, is shown in Figure 7.26.

Solving Systems of Equations in Three Variables with Cramer's Rule

Cramer's Rule can be used to solve systems of equations in three variables if the determinant D of the coefficients of the variables does not equal 0.

> **Cramer's Rule**
>
> The solution to $\begin{cases} a_1x + b_1y + c_1z = d_1 \\ a_2x + b_2y + c_2z = d_2 \\ a_3x + b_3y + c_3z = d_3 \end{cases}$ is $x = \dfrac{D_x}{D}$, $y = \dfrac{D_y}{D}$, and $z = \dfrac{D_z}{D}$, where
>
> $D = \begin{vmatrix} a_1 & b_1 & c_1 \\ a_2 & b_2 & c_2 \\ a_3 & b_3 & c_3 \end{vmatrix}$ The coefficients of the variables x, y, and z
>
> $D_x = \begin{vmatrix} d_1 & b_1 & c_1 \\ d_2 & b_2 & c_2 \\ d_3 & b_3 & c_3 \end{vmatrix}$ Replace x-coefficients with the constants at the right of the three equations.

(continued)

594 Chapter 7 Systems of Equations and Matrices

$$D_y = \begin{vmatrix} a_1 & d_1 & c_1 \\ a_2 & d_2 & c_2 \\ a_3 & d_3 & c_3 \end{vmatrix}$$ Replace y-coefficients with the constants at the right of the three equations.

$$D_z = \begin{vmatrix} a_1 & b_1 & d_1 \\ a_2 & b_2 & d_2 \\ a_3 & b_3 & d_3 \end{vmatrix}$$ Replace z-coefficients with the constants at the right of the three equations.

EXAMPLE 4 ▶ Using Cramer's Rule to Solve a System in Three Variables

Use Cramer's Rule to solve the system

$$\begin{cases} 2x - y - z = 2 \\ 3x - y + z = -3 \\ x + y - z = 7 \end{cases}$$

SOLUTION

First we will evaluate the determinant of the coefficients.

$$D = \begin{vmatrix} 2 & -1 & -1 \\ 3 & -1 & 1 \\ 1 & 1 & -1 \end{vmatrix} = 2\begin{vmatrix} -1 & 1 \\ 1 & -1 \end{vmatrix} - 3\begin{vmatrix} -1 & -1 \\ 1 & -1 \end{vmatrix} + 1\begin{vmatrix} -1 & -1 \\ -1 & 1 \end{vmatrix}$$

$$= 2((-1)(-1) - (1)(1)) - 3((-1)(-1) - (1)(-1)) + 1((-1)(1) - (-1)(-1))$$

$$= 2(1 - 1) - 3(1 + 1) + 1(-1 - 1)$$

$$= 2(0) - 3(2) + 1(-2)$$

$$= -8$$

$$D_x = \begin{vmatrix} 2 & -1 & -1 \\ -3 & -1 & 1 \\ 7 & 1 & -1 \end{vmatrix} = 2\begin{vmatrix} -1 & 1 \\ 1 & -1 \end{vmatrix} - (-3)\begin{vmatrix} -1 & -1 \\ 1 & -1 \end{vmatrix} + 7\begin{vmatrix} -1 & -1 \\ -1 & 1 \end{vmatrix}$$

$$= 2(0) + 3(2) + 7(-2)$$

$$= -8$$

$$D_y = \begin{vmatrix} 2 & 2 & -1 \\ 3 & -3 & 1 \\ 1 & 7 & -1 \end{vmatrix} = 2\begin{vmatrix} -3 & 1 \\ 7 & -1 \end{vmatrix} - 3\begin{vmatrix} 2 & -1 \\ 7 & -1 \end{vmatrix} + 1\begin{vmatrix} 2 & -1 \\ -3 & 1 \end{vmatrix}$$

$$= 2(-4) - 3(5) + 1(-1)$$

$$= -24$$

$$D_z = \begin{vmatrix} 2 & -1 & 2 \\ 3 & -1 & -3 \\ 1 & 1 & 7 \end{vmatrix} = 2\begin{vmatrix} -1 & -3 \\ 1 & 7 \end{vmatrix} - 3\begin{vmatrix} -1 & 2 \\ 1 & 7 \end{vmatrix} + 1\begin{vmatrix} -1 & 2 \\ -1 & -3 \end{vmatrix}$$

$$= 2(-4) - 3(-9) + 1(5)$$

$$= 24$$

Thus,
$$x = \frac{D_x}{D} = \frac{-8}{-8} = 1$$
$$y = \frac{D_y}{D} = \frac{-24}{-8} = 3$$
$$z = \frac{D_z}{D} = \frac{24}{-8} = -3$$

EXAMPLE 5 ▶ Business Properties

A bank lends $285,000 to a development company to purchase three business properties. One of the properties costs $45,000 more than another, and the third costs twice the sum of these two properties. What is the cost of each property?

SOLUTION

If we let x, y, and z represent the cost of the three properties, respectively, then we have the following equations:

$$x + y + z = 285{,}000$$
$$x = 45{,}000 + y$$
$$z = 2(x + y)$$

The system for these equations is

$$\begin{cases} x + y + z = 285{,}000 \\ x - y = 45{,}000 \\ 2x + 2y - z = 0 \end{cases}$$

The determinant of the coefficients is

$$\begin{vmatrix} 1 & 1 & 1 \\ 1 & -1 & 0 \\ 2 & 2 & -1 \end{vmatrix} = 1\begin{vmatrix} -1 & 0 \\ 2 & -1 \end{vmatrix} - 1\begin{vmatrix} 1 & 1 \\ 2 & -1 \end{vmatrix} + 2\begin{vmatrix} 1 & 1 \\ -1 & 0 \end{vmatrix}$$

$$= 1(1 - 0) - 1(-1 - 2) + 2(0 + 1)$$
$$= 1 + 3 + 2$$
$$= 6$$

Then

$$D_x = \begin{vmatrix} 285{,}000 & 1 & 1 \\ 45{,}000 & -1 & 0 \\ 0 & 2 & -1 \end{vmatrix} = 285{,}000\begin{vmatrix} -1 & 0 \\ 2 & -1 \end{vmatrix} - 45{,}000\begin{vmatrix} 1 & 1 \\ 2 & -1 \end{vmatrix} + 0\begin{vmatrix} 1 & 1 \\ -1 & 0 \end{vmatrix}$$

$$= 285{,}000(1) - 45{,}000(-3) + 0$$
$$= 420{,}000$$

$$D_y = \begin{vmatrix} 1 & 285{,}000 & 1 \\ 1 & 45{,}000 & 0 \\ 2 & 0 & -1 \end{vmatrix} = 1\begin{vmatrix} 45{,}000 & 0 \\ 0 & -1 \end{vmatrix} - 1\begin{vmatrix} 285{,}000 & 1 \\ 0 & -1 \end{vmatrix} + 2\begin{vmatrix} 285{,}000 & 1 \\ 45{,}000 & 0 \end{vmatrix}$$

$$= 1(-45{,}000) - 1(-285{,}000) + 2(-45{,}000)$$
$$= 150{,}000$$

$$D_z = \begin{vmatrix} 1 & 1 & 285{,}000 \\ 1 & -1 & 45{,}000 \\ 2 & 2 & 0 \end{vmatrix} = 1\begin{vmatrix} -1 & 45{,}000 \\ 2 & 0 \end{vmatrix} - 1\begin{vmatrix} 1 & 285{,}000 \\ 2 & 0 \end{vmatrix} + 2\begin{vmatrix} 1 & 285{,}000 \\ -1 & 45{,}000 \end{vmatrix}$$

$$= 1(-90{,}000) - 1(-570{,}000) + 2(330{,}000)$$

$$= 1{,}140{,}000$$

So

$$x = \frac{D_x}{D} = \frac{420{,}000}{6} = 70{,}000$$

$$y = \frac{D_y}{D} = \frac{150{,}000}{6} = 25{,}000$$

$$z = \frac{D_z}{D} = \frac{1{,}140{,}000}{6} = 190{,}000$$

Thus, the first property costs $70,000, the second costs $25,000, and the third costs $190,000.

Cramer's Rule with Inconsistent and Dependent Systems

If D, the determinant of the coefficients of the variables of a system of three equations in three variables, is 0, then the system either is inconsistent or contains dependent equations. In particular,

1. If $D = 0$ and at least one of the determinants in the numerator (D_x, D_y, or D_z) is not 0, then the system is inconsistent, and there is no solution.

2. If $D = 0$ and all the determinants in the numerators (D_x, D_y, and D_z) are 0, then the equations in the system are dependent, and there are infinitely many solutions. These solutions can be found using different methods.

SKILLS CHECK 7.5

Answers that are not seen can be found in the answer section at the back of the text.

In Exercises 1–8, evaluate each determinant.

1. $\begin{vmatrix} 3 & 2 \\ 4 & 1 \end{vmatrix}$ -5

2. $\begin{vmatrix} -2 & -1 \\ -3 & 4 \end{vmatrix}$ -11

3. $\begin{vmatrix} 5 & 1 \\ 5 & 1 \end{vmatrix}$ 0

4. $\begin{vmatrix} \frac{1}{2} & \frac{2}{3} \\ \frac{3}{4} & \frac{1}{6} \end{vmatrix}$ $\frac{-5}{12}$

5. $\begin{vmatrix} 3 & -2 & 1 \\ -1 & 0 & 2 \\ 0 & 1 & 1 \end{vmatrix}$ -9

6. $\begin{vmatrix} 1 & 2 & 2 \\ 3 & 0 & 1 \\ -1 & 0 & -2 \end{vmatrix}$ 10

7. $\begin{vmatrix} 0 & -1 & 2 \\ 3 & 1 & -1 \\ 4 & -1 & 3 \end{vmatrix}$ -1

8. $\begin{vmatrix} 1 & 2 & 3 \\ -1 & 5 & 6 \\ -1 & 3 & 3 \end{vmatrix}$ -3

A matrix has an inverse if and only if its determinant does not equal 0. In Exercises 9–12, determine whether the matrix has an inverse.

9. $\begin{bmatrix} 2 & 3 \\ -1 & 4 \end{bmatrix}$ Yes

10. $\begin{bmatrix} 3 & 3 \\ 1 & -1 \end{bmatrix}$ Yes

11. $\begin{bmatrix} 1 & 3 & -2 \\ 2 & -1 & 5 \\ 3 & 2 & 3 \end{bmatrix}$ No

12. $\begin{bmatrix} 4 & 1 & 0 \\ 0 & 5 & 0 \\ 0 & 0 & -1 \end{bmatrix}$ Yes

In Exercises 13–24, use Cramer's Rule to solve each system, if possible. If it is not possible, state whether the system is inconsistent or has infinitely many solutions.

13. $\begin{cases} x + 2y = 4 \\ 3x + 4y = 10 \end{cases}$ (2, 1)

14. $\begin{cases} 3x - 4y = 11 \\ 2x + 3y = -4 \end{cases}$ (1, −2)

15. $\begin{cases} 2x - y = 4 \\ 3x + y = 5 \end{cases}$
$(1.8, -0.4)$

16. $\begin{cases} 5x - 2y = 6 \\ 3x + 3y = 12 \end{cases}$
$(2, 2)$

17. $\begin{cases} 2x - y = 2 \\ -4x + 2y = 1 \end{cases}$
Inconsistent

18. $\begin{cases} 3x + 2y = 4 \\ 6x + 4y = 3 \end{cases}$
Inconsistent

19. $\begin{cases} 2x - 4y = 5 \\ 3x - 6y = 7.5 \end{cases}$
Infinitely many solutions

20. $\begin{cases} x - 2y = 3 \\ -2x + 4y = -6 \end{cases}$
Infinitely many solutions

21. $\begin{cases} x + y + z = 3 \\ 2x + y + z = 4 \\ 2x + 2y + z = 5 \end{cases}$
$(1, 1, 1)$

22. $\begin{cases} 2x - y - 2z = 2 \\ 3x - y + z = -3 \\ x + y - z = 7 \end{cases}$
$(1, 4, -2)$

23. $\begin{cases} x + y + 2z = 8 \\ 2x + y + z = 7 \\ 2x + 2y + z = 10 \end{cases}$
$(1, 3, 2)$

24. $\begin{cases} x - 2y + z = 0 \\ 2x + y - 2z = 2 \\ 3x + 2y - 3z = 2 \end{cases}$
$(-1, -2, -3)$

In Exercises 25–28, solve for x.

25. $\begin{vmatrix} -2 & -11 \\ x & 6 \end{vmatrix} = 32$ 4

26. $\begin{vmatrix} 5 & x \\ -1 & -3 \end{vmatrix} = 5$ 20

27. $\begin{vmatrix} x + 3 & -6 \\ x - 2 & -4 \end{vmatrix} = 18$ 21

28. $\begin{vmatrix} 2 & x & 1 \\ -3 & 1 & 0 \\ 2 & 1 & 4 \end{vmatrix} = 39$ 3

EXERCISES 7.5

Answers that are not seen can be found in the answer section at the back of the text.

In Exercises 29–38, set up each system of equations and then solve it using Cramer's Rule.

29. **Medication** Medication A is given every 4 hours, and medication B is given twice a day. For patient I, the total intake of the two medications is required to be 50.6 mg per day, and for patient II, the total intake is required to be 92 mg per day. The ratio of the dosage of A to the dosage of B is always 5 to 8.

    a. Find the dosage for each administration of each medication for patient I.
    A = 5.5 mg, B = 8.8 mg for patient I
    b. Find the dosage for each administration of each medication for patient II.
    A = 10 mg, B = 16 mg for patient II

30. **Investment** One safe investment pays 10% per year, and a more risky investment pays 18% per year. A woman has $145,000 to invest and would like to have an income of $20,000 per year from her investments. How much should she invest at each rate?
    $68,750 at 18% and $76,250 at 10%

31. **Manufacturing** A manufacturer of table saws has three models—Deluxe, Premium, and Ultimate—that must be painted, assembled, and packaged for shipping. The table gives the number of hours required for each of these operations for each type of table saw. If the manufacturer has 96 hours available for painting, 156 hours for assembling, and 37 hours for packaging, how many of each type of saw can be produced each day? 2 Deluxe, 8 Premium, 32 Ultimate

	Deluxe	Premium	Ultimate
Painting	1.6	2	2.4
Assembling	2	3	4
Packaging	0.5	0.5	1

32. **Ticket Pricing** A theater has 3600 seats, with tickets costing $40, $70, and $100, for three different seating sections: C, B, and A. The owner wants to earn $192,000 from a full house and wants to have twice as many $40 seats as the sum of all other seats. How many seats should he have in each section?
    2400 $40 seats, 800 $70 seats, 400 $100 seats

33. **Rental Cars** A car rental company has $3,160,000 available to purchase large, midsize, and compact cars. It can purchase compact cars for $18,000 each, midsize cars for $28,000 each, and large cars for $48,000 each and can rent them for $50, $70, and $90 per day, respectively. If the company's goal is to buy 120 cars and earn $7600 in daily rental, how many of each car type should it buy?
    60 compact, 40 midsize, and 20 large

34. **Investment** A trust account manager has $1,000,000 to be invested in three different accounts. The accounts pay 6%, 8%, and 10%, respectively, and the goal is to earn $86,000 with the amount invested at 10% equal to the sum of the other two investments. To accomplish this, assume that x dollars are invested at 6%, y dollars at 8%, and z dollars at 10%. Find out how much should be invested in each account to satisfy the conditions.
    $200,000 at 6%, $300,000 at 8%, $500,000 at 10%

35. **Nutrition** Three different bacterial species are cultured in one dish and feed on three nutrients. Each individual of species I consumes 1 unit of each of the first and second nutrients and 2 units of the third nutrient. Each individual of species II consumes 2 units of the first nutrient and 2 units of the third nutrient. Each individual of species III consumes 2 units of the first nutrient, 3 units of the second nutrient, and 5 units of the third nutrient. If the culture is given 5100 units of the first nutrient, 6900 units of the

second nutrient, and 12,000 units of the third nutrient, how many of each species can be supported so all of the nutrients are consumed?

36. **Investment** The members of an investment club set a goal of earning 15% on the money they earn in stocks. The members are considering purchasing four possible stocks. Their cost per share (in dollars) and their projected growth per share (in dollars) are summarized in the following table.

Stocks	Oil	Bank	Computer	Retail
Cost/share	100	40	30	20
Growth/share	10.00	3.60	6.00	2.40

If they have $102,300 to invest, how many shares of each stock should they buy to meet their goal?

37. **Investment** A woman has $800,000 to invest, and she is considering buying shares in three investment properties. One property earns 9%, one earns 8%, and one earns 7.5%. The total annual earnings on the three properties is $67,400. If she invests as much in the 9% property as the sum of her investments in the other two properties, how much did she invest in each property?
$400,000 at 9%, $280,000 at 8%, $120,000 at 7.5%

38. **Social Services** A social services agency provides services to three types of clients: A, B, and C. The agency has $800,000 available for emergency food and shelter and has $660,000 available for counseling. Type A clients require an average of $1200 for emergencies and $800 for counseling, type B clients require an average of $800 for emergencies and $500 for counseling, and type C clients require an average of $400 for emergencies and $600 for counseling. If a total of 1000 clients can be served, how many of each type of client can be served?
400 type A clients, 200 type B clients, 400 type C clients

7.6 Systems of Nonlinear Equations

KEY OBJECTIVES

- Solve systems of nonlinear equations algebraically
- Solve systems of nonlinear equations graphically

SECTION PREVIEW Market Equilibrium

The supply function for a product is given by $p = q^2 + 804$, and the demand function for this product is $p = 1200 - 60q$, where p is the price in dollars and q is the number of hundreds of units. Suppose we are interested in finding the price that gives market equilibrium.

To answer this question, we create a system of equations, one of which is nonlinear, and solve the system. In this section, we use algebraic and graphical methods to solve systems of nonlinear equations.

We have solved systems of linear equations in two variables by substitution, by elimination, and by graphing. To solve systems of nonlinear equations, we can use these same methods. A system of nonlinear equations can have any number of solutions or no solution. ∎

Algebraic Solution of Nonlinear Systems

EXAMPLE 1 ▶ Solving a Nonlinear System Algebraically

Use substitution to solve the system

$$\begin{cases} x^2 - x - 3y = 0 \\ x - y = -4 \end{cases}$$

and check the solution.

SOLUTION

To solve the system

$$\begin{cases} x^2 - x - 3y = 0 \quad (1) \\ x - y = -4 \quad (2) \end{cases}$$

by substitution, notice that Equation (2) can easily be solved for y:

$$x - y = -4$$
$$y = x + 4$$

Substituting this expression for y in Equation (1) gives

$$x^2 - x - 3y = 0$$
$$x^2 - x - 3(x + 4) = 0$$
$$x^2 - x - 3x - 12 = 0$$
$$x^2 - 4x - 12 = 0$$
$$(x - 6)(x + 2) = 0$$
$$x = 6 \quad \text{or} \quad x = -2$$

Using these values for x in $y = x + 4$ gives the solutions $(6, 10)$ and $(-2, 2)$.

To check the solution, we substitute each ordered pair into both of the original equations:

Check $(6, 10)$: $\quad x^2 - x - 3y = 0 \qquad x - y = -4$
$\qquad\qquad\qquad (6)^2 - (6) - 3(10) = 0 \quad 6 - 10 = -4$
$\qquad\qquad\qquad\qquad 36 - 6 - 30 = 0 \qquad -4 = -4$
$\qquad\qquad\qquad\qquad\qquad 0 = 0$

Check $(-2, 2)$: $\quad x^2 - x - 3y = 0 \qquad x - y = -4$
$\qquad\qquad\qquad (-2)^2 - (-2) - 3(2) = 0 \quad -2 - 2 = -4$
$\qquad\qquad\qquad\qquad 4 + 2 - 6 = 0 \qquad\qquad -4 = -4$
$\qquad\qquad\qquad\qquad\qquad 0 = 0$

Both solutions check in each equation of the original system.

EXAMPLE 2 ▶ **Solving a Nonlinear System Algebraically**

Solve the system algebraically, if $y \neq 0$:

$$\begin{cases} x + \dfrac{x}{y} = 6 \\ xy + y = 10 \end{cases}$$

SOLUTION

To solve the system

$$\begin{cases} x + \dfrac{x}{y} = 6 & (1) \\ xy + y = 10 & (2) \end{cases}$$

algebraically, we begin by multiplying Equation (1) by y:

$$xy + x = 6y \qquad (3)$$

Next we solve Equation (2) for x:

$$xy + y = 10$$
$$xy = 10 - y$$
$$x = \frac{10 - y}{y}$$

Substituting this expression for x in Equation (3) gives

$$\frac{10-y}{y}(y) + \frac{10-y}{y} = 6y$$

$$10 - y + \frac{10-y}{y} = 6y$$

Now we multiply both sides of this equation by y and solve the quadratic equation that results:

$$10 - y + \frac{10-y}{y} = 6y$$

$$10y - y^2 + 10 - y = 6y^2$$

$$0 = 7y^2 - 9y - 10$$

$$0 = (y-2)(7y+5)$$

$$y = 2 \quad \text{or} \quad y = -\frac{5}{7}$$

If $y = 2$, then

$$x = \frac{10-2}{2} = \frac{8}{2} = 4$$

and if $y = -\frac{5}{7}$, then

$$x = \frac{10 + \frac{5}{7}}{-\frac{5}{7}} = -\frac{75}{5} = -15$$

Thus, the solutions are $(4, 2)$ and $\left(-15, -\frac{5}{7}\right)$.

Both solutions check in the original system of equations.

EXAMPLE 3 ▶ Market Equilibrium

The supply function for a product is given by $p = q^2 + 804$, and the demand function for this product is $p = 1200 - 60q$, where p is the price in dollars and q is the number of hundreds of units. Find the price that gives market equilibrium.

SOLUTION

Market equilibrium occurs when the supply price (and quantity) equals the demand price (and quantity). Thus, the solution to the system

$$\begin{cases} p = q^2 + 804 & (1) \\ p = 1200 - 60q & (2) \end{cases}$$

gives the equilibrium quantity and price.

Substituting for p (from Equation (1)) into Equation (2) and solving gives

$$q^2 + 804 = 1200 - 60q$$

$$q^2 + 60q - 396 = 0$$

$$(q - 6)(q + 66) = 0$$

$$q = 6 \quad \text{or} \quad q = -66$$

7.6 Systems of Nonlinear Equations   601

The negative value of q has no meaning in this application, so we have $q = 6$ and $p = 6^2 + 804 = 840$. Thus, market equilibrium occurs when the price is $840 and the quantity is 600 units.

Graphical Solution of Nonlinear Systems

As with systems of linear equations in two variables, we can use graphical methods to check and to solve systems of nonlinear equations.

EXAMPLE 4 ▶ Solving a Nonlinear System Graphically

Solve the system

$$\begin{cases} y = -x^2 - 2x + 8 \\ 2x + y = 4 \end{cases}$$

graphically.

SOLUTION

To solve the system

$$\begin{cases} y = -x^2 - 2x + 8 & (1) \\ 2x + y = 4 & (2) \end{cases}$$

graphically, we first solve Equation (2) for y, getting $y = -2x + 4$. Graphing the two equations on the same axes gives the graph in Figure 7.27.

The points of intersection are $(2, 0)$ and $(-2, 8)$, as shown in Figures 7.28(a) and (b).

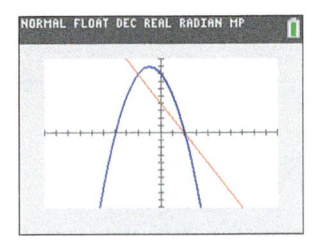

Figure 7.27

(a)   (b)

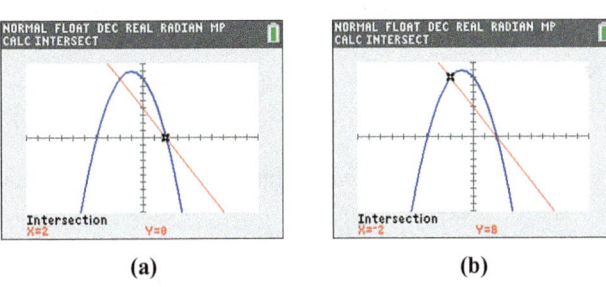

Figure 7.28

Thus, the solutions to the system are $(2, 0)$ and $(-2, 8)$.

EXAMPLE 5 ▶ Constructing a Box

An open box with a square base is to be constructed so that the area of the material used is 300 cm² and the volume of the box is 500 cm³. What are the dimensions of the box?

SOLUTION

Let x represent the length of a side of the base of the box and let y represent the height of the box (Figure 7.29). The area of the base of the box is x^2, and the area of each of the four sides is xy, so the area of the material is

$$x^2 + 4xy = 300$$

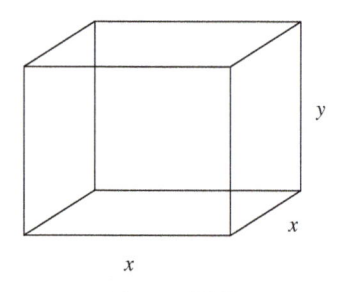

Figure 7.29

The volume of the box is $x^2 y = 500$. So the system whose solution we seek is

$$\begin{cases} x^2 + 4xy = 300 & (1) \\ x^2 y = 500 & (2) \end{cases}$$

Solving Equation (2) for y and substituting in Equation (1) gives

$$y = \frac{500}{x^2}$$

$$x^2 + 4x\left(\frac{500}{x^2}\right) = 300$$

$$x^2 + \left(\frac{2000}{x}\right) = 300$$

The resulting equation is difficult to solve algebraically, so we solve the original system by graphing. Solving Equation (1) and Equation (2) for y gives

$$y = \frac{300 - x^2}{4x} \quad \text{and} \quad y = \frac{500}{x^2}$$

Figure 7.30 shows the graphs of these equations on the same axes and the point of intersection.

Thus, $x = 10$ cm gives the length of each side of the base of the box, and $y = 5$ cm is the height of the box.

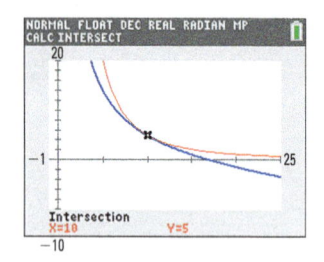

Figure 7.30

EXAMPLE 6 ▶ Oil Production

The following functions, with x equal to the number of years after 2010 and y equal to the number of millions of barrels of oil per day, give forecasts of oil production for the United States and Saudi Arabia from 2011 to 2025. Use technology to find the years in which U.S. production will equal that of Saudi Arabia, according to this model.

United States: $\quad y = -0.0332x^2 + 0.795x + 7.130$

Saudi Arabia: $\quad y = 0.00194x^3 - 0.0283x^2 - 0.0153x + 11.512$

(*Source*: International Energy Agency)

SOLUTION

Entering the equations of the forecast models and intersecting the graphs gives two points of intersection, $x = 6.22$ and $x = 15.18$. See Figure 7.31(a) and (b).

(a)   (b)

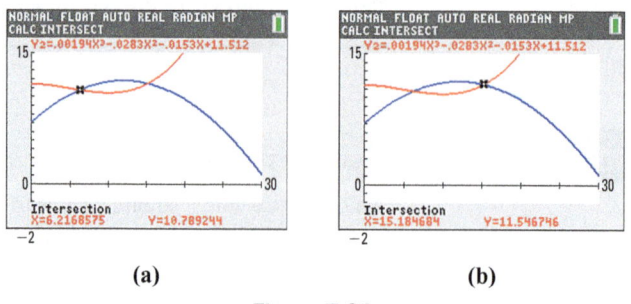

Figure 7.31

These values of x correspond to the number of years after 2010. Thus, according to this model, U.S. oil production will equal that of Saudi Arabia in 2017 (with 10.8 million barrels per day) and in 2026 (with 11.5 million barrels per day).

SKILLS CHECK 7.6

Answers that are not seen can be found in the answer section at the back of the text.

In Exercises 1–8, solve the system algebraically.

1. $\begin{cases} x^2 - y = 0 \\ 3x + y = 0 \end{cases}$
   $(0, 0), (-3, 9)$

2. $\begin{cases} x^2 - 2y = 0 \\ 2y - x = 0 \end{cases}$ $(0, 0), \left(1, \dfrac{1}{2}\right)$

3. $\begin{cases} x^2 - 3y = 4 \\ 2x + 3y = 4 \end{cases}$
   $(2, 0), (-4, 4)$

4. $\begin{cases} x^2 - 5y = 2 \\ 3x + 5y = 8 \end{cases}$ $\left(2, \dfrac{2}{5}\right), \left(-5, \dfrac{23}{5}\right)$

5. $\begin{cases} x^2 + y^2 = 80 \\ y = 2x \end{cases}$
   $(4, 8), (-4, -8)$

6. $\begin{cases} x^2 + y^2 = 72 \\ x + y = 0 \end{cases}$
   $(6, -6), (-6, 6)$

7. $\begin{cases} x + y = 8 \\ xy = 12 \end{cases}$
   $(2, 6), (6, 2)$

8. $\begin{cases} 2xy + y = 36 \\ 2x - 3y = 2 \end{cases}$
   $\left(\dfrac{11}{2}, 3\right), (-5, -4)$

In Exercises 9–16, solve the system algebraically or graphically.

9. $\begin{cases} x^2 + 5x - y = 6 \\ 2x - y + 4 = 0 \end{cases}$
   $(2, 8), (-5, -6)$

10. $\begin{cases} x^2 - y - 8x = 6 \\ y + 10x = 18 \end{cases}$
    $(4, -22), (-6, 78)$

11. $\begin{cases} 2x^2 - 2y + 7x = 19 \\ 2y - x = 61 \end{cases}$ $(5, 33), (-8, 26.5)$

12. $\begin{cases} 2x^2 + 4x - 2y = 24 \\ 2x - y + 37 = 0 \end{cases}$ $(-7, 23), (7, 51)$

13. $\begin{cases} x^2 + 4y = 28 \\ \sqrt{x} - y = -1 \end{cases}$
    $(4, 3)$

14. $\begin{cases} x^2 - y = 0 \\ 2x + \sqrt{y} = 4 \end{cases}$ $\left(\dfrac{4}{3}, \dfrac{16}{9}\right)$

15. $\begin{cases} 4xy + x = 10 \\ x - 4y = -8 \end{cases}$

16. $\begin{cases} xy + y = 4 \\ 2x - 3y = -4 \end{cases}$

In Exercises 17–20, solve the system graphically.

17. $\begin{cases} y = x^3 - 7x^2 + 10x \\ y = 70x - 10x^2 - 100 \end{cases}$ $(2, 0), (5, 0), (-10, -1800)$

18. $\begin{cases} y = x^3 - 14x^2 + 40x \\ y = 70x - 5x^2 - 200 \end{cases}$ $(-5, -675), (4, 0), (10, 0)$

19. $\begin{cases} x^3 + x^2 + y = 15 \\ 2x^2 + y = 11 \end{cases}$
    $(2, 3)$

20. $\begin{cases} x^3 - 3x + 2y = 2 \\ x^2 + x + y = 4 \end{cases}$
    $(-2, 2), (1, 2), (3, -8)$

21. a. Algebraically eliminate y from the system in Exercise 19. $x^3 - x^2 - 4 = 0$

    b. Use synthetic division to confirm that there is only one real solution to the system.

22. a. Algebraically eliminate y from the system in Exercise 20. $x^3 - 2x^2 - 5x + 6 = 0$

    b. Use synthetic division to confirm that the system has three real solutions.

23. Use graphical methods to find two solutions to the system

    $\begin{cases} x^2 + 4y = 17 \\ x^2 y - 5x = 3 \end{cases}$

    Possible answers: $(3, 2), (2, 3.25), (-0.44, 4.20), (-4.56, -0.95)$

24. Use graphical or algebraic methods to find two solutions to the system

    $\begin{cases} xy + 4x = 10 \\ \dfrac{y}{x} + y = \dfrac{3}{2} \end{cases}$ $(2, 1), (-0.\overline{90}, -15)$

EXERCISES 7.6

Answers that are not seen can be found in the answer section at the back of the text.

25. **Supply and Demand** The supply function for a product is given by $p = q^2 + 2q + 122$ and the demand function for this product is $p = 650 - 30q$, where p is the price in dollars and q is the number of hundreds of units. Find the price that gives market equilibrium and the equilibrium quantity. $290; 1200 units

26. **Supply and Demand** The supply function for a product is given by $p = q^2 + 500$ and the demand function for this product is $p = 1124 - 40q$, where p is the price in dollars and q is the number of hundreds of units. Find the price that gives market equilibrium and the equilibrium quantity. $644; 1200 units

27. **Supply and Demand** The supply function for a product is given by $p = 0.1q^2 + 50q + 1027.50$ and the demand function for this product is $p = 6000 - 20q$, where p is the price in dollars and q is the number of hundreds of units. Find the price that gives market equilibrium and the equilibrium quantity. $4700; 6500 units

28. **Supply and Demand** The supply function for a product is given by $p = q^3 + 800q + 6000$ and the demand function for this product is $p = 12{,}200 - 10q - q^2$, where p is the price in dollars and q is the number of hundreds of units. Find the price that gives market equilibrium and the equilibrium quantity. $12,077.58; 714 units

29. Break-Even The weekly total cost function for a piece of equipment is $C(x) = 2000x + 18{,}000 + 60x^2$ dollars, and the weekly revenue is $R(x) = 4620x - 12x^2 - x^3$, where x is the number of thousands of units. The x-values that are the solutions to the system

$$\begin{cases} y = 2000x + 18{,}000 + 60x^2 \\ y = 4620x - 12x^2 - x^3 \end{cases}$$

are the quantities that give break-even. Find these quantities. 18,000 units and 10,000 units

30. Break-Even The daily total cost function for a product is $C(x) = 400x + 4800$ dollars and the daily revenue is $R(x) = 600x + 4x^2 - 0.1x^3$, where x is the number of units. Use graphical methods to find the numbers of units that give break-even. 20 units and 60 units

31. Break-Even The monthly total cost function for a product is $C(x) = 400x + 0.1x^2 + 17{,}901$ dollars and the monthly revenue is $R(x) = 1000x - 0.01x^2$, where x is the number of units. Find the numbers of units that give break-even for the product. (*Hint*: Use graphical methods to find one solution less than 100 units and one greater than 5000 units.) 30 units and 5425 units

32. Break-Even The weekly total cost function for a product is $C(x) = 3x^2 + 1228$ dollars and the weekly revenue is $R(x) = 177.50x$, where x is the number of hundreds of units. Use graphical methods to find the numbers of units that give break-even. 800 units and 5117 units

33. Constructing a Box A rectangular piece of cardboard with area 180 square inches is made into an open box by cutting a 2-inch square from each corner and turning up the sides. If the box is to have a volume of 176 cubic inches, what are the dimensions of the original piece of cardboard? (See the figure below.) 15 in. by 12 in.

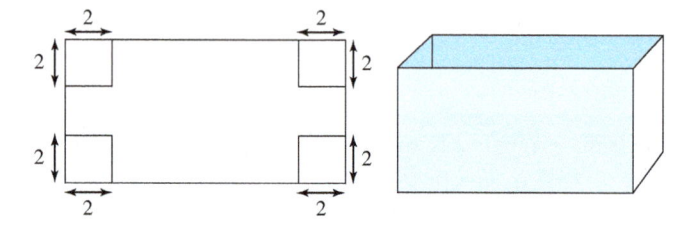

34. Constructing a Box An open box is constructed from a rectangular piece of tin by cutting a 4-centimeter square from each corner and turning up the sides. If the box has a volume of 768 cubic centimeters and the area of the original piece of tin was 512 square centimeters, what are the dimensions of the original piece of tin? 16 cm by 32 cm

35. Dimensions of a Box An open box has a rectangular base and its height is equal to the length of the shortest side of the base. What dimensions give a volume of 2000 cubic centimeters using 800 square centimeters of material? 10 cm by 10 cm by 20 cm

36. Constructing a Box A box with a top is to be constructed with the width of its base equal to half of the length of its base. If it is to have volume 6000 cubic centimeters and uses 2200 square centimeters of material, what are the dimensions of the box?

37. Trust Fund A father gives a trust fund to his daughter by depositing \$50,000 into an account paying 10%, compounded annually. This account grows according to $y = 50{,}000(1.10^t)$. He will give her \$12,968.72 per year from another account until she has received in total the value to which the trust fund has grown. For how many years will she receive the \$12,968.72? 10 yr

38. Constructing a Box An open box with a square base is to be constructed so that the area of the material used is 500 cm^2 and the volume of the box is 500 cm^3. What are the dimensions of the box?
Two possible boxes: 20 cm by 20 cm by 1.25 cm or approximately 4.14 cm by 4.14 cm by 29.14 cm

chapter 7 SUMMARY

In this chapter, we studied systems of equations and their applications. We first solved systems of linear equations in three variables by using left-to-right elimination. Matrices were introduced to help solve systems of equations, and other applications of matrices were investigated. We also solved systems of nonlinear equations by algebraic and by graphical methods.

Key Concepts and Formulas

7.1 Systems of Linear Equations in Three Variables

Left-to-right elimination method	An extension of the elimination method used to solve systems of two equations in two variables.
Possible solutions to a system of linear equations in three variables	
• Unique solution	Planes intersect in a single point.
• No solution	Planes have no common intersection.
• Many solutions	Planes intersect in a line.
Modeling systems of equations	Solution of real problems sometimes requires us to create three or more equations whose simultaneous solution is the solution to the problem.

7.2 Matrix Solution of Systems of Linear Equations

Matrix representation of systems of equations	Systems of linear equations can be represented using matrices, and the solution can be found easily.
Matrices	
• Dimension	The numbers of rows and columns of a matrix.
• Square matrix	Equal number of rows and columns.
• Coefficient matrix	Coefficients of the variables in a linear system placed in a matrix.
• Augmented matrix	A constant column is added to a coefficient matrix.
• Identity matrix	A square matrix that has 1s down its diagonal and 0s everywhere else.
• Row-echelon form	The coefficient part has all 1s or 0s on its diagonal and all 0s below its diagonal.
• Reduced row-echelon form	The coefficient part has 1s or 0s on its diagonal and 0s elsewhere.
Matrix row operations	1. Interchange two rows of the matrix. 2. Multiply a row by a nonzero constant. 3. Add a multiple of one row to another row.
Gauss–Jordan elimination method	Row operations are used to reduce a matrix to reduced row-echelon form. If the coefficient part reduces to the identity matrix, the system has a unique solution.
Using technology to find reduced row-echelon form	Graphing utilities and spreadsheets can be used to find the reduced row-echelon form.
Unique/nonunique solutions	Systems of equations can have a unique solution, no solution, or infinitely many solutions.
• Unique solution	The reduced coefficient matrix contains the identity matrix.
• No solution	A row of the reduced coefficient matrix contains all 0s and the augment of that row contains a nonzero number.
• Infinite number of solutions	If neither of the above occurs.

7.3 Matrix Operations

Addition of matrices	Matrices of the same dimension are added by adding the corresponding entries of the two matrices.
Zero matrix	Each element is zero.
Negative of a matrix	The sum of a matrix and its negative is the zero matrix.
Subtraction of matrices	If the matrices M and N are the same dimension, the difference $M - N$ is found by subtracting the elements of N from the corresponding elements of M.
Multiplication of matrices by a number	Multiplying a matrix A by a real number c results in a matrix in which each entry in matrix A is multiplied by the number c.
Matrix multiplication	The product of an $m \times n$ matrix A and an $n \times k$ matrix B is an $m \times k$ matrix $C = AB$ with the element in the ith row and jth column of matrix C given by the product of the ith row of A and the jth column of B.

7.4 Inverse Matrices; Matrix Equations

Inverse matrices	Two square matrices, A and B, are called inverses of each other if $AB = I$ and $BA = I$.
Finding the inverse of a square matrix	Placing the same dimension identity matrix in its augment and reducing the matrix to its reduced row-echelon form will give the inverse matrix in the augment if the inverse exists.
Inverses and technology	Computer software, spreadsheets, and calculators can be used to find the inverse of a matrix, if it exists.
Matrix equations	Multiplying both sides of the matrix equation $AX = C$ on the left by A^{-1}, if it exists, gives the solution to the system that the matrix equation represents.
Encoding messages and decoding messages	Matrix multiplication can be used to encode messages. The inverse of an encoding matrix can be used to decode messages.

7.5 Determinants and Cramer's Rule

Finding determinants of matrices	Every square matrix has a number associated with it called the **determinant**. The determinant for a 2×2 matrix is defined as the difference of the diagonal products. $$\begin{vmatrix} a & b \\ c & d \end{vmatrix} = ad - bc$$
Solving systems of matrices with Cramer's Rule	Cramer's Rule for a 2×2 system $$\text{If } \begin{cases} ax + by = u \\ cx + dy = w \end{cases}, \text{ then } x = \frac{\begin{vmatrix} u & b \\ w & d \end{vmatrix}}{\begin{vmatrix} a & b \\ c & d \end{vmatrix}} \text{ and } y = \frac{\begin{vmatrix} a & u \\ c & w \end{vmatrix}}{\begin{vmatrix} a & b \\ c & d \end{vmatrix}}, \text{ provided}$$ $$D = \begin{vmatrix} a & b \\ c & d \end{vmatrix} = ad - bc \neq 0.$$

Chapter 7 Skills Check

Cramer's Rule with inconsistent and dependent systems	If D, the determinant of the coefficients of the variables of a system of three equations in three variables, is 0, then the system either is inconsistent or contains dependent equations. In particular,
	1. If $D = 0$ and at least one of the determinants in the numerator (D_x, D_y, or D_z) is not 0, then the system is inconsistent, and there is no solution.
	2. If $D = 0$ and all the determinants in the numerators (D_x, D_y, and D_z) are 0, then the equations in the system are dependent, and there are infinitely many solutions. These solutions can be found using different methods.

7.6 Systems of Nonlinear Equations

Systems of nonlinear equations	A system of nonlinear equations contains at least one equation that is nonlinear. These systems can have any number of solutions or no solution.
Algebraic solution	Systems of nonlinear equations can be solved algebraically by substitution or elimination.
Graphical solution	Solutions to systems of nonlinear equations can be checked with graphical methods, or, if solving the system algebraically is difficult, graphical methods may be used to solve the system.

chapter 7 SKILLS CHECK

Answers that are not seen can be found in the answer section at the back of the text.

In Exercises 1–10, solve the systems of linear equations, if possible, by (a) left-to-right elimination and (b) using technology.

1. $\begin{cases} 2x - 3y + z = 2 \\ 3x + 2y - z = 6 \\ x - 4y + 2z = 2 \end{cases}$ $x = 2, y = 2, z = 4$

2. $\begin{cases} 3x - 2y - 4z = 9 \\ x + 3y + 2z = -1 \\ 2x + 4y + 4z = 2 \end{cases}$ $x = 3, y = -2, z = 1$

3. $\begin{cases} 3x + 2y - z = 6 \\ 2x - 4y - 2z = 0 \\ 5x + 3y + 6z = 2 \end{cases}$ $x = 1, y = 1, z = -1$

4. $\begin{cases} 3x + 6y + 9z = 27 \\ 2x + 3y - z = -2 \\ 4x + 5y + z = 6 \end{cases}$ $x = 2, y = -1, z = 3$

5. $\begin{cases} x + 2y - 2z = 1 \\ 2x - y + 5z = 15 \\ 3x - 4y + z = 7 \end{cases}$ $x = 3, y = 1, z = 2$

6. $\begin{cases} -6x + 4y - 2z = 4 \\ 3x - 2y + 5z = -6 \\ x - 4y + z = -8 \end{cases}$ $x = 1, y = 2, z = -1$

7. $\begin{cases} 2x + 5y + 8z = 30 \\ 18x + 42y + 18z = 60 \end{cases}$ Infinitely many solutions of the form $x = 41z - 160,$ $y = -18z + 70,$ $z = z$

8. $\begin{cases} 9x + 21y + 15z = 60 \\ 2x + 5y + 8z = 30 \\ x + 2y - 3z = -10 \end{cases}$ Infinitely many solutions of the form $x = 31z - 110, y = 50 - 14z, z = z$

9. $\begin{cases} x + 3y + 2z = 5 \\ 9x + 12y + 15z = 6 \\ 2x + y + 3z = -10 \end{cases}$ Inconsistent system, no solution

10. $\begin{cases} 5x + 7y + 10z = 6 \\ 2x + 5y + 6z = 1 \\ 3x + 2y + 4z = 6 \end{cases}$ Inconsistent system, no solution

In Exercises 11 and 12, solve the systems of linear equations using technology.

11. $\begin{cases} 3x + 2y - z + w = 12 \\ x - 4y + 3z - w = -18 \\ x + y + 3z + 2w = 0 \\ 2x - y + 3z - 3w = -10 \end{cases}$ $x = 1, y = 3, z = -2, w = 1$

12. $\begin{cases} 2x + y - 3z + 4w = 7 \\ x - 2y + z - 2w = 0 \\ 3x + y + 4z + w = -2 \\ x + 3y + 2z + 2w = -1 \end{cases}$ $x = 2, y = 1, z = -2, w = -1$

Perform the matrix operations, if possible, in Exercises 13–21, with

$$A = \begin{bmatrix} 1 & 3 & -3 \\ 2 & 4 & 1 \\ -1 & 3 & 2 \end{bmatrix} \quad B = \begin{bmatrix} 1 & 2 & 1 \\ 2 & -1 & 3 \end{bmatrix}$$

$$C = \begin{bmatrix} 2 & 3 \\ -1 & 2 \\ 3 & -2 \end{bmatrix} \quad D = \begin{bmatrix} -2 & 3 & 1 \\ -3 & 2 & 2 \end{bmatrix}$$

13. $B + D$
14. $D - B$
15. $5C$
16. AB Not possible
17. BA
18. CD
19. DC
20. $A^2 = A \cdot A$
21. A^{-1}

Find the inverse of each matrix in Exercises 22–24.

22. $\begin{bmatrix} 2 & 1 & 1 \\ 1 & 2 & -1 \\ 2 & 2 & 1 \end{bmatrix}$

23. $\begin{bmatrix} 1 & 2 & 1 \\ 1 & 1 & 0 \\ 0 & 2 & 1 \end{bmatrix}$

24. $\begin{bmatrix} -1 & 0 & 1 & 1 \\ 1 & 1 & 2 & 1 \\ 2 & -1 & 0 & 1 \\ 0 & 0 & 0 & 1 \end{bmatrix}$ $\begin{bmatrix} -0.4 & 0.2 & 0.2 & 0 \\ -0.8 & 0.4 & -0.6 & 1 \\ 0.6 & 0.2 & 0.2 & -1 \\ 0 & 0 & 0 & 1 \end{bmatrix}$

In Exercises 25–28, solve the system of equations by using the inverse of the coefficient matrix.

25. $\begin{cases} x + y - 3z = 8 \\ 2x + 4y + z = 15 \\ -x + 3y + 2z = 5 \end{cases}$ $x = 2, y = 3, z = -1$

26. $\begin{cases} 2x + 3y + z = 20 \\ 3x + 4y - z = 40 \\ 2x + 5y + z = 60 \end{cases}$ $x = -16, y = 20, z = -8$

27. $\begin{cases} -x_1 + x_3 + x_4 = 6 \\ x_2 + x_3 + x_4 = 12 \\ 2x_1 + 5x_2 + x_3 = 20 \\ 3x_2 + x_4 = 24 \end{cases}$ $x_1 = 2, x_2 = 4, x_3 = -4, x_4 = 12$

28. $\begin{cases} 2x_1 + 2x_2 + x_4 = 4 \\ 2x_1 + x_2 + 2x_4 = 12 \\ x_1 + x_2 + x_3 + x_4 = 4 \\ x_2 + x_4 = 8 \end{cases}$ $x_1 = -2, x_2 = 0, x_3 = -2, x_4 = 8$

In Exercises 29 and 30, find the determinant.

29. $\begin{vmatrix} -1 & 4 \\ 2 & -3 \end{vmatrix}$ -5

30. $\begin{vmatrix} 2 & 0 & -3 \\ -1 & 4 & 3 \\ 0 & 1 & -2 \end{vmatrix}$ -19

In Exercises 31 and 32, solve the system of equations using Cramer's Rule, if possible.

31. $\begin{cases} 2x - 3y = 7 \\ -x + y = 4 \end{cases}$ $(-19, -15)$

32. $\begin{cases} -x + 3y - 2z = 4 \\ -x + 2y + 2z = 5 \\ x - 2y + z = 6 \end{cases}$ $\left(\dfrac{89}{3}, \dfrac{41}{3}, \dfrac{11}{3}\right)$

33. Solve the nonlinear system

$$\begin{cases} x^2 - y = x \\ 4x - y = 4 \end{cases}$$ $x = 1, y = 0; x = 4, y = 12$

34. Find two solutions to the system

$$\begin{cases} x^2 y = 2000 \\ x^2 + 2y = 140 \end{cases}$$

Possible answers: $x = 10, y = 20; x = -10, y = 20;$ $x \approx 6.32, y = 50; x \approx -6.32, y = 50$

chapter 7 REVIEW

Answers that are not seen can be found in the answer section at the back of the text.

35. **Pricing** A concert promoter needs to make $200,000 from the sale of 4000 tickets. The promoter charges $40 for some tickets, $60 for some tickets, and $100 for the others. If the number of $60 tickets is $\dfrac{1}{4}$ the sum of the remaining tickets, how many tickets of each type should be sold?
2800 at $40, 800 at $60, 400 at $100

36. **Medication** Medication A is given six times per day, and medication B is given twice per day. Medication C is given once every two days, and its dosage is $\dfrac{1}{2}$ the sum of the dosages of the other two medications. For a certain patient, the total intake of the three medications is limited to 28.7 milligrams on days when all three medications are given. If the ratio of the dosage of medication A to the dosage of medication B is 2 to 3, how many milligrams are in each dosage?
2.8 mg of A, 4.2 mg of B, 3.5 mg of C

37. **Rental Income** A woman has $750,000 invested in three rental properties. One yields an annual return of 12% on her investment, and the second returns 15% per year on her investment. The third investment returns 10% and is limited to $\dfrac{1}{2}$ the sum of the other two investments. Her total annual return from the three investments is $89,500. How much is invested in each property?
$350,000 at 12%, $150,000 at 15%, $250,000 at 10%

38. *Investment* A woman invests $360,000 in three different mutual funds. One averages 12% per year, and another averages 16% per year. A third fund averages 8% per year, and because it has much less risk, she invests twice as much in it as in the sum of the other two funds. To realize an annual return of $35,200, how much should she invest in each fund?
$80,000 at 12%, $40,000 at 16%, $240,000 at 8%

39. *Loans* A bank lends $1,180,000 to a development company to purchase three business properties. If one of the properties costs $75,000 more than the other and the third costs three times the sum of these two properties, find the cost of each property.
$110,000; $185,000; $885,000

40. *Investment* A brokerage house offers three stock portfolios for its clients. Portfolio I consists of 10 blocks of common stock, 2 municipal bonds, and 3 blocks of preferred stock. Portfolio II consists of 12 blocks of common stock, 8 municipal bonds, and 5 blocks of preferred stock. Portfolio III consists of 10 blocks of common stock, 4 municipal bonds, and 8 blocks of preferred stock. If a client wants to combine these portfolios so that she has 290 blocks of common stock, 138 municipal bonds, and 161 blocks of preferred stock, how many units of each portfolio does she need? To answer this question, let x equal the number of units of portfolio I, y equal the number of units of portfolio II, and z equal the number of units of portfolio III. 5 units of I, 10 units of II, 12 units of III

41. *Property* A realty partnership purchases three business properties at a total cost of $375,000. One property costs $50,000 more than the second, and the third property costs half the sum of these two properties. Find the cost of each property.
$150,000; $100,000; $125,000

42. *Nutrition* A nutritionist wants to create a diet that uses a combination of three foods: I, II, and III. Each of these foods contains three additives: A, B, and C. The following table gives the percent of each additive that is present in each food. If the diet being used requires 12.5 grams per day of A, 9.1 grams of B, and 9.6 grams of C, find the number of grams of each food that should be used each day for this diet. 20 g of I, 30 g of II, 40 g of III

	Food I	Food II	Food III
Additive A (%)	10	11	18
Additive B (%)	12	9	10
Additive C (%)	14	12	8

43. *Transportation* A delivery service has three types of aircraft, each of which carries three types of cargo. The payload of each type is summarized in the table. Suppose on a given day the airline must move 2200 next-day delivery letters, 3860 two-day delivery letters, and 920 units of air freight. How many aircraft of each type should be scheduled?
3 passenger, 4 transport, 4 jumbo

	Passenger	Transport	Jumbo
Next-Day Letters	200	200	200
Two-Day Letters	300	40	700
Air Freight	40	130	70

44. *Finance* A financial planner promises a long-term return of 12% from a combination of three mutual funds. The cost per share and average annual return per share for each mutual fund are given in the table below. If the annual returns are accurate, how many shares of each mutual fund will return an average of 12% on a total investment of $210,000?

	Tech Fund	Balanced Fund	Utility Fund
Cost per Share ($)	180	210	120
Annual Return per Share ($)	18	42	18

45. *Transportation* Marshall Trucking Company has an order for delivery of three products: A, B, and C. The following table gives the volume in cubic feet, the weight in pounds, and the value for insurance in dollars per unit of each of the products. If one of the company's trucks can carry 9260 cubic feet and 12,000 pounds and is insured to carry $52,600, how many units of each product can be carried on the truck?

	Product A	Product B	Product C
Volume (cu ft)	25	30	40
Weight (lb)	30	36	60
Value ($)	150	180	200

46. *Nutrition* A biologist is growing three types of slugs (A, B, and C) in the same laboratory environment. Each day, the slugs are given different nutrients (I, II, and III). Each type A slug requires 2 units of I, 6 units of II, and 2 units of III per day. Each type B slug requires 2 units of I, 8 units of II, and 4 units of III per day. Each type C slug requires 4 units of I, 20 units of II, and 12 units of III per day. If the daily mixture contains 4000 units of I, 16,000 units of II, and 8000 units of III, find the number of slugs of each type that can be supported.

47. *Trade Balances* The tables on the next page give the actual and projected U.S. dollar value (in billions) of U.S. exports and imports of pipeline natural gas with Canada and Mexico for selected years.

  a. Form the matrix A that contains the values of the exports to these countries.

  b. Form the matrix B that contains the values of the imports to these countries.

c. Find a matrix T that describes the balance of trade with these countries.

d. Are the trade balances (exports − imports) with all these countries projected to improve? Yes

Exports

	2015	2020	2025
Canada	3.15	5.76	6.53
Mexico	2.99	5.87	7.83
Others	0.10	2.25	6.64

Imports

	2015	2020	2025
Canada	10.13	12.05	11.39
Mexico	0.20	0.21	0.11
Others	0.63	0.89	0.92

(Source: U.S. Energy Information Administration)

48. Investment An investor wants to place a total of $800,000 in two or more of three investments with different degrees of risk. One investment pays 6%, one pays 7%, and one pays 10%, and the investor wants to earn $71,600 from the investments.

a. Write a system of two equations to represent the problem, representing the amount in the 6% investment by x, the amount in the 7% investment by y, and the amount in the 10% investment by z.

b. Solve the system.

c. How much should the investor put in each investment if she wants to put more in the 6% investment than in the 7% investment, because the 6% investment has a lower risk?

d. How much should she put in each investment if she wants to put more in the 7% investment than in the 6% investment?

49. Income and SAT Scores The following table gives the average critical reading and math SAT scores for certain family income levels.

a. Use linear regression to find a linear equation that gives the critical reading SAT score as a function of family income. $y = 0.586x + 450.086$, x in thousands of $

b. Use linear regression to find a linear equation that gives the math SAT score as a function of family income. $y = 0.578x + 464.216$, x in thousands of $

c. Interpret the slopes of the equations from parts (a) and (b) as rates of change.

Family Income ($ thousands)	Critical Reading SAT	Math SAT
10	434	457
30	462	475
50	488	497
70	503	512
90	517	528
110	525	538
130	529	542
150	536	550
180	542	554
200	563	579

(Source: collegeboard.com)

50. Medication Suppose combining x cubic centimeters (cc) of a 20% solution of a medication and y cc of a 5% solution of the medication gives 10 cc of a 15.5% solution.

a. Write the equation that gives the total amount of solution in terms of x and y. $x + y = 10$

b. Write the equation that gives the total amount of medication in the solution.
$0.20x + 0.05y = 0.155(10)$

c. Use substitution to find the amount of each solution that is needed. 7 cc of 20%, 3 cc of 5%

51. Supply and Demand The supply function for a product is given by $p = q + 578$ and the demand function for this product is $p = 396 + q^2$, where p is the price in dollars and q is the number of units. Find the price that gives market equilibrium and the equilibrium quantity. $592; 14 units

52. Break-Even The daily total cost function for a product is $C(x) = 2500x + x^2 + 27{,}540$ dollars and the daily revenue is $R(x) = 3899x - 0.1x^2$, where x is the number of units. Find the numbers of units that give break-even. 20 units and approximately 1252 units

Group Activities
▶ EXTENDED APPLICATIONS

1. Salaries

Average Salaries of College Professors

		Men			Women		
Academic Rank		Public	Private-Independent	Church-Related	Public	Private-Independent	Church-Related
Doctoral	Professor	108,481	138,921	121,312	98,552	127,542	109,720
	Associate	76,030	89,936	82,941	70,764	83,148	76,814
	Assistant	65,498	78,079	70,227	60,155	71,207	65,882
	Instructor	43,352	50,418	55,513	42,012	49,246	53,549
	Lecturer	51,182	59,443	52,257	46,340	52,047	46,476
Master's	Professor	82,834	93,020	84,493	79,493	86,476	78,355
	Associate	66,102	71,130	66,448	63,630	66,881	63,338
	Assistant	56,225	57,610	55,107	53,907	55,065	52,467
	Instructor	41,495	46,444	45,997	40,820	44,315	44,174
	Lecturer	46,030	51,037	49,064	44,018	43,907	45,631
Baccalaureate	Professor	78,390	91,563	69,969	73,372	87,570	66,427
	Associate	63,658	67,210	56,752	61,363	66,444	55,315
	Assistant	52,960	55,180	47,853	50,714	54,011	47,085
	Instructor	40,844	45,139	40,966	41,183	40,705	40,181
	Lecturer	47,743	58,064	42,095	45,565	54,343	39,233

(*Source*: American Association of University Professors)

The table above gives the average salaries of male and female college professors who teach at three types of institutions at three different levels of instruction and at five job designations. To investigate the relationship between salaries of males and females, do the following:

1. Create matrix A and matrix B, which contain the salaries of male and female professors, respectively, in each type of school and at each level of instruction.
2. Subtract the matrix of male professors' salaries from the matrix of female professors' salaries.
3. What does this show about female salaries versus male salaries in every category? Based on the data, discuss gender bias in educational institutions.
4. Create a scatter plot with male professors' salaries as the input and female professors' salaries as the output.
5. Find a function that gives a relationship between the salaries of male and female professors from part 4.
6. Graph the scatter plot from part 4 and the function from part 5 on the same axes.
7. The graph of the function $y = x$ would represent the relationship between male and female professors' salaries if they were equal. Graph the function from part 5 and the function $y = x$ on the same axes.
8. Use the two graphs in part 7 to determine how female professors' salaries compare to male professors' salaries as male salaries rise.
9. Solve the two equations whose graphs are discussed in part 8 to determine at what salary level male salaries become greater than female salaries.

2. Parts Listing

Wingo Playgrounds must maintain numerous parts in inventory for its Fun-in-the-Sun swing set. The swing set has 4 legs and a top connecting the legs. Each of the legs is made from 1 nine-foot pipe connected with 1 brace and 2 bolts, and the top is made of 1 nine-foot pipe, 2 clamps, and 6 bolts. The parts listing for these swing sets can be described by the following matrix.

$$P = \begin{bmatrix} 0 & 0 & 0 & 0 & 0 & 0 & 0 \\ 4 & 0 & 0 & 0 & 0 & 0 & 0 \\ 1 & 0 & 0 & 0 & 0 & 0 & 0 \\ 0 & 1 & 1 & 0 & 0 & 0 & 0 \\ 0 & 0 & 2 & 0 & 0 & 0 & 0 \\ 0 & 1 & 0 & 0 & 0 & 0 & 0 \\ 0 & 2 & 6 & 0 & 0 & 0 & 0 \end{bmatrix} \begin{array}{l} \text{Swing sets} \\ \text{Legs} \\ \text{Top} \\ \text{Pipes} \\ \text{Clamps} \\ \text{Braces} \\ \text{Bolts} \end{array}$$

with columns labeled SS, L, T, P, C, Bc, B.

Each column indicates how many of each part are required to produce the part shown at the top of the column. For example, column 1 indicates that to produce a swing set requires 4 legs and 1 top, and the second column indicates that each leg (L) is constructed from 1 pipe, 1 brace, and 2 bolts.

Suppose an order is received for 8 complete swing sets plus the following spare parts: 2 legs, 1 top, 2 pipes, 4 clamps, 4 braces, and 8 bolts. We indicate the number of each part type that must be supplied to fill this order by the matrix

$$X = \begin{bmatrix} x_1 \\ x_2 \\ x_3 \\ x_4 \\ x_5 \\ x_6 \\ x_7 \end{bmatrix}$$

To find the numbers of each part that are needed, do the following:

1. Form the 7×1 matrix D that contains the numbers of items to be delivered to fill the order for the 8 swing sets and the spare parts. Some parts of each type are needed to produce other parts of the swing set. The matrix PX gives the number of parts of each type used to produce other parts, and the matrix $X - PX$ gives the number of parts of each type that are available to be delivered. That is,

$$X - PX = D$$

where D is the number of parts of each type that are available for delivery.

2. The matrix equation $X - PX = D$ can be rewritten in the form $(I - P)X = D$, where I is the 7×7 identity matrix. Find the matrix $I - P$ that satisfies this equation.

3. Find the inverse of the matrix $I - P$.

4. Multiply both sides of the matrix equation $(I - P)X = D$ by the inverse of $I - P$. This gives the matrix X that contains the numbers of parts of each type that are needed.

5. The primary assembly parts are the pipes, clamps, braces, and bolts. What is the total number of primary parts of each type needed to fill the order?

8

Special Topics in Algebra

To determine how many cars of different types it must buy and rent to maximize its profit, a car rental agency must consider the constraints on its purchases. The constraints include how much money it has to invest in cars and how many cars it needs. These constraints can be written as inequalities and "pictured" by graphs of systems of inequalities. The process of finding the optimal solution to a problem subject to the constraint inequalities, called linear programming, is examined in this chapter. In addition, we use sequences and series to solve several types of financial problems, including future values of lump sum investments and annuities. We also discuss the Binomial Theorem and introduce the geometric properties of conic sections.

sections	objectives	applications
8.1 Systems of Inequalities	Solve linear inequalities in two variables; solve systems of linear inequalities in two variables; solve systems of nonlinear inequalities in two variables	Auto purchases, car rental, advertising
8.2 Linear Programming: Graphical Methods	Optimize a function subject to constraints with linear programming	Maximizing car rental profit, cost minimization, profit maximization, manufacturing
8.3 Sequences and Discrete Functions	Find terms of sequences, arithmetic sequences, and geometric sequences	Football contracts, depreciation, rebounding ball, compound interest
8.4 Series	Find the sum of a finite series; find the sum of the first n terms of an infinite series, an arithmetic series, and a geometric series; find the sum of an infinite geometric series	Football contracts, depreciation, annuities
8.5 The Binomial Theorem	Evaluate factorials; use Pascal's triangle; find coefficients of terms in a binomial expansion; use the Binomial Theorem to expand binomials; find a particular term of a binomial expansion	
8.6 Conic Sections: Circles and Parabolas	Use the distance formula; use the midpoint formula; write the equation of a circle; locate the center and radius of a circle; graph circles; write the equation of a parabola given the vertex and focus; graph parabolas	Parabolic cables on a bridge, satellite dishes
8.7 Conic Sections: Ellipses and Hyperbolas	Graph horizontal ellipses; graph vertical ellipses; write the standard form of ellipses; graph horizontal hyperbolas; graph vertical hyperbolas; write the standard form of hyperbolas; identify the asymptotes of hyperbolas; graph rectangular hyperbolas	Planets, cones, bridge arches, Roman Colosseum, Ohm's law, Boyle's law

8.1 Systems of Inequalities

KEY OBJECTIVES

- Solve linear inequalities in two variables
- Graph *solution regions* to linear inequalities in two variables
- Solve systems of nonlinear inequalities in two variables

SECTION PREVIEW Car Rental

A rental agency has a maximum of $1,260,000 to invest in the purchase of at most 71 new cars of two different types: compact and midsize. The cost per compact car is $15,000, and the cost per midsize car is $28,000. The number of cars of each type is limited (constrained) by the budget available and the number of cars needed. To get a "picture" of the constraints on this rental agency, we can graph the inequalities determined by the maximum amount of money and the limit on the number of cars purchased. Because these inequalities deal with two types of cars, they will contain two variables. For example, if we denote the number of compact cars by x and the number of midsize cars by y, the statement that the agency wants to purchase at most 71 cars can be written as the inequality

$$x + y \leq 71, \quad \text{where } x \text{ and } y \text{ are integers}$$

The collection of all **constraint** inequalities can be expressed by a **system of inequalities** in two variables. Finding the values that satisfy all these constraints at the same time is called solving the system of inequalities. In this section, we will graph solutions of linear and nonlinear inequalities in two variables. ■

Linear Inequalities in Two Variables

We solved linear inequalities in one variable in Section 2.4. We now consider linear inequalities in two variables. For example, consider the inequality $y > x$. The solutions to this inequality are the ordered pairs (x, y) that satisfy the statement $y > x$. The graph of $y > x$ consists of all points above the line $y = x$ (the shaded area in Figure 8.1). The line $y = x$ is dashed in Figure 8.1 because the inequality does not include $y = x$. This (dashed) line divides the plane into two **half-planes**, $y > x$ and $y < x$. We can determine which half-plane is the solution to the inequality by selecting any point not on the line. If the coordinates of this **test point** satisfy the inequality, then the half-plane containing that point is the graph of the solution. If the test point does not satisfy the inequality, then the other half-plane is the graph of the solution. For example, the point $(2, 4)$ is above the line $y = x$, and $x = 2, y = 4$ satisfies the inequality $y > x$, so the region above the line is the graph of the solution of the inequality. Note that any other point on the half-plane above the line also satisfies the inequality.

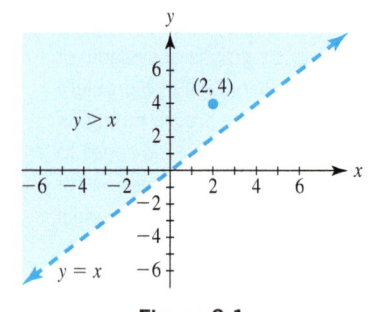

Figure 8.1

EXAMPLE 1 ▶ Graph of an Inequality

Graph the solution of the inequality $6x - 3y \leq 15$.

SOLUTION

First we graph the line $6x - 3y = 15$ (which we can also write as $y = 2x - 5$) as a solid line, because the inequality symbol "$\leq$" indicates that points on the line are part of the solution (Figure 8.2(a)). Next we pick a test point that is not on the line. If we use

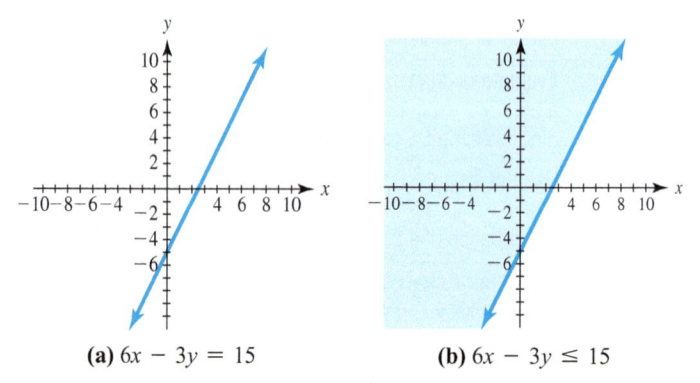

(a) $6x - 3y = 15$ (b) $6x - 3y \leq 15$

Figure 8.2

8.1 Systems of Inequalities   615

$(0, 0)$, we get $6(0) - 3(0) \leq 15$, or $0 \leq 15$, which is true, so the inequality is satisfied. Thus, the half-plane that contains the point $(0, 0)$ and the line $6x - 3y = 15$ is the graph of the solution, which we call the **solution region** of the inequality.

(a)

(b)

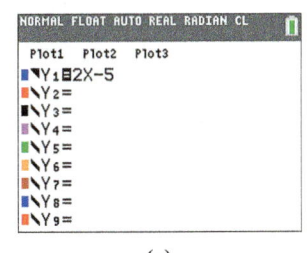

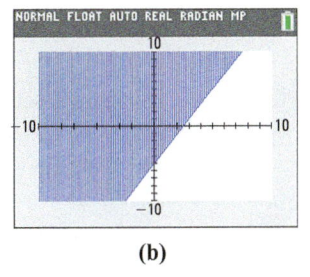

Figure 8.3

Technology Note

Graphing calculators can also be used to create shaded regions representing inequalities. To graph the solution region of the inequality $y \geq 2x - 5$, we enter $2x - 5$ to the right of Y1= in the equation editor, move to the icon at the left of Y1=, change the icon from \ to the icon showing a shaded area above a line (see Figure 8.3(a)), and press **ENTER**. Figure 8.3(b) shows the line $y = 2x - 5$ and the region above it.

EXAMPLE 2 ▶ **Car Rental Agency**

One of the constraint inequalities for the car rental company application given in the Section Preview is that the sum of x and y (the numbers of compact and midsize cars, respectively) is at most 71. This constraint can be described by the inequality

$$x + y \leq 71$$

where x and y are integers. Find the graphical solution of this inequality.

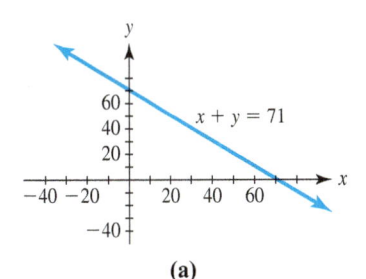

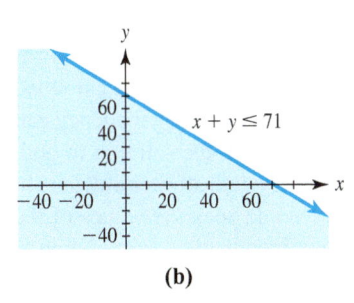

SOLUTION

The inequality $x + y \leq 71$, where x and y are integers, has many integer solutions. The inequality is satisfied if x is any integer from 0 to 71 and y is any integer less than or equal to $71 - x$. For example, some of the solutions are $(0, 71)$, $(1, 70)$, $(60, 5)$, and $(43, 20)$. As we stated in Chapter 2, problems are frequently easier to solve if we treat the data as continuous rather than discrete. With this in mind, we graph this inequality as a region on a coordinate plane rather than attempting to write all of the possible integer solutions.

We create the graph of the inequality by first graphing the equation

$$x + y = 71, \quad \text{or, equivalently,} \quad y = 71 - x$$

(Figure 8.4(a)). The graph of this line divides the coordinate plane into two half-planes, $y > 71 - x$ and $y < 71 - x$. We can find the half-plane that represents the graph of $y < 71 - x$ by testing the coordinates of a point and determining whether it satisfies $y < 71 - x$. For example, the point $(0, 0)$ is in the half-plane below the line, and $x = 0, y = 0$ satisfies the inequality $y < 71 - x$ because $0 < 71 - 0$. Thus, the half-plane below the line satisfies the inequality. If we shade the half-plane that satisfies $y < 71 - x$, the line and the shaded region constitute the graph of $x + y \leq 71$ (Figure 8.4(b)). This is the solution region for the inequality.

Note that the number of cars satisfying this constraint has other limitations. In the context of this application, the number of cars cannot be negative, so the values of x and y must be nonnegative. We can show the solution to the inequality for nonnegative x and y by using a window with $x \geq 0$ and $y \geq 0$. Figure 8.5 shows the solution of $x + y \leq 71$ for $x \geq 0$ and $y \geq 0$.

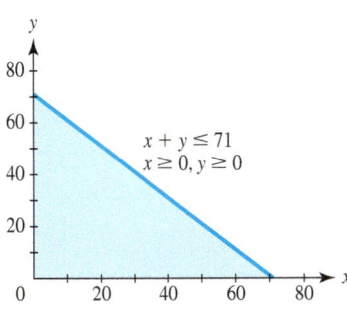

All points in this region with integer coordinates satisfy the stated conditions of this example.

(a)

(b)

Figure 8.4

Figure 8.5

We now investigate the second inequality in the car rental problem.

EXAMPLE 3 ▶ Rental Cars

The car rental agency in the Section Preview has a maximum of $1,260,000 to invest, with x compact cars costing $15,000 each and y midsized cars costing $28,000 each.

a. Write the inequality representing this information.

b. Graph this inequality on a coordinate plane.

SOLUTION

a. If x represents the number of compact cars, the cost of all the compact cars purchased is $15,000x$ dollars. Similarly, the cost of all the midsized cars purchased is $28,000y$ dollars. Because the total amount available to spend is 1,260,000 dollars, x and y satisfy

$$15,000x + 28,000y \leq 1,260,000$$

b. To graph the solutions to this inequality, we first graph the equation

$$15,000x + 28,000y = 1,260,000$$

or, equivalently,

$$y = \frac{1,260,000 - 15,000x}{28,000} = \frac{1260 - 15x}{28}$$

The graph is shown in Figure 8.6.

To decide which of the two half-planes determined by the line satisfies the inequality, we test points in one half-plane. The point $(0, 0)$ is in the half-plane below the line, and it satisfies the inequality because $15,000(0) + 28,000(0) < 1,260,000$. Thus, the half-plane containing $(0, 0)$, along with the line, forms the solution region. Because x and y must be nonnegative in this application, the graph of the inequality is shown only in the first quadrant (Figure 8.7). All points in this region with integer coordinates satisfy the stated conditions of this example.

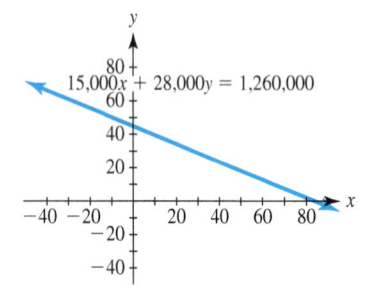

Figure 8.6

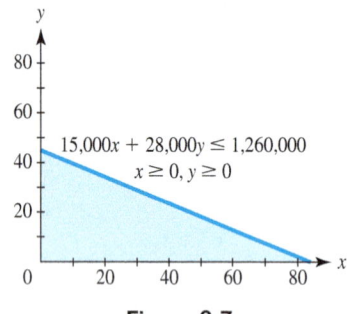

Figure 8.7

Systems of Inequalities in Two Variables

We now have pictures of the limitations (constraints) on the car purchases of the car rental agency, but how are they related? We know that the number of cars the rental agency can buy is limited by *both* inequalities in Examples 2 and 3, so we seek the solution to both inequalities *simultaneously*. That is, we seek the solution to a **system of inequalities**. In general, if we have two or more inequalities in two variables and seek the values of the variables that satisfy all inequalities, we are solving a system of inequalities. The solution to the system can be found by finding the intersection of the solution sets of the inequalities.

EXAMPLE 4 ▶ System of Inequalities

Solve the system of inequalities $\begin{cases} 2x - 4y \geq 12 \\ x + 3y > -4 \end{cases}$.

SOLUTION

To find the solution, we first solve the inequalities for y.

$$y \leq \frac{1}{2}x - 3 \quad (1)$$

$$y > -\frac{1}{3}x - \frac{4}{3} \quad (2)$$

The *borders* of the inequality region are graphed as the solid line with equation $y = \frac{1}{2}x - 3$ and the dashed line with equation $y = -\frac{1}{3}x - \frac{4}{3}$. One of these borders is a solid line because $y = \frac{1}{2}x - 3$ is part of Inequality (1), and one is a dashed line because $y = -\frac{1}{3}x - \frac{4}{3}$ is not part of Inequality (2) (Figure 8.8(a)). We find a *corner* of the solution region by finding the intersection of the two lines. We can find this point of intersection by algebraically or graphically solving the equations simultaneously. Using substitution and solving gives the point of intersection:

$$\frac{1}{2}x - 3 = -\frac{1}{3}x - \frac{4}{3}$$
$$3x - 18 = -2x - 8$$
$$5x = 10$$
$$x = 2$$
$$y = \frac{1}{2}(2) - 3 = -2$$

The two border lines divide the plane into four regions, one of which is the solution region. We can determine what region satisfies both inequalities by choosing a test point in each of the four regions. For example:

- $(0, 0)$ and $(1, -2)$ do not satisfy Inequality (1).
- $(2, -4)$ does not satisfy Inequality (2).
- $(6, -1)$ satisfies both inequalities, so the region that contains this point is the solution region.

Figure 8.8(b) shows the solution region for the system of inequalities, with its corner at $(2, -2)$. Note that the graph of $y = -\frac{1}{3}x - \frac{4}{3}$ is not part of the solution.

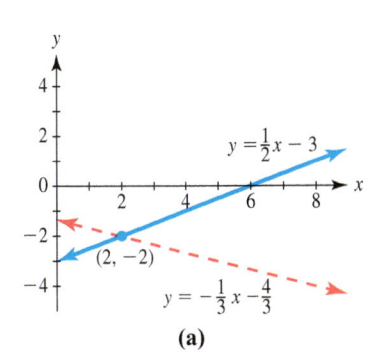

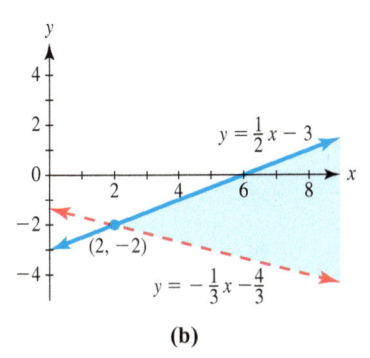

Figure 8.8

We now turn our attention to finding the values of x and y that satisfy both the constraints on the car rental agency problem in the Section Preview.

EXAMPLE 5 ▶ Car Rental Agency

The inequalities that satisfy the conditions given in the car rental problem in the Section Preview are called the **constraint inequalities** for the problem. They form the system

$$\begin{cases} y \leq 71 - x \\ y \leq \dfrac{1260 - 15x}{28} \\ x \geq 0 \\ y \geq 0 \end{cases}$$

a. Graph the solution to the system.

b. Find the coordinates of the points where the borders intersect and the points where the region intersects the x- and y-axes.

SOLUTION

a. We graphed the solutions to the first two inequalities discussed for the car rental agency application in Examples 2 and 3. If we graph them both in the first quadrant, the region where the two solution regions overlap is where both inequalities are true simultaneously (the coordinates of the points satisfy both inequalities) (Figure 8.9).

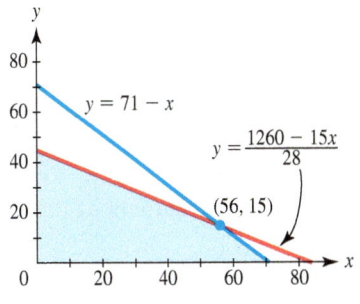

Figure 8.9

Any point in the solution region with integer coordinates is a solution to the application. For example, $x = 50, y = 8$ is a solution, as is $x = 6, y = 40$.

b. The region is bordered on one side by the *line*

$$y = 71 - x$$

and on the other side by the *line*

$$y = \frac{1260 - 15x}{28}$$

A **corner** of the graph of the solution set occurs at the *point* where these two lines intersect. We find this point of intersection graphically as shown in Figure 8.9 or by solving the two equations simultaneously.

$$\begin{cases} y = 71 - x \\ y = \dfrac{1260 - 15x}{28} \end{cases}$$

Substitution gives

$$71 - x = \frac{1260 - 15x}{28}$$

$$1988 - 28x = 1260 - 15x$$

$$728 = 13x$$

$$x = 56$$

$$y = 71 - 56 = 15$$

Thus, this point of intersection of the borders of the solution region, or the *corner*, occurs at $x = 56, y = 15$. This corresponds to purchasing 56 compact cars and 15 midsized cars and is one of many possible solutions to the rental application.

Other corners occur where the region intersects the axes:

- At the origin ($x = 0, y = 0$, representing no cars purchased)
- Where $y = 71 - x$ intersects the *x*-axis ($x = 71, y = 0$, representing 71 compact and no midsize cars)
- Where $y = \dfrac{1260 - 15x}{28}$ intersects the *y*-axis ($x = 0, y = 45$, representing 0 compact and 45 midsize cars)

(a)

(b)

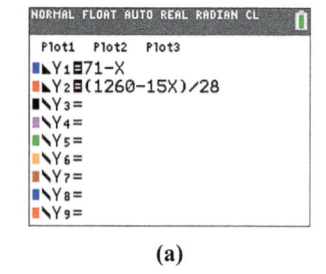

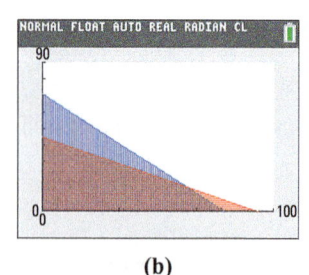

Figure 8.10

Technology Note

We can also use a graphing calculator to graph the solution to the system in Example 5. We enter the inequalities as equations in Y1 and Y2, with the icons to the left of Y1 and Y2 changed from \ to an icon denoting a shaded area below a line (see Figure 8.10(a)). The intersection of the two regions is the solution region, which is the area shaded by the intersection of blue lines and red lines in Figure 8.10(b). In Section 8.2, we will discuss how to find the corners of this region with a graphing calculator.

EXAMPLE 6 ▶ Advertising

A candidate for mayor of a city wishes to use a combination of radio and television advertisements in her campaign. Research has shown that each 1-minute spot on television reaches 0.09 million people and each 1-minute spot on radio reaches 0.006 million. The candidate believes that she must reach at least 2.16 million people, and she can buy a total of no more than 80 minutes of advertising time.

a. Write the inequalities that describe her advertising needs.

b. Graph the region determined by these constraint inequalities.

c. Interpret the solution region in the context of this problem.

SOLUTION

a. If we represent by x the number of minutes of television time, the total number of people reached by television is $0.09x$ million, and if we represent by y the number of minutes of radio time, the total number reached by radio is $0.006y$ million. Thus, one condition that must be satisfied is given by the inequality

$$0.09x + 0.006y \geq 2.16$$

In addition, the statement that the total number of minutes of advertising can be no more than 80 minutes can be written as the inequality

$$x + y \leq 80$$

Neither x nor y can be negative in this application, so we also have

$$x \geq 0$$
$$y \geq 0$$

b. Graphing the equations corresponding to these inequalities and shading the regions satisfying the inequalities give the graph of the solution (Figure 8.11).

One corner of the region occurs where the boundary lines intersect. We can write the equations of the boundary lines as

$$0.09x + 0.006y = 2.16 \quad \text{and} \quad y = 80 - x$$

Their y-values are equal where the lines intersect, so we substitute for y in the first equation and solve.

$$0.09x + 0.006(80 - x) = 2.16$$
$$0.09x + 0.48 - 0.006x = 2.16$$
$$0.084x = 1.68$$
$$x = 20$$
$$y = 80 - 20 = 60$$

The other corners occur where the region intersects the axes:

- $0.09x + 0.006y = 2.16$ intersects the x-axis at $x = 24, y = 0$.
- $x + y = 80$ intersects the x-axis at $x = 80, y = 0$.

c. Each point in this region represents a combination of television and radio time that the candidate for mayor can use in her campaign. For example, she could use 20 minutes of television time and 60 minutes of radio time or 40 minutes of television time and 20 minutes of radio time. No point outside this region can represent a number of minutes of television and radio advertising for this candidate because of the constraints given in the problem.

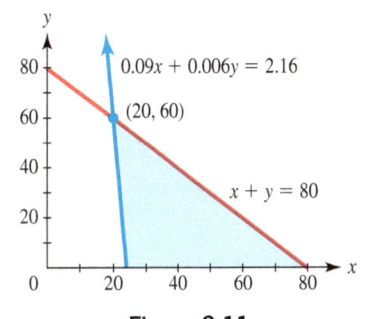

Figure 8.11

620 Chapter 8 Special Topics in Algebra

Systems of Nonlinear Inequalities

As with systems of linear inequalities in two variables, the solutions of systems of nonlinear inequalities in two variables are regions in a plane. For example, the solution of the system

$$\begin{cases} y \leq 8 - 2x - x^2 \\ y \geq 4 - 2x \end{cases}$$

is the area below the graph of $y = 8 - 2x - x^2$ and above the graph of $y = 4 - 2x$, as well as any points on the graphs of $y = 4 - 2x$ and $y = 8 - 2x - x^2$ between $(-2, 8)$ and $(2, 0)$ (Figure 8.12). The points of intersection of the two border curves are $(2, 0)$ and $(-2, 8)$.

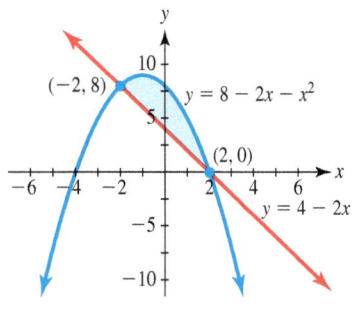

Figure 8.12

EXAMPLE 7 ▶ Nonlinear System of Inequalities

Graphically solve the system

$$\begin{cases} x^2 - 10x + y \leq 0 \\ -x^2 + 8x + y \geq 28 \end{cases}$$

SOLUTION

To solve the system

$$\begin{cases} x^2 - 10x + y \leq 0 \\ -x^2 + 8x + y \geq 28 \end{cases}$$

graphically, we first solve the inequalities for y:

$$\begin{cases} y \leq 10x - x^2 \\ y \geq x^2 - 8x + 28 \end{cases}$$

Graphing the associated equations on the same axes gives the graphs in Figure 8.13(a).

(a)   (b)

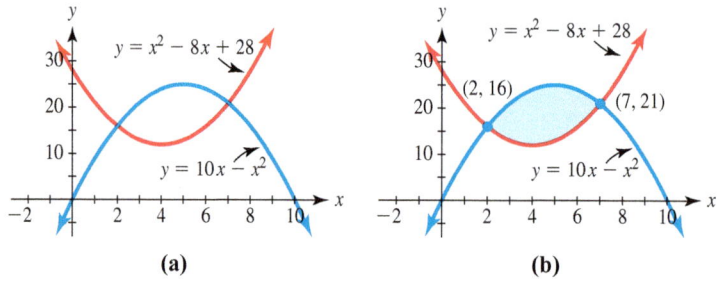

Figure 8.13

The points of intersection of the two border curves are $(2, 16)$ and $(7, 21)$. We can show this algebraically as follows:

$$x^2 - 8x + 28 = 10x - x^2$$
$$2x^2 - 18x + 28 = 0$$
$$2(x - 2)(x - 7) = 0$$
$$x = 2 \quad \text{or} \quad x = 7$$
$$y = 16 \qquad\qquad y = 21$$

The region that we seek is the portion of the graph that is above the graph of $y = x^2 - 8x + 28$ and below the graph of $y = 10x - x^2$, as well as any points on the graphs of these two functions between $(2, 16)$ and $(7, 21)$, as shown in Figure 8.13(b).

EXAMPLE 8 ▶ Nonlinear System of Inequalities

Graphically solve the system

$$\begin{cases} y > x^3 - 3x \\ y < x \\ x > -2.5 \\ x < 2.5 \end{cases}$$

SOLUTION

To solve the system

$$\begin{cases} y > x^3 - 3x \\ y < x \\ x > -2.5 \\ x < 2.5 \end{cases}$$

we first graph the related equations (Figure 8.14(a)).

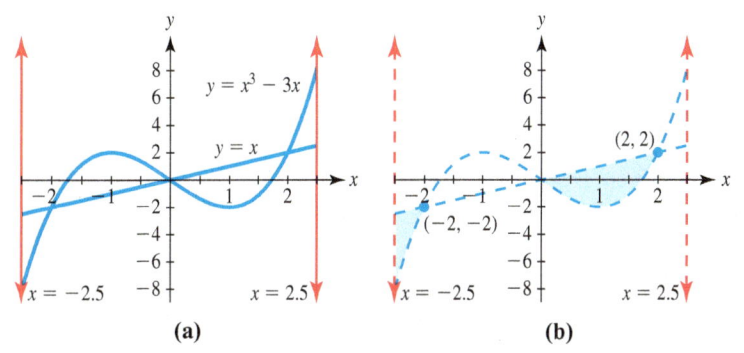

Figure 8.14

One region is bounded on the left by $x = -2.5$ and the two curves, with a corner at $(-2, -2)$. The second region is bounded by the two curves, with corners at $(0, 0)$ and $(2, 2)$. These corner points can also be found algebraically as follows.

$$x^3 - 3x = x$$
$$x^3 - 4x = 0$$
$$x(x - 2)(x + 2) = 0$$
$$x = 0 \quad \text{or} \quad x = 2 \quad \text{or} \quad x = -2$$
$$y = 0 \qquad\quad y = 2 \qquad\quad y = -2$$

The solution to the system of inequalities consists of the two regions that are above the graph of $y = x^3 - 3x$ and below the graph of $y = x$ in the interval $[-2.5, 2.5]$. This solution is shown in Figure 8.14(b).

SKILLS CHECK 8.1

Answers that are not seen can be found in the answer section at the back of the text.

In Exercises 1–6, graph each inequality.

1. $y \leq 5x - 4$
2. $y > 3x + 2$
3. $6x - 3y \geq 12$
4. $\dfrac{x}{2} + \dfrac{y}{3} \leq 6$
5. $4x + 5y \leq 20$
6. $3x - 5y \geq 30$

In Exercises 7–10, match each solution of a system of inequalities with the correct graph.

7. $\begin{cases} y > 3x - 4 \\ y < 2x + 3 \end{cases}$ C

8. $\begin{cases} y \geq x + 1 \\ y \leq 2x + 1 \\ x \geq 0 \\ y \geq 0 \end{cases}$ A

622 Chapter 8 Special Topics in Algebra

9. $\begin{cases} 3x + y \geq 5 \\ x + 3y \geq 6 \\ x \geq 0 \\ y \geq 0 \end{cases}$ D

10. $\begin{cases} x + 2y \leq 20 \\ 3x + 2y \leq 42 \\ x \geq 0 \\ y \geq 0 \end{cases}$ B

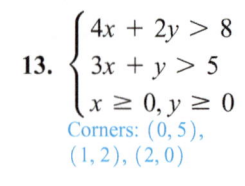

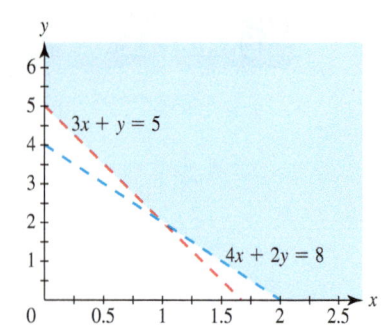

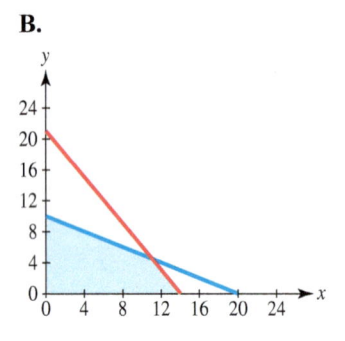

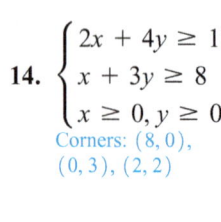

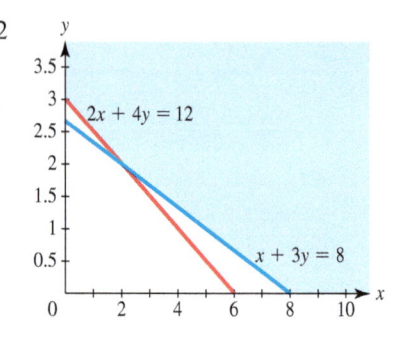

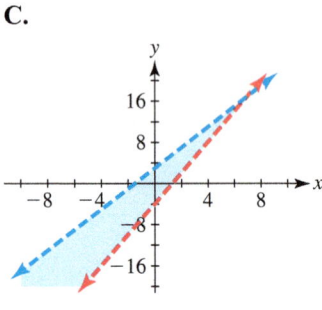

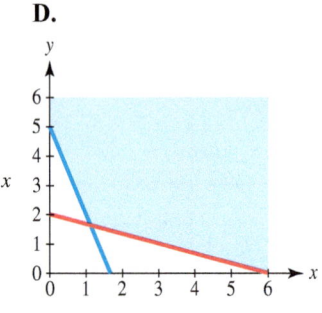

13. $\begin{cases} 4x + 2y > 8 \\ 3x + y > 5 \\ x \geq 0, y \geq 0 \end{cases}$
Corners: $(0, 5)$, $(1, 2)$, $(2, 0)$

A.

B.

C.

D.

14. $\begin{cases} 2x + 4y \geq 12 \\ x + 3y \geq 8 \\ x \geq 0, y \geq 0 \end{cases}$
Corners: $(8, 0)$, $(0, 3)$, $(2, 2)$

In Exercises 11–16, the graph of the boundary equations for each system of inequalities is shown with the system. Locate the solution region, and identify it by finding the corners.

15. $\begin{cases} 2x + 6y \geq 12 \\ 3x + y \geq 5 \\ x + 2y \geq 5 \\ x \geq 0, y \geq 0 \end{cases}$
Corners: $(0, 5)$, $(1, 2)$, $(3, 1)$, $(6, 0)$

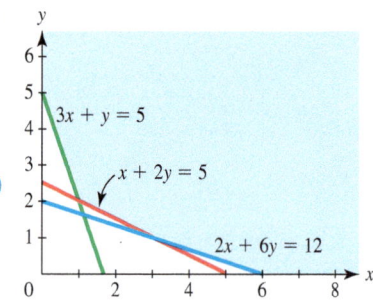

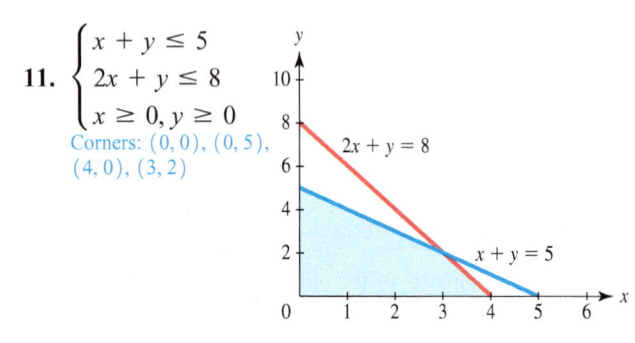

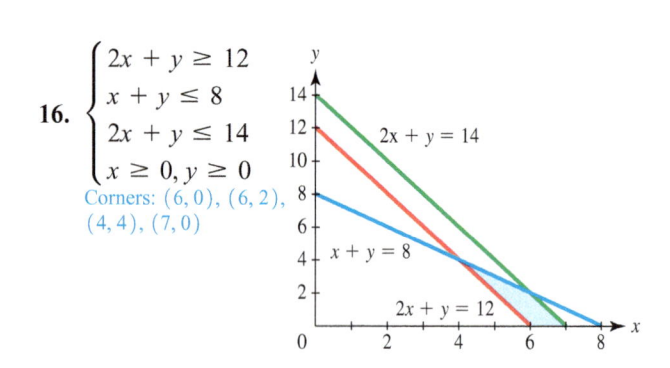

11. $\begin{cases} x + y \leq 5 \\ 2x + y \leq 8 \\ x \geq 0, y \geq 0 \end{cases}$
Corners: $(0, 0)$, $(0, 5)$, $(4, 0)$, $(3, 2)$

16. $\begin{cases} 2x + y \geq 12 \\ x + y \leq 8 \\ 2x + y \leq 14 \\ x \geq 0, y \geq 0 \end{cases}$
Corners: $(6, 0)$, $(6, 2)$, $(4, 4)$, $(7, 0)$

12. $\begin{cases} 2x + y \leq 12 \\ x + 5y \leq 15 \\ x \geq 0, y \geq 0 \end{cases}$
Corners: $(0, 3)$, $(0, 0)$, $(5, 2)$, $(6, 0)$

For each system of inequalities in Exercises 17–22, graph the solution region and identify the corners of the region.

17. $\begin{cases} y \leq 8 - 3x \\ y \leq 2x + 3 \\ y > 3 \end{cases}$

18. $\begin{cases} 2x + y \geq 10 \\ 3x + 2y \geq 17 \\ x + 2y \geq 7 \end{cases}$

19. $\begin{cases} 2x + y < 5 \\ 2x - y > -1 \\ x \geq 0, y \geq 0 \end{cases}$ 20. $\begin{cases} y < 2x \\ y > x + 2 \\ x \geq 0, y \geq 0 \end{cases}$

21. $\begin{cases} x + 2y \geq 4 \\ x + y \leq 5 \\ 2x + y \leq 8 \\ x \geq 0, y \geq 0 \end{cases}$ 22. $\begin{cases} x + y < 4 \\ x + 2y < 6 \\ 2x + y < 7 \\ x \geq 0, y \geq 0 \end{cases}$

In Exercises 23–26, graph the solution of the system of nonlinear inequalities.

23. $\begin{cases} x^2 - 3y < 4 \\ 2x + 3y < 4 \end{cases}$ 24. $\begin{cases} x + y \leq 8 \\ xy \geq 12 \end{cases}$

25. $\begin{cases} x^2 - y - 8x \leq -6 \\ y + 9x \leq 18 \end{cases}$ 26. $\begin{cases} x^2 - y > 0 \\ y > (4 - 2x)^2 \end{cases}$

EXERCISES 8.1

Answers that are not seen can be found in the answer section at the back of the text.

27. Manufacturing A company manufactures two types of leaf blowers: an electric Turbo model and a gas-powered Tornado model. The company's production plan calls for the production of at least 780 blowers per month.

   a. Write the inequality that describes the production plan, if x represents the number of Turbo blowers and y represents the number of Tornado blowers. $x + y \geq 780$

   b. Graph the region determined by this inequality in the context of the application.

28. Manufacturing A company manufactures two types of leaf blowers: an electric Turbo model and a gas-powered Tornado model. It costs $78 to produce each of the x Turbo models and $117 to produce each of the y Tornado models, and the company has at most $76,050 per month to use for production.

   a. Write the inequality that describes this constraint on production. $78x + 117y \leq 76{,}050$

   b. Graph the region determined by this inequality in the context of the application.

29. Sales Trix Auto Sales sells used cars. To promote its cars to target groups, it advertises with x 1-minute spots on cable television, at a cost of $240 per minute, and y 1-minute spots on radio, at a cost of $150 per minute. Suppose the company has at most $36,000 to spend on advertising.

   a. Write the inequality that describes this constraint on advertising. $240x + 150y \leq 36{,}000$

   b. Graph the region determined by this inequality in the context of the application.

30. Sales Trix Auto Sales sells used cars. To promote its cars to target groups, it advertises with x 1-minute spots on cable television and y 1-minute spots on radio. Research shows that it sells one vehicle for every 4 minutes of cable television advertising and it sells one vehicle for each 10 minutes of radio advertising.

   a. $\frac{1}{4}x + \frac{1}{10}y \geq 33$ Write the inequality that describes the requirement that at least 33 cars be sold from this advertising.

   b. Graph the region determined by this inequality in the context of the application.

31. Politics A political candidate wishes to use a combination of x television and y radio advertisements in his campaign. Each 1-minute ad on television reaches 0.12 million eligible voters, and each 1-minute ad on radio reaches 0.009 million eligible voters. The candidate feels that he must reach at least 7.56 million eligible voters and that he must buy at least 100 minutes of advertising.

   a. Write the inequalities that describe these advertising requirements.

   b. Graph the region determined by these inequalities in the context of the application.

32. Housing A contractor builds two models of homes: the Van Buren and the Jefferson. The Van Buren requires 200 worker-days of labor and $240,000 in capital, and the Jefferson requires 500 worker-days of labor and $300,000 in capital. The contractor has a total of 5000 worker-days and $3,600,000 in capital available per month. Let x represent the number of Van Buren models and y represent the number of Jefferson models, and graph the region that satisfies these inequalities.

33. Manufacturing A company manufactures two types of leaf blowers: an electric Turbo model and a gas-powered Tornado model. The company's production plan calls for the production of at least 780 blowers per month. It costs $78 to produce each Turbo model and $117 to manufacture each Tornado model, and the company has at most $76,050 per month to use for production. Let x represent the number of Turbo models and y represent the number of Tornado models, and graph the region that satisfies these constraints.

34. Sales Trix Auto Sales sells used cars. To promote its cars to target groups, it advertises with x 1-minute spots on cable television, at a cost of $240 per minute, and y 1-minute spots on radio, at a cost of $150 per minute. Research shows that it sells one vehicle for every 4 minutes of cable television advertising (that is, $\frac{1}{4}$ car per minute) and it sells one vehicle for each 10 minutes of radio advertising. Suppose the company has at most $36,000 to spend on advertising and sells at least 33 cars per month from this advertising. Let x represent the number of cable TV minutes and y represent the number of radio minutes, and graph the region that satisfies these constraints. That is, graph the solution to this system of inequalities.

35. **Production** A firm produces three different-size television sets on two assembly lines. The following table summarizes the production capacity of each assembly line and the minimum number of each size TV needed to fill orders.

    a. Write the inequalities that are the constraints.

    b. Graph the region that satisfies the inequalities, and identify the corners.

	Assembly Line 1	Assembly Line 2	Number Ordered
19-in. TV	80 per day	40 per day	3200
25-in. TV	20 per day	20 per day	1000
35-in. TV	100 per day	40 per day	3400

36. **Advertising** Tire Town is developing an advertising campaign. The following table indicates the cost per ad package in newspapers and the cost per ad package on radio, as well as the number of ads in each type of ad package.

	Newspaper	Radio
Cost per Ad Package ($)	1000	3000
Ads per Package	18	36

    The owner of the company can spend no more than $18,000 per month, and he wants to have at least 252 ads per month.

    a. Write the inequalities that are the constraints.

    b. Graph the region that satisfies the inequalities, and identify the corners.

37. **Nutrition** A privately owned lake contains two types of fish: bass and trout. The owner provides two types of food, A and B, for these fish. Trout require 4 units of food A and 6 units of food B, and bass require 10 units of food A and 7 units of food B. If the owner has 1600 units of each food, graph the region that satisfies the constraints.

38. **Manufacturing** Evergreen Company produces two types of printers: the inkjet and the laserjet. It takes 2 hours to make the inkjet and 6 hours to make the laserjet. The company can make at most 120 printers per day and has 400 labor-hours available per day.

    a. Write the inequalities that describe this application.

    b. Graph the region that satisfies the inequalities, and identify the corners of the region.

39. **Manufacturing** Easyboy manufactures two types of chairs: Standard and Deluxe. Each Standard chair requires 4 hours to construct and finish, and each Deluxe chair requires 6 hours to construct and finish. Upholstering takes 2 hours for a Standard chair and 6 hours for a Deluxe chair. There are 480 hours available each day for construction and finishing, and there are 300 hours available per day for upholstering.

    a. Write the inequalities that describe the application.

    b. Graph the solution of the system of inequalities, and identify the corners of the region.

40. **Manufacturing** Safeco Company produces two types of chainsaws: the Safecut and the Safecut Deluxe. The Safecut model requires 2 hours to assemble and 1 hour to paint, and the Deluxe model requires 3 hours to assemble and $\frac{1}{2}$ hour to paint. The daily maximum number of hours available for assembly is 36, and the daily maximum number of hours available for painting is 12.

    a. Write the inequalities that describe the application.

    b. Graph the solution of the system of inequalities, and identify the corners of the region.

41. **Manufacturing** A company manufactures commercial and domestic heating systems at two plant sites. It can produce no more than 1400 units per month, and it needs to fill orders of at least 500 commercial units and 750 domestic units.

    a. Write the system of inequalities that describes the constraints on production for these orders.

    b. Graph the solution set of this system of inequalities.

42. **Constructing a Box** An open box with a square base is to be constructed so that the area of the material used is at most 500 cm^2. What dimensions of the base will yield a box that gives a volume of at least 500 cm^3? If $4.14 \leq x \leq 20$, the box will have a volume of at least 500 cubic centimeters.

8.2 Linear Programming: Graphical Methods

KEY OBJECTIVE

- Find the optimal values of a function subject to constraints

SECTION PREVIEW Maximizing Profit

A rental agency has a maximum of $1,260,000 to invest in the purchase of at most 71 new cars of two different types, compact and midsize. The cost per compact car is $15,000, and the cost per midsize car is $28,000. The number of cars of each type is limited (constrained) by the budget available and the number of cars needed. We saw in the previous section that purchases of the rental agency were constrained by the number of cars needed and the amount of money available for purchasing the cars. The constraints were given by a system of inequalities, and we found the solution region determined by the constraints.

If the average yearly profit is $6000 for each compact car and $9000 for each midsize car, buying a different number of each type will affect the profit. How do we find the correct number of each type so that the profit is maximized under these conditions? To answer questions of this type without knowledge of special techniques, people must use the "guess-and-check" method repeatedly to arrive at a solution. And when the number of conditions that enter into a problem of this type increases, guessing is a very unsatisfactory solution method. We can answer these questions with a mathematical technique called **linear programming**. Linear programming is a valuable tool in management; in fact, 85% of all Fortune 500 firms use linear programming. In this section, we use graphical methods to find the **optimal solutions** to linear programming problems such as the one above. ■

Linear Programming

In any linear programming problem, we seek the point or points in the region determined by the constraint inequalities that give optimal (*maximum* or *minimum*) values for the **objective function** (the function we seek to optimize).* Any point in the region determined by the constraints is called a **feasible solution** to the linear programming problem, and the region itself is called the **feasible region**.

Note that the constraints in our car rental application above can be written as the system of inequalities

$$\begin{cases} x + y \leq 71 \\ 15x + 28y \leq 1260 \\ x \geq 0 \\ y \geq 0 \end{cases}$$

The solution set of the inequalities for the problem, the feasible region, is shown as the shaded region in Figure 8.15. Any point inside the shaded region or on its boundary satisfies the constraints, so it is a feasible solution. In the context of this application, only feasible solutions with integer coordinates can be solutions to the problem.

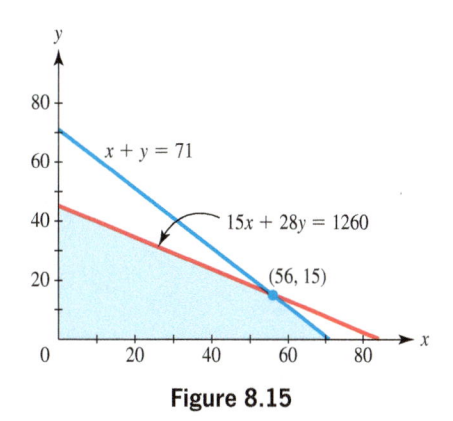

Figure 8.15

---

*The region determined by the constraints must be convex for the optimal solution to exist. A convex region is one such that for any two points in the region, the segment joining those points lies entirely within the region. We restrict our discussion to convex regions.

626 Chapter 8 Special Topics in Algebra

We will see that it is important to find the corners of feasible regions.

> **Technology Note**
> We can use graphing calculators to graph the feasible region and then use the TRACE, zero (accessed by 2ND TRACE), and intersect (accessed by 2ND TRACE) commands to find the corners. Figure 8.16 shows the calculator graph of the feasible region shown in Figure 8.15, with the corners $(0, 45)$, $(71, 0)$, and $(56, 15)$ identified.

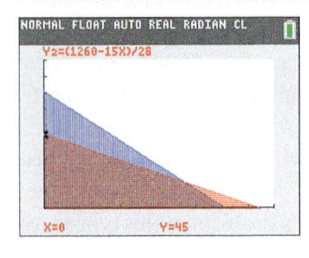

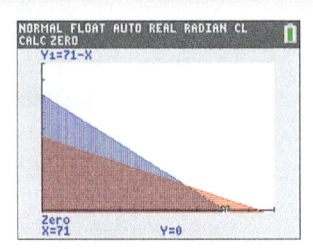

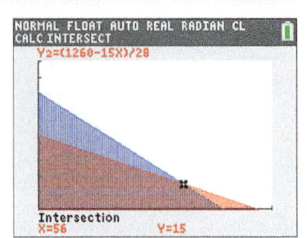

Figure 8.16

Because the profit P is 6 thousand dollars for each of the x compact cars and 9 thousand dollars for each of the y midsize cars, we can model the profit function as

$$P = 6x + 9y, \quad \text{where } P \text{ is in thousands of dollars}$$

In this application, we seek to find the maximum profit subject to the constraints listed above. Because we seek to maximize the profit subject to the constraints, we evaluate the objective function

$$P = 6x + 9y$$

for points (x, y) in the feasible region. For example, at $(20, 19)$, $P = 291$, at $(50, 10)$, $P = 390$, and at $(50, 15)$, $P = 435$. But there are a large number of points to test, so we look at a different way to find the maximum value of P. We can graph $P = 6x + 9y$ for different values of P by first writing the equation in the form

$$y = -\frac{2}{3}x + \frac{P}{9}$$

Then, for each possible value of P, the graph is a line with slope $-\frac{2}{3}$ and y-intercept $\frac{P}{9}$. Figure 8.17 shows the graphs of the feasible region and objective functions for $P = 360, 471,$ and 495. Note that the graphs are parallel $\left(\text{with slope } -\frac{2}{3}\right)$, and that larger values of P coincide with larger y-intercepts $\left(\frac{P}{9}\right)$. Observe that if $P = 471$, the graph of $y = -\frac{2}{3}x + \frac{P}{9}$ intersects the feasible region and that values of P less than 471 will give lines that pass through the feasible region, but each of these values of P

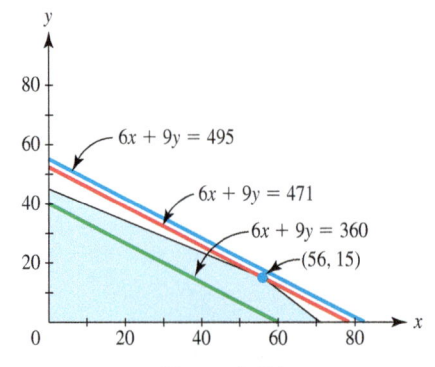

Figure 8.17

represents a smaller profit. Note also that the value $P = 495$ (or any P-value greater than 471) results in a line that "misses" the feasible region, so this value of P is not a solution to the problem. Thus, the maximum value for P, subject to the constraints, is 471 (thousand dollars). We can verify that the graph of $471 = 6x + 9y$ intersects the region at $(56, 15)$ and that $P = 6x + 9y$ will not intersect the feasible region if $P > 471$. Thus, the profit is maximized at \$471,000, when $x = 56$ and $y = 15$—that is, when 56 compact and 15 midsize cars are purchased.

Note that the values of x and y that correspond to the maximum value of P occur at $(56, 15)$, which is a corner of the feasible region. In fact, the maximum value of an objective function always occurs at a corner of the feasible region if the feasible region is closed and bounded. (A feasible region is closed and bounded if it is enclosed by and includes the lines associated with the constraints.) This gives us the basis for finding the solutions to linear programming problems using the graphical method.

Solutions to Linear Programming Problems

1. If a linear programming problem has an optimal solution, then the optimal value (maximum or minimum) of an objective function occurs at a corner of the feasible region determined by the constraints.

2. When a feasible region for a linear programming problem is closed and bounded, the objective function has a maximum and a minimum value.

3. If the objective function has the same optimal value at two corners, then it also has that optimal value at any point on the boundary line segment connecting the two corners.

4. When the feasible region is not closed and bounded, the objective function may have a maximum only, a minimum only, or neither.

Thus, for a closed and bounded region, we can find the maximum and minimum values of an objective function by evaluating the function at each of the corners of the feasible region formed by the graphical solution of the constraint inequalities.

EXAMPLE 1 ▶ Finding the Optimal Values of a Function Subject to Constraints

Find (a) the maximum value and (b) the minimum value of $C = 5x + 3y$ subject to the constraints

$$\begin{cases} x + y \leq 5 \\ 2x + y \leq 8 \\ x \geq 0, y \geq 0 \end{cases}$$

SOLUTION

The feasible region is in the first quadrant because it is bounded by $x \geq 0$ and $y \geq 0$. We can graph the remaining boundary lines with technology if we solve the first two inequalities for y.

$$\begin{cases} y \leq 5 - x \\ y \leq -2x + 8 \\ x \geq 0, y \geq 0 \end{cases}$$

The solution set for the system of inequalities (that is, the feasible region) is bounded by the lines $y = 5 - x$ and $y = -2x + 8$ and the x- and y-axes. The corners of the feasible region occur at the origin $(0, 0)$, at the x-intercept of one of the two boundary lines, at the y-intercept of one of the two boundary lines, and at the point of intersection of the lines $y = 5 - x$ and $y = -2x + 8$ (Figure 8.18).

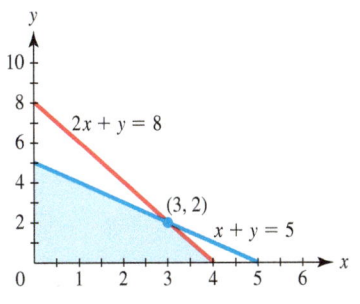

Figure 8.18

There are two y-intercept points, $(0, 5)$ and $(0, 8)$, and the one that satisfies both inequalities is $(0, 5)$. The x-intercept points are $(4, 0)$ and $(5, 0)$; the one that satisfies both inequalities is $(4, 0)$. We can find the point of intersection of the two boundary lines by solving the system of equations

$$\begin{cases} y = 5 - x \\ y = -2x + 8 \end{cases}$$

Substitution gives

$$5 - x = -2x + 8$$

Solving gives $x = 3$, and substituting gives $y = 2$. Thus, the point of intersection of these two boundary lines is $(3, 2)$. This point of intersection can also be found by using a graphing utility.

a. Any point inside the shaded region or on the boundary is a feasible solution to the problem, but the maximum possible value of

$$C = 5x + 3y$$

occurs at a corner of the feasible region. We evaluate the objective function at each of the corner points:

At $(0, 0)$: $C = 5x + 3y = 5(0) + 3(0) = 0$
At $(0, 5)$: $C = 5x + 3y = 5(0) + 3(5) = 15$
At $(4, 0)$: $C = 5x + 3y = 5(4) + 3(0) = 20$
At $(3, 2)$: $C = 5x + 3y = 5(3) + 3(2) = 21$

Thus, the maximum value of the objective function is 21 when $x = 3, y = 2$.

b. Observing the values of $C = 5x + 3y$ at each corner point of the feasible region, we see that the minimum value of the objective function is 0 at $(0, 0)$.

EXAMPLE 2 ▶ Cost Minimization

Star Manufacturing Company produces two types of Blu-ray players, which are assembled at two different locations. Plant 1 can assemble 60 units of the Star model and 80 units of the Prostar model per hour, and plant 2 can assemble 300 units of the Star model and 80 units of the Prostar model per hour. The company needs to produce at least 5400 units of the Star model and 4000 units of the Prostar model to fill an order. If it costs $2000 per hour to run plant 1 and $3000 per hour to run plant 2, how many hours should each plant spend on manufacturing Blu-ray players to minimize its cost for this order? What is the minimum cost for this order?

SOLUTION

If we let x equal the number of hours of assembly time at plant 1 and we let y equal the number of hours of assembly time at plant 2, then the cost function that we seek to minimize is

$$C = 2000x + 3000y$$

The constraints on the assembly hours follow:

Star model units: $60x + 300y \geq 5400$
Prostar model units: $80x + 80y \geq 4000$
Nonnegative hours: $x \geq 0, y \geq 0$

8.2 Linear Programming: Graphical Methods   **629**

We can solve the first two inequalities for y, write these constraints as a system of inequalities, and graph the feasible region (Figure 8.19):

$$\begin{cases} y \geq 18 - 0.2x \\ y \geq 50 - x \\ x \geq 0, y \geq 0 \end{cases}$$

The two boundary lines intersect where $18 - 0.2x = 50 - x$, at $x = 40$, which then gives $y = 10$. The boundary lines of the feasible region intersect the axes at $(0, 50)$ and at $(90, 0)$.

Thus, the corners of the feasible region are

- $(0, 50)$, the y-intercept that satisfies all the inequalities.
- $(90, 0)$, the x-intercept that satisfies all the inequalities.
- $(40, 10)$, the intersection of the two boundary lines.

To find the hours of assembly time at each plant that will minimize the cost, we test the corner points:

At $(0, 50)$: $C = 2000(0) + 3000(50) = 150{,}000$
At $(90, 0)$: $C = 2000(90) + 3000(0) = 180{,}000$
At $(40, 10)$: $C = 2000(40) + 3000(10) = 110{,}000$

Thus, the cost is minimized when assembly time is 40 hours at plant 1 and 10 hours at plant 2. The minimum cost is $110,000.

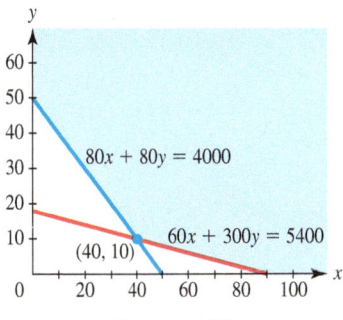

Figure 8.19

Note that the feasible region in Example 2 is not closed and bounded, and that there is no maximum value for the objective function even though there is a minimum value. It should also be noted that some applied linear programming problems require discrete solutions; for example, the optimal solution to the earlier car rental problem required a number of cars, so only nonnegative integer solutions were possible.

EXAMPLE 3 ▶ Maximizing Profit

Smoker Meat Packing Company makes two different types of hot dogs: regular and all-beef. Each pound of all-beef hot dogs requires 0.8 pound of beef and 0.2 pound of spices, and each pound of regular hot dogs requires 0.3 pound of beef and 0.2 pound of spices, with the remainder nonbeef meat products. The company has at most 1020 pounds of beef and has at most 500 pounds of spices available for hot dogs. If the profit is $0.90 on each pound of all-beef hot dogs and $1.20 on each pound of regular hot dogs, how many of each type should be produced to maximize the profit?

SOLUTION

Letting x equal the number of pounds of all-beef hot dogs and y equal the number of pounds of regular hot dogs, we seek to maximize the function $P = 0.90x + 1.20y$ subject to the constraints

$$\begin{cases} 0.8x + 0.3y \leq 1020 \\ 0.2x + 0.2y \leq 500 \\ x \geq 0, y \geq 0 \end{cases}$$

The graph of this system of inequalities is more easily found if the first two inequalities are solved for y.

$$\begin{cases} y \leq 3400 - \dfrac{8}{3}x \\ y \leq 2500 - x \\ x \geq 0, y \geq 0 \end{cases}$$

Graphing the boundary lines and testing the four regions determines the feasible region (Figure 8.20).

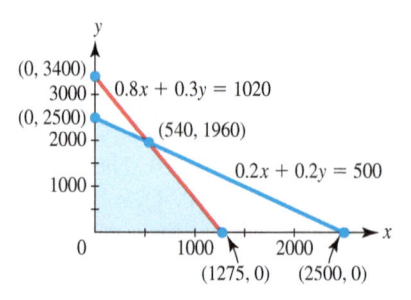

Figure 8.20

The boundary lines intersect where

$$3400 - \frac{8}{3}x = 2500 - x$$

or at $x = 540$. Substituting for x gives $y = 1960$.

The x-intercept $(1275, 0)$ and the y-intercept $(0, 2500)$ satisfy the inequalities. Thus, the corners of this region are $(0, 0)$, $(0, 2500)$, $(1275, 0)$, and $(540, 1960)$. $P = 0$ at $(0, 0)$. Testing the objective function at the other corner points gives

At $(0, 2500)$: $P = 0.90x + 1.20y = 0.90(0) + 1.20(2500) = 3000$

At $(1275, 0)$: $P = 0.90x + 1.20y = 0.90(1275) + 1.20(0) = 1147.50$

At $(540, 1960)$: $P = 0.90x + 1.20y = 0.90(540) + 1.20(1960) = 2838$

This indicates that profit will be maximized if 2500 pounds of regular hot dogs and no all-beef hot dogs are produced.

EXAMPLE 4 ▶ Manufacturing

A company manufactures air conditioning units and heat pumps at its factories in Atlanta, Georgia, and Newark, New Jersey. The Atlanta plant can produce no more than 1000 items per day, and the number of air conditioning units cannot exceed 100 more than half the number of heat pumps. The Newark plant can produce no more than 850 units per day. The profit on each air conditioning unit is $400 at the Atlanta plant and $390 at the Newark plant. The profit on each heat pump is $200 at the Atlanta plant and $215 at the Newark plant. Suppose there is an order for 500 air conditioning units and 750 heat pumps.

a. Graph the feasible region and identify the corners.

b. Find the maximum profit that can be made on this order and what production distribution will give the maximum profit.

SOLUTION

a. Because the total number of air conditioning units needed is 500, we can represent the number produced at Atlanta by x and the number produced at Newark by $(500 - x)$. Similarly, we can represent the number of heat pumps produced at Atlanta by y and the number produced at Newark by $(750 - y)$. Table 8.1 summarizes the constraints.

Table 8.1

	Air Conditioning Units	Heat Pumps	Total
Atlanta	x	+ y	≤1000
Newark	500 − x	+ (750 − y)	≤850
Other Constraints	x		≤0.5y + 100
	x		≥0
		y	≥0
Profit	400x + 390(500 − x)	+ 200y + 215(750 − y)	

Using this information gives the following system:

$$\begin{cases} x + y \leq 1000 & (1) \\ (500 - x) + (750 - y) \leq 850 & (2) \\ x \leq 0.5y + 100 & (3) \\ x \geq 0, y \geq 0 \end{cases}$$

We solve each of these inequalities for y and graph the solution set (Figure 8.21):

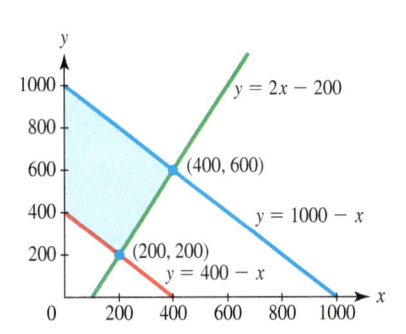

$$\begin{cases} y \leq 1000 - x & (1) \\ y \geq 400 - x & (2) \\ y \geq 2x - 200 & (3) \\ x \geq 0, y \geq 0 \end{cases}$$

Figure 8.21

The intersections of the boundary lines associated with the inequalities follow.

- Inequalities (1) and (3): where $1000 - x = 2x - 200$, at $x = 400$, $y = 600$.
- Inequalities (2) and (3): where $400 - x = 2x - 200$, at $x = 200$, $y = 200$.
- Inequalities (1) and (2): where $1000 - x = 400 - x$, which has no solution, so they do not intersect.

The other corners are on the y-axis, at $(0, 400)$ and $(0, 1000)$.

b. The objective function gives the profit for the products.

$$P = 400x + 390(500 - x) + 200y + 215(750 - y)$$
$$= 10x - 15y + 356{,}250$$

Testing the profit function at the corners of the feasible region determines where the profit is maximized.

$$\begin{aligned} \text{At } (0, 400): & \quad P = 350{,}250 \\ \text{At } (200, 200): & \quad P = 355{,}250 \\ \text{At } (0, 1000): & \quad P = 341{,}250 \\ \text{At } (400, 600): & \quad P = 351{,}250 \end{aligned}$$

Thus, the profit is maximized at $355,250 when 200 air conditioning units and 200 heat pumps are manufactured in Atlanta. The remainder are manufactured in Newark, so 300 air conditioning units and 550 heat pumps are manufactured in Newark.

Solution with Technology

As we have seen, graphing utilities can be used to graph the feasible region and to find the corners of the region satisfying the constraint inequalities. The graphical method that we have been using cannot be used if the problem involves more than two variables, and most real linear programming applications involve more than two variables. In fact, some real applications involve as many as 40 variables. Problems involving more than two variables can be solved with spreadsheet programs such as Lotus 1-2-3, Microsoft Excel, and Quatro Pro. Use of these programs to solve linear programming problems is beyond the scope of this text.

SKILLS CHECK 8.2

Answers that are not seen can be found in the answer section at the back of the text.

In Exercises 1–4, use the given feasible region determined by the constraint inequalities to find the maximum possible value and the minimum possible value of the objective function.

1. $f = 4x + 9y$ subject to the constraints
Maximum: 382 at $(10, 38)$; minimum: 0 at $(0, 0)$
$$\begin{cases} 2x + 3y \le 134 \\ x + 5y \le 200 \\ x \ge 0, y \ge 0 \end{cases}$$

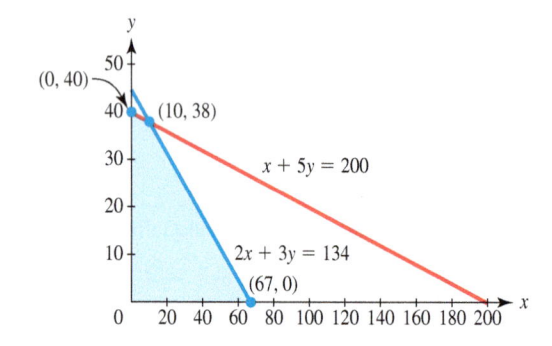

2. $f = 2x + 3y$ subject to the constraints
Maximum: 79.6 at $(14, 17.2)$; minimum: 0 at $(0, 0)$
$$\begin{cases} 6x + 5y \le 170 \\ x + 5y \le 100 \\ x \ge 0, y \ge 0 \end{cases}$$

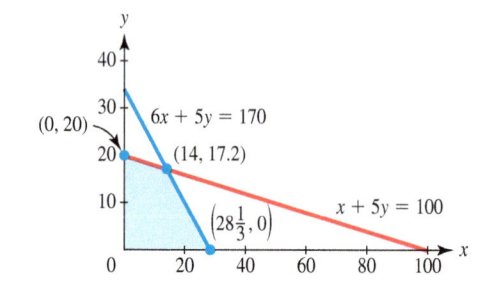

3. $f = 4x + 2y$ subject to the constraints
Maximum: 22 at $(4, 3)$; minimum: 0 at $(0, 0)$
$$\begin{cases} 3x + y \le 15 \\ x + 2y \le 10 \\ -x + y \le 2 \\ x \ge 0, y \ge 0 \end{cases}$$

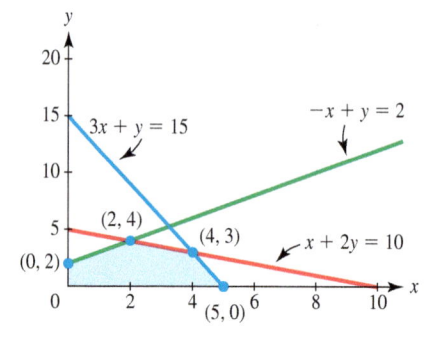

4. $f = 3x + 9y$ subject to the constraints
$$\begin{cases} 2x + y \le 12 \\ x + 3y \le 15 \\ x + y \le 7 \\ x \ge 0, y \ge 0 \end{cases}$$

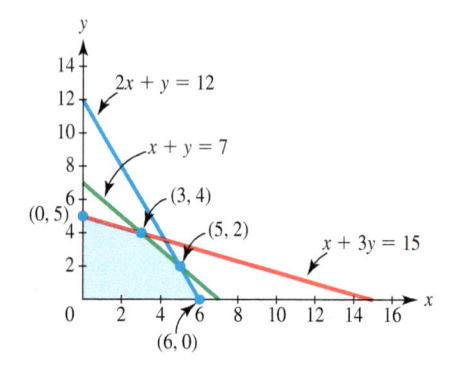

5. Perform the following steps to maximize $f = 3x + 5y$ subject to the constraints

$$\begin{cases} y \leq \dfrac{54 - 2x}{3} \\ y \leq 22 - x \\ x \geq 0, y \geq 0 \end{cases}$$

   a. Graph the region that satisfies the system of inequalities, and identify the corners of the region.

   b. Test the objective function $f = 3x + 5y$ at each of the corners of the feasible region to determine which corner gives the maximum value. Give the maximum possible value of f and the values of x and y that give the value. Maximum: 90 at $(0, 18)$

6. Perform the following steps to maximize $f = 3x + 4y$ subject to the constraints

$$\begin{cases} x + 2y \leq 16 \\ x + y \leq 10 \\ x \geq 0, y \geq 0 \end{cases}$$

   a. Graph the region that satisfies the system of inequalities, and identify the corners of the region.

   b. Test the objective function $f = 3x + 4y$ at each of the corners of the feasible region to determine which corner gives the maximum value. Give the maximum possible value of f and the values of x and y that give that value. Maximum: 36 at $(4, 6)$

7. Perform the following steps to minimize $g = 4x + 2y$ subject to the constraints

$$\begin{cases} x + 2y \geq 15 \\ x + y \geq 10 \\ x \geq 0, y \geq 0 \end{cases}$$

   a. Graph the region that satisfies the system of inequalities, and identify the corners of the region.

   b. Test the objective function $g = 4x + 2y$ at each of the corners of the feasible region to determine which corner gives the minimum value. Give the minimum possible value of g and the values of x and y that give that value. Minimum: 20 at $(0, 10)$

8. Find the maximum possible value of $f = 400x + 300y$ and the values of x and y that give that value, subject to the constraints Maximum: 2100 at $(0, 7)$

$$\begin{cases} 3x + 2y \geq 6 \\ 2x + y \leq 7 \\ x \geq 0, y \geq 0 \end{cases}$$

9. Find the maximum possible value of $f = 20x + 30y$ and the values of x and y that give that value, subject to the constraints Maximum: 140 at $(4, 2)$

$$\begin{cases} y \leq -\dfrac{1}{2}x + 4 \\ y \leq -x + 6 \\ y \leq -\dfrac{1}{3}x + 4 \\ x \geq 0, y \geq 0 \end{cases}$$

10. Find the maximum possible value of $f = 100x + 100y$ and the values of x and y that give that value, subject to the constraints Maximum: 500 at $(4, 1)$

$$\begin{cases} x + 2y \leq 6 \\ x + 4y \leq 10 \\ 2x + y \leq 9 \\ x \geq 0, y \geq 0 \end{cases}$$

11. Find the maximum possible value of $f = 80x + 160y$ and the values of x and y that give that value, subject to the constraints Maximum: 1200 at all points on the line segment from $(2, 6.5)$ to $(0, 7.5)$

$$\begin{cases} 3x + 4y \leq 32 \\ x + 2y \leq 15 \\ 2x + y \leq 18 \\ x \geq 0, y \geq 0 \end{cases}$$

12. Find the minimum possible value of $g = 30x + 40y$ and the values of x and y that give that value, subject to the constraints Minimum: 360 at $(4, 6)$

$$\begin{cases} x + 2y \geq 16 \\ x + y \geq 10 \\ x \geq 0, y \geq 0 \end{cases}$$

13. Find the minimum possible value of $g = 40x + 30y$ and the values of x and y that give that value, subject to the constraints Minimum: 140 at $(2, 2)$

$$\begin{cases} 2x + y \geq 6 \\ 4x + y \geq 8 \\ x + 2y \geq 6 \\ x \geq 0, y \geq 0 \end{cases}$$

14. Find the minimum possible value of $g = 30x + 40y$ and the values of x and y that give that value, subject to the constraints Minimum: 130 at $(3, 1)$

$$\begin{cases} x + 2y \geq 5 \\ x + y \geq 4 \\ 2x + y \geq 6 \\ x \geq 0, y \geq 0 \end{cases}$$

15. Find the minimum possible value of $g = 46x + 23y$ and the values of x and y that give that value, subject to the constraints Minimum: 161 at $(3, 1)$

$$\begin{cases} 3x + y \geq 6 \\ x + y \geq 4 \\ x + 5y \leq 8 \\ x \geq 0, y \geq 0 \end{cases}$$

16. Consider the constraints

$$\begin{cases} 2x + y \le 9 \\ 3x + y \ge 11 \\ x + y \ge 5 \\ x \ge 0, y \ge 0 \end{cases}$$

   a. Find the maximum possible value of $f = 4x + 5y$ and the values of x and y that give that value. Maximum: 33 at (2, 5)
   b. Find the minimum possible value of $g = 3x + 2y$ and the values of x and y that give that value. Minimum: 13 at (3, 2)

17. Find the minimum possible value of $g = 60x + 10y$ and the values of x and y that give that value, subject to the constraints Minimum: 70 at (0, 7)

$$\begin{cases} 3x + 2y \ge 12 \\ 2x + y \ge 7 \\ x \ge 0, y \ge 0 \end{cases}$$

18. Find the maximum possible value of $f = 10x + 10y$ and the values of x and y that give that value, subject to the constraints Maximum: 60 at (4, 2) and (2, 4) or any point on the line segment connecting (4, 2) and (2, 4)

$$\begin{cases} x + 2y \le 10 \\ x + y \le 6 \\ 2x + y \le 10 \\ x \ge 0, y \ge 0 \end{cases}$$

EXERCISES 8.2

Answers that are not seen can be found in the answer section at the back of the text.

19. **Manufacturing** A company manufactures two types of leaf blowers: an electric Turbo model and a gas-powered Tornado model. The company's production plan calls for the production of at least 780 blowers per month. It costs $78 to produce each Turbo model and $117 to manufacture each Tornado model, and the company has at most $76,050 per month to use for production. Find the number of units that should be produced to maximize profit for the company, and the maximum profit, if the profit on each Turbo model is $32 and the profit on each Tornado model is $45. (See Exercise 27 in Section 8.1.) 975 Turbo and 0 Tornado; $31,200

20. **Manufacturing** Evergreen Company produces two types of printers: the Inkjet and the Laserjet. The company can make at most 120 printers per day and has 400 labor-hours available per day. It takes 2 hours to make the Inkjet and 6 hours to make the Laserjet. If the profit on the Inkjet is $80 and the profit on the Laserjet is $120, find the maximum possible daily profit and the number of each type of printer that gives it. (See Exercise 38 in Section 8.1.) $11,200; 80 Inkjet and 40 Laserjet

21. **Manufacturing** Safeco Company produces two types of chainsaws: the Safecut and the Safecut Deluxe. The Safecut model requires 2 hours to assemble and 1 hour to paint, and the Deluxe model requires 3 hours to assemble and $\frac{1}{2}$ hour to paint. The daily maximum number of hours available for assembly is 36, and the daily maximum number of hours available for painting is 12. If the profit is $24 per unit on the Safecut model and $30 per unit on the Deluxe model, how many units of each type will maximize the daily profit and what will that profit be? (See Exercise 40 in Section 8.1.) 9 Safecut and 6 Safecut Deluxe; $396

22. **Production** Two models of riding mowers, the Lawn King and the Lawn Master, are produced on two assembly lines. Producing the Lawn King requires 2 hours on line I and 1 hour on line II. Producing the Lawn Master requires 1 hour on line I and 3 hours on line II. The number of hours available for production is limited to 60 on line I and 40 on line II. If there is $150 profit per mower on the Lawn King and $200 profit per mower on the Lawn Master, producing how many of each model maximizes the profit? What is the maximum possible profit? 28 Lawn King and 4 Lawn Master; $5000

23. **Sales** Trix Auto Sales sells used cars. To promote its cars to target groups, it advertises with x 1-minute spots per month on cable television, at a cost of $240 per minute, and y 1-minute spots per month on radio, at a cost of $150 per minute. Research shows that it sells one vehicle for every 4 minutes of cable television advertising (that is, $\frac{1}{4}$ car per minute) and it sells one vehicle for each 10 minutes of radio advertising. Suppose the company has at most $36,000 per month to spend on advertising and sells at least 33 cars per month from this advertising. Graph the region that satisfies these two inequalities. That is, graph the solution to this system of inequalities. (See Exercise 34 in Section 8.1.)

a. If the profit on cars advertised on television averages $500 and the profit on cars advertised on radio averages $550, how many minutes per month of advertising should be spent on television advertising and how many should be spent on radio advertising? 100 minutes on TV and 80 minutes on radio

b. What is the maximum possible profit? $94,000

24. *Advertising* Tire Town is developing an advertising campaign. The table below indicates the cost per ad package in newspapers and the cost per ad package on radio, as well as the number of ads in each type of ad package.

	Newspaper	Radio
Cost per Ad Package ($)	1000	3000
Ads per Package	18	36

The owner of the company can spend no more than $18,000 per month, and he wants to have at least 252 ads per month. If each newspaper ad package reaches 6000 people over 20 years of age and each radio ad reaches 8000 of these people, how many newspaper ad packages and radio ad packages should he buy to maximize the number of people over age 20 reached? How many people will these ads reach? (See Exercise 36 in Section 8.1.)
18 newspaper and 0 radio; 108,000 people

25. *Production* A firm produces three different-size television sets on two assembly lines. The table below summarizes the production capacity of each assembly line, the number of each size TV ordered by a retailer, and the daily operating costs for each assembly line. How many days should each assembly line run to fill this order with minimum cost? What is the minimum cost? (See Exercise 35 in Section 8.1.)
Assembly line 1 for 50 days and assembly line 2 for 0 days; $1,000,000

	Assembly Line 1	Assembly Line 2	Number Ordered
19-in. TV	80 per day	40 per day	3200
27-in. TV	20 per day	20 per day	1000
35-in. TV	100 per day	40 per day	3400
Daily Cost ($)	20,000	40,000	

26. *Nutrition* A privately owned lake contains two types of fish: bass and trout. The owner provides two types of food, A and B, for these fish. Trout require 4 units of food A and 5 units of food B, and bass require 10 units of food A and 4 units of food B. If the owner has 1600 units of food A and 1000 units of food B, find the maximum number of fish that the lake can support. 223 fish

27. *Housing* A contractor builds two models of homes: the Van Buren and the Jefferson. The Van Buren requires 200 worker-days of labor and $240,000 in capital, and the Jefferson requires 500 worker-days of labor and $300,000 in capital. The contractor has a total of 5000 worker-days and $3,600,000 in capital available per month. The profit is $60,000 on the Van Buren and $75,000 on the Jefferson. Building how many of each model will maximize the monthly profit? What is the maximum possible profit? (See Exercise 32 in Section 8.1.)

28. *Manufacturing* Easyboy manufactures two types of chairs: Standard and Deluxe. Each Standard chair requires 4 hours to construct and finish, and each Deluxe chair requires 6 hours to construct and finish. Upholstering takes 2 hours for a Standard chair and 6 hours for a Deluxe chair. There are 480 hours available each day for construction and finishing, and there are 300 hours available per day for upholstering. Suppose the revenue is $178 for each Standard chair and $267 for each Deluxe chair. (See Exercise 39 in Section 8.1.)

a. What is the maximum possible daily revenue? $21,360

b. How many of each type should be produced each day to maximize the daily revenue?

29. *Production* Ace Company produces three models of DVRs, the Ace, the Ace Plus, and the Ace Deluxe, at two facilities, A and B. The company has orders for at least 4000 of the Ace, at least 1800 of the Ace Plus, and at least 2400 of the Ace Deluxe. The weekly production capacity of each model at each facility and the cost per week to operate each facility are given in the table that follows. For how many weeks should each facility operate to minimize the cost of filling the orders? What is the minimum cost?
Facility A, 8 weeks; facility B, 2 weeks; $160,000

	A	B
Ace	400	400
Ace Plus	300	100
Ace Deluxe	200	400
Cost per Week ($)	15,000	20,000

30. *Politics* A political candidate wishes to use a combination of television and radio advertisements in his campaign. Each 1-minute ad on television reaches 0.12 million eligible voters and each 1-minute ad on radio reaches 0.009 million eligible voters. The candidate believes that he must reach at least 7.56 million eligible voters and that he must buy at least 100 minutes of advertisements. If television ads cost $1000 per minute and radio ads cost $200 per minute, how many minutes of radio and television advertising does he need to minimize costs? What is the minimum cost? (See Exercise 31 in Section 8.1.)
40 minutes of radio, 60 minutes of television; $68,000

31. **Nutrition** In a hospital ward, patients are grouped into two general nutritional categories depending on the amount of solid food in their diet, and are provided two different diets with different amounts of solid foods and detrimental substances. The following table gives the patient groups, the weekly diet requirements for each group, and the amount of detrimental substance in each diet. How many servings of each diet will satisfy the nutritional requirements and minimize detrimental substances? What is the minimum amount of detrimental substance?
4 servings of diet A and 10 servings of diet B; 0.71 oz

	Diet A (oz/serving)	Diet B (oz/serving)	Minimum Daily Requirements (oz)
Group 1	2	1	18
Group 2	4	1	26
Detrimental Substance	0.09	0.035	

32. **Manufacturing** Kitchen Pride manufactures toasters and can openers, which are assembled on two different lines. Line 1 can assemble 15 toasters and 20 can openers per hour, and line 2 can assemble 75 toasters and 20 can openers per hour. The company needs to produce at least 540 toasters and 400 can openers. If it costs $300 per hour to run line 1 and $600 per hour to run line 2, how many hours should each line be run to fill all the orders at the minimum cost? What is the minimum cost? Assembly line 1 for 16 hours and assembly line 2 for 4 hours; $7200

8.3 Sequences and Discrete Functions

KEY OBJECTIVES
- Find terms of sequences
- Find terms of arithmetic sequences
- Find terms of geometric sequences

SECTION PREVIEW Football Contracts

Suppose a football player is offered a chance to sign a contract for 18 games with one of the following salary plans:

Plan A: $10,000 for the first game with a $10,000 raise for each game thereafter

Plan B: $2 for the first game with his salary doubled for each game thereafter

Which salary plan should he accept if he wants to make the most money for his last game?
   The payments for the games form the sequences

Plan A: 10,000, 20,000, 30,000, . . .

Plan B: 2, 4, 8, 16, . . .

In this section, we answer this question after we have developed formulas that apply to sequences, and we use sequences to solve applied problems. ■

Sequences

In Chapter 5 we found the future values of investments at simple and compound interest. Consider a $5000 investment that pays 1% simple interest for each of 6 months, shown in Table 8.2.
   These future values are outputs that result when the inputs are positive integers that correspond to the number of months of the investment. Outputs (such as these future values) that result uniquely from the positive integer inputs define a function whose domain is a set of positive integers. Such a function is called a **sequence**, and the ordered outputs corresponding to the integer inputs are called the **terms of the sequence**.

Table 8.2

Month	Interest, $I = Prt$ ($)	Future Value of the Investment ($)
1	$(5000)(0.01)(1) = 50$	$5000 + 50 = 5050$
2	$(5000)(0.01)(1) = 50$	$5050 + 50 = 5100$
3	$(5000)(0.01)(1) = 50$	$5100 + 50 = 5150$
4	$(5000)(0.01)(1) = 50$	$5150 + 50 = 5200$
5	$(5000)(0.01)(1) = 50$	$5200 + 50 = 5250$
6	$(5000)(0.01)(1) = 50$	$5250 + 50 = 5300$

Sequences have the same properties as other functions that we have studied, *except* that the domains of sequences are positive integers, and so sequences are **discrete functions**. Rather than denoting a functional output with y, we denote the output of the sequence f that corresponds to input n with $f(n) = a_n$, where n is a positive integer.

Sequence

The functional values $a_1, a_2, a_3, \ldots$ of a sequence are called the terms of the sequence, with a_1 the first term, a_2 the second term, and so on. The general term (or nth term) is denoted by a_n.

If the domain of a sequence is the set of all positive integers, the outputs form an **infinite sequence**. If the domain is the set of positive integers from 1 to n, the outputs form a **finite sequence**. Sequences are important because they permit us to apply discrete functions to real problems instead of having to approximate them with continuous functions.

By looking at the future values in Table 8.2, we can see that the future value of $5000 at the end of each month is given by

$$a_1 = f(1) = 5000 + 50(1) = 5050$$
$$a_2 = f(2) = 5000 + 50(2) = 5100$$
$$a_3 = f(3) = 5000 + 50(3) = 5150$$
$$a_4 = f(4) = 5000 + 50(4) = 5200$$
$$a_5 = f(5) = 5000 + 50(5) = 5250$$
$$a_6 = f(6) = 5000 + 50(6) = 5300$$

These six terms can be written in the form 5050, 5100, 5150, 5200, 5250, 5300. By observing this pattern, we write the general term as $a_n = 5000 + 50n$, which represents

$$f(n) = 5000 + 50n$$

As with other functions, we can represent sequences in tables and on graphs. Figure 8.22 shows the graph of the function for these values of n.

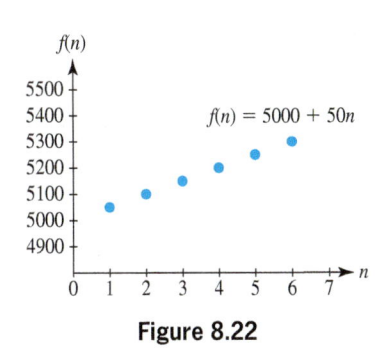

Figure 8.22

EXAMPLE 1 ▶ Finding Terms of Sequences

Find the first four terms of the sequence (for $n = 1, 2, 3, 4$) defined by

a. $a_n = \dfrac{8}{n}$

b. $b_n = (-1)^n n(n+1)$

SOLUTION

a. $a_1 = \dfrac{8}{1} = 8$, $a_2 = \dfrac{8}{2} = 4$, $a_3 = \dfrac{8}{3}$, $a_4 = \dfrac{8}{4} = 2$

We write these terms in the form $8, 4, \dfrac{8}{3}, 2$.

b. $b_1 = (-1)^1(1)(1+1) = -2$, $b_2 = (-1)^2(2)(2+1) = 6$,
$b_3 = (-1)^3(3)(3+1) = -12$, $b_4 = (-1)^4(4)(4+1) = 20$

Notice that $(-1)^n$ causes the signs of the terms to alternate. The terms are $-2, 6, -12, 20$.

Technology Note

We can find n terms of a sequence with a calculator if we set the calculator to sequence mode. Figure 8.23(a) shows the first four terms defined in Example 1(a), in decimal and fractional form. The graph of the sequence and the table of values for the sequence are shown in Figures 8.23(b) and 8.23(c), respectively.*

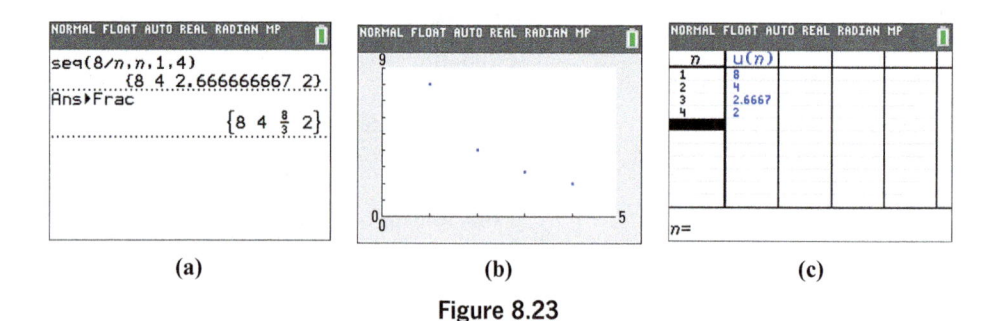

(a)   (b)   (c)

Figure 8.23

EXAMPLE 2 ▶ Depreciation

For tax purposes, a firm depreciates its $900,000 building over 30 years by the straight-line method, which depreciates the value of the building by $\dfrac{900{,}000}{30} = 30{,}000$ dollars each year. Write a sequence that gives the value of the building at the end of each of the first 5 years.

SOLUTION

The value would be reduced each year by 30,000 dollars, so the first five terms of the sequence are

$$870{,}000,\ 840{,}000,\ 810{,}000,\ 780{,}000,\ 750{,}000$$

Arithmetic Sequences

The future values of the investment described in Table 8.2 are the first six terms of the sequence

$$5050,\ 5100,\ 5150,\ 5200,\ 5250,\ 5300,\ \ldots$$

---

*For more details, see Appendix A, page 699.

We can define this sequence **recursively**, with each term after the first defined from the previous term, as

$$a_1 = 5050, a_n = a_{n-1} + 50, \quad \text{for } n > 1$$

This sequence is an example of a special sequence called an **arithmetic sequence**. In such a sequence, each term after the first term is found by adding a constant to the preceding term. Thus, we have the following definition.

Arithmetic Sequence

A sequence is called an arithmetic sequence if there exists a number d, called the **common difference**, such that

$$a_n = a_{n-1} + d, \quad \text{for } n > 1$$

EXAMPLE 3 ▶ Arithmetic Sequences

Write the next three terms of the arithmetic sequences:

a. 1, 4, 7, 10, . . . **b.** 11, 9, 7, . . . **c.** $\frac{1}{2}, \frac{2}{3}, \frac{5}{6}, \ldots$

SOLUTION

a. The common difference that gives each term from the previous one is 3, so the next three terms are 13, 16, 19.

b. The common difference is -2, so the next three terms are 5, 3, 1.

c. The common difference is $\frac{1}{6}$, so the next three terms are $1, \frac{7}{6}, \frac{4}{3}$.

Note that the differences of terms of an arithmetic sequence are constant, so the function defining the sequence is linear, with its rate of change equal to the common difference. That is, an arithmetic sequence is really a linear function whose domain is restricted to a subset of the positive integers. Because each term after the first in an arithmetic sequence is obtained by adding d to the preceding term, the second term is $a_1 + d$, the third term is $(a_1 + d) + d = a_1 + 2d$, the fourth term is $a_1 + 3d, \ldots$, and the nth term is $a_1 + (n - 1)d$. Thus, we have a formula for the nth term of an arithmetic sequence.

nth Term of an Arithmetic Sequence

The nth term of an arithmetic sequence is given by

$$a_n = a_1 + (n - 1)d$$

where a_1 is the first term of the sequence, n is the number of the term, and d is the common difference between consecutive terms.

Thus, the 25th term of the arithmetic sequence 1, 4, 7, 10, . . . (see Example 3) is

$$a_{25} = 1 + (25 - 1)3 = 73$$

EXAMPLE 4 ▶ Depreciation

For tax purposes, a firm depreciates its $900,000 building over 30 years by the straight-line method, which depreciates the value of the building by $\frac{900,000}{30} = 30,000$ dollars each year. What is the value of the building after 12 years?

SOLUTION

The description indicates that the value of the building is 900,000 dollars at the beginning, that its value is $900,000 - 30,000 = 870,000$ dollars after 1 year, and that it decreases by 30,000 dollars each year for 29 additional years. Because the value of the building is reduced by the same amount each year, after n years the value is given by the arithmetic sequence with first term 870,000, common difference $-30,000$, and nth term

$$a_n = 870,000 + (n - 1)(-30,000), \quad \text{for } n = 1, 2, 3, \ldots, 30$$

Thus, the value of the building at the end of 12 years is

$$a_{12} = 870,000 + (12 - 1)(-30,000) = 540,000 \text{ dollars}$$

EXAMPLE 5 ▶ Football Contract Plan A

Recall from the Section Preview that the first payment plan offered to the football player was

$$\text{Plan A: } 10,000, 20,000, 30,000, \ldots$$

a. Do the payments form an arithmetic sequence?

b. What is the 18th payment under this payment plan?

c. If the contract is extended into the postseason, what is the 20th payment under this payment plan?

SOLUTION

a. The payments form a finite sequence with 18 terms. The common difference between terms is 10,000, so the sequence is an arithmetic sequence.

b. Because this is an arithmetic sequence with first term 10,000 and common difference $d = 10,000$, the 18th payment is

$$a_{18} = 10,000 + (18 - 1)10,000 = 180,000 \text{ dollars}$$

c. The 20th payment is

$$a_{20} = 10,000 + (20 - 1)10,000 = 200,000 \text{ dollars}$$

Geometric Sequences

Notice in the Section Preview that Plan B pays the football player $2, $4, $8, ... It is not an arithmetic sequence, because there is no constant difference between the terms. Each successive term of this sequence is doubled (that is, multiplied by 2). This is an example of another special sequence, where each term is found by multiplying the previous term by the same number. This sequence is called a **geometric sequence**.

Geometric Sequence

A sequence is called a geometric sequence if there exists a number r, called the **common ratio**, such that

$$a_n = ra_{n-1}, \quad \text{for } n > 1$$

EXAMPLE 6 ▶ Geometric Sequences

Write the next three terms of the geometric sequences:

a. 1, 3, 9, ...   **b.** 64, 16, 4, ...   **c.** 2, −4, 8, ...

SOLUTION

a. The common ratio that gives each term from the previous one is 3, so the next three terms are 27, 81, 243.

b. The common ratio is $\frac{1}{4}$, so the next three terms are $1, \frac{1}{4}, \frac{1}{16}$.

c. The common ratio is −2, so the next three terms are −16, 32, −64.

Note that there is a **constant percent change** in the terms of a geometric sequence, so the function defining the sequence is exponential, and a geometric sequence is really an exponential function with its domain restricted to the positive integers. Because each term after the first in a geometric sequence is obtained by multiplying r times the preceding term, the second term is $a_1 r$, the third term is $a_1 r \cdot r = a_1 r^2$, the fourth term is $a_1 r^3, \ldots$, and the nth term is $a_1 r^{n-1}$. Thus, we have a formula for the nth term of a geometric sequence.

> ### nth Term of a Geometric Sequence
> The nth term of a geometric sequence is given by
> $$a_n = a_1 r^{n-1}$$
> where a_1 is the first term of the sequence, n is the number of the term, and r is the common ratio of consecutive terms.

Thus, the 25th term of the geometric sequence 2, −4, 8, ... (see Example 6) is

$$a_{25} = 2(-2)^{25-1} = 33{,}554{,}432$$

EXAMPLE 7 ▶ Rebounding

A ball is dropped from a height of 100 feet and rebounds $\frac{2}{5}$ of the height from which it falls every time it hits the ground. How high will it bounce after it hits the ground the fourth time?

SOLUTION

The first rebound is $\frac{2}{5}$ of $100 = 40$ feet, the second is $\frac{2}{5}$ of $40 = 16$ feet, and the rebounds form a geometric sequence with first term 40 and common ratio $\frac{2}{5}$.

Thus, the fourth rebound is

$$40\left(\frac{2}{5}\right)^{4-1} = 2.56 \text{ feet}$$

We found the future value of money invested at compound interest in Chapter 5 by using exponential functions. Because geometric sequences are really exponential functions with domains restricted to the positive integers, the future value of an investment with interest compounded over a number of discrete periods can be found using a geometric sequence.

EXAMPLE 8 ▶ Compound Interest

The future value of $1000 invested for 3 years at 6%, compounded annually, can be found using the simple interest formula $S = P + Prt$, as follows.

$$\begin{aligned}
\text{Year 1:} \quad S &= 1000 + 1000(0.06)(1) \\
&= 1000(1 + 0.06) \\
&= 1000(1.06) \text{ dollars} \\
\text{Year 2:} \quad S &= 1000(1.06) + [1000(1.06)](0.06)(1) \\
&= 1000(1.06)(1 + 0.06) \\
&= 1000(1.06)^2 \text{ dollars} \\
\text{Year 3:} \quad S &= 1000(1.06)^2 + [1000(1.06)^2](0.06)(1) \\
&= 1000(1.06)^2(1 + 0.06) \\
&= 1000(1.06)^3 \text{ dollars}
\end{aligned}$$

a. Do the future values of this investment form a geometric sequence?

b. What is the value of this investment in 25 years?

SOLUTION

a. The future values of the investment form a sequence with first term $1000(1.06) = 1060$ and common ratio 1.06, so the sequence is a geometric sequence.

b. The future value of the investment is the 25th term of the geometric sequence:

$$1060(1.06)^{25-1} = 4291.87 \text{ dollars}$$

EXAMPLE 9 ▶ Football Contract Plan B

Recall from the Section Preview that the second payment plan offered to the football player was

$$\text{Plan B: } 2, 4, 8, \ldots.$$

a. Do the payments form a geometric sequence?

b. What is the 18th payment under this payment plan?

SOLUTION

a. The payments form a finite sequence with 18 terms. The common ratio between terms is 2 because each payment is doubled. Thus, the sequence is a geometric sequence.

b. Because this is a geometric sequence with first term 2 and common ratio $r = 2$, the 18th payment is

$$a_{18} = 2 \cdot 2^{18-1} = 262{,}144 \text{ dollars}$$

Thus, the 18th payment ($262,144) from Plan B is significantly larger than the 18th payment ($180,000) from Plan A.

But does the fact that the 18th payment from Plan B is much larger than that from Plan A mean that the total payment from Plan B is larger than that from Plan A? We answer this question in the next section.

SKILLS CHECK 8.3

1. Find the first 6 terms of the sequence defined by $f(n) = 2n + 3$. 5, 7, 9, 11, 13, 15

2. Find the first 4 terms of the sequence defined by $f(n) = \dfrac{1}{2n} + n$. $\dfrac{3}{2}, \dfrac{9}{4}, \dfrac{19}{6}, \dfrac{33}{8}$

3. Find the first 5 terms of the sequence defined by $a_n = \dfrac{10}{n}$. $10, 5, \dfrac{10}{3}, \dfrac{5}{2}, 2$

4. Find the first 5 terms of the sequence defined by $a_n = (-1)^n(2n)$. −2, 4, −6, 8, −10

5. Find the next 3 terms of the arithmetic sequence 1, 3, 5, 7, ... 9, 11, 13

6. Find the next 3 terms of the arithmetic sequence 2, 5, 8, ... 11, 14, 17

7. Find the eighth term of the arithmetic sequence with first term −3 and common difference 4. 25

8. Find the 40th term of the arithmetic sequence with first term 5 and common difference 15. 590

9. Write 4 additional terms of the geometric sequence 3, 6, 12, ... 24, 48, 96, 192

10. Write 4 additional terms of the geometric sequence 8, 20, 50, ... 125, 312.5, 781.25, 1953.125

11. Find the sixth term of the geometric sequence with first term 10 and common ratio 3. 2430

12. Find the tenth term of the geometric sequence with first term 48 and common ratio $-\dfrac{1}{2}$. $\dfrac{-3}{32}$

13. Find the first 4 terms of the sequence with first term 5 and nth term $a_n = a_{n-1} - 2$. 5, 3, 1, −1

14. Find the first 6 terms of the sequence with first term 8 and nth term $a_n = a_{n-1} + 3$. 8, 11, 14, 17, 20, 23

15. Find the first 4 terms of the sequence with first term 2 and nth term $a_n = 2a_{n-1} + 3$. 2, 7, 17, 37

16. Find the first 5 terms of the sequence with first term 26 and nth term $a_n = \dfrac{a_{n-1} + 4}{2}$. 26, 15, 9.5, 6.75, 5.375

EXERCISES 8.3

Answers that are not seen can be found in the answer section at the back of the text.

17. **Salaries** Suppose you are offered a job with a relatively low starting salary but with a $1500 raise for each of the next 7 years. How much more than your starting salary would you be making in the eighth year? $10,500

18. **Depreciation** A new car costing $35,000 is purchased for business and is depreciated with the straight-line depreciation method over a 5-year period, which means that it is depreciated by the same amount each year. Write a sequence that gives the value after depreciation for each of the 5 years. $28,000, $21,000, $14,000, $7000, $0

19. **Landscaping** Grading equipment used for landscaping costs $300 plus $60 per hour or part of an hour thereafter.

   a. Write an expression for the cost of a job lasting n hours. $300 + 60n$

   b. If the answer to part (a) is the nth term of a sequence, write the first 6 terms of this sequence. 360, 420, 480, 540, 600, 660

20. **Profit** A new firm loses $2000 in its first month, but its profit increases by $400 in each succeeding month for the rest of the year. What is its profit in the 12th month? $2400

21. **Salaries** Suppose you are offered two identical jobs, one paying a starting salary of $40,000 with yearly raises of $2000 and a second one paying a starting salary of $36,000 with yearly raises of $2400.

   a. Which job will be paying more in 6 years? By how much? First by $2000

   b. Which job will be paying more in 11 years? By how much? They are the same.

   c. How much more will the second job pay in the 13th year? $800

22. **Phone Calls** Suppose a long-distance call costs 99¢ for the first minute plus 25¢ for each additional minute.

   a. Write the cost of a call lasting n minutes. $0.25n + 0.74$

   b. If the answer to part (a) is the nth term of a sequence, write the first 6 terms of this sequence. 0.99, 1.24, 1.49, 1.74, 1.99, 2.24

23. **Interest** If $1000 is invested at 5% interest, compounded annually, write a sequence that gives the amount in the account at the end of each of the first 4 years. $1050, $1102.50, $1157.63, $1215.51

24. **Interest** If $10,000 is invested at 6% interest, compounded annually, write a sequence that gives the amount in the account at the end of each of the first 3 years. $10,600, $11,236, $11,910.16

25. **Depreciation** A new car costing $50,000 depreciates by 20% of its original value each year.
    a. What is the value of the car at the end of the third year? $20,000
    b. Write an expression that gives the value at the end of the nth year. $50,000 - 10,000n$
    c. Write the first 5 terms of the sequence of the values after depreciation. $40,000, $30,000, $20,000, $10,000, $0

26. **Salaries** If you accept a job paying $32,000 for the first year, with a guaranteed 8% raise each year, how much will you earn in your fifth year? $43,535.65

27. **Bacteria** The size of a certain bacteria culture doubles each hour. If the number of bacteria present initially is 5000, how many will be present at the end of 6 hours? 320,000

28. **Bacteria** If a bacteria culture increases by 20% every hour and 2000 are present initially, how many will be present at the end of 10 hours? 12,383

29. **Animal Fossils** In the 1980s, a number of spherical fossils that were segmented on their surfaces were found in China. Some scientists believe these fossils were embryos of early ancestors of the first animals on earth, because they had a key characteristic of animal embryos: They "remained the same size while undergoing successive rounds of cell division." That is, different fossils consisted of different numbers of cells: one, two, four, eight, and so on.
    (*Source*: *Discover*, December 2012)
    a. What type of sequence do the numbers of cells in these fossils follow? What is the common ratio? geometric; 2
    b. Some of the fossils had more than 1000 cells. What is the smallest number in the sequence from part (a) that is larger than 1000? $2^{10} = 1024$
    c. How many cell divisions are necessary to get more than 500 cells? 9

30. **Bouncing Ball** A ball is dropped from 64 feet. If it rebounds $\frac{3}{4}$ of the height from which it falls every time it hits the ground, how high will it bounce after it hits the ground for the fourth time? $20\frac{1}{4}$ feet

31. **Profit** If changing market conditions cause a company earning a profit of $8,000,000 this year to project decreases of 2% of its profit in each of the next 5 years, what profit does it project 5 years from now? $7,231,366.37

32. **Pumps** A pump removes $\frac{1}{3}$ of the water in a container with every stroke. What amount of water is removed on the fifth stroke if the container originally had 81 cm^3 of water? $\frac{16}{3}$ cm^3

33. **Salaries** If you begin a job making $54,000 and are given a $3600 raise each year, use numerical methods to find in how many years your salary will double. 15 years

34. **Interest** If $2500 is invested at 8% interest, compounded annually, how long will it take the account to reach $5829.10? 11 years

35. **Compound Interest** Find the future value of $10,000 invested for 10 years at 8%, compounded daily. $22,253.46

36. **Compound Interest** Find the future value of $10,000 invested for 10 years at 8%, compounded annually. $21,589.25

37. **Rabbit Breeding** The number of pairs of rabbits in the population during the first 6 months can be written in the Fibonacci sequence

    $$1, 1, 2, 3, 5, 8, \ldots$$

    in which each term after the second is the sum of the two previous terms. Find the number of pairs of rabbits for the next 4 months. 13, 21, 34, 55

38. **Bee Ancestry** A female bee hatches from a fertilized egg, whereas a male bee hatches from an unfertilized egg. Thus, a female bee has a male parent and a female parent, whereas a male bee has only a female parent. Therefore, the number of ancestors of a male bee follow the Fibonacci sequence

    $$1, 2, 3, 5, 8, 13, \ldots$$

    a. Observe the pattern and write three more terms of the sequence. 21, 34, 55
    b. What do the 1, the 2, and the 3, respectively, represent for a given male bee? Mother, grandparents, great-grandparents

39. Credit Card Debt Kirsten has a $10,000 credit card debt. Each month she pays 1% interest plus a payment of 10% of the monthly balance. If she makes no new purchases, write a sequence that gives the payments for the first 4 months. $1100, $990, $891, $801.90

40. Bouncing Ball A ball is dropped from 128 feet. It rebounds $\frac{1}{4}$ of the height from which it falls every time it hits the ground. How high will it bounce after it hits the ground for the fourth time? $\frac{1}{2}$ foot

8.4 Series

KEY OBJECTIVES

- Find the sum of a finite series
- Evaluate sums written in sigma notation
- Find the sum of the first n terms of an arithmetic sequence
- Find the sum of the first n terms of a geometric sequence
- Find the sum of an infinite geometric series

SECTION PREVIEW Football Contracts

In Section 8.3, we discussed a football player's opportunity to sign a contract for 18 games for either

Plan A: $10,000 for the first game with a $10,000 raise for each game thereafter

Plan B: $2 for the first game with his salary doubled for each game thereafter

Which salary plan should he accept if he wants to make the most money?
The payments for the games form the sequences:

Plan A: 10,000, 20,000, 30,000, . . .

Plan B: 2, 4, 8, 16, . . .

We found that Plan B paid more than Plan A for the 18th game, but we did not determine which plan pays more for all 18 games. To answer this question, we need to find the *sum* of the first 18 terms of these sequences. The sum of the terms of a sequence is called a series. Series are useful in computing finite and infinite sums and in finding depreciation and future values of annuities. ∎

Finite and Infinite Series

Suppose a firm loses $2000 in its first month of operation, but its profit increases by $500 in each succeeding month for the remainder of its first year. What is its profit for the year? The description indicates that the profit is $-$$2000 for the first month and that it increases by $500 each month for 11 months. The monthly profits form the sequence that begins

$$-2000, -1500, -1000, \ldots$$

The profit for the 12 months is the sum of these terms:

$$(-2000) + (-1500) + (-1000) + (-500) + 0 + 500 + 1000 + 1500 + 2000$$
$$+ 2500 + 3000 + 3500 = 9000$$

The total profit for the year is this sum, $9000.

The sum of the 12 terms of this sequence can be referred to as a **finite series**. In general, a **series** is defined as the sum of the terms of a sequence. If there are infinitely many terms, the series is an **infinite series**.

> **Series**
> A finite series is defined by
> $$a_1 + a_2 + a_3 + \cdots + a_n$$
> where $a_1, a_2, \ldots a_n$ are terms of a sequence.
> An infinite series is defined by
> $$a_1 + a_2 + a_3 + \cdots + a_n + \cdots$$

We can use the Greek letter Σ (sigma) to express the sum of numbers or expressions. For example, we can write

$$\sum_{i=1}^{3} i \quad \text{to denote} \quad 1 + 2 + 3$$

and we can write the general finite series in the form

$$\sum_{i=1}^{n} a_i = a_1 + a_2 + a_3 + \cdots + a_n$$

The symbol $\sum_{i=1}^{n} a_i$ may be read as "The sum of a_i as i goes from 1 to n one unit at a time."

The general infinite series can be written in the form

$$\sum_{i=1}^{\infty} a_i = a_1 + a_2 + a_3 + \cdots + a_n + \cdots$$

where ∞ indicates that there is no end to the number of terms.

EXAMPLE 1 ▶ Finite Series

Find the sum of the first six terms of the sequence $6, 3, \dfrac{3}{2}, \ldots$.

SOLUTION

Each succeeding term of this sequence is found by multiplying the preceding term by $\dfrac{1}{2}$, so the sequence is

$$6, 3, \frac{3}{2}, \frac{3}{4}, \frac{3}{8}, \frac{3}{16}, \ldots$$

and the sum of the first six terms is $6 + 3 + \dfrac{3}{2} + \dfrac{3}{4} + \dfrac{3}{8} + \dfrac{3}{16} = \dfrac{189}{16}$.

Arithmetic Series

The sum of the payments to the football player in the Section Preview for the first three games under Plan A is

$$10{,}000 + 20{,}000 + 30{,}000 = 60{,}000 \text{ dollars}$$

and the sum of the payments for the first three games under Plan B is

$$2 + 4 + 8 = 14 \text{ dollars}$$

so the player would be well advised to take Plan A if there were only three games. To find the sum of the payments for 18 games and to answer other questions, it would be more efficient to have formulas to use in finding the sums of terms of sequences. We begin by finding a formula for the sum of n terms of an arithmetic sequence.

Given the first term a_1 in an arithmetic sequence with common difference d, we use s_n to represent the sum of the first n terms and write it as follows:

$$s_n = a_1 + (a_1 + d) + (a_1 + 2d) + \cdots + (a_n - 2d) + (a_n - d) + a_n$$

Writing this sum in reverse order gives

$$s_n = a_n + (a_n - d) + (a_n - 2d) + \cdots + (a_1 + 2d) + (a_1 + d) + a_1$$

Adding these two equations term by term gives

$$s_n + s_n = (a_1 + a_n) + (a_1 + a_n) + (a_1 + a_n) + \cdots + (a_1 + a_n)$$
$$+ (a_1 + a_n) + (a_1 + a_n)$$

so $2s_n = n(a_1 + a_n)$ because there are n terms, and we have $s_n = \dfrac{n(a_1 + a_n)}{2}$.

Sum of n Terms of an Arithmetic Sequence

The sum of the first n terms of an arithmetic sequence is given by

$$s_n = \frac{n(a_1 + a_n)}{2}$$

where a_1 is the first term of the sequence and a_n is the nth term.

Recall that Plan A of the football player's contract began with $a_1 = 10{,}000$ and had 18th payment $a_{18} = 180{,}000$. Thus, the sum of the 18 payments is

$$s_{18} = \frac{18(10{,}000 + 180{,}000)}{2} = 1{,}710{,}000 \text{ dollars}$$

EXAMPLE 2 ▶ Depreciation

An automobile valued at $24,000 is depreciated over 5 years with the *sum-of-the-years'-digits* depreciation method. Under this method, annual depreciation is found by multiplying the value of the property by a fraction whose denominator is the sum of the 5 years' digits and whose numerator is 5 for the first year's depreciation, 4 for the second year's, 3 for the third, and so on.

a. Write a sequence that represents the parts of the automobile value that are depreciated in each year, and show that the sum of the terms is 100%.

b. Find the yearly depreciation and show that the sum of the depreciations is $24,000.

SOLUTION

a. The sum of the 5 years' digits is $1 + 2 + 3 + 4 + 5 = 15$, so the parts of the value that are depreciated for each of the 5 years are the numbers

$$\frac{5}{15}, \frac{4}{15}, \frac{3}{15}, \frac{2}{15}, \frac{1}{15}$$

This is an arithmetic sequence with first term $\dfrac{5}{15}$, common difference $-\dfrac{1}{15}$, and fifth term $\dfrac{1}{15}$. Thus, the sum of the terms is

$$s_5 = \frac{5\left(\dfrac{5}{15} + \dfrac{1}{15}\right)}{2} = \frac{5}{2} \cdot \frac{6}{15} = 1 = 100\%$$

b. The respective annual depreciations for the automobile are

$$\frac{5}{15} \cdot 24{,}000 = 8000 \qquad \frac{4}{15} \cdot 24{,}000 = 6400 \qquad \frac{3}{15} \cdot 24{,}000 = 4800$$

$$\frac{2}{15} \cdot 24{,}000 = 3200 \qquad \frac{1}{15} \cdot 24{,}000 = 1600$$

The depreciations form an arithmetic sequence with difference -1600, first term 8000, and fifth term 1600, so the sum of these depreciations is

$$s_n = \frac{5(8000 + 1600)}{2} = 24{,}000 \text{ dollars}$$

Note that the sum is the original value of the automobile, so it is totally depreciated in 5 years.

Geometric Series

In Section 5.6, we found the future value of the annuity with payments of $1000 at the end of each of 5 years, with interest at 10%, compounded annually, by finding the sum

$$S = 1000 + 1000(1.10)^1 + 1000(1.10)^2 + 1000(1.10)^3 + 1000(1.10)^4$$
$$= 6105.10 \text{ dollars}$$

We can see that this is a finite geometric series with first term 1000 and common ratio 1.10. We can develop a formula to find the sum of this and any other finite geometric series.

The sum of the first n terms of a geometric sequence with first term a_1 and common ratio r is

$$s_n = a_1 + a_1 r + a_1 r^2 + \cdots + a_1 r^{n-1}$$

To find a formula for this sum, we multiply it by r to get a new expression and then find the difference of the two expressions.

$$s_n = a_1 + a_1 r + a_1 r^2 + a_1 r^3 + \cdots + a_1 r^{n-1}$$
$$r s_n = a_1 r + a_1 r^2 + a_1 r^3 + \cdots + a_1 r^{n-1} + a_1 r^n$$
$$s_n - r s_n = a_1 - a_1 r^n$$

Rewriting this difference and solving for s_n gives a formula for the sum.

$$s_n(1 - r) = a_1(1 - r^n)$$
$$s_n = \frac{a_1(1 - r^n)}{(1 - r)}$$

> **Sum of n Terms of a Geometric Sequence**
>
> The sum of the first n terms of a geometric sequence is
>
> $$s_n = \frac{a_1(1 - r^n)}{1 - r}, \quad r \neq 1$$
>
> where a_1 is the first term of the sequence and r is the common ratio.

EXAMPLE 3 ▶ Sums of Terms of Geometric Sequences

a. Find the sum of the first 10 terms of the geometric sequence with first term 4 and common ratio -3.

b. Find the sum of the first 8 terms of the sequence $2, 1, \frac{1}{2}, \ldots$.

SOLUTION

a. The sum of the first 10 terms of the geometric sequence with first term 4 and common ratio -3 is given by

$$s_{10} = \frac{a_1(1-r^{10})}{1-r} = \frac{4(1-(-3)^{10})}{1-(-3)} = -59{,}048$$

b. This sequence is geometric, with first term 2 and common ratio $\frac{1}{2}$, so the sum of the first 8 terms is

$$s_8 = \frac{a_1(1-r^8)}{1-r} = \frac{2\left(1-\left(\frac{1}{2}\right)^8\right)}{1-\left(\frac{1}{2}\right)} = 4\left(1-\frac{1}{256}\right) = \frac{255}{64}$$

EXAMPLE 4 ▶ Annuities

Show that if regular payments of $1000 are made at the end of each year for 5 years into an account that pays interest at 10% per year, compounded annually, the future value of this annuity is $6105.10.

SOLUTION

As we saw on the previous page, the future values of the individual payments into this annuity form a geometric sequence with first term 1000 and common ratio 1.10, so the sum of the first five terms is

$$s_5 = \frac{1000(1-1.10^5)}{1-1.10} = 6105.10$$

We found that the total payment to the football player for 18 games under Plan A is $1,710,000. The payments under Plan B form a geometric sequence with first term 2 and common ratio 2, so the sum of the payments under Plan B is

$$s_{18} = \frac{2(1-2^{18})}{1-2} = 524{,}286$$

Thus, total earnings for the 18 games for Plan A are more than three times those for Plan B. However, if the season has 20 games, the sum of the payments for Plan A is

$$s_{20} = \frac{20(10{,}000 + 200{,}000)}{2} = 2{,}100{,}000 \text{ dollars}$$

and the sum of the payments for Plan B is

$$s_{20} = \frac{2(1-2^{20})}{1-2} = 2{,}097{,}150 \text{ dollars}$$

In this case, the plans are nearly equal.

Infinite Geometric Series

If we add the terms of an infinite geometric sequence, we have an **infinite geometric series**. It is perhaps surprising that we can add an infinite number of positive numbers and get a finite sum. To see that this is possible, consider a football team that has only 5 yards remaining to score a touchdown. How many penalties on the opposing team will it take for the sum of the penalties to be 5 yards (giving them a touchdown), if each penalty is "half the distance to the goal"? The answer is that the team can never reach the goal line because every penalty is half the distance to the goal, which always leaves some distance remaining to travel. Thus, the sum of any number of penalties is less than 5 yards. To see this, let's find the sum of the first 10 such penalties. The first few penalties are $\frac{5}{2}, \frac{5}{4}$, and $\frac{5}{8}$, so the sequence of penalties has first term $\frac{5}{2}$ and common ratio $\frac{1}{2}$ and the sum of the first n penalties is

$$s_n = \frac{a_1(1-r^n)}{1-r} = \frac{\left(\frac{5}{2}\right)\left[1-\left(\frac{1}{2}\right)^n\right]}{1-\frac{1}{2}}$$

The sum of the first 10 penalties is found by letting $n = 10$, getting

$$\frac{\frac{5}{2}\left[1-\left(\frac{1}{2}\right)^{10}\right]}{1-\frac{1}{2}} \approx 4.9951$$

We can see that this sum is less than 5 yards, but close to 5 yards. Notice that as n gets larger, $\left(\frac{1}{2}\right)^n$ gets smaller, approaching 0, and the sum approaches $\frac{5/2(1-0)}{1-(1/2)} = 5$. Thus, the sum of an *infinite number* of penalties is the finite number 5 yards.

In general, if $|r| < 1$, it can be shown that r^n approaches 0 as n gets large without bound. Thus, $1 - r^n$ approaches 1, and $\frac{a_1(1-r^n)}{1-r}$ approaches $\frac{a_1}{1-r}$.

Sum of an Infinite Geometric Series

The sum of an infinite geometric series with common ratio r, $|r| < 1$, is

$$S = \frac{a_1}{1-r}$$

where a_1 is the first term.
    If $|r| \geq 1$, the sum does not exist (is not a finite number).

We can use this formula to compute the sum of an infinite number of "half the distance to the goal" penalties from the 5-yard line.

$$S = \frac{\frac{5}{2}}{1-\frac{1}{2}} = 5$$

EXAMPLE 5 ▶ Infinite Geometric Series

Find the sum represented by the following series, if possible.

a. $81 + 9 + 1 + \cdots$

b. $\sum_{i=1}^{\infty} \left(\frac{2}{3}\right)^i$

SOLUTION

a. This is an infinite geometric series with first term $a_1 = 81$ and common ratio $r = \frac{1}{9} < 1$, so the sum is

$$S = \frac{81}{1 - \frac{1}{9}} = \frac{729}{8}$$

b. This is the sigma notation for an infinite series. Each new term results in one higher power of $\frac{2}{3}$ (because i increases by one), so the common ratio is $\frac{2}{3} < 1$. The first term is $\frac{2}{3}$ (the value when $i = 1$), so the sum is

$$S = \frac{\frac{2}{3}}{1 - \frac{2}{3}} = 2$$

SKILLS CHECK 8.4

Answers that are not seen can be found in the answer section at the back of the text.

1. Find the sum of the first 6 terms of the geometric sequence 9, 3, 1, ... $\frac{364}{27}$

2. Find the sum of the first 7 terms of the arithmetic sequence 1, 3, 5, ... 49

3. Find the sum of the first 10 terms of the arithmetic sequence 7, 10, 13, ... 205

4. Find the sum of the first 20 terms of the arithmetic sequence 5, 12, 19, ... 1430

5. Find the sum of the first 15 terms of the arithmetic sequence with first term -4 and common difference 2. 150

6. Find the sum of the first 10 terms of the arithmetic sequence with first term 50 and common difference 3. 635

7. Find the sum of the first 15 terms of the geometric sequence with first term 3 and common ratio 2. 98,301

8. Find the sum of the first 12 terms of the geometric sequence with first term 48 and common ratio $\frac{1}{2}$. $\frac{12,285}{128}$

9. Find the sum of the first 10 terms of the geometric sequence 5, 10, 20, ... 5115

10. Find the sum of the first 15 terms of the geometric sequence 100, 50, 25, ... $\frac{819,175}{4096}$

11. Find the sum represented by the series $1024 + 256 + 64 + 16 + \cdots$, if possible. $\frac{4096}{3}$

12. Find the sum represented by the series $64 + 32 + 16 + \cdots$, if possible. 128

In Exercises 13–16, evaluate the sums.

13. $\sum_{i=1}^{6} 2^i$ 126

14. $\sum_{i=1}^{5} 4^i$ 1364

15. $\sum_{i=1}^{4} \frac{1+i}{i}$ $\frac{73}{12}$

16. $\sum_{i=1}^{5} \left(\frac{1}{2}\right)^i$ $\frac{31}{32}$

17. Find the sum of the series $\sum_{i=1}^{\infty} \left(\frac{3}{4}\right)^i$, if possible. 3

18. Find the sum $\sum_{i=1}^{\infty} \left(\frac{5}{6}\right)^i$, if possible. 5

19. Find the sum $\sum_{i=1}^{\infty} \left(\frac{4}{3}\right)^i$, if possible. Not possible, infinite

20. Find the following sum, if possible. 3200

$$800 + 600 + 450 + \cdots$$

EXERCISES 8.4

Answers that are not seen can be found in the answer section at the back of the text.

21. **Profit** A new firm loses $2000 in its first month, but its profit increases by $400 in each succeeding month for the rest of the year. What is the sum of its profit for the first year? $2400

22. **Salaries** Suppose you are offered a job with a relatively low starting salary but with a bonus that is $1500 for the first year and increases by $1500 each year after that. What is the total amount of your bonuses over the next 7 years? $42,000

23. **Bee Ancestry** A female bee hatches from a fertilized egg, whereas a male bee hatches from an unfertilized egg. Thus, a female bee has a male parent and a female parent, whereas a male bee has only a female parent. Therefore, the numbers of ancestors of a male bee follow the Fibonacci sequence

    $$1, 2, 3, 5, 8, 13, \ldots$$

    Use this sequence to determine the total number of ancestors of a male bee back four generations. 11

24. **Salaries** Suppose you are offered two identical jobs, one paying a starting salary of $40,000 with yearly raises of $2000 and one paying a starting salary of $36,000 with yearly raises of $3000.

    a. Which job will pay more over the first 3 years? How much more? Job 1; $9000 more

    b. Which job will pay more over the first 12 years? How much more? Job 2; $18,000 more

    c. What factor is important in deciding which job to take? How long you plan to stay with the job

25. **Clocks** A grandfather clock strikes a chime indicating each hour of the day, so it chimes 3 times at 3:00, 10 times at 10:00, and so on.

    a. How many times will it chime in a 12-hour period? 78

    b. How many times will it chime in a 24-hour day? 156

26. **Depreciation** Under the sum-of-the-year's-digits depreciation method, annual depreciation for 4 years is found by multiplying the value of the property by a fraction whose denominator is the sum of the 4 years' digits and whose numerator is 4 for the first year's depreciation, 3 for the second year's, 2 for the third year's, and so on.

    a. Write a sequence that gives the depreciation for each year on a truck with value $36,000.
    $14,400, $10,800, $7200, $3600

    b. Show that the sum of the depreciations for the 4 years adds to 100% of the original value.
    $14,400 + 10,800 + 7200 + 3600 = \$36,000$

27. **Profit** Suppose a new business makes a profit of $2000 in its first month, and its profit increases by 10% in each of the next 11 months. How much profit did it earn in its first year? $42,768.57

28. **Profit** If changing market conditions cause a company earning a profit of $8,000,000 this year to project decreases of 2% of its profit in each of the next 5 years, what is the sum of the yearly profits that it projects for the next 5 years? $37,663,047.65

29. **Pumps** A pump removes $\frac{1}{3}$ of the water in a container with every stroke. After 4 strokes, what amount of water is still in a container that originally had 81 cm³ of water? 16 cm³

30. **Salaries** If you accept a job paying $32,000 for the first year with a guaranteed 8% raise each year, how much will you earn in your first 5 years? $187,731.23

31. **Email Chain Letters** Suppose you receive an email chain letter with 5 names on it, and to keep the chain unbroken, you mail a dollar to the person whose name is at the top of the list, cross out the top name, add your name to the bottom, and mail it to 5 friends.

    a. How many friends will receive your letter? 5

    b. If your friends cross the second person from the original list off the list and mail out 5 letters each, how many people will receive letters from your friends? 25

    c. How many "descendants" of your letters will receive email letters in the next two levels if no one breaks the chain? 125; 625

    d. Describe the type of sequence that gives the number of people receiving the letter at each level.
    Geometric, with $r = 5$

32. **Email Chain Letters** Suppose the email chain letter in Exercise 31 were to go through 12 unbroken levels.

    a. How many people would receive a letter asking for money in the 12th mailing? 244,140,625

    b. What is the significance of this number? Approximately 4 out of 5 people in the U.S. would receive a letter asking for money.

33. **Credit Card Debt** Kirsten has a $10,000 credit card debt. She makes a minimum payment of 10% of the balance at the end of each month. If she makes no new purchases, how much will she owe at the end of 1 year? $2824.30

34. **Credit Card Debt** Kirsten has a $10,000 credit card debt. She makes a minimum payment of 10% of the balance at the end of each month. If she makes no new purchases, in how many months will she have half of her debt repaid? About 7 months

35. *Bouncing Ball* A ball is dropped from 128 feet. It rebounds $\frac{1}{4}$ of the height from which it falls every time it hits the ground. How far will it have traveled up and down when it hits the ground for the fifth time? 213 feet

36. *Bouncing Ball* A ball is dropped from 64 feet. If it rebounds $\frac{3}{4}$ of the height from which it falls every time it hits the ground, how far will it have traveled up and down when it hits the ground for the fourth time? 286 feet

37. *Depreciation* A new car costing $35,000 depreciates by 16% of its current value each year.

    a. Write an expression that gives the sum of the depreciations for the first n years. $35{,}000(1 - 0.84^n)$

    b. What is the value of the car at the end of the nth year? $35{,}000(0.84)^n$

38. *Loans* An interest-free loan of $15,000 requires monthly payments of 8% of the unpaid balance. What is the unpaid balance after

    a. 1 year? $5515.00

    b. 2 years? $2027.68

39. *Annuities* Find the 8-year future value of an annuity with a contribution of $100 at the end of each month into an account that pays 12% per year, compounded monthly. $15,992.73

40. *Annuities* Find the 10-year future value of an ordinary annuity with a contribution of $300 per quarter into an account that pays 8% per year, compounded quarterly. $18,120.59

41. *Present Value of Annuities* The present value of an annuity with n payments of R dollars, at interest rate i per period, is

$$A = R(1 + i)^{-1} + R(1 + i)^{-2} + R(1 + i)^{-3} + \cdots + R(1 + i)^{-(n-1)} + R(1 + i)^{-n}$$

Treat this as the sum of a geometric series with n terms, $a_1 = R(1 + i)^{-n}$, and $r = (1 + i)$, and show that the sum is

$$A = R\left[\frac{1 - (1 + i)^{-n}}{i}\right]$$

8.5 The Binomial Theorem

KEY OBJECTIVES

- Evaluate factorials
- Use Pascal's triangle
- Find coefficients of terms in a binomial expansion
- Use the Binomial Theorem to expand binomials
- Find a particular term of a binomial expansion

SECTION PREVIEW Binomial Expansions

We introduced some binomial expansions in the Chapter 6 Toolbox. We restate them here, along with some additional expansions.

$$(x + y)^0 = 1$$
$$(x + y)^1 = x + y$$
$$(x + y)^2 = x^2 + 2xy + y^2$$
$$(x + y)^3 = x^3 + 3x^2y + 3xy^2 + y^3$$
$$(x + y)^4 = x^4 + 4x^3y + 6x^2y^2 + 4xy^3 + y^4$$
$$(x + y)^5 = x^5 + 5x^4y + 10x^3y^2 + 10x^2y^3 + 5xy^4 + y^5$$

In this section, we will show how the expansions of powers of binomials can be generalized, with the Binomial Theorem giving the formula for these expansions. The Binomial Theorem uses factorial notation, which is also introduced in this section. ∎

Binomial Coefficients

By looking at the expansions above, we can observe several facts for all but the special case $(x + y)^0 = 1$.

1. The exponents of the first and last terms are the same as the power on the binomial, and they are also equal to the sum of the exponents on every term.

2. The exponent on x decreases by one on each term, and the exponent on y increases by one on each term.

3. In general, the variables in the expansion $(x + y)^n$ are

$$x^n, x^{n-1}y, x^{n-2}y^2, x^{n-3}y^3, \ldots, x^{n-k}y^k, \ldots, xy^{n-1}, y^n$$

We can also observe a special pattern for the coefficients of the terms. Writing just the coefficients gives the following pattern, which is known as **Pascal's triangle**.

$$
\begin{array}{c}
1 \\
1 \quad 1 \\
1 \quad 2 \quad 1 \\
1 \quad 3 \quad 3 \quad 1 \\
1 \quad 4 \quad 6 \quad 4 \quad 1 \\
1 \quad 5 \quad 10 \quad 10 \quad 5 \quad 1
\end{array}
$$

Observe that except for the 1's at the ends of the rows, each number is the sum of the numbers to the left and right in the row above it. For example, the 6 in the fifth row is equal to the sum of the two 3's to the left and right in the row above it, and each 4 in the fifth row is the sum of 3 and 1.

Using the observations above, we can write the expansion of $(x + y)^6$ as

$$(x + y)^6 = x^6 + 6x^5y + 15x^4y^2 + 20x^3y^3 + 15x^2y^4 + 6xy^5 + y^6$$

We can use Pascal's triangle to find the coefficients of $(x + y)^n$ for larger values of n, but we have an alternative method that is more direct, especially for large values of n. This method uses **factorial notation**, which is defined as follows.

Factorial Notation

For any positive integer n, we define n factorial as

$$n! = n(n-1)(n-2)(n-3) \cdots (3)(2)(1)$$

and we define $0! = 1$

EXAMPLE 1 ▶ Evaluating Factorials

Evaluate the following factorials.

a. $3!$ **b.** $6!$ **c.** $8!$ **d.** $0!$ **e.** $(2!)(3!)$ **f.** $\dfrac{6!}{4!\,2!}$

SOLUTION

a. $3! = 3 \cdot 2 \cdot 1 = 6$

b. $6! = 6 \cdot 5 \cdot 4 \cdot 3 \cdot 2 \cdot 1 = 720$

c. $8! = 8 \cdot 7 \cdot 6 \cdot 5 \cdot 4 \cdot 3 \cdot 2 \cdot 1 = 40{,}320$

d. $0! = 1$

e. $(2!)(3!) = (2 \cdot 1)(3 \cdot 2 \cdot 1) = 2 \cdot 6 = 12$

f. $\dfrac{6!}{4!\,2!} = \dfrac{6 \cdot 5 \cdot 4 \cdot 3 \cdot 2 \cdot 1}{4 \cdot 3 \cdot 2 \cdot 1 \cdot 2 \cdot 1} = \dfrac{6 \cdot 5}{2 \cdot 1} = 15$

We can use a graphing calculator to evaluate factorials. We type the number and then press **MATH**, move to PROB, and choose 4:!. See Figure 8.24.

Notice that $\dfrac{6!}{4!\,2!} = 15$, which is the coefficient of x^2y^4 in

$$(x+y)^6 = x^6 + 6x^5y + 15x^4y^2 + 20x^3y^3 + 15x^2y^4 + 6xy^5 + y^6$$

In fact, the coefficient of $x^{n-k}y^k$ in the expansion of $(x+y)^n$ is equal to $\dfrac{n!}{k!\,(n-k)!}$.

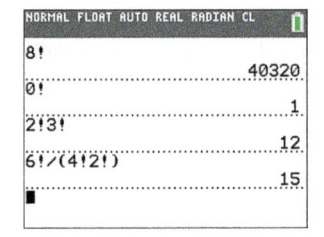

Figure 8.24

EXAMPLE 2 ▶ Finding Binomial Coefficients

a. Find the coefficient of x^2y^5 in the expansion of $(x+y)^7$.

b. Find the coefficient of x^6y^3 in the expansion of $(x+y)^9$.

SOLUTION

a. The coefficient of x^2y^5 is $\dfrac{7!}{5!\,2!} = \dfrac{7\cdot 6\cdot \cancel{5}\cdot \cancel{4}\cdot \cancel{3}\cdot \cancel{2}\cdot \cancel{1}}{\cancel{5}\cdot \cancel{4}\cdot \cancel{3}\cdot \cancel{2}\cdot \cancel{1}\cdot 2\cdot 1} = \dfrac{7\cdot 6}{2\cdot 1} = 21$.

b. The coefficient of x^6y^3 is $\dfrac{9!}{3!\,6!} = \dfrac{9\cdot 8\cdot 7\cdot \cancel{6}\cdot \cancel{5}\cdot \cancel{4}\cdot \cancel{3}\cdot \cancel{2}\cdot \cancel{1}}{3\cdot 2\cdot 1\cdot \cancel{6}\cdot \cancel{5}\cdot \cancel{4}\cdot \cancel{3}\cdot \cancel{2}\cdot \cancel{1}} = \dfrac{9\cdot 8\cdot 7}{3\cdot 2\cdot 1} = 84$.

The number $\dfrac{n!}{k!\,(n-k)!}$ is called a **binomial coefficient** and can be written as $\dbinom{n}{k}$ or as ${}_nC_k$. Thus $\dbinom{6}{4} = \dfrac{6!}{4!\,2!} = \dfrac{6\cdot 5\cdot \cancel{4}\cdot \cancel{3}\cdot \cancel{2}\cdot \cancel{1}}{\cancel{4}\cdot \cancel{3}\cdot \cancel{2}\cdot \cancel{1}\cdot 2\cdot 1} = \dfrac{6\cdot 5}{2\cdot 1} = 15$ and ${}_7C_2 = \dfrac{7!}{2!\,5!} = \dfrac{7\cdot 6}{2\cdot 1} = 21$. Note that $\dfrac{n!}{k!\,(n-k)!} = \dfrac{n!}{(n-k)!\,k!}$, so $\dbinom{n}{k} = \dbinom{n}{n-k}$ and ${}_nC_k = {}_nC_{n-k}$.

Putting these ideas together, we see that the term of the expansion of $(x+y)^n$ containing $x^{n-k}y^k$ is

$$\binom{n}{k}x^{n-k}y^k = \dfrac{n!}{k!\,(n-k)!}x^{n-k}y^k$$

and we can write the expansion with the following equation, which is called the **Binomial Theorem**.

Binomial Theorem

For any positive integer n,

$$(x+y)^n = x^n + \binom{n}{1}x^{n-1}y^1 + \binom{n}{2}x^{n-2}y^2 + \binom{n}{3}x^{n-3}y^3 + \cdots + \binom{n}{k}x^{n-k}y^k + \cdots + \binom{n}{n-1}x^1y^{n-1} + y^n$$

Because the exponent of y increases from 0 in the first term of the expansion of $(x+y)^n$ to n in the last term, we can write the expansion in sigma notation as

$$(x+y)^n = \sum_{k=0}^{n}\binom{n}{k}x^{n-k}y^k = \sum_{k=0}^{n}\dfrac{n!}{k!\,(n-k)!}x^{n-k}y^k$$

EXAMPLE 3 ▶ Using the Binomial Theorem

Expand $(x + y)^8$.

SOLUTION

We will first calculate the coefficients for the expansion with a graphing calculator by entering the power n, pressing the **MATH** key, moving to PROB, selecting nCr, and entering the exponent of y. The values for sample terms are shown in Figures 8.25(a) and (d). In the figure, we can see the *symmetry* of the coefficients, which means we need only find half of the coefficients to know what all of them are. For example, we see that $_8C_2 = {_8C_{8-2}} = {_8C_6}$ and we know that $_8C_8 = 1$ because $_8C_0 = 1$.

Using these coefficients in the Binomial Theorem gives

$$(x + y)^8 = x^8 + 8x^7y + 28x^6y^2 + 56x^5y^3 + 70x^4y^4 + 56x^3y^5 + 28x^2y^6 + 8xy^7 + y^8$$

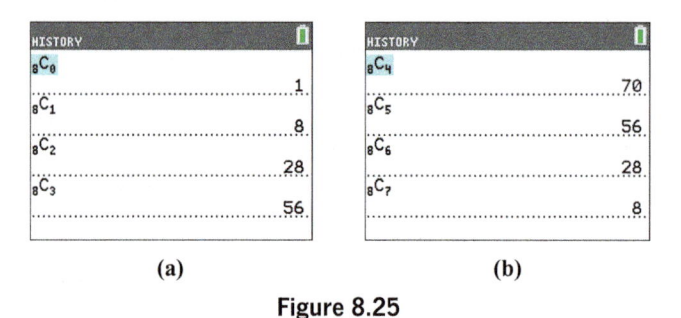

(a)   (b)

Figure 8.25

EXAMPLE 4 ▶ Finding a Particular Term in a Binomial Expansion

Find the sixth term of $(2x - y)^9$.

SOLUTION

This binomial has the form $(a + b)^9$ with $a = 2x$ and $b = -y$, so we will first find the sixth term of $(a + b)^9$. The sixth term is the term $\binom{n}{k}a^{n-k}b^k$, in which $k = 5$, because the second term has the power $k = 1$. The coefficient of this term, found with factorials or with a graphing calculator, is

$$\binom{9}{5} = {_9C_5} = 126$$

Thus the sixth term of $(a + b)^9$ is $126a^{9-5}b^5 = 126a^4b^5$. To find the sixth term of $(2x - y)^9$, we substitute $2x$ for a and $-y$ for b.

$$126(2x)^4(-y)^5 = 126(16x^4)(-y^5) = -2016x^4y^5$$

SKILLS CHECK 8.5

Evaluate the following.

1. $\dfrac{8!}{4!\,4!}$ 70
2. $\dfrac{7!}{4!\,3!}$ 35
3. $\dfrac{9!}{0!\,9!}$ 1
4. $\dfrac{7!}{7!\,0!}$ 1
5. $\dfrac{9!}{3!\,6!}$ 84
6. $\dfrac{11!}{7!\,4!}$ 330
7. $\binom{12}{2}$ 66
8. $\binom{13}{10}$ 286
9. $\binom{10}{3}$ 120
10. $\binom{9}{7}$ 36
11. $_{11}C_8$ 165
12. $_{14}C_{12}$ 91

Expand the following binomial powers.

13. $(a+b)^4$
14. $(m+n)^6$
15. $(x-y)^4$
16. $(x-y)^5$
17. $(x+2y)^6$
18. $(2x+y)^4$
19. $(3x+4y)^5$
20. $(5x+6y)^4$
21. $(3a+b)^4$
22. $(2r-3s)^5$
23. $(x+y)^7$
24. $(x-y)^7$

Write the indicated term of each binomial expansion.

25. The fifth term of $(m+n)^6$ $15m^2n^4$
26. The fourth term of $(a+b)^6$ $20a^3b^3$
27. The eighth term of $(x+y)^{10}$ $120x^3y^7$
28. The ninth term of $(6x+y)^{11}$ $35{,}640x^3y^8$
29. The third term of $(3m-n)^5$ $270m^3n^2$
30. The fourth term of $(r-3s)^6$ $-540r^3s^3$
31. The fourteenth term of $(m-n)^{16}$ $-560m^3n^{13}$
32. The twelfth term of $(2r+s)^{14}$ $2912r^3s^{11}$
33. The middle term of $(x+y)^{10}$ $252x^5y^5$
34. The middle term of $(x+y)^{12}$ $924x^6y^6$

8.6 Conic Sections: Circles and Parabolas

KEY OBJECTIVES

- Use the distance formula
- Use the midpoint formula
- Write the equation of a circle
- Locate the center and radius of a circle
- Graph circles
- Write the equation of a parabola given the vertex and focus
- Graph parabolas

SECTION PREVIEW Circles and Parabolas

In earlier sections, we have discussed the algebraic properties of **circles** and **parabolas**. We will study the geometric properties of these two **conic sections** in this section and of two others, the **ellipse** and the **hyperbola**, in the next section. These special curves are called conic sections because they can be formed by the intersection of a plane and a (double-napped) right circular cone. The intersection curves are illustrated in Figure 8.26. Note that if the plane passes through the vertex of the cone, the intersection is a point or lines, and conic sections are not formed. The resulting point or lines formed by the intersection are called *degenerate conics*.

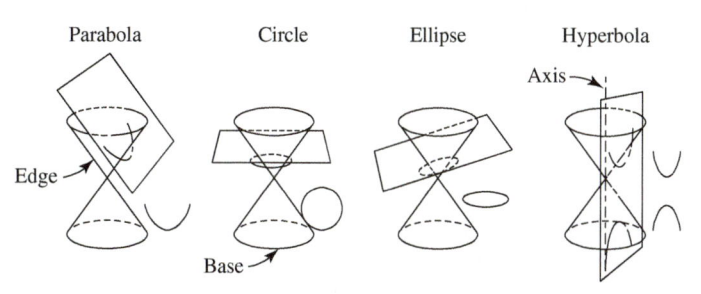

Figure 8.26 Conic Sections

The geometric properties of these conic sections, which we will discuss further in this chapter, were discovered by the Greeks around 200 BC. New applications of conic sections were developed in the 17th century. Using data from Tyco Brahe's 16th-century observations of planetary movement, Johannes Kepler determined that planets move in elliptical orbits about the sun. We now know that many comets have parabolic paths, and other applications of parabolas include parabolic searchlights, bridge cables, and bridge arches.

Conic sections can be defined as intersections of planes and cones as the Greeks defined them, as the graphs of the **second-degree equation**

$$Ax^2 + Bxy + Cy^2 + Dx + Ey + F = 0$$

or as sets of all points satisfying specific geometric properties. We will use this last definition in our discussion of conic sections in this section and the next. ■

Distance and Midpoint Formulas

Our definitions of the conic sections use the distance between two points, so we will develop the formula used to find this distance. The notation (x_a, y_b) is used to denote a fixed but arbitrary point in the coordinate plane, so the points $P_1(x_1, y_1)$ and $P_2(x_2, y_1)$ have the same y-coordinates but different x-coordinates, while the points $P_2(x_2, y_1)$ and $P_3(x_2, y_2)$ have the same x-coordinates but different y-coordinates. (See Figure 8.27.)

From our definition of distance on a number line, the distance between P_1 and P_2 on the horizontal line is $|x_2 - x_1|$, and the distance between P_2 and P_3 on the vertical line is $|y_2 - y_1|$. By the Pythagorean Theorem, $c^2 = a^2 + b^2$, the distance between P_1 and P_3 is

$$d = \sqrt{|x_2 - x_1|^2 + |y_2 - y_1|^2}$$

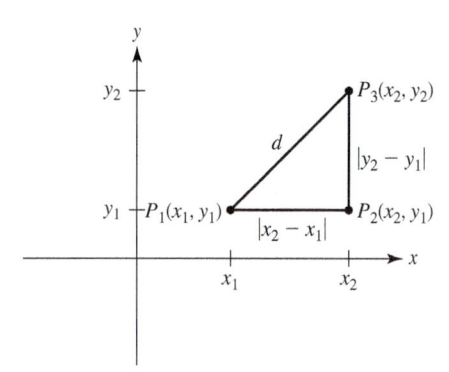

Figure 8.27

But since $|a|^2 = a^2$ for every real number a, we have the following formula for the distance between two points.

Distance Formula

The distance between the points $P_1(x_1, y_1)$ and $P_2(x_2, y_2)$ in a plane is

$$d = \sqrt{(x_2 - x_1)^2 + (y_2 - y_1)^2}$$

EXAMPLE 1 ▶ Using the Distance Formula

a. Find the distance from $(-3, 2)$ to $(1, -4)$.

b. Use the distance formula to show that the points $A(-3, 2)$, $B(1, -4)$, and $C(3, -7)$ lie on a line.

SOLUTION

a. $d = \sqrt{[1-(-3)]^2 + (-4-2)^2} = \sqrt{16+36} = \sqrt{52} = 2\sqrt{13}$

b. The three points lie on the same line if the sum of the distances between two pairs of points equals the distance between the third pair of points. The distances are

$$\overline{AB} = \sqrt{(1+3)^2 + (-4-2)^2} = \sqrt{52} = 2\sqrt{13}$$
$$\overline{BC} = \sqrt{(3-1)^2 + (-7+4)^2} = \sqrt{13}$$
$$\overline{AC} = \sqrt{(3+3)^2 + (-7-2)^2} = \sqrt{117} = 3\sqrt{13}$$

$\overline{AB} + \overline{BC} = \overline{AC}$, so the points lie on a line.

The midpoint formula is also useful in the study of conic sections. For instance, the center of a circle can be found by the midpoint formula if the endpoints of one of its diameters are known.

Midpoint Formula

The midpoint M of the line segment joining $P_1(x_1, y_1)$ and $P_2(x_2, y_2)$ is

$$M\left(\frac{x_1 + x_2}{2}, \frac{y_1 + y_2}{2}\right)$$

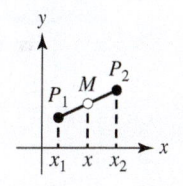

For example, if the endpoints of a diameter of a circle are $(3, 6)$ and $(5, -2)$, the center is at the midpoint of the diameter, $M\left(\dfrac{3+5}{2}, \dfrac{6+(-2)}{2}\right) = M(4, 2)$.

Circles

Each of the conic sections is defined as the set of all points in a plane satisfying a given geometric definition. We define the **circle** as the set of all points in a plane that are equidistant from a fixed point called the center. Recall the distance formula, which states that the distance between the point $P_1(x_1, y_1)$ and the point $P_2(x_2, y_2)$ in a plane is

$$d = \sqrt{(x_2 - x_1)^2 + (y_2 - y_1)^2}$$

Thus any point (x, y) on a circle of radius r with center at (h, k) satisfies the equation

$$r = \sqrt{(x-h)^2 + (y-k)^2}$$

Squaring both sides of this equation gives the equation of the circle of radius r with center at (h, k).

$$r^2 = (x-h)^2 + (y-k)^2$$

Standard Form of the Equation of a Circle

The equation

$$(x-h)^2 + (y-k)^2 = r^2$$

is the standard form of an equation of a circle with center at (h, k) and radius r.

Figures 8.28 and 8.29 show the graphs of $x^2 + y^2 = 3^2$ and $(x + 2)^2 + (y - 3)^2 = 4^2$, respectively. Note that the center of the first circle is at the origin, and the center of the second is at $(-2, 3)$.

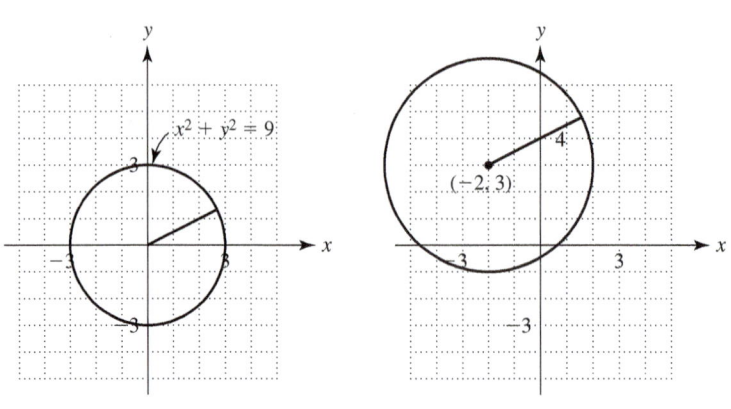

Figure 8.28

Figure 8.29

EXAMPLE 2 ▶ Finding the Equation of a Circle

Write the equation of the circle with center at $(-3, 4)$ and radius 5.

SOLUTION

The standard form of the equation is

$$[x - (-3)]^2 + (y - 4)^2 = 5^2 \quad \text{or} \quad (x + 3)^2 + (y - 4)^2 = 25$$

EXAMPLE 3 ▶ Finding the Equation of a Circle

Write the equation of the circle centered at $(3, -1)$ and passing through $(4, 6)$.

SOLUTION

The radius is the distance between $(3, -1)$ and $(4, 6)$:

$$r = \sqrt{(4 - 3)^2 + [6 - (-1)]^2} = \sqrt{1 + 49} = \sqrt{50}$$

Thus, substituting in $(x - h)^2 + (y - k)^2 = r^2$ gives the equation

$$(x - 3)^2 + (y + 1)^2 = 50$$

The *general form* of the equation of a circle is

$$Ax^2 + Ay^2 + Dx + Ey + F = 0$$

By completing the square on this equation, we can convert it to the standard form

$$(x - h)^2 + (y - k)^2 = r^2$$

EXAMPLE 4 ▶ Converting the Equation of a Circle to Standard Form

Write $x^2 + y^2 + 4x - 6y - 3 = 0$ in standard form and locate its center and radius.

SOLUTION

The equation is equivalent to

$$(x^2 + 4x) + (y^2 - 6y) = 3$$

8.6 Conic Sections: Circles and Parabolas

We add 4 to complete the square on the x portion and 9 to complete the square on the y portion, and we add 13 on the right side of the equation to preserve the equality:

$$(x^2 + 4x + 4) + (y^2 - 6y + 9) = 3 + 4 + 9$$

Rewriting the trinomials as binomials squared gives

$$(x + 2)^2 + (y - 3)^2 = 16$$

The center is at $(-2, 3)$ and the radius is 4. (See Figure 8.30.)

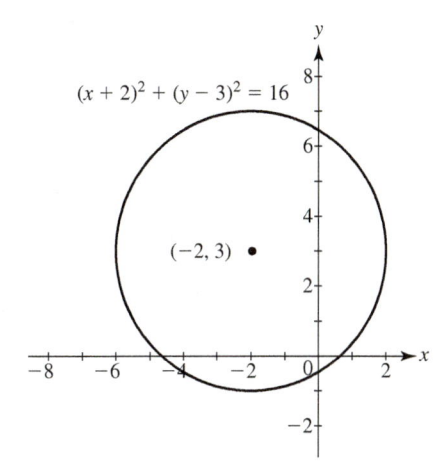

Figure 8.30

Sometimes we encounter functions of the form

$$y = \sqrt{4 - x^2}$$

If we square both sides of this equation, we get the equation

$$y^2 = 4 - x^2 \quad \text{or} \quad x^2 + y^2 = 4$$

whose graph is a circle with center at the origin and radius 2. Because $y = \sqrt{4 - x^2}$ gives only nonnegative values for y, its graph is the upper half of the circle. This graph is the solid curve shown in Figure 8.31(a).

(a)

(b)

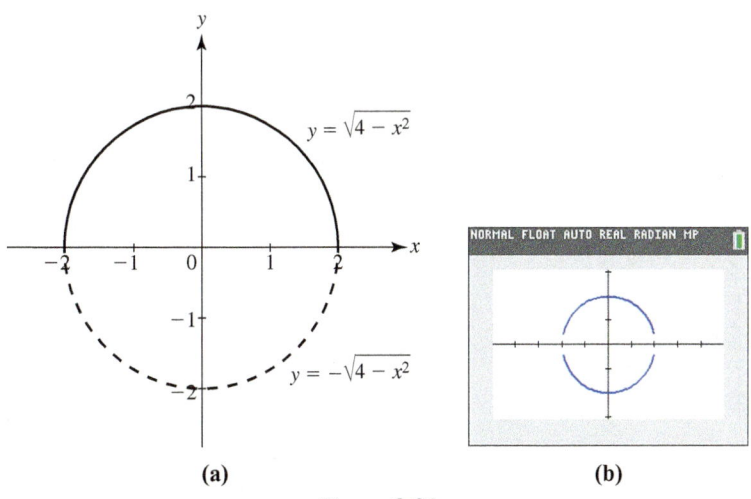

Figure 8.31

We can graph the equation $x^2 + y^2 = 4$ with technology by graphing the two functions $y = \sqrt{4 - x^2}$ and $y = -\sqrt{4 - x^2}$ on the same axes, as shown Figure 8.31(b).

Note that if the conversion of a quadratic equation from general form to standard form results in an equation of the form $(x - h)^2 + (y - k)^2 = 0$, its graph is a circle of radius 0 (the point (h, k)).

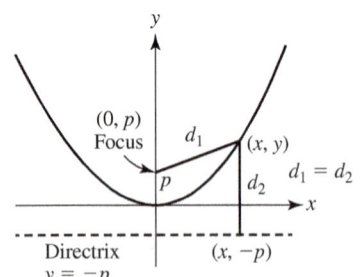

Parabolas

We have seen that a function of the form $y = ax^2 + bx + c$ ($a \neq 0$) has as its graph a parabola that opens upward if $a > 0$, and opens downward if $a < 0$. The following geometric definition applies to parabolas opening in any direction.

> A **parabola** is the set of all points in a plane that are equidistant from a fixed point called the **focus** and a fixed line called the **directrix**. The midpoint between the focus and the directrix is the **vertex**, and the line passing through the vertex and the focus is the **axis** of the parabola. See Figure 8.32.

We can use this definition to derive the standard equation of a parabola with the vertex at the origin. See Exercises 63 and 64.

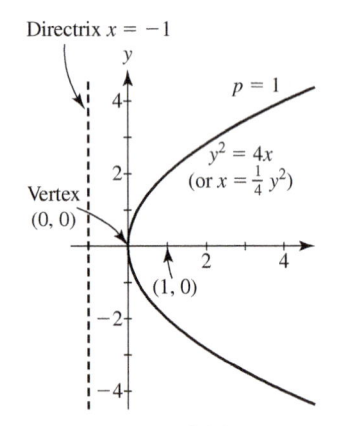

Standard Equation of a Parabola with Vertex at Origin

The parabola with vertex at $(0, 0)$, focus at $(0, p)$, and directrix $y = -p$ has equation

$$x^2 = 4py \quad \text{(Its axis is vertical; see Figure 8.32.)}$$

The parabola with vertex at $(0, 0)$, focus at $(p, 0)$, and directrix $x = -p$ has equation

$$y^2 = 4px \quad \text{(Its axis is horizontal; see Figure 8.33.)}$$

Note that the equation for a parabola with a vertical axis and vertex $(0, 0)$ can be solved for y, yielding the function

$$y = \frac{1}{4p}x^2$$

and the parabola can be graphed using technology as well as by hand. It will open up if $p > 0$ and down if $p < 0$. The graph of the equation $y^2 = 4px$ is a parabola with a horizontal axis, vertex at $(0, 0)$, and focus at $(p, 0)$. It opens to the right if $p > 0$ and to the left if $p < 0$. The equation $y^2 = 4x$ has the form $y^2 = 4px$ with $p = 1$. Thus the graph is a horizontal parabola that opens to the right. The focus is located at $(1, 0)$, and the directrix is $x = -1$. See Figure 8.33.

The equation $y^2 = 4px$ can also be written in the form

$$y = \sqrt{4px} \quad \text{and} \quad y = -\sqrt{4px}$$

The first function gives the upper half of a parabola opening to the right or left, and the second function gives the bottom half. Graphing both functions with technology or by hand on the same axes gives the parabola that opens right or left. See the graphs of $y = \sqrt{4x}$ and $y = -\sqrt{4x}$ in Figure 8.34. Graphing these two functions on the same axes gives the graph of $y^2 = 4x$.

The graphs of equations of the form $x^2 = 4py$ or $y^2 = 4px$ are parabolas with vertices at $(0, 0)$. If the graph of a parabola is the same as the graph of one of these equations except that its vertex is shifted from $(0, 0)$ to (h, k), then its equation is the same except that x is replaced by $x - h$, and y is replaced by $y - k$.

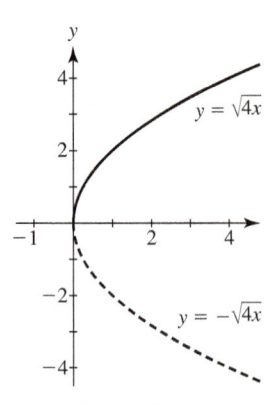

8.6 Conic Sections: Circles and Parabolas

EXAMPLE 5 ▶ Finding the Equation of a Parabola

Write the equation of the parabola with vertex at $(-1, 3)$ and focus at $(-3, 3)$.

SOLUTION

The axis contains the vertex at $(-1, 3)$ and focus at $(-3, 3)$, so it is horizontal. The distance from the vertex to the focus is $-3 - (-1) = -2$, so $p = -2$. Thus the standard form of the equation of the parabola is

$$(y - 3)^2 = 4(-2)[x - (-1)] \quad \text{or} \quad (y - 3)^2 = -8(x + 1)$$

EXAMPLE 6 ▶ Graphing the Equation of a Parabola

Graph the equation $y^2 + 4x - 6y + 5 = 0$.

SOLUTION

There is a y^2-term and no x^2-term, so the graph is a parabola that opens to the right or left, with standard form $(y - k)^2 = 4p(x - h)$. We write the equation in this form by completing the square as follows.

$$y^2 - 6y = -4x - 5$$
$$y^2 - 6y + 9 = -5 - 4x + 9$$
$$(y - 3)^2 = -4x + 4$$
$$(y - 3)^2 = -4(x - 1)$$

Hence the vertex point is $(1, 3)$. If we select y-values near the vertex, calculate the x-values, and plot the points, we obtain the graph shown in Figure 8.35.

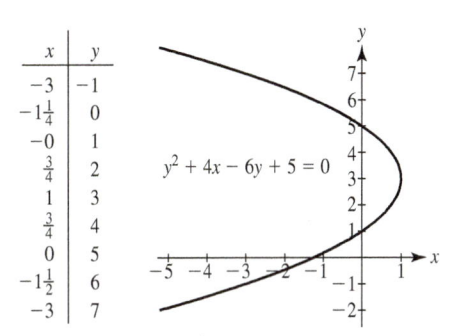

x	y
-3	-1
$-1\frac{1}{4}$	0
-0	1
$\frac{3}{4}$	2
1	3
$\frac{3}{4}$	4
0	5
$-1\frac{1}{2}$	6
-3	7

Figure 8.35

The definition of a parabola is useful in many applications of the parabolic shape. Figure 8.36 shows the parabolic cable on a bridge. Parabolic dishes have the property that entering rays are reflected to the focus and that rays initiated from the focus are reflected outward in parallel lines. These parabolic dishes have applications to satellite transmission, sideline microphones for sporting events, and spotlights.

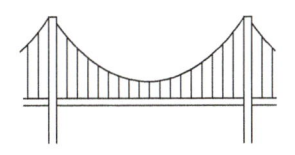

Figure 8.36

EXAMPLE 7 ▶ Design of a Satellite Dish

A satellite dish is a paraboloid formed by rotating a parabola about its axis. Its design results in incoming signals parallel to the axis being reflected to the receiver at the focus of the parabola. If the dish is 32 inches across and 4 inches deep, where should the receiver be placed?

SOLUTION

We can represent a cross section of the dish with a parabola that has its vertex at $(0, 0)$ and contains the points $(16, 4)$ and $(-16, 4)$. This parabola has its axis along the y-axis, and we can use the equation $x^2 = 4py$ and one of the points to find p.

$$16^2 = 4p(4)$$
$$256 = 16p$$
$$p = 16$$

Thus the focus is at $(0, 16)$, and the receiver should be 16 inches from the vertex of the dish.

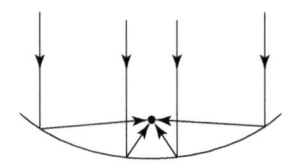

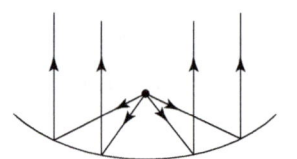

SKILLS CHECK 8.6

Answers that are not seen can be found in the answer section at the back of the text.

1. **a.** Write the equation of the circle with center at $(-1, 2)$ and radius 3.
   b. Graph the circle.

2. Write the equation of the circle with center at $(2, 5)$ and radius $\sqrt{5}$, and graph the circle.

In Exercises 3–8, find the radius and center of a circle with the given equation.

3. $x^2 + y^2 = 36$ $C(0, 0); r = 6$
4. $x^2 + y^2 = 49$ $r = 7; (0, 0)$
5. $(x - 2)^2 + (y + 1)^2 = 14$ $C(2, -1); r = \sqrt{14}$
6. $(x + 1)^2 + (y - 1)^2 = 1$ $r = 1; (-1, 1)$
7. $(x - 0.5)^2 + (y + 3.2)^2 = 7$ $C(0.5, -3.2); r = \sqrt{7}$
8. $(x + 1.3)^2 + \left(y - \dfrac{3}{2}\right)^2 = \dfrac{9}{16}$ $r = \dfrac{3}{4}; (-1.3, 1.5)$

In Exercises 9–26, graph the equation.

9. $x^2 + y^2 - 49 = 0$
10. $x^2 + y^2 - 18 = 0$
11. $x^2 + y^2 - 6 = 8$
12. $x^2 + (y - 2)^2 = 20$
13. $(x + 1)^2 + (y + 3)^2 = 1$
14. $(x - 4)^2 + (y - 1)^2 = \dfrac{1}{4}$
15. $\left(x + \dfrac{1}{2}\right)^2 + (y - 2)^2 = 6$
16. $(x + 2)^2 + y^2 = 1$
17. $x^2 + (y + 4)^2 = 16$
18. $(x - 6)^2 + (y - 6)^2 = 12$
19. $x^2 + y^2 + 4x - 2y - 4 = 0$
20. $x^2 + y^2 - 6x + 10y - 2 = 0$
21. $x^2 + y^2 - x - 4y + 4 = 0$
22. $4x^2 + 4y^2 + 12x - 8y = 3$
23. $x^2 + y^2 - y = 0$
24. $x^2 - x + y^2 - 2y = 0$
25. $y = \sqrt{9 - x^2}$
26. $y = \sqrt{16 - x^2}$

In Exercises 27–32, write the equation of the circle with the given center and radius.

27. $(4, 0); 4$
28. $(-2, 3); 5$
29. $(-3, -1); 2$
30. $(0, -3); 13$
31. $(5, -2); 1/2$
32. $(1, -1); \sqrt{3}$

In Exercises 33–36, write the equation of the shape described.

33. The upper semicircle centered at $(0, 0)$ with radius 3
    $y = \sqrt{9 - x^2}$
34. The lower semicircle centered at $(0, 0)$ with radius 5
    $y = -\sqrt{25 - x^2}$
35. The left semicircle centered at $(0, -1)$ with radius 1
    $x = -\sqrt{-y^2 - 2y}$
36. The right semicircle centered at $(0, 0)$ with radius 6
    $x = \sqrt{36 - y^2}$

37. Write the equation of the circle that has the ends of a diameter at $(1, 6)$ and $(4, -1)$.

38. Write the equation of the circle centered at $(2, -3)$ and passing through the point $(1, 1)$. $(x - 2)^2 + (y + 3)^2 = 17$

39. Write the equation of the circle that is tangent to the x-axis at the point $(-4, 0)$ and passes through the point $(0, 2)$. $(x + 4)^2 + (y - 5)^2 = 25$

40. Write the equation of the circle of radius 5 that is tangent to both axes and passes through the point $(-8, 1)$. $(x + 5)^2 + (y - 5)^2 = 25$

In Exercises 41–60, graph each equation.

41. $y^2 - 12x = 0$
42. $y^2 - 8x = 0$
43. $y^2 + 4x = 0$
44. $y^2 + 8x = 0$
45. $x^2 - 4y = 0$
46. $x^2 - 8y = 0$
47. $x^2 + 12y = 0$
48. $x^2 + 4y = 0$
49. $y = \sqrt{x - 3}$
50. $y = \sqrt{2x - 2}$
51. $y = \sqrt{3 - x}$
52. $y = \sqrt{2 - 3x}$
53. $y^2 - 6y + 4x + 5 = 0$
54. $x^2 + 10x + 8y + 49 = 0$
55. $4x^2 + 12x + 32y + 25 = 0$
56. $4y^2 - 4y + 24x + 37 = 0$
57. $y = x^2 + 10x + 25$
58. $y = x^2 + x$
59. $y = 3 + x - 2x^2$
60. $y = 9 - x^2$

EXERCISES 8.6

Answers that are not seen can be found in the answer section at the back of the text.

61. **Bridges** A cable of a suspension bridge is suspended between two towers that are 300 feet apart and 80 feet above the road. The cable is in the shape of a parabola whose low point is 20 feet above the road. Find the equation of this cable if the point $(0, 0)$ is the point on the road directly below the low point of the cable. $x^2 = 375(y - 20)$

62. **Satellite Dishes** A satellite dish is 20 feet across and 5 feet deep. How far from the vertex of the dish should the receiver be placed to receive radio signals coming in parallel to the axis of the dish? 5 feet

63. **Parabolas** The points on a parabola satisfy the equation $\sqrt{(x - 0)^2 + (y - p)^2} = y + p$ because the distance from the focus $(0, p)$ to any point on the curve, (x, y), must equal the distance from (x, y) to the directrix $y = -p$. Show that this equation can be simplified to obtain the equation $x^2 = 4py$.

64. **Parabolas** Use the definition of a parabola to show that the equation of a parabola with vertex at $(0, 0)$, focus at $(p, 0)$, and directrix $x = -p$ can be written in the form $y^2 = 4px$.

8.7 Conic Sections: Ellipses and Hyperbolas

KEY OBJECTIVES

- Graph horizontal ellipses
- Graph vertical ellipses
- Write the equation of an ellipse in standard form
- Graph horizontal hyperbolas
- Graph vertical hyperbolas
- Write the equation of a hyperbola in standard form
- Identify the asymptotes of hyperbolas
- Graph rectangular hyperbolas

SECTION PREVIEW Ellipses and Hyperbolas

As we mentioned in the previous section, Johannes Kepler determined that planets moved in elliptical orbits about the Sun. We now know that objects in space can move in elliptical or hyperbolic paths, as well as parabolic paths. In particular, we know there are more than 1000 comets, many of which have elliptical paths, and many which have hyperbolic paths. Other applications include elliptical whispering chambers and hyperbolic antennas for radar searches. ■

Ellipses

An **ellipse** is defined as the set of all points in a plane such that the sum of their distances from two fixed points (called **foci**) is a constant (see Figure 8.37). We can see how this definition results in an ellipse by placing two thumbtacks on a corkboard, fastening the ends of a fixed length of string to the tacks, and moving a pencil against the string to sketch the ellipse.

666　Chapter 8　Special Topics in Algebra

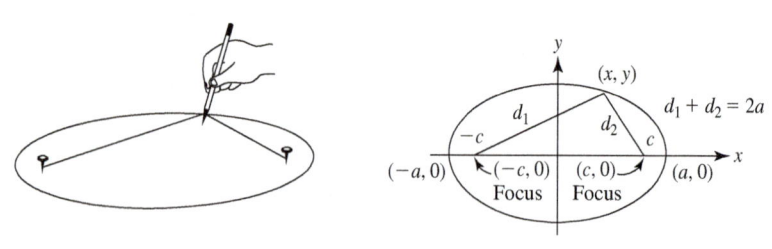

Figure 8.37

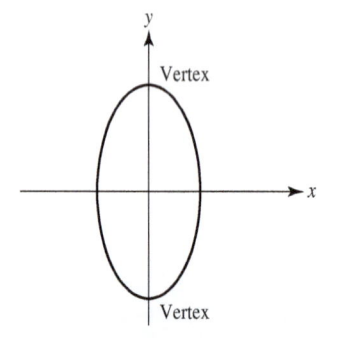

Figure 8.38

The line segment through the foci that ends at its intersection with the ellipse is the **major axis**. For the ellipse in Figure 8.37, the points of intersection are the **vertices** $(a, 0)$ and $(-a, 0)$, and the length of the major axis is $2a$. The line segment through the center perpendicular to the major axis and ending at its intersection with the ellipse is the **minor axis**. The length of the minor axis is $2b$, with $a > b > 0$. The major and minor axes are axes of symmetry for the ellipse. An ellipse with a horizontal major axis is a **horizontal ellipse** (see Figure 8.37), and an ellipse with a vertical major axis is a **vertical ellipse** (see Figure 8.38).

If the foci are on the x-axis (or on the y-axis) equidistant from the origin, we can use the distance formula and the definition of an ellipse to find the following. (See Exercise 53.)

Standard Forms of Equations of Ellipses

The ellipse centered at the origin with vertices at $(a, 0)$ and $(-a, 0)$, foci at $(c, 0)$ and $(-c, 0)$, and endpoints of the minor axis at $(0, b)$ and $(0, -b)$ has the equation

$$\frac{x^2}{a^2} + \frac{y^2}{b^2} = 1 \qquad \text{(See Figure 8.37.)}$$

The ellipse centered at the origin with vertices at $(0, a)$ and $(0, -a)$, foci at $(0, c)$ and $(0, -c)$, and endpoints of the minor axis at $(b, 0)$ and $(-b, 0)$ has the equation

$$\frac{y^2}{a^2} + \frac{x^2}{b^2} = 1 \qquad \text{(See Figure 8.38.)}$$

EXAMPLE 1 ▶ **Graphing the Equation of an Ellipse**

Graph the equation $9x^2 + 25y^2 = 225$.

SOLUTION

This equation can be written in the form

$$\frac{x^2}{25} + \frac{y^2}{9} = 1$$

so we see that it is an ellipse centered at the origin. Since $a^2 = 25$ and $b^2 = 9$, the ellipse intersects the x-axis at 5 and -5, and it intersects the y-axis at 3 and -3. The sketch is shown in Figure 8.39.

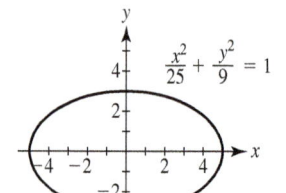

Figure 8.39

The graph of

$$\frac{x^2}{4} + \frac{y^2}{16} = 1$$

is a vertical ellipse with center at the origin and with the ends of the longer (major) axis 4 units up and 4 units down from the center. The shorter (minor) axis ends 2 units to the left and 2 units to the right of the center. The graph is shown in Figure 8.40.

The graph of

$$\frac{(x-1)^2}{4} + \frac{(y+2)^2}{16} = 1$$

is the graph in Figure 8.40 shifted to the right 1 unit and down 2 units. The center is shifted from $(0, 0)$ to $(1, -2)$; the *lengths* of the axes are unchanged (see Figure 8.41).

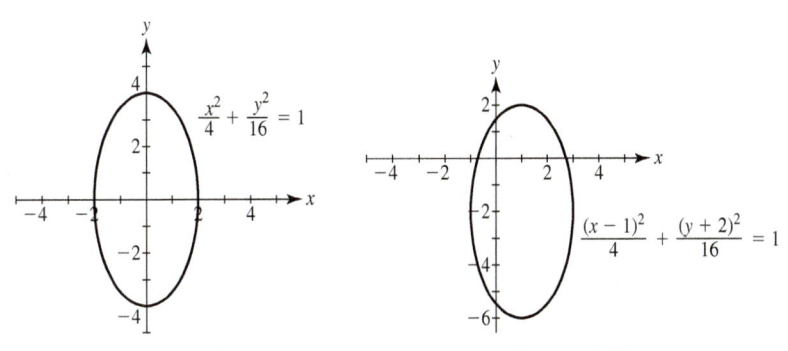

Figure 8.40

Figure 8.41

In general, we have the following.

Horizontal and Vertical Shifts of Ellipses

The graph of the equation

$$\frac{(x-h)^2}{a^2} + \frac{(y-k)^2}{b^2} = 1$$

is the graph of the equation $\frac{x^2}{a^2} + \frac{y^2}{b^2} = 1$ with the center shifted h units horizontally and k units vertically.

The *general form* of the equation of an ellipse with its major axis parallel to the x-axis or the y-axis is

$$Ax^2 + Cy^2 + Dx + Ey + F = 0 \; (A \neq C, AC > 0)$$

As with the circle, we can complete the square to write the equation in its *standard form*.

EXAMPLE 2 ▶ Converting the Equation of an Ellipse to Standard Form

Write the standard form and sketch the graph of $4x^2 + y^2 + 16x - 2y + 1 = 0$.

SOLUTION

From the coefficients of x^2 and y^2 we see that this equation represents an ellipse. We obtain the standard form as follows.

$$4x^2 + y^2 + 16x - 2y + 1 = 0$$
$$4x^2 + 16x + y^2 - 2y = -1 \qquad \text{Grouping}$$
$$4(x^2 + 4x + 4) + (y^2 - 2y + 1) = -1 + 16 + 1 \qquad \text{Factoring and completing squares}$$

(Note that adding 4 inside the first parentheses actually adds 4 · 4 to the left side, so 16 must also be added to the right side.)

$$4(x+2)^2 + (y-1)^2 = 16 \quad \text{Factoring}$$

$$\frac{(x+2)^2}{4} + \frac{(y-1)^2}{16} = 1 \quad \text{Dividing both sides by 16}$$

Thus the center is $(-2, 1)$; the major axis ends 4 units above and 4 units below the center. The minor axis ends 2 units to the left and 2 units to the right of the center (see Figure 8.42).

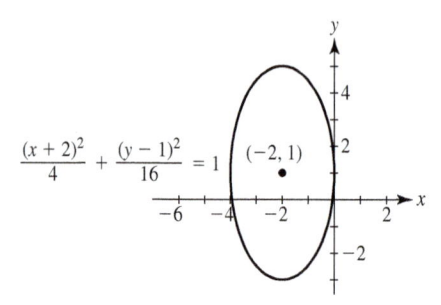

Figure 8.42

As with semicircles, equations of the top half of ellipses with the form

$$y = \sqrt{b^2 - \frac{b^2}{a^2}x^2}$$

can often be graphed with technology or by squaring both sides of the equation. In particular, squaring both sides of

$$y = \sqrt{9 - \frac{9}{25}x^2}$$

gives

$$y^2 = 9 - \frac{9}{25}x^2$$

$$\frac{9}{25}x^2 + y^2 = 9 \quad \text{or} \quad \frac{x^2}{25} + \frac{y^2}{9} = 1$$

Thus the graph of

$$y = \sqrt{9 - \frac{9}{25}x^2}$$

is the upper half of this ellipse, because only nonnegative values of y are possible (see Figure 8.43).

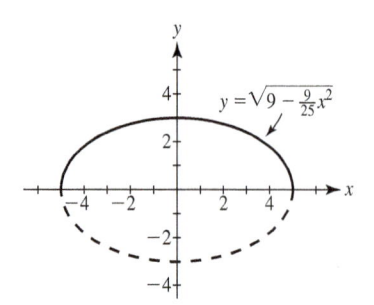

Figure 8.43

Hyperbolas

A hyperbola is defined as the set of all points in a plane such that the difference of their distances from two fixed points (called foci) is a constant. If the foci are on the x-axis equidistant from the origin, the distance formula can be used to develop the equation of this **horizontal hyperbola**, which intercepts the x-axis at its vertices a and $-a$ (see Figure 8.44).

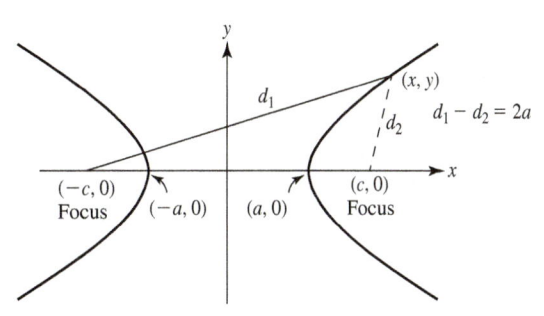

Figure 8.44 shows the graph of the horizontal hyperbola with equation

$$\frac{x^2}{a^2} - \frac{y^2}{b^2} = 1 \text{ with } b^2 = c^2 - a^2$$

A similar, but different, equation results in a **vertical hyperbola** (in which the curve intersects the y-axis). A comparison of horizontal and vertical hyperbolas follows.

Standard Forms of Equations of Hyperbolas

The horizontal hyperbola centered at the origin with x-intercepts at $(a, 0)$ and $(-a, 0)$ and foci at $(c, 0)$ and $(-c, 0)$ has the equation

$$\frac{x^2}{a^2} - \frac{y^2}{b^2} = 1$$

with $b^2 = c^2 - a^2$.

The vertical hyperbola centered at the origin with x-intercepts at $(0, a)$ and $(0, -a,)$ and foci at $(0, c)$ and $(0, -c)$ has the equation

$$\frac{y^2}{a^2} - \frac{x^2}{b^2} = 1$$

with $b^2 = c^2 - a^2$.

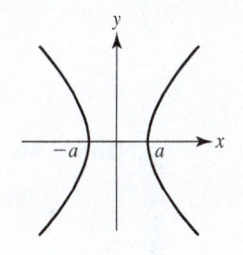

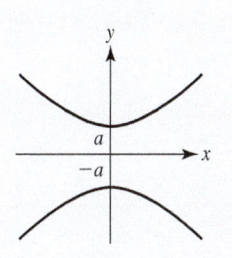

Figure 8.45 also shows the graph of $\frac{x^2}{a^2} - \frac{y^2}{b^2} = 1$. Note that the graph does not intersect the y-axis and that the curve approaches, but does not touch, the *asymptotes* (the dashed lines in Figure 8.45). The equations for the asymptotes are found by solving for y, which yields

$$y = \pm \frac{b}{a}\sqrt{x^2 - a^2}$$

and observing that x gets much larger than a as $|x| \to \infty$, so $\sqrt{x^2 - a^2} \to \sqrt{x^2}$. Hence y approaches $\pm \left(\frac{b}{a}\right)x$ as $|x| \to \infty$. Thus, if the hyperbola is centered at the origin, the asymptotes are lines through the origin with slopes b/a and $-b/a$, respectively.

Figure 8.45

EXAMPLE 3 ▶ Graphing the Equation of a Hyperbola

Graph the equation $9x^2 - 4y^2 = 36$.

SOLUTION

Dividing both sides of the equation by 36 gives the equation in the standard form:

$$\frac{x^2}{4} - \frac{y^2}{9} = 1$$

We see that its graph is a horizontal hyperbola with

Center: $(0, 0)$

Vertices: 2 units to left and 2 units to the right of center: $(-2, 0), (2, 0)$

Asymptotes: Through $(0, 0)$ with slopes $3/2$ and $-3/2$

Graph: See Figure 8.46.

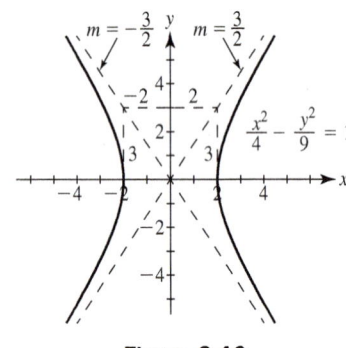

Figure 8.46

The standard forms of the equations of hyperbolas that are shifted horizontally and vertically follow.

Horizontal and Vertical Shifts of Hyperbolas

The graph of the equation

$$\frac{(x-h)^2}{a^2} - \frac{(y-k)^2}{b^2} = 1$$

is the graph of the equation $\frac{x^2}{a^2} - \frac{y^2}{b^2} = 1$ with the center shifted h units horizontally and k units vertically.

The graph of the equation

$$\frac{(y-k)^2}{a^2} - \frac{(x-h)^2}{b^2} = 1$$

is the graph of the equation $\frac{y^2}{a^2} - \frac{x^2}{b^2} = 1$ with the center shifted h units horizontally and k units vertically.

The graph of $\frac{(x-1)^2}{4} - \frac{(y+2)^2}{9} = 1$ is the graph of $\frac{x^2}{4} - \frac{y^2}{9} = 1$ shifted to the right 1 unit and down 2 units. The center is shifted from $(0, 0)$ to $(1, -2)$; the vertices are still 2 units to the left and to the right of the center, and the asymptotes still pass through the center with slopes 3/2 and −3/2. This graph is shown in Figure 8.47.

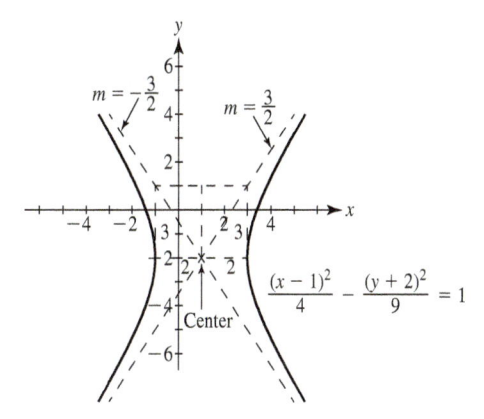

Figure 8.47

For graphing purposes, we can convert the general form of a hyperbola,

$$Ax^2 + Cy^2 + Dx + Ey + F = 0 \quad (AC < 0)$$

to its standard form by completing the square.

EXAMPLE 4 ▶ Converting the Equation of a Hyperbola to Standard Form

Write the standard form and sketch the graph of

$$x^2 - 4y^2 - 2x - 8y + 1 = 0$$

SOLUTION

Note from the coefficients of x^2 and y^2 that this equation represents a hyperbola. We obtain the standard form as follows.

$$(x^2 - 2x) - 4(y^2 + 2y) = -1 \quad \text{Grouping}$$

$$(x^2 - 2x + 1) - 4(y^2 + 2y + 1) = -1 + 1 - 4 \quad \text{Completing squares}$$

$$(x - 1)^2 - 4(y + 1)^2 = -4$$

$$\frac{(x - 1)^2}{-4} + \frac{(y + 1)^2}{1} = 1$$

$$\frac{(y + 1)^2}{1} - \frac{(x - 1)^2}{4} = 1 \quad \text{Standard form}$$

This is the standard form for the equation of a vertical hyperbola with

Center: $(1, -1)$

Vertices: 1 unit above and 1 unit below center: $(1, 0)$, $(1, -2)$

Asymptotes: Through center with slopes $\pm 1/2$

Graph: See Figure 8.48.

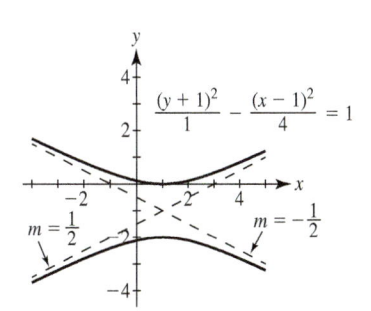

Figure 8.48

Rectangular Hyperbolas

Although we have not discussed quadratic equations containing xy terms, there is one other important equation form that has the hyperbola as its graph. This equation has the form $xy = c$, and its graph is called a **rectangular hyperbola**. If $c > 0$, the graph has the form shown in Figure 8.49(a), and if $c < 0$, the graph has the form shown in Figure 8.49(b).

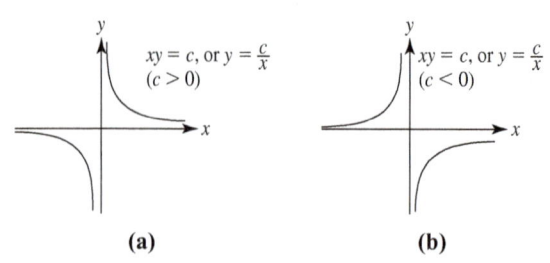

Figure 8.49

SKILLS CHECK 8.7

Answers that are not seen can be found in the answer section at the back of the text.

In Exercises 1–16, sketch the graph of each equation. In Exercises 1–12, give the center and the endpoints of the major axis.

1. $\dfrac{x^2}{9} + \dfrac{y^2}{4} = 1$
2. $\dfrac{x^2}{9} + \dfrac{y^2}{5} = 1$
3. $4x^2 + y^2 = 16$
4. $6x^2 + 49y^2 - 294 = 0$
5. $x^2 + 4y^2 = 1$
6. $2x^2 + 3y^2 = 6$
7. $\dfrac{(x+2)^2}{4} + \dfrac{(y-1)^2}{16} = 1$
8. $\dfrac{(x-1)^2}{9} + y^2 = 1$
9. $x^2 + 4y^2 + 4x - 8y + 4 = 0$
10. $2x^2 + y^2 - 4x + 6y + 7 = 0$
11. $9x^2 + y^2 + 6y = 0$
12. $4x^2 + 9y^2 - 24x = 0$
13. $y = \dfrac{\sqrt{1-4x^2}}{3}$
14. $y = -\sqrt{4-9x^2}$
15. $x = \dfrac{-2\sqrt{225-y^2}}{3}$
16. $x = 2\sqrt{1-y^2}$

In Exercises 17–20, determine whether the graph of each equation is a circle or an ellipse.

17. $\dfrac{y^2}{16} + \dfrac{x^2}{25} = 1$ Ellipse
18. $\dfrac{x^2}{50} + \dfrac{y^2}{50} = 1$ Circle
19. $9x^2 + 9y^2 + 18x - 36y - 180 = 0$ Circle
20. $4x^2 + 2y^2 + 16x - 8y - 82 = 0$ Ellipse
21. Sketch the graph of the ellipse with center at $(0, 0)$, vertical major axis 10 units long, and minor axis 2 units long.
22. Sketch the graph of the ellipse that has a minor axis half the length of its major axis and that has major axis endpoints at $(3, 16)$ and $(3, -12)$.

In Exercises 23–46, sketch the graph of each equation. Give the center and the slopes of the asymptotes in Exercises 23–34.

23. $\dfrac{x^2}{16} - \dfrac{y^2}{25} = 1$
24. $\dfrac{x^2}{25} - \dfrac{y^2}{9} = 1$
25. $y^2 - x^2 = 1$
26. $5y^2 - x^2 - 20 = 0$
27. $y^2 - 4x^2 = 4$
28. $y^2 - 4x^2 - 6 = 16$
29. $\dfrac{(x+1)^2}{36} - \dfrac{(y+4)^2}{36} = 1$
30. $\dfrac{(y+3)^2}{4} - \dfrac{(x-0.5)^2}{9} = 1$
31. $x^2 - 4y^2 - 24y - 40 = 0$
32. $4x^2 - 4y^2 + 36x + 77 = 0$
33. $5x^2 - 4y^2 + 20x - 16y + 24 = 0$
34. $x^2 - 4y^2 - 8x + 56y - 164 = 0$
35. $y = -\sqrt{1+x^2}$
36. $y = \sqrt{4+9x^2}$
37. $x = \dfrac{\sqrt{9+y^2}}{3}$
38. $x = \dfrac{-2\sqrt{25+y^2}}{5}$
39. $xy = 16$
40. $xy = 2$
41. $xy = -\dfrac{1}{2}$
42. $xy = -9$
43. $(x-3)y = 4$
44. $(x+1)(y-1) = 9$
45. The left half of the hyperbola $x^2 - y^2 = 1$
46. The upper half of the hyperbola $y^2 - 9x^2 = 36$

EXERCISES 8.7

Answers that are not seen can be found in the answer section at the back of the text.

47. Bridges An arch under a bridge is the upper half of an ellipse with a width of 100 feet and a height of 16 feet at its highest point. What is the height of the arch 19 feet from its center? 14.8 feet

48. Bridges An arch of a bridge across a stream is in the shape of half an ellipse. If the span is 50 feet wide and 12 feet high, write the equation of an ellipse whose upper half represents this arch.

49. Halley's Comet The orbit of Halley's Comet is an ellipse 36.18 AU (astronomical units) long and 9.12 AU wide. Write the equation of an ellipse that represents this path.

50. Roman Colosseum The outside of the Roman Colosseum is an ellipse with a major axis of length 620 feet and a minor axis of length 513 feet. Write the equation of this ellipse.

51. Ohm's Law Ohm's law states that $E = IR$, where E is the voltage in a circuit, I is the current, and R is the resistance. In a circuit where voltage is a constant 120 volts, sketch the graph of current as a function of resistance.

52. Boyle's Law Boyle's law states that for a perfect gas at a constant temperature, the product of its pressure P and its volume V is a constant. Sketch the graph of pressure as a function of volume if their product is 24.

53. Ellipses The ellipse in Figure 8.37 is the set of all points such that the sum of the distances from the foci, $(-c, 0)$ and $(c, 0)$, to any point (x, y) on the ellipse is a constant, $2a$. Thus any point (x, y) on this ellipse satisfies the equation

$$\sqrt{(x+c)^2 + y^2} + \sqrt{(x-c)^2 + y^2} = 2a$$

Show that if we let $b^2 = a^2 - c^2$, this equation can be written in the form

$$\frac{x^2}{a^2} + \frac{y^2}{b^2} = 1$$

54. Hyperbolas The hyperbola in Figure 8.44 is the set of all points such that the difference of the distances from the foci, $(-c, 0)$ and $(c, 0)$, to any point (x, y) on the hyperbola is a constant, $\pm 2a$. Thus any point (x, y) on this hyperbola satisfies the equation

$$\sqrt{(x+c)^2 + y^2} - \sqrt{(x-c)^2 + y^2} = 2a$$

Show that if we let $b^2 = c^2 - a^2$, this equation can be written in the form

$$\frac{x^2}{a^2} - \frac{y^2}{b^2} = 1$$

chapter 8 SUMMARY

In this chapter, we graphed linear inequalities in two variables and systems of linear inequalities in two variables as regions in the two-dimensional plane. We discussed an important application of linear inequalities, linear programming, where we seek a point or points in a region determined by constraints that give an optimal value for an objective function.

We also discussed sequences, which are special discrete functions whose domains are sets of positive integers. Sequences can be used to solve numerous problems involving discrete inputs, including investment problems. We also discussed series, which are the sums of terms of a sequence. Series are useful in finding the future value of annuities if the interest is compounded over discrete time periods.

Finally, we discussed the Binomial Theorem and conic sections (circles, parabolas, ellipses, and hyperbolas).

Key Concepts and Formulas

8.1 Systems of Inequalities

Linear inequalities in two variables	The solution to a linear inequality in two variables is the set of ordered pairs that satisfy the inequality.
Systems of linear inequalities	The solution to a system of linear inequalities in two or more variables consists of the values of the variables that satisfy all the inequalities.
Borders	The lines that bound the solution region are called the borders of the system of linear inequalities.
Solution region	The region that includes the half-plane containing the ordered pairs that satisfy an inequality is called the solution region. The border line is included in the solution region if the inequality includes an equal sign.
Corner	A corner of the graph of a solution to a system of inequalities occurs at a point where boundary lines intersect.
Systems of nonlinear inequalities	As with systems of linear inequalities in two variables, the solutions of systems of nonlinear inequalities in two variables are regions in a plane.

8.2 Linear Programming: Graphical Methods

Linear programming	Linear programming is a technique that can be used to solve problems when the constraints on the variables can be expressed as linear inequalities and a linear objective function is to be maximized or minimized.
Constraints	The inequalities that limit the values of the variables in a linear programming application are called the constraints.
Feasible region	The constraints of a linear programming application form a feasible region in which the solution lies.
Feasible solution	Any point in the region determined by the constraints is called a feasible solution.
Optimal value	A maximum or minimum value of an objective function is called an optimal value.
Solving a linear programming problem	1. If a linear programming problem has a solution, then the optimal value (maximum or minimum) of an objective function occurs at a corner of the feasible region determined by the constraints. 2. When a feasible region for a linear programming problem is closed and bounded, the objective function has a maximum and a minimum value. 3. When the feasible region is not closed and bounded, the objective function may have a maximum only, a minimum only, or neither.

8.3 Sequences and Discrete Functions

Sequence	A sequence is a function whose outputs result uniquely from positive integer inputs. Because the domains of sequences are positive integers, sequences are discrete functions.
Terms of the sequence	The ordered outputs corresponding to the integer inputs of a sequence are called the terms of the sequence.
Infinite sequence	If the domain of a sequence is the set of all positive integers, the outputs form an infinite sequence.

Finite sequence	If the domain of a sequence is a set of positive integers from 1 to n, the outputs form a finite sequence.
Arithmetic sequence	A sequence is called an arithmetic sequence if there exists a number d, called the common difference, such that $a_n = a_{n-1} + d$ for $n > 1$.
nth term of an arithmetic sequence	The nth term of an arithmetic sequence is given by $a_n = a_1 + (n-1)d$, where a_1 is the first term of the sequence, n is the number of the term, and d is the common difference between the terms.
Geometric sequence	A sequence is called a geometric sequence if there exists a number r, the common ratio, such that $a_n = r a_{n-1}$ for $n > 1$.
nth term of a geometric sequence	The nth term of a geometric sequence is given by $a_n = a_1 r^{n-1}$, where a_1 is the first term of the sequence, n is the number of the term, and r is the common ratio of consecutive terms.

8.4 Series

Series	A series is the sum of the terms of a sequence.
Finite series	A finite series is defined by $a_1 + a_2 + a_3 + \cdots + a_n$, where $a_1, a_2, \ldots a_n$ are the terms of a sequence.
Infinite series	An infinite series is defined by $a_1 + a_2 + a_3 + \cdots + a_n + \cdots$.
Sum of n terms of an arithmetic sequence	The sum of the first n terms of an arithmetic sequence is given by the formula $s_n = \dfrac{n(a_1 + a_n)}{2}$, where a_1 is the first term of the sequence and a_n is the nth term.
Sum of n terms of a geometric sequence	The sum of the first n terms of a geometric sequence is $s_n = \dfrac{a_1(1 - r^n)}{1 - r}$, where a_1 is the first term of the sequence and r is the common ratio, where $r \neq 1$.
Infinite geometric series	An infinite geometric series sums to $S = \dfrac{a_1}{1 - r}$, where a_1 is the first term and r is the common ratio, if $|r| < 1$. If $|r| \geq 1$, the sum does not exist (is not a finite number).

8.5 The Binomial Theorem

Pascal's triangle	Pascal's triangle can be used to find the coefficients of $(x + y)^n$ for larger values of n.
Factorial notation	For any positive integer n, we define n factorial as $$n! = n(n-1)(n-2)(n-3) \cdots (3)(2)(1)$$ and we define $0! = 1$.
Binomial coefficients	The coefficient of $x^{n-k}y^k$ in the expansion of $(x + y)^n$ is equal to $\dfrac{n!}{k!(n-k)!}$.
Binomial theorem	For any positive integer n, $$(x + y)^n = x^n + \binom{n}{1}x^{n-1}y^1 + \binom{n}{2}x^{n-2}y^2 + \binom{n}{3}x^{n-3}y^3 + \cdots$$ $$+ \binom{n}{k}x^{n-k}y^k + \cdots + \binom{n}{n-1}x^1 y^{n-1} + y^n$$

8.6 Conic Sections: Circles and Parabolas

Conic sections	Conic sections can be defined as intersections of planes and cones as the Greeks defined them, as the graphs of the second-degree equation $Ax^2 + Bxy + Cy^2 + Dx + Ey + F = 0$, or as sets of all points satisfying specific geometric properties.
The distance formula	The distance between the points $P_1(x_1, y_1)$ and $P_2(x_2, y_2)$ in a plane is $d = \sqrt{(x_2 - x_1)^2 + (y_2 - y_1)^2}$.
The midpoint formula	The midpoint M of the line segment joining $P_1(x_1, y_1)$ and $P_2(x_2, y_2)$ is $M\left(\dfrac{x_1 + x_2}{2}, \dfrac{y_1 + y_2}{2}\right)$.
Equation of a circle	The equation $(x - h)^2 + (y - k)^2 = r^2$ is the standard form of an equation of a circle with center at (h, k) and radius r.
Parabolas	A parabola is the set of all points in a plane that are equidistant from a fixed point called the focus and a fixed line called the directrix. The midpoint between the focus and the directrix is the vertex, and the line passing through the vertex and the focus is the axis of the parabola.
Equation of a parabola	The parabola with vertex at $(0, p)$ and directrix $y = -p$ has equation $x^2 = 4py$, and its axis is vertical. The parabola with vertex at $(p, 0)$ and directrix $x = -p$ has equation $y^2 = 4px$, and its axis is horizontal.
Shifted graphs of parabolas	The graphs of equations of the form $x^2 = 4py$ or $y^2 = 4px$ are parabolas with vertices at $(0, 0)$. If a parabola is shifted, its graph is the same as the graph of one of these equations except that its vertex is shifted from $(0, 0)$ to (h, k), and its equation is the same except that x is replaced by $x - h$, and y is replaced by $y - k$.

8.7 Conic Sections: Ellipses and Hyperbolas

Ellipses	An ellipse is defined as the set of all points in a plane such that the sum of their distances from two fixed points (called foci) is a constant.
Horizontal ellipse	The ellipse centered at the origin with vertices at $(a, 0)$ and $(-a, 0)$, foci at $(c, 0)$ and $(-c, 0)$, and endpoints of the minor axis at $(0, b)$ and $(0, -b)$ has the equation $\dfrac{x^2}{a^2} + \dfrac{y^2}{b^2} = 1$.
Vertical ellipse	The ellipse centered at the origin with vertices at $(0, a)$ and $(0, -a)$, foci at $(0, c)$ and $(0, -c)$, and endpoints of the minor axis at $(b, 0)$ and $(-b, 0)$ has the equation $\dfrac{y^2}{a^2} + \dfrac{x^2}{b^2} = 1$.
Shifted graphs of ellipses	The graph of the equation $$\frac{(x - h)^2}{a^2} + \frac{(y - k)^2}{b^2} = 1$$ is the graph of the equation $\dfrac{x^2}{a^2} + \dfrac{y^2}{b^2} = 1$ with the center shifted h units horizontally and k units vertically.

Hyperbolas	A hyperbola is defined as the set of all points in a plane such that the difference of their distances from two fixed points (called foci) is a constant.
Shifted graphs of hyperbolas	The graph of the equation $\dfrac{(x-h)^2}{a^2} - \dfrac{(y-k)^2}{b^2} = 1$ is the graph of the equation $\dfrac{x^2}{a^2} - \dfrac{y^2}{b^2} = 1$ with the center shifted h units horizontally and k units vertically. The graph of the equation $\dfrac{(y-k)^2}{a^2} - \dfrac{(x-h)^2}{b^2} = 1$ is the graph of the equation $\dfrac{y^2}{a^2} - \dfrac{x^2}{b^2} = 1$ with the center shifted h units horizontally and k units vertically.

chapter 8 SKILLS CHECK

Answers that are not seen can be found in the answer section at the back of the text.

Graph the inequalities in Exercises 1 and 2.

1. $5x + 2y \leq 10$
2. $5x - 4y > 12$

Graph the solutions to the systems of inequalities in Exercises 3–6. Identify the corners of the solution regions.

3. $\begin{cases} 2x + y \leq 3 \\ x + y \leq 2 \\ x \geq 0, y \geq 0 \end{cases}$

4. $\begin{cases} 3x + 2y \leq 6 \\ 3x + 6y \leq 12 \\ x \geq 0, y \geq 0 \end{cases}$

5. $\begin{cases} 2x + y \leq 30 \\ x + y \leq 19 \\ x + 2y \leq 30 \\ x \geq 0, y \geq 0 \end{cases}$

6. $\begin{cases} 2x + y \leq 10 \\ x + 2y \leq 11 \\ x \geq 0, y \geq 0 \end{cases}$

Graph the solutions to the systems of nonlinear inequalities in Exercises 7 and 8.

7. $\begin{cases} 15x - x^2 - y \geq 0 \\ y - \dfrac{44x + 60}{x} \geq 0 \\ x \geq 0, y \geq 0 \end{cases}$

8. $\begin{cases} x^3 - 26x + 100 - y \geq 0 \\ 19x^2 - 20 - y \leq 0 \\ x \geq 0, y \geq 0 \end{cases}$

9. Minimize $g = 3x + 4y$ subject to the constraints   Minimum: 14 at (2, 2)
$\begin{cases} 3x + y \geq 8 \\ 2x + 5y \geq 14 \\ x \geq 0, y \geq 0 \end{cases}$

10. Maximize the function $f = 7x + 12y$ subject to the constraints Maximum: 168 at (12, 7)
$\begin{cases} 7x + 3y \leq 105 \\ 2x + 5y \leq 59 \\ x + 7y \leq 70 \\ x \geq 0, y \geq 0 \end{cases}$

11. Find the maximum value of $f = 3x + 5y$ and the values of x and y that give that value, subject to the constraints Maximum: 79 at (8, 11)
$\begin{cases} 2x + y \leq 30 \\ x + y \leq 19 \\ x + 2y \leq 30 \\ x \geq 0, y \geq 0 \end{cases}$

Determine whether the sequences in Exercises 12–14 are arithmetic or geometric. Then state the value of either r or d.

12. $\dfrac{1}{9}, \dfrac{2}{3}, 4, 24, \ldots$ Geometric; $r = 6$

13. $4, 16, 28, \ldots$ Arithmetic; $d = 12$

14. $16, -12, 9, \dfrac{-27}{4}$ Geometric; $r = -\dfrac{3}{4}$

15. Find the fifth term of the geometric sequence with first term 64 and eighth term $\dfrac{1}{2}$. 4

16. Find the sixth term of the geometric sequence $\dfrac{1}{9}, \dfrac{1}{3}, 1, \ldots$ 27

17. Find the sum of the first 10 terms of the geometric sequence with first term 5 and common ratio -2. -1705

18. Find the sum of the first 12 terms of the sequence $3, 6, 9, \ldots$. 234

19. Find $\sum_{i=1}^{\infty} \left(\frac{4}{5}\right)^i$, if possible. 4

20. Find the sum of the first 5 terms of the arithmetic sequence with first term 20 and common difference -3. 70

21. Find the sum of the first 7 terms of the geometric sequence $40, 120, 360, \ldots$. $43,720$

22. Find the sum represented by the series $4096 + 1024 + 256 + 64 + \ldots$, if possible. $16,384/3$

23. Find the sum $\sum_{i=1}^{6} \left(\frac{1}{3}\right)^i$. $364/729$

24. Find the sum $\sum_{i=1}^{\infty} \left(\frac{5}{3}\right)^i$, if possible. Not possible; infinite

25. Evaluate $\frac{6!}{5!3!}$. 1

26. Evaluate $\binom{9}{4}$. 126

27. Evaluate $_{10}C_7$. 120

28. Expand the binomial power: $(x + 2y)^4$ $x^4 + 8x^3y + 24x^2y^2 + 32xy^3 + 16y^4$

29. Expand the binomial power: $(2a - 3b)^6$

30. Write the fourth term of $(m - n)^7$. $-35m^4n^3$

31. Write the sixth term of $(2x + 3y)^9$. $489,888x^4y^5$

32. Write the equation of a circle with center at $(2, -1)$ and radius 3. Graph the circle. $(x - 2)^2 + (y + 1)^2 = 9$

33. Find the radius and center of the circle $x^2 - 6x + y^2 - 2y = 15$. $r = 5$, center at $(3, 1)$

34. Write the equation of the lower semicircle centered at $(0, 0)$ with radius 6. $y = -\sqrt{36 - x^2}$

35. Graph the equation $y^2 + 6x = 0$.

36. Graph the equation $y = x^2 - 8x + 16$.

37. Graph the equation $4y^2 - 4y + 12x + 37 = 0$.

38. Graph the equation $\frac{x^2}{4} + \frac{y^2}{9} = 1$, and give the center and the endpoints of the major axis.

39. Graph the equation $\frac{(x-2)^2}{16} + \frac{(y+1)^2}{4} = 1$, and give the center and the endpoints of the major axis.

40. Sketch the graph of $y = -\sqrt{9 - 4x^2}$.

41. Sketch the graph of the equation $\frac{x^2}{9} - \frac{y^2}{25} = 1$.

42. Sketch the graph of the equation $16y^2 - 9x^2 = 64$.

43. Sketch the graph of the equation $xy = -12$.

chapter 8 REVIEW

Answers that are not seen can be found in the answer section at the back of the text.

44. **Manufacturing** Ace Manufacturing produces two types of DVD players: the Deluxe model and the Superior model, which also plays videotapes. Each Deluxe model requires 3 hours to assemble and $40 for assembly parts, and each Superior model requires 2 hours to assemble and $60 for assembly parts. The company is limited to 1800 assembly hours and $36,000 for assembly parts each month. If the profit on the Deluxe model is $30 and the profit on the Superior model is $40, how many of each model should be produced to maximize profit? What is the maximum profit? $25,200 profit at 360 units of each

45. **Manufacturing** A company manufactures two grades of steel, cold steel and stainless steel, at the Pottstown and Ethica factories. The table gives the daily cost of operation and the daily production capabilities of each of the factories, as well as the number of units of each type of steel that is required to fill an order.

a. How many days should each factory operate to fill the order at minimum cost? Pottstown, 20 days; Ethica, 30 days

b. What is the minimum cost? $1,120,000

	Units/Day at Pottstown	Units/Day at Ethica	Required for Order
Units of Cold Steel	20	40	At least 1600
Units of Stainless Steel	60	40	At least 2400
Cost/Day for Operation ($)	20,000	24,000	

46. **Nutrition** A laboratory wishes to purchase two different feeds, A and B, for its animals. Feed A has 2 units of carbohydrates per pound and 4 units of protein per pound, and feed B has 8 units of carbohydrates per pound and 2 units of protein per pound. Feed A costs $1.40 per pound, and feed B costs $1.60 per pound. If at least 80 units of carbohydrates and 132 units of protein are required, how many pounds of each feed are required to minimize the cost? What is the minimum cost? Minimum cost is $48 with 32 units of feed A and 2 units of feed B

47. **Manufacturing** A company manufactures leaf blowers and weed wackers. One line can produce 260 leaf blowers per day, and another line can produce 240 weed wackers per day. The combined number of leaf blowers and weed wackers that shipping can handle is 460 per day. How many of each should be produced daily to maximize company profit if the profit is $5 on the leaf blowers and $10 on the weed wackers? 220 leaf blowers and 240 weed wackers

48. **Profits** A woman has a building with 60 two-bedroom and 40 three-bedroom apartments available to rent to students. She has set the rent at $800 per month for the two-bedroom units and $1150 for the three-bedroom units. She must rent to one student per bedroom, and zoning laws limit her to at most 180 students in this building. How many of each type of apartment should she rent to maximize her revenue? 60 two-bedroom and 20 three-bedroom

49. **Loans** A finance company has at most $30 million available for auto and home equity loans. The auto loans have an annual return rate of 8% to the company, and the home equity loans have an annual rate of return of 7% to the company. If there must be at least twice as many auto loans as home equity loans, how much should be put into each type of loan to maximize the profit to the finance company? $30 million in auto loans and no home equity loans

50. **Salaries** Suppose you are offered two identical jobs, job 1 paying a starting salary of $20,000 with yearly raises of $1000 at the end of each year and job 2 paying a starting salary of $18,000 with yearly increases of $1600 at the end of each year.
    a. Which job will pay more in the fifth year on the job? Job 2
    b. Which job will pay more over the first 5 years? Job 1

51. **Drug in the Bloodstream** Suppose a 400-milligram dose of heart medicine is taken daily. During each 24-hour period, the body eliminates 40% of the drug (so that 60% remains in the body). Thus, the amount of drug in the body just after the 21 doses, over 21 days, is given by
$$400 + 400(0.6) + 400(0.6)^2 + 400(0.6)^3 + \cdots + 400(0.6)^{19} + 400(0.6)^{20}$$
Find the level of the drug in the bloodstream at this time. Approximately 999.98 mg

52. **Chess Legend** Legend has it that when the king of Persia offered to reward the inventor of chess with any prize he wanted, the inventor asked for one grain of wheat on the first square of the chessboard, with the number of grains doubled on each square thereafter for the remaining 63 squares.
    a. How many grains of wheat would there be on the 64th square? 9.22337×10^{18} grains
    b. What is the total number of grains of wheat for all 64 squares? (This is enough wheat to cover Alaska more than 3 inches deep in wheat.) 1.84467×10^{19} grains

53. **Compound Interest** Find the future value of $20,000 invested for 5 years at 6%, compounded annually. $26,764.51

54. **Annuities** Find the 5-year future value of an ordinary annuity with a contribution of $300 per month into an account that pays 12% per year, compounded monthly. $24,500.90

55. **Profit** The weekly total cost function for a product is $C(x) = 3x^2 + 1228$ dollars and the weekly revenue is $R(x) = 177.50x$, where x is the number of hundreds of units. Use graphical methods to find the number of units that gives profit for the product. Between 800 and 5200 units

56. **Bridges** The towers of a suspension bridge are 800 feet apart and rise 160 feet above the ground. The cable between the towers has the shape of a parabola, and the cable just touches the sides of the road midway between the towers. What is the height of the cable 100 feet from a tower? 90 feet

57. **Archway** An archway is the upper half of an ellipse and has a height of 20 feet and a width of 50 feet. Can a truck 14 feet high and 10 feet wide drive under the archway without going into the other lane? Yes

58. **Construction** Two houses are designed so that they are shaped and positioned like the branches of the hyperbola whose equation is $625y^2 - 900x^2 = 562,500$, where x and y are in yards. How far apart are the houses at their closest point? 60 yards

Group Activities
▶ EXTENDED APPLICATIONS

Salaries

In the 1970s, a local teachers' organization in Pennsylvania asked for a $1000 raise at the end of each year for the next 3 years. When the school board said it could not afford to do this, the teachers' organization then said that the teachers would accept a $300 raise at the end of each 6-month period. The school board initially responded favorably because the members thought it would save them $400 per year per teacher.

As an employee, would you rather be given a raise of $1000 at the end of each year (Plan I) or be given a $300 raise at the end of each 6-month period (Plan II)? To answer the question, complete the table for an employee whose base salary is $40,000 (or $20,000 for each 6-month period), and then answer the questions below.

1. Find the total additional money earned for the first 3 years from the raises of Plan I.
2. Find the total additional money earned for the first 3 years from the raises of Plan II.
3. Which plan gives more money from the raises, and how much more? Which plan is better for the employee?
4. Find the total additional money earned (per employee) from raises for each of Plan I and Plan II if they are extended to 4 years. Which plan gives more money in raises, and how much more? Which plan is better for the employee?
5. Did the school board make a mistake in thinking that the $300 raise every 6 months instead of $1000 every year would save money? If so, what was the mistake?
6. If there were 200 teachers in the school district, how much extra would Plan II cost over Plan I for the 4 years?

Year	Period (in months)	Salary Received per 6-Month Period ($) Plan I	Salary Received per 6-Month Period ($) Plan II
1	0–6	20,000	20,000
	6–12	20,000	20,300
2	12–18	20,500	20,600
	18–24	20,500	20,900
3	24–30		
	30–36		
Total for 3 years			

Appendix A

Basic Graphing Calculator Guide

Operating the TI-84 Plus and TI-84 Plus C Silver Edition

Turning the Calculator On and Off

ON — Turns the calculator on

2ND **ON** — Turns the calculator off

Adjusting the Display Contrast

2ND △ — Increases the contrast (darkens the screen)

2ND ▽ — Decreases the contrast (lightens the screen)

Note: If the display begins to dim (especially during calculations), then batteries are low and you should replace the batteries or charge the calculator.

The TI-84 Plus keyboards are divided into four zones: graphing keys, editing keys, advanced function keys, and scientific calculator keys (Figure 1).

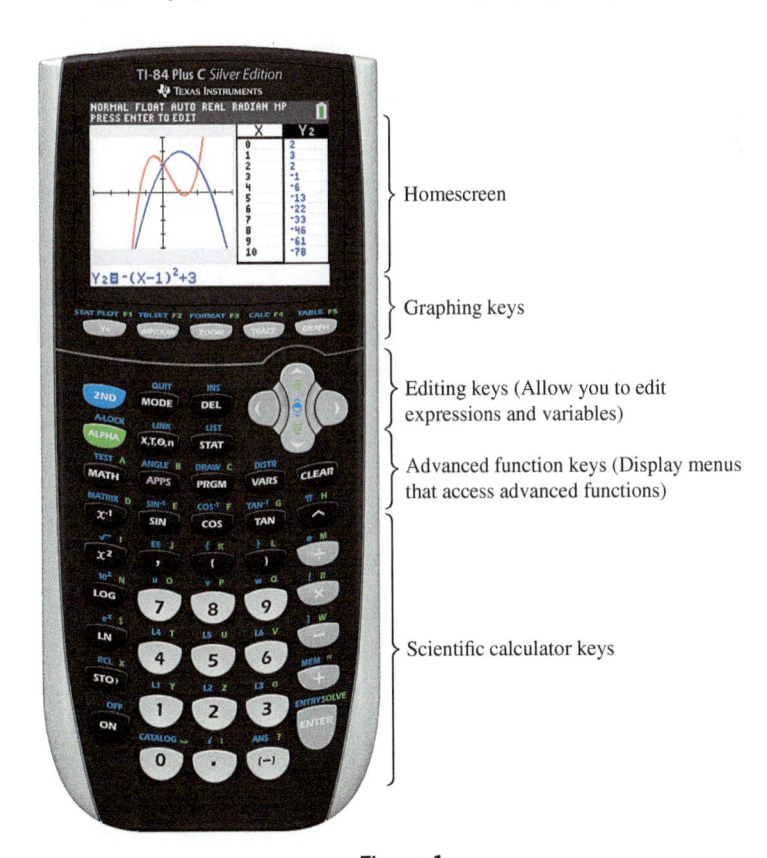

Figure 1

Keystrokes on the TI-84 Plus, and TI-84 Plus C Silver Edition

Keys	Description
ENTER	Executes commands or performs a calculation
2ND	Pressing the **2ND** key *before* another key accesses the character located above the key and printed in blue
ALPHA	Pressing the **ALPHA** key *before* another key accesses the character located above the key and printed in green
2ND **ALPHA**	Locks in the ALPHA keyboard
CLEAR	Pressing **CLEAR** once while typing on a line clears the line
	Pressing **CLEAR** after pressing **ENTER** clears the screen
2ND **MODE**	Returns to the homescreen
DEL	Deletes the character at the cursor
2ND **DEL**	Inserts characters at the underline cursor
X,T,θ,n	Enters an X in Function mode, a T in Parametric mode, a θ in Polar mode, or an n in Sequence mode
STO)	Stores a value to a variable
^	Raises to an exponent
2ND **^**	The number π
(-)	Negative symbol
MATH **1**	Converts a rational number to a fraction
MATH **)** 1: abs(	Computes the absolute value of the number or expression in parentheses
2ND **ENTER**	Recalls the last entry
ALPHA **.**	Used to enter more than one expression on a line
2ND **(-)**	Recalls the most recent answer to a calculation
x²	Squares a number or an expression
x⁻¹	Inverse; can be used with a real number or a matrix
2ND **x²**	Computes the square root of a number or an expression in parentheses
2ND **LN**	Returns the constant *e* raised to a power
ALPHA **0**	Space
2ND **(**	Moves the cursor to the beginning of an expression
2ND **)**	Moves the cursor to the end of an expression

Creating Scatter Plots

Clearing Lists

Press **STAT** and under EDIT press 4:ClrList (Figure 2). Press **2ND** **1** (L1) **ENTER** to clear List 1. Repeat the line by pressing **2ND** **ENTER** and moving the cursor to L1. Press **2ND** **2** (L2) and **ENTER** to clear List 2 (Figure 3).

Another way to clear a list in the STAT menu is, after accessing the list, to press the up arrow until the name of the list is highlighted and then press **CLEAR** and **ENTER**. *Caution:* Do not press the **DEL** key. This will not actually delete the list, but hide it from view. To view the list, press **2ND** **DEL** (INS) and type the list name.

Entering Data into Lists

Press **STAT** and under EDIT press 1:Edit. This will bring you to the screen where you enter data into lists.

Enter the *x*-values (input) in the column headed L1 and the corresponding *y*-values (output) in the column headed L2 (Figure 4).

Create a Scatter Plot of Data Points

Go to the Y= menu and turn off or clear any functions entered there. To turn off a function, move the cursor over the = sign and press **ENTER**.

Press **2ND** **Y=**, 1:Plot1, highlight On, and then highlight the first graph type (Scatter Plot). Enter Xlist:L1, Ylist:L2, and pick the point plot mark and color you want (Figure 5).

Choose an appropriate window for the graph and press **GRAPH** or **ZOOM**, 9:ZoomStat to plot the data points (Figure 6).

Tracing Along the Plot

Press **TRACE** and the right arrow to move from point to point. The *x*- and *y*-coordinates will be displayed at the bottom of the screen.

In the upper left-hand corner of the screen, Plot1:L1,L2 will be displayed. This tells you that you are tracing along the scatter plot (Figure 7).

Turning Off Stat Plot

After creating a scatter plot, turn Stat Plot1 off by pressing **Y=** moving up to Plot1 with the cursor, and pressing **ENTER**. Stat Plot1 can be turned on by pressing **ENTER** with the cursor on Plot1. It will be highlighted when it is on.

Graphing Equations

Setting Windows

The window defines the highest and lowest values of *x* and *y* on the graph of the function that will be shown on the screen. To set the window manually, press the **WINDOW** key and enter the values that you want (Figure 8).

The values that define the viewing window can also be set by using **ZOOM** keys (Figure 9). Frequently the standard window (**ZOOM** 6) is appropriate. The standard window sets values Xmin = −10, Xmax = 10, Ymin = −10, Ymax = 10. Often a decimal or integer viewing window (**ZOOM** 4 or **ZOOM** 8) gives a better representation of the graph.

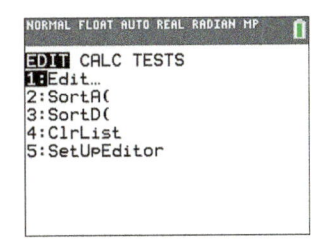

Figure 2

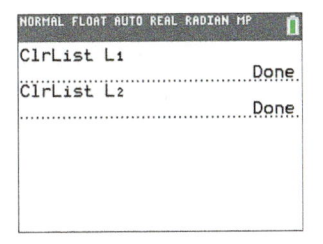

Figure 3

Figure 4

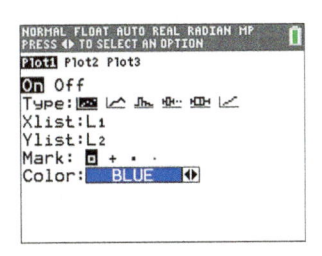

Figure 5

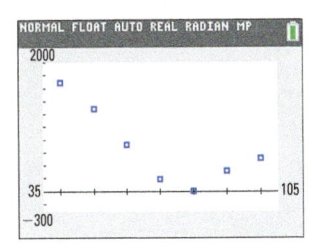

Figure 6

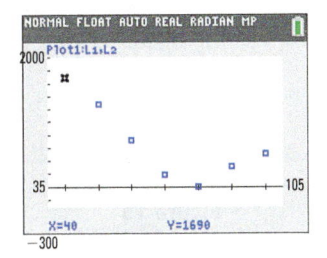

Figure 7

Figure 8

Figure 9

Figure 10 The function $y = x^3 - 3x^2 - 13$ is entered in Y1.

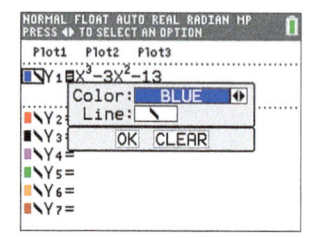

Figure 11

The window should be set so that the important parts of the graph are shown and the unseen parts are suggested. Such a graph is called complete.

Graphing Equations

To graph an equation in the variables x and y, first solve the equation for y in terms of x. If the equation has variables other than x and y, solve for the dependent variable and replace the independent variable with x.

Press the **Y=** key to access the function entry screen. You can input up to ten functions in the **Y=** menu (Figure 10). Enter the equation, using the **X,T,θ,n** key to input x. Use the subtraction key **−** between terms. Use the negation key **(−)** when the first term is negative. Use parentheses as needed so that what is entered agrees with the order of operations (Figure 10).

To change the color of the graph, use the left arrow **◁** to move to the left of Y$_1$, press **ENTER**, and choose the color and/or graph style you want. Colors are changed by using the right arrow **▷** and **ENTER** (Figure 11).

The = sign is highlighted to show that Y1 is *on* and ready to be graphed. If the = sign is not highlighted, the equation will remain but its graph is "turned off" and will not appear when **GRAPH** is pressed. (The graph is "turned on" by repeating the process.)

To erase an equation, press **CLEAR**. To return to the homescreen, press **2ND** **MODE** (QUIT).

Determine an appropriate viewing window.

Pressing **GRAPH** or a **ZOOM** key will activate the graph (Figure 12).

Finding Function Values

Using TRACE, VALUE Directly on the Graph

Enter the function to be evaluated in Y1. Choose a window that contains the x-value whose y-value you seek (Figure 13).

Turn all Stat Plots off.

Press **TRACE** and then enter the selected x-value followed by **ENTER**. The cursor will move to the selected value and give the resulting y-value if the selected x-value is in the window.

If the selected x-value is not in the window, ERROR:INVALID will occur.

If the x-value is in the window, the y-value will be given even if it is not in the window (Figures 14 and 15).

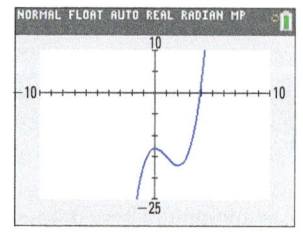

Figure 12 The graph of the function using the window $[-10, 10]$ by $[-25, 10]$.

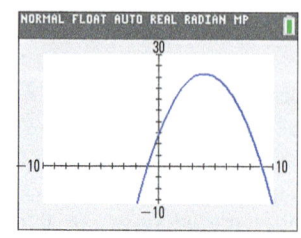

Figure 13 To evaluate $y = -x^2 + 8x + 9$ when $x = 3$ and when $x = -5$, graph the function using the window $[-10, 10]$ by $[-10, 30]$.

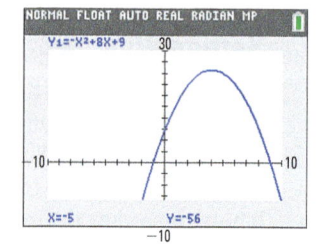

Figure 14

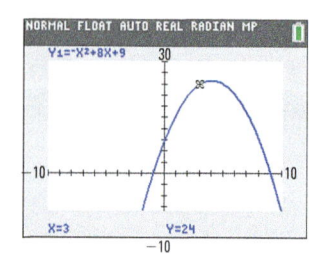

Figure 15

Basic Graphing Calculator Guide   685

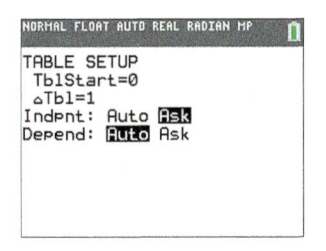

Figure 16

Figure 17

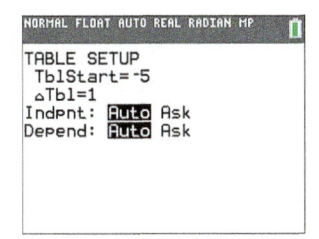

Figure 18

Figure 19

Figure 20

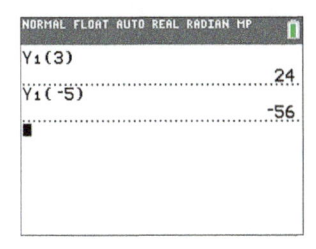

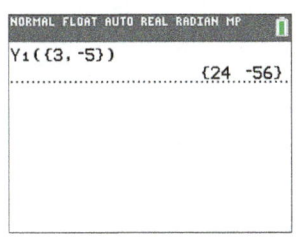

Figure 21

Figure 22

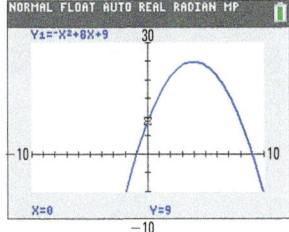

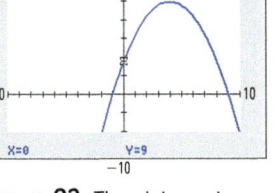

Figure 23 The y-intercept of the graph of the function $y = -x^2 + 8x + 9$ is 9.

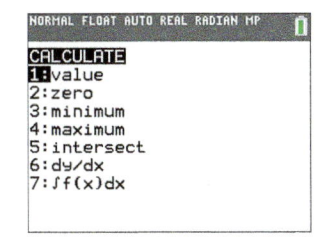

Figure 24

Using the TABLE ASK Feature

Enter the function with the `Y=` key. (*Note*: The = sign must be highlighted.) Press **2ND** `WINDOW` (TBLSET), move the cursor to Ask opposite Indpnt:, and press **ENTER** (Figure 16). This allows you to input specific values for *x*. Pressing **DEL** will clear entries in the table.

Then press **2ND** `GRAPH` (TABLE) and enter the specific values (Figure 17).

Making a Table of Values with Uniform Inputs

Press **2ND** `WINDOW` (TBLSET), enter an initial *x*-value in the table (TblStart), and enter the desired change in the *x*-value in the table (ΔTbl). Select Auto for Indpnt and Depend (Figure 18).

Enter **2ND** `GRAPH` (TABLE) to get the list of *x*-values and the corresponding *y*-values. The value of the function at the given value of *x* can be read from the table (Figure 19).

Use the up or down arrow to find the *x*-values where the function is to be evaluated (Figure 20).

Using Y-VARS

Use the `Y=` key to store $Y1 = f(x)$. Press **2ND** (QUIT).

Press **VARS**, Y-VARS 1,1 to display Y1 on the homescreen. Then press **(**, the *x*-value, **)**, and **ENTER** (Figure 21).

Alternatively, you can enter the *x*-values needed as follows: Y1 ({value 1, value 2, etc.}) **ENTER** (Figure 22). Values of the function will be displayed.

Finding Intercepts of Graphs

Solve the equation for *y*. Enter the equation with the `Y=` key.

Finding the y-Intercept

Press `TRACE` and enter the value 0. The resulting *y*-value is the *y*-intercept of the graph (Figure 23).

Finding the x-Intercept(s)

Set the window so that the intercepts to be located can be seen. The graph of a linear equation will cross the *x*-axis at most one time; the graph of a quadratic equation will cross the *x*-axis at most two times; etc.

To find the point(s) where the graph crosses the *x*-axis, press **2ND** `TRACE` to access the CALC menu. Select 2:zero (Figure 24).

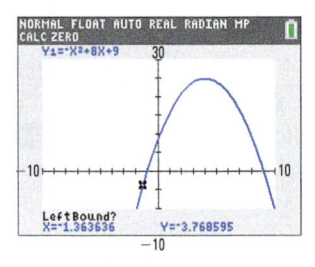

Answer the question "Left Bound?" with **ENTER** after moving the cursor close to and to the left of an *x*-intercept (Figure 25).

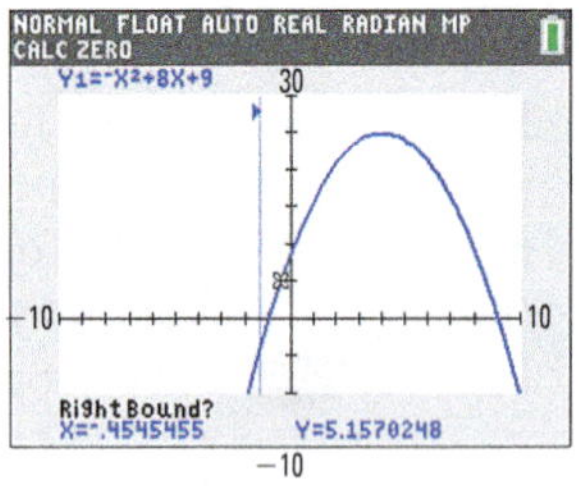

Answer the question "Right Bound?" with **ENTER** after moving the cursor close to and to the right of this *x*-intercept (Figure 26).

To answer the question "Guess?" press **ENTER**.

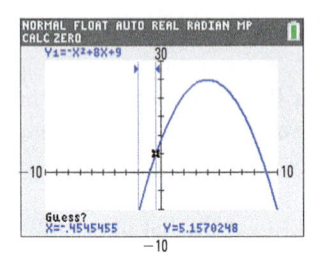

Note that vertical dotted lines show where the left bound and the right bound have been marked (Figure 27).

The coordinates of the *x*-intercept will be displayed (Figure 28). Repeat to get all *x*-intercepts (Figure 29). (Some of these values may be approximate.)

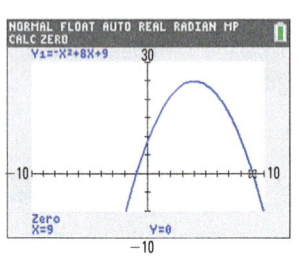

Finding Maxima and Minima

To locate a local maximum on the graph of an equation, graph the function using a window that shows all the turning points (Figure 30).

Press **2ND** **TRACE** to access the CALC menu. Select 4:maximum. The calculator will ask the question "Left Bound?" (Figure 31). Move the cursor close to and to the left of the maximum and press **ENTER**. Next the calculator will ask the question "Right Bound?" (Figure 32). Move the cursor close to and to the right of the maximum and press **ENTER**. Note that vertical dotted lines show where the left bound and the right bound have been marked. When the calculator asks "Guess?" move the cursor close to the maximum and press **ENTER** (Figure 33). The coordinates of the maximum will be displayed (Figure 34).

Note: When selecting a left bound or a right bound, you may also type in a value of *x* if you wish.

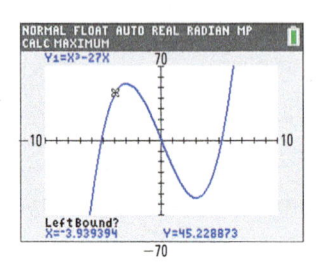

To locate a local minimum on the graph of an equation, press **2ND** **TRACE** to access the CALC menu. Select 3:minimum, and follow the same steps as above (Figure 35).

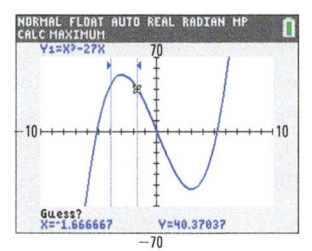

Figure 25

Figure 26

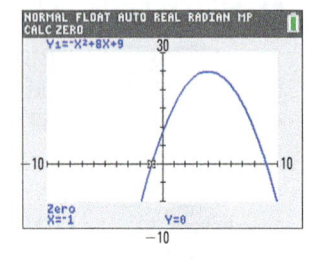

Figure 27

Figure 28

Figure 29

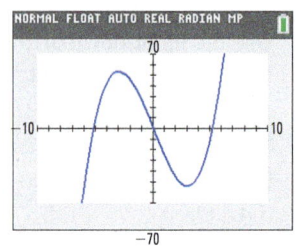

Figure 30 The graph of the function $y = x^3 - 27x$ on the window $[-10, 10]$ by $[-70, 70]$

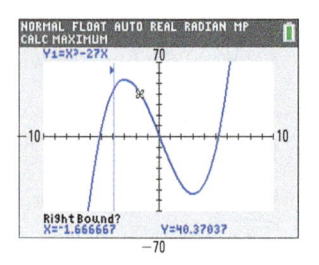

Figure 31

Figure 32

Figure 33

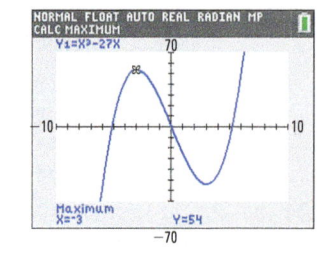

Figure 34

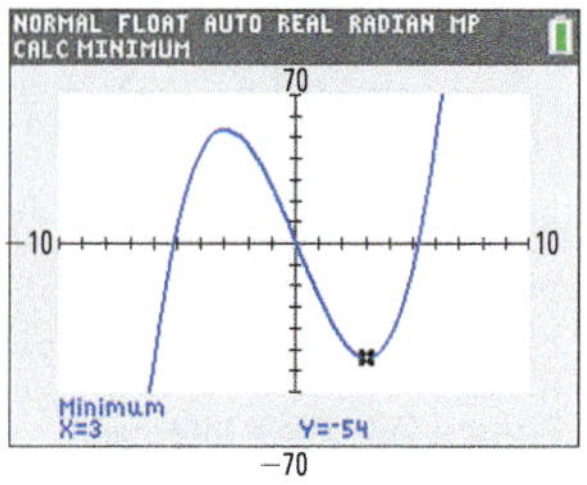

Figure 35

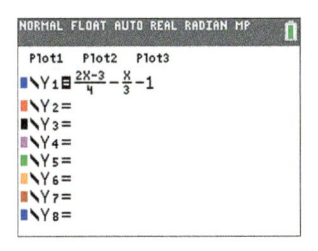

Figure 36

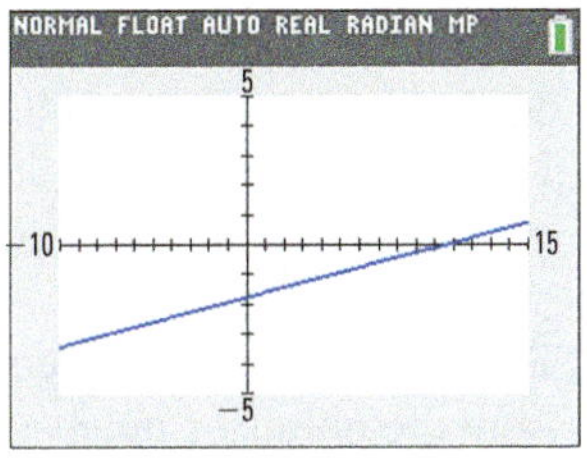

Figure 37

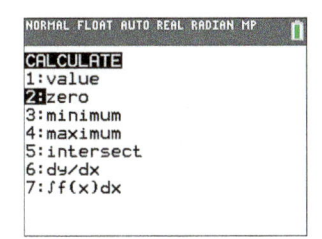

Figure 38

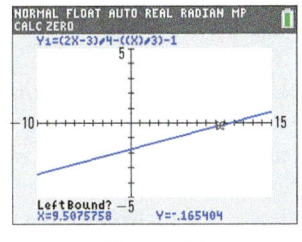

Figure 39

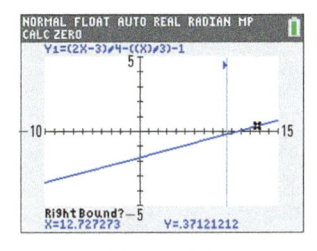

Figure 40

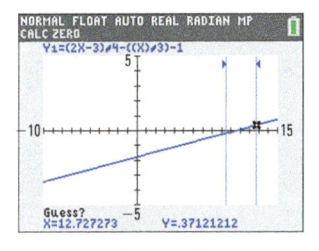

Figure 41

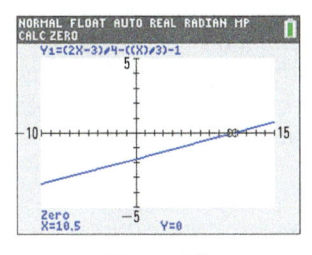

Figure 42

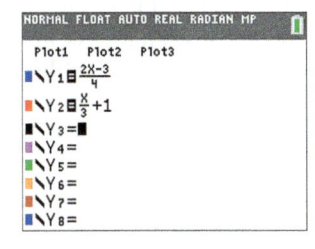

Figure 43

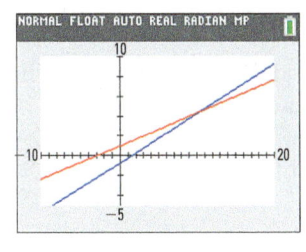

Figure 44

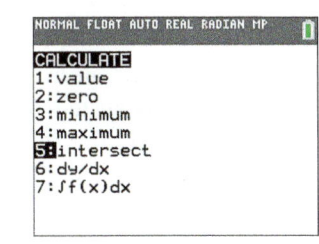

Figure 45

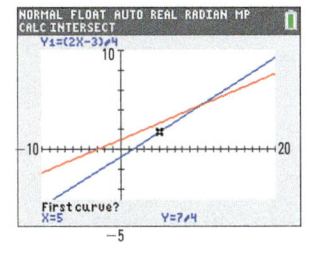

Figure 46

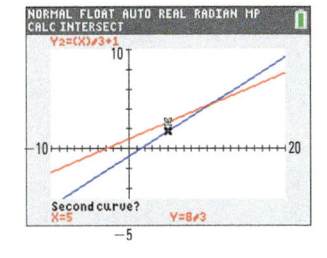

Figure 47

Basic Graphing Calculator Guide   **687**

Solving Equations Graphically

The x-Intercept Method

To solve an equation graphically using the x-intercept method, first rewrite the equation with 0 on one side of the equation. Using the Y= key, enter the nonzero side of the equation equal to Y1. Then graph the equation with a window that shows all points where the graph crosses the x-axis.

For example, to solve the equation

$$\frac{2x-3}{4} = \frac{x}{3} + 1$$

enter

$$Y1 = \frac{2x-3}{4} - \frac{x}{3} - 1$$

and graph using the window $[-10, 15]$ by $[-5, 5]$ (Figures 36 and 37).

The graph will intersect the x-axis where $y = 0$—that is, when x is a solution to the original equation. To find the point(s) where the graph crosses the x-axis and the equation has solutions, press 2ND TRACE to access the CALC menu (Figure 38). Select 2:zero.

Select "Left Bound?" (Figure 39), "Right Bound?" (Figure 40), and "Guess?" (Figure 41). The coordinates of the x-intercept will be displayed.

The x-intercept is the solution to the original equation (Figure 42).

Repeat to get all x-intercepts (and solutions).

The Intersection Method

To solve an equation graphically using the intersection method, under the Y= menu, assign the left side of the equation to Y1 and the right side of the equation to Y2. Then graph the equations using a window that contains the point(s) of intersection of the graphs.

To solve the equation

$$\frac{2x-3}{4} = \frac{x}{3} + 1$$

enter the left and right sides as shown in Figure 43 and graph using the window $[-10, 20]$ by $[-5, 10]$ (Figure 44).

Press 2ND TRACE to access the CALC menu (Figure 45). Select 5:intersect.

Answer the question "First curve?" by pressing ENTER (Figure 46) and "Second curve?" by pressing ENTER (Figure 47). (Or press the down arrow to move to one of the two curves.)

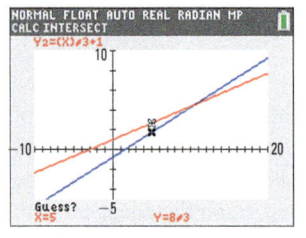

Figure 48

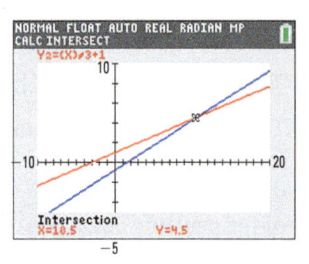

Figure 49 The solution to the equation is $x = 10.5$.

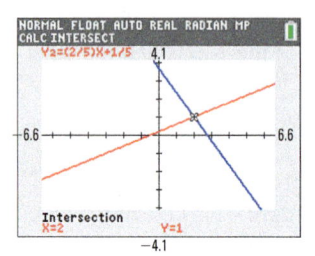

Figure 50

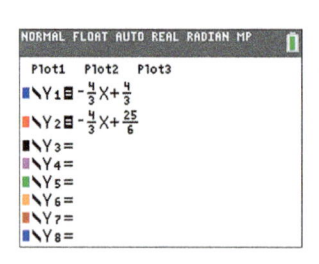

Figure 51

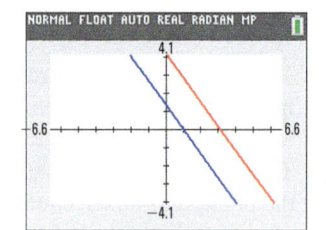

Figure 52

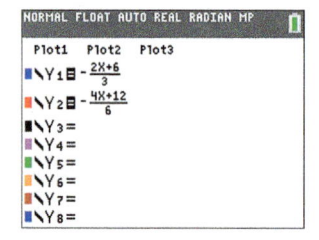

Figure 53

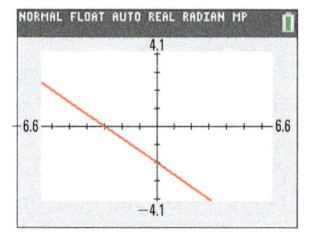

Figure 54

To the question "Guess?" move the cursor close to the desired point of intersection and press **ENTER** (Figure 48). The coordinates of the point of intersection will be displayed. Repeat to get all points of intersection.

The solution(s) to the equation will be the values of x from the points of intersection (Figure 49).

Solving Systems of Equations

To find the solution of a system of equations graphically, first solve each equation for y and use the **Y=** key with Y1 and Y2 to enter the equations. Graph the equation with an appropriate window.

For example, to solve the system

$$\begin{cases} 4x + 3y = 11 \\ 2x - 5y = -1 \end{cases}$$

graphically, we solve for y:

$$y_1 = \frac{11}{3} - \left(\frac{4}{3}\right)x$$

$$y_2 = \left(\frac{2}{5}\right)x + \frac{1}{5}$$

We graph using **ZOOM** 4:ZDecimal and then Intersect under the CALC menu to find the point of intersection. If the two lines intersect in one point, the coordinates give the x- and y-values of the solution (Figure 50). The solution of the system above is $x = 2$, $y = 1$. If the two lines are parallel, there is no solution; the system of equations is inconsistent.

For example, to solve the system

$$\begin{cases} 4x + 3y = 4 \\ 8x + 6y = 25 \end{cases}$$

we solve for y (Figure 51) and obtain the graph in Figure 52. This system has no solution. Note that if the lines are parallel, then when we solve for y the equations will show that the lines have the same slope and different y-intercepts.

If the two graphs of the equations give only one line, every point on the line gives a solution to the system and the system is dependent. For example, to solve the system

$$\begin{cases} 2x + 3y = -6 \\ 4x + 6y = -12 \end{cases}$$

we solve for y (Figure 53) and obtain the graph in Figure 54.

This system has many solutions and is dependent. Note that the two graphs will be the same graph if, when we solve for y to use the graphing calculator, the equations are equivalent.

Basic Graphing Calculator Guide 689

Solving Inequalities Graphically

To solve a linear inequality graphically, first rewrite the inequality with 0 on the right side and simplify. Then, under the Y= menu, assign the left side of the inequality to Y1, so that Y1 = $f(x)$. For example, to solve $3x > 6 + 5x$, rewrite the inequality as $3x - 5x - 6 > 0$ or $-2x - 6 > 0$. Set Y1 = $-2x - 6$.

Graph this equation so that the point where the graph crosses the *x*-axis is visible. Note that the graph will cross the axis in at most one point because the function is of degree 1. (Using ZOOM OUT can help you find this point.)

Using ZOOM 4, the graph of Y1 = $-2x - 6$ is shown in Figure 55.

Use the ZERO command under the CALC menu to find the *x*-value where the graph crosses the *x*-axis (the *x*-intercept). This value can also be found by finding the solution to $0 = f(x)$ algebraically.

The *x*-intercept of the graph of Y1 = $-2x - 6$ is $x = -3$ (Figure 56).

Observe the inequality as written with 0 on the right side. If the inequality is $<$, the solution to the original inequality is the interval (bounded by the *x*-intercept) where the graph is below the *x*-axis. If the inequality is $>$, the solution to the original inequality is the interval (bounded by the *x*-intercept) where the graph is above the *x*-axis.

The solution to $-2x - 6 > 0$, and thus to the original inequality $3x > 6 + 5x$, is $x < -3$.

The region above the *x*-axis and under the graph can be shaded by pressing 2ND PRGM (DRAW), 7:Shade(, and then entering (0, Y1) to shade on the homescreen (Figure 57).

The *x*-interval where the shading occurs is the solution.

Piecewise-Defined Functions

A piecewise-defined function is defined differently over two or more intervals.

To graph a piecewise-defined function

$$y = \begin{cases} f(x) & \text{if } x \leq a \\ g(x) & \text{if } x > a \end{cases}$$

press Y= and enter

$$Y1 = (f(x))/(x \leq a)$$

and

$$Y2 = (g(x))/(x > a)$$

Note that the inequality symbols are found by pressing 2ND MATH to access the TEST menu (Figure 58).

To illustrate, we will enter the function

$$f(x) = \begin{cases} x + 7 & \text{if } x \leq -5 \\ -x + 2 & \text{if } x > -5 \end{cases}$$

(Figure 59) and graph using an appropriate window (Figure 60).

Evaluating a piecewise-defined function at a given value of *x* requires that the correct equation ("piece") be selected. For example, $f(-6)$ and $f(3)$ are shown in Figures 61 and 62, respectively.

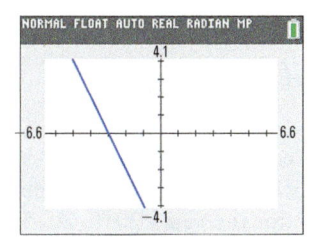

Figure 55

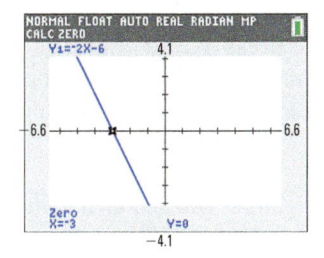

Figure 56

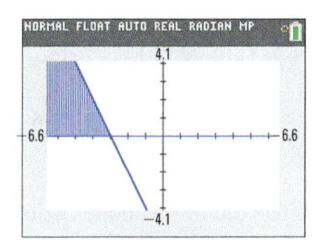

Figure 57

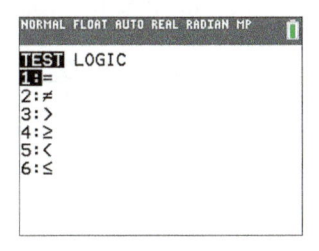

Figure 58

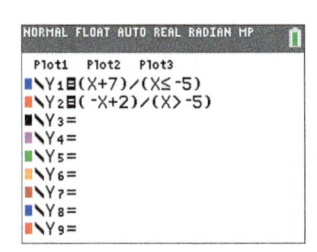

Figure 59

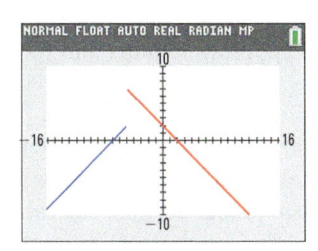

Figure 60

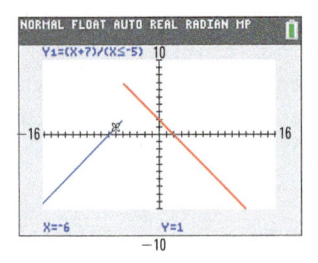

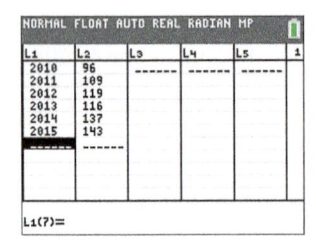

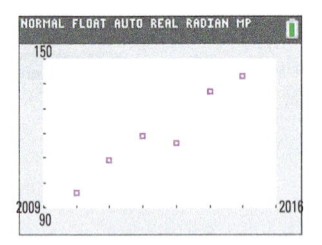

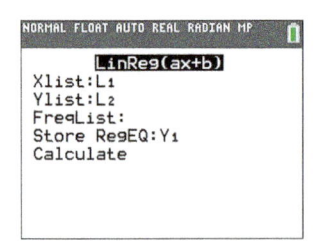

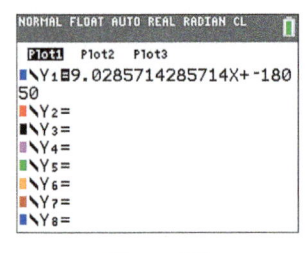

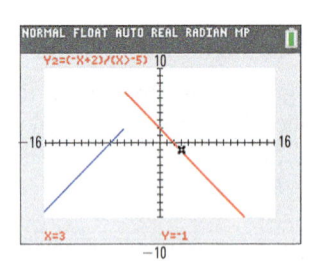

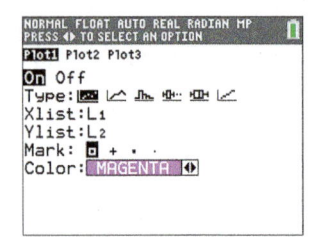

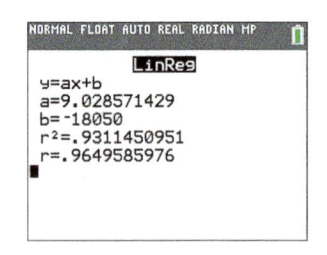

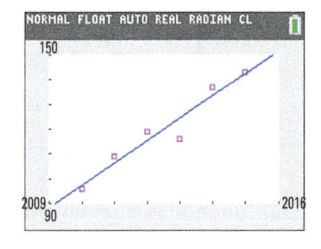

Modeling—Regression Equations

Finding an Equation That Models a Set of Data Points

To illustrate, we will find the equation of the line that is the best fit for the average daily numbers of inmates in the Springfield Detention Center for the years 2010–2015, given by the data points (2010, 96), (2011, 109), (2012, 119), (2013, 116), (2014, 137), (2015, 143).

Press **STAT** and under EDIT press 1:Edit. Enter the x-values (inputs) in the column headed L1 and the corresponding y-values (outputs) in the column headed L2 (Figure 63).

Press **2ND** **Y=** 1:Plot 1, highlight On, and then highlight the first graph type. Enter Xlist:L1, Ylist:L2, and pick the point plot mark you want (Figure 64).

Press **GRAPH** with an appropriate window or **ZOOM**, 9:ZoomStat to plot the data points (Figure 65).

Observe the point plots to determine what type of function would best model the data.

The graph looks like a line, so use the linear model, with LinReg.

Press **STAT**, move to CALC, and select the function type to be used to model the data (Figure 66). Enter L1 for Xlist, L2 for Ylist, and for Store RegEQ press the **VARS** key, go over to Y-VARS, and select 1:Function and 1:Y1 (Figure 67). Press **ENTER** twice. The coefficients of the equation will appear on the screen (Figure 68), and the linear regression equation will appear as Y1 on the Y= screen (see Figure 69).

To see how well the equation models the data, press **GRAPH**. If the graph does not fit the points well, another type function may be used to model the data.

For some models, r is a diagnostic value that gives the **correlation coefficient**, which determines the strength of the relationship between the independent and dependent variables. To display this value, select DiagnosticOn after pressing **2ND** **0** (CATALOG) (Figure 68).

Report the equation in a way that makes sense in the context of the problem, with the appropriate units and the variables identified.

The equation that models the average daily number of inmates in the Springfield Detention Center is $y = 9.0286x - 18{,}050$ (rounded to four decimal places), where x is the year and y is the number of inmates (Figures 69 and 70).

Basic Graphing Calculator Guide   691

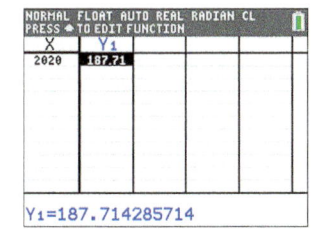

Figure 71 Using TABLE

Figure 72 Using Y-VARS

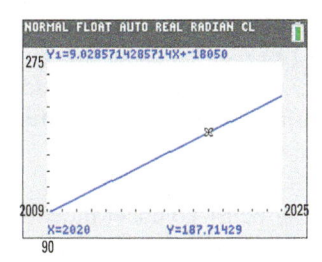

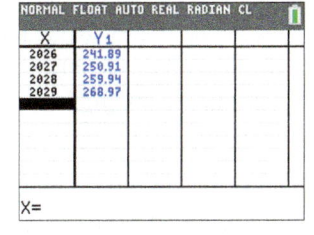

Figure 73 Using TRACE

Figure 74

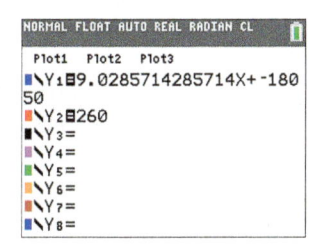

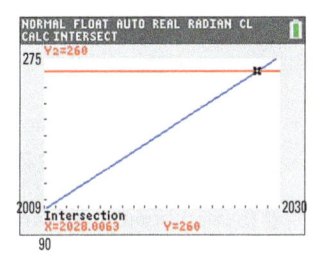

Figure 75

Figure 76

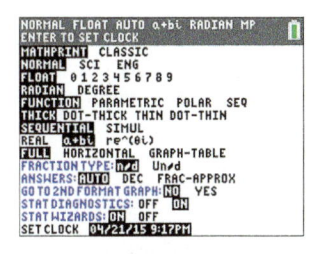

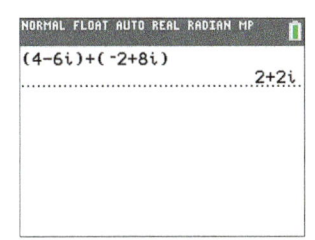

Figure 77

Figure 78

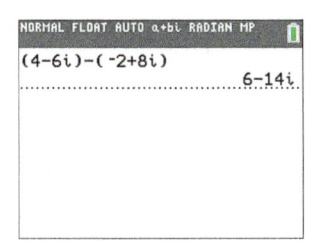

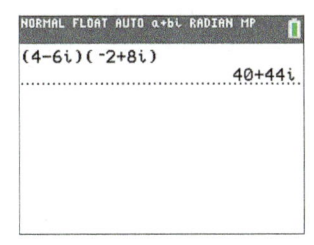

Figure 79

Figure 80

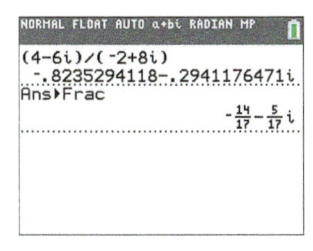

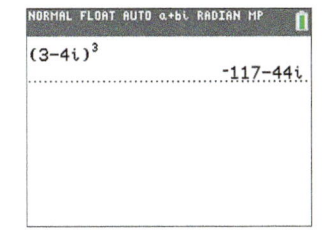

Figure 81

Figure 82

Using a Model to Find an Output

To use the model to find output values inside the data range (*interpolation*) or outside the data range (*extrapolation*), evaluate the function at the desired input value. This may be done using TABLE (Figure 71), Y-VARS (Figure 72), or TRACE (Figure 73). (Note that you may have to change the window to see the *x*-value to which you are tracing.)

To predict the average daily inmate population in the Beaufort County jail in 2020, we compute Y1 (2020). According to the model, approximately 188 inmates were, on average, in Springfield jail in 2020.

When TRACE is used, the window must be changed so that $x = 2020$ is visible (Figure 73).

Using a Model to Find an Input

To use the model to estimate an input for a given output, set the model function equal to Y1 and the desired output equal to Y2 and solve the resulting equations for the input variable. An approximate solution can be found using TABLE or 2ND TRACE (CALC), 5:Intersect.

To use the model to determine in what year the population will be 260, we must solve the equations $Y1 = f(x)$, $Y2 = 260$. TABLE may be used to find the *x*-value when *y* is approximately 260 (Figure 74).

We can also solve the equation $Y1 = Y2$ using the intersect method (Figures 75 and 76).

According to the model, the population of the jail will reach 260 in the year 2028.

Complex Numbers

Calculations that involve complex numbers of the form $a + bi$ will be displayed as complex numbers only if MODE is set to a + bi (Figure 77).

Complex numbers can be added (Figure 78), subtracted (Figure 79), multiplied (Figure 80), divided (Figure 81), and raised to powers (Figure 82). The number *i* is accessed by pressing 2ND .

Operations with Functions

Combinations of Functions

To find the graphs of combinations of two functions $f(x)$ and $g(x)$, enter $f(x)$ as Y1 and $g(x)$ as Y2 under the Y= menu. Consider the example $f(x) = 4x - 8$ and $g(x) = x^2$. To graph $(f + g)(x)$, enter Y1 + Y2 as Y3

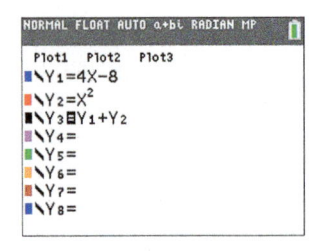

Figure 83

Figure 84

under the Y= menu (Figure 83). Place the cursor on the = sign beside Y1 and press **ENTER** to turn off the graph of Y1. Repeat with Y2. Press **GRAPH** with an appropriate window (Figure 84).

To graph $(f - g)(x)$, enter Y1 − Y2 as Y3 under the Y= menu (Figure 85). Place the cursor on the = sign beside Y1 and press **ENTER** to turn off the graph of Y1. Repeat with Y2. Press **GRAPH** (Figure 86).

Figure 85

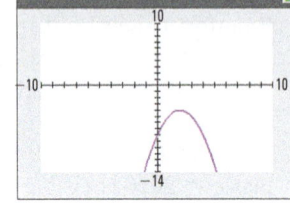

Figure 86

To graph $(f*g)(x)$, enter Y1*Y2 as Y3 under the Y= menu (Figure 87). Place the cursor on the = sign beside Y1 and press **ENTER** to turn off the graph of Y1. Repeat with Y2. Press **GRAPH** (Figure 88).

To graph $(f/g)(x)$, enter Y1/Y2 as Y3 under the Y= menu (Figure 89). Place the cursor on the = sign beside Y1 and press **ENTER** to turn off the graph of Y1. Repeat with Y2. Press **GRAPH** with an appropriate window (Figure 90).

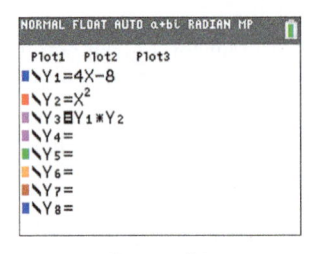

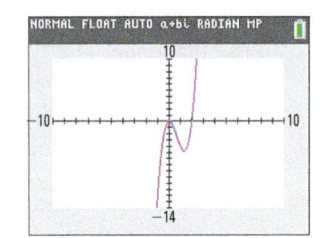

Figure 87

Figure 88

To evaluate $f + g, f - g, f*g$, or f/g at a specified value of x, enter Y1, the value of x enclosed in parentheses, the operation to be performed, Y2, and the value of x enclosed in parentheses, and press **ENTER** (Figure 91). Or, if the combination of functions is entered as Y3, enter Y3 and the value of x enclosed in parentheses (Figure 92), or use TABLE.

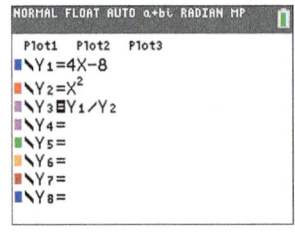

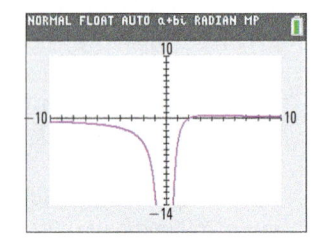

Figure 89

Figure 90

The value of $(f + g)(3)$ is shown in Figures 91 and 92.

Note: Entering $(Y1 + Y2)(3)$ does not produce the correct result (Figure 93).

Composition of Functions

To graph the composition of two functions $f(x)$ and $g(x)$, enter $f(x)$ as Y1 and $g(x)$ as Y2 under the Y= menu. Consider the example $f(x) = 4x - 8$ and $g(x) = x^2$.

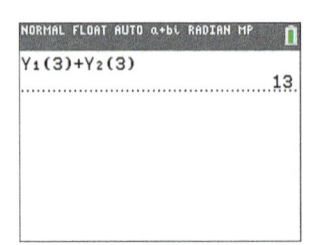

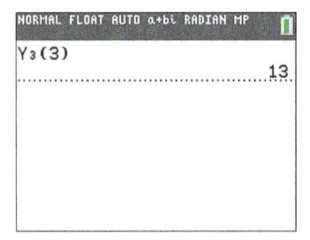

Figure 91

Figure 92

To graph $(f \circ g)(x) = f(g(x))$, enter Y1(Y2) as Y3 under the Y= menu (Figure 94). Place the cursor on the = sign beside Y1 and press **ENTER** to turn off the graph of Y1. Repeat with Y2. Press **GRAPH** with an appropriate window (Figure 95).

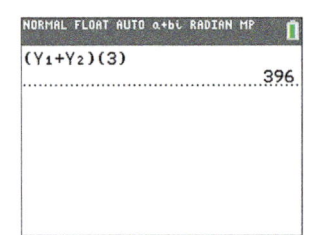

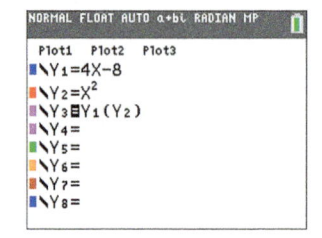

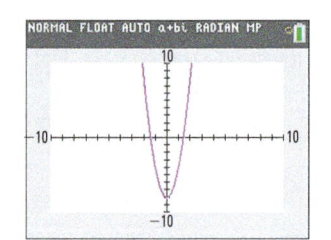

Figure 93

Figure 94

Figure 95

Basic Graphing Calculator Guide   693

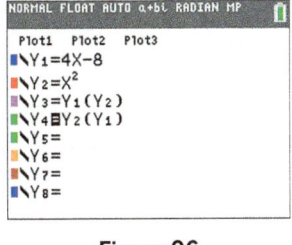

Figure 96

Figure 97

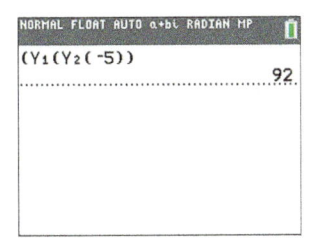

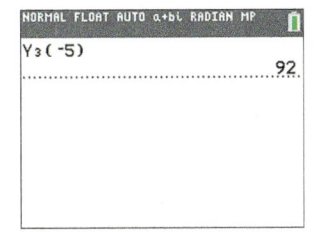

Figure 98

Figure 99

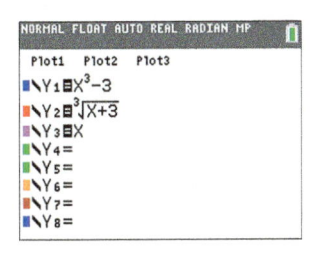

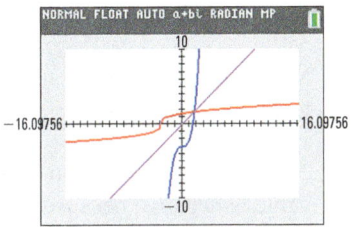

Figure 100

Figure 101

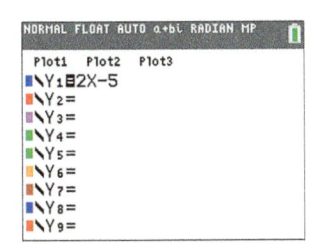

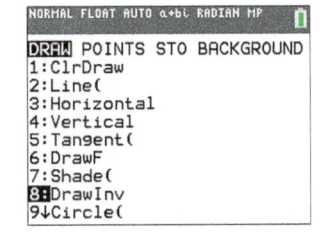

Figure 102

Figure 103

Figure 104

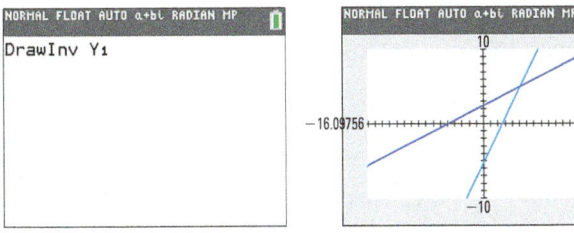

Figure 105

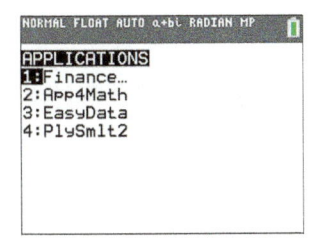

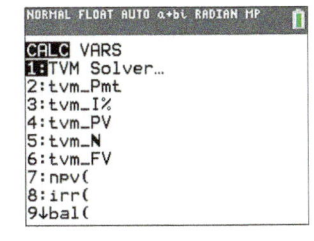

Figure 106

Figure 107

To graph $(g \circ f)(x)$, enter Y2(Y1) as Y4 under the Y= menu (Figure 96). Turn off the graphs of Y1, Y2, and Y3. Press GRAPH with an appropriate window (Figure 97).

To evaluate $(f \circ g)(x)$ at a specified value of x, enter Y1(Y2(, the value of x, and two closing parentheses (Figure 98), and press ENTER. Or, if the combination of functions is entered as Y3, enter Y3 and the value of x enclosed in parentheses (Figure 99). The value of $(f \circ g)(-5) = f(g(-5))$ is shown in Figures 98 and 99.

Inverse Functions

To show that the graphs of two inverse functions are symmetrical about the line $y = x$, graph the functions on a square window. Enter $f(x)$ as Y1 and the inverse function $g(x)$ as Y2, and press ENTER. Press GRAPH with an appropriate window and different colored curves. Enter Y3 = x under the Y= menu and graph using ZOOM, 5:Zsquare. For example, the graphs of $f(x) = x^3 - 3$ and its inverse $g(x) = \sqrt[3]{x + 3}$ are symmetric about the line $y = x$, as shown in Figures 100 and 101.

To graph a function $f(x)$ and its inverse, under the Y= menu, enter $f(x)$ as Y1 (Figure 102). Choose a square window. Press 2ND PRGM (DRAW) (Figure 103), 8:DrawInv (Figure 104), press VARS, move to Y-VARS, and press ENTER three times. The graph of $f(x)$ and its inverse will be displayed.

To clear the graph of the inverse, press 2ND PRGM (DRAW), 1:ClrDraw. The graphs of $f(x) = 2x - 5$ and its inverse are shown on a square window in Figure 105.

Finance

Future Value of an Investment

The TVM Solver displays the time-value-of-money variables. Given four TVM variables, the TVM Solver solves for the fifth variable.

To find the future value of a present amount, press the APPS key, select 1:Finance and next select 1:TVM Solver (Figures 106 and 107). Then set the variables as follows:

Set N = the total number of compounding periods.

Set I% = the annual percentage rate.

Set PV = lump-sum investment (with a negative sign because the sum is leaving the investor and going into the investment).

Set PMT = 0 because no additional money is being invested.

Set both P/Y (payments/year) and C/Y (compounding periods per year) equal to the number of compounding periods per year.

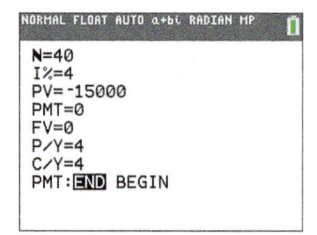

```
NORMAL FLOAT AUTO a+bi RADIAN MP
N=40
I%=4
PV=-15000
PMT=0
FV=0
P/Y=4
C/Y=4
PMT:END BEGIN
```

Figure 108

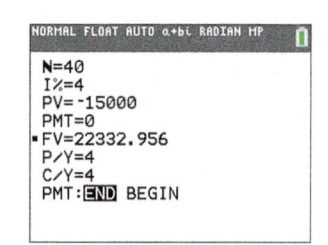

```
NORMAL FLOAT AUTO a+bi RADIAN MP
N=40
I%=4
PV=-15000
PMT=0
•FV=22332.956
P/Y=4
C/Y=4
PMT:END BEGIN
```

Figure 109

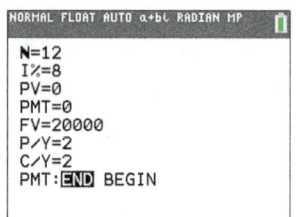

Highlight END, move the cursor to FV, and press **ALPHA** **ENTER** (SOLVE) to get the future value.

For example, to find the future value of $15,000 invested at 4% compounded quarterly for 10 years, use the settings in Figure 108. We use N = 10 years × 4 quarters per year, I% = 4, PV = −15000 to indicate the amount that is being invested, PMT = 0 because no additional money is being invested, and P/Y and C/Y = 4 compounding periods per year.

The future value, FV in Figure 109, is $22,332.96.

Present Value of an Investment

Use the same steps as above, except set FV = the desired future value, and solve for PV.

```
NORMAL FLOAT AUTO a+bi RADIAN MP
N=12
I%=8
PV=0
PMT=0
FV=20000
P/Y=2
C/Y=2
PMT:END BEGIN
```

Figure 110

```
NORMAL FLOAT AUTO a+bi RADIAN MP
N=12
I%=8
•PV=-12491.94099
PMT=0
FV=20000
P/Y=2
C/Y=2
PMT:END BEGIN
```

Figure 111

To find the present value of $20,000 at 8% compounded semiannually for 6 years, use the settings in Figure 110.

Move the cursor to PV, and press **ALPHA** **ENTER** (SOLVE) to get the present value.

The present value, PV in Figure 111, is $12,491.94. This value is negative because it is the investment leaving the investor.

Future Value of an Ordinary Annuity

To find the 5-year future value of an ordinary annuity with a contribution of $500 per quarter into an account that pays 8% per year compounded quarterly, press the **APPS** key, select **1:Finance**, and next select **1:TVM Solver**.

Set N = the total number of compounding periods = 5*4 = 20.

Set I% = the annual percentage rate = 8.

Set PV = lump-sum investment = 0.

Set PMT = quarterly payment = −500 (negative because it is leaving the investor).

Set both P/Y (payments/year) and C/Y (compounding periods per year) equal to the number of compounding periods per year = 4 Highlight END.

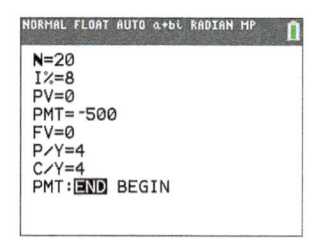

```
NORMAL FLOAT AUTO a+bi RADIAN MP
N=20
I%=8
PV=0
PMT=-500
FV=0
P/Y=4
C/Y=4
PMT:END BEGIN
```

Figure 112

See the settings in Figure 112.

Move the cursor to FV, and press **ALPHA** **ENTER** (SOLVE) to get the future value.

The future value, FV in Figure 113 on the next page, is $12,148.68.

Present Value of an Ordinary Annuity

We can find the present value of an annuity by using the same steps used to find future value, with the inputs for PV and FV changed.

To find the amount to put in an annuity that will provide $2000 at the end of each month for 20 years at 6% compounded monthly, press the APPS key, select **1:Finance**, and next select **1:TVM Solver.**

Set N = the total number of compounding periods = 240.

Set I% = the annual percentage rate = 6.

Set FV = value after all payments are made = 0.

Set PMT = quarterly payment = 2000.

Set both P/Y (payments/year) and C/Y (compounding periods per year) equal to the number of compounding periods per year = 12. Highlight END.

See the settings in Figure 114.

Move the cursor to PV, and press ALPHA ENTER (SOLVE) to get the present value.

The amount needed to fund the annuity is $279,161.54, found in Figure 115.

Payments to Amortize a Loan

To find the size of periodic payments to amortize a loan:

1. Press the APPS key and select Finance; press ENTER.
2. Select TVM Solver; press ENTER.
3. Set N = the total number of periods; set I% = the APR.
4. Set PV = loan value and set both P/Y and C/Y = the number of periods per year. END should be highlighted.
5. Set FV = 0.
6. Put the cursor on PMT and press ALPHA ENTER to get the payment.

For example, to repay a loan of $180,000 in 300 monthly payments with annual interest 9%, each payment must be $1510.55 See Figures 116 and 117.

Matrices

Entering Data into Matrices

To enter data into matrices, press 2ND x^{-1} to get the MATRIX menu (Figure 118). Move the cursor to EDIT. Select the number of the matrix into which the data are to be entered. Enter the dimensions of the matrix, and enter the value for each entry of the matrix (Figure 119). Press ENTER after each entry.

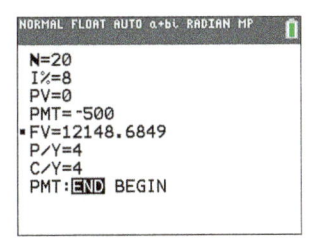

Figure 113

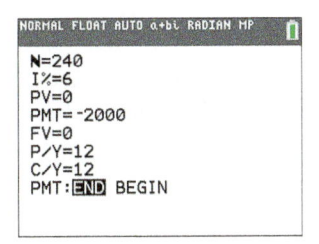

Figure 114

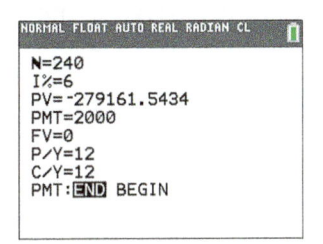

Figure 115

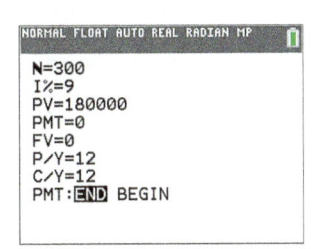

Figure 116

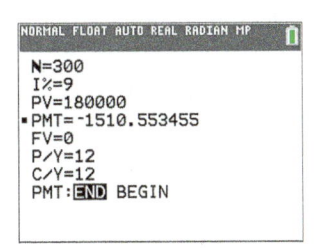

Figure 117

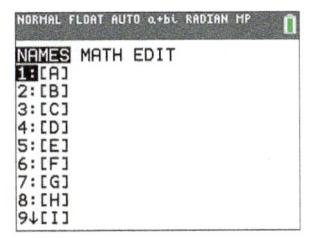

Figure 118

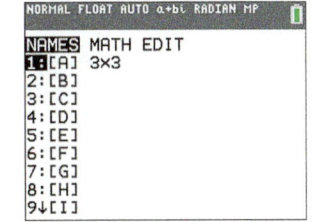

Figure 119

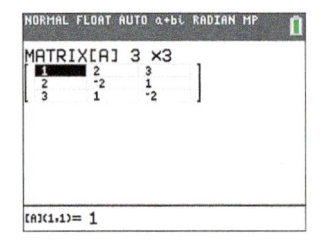

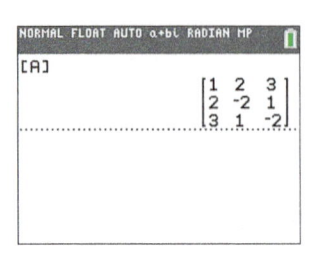

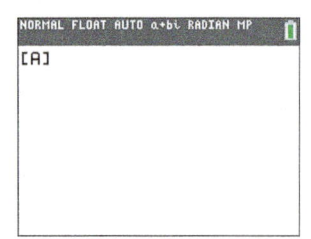

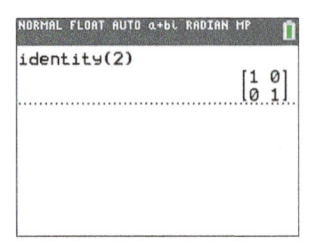

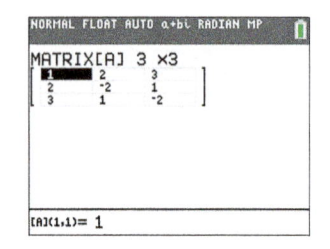

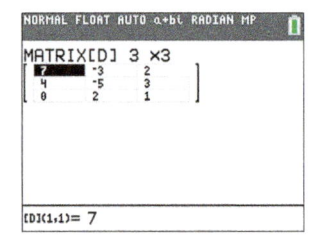

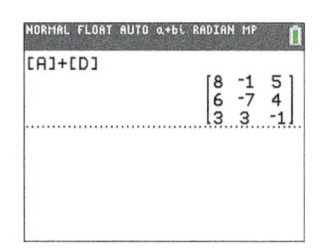

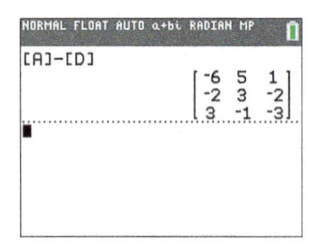

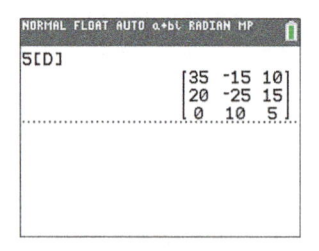

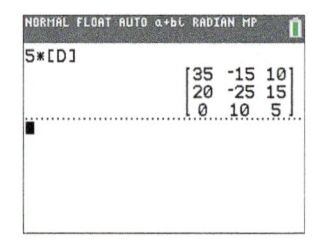

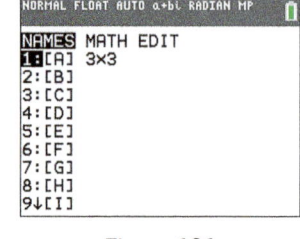

For example, to enter as [A] the matrix

$$\begin{bmatrix} 1 & 2 & 3 \\ 2 & -2 & 1 \\ 3 & 1 & -2 \end{bmatrix}$$

enter 3's to set the dimension and enter the elements of the matrix (Figure 120).

To perform operations with the matrix or leave the editor, first press **2ND** **MODE** (QUIT).

To view the matrix, access MATRIX, enter the name of the matrix (Figure 121), and press **ENTER** (Figure 122).

The Identity Matrix

To display an identity matrix of order n (an $n \times n$ matrix consisting of 1's on the main diagonal and 0's elsewhere), access MATRIX, move to MATH, select 5:identity(, and enter the order of the identity matrix desired (Figure 123). The identity matrix of order 2 is shown in Figure 124.

An identity matrix can also be created by entering the numbers directly with MATRIX and EDIT.

Adding and Subtracting Matrices

To find the sum of two matrices [A] and [D], enter the values of the elements of [A] using MATRIX and EDIT (Figure 125). Press **2ND** **MODE** (QUIT). Enter the values of the elements of [D] using MATRIX and EDIT (Figure 126). Press **2ND** **MODE** (QUIT).

Use MATRIX and NAMES to enter $[A] + [D]$, and press **ENTER**. If the matrices have the same dimensions, they can be added (or subtracted). If they do not have the same dimensions, an error message will occur.

For example, Figure 127 shows the sum

$$\begin{bmatrix} 1 & 2 & 3 \\ 2 & -2 & 1 \\ 3 & 1 & -2 \end{bmatrix} + \begin{bmatrix} 7 & -3 & 2 \\ 4 & -5 & 3 \\ 0 & 2 & 1 \end{bmatrix}$$

To find the difference of the matrices [A] and [D], enter $[A] - [D]$ and press **ENTER** (Figure 128).

Multiplying a Matrix by a Real Number

We can multiply a matrix [D] by a real number (scalar) k by entering $k[D]$ (or $k*[D]$). In Figures 129 and 130, we multiply the matrix [D] by 5.

Multiplying Two Matrices

To find the product of matrices $[C][A]$, access MATRIX, move to EDIT, select 1: [A], enter the dimensions of [A] (Figure 131), and enter the elements of [A]

Basic Graphing Calculator Guide   **697**

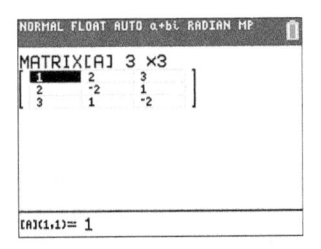

Figure 132

Figure 133

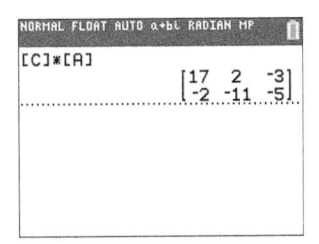

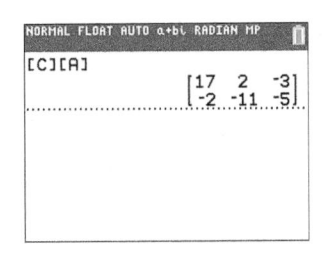

Figure 134

Figure 135

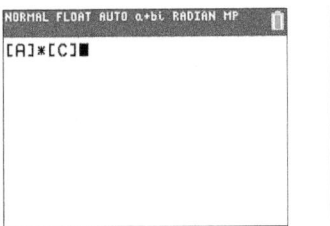

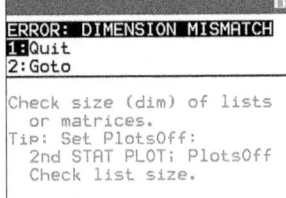

Figure 136

Figure 137

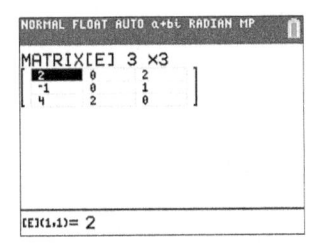

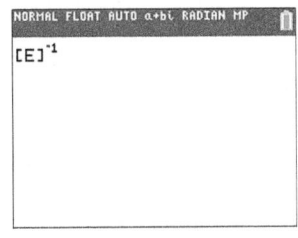

Figure 138

Figure 139

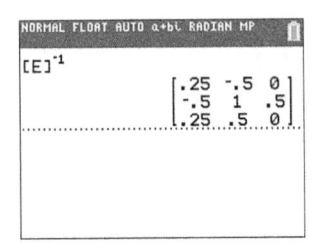

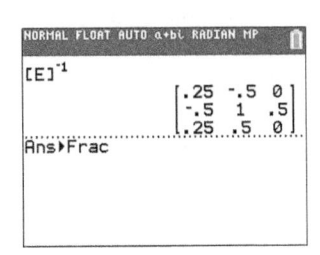

Figure 140

Figure 141

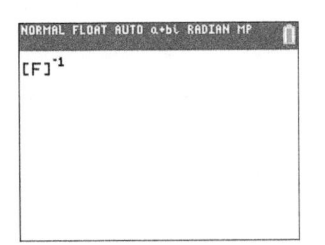

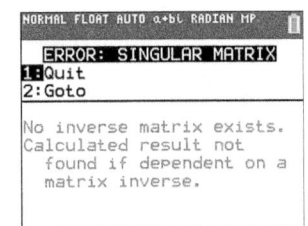

Figure 142

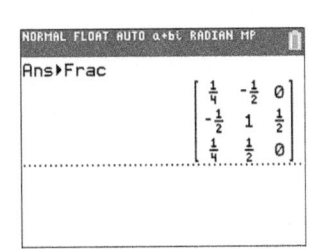

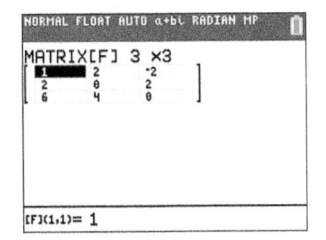

Figure 143

Figure 144

Figure 145

(Figure 132). Press **2ND** **MODE** (QUIT). Do the same for matrix [C] (Figure 133) and access **2ND** **MODE** (QUIT). Access MATRIX [C], *, MATRIX [A], and **ENTER** (Figure 134). Or press MATRIX [C], MATRIX [A], and **ENTER** (Figure 135).

The product

$$\begin{bmatrix} 1 & 2 & 4 \\ -3 & 2 & -1 \end{bmatrix} \begin{bmatrix} 1 & 2 & 3 \\ 2 & -2 & 1 \\ 3 & 1 & -2 \end{bmatrix}$$

is shown in Figures 134 and 135.

Note that $[A][C]$ does not always equal $[C][A]$. The product $[A][C]$ may be the same as $[C][A]$, may be different from $[C][A]$, or may not exist.

In Figures 136 and 137, $[A][C]$ cannot be computed because the matrices have dimensions such that they cannot be multiplied.

Finding the Inverse of a Matrix

To find the inverse of a matrix, enter the elements of the matrix using MATRIX and EDIT (Figure 138). Press **2ND** **MODE** (QUIT). Access MATRIX, enter the name of the matrix, and then press the **x⁻¹** key and **ENTER** (Figure 139).

Figure 140 shows the inverse of

$$E = \begin{bmatrix} 2 & 0 & 2 \\ -1 & 0 & 1 \\ 4 & 2 & 0 \end{bmatrix}$$

To see the entries as fractions, press **MATH**, 1:Frac, and **ENTER** (Figures 141 and 142).

Not all matrices have inverses. Matrices that do not have inverses are called singular matrices (Figures 143–145).

Solving Systems—Reduced Echelon Form

To solve a system of linear equations by using rref under the MATRIX MATH menu, create an augmented matrix [A] with the coefficient matrix augmented by the constants (Figure 146). Use the MATRIX menu to produce a reduced row echelon form of [A], as follows:

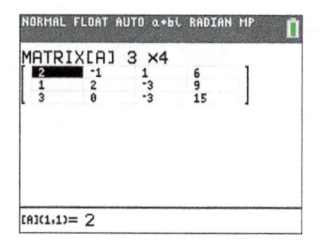

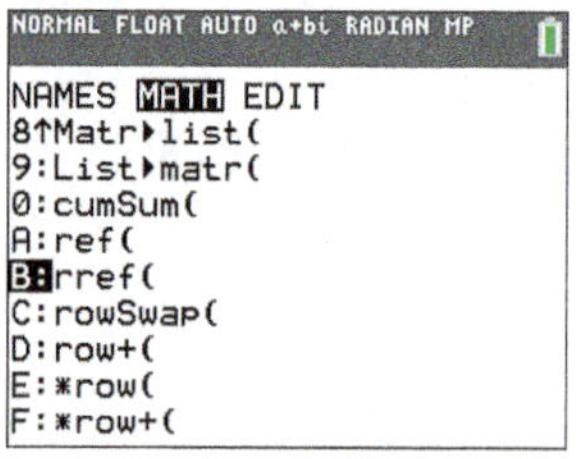

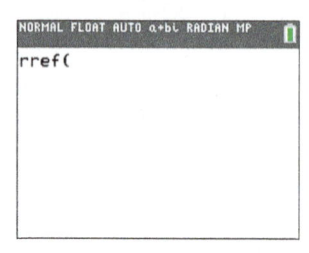

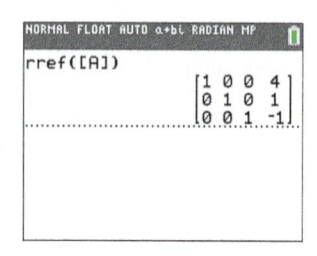

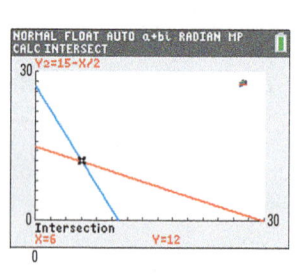

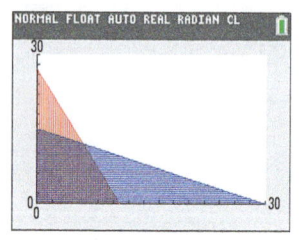

1. Access MATRIX and move to the right to MATH (Figure 147).
2. Scroll down to B:rref(and press **ENTER**, or press **ALPHA** B (Figure 148). Access MATRIX, 1: [A] to get rref([A]). Press **ENTER** to get the reduced echelon form.

The system
$$\begin{cases} 2x - y + z = 6 \\ x + 2y - 3z = 9 \\ 3x - 3z = 15 \end{cases}$$
is solved in Figure 149.

If each row in the coefficient matrix (the first three columns) contains a 1 with the other elements 0's, the solution is unique and the number in column 4 of a row is the value of the variable corresponding to a 1 in that row. The solution to the system above is unique: $x = 4$, $y = 1$, and $z = -1$.

If the bottom row contains all zeros, the system has many solutions. The values for the first two variables are found as functions of the third.

If there is a nonzero element in the augment of row 3 and zeros elsewhere in row 3, there is no solution to the system.

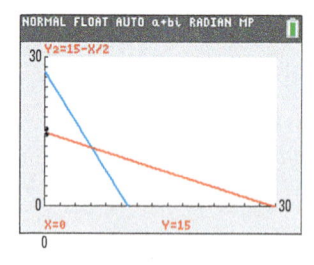

Linear Programming

To solve a linear programming problem involving two constraints graphically, write the inequalities as equations, solved for y. Graph the equations. The inequalities $x \geq 0$, $y \geq 0$ limit the graph to Quadrant I, so choose a window with Xmin = 0 and Ymin = 0. Use **TRACE** or intersect (accessed by pressing **2ND** **TRACE**) to find the corners of the region, where the borders intersect.

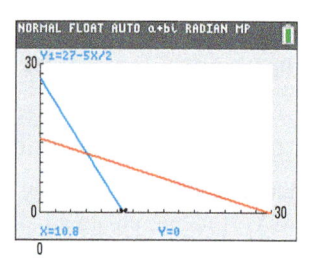

For example, to find the region defined by the inequalities
$$5x + 2y \leq 54$$
$$2x + 4y \leq 60$$
$$x \geq 0, y \geq 0$$
write $y = 27 - 5x/2$ and $y = 15 - x/2$, graph, and find the intersection points as shown in Figures 150 to 152.

The corners of the region determined by the inequalities are $(0, 15)$, $(6, 12)$, and $(10.8, 0)$.

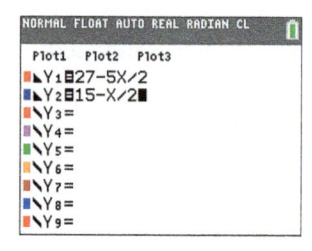

Change the icons to the left of Y1= and Y2= to show regions below the lines (Figure 153). Then graph these inequalities. The cross-hatched region is the solution region (Figure 154).

Evaluating the objective function $P = 5x + 11y$ at the coordinates of each of the corners determines where the objective function is maximized or minimized.

At $(0, 15)$, $f = 165$

At $(6, 12)$, $f = 162$

At $(10.8, 0)$, $f = 54$

The maximum value of f is 165 at $x = 0$, $y = 15$.

Sequences and Series

Evaluating a Sequence

To evaluate a sequence for different values of n, press **MODE** and highlight Seq (Figure 155). Press **ENTER** and **2ND** **MODE** (QUIT). Store the formula for the sequence (in quotes) in u, using **STO** u. (Press the **X,T,θ,n** key to enter n for the formula, and **2ND** 7 to get u.) Press **ENTER**.

For example, to evaluate the sequence with nth term $n^2 + 1$ at $n = 1, 3, 5,$ and 9, we store the formula as shown in Figure 156.

Enter u($\{a, b, c, \ldots\}$) to evaluate the sequence at $a, b, c, \ldots$, and press **ENTER** (Figure 157).

To generate a sequence after the formula is defined, enter u(nstart, nstop, step) and press **ENTER**. For the sequence formula above, evaluate every third term beginning with the second term and ending with the eleventh term (Figure 158).

Finding the nth Term of an Arithmetic Sequence

To find the nth term of an arithmetic sequence with first term a and common difference d, press **MODE** and highlight Seq (Figure 159). Press **ENTER** and press **2ND** **MODE** (QUIT).

Press **Y=**. At u(n) =, enter the formula for the nth term of an arithmetic sequence, using the **X,T,θ,n** key to enter n. The formula is $a + (n - 1)d$, where a is the first term and d is the common difference.

For example, to find the 12th term of the arithmetic sequence with first term 10 and common difference 5, substitute 10 for a and 5 for d, as shown in Figure 160, to get u(n) = $10 + (n - 1)5$.

Press **2ND** **MODE** (QUIT). To find the nth term of the sequence, press **2ND** 7, followed by the value of n in parentheses, to get u(n); then press **ENTER** (Figure 161).

Additional terms can be found in the same manner (Figure 162).

Finding the Sum of an Arithmetic Sequence

To find the sum of the first n terms of an arithmetic sequence, press **MODE** and highlight Seq (Figure 163). Press **ENTER** and press **2ND** **MODE** (QUIT).

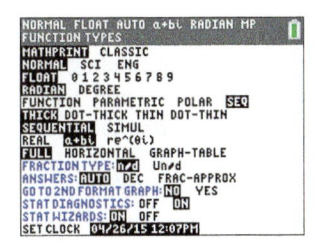

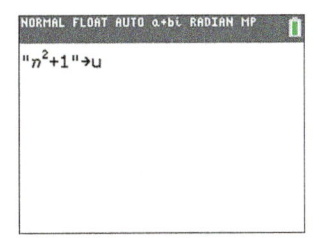

Figure 155

Figure 156

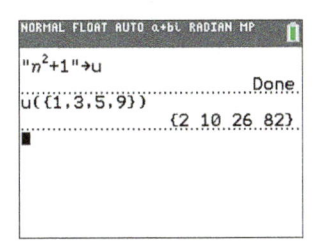

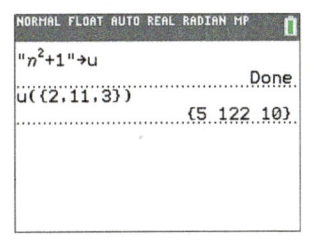

Figure 157

Figure 158

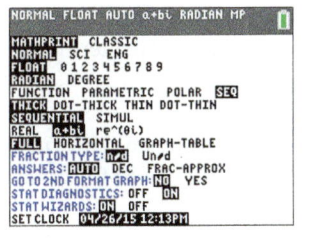

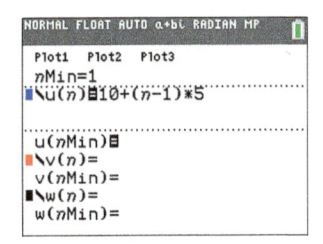

Figure 159

Figure 160

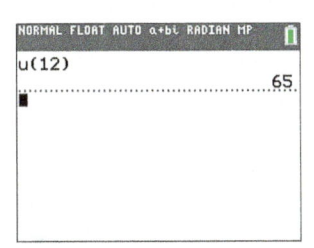

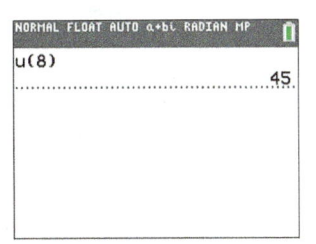

Figure 161 The 12th term

Figure 162 The 8th term

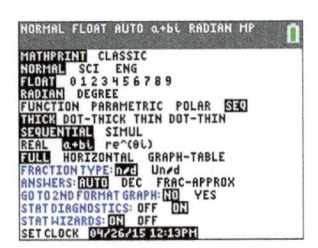

Figure 163

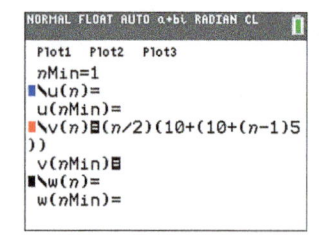

Figure 164

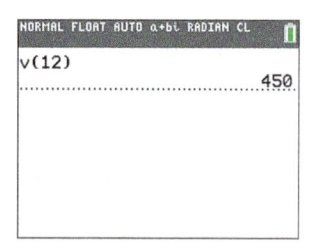

Figure 165 The sum of the first 12 terms

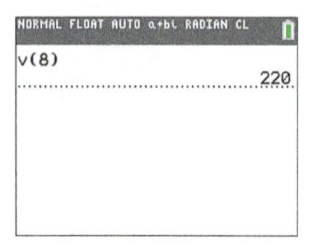

Figure 166 The sum of the first 8 terms

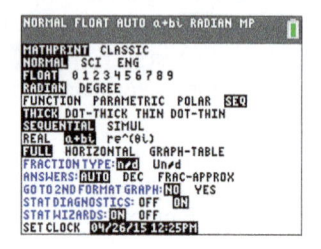

Figure 167

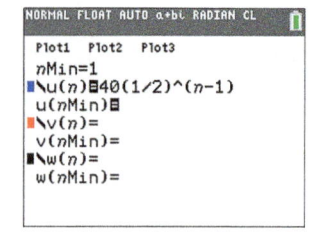

Figure 168

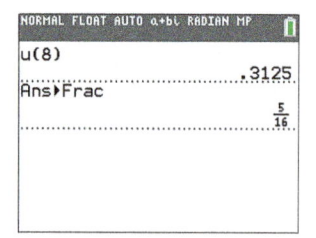

Figure 169

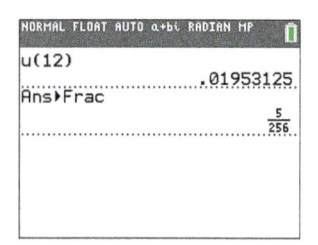

Figure 170

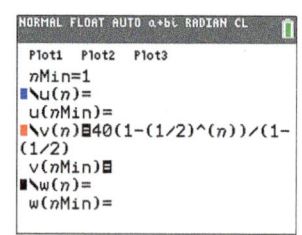

Figure 171

Figure 172 The sum of the first 12 terms

Figure 173 The sum of the first 8 terms

Press $Y=$. At $v(n) =$, enter the formula for the sum of the first n terms of an arithmetic sequence, using the X,T,θ,n key to enter n. The formula is $(n/2)(a + (a + (n-1)d))$, where a is the first term and d is the common difference (Figure 164).

Press $2ND$ $MODE$ (QUIT). To find the sum of the first n terms of the sequence, press $2ND$ 8, followed by the value of n in parentheses, to get $v(n)$; then press $ENTER$. For example, the sum of the first 12 terms of the arithmetic sequence with first term 10 and common difference 5 is shown in Figure 165.

Other sums can be found in the same manner (Figure 166).

Finding the *n*th Term and Sum of a Geometric Sequence

To find the *n*th term of a geometric sequence with first term a and common ratio r, press $MODE$ and highlight Seq (Figure 167). Press $ENTER$ and $2ND$ $MODE$ (QUIT).

Press $Y=$. At $u(n) =$, enter the formula for the *n*th term of a geometric sequence, using the X,T,θ,n key to enter n. The formula is ar^{n-1}, where a is the first term and r is the common ratio.

For example, to find the geometric sequence with first term 40 and common ratio $1/2$, we substitute 40 for a and $(1/2)$ for r (Figure 168).

Press $2ND$ $MODE$ (QUIT). To find the *n*th term of the sequence, press $2ND$ 7, followed by the value of n in parentheses, to get $u(n)$; then press $ENTER$. The 8th and 12th terms of the geometric sequence above are shown in Figures 169 and 170, respectively. To get a fractional answer, press $MATH$, 1:Frac. Additional terms can be found in the same manner.

To find the sum of the first n terms of a geometric sequence, press $Y=$. At $v(n) =$, enter the formula for the sum of the first n terms of a geometric sequence, using the X,T,θ,n key to enter n. The formula is $a(1 - r^n)/(1 - r)$, where a is the first term and r is the common ratio. Press $2ND$ $MODE$ (QUIT). To find the sum of the first n terms of the sequence, press $2ND$ 8, followed by the value of n in parentheses, to get $v(n)$; then press $ENTER$.

For example, to find the sum of the first 12 terms of the geometric sequence with first term 40 and common ratio $1/2$, substitute 40 for a and $(1/2)$ for r (Figure 171).

To get a fractional answer, press $MATH$, 1:Frac (Figure 172).

Other sums can be found in the same manner (Figure 173).

Appendix B

Basic Guide to Excel

This appendix describes the basic features and operation of Excel. More details of specific features can be found in the *Graphing Calculator and Excel® Manual*.

The Excel steps shown in this guide are for Excel 2013.

Some other versions of Excel may use a few slightly different menus and displays. To find information about how to use these menus and displays, use Excel Help. In Excel 2013, click the ⓘ icon in the upper right corner of the screen to get Excel Help.

Excel Worksheet

When you start up Excel by using the instructions for your software and computer, the screen in Figure 1 will appear. The components of the **spreadsheet** are shown, and the grid shown is called a **worksheet**. By clicking on the tabs at the bottom, you can move to other worksheets.

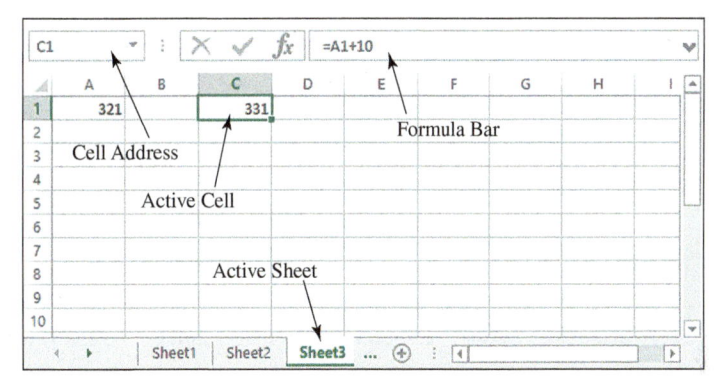

Figure 1

Addresses and Operations

Notice the letters at the top of the columns and the numbers identifying the rows. The cell addresses are given by the column and row; for example, the first cell has address A1. You can move from one cell to another with arrow keys, or you can select a cell with a mouse click. After you enter an entry in a cell, press ENTER to accept the entry. To edit the contents of a cell, make the edits in the formula bar at the top. To delete the contents, press the delete key.

File operations such as "open a new file," "saving a file," and "printing a file" are similar to those in Word. For example, <CTRL>S saves a file. You can also format a cell entry by selecting it and using menus similar to those in Word.

Working with Cells

Cell entries, rows containing entries, and columns containing entries can be copied and pasted with the same commands as in Word. For example, a highlighted cell can be copied with <CTRL>C. Sometimes an entry exceeds the width of the cell containing it, especially if it is text. To widen the cells in a column, move the mouse to the right side of the column heading until you see the symbol ↔; then hold down the left button and move the mouse to the right (moving to the left makes the cells narrower). If entering a number results in #####, the number is too long for the cell, and the cell should be widened.

You can work with a cell or with a range of cells. To select a range of cells,

1. Click on the beginning of the range of cells, hold down the left mouse button, and drag to the end of the desired range.
2. Release the mouse button. The range of cells will be highlighted. If the range of cells were from A2 through B6, the range would be indicated by A2:B6.

Creating Tables of Numbers in Excel

Entering Independent Values

Type the column heading, such as x, in cell A1, and enter each data value in a cell of the column (Figure 2).

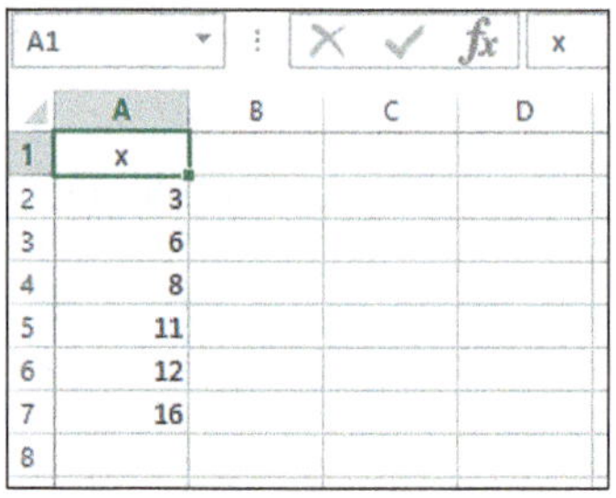

Figure 2

Fill-Down Method for Values That Change by a Constant Increment

1. Type in the first two numbers (in C2 and C3, for example).
2. Select the cells C2 and C3 (C2:C3) (Figure 3).

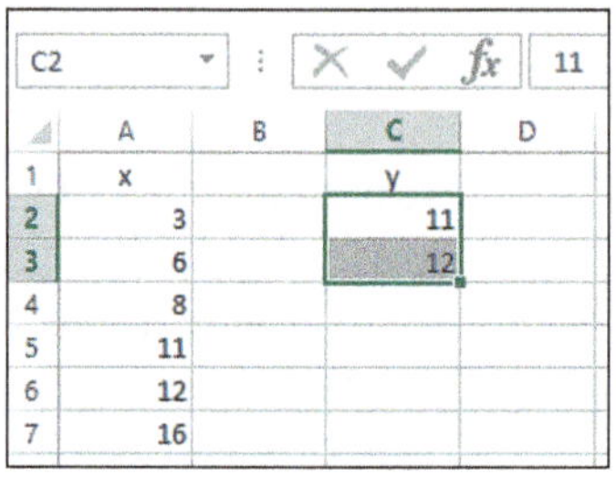

Figure 3

3. Move the mouse to the lower right corner until the mouse becomes a thin + sign.
4. Drag the mouse down to the last cell where data are required (Figure 4). Press the right arrow to remove the highlight.

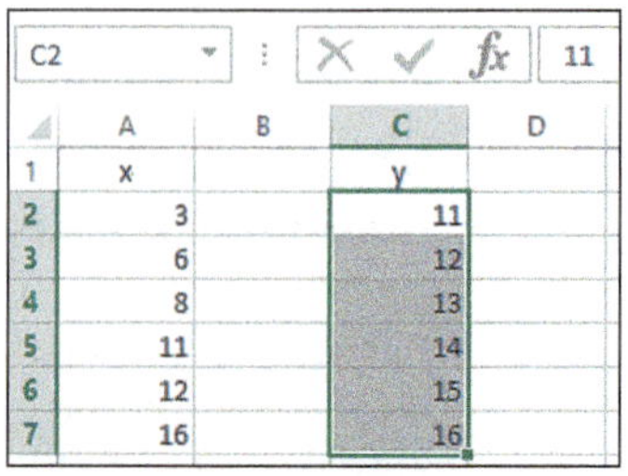

Figure 4

Evaluating a Function

1. Put headings on the two columns (x and $f(x)$, for example).
2. Begin a formula (in cell B2, for example) by entering =; then enter the function performing operations with data in cells by entering the operation side of the formula with appropriate cell address(es) representing the variable(s) (Figure 5). Use * to indicate multiplication and ^ to indicate power.
3. Enter the value(s) of the variable(s) for which the function is to be evaluated and press ENTER (Figure 6).

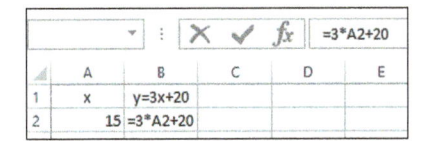

	A	B	C	D	E
1	x	y=3x+20			
2	15	=3*A2+20			

Figure 5

	A	B	C	D	E
1	x	y=3x+20			
2	15	65			

Figure 6

4. Changing the input values will give different function values (Figure 7). Press ENTER to complete the process (Figure 8).

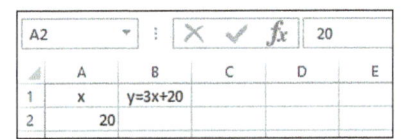

A2 : fx 20

	A	B	C	D	E
1	x	y=3x+20			
2	20				

Figure 7

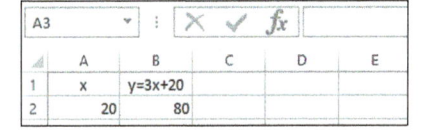

A3 : fx

	A	B	C	D	E
1	x	y=3x+20			
2	20	80			

Figure 8

Creating a Table for a Function Using Fill Down

1. Put headings on the two columns (x and $f(x)$, for example).
2. Fill the inputs (x-values) as described on page 702.
3. Enter the formula for the function as described above.
4. Select the cell containing the formula for the function (B2, for example) (Figure 9).
5. Move the mouse to the lower right corner until there is a thin + sign.
6. Drag the mouse down to the last cell where the formula is required, and the values will be displayed (Figure 10). Using an arrow to move to a cell to the right will remove the highlight from the outputs.

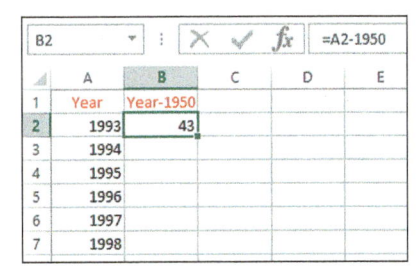

B2 : fx =A2-1950

	A	B	C	D	E
1	Year	Year-1950			
2	1993	43			
3	1994				
4	1995				
5	1996				
6	1997				
7	1998				

Figure 9

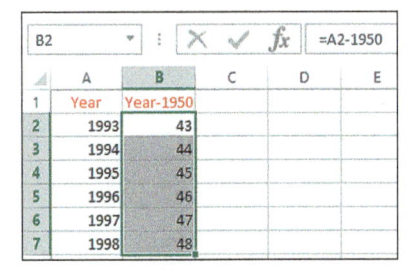

B2 : fx =A2-1950

	A	B	C	D	E
1	Year	Year-1950			
2	1993	43			
3	1994	44			
4	1995	45			
5	1996	46			
6	1997	47			
7	1998	48			

Figure 10

The worksheet in Figure 11 has headings typed in, numbers entered with Fill Down in column A, inputs entered in column B, the numbers in column D created with Fill Down using the formula A5 − 1950, and column B copied in column E.

704   Appendix B

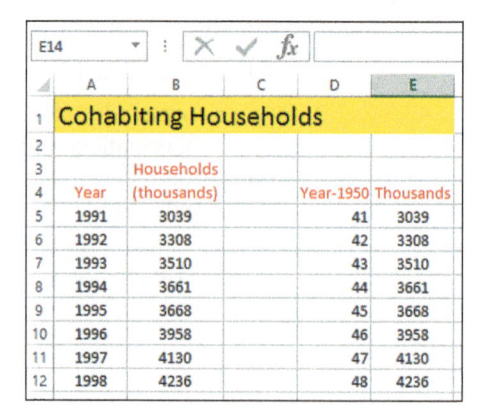

Figure 11

Graphing a Function of a Single Variable ($y = f(x)$, for example)

1. Use the function to create a table containing values for x and $f(x)$.
2. Highlight the two columns containing the values of x and $f(x)$ (Figure 12).

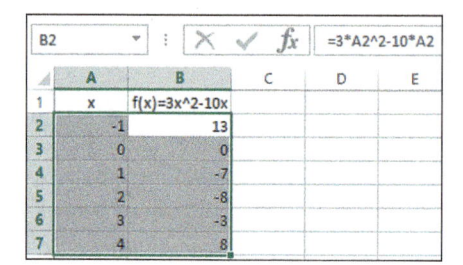

Figure 12

3. Select the Insert tab.
4. Select the scatter plot icon in the Charts group to get the Scatter menu (Figure 13).

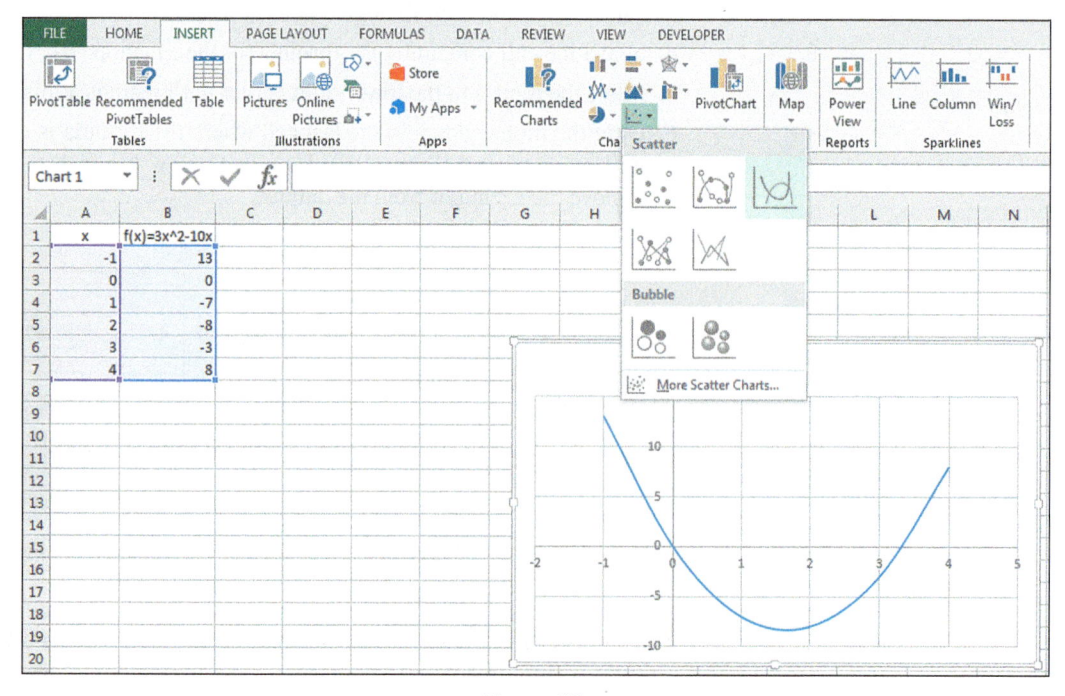

Figure 13

5. Select the smooth curve option under Scatter, and the graph will be as shown in Figure 14.
6. To add or change titles or legends, double click the item to be changed or use the subgroups under Chart Tools.

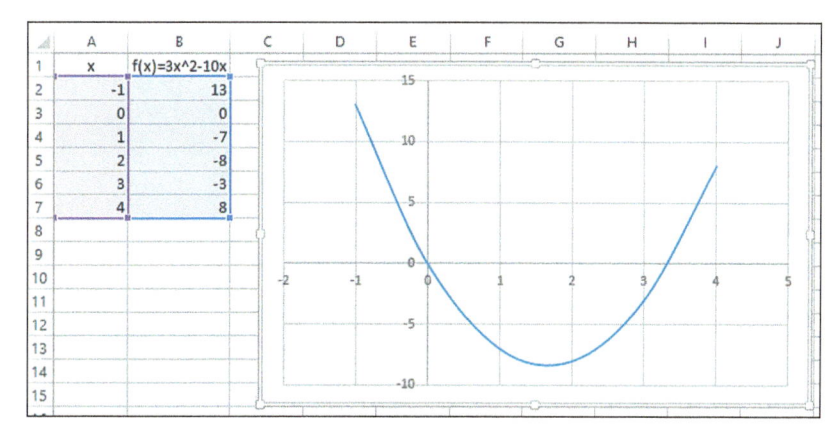

Figure 14

Graphing More Than One Function

1. Create a table with an input column and one column for each function, with multiple function headings.
2. Highlight the columns containing the variable and function values (Figure 15).
3. Proceed with the same steps as are used to graph a single function (Figure 16).

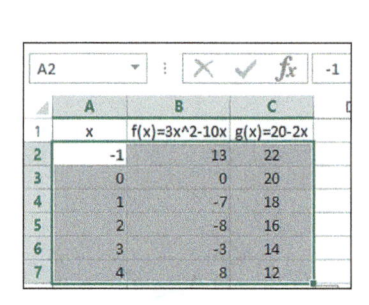

Figure 15

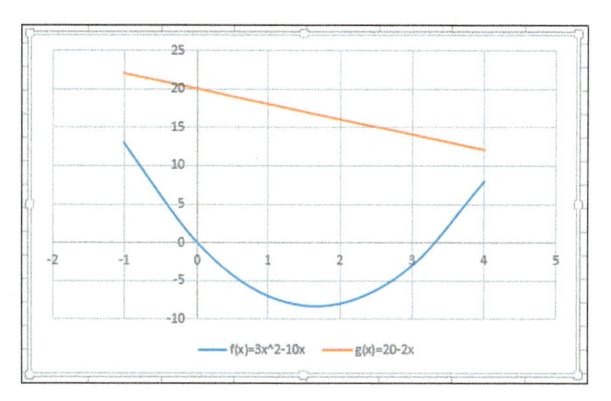

Figure 16

Changing Graphing Windows

1. To change the x-scale,
    a. Double click on the x-axis, and choose Axis Options. This opens a dialog box with Axis Options as the first option. Once there, choose the bar chart icon and open the Axis Options dropdown menu.
    b. Change the minimum and maximum values to the desired values.
2. To change the y-scale, double click on the y-axis and proceed as for the x-axis.
3. For shading options, double click on the plot area to open Format Plot Area options. It opens as a pane on the right-hand side of the screen.
4. To change the Chart Title, double click on it.

Graphing Discontinuous Functions

An Excel graph will connect all points corresponding to values in the table, so if the function you are graphing is discontinuous for some x-value a, enter x-values near this value and leave (or make) the corresponding $f(a)$ cell blank. Then graph the function using the steps described on page 704 (Figure 17).

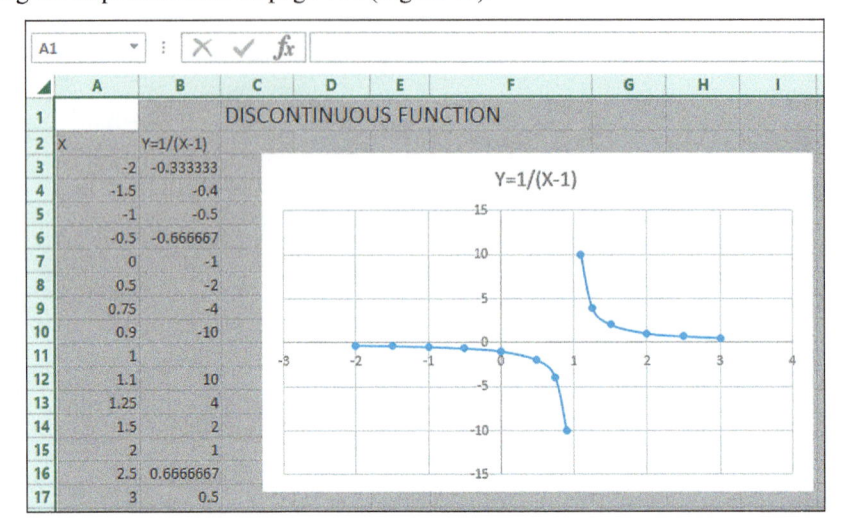

Figure 17

Finding a Minimum or Maximum of a Function with Solver

Finding the Minimum of a Function (if it has a minimum)

1. Enter the inputs (x-values) in column A and the functional values in column B, and use them to graph the function over an interval that shows a minimum (Figure 18). (If a minimum exists but does not show, add more points.)

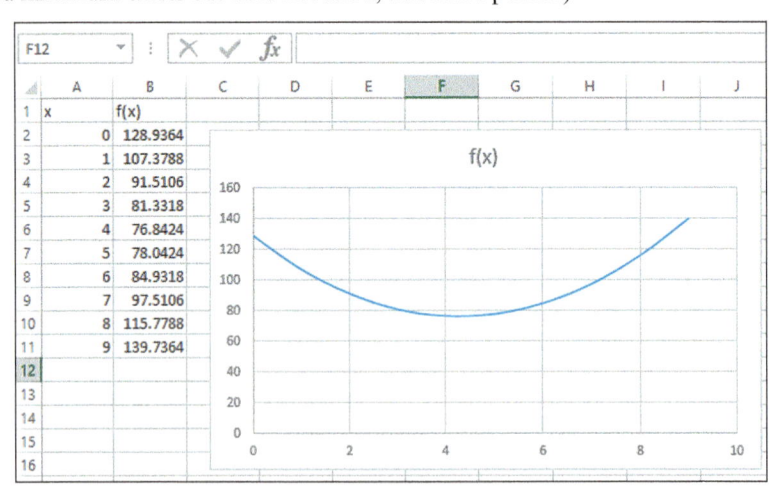

Figure 18

2. Use any value for x in A2 and the function formula in B2 (Figure 19).

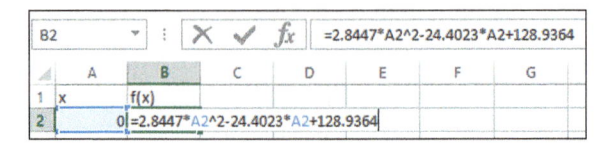

Figure 19

3. Select the Data tab, and select ?Solver from the Analysis group. [Note: Solver is an Add-In. To install it, select the File tab and choose Options. Choose Add-Ins, and in the Manage Box, select Excel Add-Ins and click Go. In the Add-Ins Available box, select Solver Add-In and click OK. If Solver Add-In isn't available, choose Browse to find it.]
4. Click on Min in the Solver dialog box.
5. Set the Objective Cell to B2 by choosing the box and clicking on B2. Then set the Changing Cells to A2 with a similar process (Figure 20).

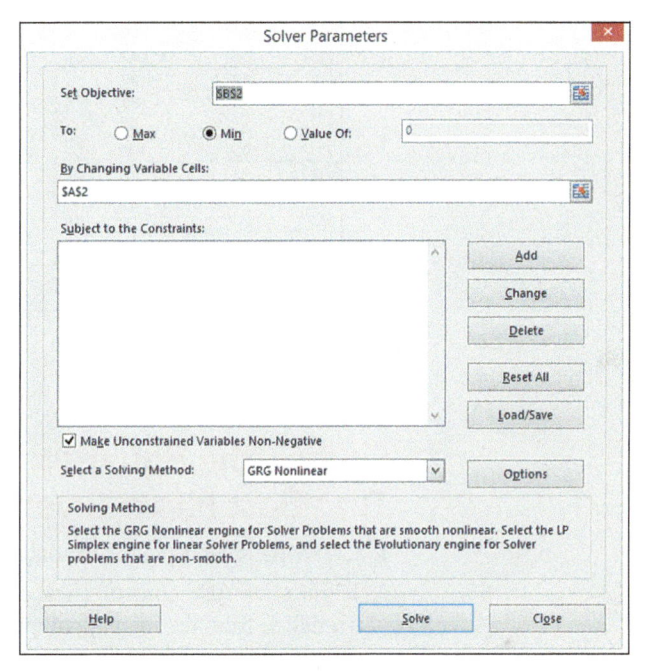

Figure 20

6. Click Solve in the dialog box, and the minimum will show in B2. Click OK to save the solution (Figure 21).

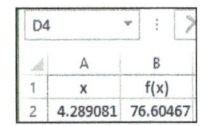

Figure 21

Finding the Maximum of a Function (if it has a maximum)

Use the process above, but in step 4 click on Max in the Solver dialog box.

Scatter Plots of Data

1. Enter the inputs (x-values) in column A and the outputs (y-values) in column B.
2. Highlight the two columns.
3. Select the Insert tab and select the scatter plot icon in the Charts group to get the Scatter menu.

4. Select the points option under Scatter. The graph will appear (Figure 22).
5. To add or change titles or legends, double click the item to be changed or use the subgroups under Chart Tools.

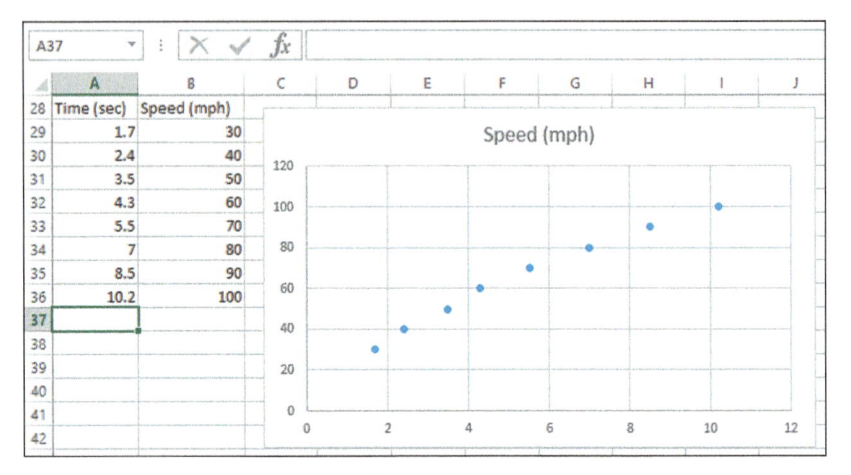

Figure 22

Finding the Equation of the Line or Curve That Best Fits a Given Set of Data Points

1. Create the scatter plot of the data in the worksheet.
2. Right click (use control-click on Mac) on one of the data points and select Add Trendline from the menu that appears (Figure 23).

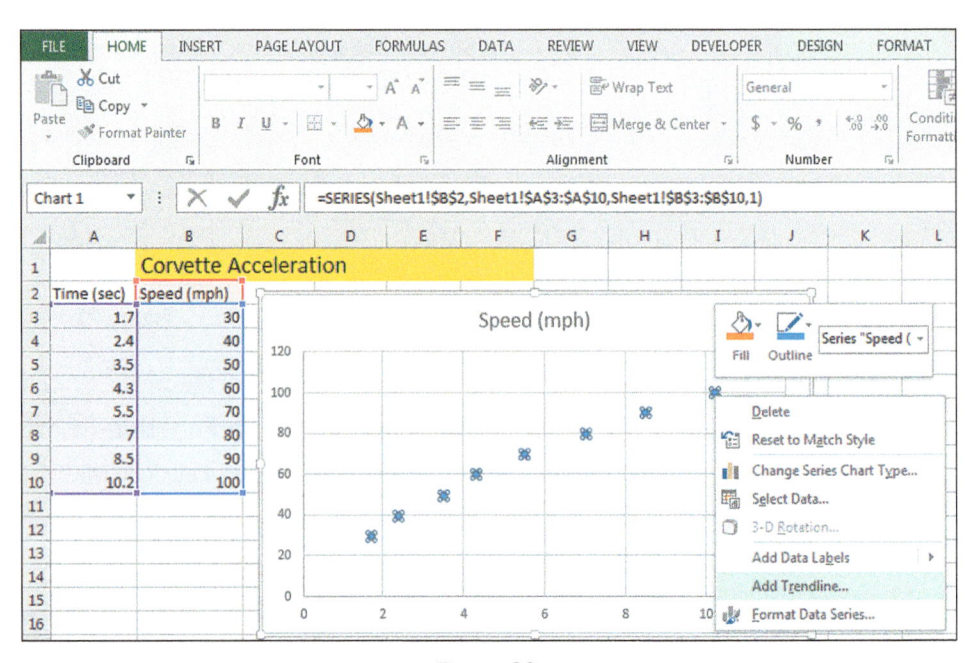

Figure 23

3. Select the desired regression type and check Display Equation on chart (Figure 24). [Note: If Polynomial is selected, choose the appropriate Order (degree).]
4. Close the box, and the function and its graph will appear (Figure 25).

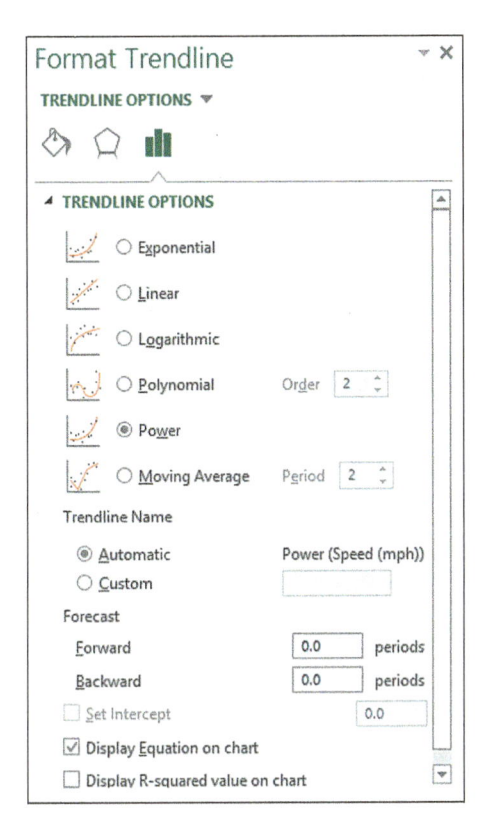

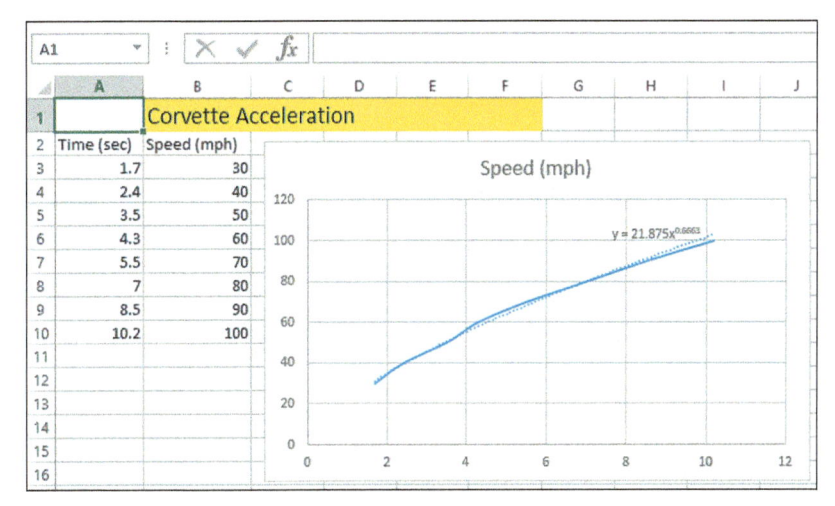

Figure 24

Figure 25

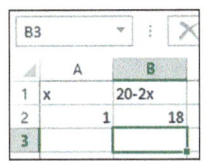

Figure 26

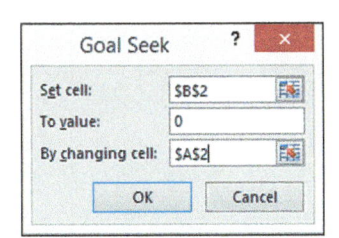

Figure 27

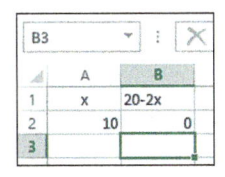

Figure 28

Solving a Linear Equation of the Form $f(x) = 0$ Using the x-Intercept Method

The solution will also be the x-intercept of the graph and the zero of the function.

1. Enter an x-value in A2 and the function formula in B2, as described on page 706 (Figure 26).
2. Highlight B2.
3. Select the Data tab.
4. Select What-if Analysis in the Data Tools group.
5. Select Goal Seek under What-if Analysis.
6. In the Goal Seek dialog box,
    a. Click the Set cell box and click on the B2 cell.
    b. Enter 0 in the To value box.
    c. Click the By changing box and click on the A2 cell (Figure 27).
7. Click OK to see the Goal Status box.
8. Click OK again. The solution will be in A2 and 0 will be in B2 (Figure 28). The solution may be approximate.

Solving a Quadratic Equation of the Form $f(x) = 0$ Using the x-Intercept Method

The solutions will also be the x-intercepts of the graph and the zeros of the function. Because the graph of a quadratic function $f(x) = ax^2 + bx + c$ can have two intercepts, it is wise to graph the function using an x-interval with the x-coordinate $\left(h = \dfrac{-b}{2a}\right)$ near the center.

1. Enter x-values centered around h in column A and use the function formula to find the values of $f(x)$ in column B, as described on page 703 (Figure 29).

2. Graph the function as described on page 704, and observe where the function values are at or near 0.

3. Select the Data tab and choose What-If Analysis in Data Tools. Select Goal Seek, enter the address of a cell with a function value in column B at or near 0, enter 0 in To value, and enter the corresponding cell in column A in By changing cell (Figure 30). Click OK to find the x-value of the solution in column A. The solution may be approximate.

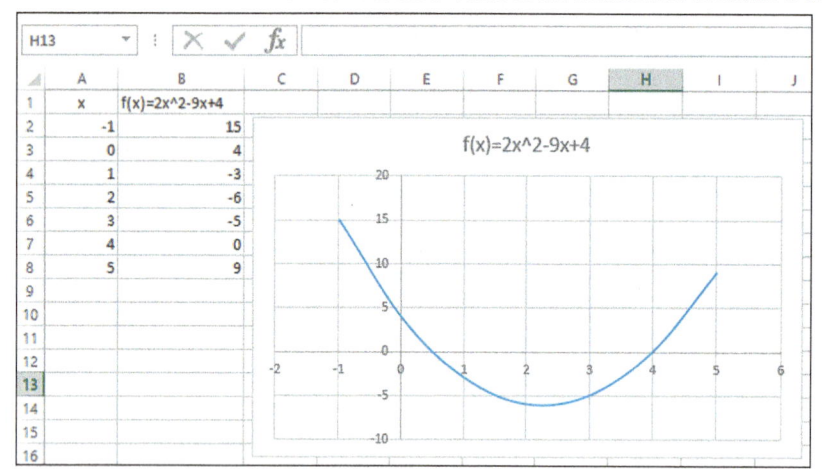

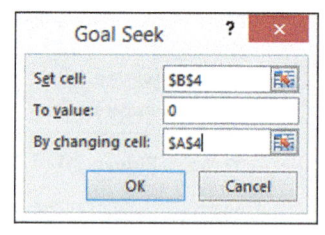

Figure 29

Figure 30

4. One solution is $x = 0.5$ (Figure 31).*

5. After finding the first solution, repeat the process using a second function value at or near 0.

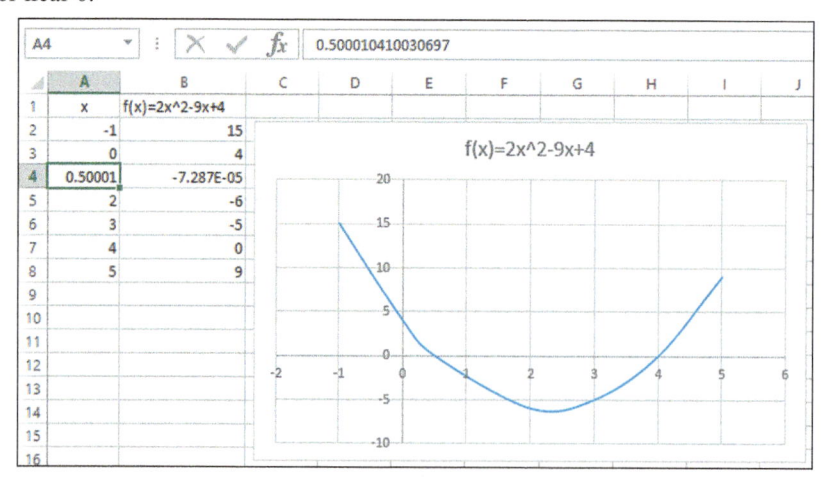

Figure 31

---

*__Notes:__ (1) The solution $x = 0.50001$, giving $f(x) = -7.287\text{E}-05$, is an approximation of the exact solution $x = 0.5$, which gives $f(x) = 0$. (2) Solutions of other equations (logarithmic, etc.) can be found with similar solution methods.

Solving an Equation of the Form $f(x) = g(x)$ Using the Intersection Method

If One Solution Is Sought

1. Enter a value for the input variable (*x*, for example) in cell A2 and the formula for each of the two sides of the equation in cells B2 and C2, respectively.
2. Enter "=B2−C2" in cell D2 (Figure 32).
3. Select the Data tab and choose What-If Analysis in Data Tools. Select Goal Seek to find *x* when B2−C2=0.
4. In the dialog box,
   a. Click the Set cell box and click on the D2 cell.
   b. Enter 0 in the To value box.
   c. Click the By changing cell box and click on the A2 cell (Figure 33).

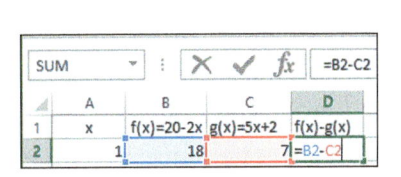

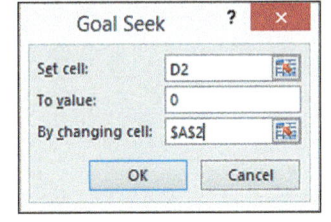

Figure 32

Figure 33

5. Click OK to find the *x*-value of the solution in cell A2 (Figure 34). The solution may be approximate.

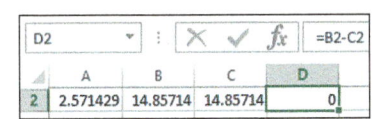

Figure 34

If Multiple Solutions Are Sought (as in the case of a quadratic equation)

1. Proceed as in steps 1 and 2 above.
2. Enter values for $f(x)$ and for $g(x)$, and use them to graph the function over an interval that shows points of intersection. Input values near those of these points will give column D values that are at or near 0. Entering *x*-values with $h = \frac{-b}{2a}$ near the center is useful when solving quadratic equations.
3. Select the Data tab and choose What-If Analysis in Data Tools. Use Goal Seek to enter in D2 the address of a cell with a function value at or near 0, and complete the process as above.
4. After finding the first solution, repeat the process using a second function value at or near 0.

Investments

Finding the Future Value of a Lump-Sum Investment

To find the future value of a lump-sum investment:

1. Type the headings Principal, Number of Periods, Payment, Annual Rate, Periods per Year, and Periodic Rate into row 1, cells A1 through F1, and enter their values (with interest rates as decimals) in row 2. Enter the investment under Principal as a negative number, and divide the annual rate by the periods per year to get the periodic rate.

2. Type the heading Future Value in cell A4.

3. In cell B4, type the formula =fv(F2,B2,C2,A2, 0) and press ENTER to compute the future value.

The future value of $10,000 invested at 8% per year for 10 years, compounded daily, is shown in Cell B4 of Figure 35.

	A	B	C	D	E	F
1	Principal	Number of periods	Payment	Annual Rate	Periods per year	
2	-10,000	3650	0	0.08	365	0.000219178
3						
4	Future Value	$22,253.46				

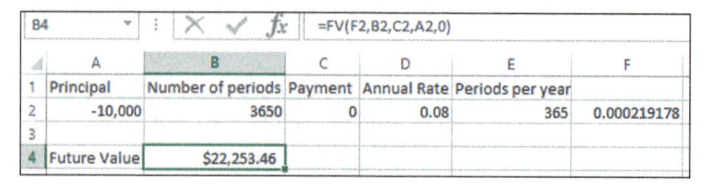

B4 =FV(F2,B2,C2,A2,0)

Figure 35

Finding the Present Value of an Investment

To find the present value (investment) that gives the given future value:

1. Type the headings Future Value, Number of Periods, Payment, Annual Rate, Periods per Year, and Periodic Rate into row 1, cells A1 through F1, and enter their values (with interest rates as decimals) in row 2. Enter the investment as a negative number, and divide the annual rate by the periods per year to get the periodic rate.

2. Type the heading Present Value in cell A4.

3. In cell B4, type the formula =pv(F2,B2,C2,A2, 0) and press ENTER to compute the present value.

The lump sum that must be invested at 10% per year, compounded semiannually, to get $15,000 in 7 years is shown as the present value in cell B4 of Figure 36. This present value will be negative because it leaves the investor.

	A	B	C	D	E	F
1	Future Value	Number of Periods	Payment	Annual Rate	Periods per Year	Periodic Rate
2	15,000	14	0	0.1	2	0.05
3						
4	Present Value	-7,576.02				

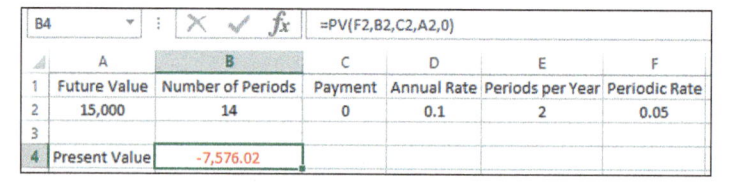

B4 =PV(F2,B2,C2,A2,0)

Figure 36

Annuities

Finding the Future Value of an Ordinary Annuity

To find the future value of an ordinary annuity:

1. Type the headings Principal, Number of Periods, Payment, Annual Rate, Periods per Year, and Periodic Rate into row 1, cells A1 through F1, and enter their values (with interest rates as decimals) in row 2. Enter the periodic investment under payment as a negative number, and divide the annual rate by the periods per year to get the periodic rate.

2. Type the heading Future Value in cell A4.

3. In cell B4, type the formula =fv(F2,B2,C2,A2, 0) and press ENTER to compute the future value.

The future value of an ordinary annuity with a contribution of $500 per quarter into an account that pays 8% per year, compounded quarterly, is shown in Cell B4 of Figure 37.

	A	B	C	D	E	F
1	Principal	Number of Periods	Payment	Annual Rate	Periods per year	Periodic Rate
2	0	20	-500	0.08	4	0.02
3						
4	Future Value	$12,148.68				

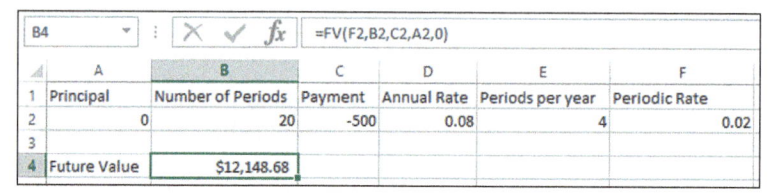

B4 =FV(F2,B2,C2,A2,0)

Figure 37

Finding the Present Value of an Ordinary Annuity

To find the present value of an ordinary annuity:

1. Type the headings Principal, Number of Periods, Payment, Annual Rate, Periods per Year, and Periodic Rate into row 1, cells A1 through F1, and enter their values (with interest rates as decimals) in row 2. Divide the annual rate by the periods per year to get the periodic rate.
2. Type the heading Present Value in cell A4.
3. In cell B4, type the formula =pv(F2,B2,C2,A2, 0) and press ENTER to compute the present value. The present value is negative because the money leaves the investor to start the annuity.

The lump sum necessary to establish an ordinary annuity that provides $2000 at the end of each month for 20 years, if it earns 6%, compounded monthly, is shown in cell B4 of Figure 38.

	A	B	C	D	E	F
1	Future Value	Number of Periods	Payment	Annual Rate	Periods per Year	Periodic Rate
2	0	240	2000	0.06	12	0.005
3						
4	Present Value	-279,161.54				

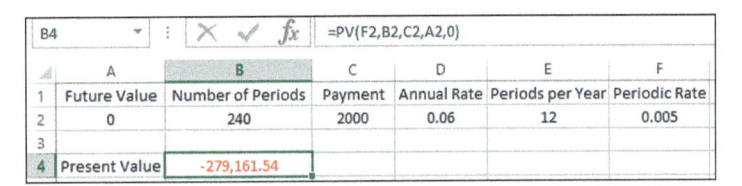

B4 =PV(F2,B2,C2,A2,0)

Figure 38

Payments to Amortize a Loan

To find the periodic payment to pay off a loan:

1. Type the headings in row 1 and their values (with the interest rate as a decimal) in row 2.
2. Enter the formula =(D2/E2) in F2 to compute the rate per period.
3. Type the heading Payment in A4.
4. In cell B4, type the formula =PMT(F2,B2,A2,C2,0) to compute the payment.

The spreadsheet in Figure 39 gives the annual payment of a loan of $10,000 over 5 years when interest is 10% per year.

The red value indicates a payment out.

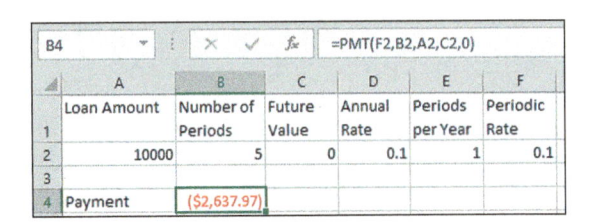

	A	B	C	D	E	F
1	Loan Amount	Number of Periods	Future Value	Annual Rate	Periods per Year	Periodic Rate
2	10000	5	0	0.1	1	0.1
3						
4	Payment	($2,637.97)				

B4 =PMT(F2,B2,A2,C2,0)

Figure 39

Solving a System of Two Linear Equations in Two Variables

1. Write the two equations as linear functions in the form $y = mx + b$.
2. Enter a value for the input variable (x) in cell A2 and the formula for each of the two equations in cells B2 and C2, respectively.
3. Enter "=B2−C2" in cell D2 (Figure 40).
4. Select the Data tab and choose What-If Analysis in Data Tools. Select Goal Seek to find x when B2−C2=0, as on page 711.

5. In the dialog box,
   a. Click the Set cell box and click on the D2 cell.
   b. Enter 0 in the To value box.
   c. Click the By changing cell box and click on the A2 cell.
6. Click OK in the Goal Seek dialog box, getting the solution.
7. The *x*-value of the solution will be in cell A2, and the *y*-value will be in both B2 and C2 (Figure 41).

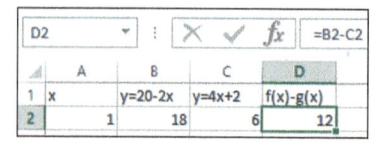

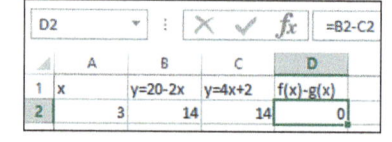

Figure 40

Figure 41

Matrices

Excel is useful to add, subtract, multiply, and find inverses of matrices. To perform operations with a matrix, enter each of its elements in a cell of a worksheet. Addition and subtraction of matrices are intuitive with Excel, but finding products and inverses requires special commands.

Adding and Subtracting Matrices (steps for two 3 × 3 matrices)

1. Type a name A in A1 to identify the first matrix.
2. Enter the elements of matrix A in the cells B1:D3.
3. Type a name B in A5 to identify the second matrix.
4. Enter the elements of matrix B in the cells B5:D7.
5. Type a name A+B in A9 to indicate the matrix sum.
6. Type the formula "=B1+B5" in B9 and press ENTER (Figure 42).
7. Use Fill Across to copy this formula across the row to C9 and D9.
8. Use Fill Down to copy the row B9:D9 to B11:D11, which gives the sum (Figure 43).

Figure 42

Figure 43

9. To subtract the matrices, change the formula in B9 to "=B1−B5" (Figure 44) and proceed as with addition (Figure 45).

Basic Guide to Excel 715

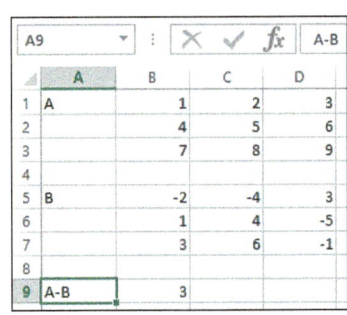

Figure 44

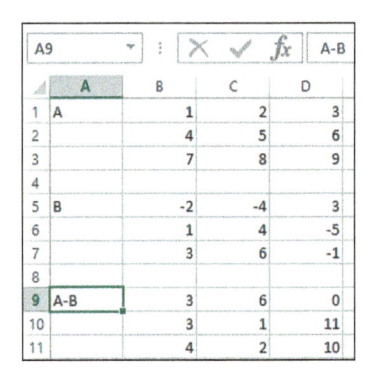

Figure 45

Finding the Product of Two Matrices (steps for two 3 × 3 matrices)

1. Enter the names and elements of the matrices as described on page 714.
2. Enter the name A×B in A9 to indicate the matrix product (Figure 46).
3. Select a range of cells that is the correct size to contain the product (B9:D11 in this case).
4. Type "=mmult(" in the formula bar, and then select the cells containing the elements of matrix A (B1:D3 in this case) (Figure 47).

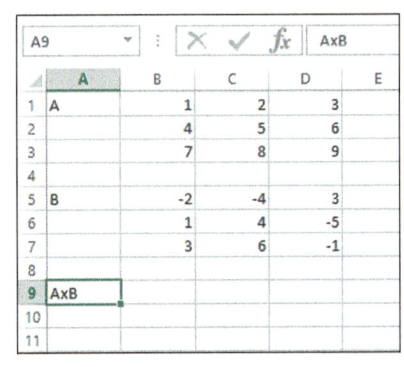

Figure 46

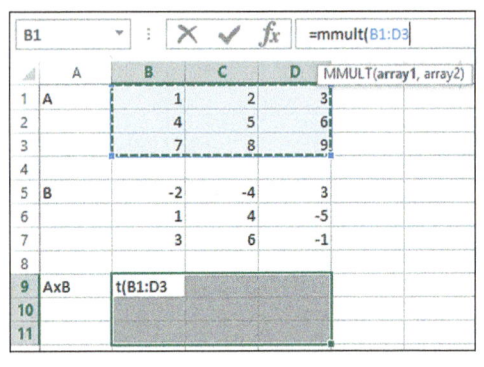

Figure 47

5. Staying in the formula bar, type a comma, select the matrix B elements (B5:D7 in this case), and close the parentheses.
6. Hold the CTRL and SHIFT keys down and press ENTER, which will give the product (Figure 48).

Figure 48

Finding the Inverse of a Matrix
(steps for a 3 × 3 matrix)

1. Enter the name A in A1 and the elements of the matrix in B1:D3, as described on page 714.
2. Enter the name "Inverse(A)" in A5 and select a range of cells that is the correct size to contain the inverse (B5:D7 in this case).
3. In the formula bar, enter "=minverse(", select matrix A (B1:D3), and close the parentheses (Figure 49).
4. Hold the CTRL and SHIFT keys down and press ENTER, which will give the inverse (Figure 50).

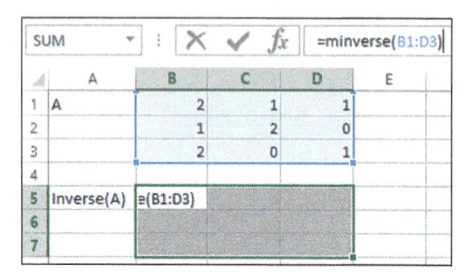

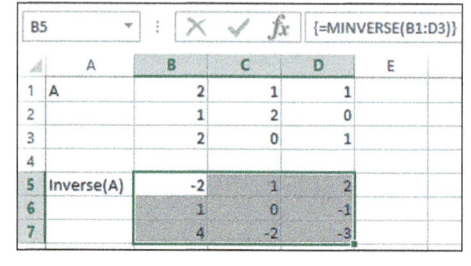

Figure 49

Figure 50

Solving Systems of Linear Equations with Matrix Inverses

A system of linear equations can be solved by multiplying the matrix containing the augment by the inverse of the coefficient matrix. Following are the steps used to solve

$$\begin{cases} 2x + y + z = 8 \\ x + 2y = 6 \\ 2x + z = 5 \end{cases}$$

1. Enter the coefficient matrix A in B1:D3.
2. Compute the inverse of A in B5:D7, as described above (Figure 50).
3. Enter B in cell A9 and enter the matrix containing the augment in B9:B11.
4. Enter X in A13, Y in A14, and Z in A15 and select the cells B13:B15.
5. In the formula bar, type "=mmult(", select matrix inverse(A) in B5:D7, type a comma, select matrix B in B9:B11, and close the parentheses (Figure 51).
6. HOLD the CTRL and SHIFT keys down and press ENTER, which will give the solution.
7. Matrix X gives the solution $x = 0$, $y = 3$, $z = 5$ (Figure 52).

Figure 51

Figure 52

Photo Credits

Cover: ssuaphotos/Shutterstock

Chapter 1 **p. 1,** kaykhoon/Shutterstock **p. 12,** Mark Van Scyoc/Shutterstock **p. 31,** Darren Brode/Shutterstock **p. 62,** alice-photo/Shutterstock

Chapter 2 **p. 88,** Karin Hildebrand Lau/Shutterstock **p. 108,** Marc Henauer/123RF **p. 138,** CandyBox Images/Shutterstock

Chapter 3 **p. 161,** PR Image Factory/Shutterstock **p. 170,** Claudio Divizia/Shutterstock **p. 186,** Rawpixel.com/Shutterstock **p. 243,** Temistocle Lucarelli/123RF

Chapter 4 **p. 255,** Blend Images/Shutterstock **p. 316,** rido/123RF

Chapter 5 **p. 325,** rblfmr/Shutterstock **p. 332,** 3d_man/Shutterstock **p. 361,** podorojniy/Shutterstock **p. 362,** Artwork by Liz Bradford/OXHIP, LLC **p. 426,** Odua Images/Shutterstock

Chapter 6 **p. 440,** Adam Calaitzis/Shutterstock

Chapter 7 **p. 533,** KeepWatch/Shutterstock

Chapter 8 **p. 613,** Alex Kravtsov/Shutterstock **p. 644,** Subbotina Anna/Shutterstock

Figure 1 on page 681 and graphing calculator screenshots from Texas Instruments. Courtesy of Texas Instruments Incorporated.

Additional Answers

Chapter 1 Functions, Graphs, and Models; Linear Functions

Toolbox Exercises

34. 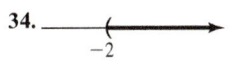 (graph: open circle at -2, arrow right)

35.  (graph: closed circle at 2, open circle at 5)

36.  (graph: arrow left from open circle at 3)

58. 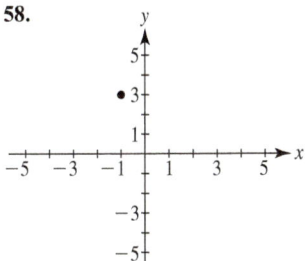 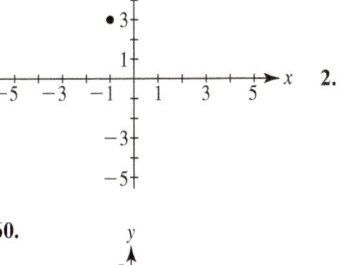 (graph with point at (0, 3))

59. 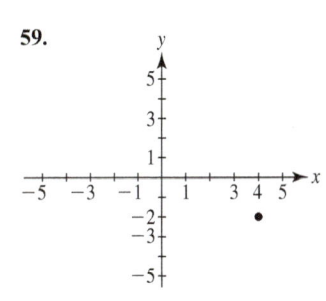 (graph)

60. 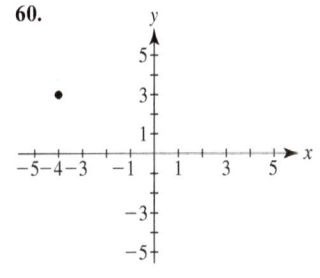 (graph)

61. 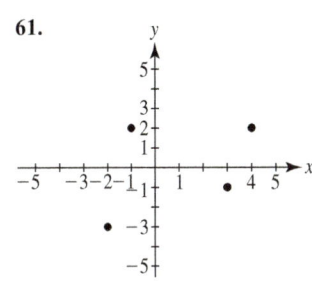 (graph)

62. 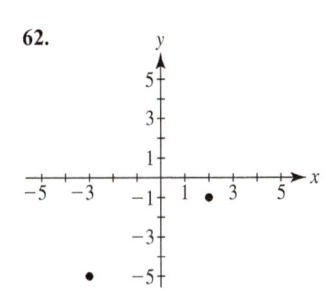 (graph)

1.1 Exercises

41. b. $\{2014, 2015, 2016, 2017, 2018, 2019, 2020, 2021, 2022, 2023, 2024\}$ is the domain; $\{207.3, 240.3, 290.1, 349.3, 414.0, 454.2, 475.1, 497.0, 515.7, 532.2, 544.8\}$ is the range **42.** m is the number of minutes after the outage, T is the temperature; there is one temperature for each minute, shown by the vertical line test. **55. b.** 7.7; in 2030, 7.7 million U.S. citizens age 65 and older are expected to have Alzheimer's disease.

69. c.

x	y
10	6,800
15	10,800
20	11,200
21	10,584
19	11,552
18	11,664
17	11,560

Testing values in the table shows a maximum volume of 11,664 cubic inches when $x = 18$ inches; the dimensions of the box are 18 in. by 18 in. by 36 in.

1.2 Skills Check

1. a. 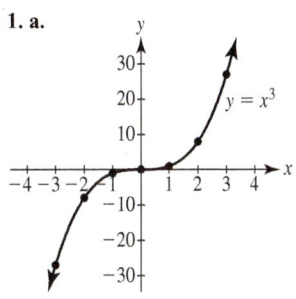 (graph of $y = x^3$) **b.** 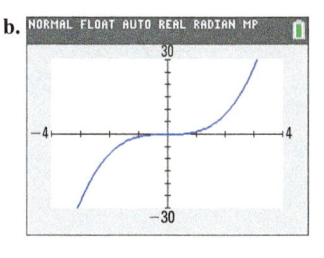 (calculator screen: NORMAL FLOAT AUTO REAL RADIAN MP, window $[-4, 4]$ by $[-30, 30]$)

2. a. 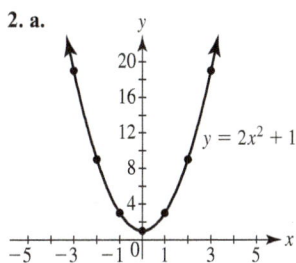 (graph of $y = 2x^2 + 1$) **b.** 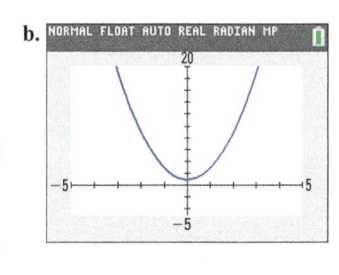 (calculator screen: NORMAL FLOAT AUTO REAL RADIAN MP, window $[-5, 5]$ by $[-5, 20]$)

3. Answers vary.

x	$f(x)$
-2	-7
-1	-4
0	-1
1	2
2	5

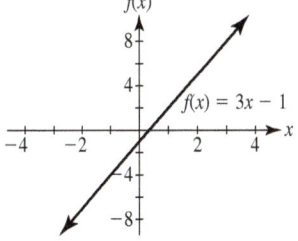

(graph of $f(x) = 3x - 1$)

4. 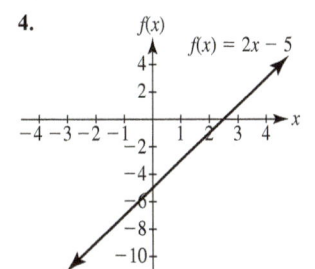 (graph of $f(x) = 2x - 5$)

Answers vary.

x	$f(x)$
-2	-9
-1	-7
0	-5
1	-3
2	-1

5. Answers vary.

x	$f(x)$
-4	8
-2	2
0	0
2	2
4	8

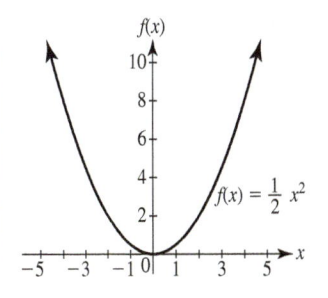

(graph of $f(x) = \frac{1}{2}x^2$)

A-1

Additional Answers

6.

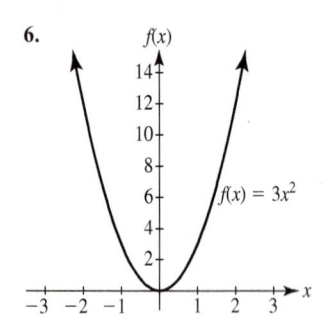

x	y
−2	12
−1	3
0	0
1	3
2	12

Answers vary. Graph of $f(x) = 3x^2$.

7. Answers vary.

x	f(x)
−2	−1/4
−1	−1/3
0	−1/2
1	−1
2	undefined
3	1
4	1/2

Graph of $f(x) = \dfrac{1}{x-2}$.

8. $f(x) = \dfrac{x}{x+3}$

Answers vary.

x	f(x)
−6	2
−2	−2
−1	−0.5
0	0
1	0.25
3	0.5

9.

10.

11.

12.

13.

14.

15. a. **b.** View (b) is better.

16. a. **b.** View (b) is better.

17. a. **b.** View (b) is better.

18. a. **b.** View (b) is better.

19. Letting y vary from −5 to 100 gives one view.

Turning point coordinates: $(0, 50)$

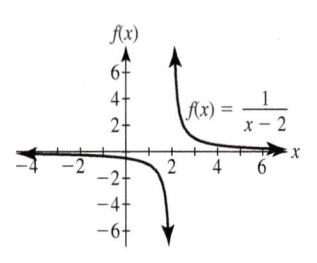

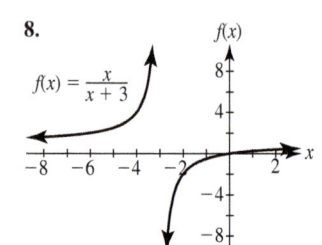

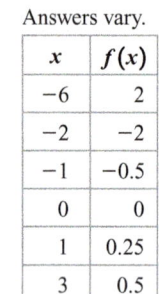

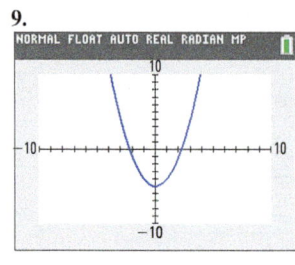

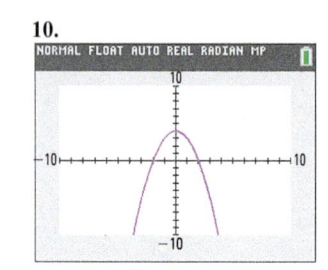

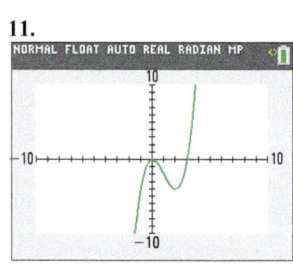

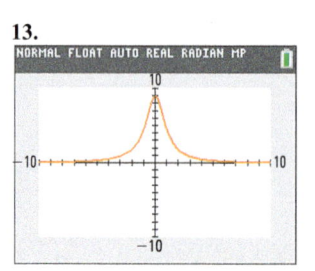

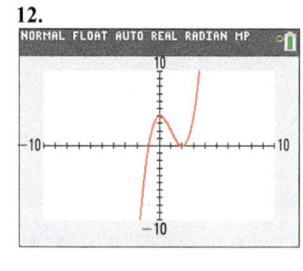

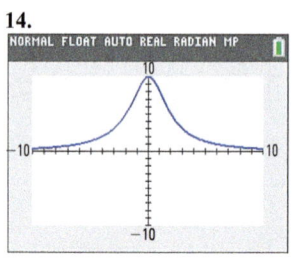

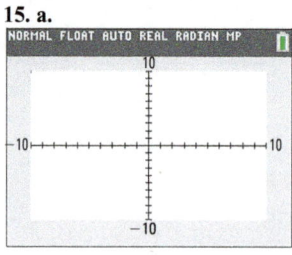

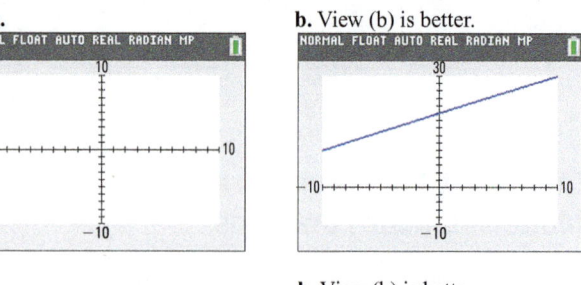

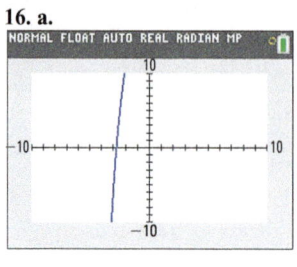

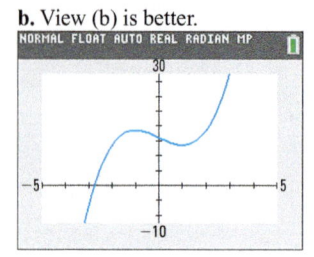

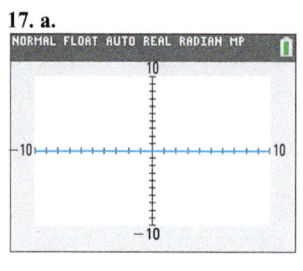

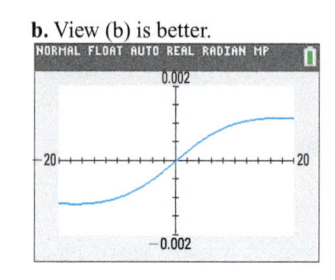

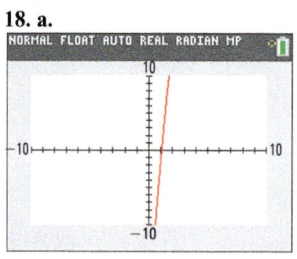

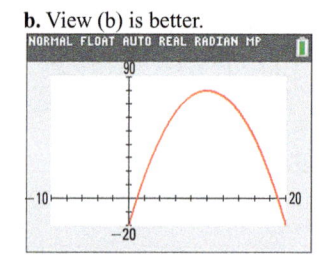

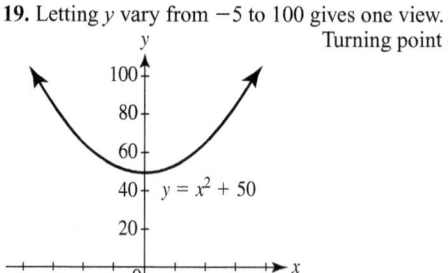

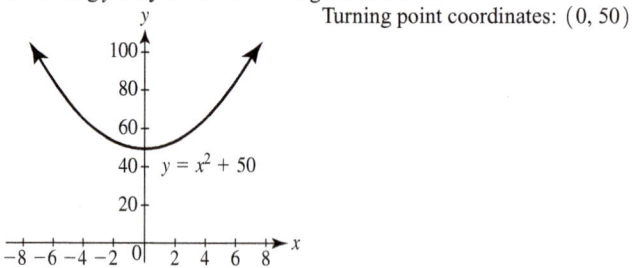

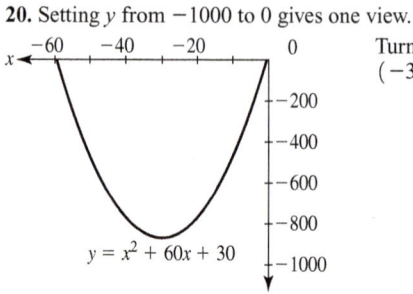

Graph of $y = x^2 + 50$.

20. Setting y from −1000 to 0 gives one view.

Turning point coordinates: $(-30, -870)$

Graph of $y = x^2 + 60x + 30$.

21. Setting y from -200 to 300 gives one view.

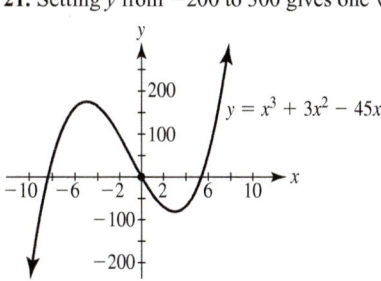

Turning point coordinates: $(-5, 175)$, $(3, -81)$

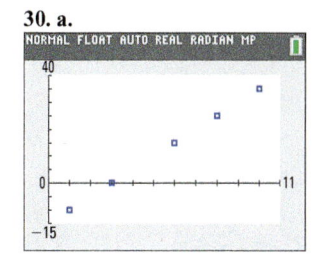

$y = x^3 + 3x^2 - 45x$

22. Setting y from -30 to 30 gives one view.

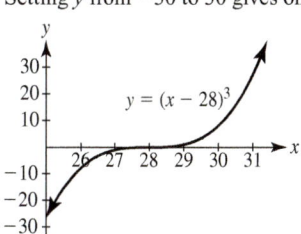

No turning point

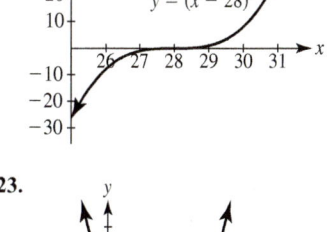

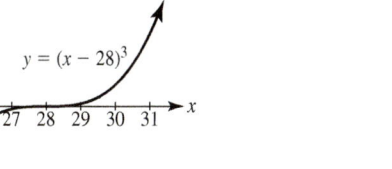

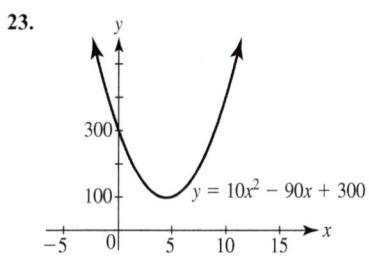

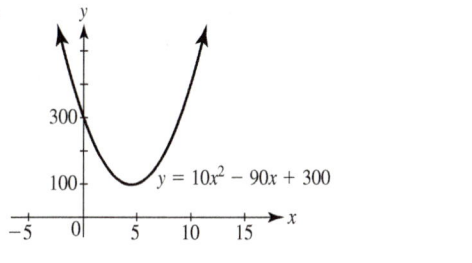

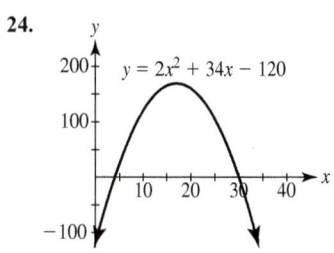

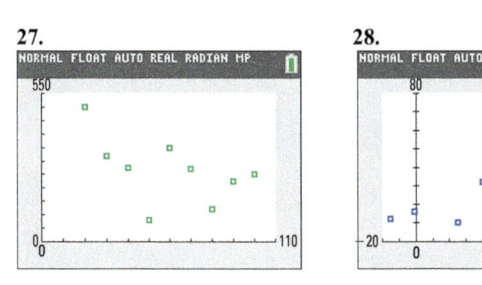

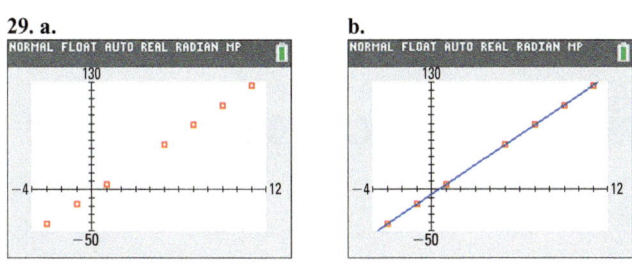

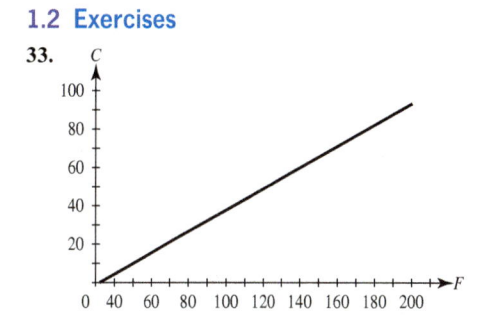

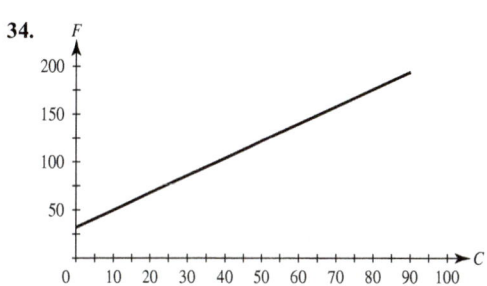

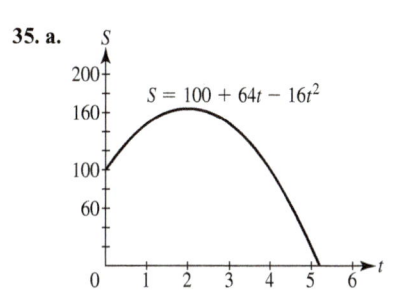

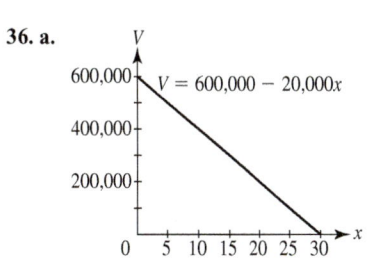

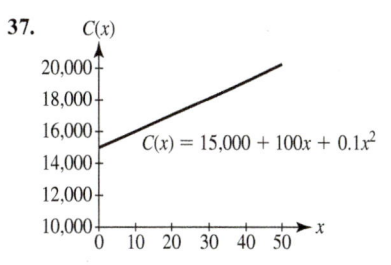

$y = (x - 28)^3$

23. $y = 10x^2 - 90x + 300$

24. $y = 2x^2 + 34x - 120$

27.

28.

29. a.

b.

30. a.

b.

1.2 Exercises

33.

34.

35. a. $S = 100 + 64t - 16t^2$

36. a. $V = 600,000 - 20,000x$

37. $C(x) = 15,000 + 100x + 0.1x^2$

A-4 Additional Answers

38.

41. a.

42. a.

43. d.

45. a.

46. $P(x) = 200x - 0.01x^2 - 5000$

47. $P(x) = 1500x - 8000 - 0.01x^2$

48. a.

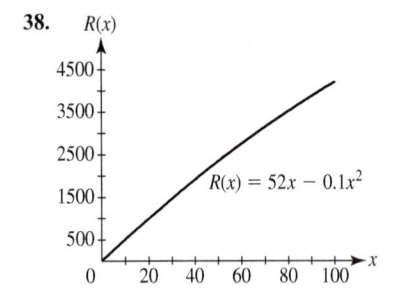

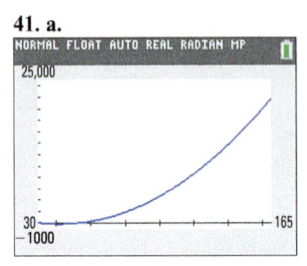

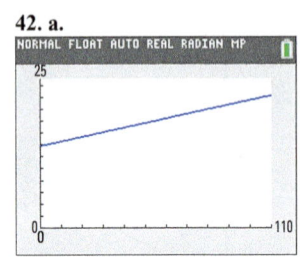

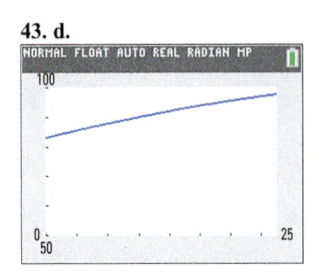

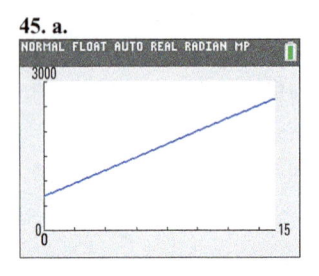

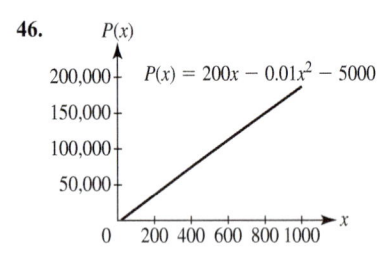

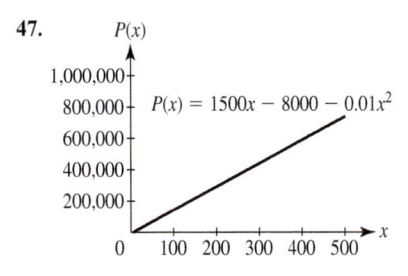

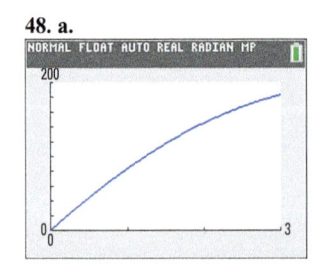

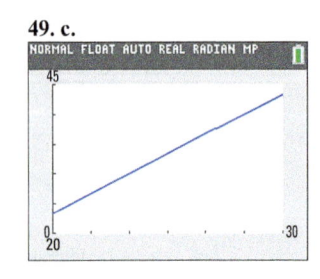

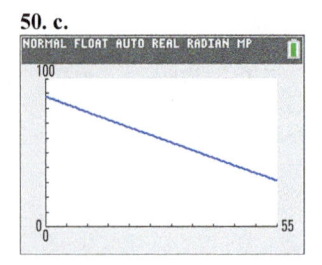

49. c.

50. c.

51. b.

Years after 2000	Population (millions)
0	275.3
10	299.9
20	324.9
30	351.1
40	377.4
50	403.7
60	432.0

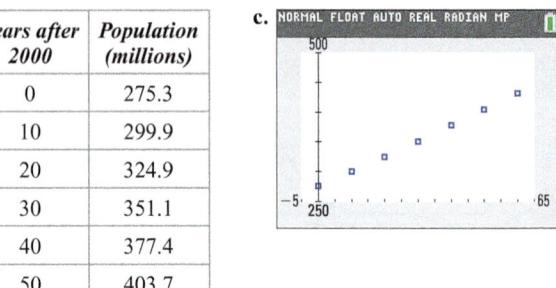

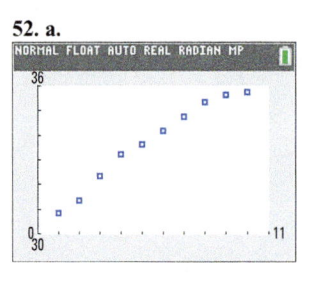

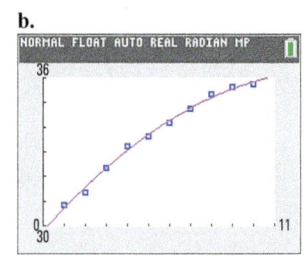

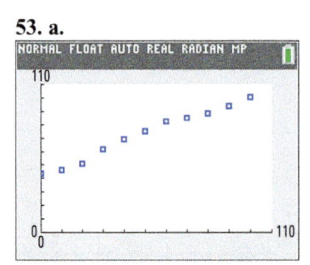

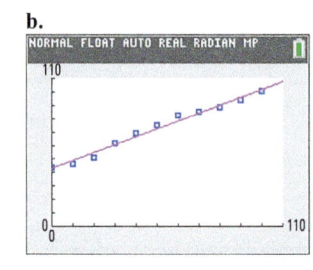

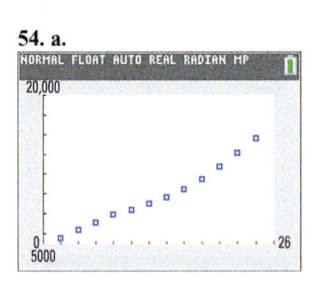

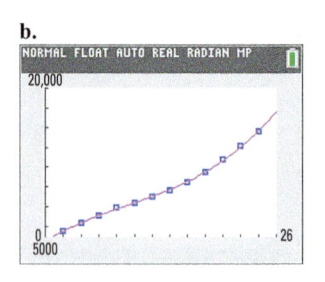

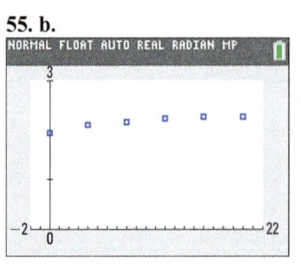

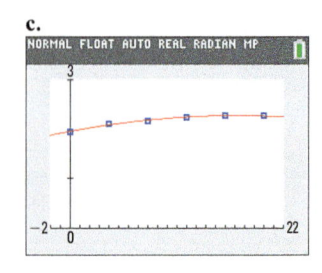

c.

52. a. **b.**

53. a. **b.**

54. a. **b.**

55. b. **c.**

(Graph 38 shows $R(x) = 52x - 0.1x^2$)

56. a.

b.

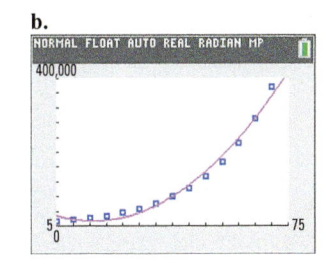

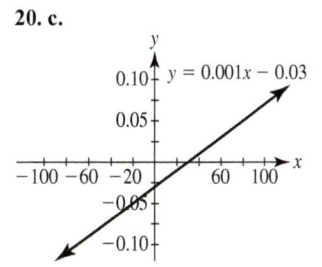

20. c. $y = 0.001x - 0.03$

1.3 Skills Check

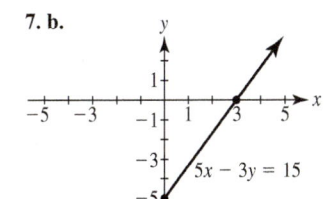

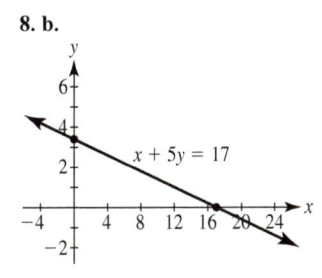

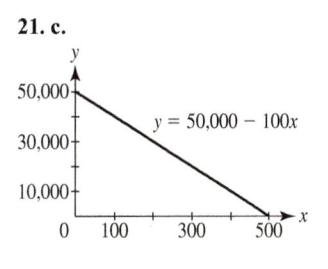

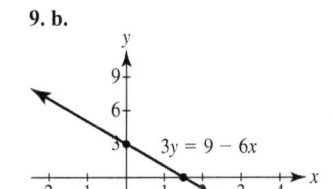

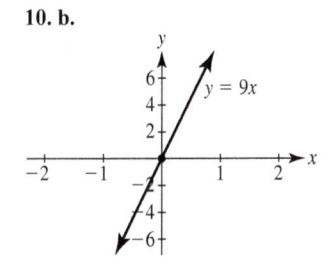

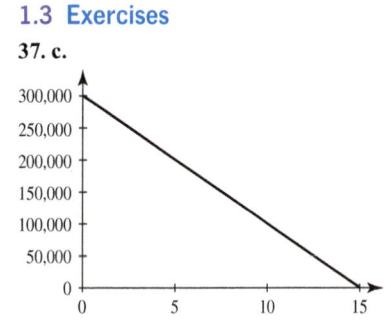

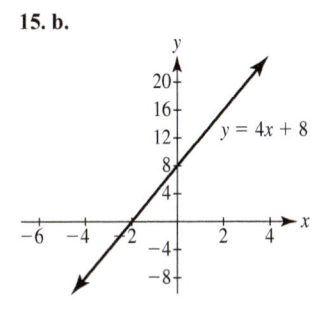

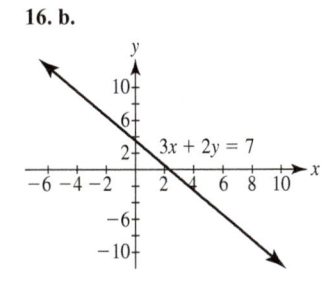

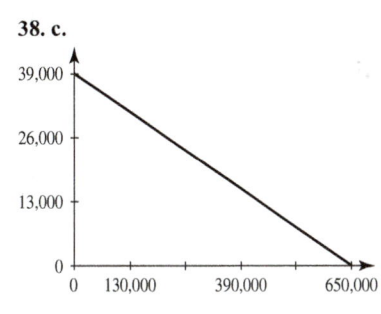

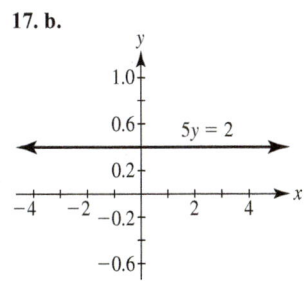

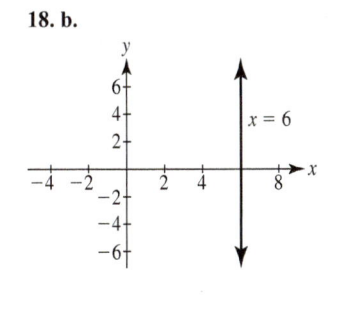

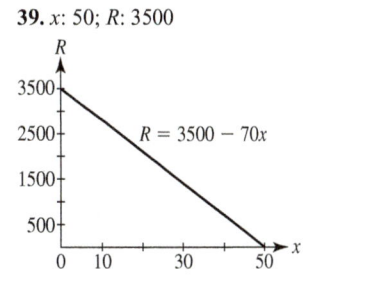

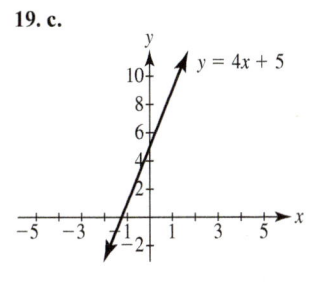

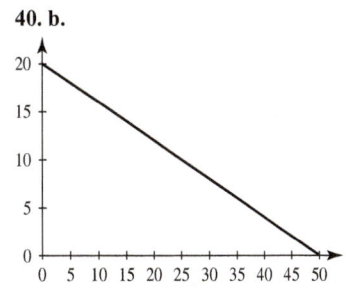

7. b. $5x - 3y = 15$

8. b. $x + 5y = 17$

21. c. $y = 50,000 - 100x$

9. b. $3y = 9 - 6x$

10. b. $y = 9x$

1.3 Exercises

37. c.

15. b. $y = 4x + 8$

16. b. $3x + 2y = 7$

38. c.

17. b. $5y = 2$

18. b. $x = 6$

39. x: 50; R: 3500

$R = 3500 - 70x$

19. c. $y = 4x + 5$

40. b.

Additional Answers

42. a.

d.

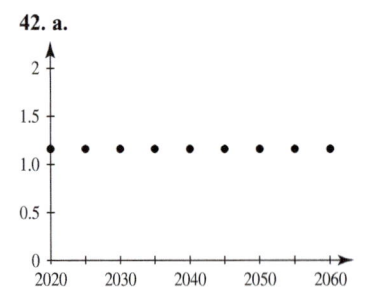

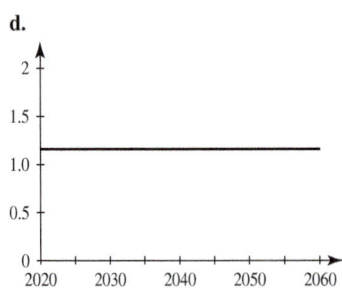

48. b. For every one-unit increase in depth, there is a corresponding increase in pressure of $\frac{6}{11}$ pound per square inch.   **50. c.** The percent of U.S. sales that are electric or plug-in hybrid electric vehicles is projected to increase by 0.77 percentage point each year.
51. b. The population of China aged 15 years and older is projected to increase by 9.32 million per year.

1.4 Exercises

68. b. The y-intercept is 500. If no clients from the first group are served, then 500 clients from the second group can be served. The slope is -1.5. For each one-person increase in the number of clients served from the first group, there is a corresponding decrease of 1.5 clients served from the second group.

Chapter 1 Skills Check

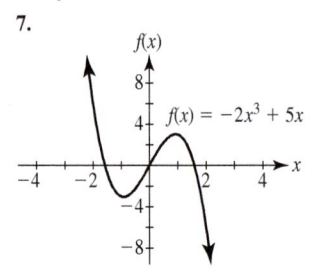

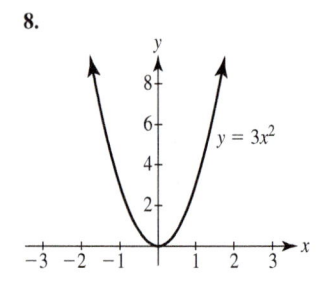

7. $f(x) = -2x^3 + 5x$

8. $y = 3x^2$

9. Standard window:

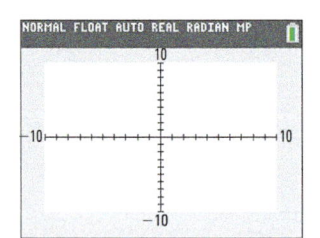

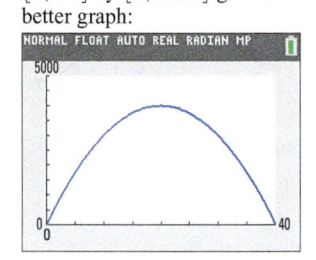

$[0, 40]$ by $[0, 5000]$ gives a better graph:

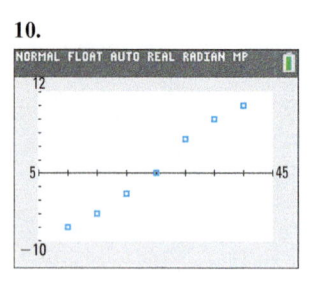

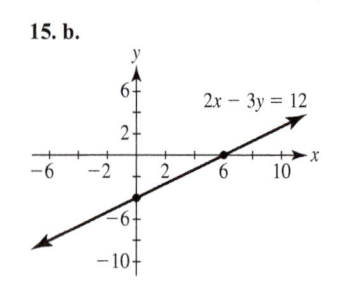

10.

Chapter 1 Review

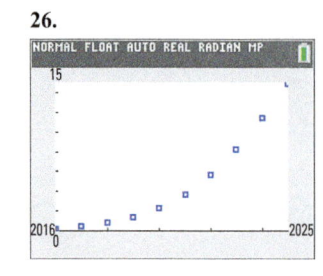

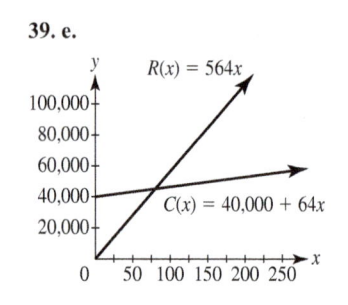

26.

15. b. $2x - 3y = 12$

39. e. $R(x) = 564x$; $C(x) = 40{,}000 + 64x$

Chapter 2 Linear Models, Equations, and Inequalities

2.1 Skills Check

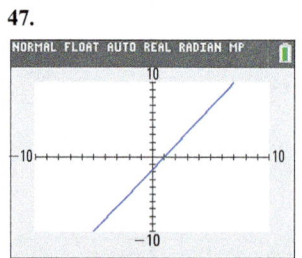

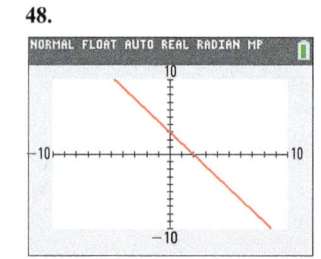

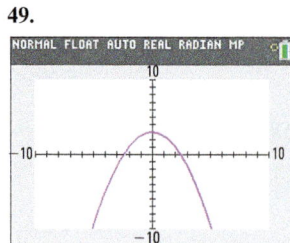

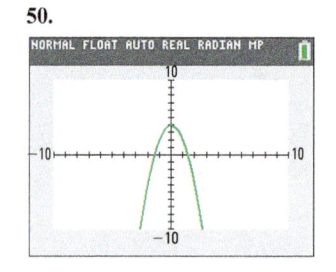

47.

48.

49.

50.

2.2 Skills Check

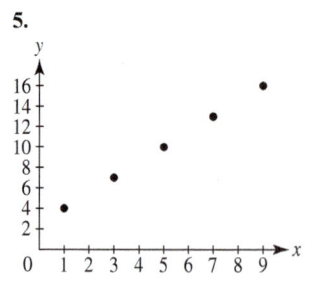

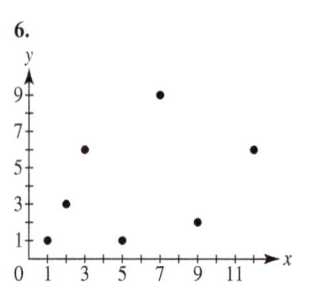

5.

6.

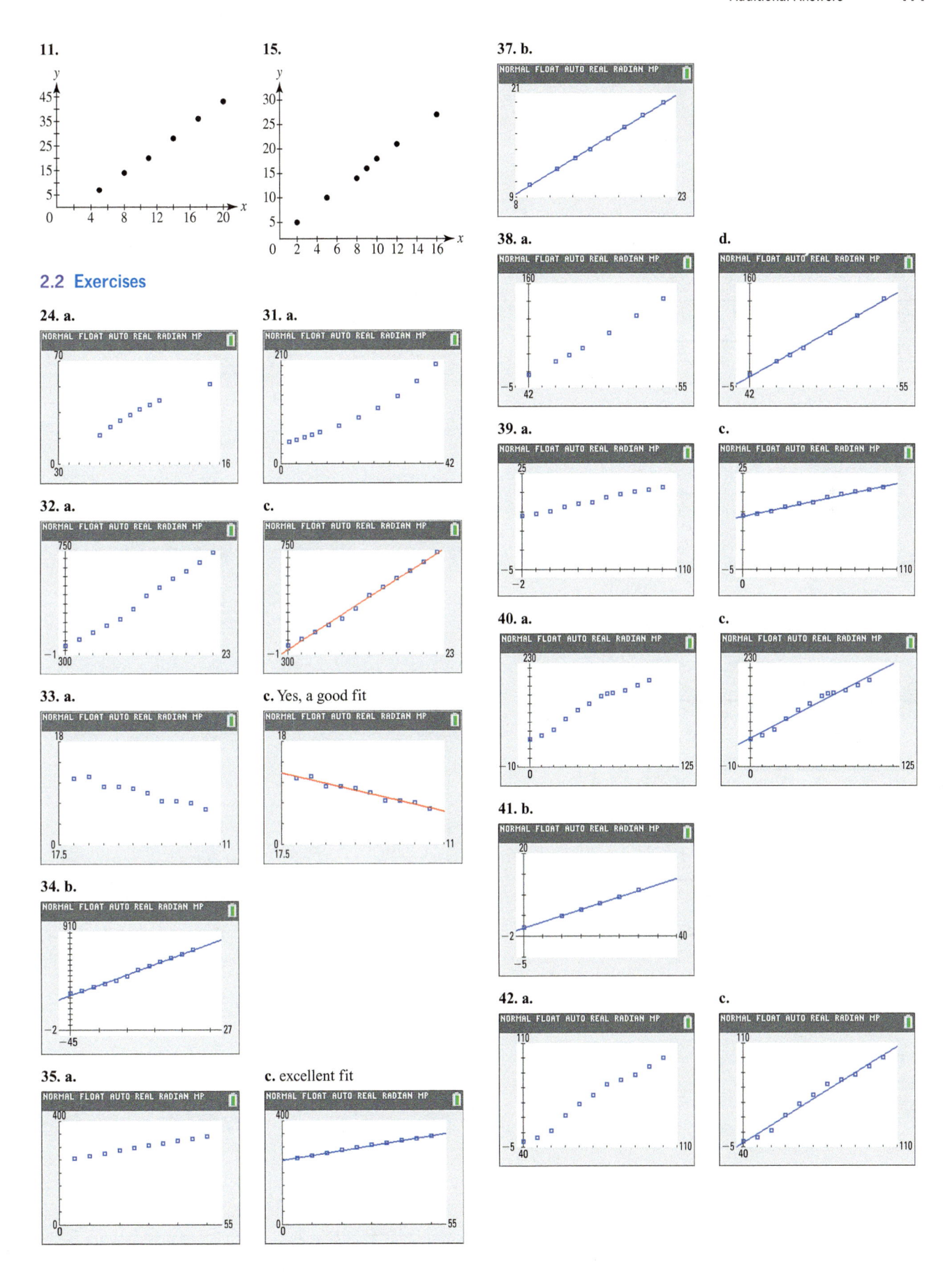

11.

15.

2.2 Exercises

24. a.

31. a.

32. a.

c.

33. a.

c. Yes, a good fit

34. b.

35. a.

c. excellent fit

37. b.

38. a.

d.

39. a.

c.

40. a.

c.

41. b.

42. a.

c.

Additional Answers

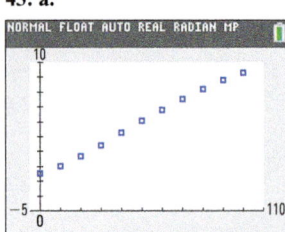

43. a. [calculator screen: scatter plot] **c.** [calculator screen: scatter plot with regression line]

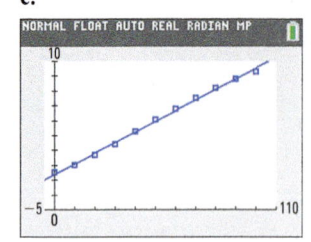

2.4 Skills Check

1. 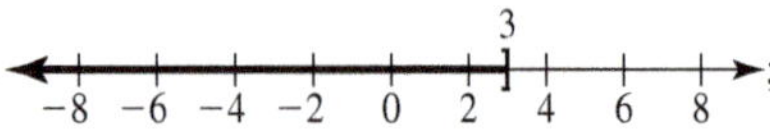 [number line with closed circle at 3, shaded left]; $(-\infty, 3]$
2. 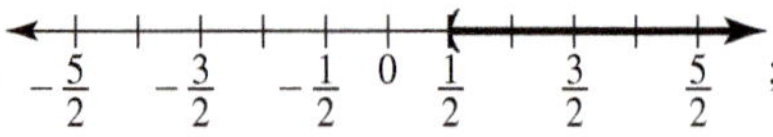 [number line with open circle at $\frac{1}{2}$, shaded right]; $\left(\frac{1}{2}, \infty\right)$
3. 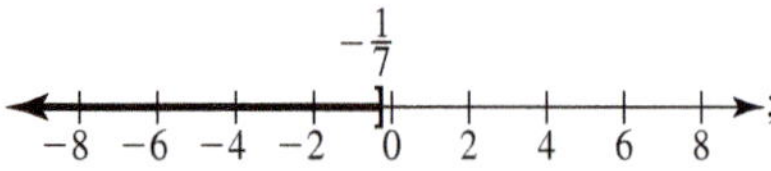 [number line with closed circle at $-\frac{1}{7}$, shaded left]; $\left(-\infty, -\frac{1}{7}\right]$
4. 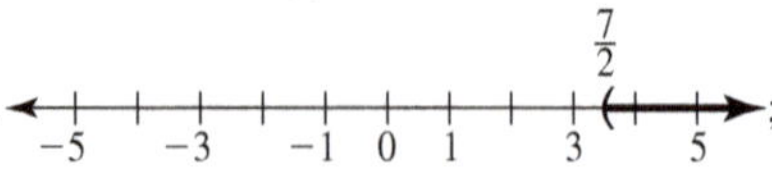 [number line with open circle at $\frac{7}{2}$, shaded right]; $\left(\frac{7}{2}, \infty\right)$
5. 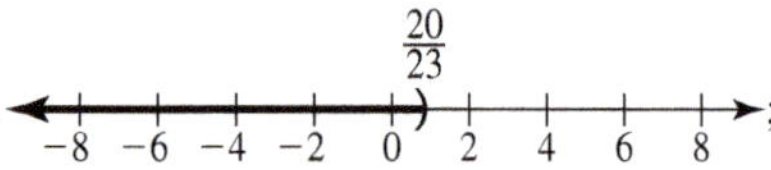 [number line with open circle at $\frac{20}{23}$, shaded left]; $\left(-\infty, \frac{20}{23}\right)$
6. 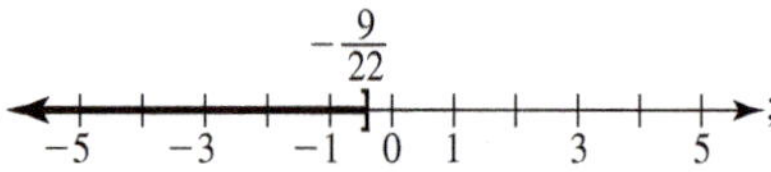 [number line with closed circle at $-\frac{9}{22}$, shaded left]; $\left(-\infty, \frac{-9}{22}\right]$
7. 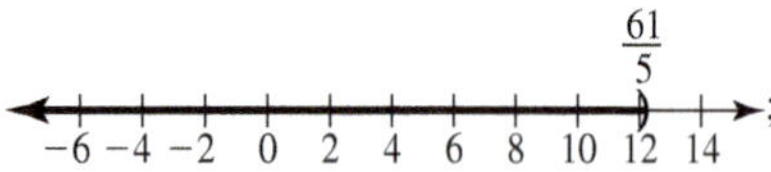 [number line with open circle at $\frac{61}{5}$, shaded left]; $\left(-\infty, \frac{61}{5}\right)$
8. 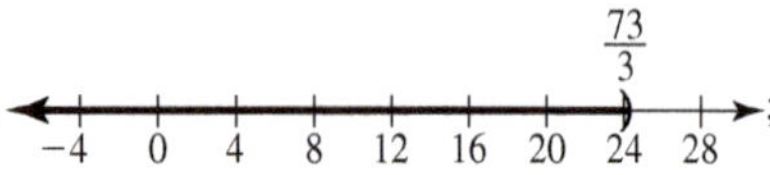 [number line with closed circle at $\frac{73}{3}$, shaded left]; $\left(-\infty, \frac{73}{3}\right]$
9. 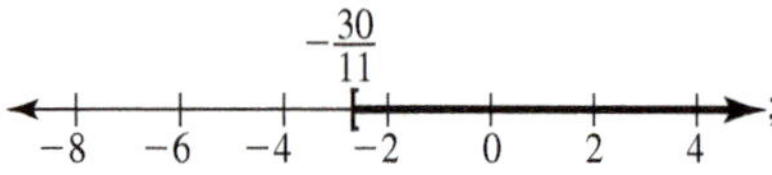 [number line with closed circle at $-\frac{30}{11}$, shaded right]; $\left[-\frac{30}{11}, \infty\right)$
10. 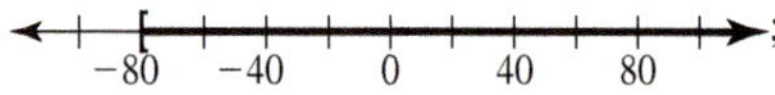 [number line with closed circle at -80, shaded right]; $[-80, \infty)$
11. 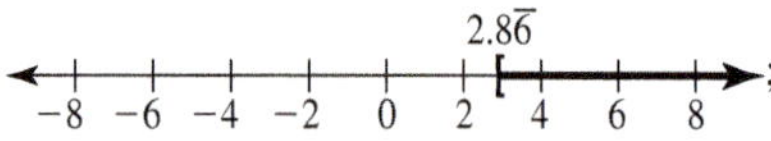 [number line with closed circle at $2.8\overline{6}$, shaded right]; $[2.8\overline{6}, \infty)$
12. 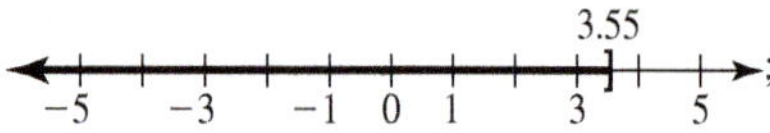 [number line with closed circle at 3.55, shaded left]; $(-\infty, 3.55]$

2.4 Exercises

44. b. [calculator screen: CALC INTERSECT, Intersection X=?, Y=250]

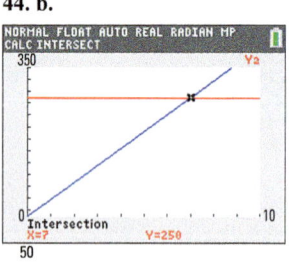

Chapter 2 Skills Check

9. [calculator screen: line graph] **10.** [calculator screen: scatter plot]

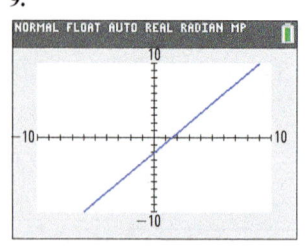

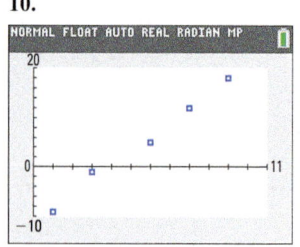

12. [calculator screen: scatter plot with regression line]

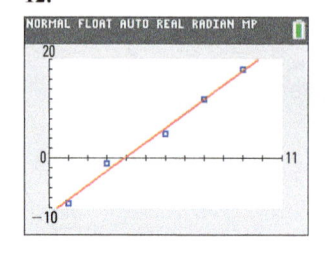

Chapter 2 Review

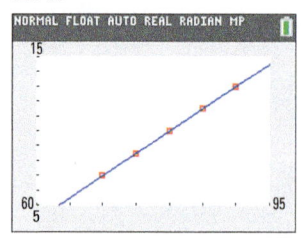

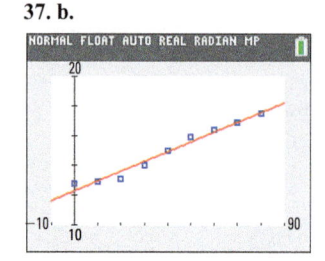

31. b. [calculator screen: scatter plot with line] **36. b.** [calculator screen: scatter plot with line]

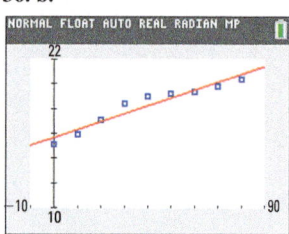

37. b. [calculator screen: scatter plot with line] **38. b.** [calculator screen: scatter plot with line]

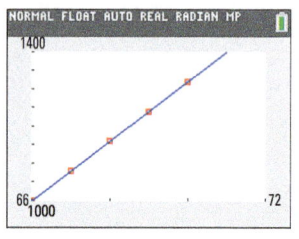

39. b. [calculator screen: scatter plot with line]

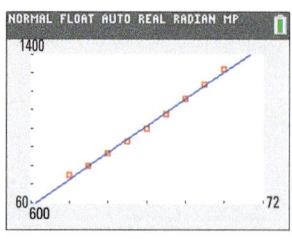

40. a. [calculator screen: scatter plot] **c.** [calculator screen: scatter plot with line]

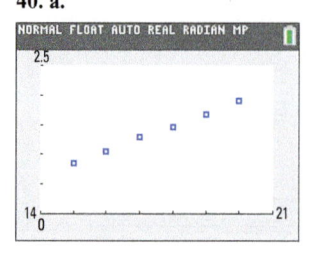

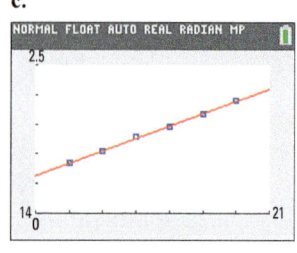

Chapter 3 Quadratic, Piecewise-Defined, and Power Functions

3.1 Skills Check

7. a. graph of $y = 2x^2 - 8x + 6$

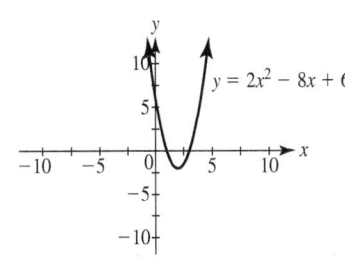

8. a. graph of $f(x) = x^2 + 4x + 4$

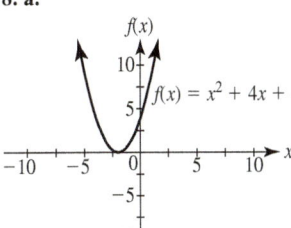

9. a. graph of $g(x) = -5x^2 - 6x + 8$

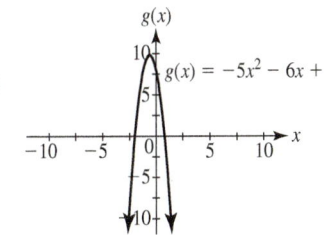

10. a. graph of $h(x) = -2x^2 - 4x + 6$

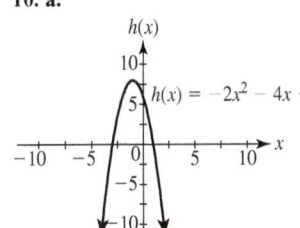

11. a. graph of $y = x^2 + 8x + 19$

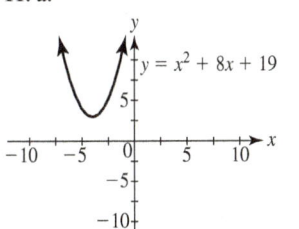

12. a. graph of $y = x^2 - 4x + 5$

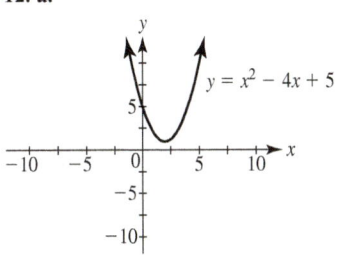

13. a. graph of $y = 0.01x^2 - 8x$

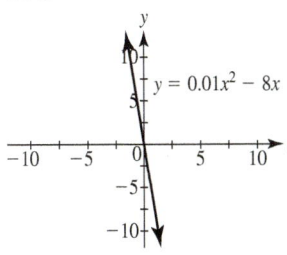

14. a. graph of $y = 0.1x^2 + 8x + 2$

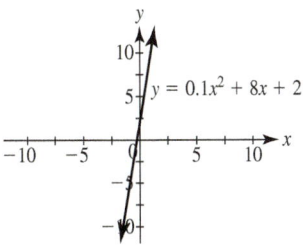

21. b. graph of $y = (x - 1)^2 + 3$

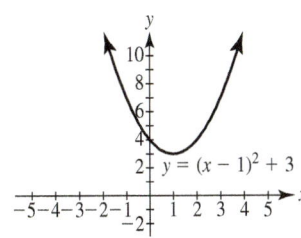

22. b. graph of $y = (x + 10)^2 - 6$

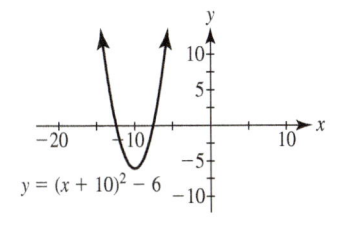

23. b. graph of $y = (x + 8)^2 + 8$

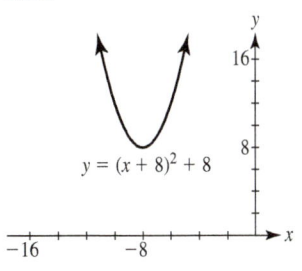

24. b. graph of $y = (x - 12)^2 + 1$

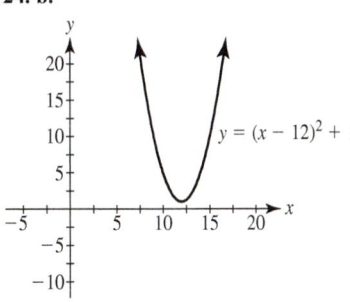

25. b. graph of $f(x) = 2(x - 4)^2 - 6$

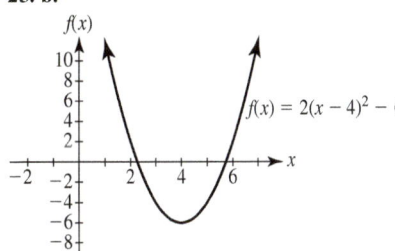

26. b. graph of $f(x) = -0.5(x - 2)^2 + 1$

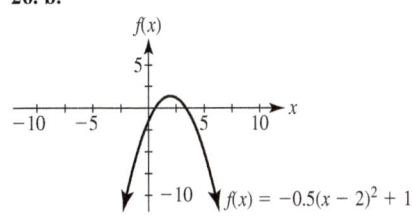

27. b. graph of $y = 12x - 3x^2$

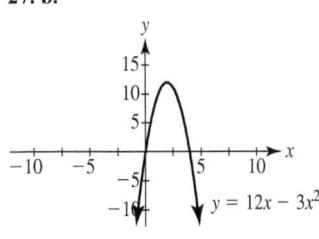

28. b. graph of $y = 3x + 18x^2$

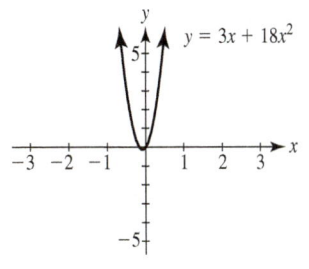

Additional Answers

29. b. Graph of $y = 3x^2 + 18x - 3$

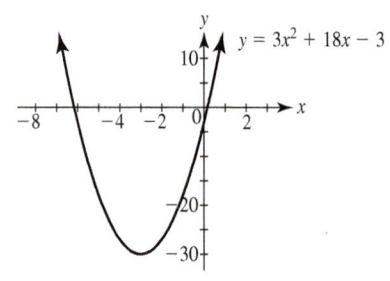

30. b. Graph of $y = 5x^2 + 75x + 8$

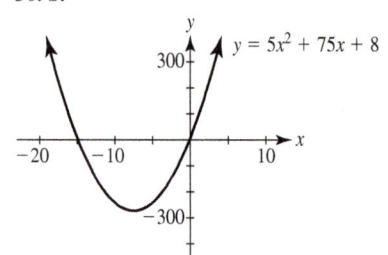

31. b. Graph of $y = 2x^2 - 40x + 10$

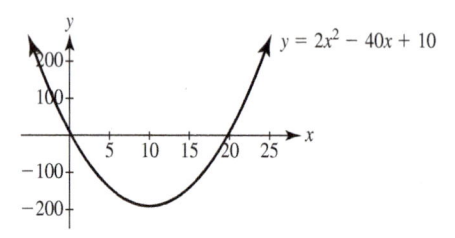

32. b. Graph of $y = -3x^2 - 66x + 12$

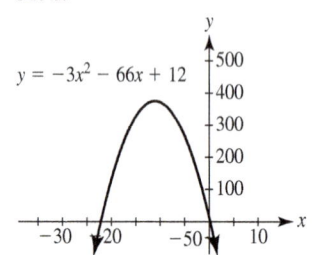

33. b. Graph of $y = -0.2x^2 - 32x + 2$

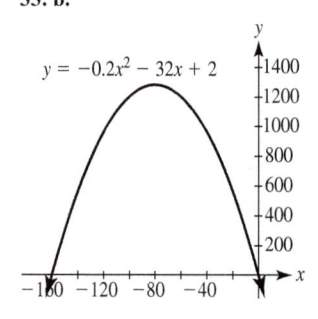

34. b. Graph of $y = 0.3x^2 + 12x - 8$

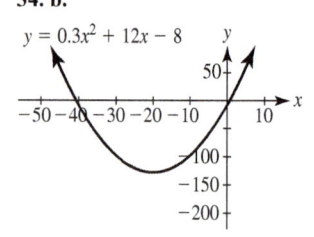

35. Graph of $y = x^2 + 24x + 144$

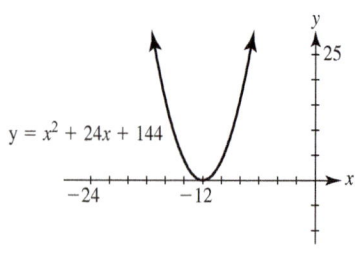

36. Graph of $y = x^2 - 36x + 324$

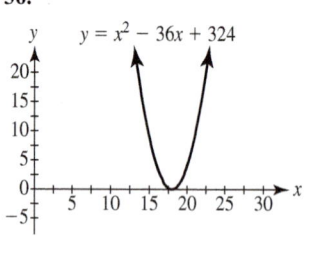

37. Graph of $y = -x^2 - 100x + 1600$

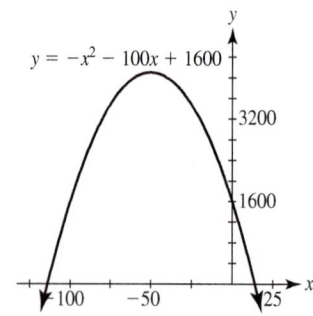

38. Graph of $y = -x^2 - 80x - 2000$

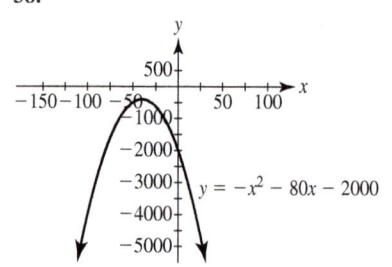

39. Graph of $y = 2x^2 + 10x - 600$

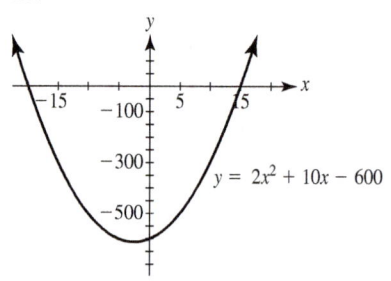

40. Graph of $y = 2x^2 - 75x - 450$

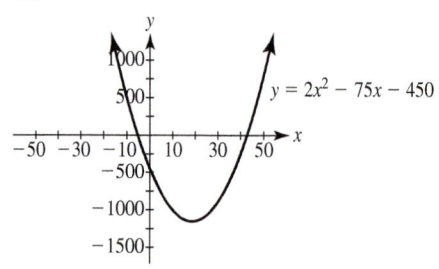

3.1 Exercises

47. a. Graph of $P = 32x - 0.01x^2 - 1000$

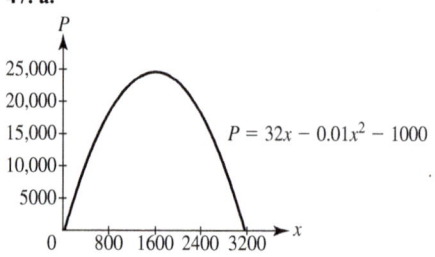

48. a.

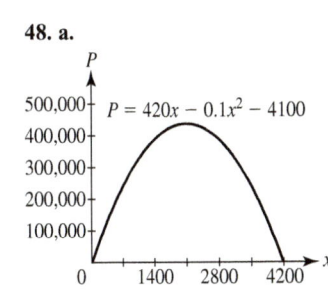

Graph of $P = 420x - 0.1x^2 - 4100$

49. a.

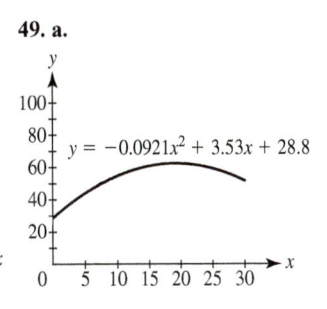

Graph of $y = -0.0921x^2 + 3.53x + 28.8$

50. a.

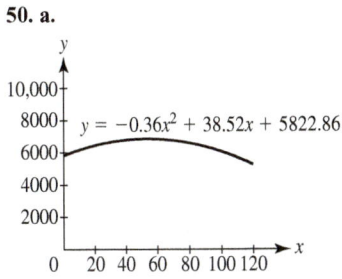

Graph of $y = -0.36x^2 + 38.52x + 5822.86$

51. a.

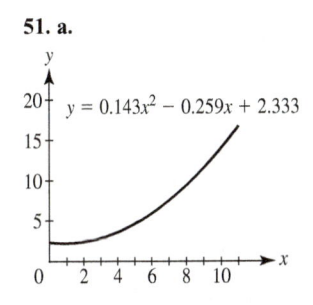

Graph of $y = 0.143x^2 - 0.259x + 2.333$

54. a.

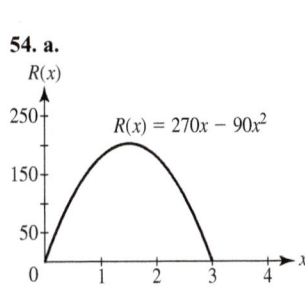

Graph of $R(x) = 270x - 90x^2$

55. c.

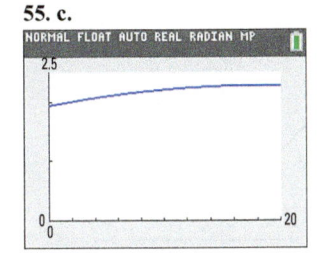

Calculator graph (NORMAL FLOAT AUTO REAL RADIAN MP), window 0 to 20, 0 to 2.5

56.

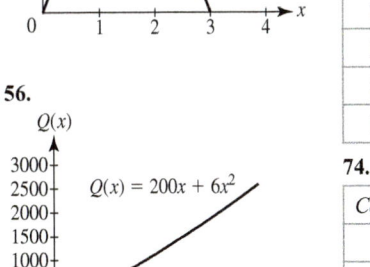

Graph of $Q(x) = 200x + 6x^2$

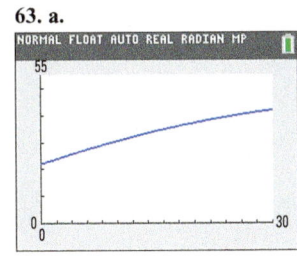

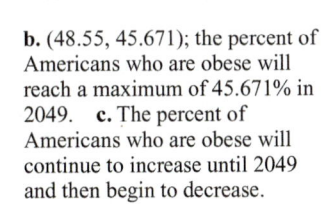

b. (48.55, 45.671); the percent of Americans who are obese will reach a maximum of 45.671% in 2049. **c.** The percent of Americans who are obese will continue to increase until 2049 and then begin to decrease.

63. a.

Calculator graph (NORMAL FLOAT AUTO REAL RADIAN MP), window 0 to 30, 0 to 55

65. a.

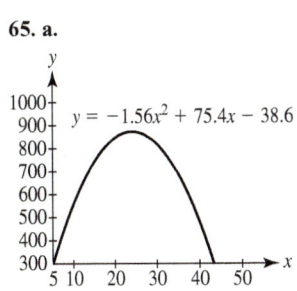

Graph of $y = -1.56x^2 + 75.4x - 38.6$

67. a.

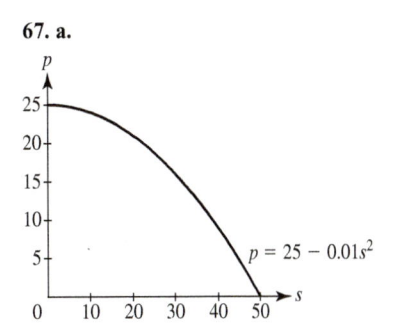

Graph of $p = 25 - 0.01s^2$

68. a.

Graph of $S = 1000x - x^2$

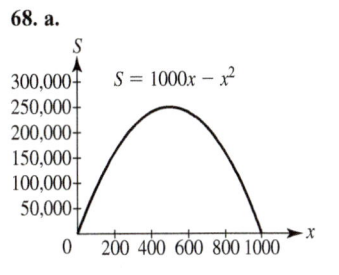

d. The sensitivity to the drug drops to zero when the dosage is 1000. This could indicate that the person is overdosed on the drug and no longer able to detect sensitivity to the drug. **69.** The t-intercepts are 3.5 and -3.75. This means that the ball will strike the pool in 3.5 seconds; -3.75 is meaningless.

73. a.

Rent ($)	Number of Apartments	Revenue ($)
1200	100	120,000
1240	98	121,520
1280	96	122,880
1320	94	124,080

74. a.

Cost per Person ($)	Number of Skaters	Total Revenue ($)
11.50	66	759
11.00	72	792
10.50	78	819

3.2 Exercises

71. b. When x is zero, there is no amount of drug in a person's system, and therefore no sensitivity to the drug. When x is 100 mL, the amount of drug in a person's system is so high that the person may be overdosed on the drug and therefore have no sensitivity to the drug.

85. a.

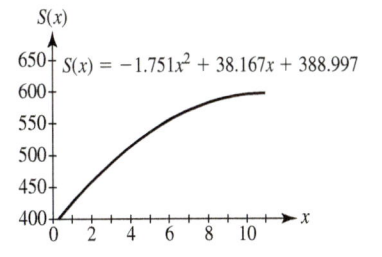

Graph of $S(x) = -1.751x^2 + 38.167x + 388.997$

3.3 Skills Check

15. graph of $y = x^{3.5}$

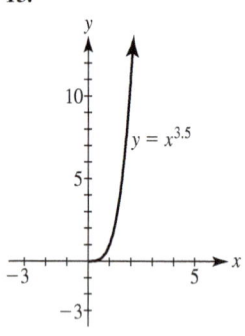

16. graph of $y = x^{-1.5}$

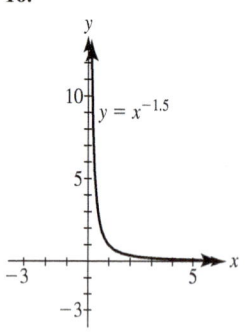

17. graph of $y = 2\sqrt[3]{x}$

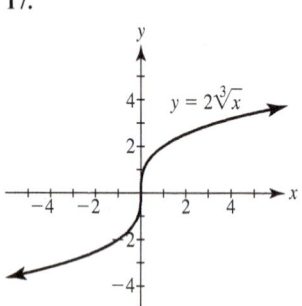

18. graph of $f(x) = -3\sqrt[4]{x}$

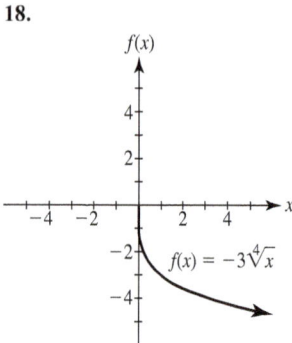

3.3 Exercises

35. a. graph of $Q = 489L^{0.6}$

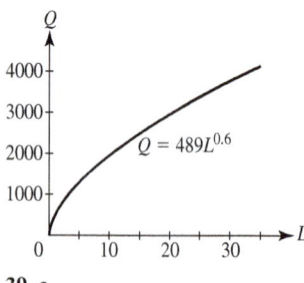

37. b. graph of $y = 66.164x^{-0.094}$

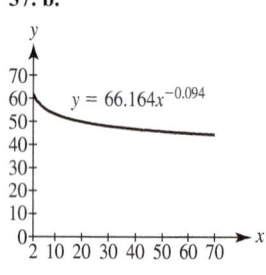

39. a. graph of $P = 1200x^{5/2}$

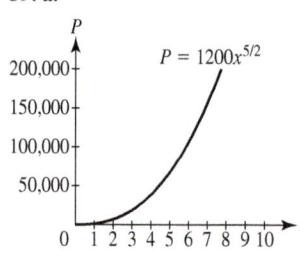

40. a. graph of $y = 1.595x^{-0.343}$

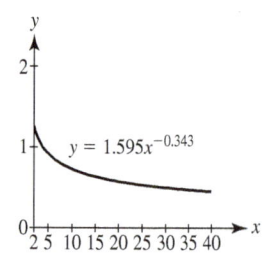

41. b. graph of $y = 2.15x^{2.12}$

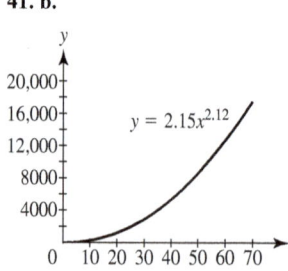

43. a. graph of $y = 0.132x^{1.921}$

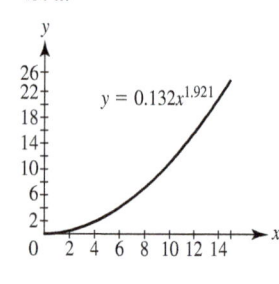

44. a. graph of $f(x) = 27.334\sqrt[10]{x^3}$

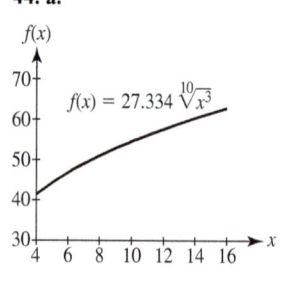

45. a. graph of $f(x) = 7.53\sqrt{x^3}$

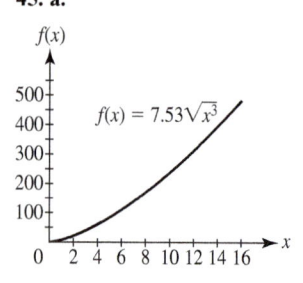

3.4 Skills Check

5. graph of $y = \begin{cases} -1 \text{ if } x < 0 \\ 1 \text{ if } x \geq 0 \end{cases}$

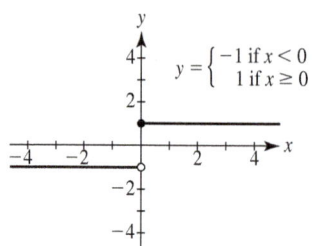

6. graph of $y = \begin{cases} 2 \text{ if } x \geq 2 \\ 6 \text{ if } x < 2 \end{cases}$

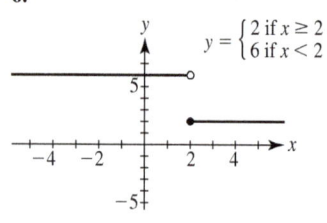

7. a. step function graph

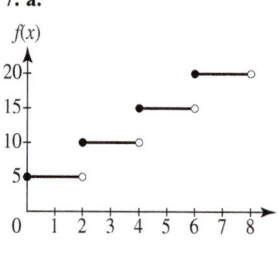

8. a. step function graph

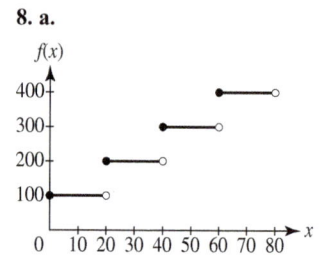

9. a. piecewise function graph

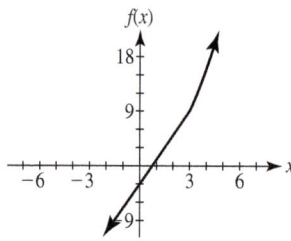

10. a. piecewise function graph

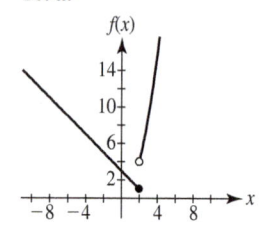

11. b. step function graph

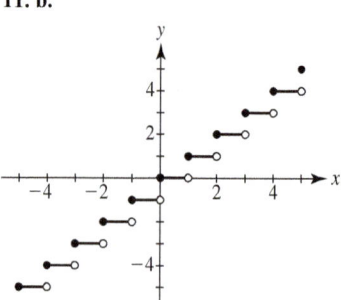

12. a. The graph seems to be shifted up.

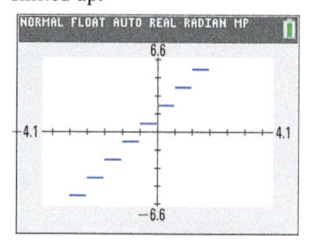

b. The graph seems to be shifted to the left.

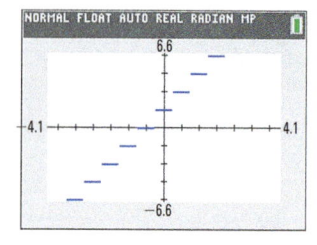

13. a.

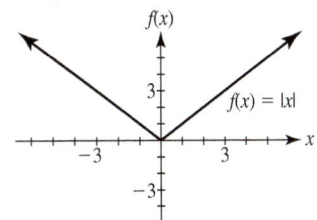

14. a.

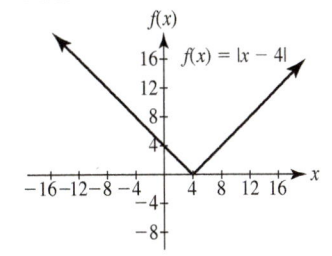

30. c.

15.

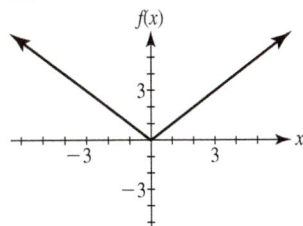

16.

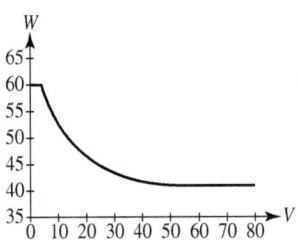

31. a. $F(x) = \begin{cases} 0, & \text{if } 40 \le x \le 72 \\ 1, & \text{if } 73 \le x \le 112 \\ 2, & \text{if } 113 \le x \le 157 \\ 3, & \text{if } 158 \le x \le 206 \\ 4, & \text{if } 207 \le x \le 260 \\ 5, & \text{if } 261 \le x \le 318 \end{cases}$

b.

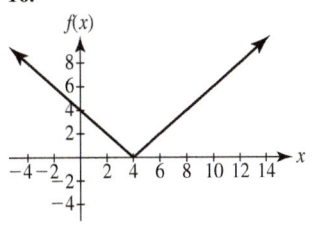

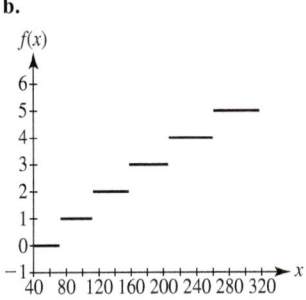

17.

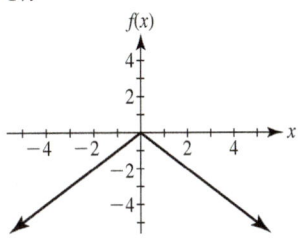

3.4 Exercises

25. $y = \begin{cases} 2.75 & \text{if } 0 < x \le 1 \\ 3.27 & \text{if } 1 < x \le 2 \\ 3.79 & \text{if } 2 < x \le 3 \\ 4.31 & \text{if } 3 < x \le 4 \\ 4.83 & \text{if } 4 < x \le 5 \end{cases}$

26. a. $f(x) = \begin{cases} 7.10 + 0.06747x & \text{if } 0 \le x \le 1200 \\ 88.06 + 0.05788(x - 1200) & \text{if } x > 1200 \end{cases}$

27. a.
$T(x) = \begin{cases} 8907 + 0.22(x - 77{,}400) & \text{if } 77{,}401 \le x \le 165{,}000 \\ 28{,}179 + 0.24(x - 165{,}000) & \text{if } 165{,}001 \le x \le 315{,}000 \\ 64{,}179 + 0.32(x - 315{,}000) & \text{if } 315{,}001 \le x \le 400{,}000 \\ 91{,}379 + 0.35(x - 400{,}000) & \text{if } 400{,}001 \le x \le 600{,}000 \end{cases}$

28. a. $P(x) = \begin{cases} 100 & \text{if } 0 < x \le 1 \\ 115 & \text{if } 1 < x \le 2 \\ 130 & \text{if } 2 < x \le 3 \\ 145 & \text{if } 3 < x \le 4 \end{cases}$

29. b.

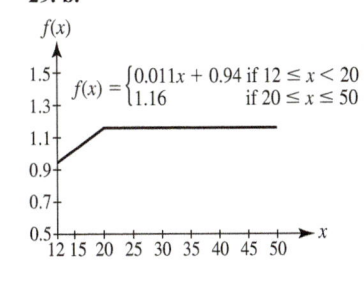

$f(x) = \begin{cases} 0.011x + 0.94 & \text{if } 12 \le x < 20 \\ 1.16 & \text{if } 20 \le x \le 50 \end{cases}$

32. a. $L(x) = \begin{cases} 2.61 & \text{if } 0 \le x \le 1 \\ 2.61 + 0.49(x - 1) & \text{if } 1 < x \le 70 \end{cases}$

b.

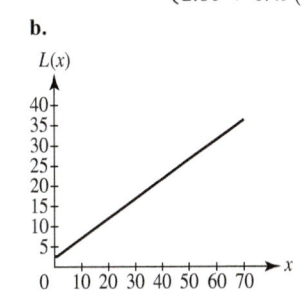

33. a. $f(x) = \begin{cases} 2.66 & \text{if } x \le 4 \\ 3.18 & \text{if } 4 < x \le 8 \\ 3.82 & \text{if } 8 < x \le 12 \\ 4.94 & \text{if } 12 < x < 16 \end{cases}$

b.

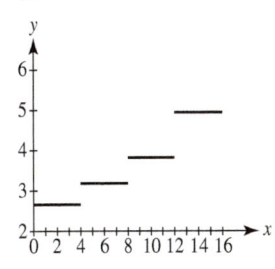

34. a. $C(x) = \begin{cases} 0.5658x & \text{if } x \le 650 \\ 36.78 + 04853(x - 650) & \text{if } 650 < x \le 1000 \\ 53.77 + 0.04764(x - 1000) & \text{if } x > 1000 \end{cases}$

A-14 Additional Answers

d.

[Graph of C(x) showing a linear function from 0 to 2000 on x-axis, 0 to 140 on y-axis]

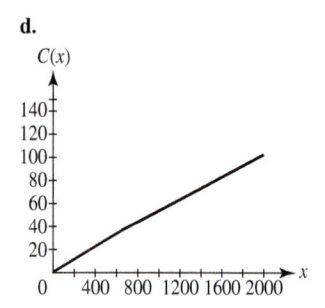

35. a. $H(x) = \begin{cases} 1.3x^2 + 0.57x + 104 & \text{if } 0 \leq x \leq 6 \\ -31.6x + 318 & \text{if } 6 < x \leq 9 \\ 6.23x - 24.9 & \text{if } 9 < x \leq 19 \end{cases}$

3.5 Skills Check

14. a.

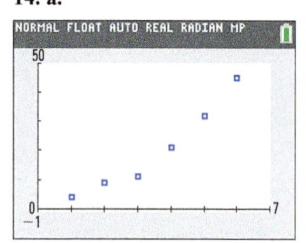

18. a.

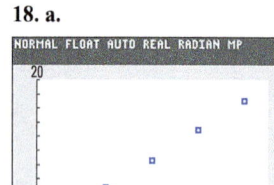

3.5 Exercises

23. b. Excellent fit

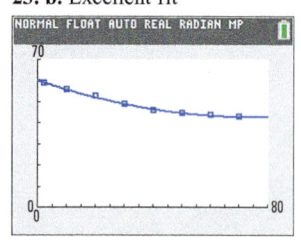

24. a. Quadratic should fit 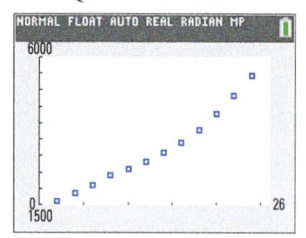   **c.** Very good fit

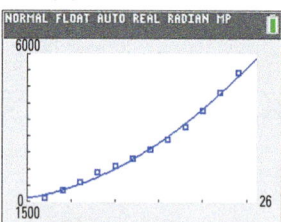

25. a. Quadratic is reasonable 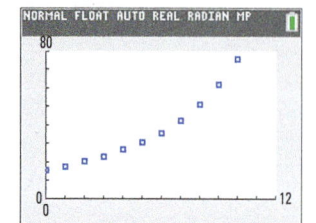   **c.** Excellent fit

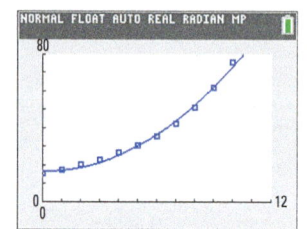

26. a.

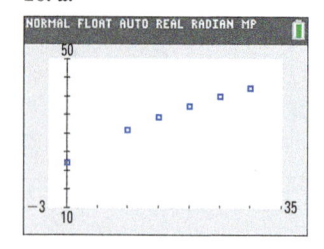

28. a. 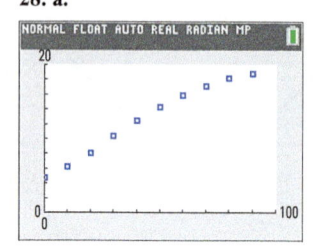   **d.**

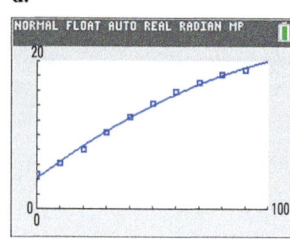

29. b. 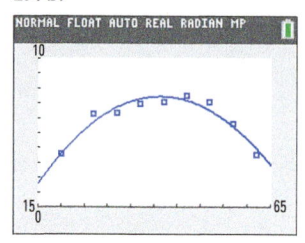   **30. a.**

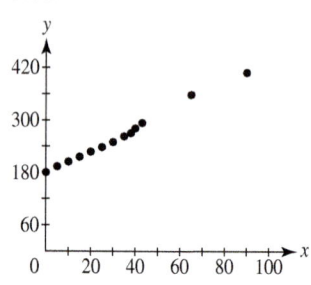

31. a. 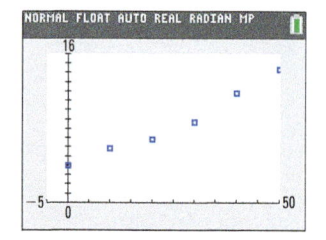   **c.**

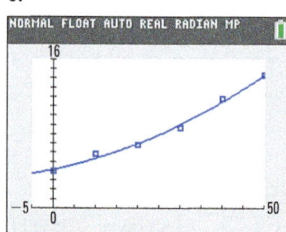

33. a. 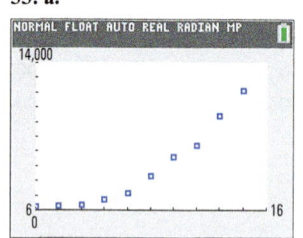   **c.** Very good fit

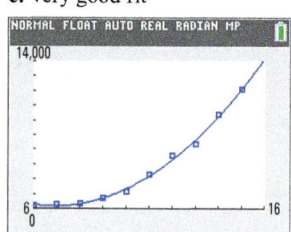

36. b. Very good fit 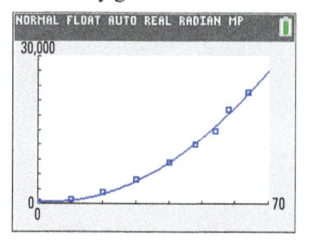   **37. b.**

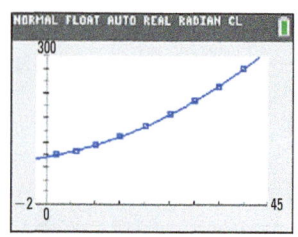

38. a. I. II.

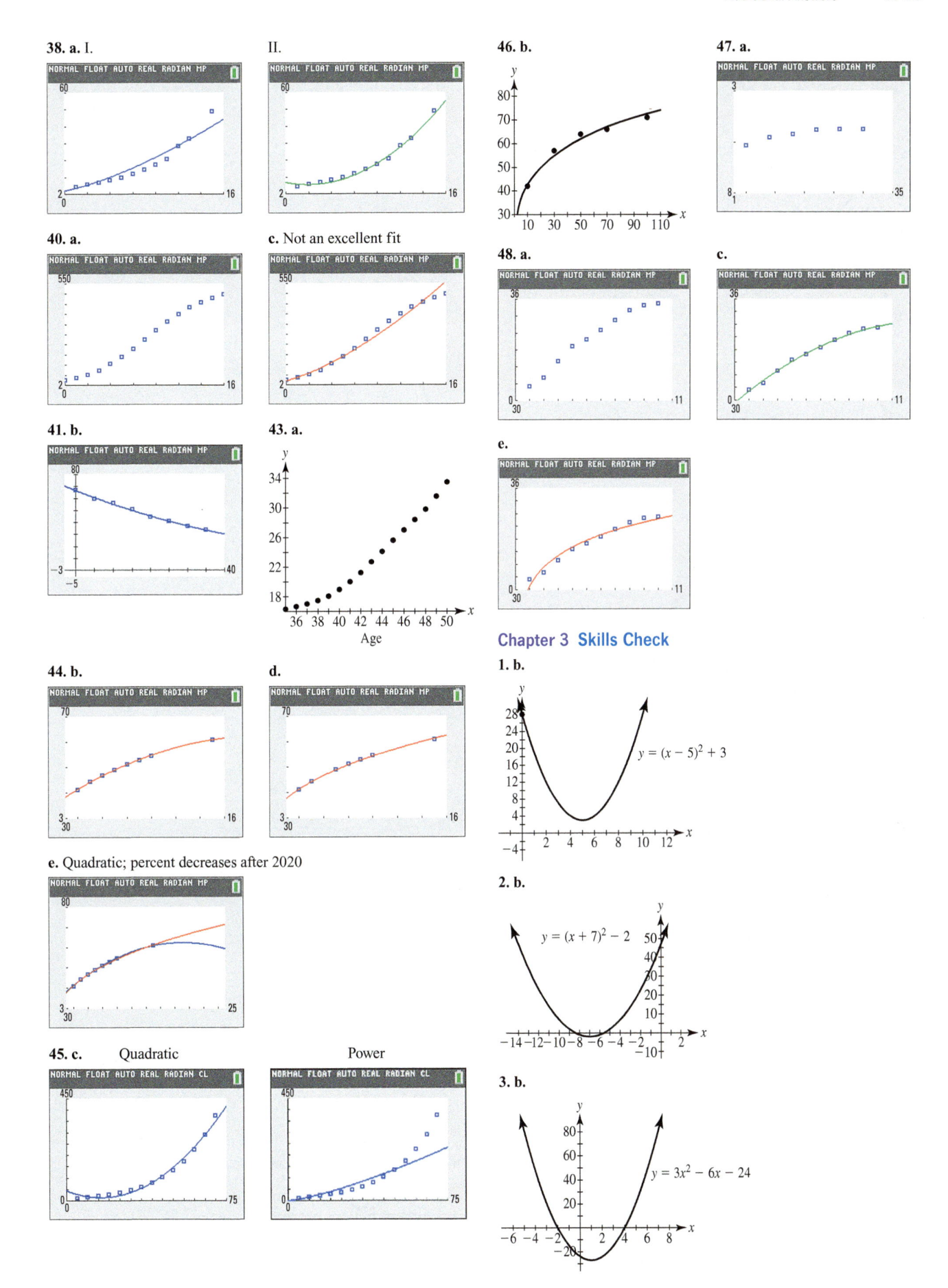

40. a. c. Not an excellent fit

41. b. **43. a.**

44. b. d.

e. Quadratic; percent decreases after 2020

45. c. Quadratic Power

46. b.

47. a.

48. a. c.

e.

Chapter 3 Skills Check

1. b. $y = (x - 5)^2 + 3$

2. b. $y = (x + 7)^2 - 2$

3. b. $y = 3x^2 - 6x - 24$

Additional Answers

4. b.

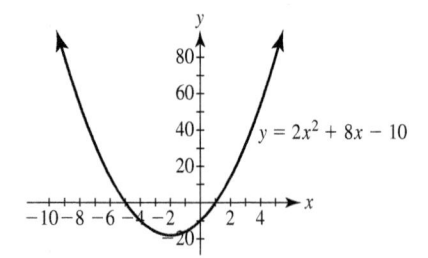

Graph of $y = 2x^2 + 8x - 10$

5. b.

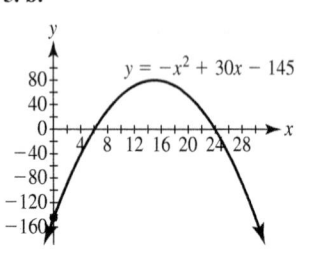

Graph of $y = -x^2 + 30x - 145$

6. b.

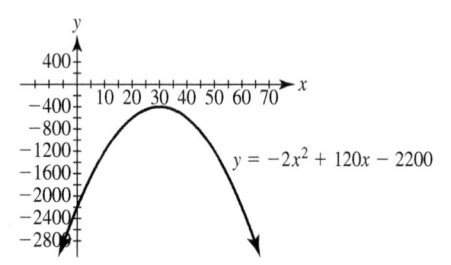

Graph of $y = -2x^2 + 120x - 2200$

7. b.

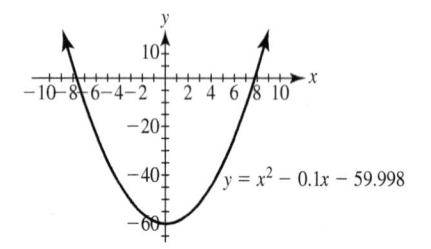

Graph of $y = x^2 - 0.1x - 59.998$

8. b.

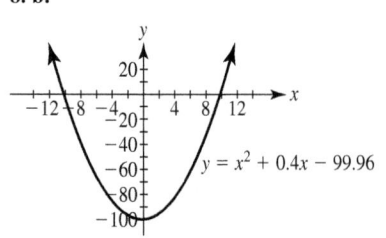

Graph of $y = x^2 + 0.4x - 99.96$

23.

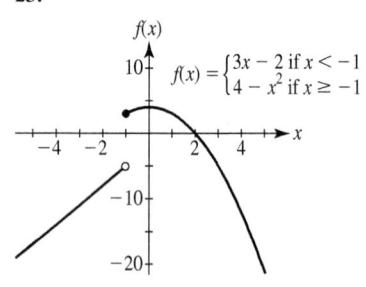

Graph of $f(x) = \begin{cases} 3x - 2 & \text{if } x < -1 \\ 4 - x^2 & \text{if } x \geq -1 \end{cases}$

24.

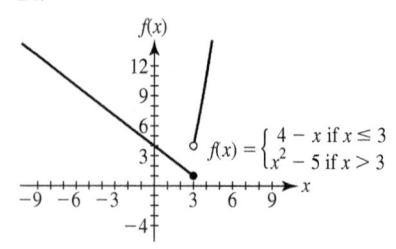

Graph of $f(x) = \begin{cases} 4 - x & \text{if } x \leq 3 \\ x^2 - 5 & \text{if } x > 3 \end{cases}$

25.

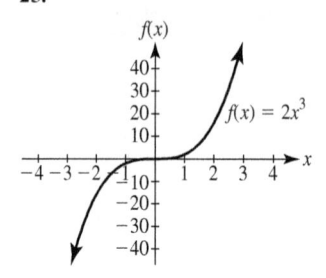

Graph of $f(x) = 2x^3$

26.

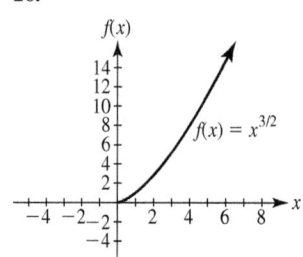

Graph of $f(x) = x^{3/2}$

27.

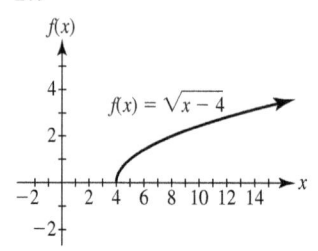

Graph of $f(x) = \sqrt{x - 4}$

28.

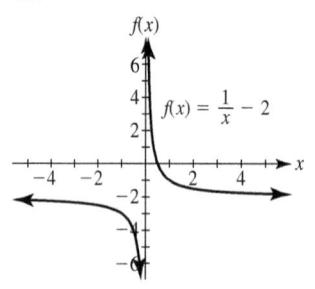

Graph of $f(x) = \dfrac{1}{x} - 2$

29.

Graph of $y = x^{4/5}$

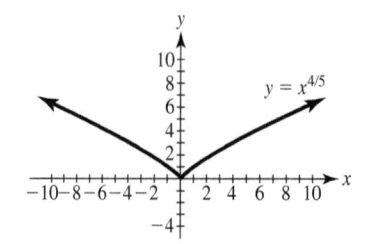

30.

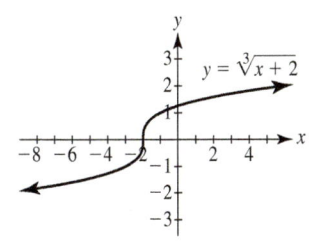

Chapter 3 Review

53. a. $y = \begin{cases} 0.08x^2 - 2.64x + 22.35 & \text{if } 15 \leq x \leq 45 \\ -0.525x + 89.82 & \text{if } 45 < x \leq 110 \end{cases}$

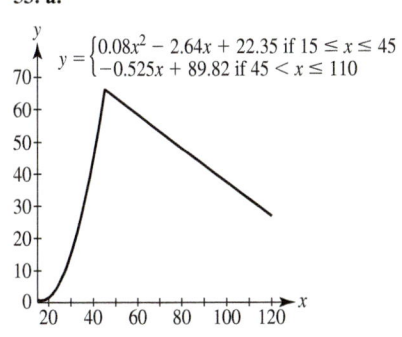

62. a.

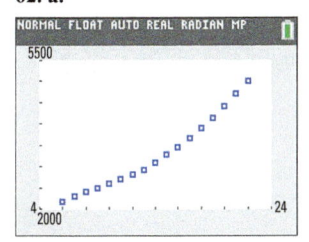

c.

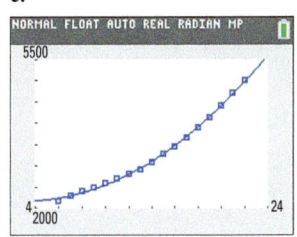

63. a.

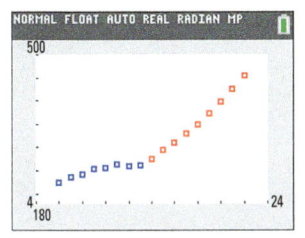

d. $y = \begin{cases} -0.99x^2 + 24.27x + 114.11, & \text{if } 6 \leq x \leq 13 \\ 0.93x^2 - 11.55x + 256.47, & \text{if } 13 < x \leq 22 \end{cases}$

64. b.

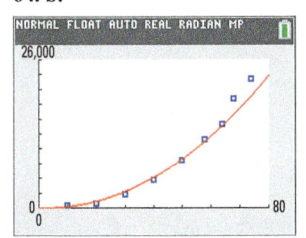

65. c.

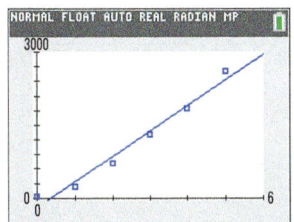

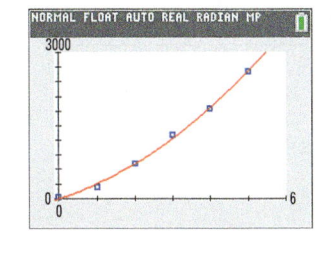

66. b. Not an excellent fit

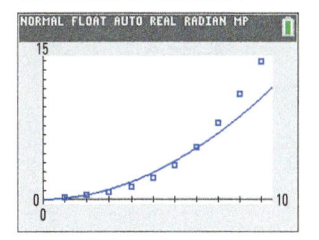

67. a.

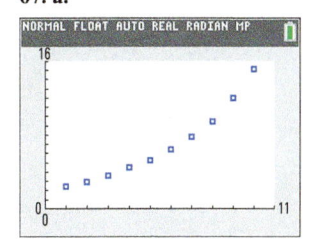

c.

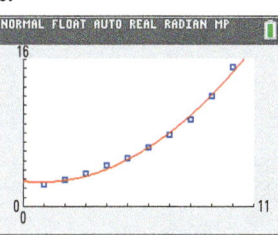

68. d. $H(x) = \begin{cases} 1.3x^2 + 0.57x + 104 & \text{if } 0 \leq x \leq 6 \\ -31.6x + 318 & \text{if } 6 < x \leq 9 \\ 6.226x - 24.888 & \text{if } 9 < x \leq 19 \end{cases}$

Chapter 4 Additional Topics with Functions

Toolbox Exercises

13. $f(x) = x^3$

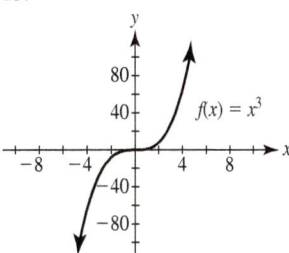

14. $f(x) = \sqrt{x}$

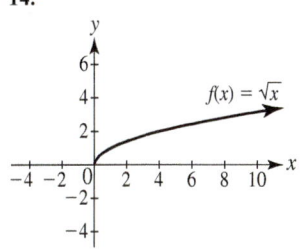

15. $f(x) = 8\sqrt[6]{x}$

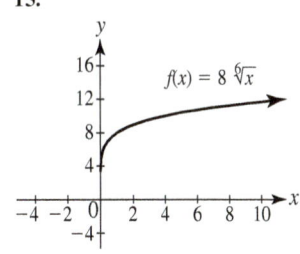

16. $f(x) = -\dfrac{4}{x^3}$

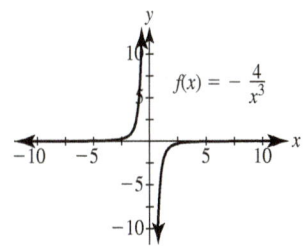

4.1 Skills Check

1. a. $y = x^3$; $y = x^3 + 5$

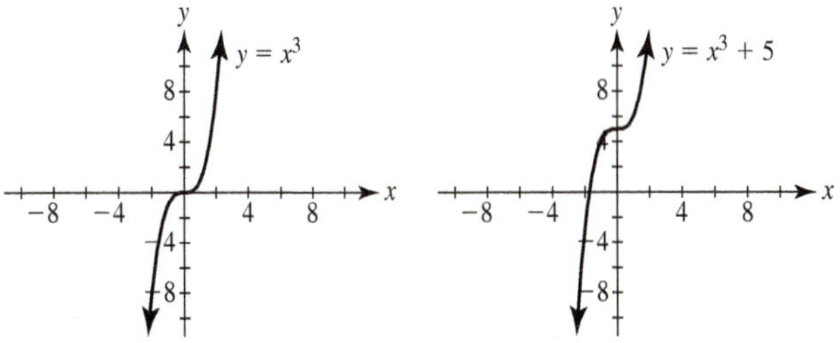

A-18 Additional Answers

2. a. Graphs of $y = x^2$ and $y = x^2 + 3$

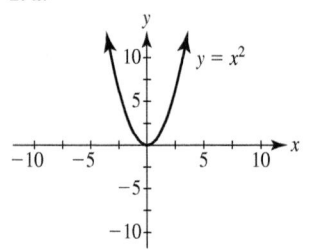

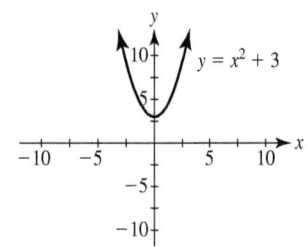

3. a. Graphs of $y = \sqrt{x}$ and $y = \sqrt{x-4}$

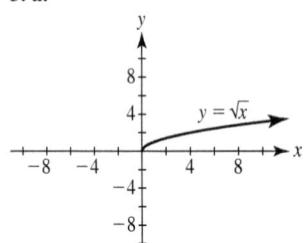

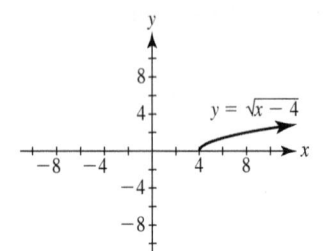

4. a. Graphs of $y = x^2$ and $y = (x+2)^2$

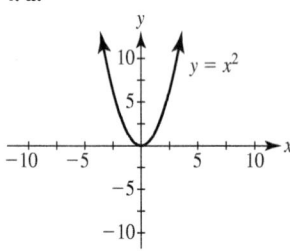

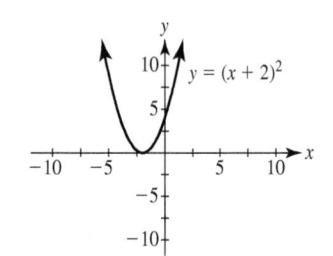

5. a. Graphs of $y = \sqrt[3]{x}$ and $y = \sqrt[3]{x-2} - 1$

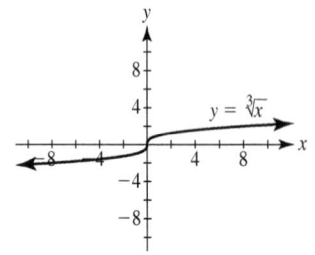

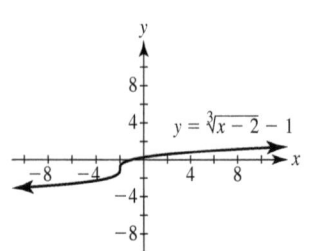

6. a. Graphs of $y = x^3$ and $y = (x-5)^3 - 3$

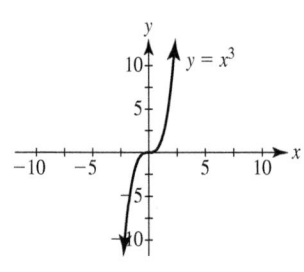

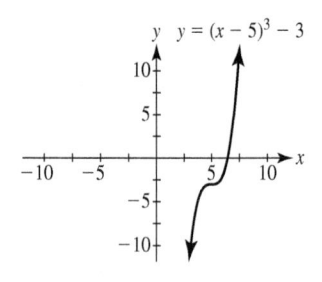

7. a. Graphs of $y = |x|$ and $y = |x-2| + 1$

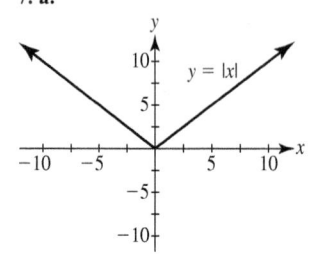

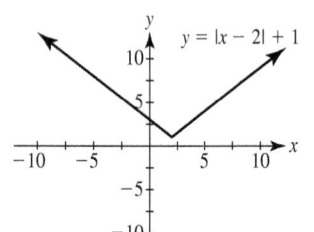

8. a. Graphs of $y = |x|$ and $y = |x+3| - 4$

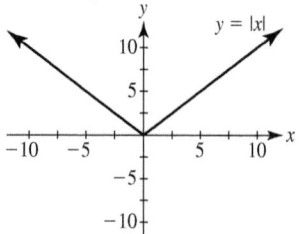

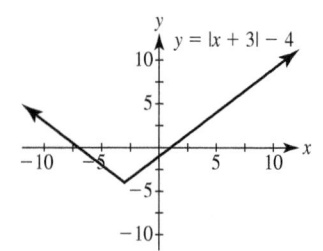

9. a. Graphs of $y = x^2$ and $y = -x^2 + 5$

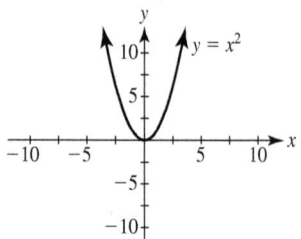

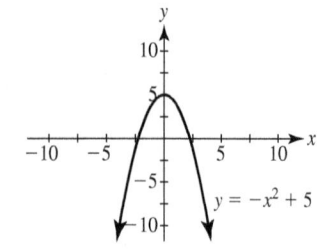

10. a. Graphs of $y = \sqrt{x}$ and $y = -\sqrt{x-2}$

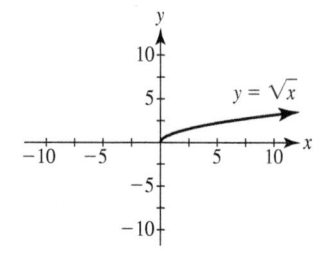

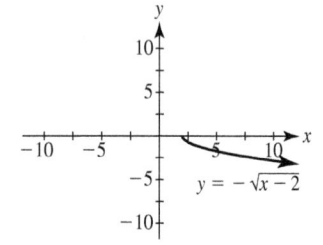

11. a. Graphs of $y = \dfrac{1}{x}$ and $y = \dfrac{1}{x} - 3$

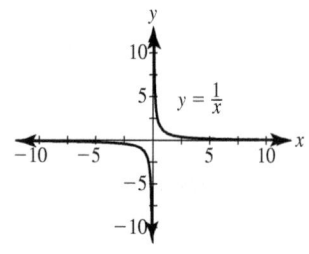

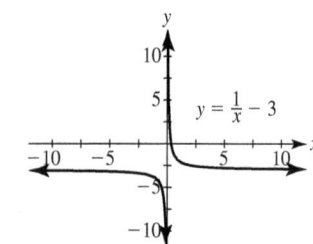

12. a. Graphs of $y = \dfrac{1}{x}$ and $y = \dfrac{2}{x-1}$

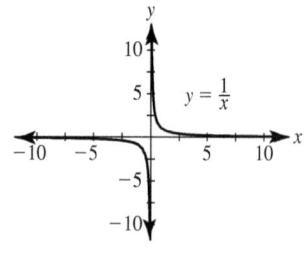

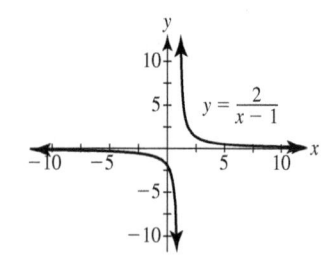

13. a. Graphs of $y = x^2$ and $y = \dfrac{1}{3}x^2$

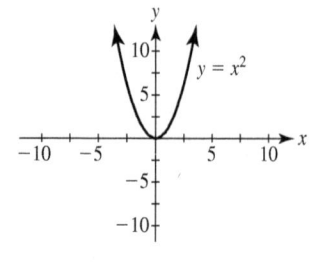

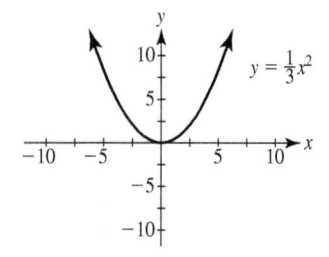

14. a.

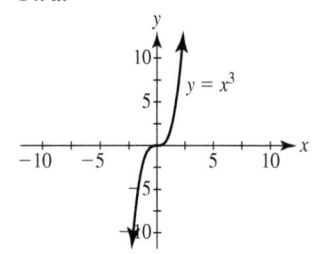

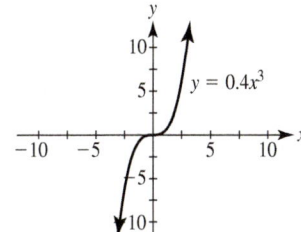

15. a.

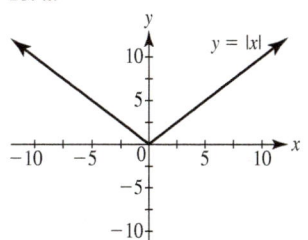

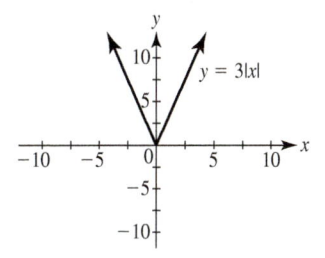

16. a.

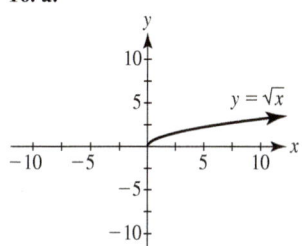

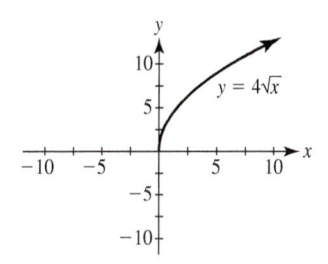

23. $f(x) = \sqrt{x+6} + 3$

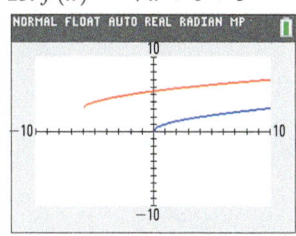

24. $f(x) = (x-5)^2 + 2$

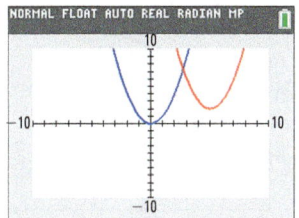

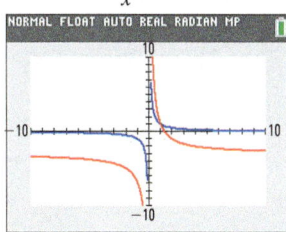

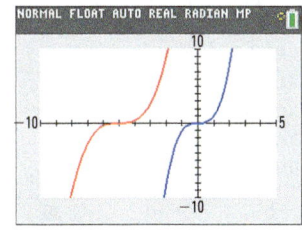

25. $f(x) = \dfrac{4}{x} - 3$

26. $f(x) = \dfrac{1}{3}(x+5)^3$

27. $f(x) = -\sqrt{x} + 2$

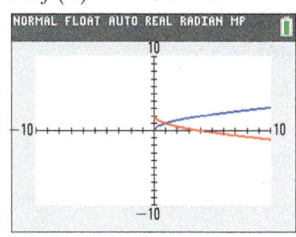

28. $f(x) = -\sqrt[3]{x-1}$

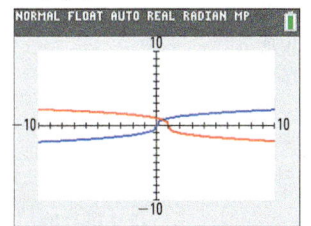

55.

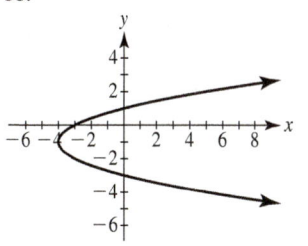

4.1 Exercises

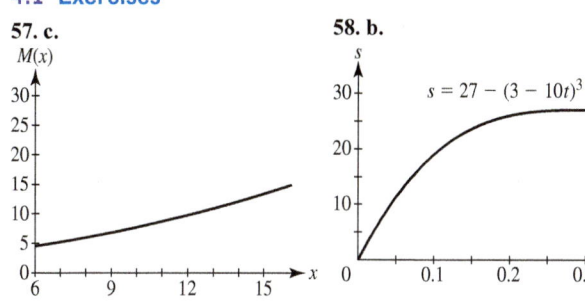

57. c.

58. b. $s = 27 - (3 - 10t)^3$

62. a. Shift 100 units right, reflect about the p-axis, and vertical stretch using a factor of 10,500

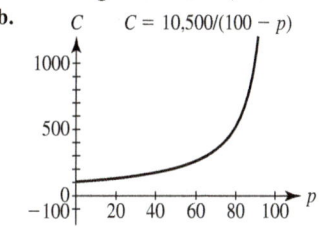

b. $C = 10{,}500/(100 - p)$

63. a. Shift 10 units left, shift 1 unit down, reflect about the x-axis, and vertical stretch using a factor of 1000.

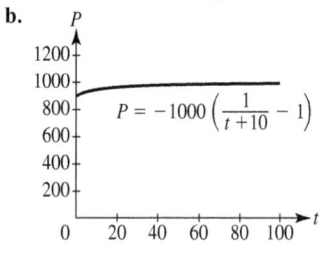

b. $P = -1000\left(\dfrac{1}{t+10} - 1\right)$

64. c. Reflect about the y-axis, shift 110 units to the right, and vertically stretch using a factor of 4700.

4.2 Skills Check

1. a. $2x - 1$ **b.** $4x - 9$ **c.** $-3x^2 + 17x - 20$
d. $\dfrac{3x - 5}{4 - x}$ **e.** All real numbers except 4

2. a. $x + 2$ **b.** $3x - 8$ **c.** $-2x^2 + 13x - 15$
d. $\dfrac{2x - 3}{5 - x}$ **e.** All real numbers except 5 **3. a.** $x^2 - x + 1$
b. $x^2 - 3x - 1$ **c.** $x^3 - x^2 - 2x$ **d.** $\dfrac{x^2 - 2x}{1 + x}$ **e.** All real numbers except -1 **4. a.** $2x^2 + x + 1$ **b.** $2x^2 - 3x - 1$
c. $4x^3 - x$ **d.** $\dfrac{2x^2 - x}{2x + 1}$ **e.** All real numbers except $-\dfrac{1}{2}$

5. a. $\dfrac{x^2 + x + 5}{5x}$ **b.** $\dfrac{-x^2 - x + 5}{5x}$ **c.** $\dfrac{x + 1}{5x}$
d. $\dfrac{5}{x(x + 1)}$ **e.** All real numbers except 0 and -1

Additional Answers

6. a. $\dfrac{x^2 - 2x + 3}{3x}$ **b.** $\dfrac{x^2 - 2x - 3}{3x}$ **c.** $\dfrac{x - 2}{3x}$

d. $\dfrac{x^2 - 2x}{3}$ **e.** All real numbers except 0 **7. a.** $\sqrt{x} + 1 - x^2$

b. $\sqrt{x} - 1 + x^2$ **c.** $\sqrt{x}(1 - x^2)$ **d.** $\dfrac{\sqrt{x}}{1 - x^2}$

e. All real numbers $x \geq 0$ except 1 **8. a.** $x^3 + \sqrt{x + 3}$

b. $x^3 - \sqrt{x + 3}$ **c.** $x^3\sqrt{x + 3}$ **d.** $\dfrac{x^3}{\sqrt{x + 3}}$ **e.** $(-3, \infty)$

13. a. $\dfrac{1}{x^2}$ **b.** $\dfrac{1}{x^2}$ **14. a.** $\dfrac{8}{x^3}$ **b.** $\dfrac{2}{x^3}$

15. a. $\sqrt{2x - 8}$ **b.** $2\sqrt{x - 1} - 7$
16. a. $\sqrt{8 - x}$ **b.** $\sqrt{3 - x} - 5$

4.2 Exercises

31. a. $\overline{C}(x)$ is $C(x) = 50{,}000 + 105x$ divided by $f(x) = x$; that is, $\overline{C}(x) = \dfrac{C(x)}{x}$.

34. a. $\overline{C}(x) = \dfrac{2.15x + 2350}{x}$

40. c. $(0, -87{,}500)$ represents the fixed costs for manufacturing and selling the computers. If no computers are manufactured and sold, the company will lose $87,500 per month.

52. b. $C = 59.914 - 2.35s - 20.14\sqrt{s}$

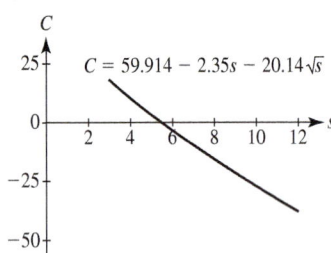

54. a. $\overline{C}(x) = \dfrac{100{,}000}{x} + 150$

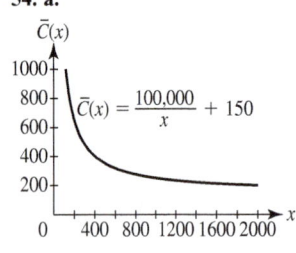

4.3 Skills Check

1. Domain → Range (f^{-1}): 3→2, 4→5, 1→3, 8→7

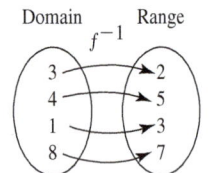

26. $g^{-1}(x)$, $g(x) = \sqrt{x}$

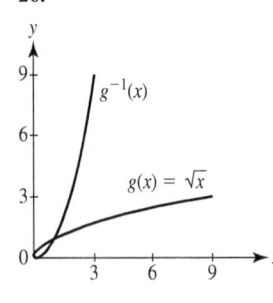

27. $g^{-1}(x) = x^3$, $g(x) = \sqrt[3]{x}$

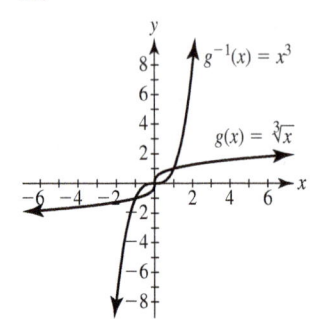

30. $f(x)$, $y = x$, $f^{-1}(x)$

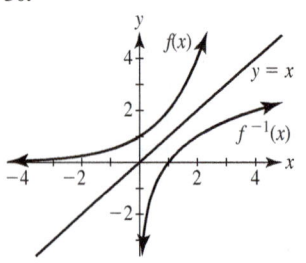

31. $f(x)$, $f^{-1}(x)$

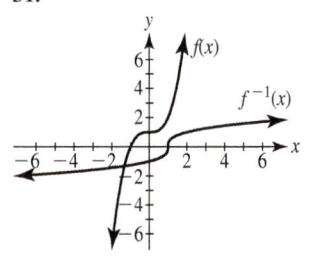

4.3 Exercises

50. a. $P(x) = \begin{cases} 0.55 & \text{if } 0 < x \leq 1 \\ 0.70 & \text{if } 1 < x \leq 2 \\ 0.85 & \text{if } 2 < x \leq 3 \end{cases}$

Chapter 4 Skills Check

3. a. $f(x) = \sqrt{x}$, $g(x) = \sqrt{x + 2} - 3$

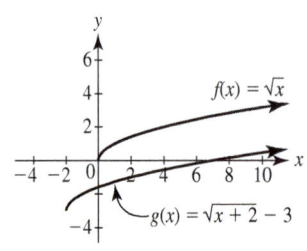

24. NORMAL FLOAT AUTO REAL RADIAN MP

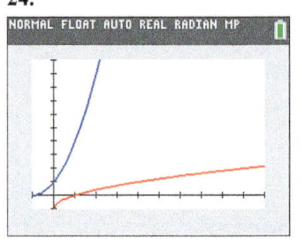

Chapter 4 Review

37. a. $s = 64 - (4 - 10t)^3$

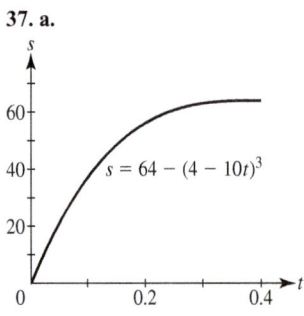

38. c. Stretched vertically using a factor of 16, and reflected across the x-axis, then shifted 4 units right and 380 units up.

42. b. $\overline{C}(x)$

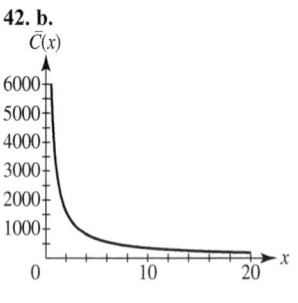

43. a. $\overline{C}(x) = \dfrac{50{,}000}{x} + 120$

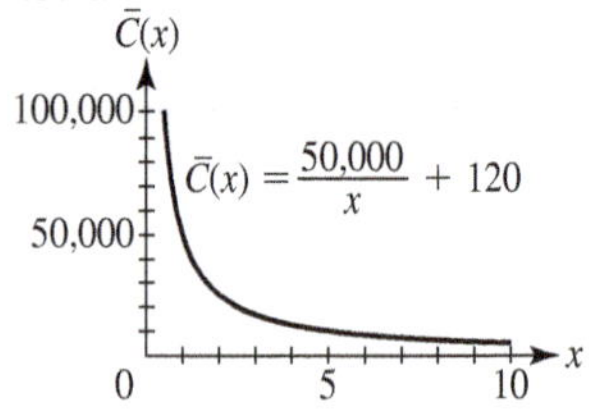

44. b. $p = \dfrac{300}{q} - 20$, $p = \dfrac{180 + q}{6}$

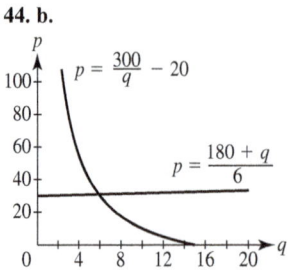

45. b. $p = \dfrac{2555}{q + 5}$, $p = 58 + \dfrac{q}{2}$

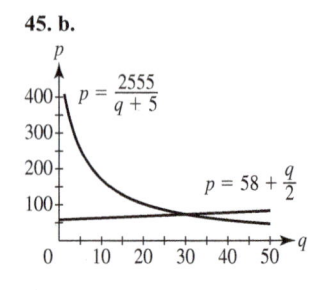

47. a. $W(M(x)) = 0.781x + 20.2$; $W(M(x))$ gives the number of millions of women in the workforce as a function of the number of millions of men in the workforce.

Chapter 5 Exponential and Logarithmic Functions

5.1 Skills Check

3. a. Graph of $f(x) = e^x$

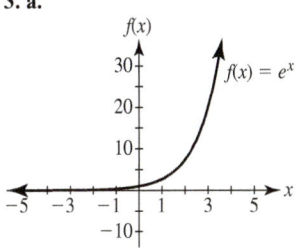

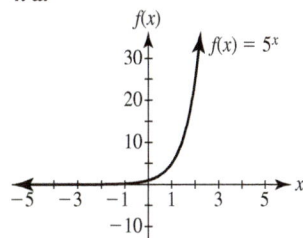

4. a. Graph of $f(x) = 5^x$

5. Graph of $y = 3^x$

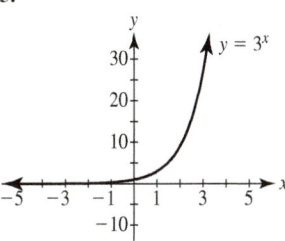

6. Graph of $y = 5e^{-x}$

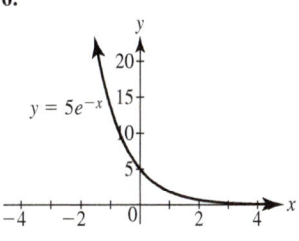

7. Graph of $y = 5^{-x}$

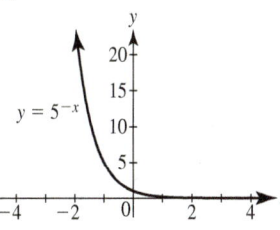

8. Graph of $y = 3^{-2x}$

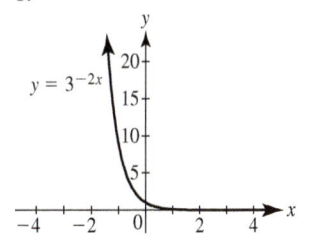

9. Graph of $y = 3^x + 5$

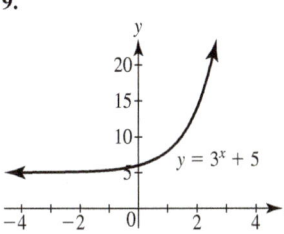

10. Graph of $y = 2(1.5)^{-x}$

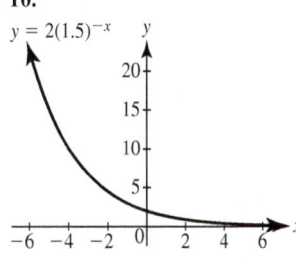

11. Graph of $y = 3^{(x-2)} - 4$

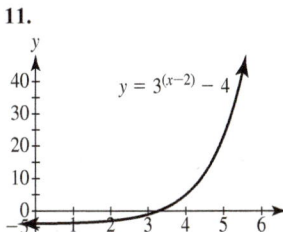

12. Graph of $y = 3^{(x-1)} + 2$

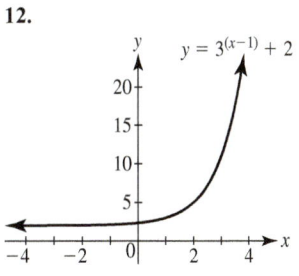

19. Vertical shift, 2 units up; graph of $y = 4^x + 2$

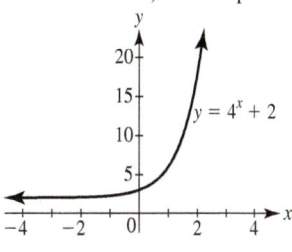

20. Horizontal shift, 1 unit right; graph of $y = 4^{x-1}$

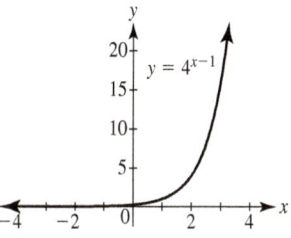

21. Reflection across the y-axis; graph of $y = 4^{-x}$

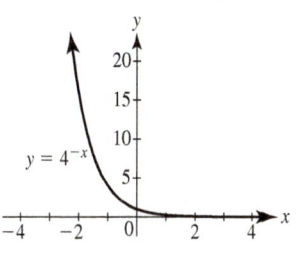

22. Reflection across the x-axis; graph of $y = -4^x$

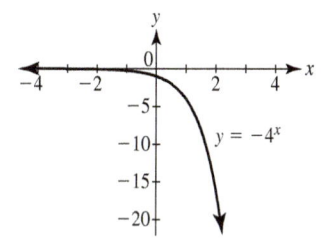

23. Vertical stretch by a factor of 3; graph of $y = 3(4^x)$

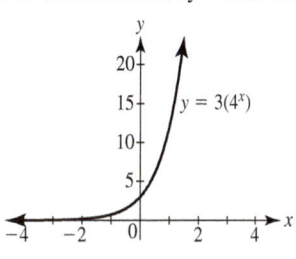

24. Vertical stretch by a factor of 3, shift right 2 units, and shift down 3 units; graph of $y = 3 \cdot 4^{(x-2)} - 3$

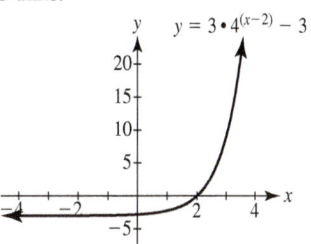

25. The graph of $y = 3 \cdot 4^{(x-2)} - 3$ is the graph of $y = 3(4^x)$ shifted to the right 2 units and shifted down 3 units.

27. Graphs of $y = e^x$ and $y = e^{x-3}$

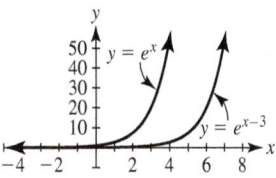

28. Graphs of $y = e^x$ and $y = e^x - 4$

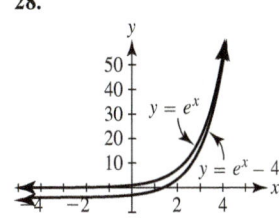

29. Graphs of $y = e^x$ and $y = \frac{1}{2}e^x$

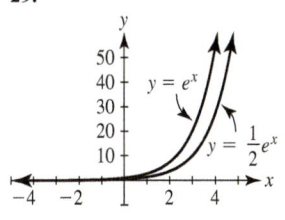

A-22 Additional Answers

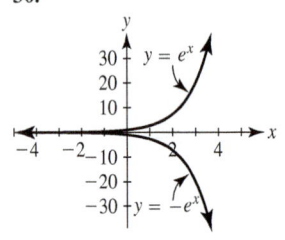

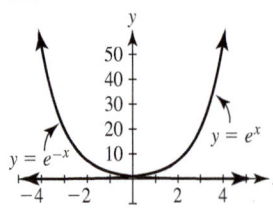

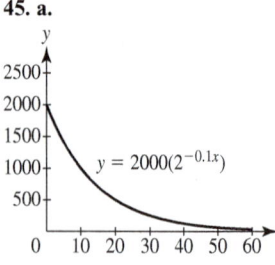

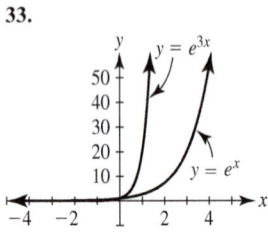

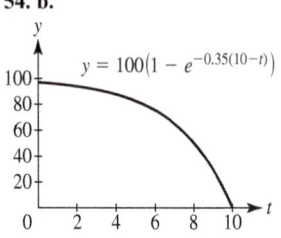

30. Graph of $y = e^x$ and $y = -e^x$

31. Graph of $y = e^{-x}$ and $y = e^x$

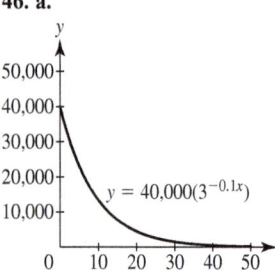

32. Graph of $y = e^{x+2} + 1$ and $y = e^x$

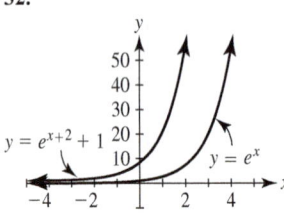

33. Graph of $y = e^{3x}$ and $y = e^x$

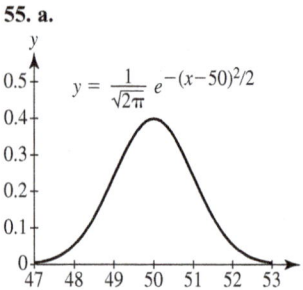

34. Graph of $y = e^x$ and $y = e^{\frac{1}{3}x}$

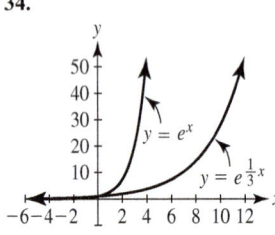

35. a. Graph of $f(x) = 12e^{-0.2x}$

45. a. Graph of $y = 2000(2^{-0.1x})$

46. a. Graph of $y = 40{,}000(3^{-0.1x})$

54. b. Graph of $y = 100(1 - e^{-0.35(10-t)})$

55. a. Graph of $y = \frac{1}{\sqrt{2\pi}} e^{-(x-50)^2/2}$

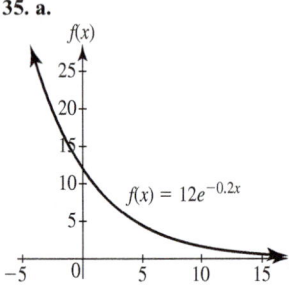

56. Graph of $y = \frac{1}{\sqrt{2\pi}} e^{-(x-100)^2/20}$

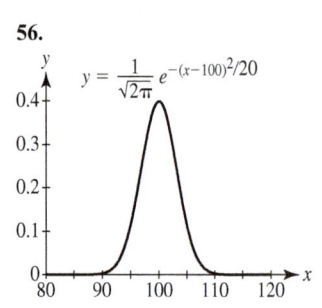

5.1 Exercises

39. a. Graph of $S = 80{,}000(1.05^t)$

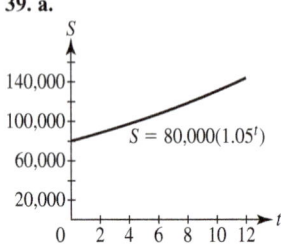

40. a. Graph of $S = 56{,}000(1.09^t)$

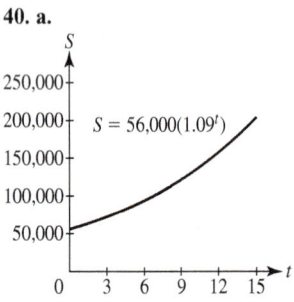

41. a. Graph of $S = 8000e^{0.08t}$

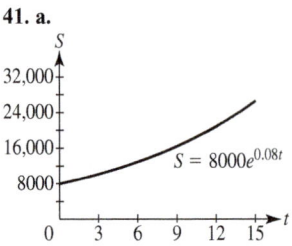

42. a. Graph of $S = 35{,}000e^{0.09t}$

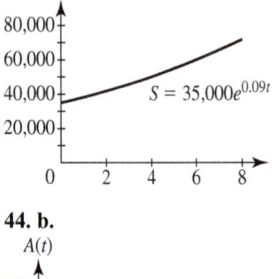

43. b. Graph of $A(t) = 500e^{-0.02828t}$

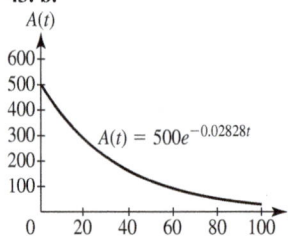

44. b. Graph of $A(t) = 100e^{-0.00002876t}$

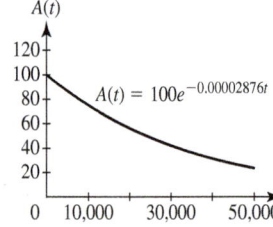

5.2 Skills Check

14. a. Graph of $y = \log_3 x$

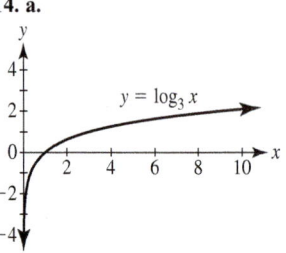

b. Graph of $y = \log_5 x$

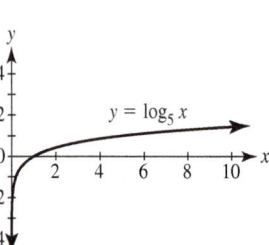

15. Graph of $y = 2 \ln x$

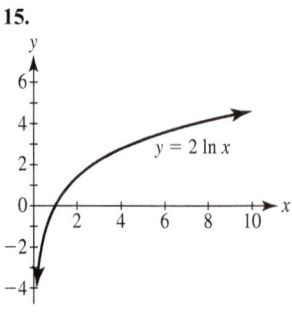

16. Graph of $y = 4 \log x$

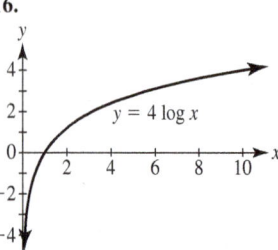

Additional Answers A-23

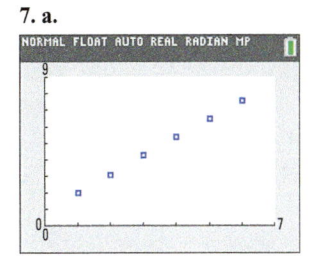

17. $y = \log(x+1) + 2$

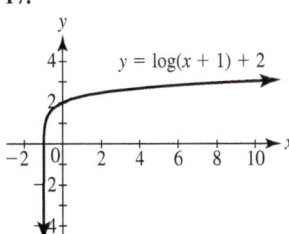

18. $y = \ln(x-2) - 3$

5.4 Skills Check

7. a.

10.

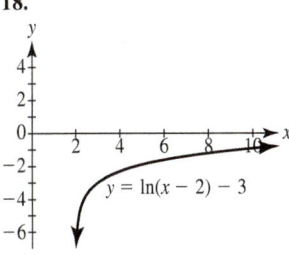

19. b. $y = 4^x$, $y = x$, $y = \log_4 x$

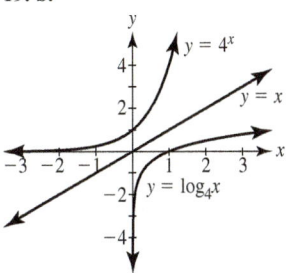

20. b. $y = 3^x$, $y = x$, $y = \log_3 x$

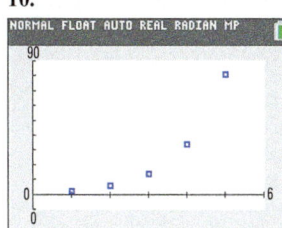

13. a.

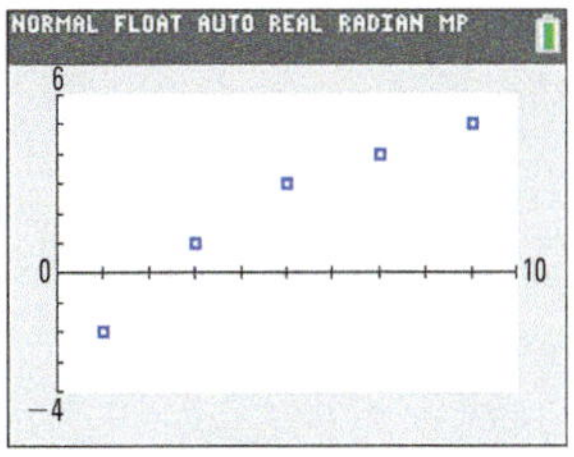

14. c. Logarithmic   Linear

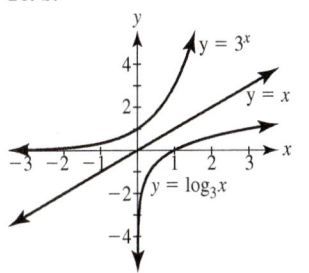

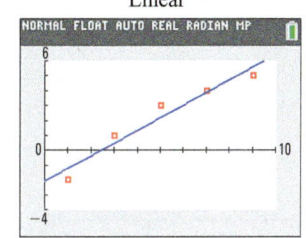

33. $\dfrac{1}{3}\log_3(4x+1) - \log_3(4) - 2\log_3(x)$

34. $\dfrac{1}{3}\log_3(3x-1) - \log_3 5 - 2\log_3 x$

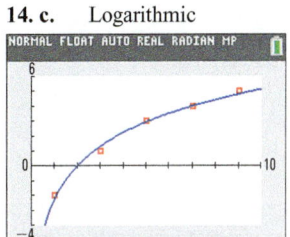

5.2 Exercises

41. c.

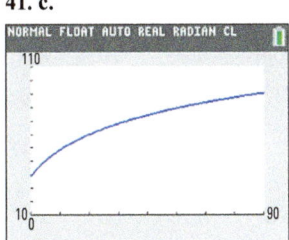

15. a.

5.3 Exercises

62. $2P = P(1.10)^n$
$2 = 1.10^n$
$n = \log_{1.10} 2$

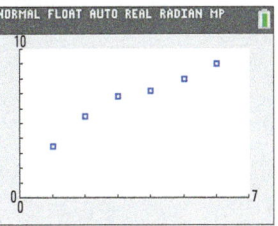

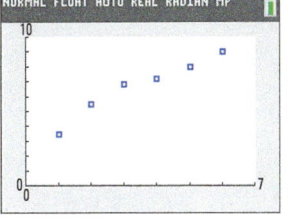

5.4 Exercises

23. b.

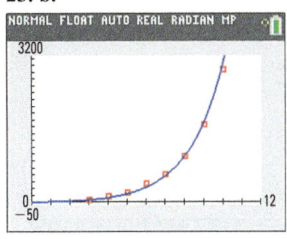

72. a.

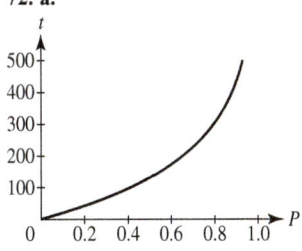

82. b.

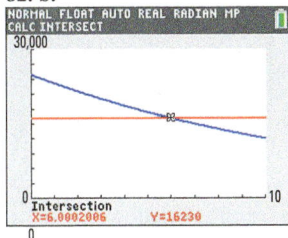

24. c.

83. b. $S = 600e^{-0.05x}$
(16.0, 269.6)
$S = 269.60$

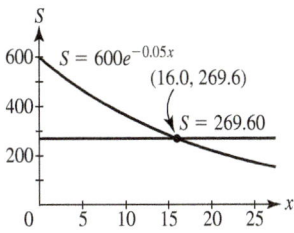

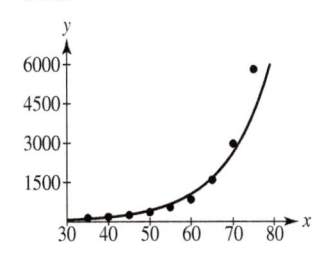

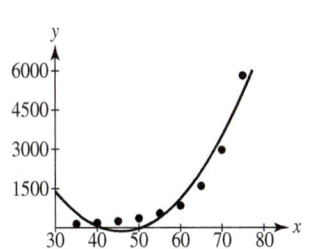

A-24 Additional Answers

25. b.

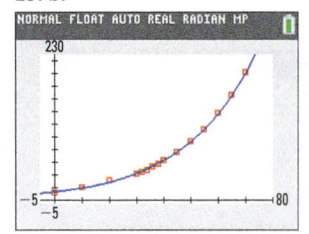

26. a.

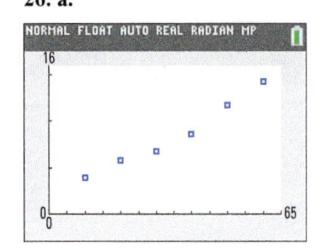

d.

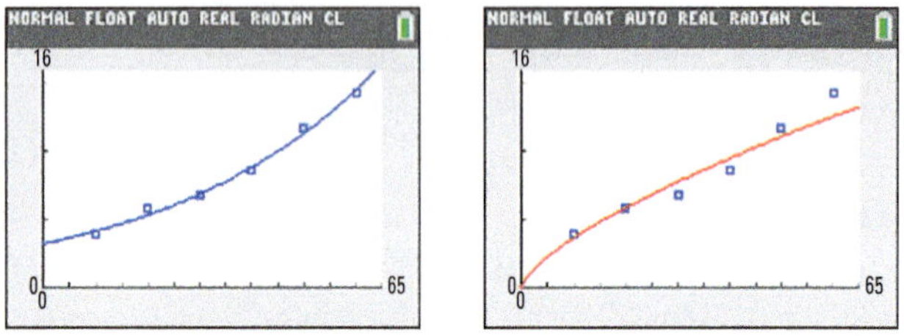

27. c. Logarithmic 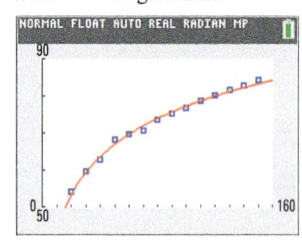   Quadratic

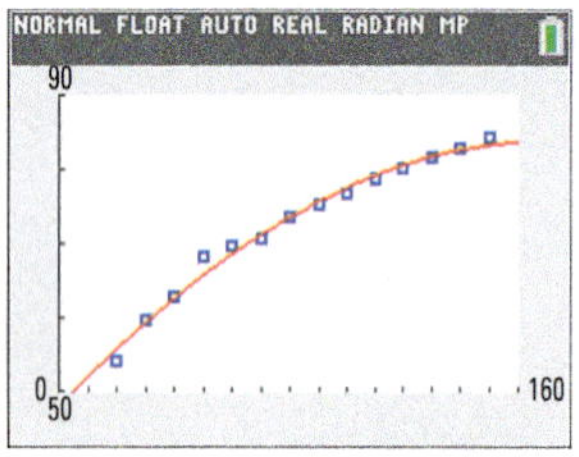

28. b.

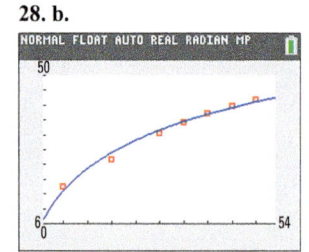

29. c. $y = 3.885 - 0.733 \ln x$ $y = 2.572(0.982^x)$

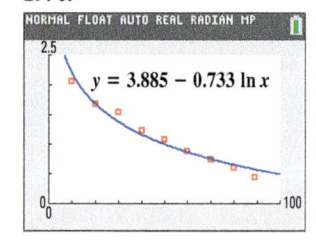

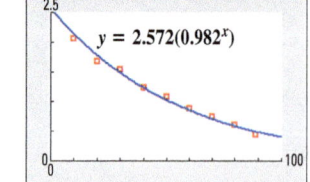

30. a.

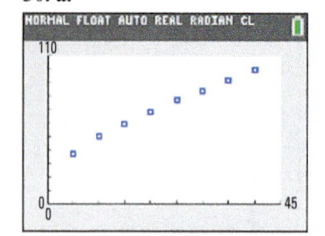

31. a. 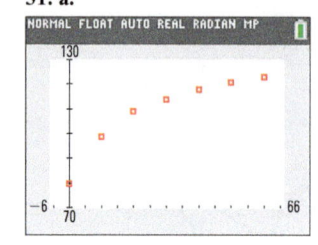   **c.**

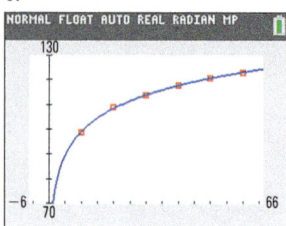

32. a. 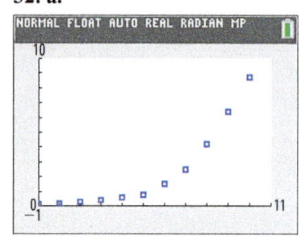   **c.**

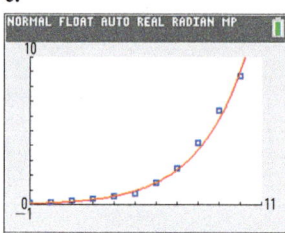

36. b.

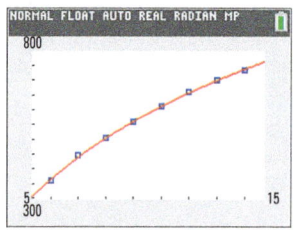

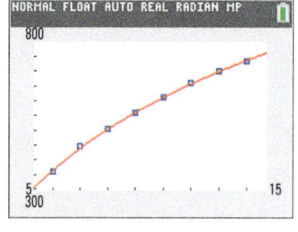

5.5 Exercises

17. a. $S = 3300(1.10)^x$

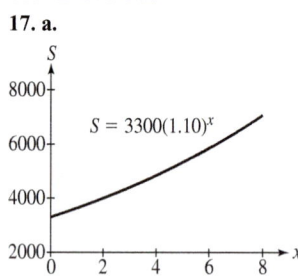

18. a. $y_1 = 5500(1.03)^{4x}$

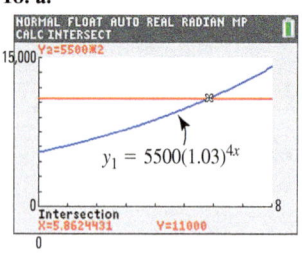

5.7 Skills Check

5. a. $f(x) = \dfrac{100}{1 + 3e^{-x}}$

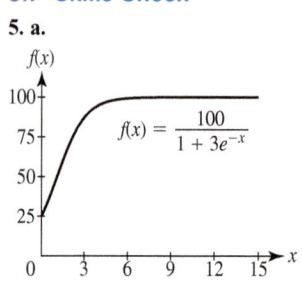

6. a. $f(t) = \dfrac{1000}{1 + 9e^{-0.9t}}$

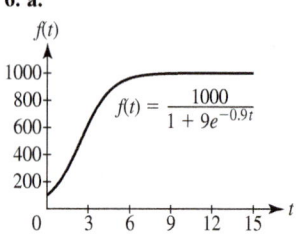

7. a. $y = 100(0.05)^{0.3x}$

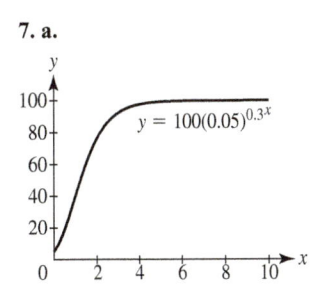

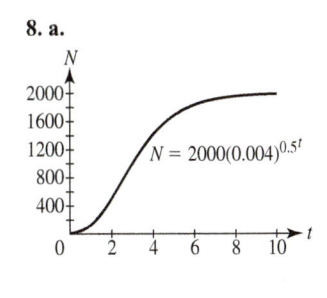

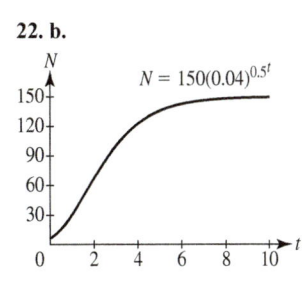

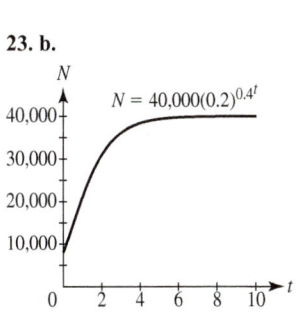

8. a. $N = 2000(0.004)^{0.5t}$

22. b. $N = 150(0.04)^{0.5t}$

23. b. $N = 40{,}000(0.2)^{0.4t}$

5.7 Exercises

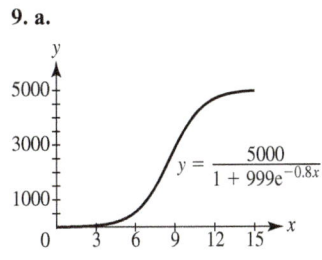

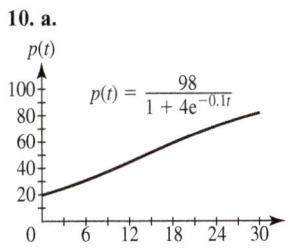

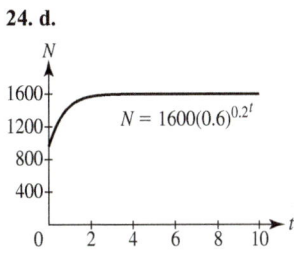

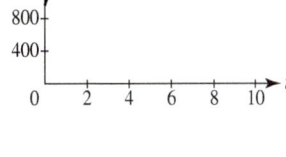

9. a. $y = \dfrac{5000}{1 + 999e^{-0.8x}}$

10. a. $p(t) = \dfrac{98}{1 + 4e^{-0.1t}}$

24. d. $N = 1600(0.6)^{0.2t}$

11. a. $y = \dfrac{89.786}{1 + 4.6531e^{-0.8256x}}$

12. a. $y = \dfrac{83.84}{1 + 13.9233e^{-0.9248x}}$

Chapter 5 Skills Check

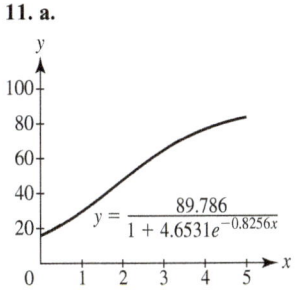

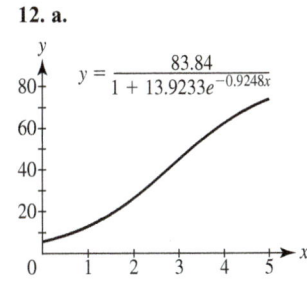

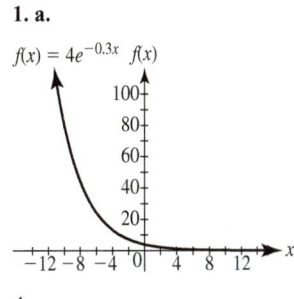

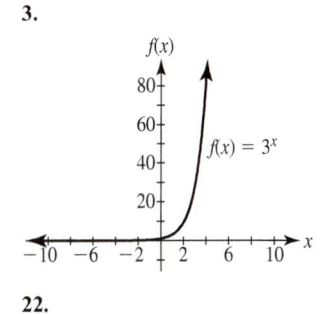

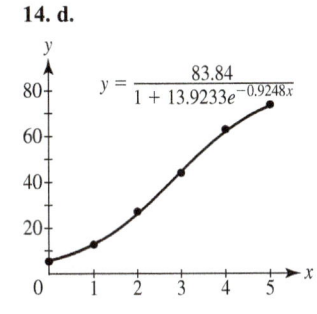

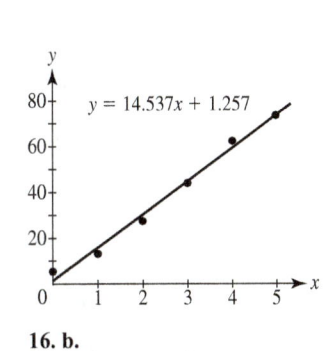

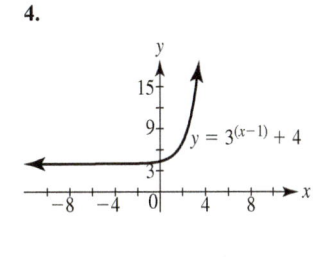

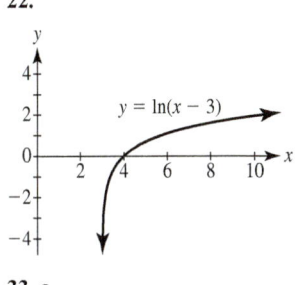

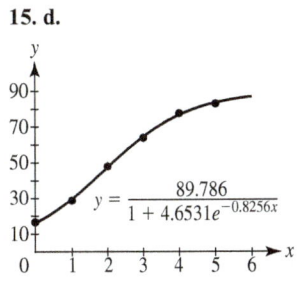

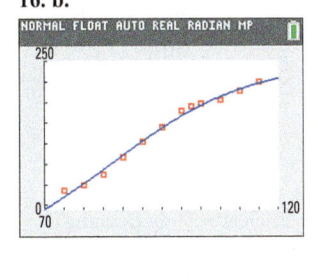

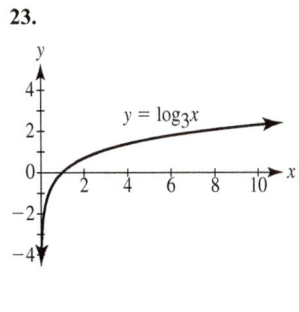

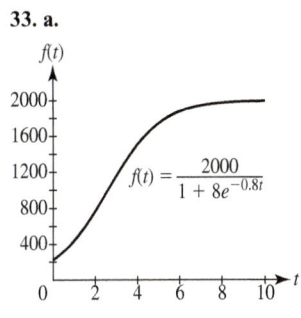

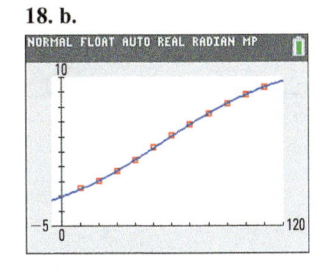

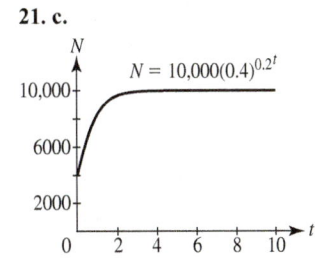

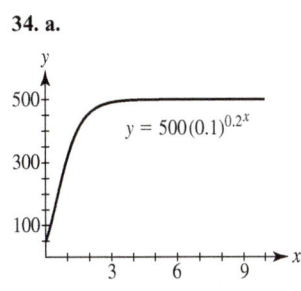

1. a. $f(x) = 4e^{-0.3x}$

3. $f(x) = 3^x$

14. d. $y = \dfrac{83.84}{1 + 13.9233e^{-0.9248x}}$; $y = 14.537x + 1.257$

4. $y = 3^{(x-1)} + 4$

22. $y = \ln(x - 3)$

15. d. $y = \dfrac{89.786}{1 + 4.6531e^{-0.8256x}}$

16. b. NORMAL FLOAT AUTO REAL RADIAN MP

23. $y = \log_3 x$

33. a. $f(t) = \dfrac{2000}{1 + 8e^{-0.8t}}$

18. b. NORMAL FLOAT AUTO REAL RADIAN MP

21. c. $N = 10{,}000(0.4)^{0.2t}$

34. a. $y = 500(0.1)^{0.2x}$

Chapter 5 Review

35. b.

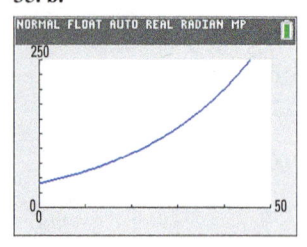

44. b. $P(x) = 10(1.26^x) - 2x - 50$; point $(9.9596, 30)$; $y = 30$

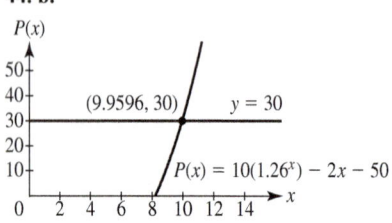

52. b.

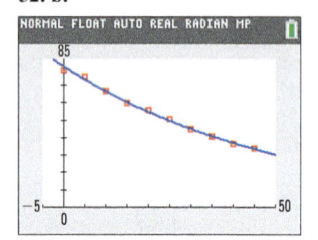

53. b.

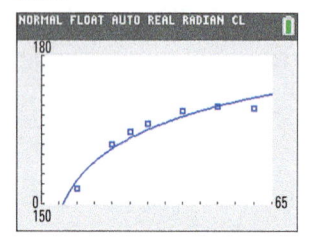

56. a.

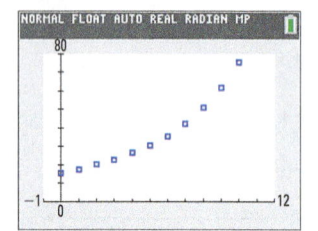

c.

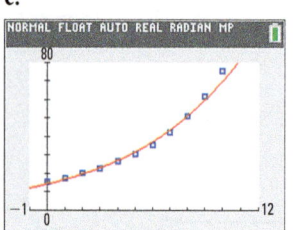

65. b.

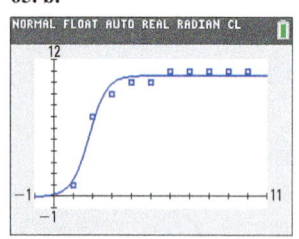

Chapter 6 Higher-Degree Polynomial and Rational Functions

6.1 Skills Check

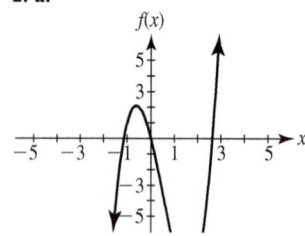

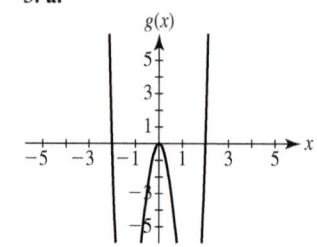

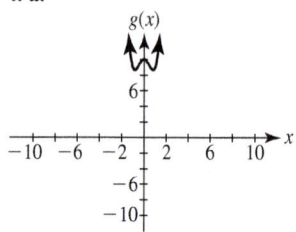

1. a. $h(x)$ graph **b.** $h(x)$ graph

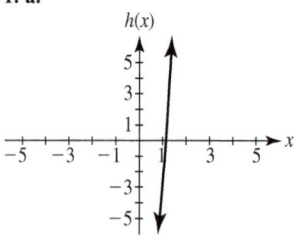

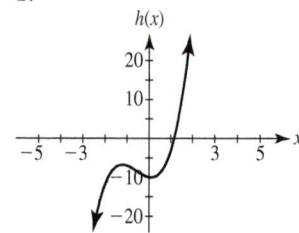

2. a. $f(x)$ graph **b.** $f(x)$ graph

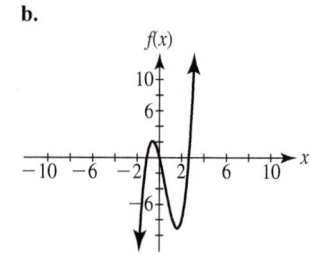

3. a. $g(x)$ graph **b.** $g(x)$ graph

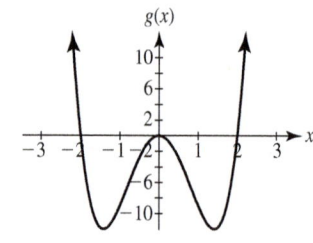

4. a. $g(x)$ graph **b.** $g(x)$ graph

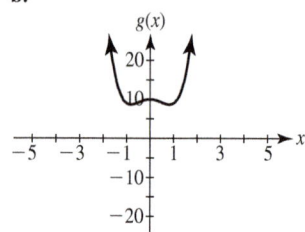

17. a. Degree 3; leading coefficient 2
b. Opening up to right and down to left
c. $f(x)$ graph

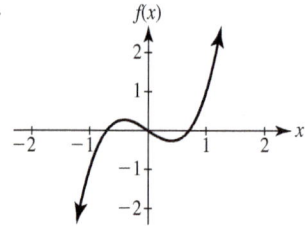

18. a. Degree 4; leading coefficient 0.3 **b.** Both ends opening up
c. $g(x)$ graph

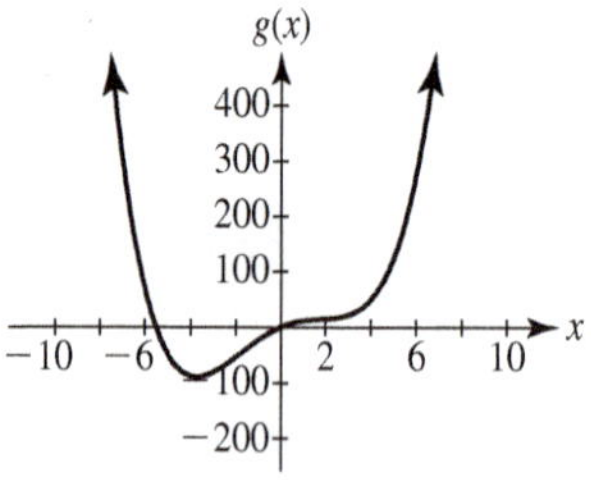

19. a. Degree 3; leading coefficient -2 **b.** Opening down to right and up to left
c. $f(x)$ graph

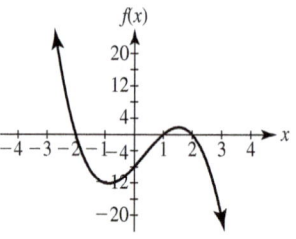

20. a. Degree 4; leading coefficient -3 **b.** Both ends opening down
c. $g(x)$ graph

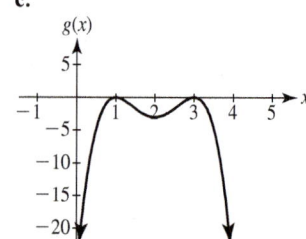

21. a.

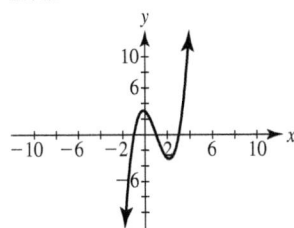

22. a.

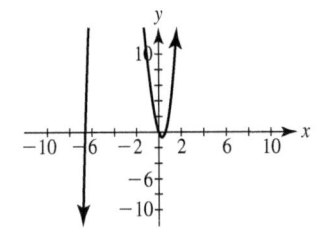

23. a.

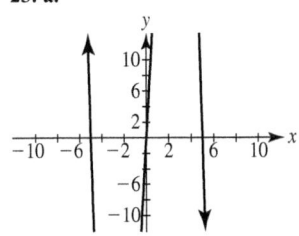

b.

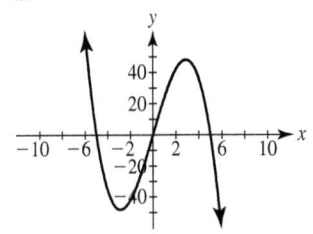

24. a.

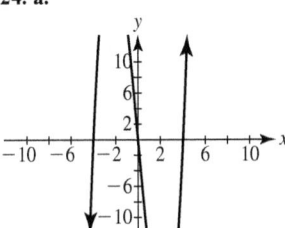

b.

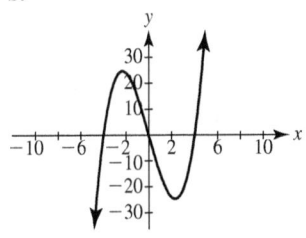

25. a.

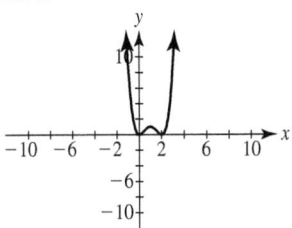

b.

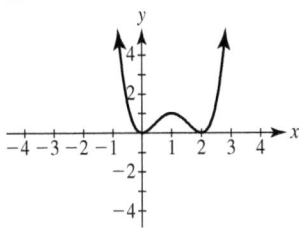

26. a.

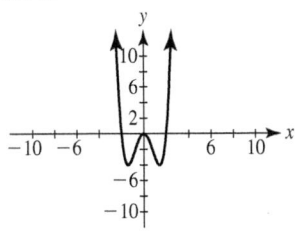

27. a.

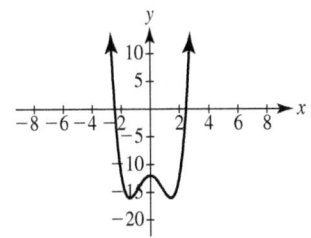

28. a.

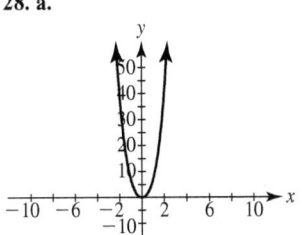

29. Answers will vary. One such graph is the graph of $f(x) = -4x^3 + 4$, shown here.

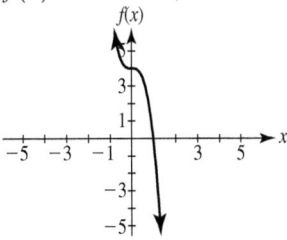

30. Answers will vary. One such graph is the graph of $f(x) = x^3 - 5x - 1$, shown here.

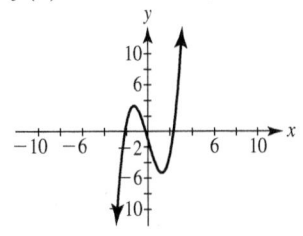

31. Answers will vary. One such graph is the graph of $f(x) = x^4 - 3x^2 - 4$, shown to the right.

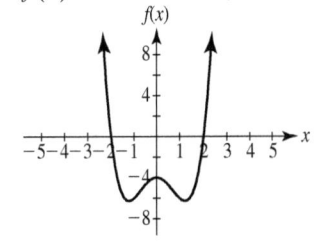

32. Answers will vary. One such graph is the graph of $f(x) = -x^3 + 3x^2 + x - 3$, shown here.

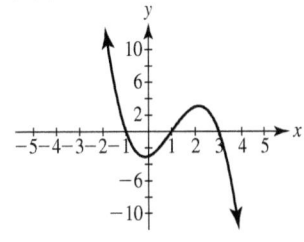

33. a.

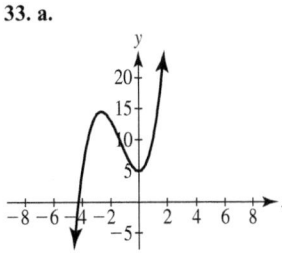

34. a.

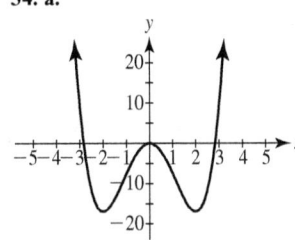

35.

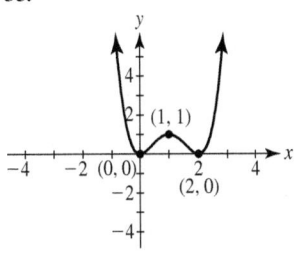

Additional Answers

36. a.

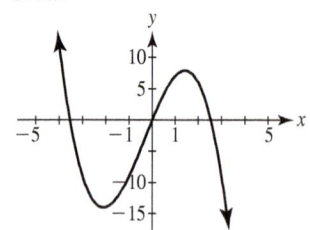

b.

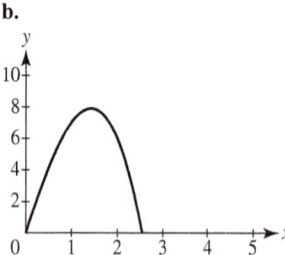

43. a.

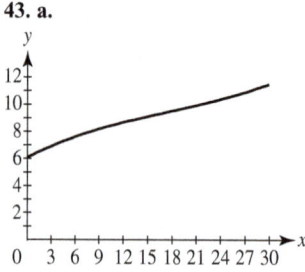

44. a.

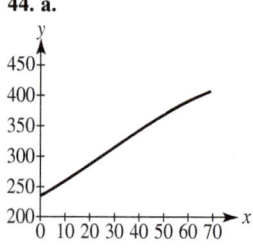

6.1 Exercises

37. a.

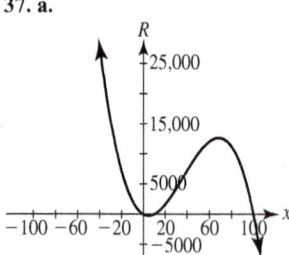

c.

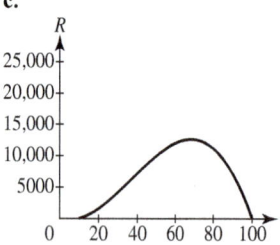

45. a.

38. a.

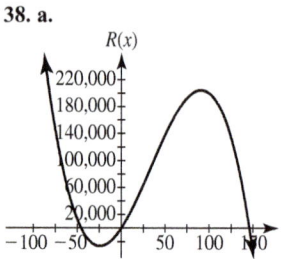

c.

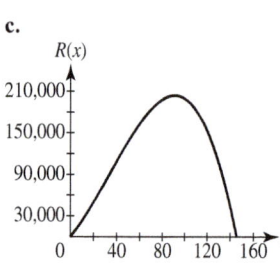

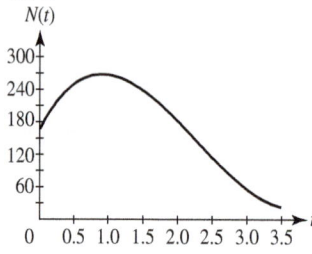

46. b. The local maximum is approximately at $(46.062, 7.376)$. This and looking at the graph indicates that the maximum number of full-time workers, 7.4 million, occurs at age 46. **c.** Approximately 6.27 at $x = 27.8$. This and looking at the graph indicates that there is a low point in the number of workers at age 27, but that it is not the age where the lowest number occurs according to the data.

47. a.

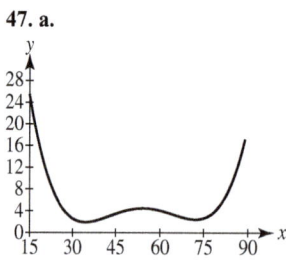

39. a.

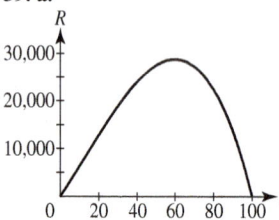

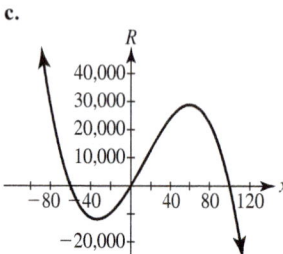

c.

48. a.

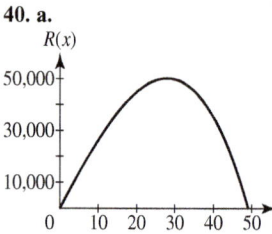

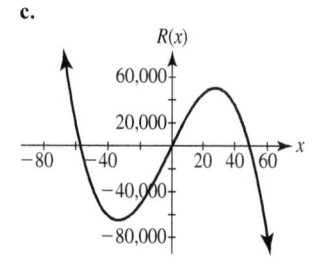

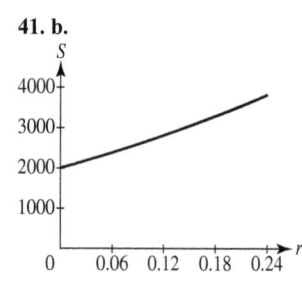

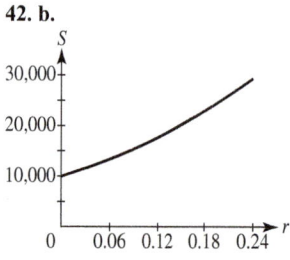

40. a.

c.

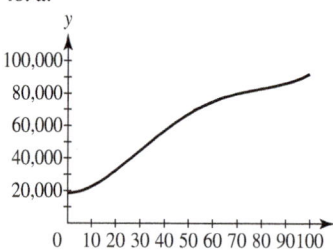

6.2 Skills Check

5. a.

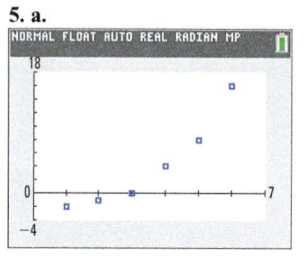

6. a.

41. b.

42. b.

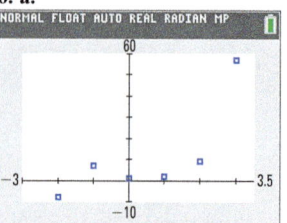

Additional Answers   **A-29**

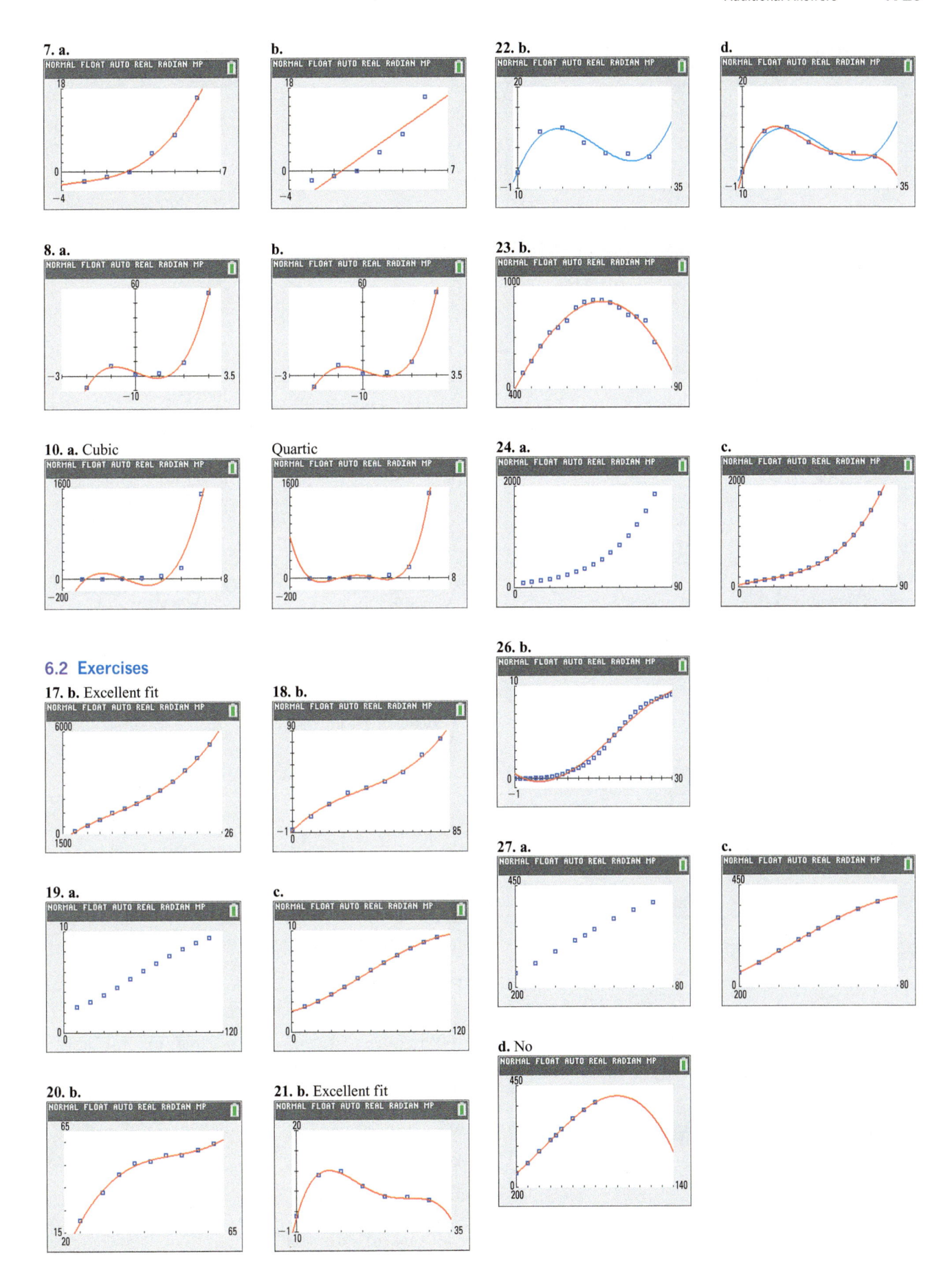

7. a.

b.

22. b.

d.

8. a.

b.

23. b.

10. a. Cubic

Quartic

24. a.

c.

6.2 Exercises

17. b. Excellent fit

18. b.

26. b.

19. a.

c.

27. a.

c.

d. No

20. b.

21. b. Excellent fit

A-30 Additional Answers

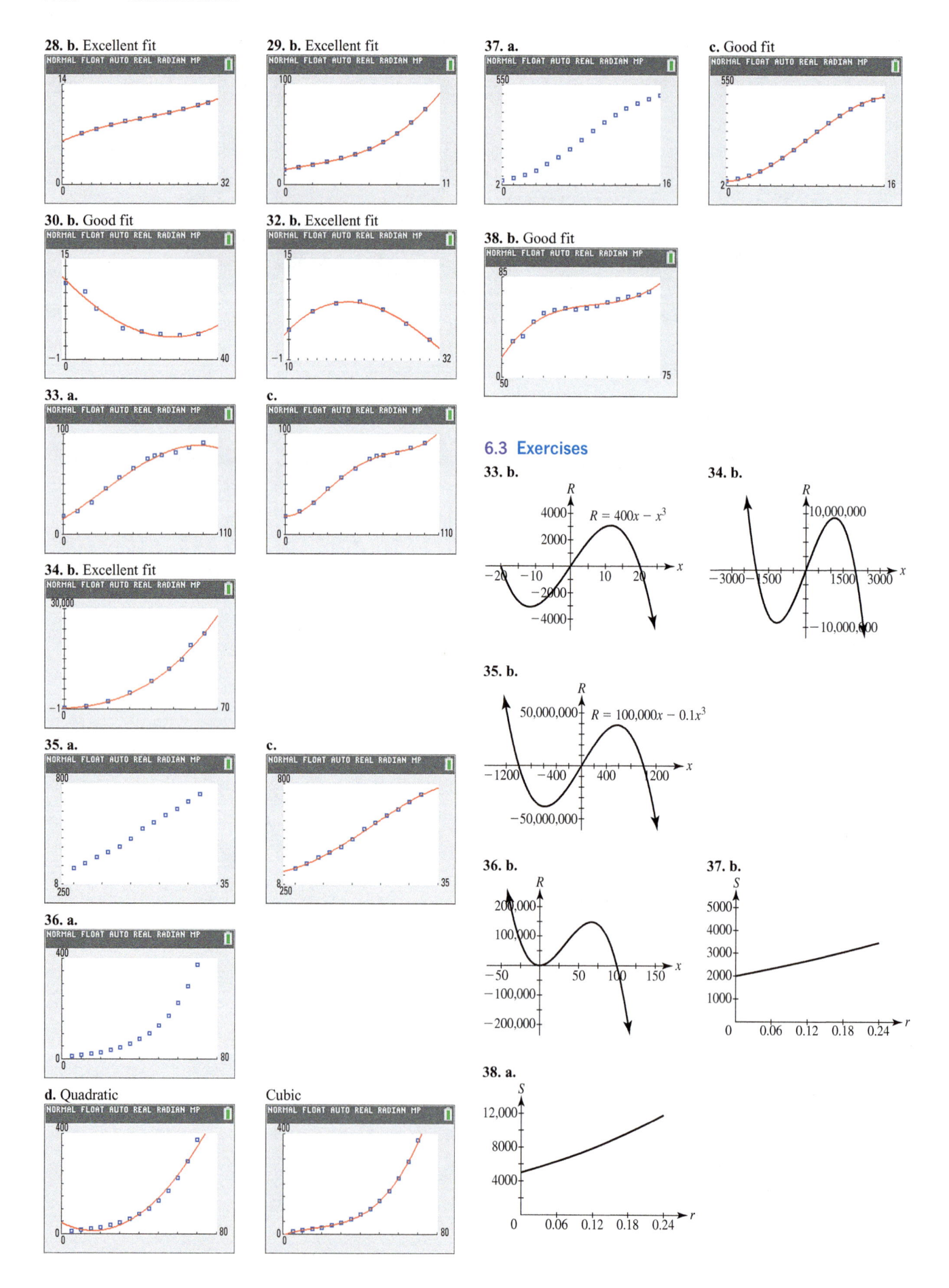

28. b. Excellent fit

29. b. Excellent fit

37. a.

c. Good fit

30. b. Good fit

32. b. Excellent fit

38. b. Good fit

33. a.

c.

6.3 Exercises

33. b. $R = 400x - x^3$

34. b.

34. b. Excellent fit

35. b. $R = 100{,}000x - 0.1x^3$

35. a.

c.

36. a.

36. b.

37. b.

38. a.

d. Quadratic

Cubic

40. b. For the values calculated in part (a) no box can be formed. The calculated values of x yield no sides that can be folded up to form the box.

d.

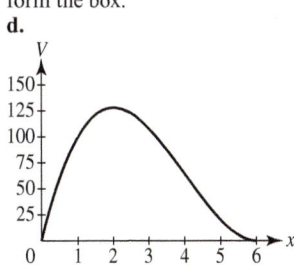

6.4 Exercises

33. a.

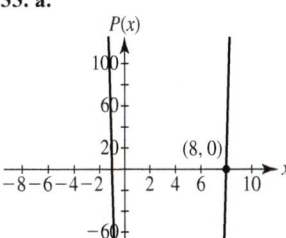

34. a.

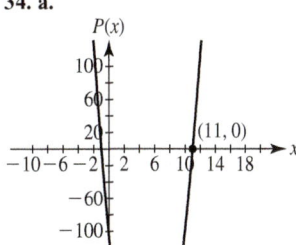

6.5 Skills Check

15. c.

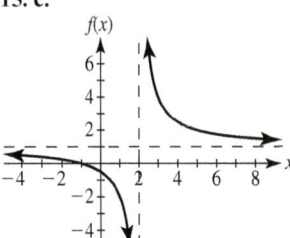

16. c.

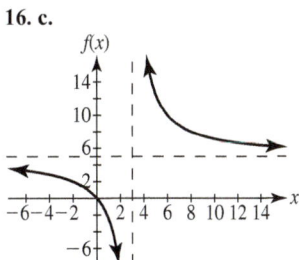

17. c.

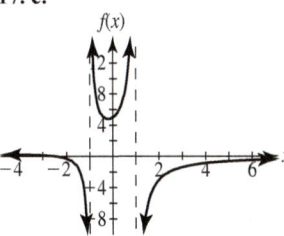

18. c.

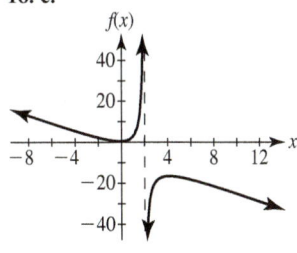

19.

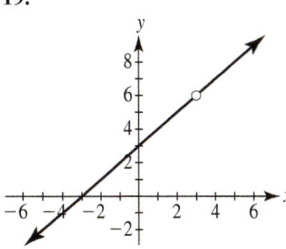

20.

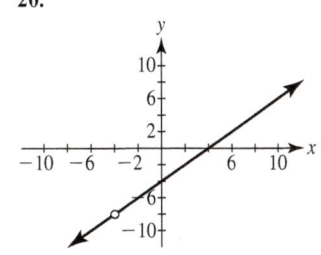

21.

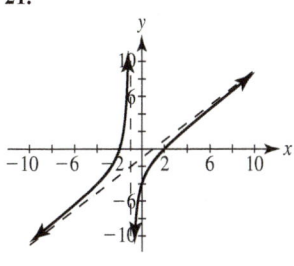

22.

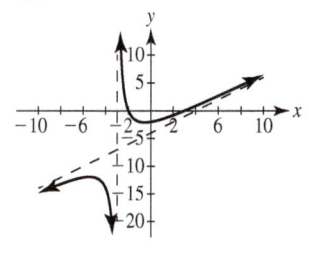

23.

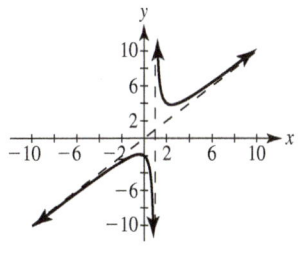

24.

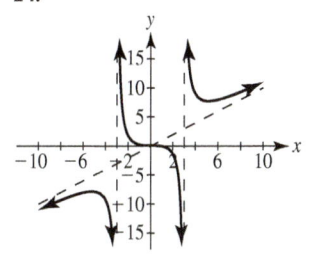

25.

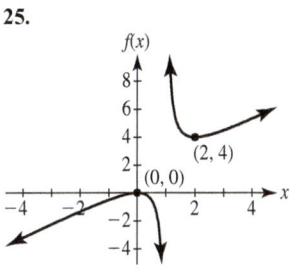

26.

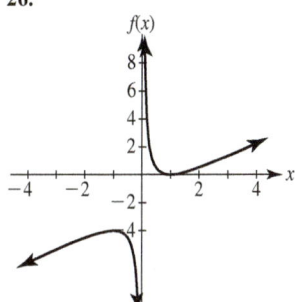

27.

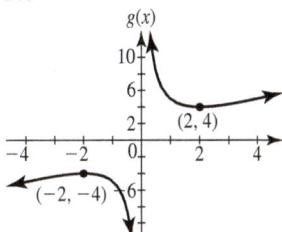

28.

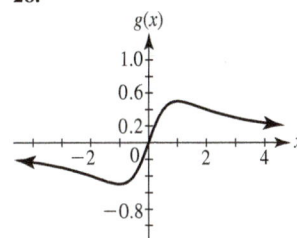

29. a.

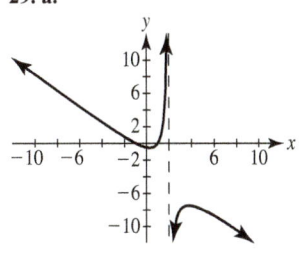

30. a.

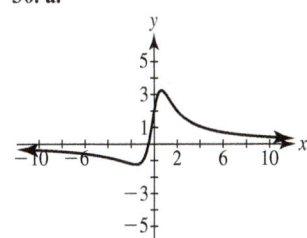

31. a.

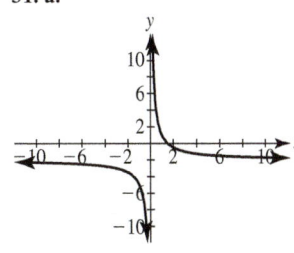

32. a.

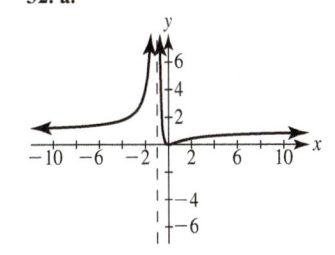

Additional Answers

6.5 Exercises

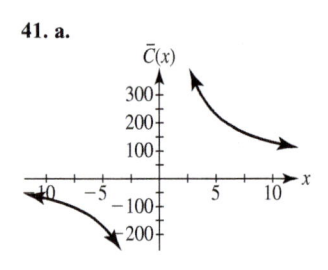

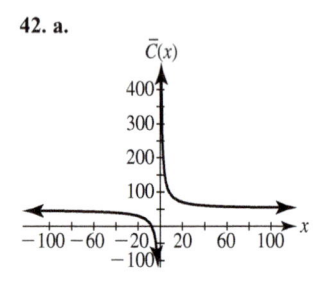

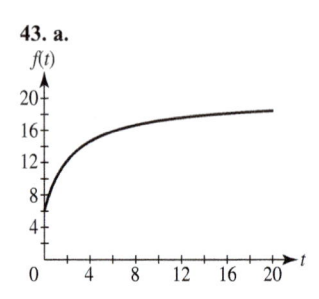

40. a. [graph of f(t)] **b.** [graph of f(t)]

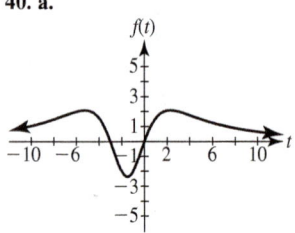

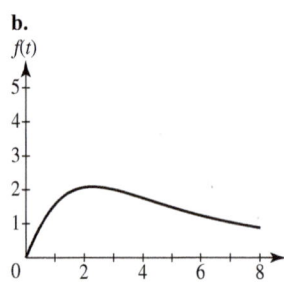

41. a. [graph of C̄(x)] **b.** [graph of C̄(x)]

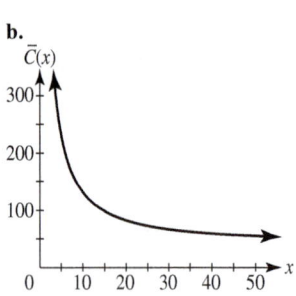

42. a. [graph of C̄(x)] **b.** [graph of C̄(x)]

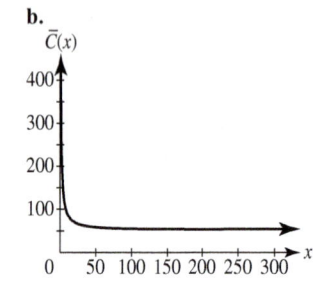

43. a. [graph of f(t)] **44. a.** [graph of C(t)]

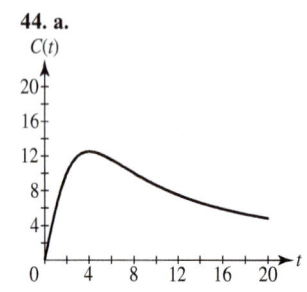

d. Concentration increases over the first 4 hours and decreases after 4 hours, approaching 0% as t approaches ∞.

47. a. [graph of V vs p]

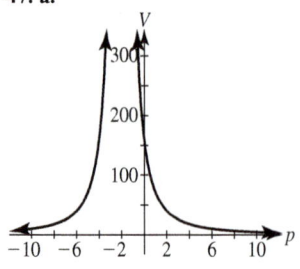

b.

Price/Unit ($)	5	20	50	100	200	500
Weekly Sales Volume	13,061	1322	237	62	16	3

49. a. [graph of p vs q]

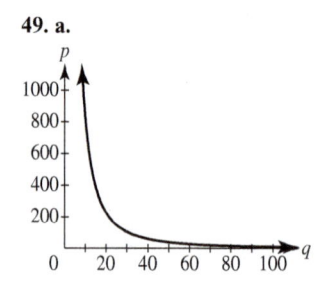

c.

Weekly Expenses ($)	Weekly Sales
0	0
50	14,814.81
100	15,384.62
200	15,686.27
300	15,789.47
500	15,873.02

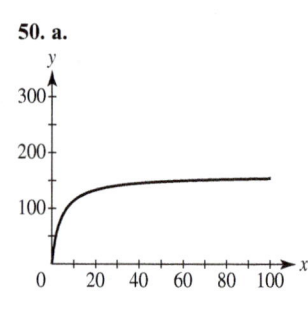

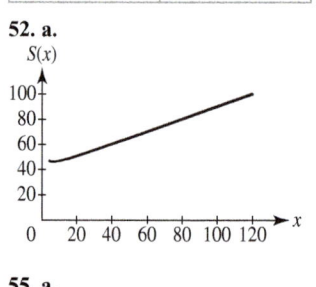

52. a. [graph of S(x)]

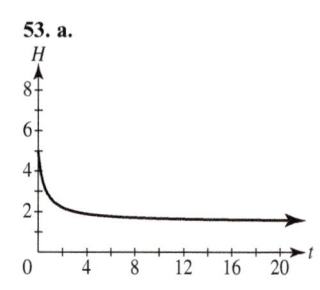

55. a. [graph of y vs x]

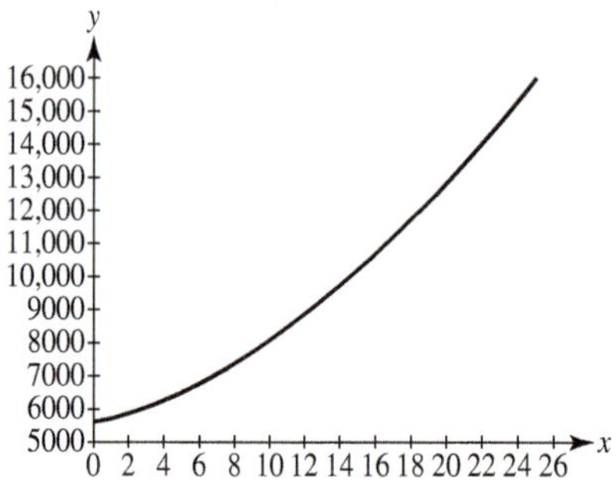

56. c. $L = \dfrac{51,200}{y} + 2y$

50. a. [graph of y vs x]

53. a. [graph of H vs t]

d. [graph of L vs y]

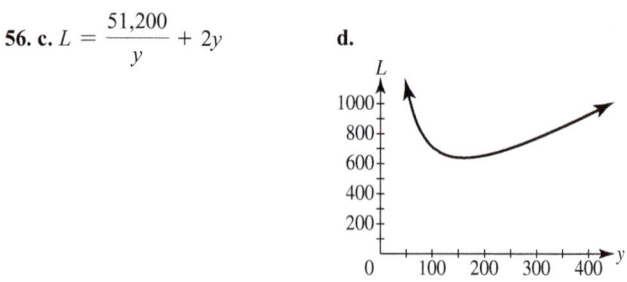

Chapter 6 Skills Check

3. [graph of y vs x]

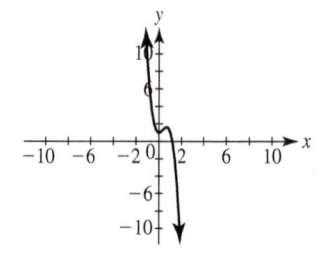

4. a. **b.**

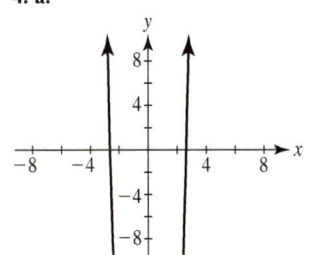

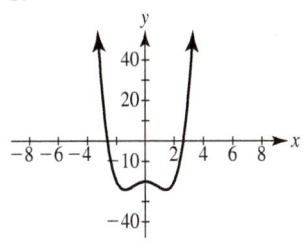

5. a.

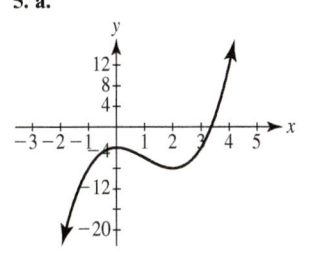

6. a. 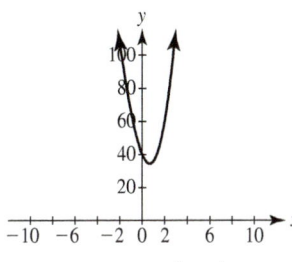 **c.**

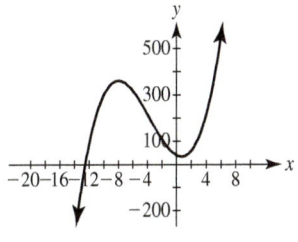

21. a. y-intercept: $\left(0, \frac{1}{2}\right)$; x-intercepts: $(-1, 0)$, $(1, 0)$

d.

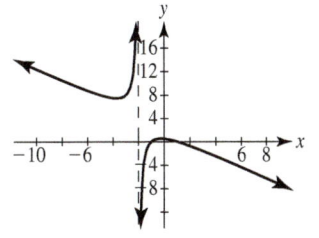

22. a. x-intercept $\left(\frac{2}{3}, 0\right)$; y-intercept $\left(0, \frac{2}{3}\right)$

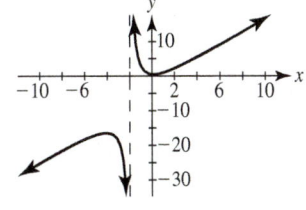

c. **23.**

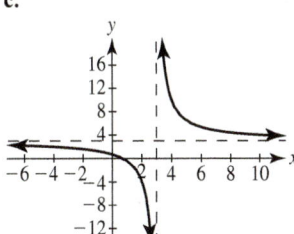

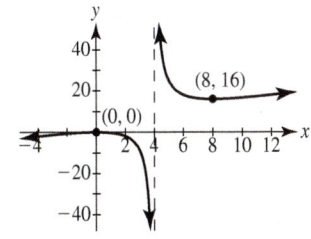

24. 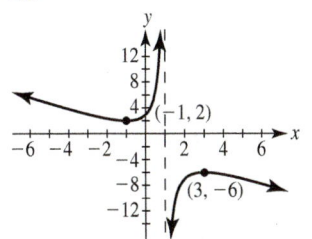 **25. a.**

Chapter 6 Review

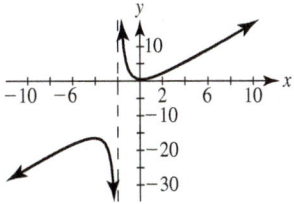

33. a. **b.**

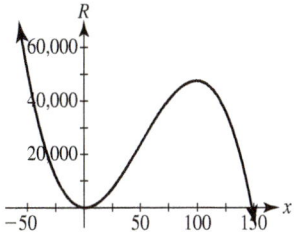

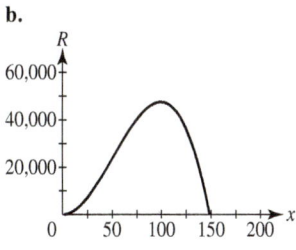

35. b. 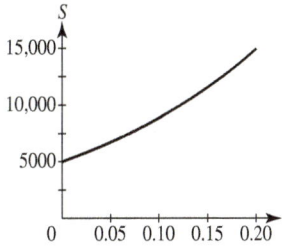 **37. a.**

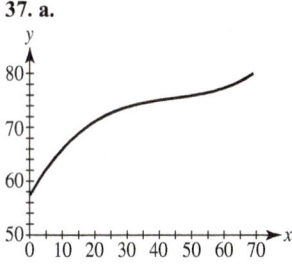

38. a. 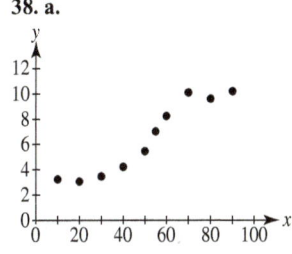 **c.**

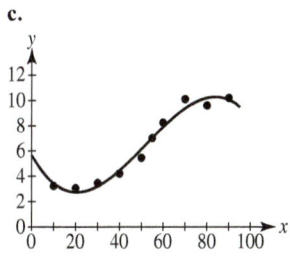

39. a. 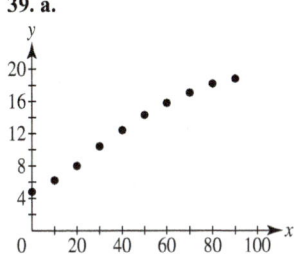 **d.** Good fit

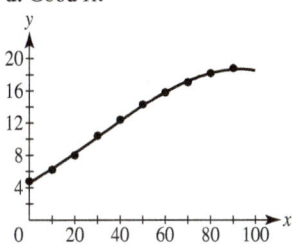

42. a. 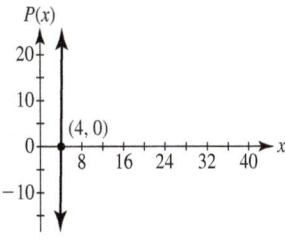 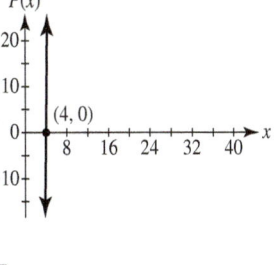 **46. a.**

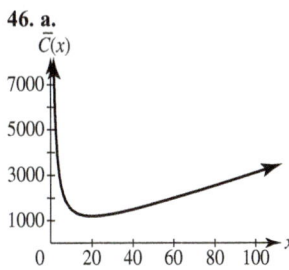

47. $p = \dfrac{30{,}000 - 20q}{q}$ **49. a.**

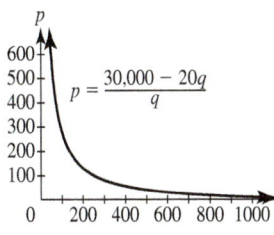

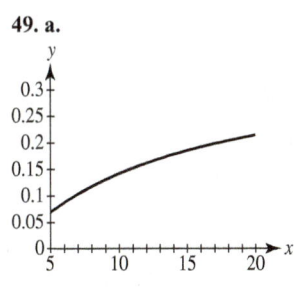

51. a.

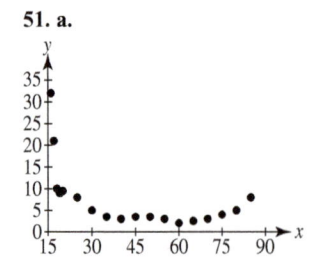

c.

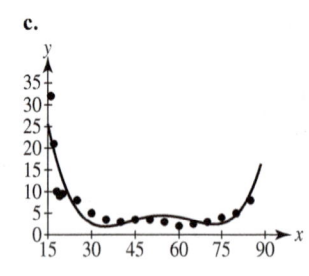

25. Infinitely many solutions of the form
$x = \frac{32}{19} - \frac{22z}{19}, y = \frac{10}{19} - \frac{14z}{19}, z = z$

28. Infinitely many solutions of the form
$x = 3y + \frac{2}{3}, z = \frac{17}{3}, y = y$

29. Infinitely many solutions of the form
$x = \frac{23}{14} + \frac{z}{2}, y = -\frac{4}{7} + z, z = z$

30. Infinitely many solutions of the form
$x = \frac{1}{13}z + \frac{18}{13}, y = \frac{5}{13}z - \frac{1}{13}, z = z$

Chapter 7 Systems of Equations and Matrices

7.1 Skills Check

17. Infinitely many solutions of the form $x = 5z - 14, y = 2z - 4, z = z$ **18.** Infinitely many solutions of the form $x = 4z - 8, y = 6 - 3z, z = z$ **19.** Infinitely many solutions of the form $x = 20 - 10z, y = 16 - 8z, z = z$

7.1 Exercises

34. a. $\begin{cases} 50x + 108y + 127z = 393 \\ 0.1y + 5.5z = 5.7 \\ 22x + 25.5y + 18.0z = 91.0 \end{cases}$

36. a. $\begin{cases} x + y + z = 200{,}000 \\ z = 45{,}000 + (x + y) \end{cases}$

38. Let $x =$ the number of units of product A, $y =$ the number of units of product B, and $z =$ the number of units of product C. Then $x = 175 - 5z, y = 3.5z - 10, z = z$, with x, y, z positive integers, $y \geq 0 \Rightarrow z \geq \frac{10}{3.5}$ and $x \geq 0 \Rightarrow z \leq 35$, so $3 \leq z \leq 35$, z must be even so y is an integer.
Sample solutions are

Product C	4	6	...	34
Product A	155	145	...	5
Product B	4	11	...	109

39. Food I $= 132 - (4.4)$ food III, food II $= (3.2)$ food III $- 64$, where $20 \leq$ food III ≤ 30 g
40. a. Let $x =$ amount invested in the 8% account, $y =$ amount invested in the 7.5% account, and $z =$ amount invested in the 10% account. Strategies with $x = 3{,}000{,}000 - 5z, y = 4z - 1{,}600{,}000$, where $400{,}000 \leq z \leq 600{,}000$, give earnings of $120{,}000.

7.2 Skills Check

1. $\begin{bmatrix} 1 & 1 & -1 & | & 4 \\ 1 & -2 & -1 & | & -2 \\ 2 & 2 & 1 & | & 11 \end{bmatrix}$ **2.** $\begin{bmatrix} 3 & -4 & 6 & | & 10 \\ 2 & -4 & -5 & | & -14 \\ 1 & 2 & -3 & | & 0 \end{bmatrix}$

3. $\begin{bmatrix} 5 & -3 & 2 & | & 12 \\ 3 & 6 & -9 & | & 4 \\ 2 & 3 & -4 & | & 9 \end{bmatrix}$ **4.** $\begin{bmatrix} 1 & -3 & 4 & | & 7 \\ 2 & 2 & -3 & | & -3 \\ 1 & -3 & 1 & | & -2 \end{bmatrix}$

12. Infinitely many solutions of the form
$x = 1 - 2z, y = 5 - 3z, z = z$
13. Infinitely many solutions of the form
$x = 2 - 3z, y = 5z + 5, z = z$
14. Infinitely many solutions of the form
$x = z + 3, y = -2 - 2z, z = z$
24. Infinitely many solutions of the form
$x = z - 2, y = 3 - 2z, z = z$

7.2 Exercises

35. a. $\begin{cases} 15x + 10y + 5z = 100 \\ y = 2x \\ z = 3x \end{cases}$

36. a. Let $x =$ the number of true-false questions, $y =$ the number of multiple-choice questions, and $z =$ the number of essay questions.
Then $\begin{cases} x + y + z = 35 \\ y - 2z = 0 \\ x - 2y = 0 \end{cases}$

42. a. Let $x =$ the amount invested at 7.5%, $y =$ the amount invested at 10%, and $z =$ the amount invested at 8%.
$\begin{cases} x + y + z = 235{,}000 \\ 0.075x + 0.10y + 0.08z = 18{,}000 \end{cases}$

44. The solution to the system is $x = 3 - z, y = 4 - z, z = z$. The table represents all the potential investment choices based on units per plan.

Plan I (x)	Plan II (y)	Plan III (z)
3	4	0
2	3	1
1	2	2
0	1	3

45. a. $\begin{cases} x + y + z = 4 \\ 40{,}000x + 30{,}000y + 20{,}000z = 100{,}000 \end{cases}$

d. 2 at $30,000, 2 at $20,000, and 0 at $40,000 or 1 at $40,000, 0 at $30,000, and 3 at $20,000

46. a. $\begin{cases} x + y + z = 400{,}000 \\ 0.08x + 0.10y + 0.12z = 42{,}400 \end{cases}$

47. a. $\begin{cases} x + y + z = 500{,}000 \\ .04x + .08y + .10z = 35{,}000 \end{cases}$

b. $x = 0.5z + 125{,}000$ and $y = 375{,}000 - 1.5z$ with nonnegative values of x, y, and z **c.** $225,000 at 8%, $100,000 at 10%, and $175,000 at 4%

48. a. $\begin{cases} x + y + z = 700{,}000 \\ .04x + .08y + .12z = 42{,}000 \end{cases}$

b. $x = z + 350{,}000$ and $y = 350{,}000 - 2z$ with nonnegative values of x, y, and z.   **c.** The investor should invest $525,000 at 4%, $0 at 8%, and $175,000 at 12%
49. Traffic from intersection A to intersection B is 650 less than the traffic from intersection D to intersection A. Traffic from intersection B to intersection C is 100 less than the traffic from intersection D to intersection A. Traffic from intersection C to intersection D is 1200 plus the traffic from intersection D to intersection A.
50. b. Traffic from intersection A to intersection B is 130 plus the traffic from intersection D to intersection A. Traffic from intersection B to intersection C is 50 plus the traffic from intersection D to intersection A. Traffic from intersection C to intersection D is 100 plus the traffic from intersection D to intersection A.

7.3 Skills Check

2. AC, AD, AE
BA, BC, BD, BE
CB, CF
DA, DC, DE
EA, EC, ED
FA, FC, FD, FE
AA, DD, EE

3. $\begin{bmatrix} 3 & 6 & -1 \\ 6 & 5 & 3 \\ -3 & 8 & 7 \end{bmatrix}$

4. a. $D + E = \begin{bmatrix} 11 & 5 & -6 \\ -2 & 4 & 4 \\ 9 & 1 & 0 \end{bmatrix}$,

$E + D = \begin{bmatrix} 11 & 5 & -6 \\ -2 & 4 & 4 \\ 9 & 1 & 0 \end{bmatrix}$

5. $\begin{bmatrix} 3 & 9 & -6 \\ 9 & 3 & 12 \\ -15 & 9 & 18 \end{bmatrix}$ **7.** $\begin{bmatrix} 0 & -6 & 10 \\ -6 & 4 & -18 \\ 24 & -2 & -22 \end{bmatrix}$

9. a. $AD = \begin{bmatrix} 7 & 5 & -4 \\ 17 & 33 & 6 \\ 11 & 27 & -2 \end{bmatrix}$, $DA = \begin{bmatrix} 6 & 12 & 14 \\ 20 & 10 & 4 \\ 12 & 14 & 22 \end{bmatrix}$

10. a. $BC = \begin{bmatrix} 1 & 8 \\ 19 & 7 \end{bmatrix}$, $CB = \begin{bmatrix} 11 & 7 & 11 \\ 7 & 4 & 2 \\ 3 & 1 & -7 \end{bmatrix}$

11. a. $DE = ED = \begin{bmatrix} 10 & 0 & 0 \\ 0 & 10 & 0 \\ 0 & 0 & 10 \end{bmatrix}$

15. $\begin{bmatrix} 3a - 2 & 3b - 4 \\ 3c - 6 & 3d - 8 \end{bmatrix}$

21. $AB = \begin{bmatrix} a + 3b + 5c & 2a + 4b + 6c \\ d + 3e + 5f & 2d + 4e + 6f \end{bmatrix}$;

$BA = \begin{bmatrix} a + 2d & b + 2e & c + 2f \\ 3a + 4d & 3b + 4e & 3c + 4f \\ 5a + 6d & 5b + 6e & 5c + 6f \end{bmatrix}$

22. $EF = \begin{bmatrix} ae + bg & af + bh \\ ce + dg & cf + dh \end{bmatrix}$; $FE = \begin{bmatrix} ea + fc & eb + fd \\ ga + hc & gb + hd \end{bmatrix}$

24. $AB = \begin{bmatrix} 0 & 14 & 6 \\ 13 & 3 & 5 \\ -10 & 2 & -2 \end{bmatrix}$; $BA = \begin{bmatrix} 6 & 18 \\ 6 & -5 \end{bmatrix}$

25. $AB = \begin{bmatrix} -2 & 0 \\ 5 & 19 \end{bmatrix}$; $BA = \begin{bmatrix} 6 & 1 & 10 \\ 10 & 11 & 14 \\ 1 & 6 & 0 \end{bmatrix}$

26. a. $AB = \begin{bmatrix} 1 & 0 & 0 \\ 0 & 1 & 0 \\ 0 & 0 & 1 \end{bmatrix}$; $BA = \begin{bmatrix} 1 & 0 & 0 \\ 0 & 1 & 0 \\ 0 & 0 & 1 \end{bmatrix}$

7.3 Exercises

27. a. $B = \begin{bmatrix} 1341.3 & 148.7 & 93.3 \\ 73.8 & 69.4 & 69.2 \\ 1.56 & 2.16 & 3.05 \\ 19.6 & 41.8 & 20.9 \end{bmatrix}$

$A = \begin{bmatrix} 1211.5 & 192.4 & 165.5 \\ 80.8 & 79.0 & 78.4 \\ 1.88 & 1.68 & 1.95 \\ 8.3 & 12.5 & 10.2 \end{bmatrix}$

b. $C = \begin{bmatrix} -129.8 & 43.7 & 72.2 \\ 7 & 9.6 & 9.2 \\ 0.32 & -0.48 & -1.10 \\ -11.3 & -29.3 & -10.7 \end{bmatrix}$

c. The negative entries in row 4 indicate that the infant mortality rates of all three countries are projected to decline, which is a positive change for the countries.

28. a.

	Under Age 5	5 – 13	14 – 17	18 – 64	Over Age 64
2015	475	796	388	−686	−5613
2020	492	669	374	41	−6029
2025	500	909	379	842	−6644
2030	502	932	425	1671	−7357

The matrix shows the projected differences between male and female populations for different ages and years.
b. For ages 18 – 64 in 2015, and for those over age 64 in all years 2015 – 2030

29. a. $A = \begin{bmatrix} 210 & 140 & 50 & 0 \\ 240 & 200 & 30 & 0 \\ 260 & 270 & 60 & 0 \\ 230 & 300 & 80 & 0 \\ 210 & 360 & 100 & 0 \\ 210 & 330 & 120 & 0 \end{bmatrix}$

$B = \begin{bmatrix} 180 & 350 & 170 & 0 \\ 160 & 340 & 210 & 0 \\ 140 & 330 & 260 & 90 \\ 120 & 330 & 300 & 160 \\ 100 & 310 & 360 & 290 \\ 40 & 270 & 390 & 560 \end{bmatrix}$

b. $C = B - A = \begin{bmatrix} -30 & 210 & 120 & 0 \\ -80 & 140 & 180 & 0 \\ -120 & 60 & 200 & 90 \\ -110 & 30 & 220 & 160 \\ -110 & -50 & 260 & 290 \\ -170 & -60 & 270 & 560 \end{bmatrix}$

30. a. $A = \begin{bmatrix} 74,128 & 82,304 \\ 120,073 & 136,310 \\ 83,861 & 100,013 \\ 56,441 & 98,164 \end{bmatrix}$; $B = \begin{bmatrix} 2445 & 3254 \\ 20,704 & 25,169 \\ 16,665 & 24,520 \\ 8079 & 25,288 \end{bmatrix}$

b. $C = \begin{bmatrix} 71,683 & 79,050 \\ 99,369 & 111,141 \\ 67,196 & 75,493 \\ 48,362 & 72,876 \end{bmatrix}$

31. a. $C = \begin{bmatrix} 37.04 & 37.54 & 36.87 \\ 25.86 & 26.77 & 27.28 \\ 34.82 & 36.73 & 38.19 \end{bmatrix}$

b. $M = \begin{bmatrix} 3.09 & 3.13 & 3.07 \\ 2.16 & 2.23 & 2.27 \\ 2.90 & 3.06 & 3.18 \end{bmatrix}$

32. a. $A = \begin{bmatrix} 5.13 & 6.42 & 12.83 & 25.67 \\ 6.53 & 8.17 & 16.33 & 32.67 \\ 8.63 & 10.79 & 21.58 & 43.17 \\ 13.77 & 17.21 & 34.42 & 68.83 \\ 22.75 & 28.44 & 56.88 & 113.75 \end{bmatrix}$

b. $B = \begin{bmatrix} 5.64 & 7.06 & 14.11 & 28.24 \\ 7.18 & 8.99 & 17.96 & 35.94 \\ 9.49 & 11.87 & 23.74 & 47.49 \\ 15.15 & 18.93 & 37.86 & 75.71 \\ 25.03 & 31.28 & 62.57 & 125.13 \end{bmatrix}$

33. $BA = \begin{bmatrix} 1210 \\ 980 \\ 594 \end{bmatrix} \begin{matrix} \text{Cost} \\ \text{Singles} \\ \text{Males 35–55} \\ \text{Females 65+} \end{matrix}$

34. a. $\begin{matrix} & \text{Men} & \text{Women} \\ \text{Robes} \\ \text{Hoods} \end{matrix} \begin{bmatrix} 3365 & 1745 \\ 1300 & 740 \end{bmatrix}$

35. a. $\begin{matrix} & \text{DeTuris} & \text{Marriott} \\ \text{Dept. A} \\ \text{Dept. B} \end{matrix} \begin{bmatrix} 54,000 & 49,600 \\ 42,000 & 42,400 \end{bmatrix}$

36. $\begin{bmatrix} 5 & 20 & 0 & 10 \\ 10 & 9 & 0 & 0 \\ 5 & 10 & 10 & 10 \end{bmatrix} \begin{bmatrix} 15 \\ 12 \\ 14 \\ 30 \end{bmatrix} \begin{matrix} \text{Style A} \\ \text{Style B} \\ \text{Style C} \end{matrix} \begin{bmatrix} \text{Manufacturing} \\ \text{Cost} \\ 615 \\ 258 \\ 635 \end{bmatrix}$

39. a. $W = \begin{bmatrix} 1033 & 772 & 1299 & 728 \\ 826 & 709 & 1017 & 631 \end{bmatrix}$

b. $G = \begin{bmatrix} 1094.98 & 818.32 & 1376.94 & 771.68 \\ 892.08 & 765.72 & 1098.36 & 681.48 \end{bmatrix}$

40. a. $A = \begin{bmatrix} 23.2 & 2.5 \\ 24.1 & 14.2 \\ 35.1 & 3.1 \\ 40.3 & 19.9 \\ 33.1 & 2.9 \\ 49.4 & 3.6 \end{bmatrix}$ $B = \begin{bmatrix} 34.7 & 1.6 \\ 16.3 & 8.6 \\ 45.3 & 2.5 \\ 31.3 & 15.9 \\ 27.5 & 1.2 \\ 43.3 & 2.4 \end{bmatrix}$

b. $C = \begin{bmatrix} 11.5 & -0.9 \\ -7.8 & -5.6 \\ 10.2 & -0.6 \\ -9 & -4 \\ -5.6 & -1.7 \\ -6.1 & -1.2 \end{bmatrix}$

7.4 Skills Check

1. a. $AB = BA = \begin{bmatrix} 1 & 0 \\ 0 & 1 \end{bmatrix}$ **2. a.** $AB = BA = \begin{bmatrix} 1 & 0 & 0 \\ 0 & 1 & 0 \\ 0 & 0 & 1 \end{bmatrix}$

3. $AB = BA = \begin{bmatrix} 1 & 0 & 0 \\ 0 & 1 & 0 \\ 0 & 0 & 1 \end{bmatrix}$

4. $CD = DC = \begin{bmatrix} 1 & 0 & 0 \\ 0 & 1 & 0 \\ 0 & 0 & 1 \end{bmatrix}$; thus C and D are inverses.

5. $A^{-1} = \begin{bmatrix} 7 & -3 \\ -2 & 1 \end{bmatrix}$ **6.** $A^{-1} = \begin{bmatrix} \frac{5}{2} & -2 \\ -1 & 1 \end{bmatrix}$

8. $A^{-1} = \begin{bmatrix} \frac{1}{2} & \frac{1}{4} & \frac{1}{2} \\ \frac{1}{4} & -\frac{3}{8} & \frac{1}{4} \\ 1 & -\frac{1}{2} & 0 \end{bmatrix}$ **9.** $A^{-1} = \begin{bmatrix} -\frac{1}{3} & -1 & \frac{1}{3} \\ 1 & 1 & 0 \\ -\frac{1}{3} & 0 & \frac{1}{3} \end{bmatrix}$

10. $C^{-1} = \begin{bmatrix} 0 & 1 & 0 \\ 1 & 0 & 1 \\ 0 & 0 & 1 \end{bmatrix}$ **11.** $A^{-1} = \begin{bmatrix} 0.9 & 0.2 & -0.7 \\ -0.5 & 0 & 0.5 \\ 0.7 & -0.4 & -0.1 \end{bmatrix}$

16. $X = \begin{bmatrix} 7 \\ 2 \\ 6 \end{bmatrix}$

7.4 Exercises

29. a. $\begin{cases} x + y + z = 400{,}000 \\ x + y - z = -100{,}000 \\ x - \frac{1}{2}y = 0 \end{cases}$

7.5 Exercises

35. Species III = any amount (between 1800 and 2300)
Species I = 6900 − 3(Species III)
Species II = 1/2 (Species III) − 900
36. Oil = 6138 − 3.3C − 0.6R
Bank = 7.5C + R − 12,787.50, where R and C are any nonnegative amounts for which Oil and Bank are also nonnegative.

7.6 Skills Check

15. $\left(1, \frac{9}{4}\right), \left(-10, -\frac{1}{2}\right)$ **16.** $\left(-4, -\frac{4}{3}\right), (1, 2)$

21. b. Synthetic division by $x - 2$ gives $x^2 + x + 2$, which has no real zeros; thus, 2 is the only real solution to the system.
22. b. Synthetic division by $x + 2$ gives
$x^2 - 4x + 3 = (x - 3)(x - 1)$ which has $x = 3$ and $x = 1$ as solutions. Thus the solutions are $(-2, 2), (1, 2),$ and $(3, -8)$.

7.6 Exercises

36. Two possible boxes: 10 cm by 20 cm by 30 cm or approximately 33.59 cm by 16.79 cm by 10.64 cm

Chapter 7 Skills Check

13. $\begin{bmatrix} -1 & 5 & 2 \\ -1 & 1 & 5 \end{bmatrix}$ **14.** $\begin{bmatrix} -3 & 1 & 0 \\ -5 & 3 & -1 \end{bmatrix}$ **15.** $\begin{bmatrix} 10 & 15 \\ -5 & 10 \\ 15 & -10 \end{bmatrix}$

17. $\begin{bmatrix} 4 & 14 & 1 \\ -3 & 11 & -1 \end{bmatrix}$ **18.** $\begin{bmatrix} -13 & 12 & 8 \\ -4 & 1 & 3 \\ 0 & 5 & -1 \end{bmatrix}$

19. $\begin{bmatrix} -4 & -2 \\ -2 & -9 \end{bmatrix}$ **20.** $\begin{bmatrix} 10 & 6 & -6 \\ 9 & 25 & 0 \\ 3 & 15 & 10 \end{bmatrix}$

21. $\begin{bmatrix} -\frac{1}{8} & \frac{3}{8} & -\frac{3}{8} \\ \frac{1}{8} & \frac{1}{40} & \frac{7}{40} \\ -\frac{1}{4} & \frac{3}{20} & \frac{1}{20} \end{bmatrix}$ **22.** $\begin{bmatrix} \frac{4}{3} & \frac{1}{3} & -1 \\ -1 & 0 & 1 \\ -\frac{2}{3} & -\frac{2}{3} & 1 \end{bmatrix}$

23. $\begin{bmatrix} 1 & 0 & -1 \\ -1 & 1 & 1 \\ 2 & -2 & -1 \end{bmatrix}$

Chapter 7 Review

44. If $x =$ tech, $y =$ balanced, $z =$ utility,
$x = \dfrac{2800 - z}{3}$, $y = 200 - \left(\dfrac{2}{7}\right)z$, $0 \leq z \leq 700$

45. If $x =$ number of units of product A,
$y =$ number of units of product B,
$z =$ number of units of product C, $x = 252 - 1.2y$,
$0 \leq y \leq 210$, $z = 74$

46. $x = 2z$, $y = 2000 - 4z$ for $0 \leq z \leq 500$; $A =$ twice number of C, $B = 2000 - 4$ times of C, $0 \leq C \leq 500$

47. a. $A = \begin{bmatrix} 3.15 & 5.76 & 6.53 \\ 2.99 & 5.87 & 7.83 \\ 0.10 & 2.25 & 6.64 \end{bmatrix}$ **b.** $B = \begin{bmatrix} 10.13 & 12.05 & 11.39 \\ 0.20 & 0.21 & 0.11 \\ 0.63 & 0.89 & 0.92 \end{bmatrix}$

c. $T = \begin{bmatrix} -6.98 & -6.29 & -4.86 \\ 2.79 & 5.66 & 7.72 \\ -0.53 & 1.36 & 5.72 \end{bmatrix}$

48. a. $\begin{cases} x + y + z = 800{,}000 \\ .06x + .07y + .10z = 71{,}600 \end{cases}$

b. $x = 3z - 1{,}560{,}000$ and $y = 2{,}360{,}000 - 4z$ with x, y, and z nonnegative; $z \geq 520{,}000$ and $z \leq 590{,}000$

c. $210{,}000 at 6%, $0 at 7%, and $590{,}000 at 10%

d. $280{,}000 at 7%, $0 at 6%, and $520{,}000 at 10%

49. c. For each thousand-dollar increase in family income, the critical reading SAT score increases by 0.586 point and the math SAT score increases by 0.578 point.

Chapter 8 Special Topics in Algebra

8.1 Skills Check

1. $y \leq 5x - 4$

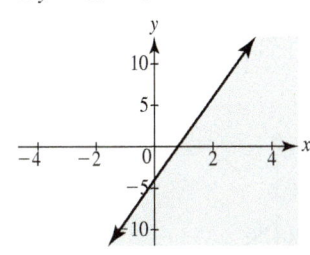

2. $y > 3x + 2$

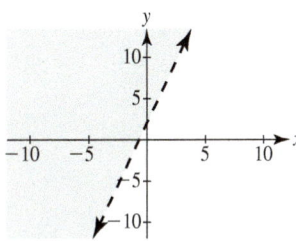

3. $6x - 3y \geq 12$

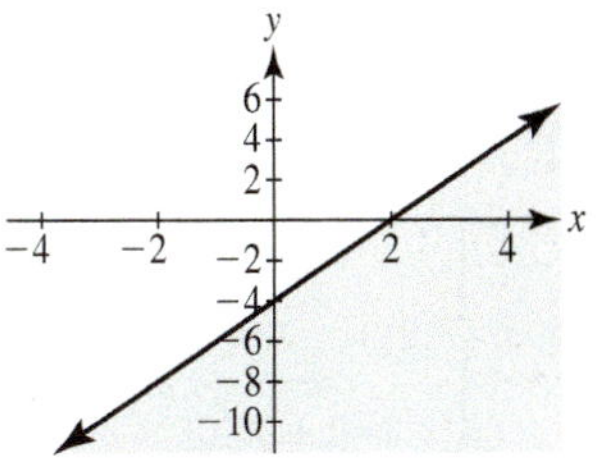

4. $\dfrac{x}{2} + \dfrac{y}{3} \leq 6$

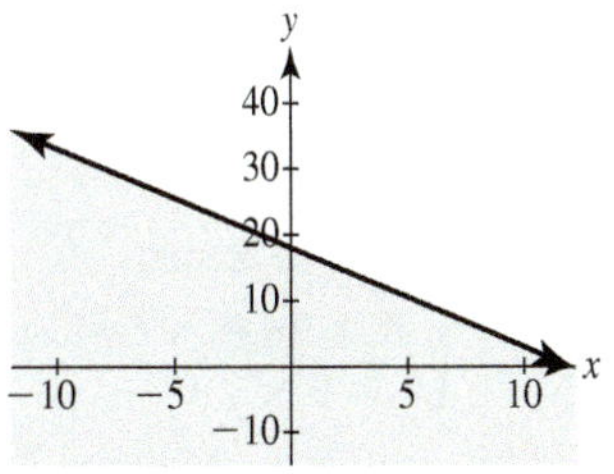

5. $4x + 5y \leq 20$

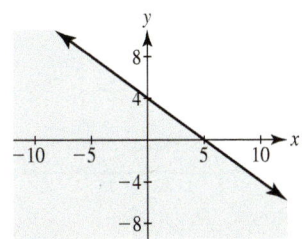

6. $3x - 5y \geq 30$

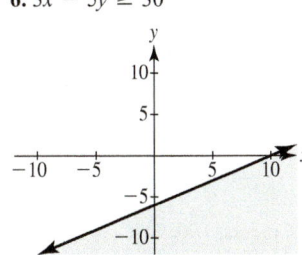

17. $\begin{cases} y \leq 8 - 3x \\ y \leq 2x + 3 \\ y > 3 \end{cases}$

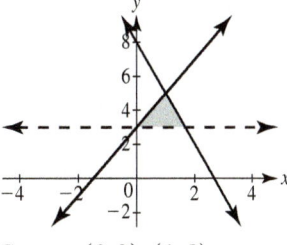

Corners: $(0, 3)$, $(1, 5)$, $(5/3, 3)$

18. $\begin{cases} 2x + y \geq 10 \\ 3x + 2y \geq 17 \\ x + 2y \geq 7 \end{cases}$

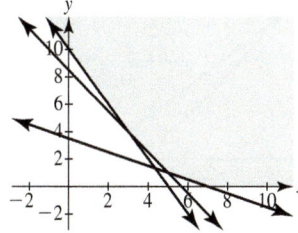

Corners: $(0, 10)$, $(3, 4)$, $(5, 1)$, $(7, 0)$

19. $\begin{cases} 2x + y < 5 \\ 2x - y > -1 \\ x \geq 0, y \geq 0 \end{cases}$

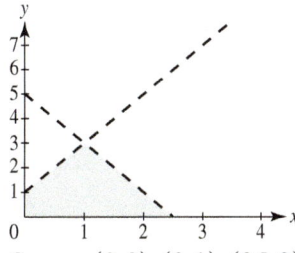

Corners: $(0, 0)$, $(0, 1)$, $(2.5, 0)$, $(1, 3)$

20. $\begin{cases} y < 2x \\ y > x + 2 \\ x \geq 0, y \geq 0 \end{cases}$

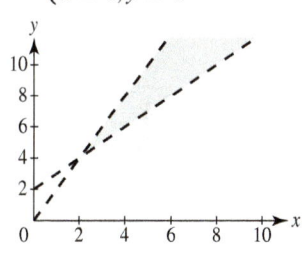

Corners: $(2, 4)$

21. $\begin{cases} x + 2y \geq 4 \\ x + y \leq 5 \\ 2x + y \leq 8 \\ x \geq 0, y \geq 0 \end{cases}$

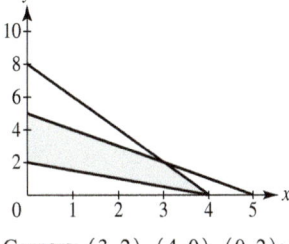

Corners: $(3, 2)$, $(4, 0)$, $(0, 2)$, $(0, 5)$

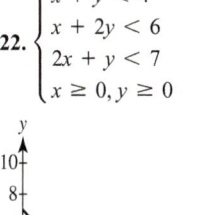

22. $\begin{cases} x + y < 4 \\ x + 2y < 6 \\ 2x + y < 7 \\ x \geq 0, y \geq 0 \end{cases}$

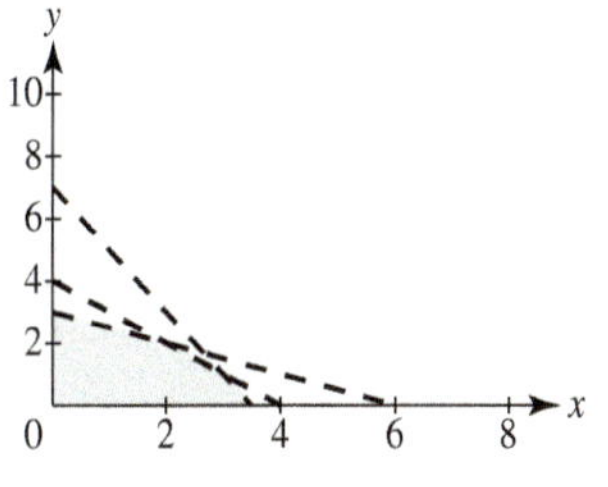

Corners: $(0, 0)$, $(0, 3)$, $(3.5, 0)$, $(2, 2)$, $(3, 1)$

23. $\begin{cases} x^2 - 3y < 4 \\ 2x + 3y < 4 \end{cases}$

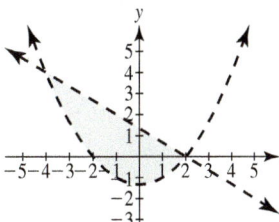

Corners: $(-4, 4)$, $(2, 0)$

24. $\begin{cases} x + y \leq 8 \\ xy \geq 12 \end{cases}$

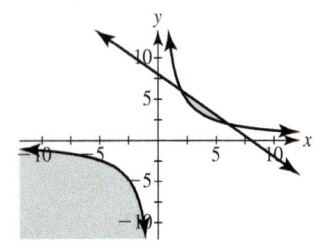

Corners: $(2, 6)$, $(6, 2)$

A-38 Additional Answers

25. $\begin{cases} x^2 - y - 8x \leq -6 \\ y + 9x \leq 18 \end{cases}$

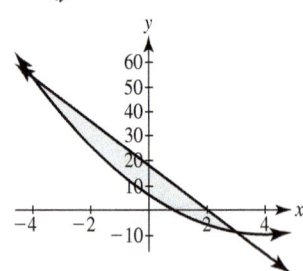

Corners: $(-4, 54)$, $(3, -9)$

26. $\begin{cases} x^2 - y > 0 \\ y > (4 - 2x)^2 \end{cases}$

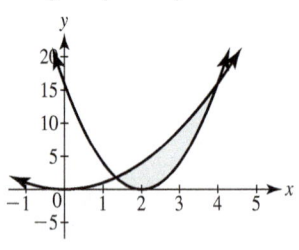

Corners: $(1.\overline{3}, 1.\overline{7})$, $(4, 16)$

33. $\begin{cases} x + y \geq 780 \\ 78x + 117y \leq 76{,}050 \\ x \geq 0, y \geq 0 \end{cases}$

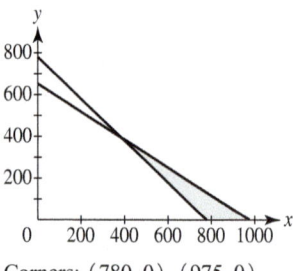

Corners: $(780, 0)$, $(975, 0)$, $(390, 390)$

34. $\begin{cases} 240x + 150y \leq 36{,}000 \\ \frac{1}{4}x + \frac{1}{10}y \geq 33 \\ x \geq 0, y \geq 0 \end{cases}$

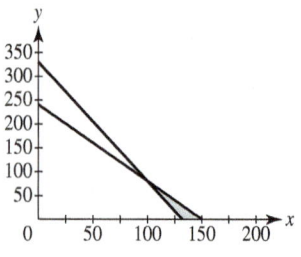

Corners: $(100, 80)$, $(150, 0)$, $(132, 0)$

8.1 Exercises

27. b.

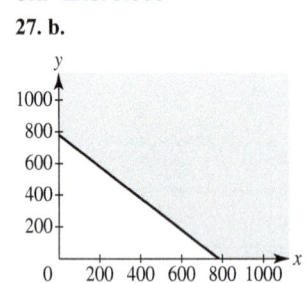

28. b.

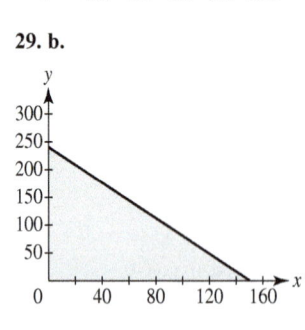

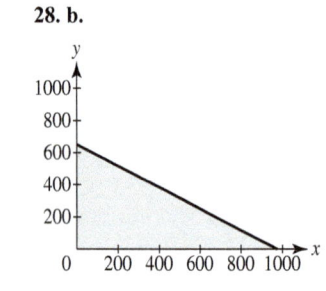

35. a. $\begin{cases} 80x + 40y \geq 3200 \\ 20x + 20y \geq 1000 \\ 100x + 40y \geq 3400 \\ x \geq 0, y \geq 0 \end{cases}$

b. Corners: $(30, 20)$, $(0, 85)$, $(50, 0)$, $(10, 60)$

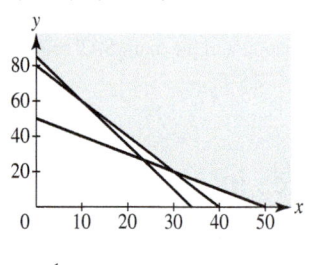

36. a. $\begin{cases} 1000x + 3000y \leq 18{,}000 \\ 18x + 36y \geq 252 \\ x \geq 0, y \geq 0 \end{cases}$

b.

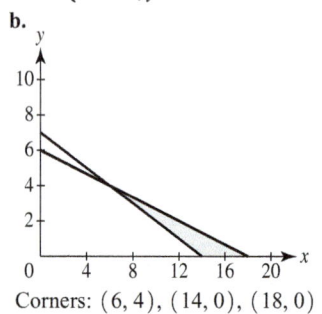

Corners: $(6, 4)$, $(14, 0)$, $(18, 0)$

29. b.

30. b.

37. $\begin{cases} 4x + 10y \leq 1600 \\ 6x + 7y \leq 1600 \\ x \geq 0, y \geq 0 \end{cases}$

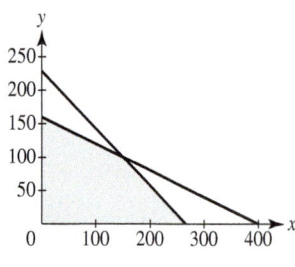

Corners: $(0, 0)$, $(0, 160)$, $(150, 100)$, $(266\frac{2}{3}, 0)$

38. a. $\begin{cases} x + y \leq 120 \\ 2x + 6y \leq 400 \\ x \geq 0, y \geq 0 \end{cases}$

b.

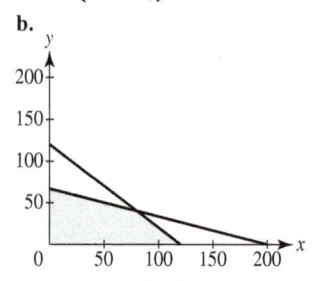

Corners: $(0, 0)$, $(80, 40)$, $(120, 0)$, $(0, 66.\overline{6})$

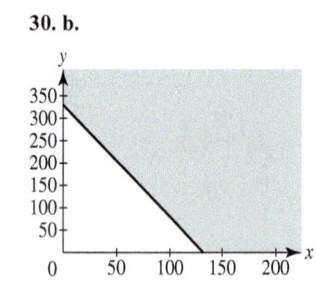

31. a. $\begin{cases} 0.12x + 0.009y \geq 7.56 \\ x + y \geq 100 \\ x \geq 0, y \geq 0 \end{cases}$

b. Corners: $(0, 840)$, $(60, 40)$, $(100, 0)$

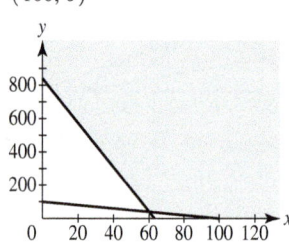

39. a. $\begin{cases} 4x + 6y \leq 480 \\ 2x + 6y \leq 300 \\ x \geq 0, y \geq 0 \end{cases}$

b.

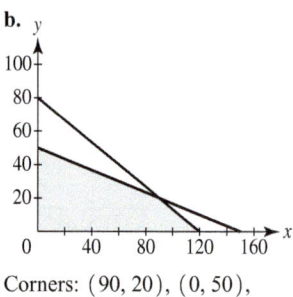

Corners: $(90, 20)$, $(0, 50)$, $(120, 0)$, $(0, 0)$

40. a. $\begin{cases} 2x + 3y \leq 36 \\ 1x + \frac{1}{2}y \leq 12 \\ x \geq 0, y \geq 0 \end{cases}$

b.

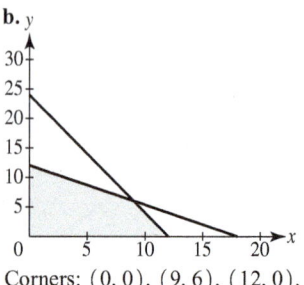

Corners: $(0, 0)$, $(9, 6)$, $(12, 0)$, $(0, 12)$

32. $\begin{cases} 200x + 500y \leq 5000 \\ 240{,}000x + 300{,}000y \leq 3{,}600{,}000 \\ x \geq 0, y \geq 0 \end{cases}$

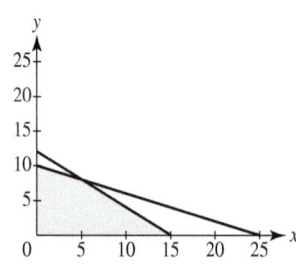

Corners: $(5, 8)$, $(0, 0)$, $(0, 10)$, $(15, 0)$

41. a. $\begin{cases} x + y \leq 1400 \\ x \geq 500 \\ y \geq 750 \\ x \geq 0, y \geq 0 \end{cases}$

b.

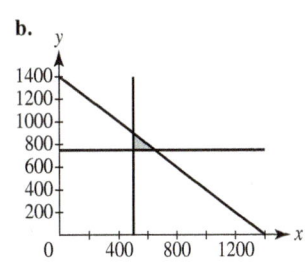

Corners: $(500, 750)$, $(500, 900)$, $(650, 750)$

8.2 Skills Check

4. Maximum: 45 at $(0, 5)$ and $(3, 4)$ or at any point on the line segment connecting $(0, 5)$ and $(3, 4)$; minimum: 0 at $(0, 0)$.

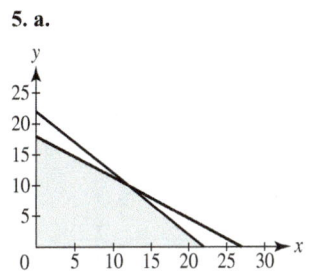

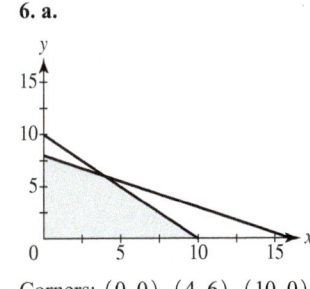

5. a.

Corners: $(0, 0)$, $(0, 18)$, $(22, 0)$, $(12, 10)$

6. a.

Corners: $(0, 0)$, $(4, 6)$, $(10, 0)$, $(0, 8)$

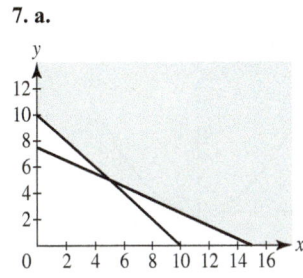

7. a.

Corners: $(0, 10)$, $(5, 5)$, $(15, 0)$

8.2 Exercises

27. 15 Van Buren and 0 Jefferson models, 5 Van Buren and 8 Jefferson models, or 10 Van Buren and 4 Jefferson models; $900,000

28. b. 90 standard and 20 deluxe, 120 standard and 0 deluxe, or other types corresponding to integer values of x and y on the line segment connecting $(90, 20)$ and $(120, 0)$; for example: $(93, 18)$, $(96, 16)$, $(99, 14)$, $(102, 12)$, ..., $(117, 2)$

8.4 Exercises

41. kth term: $R(1 + i)^{-n}(1 + i)^{k-1}$

$s_n = \dfrac{R(1+i)^{-n}(1 - (1+i)^n)}{1 - (1+i)}$

$= R\left[\dfrac{(1+i)^{-n} - (1+i)^0}{-i}\right]$

$= R\left[\dfrac{(1+i)^{-n} - 1}{-i}\right]$

$= R\left[\dfrac{1 - (1+i)^{-n}}{i}\right]$

8.5 Skills Check

13. $a^4 + 4a^3b + 6a^2b^2 + 4ab^3 + b^4$
14. $m^6 + 6m^5n + 15m^4n^2 + 20m^3n^3 + 15m^2n^4 + 6mn^5 + n^6$
15. $x^4 - 4x^3y + 6x^2y^2 - 4xy^3 + y^4$
16. $x^5 - 5x^4y + 10x^3y^2 - 10x^2y^3 + 5xy^4 - y^5$
17. $x^6 + 12x^5y + 60x^4y^2 + 160x^3y^3 + 240x^2y^4 + 192xy^5 + 64y^6$
18. $16x^4 + 32x^3y + 24x^2y^2 + 8xy^3 + y^4$
19. $243x^5 + 1620x^4y + 4320x^3y^2 + 5760x^2y^3 + 3840xy^4 + 1024y^5$
20. $625x^4 + 3000x^3y + 5400x^2y^2 + 4320xy^3 + 1296y^4$
21. $81a^4 + 108a^3b + 54a^2b^2 + 12ab^3 + b^4$
22. $32r^5 - 240r^4s + 720r^3s^2 - 1080r^2s^3 + 810rs^4 - 243s^5$
23. $x^7 + 7x^6y + 21x^5y^2 + 35x^4y^3 + 35x^3y^4 + 21x^2y^5 + 7xy^6 + y^7$
24. $x^7 - 7x^6y + 21x^5y^2 - 35x^4y^3 + 35x^3y^4 - 21x^2y^5 + 7xy^6 - y^7$

8.6 Skills Check

1. a. $(x + 1)^2 + (y - 2)^2 = 9$

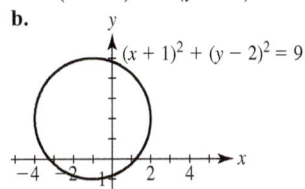

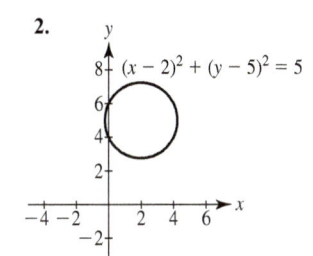

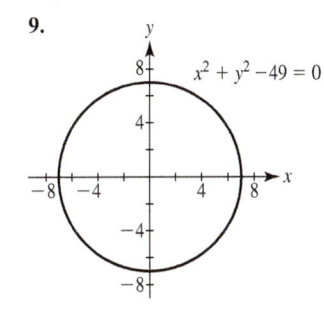

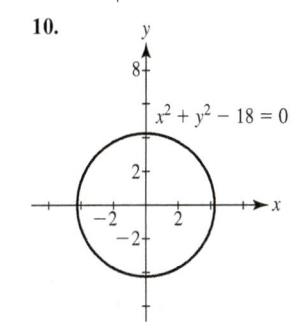

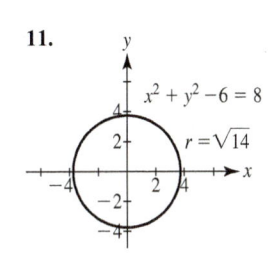

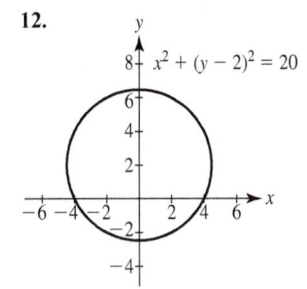

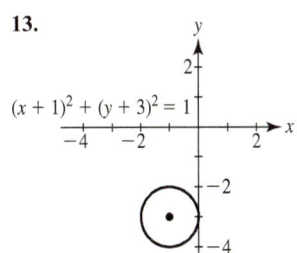

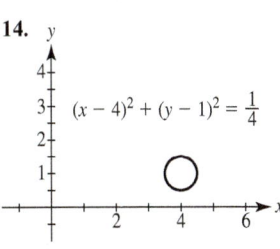

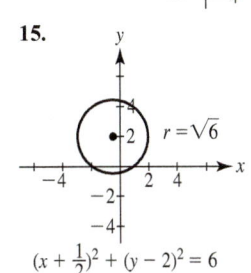

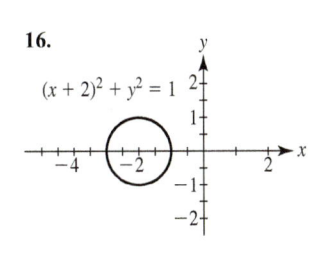

b. $(x + 1)^2 + (y - 2)^2 = 9$

2. $(x - 2)^2 + (y - 5)^2 = 5$

9. $x^2 + y^2 - 49 = 0$

10. $x^2 + y^2 - 18 = 0$

11. $x^2 + y^2 - 6 = 8$, $r = \sqrt{14}$

12. $x^2 + (y - 2)^2 = 20$

13. $(x + 1)^2 + (y + 3)^2 = 1$

14. $(x - 4)^2 + (y - 1)^2 = \dfrac{1}{4}$

15. $\left(x + \dfrac{1}{2}\right)^2 + (y - 2)^2 = 6$, $r = \sqrt{6}$

16. $(x + 2)^2 + y^2 = 1$

A-40 Additional Answers

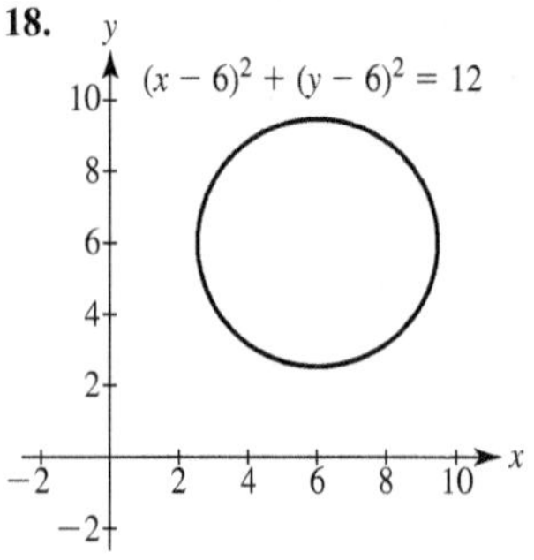

17. Graph of $x^2 + (y+4)^2 = 16$, center $(0, -4)$

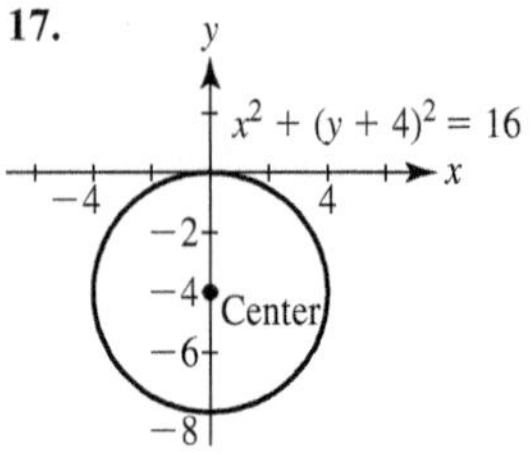

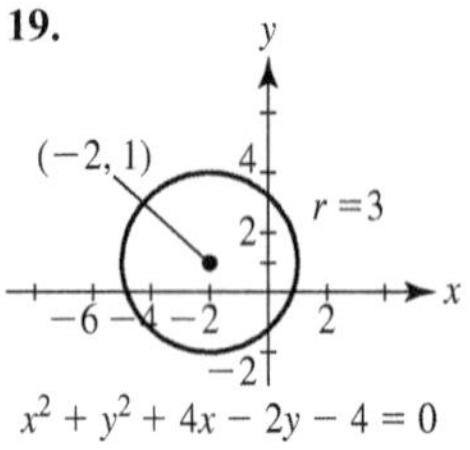

18. Graph of $(x-6)^2 + (y-6)^2 = 12$

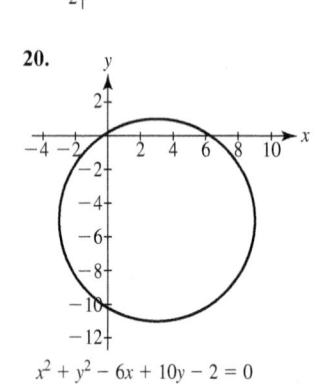

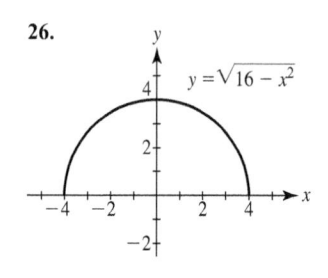

27. $x^2 - 8x + y^2 = 0$
28. $x^2 + 4x + y^2 - 6y - 12 = 0$
29. $x^2 + 6x + y^2 + 2y + 6 = 0$
30. $x^2 + y^2 + 6y - 160 = 0$
31. $4x^2 - 40x + 4y^2 + 16y + 115 = 0$
32. $x^2 - 2x + y^2 + 2y - 1 = 0$
37. $\left(x - \dfrac{5}{2}\right)^2 + \left(y - \dfrac{5}{2}\right)^2 = \dfrac{58}{4}$

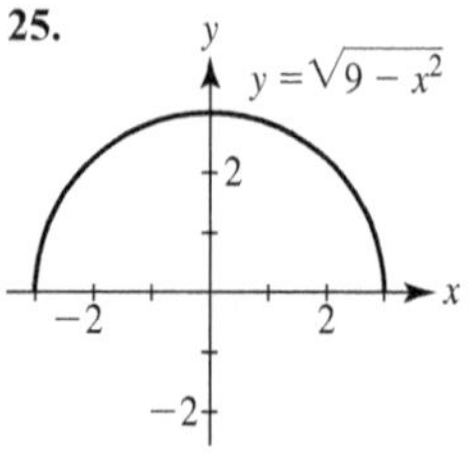

19. Graph with center $(-2, 1)$, $r = 3$; $x^2 + y^2 + 4x - 2y - 4 = 0$

20. Graph of $x^2 + y^2 - 6x + 10y - 2 = 0$

41. Graph of $y^2 = 12x$

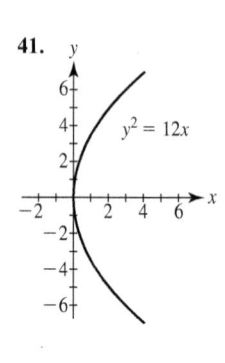

42. Graph of $y^2 - 8x = 0$

21. Graph with Center $\left(\dfrac{1}{2}, 2\right)$, $r = \dfrac{1}{2}$; $x^2 + y^2 - x - 4y + 4 = 0$

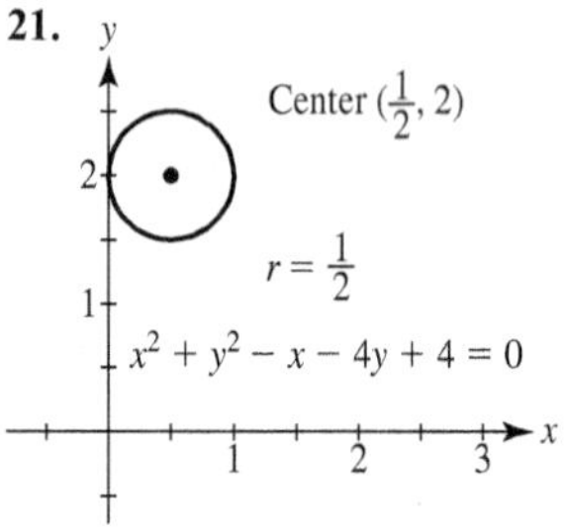

22. Graph of $4x^2 + 4y^2 + 12x - 8y = 3$

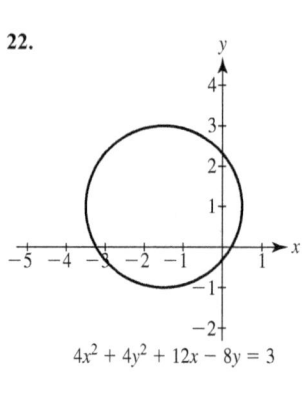

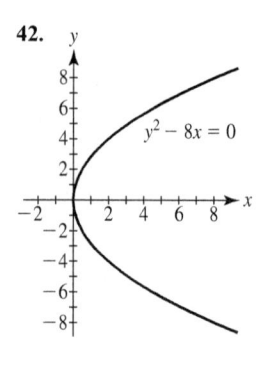

43. Graph of $y^2 + 4x = 0$

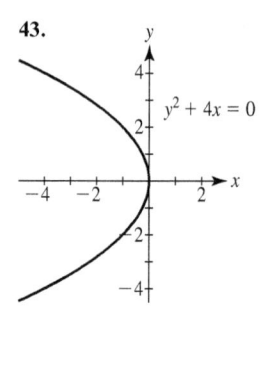

44. Graph of $y^2 + 8x = 0$

23. Graph with Center $\left(0, \dfrac{1}{2}\right)$, $r = \dfrac{1}{2}$; $x^2 + y^2 - y = 0$

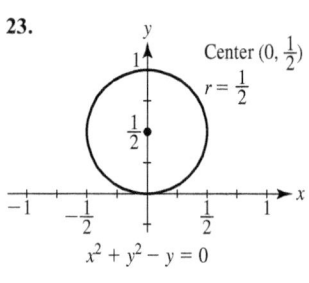

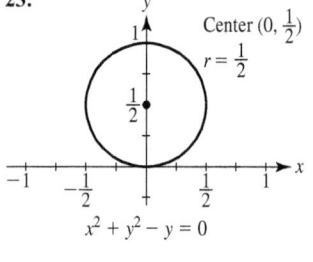

24. Graph of $x^2 - x + y^2 - 2y = 0$

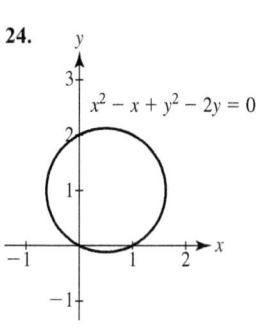

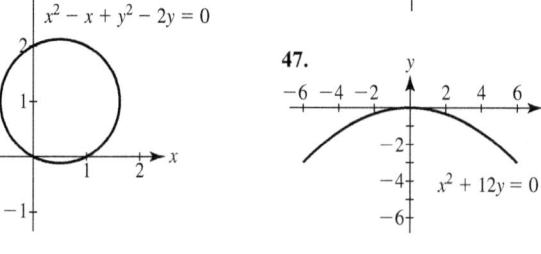

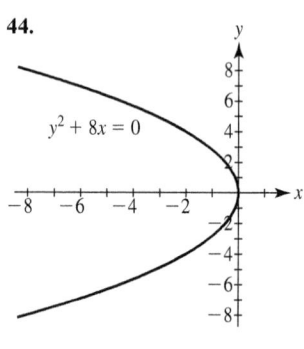

45. Graph of $x^2 - 4y = 0$

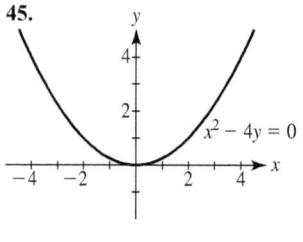

46. Graph of $x^2 - 8y = 0$

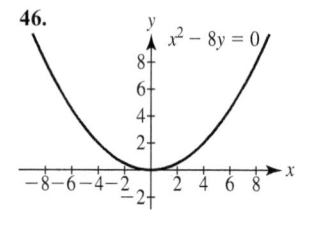

47. Graph of $x^2 + 12y = 0$

48. Graph of $x^2 + 4y = 0$

25. Graph of $y = \sqrt{9 - x^2}$

26. Graph of $y = \sqrt{16 - x^2}$

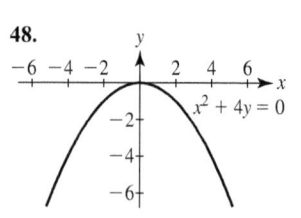

49. Graph of $y = \sqrt{x - 3}$

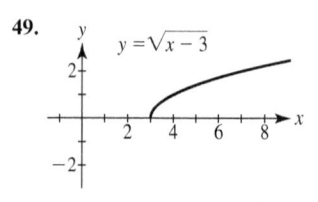

50. Graph of $y = \sqrt{2x - 2}$

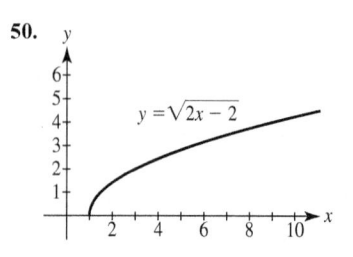

51. $y = \sqrt{3-x}$

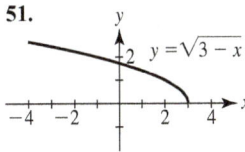

52. $y = \sqrt{2-3x}$

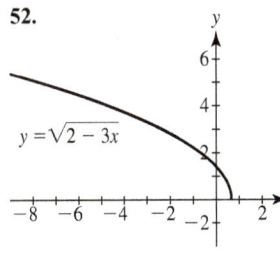

53. $y^2 - 6y + 4x + 5 = 0$, Vertex $(1, 3)$

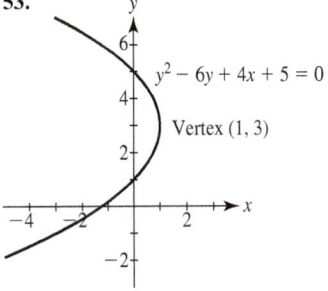

54. $x^2 + 10x + 8y + 49 = 0$

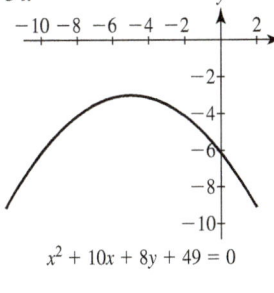

55. Vertex $\left(-\frac{3}{2}, -\frac{1}{2}\right)$, $y = -\frac{4}{32}x^2 - \frac{12}{32}x - \frac{25}{32}$

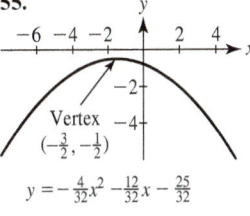

56. $4y^2 - 4y + 24x + 37 = 0$

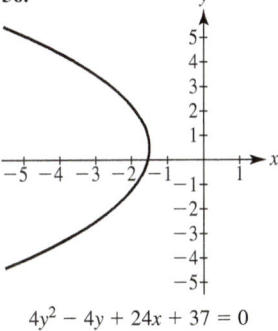

57. $y = x^2 + 10x + 25$, Vertex $(-5, 0)$

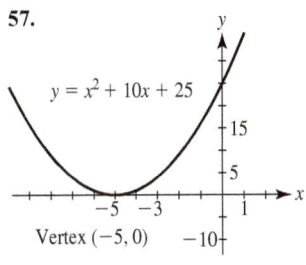

58. $y = x^2 + x$

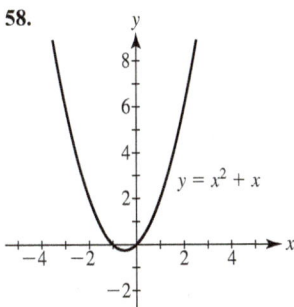

59. $y = 3 + x - 2x^2$, Vertex $\left(\frac{1}{4}, 3\frac{1}{8}\right)$

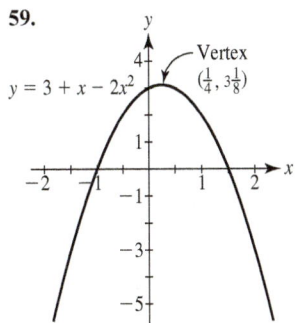

60. $y = 9 - x^2$

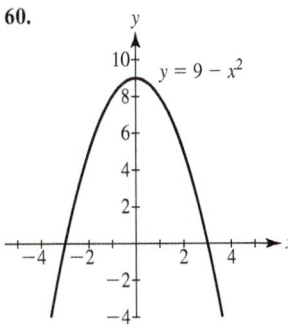

8.7 Skills Check

1. $(0, 0)$; $(3, 0)$, $(-3, 0)$; $\frac{x^2}{9} + \frac{y^2}{4} = 1$

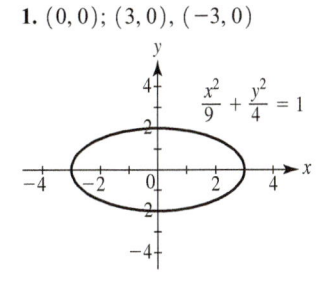

2. $(0, 0)$; $(-3, 0)$, $(3, 0)$; $\frac{x^2}{9} + \frac{y^2}{5} = 1$

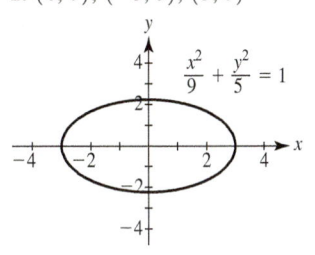

3. $(0, 0)$; $(0, -4)$, $(0, 4)$; $4x^2 + y^2 = 16$

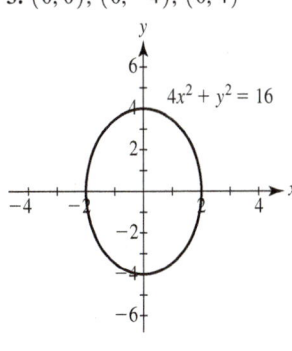

4. $(0, 0)$; $(0, -7)$, $(0, 7)$; $6x^2 + 49y^2 - 294 = 0$

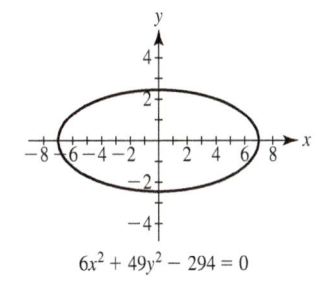

5. $(0, 0)$; $(1, 0)$, $(-1, 0)$; $x^2 + 4y^2 = 1$

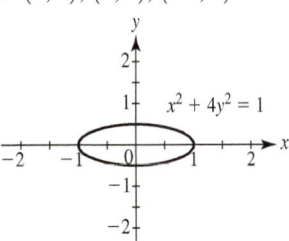

6. $(0, 0)$; $(-\sqrt{3}, 0)$, $(\sqrt{3}, 0)$; $2x^2 + 3y^2 = 6$

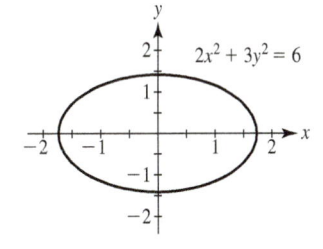

7. $(-2, 1)$; $(-2, 5)$, $(-2, -3)$; $\frac{(x+2)^2}{4} + \frac{(y-1)^2}{16} = 1$

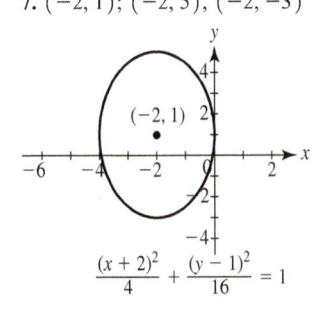

8. $(1, 0)$; $(-2, 0)$, $(4, 0)$; $\frac{(x-1)^2}{9} + y^2 = 1$

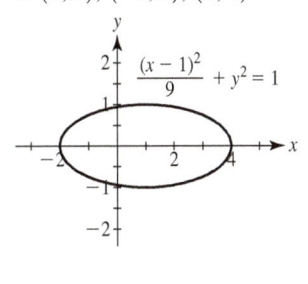

9. $(-2, 1)$; $(0, 1)$, $(-4, 1)$; $\frac{(x+2)^2}{4} + (y-1)^2 = 1$

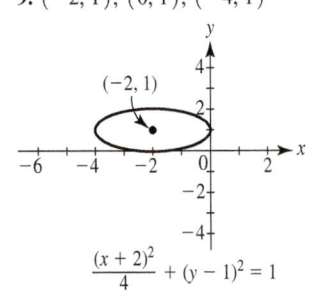

10. $(1, -3)$; $(1, -1)$, $(1, -5)$; $2x^2 + y^2 - 4x + 6y + 7 = 0$

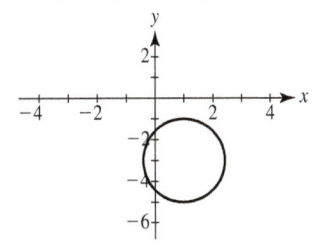

11. $(0, -3)$; $(0, 0)$, $(0, -6)$

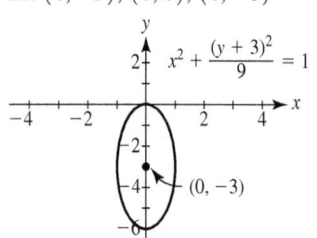

$x^2 + \dfrac{(y+3)^2}{9} = 1$

12. $(3, 0)$; $(0, 0)$, $(6, 0)$

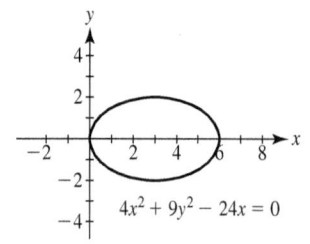

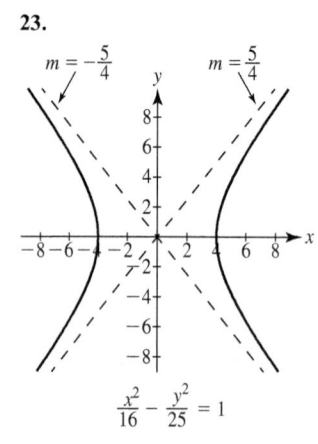

$4x^2 + 9y^2 - 24x = 0$

23. $\dfrac{x^2}{16} - \dfrac{y^2}{25} = 1$; $m = -\dfrac{5}{4}$, $m = \dfrac{5}{4}$

24. $(0, 0)$; $\pm 3/5$

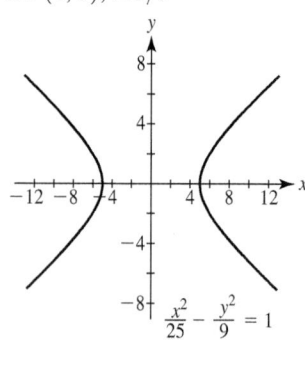

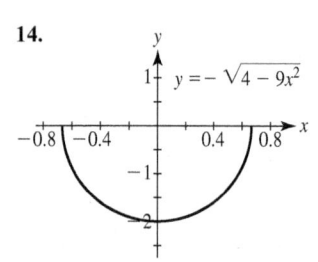

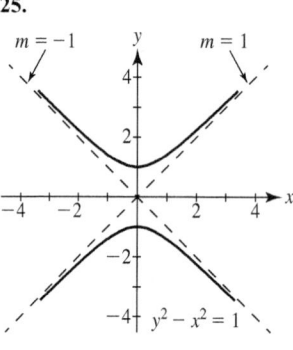

$\dfrac{x^2}{25} - \dfrac{y^2}{9} = 1$

13. $y = \dfrac{\sqrt{1-4x^2}}{3}$

14. $y = -\sqrt{4 - 9x^2}$

25. $y^2 - x^2 = 1$; $m = -1$, $m = 1$

26. $(0, 0)$; $\pm 1/\sqrt{5}$

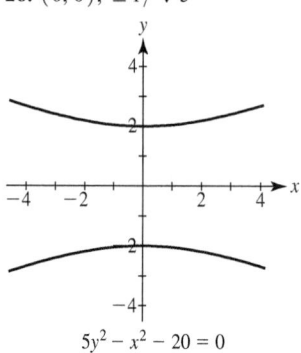

$5y^2 - x^2 - 20 = 0$

15. $x = \dfrac{-2\sqrt{225 - y^2}}{3}$

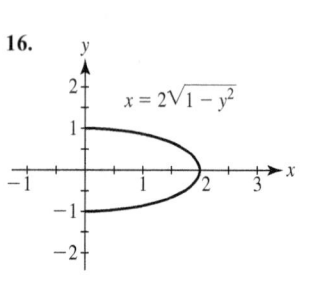

16. $x = 2\sqrt{1 - y^2}$

27. $y^2 - 4x^2 = 4$; $m = -2$, $m = 2$

28. $(0, 0)$; ± 2

$y^2 - 4x^2 - 6 = 16$

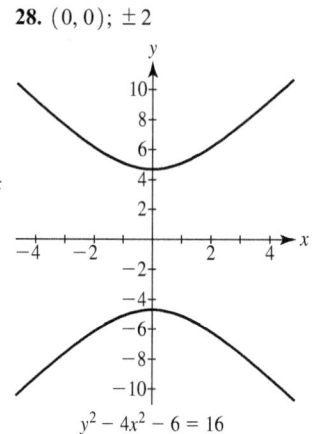

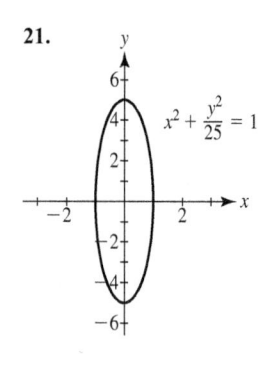

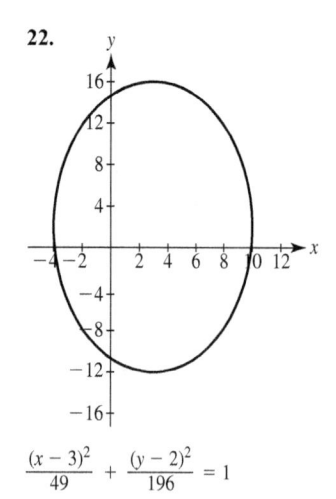

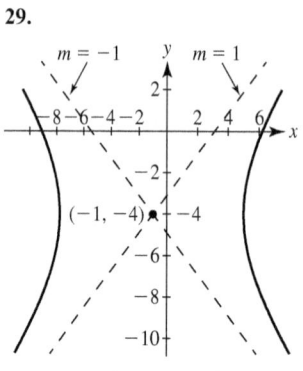

21. $x^2 + \dfrac{y^2}{25} = 1$

22. $\dfrac{(x-3)^2}{49} + \dfrac{(y-2)^2}{196} = 1$

29. $\dfrac{(x+1)^2}{36} - \dfrac{(y+4)^2}{36} = 1$; $(-1, -4)$; $m = -1$, $m = 1$

30. $(0.5, -3)$; $\pm 2/3$

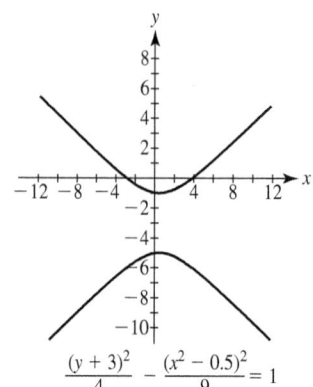

$\dfrac{(y+3)^2}{4} - \dfrac{(x - 0.5)^2}{9} = 1$

Additional Answers   **A-43**

31.
$\frac{x^2}{4} - (y+3)^2 = 1$
$m = \frac{1}{2}$, $m = -\frac{1}{2}$

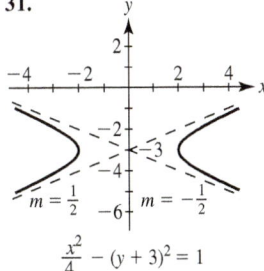

32. $(-9/2, 0)$; ± 1
$4x^2 - 4y^2 + 36x + 77 = 0$

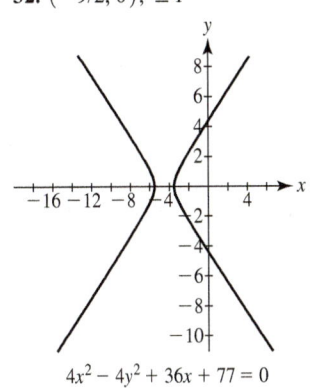

33. $m = -\frac{\sqrt{5}}{2}$, $m = \frac{\sqrt{5}}{2}$
$(-2, -2)$
$\frac{(y+2)^2}{5} - \frac{(x+2)^2}{4} = 1$

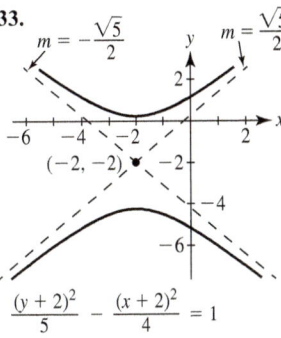

34. $(4, 7)$; $\pm 1/2$
$x^2 - 4y^2 - 8x + 56y - 164 = 0$

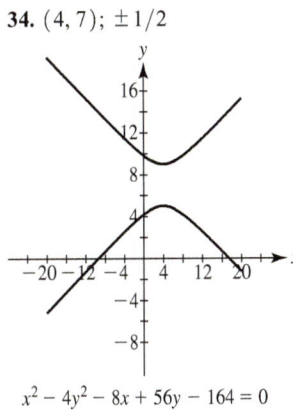

35. The lower branch of the graph of the equation in Exercise 25.

36. $y = \sqrt{4 + 9x^2}$

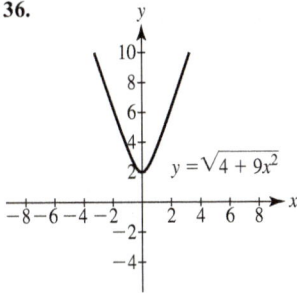

37. $m = -3$, $m = 3$
$x = \frac{\sqrt{9+y^2}}{3}$

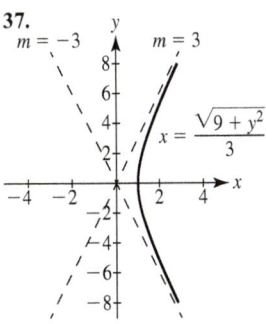

38. $x = \frac{-2\sqrt{25+y^2}}{5}$

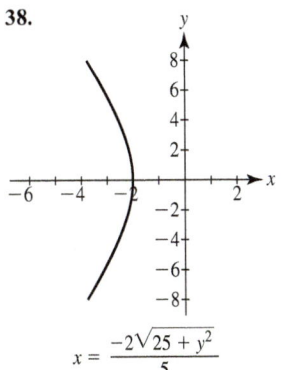

39. $xy = 16$, $x = 0$, $y = 0$

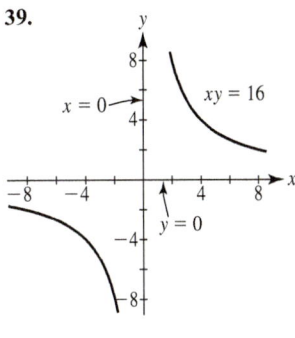

40. $xy = 2$

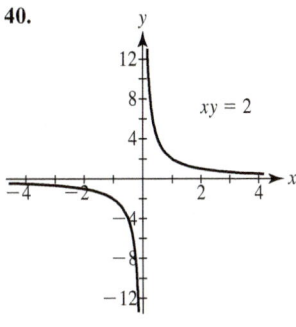

41. $xy = -\frac{1}{2}$, $x = 0$, $y = 0$

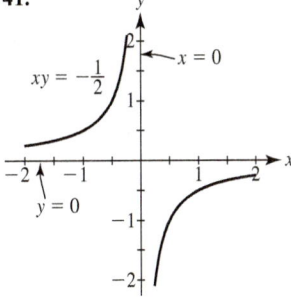

42. $xy = -9$

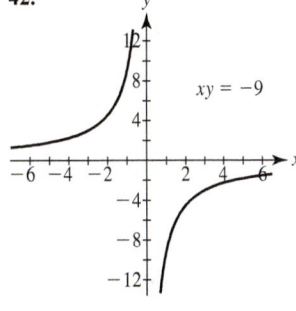

43. $(x-3)y = 4$, $(3, 0)$, $y = 0$, $x = 3$

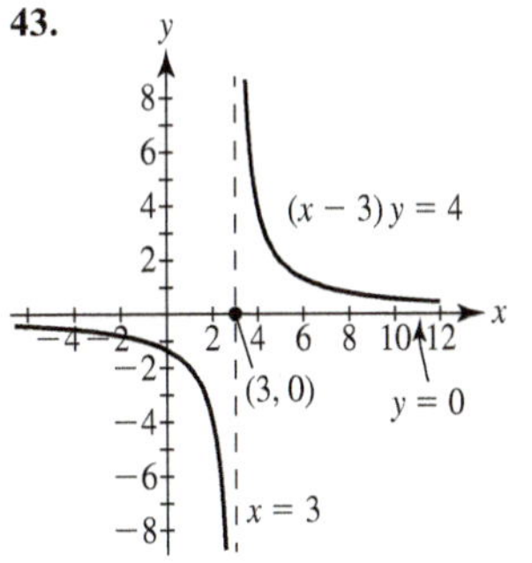

44. $(x+1)(y-1) = 9$

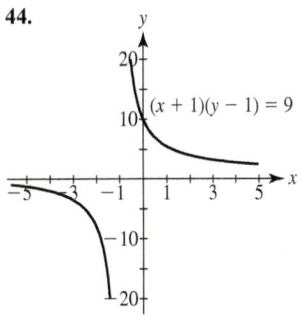

45. $x = -\sqrt{1+y^2}$

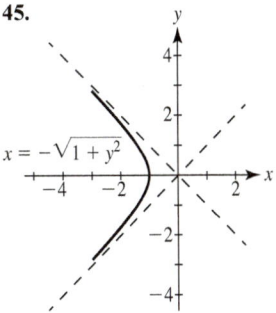

46. $y^2 - 9x^2 = 36$

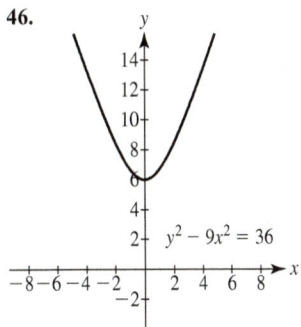

8.7 Exercises

48. $y = \dfrac{\sqrt{90{,}000 - 144x^2}}{25}$

49. $\dfrac{x^2}{327.25} + \dfrac{y^2}{20.79} = 1$

50. $\dfrac{x^2}{96{,}100} + \dfrac{y^2}{65{,}792.25} = 1$

51.

IR = 120
(I > 0, R > 0)

52.

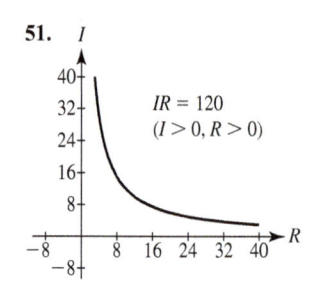

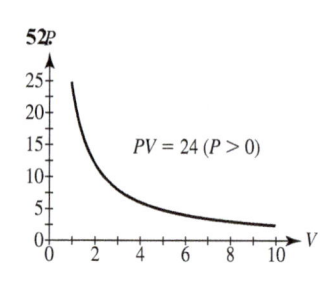

PV = 24 (P > 0)

7. $\begin{cases} 15x - x^2 - y \geq 0 \\ y - \dfrac{44x + 60}{x} \geq 0 \\ x \geq 0, y \geq 0 \end{cases}$

(6, 54), (10, 50)

8. $\begin{cases} x^3 - 26x + 100 - y \geq 0 \\ 19x^2 - 20 - y \leq 0 \\ x \geq 0, y \geq 0 \end{cases}$

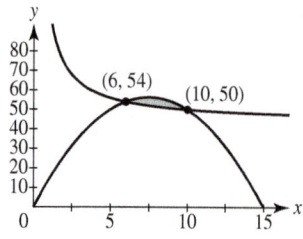

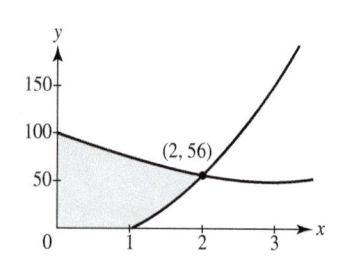

(2, 56)

Chapter 8 Skills Check

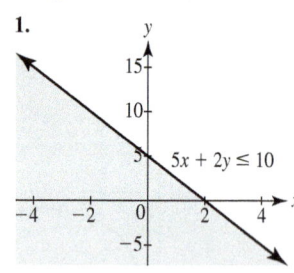

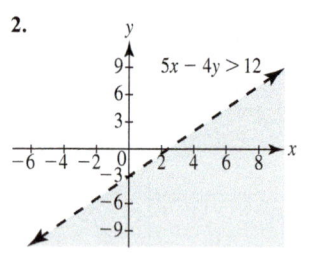

1. $5x + 2y \leq 10$

2. $5x - 4y > 12$

29. $64a^6 - 576a^5b + 2160a^4b^2 - 4320a^3b^3 + 4860a^2b^4 - 2916ab^5 + 729b^6$

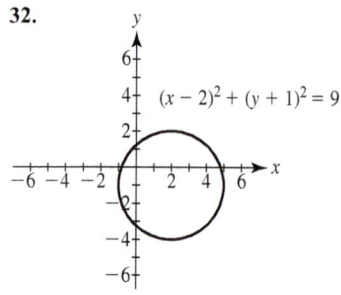

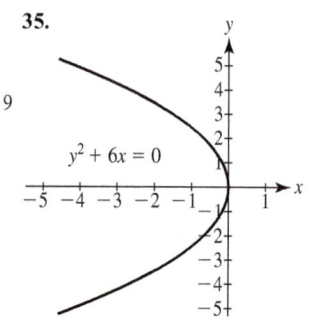

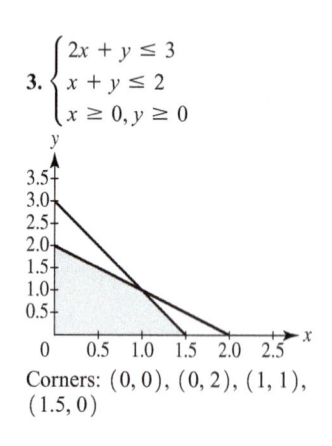

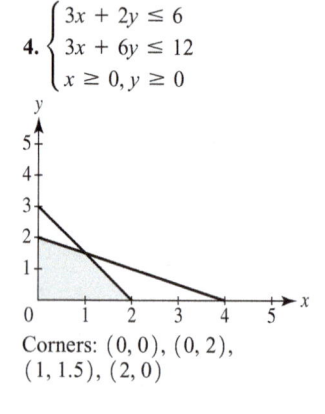

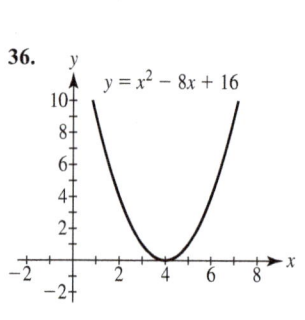

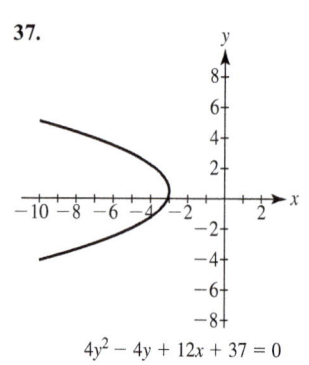

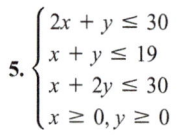

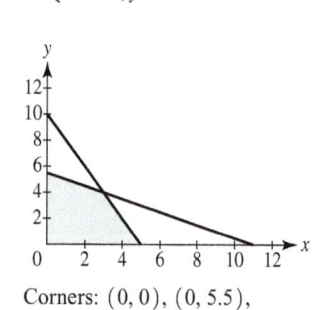

32. $(x - 2)^2 + (y + 1)^2 = 9$

35. $y^2 + 6x = 0$

3. $\begin{cases} 2x + y \leq 3 \\ x + y \leq 2 \\ x \geq 0, y \geq 0 \end{cases}$

4. $\begin{cases} 3x + 2y \leq 6 \\ 3x + 6y \leq 12 \\ x \geq 0, y \geq 0 \end{cases}$

Corners: $(0, 0), (0, 2), (1, 1), (1.5, 0)$

Corners: $(0, 0), (0, 2), (1, 1.5), (2, 0)$

36. $y = x^2 - 8x + 16$

37. $4y^2 - 4y + 12x + 37 = 0$

5. $\begin{cases} 2x + y \leq 30 \\ x + y \leq 19 \\ x + 2y \leq 30 \\ x \geq 0, y \geq 0 \end{cases}$

6. $\begin{cases} 2x + y \leq 10 \\ x + 2y \leq 11 \\ x \geq 0, y \geq 0 \end{cases}$

38. Center at $(0, 0)$, endpoints of major axis $(0, 3)$ and $(0, -3)$

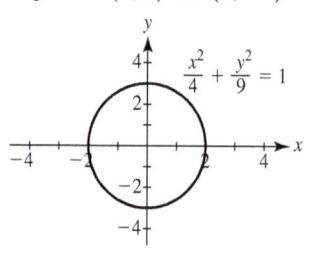

$\dfrac{x^2}{4} + \dfrac{y^2}{9} = 1$

39. Center at $(2, -1)$, endpoints of major axis $(-2, -1)$ and $(6, -1)$

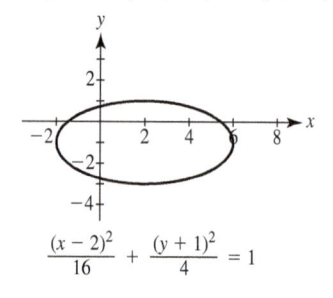

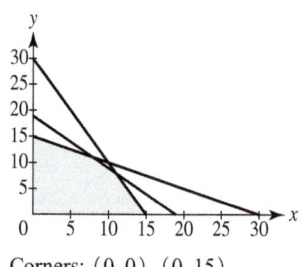

$\dfrac{(x - 2)^2}{16} + \dfrac{(y + 1)^2}{4} = 1$

Corners: $(0, 0), (0, 15), (15, 0), (11, 8), (8, 11)$

Corners: $(0, 0), (0, 5.5), (3, 4), (5, 0)$

40. $y = -\sqrt{9 - 4x^2}$

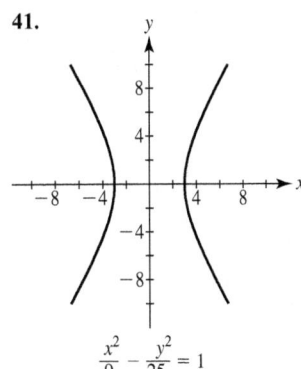

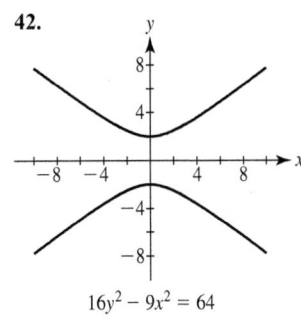

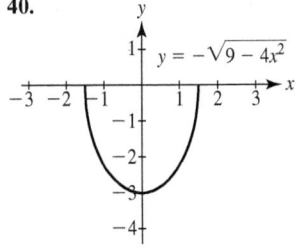

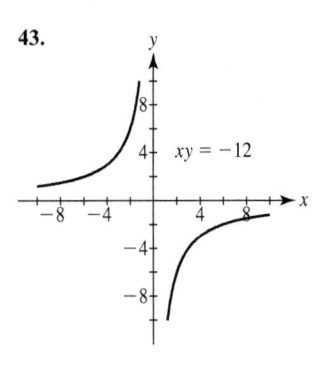

41. $\dfrac{x^2}{9} - \dfrac{y^2}{25} = 1$

42. $16y^2 - 9x^2 = 64$

43. $xy = -12$

Index

Absolute maximum point, 453
Absolute minimum point, 453
Absolute value(s), 163
   inequalities involving, 313
Absolute-value equations, 220–221
Absolute-value functions, 215, 220
Addition. *See also* Sums
   Associative Property of, 5
   Commutative Property of, 5
   of complex numbers, 448
   of fractions, 89
   of functions, 279
   of matrices, 561–562
      with Excel, 714–715
      on graphing calculators, 696
   of polynomials, 166
   of radicals, 328
   of rational expressions, 444–445
   of real numbers, 4–5
Addition Property
   of equations, 8, 90
   of inequalities, 93
Additive identity, 5
Additive inverse, 5
Algebra, Fundamental Theorem of, 497–498
Algebraic expressions, 7
   evaluating, 7
   simplifying, 8
   terms of, 7
Aligned inputs, 21–22
Amortization, 412–414
   on graphing calculators, 695
   using Excel, 713
Amortization formula, 412–413
Annual compounding, 394, 396, 400–401
Annual interest rate, 396, 412, 414
Annual percentage rate (APR), 414
Annuities, 406–411
   future value of, 406–409
      on graphing calculators, 694
      with spreadsheet, 408, 712
   geometric sequences and, 649
   present value of, 409–411
      on graphing calculators, 695
      with spreadsheet, 411, 713
Approximately linear data, 71–72
Arguments, of logarithms, 347
Arithmetic sequences, 638–640
   common difference and, 639
   definition of, 639
   nth term of, 639
Arithmetic series, 646–648
   sum of n terms of, 647–648
Associative Property of Addition, 5
Associative Property of Multiplication, 5

Asymptotes, 208–209, 333
   horizontal
      of an exponential function, 233
      of a logistic function, 417, 419
      of a power function, 208, 209
      of a rational function, 503–504
   of a rational function, 502, 503–504
   vertical
      of a power function, 208
      of a rational function, 502
Augment, of a matrix, 548
Augmented matrices, 548
Average cost, 281–282, 501, 504–505, 517–518, 520
Average rate of change, 68–69
Axes
   of an ellipse, 666
   coordinate, 9
      reflections of graphs across, 267
   of a parabola, 662
   of symmetry, of a parabola, 173, 174
   x-axis, 9
      reflection of graphs across, 335
      symmetry of graphs with respect to, 271
   y-axis, 9
      reflection of graphs across, 335
      symmetry of graphs with respect to, 270

Babbage, Charles, 589
Back substitution, 538
Best-fit line, 113
   finding equation of, using Excel, 708–709
Binomial(s), 165–169
Binomial coefficients, 653–656
Binomial expansions, finding a particular term in, 656
Binomial Theorem, 653–657
   binomial coefficients and, 653–656
   using, 656
Borders, of inequality region, 617
Brahe, Tyco, 658
Break-even analysis, 484, 493–494

Calculators. *See* Graphing calculators; Technology
Cartesian coordinate system, 9–10
Cells, in Excel, 701–702
Change-of-base formula, 365–367
Circles, 659–662
   standard form of equation of, 659
      converting from general form to, 660–662
Closed intervals, 6
Coefficient(s)
   binomial, 653–656

correlation, 121
   on graphing calculators, 690
   leading, of a polynomial, 165, 441
   numerical, 7
   rounding of, 94–95
Coefficient matrices, 548
Column matrices, product of a row matrix and, 566
Combinations of functions, 285–286
   on graphing calculators, 691–692
Combined variation, 510–511
Common denominators, subtracting fractions with, 89
Common difference, arithmetic sequences and, 639
Common logarithms, 349–352
Common ratio, geometric sequences and, 640
Commutative Property
   of addition, 5
   of multiplication, 5
Complete graphs, 32–33, 34
Completing the square, 193–194
Complex coordinate system, 170
Complex fractions, simplification of, 445–447
Complex number(s), 169–171
   addition of, 448
   definition of, 170
   division of, 449–450
   on graphing calculators, 691
   multiplication of, 449
   operations with, 448–450
   simplifying, 170–171
   in standard form, 170
   subtraction of, 449
Complex number system, solutions to polynomial equations in, 497–498
Complex zeros, 497
Composite functions, 283
Composition of functions, 282–286
   on graphing calculators, 692–693
Compounding, 394–406
   annual, 394, 396, 400–401
   continuous, 398–401
   future value and, 394, 395–398, 400
   periodic, 396–397
   present value and, 394, 401–402
Compressing graphs, 266–267, 268, 335, 336–337
Concave down parabolas, 174
Concave up parabolas, 174
Conditional equations, 90
Conic sections, 657–665. *See also* Circles; Ellipses; Hyperbolas; Parabolas
   degenerate, 657

I-1

distance formula and, 658–659
midpoint formula and, 659
Constant(s)
  literal, 7
  of proportionality (variation), 105, 534
  rounding of, 94–95
Constant functions, 56–57
Constant percent changes
  geometric sequences and, 641
  of a set of data, 380
Constant rate of change, 54
Constant terms, of an algebraic
    expression, 7
Constraint inequalities, 617
  systems of inequalities and, 614
Continuous compounding, 398–401
  annual compounding vs., 400–401
  future value and, 400
  number e and, 398–401
Continuous functions, 111
Contradictions, 90
Contrast, on graphing calculator display,
    adjusting, 681
Coordinate axes, 9
  reflections of graphs across, 267
Corners, of solution region, 617, 618
Correlation coefficient, 121
  on graphing calculators, 690
Cost(s), 279–280
  average, 281–282, 501, 504–505,
    517–518, 520
  marginal, 56
  total, 55
  variable, 55
Cost-benefit analysis, 502–503
Cramer's Rule
  with dependent systems, 596
  with inconsistent systems, 596
  solving systems of linear equations
    in three variables using, 593–596
  solving systems of linear equations in
    two variables using, 589–591
Cubic equations
  Rational Solutions Test and, 495
  solving with synthetic division, 491–492
Cubic functions, 452–455
  graphs of, 453–454
  modeling with, 464–466, 468–470
    with spreadsheet, 465
Cubing function, 203–204

Data, definition of, 12
Data points
  fitting lines to. *See* Linear regression
  graphing with graphing calculators, 39–40
Decay functions, logistic, 419–421
Decay models, exponential, 339–340, 383
  finding rates of change in, 380–381
Decoding messages, 580–581
Decreasing functions, 174
Degenerate conics, 657

Degrees
  of a polynomial, 165, 441
  of a term of a polynomial, 165
Demand, 130
Denominators
  common, subtracting fractions
    with, 89
  least common, 444
  rationalization of, 328–329
  unlike, adding and subtracting fractions
    with, 89
Dependent systems, 135–137, 542
  Cramer's Rule with, 596
  of linear equations in three variables,
    553–555
Dependent variable, 14
Depreciation, straight-line, 62
Determinants
  minors of, 592
  of a 3×3 matrix, 591–593
  of a 2×2 matrix, 589
Difference(s). *See also* Subtraction
  second, 230
  third and fourth, 470–471
  of two squares, factoring, 168
Difference quotient, 69–71
Dimension, of a matrix, 548
Direct variation (proportionality), 105, 212
Directrix, of a parabola, 662
Discontinuous functions, graphing using
    Excel, 706
Discrete functions, 111
  sequences as, 637
Discriminant, of a quadratic equation, 196
Disjoint sets, 3
Display contrast, on graphing calculators,
    adjusting, 681
Distance formula, 658–659
Distributive Property of Multiplication
    over Addition, 5, 8
  with polynomials, 166
Division
  of complex numbers, 449–450
  of fractions, 89
  of functions, 279
  of polynomials, 167, 447–448
    long, 447–448
    by synthetic division, 490–492
  of radicals, 328
  of rational expressions, 443–444
  of real numbers, 5
  synthetic, 490–492
  by zero, 5
Division Property of Equations, 9, 90
Domain of a function, 13–15
  limited, inverse functions on, 297–298
Double inequalities, 6, 147–148

e (irrational number), 340–342
Echelon forms
  of matrices, 548–550

  of three-variable systems of
    equations, 538
Elimination method
  Gauss-Jordan, 550–551
  left-to-right, 538–540
  for systems of linear equations, 133
Ellipses, 665–668
  equations of
    converting to standard form, 667–668
    graphing, 666–667
    standard forms of, 666
  foci of, 665
  horizontal, 666
  horizontal and vertical shifts of, 667
  major axis of, 666
  minor axis of, 666
  vertical, 666
Encoding/decoding messages, 578–581
Equal sets, 2
Equations
  absolute-value, 220–221
  Addition Property of, 8, 90
  of a circle
    general form of, 660–662
    standard form of, 659, 660–662
  conditional, 90
  containing rational powers, 305
  cubic
    Rational Solutions Test and, 495
    solving with synthetic division,
      491–492
  Division Property of, 9, 90
  of ellipses
    general form of, 557
    standard form of, 666
  equivalent, 8, 90
  exponential. *See* Exponential equations
  graphing with graphing calculators,
    683–684
  of hyperbolas
    converting to standard form, 671
    graphing, 670
    standard forms of, 669
  linear. *See* Linear equations
  literal, 103–104
  matrix, 581–584
    solution with inverses, 582–583
    technology and, 583–584
  Multiplication Property of, 9, 90
  nonlinear, systems of. *See* Systems of
    nonlinear equations
  polynomial. *See* Polynomial equations
  properties of, 90–92
  quadratic, solving using Excel, 710
  in quadratic form, 306–307
  quartic, Rational Solutions Test and,
    495–497
  radical, 303–305
  rational
    algebraic solution of, 507–509
    graphical solution of, 509

Equations (continued)
  regression, on graphing calculators, 690–691
  second-degree, graphs of, 658
  Substitution Property of, 90
  Subtraction Property of, 9, 90
Equilibrium price, 130
Equivalent equations, 8, 90
Equivalent inequalities, 142
Even functions, 270
Excel, 701–716. *See also* Spreadsheets; Technology
  addresses and, 701
  cells in, 701–702
  changing graphing windows in, 705
  Fill Down function in, 703–704
  finding maximum or minimum values with Solver using, 706–707
  finding the equation of the best fit line using, 708–709
  future value using, 711–712
  graphing a function of a single variable using, 704–705
  graphing discontinuous functions using, 706
  graphing multiple functions using, 705
  Help in, 701
  loan amortization using, 713
  matrices using, 714–716
  operations using, 701
  present value using, 712, 713
  scatter plots in, 707–708
  solving a system of two linear equations in two variables using, 713–714
  solving linear equations using the x-intercept method with, 709
  solving quadratic equations using the intercept method with, 711
  solving quadratic equations using the x-intercept method with, 710
  tables of numbers in, creating, 702–704
    using Fill Down, 703–704
  worksheet in, 701–702
Exponent(s)
  integer, 162–163
  logarithms as, 346
  negative, 162–163
    power functions with, 208–209
  properties of, 162, 326–327
  rational, 164–165
  real number, 327
  zero, 162–163
Exponential equations
  change-of-base formula and, 365–367
  solving by converting to exponential form, 369–371
  solving using logarithmic forms, 363–365
  solving using logarithmic properties, 367–369

Exponential forms, 346–347
  converting from logarithmic to, 347
  converting to logarithmic form from, 346–347
  solving exponential equations by converting to, 369–371
Exponential functions, 332–345
  definition of, 333
  graphs of, 334–337
    transformations of, 335–337
  initial value of, 382
  investing and, 394–406
  modeling with, 377–380
  number e and, 340–342
Exponential inequalities, 371–372
Exponential models, 382–383
  decay, 339–340, 383
    finding rates of change in, 380–381
  definition of, 382
  fitting to data using technology, 384
  growth, 337–339, 383
    finding rates of change in, 380–381
  linear regression for creating, 387–388
Extrapolation, 120

Factor(s), of polynomials, 167
Factorial(s), evaluating, 654–655
Factorial notation, 654
Factoring
  by grouping, 482–483
  of polynomials, 167–169
    completely, 168–169
    higher-degree, 441–442
  of quadratic functions, 186–202
    combining graphing with, 189–190
  solving polynomial equations by, 479–482
  of special products, 168
Feasible regions, linear programming and, 625–627
Feasible solutions, linear programming and, 625
Finite sequences, 637
First differences, constant, 110
Focus(i)
  of ellipses, 666
  of a parabola, 662
FOIL method, 166
Fourth differences, 470–471
Fractal images, 170
Fractions
  addition of, 89
  complex, simplification of, 445–447
  division of, 89
  least common denominator of, 444
  multiplication of, 89
  subtraction of, 89
Function(s), 12–20, 255–324
  absolute-value, 215, 220
  combinations of, 285–286
    on graphing calculators, 691–692

  composite, 283
  composition of, 282–286
    on graphing calculators, 692–693
  constant, 56–57
  cubic, 452–455
    graphs of, 453–454
    modeling with, 464–466, 468–470
  cubing, 203–204
  decay, logistic, 419–421
  decreasing, 174
  definition of, 13
  definitions related to, 12–13
  discontinuous, graphing using Excel, 706
  discrete, sequences as, 637
  domain of, 13–15
    limited, inverse functions on, 297–298
  even, 270
  exponential. *See* Exponential functions
  finding values using graphing calculators, 684–685
  Gompertz, 422–423
  graphs of, 31–44. *See also* Graphs/graphing
  increasing, 174
  inverse. *See* Inverse functions
  linear. *See* Linear functions
  logarithmic. *See* Logarithmic functions
  logistic, 417–421
  odd, 270–271
  one-to-one, 292–297
    horizontal line test and, 293
  operations with, 279–282
  piecewise-defined. *See* Piecewise-defined functions
  polynomial. *See* Higher-degree polynomial functions; Polynomial functions
  power. *See* Power functions
  quadratic (second-degree). *See* Parabolas; Quadratic functions
  quartic, 455
    modeling with, 467–470
  range of, 13–15
  rational. *See* Rational functions
  relationships among x-intercepts, zeros, and solutions of, 100
  root, 209–212
  special, 258–259
  squaring, 203–204
  tests for, 15–17
  zeros of, 100
Function notation, 18–20
Fundamental Principle of Rational Expressions, 442
Fundamental Theorem of Algebra, 497–498
Future value, 394, 395–398, 400
  of an annuity, 406–409

on graphing calculators, 694
  spreadsheet for finding, 408, 712
 of an investment
  with annual compounding, 396, 484–485
  with continuous compounding, 400
  on graphing calculators, 693–694
  with periodic compounding, 396–397
  with spreadsheet, 398, 711–712

Gauss-Jordan elimination, 550–551
General form
 of the equation of a circle, converting to standard form, 660–662
 of the equation of an ellipse, 557
 of linear equations, 67–68
 of a polynomial, 165
Geometric sequences, 640–642
 common ratio and, 640
 constant percent change and, 641
 definition of, 640
 nth term of, 641
Geometric series, 648–651
 infinite, 649–651
  sum of, 650
 sum of n terms of, 648–649
Gompertz functions, 422–423
Goodness of fit, 120–121
Graphing calculators, 681–700. *See also* Technology
 binomial coefficients using, 656
 checking reasonableness of a solution using, 100
 combinations of functions using, 691–692
 complex numbers on, 691
 composition of functions using, 692–693
 display contrast adjustment for, 681
 evaluating determinants using, 593
 evaluating factorials using, 655
 function values on, 684–685
 future value using, 397–398, 693–695
 goodness of fit using, 120–121
 GRAPH key on, 685
 graphing with, 33–36
  of data points, 39–40
  of equations, 683–684
  of inequalities, 615
  intercepts and, 48, 685–686
  of polynomial functions, 454, 455
  of quadratic functions, 178
  of solutions to a system of inequalities, 618
 greatest integer function using, 219–220
 intersection method using, 687–688
 inverse functions on, 693
 keystrokes on, 682
 linear programming using, 698–699

linear regression using, 116–119
loan amortization using, 414, 695
matrices on, 695–698
maxima and minima on, 686
modeling data using, 114–115
modeling with regression equations using, 690–691
piecewise-defined functions on, 689
rounding using, 115
scatter plots on, 683
sequences and, 638, 699–700
solving a system of linear equations using, 552
solving equations graphically using, 687, 689
solving linear equations graphically using, 101–103, 104
solving systems of equations using, 688
symmetry and, 273
TABLE ASK feature of, 685
TABLE key on, 685
TRACE key on, 684–685
turning on and off, 681
viewing window of, 31, 33–34, 37–38
WINDOW key on, 685
Y= key on, 685
Y-VARS feature of, 685
Graphs/graphing
 asymptotes of. *See* Asymptotes
 complete, 32–33, 34
 of the equation of an ellipse, 666–667
 Excel for. *See* Excel
 of exponential functions, 334–337
  transformations of, 335–337
 finding intercepts using graphing calculators, 685–686
 of functions, 14–15, 31–44
  complete graph and, 32–33, 34
  data points and, 39–40
  by plotting points, 31–32
  reflections of, 267–269
  shifts of, 262–266
  stretching and compressing, 266–267
  symmetric, 270–273
  with technology, 33–37
 with graphing calculators. *See* Graphing calculators
 of a hyperbola, 670
 of inequalities, 614–615
 of inverse functions, 295–296
 of linear equations, 101–103
 local extrema points of, 453
 of logarithmic functions, 348–349
 of a parabola, 663
 of piecewise-defined functions, 216–217
 of rational functions, 501–505

spreadsheets for. *See* Excel; Spreadsheets
symmetry of, 270–273
 with respect to origin, 270–272
 with respect to x-axis, 271
 with respect to y-axis, 270
 of systems of linear equations, in two variables, 129–130
 transformations of. *See* Transformations of graphs
Greatest common factor (gcf), 167–168
Greatest integer function, 215, 219–220
Grouping, factoring by
 of higher-degree polynomial functions, 482–483
 of polynomials, 168–169
Growth, rate of, 383
Growth functions, logistic, 417–419
Growth models, exponential, 337–339, 383
 finding rates of change in, 380–381

Half-open intervals, 6
Higher-degree polynomial(s), factoring, 441–442
Higher-degree polynomial functions, 451–463
 cubic, 452–455, 464–466
 factoring, 482–483
 quartic, 455, 467–468
Horizontal asymptotes
 of an exponential function, 333
 of a logistic function, 417, 419
 of a power function, 208, 209
 of a rational function, 503–504
Horizontal ellipses, 666
Horizontal hyperbolas, 668–669
Horizontal line(s), equations of, 66–67, 68
Horizontal line test, 293
Horizontal shifts
 of ellipses, 667
 of graphs, 264, 268, 335
 of hyperbolas, 670–671
Horizontal stretch/compression of graphs, 335, 336–337
Hyperbolas, 668–672
 equations of
  converting to standard form, 671
  graphing, 670
  standard forms of, 669
 horizontal, 668–669
 horizontal and vertical shifts of, 670–671
 rectangular, 671–672
 vertical, 669

Identities, 90
 additive, 5
 multiplicative, 5
Identity function, 57

Identity matrices, 548, 570
   on graphing calculators, 696
Imaginary numbers, 170
Inconsistent systems, 135–137, 542
   Cramer's Rule with, 596
   of linear equations in three variables, 552, 555–556
Increasing functions, 174
Independent variable, 14
Index (root), 163
Inequalities
   Addition Property of, 93
   constraint, 617
      systems of inequalities and, 614
   double, 6, 147–148
   equivalent, 142
   exponential, 371–372
   graph of, 614–615
   involving absolute values, 313
   linear. *See* Linear inequalities
   logarithmic, 371–372
   Multiplication Property of, 93
   on the number line, 5–6
   polynomial, 518–519
   power, 312
   properties of, 92–94
   quadratic, 307–312
      algebraic solution of, 308, 310
      graphical solution of, 309, 310–311
   rational, 519–521
   solution region of, 615
   Substitution Property of, 93
   Subtraction Property of, 93
   systems of, 614–624
      linear, in two variables, 616–619
      nonlinear, 620–621
Inequality notation, 6
Infinite geometric series, 649–651
   sum of, 650
Infinite sequences, 637
Initial value, of exponential functions, 382
Inputs, 13
   aligned, 21–22
   finding using a model, on graphing calculators, 691
   uniform, 110
Integer exponents, 162–163
Intercepts. *See also* x-intercept(s); x-intercept method; y-intercepts
   slope-intercept form of linear equations and, 63, 68
Interest, compound. *See* Compounding
Interest rate, annual (nominal), 396, 412, 414
Interpolation, 120
Intersection method
   for linear equations, 102–103
   for linear inequalities, 144–145
   for systems of linear equations, 129–130
   using Excel, 711
   using graphing calculators, 687–688
Intersection of sets, 3
Interval(s)
   closed, 6
   half-open, 6
   on the number line, 5
   open, 6
Interval notation, 6
Inverse(s), additive and multiplicative, 5
Inverse functions, 290–292, 294–298
   on graphing calculators, 693
   graphs of, 295–296
   on limited domains, 297–298
Inverse matrices, 575–578
   definition of, 575
   finding using technology, 577–578, 697, 716
   solving matrix equations with, 582–583
   of a square matrix, 576
   of a 3 × 3 matrix, 577
Inverse variation, 509–510
Investment(s)
   future value of. *See* Future value
   present value of, 394, 401–402
      with Excel, 712
      on graphing calculators, 694
Investment models, 403
Irrational numbers, 4

Joint variation, 511

Kepler, Johannes, 207, 658, 665
Keystrokes, on graphing calculators, 682

Leading coefficient, of a polynomial, 165, 441
Leading term, of a polynomial, 165
Least common denominator (LCD), 444
Least squares line, 113
Least squares method. *See* Linear regression
Left-to-right elimination method, 538–540
Like terms, combining, 7–8
Limiting value, of logistic function, 417
Line(s)
   best-fit, 113
      finding equation of, using Excel, 708–709
   horizontal, equations of, 66–67, 68
   horizontal line test and, 293
   least squares, 113
   number
      inequalities on, 5–6
      intervals on, 5
      real, 4
   orientation of, slope and, 51–52
   parallel, equations of, 66–67
   perpendicular, equations of, 67–68
   vertical, equations of, 66–67, 68
   vertical line test and, 17–18
Linear equations, 62–78
   algebraic solution of, 97–99
   approximately linear data and, 71–72
   average rate of change and, 68–69
   difference quotient and, 69–71
   functional form of, 104
   general form of, 67–68
   graphical solution of, 101–103
   of horizontal lines, 66–67, 68
   literal, 103–104
   in one variable, solving, 98
   of parallel lines, 67–68
   of perpendicular lines, 67–68
   point-slope form of, 63–66, 68
   slope-intercept form of, 63, 68
   solving, 8–9
   using spreadsheets, 103, 709
   systems of. *See* Systems of linear equations; Systems of nonlinear equations
   in three variables, 535–536
   of vertical lines, 66–67, 68
Linear functions, 45–62, 256
   constant, 56–57
   constant rate of change and, 54
   continuous, 111
   definition of, 45
   discrete, 111
   exact vs. approximate, 110–113
   identity, 57
   intercepts of, 46–49, 52–53
   slope of, 49–53
Linear inequalities, 141–151
   algebraic solution of, 142–144
   definition of, 142
   graphical solution of, 144–146
   in two variables, 614–616
   systems of, 616–619
Linear models, quadratic models compared with, 230
Linear programming, 625–636
   feasible regions and, 625–627
   feasible solution and, 625
   objective function and, 625
   optimal solutions and, 625
   optimal values of a function subject to constraints and, 627–628
   solutions to problems in, 627
   technology for, 632, 698–699
Linear regression, 113–119
   applying models and, 120
   creating exponential models using, 387–388
   goodness of fit and, 120–121
   with spreadsheet, 119
Linear regression line, 113
Linear systems of inequalities, in two variables, 616–619
Literal constants, 7

Literal equations, 103–104
Loan repayment, 412–414
 amortization formula and, 412–413
 with Excel, 713
 on graphing calculators, 695
Local extrema points, 453
Local maximum point, 453
Local minimum point, 453
Logarithm(s)
 argument of, 347
 base 10, 350
 common, 349–352
 as exponents, 346
 natural, 353–354
 properties of, 354–356
  solving exponential equations using, 367–369
 rewriting, 355–356
Logarithmic forms, solving exponential equations using, 363–365
Logarithmic functions, 345–349
 converting from exponential to logarithmic form and, 346–347
 converting from logarithmic to exponential form and, 347
 definition of, 346
 finding using spreadsheets, 387
 graphs of, 348–349
  with spreadsheets, 350
  transformations of, 352
Logarithmic inequalities, 371–372
Logarithmic models, 385–387
 fitting to data using technology, 386
Logistic functions, 417–421
 decay, 419–421
 growth, 417–419
Long division, of polynomials, 447–448

Major axis, of an ellipse, 666
Marginal cost, 56
Marginal profit, 56
Marginal revenue, 56
Mathematical models. *See* Models/modeling
Matrices, 547–598
 addition of, 561–562
  with Excel, 714–715
  on graphing calculators, 696
 augment of, 548
 augmented, 548
 coefficient, 548
 Cramer's Rule and, 589–591
 dimension of, 548
 echelon forms of, 548–550
 encoding/decoding messages and, 578–581
 Gauss-Jordan elimination and, 550–551
 on graphing calculators, 695–698
 identity, 548, 570
  on graphing calculators, 696
 inverse. *See* Inverse matrices

matrix representation of systems of equations and, 548
 multiplication of, 564–570
  by a matrix, 565–568, 696–697
  by a number, 564–565, 696–697
  with technology, 568–570, 696–697, 715
 negatives of, 562
 reduced row-echelon form of, 550
 row operations and, 549
 row-echelon form of, 549
  on graphing calculators, 697–698
 square, 548
 subtraction of, 562–564
  on graphing calculators, 696
  with spreadsheet, 564, 714–715
 3 × 3, determinant of, 591–593
 2 × 2, determinant of, 589
 zero, 562
Matrix equations, 581–584
 solution with inverses, 582–583
 technology and, 583–584
Maxima (maximum points)
 absolute, 453
 finding using Excel, 706–707
 finding using graphing calculators, 686
 local, 453
 of a parabola, 174
Midpoint formula, 659
Minima (minimum points)
 absolute, 453
 finding using Excel, 706–707
 finding using graphing calculators, 686
 local, 453
 of a parabola, 174
Minor(s), of determinants, 592
Minor axis, of an ellipse, 666
Models/modeling, 12, 20–22, 114–115
 aligned inputs for, 21–22
 applying, 120
 comparison of, 384
 with cubic functions, 464–466, 468–470
  with spreadsheet, 465
 exponential, 382–383
  decay, 339–340, 380–381, 383
  definition of, 382
  finding rates of change in, 380–381
  growth, 337–339, 380–381, 383
 investment, 403
 linear and quadratic models compared and, 230
 linear regression and. *See* Linear regression
 logarithmic, 385–387
  fitting to data using technology, 386
 power, quadratic models compared with, 234–235
 with power functions, 231–233
  with spreadsheet, 233
 quadratic, 227–230

 linear models compared with, 230
 power models compared with, 234–235
  with spreadsheet, 229–230
 of quadratic functions, from three points on its graph, 226–227
 with quartic functions, 467–470
 with regression equations, on graphing calculators, 690–691
 of systems of linear equations, 134–135
  in three variables, 540–541
 third and fourth differences and, 470–471
Monomials, 165
 dividing polynomials by, 167
Mortgages, 411, 413–414
Multiplication. *See also* Product(s)
 Associative Property of, 5
 Commutative Property of, 5
 of complex numbers, 449
 of fractions, 89
 of functions, 279
 of matrices, 564–570
  on graphing calculators, 696–697
  by a matrix, 565–568
  by a number, 564–565
  with technology, 568–570
 over Addition, Distributive Property of, 5, 8, 166
 of polynomials, 166–167
 of radicals, 328
 of rational expressions, 443
 of real numbers, 5
 by zero, 5
Multiplication Property
 of equations, 9, 90
 of inequalities, 93
Multiplicative identity, 5
Multiplicative inverse, 5
Multiplicity, solutions with, 196, 492

Natural logarithms, 353–354
Negative exponents, 162–163
 power functions with, 208–209
Nominal interest rate, 396, 412, 414
Nonlinear systems of equations. *See* Systems of nonlinear equations
Nonlinear systems of inequalities, 620–621
Notation
 factorial, 654
 function, 18–20
 inequality, 6
 interval, 6
 radical, 165
 scientific, 329–330
nth power, direct variation as, 212
nth term
 of an arithmetic sequence, 639
 of a geometric sequence, 641

Number(s)
 complex. *See* Complex number(s)
 imaginary, 170
 real. *See* Real number(s)
Number e, 340–342
 continuous compounding and, 398–401
Number line
 inequalities on, 5–6
 intervals on, 5
 real, 4
Numerical coefficients, 7

Objective function, linear programming and, 625
Odd functions, 270–271
One-to-one functions, 292–298
 horizontal line test and, 293
Open intervals, 6
Optimal solutions, linear programming and, 625
Ordered pairs, 9–10
Orientation of a line, slope and, 51–52
Origin
 of coordinate system, 9
 symmetry of graphs with respect to, 270
Outputs, 13
 of composite functions, 284–285
 finding using a model, on graphing calculators, 691
 first differences of, 110
 second differences of, 230
 third and fourth differences of, 470–471

Parabolas, 180, 662–664. *See also* Quadratic functions
 axis of, 662
 axis of symmetry of, 173, 174
 combining graphing and factoring of quadratic equations and, 189–190
 concave down, 174
 concave up, 174
 definition of, 662
 directrix of, 662
 focus of, 662
 graphing the equation of, 663
 maximum point of, 174
 minimum point of, 174
 standard equation of, 662–663
 vertex of, 173, 662
Parallel lines, equations of, 66–67
Parentheses, removing, 8
Pascal's triangle, 654
Perfect squares, factoring, 168
Perpendicular lines, equations of, 67–68
Piecewise-defined functions, 215–225, 257
 absolute-value functions as, 215, 220
 graphing, 216–217
 on graphing calculators, 689
 greatest integer function as, 215, 219–220

Planes, 535–536
Point-plotting method, 31–32
Point-slope form, of linear equations, 63–66, 68
Polynomial(s), 165–169, 441–442
 degree of, 165, 441
 division of, 447–448
  by synthetic division, 490–492
 factoring, 167–169
  completely, 168–169
  of higher-degree polynomials, 441–442
 factors of, 167
 general form of, 165
 higher-degree, factoring, 441–442
 leading coefficient of, 165, 441
 operations with, 166–167
 in quadratic form, 441–442
 in rational expressions. *See* Rational expressions
 in x, 165
Polynomial equations, 479–500
 estimating solutions with technology and, 485–486
 factoring by grouping and, 482–483
 graphical and algebraic methods for solving, 492–494
 Rational Solutions Test and, 495–497
 root method for solving, 483–485
 solution in the complex number system, 497–498
 solving by factoring, 479–482
Polynomial functions
 graphs of, 454, 455, 456–458
 higher-degree, 451–463
  cubic, 452–455, 464–466
  factoring, 482–483
  quartic, 455, 467–468
Polynomial inequalities, 518–519
Power functions, 164, 203–209, 210–211, 257
 definition of, 203
 modeling with, 231–233
  with spreadsheet, 233
 with negative exponents, 208–209
 root functions as. *See* Root functions
 special, 258
Power inequalities, 312
Power models, quadratic models compared with, 234–235
Present value
 of an annuity, 409–411
  on graphing calculators, 695
  with spreadsheet, 411, 713
 of an investment, 394, 401–402
  with Excel, 712
  on graphing calculators, 694
 definition of, 401
Principal square root, 163
Product(s). *See also* Multiplication
 of a number and a matrix, 564–565

 of a row matrix and a column matrix, 566
 of two matrices, 566–568
Product Rule for Radicals, 163–164
Profit, 55–56, 279–280
 marginal, 56
Proportionality, 534–535. *See also* Variation
 constant of, 105, 534
 direct, 105
  as the nth power, 212
 inverse, 509–510
 proportional pairs and, 534
 proportional triples and, 535
Pure imaginary numbers, 170

Quadrants, 9
Quadratic equations, solving, using spreadsheets, 191–192, 710
Quadratic form
 equations in, 306–307
 polynomials in, 441–442
Quadratic formula, 194–196
Quadratic functions, 186–202, 256. *See also* Parabolas
 aids for solving, 197
 combined graphical and factoring solution of, 189–190
 completing the square and, 193–194
 with complex solutions, 197–198
 discriminant of, 196
 equation of, 226–227
 factoring of, 186–202
  combining graphing with, 189–190
 graphical solution of, 188–189, 190–192
  factoring solution combined with, 189–190
 graphing using spreadsheets, 176
 modeling of, from three points on its graph, 226–227
 modeling with, 227–230
  linear models compared with, 230
  with spreadsheet, 229–230
 numerical solution of, 191
 quadratic formula and, 194–196
 spreadsheet solution of, 191–192
 square root method for solving, 192–193, 197
 vertex form of, 179–181
Quadratic inequalities, 307–312
 algebraic solution of, 308, 310
 graphical solution of, 309, 310–311
Quadratic models, power models compared with, 234–235
Quartic equations, Rational Solutions Test and, 495–497
Quartic functions, 455
 modeling with, 467–470
Quotients. *See also* Division
 difference, 69–71

Radical(s), 163–165
  addition of, 328
  division of, 328
  multiplication of, 328
  Product Rule for, 163–164
  in simplified form, 163–164
  subtraction of, 328
Radical equations, 303–305
Radical notation, 165
Radicand, 163
Range of a function, 13–15
Rate of growth, 383
Rational equations
  algebraic solution of, 507–509
  graphical solution of, 509
Rational exponents, 164–165
Rational expressions, 442–445
  addition of, 444–445
  division of, 443–444
  Fundamental Principle of, 442
  multiplication of, 443
  simplification of, 442–443
  subtraction of, 444, 445
Rational functions
  definition of, 501
  graphs of, 501–505
  horizontal asymptotes of, 503–504
  slant asymptotes of, 505–507
  vertical asymptotes of, 502
Rational inequalities, 519–521
Rational numbers, 3, 4
Rational powers, equations containing, 305
Rational Solutions Test, 495–497
Rationalization, of denominators, 328–329
Real number(s), 3–5
  calculating with, 4–5
  integers, 4
  irrational, 4
  natural, 4
  properties of, 5
  rational, 3, 4
Real number exponents, 327
Real number line, 4
Real number system, 3
Reciprocals, 5
Rectangular coordinate(s), 10
Rectangular coordinate system, 9–10
Rectangular hyperbolas, 671–672
Reduced row-echelon form, of matrices, 550
  on graphing calculators, 697–698
Reflections of graphs, 267–269, 335
  across x-axis, 335
  across y-axis, 335
Regression equations, on graphing calculators, 690–691
Revenue, 55–56, 279–280
  marginal, 56
Richter scale, 356–358
Root functions, 209–212
Root method, solving polynomial equations using, 483–485

Rounding
  of coefficients, 94–95
  of constants, 94–95
  using graphing calculators, 115
Row matrices, product of a column matrix and, 566
Row operations, matrices and, 549
Row-echelon form, of matrices, 549

Scatter plots, 13
  on Excel, 707–708
  on graphing calculators, 683
Scientific notation, 329–330
Second differences, 230
Second-degree equations, graphs of, 658
Second-degree functions. See Parabolas; Quadratic functions
Sequences, 636–642
  arithmetic, 638–640
    common difference and, 639
    definition of, 639
    nth term of, 639, 699
    sum of, finding with graphing calculators, 699–700
  definition of, 637
  as discrete functions, 637
  finite, 637
  geometric, 640–642
    common ratio and, 640
    constant percent change and, 641
    definition of, 640
    nth term of, 641, 700
    sum of, finding with graphing calculators, 700
  infinite, 637
  technology and, 638, 699–700
  terms of, 636, 637–638
Series, 645–653
  arithmetic, 646–648
    sum of n terms of, 647–648
  defined, 645
  finite, 645–646
  geometric, 648–651
    infinite, 649–651
    sum of n terms of, 648–649
  infinite, 645
Sets, 2–3
  relations between, 2–3
Shifts of graphs, 262–266
  horizontal, 264, 268, 335
  vertical, 263, 268, 335
Significant digits, 95
Simplification
  of complex fractions, 445–447
  of radicals, 163–164
  of rational expressions, 442–443
Slant asymptotes, of rational functions, 505–507
Slope
  constant rate of change and, 54
  of a line, 49–52

  orientation and, 51–52
  y-intercept and, 52–53
Slope-intercept form, of linear equations, 63, 68
Solution(s)
  of an equation, 8–9
  complex, quadratic equations with, 197
  feasible, linear programming and, 625
  with multiplicity, 196, 492
  of systems of linear equations, 128, 129, 536
Solution region, of an inequality, 615
Special products, factoring, 168
Spreadsheets. See also Excel; Technology
  finding future value of an annuity using, 408
  finding logarithmic functions using, 387
  finding present value of an annuity using, 411
  future value of an investment using, 398, 711–712
  graphing logarithmic functions using, 350
  graphing quadratic functions using, 176
  inverse matrices and, 578
  linear regression using, 119
  matrix equations and, 584
  matrix multiplication using, 569–570
  matrix subtraction using, 564
  modeling with cubic functions using, 465
  modeling with power functions using, 233
  modeling with quadratic functions using, 229–230
  solving linear equations using, 103, 709
  solving quadratic equations using, 191–192, 710
Square(s)
  difference of, factoring, 168
  perfect, factoring, 168
Square matrices, 548
  inverses of, 576–577
Square root(s), principal, 163
Square root method, for quadratic equations, 192–193, 197
Squaring function, 203–204
Standard form
  complex numbers in, 170
  of the equation of a circle, 659
    converting from general form to, 660–662
  of the equation of a parabola, 662–663
  of the equation of an ellipse, 666
Straight-line depreciation, 62
Stretching graphs, 266–267, 268, 335, 336–337
Subscripts, 10
Subsets, 2
Substitution, back, 538

Substitution method, for systems of linear equations, 130–132
Substitution Property
  of equations, 90
  of inequalities, 93
Subtraction. *See also* Difference(s)
  of complex numbers, 449
  of fractions, 89
  of functions, 279
  of matrices, 562–564
    with Excel, 714–715
    on graphing calculators, 696
  of polynomials, 166
  of radicals, 328
  of rational expressions, 444, 445
  of real numbers, 5
Subtraction Property
  of equations, 9, 90
  of inequalities, 93
Sums. *See also* Addition
  of *n* terms of a geometric series, 648–649
  of *n* terms of an arithmetic series, 647–648
  of squared errors (SSE), 113n
Supply, 130
Symmetry of graphs, 270–273
  with respect to origin, 270–272
  with respect to *x*-axis, 271
  with respect to *y*-axis, 270
Synthetic division, 490–492
Systems of equations
  linear. *See* Systems of linear equations
  nonlinear, 598–601
    algebraic solution of, 598–601
    graphical solution of, 601–602
    solving using graphing calculators, 688
Systems of inequalities, 614–624
  linear, in two variables, 616–619
  nonlinear, 620–621
Systems of linear equations, 128
  matrix solutions of, 547–560
    of dependent systems, 553–555
    echelon forms of matrices and, 548–550
    Gauss-Jordan elimination and, 550–551
    of inconsistent systems, 555–556
    matrix representation of systems of equations and, 548
    nonunique, 552–553
    technology for, 552
  in three variables, 537–547
    back substitution and, 538
    Cramer's Rule for solving, 593–596
    dependent, 542, 553–555
    in echelon form, 538
    inconsistent, 542, 552, 555–556
    left-to-right elimination and, 538–540
    modeling, 540–541
    nonunique solutions to, 541–544

  operations on, 539
  solutions of, 536
  in two variables, 128–141
    Cramer's Rule for solving, 591
    dependent, 135–137
    elimination method for solving, 133
    graphical solution of, 129–130
    inconsistent, 135–137
    modeling, 134–135
    with nonunique solutions, 135–136
    solving with Excel, 713–714
    substitution method for solving, 130–132
Systems of nonlinear equations, 598–601
  algebraic solution of, 598–601
  graphical solution of, 601–602

Technology. *See also* Excel; Graphing calculators; Spreadsheets
  estimating solutions with, 485–486
  fitting exponential models to data using, 384
  fitting logarithmic models to data using, 386
  inverse matrices and, 577–578
  for linear programming, 632
  matrix equations and, 583–584
  matrix multiplication using, 568–570
Terms
  of an algebraic expression, 7
  in binomial expansions, finding, 656
  constant, of an algebraic expression, 7
  like, combining, 7–8
  *n*th
    of an arithmetic sequence, 639
    of a geometric sequence, 641
  of a polynomial
    degree of, 165
    leading, 165
  of sequences, 636, 637–638
Third differences, 470–471
TI-84 Plus/TI-84 Plus C Silver Edition. *See* Graphing calculators
Total cost, 55
Transformations of graphs
  compressing as, 266–267, 268, 336–337
  of exponential functions, 335–337
  of logarithmic functions, 352
  reflections as, 267–269, 335
  shifts as, 262–266, 335
    horizontal, 263, 268, 335, 667, 670–671
    vertical, 263, 268, 335, 667, 670–671
  stretching as, 266–267, 268, 336–337
Trinomials, 165
Turning points, 32, 33, 453

Uniform inputs, 110
Union of sets, 3
Unlike denominators, adding and subtracting fractions with, 89

Upper limit, of logistic function, 417

Variable(s), 7
  independent and dependent, 14
  specified, solving equations for, 103–104
Variable costs, 55
Variation. *See also* Proportionality
  combined, 510–511
  constant of, 105
  direct, 105, 212
  inverse, 509–510
  joint, 511
Vertex, of a parabola, 173, 662
Vertex form, of a quadratic function, 179–181
Vertical asymptotes
  of a power function, 208
  of a rational function, 502
Vertical compression of graphs, 266
Vertical ellipses, 666
Vertical hyperbolas, 669
Vertical line(s), equations of, 66–67, 68
Vertical line test, 17–18
Vertical shifts
  of ellipses, 667
  of graphs, 263, 268, 335
  of hyperbolas, 670–671
Vertical stretch/compression of graphs, 266, 335
Viewing window, of graphing calculator, 31, 33–34

x-axis, 9
  reflection of graphs across, 335
  symmetry of graphs with respect to, 271
x-intercept(s), 46–49
  finding using graphing calculators, 685–686
x-intercept method
  for linear equations, 101–102
  for linear inequalities, 145–146
  using Excel, 709–710
  using graphing calculators, 687

y-axis, 9
  reflection of graphs across, 335
  symmetry of graphs with respect to, 270
y-intercepts, 46–49
  finding using graphing calculators, 685
  slope and, 52–53

Zero(s)
  complex, 497
  division by, 5
  of functions, 100
  multiplication by, 5
Zero exponents, 162–163
Zero product property, 186–187

Pricing, 139–140, 159
Purchasing power, 214, 234–235, 316, 344, 374, 434
Real estate inflation, 344, 374
Rule of 72, 375
Supply, 138, 140–141, 276–277, 301, 322, 360, 373, 374, 523, 603, 610
Trade balances, 609–610
U.S. gross domestic product, 242
U.S. national consumption, 125–126, 277–278

Education

College enrollment, 489
College tuition, 405, 415
Dean's list, 415
Degrees earned, 321
Doctorates, 151
Gender and PhDs, 290
Grades, 108, 147–149, 150
IQ measure, 345
Learning rate, 59
Normal curve, 345
Public school enrollment, 71–72
Reading tests, 107, 151
SAT scores, 150, 610
Student loans, 43, 59, 128
Test reliability, 30
Test scores, 27
Testing, 558

Environment

Apparent temperature, 146
Carbon dioxide emissions, 124, 125, 477–478
Climate change, 371, 374, 531
Deforestation, 376
Earthquakes, 356–358, 360–361, 433
Fujita scale, 224
Japan tsunami, 361
Pollution, 124, 125, 268–269, 277, 278, 477–478
Richter scale, 356–358
Seawater pressure, 60
Sulfur emissions, 476, 500
Temperature-humidity index, 107
Wind and pollution, 185, 200
Wind-chill factor, 218–219, 223–224, 238, 290, 316, 517

Finance

Annual compounding, 404
Annuities, 376, 406, 407–411, 415, 436, 649, 653, 679
Apartment rental, 185
Appliance repair, 63
Auto leasing, 415
Auto loans, 415
Break-even, 129–130, 138, 150, 159, 185, 200, 249, 489, 493–494, 499, 530, 604, 610
Business loans, 416, 586–587, 588, 595–596, 609
Car financing, 27, 84, 156, 415

Checkbook balance, 301
Compound interest, 404, 642, 643–644, 679
Continuous compounding, 343–344, 404
Cost-benefit, 30, 35–36, 278, 288, 372, 502–503, 515, 531
Credit card debt, 99, 645, 652
Depreciation, 26, 42, 59, 61, 62, 64–65, 74, 75, 85, 107, 149, 157, 638, 640, 643, 644, 647–648, 652, 653
Doubling time, 360, 364, 374, 375, 376, 404, 405
Down payments, 414–415
Earnings per share, 158
Future value, 123, 395–396, 397–398, 404, 405, 407–409, 414, 415, 484–485, 488, 522–523
Home appraisal, 151
Insurance payments, 415
Insurance premiums, 151, 241–242, 250, 377–378, 390, 573
Interest rates, 406, 437
Investment, 77, 92, 107, 109, 134, 136, 140, 156, 159, 215, 290, 296–297, 300, 312, 341, 343, 360–361, 367, 374, 394, 404, 405–406, 433, 434, 436, 437, 462, 517, 529, 530, 545–546, 547, 551, 558, 559, 587, 597, 598, 609, 610
IRAs, 414
Land cost, 109
Life insurance, 28, 60, 75
Life-income plan, 156–157
Loan amortization, 436
Loan amounts, 586–587, 588, 595–596, 609
Loan balance, 48–49, 53
Loan repayment, 360, 412–414, 415–416
Loans, 546, 653, 679
Mortgages, 27, 240, 277, 411, 413–414, 415, 416
Mutual funds, 403, 609
Present value, 401–402, 405, 409–411, 436, 653
Rental income, 159, 608
Retirement plans, 75, 110–111, 405, 415
Rink rental, 185–186
Simple interest, 103
Stock market, 99, 150
Stock prices, 139
Student loans, 43, 59, 128
Trust funds, 405, 604

Geometry

Archway, 679
Area of a rectangle, 184
Bridge design, 665, 673, 679
Circles, 109
Constructing a box, 488, 519, 522, 530, 601–602, 604, 624
Dimensions of a box, 604
Ellipses, 673
Halley's Comet orbit, 673
House construction, 679

Hyperbolas, 673
Maximum volume of a box, 480–481
Parabolas, 665
Roman Colosseum, 673
Satellite dish design, 663, 665
Surface area, 301
Volume of a cube, 301
Volume of a pyramid, 239, 511
Volume of a sphere, 301

Government

Federal tax per capita, 43
Income taxes, 160, 223, 249, 250, 301
Politics, 574, 586, 623, 635
Postage, 216, 223, 224, 302
Postal rates, 223
Sales tax, 108
Social Security, 29, 75, 123, 141, 150, 158, 301, 473
Social services, 78, 141, 558–559, 598
Supplemental Nutrition Assistance Program, 467–468

Health. *See also* Medical

Alcohol use, 140
Blood alcohol percent, 12, 21, 76, 105, 149–150
Body mass index, 86, 109, 510–511
Body temperature, 143–144
Body-heat loss, 30, 200–201, 300–301
Calories, 109
Cardiac health, 123
China's tobacco smoking, 150
Drug testing, 43
Health care spending, 434, 516
Hearing, 434
Heart rate, 74, 84–85, 108, 158–159
Life expectancy, 42, 59, 84, 108, 126, 157–158, 360, 374, 391, 419–421, 426, 560–561, 563–564
Medicaid, 436
National health care, 127, 151, 237, 473, 488–489
National health expenditures, 44, 190–191, 202, 250
Nutrition, 140, 546, 547, 558, 597–598, 609, 624, 635, 636, 679
Obesity, 26, 43, 76, 126–127, 184, 237–238, 311–312, 360, 390, 391–392
Sleep, 74
Spread of a disease, 424, 427, 437
Tobacco smoking, 150, 289, 575
Uninsured Americans, 421
Velocity of blood, 200, 269, 298

Labor

Aging workers, 37–38, 238, 463–465
Annual wage, 107, 124, 151, 241, 391, 433
China's labor pool, 201, 239–240, 316, 474, 485
Civilian labor force, 84, 107, 126, 128, 132, 322, 489

Index of Applications

Automotive
Auto leasing, 415
Auto noise, 242
Car financing, 27, 84, 156, 415
Car sales profit, 151
Corvette acceleration, 435
Drinking and driving, 75
EV/PHEV sales, 31, 35, 61, 425, 436
Fuel economy, 393
Fuel purchases, 85, 156
HID headlights, 150
Rental cars, 545, 558, 597, 614, 615–616, 617–618
Risky drivers, 463, 531
Unconventional vehicle sales, 277, 280–281

Biology
Bacterial growth, 383, 644
Bee ancestry, 644, 652
Crickets, 61
Deer population, 422
Growth of a colony of bees, 394
Home range of animals, 249
Paramecia, 332–333
Photosynthesis, 184, 481–482, 529
Rabbit breeding, 644
Weight of fish, 301
Wildlife management, 108–109
Wingspan of birds, 206–207

Business
Advertising, 59, 97, 102–103, 427, 508–509, 514, 515, 516, 522, 567–568, 573, 619, 624, 635
Average cost, 277, 281–282, 288, 290, 321, 501, 504–505, 513–514, 517–518, 520, 522, 530–531
Barcodes, 26
Bookcase production, 587
Business loans, 416
Business sale, 415
China's manufacturing, 139
China's shale-natural gas production, 374, 390
Cloud revenue, 215, 228–230, 241, 277, 301, 316, 322, 478
Company growth, 423, 426–427, 437
Competition, 574, 586
Cost, 29–30, 42, 55–56, 61, 75, 279–280, 287–288, 374, 488, 522, 574
Cost minimization, 180, 628–629
Crude oil production, 44, 85, 156, 184, 242–243, 451, 458, 474, 602
Daily revenue, 461–462
Depreciation, 26, 42, 59, 61, 62, 64–65, 74, 75, 85, 107, 149, 157, 638, 640, 643, 644, 647–648, 652, 653
Discount prices, 290
Drone services revenue, 392–393
E-commerce, 77, 185, 316, 393–394, 531
Electric charges, 223, 224–225
Electronic component production, 288
Energy consumption, 44, 124–125, 185, 243
Global biometrics revenue, 183, 249, 251–252
Housing construction, 225, 623, 625
Industrial shipments, 18–19, 462, 476, 530
Lithium demand, 59, 107, 202
Loans, 416, 586–587, 588, 595–596, 609
Machine assembly, 517
Manufacturing, 29, 477, 489, 500, 537–538, 540–541, 546–547, 558, 569–570, 574, 583–584, 597, 623, 624, 630–632, 634, 636, 678, 679
Marginal cost, 56, 61
Marginal profit, 56, 85, 156
Marginal revenue, 56, 61–62
Market share, 376
Marketing, 156
Mobile home sales, 434
Oil imports, 124
Parts listing, 612
Pet industry, 321
Phone call costs, 643
Price increases, 565
Pricing, 139–140, 159, 188–189, 545, 587, 608
Printer production cost, 288
Printing, 531
Production, 184, 212, 214, 624, 634, 635
Productivity, 514
Profit, 30, 43, 55–56, 61, 62, 75, 85, 87, 107, 143, 151, 157, 180–181, 183, 184, 248, 249, 264–265, 278, 279–280, 287, 289, 302, 315, 322, 463, 488, 643, 644, 652, 679
Profit maximization, 248, 625, 629–630
Property costs, 609
Purchasing, 543–544, 559
Revenue, 29, 42, 55–56, 85, 173, 184, 279–280, 287–288, 461, 462, 487–488, 499, 522, 523, 529, 531
Revenue maximization, 176–177
Sale of a practice, 415
Sales, 185, 202, 316, 393–394, 531, 623–624, 634–635
Sales and advertising, 508–509, 514, 515, 516, 522
Sales and training, 515–516, 531
Sales commissions, 108
Sales decay, 340, 343, 344, 371–372, 373, 376, 381, 389, 433, 434
Sales growth, 426, 427, 437
Sales volume, 515
Service calls, 219
Shipping restrictions, 30–31
3-D printing market, 240
Ticket pricing, 545, 558, 597
Total cost, 55, 87, 289
Total revenue, 87
Utilities, 30
Venture capital, 587
Worldwide 5G shipments, 77

Chemistry
Carbon-14 dating, 341–342, 345, 362, 364–365, 376, 434
Decomposition of chlorine in a pool, 394
pH scale, 351, 361–362
Radioactive decay, 344, 374

Consumer
Auto leasing, 415
Cell phone subscribers, 108, 186, 195–196, 201, 202, 249, 251, 316, 475
Credit card debt, 99, 645, 652
Energy use, 43, 434–435, 573
Game show question, 107
IRAs, 414
Mortgages, 27, 240, 277, 411, 413–414, 415, 416
Phone bills, 74
Postal rates, 223
Residential power costs, 215, 217
Retirement plans, 75, 110–111, 405, 415
Utility charges, 74

Economics
Big data market size, 109–110, 116–117
China's GDP vs. U.S. GDP, 253–254, 316
Consumer price index, 75, 240, 378–380, 474–475
Cost-benefit, 30, 35–36, 278, 288, 372, 502–503, 515, 531
Currency conversion, 300, 302
Demand, 30, 138, 140–141, 276–277, 322, 360, 373, 374, 515, 523, 531, 603, 610
Disposable income, 60, 76, 107, 124, 369
Earnings, 26
E-commerce in ASEAN region, 77
Energy use, 241
Exchange rates, 290
Global payment market, 316
Global telecom spending, 61
Gross domestic product, 44, 315, 478
Housing starts, 225, 252
Income by age, 473–474
Inflation, 337–338, 344, 382–383, 389
Market analysis, 141, 159
Market equilibrium, 131, 138, 139, 141, 159, 201, 598, 600–601
Personal income, 214, 240, 251, 277, 322, 338–339, 389–390, 477